PO. 669071, Rel 15
Item 1
S. Eckard

THE CHEMICAL THERMODYNAMICS
OF ORGANIC COMPOUNDS

"In this work, when it shall be found that much is omitted, let it not be forgotten that much likewise is performed."

Dr. Samuel Johnson
(on completion of his
Dictionary, 1755)

THE CHEMICAL THERMODYNAMICS OF ORGANIC COMPOUNDS

DANIEL R. STULL
Thermal Research Laboratory
Dow Chemical Company
Midland, Michigan

EDGAR F. WESTRUM, Jr.
Professor of Chemistry
University of Michigan
Ann Arbor, Michigan

GERARD C. SINKE
Thermal Research Laboratory
Dow Chemical Company
Midland, Michigan

John Wiley & Sons, Inc.
New York · London · Sydney · Toronto

LEGAL NOTICE

Neither the authors nor the publisher nor any person or organization acting either on behalf of them or otherwise in the compilation, publication, or distribution of the information in this book

(1) makes any warranty or representation expressed or implied with respect to the accuracy, completeness, or usefulness of the information which it contains, or that the use of any information, apparatus, method, or process disclosed may not infringe privately owned rights; or

(2) assumes any liabilities with respect to the use of, or for damage resulting from the use of, any information, apparatus, method, etc.

Copyright © 1969 by John Wiley & Sons, Inc.

All rights reserved. No part of this book may
be reproduced by any means, nor transmitted, nor translated into a machine language without the written permission of the publisher.

Library of Congress Catalog Card Number: 68-9250
SBN 471 83490 4
Printed in the United States of America
10 9 8 7 6 5 4 3 2

FOREWORD

The critical evaluation of published data, culminating in the compilation of tables of selected values of physical and thermodynamic properties of chemical substances, constitutes the viable end product of experimental thermodynamics for practical and industrial purposes. Before 1939 most compilations in common use were the result of individual enterprise, as instanced by the extensive review of heats of combustion of organic substances by M. S. Kharasch in 1929 and by the book *Thermochemistry of Chemical Substances* by F. R. Bichowsky and F. D. Rossini, published in 1936. Even the well-known Landolt-Börnstein Tabellen, although produced as a result of organized collective effort, were based on individual contributions. Several scientific groups are currently engaged in the systematic compilation of thermodynamic tables, such as those at the National Bureau of Standards, at Texas A & M University, at The Dow Chemical Company (Midland, Michigan), at the Academy of Sciences, Moscow, and at Imperial College, London. Full-time data compilation by specialist groups is essentially a postwar development, necessitated by the increasing rate of output of new data in recent years and encouraged by the advent of computer aids.

American Chemical Society Monograph No. 60, by G. S. Parks and H. M. Huffman, *The Free Energies of Some Organic Compounds*, one of the most influential of earlier compilations, aimed at presenting thermodynamics and equilibria in a practical manner as well as providing the reader with a reference book on the thermodynamic properties of the more common organic chemicals. The monograph went out of print in 1938 and the authors subsequently began the task of revision, but the war intervened and little progress was made. Huffman, Parks, and Stull discussed the possibility of a complete rewrite in 1948 at the Portland, Oregon, A.C.S. meeting. All agreed that a new book was needed, but both original authors felt unable to give a sufficient amount of their time to it and suggested that Stull accept full responsibility for the updating of the original monograph.

In June 1952 Sinke joined the Thermal Laboratory of the Dow Chemical Company at Midland, and later he accepted an invitation from Stull to collaborate with him in the preparation of the book. The primary task was to collect and evaluate all relevant data and reduce them to convenient tabular form. Previously, toward the end of 1948, Stull had developed machine computer techniques specifically for the compilation of thermodynamic properties, which he described at the 118th A.C.S. meeting in Chicago (reported in *Chemical and Engineering News*, September 25, 1950). These techniques were thereafter used and developed in the day-to-day operations of the Dow Thermal Laboratory and also served in the preparation of the book *Thermodynamic Properties of the Elements* by Stull and Sinke, published by the American Chemical Society in 1956 as No. 18 in the *Advances in Chemistry* series. In 1959 the Dow Thermal Laboratory was commissioned to prepare the Joint Army-Navy-Air Force Thermodynamic Tables (JANAF Tables), by which time Stull and Sinke had fully developed the computer techniques necessary for the preparation of such extensive tables and of the tables in this book. The reader is given complete tabular data, as opposed to generating equations.

In May 1961 Stull invited Westrum of the University of Michigan to collaborate in the preparation of the book, but completion of the task required six more years. Data compilation today is a frustrating occupation; all too often new data appear that demand instant reassessment of completed work, so that the end of the road is never reached.

The book follows the philosophy of A.C.S. Monograph 60. It goes further than its parent in appealing to the industrial chemist and is perhaps the clearest available exposition on *how* to apply thermodynamics to practical problems in the laboratory and in the plant. The Dow Thermal Laboratory for many years has served industrial needs of the company profitably, and, in an interesting manner, some of its experience and much of its enthusiasm happily have entered the pages of Stull, Westrum, and Sinke.

The authors are to be complimented on their fine work and for their perseverance over many years in developing comprehensive tables, despite the need for continual revisions. Acknowledgment is surely due to the Dow Chemical Company for its early and continuing support of the Thermal Laboratory in all its aspects. Thermodynamicists and thermochemists will welcome this publication, which will find its place in teaching and, as a reference book, in laboratories throughout the world.

University of Manchester
September 1967

HENRY A. SKINNER,

Chairman, Commission on
Thermodynamics and Thermochemistry
International Union of
Pure and Applied Chemistry

PREFACE

The classical publication *Thermodynamics and the Free Energy of Substances*, written in 1923 by Lewis and Randall, is generally conceded to be the first complete mathematical formulation of chemical thermodynamics. A generation of students studied this famous text, and they are convinced of the usefulness of these relationships in technical applications. One of two significant developments since 1923 has been the experimental verification of the third law, spearheaded by W. F. Giauque and his students. The other is the development of statistical mechanical methods for computing the thermodynamic properties of an ideal gas from first principles. There is now no question that free energy values based on thermal data and statistical calculations can be used reliably to predict reaction equilibria.

Nevertheless, two factors have prevented full use of this powerful tool in the field of organic chemistry. Because of the employment of inaccurate information or failure to recognize the limitations of thermodynamics, some conclusions have been reached that are not in accord with experimental results and have led some to mistrust the predictions based on thermodynamics. The other factor that has impeded the application of thermodynamic reasoning to problems of organic chemistry is the lack of easily accessible numerical values of free energies. Moreover, the data of chemical thermodynamics are so widely scattered throughout the technical literature that compilations are an indispensable aid to the practicing thermochemist. The first authors to note this lack of data for organic compounds were G. S. Parks and H. M. Huffman, who introduced their pioneering monograph, *The Free Energies of Organic Compounds*, in 1932. This volume was received favorably by academic and industrial chemists alike and was used as seminar material in numerous universities. It struck resolutely at the two disparaging factors mentioned above and possessed a high degree of utility. About the time the book went out of print war clouds were gathering and the authors turned to other tasks.

A few limited attempts have been made to summarize the thermodynamic and thermochemical data for organic substances. Rossini, Pitzer, Arnett, Braun, and Pimentel (1248) Rossini et al. (1250), and Zwolinski et al. (1653) have provided continuing attention to thermodynamic data on hydrocarbons. Ribaud (1222) has tabulated some Gibbs energy functions (based on degrees Kelvin) for saturated hydrocarbons and other organic substances. Westrum and McCullough (1598) have summarized work on the thermodynamics of the organic solid state; Scott and McCullough (1316) have tabulated data for gaseous organic sulfur compounds; Kharasch (744) has provided data on heats of combustion; and Auer (52) has compiled 25°C data on the enthalpies of combustion, heat capacities, and entropies of organic and metal organic compounds. Moreover, Chermin [cf. (223) et prec.] and Kobe (784) have treated the thermodynamics of certain classes of hydrocarbons and their derivatives. However, no broad review or tabulation covering the increasing wealth of thermodynamic data on organic compounds has appeared since the pioneer work of Parks and Huffman (1105) in 1932.

By contrast, chemical thermodynamic data have been tabulated for the elements by Stull and Sinke (1437), for metals and alloys by Hultgren, Orr, Anderson, and Kelley (660), supplemented by Hultgren, Orr, and Kelley (661), and for inorganic substances by Kelley (736, 737), Kelley and King (738), Bichowsky and Rossini (125), Coughlin (278), Kubaschewski and Evans (812), and, as "Circular 500," by Rossini, Wagman, Evans, Levine, and Jaffe (1249). "Circular 500" has been revised and extended in part by Technical Notes issued by Wagman, Evans, Halow, Parker, Bailey, and Schumm (1560, 1561). Other tabulations for more or less restricted groups of inorganic substances have been compiled by Wagman and Evans (1559), Zwolinski et al. (1654), Brewer, Bromley, Gilles, and Lofgren (167), Ribaud (1222), Zeise (1643), Margrave (930), Katz and Margrave (726), Stull et al. (1435), Auer (52), Hilsenrath et al. (604), Glushko, Gurvich, Khachkuruzov, Veits, and Medvedev (499, 500), Gerasimov, Krestovnikov, and Shakov (467–470), and Schick (1282, 1283). In addition, preliminary tabulations to meet specific and urgent needs have been issued by government agencies and other organizations.

Because progress in the scientific and technological aspects of organic chemistry is increasingly dependent on basic information on the properties of substances, the lack of a unifying treatment of organic chemical thermodynamics that is both comprehensive and systematic has unquestionably delayed more rapid advances. It is the purpose of this book to rectify this conspicuous neglect.

By 1940 research workers at the Dow Chemical Company had recognized the need for quantitative measurements of thermal properties of chemical compounds. The laboratory established for this purpose has successfully dealt with many of the day-to-day problems by application of Gibbs energy data. Problems of organic compounds other than hydrocarbons required continual searching of the literature. Thus this book has been born of industrial necessity and is intended to provide the organic chemist with a single exhaustive survey of thermodynamic data pertinent to his concerns. Industrial organic chemists and chemical engineers outside the petroleum industry appear to have made too little application of thermodynamic data in solving their problems. This situation is changing, however, and many of the newer industrial organic chemists are beginning to seek those synthetic routes that are the most favorable energetically because the processes that require the least energy will probably be most economical and will therefore have long-term utility. Moreover, as energy is becoming more expensive daily, we are rapidly approaching the time when the chemical engineer must be as concerned about the energy balance in his process as he is about the mass balance.

We have therefore made a complete search of the literature up to January 1, 1966. For the ideal gas state we have tabulated the heat capacity, entropy, enthalpy, and Gibbs energy of formation for 741 pure organic compounds from 298 to 1000°K. The entropy, enthalpy, and Gibbs energy of formation are presented as fully as possible for approximately 4400 organic compounds in the ideal gaseous and condensed states at 298°K. Few organic compounds are stable above 1000°K (727°C), and so tabulations of organic compounds have been limited to the temperature range from 298.15 to 1000°K. We have pointed out some of the regularities existing between hydrocarbons and the other classes of organic compounds, which are useful in the estimation of thermodynamic properties of compounds that have not been experimentally measured.

We assume that users of this book have some knowledge of thermodynamics and mathematics. However, sufficient review of relevant elements of thermodynamics and discussion of calculational methods are included in Section I to assist the laboratory worker in utilizing the data tables provided herein in the solution of his problems. The organization is intended to provide ready reference to practical calculations of particular types and is meant to be for supplementary reading rather than for initial exposure. Although many troublesome matters are discussed, the brief scope of this review can hardly plumb the depths and subtle aspects of interest to thermodynamic specialists. Brief descriptions of calorimetric measurements are provided to acquaint the organic experimentalist with the measurement of thermal data, the primary source of the information tabulated in Sections II and III. Chapter 6 introduces the reader to the methods used to estimate thermodynamic properties of compounds that have not yet been measured but are essential to his application.

Still more recently, academically oriented organic chemists have come to appreciate the significant contribution made by chemical thermodynamic data to the understanding of molecular structures, molecular freedom in crystals, stabilities of organic molecules, strain energies, and the manifold energetic restrictions on transitional and transformational perturbation in structures. Such applications also can be greatly facilitated by the compiled experimental values contained in this volume. Organic chemists in general will therefore find much of interest and relevance here.

Finally, it should be noted that many academic thermal studies involve molecules of no immediate or present direct industrial interest. It is hoped that a further dividend from the use of this book will be the recognition of extant serious gaps in chemical thermodynamic data and the stimulation of measurements of thermal and thermochemical properties of technically important compounds.

We have made an honest attempt to avoid errors, but in an effort of this magnitude some may have escaped our notice. Consequently we hope that errors of commission and omission will be reported to us. We solicit the patience and understanding of the reader.

Midland, Michigan DANIEL R. STULL
Ann Arbor, Michigan EDGAR F. WESTRUM, JR.
  GERARD C. SINKE

ACKNOWLEDGMENTS

One of the pleasantest aspects of writing a book is the opportunity it affords to thank one's friends for their kindness and gracious help. Had it not been for the sound decision of The Dow Chemical Company to develop a laboratory for studying the energy relationships of molecules, the two industrial authors would have had neither the background nor the inclination to write this book. The abiding support of J. J. Grebe, W. R. Veazey, A. W. Beshgetoor, G. F. Dressel, R. H. Boundy, L. C. Chamberlain, C. D. Alstad, and F. A. Landee through the systolic and diastolic pulses of business have helped to make the Dow Thermal Research Laboratory a genuine reality. Thanks are due The Dow Chemical Company for the use of its library and computational facilities and for permission to undertake and publish this synoptic review.

The other author (E.F.W.) acknowledges his appreciation of the support by the Division of Research of the United States Atomic Energy Commission in chemical thermodynamic research, from which his interest and impetus in the present compilation developed.

Together we voice our appreciation for the patience of our colleagues, students, and especially our wives, who were, on occasion, deprived of merited attention by our compiling and editing labors. We are deeply grateful to Miss Kathie Robison and Mrs. Lynne Lurie for their infinite skill and attention to detail in typing the manuscript and in organizing related matters. Mrs. Carolyn Barber and William G. Lyon contributed significantly to the scientific aspects of manuscript preparation. We are indebted also for the patience and helpfulness of the publisher. Many colleagues have furnished us with data in advance of publication, for which we give grateful thanks. The compilation of Dr. W. Auer was especially useful as a secondary source of primary references.

The following illustrations are based on diagrams in the sources named: Figure 5–7 from John Wiley & Sons, Inc.; Figures 2–6 and 2–7 from *The Journal of the American Chemical Society*; Figure 2–1 from *The Journal of Chemical Education*; Figure 2–5 from *The Journal of Chemical Physics*; Figures 7–13 and 7–14 from *The Journal of the Chemical*

Society; Figure 2-3 from *The Journal of Physical Chemistry*; Figures 7-3 through 7-9 from Reinhold Publishing Corporation; Figures 3-3 and 3-5 from *The Journal of Research of the National Bureau of Standards*; Figure 3-4 from the Parr Instrument Company; and Figure 2-13 from D. Van Nostrand Company, Inc.

D.R.S.
E.F.W.
G.C.S.

CONTENTS

SECTION I.	**BASIC THERMODYNAMIC PRINCIPLES**	**1**
Chapter 1.	**Chemical Thermodynamic Concepts**	**3**
	Introduction	3
	Basic Definitions	4
	Equation of State	8
	Energy, Work, and Heat	10
	Enthalpy	12
	Reference States	13
	Vapor Pressure	14
Chapter 2.	**Heat Capacity and Enthalpy of Transition**	**17**
	Heat Capacity	17
	The Statistico-Mechanical Evaluation of Heat Capacity of a Gas	32
	Enthalpy	50
Chapter 3.	**Thermochemistry**	**61**
	Enthalpy of Reaction	62
	Temperature Dependence of Enthalpy of Reaction	64
	Enthalpy of Formation from Enthalpy of Combustion	67
	Enthalpy of Combustion from Bomb Calorimetry	67
	Enthalpy of Combustion from Flame Calorimetry	72
	Enthalpy of Formation from Heat of Reaction	75
	Enthalpy of Formation from Equilibrium Data	78
	Enthalpy of Formation by Other Methods	86
Chapter 4.	**Evaluation of Entropy**	**87**
	Second Law of Thermodynamics	87
	Third Law of Thermodynamics	90
	Entropy from Low-Temperature Heat Capacity	92
	Statistico-Mechanical Evaluation of Entropy	93
	Evaluation of Entropy as a Function of Temperature	108
	Entropy Increment for Chemical Reaction	112

Contents

Chapter 5.	**Gibbs Energy and Chemical Equilibrium**	**115**
	Gibbs Energy	115
	Gibbs Energy and Equilibrium Composition	118
Chapter 6.	**Methods for Estimating Thermodynamic Quantities**	**140**
	Introduction	140
	Methods Involving Valence Bond Contributions	141
	Methods Involving Group Contributions	146
Chapter 7.	**Application to Industrial Problems**	**155**
	Introduction	155
	The Petroleum Industry	158
	Chemicals from Methane	165
	Styrene Manufacture	171
	Acrylonitrile Synthesis	175
	Vinyl Chloride Synthesis	178
	Thermodynamics in Methanol Synthesis	180
	Manufacture of Formaldehyde from Methanol	181
	Manufacture of Acetic Acid	188
	The Gattermann-Koch Reaction	190
	Utilization of Thermodynamics in Catalyst Selection	192

SECTION II. THERMAL AND THERMOCHEMICAL PROPERTIES IN THE IDEAL GASEOUS STATE FROM 298° TO 1000°K   **198**

Chapter 8.	**Thermodynamic Symbols, Constants, Methods of Calculation, and Data for Elements and Selected Inorganic Compounds**	**199**
	Definition of Thermodynamic Symbols	199
	Constants	202
	Machine Computation of the Tables	204
	Reference States of the Elements	208
	Thermal Properties of Some Inorganic Compounds	214
Chapter 9.	**Chemical Thermodynamics of Hydrocarbon Compounds**	**235**
	Historical Introduction	235
	Importance of the Hydrocarbons	237
	Regularity of Properties	238
	Alkane Ideal Gas Tables	243
	Alkene Ideal Gas Tables	312
	Alkadiene Ideal Gas Tables	330
	Alkyne Ideal Gas Tables	334
	Cycloalkane Ideal Gas Tables	343

	Cycloalkene Ideal Gas Tables	346
	Alkylcyclopentane Ideal Gas Tables	348
	Alkylcyclopentene Ideal Gas Tables	357
	Alkylcyclohexane Ideal Gas Tables	358
	Alkylbenzene Ideal Gas Tables	367
	Styrene Ideal Gas Tables	386
	Alkylnaphthalene Ideal Gas Tables	389
	Miscellaneous Hydrocarbon Ideal Gas Tables	400
Chapter 10.	Chemical Thermodynamics of Compounds of Carbon, Hydrogen, and Oxygen	405
	Introduction	405
	Oxygen Compound Tables	412
	Aliphatic Ether Ideal Gas Tables	412
	Cyclic Ether Ideal Gas Tables	419
	Alkanol Ideal Gas Tables	422
	Alkenol Ideal Gas Table	436
	Alkanediol Ideal Gas Table	436
	Cycloalkanol Ideal Gas Table	437
	Alkanal Ideal Gas Tables	438
	Alkanone Ideal Gas Tables	444
	Alkenone Ideal Gas Table	446
	Cycloalkanone Ideal Gas Table	447
	Alkanoic Acid and Derivative Ideal Gas Tables	448
	Alkenoic Acid Ideal Gas Tables	452
	Phenol Ideal Gas Tables	454
	Benzenecarboxylic Acid Ideal Gas Table	456
Chapter 11.	Chemical Thermodynamics of Nitrogen Compounds	457
	Introduction	457
	Primary Alkylamine Ideal Gas Tables	461
	Secondary Alkylamine Ideal Gas Tables	466
	Tertiary Alkylamine Ideal Gas Tables	467
	Cyclic Amine Ideal Gas Tables	468
	Arylamine Ideal Gas Table	472
	Nitrile Ideal Gas Tables	473
	Nitroalkane Ideal Gas Tables	477
	Alkyl Nitrite and Alkyl Nitrate Ideal Gas Tables	481
Chapter 12.	Chemical Thermodynamics of Halogen Compounds	486
	Introduction	486
	Aliphatic Fluorine Compound Ideal Gas Tables	494
	Aromatic Fluorine Compound Ideal Gas Tables	504

xvi Contents

	Aliphatic Chlorine Compound Ideal Gas Tables	508
	Aromatic Chlorine Compound Ideal Gas Tables	532
	Acyl Chloride Ideal Gas Tables	536
	Aliphatic Bromine Compound Ideal Gas Tables	537
	Aromatic Bromine Compound Ideal Gas Table	547
	Aliphatic Iodine Compound Ideal Gas Tables	549
	Aromatic Iodine Compound Ideal Gas Table	556
Chapter 13.	**Chemical Thermodynamics of Organic Sulfur Compounds**	**558**
	Introduction	558
	Monothiaalkane Ideal Gas Tables	563
	Dithiaalkane Ideal Gas Tables	600
	Thiacycloalkane Ideal Gas Tables	605
	Thiacycloalkene Ideal Gas Tables	609
	Alkanethiol Ideal Gas Tables	611
	Cycloalkanethiol Ideal Gas Table	624
	Aromatic Thiol Ideal Gas Table	625
	Miscellaneous Sulfur Compound Ideal Gas Tables	626

SECTION III. THERMAL AND THERMOCHEMICAL DATA AT 298°K — **629**

Chapter 14.	**Selected Values of Enthalpy of Formation and Entropy of Organic Compounds at 298°K**	**631**
	Introduction	631
	Summary Table	771

APPENDICES — **771**

Appendix A-1.	Glossary of Symbols	773
Appendix A-2.	Einstein Function Table for Heat Capacity	776
Appendix A-3.	Einstein Function Table for Enthalpy	777
Appendix A-4.	Einstein Function Table for Entropy	778
Appendix A-5.	Debye Function Table for Heat Capacity	779
Appendix A-6.	Debye Function Table for Energy	780
Appendix A-7.	Debye Function Table for Entropy	781
Appendix A-8.	Heat Capacity Table for Restricted Rotation	782
Appendix A-9.	Enthalpy Function Table for Restricted Rotation	783

Appendix A-10.	Entropy Function Table for Restricted Rotation	784
Appendix A-11.	Entropy Table for Decrease from Free Rotation	785
Appendix A-12.	Comparative Thermodynamic Properties of the Restricted Rotator and the Harmonic Oscillator	786
Appendix A-13.	Harmonic Oscillator Contributions to the Heat Capacity	787

REFERENCES 799

COMPOUND INDEX 847

SUBJECT INDEX 855

THE CHEMICAL THERMODYNAMICS
OF ORGANIC COMPOUNDS

SECTION I

BASIC THERMODYNAMIC PRINCIPLES

CHAPTER 1

CHEMICAL THERMODYNAMIC CONCEPTS

INTRODUCTION

Over a century ago Thomson (Lord Kelvin), Rankine, Clausius, Regnault, and a score of others labored to clarify the relationship between heat and work that had been established for the steam engine, and thereby they set forth the fundamentals of thermodynamics. The name of this science was derived from two Greek words: *thermos* (meaning hot or warm) and *dynamos* (meaning force in action). Later Berthelot, Thomsen, and their contemporaries sought a better understanding of the application of heat to chemical reactions. They wrongly believed that heats of reaction and transformation could be coupled with *sensible heat* to yield a reliable index to the tendency for chemical reaction to take place. The term *thermochemistry* came to be associated with this heat index to a chemical reaction. It remained for Gibbs, van't Hoff, Horstman, Nernst, Planck, and many others to couple entropy with heat and obtain the Gibbs energy, which does yield a reliable index to chemical affinity and reactivity. The application of Gibbs energies and enthalpies of reaction to chemical problems has come to be known as *chemical thermodynamics*. This area of science is sometimes further subdivided into the study of *thermal properties*—that is, those essentially *physical* and not involving chemical reactions—and *thermochemistry*, which by contrast involves the heat associated with *chemical* reactions and transformations.

The chemist is interested in the quantitative aspects of the changes accompanying the gain or loss of one or more forms of energy by substances. These changes may be physical, chemical, or both, and may take place under a wide variety of conditions. Thermochemical principles are useful in predicting the behavior of reacting systems moving toward a more stable state of equilibrium. Reaction equilibria based on thermal and thermochemical measurements alone may be calculated from a minimum of experimental information. Gibbs (or free) energies provide a measure of the forces driving processes and reactions spontaneously toward an equilibrium state. The rate of approach to this equilibrium state, however, depends on additional factors, such as kinetics and catalysis. Chemical thermodynamics deals with equilibrium systems and cannot predict the time required to complete the transformation. The chemist must therefore combine his knowledge of kinetics and catalysis with free energies to predict the course of a given chemical reaction.

BASIC DEFINITIONS

The application of thermodynamics involves a clear formulation of entities and boundaries between them. Thus, in a typical thermodynamic phenomenon, bodies of matter interact, chemically or otherwise, with other bodies; there are energy interchanges and changes in the properties of the bodies involved. It is convenient to divide all of the matter that is in any way involved into two parts: one of these is designated the *system*, and the other, the *surroundings*. The system is then a definite kind and quantity of matter that is under discussion. The surroundings comprise any other bodies with which the system interacts in the course of the discussion. If, for example, an esterification reaction in a sealed flask is considered, it is convenient to regard the reaction mass as the system and the flask and bath as the surroundings.

In any particular discussion there may be a certain arbitrariness in just what is chosen as the system. Hence a precise statement of exactly which system is under discussion is important. For example, in the discussion of a calorimetric experiment, we might or might not wish to count the calorimeter itself as part of the system; but it would be important to be clear as to which was being done. A *homogeneous* system is the same throughout and is referred to as a single phase, while a *heterogeneous* system contains more than one phase, with definite boundaries between the separate phases. The physical properties of the system that are proportional to the mass are known as *extensive* properties, such as volume and heat capacity. Physical properties of the system that are independent of mass and characteristic of the substance, such as density and temperature, are known as *intensive* properties. Specifications of the extensive and intensive properties describe the *state* of the system. At the beginning of some transformation the system will be in a given condition, known as the *initial* state, while after the transformation is finished the system will be in a different condition, known as the *final* state. Chemical transformation of matter is described as a *reaction*, while a physical transformation is called a *process*. To avoid ambiguity, the practice initiated by Lewis and Randall (861) is used. Each reaction or process is explicitly described; for example,

$$C(\text{graphite}) + 2Cl_2(g) \rightarrow CCl_4(l)$$

indicates that 1 mole of carbon (graphite) reacts with 2 moles of *gaseous* chlorine to form 1 mole of *liquid* carbon tetrachloride, whereas

$$CCl_4(l) \rightarrow CCl_4(s)$$

indicates that 1 mole of *liquid* carbon tetrachloride, by the process of freezing, is transformed into a mole of *solid* or crystalline carbon tetrachloride.

Equilibrium may be said to exist in a chemical or physical thermodynamic system when its composition and properties undergo no observable change under constant external conditions or constraints even after an indefinite period of time. The system is then macroscopically in a state of rest or stable equilibrium, although microscopically a dynamic state of opposed but balanced reactions or processes exists. Any system not in a state of equilibrium must be changing continuously toward such a state with greater or less rapidity. If the rate of approach to equilibrium is so slow as to be imperceptible during the time of observation, the system may be in a state of metastable equilibrium. The presence of a suitable catalyst will demonstrate that the tendency to reach equilibrium is there. Thus a number of common organic substances are thermodynamically unstable, for example, nitroglycerine or benzene. A mixture of hydrogen and oxygen gases at room temperature does not react spontaneously to form the more stable product, water, in the absence of a catalyst, a palladium alloy foil or an electric spark. Such examples need introduce no particular concern provided that it is appreciated which reactions are proceeding at measurable speed, and therefore are subject to thermodynamic scrutiny, and which are substantially "frozen" by kinetic considerations.

Pressure is an intensive property of a system that has the dimensions of force per unit area. In this book pressure is given in terms of the standard atmosphere, which is defined as $1,013,250$ dynes cm^{-2} and is equal to the pressure of 760 mm of mercury at 0°C and a standard gravity of 980.665 cm sec^{-2}.

Volume is an extensive property of a system that has the dimensions of length cubed. Here the space required to contain 6.02308×10^{23}

molecules of an ideal gas at 1 atm total pressure at the ice point (273.15°K) is the standard molar volume and is equal to 22,414.6 cm^3.

Temperature is an intensive property of a system that is fundamental to the study of thermodynamics. The Latin verb *temperare* (to temper), meaning to bring anything to a suitable condition by mingling with something else, and the suffix *-ure*, designating the result of the action expressed by the verb, have formed the word temperature. This concept came to be associated with hotness and coldness. A quantitative definition of temperature involves the measurable effect of heat upon some system. Anders Celsius developed a scale of temperature by assigning 100 even degrees to the volume change of a liquid between the ice point, 0°C, and the steam point, 100°C.

Robert Boyle observed that the quantity of gas of volume V and pressure P at a constant temperature obeyed the relation

$$PV = \text{constant} \quad \text{or} \quad P_2 V_2 = P_1 V_1, \qquad [1\text{-}1]$$

in which the subscripts 1 and 2 refer to any two different states.

Jacques Charles found that the volume of a gas at constant pressure increases in a regular manner as the temperature is increased, or

$$V_t = V_0(1 + at) \qquad [1\text{-}2]$$

where V_0 and V_t are the volumes at 0 and t°C, and a is a constant. Insertion of experimental information leads to the approximate numerical value of 1/273 for a, and [1-2] becomes

$$\frac{V_t}{V_0} = \frac{273 + t}{273}. \qquad [1\text{-}3]$$

William Thomson (Lord Kelvin) noted that a simplification would result if a new scale of temperature were defined having the same size degree as the Celsius degree, but having its scale zero at -273°C. Equation [1-3] is then

$$\frac{V_T}{V_0} = \frac{T}{273} \quad \text{or} \quad \frac{V_2}{V_1} = \frac{T_2}{T_1}, \qquad [1\text{-}4]$$

in which T is known as the *absolute* or *Kelvin* temperature (°K).

Hence, by combining [1-1] and [1-4], the gas obeying Boyle's and Charles' rules is governed by the relation

$$PV = rT,$$

in which $P_0 V_0 / 273 = r$ is a specific gas constant characteristic for each gas.

Basic Definitions

Although actual (or real) gases only approximate this behavior, as the pressure of any gas diminishes its individual characteristics become less distinct, and at infinite attenuation (a purely hypothetical state) all gases show ideal behavior as defined by the relation

$$PV = RT, \qquad [1\text{-}5]$$

in which R is a *universal gas constant* per gram molecular weight (or mole) of the gas. T is proportional to the PV product of this ideal gas and may be shown to be identical with the *thermodynamic temperature scale*, usually referred to as the *Kelvin scale*, since Lord Kelvin deduced this temperature scale from the efficiency of a heat engine.

An ideal gas, as we have seen, is a hypothetical concept that can be approached by a real gas only at low pressure. Because a low-pressure gas is experimentally difficult to work with, a more convenient scale of temperature has been adopted by international agreement. In 1948 the Ninth General Conference on Weights and Measures (1423) agreed on the details and the definition of the *international temperature scale of 1948*, designated as degrees Celsius (or °C) and denoted by the symbol t. This international temperature scale of 1948 is a practical, working scale and is divided into two parts. From $-182.970°C$ (oxygen point) to $630.5°C$ (freezing point of antimony), the temperature is defined by the resistance $\mathscr{R}$ of a specified type of platinum resistance thermometer according to the formula

$$\mathscr{R}_t = \mathscr{R}_0[1 + At + Bt^2 + C(t-100)t^3].$$

Here $\mathscr{R}_0$ is the resistance at the ice point and A, B, and C are constants evaluated from the fixed points (c), (d), and (a) of Table 1-1, respectively. The last or quartic term in this equation is required only between the oxygen and ice points.

From $630.5°C$ to $1063.0°C$ the scale is defined by the formula

$$E = a + bt + ct^2,$$

where E is the electromotive force of a standard thermocouple of platinum and platinum-rhodium alloy with one junction at the ice point and the other at the temperature t. The constants a, b, and c are determined by measurements of E at the freezing points of antimony, silver, and gold. Since the international temperature scale and the Kelvin scale are based on different definitions, they are not identical. At $200°K$ the Kelvin scale is about $0.04°$ above the international scale, while at $660°K$ the Kelvin scale is about $0.15°$ above the international scale.

Chemical Thermodynamic Concepts

Table 1-1 Fundamental and Primary Fixed Temperature Points at 1 atm Total Pressure

Fixed Point	Temperature (°C)
(a) Temperature of equilibrium between liquid oxygen and its vapor (oxygen point)	−182.970
(b) Temperature of equilibrium between ice and air saturated water (ice point) (*fundamental fixed point*)	0.000
(c) Temperature of equilibrium between liquid water and its vapor (steam point) (*fundamental fixed point*)	100.000
(d) Temperature of equilibrium between liquid sulfur and its vapor (sulfur point)	444.600
(e) Temperature of equilibrium between solid and liquid antimony (antimony point)	630.5
(f) Temperature of equilibrium between solid and liquid silver (silver point)	960.8
(g) Temperature of equilibrium between solid and liquid gold (gold point)	1063.0

Lord Kelvin noted in 1854 that when the value of the ice point is known with sufficient accuracy the temperature scale can be uniquely defined in terms of the absolute value of the ice point rather than by selecting 100° for the difference between the ice and steam points. In 1954 the Tenth General Conference on Weights and Measures (1424) redefined the Kelvin scale by assigning 273.16°K to the triple point of water, and the ice point then became 273.15°K. For practical calculations, conversion between the two scales is

$$T(°K) = t(°C) + 273.15°.$$

EQUATION OF STATE

In the study of chemical thermodynamics the state or condition of a system may be completely defined by four observable properties—composition, pressure, temperature, and volume (in the absence of other constraints such as gravitational, magnetic, and electric fields)—of which the first three are intensive properties. For a pure, homogeneous substance the composition is already fixed. The remaining three properties are related, so that only two need be specified. For example, the pressure is some function of the temperature and volume:

$$P = f(T, V).$$

Equation of State

Here, for the various values of T and V, P is fixed by the mathematical relationship called an *equation of state*. For n moles of an ideal gas, the equation of state from [1-5] is

$$PV = nRT. \qquad [1\text{-}5]$$

By measuring the PV product of several real gases for a series of pressures at the ice point and extrapolating to zero pressure, careful experimental work indicates

$$(PV)_{P=0} = 22{,}414.6 \text{ cm}^3\text{-atm/mole}.$$

This limiting value of the PV product for an ideal gas may be substituted into [1-5], leading to

$$\frac{22{,}414.6}{273.15} = R = 82.05967 \text{ cm}^3 \text{ atm/(mole °K)}.$$

Since 1 atm is 1,013,250 dynes cm^{-2},

$$R = 83.1469 \times 10^6 \text{ ergs/(mole °K)}.$$

The *thermochemical calorie* is defined as 4.1840×10^7 ergs (1424), so that

$$R = 1.98726 \text{ cal/(mole °K)}.$$

As a first approximation, many real gases are nearly ideal at pressures of one atmosphere or less; therefore [1-5] may be used to approximate the behavior of real gases. For more exact requirements and at pressures above 1 atm the equation of state becomes more complex. The experimental compressibility of a gas may be represented by empirical equations expressed in terms of the volume or pressure of the gas.

$$PV = A_V + \frac{B_V}{V} + \frac{C_V}{V^2} + \frac{D_V}{V^4} + \frac{E_V}{V^6} + \ldots,$$

in which A_V, B_V, C_V, D_V, E_V are the first, second, third, fourth, etc., *virial coefficients*. Also

$$PV = A_P + B_P P + C_P P^2 + D_P P^4 + E_P P^6 + \ldots,$$

in which A_P, B_P, C_P, D_P, E_P are also known as virial coefficients. These so-called virial coefficients are temperature-dependent and have been modified into a formidable array of gaseous equations of state by numerous workers.

An equation of state used by many thermodynamicists is that of D. Berthelot (120),

$$PV = nRT\left[1 + \frac{9}{128}\frac{PTc}{PcT}\left(1 - 6\frac{Tc^2}{T^2}\right)\right].$$

This form of Berthelot's equation requires only knowledge of the critical temperature and pressure for its use, and gives accurate results in the vicinity of room temperature for unassociated substances at moderate pressures.

A generalized approach that has met with some success was devised by Lydersen, Greenkorn, and Hougen (894). It employs a compressibility factor Z obtained from a correlation (cf. Fig. 5-7) of the reduced temperature, $Tr = T/Tc$, and reduced pressure, $Pr = P/Pc$. This enters into the equation of state in the following way:

$$PV = nZRT.$$

ENERGY, WORK, AND HEAT

A system that can do work is said to possess *energy*. Einstein has indicated that the total energy is related to the mass of the system, but the energy changes observed in the systems considered here are equivalent to virtually unobservable changes in mass. The total energy possessed by a system is unknown, but when a system is transformed from some initial state to a final state the change in its energy can be observed. Energy residing in a system as a result of its position, composition, or condition, such as a lifted weight, an endothermic compound, or a compressed gas, is known as *potential* or stored energy. Potential energy is the product of force multiplied by distance. Energy residing in a system as a result of its motion is known as *kinetic* energy. Kinetic energy is the product of half the mass multiplied by the square of the velocity. The force that will impart to 1 g a velocity of 1 cm/sec is called a *dyne*. The energy produced by a dyne acting through 1 cm is called an *erg*. A *joule* is defined as 10^7 ergs. It is also a *watt-second*.

There are numerous forms of energy, such as mechanical, electrical, chemical, gravitational, centrifugal, and magnetic. The distribution of the energy of a system among these different forms varies according to the conditions imposed on the system. As the conditions change, the quantity of energy in a given form may change, but for any conceivable process the sum of all the energy changes is zero; that is, energy can neither be created nor destroyed, or

$$\Sigma\, dE = 0.$$

This principle, born of human experience, is the basis of the first law of thermodynamics.

When the environment exerts a constant pressure P on a system that expands against that pressure by an amount ΔV, the system does *work* W, defined by

$$W = P\,\Delta V.$$

If the external pressure varies from the initial state 1 to the final state 2, the work is then

$$W = \int_1^2 P\,dV. \qquad [1\text{-}6]$$

When work is done *by* the system on the surroundings, such work is considered to have a positive sign.

While superintending the boring of cannon at the Munich Arsenal in 1789, Benjamin Thompson (Count Rumford) observed the disappearance of mechanical energy on the generation of heat. By experimentation he found that the quantity of heat produced was proportional to the amount of energy expended. He announced that heat is a kind of motion, and whenever motion vanishes a proportional amount of heat appears in its place. He attempted to weigh heat, but concluded that all attempts to discover any effect of heat on the apparent weight of bodies were fruitless. In 1840 James Prescott Joule studied the conversion of mechanical energy into heat and found that a given amount of mechanical energy always produces the same quantity of heat.

When a system is isolated from its surroundings, it may have a different temperature from its surroundings. If the isolation is terminated, heat will flow from the hotter surroundings to the colder system, or vice versa, until both are at the same temperature. Energy transferred by virtue of a temperature difference is called *heat*, and the temperature change of a system is a measure of the heat added. However, a system may absorb heat from or release heat to its surroundings without undergoing a change in temperature; for instance, the system may undergo a physical or chemical change. Thus benzene may melt and absorb heat while its temperature remains constant at the melting point, 278.69°K. A chemical reaction may take place at a constant temperature while its heat of reaction is absorbed from or transferred to the surroundings.

This book follows the American practice of considering energy transferred from the system to its surroundings to be negative in sign. Although Clausius adopted this convention in 1850, many European thermochemists have regarded energy lost by the system as positive; consequently thermochemical literature requires careful analysis with respect to this factor.

VAPOR PRESSURE

Vapor pressure is the pressure existing in a system consisting of a vapor phase in dynamic equilibrium with a condensed phase. The condensed phase may be either solid or liquid (or both at the triple point). Vapor pressure is an important index to the behavior of the system and has been the subject of numerous measurements. Some of the various ways of measuring vapor pressure have been reviewed by Partington (1124). Considerable attention to the details of this measurement is required to produce reliable data. The vapor pressures of numerous organic compounds have been presented by Stull (1432), who listed selected values of the temperature for a series of pressures.

When a condensed phase is present in the system, the equation of state is more complex than that of a purely gaseous system. It is restricted to the saturation line and terminates at the critical point. Equations representing vapor pressure data have been reviewed by Thomson (1496). Because vapor pressure is related to enthalpies of transition, it is desirable to be able to interpolate measured vapor pressure data at temperatures of interest. Of the numerous mathematical formulas relating the temperature and pressure of the gas phase in equilibrium with the condensed phase, the equation proposed by Antoine (24) serves admirably:

$$\log_{10} P = A - \frac{B}{t+C}, \qquad [1\text{-}12]$$

in which P is the pressure, t is in degrees Celsius, and A, B, and C are constants. Experimental or tabulated values of vapor pressure data are often given at random temperatures, and the constants may be obtained by least squares fitting as described by Thomson (1496). If, as is often the situation, the data are too limited to justify such a procedure, the constant C may be estimated, and the remaining constants evaluated by least squares. Since the Antoine equation is nonlinear in the constants, this calculation is too complex for desk calculation. A simple "three-point method" will, however, serve for many purposes.

Three widely spaced valid points are selected from the data and are used to establish three equations to be solved simultaneously for A, B, and C. The task may be simplified by first converting the pressure values to their common logarithm and designating the three points as follows:

$$t_1, y_1, \qquad t_2, y_2, \qquad t_3, y_3, \qquad [1\text{-}13]$$

in which $y_i = \log P_i$. C can be obtained by direct substitution into the

By definition, the quantity of energy a system contains is related to the mass and temperature of the system. Early calorimetrists measured heat by absorbing it in a fluid and observing the temperature rise. This led to the original definition of the calorie: the quantity of heat required to raise the temperature of 1 g of water 1°C. Later developments disclosed the fact that the size of the calorie depends on the temperature at which it is measured. Modern calorimetry is based on duplicating the temperature rise produced chemically with a measured amount of electro-mechanical work (that is, by a measured electrical current flowing for a measured period of time). An excellent exposition of the thermodynamics of modern calorimetry has been made by McGlashan (976). The Ninth General Conference on Weights and Measures of 1948 (1423) adopted the absolute joule (10^7 ergs) as the basic energy unit, with 1 cal defined equal to 4.1840 absolute joules. One absolute joule is obtained when one absolute ampere flows through one absolute ohm for one mean solar second. (By Ohm's law there will be one absolute volt of potential across the above resistor.)

For systems composed of organic compounds, the energy increment E of a process or reaction from state 1 to state 2 may be expressed as a sum of heat Q and work W, provided that other factors, such as surface energy and magnetic or gravitational effects, may be neglected. Thus the first law may be postulated.

$$\Delta E = E_2 - E_1 = Q - W. \qquad [1\text{-}7]$$

ENTHALPY

For a transformation taking place at constant pressure, the work according to [1-6] is

$$P(V_2 - V_1) = PV_2 - PV_1.$$

Therefore

$$Q = E_2 - E_1 + (PV_2 - PV_1) = (E_2 + PV_2) - (E_1 + PV_1). \qquad [1\text{-}8]$$

For convenience we define enthalpy H as

$$H = E + PV. \qquad [1\text{-}9]$$

Equation [1-8] may then be simplified to

$$Q = H_2 - H_1 = \Delta H, \qquad [1\text{-}10]$$

in which ΔH, often loosely designated the "heat" of reaction, is more appropriately called the *enthalpy increment*. Consequently it is apparent that a purely chemical process at a constant volume (which therefore does no PV work) has in its heat, Q, an exact measure of the energy increment, ΔE, as shown by [1-8]. The Q for a process or reaction that takes place at constant pressure provides a direct measure of the enthalpy increment, ΔH, by [1-10].

REFERENCE STATES

Although it is impractical to endeavor to measure the absolute value of enthalpy or energy of a substance, differences in the enthalpy between two temperatures can be measured, for example,

$$H_{298} - H_0 = \int_0^{298} dH = \Delta H \Big|_0^{298}, \qquad [1\text{-}11]$$

and the value of the integral can be calculated, but H_0 remains unknown. Although the zero point energy or enthalpy is important for some applications, it is unimportant to the thermochemist. There is need for a datum level of enthalpy on which to base practical calculations. In view of the lack of heat capacity data below room temperature for many substances, it is convenient to select 298.15°K for this arbitrary datum zero. The practice of using subscript numerals (as in [1-11]) to designate the temperatures pertaining to the states involved is used.

In discussing processes and reactions it is important that the state of the material be clearly described. A material in its *reference state* is in the state agreed on for discussion and no other. All other states may be compared with the reference state. For elements the reference state is the condensed state up to the temperature at which the vapor pressure of the element reaches 1 atm, and the ideal gas state above this temperature. Reference states for the elements are given in Chapter 8. There is a decided lack of enthalpy data on organic liquids, whereas those on organic gases are much better known. In view of this and the additional fact that organic compounds reach 1 atm total pressure by 500°K or less, much confusion can be avoided by selecting the ideal gas state as the reference state for all organic compounds. In a few cases, such as for compounds that decompose before attaining an appreciable vapor pressure, e.g., sugar, a condensed state is used and noted. With these exceptions the reference state of an organic compound in this book is the ideal gas state at 1 atm total pressure. The superscript ° indicates materials in their reference state.

following equation:

$$\left(\frac{y_3-y_2}{y_2-y_1}\right)\left(\frac{t_2-t_1}{t_3-t_2}\right) = 1 - \left(\frac{t_3-t_1}{t_3+C}\right). \qquad [1\text{-}14]$$

B is next obtained from

$$B = \left(\frac{y_3-y_1}{t_3-t_1}\right)(t_1+C)(t_3+C), \qquad [1\text{-}15]$$

and, finally, A is found readily from

$$A = y_2 + \frac{B}{t_2+C}, \qquad [1\text{-}16]$$

Pennington, Scott, Finke, McCullough, Messerly, Hossenlopp, and Waddington (1136) have measured the vapor pressure of 1-propanethiol from 24° to 102°C. From three points,

Number	t (°C)	P (mm)	$y = \log P$
1	24.28	149.41	2.17438
2	62.14	633.99	2.80208
3	102.09	2026.0	3.30664,

[1-14], [1-15], and [1-16] are used to compute the Antoine constants:

$$\left(\frac{3.30664-2.80208}{2.80208-2.17438}\right)\left(\frac{62.14-24.28}{102.09-62.14}\right) = 1 - \left(\frac{102.09-24.28}{102.09+C}\right).$$

Therefore

$$C = 223.48$$

$$B = \left(\frac{3.30664-2.17438}{102.09-24.28}\right)(24.28+223.48)(102.09+223.48)$$

$$= 1173.81$$

$$A = 2.80208 + \frac{1173.81}{62.14+223.48} = 6.91177.$$

To calculate the temperature of the normal boiling point, [1-12] may be rearranged to

$$t = \left(\frac{B}{A-\log_{10}P}\right) - C$$

and at 760 mm pressure

$$t = \left(\frac{1173.81}{6.91177 - 2.88081}\right) - 223.48 = 67.72°C.$$

Their *observed* normal boiling point is 67.719°C.

CHAPTER 2

HEAT CAPACITY AND ENTHALPY OF TRANSITION

HEAT CAPACITY

If a quantity of heat Q is imparted to a mole of substance to increase its temperature from T_1 to T_2, the mean heat capacity $\bar{C}$ for the temperature interval is

$$\bar{C} = \frac{Q}{T_2 - T_1} = \frac{Q}{\Delta T},$$

whereas the heat capacity C is defined by the limit

$$\lim_{\Delta T \to 0} \frac{Q}{\Delta T} = \frac{\delta Q}{dT} = C.$$

If this process takes place at constant volume, no expansion work is done and the infinitesimal quantity of heat δQ is equal to the infinitesimal energy increment dE; therefore

$$\left(\frac{\partial E}{\partial T}\right)_V = Cv. \qquad [2\text{-}1]$$

Cv is defined as the *heat capacity at constant volume*.

For heat addition at constant pressure, increase of temperature also usually causes a change in volume; consequently PV work is done; on application of the first law,

$$\delta Q = dE + P\,dV,$$

and the heat capacity at constant pressure is given by the expression

$$\left(\frac{\partial E}{\partial T}\right)_P + P\left(\frac{\partial V}{\partial T}\right)_P = Cp. \qquad [2\text{-}2]$$

But noting that for constant pressure

$$P\left(\frac{\partial V}{\partial T}\right)_P = \left(\frac{\partial (PV)}{\partial T}\right)_P,$$

[2-2] becomes

$$\left(\frac{\partial (E+PV)}{\partial T}\right)_P = \left(\frac{\partial H}{\partial T}\right)_P = Cp. \qquad [2\text{-}3]$$

Cp is simply the *heat capacity at constant pressure*. The difference between Cp and Cv may be calculated by subtracting [2-1] from [2-3]:

$$Cp - Cv = \left(\frac{\partial H}{\partial T}\right)_P - \left(\frac{\partial E}{\partial T}\right)_V$$

$$= \left[\left(\frac{\partial E}{\partial T}\right)_P + \left(\frac{\partial (PV)}{\partial T}\right)_P\right] - \left(\frac{\partial E}{\partial T}\right)_V.$$

Since

$$\left(\frac{\partial E}{\partial T}\right)_P = \left(\frac{\partial E}{\partial T}\right)_V + \left(\frac{\partial E}{\partial V}\right)_T \left(\frac{\partial V}{\partial T}\right)_P,$$

$$Cp - Cv = \left[P + \left(\frac{\partial E}{\partial V}\right)_T\right] \left(\frac{\partial V}{\partial T}\right)_P. \qquad [2\text{-}4]$$

For an ideal gas

$$\left(\frac{\partial E}{\partial V}\right)_T = 0 \quad \text{and} \quad \left(\frac{\partial V}{\partial T}\right)_P = \frac{R}{P};$$

consequently

$$Cp - Cv = R. \qquad [2\text{-}5]$$

Because theoretical expressions for heat capacities of condensed phases are generally derived from considerations of the energy of the system, Cv is involved. On the other hand, measurements of the heat capacity are usually carried out under conditions approximating constant pressure, thus giving values of Cp. Most chemical processes and reactions are performed under constraint of constant pressure rather than of constant volume; as a result the chemist usually employs heat capacities at constant pressure.

Division of the heat capacity of a substance by its gram molecular weight M yields the corresponding *specific heat* c of the substance; for example, $Cp/M = cp$.

Direct experimental determinations are the primary source of heat capacity values, but theoretical methods are also applied in favorable cases for the evaluation of thermal data of organic substances. Consequently several of the most important aspects of both approaches are described in the following sections.

Experimental Determination of Low-Temperature Heat Capacities of Condensed Phases

Although measurements of the heat capacity at low temperatures had been roughly made for half a century, it remained for Nernst and his collaborators in 1910 (1055) to develop and exploit the first successful low-temperature vacuum calorimeter. Since that time there has been a steady improvement in low-temperature calorimetry. The adiabatic low-temperature calorimetric cryostat designed and constructed by Westrum (1594) is typical of modern, high-precision apparatus operated either completely manually or with automatic adiabatic shield control. Since many heat capacities of organic substances have been obtained with such a cryostat, it has been selected for a detailed portrayal. It is a modernization of the Ruehrwein and Huffman (1264) design of the prototype version of Nernst (1054).

The cryostat is shown schematically in Figure 2-1. Here A and B represent chromium-plated copper tanks for liquid nitrogen and liquid helium, respectively, which serve as thermal sinks. C, D, and E are thin-walled, chromium-plated copper radiation shields, which serve not only to conserve refrigerants, but also to generate zones of uniform and progressively lower temperature. For studies above 90°K liquid nitrogen is used as refrigerant in both tanks; temperatures as low as 50°K may be achieved by evacuating tank B and solidifying the nitrogen in it. For the range 4–50°K liquid helium is used as refrigerant in tank B. Temperatures approaching 1°K may be achieved by pumping on the liquid helium. Suitable modification in the fabrication of the apparatus will permit its operation up to 600°K. Thin-walled stainless steel tubes attached to the tanks serve for filling, venting, and pumping as well as for support. The cover plate is readily sealed to the outer brass vacuum can by an O-ring gasket (G) inside the bolt circle. A high-speed oil diffusion pump connected at M is backed by a mechanical vacuum pump for evacuating the can.

A light-weight, chromium-plated, copper adiabatic shield (F) surrounds the calorimeter. The ends and cylindrical portion of the adiabatic shield are each provided with a bifilarly wound constantan heater covering the entire outer surface. A 50-cm shallow helical groove around the shield permits equilibration of the bundle of lead wires under the heater

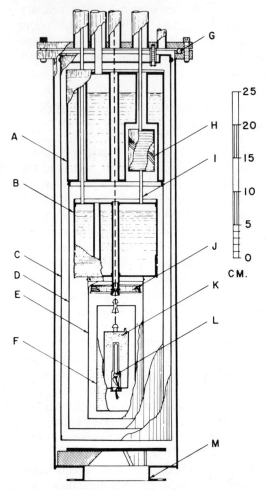

Figure 2-1 Cross-sectional schematic diagram of the Mark II cryostat: A, liquid nitrogen tank; B, liquid helium tank; C, D, and E, radiation shields; F, adiabatic shield; G, O-ring gasket sealing brass vacuum jacket; H, effluent helium vapor exchanger ("economizer"); I, helium exit tube; J, ring for adjusting temperature of leads; K, calorimeter assembly; L, platinum resistance thermometer; M, connection to vacuum diffusion pump. From Westrum (1594).

windings. The wires are all continuous to the outside and are thermally anchored by baked Formvar enamel to the nitrogen tank, the economizer (H) (which serves as a heat exchanger by using the heat capacity of cold effluent helium gas to absorb much of the heat conducted down the leads and thus conserve liquid helium), the helium tank, and the ring (J). The temperature of the lead bundle is tempered by a ring and finally adjusted to the temperature of the calorimeter by the adiabatic shield, which is maintained within $\pm 0.002°K$ of the calorimeter temperature. Copper-constantan thermocouples monitor the difference in temperature between calorimeter and shield and between shield and ring, and actuate three separate channels of recording electronic circuitry (corresponding to the three portions of the adiabatic shield) provided with proportional, rate, and reset control actions for establishment of adiabaticity. Alternatively, manual monitoring and control of the adiabatic shield may be employed.

The cryostat thus provides a way of maintaining a sample of chemical substance within a calorimeter vessel at any desired temperature under such conditions that no heat is exchanged with the environment except that introduced by an electrical heater during the measurement. The calorimeter vessel containing the sample under investigation is suspended in isolation by a braided silk line within the adiabatic shield. The calorimeter and adiabatic shield may be mechanically brought into direct thermal contact through carefully machined cones with a refrigerant tank, and thus be cooled to the desired operating temperature and again isolated by lowering it. This system of cooling the calorimeter does not require a transfer gas and permits cycling of the sample temperature in the neighborhood of a transition region to produce alternate cooling and heating curves at will.

The calorimeter assembly (K) is composed of a sample container typically of gold-plated copper and of size varying from 2 to 100 cc, depending on the heat capacity and availability of the substance to be studied. Radial vanes plus a small pressure of helium gas in the sample space tend to promote rapid thermal equilibration after energy input.

The equilibrium temperature of the sample is determined with a platinum-encapsulated, platinum-resistance thermometer (L) mounted axially within the calorimeter vessel. As the temperature of the calorimeter is increased by electrical energy input to the heating element installed within the vessel, that of the adiabatic shield is made to follow it so closely that no unmeasured energy enters or leaves the calorimeter. After the input the new equilibrium temperature is determined. Under normal conditions for typical substances several such runs may be made per hour. Especially during phase and other transformations, equilibration periods of many hours may occasionally be required. Here the precise automatic control of the adiabatic shield is indispensable in making reliable heat capacity determinations. The heat capacity of the sample may then be calculated from its mass, the observed temperature increment, the measured energy input, and the previously determined heat capacity of the empty calorimeter-heater-thermometer system. A precision of the order of a few hundredths of 1% may be obtained over most of the temperature range. Accuracy is assured by ultimately referring all determinations of mass, time, temperature, resistance, and potential to calibrations performed by the National Bureau of Standards and by the measurement of heat capacity standards provided by the Calorimetry Conference (487).

Careful error analysis, together with the study of standard samples and interlaboratory comparisons, indicates that the heat capacities obtained with such a calorimetric cryostat will provide thermodynamic functions such as entropies with a probable error of $\pm 0.1\%$ above $100°K$.

An extensive discussion of the current development of adiabatic cryogenic calorimetry has been developed by Westrum, Furukawa, and McCullough (1597), and a corresponding treatment of the isothermal cryogenic approach has been made by Stout (1427).

Useful calorimetric apparatuses for measurement of specific heats at room temperatures and above have been described by Williams (1614), Corruccini and Ginnings (274), Aston, Fink, Tooke, and Cines (36), Stow and Elliott (1429), West and Ginnings (1591), and Westrum and Trowbridge (1599). A summary of adiabatic calorimetry in the intermediate temperature range has been made by West and Westrum (1592).

Even in the intermediate temperature range, $300-700°K$, the adiabatic method of calorimetry has many advantages over the "method of mixtures" (452) for determination of thermal properties of organic substances, which often involve metastable states and irreversible transitions on heating or do not give a thermodynamically reproducible state on quenching. With automatic adiabatic shield control, the study of transitions is possible even when thermal equilibration and hysteresis require observations over periods of many hours. A convenient adiabatic calorimetric system, described by Westrum and Trowbridge (1599) as suitable for precise measurements of the heat capacities of condensed phases together with the associated enthalpies of transition and fusion from $300°$ to $600°K$, is depicted in Figure 2-2. This device, involving isolation of the calorimeter within reflective chromium-plated radiation shields, is similar in principle and operation to the adiabatic cryogenic calorimeter already described. The calorimeter is machined from silver and contains an axial well with a $250\text{-}\Omega$ heater and an encapsulated platinum resistance thermometer, held firmly within the tapered bore of the heater sleeve by a beryllium copper collet. The outer surface of the heater sleeve is also tapered and snugly clamped into the conical well (G) by a threaded section. Six vertical vanes machined as an integral part of the well assembly aid in the establishment of thermal equilibrium. To facilitate

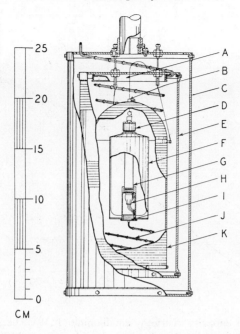

Figure 2-2 Cross-sectional schematic diagram of an intermediate-range adiabatic calorimeter (1599): A, guard shield ring; B, calorimeter suspension collar; C, primary radiation shield; D, calorimeter closure assembly; E, guard shield; F, calorimeter vessel; G, thermometer heater assembly; H, thermometer; I, thermal equilibration spool; J, lead bundle; K, adiabatic shield.

loading of samples, a screw-type closure (D), in which a sharp circular edge is forced against an annealed gold gasket, provides a vacuum-tight seal. This closure is effected while the calorimeter is within a vacuum system, thus permitting addition of helium gas to the sample space to facilitate establishment of thermal equilibrium. With this apparatus measurements are made on an intermittent basis, with time allowed for the establishment of thermal equilibrium after each input of electrical energy.

As an indication of the precision attained by this type of calorimeter, data on triethylenediamine (1,4-diazabicyclo[2,2,2]octane) obtained by Chang and Westrum (210) over the low-temperature region and by Trowbridge and Westrum (1519) into the liquid region are shown in Figure 2-3. In this substance, composed of globular molecules, the low-temperature crystal II state transforms to the plastically crystalline (crystal I) state at 351.08°K with an enthalpy of transition of 2524.0 cal/mole, and melts at 432.98°K with an enthalpy increment of 1776.5 cal/mole.

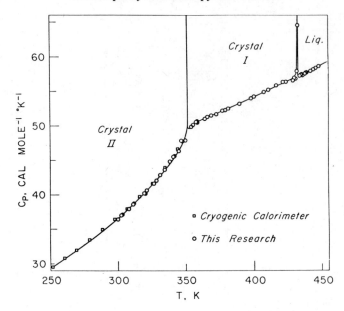

Figure 2-3 Heat capacity of triethylenediamine (1519). The squares represent the data taken with an adiabatic cryogenic calorimeter (210), and the circles, the data taken with the intermediate-range calorimeter shown in Figure 2-2.

The similarity of crystal I and liquid phases is unusual and suggests that the plastically crystalline phase already possesses most of the degrees of freedom of the liquid.

Few studies have been reported on organic molecular crystals at elevated temperature, in part because of their relatively low melting points and their tendency to decompose or isomerize. The technique of calorimetry in the range above about 700°K has been reviewed by Douglas and King (338).

Theoretical Evaluation of the Heat Capacities of Solids

In 1907 Einstein (375) treated a monatomic crystalline solid of n atoms as a system of $3n$ independent oscillators vibrating in three mutually perpendicular directions with a single characteristic simple harmonic frequency v. He derived the heat capacity at constant volume of such a system as

$$Cv = \frac{3Ru^2 e^u}{(e^u - 1)^2}, \qquad [2\text{-}6]$$

where the quantity $u = hv/kT = \theta_E/T$. Here h is Planck's constant and

Heat Capacity

k is Boltzmann's constant or the gas constant per atom. θ_E is a characteristic constant (the Einstein temperature) for a given substance. Equation [2-6] gives $Cv = 0$ when $T = 0$ and reaches a maximum value of $3R$ when $T = \infty$.

About 1910 Nernst and his collaborators measured heat capacities at very low temperatures and reported that the heat capacity did indeed approach zero at 0°K but deviated rather significantly from the temperature dependence predicted by [2-6].

In 1912 Debye (315) and, independently, Born and von Kármán (147) sought to derive a theoretical expression in better accord with measured low-temperature heat capacities. Debye considered the monatomic isotropic crystalline state to be an elastic continuum having a spectrum of $3n$ oscillations with a characteristic distribution terminating at a maximum frequency, ν_{max}. By applying vibration theory to the problem, Debye derived the heat capacity of a continuous elastic solid at constant volume

$$Cv = 9R\left[\frac{4}{x^3}\int_0^x \frac{u^3}{e^u - 1}du - \frac{x}{e^x - 1}\right], \qquad [2\text{-}7]$$

where $x = h\nu_{max}/kT = \theta_D/T$ and $u = h\nu/kT$. Equation [2-7] may also be expressed in the compact form

$$Cv = f_D\left(\frac{T}{\theta_D}\right), \qquad [2\text{-}8]$$

where f_D is the Debye heat capacity function applicable to all substances. The parameter θ_D is a characteristic constant (the Debye temperature) for a given substance. The Debye function closely represents the experimental heat capacity data for a large number of simple substances but is limited to simple molecules, for which the required numerical values of the vibrational frequencies are available. As yet, incomplete data concerning the vibrational frequencies of polyatomic molecules prevent calculation of their crystalline heat capacities from first principles.

Both the Einstein heat capacity equation [2-6] and the Debye heat capacity equation [2-7] contain the universal gas constant R and were developed for monatomic solids vibrating in three dimensions or having three degrees of vibrational freedom. Tables A-2 and A-5 in the Appendix contain tabulated values of the Einstein and Debye heat capacity functions divided by R for one degree of freedom. These tables are not invalidated by changing experimental values of R and may be multiplied by the number of degrees of freedom required by a given problem. Figure 2-4 shows a comparison of these two functions for various characteristic vibrational frequencies.

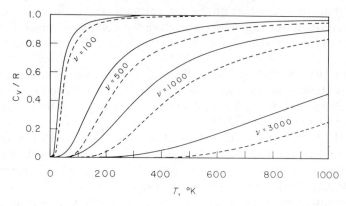

Figure 2-4 Comparison of Debye (solid lines) and Einstein (dashed lines) functions for characteristic frequencies.

Since the difference between Cp and Cv for a molecular solid becomes negligible in the vicinity of 30°K, the Debye heat capacity function can be used to extrapolate measured values of Cp to absolute zero. Kelley and King (738) discussed the extrapolation of observed Cp values for inorganic compounds and presented a trial-and-error method using an appropriate combination of Einstein and Debye functions for the purpose. For organic compounds many workers have obtained an adequate extrapolation using Debye functions only. Putnam and Kilpatrick (1208) measured the low-temperature heat capacity of 1,2,4-trimethylbenzene and reported the data given in Table 2-1. By trial-and-error they found

Table 2-1 Extrapolation of Measured Heat Capacity to 0°K for 1,2,4-Trimethylbenzene[a]

$T(°K)$	Experimental	$5D\left(\dfrac{103\cdot 7}{T}\right)$	$3D\left(\dfrac{80\cdot 5}{T}\right) + 6E\left(\dfrac{155}{T}\right)$
14.77	1.901	1.883	1.934
15.80	2.188	2.190	2.194
16.96	2.518	2.541	2.491
18.40	2.914	2.977	2.863
20.36	3.472	3.556	3.382
22.82	4.165	4.234	4.050
25.46	4.900	4.890	4.771
28.16	5.626	5.483	5.577
30.91	6.360	6.002	6.371

[a] Data of Putnam and Kilpatrick (1208), in cal/(mole °K).

that, their measured values could be extrapolated to 0°K by assuming five degrees of freedom and $\theta_D = 103.7$ on fitting to their first six values. Such empirical use of the Debye function with more than three degrees of freedom is commonplace in the literature. Alternatively, the extrapolation may be made by using a Debye function with three degrees of freedom for the lattice vibrations, together with the appropriate number of Einstein functions to represent the vibrational and torsional oscillation of the molecules. Hence the data of Putnam and Kilpatrick may also be fitted by

$$3D\left(\frac{80.5}{T}\right) + 6E\left(\frac{155}{T}\right).$$

This combination of Debye and Einstein functions will adequately fit all the measured values up through 30.91°K, as may be seen in Figure 2-5. Such an approach has been computerized by Phillips and Klimpel (1142).

Alternatively, especially for more recent data extending below 10°K, the extrapolation may be performed by use of a plot of Cp/T versus T^2 as described by Pitzer and Brewer (1158) and detailed further by Westrum, Furukawa, and McCullough (1597).

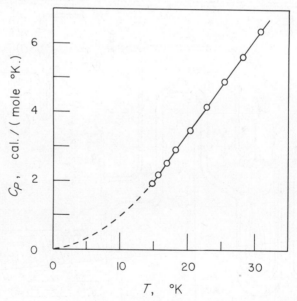

Figure 2-5 Extrapolation of the heat capacity of 1,2,4-trimethylbenzene. Data (circles) from Putnam and Kilpatrick (1208) and combined Einstein and Debye functions (solid and dashed curve). See the text and Table 2-1.

Heat Capacities of Liquids

Since there is no adequate theory of the liquid state, there is no satisfactory method of calculating the heat capacity of a liquid. Theoretical discussions tend to be somewhat contradictory [cf. Moelwyn-Hughes (1017, 1018) and Rowlinson (1257)]. Because there are more degrees of freedom in the liquid state than in the solid state, the liquid heat capacity tends to be somewhat higher than the heat capacity of the solid. However, direct measurements must be relied on for definitive values.

Experimental Determination of Heat Capacities of Gases and Enthalpies of Vaporization

An apparatus designed to determine both enthalpy of vaporization of a liquid and vapor heat capacity has been described by Waddington, Todd, and Huffman (1552) and is shown in a somewhat simplified representation in Figure 2-6. A liquid sample 150–250 cc in volume is boiled steadily

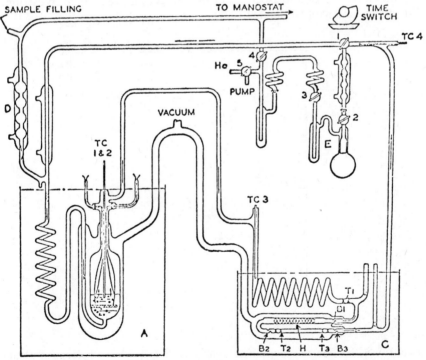

Figure 2-6 Apparatus for measurement of enthalpy of vaporization and vapor heat capacity. From Waddington, Todd, and Huffman (1552).

in the vacuum-jacketed vaporizer (A) by electrical energy from storage batteries. The heater is wound on a Pyrex spiral fused to the inner wall of the vaporizer and contains current and potential leads covered with glass, reaching from below the liquid level to the exterior. Electrically heated flow lines prevent condensation, and the vapor, after passing through the gas calorimeter (C), is condensed, heated to bath temperature, and returned to the vaporizer. Maintenance of the vaporizer bath about 0.2°K above that of the boiling liquid and constant within ±0.003°K makes heat flow negligible in the silvered and evacuated vaporizer jacket. Thermocouples (TC_1 and TC_2) are used to measure the temperature of the boiling liquid and the vapor. The remainder of the system is filled with helium, the pressure of which is controlled to within ±0.1 mm with a carefully constructed manostat.

The flow calorimeter (C) is located in a second bath controlled within ±0.002°K. The heating element on the flow line from the vaporizer is adjusted so that a thermocouple (TC_3) indicates a vapor temperature within ±2.0°K of bath temperature. After passing through a glass helix, the vapor passes over a platinum resistance thermometer (T_1), where the vapor temperature is compared with the bath temperature measured by another platinum resistance thermometer (not shown). Next the vapor enters the calorimeter, made of a thin-walled, Pyrex U-tube in a silvered and evacuated jacket, and passes over the heater (H) and two platinum resistance thermometers (T_2 and T_3). Current and potential leads from the heater leave the system against the gas flow, while similar leads from the two thermometers leave the system with the gas flow. Radiation losses are minimized by baffles (B_1 and B_3), while another baffle (B_2) promotes mixing subsequent to the heating of the vapor. A heated exit vapor line conducts the vapor to a three-way stopcock (1), the temperature of which is measured by TC_1 and maintained about 20°K above the boiling point. This stopcock may recycle the vapor to a boiler (A), or it may direct the vapor to a condenser and a double-trap sample-collecting system. The plug of stopcock 1 is attached to a stainless steel sector, which dips into a pool of mercury only for the period during which vapor flows into the trap, and thereby completes an electrical timing circuit. By this device the sampling interval of 5 or 6 min is measured with an estimated accuracy of ±0.05 sec. The traps are removable for determination of the weight of sample collected.

The five platinum resistance thermometers are of the coiled filament, strain-free type (995) and are calibrated by comparison with another thermometer of this type, calibrated and certified by the National Bureau of Standards. A White double potentiometer with associated calibrated resistors, standard cell, and high-sensitivity galvanometer is employed to measure the currents through the heaters and the potentials across them.

By weighing a sample trapped during a timed interval, the flow rate is measured. The measured values of the current and voltage in the heater during this interval of time give the electrical energy delivered to the vaporizer. It is necessary to make four corrections to obtain the energy actually used in vaporizing the sample: (a) a 0–20 cal adjustment for variation in the boiling temperature by $\pm 0.05°K$, as indicated by TC_2, (b) a 10–35 cal adjustment for variation in the temperature gradient between heater and liquid surface in the boiler, (c) an adjustment of about 25 cal for vaporization by heat developed in the current leads beyond the junctions of the potential leads, and (d) an adjustment for vaporization into the additional vapor space provided by liquid removed from the boiler. Establishment of the proportionality between power input and mass rate of flow at a given boiling temperature permits use of boiler power input to indicate the flow rate in the heat capacity measurements that follow. At each temperature this proportionality is independent of flow rate and of the duration of the flow measurement.

Observations of the temperature produced in the gas calorimeter by measured power inputs to the calorimetric heater are made at four or more flow rates to measure the apparent heat capacities of the gas sample. Independent values of the apparent heat capacity at each flow rate are obtained from resistance thermometers T_2 and T_3. Plotting the reciprocal of the flow rate against the apparent heat capacity gives a linear relationship within about $\pm 0.05\%$ (see Figure 2-7). Extrapolation of the reciprocal to zero value corrects for the heat losses, which even at the highest temperature and lowest flow rate are less than 2%.

With vapor flowing the temperatures of resistance thermometers T_1, T_2, and T_3 are measured with heater H off; then measured power is supplied to H, and after a steady state has been reached the temperatures of resistance thermometers T_1, T_2, and T_3 are again measured. Slight irregularities in the flow rate cause the temperatures to vary from ± 0.003 to $\pm 0.008°$. Temperature increments (for instance, $T_{2,\text{final}} - T_{2,\text{initial}}$) for each power input are observed on each thermometer, and tend to minimize errors from differences in the thermometer calibration, as well as Joule-Thomson cooling caused by the pressure drop through the calorimeter. Adequate adjustment is applied for the temperature coefficient of the Joule-Thomson cooling over the 8° temperature increment used. This effect is about 0.03% of the total. If the mean temperature of a series of measurements is the same, experiments demonstrate that the apparent heat capacities are independent of the magnitude of the ΔT used. Under experimental conditions the pressure drop between the vaporizer and the calorimeter varies from 1 to 35 mm and attains its maximum

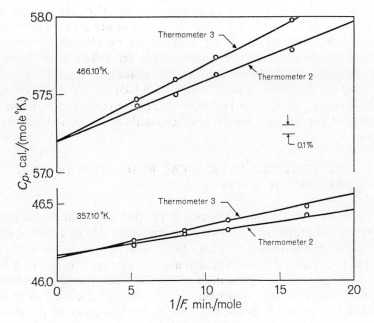

Figure 2-7 The apparent heat capacity of heptane as a function of flow rate. From Waddington, Todd, and Huffman (1552).

value for the highest flow rate and the lowest operating pressure. Correction [which does not exceed 0.1% and which involves a measured value of the quantity $(\Delta C p/\Delta P)_T$] is applied for this pressure drop. Values of the heat capacity identical to within 0.05% on the average are obtained using ΔT's based on either thermometer T_2 or T_3.

Table 2-2 Heat Capacity of Gaseous Heptane[a]

$T(°K) \rightarrow$ $P(\text{mm})$ $\downarrow$	357.10	373.15	400.40	434.35	466.10
195.5	46.140	47.805	50.555	53.965	57.075
391.2	46.515	48.095			
597.7		48.410	50.930	54.200	57.215
760.0	47.22	48.66	51.08	54.30	57.27
0.0	45.77	47.51	50.37	53.85	57.00

[a] Data of Waddington, Todd, and Huffman (1552), in cal/(mole °K).

Correction of Cp at 1 atm to Cp at zero pressure. $Cp°$ (that is, to the ideal vapor state), may be obtained from the experimental data at two or more pressures representing the same temperature by plotting Cp versus P and extrapolating to zero pressure. This procedure gives $Cp°$ and $(\Delta Cp/\Delta P)_T$ and serves to define an equation of state.

Typical experimental gaseous heat capacity data on n-heptane obtained by Waddington, Todd, and Huffman (1552) are presented in Table 2-2. Enthalpy of vaporization results are discussed and correlated later in this chapter.

THE STATISTICO-MECHANICAL EVALUATION OF HEAT CAPACITY OF A GAS

The function of statistical mechanics in providing a theoretical approach to the thermal properties of gases is analogous to that of actuarial statistics. Although the longevity of an individual cannot be foretold, it is possible to predict in a statistical sense the average lifespan of a large number of individuals. Application of the powerful statistico-mechanical technique requires, first, information as to whether classical or quantal mechanics is required to describe the various *degrees of freedom* or characteristic modes of partitioning energy of molecules, and, second, a means of averaging or computing the distribution of the energy among these modes. Although quantal mechanics provides the best description of the energetics of molecules, if the spacing of the discrete energy levels of molecules is small compared to kT, the use of classical mechanics provides an adequate basis for the evaluation of thermodynamic properties in several instances when these energy levels are fully excited. Fortunately, it is possible to simplify the treatment by taking advantage of the fact that the coordinates and momenta of the various degrees of freedom are frequently separable from the others, and by idealizing our model of the molecule in the same fashion characteristic of the mechanistic approach of the kinetic theory of gases.

Use may thus be made of the classical mechanical principle of *equipartition of energy*, which may be formulated as: The average energy associated with each single variable, whether coordinate or momentum, that contributes a *quadratic term* (or *squared term*) to the energy is $RT/2$ per mole. This conclusion is rigorously derived in treatments of statistical mechanics [e.g., Aston (30), Tolman (1513), Mayer and Mayer (942), or Fowler and Guggenheim (436)]. The heat capacity contribution by differentiation is

$$Cv = \left(\frac{\partial E}{\partial T}\right)_V = \frac{R}{2}. \qquad [2\text{-}9]$$

The Statistico-Mechanical Evaluation of Heat Capacity of Gas

The kinetic energy component of the translational motion of a molecule in each coordinate direction may be expressed in the familiar form, $E_{tr}/n = 1/2\, m(v_i)^2$, or, alternatively, as a quadratic function of the corresponding momentum $p_i = mv_i$, for example $E_R/n = (p_i)^2/2m$; hence the mean molal value of the kinetic energy in that direction is $RT/2$. Since the degrees of freedom of the monatomic gaseous molecule correspond to the number of coordinates needing specification to locate the center of mass of the molecule in space (x, y, and z in Cartesian coordinates or r, θ, and ϕ in polar coordinates), they are three in number and are totally kinetic in nature. Consequently the molal energy is $3RT/2$ and the molal heat capacity $Cv = 3R/2$. The heat capacity at constant pressure for such an ideal gas is thus

$$Cp = \frac{3R}{2} + R = \frac{5R}{2}.$$

These theoretical heat capacities are in accord with those observed experimentally to within a few degrees of 0°K for helium and thus verify the validity of the classical approach above the temperatures at which quantal effects become manifest.

Rotational energy is also entirely kinetic in character; that is,

$$E_{or} = \frac{(p_{or})^2}{2I},$$

in which p_{or} is the angular momentum and I the moment of inertia about the axis of rotation. The subscript "or" represents over-all rotation. Because the energy of rotation is also a quadratic function of the momentum, each type of rotation of the entire molecule will also contribute $RT/2$ cal/mole to the total energy.

The energy of a simple harmonic oscillator, however, includes the sum of two squared terms, one involving the momentum p_x, and the other the related displacement coordinate x; that is,

$$E_{vib} = \frac{(p_x)^2}{2m} + \frac{fx^2}{2},$$

in which f is a force constant. The first term thus involves vibrational kinetic energy, and the second term is the potential energy. Hence each vibrational mode or degree of freedom contributes twice $RT/2$, or RT, per mole to the total energy.

For a nonlinear polyatomic molecule containing n atoms, the total number of degrees of freedom required to localize the molecule will be $3n$. Three of these, corresponding to motion of the center of gravity of the molecule, are translational; three, corresponding to orientation of the molecule as a whole, are rotational; and the remainder, $3n-6$, are vibrational. Hence the total heat capacity at constant volume according to the equipartition principle is $(3n-3)R$. For a linear molecule one less rotational degree of freedom and one more (i.e., $3n-5$) vibrational degree of freedom are involved, corresponding to a heat capacity at constant volume of $(3n-5/2)R$, or $7R/2$, for a diatomic molecule. Heat capacity data on gases reveal that these values are achieved only at high temperatures; at lower temperatures smaller values are obtained. Neglect of the quantization of energy is the prime reason for the deviation; this is especially serious for the vibrational energies. In addition it should be remarked that deviations from exact quadratic functional form for the energy must inevitably vitiate the principle of equipartition even at high temperatures. For example, oscillations that are not strictly harmonic in character will also cause deviations. Finally, the presence of partially restricted internal rotation must also be taken into account.

The procedure for evaluation of the heat capacities is revealed by a series of examples for increasingly more complex molecules. The use of quantal evaluation of the vibrational contribution from spectral data will be introduced as necessary to give values in accord with experiment. A diatomic molecule, shown in Figure 2-8, behaves like two mass points held together with springlike forces. Each atom has three degrees of freedom, or a total of six degrees for the diatomic molecule. Since the molecule (idealized to rigidity) has three degrees of freedom of translation, each of its three translational contributions to the heat capacity is the same as given above in [2-9]. In addition, the molecule can rotate as a whole about the y and z axes, but the rotational inertia along the x axis is negligibly small. Thus the diatomic molecule has two degrees of rotational freedom.

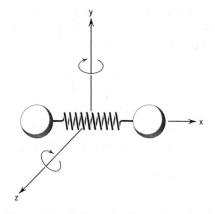

Figure 2-8 Schematic diatomic molecule.

The remaining degree of freedom is one of vibration along the x axis. Application of the quantal theory to this problem led to an expression for the contribution of a vibrational mode to the heat capacity having the same form as that discussed earlier for solids, that is, [2-6], except that now only a single vibrational mode is considered at a time:

$$C_{\text{vib}} = \frac{u^2 e^u}{(e^u - 1)^2} R, \qquad [2\text{-}10]$$

in which $u = h\nu/kT = hc\omega/kT = 1.4386\omega/T$. It is customary to express frequency (units, sec^{-1}) in terms of wave numbers, and ω (i.e., ν/c), in cm^{-1}. Tabulation of $C°/R$ contributions of simple harmonic vibrator frequencies (in cm^{-1}) calculated from the Einstein heat capacity equation [2-6] is given in Appendix Table A-2.

As a further illustration, the heat capacity of gaseous methyl bromide is calculated. The structure of this pentatomic, nonlinear molecule, CH_3Br, is represented in Figure 2-9. The three translational and three rotational degrees of freedom each contribute $3R/2$ cal/(mole °K) to the heat capacity at constant volume, and the remaining $3 \times 5 - 6 = 9$ degrees of vibrational freedom are evaluated from spectral data. Weissman, Bernstein, Rosser, Meister, and Cleveland (1584) studied the infrared spectrum from 400 to 4000 cm^{-1} and identified the fundamental frequencies (in cm^{-1}) and their degeneracies (in brackets) as follows: 611 [1], 955 [2], 1305 [1], 1444 [2], 2972 [1], 3057 [2]. A vibrational degree of freedom is said to have a degeneracy g_i, if g_i fundamental modes oscillate with the same frequency.

Methyl bromide has three doubly degenerate frequencies; therefore the heat capacity contribution from these three frequencies must be doubled. The vibrational contribution to the heat capacity is found for

Figure 2-9 Schematic view and orthographic projection of the bromomethane molecule.

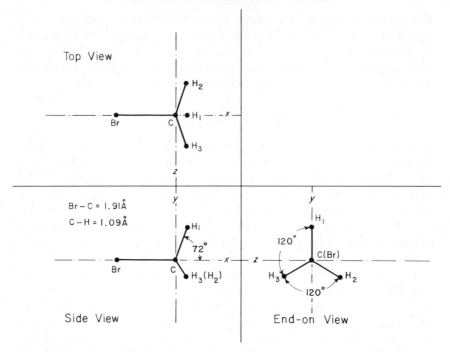

Figure 2-9 Schematic view and orthographic projection of the bromomethane molecule.

each of the above frequencies from Table A-2 in the Appendix, either directly or by interpolation. At 300°K the contributions are

ω_i (cm^{-1})	g_i	$\dfrac{C°}{R}$
611	1	0.5116
955	2	0.4392
1305	1	0.0762
1444	2	0.0948
2972	1	0.0001
3057	2	0.0002

Vibrational contribution	1.122
Translational and over-all rotational contribution	3.000
Conversion, Cv to Cp	1.000

Total calculated $Cp°_{300}/R$ of ideal gaseous CH_3Br = 5.122
or 5.122 × 1.9873 = 10.18 cal/(mole °K).

In addition to the translational, over-all rotational, and vibrational degrees of freedom considered above, an additional mode arises from internal rotation; that is, the rotation of one portion of the molecule about a valence bond joining that portion to the rest of the molecule. When the energy of the rotator is much lower than the energy of the hindering barrier, no rotation takes place, but the bond acts as a torsional vibrator. As the temperature increases, the amplitude of the torsional vibration increases until it approximates the value of the barrier, whereupon hindered rotation takes place. As the temperature and energy increase, the freely rotating state is approached. This is represented graphically in Figure 2-10, taken from Herzberg (589). For small barriers the heat capacity contribution rises sharply above that of free rotation and then declines toward it. Moreover, as the value of the barrier increases, the peak heat capacity contribution exceeds the maximum contribution of a harmonic vibrator. The value of the barrier may be obtained from observations of the gaseous heat capacity at one temperature, or from an experimental entropy at some temperature. Once the value of the barrier is known, its contribution to the thermodynamic properties at any temperature can be calculated. No perfectly general method of predicting potential barriers hindering rotation has been developed, but some useful correlations exist, as is evident in Table 2-3. The thermodynamic technique is rather insensitive to small barriers, but effective for high ones.

Kistiakowsky and Rice (758) measured the gaseous heat capacity of dimethylacetylene, shown in Figure 2-11. Since the methyl hydrogens are well separated in this molecule, one methyl group should rotate freely

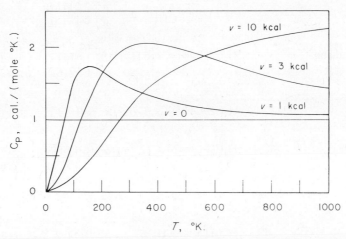

Figure 2-10 Heat capacity contribution as a function of barrier height for hindered rotation of a methyl group. From Herzberg (589).

Table 2-3 Potential Energy Barriers to Internal Rotation

Substance	Structure and Bond		Barrier (kcal)[a]	References
Hydrocarbons				
Ethane	$H_3C\stackrel{a}{-}CH_3$	a	2.88 (T)	(1157)
		a	3.03 (S)	(866)
Propane	$H_3C\stackrel{a}{-}CH_2\stackrel{a}{-}CH_3$	a	3.4 (T)	(1157)
		a	3.3 (T)	(31)
2-Methylpropane	$H_3C\stackrel{a}{-}CH\stackrel{a}{-}CH_3$	a	3.87 (T)	(38)
	$\|a$	a	3.62 (T)	(1164)
	CH_3	a	3.9 (M)	(873)
2,2-Dimethylpropane	$\begin{array}{c}CH_3\\\|a\\H_3C\stackrel{a}{-}C\stackrel{a}{-}CH_3\\a\|\\CH_3\end{array}$	a	4.54 (T)	(29)
		a	4.30 (T)	(1164)
Propylene	$H_3C\stackrel{a}{-}CH=CH_2$	a	1.95 (T)	(751)
		a	1.98 (M)	(870)
		a	2.0 (S)	(410)
2-Methylpropene	$H_3C\stackrel{a}{-}C=CH_2$	a	2.35 (T)	(751)
	$\|a$	a	2.21 (M)	(850)
	CH_3	a	2.12 (S)	(408)
1,2-Butadiene	$H_3C\stackrel{a}{-}CH=C=CH_2$	a	1.65 (T)	(44)
		a	1.59 (M)	(871)
1-Butyne	$H_3C\stackrel{a}{-}CH_2-C\equiv CH$	a	3.0 (T)	(39)
2-Butyne	$H_3C\stackrel{a}{-}C\equiv C\stackrel{a}{-}CH_3$	a	<0.5 (T)	(758)
		a	<0.5 (T)	(1638)
		a	<0.1 (S)	(181, 756, 800)
Toluene	$H_3C\stackrel{a}{-}C_6H_5$	a	0 (T)	(1314)

Table 2-3 *(Continued)*

Substance	Structure and Bond		Barrier(kcal)[a]	References
o-Xylene	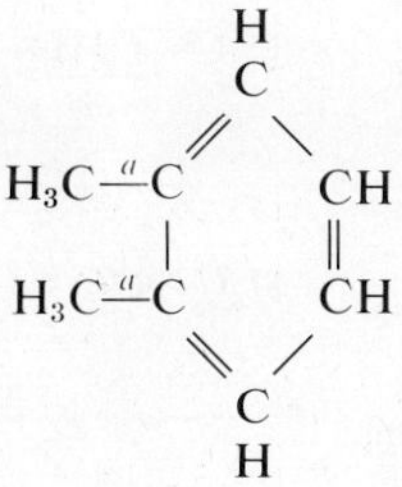 $H_3C\overset{a}{-}C\begin{smallmatrix}H\\\\C\end{smallmatrix}CH$ (ortho-xylene ring structure) $H_3C\overset{a}{-}C\;\;\;\;\;CH$	a	1.8 (T)	(577)

Oxygen compounds

Substance	Structure and Bond		Barrier(kcal)[a]	References
Methyl alcohol	$H_3C\overset{a}{-}OH$	a	1.1 (T)	(674)
		a	1.1 (M)	(673, 1455)
Ethyl alcohol	$H_3C\overset{a}{-}CH_2\overset{b}{-}OH$	a	3.3 (T)	(536)
		b	0.8 (T)	(536)
Isopropyl alcohol	$H_3C\overset{a}{-}CH\overset{b}{-}OH$ $\;\;\;\;\;\;\;\;\;\overset{a}{|}$ $\;\;\;\;\;\;\;CH_3$	a	4.0 (T)	(538)
		b	0.8 (T)	(538)
$tert$-Butyl alcohol	$\;\;\;\;\;\;\;\;\overset{b}{OH}$ $H_3C\overset{a}{-}C\overset{a}{-}CH_3$ $\;\;\;\;\;\;\overset{a}{|}$ $\;\;\;\;\;CH_3$	a	3.8 (T)	(124)
		b	0.9 (T)	(124)
Methyl ether	$H_3C\overset{a}{-}O\overset{a}{-}CH_3$	a	2.7 (T)	(31)
		a	2.72 (M)	(722)
		a	2.63 (S)	(408)
Propylene oxide	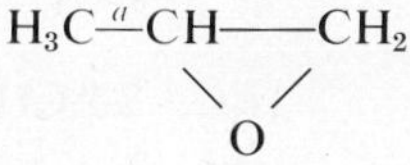 $H_3C\overset{a}{-}CH\!\!-\!\!-\!\!-\!\!CH_2$ $\;\;\;\;\;\;\;\;\;\;\;\backslash\;\;/$ $\;\;\;\;\;\;\;\;\;\;\;\;O$	a	2.56 (M)	(588)
		a	2.52 (S)	(410)

[a] (T), (S), and (M) indicate values obtained from thermal data, far infrared spectroscopy, and microwave spectroscopy, respectively. The sources of the original thermal data have been cited; however, some values of barrier heights have been adjusted to accommodate improved vibrational assignments or moments of inertia.

Table 2-3 *(Continued)*

Substance	Structure and Bond		Barrier(kcal)[a]	References
Acetaldehyde	$H_3C\overset{a}{-}CHO$	a	1.0 (T)	(1167)
		a	1.17 (M)	(407)
		a	1.18 (S)	(587)
Acetone	$H_3C\overset{a}{-}CO\overset{a}{-}CH_3$	a	1.0 (T)	(1135)
		a	0.78 (M)	(1456)
		a	0.83 (S)	(408)
Acetic acid	$H_3C\overset{a}{-}CO_2H$	a	2.5 (T)	(1587)
		a	0.5 (M)	(1463)
Phenol	$HO\overset{a}{-}C_6H_5$	a	3.4 (S)	(392)
		a	3.2 (M)	(788)
Sulfur compounds				
Methanethiol	$H_3C\overset{a}{-}SH$	a	1.27 (T)	(1271, 1316)
		a	1.27 (M)	(790)
Ethanethiol	$H_3C\overset{a}{-}CH_2\overset{b}{-}SH$	a	3.31 (T)	(962)
		b	1.64 (T)	(962)
2-Propanethiol	$H_3C\overset{a}{-}CH\overset{b}{-}SH$ $\quad\;\;\overset{a}{\mid}$ $\quad\;\;CH_3$	a b	3.95 (T) 1.39 (T)	(955) (955)
2-Methyl-2-propanethiol	$\quad\;\;SH$ $\quad\;\;\overset{b}{\mid}$ $H_3C\overset{a}{-}C\overset{a}{-}CH_3$ $\quad\;\;\overset{a}{\mid}$ $\quad\;\;CH_3$	a b	5.10 (T) 1.36 (T)	(963) (963)
Methyl sulfide	$H_3C\overset{a}{-}S\overset{a}{-}CH_3$	a	2.10 (T)	(958)
		a	2.09 (S)	(408)
		a	2.18 (M)	(354)
		a	2.13 (M)	(1143)
Benzenethiol	$HS\overset{a}{-}C_6H_5$	a	0 (T)	(1319)
Nitrogen compounds				
Methylamine	$H_3C\overset{a}{-}NH_2$	a	1.98 (M)	(1346)

Table 2-3 *(Continued)*

Substance	Structure and Bond		Barrier (kcal)[a]	References
Dimethylamine	$H_3C\stackrel{a}{-}NH\stackrel{a}{-}CH_3$	a	3.46 (T)	(32)
		a	3.28 (S)	(408)
Trimethylamine	$H_3C\stackrel{a}{-}N\stackrel{a}{-}CH_3$	a	4.27 (T)	(42)
	$\quad\quad\mid a$	a	4.40 (M)	(872)
	$\quad\quad CH_3$	a	4.41 (S)	(408)
N-Methylmethylenimine	$H_3C\stackrel{a}{-}N{=}CH_2$	a	1.97 (M)	(1277)
Nitromethane	$H_3C\stackrel{a}{-}NO_2$	a	0 (T)	(964)
		a	0 (M)	(1466)
Methyl nitrate	$H_3C\stackrel{a}{-}O\stackrel{b}{-}NO_2$	a	0 (T)	(524)
		a	2.3 (M)	(330)
		b	0 (T)	(524)
		b	9.1 (M)	(330)
2-Methylpyridine	$H_3C\stackrel{a}{-}C_5H_4N$	a	0 (T)	(1315)
3-Methylpyridine	$H_3C\stackrel{a}{-}C_5H_4N$	a	0 (T)	(1310)
Aniline	$H_2N\stackrel{a}{-}C_6H_5$	a	3.43 (T)	(579)
		a	3.54 (S)	(391)
Halogen compounds				
Ethyl fluoride	$H_3C\stackrel{a}{-}CH_2F$	a	3.33 (S)	(1274)
		a	3.30 (M)	(586)
1,1-Difluoroethane	$H_3C\stackrel{a}{-}CHF_2$	a	3.18 (M)	(586)
		a	3.21 (S)	(407)
1,1,1-Trifluoroethane	$H_3C\stackrel{a}{-}CF_3$	a	3.34 (T)	(1402)
		a	3.20 (T)	(1270)
		a	3.04 (S)	(204)

[a] (T), (S), and (M) indicate values obtained from thermal data, far infrared spectroscopy, and microwave spectroscopy, respectively. The sources of the original thermal data have been cited; however, some values of barrier heights have been adjusted to accommodate improved vibrational assignments or moments of inertia.

Table 2-3 *(Continued)*

Substance	Structure and Bond	Barrier (kcal)[a]	References
Hexafluoroethane	$F_3C\overset{a}{-}CF_3$	a 3.96 (T)	(673)
Benzotrifluoride	$F_3C\overset{a}{-}C_6H_5$	a 0 (T)	(1301)
Ethyl chloride	$H_3C\overset{a}{-}CH_2Cl$	a 3.7 (T)	(518)
		a 3.69 (M)	(1296)
		a 3.69 (S)	(410)
1,1-Dichloroethane	$H_3C\overset{a}{-}CHCl_2$	a 3.75 (T)	(862)
		a 3.75 (S)	(302)
1,1,1-Trichloroethane	$H_3C\overset{a}{-}CCl_3$	a 2.7 (T)	(1261)
		a 2.8 (S)	(377)
Ethyl bromide	$H_3C\overset{a}{-}CH_2Br$	a 3.68 (M)	(430)
Ethyl iodide	$H_3C\overset{a}{-}CH_2I$	a 3.2 (M)	(723)

[a] (T), (S), and (M) indicate values obtained from thermal data, far infrared spectroscopy, and microwave spectroscopy, respectively. The sources of the original thermal data have been cited; however, some values of barrier heights have been adjusted to accommodate improved vibrational assignments or moments of inertia.

relative to the other. The fundamental frequencies (in cm^{-1}) and degeneracies (in brackets) given by Kistiakowsky and Rice are: 213 [2], 371 [2], 725 [1], 1029 [2], 1050 [2], 1126 [1], 1380 [2], 1448 [2], 1468 [2], 2270 [1], 2916 [1], 2966 [2], 2976 [3]. This molecule contains 10 atoms and is not linear, so that it will have $3 \times 10 - 6 = 24$ vibrational degrees

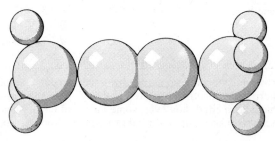

Figure 2-11   Schematic view of dimethylacetylene molecule.

of freedom. In the absence of a potential energy barrier each degree of *free* internal rotation has energy in the amount of $E_{ir} = RT/2$ and would contribute $R/2 = 0.994$ cal/(mole °K) to the heat capacity. The translational and over-all rotational portion of the heat capacity of dimethylacetylene is $3R$. At 336.07°K the contributions are

ω_i (cm^{-1})	g_i	$\dfrac{C^\circ}{R}$
213	2	1.8674
317	2	1.6287
725	1	0.4732
1029	2	0.4862
1050	2	0.5614
1126	1	0.1905
1380	2	0.1905
1448	2	0.1569
1468	2	0.1479
2270	1	0.0057
2916	1	0.0006
2966	2	0.0010
2976	3	0.0014

Vibrational contribution	5.711
Translational and over-all rotational contribution	3.000
Conversion, Cv to Cp	1.000
Internal rotation	0.500
	10.211

The total calculated $Cp^\circ_{336.07}$ of ideal gaseous dimethylacetylene is $10.211 \times 1.9873 = 20.29$ cal/(mole °K). This value is in agreement with the value $Cp^\circ = 20.21$ cal/(mole °K) (within experimental error) measured by Kistiakowsky and Rice and thus indicates that the methyl groups undergo essentially free rotation.

Vanderkooi and De Vries (1527) measured the heat capacity of gaseous 1,1,1-trifluoroethane, shown in Figure 2-12, and at 300°K found $Cv = 16.83 \pm 0.18$ cal/(mole °K). Nielsen, Claassen, and Smith (1070) studied the spectra and assigned the following fundamental vibrational frequencies (in cm^{-1}) and degeneracies (in brackets) to this molecule: 238 [1], 365 [2], 541 [2], 603 [1], 830 [1], 969 [2], 1232 [2], 1278 [1], 1408 [1], 1443 [2], 2978 [1], 3036 [2]. This nonlinear molecule contains 8 atoms

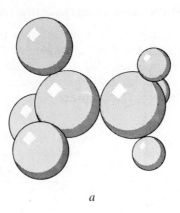

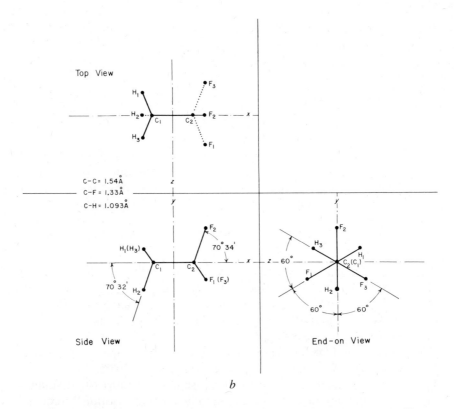

Figure 2-12 Schematic view and orthographic projection of the 1,1,1-trifluoroethane molecule.

and has $3 \times 8 - 6 = 18$ vibrational degrees of freedom. Nielsen, Claassen, and Smith refer to the frequency 238 [1] as the torsional frequency associated with the C—C bond, and in this calculation it will be treated and discussed last. At 300°K the contributions to the heat capacity are

ω_i (cm^{-1})	g_i	$\dfrac{C°}{R}$
365	2	1.5590
541	2	1.1788
603	1	0.5265
830	1	0.3074
969	2	0.4200
1232	2	0.1908
1278	1	0.0825
1408	1	0.0527
1443	2	0.0948
2978	1	0.0001
3036	2	0.0002

Vibrational (except torsional) contribution	4.413
Translational and over-all rotational contribution	3.000
Total	7.413

Thus $Cv = 7.413 \times 1.9872 = 14.731$ cal/(mole °K). Hence the observed value less the calculated contribution is $16.83 - 14.73 = 2.10$ cal/(mole °K) and may be taken as the contribution of the central bond. This value is considerably greater than the 1.782 cal/(mole °K) calculated for a torsional frequency of 238 cm^{-1}, is over twice that contributed by a free internal rotation, and is the result of forces acting between the $-CF_3$ and $-CH_3$ groups.

As the methyl group rotates relative to the trifluoromethyl group, it may be assumed to be repulsed by a cosine potential barrier V with maxima every 120°, as shown in Figure 2-13. This is a case of what Pitzer and Gwinn (1162) designated a "symmetrical top attached to a rigid frame," and for which they presented a general quantal treatment of the problem for a wide range of barrier values and computed the contributions of such a hindered rotator to the thermodynamic functions.

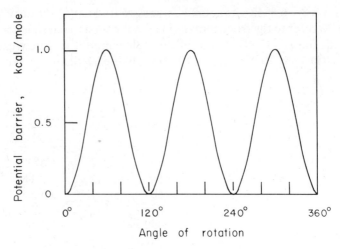

Figure 2-13 Potential energy barrier as a function of angle for an internal rotator.

Their values are reproduced in the Appendix in Tables A-8 through A-11 in dimensionless ratios (V/RT), and

$$\frac{1}{Q_f} = \frac{h\sigma_r}{(8\pi^3 I_r kT)^{1/2}} = \frac{n}{2.7935 \times 10^{19}(I_r T)^{1/2}}, \quad [2\text{-}11]$$

where Q_f is the partition function for a rotator of symmetry number σ_r, n is the number of potential energy maxima occurring during one revolution of the rotating group, and I_r is the reduced moment of inertia of the two rotating groups A and B. If groups A and B are symmetrical about the axis of rotation, then

$$\frac{1}{I_r} = \frac{1}{I_A} + \frac{1}{I_B}.$$

I is the moment of inertia and is the sum of the products of the rotating masses, m_i, multiplied by the square of their perpendicular distances, d_i, from the axis of rotation,

$$I = \Sigma\, m_i(d_i)^2. \quad [2\text{-}12]$$

It is customary to calculate the moment of inertia using interatomic distances in Ångström units (1 Å = 10^{-8} cm), and atomic weights as grams per molecule. Since a methyl group has tetrahedral angles (109°28′)

and a C−H distance of 1.08 Å, the moment of inertia for the methyl group about the axis of internal rotation will be

$$I_{CH_3} = \frac{3 \times 1.008(1.08 \sin 71°)^2}{6.023 \times 10^{23} \times 10^{16}} = 5.232 \times 10^{-40} \text{ g-cm}^2.$$

Brandt and Livingston (161) made electron diffraction measurements on 1,1,1-trifluoroethane and reported the C−F distance as 1.33 Å and ∠CCF = 111.5°, so that the moment of inertia of the trifluoromethyl group is

$$I_{CF_3} = \frac{3 \times 19.00(1.33 \sin 68.5°)^2}{6.023 \times 10^{23} \times 10^{16}} = 144.90 \times 10^{-40} \text{ g-cm}^2,$$

giving a reduced moment of inertia

$$I_r = \left[\frac{1}{5.232 \times 10^{-40}} + \frac{1}{144.90 \times 10^{-40}}\right]^{-1} = 5.050 \times 10^{-40} \text{ g-cm}^2.$$

Substituting these values into [2-11] gives for 300°K

$$\frac{1}{Q_f} = \frac{3}{2.7935 \times 10^{19}(5.050 \times 10^{-40} \times 300)^{1/2}} = 0.2759.$$

Reference to heat capacity Table A-8 indicates $V/RT = 3.76$ or 5.91 for the 2.10 cal/(mole °K) measured value for the contribution from hindered internal rotation, leading to a potential barrier of 2242 or 3523 cal/mole, respectively, for 3.76 and 5.91. More information is required to make a choice. This additional information is provided by the low-temperature heat capacity measurements of Russell, Golding, and Yost (1270), who reported a barrier of 3450 cal/mole. These independent measurements are in good agreement with each other, as well as with the barriers of 3290 cal/mole reported by Nielsen, Claassen, and Smith (1070) and of 3250 cal/mole reported by Thompson and Temple (1493).

The more complicated problem of calculating the thermodynamic properties of unsymmetrical tops attached to a rigid frame has been treated by Pitzer (1156) and extends the use of the tables described above for these cases. Kilpatrick and Pitzer (752) developed procedures for the general case of one or more tops of any symmetry type attached to a frame. Aston (31) reviewed calculational procedures for molecules with energy differences between rotational isomers, and Scott and McCullough (1317) presented tables of contributions to the thermodynamic functions for a limited number of such cases.

Recently two other useful techniques have been developed for the determination of barriers to internal rotation. Lin and Swalen (875) reviewed the application of microwave spectroscopy to internal rotation problems, while Fateley and Miller (407, 408, 410) presented the theory and practice of evaluating barriers from far infrared spectroscopy.

Wulff (1632) described a method by which barrier heights in potential functions hindering the internal rotation of methyl groups, or spectroscopically inactive vibrational frequencies, can be obtained solely from thermal data for the solid phase. He showed that barrier heights derived by this method are in accord with those estimated from comparison of gas phase entropies, but can be determined with greater precision than such values.

A limited survey of potential barriers is given in Table 2-3. Many of the examples listed were investigated by thermal as well as spectroscopic techniques. With the pronounced exceptions of acetic acid and methyl nitrate, the agreement is very good.

Other methods for calculating the contribution of a hindered rotator to the thermodynamic functions have been devised by Price (1186), Wilson (1620), and Halford (563).

Estimation of the heat capacity of gases by empirical and semiempirical methods where direct measurements or spectral data are lacking is described in Chapter 6.

Heat Capacity Equations

Since heat capacity data are basic for many types of thermal and thermochemical calculations, it is important to present this information in the most convenient manner. The temperature dependence of the heat capacity is complex, and so there is no "best" mathematical equation that will express the data. If the temperature range is restricted, the heat capacity of any phase may be represented adequately by

$$Cp = a + bT + cT^2. \qquad [2\text{-}13]$$

in which a, b, and c are empirical constants. A rapid solution for the coefficients of a parabolic equation such as [2-13] through any three points may be made by designating the three points: (T_1, C_{p_1}), (T_2, C_{p_2}), and (T_3, C_{p_3}). Then

$$\frac{Cp_1}{(T_1-T_2)(T_1-T_3)} + \frac{Cp_2}{(T_2-T_1)(T_2-T_3)} + \frac{Cp_3}{(T_3-T_2)(T_3-T_1)} = c,$$

$$\frac{Cp_1 - Cp_2}{T_1 - T_2}[-(T_1+T_2) \times c] = b,$$

and
$$(Cp_1 - bT_1) - c(T_1)^2 = a.$$

Taking data on butane from Table 2-4, one finds

$$T_1 = 300, \quad Cp_1 = 23.40,$$
$$T_2 = 400, \quad Cp_2 = 29.60,$$
$$T_3 = 500, \quad Cp_3 = 35.34.$$

$$\frac{23.40}{100 \times 200} + \frac{29.60}{100 \times (-100)} + \frac{35.34}{100 \times 200} = -0.000023 = c,$$

$$\frac{23.40 - 29.60}{300 - 400} \times [-(300 + 400)(-0.000023)] = 0.0781 = b,$$

$$(23.40 - 23.43) + 2.07 = 2.04 = a.$$

Thus
$$Cp = 2.04 + 0.0781\,T - 0.000023\,T^2.$$

Check (at 400°K):

$$Cp = a + bT + cT^2 = 2.04 + 31.24 - 3.68 = 29.60 \text{ cal/(mole °K)}.$$

The use of such a parabolic fit is appropriate for interpolation (especially of smoothed data presented at rounded temperatures) as well as for integration. Both operations may be thus readily performed on a desk calculator. On theoretical grounds Bernstein (116) has shown that gaseous heat capacities may be advantageously represented by

$$Cp = a' + b'\,T^{-1} + c'\,T^{-2}, \qquad [2\text{-}14]$$

in which a', b', and c' are different empirical constants. An alternative three-term equation,

$$Cp = a'' + b''T + c''T^{-2}, \qquad [2\text{-}15]$$

has been found to be very convenient for all phases of matter, especially above 300°K. The need to add, to subtract, and to integrate these equations makes uniform, simple mathematical equations desirable. Moreover, certain terms are found to be more convenient than others in the eventual use of these equations in Gibbs energy calculations. In this book heat capacity

Table 2-4 Heat Capacity of n-Butane[a]

T(°K)	Cp
298.15	23.29
300	23.40
400	29.60
500	35.34
600	40.30

[a]Data of Rossini, Pitzer, Arnett, Braun, and Pimental (1248), in calories per (mole degrees Kelvin).

ENTHALPY

Enthalpy, designated by the letter H and defined by the relation $H = E + PV$, is employed for the simplification that it provides in many equations and problems. Like energy it is a thermodynamic function of the state of the system and is ambiguous to the extent of an (unknown) additive constant. Although various names are frequently ascribed to H, those such as "heat content" or "total heat" are objectionable semantically, inasmuch as the quantity H cannot be identified with Q, the *heat* of thermodynamics; even ΔH is, in general, not equal to Q, nor does "the amount of heat" in a body or even its change make thermodynamic sense. As a physical concept it differs from energy only, as shown by the defining equation, in that it includes the energy that the system has as a consequence of occupying space V in a fluid (for example, the ambient atmosphere) at pressure P. Since many chemical processes are carried out under the constraint of the essentially constant atmospheric pressure, enthalpy is of great convenience in chemical science and technology.

For a substance the enthalpy (like the energy) is a single-valued, monotonically increasing function of temperature, continuous within the region of existence of a particular phase. The total quantity of heat required to raise the system from temperature T_1 to T_2 under constraint of constant pressure may be calculated by integration of [2-3],

$$H_2 - H_1 = \int_{T_1}^{T_2} Cp \, dT, \qquad [2\text{-}16]$$

and is known as the enthalpy increment of the system. If the temperature range of interest includes changes of phase, the enthalpies of transition must be included in the above equation. Thus the enthalpy from $0°$ to $1000°K$ is represented by

$$H_{1000} - H_0 = \int_0^{Tt} Cp(s, \text{II}) \, dT + \Delta Ht + \int_{Tt}^{Tm} Cp(s, \text{I}) \, dT + \Delta Hm$$
$$+ \int_{Tm}^{Tb} Cp(l) \, dT + \Delta Hv + \int_{Tb}^{1000} Cp(g) \, dT.$$

Even though H_0 is not known, the enthalpy increment can be computed if the heat capacity and enthalpies of transition over the temperature

Enthalpy

range of interest are known. Data for the evaluation of enthalpies below 298.15°K often are not available, and so throughout this volume the reference temperature for enthalpy is taken as 298°K.

For convenience in calculating enthalpy values heat capacity data are often presented by analytical expressions such as [2-13], [2-14], or [2-15], or in tabular form. The intended ultimate use of such information usually dictates the form selected. Integration of heat capacities expressed in the form of [2-13] by means of [2-16] leads to

$$H_{T_2} - H_{T_1} = a(T_2 - T_1) + \frac{b(T_2^2 - T_1^2)}{2} + \frac{c(T_2^3 - T_1^3)}{3}. \quad [2\text{-}17]$$

The heat capacity of liquid ethanol is well represented by

$$Cp = -10.83 + 0.118T + 0.000025T^2 \text{ cal/(mole °K)}.$$

The enthalpy increment of ethanol from 300° to 350°K, obtained by substitution into [2-17], is

$$H_{350} - H_{300} = -10.83[350 - 300] + 0.059[(350)^2 - (300)^2]$$
$$+ 0.0000083[(350)^3 - (300)^3] = 1508 \text{ cal/mole}.$$

Similar treatment of the heat capacity equation [2-15] yields

$$H_{T_2} - H_{T_1} = a''(T_2 - T_1) + b'' \frac{[(T_2)^2 - (T_1)^2]}{2} - c''[(T_2)^{-1} - (T_1)^{-1}].$$

Experience has shown that thermodynamic information is conveniently handled in the tabular form utilized in the presentation of the compiled data in later chapters of this book. Thus heat capacity values obtained from smoothed curves through experimental points or calculated from analytical expressions are tabulated at selected temperatures for ready accessibility or use in high-speed calculations and other purposes. However, integration of a series of heat capacity values at evenly spaced temperatures is conveniently accomplished even with a desk calculator using Simpson's parabolic rule. Heat capacity values $Cp_0, Cp_1, Cp_2, \ldots$ at temperatures $T_0, T_1, T_2, \ldots$ (each separated by equal ΔT's, as shown in Figure 2-14) can be integrated by passing second-degree parabolic arcs through sets of three successive points. In general, for $n+1$ points (where n is even),

$$H_{T_n} - H_{T_0} = \frac{\Delta T}{3} \left[Cp_0 + 4 \sum_{i=1,3,5,\ldots}^{i=n-1} Cp_i + 2 \sum_{i=2,4,6,\ldots}^{i=n-2} Cp_i + Cp_n \right]. \quad [2\text{-}18]$$

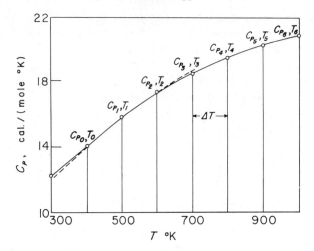

Figure 2-14 Application of Simpson's parabolic rule to heat capacity integration.

For the enthalpy of two adjacent ΔT intervals

$$H_{T_n} - H_{T_{n-2}} = \frac{\Delta T}{3}\left[Cp_{n-2} + 4Cp_{n-1} + Cp_n\right], \qquad [2\text{-}19]$$

and beginning with T_0, for example,

$$H_{T_2} - H_{T_0} = \frac{\Delta T}{3}(Cp_0 + 4Cp_1 + Cp_2). \qquad [2\text{-}20]$$

Usually one portion of the gaseous heat capacity curve is approximately linear, and ΔH for that temperature interval ΔT can be calculated directly. For example, $(H_{T_1} - H_{T_0})$ can be calculated as a linear interval, and by difference from [2-20] $(H_{T_2} - H_{T_1})$ can be obtained, and so on. From the heat capacity of butane given in Table 2-4 enthalpy increments may be conveniently evaluated. The enthalpy increment from 298.15° to 300°K is 1.85 (23.29 + 23.40)/2 = 43 cal/mole, and, assuming that the heat capacity from 300° to 400°K is linear, $H_{400} - H_{300} = 100$ (23.40 + 29.60)/2 = 2650 cal/mole. Substitution into [2-19] gives for the double interval

$$H_{500} - H_{300} = 100\frac{[23.40 + 4(29.60) + 35.34]}{3} = 5907 \text{ cal/mole};$$

hence $H_{500} - H_{400} = 5907 - 2650 = 3257$ cal/mole. The next double interval is

$$H_{600} - H_{400} = 100 \frac{[29.60 + 4(35.34) + 40.30]}{3} = 7042 \text{ cal/mole},$$

and $H_{600} - H_{500} = 7042 - 3257 = 3785$ cal/mole.

Decreasing the size of the interval ΔT improves the accuracy of this method. If the interval from 300° to 400°K had not been linear, its value could have been computed using 50° intervals. This would have required a single extra tabular value at 350°K.

For convenience in interpolation the *enthalpy function*, $(H_T^\circ - H_0^\circ)/T$ or $(H_T^\circ - H_{298}^\circ)/T$, is frequently tabulated rather than the enthalpy increment.

Further development of the concept of the enthalpy increment of a chemical reaction will be found in the next chapter.

Enthalpy of Transition

As the temperature of a crystalline solid is raised, the vibrational amplitudes of the atoms increase in magnitude. Ultimately a particular crystal lattice structure becomes unstable and is transformed to a more stable lattice with absorption of heat at some more or less definite temperature. This temperature is known as the transition temperature, Tt, and the heat absorbed at constant pressure is designated the enthalpy of the transition, ΔHt. No generally adequate theory for predicting either Tt or ΔHt is known. More than a third of the organic compounds subjected to detailed thermal investigation have revealed the existence of some type of transformation. Deffet (319) listed nearly 1200 organic substances in which polymorphism had been detected by various experimental techniques. Polymorphic forms of substances are differentiated by (a) crystal structures or arrangement of molecules on a lattice and (b) the conformation and orientation of molecules at lattice sites. Both the molecular and the macroscopic aspects of polymorphism in organic compounds have been treated by Westrum and McCullough (1598), together with an enumeration of the types of transformations encountered.

Most crystalline solids may be fused at a definite temperature because the thermal energy exceeds the binding energy holding the crystal together; consequently the solid is converted to liquid. At atmospheric pressure this temperature is designated the melting point, Tm, and the associated heat as the enthalpy of melting, ΔHm. If the crystal is melted in equilibrium with only the liquid and vapor phases of the pure substance at saturation pressure, the equilibrium temperature is known as the triple point, T_{tp}.

Although enthalpies of transition and melting may be directly determined by adiabatic calorimetric procedures and are routinely done in studies on which the third-law entropy data of Chapter 4 are based, alternative methods often provide useful values. For example, the enthalpy of melting may be evaluated from depression of the melting point of a substance by addition of a solute miscible with the liquid but not the crystalline phase of the substance.

This depends on the relation between the mole fraction of solute (N_2) in an ideal binary solution, the enthalpy of melting (ΔHm), and the melting temperatures of the pure solvent (Tm) and of the solution (T), that is,

$$\log N_1 = \frac{\Delta Hm}{2.3026R}\left(\frac{1}{Tm}-\frac{1}{T}\right), \qquad [2\text{-}21]$$

which may readily be obtained from thermodynamic consideration of the equilibrium conditions. For example, pure monomeric styrene melts at 242.52°K, and when diluted with 0.0375 mole fraction of benzene forms an ideal solution that freezes at 240.84°K. Substitution into [2-21] gives

$$-0.01660 = \frac{\Delta Hm}{4.576}\left(\frac{1}{242.52}-\frac{1}{240.84}\right).$$

Solving, ΔHm is 2640 cal/mole, in good agreement with the calorimetric value of 2630 cal/mole of Brickwedde (168). In this case the agreement between ΔHm obtained from the ideal binary solution and the calorimetric ΔHm is good; but if the system studied is not ideal, the deviation between the two can be considerable. Appreciable partial miscibility of the solute in the solid phase occasions discrepancy.

Although the Kirchoff equation [3-3] developed in Chapter 3 may be employed to evaluate the temperature variation of ΔHt and ΔHm at *constant pressure*, that is,

$$\left(\frac{\partial \Delta Hm}{\partial T}\right)_P = Cp(l)-Cp(s) = \Delta Cmp,$$

for the more general case in which *both* temperature and pressure are allowed to vary, the equation

$$\frac{d\Delta Hm}{dT} \simeq \Delta Cmp + \frac{\Delta Hm}{T}$$

provides a much better approximation to the temperature dependence of the enthalpy of melting. Neglect of the last term may introduce errors as large as 40% in the temperature coefficient.

Enthalpy of Vaporization

The pressure of the vapor in equilibrium with a liquid increases with increasing temperature. When the pressure of the vapor reaches 1 atm, the liquid boils and is completely converted to vapor on absorption of the enthalpy of vaporization, ΔHv, at the normal boiling point, Tb. By confining the liquid it may be heated above Tb until the critical temperature, Tc, is reached; here the liquid and vapor states become identical, and ΔHv becomes zero. Table 2-5 lists enthalpy of vaporization data for benzene, which are also presented graphically in Figure 2-15. The striking increase in ΔHv as the system is lowered below the critical temperature is noteworthy.

The relation expressing the dynamic equilibrium existing between the vapor and the condensed phase of a pure substance was developed by Clapeyron in the form

$$\frac{dP}{dT} = \frac{\Delta Hv}{T \Delta V}, \qquad [2\text{-}22]$$

in which ΔHv is the enthalpy and ΔV is the volume increment between the vapor phase and the condensed phase. If the condensed phase is solid, the enthalpy increment is that of sublimation, ΔHs. To a first approximation at constant temperature

$$\Delta Hs = \Delta Hm + \Delta Hv. \qquad [2\text{-}23]$$

At low pressure the volume of the condensed phase may be neglected in comparison with that of the vapor, and the perfect gas law may be adopted; substitution of $V = RT/P$ into [2-22] and rearranging gives the Clausius-Clapeyron equation,

$$\frac{dP}{P \, dT} = \frac{\Delta Hv}{RT^2}. \qquad [2\text{-}24]$$

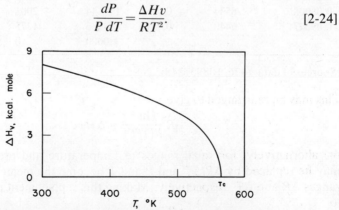

Figure 2-15 Enthalpy of vaporization of benzene.

Heat Capacity and Enthalpy of Transition

Table 2-5 Enthalpy of Vaporization of Benzene[a]
$Tc = 562.65°K$ and $Pc = 48.7$ atm.

Pr	ΔHv (cal/mole)	Tr	Pr	ΔHv (cal/mole)	Tr
0.0004801	8416	0.4736	0.1022	6345	0.7402
0.0006674	8354	0.4835	0.1144	6248	0.7501
0.0009123	8291	0.4934	0.1274	6149	0.7599
0.001229	8228	0.5032	0.1427	6049	0.7698
0.001633	8164	0.5131	0.1586	5947	0.7797
0.002142	8100	0.5230	0.1755	5844	0.7896
0.002772	8035	0.5329	0.1941	5739	0.7994
0.003550	7969	0.5427	0.2143	5632	0.8093
0.004495	7902	0.5526	0.2361	5523	0.8192
0.005649	7833	0.5625	0.2589	5411	0.8291
0.007019	7763	0.5723	0.2831	5295	0.8389
0.008651	7692	0.5822	0.3088	5174	0.8488
0.01057	7619	0.5921	0.3374	5047	0.8587
0.01282	7545	0.6020	0.3683	4914	0.8686
0.01543	7470	0.6118	0.4000	4774	0.8784
0.01844	7393	0.6217	0.4338	4627	0.8883
0.02194	7314	0.6316	0.4705	4472	0.8982
0.02584	7233	0.6415	0.5089	4308	0.9081
0.03036	7151	0.6513	0.5497	4133	0.9179
0.03528	7068	0.6612	0.5937	3944	0.9278
0.04077	6983	0.6711	0.6394	3736	0.9377
0.04709	6896	0.6810	0.6896	3502	0.9475
0.05400	6807	0.6908	0.7415	3232	0.9574
0.06201	6717	0.7007	0.7967	2911	0.9673
0.07031	6626	0.7106	0.8539	2515	0.9772
0.08008	6534	0.7204	0.9144	2001	0.9870
0.09047	6440	0.7303	0.9786	1275	0.9969
			1.0000	0	1.0000

[a]Sources of data: (426, 1089, 1248).

This may be rearranged to give

$$-R\frac{d(\ln P)}{d(1/T)} = \Delta Hv, \qquad [2\text{-}25]$$

or, alternatively, for small ranges of temperature and pressure, dP/dT may be replaced by $\Delta P/\Delta T$ and $\bar{P}$ and $\bar{T}$ become the mean values for the ranges ΔP and ΔT, respectively. Making this replacement in [2-24] leads to

$$\frac{R\bar{T}^2 \Delta P}{\bar{P} \Delta T} = \Delta Hv. \qquad [2\text{-}26]$$

Enthalpy

Timmermans (1501) reported the Tm of phenol as 313.90°K, and Biddiscombe and Martin (129) measured the vapor pressure of solid and liquid phenol. Because their liquid data were measured between 394° and 455°K, extrapolation of their Antoine equation[1-12] to the melting point gives a pressure lower than that of the solid by about 0.06 mm of mercury. Slight adjustment of their Antoine equation constants brings solid and liquid vapor pressures into coincidence at the triple point. Their solid vapor pressure equation and the revised one for the liquid vapor pressure lead to the following results:

T (°K)	P (mm Hg)	
311.15	1.0907	solid
313.15	1.2920	
314.15	1.4146	liquid
316.15	1.6320	

Substitution of these data into [2-26] gives for the sublimation at the mean temperature 312.15°K

$$\Delta Hs = \frac{1.9872 \times 0.2013 \times (312.15)^2}{1.1913 \times 2.00} = 16{,}379 \text{ cal/mole}$$

and for the vaporization at the mean temperature of 315.15°K

$$\Delta Hv = \frac{1.9872 \times 0.2174 \times (315.15)^2}{1.5233 \times 2.00} = 14{,}083 \text{ cal/mole.}$$

Substitution of these two values into [2-23] gives

$$\Delta Hm = 16{,}379 - 14{,}083 = 2296 \text{ cal/mole,}$$

which may be compared with the cryoscopic value of 2530 cal/mole of Timmermans (1503) or the calorimetric value of 2752 cal/mole of Andon, Counsell, Herington, and Martin (19).

Trouton (1518) reported a rough empirical relationship between the normal boiling point and the enthalpy of vaporization,

$$\frac{\Delta Hv}{Tb} = 21 \text{ cal/(mole °K).}$$

This applies best to nonpolar liquids, which form unassociated vapors. A general method for evaluating enthalpies of vaporization, based on the theory that molecules in the same reduced state have identical properties, has been presented by Othmer (1093). A standard reference substance (designated by an asterisk), for which the vapor pressure, enthalpy of vaporization as a function of temperature, critical temperature (Tc), and critical pressure (Pc) are known, is selected. From the relationship

$$\Delta Hv = \Delta Hv^* \times \frac{Tc \log Pr}{Tc^* \log Pr^*} \qquad [2\text{-}27]$$

the enthalpy of vaporization may be calculated for another substance at the reduced temperature, $Tr = T/Tc$. For benzene as the reference substance (cf. Table 2-5), the enthalpy of vaporization of fluorobenzene is computed at 357.89°K (the normal boiling point). Douslin, Moore, Dawson, and Waddington (343) listed the critical constants of fluorobenzene as $Tc = 560.07°K$ and $Pc = 44.91$ atm, from which $Tr = 0.6390$ and $Pr = 0.02227$. For benzene, $Tc^* = 562.65°K$, $Pc^* = 48.7$ atm; $Tr^* = 0.6390$; therefore $\Delta Hv^* = 7253$ cal/mole and $Pr^* = 0.02432$. Substitution into [2-27] gives

$$\Delta Hv = \frac{7253 \times 560.07 \times \log 0.02227}{562.65 \times \log 0.02432}$$

$$= 7391 \text{ cal/mole.}$$

This calculated value is less than 1% below the value of 7457 cal/mole measured by Scott, McCullough, Good, Messerly, Pennington, Kincheloe, Hossenlopp, Douslin, and Waddington (1318). Application of Trouton's rule gives $21 \times 357.89 = 7515$ cal/mole which is less than 1% too high.

Scott et al. (1318) reported the Antoine equation for fluorobenzene as

$$\log_{10} P(\text{mm}) = 6.95208 - \frac{1248.083}{t + 221.827}.$$

From this one may calculate

| $T(°K)$ | 356.82 | 357.89 | 358.95 |
| $P(\text{mm})$ | 735 | 760 | 785 |

Substitution into the Clausius-Clapeyron equation [2-26] gives

$$\frac{50 \times (357.89)^2 \times 1.9872}{760 \times (358.95 - 356.82)} = 7860 \text{ cal/mole.}$$

This value is more than 5% above the measured value, owing to the approximations made in the derivation of the Clausius-Clapeyron equation.

Enthalpy

A more accurate value may be computed from a modification of [2-24] derived by Haggenmacher (560) by inclusion of a compressibility factor (in brackets):

$$\Delta Hv = \frac{RT^2}{P}\left(\frac{dP}{dT}\right)\left[1 - \frac{Tc^3P}{T^3Pc}\right]^{1/2}. \qquad [2\text{-}28]$$

Differentiation of the Antoine equation [1-12] leads to the following expression with T in degrees Kelvin:

$$\frac{dP}{dT} = \frac{2.303\, BP}{(T+C-273.15)^2}. \qquad [2\text{-}29]$$

Combination of [2-28] and [2-29] yields

$$\Delta Hv = \frac{4.5757\, T^2 B}{(T+C-273.15)^2}\left[1 - \frac{Tc^3P}{T^3Pc}\right]^{1/2}. \qquad [2\text{-}30]$$

Substitution of the fluorobenzene data at 357.89°K into [2-30] gives

$$\Delta Hv = \frac{4.5757\,(357.89)^2 \times 1248.083}{(357.89 + 221.83 - 273.15)^2}\left[1 - \frac{(560.07)^3 \times 1.00}{(357.89)^3 \times 44.91}\right]^{1/2}$$
$$= 7443 \text{ cal/mole},$$

which is in excellent agreement with the measured value.

The experimental determination of the enthalpy of vaporization has been described earlier in this chapter. Table 2-6 presents values for the

Table 2-6 Enthalpy of Vaporization of Heptane[a]

| | | | ΔHv (cal/mole) | |
P(mm)	T(°K)	No. of Determinations	Experimental	Clapeyron Equation
597.7	363.63	3	7715 ± 2	7734
391.2	350.48	3	7938 ± 5	7945
195.2	331.21	4	8244 ± 2	8245

[a]Data of Waddington, Todd, and Huffman (1552).

enthalpy of vaporization of heptane measured by Waddington, Todd, and Huffman (1552) in a flow calorimeter. These results are compared with values calculated with the Clapeyron equation [2-22] using vapor volumes obtained from the equation of state and liquid volumes based on experimental densities.

Heat Capacity and Enthalpy of Transition

The total derivative with respect to temperature of the enthalpy of vaporation along the saturation line is very nearly

$$\frac{\Delta Hv}{T} = \Delta Cvp$$

for an ideal gas, as a result of an interesting cancellation of terms. The corresponding derivative for a real gas is treated by Glasstone (492).

CHAPTER 3

THERMOCHEMISTRY

The heat attending the formation of a chemical compound by direct (or hypothetical) union of its elements at constant pressure is termed the enthalpy (heat) of formation, ΔHf. Such data are useful to the engineer in plant design problems and to the rocket expert in the evaluation of jet and rocket fuels. Modern chemical technology requires accurate enthalpies of formation for the calculation of equilibrium constants of reaction. To the theoretician, enthalpies of formation are important in the investigation of bond energies, resonance energies, and the nature of the chemical bond. It is not surprising, therefore, that considerable endeavor has been directed toward the determination of enthalpies of formation over the past century.

Unfortunately, many of the available data are not sufficiently accurate for definitive thermochemical calculations. Early workers usually did not ascertain the purity of the substances in their experiments. Incomplete description of the calorimetric procedures frequently makes it impossible to apply corrections now recognized as important. Moreover, there were no generally recognized standard substances to permit comparison of the accuracy of results from different laboratories. Halogen and sulfur-containing compounds present unique problems, some of which have been adequately solved only recently. Hence these early values are difficult to evaluate and frequently can only be used as rough guides. For illustration one may compare the values selected by Parks and Huffman in 1932 (1105) with modern data (1248). For gaseous ethane, they adopted $\Delta Hf^\circ_{298} = -23,460$ cal/mole, whereas the presently accepted figure is $-20,236$ cal/mole. For gaseous propylene, they adopted $\Delta Hf^\circ_{298} = 4550$ cal/mole, compared to the present 4879 cal/mole. In the first instance there is a serious discrepancy of over 3000 cal, while in the latter the agreement is much better; yet in both cases the experimental basis was the heat of combustion data of Thomsen (1495).

With the development of the petroleum and petrochemical industries, a need for more accurate and dependable thermodynamic properties of hydrocarbons has arisen. The support of the American Petroleum Institute has made possible an extensive program for the determination of thermodynamic properties of hydrocarbons. The very accurately measured enthalpies of formation obtained by this program make an invaluable foundation to support the values for other classes of organic compounds. The regularities observed in the hydrocarbon series can be used to fill the gaps in current knowledge of oxygen, nitrogen, halogen, and sulfur-containing molecules, as well as to reveal inconsistencies and errors in published experimental results.

ENTHALPY OF REACTION

When the formation of a compound from its elements is accompanied by evolution of heat from the system to its surroundings, the enthalpy of formation, ΔHf, of that compound by the adopted convention has a negative value. Conversely, if heat is absorbed by the system from its surroundings, the enthalpy of formation from elements has a positive value. It is necessary that the temperature and the state of reactants and products be specified in order to define the reaction completely. By convention, elements in their standard reference states at all temperatures have $\Delta Hf° = 0$. Table 3-1 lists the enthalpies of formation of some important

Table 3-1 Enthalpies of Formation

Reaction	$\Delta Hf°_{298}$ (kcal/mole)	Source
$C\,(\text{graphite}) + \tfrac{1}{2}O_2(g) = CO(g)$	-26.416	(1249)
$C\,(\text{graphite}) + O_2(g) = CO_2(g)$	-94.052	(1249)
$H_2(g) + \tfrac{1}{2}O_2(g) = H_2O(l)$	-68.317	(1249)
$\tfrac{1}{2}H_2(g) + \tfrac{1}{2}Cl_2(g) = HCl(g)$	$-22.063 \Big\} -39.885$	(1249)
$HCl(g) + 600H_2O(l) = HCl \cdot 600H_2O(l)$	-17.822	(1249)
$\tfrac{1}{2}H_2(g) + \tfrac{1}{2}Br_2(l) = HBr(g)$	$-8.66 \Big\} -28.775$	(1249)
$HBr(g) + 600H_2O(l) = HBr \cdot 600H_2O(l)$	-20.115	(1249)
$\tfrac{1}{2}H_2(g) + \tfrac{1}{2}F_2(g) = HF(g)$	$-64.92 \Big\} -77.03$	(833)
$HF(g) + 50H_2O(l) = HF \cdot 50H_2O(l)$	-12.11	(833)
$S(\text{rhombic}) + \tfrac{3}{2}O_2(g) = SO_3(g)$	$-94.45 \Big\}$	(1249)
$SO_3(g) + H_2(g) + \tfrac{1}{2}O_2(g) + 115\,H_2O(l)$	$\Big\} -212.20$	(926)
$\quad = H_2SO_4 \cdot 115\,H_2O(l)$	-117.75	

compounds from their constituent elements.

In 1840 G. H. Hess (592) demonstrated that enthalpies of reaction are additive and depend only on the initial and final states of the reaction and

Enthalpy of Reaction

not on the particular sequence of steps by which the over-all reaction is achieved. Application to the evaluation of the enthalpy increment of reaction, ΔHr, may thus be made since, by definition,

$$\Delta Hr^\circ = \sum \nu_i H_i^\circ,$$

in which ν_i is the number of moles of each substance (that is, the stoichiometric coefficient) in the balanced chemical equation. It is important to observe that ν_i is positive for the reaction products and has a negative sense for reactants.* Introducing the enthalpy of formation concept to facilitate evaluation of the above equation,

$$\Delta Hr^\circ = \sum \nu_i \Delta Hf_i^\circ. \qquad [3\text{-}1]$$

Application of this equation to evaluation of the enthalpy increment of the esterification reaction at 298°K may be made from the following enthalpies of formation (in kilocalories per mole): $C_2H_5OH(l) = -66.4$, $H_3CCOOH(l) = -116.0$, $H_3CCOOC_2H_5(l) = -115.1$, and $H_2O(l) = -68.3$. Writing out the explicit equations and reversing the sense of the third and fourth,

$$4C\text{ (graphite)} + 4H_2(g) + O_2(g) = H_3CCOOC_2H_5(l)$$
$$H_2(g) + \tfrac{1}{2}O_2(g) = H_2O(l)$$
$$C_2H_5OH(l) = 2C\text{ (graphite)} + 3H_2(g) + \tfrac{1}{2}O_2(g)$$
$$H_3CCOOH(l) = 2C\text{ (graphite)} + 2H_2(g) + O_2(g)$$

$$\overline{C_2H_5OH(l) + H_3CCOOH(l) = H_3CCOOC_2H_5(l) + H_2O(l)}$$

yields
$$\Delta Hr^\circ = -115.1 - 68.3 + 66.4 + 116.0 = -1.0 \text{ kcal.}$$

---

*If mathematicians, rather than chemists, had derived the format for chemical equations, the reaction
$$C + 2H_2 = CH_4$$
would probably have been formulated as
$$CH_4 - C - 2H_2 = 0,$$
or, in a more generalized way, any chemical equation might be indicated by the formulation
$$\sum \nu_i \mathscr{S}_i = 0,$$
in which ν_i is the stoichiometric coefficient (with the sense previously indicated) and $\mathscr{S}_i$ represents a mole of the ith substance.

Alternatively, enthalpies of formation at 298°K may be obtained from enthalpy of combustion (ΔHc) data. For the reaction

$$CH_4(g) + 2O_2(g) = CO_2(g) + 2H_2O(l)$$

with all materials in their standard reference states, $\Delta Hc^{\circ}_{298} = -212.797$ kcal/mole. Insertion of ΔHf° values for $CO_2(g)$ and $H_2O(l)$ from Table 3-1 into [3-1] leads to

$$-212.797 = -94.052 + 2(-68.317) - [\Delta Hf^{\circ} CH_4 + 2(0)],$$

in which 0 represents the enthalpy of formation of oxygen. Solution of this equation gives the enthalpy of formation of methane at 298°K as -17.889 kcal/mole. The relations involved are presented graphically in Figure 3-1.

TEMPERATURE DEPENDENCE OF ENTHALPY OF REACTION

It is useful to be able to calculate the enthalpy for a reaction at different temperatures. Suppose, for example, that at temperature T_1 the enthalpy of a reaction ΔHr_{T_1} is known, but that the reaction is to be operated at a higher temperature, T_2, at which the reaction rate will be adequate. Heating the reactants from temperature T_1 to temperature T_2 makes it possible to conduct the reaction at that temperature with the accompanying enthalpy of reaction, ΔHr_{T_2}. An obvious parallel with [3-1] exists, and

$$\Delta Hr_{T_2} - \Delta Hr_{T_1} = \sum \nu_i [\Delta Hf_{T_2} - \Delta Hf_{T_1}]_i, \qquad [3\text{-}2]$$

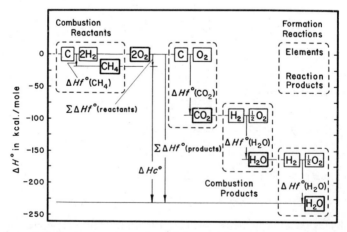

Figure 3-1 Schematic representation of the enthalpy of combustion of methane.

in which the terms in the brackets represent the increments in the enthalpies of the substances arising from their heat capacities over the temperature interval. From the definition of heat capacity at constant pressure (equation 2-3), [3-2] may be expressed as

$$\Delta Hr_{T_2} - \Delta Hr_{T_1} = \sum \nu_i \int_{T_1}^{T_2} Cp_i \, dT.$$

This may be further shortened to

$$\Delta Hr_{T_2} = \Delta Hr_{T_1} + \int_{T_1}^{T_2} \Delta Cp \, dT \qquad [3\text{-}3]$$

since $\sum \nu_i Cp_i = \Delta Cp$. Equation 3-3 was developed in 1858 by G. R. Kirchoff and has come to be known as the Kirchoff equation.

When the heat capacities of reactants and products are expressed in the form of [2-13], the ΔCp term in [3-3] is actually

$$\Delta Cp = \sum \nu_i [a + bT + cT^2]_i = \Delta a + \Delta bT + \Delta cT^2. \qquad [3\text{-}4]$$

Performing the integration indicated by [3-3] leads to

$$\Delta Hr_{T_2} = \Delta Hr_{T_1} + \Delta a(T_2 - T_1) + \frac{\Delta b}{2}(T_2^2 - T_1^2) + \frac{\Delta c}{3}(T_2^3 - T_1^3).$$

If the heat capacities of the reactants and products are expressed in the form of [2-15], the ΔCp term of [3-3] becomes

$$\Delta Cp = \Delta a'' + \Delta b'' T + \Delta c'' T^{-2}$$

and the integral of [3-3] is then

$$\Delta Hr_{T_2} = \Delta Hr_{T_1} + \Delta a''(T_2 - T_1) + \frac{\Delta b''}{2}(T_2^2 - T_1^2) - \Delta c''(T_2^{-1} - T_1^{-1}).$$

As an example of a calculation by [3-2], consider the gaseous reaction of cracking ethylbenzene to hydrogen and styrene,

$$C_6H_5\text{—}C_2H_5 = H_2 + C_6H_5\text{—}C_2H_3.$$

Numerical data for these substances are given in Table 3-2 and are used to calculate the enthalpy increment of the cracking reaction at 900°K. Substitution of the enthalpy of formation data at 298°K into [3-1] gives the enthalpy increment of the gaseous reaction taking place at that temperature,

$$\Delta Hr_{298}^\circ = 35.22 - 7.12 = 28.10 \text{ kcal/mole}.$$

Thermochemistry

Table 3-2 Ethylbenzene Cracking Data

Compound	State	$\Delta H f^\circ_{298}$ (kcal/mole)	$H^\circ_{900} - H^\circ_{298}$ (kcal/mole)	Data from
Ethylbenzene	(g)	7.12	32.82	Chapter 9
Hydrogen	(g)	0.00	4.22	Chapter 8
Styrene	(g)	35.22	30.32	Chapter 9

At 298°K the cracking reaction absorbs 28.10 kcal/mole. Utilizing the enthalpy data in Table 3-2 yields

$$\Delta H r^\circ_{900} = 28.10 + [4.22 + 30.32 - 32.82] = 29.82 \text{ kcal/mole.}$$

Hence at 900°K the cracking reaction absorbs 29.82 kcal/mole. These calculations are demonstrated graphically in Figure 3-2.

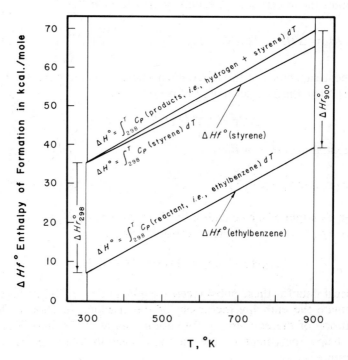

Figure 3-2 Illustration of Kirchoff's law. Note that the enthalpy of formation of hydrogen (an element in its standard state) is zero at all temperatures.

ENTHALPY OF FORMATION FROM ENTHALPY OF COMBUSTION

By far the most important method of determining enthalpies of formation of organic compounds is the measurement of the heat of combustion in oxygen. This technique, by suitable modification, is now applicable to many organic substances. The advantage of this procedure is the relative ease of attaining a complete reaction and well-defined end products for many organic compounds. An inherent disadvantage of the method is the circumstance that enthalpies of formation are derived as relatively small differences of the two enthalpies of combustion, so that small fractional errors in the measured values or relatively small contamination by impurities may result in a more significant fractional error in the enthalpy of formation than in a direct measurement.

Heats of combustion can be determined by burning the compound either at constant volume under a high pressure of oxygen in some type of bomb, or at a constant pressure (about 1 atm) in a flame. The use of flame calorimeters is limited to compounds that either are gaseous or have high vapor pressures at the calorimeter operating temperature. Since gases and volatile liquids are difficult to handle by the bomb method, the two methods are complementary.

ENTHALPY OF COMBUSTION FROM BOMB CALORIMETRY

In nearly all cases the heat released by the combustion reaction has been determined by measuring the rise in temperature of a calorimeter, consisting of a mass of stirred water in which the bomb is immersed. This technique was first thoroughly discussed and systematized by Dickinson (327). The calorimeter is surrounded by an isothermal jacket presenting a constant temperature surface to the calorimeter. Jessup and Green (690) have described a typical calorimeter, shown schematically in Figure 3-3. Details of a self-sealing oxygen bomb manufactured by the Parr Instrument Company are depicted in Figure 3-4. Some papers giving refinements and improvements in apparatus or procedure are those of Hubbard, Knowlton, and Huffman (635), Prosen (1191), Jessup (689), Prosen and Rossini (1199), Coops, Van Nes, Kentie, and Dienske (272), Pilcher and Sutton (1147), Challoner, Gundry, and Meetham (208), Aitken, Boxall, and Cook (6), Bjellerup (135, 137), and Keith and Mackle (731). Two recent monographs give extensive and detailed discussion of many phases of bomb calorimetry (1247, 1362).

Bomb calorimeters can be calibrated with electrical energy, as described

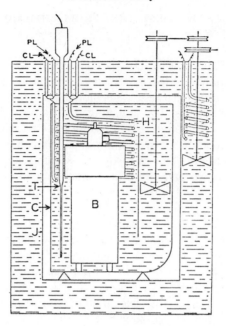

Figure 3-3 Schematic diagram of bomb calorimeter of Jessup and Green (690).

in several of the above references. Since the apparatus needed to measure the electrical energy with the required accuracy is not always available, the United States Bureau of Standards provides samples of benzoic acid that can be used for calibration. The heats of combustion of these standard samples under specified conditions are very accurately known and are certified by the Bureau. An advantage in using benzoic acid is the elimination of systematic errors, since conditions in calibration experiments are very nearly the same as in the combustion of other samples. Although other substances have been proposed as standards, none has been investigated with the necessary thoroughness.

Typical conditions for a combustion experiment are the burning of 1 g of sample in a bomb of $\frac{1}{3}$-liter capacity containing oxygen at a pressure of 30 atm and 1 g of water. The heat of this reaction differs from that in which all reactants and products are in their reference states of pure liquids and ideal gases, since some carbon dioxide is dissolved in the liquid water and the gases deviate appreciably from ideal behavior. Replacing the currently outdated "Washburn corrections" (1578) (which applied only to compounds of carbon, hydrogen, and oxygen) for adjustment to the reference state is the method of Prosen (1191) or the equation derived by Cox, Challoner, and Meetham (288) for compounds

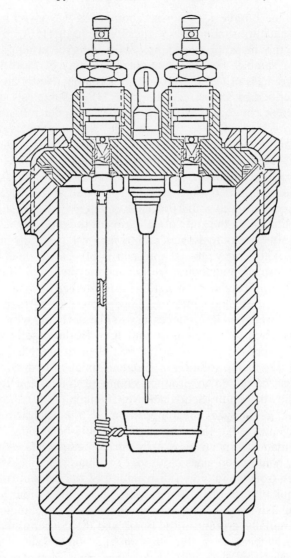

Figure 3-4 Self-sealing Parr combustion bomb.

of carbon, hydrogen, oxygen, and nitrogen, or the more detailed procedure of Hubbard, Scott, and Waddington (637), which also treats compounds containing sulfur or halogens. The outline of Hubbard et al. was necessarily incomplete, however, because parameters for some of the thermochemical corrections were unavailable. Additional information to complete the outline for fluorine compounds was given by Cox and Head (292), and for bromine compounds by Bjellerup (136).

Although the mass of the sample is usually determined by direct weighing, for samples containing impurities it may be more accurate to determine the mass of carbon dioxide after combustion by an absorption train as described by Prosen and Rossini (1199). Samples that are volatile, hygroscopic, or slowly oxidized by oxygen under pressure must be enclosed. For this purpose thin, flexible glass ampoules have been recommended by Jessup (686) and by Coops, Mulder, Dienske, and Smittenberg (269); a technique for using plastic film has also been described by Good and Scott (513) and by Mackle and Mayrick (897).

Combustion calorimetry has developed to the stage where the heat evolved can be measured with a precision of 0.01%. The over-all accuracy, however, is limited by the state of definition of the products. Compounds containing only carbon, hydrogen, and oxygen will generally burn cleanly to carbon dioxide and water. If nitrogen is present, a small amount of nitric acid is formed in aqueous solution and may be readily analyzed. An appropriate thermochemical correction for it can then be made.

Compounds containing sulfur or halogens produce corrosive combustion products, which makes it necessary to line the bomb with a relatively inert material such as platinum or tantalum. In the case of chlorine or bromine compounds, a mixture of the free element in the halogen acid is formed, and a reducing solution of arsenous oxide or hydrazine dihydrochloride must be added to produce complete conversion to the acid. Recent studies by Smith, Scott, and McCullough (1387) showed that a platinum lining catalyzes decomposition of the hydrazine dihydrochloride reagent, so that this combination should be avoided. Stationary bomb equipment has been used for chlorine compounds by supporting the reducing solution on glass cloth (635) or quartz fibers (1384), but the results suffer from nonuniform distribution of the acids formed and incomplete equilibration of the bomb gases with the solution. Sulfur compounds have also been burned by a stationary bomb technique, in which no water is initially present in the bomb and the concentrated H_2SO_4 is allowed to condense as a fine mist on the bomb walls (633). These methods became obsolete with the development of the moving bomb calorimeter described by Popov and Schirokich (1180) and the rotating bomb calorimeters described by Sunner (1442) and by Hubbard, Katz, and Waddington (633). Rapid, complete reduction of free halogen and equilibration of the solution with the gaseous products are achieved in the latter device.

Compounds containing less than 50% fluorine produce only hydrogen fluoride, but highly fluorinated compounds may also yield carbon tetrafluoride as a product. Techniques for analyzing these products and deriving enthalpies of formation have been described by Good, Scott, and Waddington (514) and by Good, Douslin, Scott, George, Lacina, Dawson, and Waddington (508).

Iodine compounds yield only elemental iodine in addition to carbon dioxide and water and can be burned in stationary bombs as well as in rotating bomb calorimeters.

To illustrate the derivation of enthalpies of formation from bomb combustion data, the results for 1,2-dichloroethane may be considered. This compound was recently burned by Sinke and Stull (1359), who reported $\Delta Hf^{\circ}_{298}(l) = -39.6 \pm 0.4$ kcal/mole after all corrections had been made. A second result is that of Smith, Bjellerup, Krook, and Westermark (1384), $\Delta Ec^{\circ}_{293}(l) = -3004.3$ cal/g. However, their correction (-3.0 cal/g) for the change in the heat of oxidation of arsenous oxide from 19.6 to 18.6 cal/meq should be modified, since Bjellerup, Sunner, and Wadsö (142) have reported a more accurate value (19.35 cal/meq) for this reaction. Hence the revised As_2O_3 correction is only -0.8 cal/g. Although the value used for benzoic acid in calibration is not specifically stated, it is reasonable to assume that the correct value was used, since for at least part of the data a sample from the U.S. Bureau of Standards was employed. Also, Smith et al. stated that a small additional term in the Washburn corrections can be estimated as -0.5 cal/g. After these two corrections have been applied, the energy of combustion is $\Delta Ec^{\circ}_{293}(l) = -3001.6$ cal/g, or, using a molecular weight of 98.968 g, $\Delta Ec^{\circ}_{293}(l) = -297.06$ kcal/mole for the reaction

$$C_2H_4Cl_2(l) + \tfrac{5}{2}O_2(g) = 2CO_2(g) + H_2O(l) + 2HCl\,(\text{in }600H_2O)(l).$$

To calculate the enthalpy of combustion at constant pressure, [1-16] and [1-27] are combined to give

$$\Delta H = \Delta E + \Delta n RT,$$

in which Δn is the increment in moles of gaseous substances during the reaction. For the present example Δn is 0.5. Thus

$$\Delta Hc^{\circ}_{293}(l) = -297.06 - 0.30 = -297.36 \text{ kcal/mole}.$$

Pitzer (1151) measured the heat capacity of liquid 1,2-dichloroethane as 30.80 cal/(mole °K), whereas Rossini, Pitzer, Arnett, Braun, and Pimentel (1248) gave 7.017 and 8.874 cal/(mole °K) for oxygen and carbon dioxide, respectively. From the data of Gucker and Schminke (545) on the heat capacity of dilute solutions of hydrogen chloride relative to pure water, the apparent molal heat capacity of hydrogen chloride (in $600H_2O$) is

-25.4 cal/(mole °K). Hence the increment in heat capacity for the combustion reaction can be calculated as $\Delta Cp = -63.3$ cal/°K. Using the Kirchoff equation [3-3],

$$\Delta Hc^\circ_{298}(l) = \Delta Hc^\circ_{293} + \Delta Cp \Delta T = -297.36 - 0.32 = -297.68 \text{ kcal/mole}.$$

Writing the appropriate values from Table 3-1 beneath the combustion equation,

$$\underset{\Delta Hf^\circ}{C_2H_4Cl_2(l)} + \underset{0}{\tfrac{5}{2}O_2(g)} = \underset{2(-94.052)}{2CO_2(g)} + \underset{(-68.317)}{H_2O(l)} + \underset{2(-39.885)}{2HCl\,(\text{in }600H_2O)},$$

and employing [3-1] yields

$$-297.68 = 2(-94.052) + 2(-39.885) + (-68.317) - (\Delta Hf^\circ C_2H_4Cl_2),$$
$$\Delta Hf^\circ_{298} C_2H_4Cl_2(l) = -38.5 \text{ kcal/mole},$$

differing by 1.1 kcal from the value found by Sinke and Stull. A value found by a different method, described later, is in excellent agreement with the work of Sinke and Stull, and it appears that the result of Smith et al. in this case is in error, possibly because of impurities in the sample.

ENTHALPY OF COMBUSTION FROM FLAME. CALORIMETRY

Enthalpies of combustion can also be determined by measuring the rise in temperature of a calorimeter consisting of a mass of stirred water surrounding a chamber in which a gaseous mixture of oxygen and the sample reacts in a flame at constant pressure. Much of the early work of Thomsen was done by this method, more recently brought to a high degree of development by Prosen, Maron, and Rossini (1198) in the measurement of the enthalpy of formation of butadiene. The reaction vessel used by Rossini et al. is shown in Figure 3-5. Since the evolution of heat in these combustion experiments is a linear function of time, the time-temperature curve can be almost exactly reproduced by electrical calibration, with a resulting high accuracy. Investigators who do not have the necessary apparatus for very accurate electrical measurements may employ the reaction of hydrogen and oxygen for calibration. Here the amount of water formed can be accurately determined by direct weighing of that formed in the calorimeter together with the small amount swept out by excess gas and absorbed in magnesium perchlorate or phosphorus pentoxide. Small corrections are necessary for (a) the heat of vaporization of the water swept out, (b) the electrical energy of the spark used for

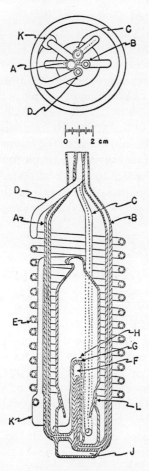

Figure 3-5 Diagram of the reaction vessel in the flame calorimeter of Prosen, Maron, and Rossini (1198).

ignition, and (c) the vaporized water remaining in the calorimeter. For further details the reader is referred to articles by Rossini (1238, 1239).

Interesting recent applications of the techniques of flame calorimetry are the determination of the enthalpy of combustion of methane in fluorine by Jessup, McCoskey, and Nelson (691) and the enthalpy of combustion of cyanogen measured by flame calorimetry by Knowlton and Prosen (769).

An example of an enthalpy of formation from flame combustion data may be cited here for comparison with other methods. Miles and Hunt (1000) measured the enthalpy of combustion of acetone vapor. Their calorimeter was calibrated by burning hydrogen in oxygen, using for water $\Delta Hf^\circ_{298}(l) = -68.318$ kcal/mole. The molecular weight of water was taken as 18.0156, which gives the enthalpy of formation as -3792.16 ± 0.19 cal/g. The most recent value for this quantity is given by Wagman, Kilpatrick, Taylor, Pitzer, and Rossini (1562) as -3792.04 ± 0.53 cal/g. Adjustment to the value of Wagman et al. lowers the enthalpy of combustion values of Miles and Hunt by 0.003%, which is negligible in view of the experimental error of $\pm 0.05\%$. The enthalpy of combustion was measured at a partial pressure close to the vapor pressure of acetone at 298°K, about 200 mm of mercury. The result found by Miles and Hunt, $\Delta Hc_{298}(g, 200 \text{ mm}) = -435.32 \pm 0.20$ kcal/mole, can be used to calculate the enthalpy of formation of acetone from the equation

$$C_3H_6O(g) + 4O_2(g) = 3CO_2(g) + 3H_2O(l).$$
$$\quad\;\;\Delta Hf\qquad\quad 0 \qquad\quad 3(-94.052) \quad\; 3(-68.317)$$

Substitution into [3-1] gives

$$-435.32 = 3(-94.052) + 3(-68.317) - [\Delta Hf^\circ_{298} C_3H_6O(g)],$$

$$\Delta Hf_{298} C_3H_6O(g, 200 \text{ mm}) = -51.787 \pm 0.20 \text{ kcal/mole}.$$

This value may be used to derive the enthalpy of formation of the liquid and of the ideal gas using the enthalpy of vaporization and gas imperfection data of Pennington and Kobe (1135). The enthalpy of vaporization at 300.42°K for acetone is given as 7.372 ± 0.002 kcal/mole. This may be corrected to 298°K by employing a ΔCp of vaporization of -12 cal/(mole °K) to give 7.399 kcal/mole. From this and the enthalpy of formation of the gas, $\Delta Hf^\circ_{298}(l) = -59.19 \pm 0.20$ kcal/mole.

To calculate the enthalpy of formation of the ideal gas the following thermodynamic relation is utilized:

$$H_{P_2} - H_{P_1} = \int_{P_1}^{P_2} \left[V - T \left(\frac{\partial V}{\partial T} \right)_P \right] dP \qquad [3\text{-}5]$$

with the equation of state taken as

$$V = \frac{RT}{P} + B.$$

The temperature dependence of B is evaluated from the data of Pennington and Kobe as

$$B = -40.4 e^{(1164/T)}.$$

Hence

$$H_{P_2} - H_{P_1} = \int_{P_1}^{P_2} \left[B - T \left(\frac{\partial B}{\partial T} \right)_P \right] dP.$$

Taking the ideal gas state as equivalent to the real gas at very low pressure yields, at 298°K,

$$H_0 - H_{200\,mm} = 50 \text{ cal/mole.}$$

Applying this correction to the value of Miles and Hunt for the gas at 200 mm yields $\Delta Hf^\circ_{298}(g) = -51.74 \pm 0.20$ kcal/mole, in which the superscript degree indicates that all products and reactants are in their standard reference states of ideal gases at 1 atm pressure, and pure liquids or solids under a pressure of 1 atm.

ENTHALPY OF FORMATION FROM HEAT OF REACTION

Several workers have used reactions other than combustion to determine the enthalpies of formation of organic compounds. Kistiakowsky and coworkers measured the heats of addition of hydrogen to carbon-carbon and carbon-oxygen bonds in a series of researches (261, 334, 335, 759, 760, 761, 762). Lacher, Emery, Bohmfalk, and Park (824) described a calorimeter used for the heats of hydrogenation of methyl, ethyl, and vinyl chlorides. Other researches by this group include the heat of addition of hydrogen bromide to propylene and cyclopropane (829, 834), to the isomeric butenes (822), and to fluoroolefins (830). Also studied were the addition of chlorine to fluoroolefins (831, 832) and of bromine to fluoroolefins (823). Kistiakowsky et al. also studied the heats of addition of chlorine and bromine to olefins (259, 260) as well as heats of hydrolysis of acid anhydrides (258). Skinner and co-workers measured heats of hydrolysis of acetyl halides (198, 1188), chlorinated acetyl chlorides (1187), benzoyl halides (197), and chloral and bromal (1189). In addition, a large number of organometallic compounds were studied.

These examples are representative of other, less extensive investigations. Because the heat of reaction is usually much smaller than that of combustion, thermochemical reaction data of less fractional precision yield comparable precision in enthalpy of formation values. To illustrate, the enthalpy of addition of chlorine to ethylene measured by Conn, Kistiakowsky, and Smith (259) is used to calculate the enthalpy of formation of 1,2-dichloroethane. For the reaction

$$C_2H_4(g) + Cl_2(g) = C_2H_4Cl_2(g)$$

$\Delta Hr_{355} = -43,653 \pm 120$ cal. Rossini, Pitzer, Arnett, Braun, and Pimentel gave heat capacity data for ethylene (1248), and Potter for chlorine (1183) while Gwinn and Pitzer (556) measured the heat capacity of gaseous 1,2-dichloroethane. The average value of ΔCp for this reaction between 298° and 355°K is calculated as 0.4 cal/(mole °K), and the enthalpy of reaction at 298°K is derived using the Kirchoff equation [3-3],

$$\Delta Hr_{298} = -43{,}653 - (0.4 \times 57) = -43{,}676 \pm 120 \text{ cal/mole}.$$

McDonald (968) found for the enthalpy of vaporization of 1,2-dichloroethane at 298°K a value of 8470 ± 20 cal/mole, and Prosen and Rossini (1204) listed the enthalpy of formation of ethylene gas at 298°K as $12{,}496 \pm 66$ cal/mole. Combining these data, one derives for 1,2-dichloroethane $\Delta Hf^\circ_{298}(l) = -39.65 \pm 0.20$ kcal/mole. This value is in excellent agreement with the combustion value of Sinke and Stull mentioned earlier. It should be noted that the enthalpy of combustion was believed accurate to 0.14%, while the estimated accuracy of the enthalpy of hydrogenation was 0.28%; yet the enthalpy of formation derived from chlorination data has only half the uncertainty of that derived from combustion data.

Dolliver, Gresham, Kistiakowsky, Smith, and Vaughan (334) also measured the enthalpy of hydrogenation of acetone to 2-propanol at 335°K. They found for the reaction

$$C_3H_6O(g, 0.1 \text{ atm}) + H_2(g, 0.9 \text{ atm}) = C_3H_8O(g, 0.1 \text{ atm}),$$

$$\Delta Hr_{355} = -13{,}343 \pm 100 \text{ cal/mole}.$$

Williamson and Harrison (1617) found the enthalpy of vaporization of 2-propanol at 353.33°K and 702 mm pressure to be 9625 ± 12 cal/mole. An equation of state for 2-propanol has been determined by Kretschmer and Wiebe (811). Ginnings and Corruccini (486) measured the enthalpy of liquid 2-propanol, and Parks, Mosely, and Peterson (1116) and Parks and Manchester (1111) the heat of combustion of liquid 2-propanol. These data will be combined to derive the enthalpy of formation of acetone.

Hydrogenation reactions of this type have a ΔCp of close to -4 cal/(mole °K), as may be verified by examples from tables in this book. The enthalpy of reaction may therefore be adjusted to 353.33°K as

$$\Delta Hr_{353.33} = -13{,}343 + (4 \times 1.67) = -13{,}336 \pm 100 \text{ cal/mole}.$$

Kretschmer and Wiebe give for 2-propanol

$$V = \frac{RT}{P} + B + DP^2,$$

Enthalpy of Formation from Heat of Reaction

in which
$$B = -300 - 0.755 e^{2483/T} \text{ ml/mole}$$

$$D = (2.70 \times 10^{-5}) - (3.16 \times 10^{-15}) e^{9215/T} \text{ ml/(mm}^2 \text{ mole)}.$$

Again, using the relation [3-5] and the above data yields

$$H_{P_2} - H_{P_1} = B(P_2 - P_1) + \frac{1}{3} D(P_2^3 - P_1^3) - T\left(\frac{\partial B}{\partial T}\right)_P$$
$$\times (P_2 - P_1) - \frac{1}{3} T \left(\frac{\partial D}{\partial T}\right)_P (P_2^3 - P_1^3).$$

Inserting the above values for B and D and the partial differentials at 353.33°K with $P_2 = 702$ mm and $P_1 = 76$ mm and converting to calories per mole, one calculates

$$H_{P_2} - H_{P_1} = -226 \text{ cal/mole}.$$

The corrections to the ideal gas state for acetone and hydrogen are less than 10 cal/mole and will be neglected. The enthalpies of acetone and hydrogen in the ideal gas states at 353.33°K can be interpolated from data in later chapters. The following reaction sequence can then be written by employing Hess' law.

Reaction		ΔHr(cal/mole)
$C_3H_6O(g, 353°) + H_2(g, 353°)$	$= C_3H_8O(g, 0.1 \text{ atm}, 353°)$	$-13,336$
$C_3H_8O(g, 0.1 \text{ atm}, 353°)$	$= C_3H_8O(g, 0.924 \text{ atm}, 353°)$	-226
$C_3H_8O(g, 0.924 \text{ atm}, 353°)$	$= C_3H_8O(l, 353°)$	$-9,625$
$C_3H_8O(l, 353°)$	$= C_3H_8O(l, 298°)$	$-2,364$
$H_2(g, 298°)$	$= H_2(g, 353°)$	381
$C_3H_6O(g, 298°)$	$= C_3H_6O(g, 353°)$	$1,054$
$C_3H_6O(g, 298°) + H_2(g, 298°)$	$= C_3H_8O(l, 298°)$	$-24,116$

Upon combining this enthalpy of reaction with the enthalpy of formation of 2-propanol, $\Delta Hf°_{298}(l) = -76,190 \pm 80$ cal/mole, one derives for acetone

$$\Delta Hf°_{298}(g) = -52,074 \text{ cal/mole}.$$

It is difficult to estimate the uncertainty, since there are many sources of data used, but ± 250 cal/mole would appear adequate. This value accords fairly well with the flame calorimetric value of Miles and Hunt (1000), $-51,740$ cal/mole.

ENTHALPY OF FORMATION FROM EQUILIBRIUM DATA

Enthalpies of formation may be derived from measured equilibrium constants by using the van't Hoff equation [5-19]. Recent examples of this procedure are the equilibrium studies on the hydrogenation of alcohols by Cubberley and Mueller (297), dehydrohalogenation of ethyl, isopropyl, and tert-butyl chlorides by Howlett (619,620) and dehalogenation of diiodoethane by Abrams and Davis (1). Three methods of treating such data can be used. These are illustrated here by employing the equilibrium data of Kolb and Burwell (791) on the dehydrogenation of 2-propanol presented in Table 3-3 and the data of Ruff and Li (1267) on the CF_4-CO_2-CF_2O equilibria. The examples have been carefully

Table 3-3 Equilibrium Data for the Gaseous Reaction 2-Propanol = Acetone + Hydrogen[a]

$T(°K)$	Kp	$1000/T$	Log Kp
Equilibria approached from excess acetone + hydrogen			
417.8	0.132	2.393	−0.8794
422.7	0.156	2.366	−0.8069
433.4	0.234	2.307	−0.6308
452.2	0.454	2.211	−0.3429
464.3	0.748	2.154	−0.1261
472.8	0.975	2.115	−0.0110
491.6	1.59	2.034	0.2014
Equilibria approached from excess 2-propanol			
416.7	0.124	2.400	−0.9066
422.7	0.154	2.366	−0.8125
436.4	0.276	2.291	−0.5591
444.8	0.356	2.248	−0.4486
452.2	0.435	2.211	−0.3615
455.7	0.525	2.194	−0.2798
464.3	0.683	2.154	−0.1656
472.8	0.959	2.115	−0.0182
491.6	1.57	2.034	0.1959

[a] Data of Kolb and Burwell (791); P expressed in atmospheres.

Enthalpy of Formation from Equilibrium Data

selected to emphasize the hazard of relying solely on the second-law method.

The first method assumes the enthalpy of reaction to be constant over the range of experimental temperatures. Thus the van't Hoff equation can be integrated as

$$\log Kp = \frac{-\overline{\Delta Hr}}{4.5758\,T} + I',$$

in which $\overline{\Delta Hr}$ is the average enthalpy of reaction over the temperature range studied and I' is a constant of integration. At these temperatures the gases are nearly ideal and, moreover, deviations from ideality largely cancel. It is therefore possible to use the equilibrium constants as measured. If the results are plotted as $\log Kp$ versus $1/T$, a straight line is obtained, as shown in Figure 3-6. The slope of this line is equal to $-\overline{\Delta Hr}/4.5758$ and, according to Kolb and Burwell, $\overline{\Delta Hr}$ for the reaction

$$C_3H_8O(g) = C_3H_6O(g) + H_2(g)$$

is thus obtained as

$$\overline{\Delta Hr}^\circ_{416-492} = 13{,}980 \text{ cal/mole}.$$

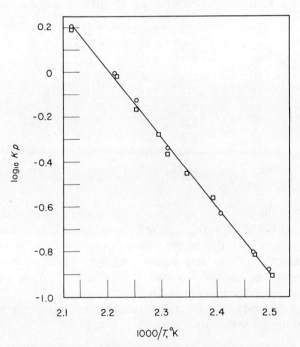

Figure 3-6 Equilibrium constant for the gaseous reaction 2-propanol = acetone + hydrogen.

Equation 3-2 can be used to reduce this result from the average temperature of 454°K to 298°K. From the appropriate tables in Section II one may calculate, by interpolation,

$$\Delta Hr^\circ_{298} = \Delta Hr^\circ_{454} [(H^\circ_{454} - H^\circ_{298}) \text{ acetone} + (H^\circ_{454} - H^\circ_{298})$$
$$\text{hydrogen} - (H^\circ_{454} - H^\circ_{298}) \text{2-propanol}],$$

$$\Delta Hr^\circ_{298} = 13,980 - [3323 + 1084 - 3958] = 13,530 \text{ cal/mole}.$$

The data of Ruff and Li for the equilibrium

$$2CF_2O(g) = CO_2(g) + CF_4(g)$$

present a more difficult case. The experimental points are given in Table 3-4 and plotted as $\log Kp$ versus $1/T$ in Figure 3-7. The points are

Table 3-4 Equilibrium Data for Gaseous Reaction
$2CF_2O = CO_2 + CF_4{}^a$

Catalyst	T(°K)	Kp	1000/T	Log Kp
Ni	573	0.00195	1.745	−2.710
Ni	773	0.00391	1.294	−2.408
Ni	923	0.00897	1.083	−2.047
Pt	1273	0.4659	0.786	−0.332
Pt	1273	0.6975	0.786	−0.156
Pt	1373	0.5952	0.729	−0.225
Pt	1473	1.805	0.679	0.256
Pt	1273	2.04	0.786	0.309

[a] Data of Ruff and Li (1267); P expressed in atmospheres.

scattered, but the straight line drawn corresponds to that of Ruff and Li and to an enthalpy of $\Delta Hr^\circ_{923-1473} = 26$ kcal/mole. This value has been used (uncritically and also in the wrong sense) both by Duus (360) and by Neugebauer (1058) in deriving the enthalpy of formation of carbonyl fluoride. As will be shown, however, it is seriously in error and gives far too much weight to the point at 923°K.

A second method is more general, taking into account the variation of ΔH with temperature by substituting [3-3] into [5-19]:

$$\frac{d \ln Kp}{dT} = \frac{\Delta H_0}{RT^2} + \frac{1}{RT^2} \int^T \Delta Cp \, dT. \qquad [3\text{-}6]$$

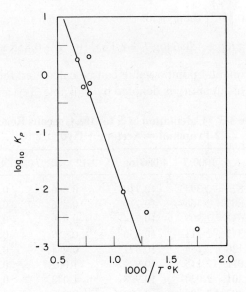

Figure 3-7 Data for the gaseous reaction $2CF_2O = CO_2 + CF_4$, pressure in atmospheres. The data and curve are those of Ruff and Li (1267).

If the heat capacities of reactants and products are expressed in the form

$$Cp = a + bT + cT^2,$$

substitution in [3-6] and integration give

$$\log Kp = \frac{-\Delta H_0}{4.5758T} + \frac{\Delta a}{1.9873} \log T + \frac{\Delta b}{9.1514} T + \frac{\Delta c}{27.454} T^2 + I.$$

Rearranging,

$$\frac{\Delta H_0}{4.5758T} - I = -\log Kp + \frac{\Delta a}{1.9873} \log T + \frac{\Delta b}{9.1514} T + \frac{\Delta c}{27.454} T^2. \quad [3\text{-}7]$$

Fitting heat capacity equations to the three substances involved in the Kolb and Burwell data, one obtains the following:

2-Propanol $Cp° = 3.06 + 68.3 \times 10^{-3}T - 23.5 \times 10^{-6}T^2$
Acetone $Cp° = 5.04 + 45.2 \times 10^{-3}T - 7.0 \times 10^{-6}T^2$
Hydrogen $Cp° = 6.14 + 3.7 \times 10^{-3}T - 4.0 \times 10^{-6}T^2.$

Adding algebraically gives

$$\Delta Cp° = 8.12 - 19.4 \times 10^{-3}T + 12.5 \times 10^{-6}T^2.$$

Substitution into [3-7] then yields

$$\Sigma = -\log Kp + 4.086 \log T - 2.12 \times 10^{-3}T + 0.455 \times 10^{-6}T^2.$$

For each experimental point a value can be calculated for the right-hand side of [3-7]. This quantity is denoted by Σ (Table 3-5) and when plotted

Table 3-5 Calculation of Σ for the Gaseous Reaction
2-Propanol = Acetone + Hydrogen[a]

$T(°K)$	Log Kp	$1000/T$	$4.086 \log T$	$2.12 \times 10^{-3}T$	$0.455\ T^2$	Σ
417.8	−0.879	2.393	10.710	0.886	0.079	10.782
422.7	−0.807	2.366	10.730	0.896	0.081	10.722
433.4	−0.631	2.307	10.744	0.919	0.085	10.571
452.2	−0.343	2.211	10.850	0.959	0.093	10.327
464.3	−0.126	2.154	10.897	0.984	0.098	10.137
472.8	−0.011	2.115	10.929	1.002	0.100	10.038
491.6	0.201	2.034	10.998	1.042	0.110	9.865
416.7	−0.907	2.400	10.705	0.883	0.079	10.808
422.7	−0.813	2.366	10.730	0.896	0.081	10.728
436.4	−0.559	2.291	10.787	0.925	0.087	10.508
444.8	−0.449	2.248	10.821	0.943	0.090	10.417
452.2	−0.362	2.211	10.850	0.959	0.093	10.345
455.7	−0.280	2.194	10.863	0.966	0.094	10.271
464.3	−0.166	2.154	10.897	0.984	0.098	10.176
472.8	−0.018	2.115	10.929	1.002	0.100	10.045
491.6	0.196	2.034	10.998	1.042	0.110	9.870

[a] Data of Kolb and Burwell (791); P expressed in atmospheres.

against $1/T$ (Fig. 3-8) gives ΔH_0 from the slope of the line. The enthalpy of reaction at any other temperature is then calculable from [3-3].

The slope of the Σ versus $1/T$ plot gives

$$\Delta H_0 = 15{,}108 \text{ cal/mole}.$$

Substituting the expression for $\Delta Cp°$ in [3-3], we obtain

$$\Delta Hr_T° = \Delta H_0 - 8.12T + 0.0097T^2 - (4.2 \times 10^{-6}T^3),$$

from which

$$\Delta Hr_{298} = 15{,}108 - 2421 + 862 - 111 = 13{,}438 \text{ cal/mole}.$$

This result is in excellent agreement with the first method. In general, the assumption of a constant ΔHr over the experimental range of tem-

Enthalpy of Formation from Equilibrium Data

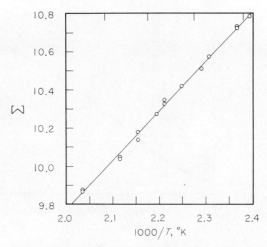

Figure 3-8 Σ plot for the gaseous reaction 2-propanol = acetone + hydrogen.

perature does not introduce serious error for organic reactions. The chief difficulty in both methods is the judgment required in drawing the "best" straight line through the experimental points.

If the thermodynamic functions—that is, heat capacity or entropy and enthalpy—of the reactants and products are known or can be estimated with reasonable accuracy, a better procedure is the employment of a third method using the Gibbs energy function. This function is defined and discussed in Chapter 5. For the present purpose the relation needed is

$$\frac{\Delta Hr^\circ_{298}}{T} = -R \ln Kp + \sum \nu_i \left(\frac{G^\circ_T - H^\circ_{298}}{T}\right)_i.$$

Free energy functions for acetone, hydrogen, and 2-propanol are given at even 100° intervals in the appropriate tables. Rather than interpolate each reactant and product, a more convenient procedure is to calculate the increment in the Gibbs energy function,

$$\Delta\left(\frac{G^\circ_T - H^\circ_{298}}{T}\right) = \sum \nu_i \left(\frac{G^\circ_T - H^\circ_{298}}{T}\right)_i,$$

at the even 100° intervals, plot it as a function of temperature, and interpolate graphically. Data for the acetone–hydrogen–2-propanol reaction are plotted in Figure 3-9. In drawing such a plot it is necessary to consider that, by definition, all Gibbs energy functions and usually the increment also will have a minimum at 298°K.

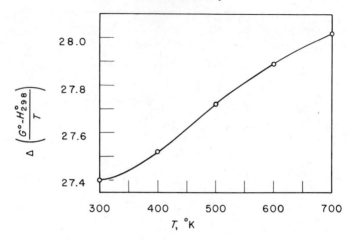

Figure 3-9 Increment in the Gibbs energy function for the gaseous reaction 2-propanol = acetone + hydrogen.

It is now possible to calculate ΔHr°_{298} for each experimental point, as given in Table 3-6. That the ΔHr°_{298} values do not show a trend with temperature indicates the absence of systematic error. Moreover, since forward and backward approaches give average values in excellent agreement, this is also good evidence that equilibrium was reached. The final average is significantly lower than that found by the other methods, and is to be preferred, since it is consistent with the thermodynamic functions, which are based on independent low temperature heat capacity, vapor heat capacity, or spectroscopic data.

From data given earlier in this chapter the enthalpy of vaporization of 2-propanol at 298°K is 10,960 cal/mole, which, when combined with the over-all average, gives for the reaction

$$C_3H_6O(g) + H_2(g) = C_3H_8O(l)$$

$\Delta Hr^\circ_{298} = -24,140$ cal/mole, in excellent agreement with the directly determined enthalpy of hydrogenation.

The equilibrium data of Ruff and Li provide a more striking illustration. In this case the Gibbs energy function changes very slowly with temperature, and linear interpolation between 100° intervals is sufficiently accurate. From the data given in Table 3-7 it is immediately obvious that the value of 26 kcal/mole derived by Ruff and Li using the first method is incompatible with the known thermodynamic functions of these compounds. The first three points (taken with a nickel catalyst) form

Enthalpy of Formation from Equilibrium Data

Table 3-6 Analysis of Thermal Equilibrium Data for the Gaseous Reaction 2-Propanol = Acetone + Hydrogen[a]

$T(°K)$	Kp	$\dfrac{\Delta Gr^{ob}}{T}$	$\dfrac{\Delta(G_T° - H_{298}°)^b}{T}$	$\Delta Hr_{298}°{}^c$
Equilibria approached from excess acetone and hydrogen				
417.8	0.132	4.02	27.55	13,190
422.7	0.156	3.69	27.56	13,209
433.4	0.234	2.89	27.58	13,206
452.2	0.454	1.57	27.62	13,200
464.3	0.748	0.58	27.65	13,107
472.8	0.975	0.05	27.67	13,106
491.6	1.59	−0.92	27.70	13,165
Average				13,169
Equilibria approached from excess 2-propanol				
416.7	0.124	4.15	27.55	13,209
422.7	0.154	3.72	27.56	13,222
436.4	0.276	2.56	27.59	13,157
444.8	0.356	2.05	27.61	13,193
452.2	0.435	1.65	27.62	13,236
455.7	0.525	1.28	27.63	13,174
464.3	0.683	0.76	27.65	13,191
472.8	0.959	0.08	27.67	13,120
491.6	1.57	−0.90	27.70	13,175
Average				13,186
Over-all Average				13,180

[a] Data of Kolb and Burwell (791); P expressed in atmospheres.
[b] In calories per (mole degrees Kelvin).
[c] In calories per mole.

one group, and the last five (taken with a platinum catalyst and at considerably higher temperatures) form another. The equilibrium constants at low temperatures are very small and difficult to measure accurately, and true equilibrium may not have been reached. The best value available from these data appears to be an average of the last five points, or

$$\Delta Hr_{298}°(g) = -12 \pm 3 \text{ kcal/mole.}$$

Table 3-7 Analysis of Thermal Equilibrium Data for the Gaseous Reaction $2CF_2O = CO_2 + CF_4$ [a]

Catalyst	T (°K)	$\dfrac{\Delta Gr^{ob}}{T}$	$\dfrac{\Delta(G_T° - H_{298}°)^b}{T}$	$\Delta Hr_{298}°$ [c]
Ni	573	12.40	−9.89	1,440
Ni	773	11.02	−9.71	1,010
Ni	923	9.37	−9.52	−140
Pt	1273	1.52	−9.24	−9,830
Pt	1273	0.71	−9.24	−10,860
Pt	1373	1.03	−9.16	−11,160
Pt	1473	−1.17	−9.09	−15,113
Pt	1273	−1.42	−9.24	−13,570

[a] Data of Ruff and Li (1267).
[b] In calories per (mole degree Kelvin).
[c] In calories per mole.

Combining this with the enthalpies of formation of carbon dioxide and carbon tetrafluoride, that for carbonyl fluoride is

$$\Delta Hf_{298}°(g) = -152.5 \pm 3 \text{ kcal/mole}.$$

This value is in good agreement with −153.2 ± 0.20 kcal/mole found by Wartenberg (1572) by hydrolysis of CF_2O. The 26 kcal/mole used by Duus (360) gave $\Delta Hf_{298}°(g) = -169$ kcal/mole, which is quite erroneous. Thus the use of the Gibbs energy function yields useful information from equilibrium data that are impossible to treat reliably by the first two methods described.

ENTHALPY OF FORMATION BY OTHER METHODS

For a few molecules special methods can be used to derive enthalpies of formation. For diatomic molecules spectroscopic data can be used to derive dissociation energies as described by Gaydon (461) and Herzberg (590). Dissociation energies combined with enthalpies of formation of gaseous atomic elements may be used to obtain enthalpies of formation of diatomic molecules. Techniques for determining free energies of free radicals have been reviewed by Szwarc (1462).

CHAPTER 4

EVALUATION OF ENTROPY

SECOND LAW OF THERMODYNAMICS

The so-called second law, another classical postulate of thermodynamics, provides a basis for the definition of another function of state, entropy. This term, selected by Clausius (231) in 1865, is based on the Greek stem τροπή (meaning *to turn about or change*) and is succinctly incorporated in Clausius' second-law statement: "The entropy of the world tends toward a maximum value." For the present purpose a mathematical expression of the second-law postulate is advantageous; it may be formulated in terms of line integrals for reversible and actual processes:

$$\int_{C_1}^2 \frac{\delta Q}{T}\bigg|_{rev} = f(1,2) > \int_{C_1}^2 \frac{\delta Q}{T}\bigg|_{actual}. \qquad [4\text{-}1]$$

Couched in verbal form, this might be interpreted: The line integral of $\delta Q/T$ for a given change in state, $1 \to 2$, over a reversible process, is dependent only on the initial and final states, and is greater than the corresponding integral over any actual path connecting these two states. Designation of a process as *reversible* implies an infinitesimal driving force (and hence reversal of the direction of the process by an infinitesimal diminution of this force). Such processes require an infinite duration to produce a finite change and hence are hypothetical limiting processes, clearly differentiated from the *actual* or *spontaneous* processes observed daily, which are characterized by finite driving force and rate (plus inefficiency, occasioning frictional dissipation of energy). Mathematically, the expression shows $1/T$ to be the *integrating factor* rendering *exact* (or integrable) the *inexact* differential, δQ. The postulate thus provides the fundamental definition of T, the thermodynamic temperature scale. Although the presence of the greater-than sign provides the basic criterion for the thermodynamic equilibrium and the distinction between reversible and irreversible processes, the equality provides the basis for the definition of a *thermodynamic function*, entropy, designated by S,

$$S_2 - S_1 = \oint_{1\text{rev}}^{2} \frac{\delta Q}{T},$$

as well as a *recipe* for its evaluation. In differential form,

$$dS_{\text{rev}} = \frac{\delta Q}{T}, \qquad [4\text{-}2]$$

in which the subscript rev provides a reminder of the restriction to reversible processes.

Entropy of Heating

For the process of heating at constant volume, $dQ_V = Cv\, dT$ and

$$dS_V = \frac{Cv\, dT}{T}.$$

For a similar process under the constraint of constant pressure $dQ_P = Cp\, dT$, or

$$dS_P = \frac{Cp\, dT}{T}. \qquad [4\text{-}3]$$

It is obvious that the units of entropy are the same as those of heat capacity, that is, calories per (mole degrees Kelvin), occasionally abbreviated, in less formal usage, eu for *entropy unit*. Integration of [4-3] over a range of temperature yields

$$S_{T_2} - S_{T_1} = \int_{T_1}^{T_2} \frac{Cp\, dT}{T}. \qquad [4\text{-}4]$$

Entropy of Mixing

Schottky and Wagner (1291) demonstrated that the entropy of mixing noninteracting components of solution, or noninteracting ideal gases (each separate gas is at the same pressure before mixing as the final pressure of the mixture), is given by the expression

$$\Delta S = -R \sum (n_i \ln N_i), \qquad [4\text{-}5]$$

in which n_i is the mole number and N_i is the mole fraction of the ith species. For the entropy of mixing a total of 1 mole of the mixture of ideal gases

$$\Delta S = -R \sum (N_i \ln N_i).$$

Hence, for a binary mixture with $N_1 = N_2 = 0.5$, $\Delta S = -R(N_1 \ln N_1 + N_2 \ln N_2) = 1.38$ cal/(mole °K).

Entropy of Phase Transition

Entropy increments associated with isothermal phase transitions, such as enantiomorphic solid transitions, melting, sublimation, or vaporization, may be readily calculated from [4-2] because of the constancy of temperature. Using fusion as an example,

$$\Delta Sm = \frac{\Delta Hm}{T}. \qquad [4\text{-}6]$$

Because of the restriction to reversible processes inherent in [4-2], evaluations of the entropy increments must be made only for conditions under which both phases involved are in equilibrium. For λ-type or Schottky transitions, in which the associated enthalpy effect is not isothermal, an appropriate integral of the heat capacity must be employed.

Effect of Pressure on Entropy

Change of pressure affects the entropy of a gas and must be taken into consideration in entropy evaluations. As is shown in Chapter 5, we may readily derive the equation

$$\left(\frac{\partial S}{\partial P}\right)_T = -\left(\frac{\partial V}{\partial T}\right)_P, \qquad [4\text{-}7]$$

which relates the dependence of entropy on pressure at constant temperature to the dependence of volume on temperature at constant pressure. Hence

$$dS = -\left(\frac{\partial V}{\partial T}\right)_P dP,$$

or, for a finite change in the entropy of a substance with compression from an initial pressure P_1 to a final pressure P_2,

$$S_2 - S_1 = -\int_{P_1}^{P_2} \left(\frac{\partial V}{\partial T}\right)_P dP. \qquad [4\text{-}8]$$

For a perfect gas, since $PV = RT$, it is evident that $(\partial V/\partial T)_P = R/P$, and substitution into [4-8] gives, on integration,

$$S_2 - S_1 = -R \ln\left(\frac{P_2}{P_1}\right). \qquad [4\text{-}9]$$

Thus compression from P_1 to P_2 diminishes the entropy by the quantity $S_2 - S_1$, for this process causes an increase in order. Since the vapor pressure of chlorobenzene at 298.15°K is 11.5 mm of mercury (1432), the entropy decrease for compression from 11.5 mm to 760 mm pressure is thus $-1.9873 \ln 760/11.5 = -8.328$ cal/(mole °K).

Treatment of an actual gas requires the use in [4-8] of the equation of state for the real gas. Such equations may be complicated [see Barrow (76) and Weltner and Pitzer (1588)]. Differentiation of the Berthelot equation of state, for example, leads to the expression

$$\left(\frac{\partial V}{\partial T}\right)_P = \frac{R}{P}\left[1 + \frac{27}{32} \times \frac{PTc^3}{PcT^3}\right],$$

which, by substitution into [4-8], gives

$$S_2 - S_1 = R\left[1 + \frac{27}{32} \times \frac{RTc^3}{PcT^3}\right] \int_{P_1}^{P_2} \frac{dP}{P}. \qquad [4\text{-}10]$$

One may figuratively take a real gas at an initial pressure P_1 and expand it to a final pressure P_2 so low that for all practical purposes it behaves as an ideal gas, and then compress the ideal gas to P_1. Under these conditions S_1 is the entropy of the real gas, while S_2 is the entropy of the ideal gas in the standard state ($S°$). From [4-10] and the corresponding equation for an ideal gas, one obtains

$$S° - S_1 = \frac{27}{32} \times \frac{RTc^3}{PcT^3} \times P_1. \qquad [4\text{-}11]$$

The quantity $S° - S_1$ represents the entropy increment between an ideal gas at P_1 and a real gas at the same pressure. It is often called the *gas imperfection correction*.

For cyclopentane Aston, Fink, and Schumann (35) reported $Tc = 520°K$ and $Pc = 44.2$ atm, and from [4-11] they calculated the value of the gas imperfection at 1 atm to be 0.084 cal/(mole °K).

THIRD LAW OF THERMODYNAMICS

Of such special interest is the entropy at the lower bound of temperature, $S_{0°K}$, for chemico-thermodynamical purposes, that an additional postulate, the third law, has been predicated. In 1906 Nernst (1053) concluded that the entropy increment for a chemical reaction at absolute zero is negligibly small. Planck (1168) reasoned in 1912 that a crystalline lattice at absolute zero possessed maximum order and hence should have zero entropy. Simon (1350) and, more recently, Wilks (1613) traced the development of this concept from its heat theorem initiation by Nernst to the present time. An extensive discussion and applications have been provided by Aston (30). For chemical thermodynamic purposes the formulation of Lewis and Randall (860) makes the essential quantum mechanical nature of the postulate evident and provides a ready application to chemical systems:

"If the entropy of each element in some crystalline state be taken as zero at the absolute zero of temperature, every substance has a finite positive entropy; but at the absolute zero of temperature the entropy may become zero, and does so become in the case of perfect crystalline substances."

In confirmation of the third law, many chemical and physical tests have been performed, and substances that do not have a null zero-point entropy have deviations from the perfectly crystalline state which are well understood. Tests involving organic substances have been discussed by Westrum and McCullough (1598). Although gases, liquids, metastable vitreous phases, substances with frozen-in disorder, and solutions are not to be expected to accord with the third law, substances in internal thermodynamic equilibrium may be expected to have no entropy at 0°K. Substances that do not form ordered crystalline solids at 0°K cannot be expected to have $S_0^\circ = 0$. Examples of such materials are *cis, trans* isomeric mixtures, optically active isomeric substances forming solid solutions, and disordered solids. An "absolute" value for the entropy of a perfectly crystalline substance may be evaluated by the use of [4-4] with 0°K as its lower bound and $S_0^\circ = 0$. Such terms as those given by [4-5] must be included as the zero-point entropy for mixtures. However, since ratios of isotopes and those of nuclear spin states do not change in ordinary chemical reactions, it is customary to neglect these contributions to the entropy (and Gibbs energy). The resulting practical entropies are those ordinarily tabulated and hence applicable to chemical thermodynamic purposes. All entropies given in this book are practical entropies.

Third-law Entropies

Absolute entropies are predicated on measurements of the heat capacity of a substance from near 0°K to the temperature of interest. Because experiments in the neighborhood of absolute zero present considerable practical difficulties, heat capacities below 4–10°K are frequently not available, and so the entropy of this region is often extrapolated by substitution of the Debye heat capacity equation [2-7] into [4-4]. Appendix

Evaluation of Entropy

Table A-7 relates θ_D to its entropy value. Table 2-1 gives measured heat capacity values for 1,2,4-trimethylbenzene fitted by assuming five degrees of freedom and $\theta_D = 103.7°$. Thus θ_D/T for 15°K is 6.91, and the table gives $S_{15} = 0.146$ cal/(mole °K) for each degree of freedom, or a total entropy at 15°K of 0.729 cal/(mole °K), in agreement with the value reported by Putnam and Kilpatrick (1208).

As noted previously, heat capacity data can also be fitted by a combination of Debye and Einstein functions, namely,

$$3D\left(\frac{80.5}{T}\right) + 6E\left(\frac{155}{T}\right).$$

These values of θ_D and θ_E may be calculated from Appendix Tables A-4 and A-7 to give the following values at 15°K [in calories per (mole degrees Kelvin)]:

$80.5/T = 5.366$, giving for $3\theta_D/T$ 0.838
$155/T = 10.333$, giving for $6\theta_E/T$ 0.004
Total calculated entropy at 15°K   0.842

This value is about 10% larger than that calculated by the more empirical method used by Putnam and Kilpatrick.

General Expression for Entropy

On the basis of the preceding discussion, the general expression for the evaluation of the standard state temperature-entropy function for a compound with an enantiomorphic transition at Tt, melting at Tm, and a normal boiling point at Tb may be written

$$S_{1000} = S_0 + \int_{0°K}^{Tb} \frac{Cp(\text{II})\,dT}{T} + \frac{\Delta Ht}{Tt} + \int_{Tt}^{Tm} \frac{Cp(\text{I})\,dT}{T} + \frac{\Delta Hm}{Tm}$$
$$+ \int_{Tm}^{Tb} \frac{Cp(l)\,dT}{T} + \frac{\Delta Hv}{Tb} + \int_{Tb}^{1000} \frac{Cp(g)\,dT}{T}. \quad [4\text{-}12]$$

ENTROPY FROM LOW-TEMPERATURE HEAT CAPACITY

Since the entropies of pure crystalline materials may be evaluated from a knowledge of the heat capacities, both cryogenic calorimeters and gas calorimeters of the types described in Chapter 2 have provided useful data for the evaluation of entropies and entropy increments of transition.

With a low-temperature calorimeter similar in type to that described, Guthrie, Scott, Hubbard, Katz, McCullough, Gross, Williamson, and

Waddington (551) measured the heat capacity of liquid and solid furan together with its enthalpies of fusion and vaporization, vapor pressure, and gaseous heat capacity. From their data, presented in Figure 4-1, the third law entropy of the liquid was calculated as shown in Table 4-1.

STATISTICO-MECHANICAL EVALUATION OF ENTROPY

Because of the experimental difficulty of measuring the heat capacities of gases, thermodynamic properties of relatively few gases were formerly available. However, application of quantal theory to the energy relationships of gaseous molecules has led to the development of reliable means for calculating these properties of simple gaseous substances. Although

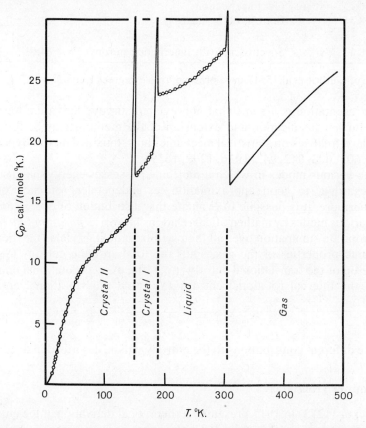

Figure 4-1 Heat capacity of furan. From Guthrie, Scott, Hubbard, Katz, McCullough, Gross, Williamson, and Waddington (551).

Evaluation of Entropy

Table 4-1 Entropy of Furan[a]

$T(°K)$		S
0–12	Debye extrapolation, $\theta = 95.5°$, 3.5 degrees of freedom	0.347
12–150.0	Crystal II, graphical $\int C_{satd}\, d\ln T$	18.076
150.0	Transition, 489.2/150.0	3.262
150.0–187.55	Crystal I, graphical $\int C_{satd}\, d\ln T$	4.047
187.55	Fusion, 908.8/187.55	4.846
187.55–298.15	Liquid, graphical $\int C_{satd}\, d\ln T$	11.638
298.15	Entropy of liquid	42.22 ± 0.08
298.15	Vaporization, 6561/298.15	22.005
298.15	Compression, $R \ln (P/760)$	−0.471
298.15	Gas imperfection	0.112
298.15	Entropy of the ideal gas	63.86 ± 0.10
298.15	Spectroscopically determined entropy	64.03

[a] Data of Guthrie et al. (551), in calories per (mole degrees Kelvin).

this presentation is a somewhat abbreviated survey, derivation together with further justification and extension of the methods may be found in standard treatises on statistical mechanics and statistical thermodynamics (436, 602, 603, 842, 942, 1036, 1292, 1619).

The various modes in which a molecule possesses energy may usually be regarded, to good approximation, as independent of each other. Furthermore, it is possible to evaluate this distribution of energy among the various modes by mathematical relations known as *partition functions*. Appropriate summation over all the partition functions yields the thermodynamic properties of the gas. Thus the total absolute entropy (S_{tot}) is made up of the translational entropy (S_{tr}), the over-all rotational entropy (S_{or}), the internal rotational entropy (S_{ir}), and the vibrational entropy (S_{vib}):

$$S_{tot} = S_{tr} + S_{or} + S_{ir} + S_{vib}. \qquad [4\text{-}13]$$

These different contributions to the entropy will be described in turn.

Entropy of Translation

Sackur (1273) in 1911 presented a theoretical derivation of the entropy of a monatomic gas that accorded with experience. Any gaseous molecule has only three degrees of translational freedom, and so the same expression gives the translational contribution to the entropy for any molecule. The translational entropy is

$$S_{tr} = \tfrac{3}{2}R \ln M + \tfrac{3}{2}R \ln T + R \ln V + \tfrac{5}{2}R + C^*,$$

where M is the molecular weight of the gas, T the absolute temperature, V the molal volume, R the ideal gas constant per mole, and C^* a constant for all substances. A year later Tetrode (1480) showed that the constant C^* had the value

$$R \ln \frac{(2\pi k)^{3/2}}{h^3 N^{5/2}} = -16.0412 \text{ cal/(mole °K)},$$

where k is the ideal gas constant per molecule, h is Planck's constant, and N is Avogadro's number. Combining these two equations yields the Sackur-Tetrode equation,

$$S_{tr} = \tfrac{3}{2}R \ln M + \tfrac{3}{2}R \ln T + R \ln V - 11.0731 \text{ cal/(mole °K)}, \quad [4\text{-}14]$$

which for an ideal gas at 298.15°K and 1 atm becomes

$$S_{tr} = 6.8637 \log M + 25.8182 \text{ cal/(mole °K)}. \quad [4\text{-}15]$$

Entropy of Rotation

To obtain thermodynamic functions it is necessary to evaluate the moments of inertia about the center of mass of the molecule, whether diatomic, linear polyatomic, or nonlinear. This calculation requires that atomic masses, interatomic distances, and valence angles be known for the molecules of interest. Much of this information comes from electron diffraction studies. Wheland (1602) collected such data through mid-1954 and listed detailed information on bond lengths and angles for well over 600 organic molecules. The values presented in Table 4-2 have been obtained by averaging the measured values for a given bond length or angle contained in the molecules catalogued by Wheland. These values may be utilized judiciously to approximate the structures of unstudied molecules.

Evaluation of the moments of inertia of linear molecules requires only the location of the center of mass and summation of the moments of inertia for each atom, since the moment of inertia for the ith atom is

$$I_i = m_i d_i^2,$$

where d_i is the distance (in centimeters) from the ith atom to the center of mass, and m_i is the mass of the atom (in grams). A linear molecule can

Table 4-2 Bond Lengths and Bond Angles[a]

Bond	Length (Å)	Bond Angles — Tetrahedral	Bond Angles — Planar	Bond Angles — Linear	Bond Angles — Other
Single bonds					
H—C	1.08	—C—H 109°	=C(H) 120°	≡C—H 180°	
H—N	1.02	—N—H 109°			=N(H) 108°
H—O	0.96				—O—H 105°
H—Si	1.48	—Si—H 109°			
H—P	1.42	—P—H 100°			=P(H) 100°
H—S	1.33				—S—H 96°
C—C (chain)	1.53	—C—C 109°	=C(C) 120°	≡C—C 180°	
C—C (aromatic)	1.39		=C(C) 120°		
C—N (chain)	1.45	—C—N 109°	=C(N) 120°	≡C—N 180°	=C=N 180°
C—N (ring)	1.34		=C(N) 120°		
C—O	1.40	—C—O 111°	=C(O) 120°	≡C—O 180°	
C—Si	1.88	—C—Si 111°			
C—P	1.90				C(P—C) 100°
C—S	1.80	S(C—S) 114°			—S—C 103°
C—F	1.35	—C—F 109°	=C(F) 120°	≡C—F 180°	
C—Cl	1.75	—C—Cl 109°	=C(Cl) 120°	≡C—Cl 180°	
C—Br	1.91	—C—Br 109°	=C(Br) 120°	≡C—Br 180°	
C—I	2.06	—C—I 109°	=C(I) 120°	≡C—I 180°	
N—N	1.47		N—N(H) 127°		
N—O	1.37				=N(O) 110°
N—S (chain)	1.75				—N(S) 107°
N—S (ring)	1.60				—N(S) 104°

Table 4-2 (*Continued*)

Bond	Length (Å)	Bond Angles			
		Tetrahedral	Planar	Linear	Other
Single bonds					
N—Cl	1.69				−N⟨Cl 110°
O—O	1.46		O—O⟨H 102°		
O—P	1.65		O—P⟨O 99°		P—O⟨P/S 128°
O—S	1.70		O—S⟨ 120°		S—O⟨S 120°
O—Cl	1.69				
S—S	2.07				S—S 105°
Double bonds					
C=C	1.34		C=C 120°	=C=C 180°	
C=N	1.22		N=C 120°	=C=N 180°	
C=O	1.22		O=C 120°	=C=O 180°	
C=S	1.61		S=C 120°	=C=S 180°	
N=N (chain)	1.26		N=N 120°		
N=N (ring)	1.39		N=N 120°		
N=O	1.22		N=N 120°		
S=O	1.44		O=S 120°		
P=O	1.54				
Triple bonds					
C≡C	1.20			C≡C— 180°	
C≡N	1.15			N≡C— 180°	
N≡N	1.11			N≡N= 180°	

[a] Averaged from Wheland (1602).

rotate about two perpendicular axes and the moments of inertia will be the same about each:

$$I_y = I_z = \sum (m_i d_i^2). \quad [4\text{-}16]$$

For nonlinear polyatomic molecules the problem is more complex, but it has been systematized by Hirschfelder (608), whose method is presented here. From the information on masses, angles, and distances a three-dimensional diagram is prepared, showing the position of each atom with respect to the three coordinate axes x, y, and z. The Cartesian coordinates x_i, y_i, and z_i are written for each atom. The product of the three principal moments of inertia is equal to the determinant:

$$I_x I_y I_z = \begin{bmatrix} A & -D & -E \\ -D & B & -F \\ -E & -F & C \end{bmatrix}$$

$$= ABC - AF^2 - CD^2 - 2DEF - BE^2,$$

where

$$A = \sum m_i(y_i^2 + z_i^2) - \frac{1}{M}\left(\sum m_i y_i\right)^2 - \frac{1}{M}\left(\sum m_i z_i\right)^2,$$

$$B = \sum m_i(x_i^2 + z_i^2) - \frac{1}{M}\left(\sum m_i x_i\right)^2 - \frac{1}{M}\left(\sum m_i z_i\right)^2,$$

$$C = \sum m_i(x_i^2 + y_i^2) - \frac{1}{M}\left(\sum m_i x_i\right)^2 - \frac{1}{M}\left(\sum m_i y_i\right)^2,$$

$$D = \sum m_i x_i y_i - \frac{1}{M}\left(\sum m_i x_i\right)\left(\sum m_i y_i\right),$$

$$E = \sum m_i x_i z_i - \frac{1}{M}\left(\sum m_i x_i\right)\left(\sum m_i z_i\right),$$

$$F = \sum m_i y_i z_i - \frac{1}{M}\left(\sum m_i y_i\right)\left(\sum m_i z_i\right),$$

$$M = \sum m_i = \text{molecular weight in grams.} \quad [4\text{-}17]$$

It is convenient to express the masses in grams and the distances in Ångström units (10^{-8} cm) and to designate such moments of inertia by primes, as I'. Consequently

$$I_x = \frac{I'_x}{6.0231 \times 10^{23} \times (10^8)^2} = \frac{I'_x}{6.0231 \times 10^{39}}$$

from which

$$I_x I_y I_z = \frac{I'_x I'_y I'_z}{218.504 \times 10^{117}}. \quad [4\text{-}18]$$

Another factor needed to calculate entropy is the symmetry number, σ. Modern usage has separated this factor into two portions, the symmetry number of the molecule as a whole, σ_w and the symmetry of a rotating portion, σ_r. The total symmetry number, σ_t, is the product of all of the symmetry numbers of the molecule:

$$\sigma_t = \sigma_w \Pi \sigma_{r_i} \quad [4\text{-}19]$$

σ_w may be defined as the number of different positions into which a rigid polyatomic molecule can be rotated and still appear unchanged. For the carbon monoxide molecule, C=O, $\sigma_w = 1$. For carbon dioxide, O=C=O, $\sigma_w = 2$. For a tetrahedral molecule such as methane, CH_4, $\sigma_w = 12$, since in each of four molecular orientations there are three separate orientations of the methyl group. For methyl chloride, H_3CCl, $\sigma_w = 3$, since for one orientation of the molecule there are three orientations of the methyl group, or 1×3.

When a portion of the molecule can rotate relative to the rest of the molecule, the symmetry of the rotating part may introduce additional indistinguishable positions giving rise to σ_r. The rigid ethane molecule, H_3C—CH_3, may be rotated end for end, giving two identical positions, and about its C—C axis it may be rotated into three identical positions in 360°, so that $\sigma_w = 2 \times 3 = 6$. In addition, one half of the molecule can be kept rigid while the other half is rotated into three additional identical positions in 360°, so that $\sigma_r = 3$. Thus $\sigma_t = 18$. For a branched hydrocarbon with n methyl groups there are 3^n different arrangements of these methyl groups, leading to $\sigma_r = 3^n$. Neopentane, C_5H_{12}, has $\sigma_w = 12$ (like methane), and three additional arrangements for each of the four methyl groups, so that $\sigma_r = 3^4$, and hence $\sigma_t = 972$.

The amine moiety (e.g., CNH_2) is pyramidal and has $\sigma_r = 1$. The C—O—H moiety is nonlinear and also has $\sigma_r = 1$. Table 4-3, listing symmetry numbers for a few aromatic molecules, is instructive.

Table 4-3 Symmetry Numbers

Molecule	σ_w	σ_r	σ_t	Reference
Benzene	12	1	12	(1477)
Toluene	1	6	6	(1477)
Ethylbenzene	1	6	6	(1477)
o-Xylene	2	9	18	(1477)
m-Xylene	2	9	18	(1477)
p-Xylene	4	9	36	(1477)
1,3,5-Trimethylbenzene	6	27	162	(1477)
Hexamethylbenzene	12	729	8748	(577)
Phenol	1	2	2	(392)
Aniline	1	2	2	(391)
Benzenethiol	1	2	2	(1319)
Biphenyl	4	2	8	(725)

For a linear gaseous molecule the entropy contribution for over-all rotation is given by

$$S_{or}^{\circ} = R + R \ln \left(\frac{8\pi^2 I k T}{h^2 \sigma_w}\right). \quad [4\text{-}20]$$

Substitution of the known constants into this equation leads to

$$S_{or}^{\circ} = 4.5758[\log(I \times 10^{39}) + \log T - \log \sigma_w] - 0.7804 \text{ cal/(mole °K)}. \quad [4\text{-}21]$$

For a nonlinear gaseous molecule the entropy contribution for over-all rotation is given by

$$S_{or}^{\circ} = 1.5 R + R \ln \left[\frac{\pi^{1/2}}{\sigma_w}\left(\frac{8\pi^2 k T}{h^2}\right)^{3/2} (I_x I_y I_z)\right]^{1/2}.$$

Evaluating the constants in this equation gives

$$S_{or}^{\circ} = 2.2879 \log (I_x I_y I_z \times 10^{117}) - 4.5758 \log \sigma_w + 6.8637 \log T$$
$$- 0.0332 \text{ cal/(mole °K)}. \quad [4\text{-}22]$$

Entropy of Internal Rotation

Besides over-all rotation, many gaseous molecules exhibit internal rotation between two or more parts of the molecule. For free internal rotation the entropy of a single rotating group is given by

$$S_{ir}^{\circ} = \tfrac{1}{2}R + R \ln (8\pi^3 k I_r T)^{1/2}.$$

Evaluating all of the constants and expressing I_r (the moment of the rotating fragment about its axis) in gram-centimeters squared gives [in calories per (mole degrees Kelvin)]

$$S_{ir}^{\circ} = 3.0348 - 4.5758 \log \sigma_r + 2.2879 \log (I_r \times 10^{38}) + 2.2879 \log T. \quad [4\text{-}23]$$

When the internal rotation is hindered, it is convenient to use the tables presented by Pitzer (1156) and by Pitzer and Gwinn (1162) and given in the Appendix as Tables A-8 through A-11. For symmetrical tops, [2-11] can be used to calculate the parameter $1/Q_f$ by setting $n_r = \sigma_r$. The decrease in entropy due to the potential barrier of V is then taken from the tables. For unsymmetrical tops with small off-balance factors, it is still possible to use the tables by calculating the contribution of the free rotator from [4-23]. The unsymmetrical top will usually be attached to a carbon, which is in turn bonded tetrahedrally to three other groups and will therefore have three potential energy maxima per revolution. As a first approximation, the maxima will be equal in height and the parameter $1/Q_f$ is calculated using [2-11] and $n_r = 3$. The decrease in entropy due to the potential V is again taken from the Pitzer and Gwinn tables. Other methods have been devised to compute the contributions of hindered internal rotators, for example that of Halford (563). This method is also presented in tabular form in Table A-12 and can be readily used.

Entropy of Vibration

Einstein derived the contribution of each harmonic oscillator to the entropy in the form

$$S_{\text{vib}} = R\left[\frac{x}{e^x - 1} - \ln(1 - e^{-x})\right] \quad [4\text{-}24]$$

on the basis of the same theoretical treatment as for [2-6]. Here $x = hc\omega/kT$, h is Planck's constant, ω is the wave number of the vibrational mode (i.e., the reciprocal of the wavelength expressed in cm^{-1}), k is Boltzmann's constant, and T is the absolute temperature. Values of this function are tabulated as Table A-4 in the Appendix. For a polyatomic molecule the summation of the contributions from each of the normal coordinate frequencies with the translational and rotational terms gives the total entropy. A few examples are presented to demonstrate the use of the foregoing equations.

The linear carbon disulfide molecule shown schematically in Figure 4-2 has a molecular weight of 76.143 g. From Table 4-2 the interatomic C=S distance is 1.61 Å. Brown and Manov (175) have listed the vibrational frequencies, expressed in cm^{-1}, as 1523 [1], 655.5 [1], 396.8 [2]. The numbers in brackets are the degeneracies g_i. The symmetry number σ_w is 2. Substitution into [4-15] gives $S_{\text{tr}} = 6.8637 \log 76.143 + 25.8182 = 38.7331$ cal/(mole °K) at 298.15°K. Application of [4-16] gives the moments of inertia about the y and z axes,

$$I_y = I_z = 2\left[\frac{32.066}{6.023 \times 10^{23}} \times (1.61 \times 10^{-8})^2\right]$$

$$= 2(5.324 \times 10^{-23})(2.592 \times 10^{-16}) = 27.60 \times 10^{-39} \text{ g-cm}^2.$$

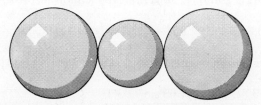

Figure 4-2 Schematic carbon disulfide molecule.

Using [4-21] at 298.15°K gives

$$S°_{or} = 4.5758 (\log 27.60 + \log 298.15 - \log 2) - 0.7804$$
$$= 15.758 \text{ cal/(mole °K)}.$$

Values of x for each of the three vibrational wave numbers at 298.15° are referred to Table A-4 of the Appendix for the corresponding entropy contribution. Thus, taking degeneracies into account,

ω_i(cm^{-1})	g_i	$x_i = \dfrac{hc\omega_i}{kT}$	$\dfrac{S°_i}{R}$	$g_i S°_i$ [cal/(mole °K)]
397	2	1.916	0.48992	1.9472
656	1	3.166	0.18248	0.3626
1523	1	7.350	0.00537	0.0106
			$S_{\text{vib}} =$	2.320
			$S_{\text{or}} =$	15.758
			$S_{\text{tr}} =$	38.733
			$S_{\text{tot}} =$	56.811

This evaluation checks a similar calculation made by Cross (295), who obtained 56.84 cal/(mole °K). Brown and Manov (175) measured the third-law entropy and obtained $S°_{319.35}$, ideal gas = 57.60 cal/(mole °K). Adjusting this value to a corresponding temperature using the mean heat capacity, $[11.01 \times (319.35 - 298.15)]/308.75 = -0.76$, yields $S°_{298.15}$, ideal gas = 56.84 cal/(mole °K), a value well within experimental error.

The methyl bromide molecule shown schematically in Figure 2-9 has a molecular weight of 94.951 g. Egan and Kemp (373) listed its interatomic distances and angles as follows: C—Br, 1.91 Å; C—H, 1.09 Å; ∠H—C—H, 111°; from which can be calculated ∠Br—C—H, 107°57'. The molecule has a symmetry number of $\sigma_w = 3$. Weissman, Bernstein, Rosser, Meister, and Cleveland (1584) reported the following fundamental frequencies (expressed in cm^{-1}) and degeneracies 618[1], 953[2], 1290[1], 1453[2], 2965[1], 3082[2]. Application of [4-15] at 298.15°K gives

$$S_{\text{tr}} = 6.8637 \log 94.951 + 25.8182 = 39.3911 \text{ cal/(mole °K)},$$

and [4-16] gives the moments of inertia about the threefold (or x) axis:

$$I_x = \frac{3 \times 1.008 \times (1.09 \sin 72°3')^2}{6.023 \times 10^{23} \times (10^8)^2} = 5.399 \times 10^{-40} \text{ g-cm}^2.$$

Calculation of the remaining moments of inertia may be accomplished by carefully plotting the molecule to scale (see Fig. 4-3). Substitution of the requisite quantities into [4–17] leads to the following arrays.

Atom	x_i	x_i^2	y_i	y_i^2	z_i	z_i^2
Br	−1.91	3.6481	0	0	0	0
	(1.09 cos 72°)		(1·09 sin 72°)			
H_1	0.319	0.1018	1.0443	1.0733	0	0
			(−1.036 sin 30°)		(−1.036 cos 30°)	
H_2	0.319	0.1018	−0.518	0.2683	−0.897	0.8046
H_3	0.319	0.1018	−0.518	0.2683	+0.897	0.8046

Atom	$m_i x_i$	$m_i y_i$	$m_i z_i$
Br	−152.6396	0	0
H_1	0.3215	1.0443	0
H_2	0.3215	−0.5221	−0.9042
H_3	0.3215	−0.5221	+0.9042
	−151.6751	0	0

Atom	$m_i x_i y_i$	$m_i y_i z_i$	$m_i x_i z_i$
Br	0	0	0
H_1	0.3331	0	0
H_2	−0.1665	−0.4683	0.2884
H_3	−0.1665	0.4683	−0.2884
	0	0	0

Atom	$m_i(x_i^2+y_i^2)$	$m_i(y_i^2+z_i^2)$	$m_i(x_i^2+z_i^2)$
Br	291.4515	0	291.5415
H_1	1.1857	1.0818	0.1026
H_2	0.3731	1.0815	0.9136
H_3	0.3731	1.0815	0.9136
	293.4734	3.2448	293.4713

$$\frac{1}{94.951} = 0.010532,$$

Evaluation of Entropy

$A = 3.2448 - 0.010532(0) - 0.010532(0) = 3.2448,$
$B = 293.4713 - 0.010532\ (151.6751)^2 - 0.010532(0) = 51.1792,$
$C = 293.4734 - 0.010532\ (151.6751)^2 - 0.010532(0) = 51.1813,$
$D = 0 - 0.010532(0)(151.6751) = 0,$
$E = 0 - 0.010532(151.6751)(0) = 0,$
$F = 0 - 0.010532(0)(0) = 0.$

$I'_x I'_y I'_z = (3.2448)(51.1792)(51.1813) - 0 - 0 - 2(0) - 0 = 8499.5,$

$$I_x I_y I_z = \frac{8499.5}{218.5 \times 10^{117}} = 38.899 \times 10^{-117},$$

$$I_y I_z = \frac{38.899 \times 10^{-117}}{5.399 \times 10^{-40}} = 7204 \times 10^{-80},$$

$I_y = I_z = (7204 \times 10^{-80})^{1/2} = 84.88 \times 10^{-40}$ g-cm^2.

Substitution of these results into [4–22] leads to

$S_{or} = 2.2879 \log 38.899 - 4.5758 \log 3 + 6.8637 \log 298.15 - 0.0332$
$\phantom{S_{or}} = 18.405$ cal/(mole °K) at 298.15°K.

The vibrational contributions to the entropy are obtained from Table A-4 of the Appendix using values of x based on the fundamental frequencies.

ω_i (cm^{-1})	g_i	$x_i = \dfrac{hc\omega_i}{kT}$	$\dfrac{S^\circ_i}{R}$	$g_i S^\circ_i$ [cal/(mole °K)]
618	1	2.982	0.21125	0.4198
953	2	4.599	0.05681	0.2258
1290	1	6.225	0.01432	0.0285
1453	2	7.012	0.00722	0.0287
2965	1	14.039	0.00001	0.0000
3082	2	14.873	0.00001	0.0000
			$S_{vib} =$	0.703
			$S_{tr} =$	39.391
			$S_{or} =$	18.405
			$S_{tot} =$	58.499

This value compares favorably with the 298.10°K calorimetric value, 58.61 ± 0.10 cal/(mole °K), of Egan and Kemp (373).

An interesting example revealing internal rotation is 1,1,1-trifluoroethane, shown schematically in Figure 2-12. The most consistent values of the three sets of interatomic distances and angles reported by Wheland (1602) are: C−C, 1.54 Å; C−F, 1.33 Å; C−H, 1.09 Å; ∠ FCF, 108.5°; ∠ HCC, 109°28′, from which can be calculated ∠ FCC, 110°26′. The molecule has a symmetry number σ_w of 3. Nielsen, Claassen, and Smith (1070) reported the following fundamental vibrational frequencies (in terms of wave numbers, cm^{-1}) and degeneracies (in brackets): 238 [1], 365 [2], 541 [2], 603 [1], 830 [1], 969 [2], 1232 [2], 1278 [1], 1408 [1], 1443 [2], 2978 [1], 3036 [2].

At 298.15°K the translational entropy is given by substitution of the molecular weight (84.046 g) into [4-15],

$$S_{tr} = 6.8637 \log 84.046 + 25.8182 = 39.0275 \text{ cal/(mole °K)}.$$

A scale plot of the molecule (see Fig. 2-12) is of assistance in computing the over-all rotational moments of inertia by the method of [4-16]. The coordinates thus deduced yield the following arrays.

Atom	x_i	x_i^2	y_i	y_i^2	z_i	z_i^2
C_1	−0.770	0.593	0	0	0	0
C_2	0.770	0.593	0	0	0	0
F_1	1.237	1.530	−0.623	0.388	1.079	1.164
F_2	1.237	1.530	1.246	1.552	0	0
F_3	1.237	1.530	−0.623	0.388	−1.079	1.164
H_1	−1.133	1.284	0.513	0.263	−0.889	0.790
H_2	−1.133	1.284	−1.027	1.055	0	0
H_3	−1.133	1.284	0.513	0.263	0.889	0.790

Atom	$m_i x_i$	$m_i y_i$	$m_i z_i$
C_1	−9.248	0	0
C_2	9.248	0	0
F_1	23.503	−11.837	20.501
F_2	23.503	23.674	0
F_3	23.503	−11.837	−20.501
H_1	−1.142	0.517	−0.896
H_2	−1.142	−1.035	0
H_3	−1.142	0.517	0.896
	67.083	0	0

Evaluation of Entropy

Atom	$m_i x_i y_i$	$m_i y_i z_i$	$m_i x_i z_i$
C_1	0	0	0
C_2	0	0	0
F_1	−14.642	−12.772	25.359
F_2	29.284	0	0
F_3	−14.642	12.772	−25.359
H_1	−0.586	−0.460	1.015
H_2	1.172	0	0
H_3	−0.586	0.460	−1.015
	0	0	0

Atom	$m_i(x_i^2+y_i^2)$	$m_i(y_i^2+z_i^2)$	$m_i(x_i^2+z_i^2)$
C_1	7.122	0	7.122
C_2	7.122	0	7.122
F_1	36.442	29.488	51.186
F_2	58.558	29.488	29.070
F_3	36.442	29.488	51.186
H_1	1.559	1.061	2.091
H_2	2.358	1.063	1.294
H_3	1.559	1.061	2.091
	151.162	91.649	151.162

$$\frac{1}{84.046} = 0.011898,$$

$A = 91.649 - 0.01190(0)^2 - 0.01190(0)^2 = 91.649,$

$B = 151.162 - 0.01190(67.083)^2 - 0.01190(0)^2 = 97.611,$

$C = 151.162 - 0.01190(67.083)^2 - 0.01190(0)^2 = 97.611,$

$D = 0 - 0.01190(67.083)(0) = 0,$

$E = 0 - 0.01190(67.083)(0) = 0,$

$F = 0 - 0.01190(0)(0) = 0.$

$I'_x I'_y I'_z = (91.649)(97.611)(97.611) - (0) - (0) - (0) - (0)$
$= 873,223,$

$$I_x I_y I_z = \frac{873,223}{218.5 \times 10^{117}} = 3996.4 \times 10^{-117}.$$

Substitution of this value into [4-22] gives (at 298.15°K)

$S_{or} = 2.2879 \log 3996.4 - 4.5758 \log 3 + 6.8637 \log 298.15 - 0.0332 =$

$= 23.0075 \text{ cal/(mole °K)}.$

As shown on page 45, the lowest vibrational frequency, 238[1], is in reality the frequency of torsional oscillation of the H_3C and CF_3 groups about the C—C bond with a barrier of 3.52 kcal/mole hindering this rotation. This type of hindered rotation has been treated by Pitzer and Gwinn (1162), and, when information is substituted into [2-11] as shown in the example on page 47,

$$\frac{1}{Q_f} = 0.2766 \quad \text{and} \quad \frac{V}{RT} = 5.85.$$

Reference to Table A-10 yields the entropy contribution for this hindered rotator, S_{hr}, of 2.022 cal/(mole °K).

The vibrational contribution to the entropy is obtained from the fundamental frequencies and degeneracies by the use of Table A-4.

ω_i (cm^{-1})	g_i	$x_i = \dfrac{hc\omega_i}{kT}$	$\dfrac{S_i^\circ}{R}$	$g_i S_i^\circ$ [cal/(mole °K)]
365	2	1.761	0.55408	2.2022
541	2	2.611	0.28331	1.1260
603	1	2.910	0.22368	0.4445
830	1	4.006	0.09266	0.1841
969	2	4.676	0.05333	0.2120
1232	2	5.946	0.01822	0.0724
1278	1	6.168	0.01506	0.0299
1408	1	6.795	0.00873	0.0173
1443	2	6.964	0.00754	0.0299
2978	1	14.372	0.00001	0.0000
3036	2	14.652	0.00001	0.0000
			$S_{vib} =$	4.318
			$S_{tr} =$	39.027
			$S_{or} =$	23.008
			$S_{hr} =$	2.022
		S_{298}° (ideal gas) $=$		68.375

Russell, Golding, and Yost (1270) determined the third-law entropy of the real gas at 224.40°K and 0.9330 atm as 63.814 ± 0.10 cal/(mole °K). Correcting this for gas imperfection (0.135), for compression to 1 atm

($R \ln 0.933 = -0.138$), and for change in temperature to 298°K as an ideal gas $\left(\int Cp \, d\ln T = 4.829 \right)$ (cf. 1070, 1527) yields $S°_{298}$ for the ideal gas = 68.640 ± 0.15 cal/(mole °K), which compares favorably with the calculated value.

EVALUATION OF ENTROPY AS A FUNCTION OF TEMPERATURE

There are two general methods for evaluating the entropy of a substance as a function of temperature. One of these rests on [4-12]; it requires the measurement of heat capacity data and its integration from near 0°K through the temperature range of interest and may concern either gaseous or condensed phases. Values obtained in this way are known as *third-law* entropies.

The other approach for gaseous substances depends on spectroscopic methods as presented on pages 93-107. Entropies calculated in this manner are often referred to as *spectroscopic* values. They are usually considered to be more reliable, provided that the basic data for the calculation are well known. Especially for simple molecules having clear-cut frequency assignments and no internal rotation, the calculated entropies are very reliable. For larger and more complex molecules, however, the reliability of the spectroscopic calculation is somewhat diminished by inadequately known factors:

1. Frequency assignment (incomplete or obscured by complexities and/or uncertainties).
2. Detailed configuration of the molecule (e.g., *gauche, skew, cis,* and *trans* forms occur, and *dextro, levo, racemic,* and *meso* forms also are encountered for optically active substances).
3. Interatomic distances and bond angles.
4. Energies and shapes of potential barriers to internal rotation (unknown or poorly defined).

When one value of the absolute entropy is obtained from either third-law or spectroscopic data, the entropy at any other temperature may readily be calculated from [4-4]. As noted earlier, the reference state for organic compounds is taken here to be the ideal gaseous state unless stated otherwise. Substitution of the gaseous heat capacity equations [2-13] and [2-15] into [4-4] gives

$$S_{T_2} = S_{T_1} + \int_{T_1}^{T_2} \left(\frac{a+bT+cT^2}{T}\right) dT$$

$$= S_{T_1} + a \ln \frac{T_2}{T_1} + b(T_2 - T_1) + \frac{c}{2}(T_2^2 - T_1^2) \quad [4\text{-}25]$$

and

$$S_{T_2} = S_{T_1} + \int_{T_1}^{T_2} \left(\frac{a''+b''T+c''T^{-2}}{T}\right) dT$$

$$= S_{T_1} + a'' \ln \frac{T_2}{T_1} + b''(T_2 - T_1) - \frac{c}{2}(T_2^{-2} - T_1^{-2}), \quad [4\text{-}26]$$

in which the constants a, b, c, a'', b'', and c'' have the same significance as in the respective heat capacity equations.

The heat capacity of liquid ethanol is given by

$$Cp = -10.83 + 0.118T + 0.000025T^2 \text{ cal/(mole °K)}.$$

From the measurements of Kelley (733) the entropy of liquid ethanol at 300°K is 38.56 cal/(mole °K). Substitution of this information into [4-24] gives at 350°K

$$S_{350} = 38.56 - 10.83 \ln \frac{350}{300} + 0.118(350-300) + 1.25$$

$$\times 10^{-5}[(350)^2 - (300)^2] = 43.20 \text{ cal/(mole °K)}.$$

On page 51 the enthalpy increment of liquid ethanol over this temperature range was calculated to be 1508 cal/mole. For temperature increments small compared to $\bar{T}$ (the mean temperature of the interval), $\Delta S \approx \Delta H/\bar{T}$ or $S_{350} - S_{300} = 1508/325 = 4.64$ cal/(mole °K). This method is seen to check the one immediately above and is a useful shortcut.

At 500°K the entropy of gaseous ethanol according to Barrow (76) is 77.40 cal/(mole °K) while the gaseous heat capacity may be represented by

$$Cp = 51.84 - 2.023 \times 10^4 T^{-1} + 2.843 \times 10^6 T^{-2} \text{ cal/(mole °K)}.$$

Use of [4-25] gives at 600°K

$$S_{600} = 77.40 + 51.84 \ln \frac{600}{500} + 2.023 \times 10^4 [(600)^{-1} - (500)^{-1}]$$

$$- 1.421 \times 10^6 [(600)^{-2} - (500)^{-2}] = 81.84 \text{ cal/(mole °K)}.$$

The above heat capacity equation leads to an enthalpy increment for gaseous ethanol over the temperature range of 2444 cal/mole. As a check $S_{600} - S_{500} \approx 2444/550 = 4.44$ cal/(mole °K).

For high-speed machine calculation it is convenient to represent the heat capacity in tabular form at evenly spaced temperatures. (See page 51 of Chapter 2.) Such values may be read from smoothed curves through experimental points or may be interpolated analytically at even values of the temperature. Simpson's parabolic rule permits rapid integration of such a series of evenly spaced values. Division of each value of the heat capacity by its temperature yields quotients designated $S_0, S_1, S_2, \ldots$, at temperatures $T_0, T_1, T_2, \ldots$, separated by evenly spaced increments ΔT, as shown in Figure 4-3. Passing a parabolic arc through three points at a time permits integration over that region. Beginning at T_0, T_1, and T_2, the increment in entropy over the two ΔT intervals equals $S_{T_2} - S_{T_0} = \Delta T \times (S_0 + 4S_1 + S_2)/3$.

If the intervals are spaced closely enough, deviation from linearity may be sufficiently small to permit direct calculation of ΔS for that interval by means of the relationship

$$S_{T_2} - S_{T_1} = \frac{H_{T_2} - H_{T_1}}{(T_2 + T_1)/2}.$$

For example, $S_{T_1} - S_{T_0}$ may be calculated in this fashion and subtracted from $S_{T_2} - S_{T_0}$ to give $S_{T_2} - S_{T_1}$, and so on; alternatively, increments of size $\Delta T/2$ may be employed.

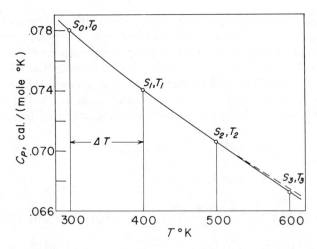

Figure 4-3 Entropy integration of Simpson's parabolic rule.

Evaluation as a Function of Temperature

The entropy of gaseous n-butane at 298.15°K is given as 74.12 cal/(mole °K). The enthalpy increment of n-butane from 298.15°K to 300°K has previously been calculated as 43 cal/mole, so that $43/299.07 = 0.14$ cal/(mole °K). Likewise the enthalpy increment from 300° to 400°K was calculated to be 2650 cal/mole, so that $2650/350 = 7.57$ cal/(mole °K).

Table 4-4 Heat Capacity and Entropy of Gaseous n-Butane[a]

T(°K)	Cp	$\dfrac{Cp}{T}$	S, Calculated	S, Literature[b]
298.15	23.29	0.07811	74.12	74.12
300	23.40	0.07800	74.26	74.27
400	29.60	0.07400	81.83	81.86
500	35.34	0.07068	89.08	89.10
600	40.30	0.06717	95.96	95.97

[a] In calories per (mole degrees Kelvin).
[b] Reported by Person and Pimentel (1140).

Utilizing heat capacity values from Table 4-4 and applying Simpson's parabolic rule to the interval from 300° to 500°K gives

$$S_{500} - S_{300} = 100 \left[\frac{0.07800 + 4(0.07400) + 0.07068}{3} \right]$$

$$= 14.822 \text{ cal/(mole °K)}.$$

Thus $(S_{500} - S_{300}) - (S_{400} - S_{300}) = (S_{500} - S_{400}) = 14.82 - 7.57 = 7.25$ cal/(mole °K). The next double interval is

$$S_{600} - S_{400} = 100 \left[\frac{0.07400 + 4(0.07068) + 0.06717}{3} \right]$$

$$= 14.129 \text{ cal/(mole °K)},$$

and $(S_{600} - S_{400}) - (S_{500} - S_{400}) = (S_{600} - S_{500}) = 14.13 - 7.25 = 6.88$ cal/(mole °K). The entropies calculated by the above process are tabulated in column 4 of Table 4-4 and compared with those obtained by Person and Pimentel (1140) in the last column.

ENTROPY INCREMENT FOR CHEMICAL REACTION

When a single substance changes from state A to state B, the change is accompanied by an entropy increment $S_B - S_A = \Delta S$. When a chemical reaction takes place there is also an entropy change, which may be generalized by the notation developed in Chapter 3 as follows:

$$\Delta Sr = \sum \nu_i S_i. \qquad [4\text{-}27]$$

In discussing processes and reactions it is important that the state of the substances involved be clearly described. For elements the standard reference state is selected as the condensed state up to the temperature at which the vapor pressure of the element reaches 1 atm, and as the ideal gas state above this temperature. Practical entropy values for the standard reference states of elements are given in Chapters 8 and 14. In conformance with the discussion given in Chapter 1, the standard reference state for organic compounds in this book has been selected as the ideal gas state at 1 atm pressure (indicated by a degree sign) at all temperatures unless otherwise noted. Other states of interest may be related to the standard state.

Spectroscopic calculations, or measurements of heat capacity from near absolute zero to 298.15°K (a convenient comparison temperature), have been used to compile the practical entropies in Table 4-5. From this information we may calculate the entropy increment for the hypothetical reaction

$$C\,(\text{graphite}) + 2H_2(g) = CH_4(g):$$

$$\Delta Sr^\circ_{298} = \Delta Sf^\circ_{298}\, CH_4(g) = 44.50 - [1.36 + 2(31.21)]$$

$$= -19.28 \text{ cal/(mole °K)}.$$

Here the superscript degree sign indicates that each substance is in its standard reference state, and the modifier f indicates that the reaction involves *formation* from elements at the indicated temperature. Note that when a system goes from a state of lower order (reactants in the above example) to one of higher order (products) its entropy diminishes.

Other reactions for which entropy changes can be calculated from the data in Table 4-5 are

$$C\,(\text{graphite}) + O_2(g) = CO_2(g):$$

$$\Delta Sf^\circ_{298} = 51.07 - [1.36 + 49.00] = 0.69 \text{ cal/(mole °K)}.$$

Note that the increase in entropy of an ordered solid (graphite) changed to

Table 4-5 Practical Molal Entropies at 298.15°K[a]

Substance	State	S°_{298}
Graphite	(s)	1.36
Hydrogen	(g)	31.21
Methane	(g)	44.50
Oxygen	(g)	49.00
Carbon dioxide	(g)	51.07
Water	(g)	45.11

[a]In calories per (mole degrees Kelvin); see Chapter 8.

a gaseous compound was not compensated for here by as large a decrease in the moles of gas as in the previous example. Also,

$$H_2(g) + \tfrac{1}{2}O_2(g) = H_2O(g):$$

$$\Delta S f^0_{298} = 45.11 - [31.21 + 24.50] = -10.60 \text{ cal/(mole °K)}$$

and

$$CH_4(g) + 2O_2(g) = CO_2(g) + 2H_2O(g):$$

$$\Delta S c^\circ_{298} = 51.07 + 2(45.11) - [44.50 + 2(49.00)] = -1.23 \text{ cal/(mole °K)}.$$

Entropy Increment of a Reaction from Its Equilibrium Constant at Two Temperatures

At constant pressure the entropy change of a reaction in terms of its equilibrium constant is given by differentiation of the relation $\Delta G^\circ = -RT \ln Kp$:

$$-\left(\frac{\partial \Delta G^\circ}{\partial T}\right)_P = \Delta S^\circ = R\left[T\left(\frac{\partial \ln Kp}{\partial T}\right)_P + \ln Kp\right].$$

Hence at the mean temperature, $\bar{T}$, of a sufficiently small interval.

$$\Delta S^\circ_{\bar{T}} = 4.575\left[\bar{T}\left(\frac{\log Kp_2 - \log Kp_1}{T_2 - T_1}\right) + \left(\frac{\log Kp_2 + \log Kp_1}{2}\right)\right] \text{ cal/(mole °K)}$$

[4-28]

Ghosh, Guha, and Roy (472) measured the equilibrium constant for the reaction involving the dehydrogenation of ethylbenzene to styrene,

$$C_6H_5CH_2CH_3(g) = C_6H_5CHCH_2(g) + H_2(g),$$

and reported $Kp_{703} = 4.95 \times 10^{-3}$ and $Kp_{733} = 1.20 \times 10^{-2}$ (pressures in atmospheres). Substitution into [4-28] gives

$$\Delta S^\circ_{718} = 4.5758 \left\{ 718 \left[\frac{(-1.9208) - (-2.3054)}{733 - 703} \right] \right.$$
$$\left. + (-2.1131) \right\} = 32.45 \text{ cal/(mole °K)}.$$

Reference to the tables of thermal properties for the above materials shows that the calculated standard entropy increments are 30.87 and 31.07 cal/(mole °K) at 700° and 800°K, respectively. In view of the experimental difficulty of measuring these equilibria at these temperatures, the agreement is satisfactory.

CHAPTER 5

GIBBS ENERGY AND CHEMICAL EQUILIBRIUM

GIBBS ENERGY

It is common experience that active, spontaneous reactions, such as combustions of hydrocarbons, are accompanied by the liberation of heat. It was natural, therefore, for such researchers as M. Berthelot (121) and Julius Thomsen (1495) to generalize that the amount of heat generated by a reaction is a measure of its "driving force." By means of thermochemical measurements and the application of Hess' law, these scientists hoped to be able to predict the feasibility of any postulated reaction without the necessity of making an actual laboratory experiment. It later became evident, however, that entropy as well as enthalpy is involved in the driving force of a reaction. A new function, originally called the *free energy* and currently termed the *Gibbs energy*, was proposed and discussed by J. W. Gibbs (479) and further developed by G. N. Lewis and M. Randall (860). This function, G, is mathematically defined by the relation

$$G = H - TS. \qquad [5\text{-}1]$$

Because both H and S are extensive properties of the system, G is also a single-valued, extensive property of the system, and its value is independent of the path traversed in passing from initial state 1 to final state 2. For an isothermal process or reaction

$$G_2 - G_1 = \Delta Gr = \Delta Hr - T\Delta Sr. \qquad [5\text{-}2]$$

This is one form of the Gibbs-Helmholtz equation, and ΔGr may be described as the *Gibbs energy increment* or the *free energy of reaction*. A similar function can be defined in terms of energy and entropy:

$$A = E - TS. \qquad [5\text{-}3]$$

A is referred to as the Helmholtz energy or the *work function*. Since constant pressure conditions are more generally encountered by chemists, G is the more useful function and will be called simply the Gibbs energy. It should be noted that in the past F has been used chiefly in the United States and Russia; in European and other countries G has been commonly used to denote the Gibbs energy (see discussion in Chapter 8).

The function ΔGr is a true index of the driving force of a reaction or process. In general a spontaneous reaction must be accompanied by a decrease in Gibbs energy of the system. When the Gibbs energy of a system is at a minimum, there is no tendency toward further change.

Since Gibbs energies involve enthalpies, absolute values of the Gibbs energy cannot be assigned to a system. Gibbs energies can only be expressed as differences between a given state and some standard reference state. Chapter 8 presents thermodynamic data for the standard reference state of each element present in organic compounds, and for selected inorganic compounds; similar data tables for the standard reference states of organic compounds are to be found in succeeding chapters. The data for the compounds contain the Gibbs energy increment accompanying the formation of the compound in its standard reference state from the elements in their standard reference states. For the compounds themselves a state of the broadest utility is desired, and the standard reference state for most of the compounds listed is the hypothetical ideal gas at 1 atm. For some compounds at very low vapor pressures, data are given for the solid or liquid state. The ideal gas standard state can in general be used directly in calculations at low pressures, and corrections for nonideality at high pressures are readily applied to the ideal state properties. The Gibbs energy of formation of a compound in its standard state of ideal gas, pure liquid, or solid at 1 atm from the elements in their standard states is called the *standard Gibbs energy of formation* and is identified by a superscript degree sign: $\Delta Gf°$.

In illustration of these principles the thermodynamic properties of the reaction

$$C_2H_5Cl(g) = C_2H_4(g) + HCl(g)$$

may be evaluated from tables of thermodynamic data in later chapters at 298°K as

$\Delta Hr° = 17.14$ kcal/mole, $\quad \Delta Sr° = 31.15$ cal/(mole °K),

$$\Delta Gr° = 7.86 \text{ kcal/mole}.$$

The reaction as written absorbs heat and also involves an increase in Gibbs energy, and from neither the Gibbs energy nor the enthalpy criteria would the reaction be expected to proceed; in other words, ethyl chloride at 298°K is stable with respect to dissociation into ethylene and hydrogen chloride. However, at 1000°K the data are

$$\Delta Hr° = 17.07 \text{ kcal/mole}, \quad \Delta Sr° = 31.50 \text{ cal/(mole °K)},$$
$$\Delta Gr° = -14.42 \text{ kcal/mole}.$$

The entropy and enthalpy increments of reaction have changed little, but the reaction now involves a considerable decrease in Gibbs energy because of the increased importance of the $T\Delta Sr$ term. Experimentally, ethyl chloride at high temperatures spontaneously dissociates to ethylene and hydrogen chloride, confirming the validity of ΔGr, rather than ΔHr, as an index to the driving force. For many reactions ΔSr is small, and for these reactions ΔHr is quite similar to ΔGr. If there is a decrease in entropy, the Gibbs energy decreases less then the enthalpy, for example, for the reaction at 298°K

$$C \text{ (graphite)} + 2H_2(g) = CH_4(g),$$
$$\Delta Hf° = -17.89 \text{ kcal/mole}, \quad \Delta Sf° = -19.29 \text{ cal/(mole °K)},$$
$$\Delta Gf° = -12.14 \text{ kcal/mole}.$$

If the entropy increment is small, $\Delta Gr \simeq \Delta Hr$. For the reaction

$$C(\text{graphite}) + O_2(g) = CO_2(g)$$
$$\Delta Hf° = -94.05 \text{ kcal/mole}, \quad \Delta Sf° = 0.68 \text{ cal/(mole °K)},$$
$$\Delta Gf° = -94.27 \text{ kcal/mole}.$$

If the entropy increases, the decrease in Gibbs energy is greater than the decrease in enthalpy; for example, for the reaction

$$C \text{ (graphite)} + \tfrac{1}{2}O_2(g) = CO(g)$$
$$\Delta Hf° = -26.42 \text{ kcal/mole}, \quad \Delta Sf° = 21.42 \text{ cal/(mole °K)},$$
$$\Delta Gf° = -32.80 \text{ kcal/mole}.$$

Thus methane is less stable and carbon monoxide more stable than would be expected from the enthalpy of reaction.

As has been stated before, factors other than the Gibbs energy change are involved in chemical reactions. Kinetic factors are of great importance; for example, the higher members of the aliphatic hydrocarbons have positive Gibbs energies of formation (see tables in Chapter 9) and should spontaneously decompose to carbon and lower members such as methane. At 298°K the reaction rate is imperceptible, and hydrocarbons such as kerosene and lubricating oils can be stored indefinitely. However, large amounts of methane are formed when these hydrocarbons are passed over heated brickwork and the reaction rate is high. Catalysts and increased temperatures are frequently necessary to allow organic reactions to proceed at practical rates.

It is tempting to the novice to oversimplify the $\Delta Gr°$ criterion and regard reactions with positive standard Gibbs energy increments as impracticable. Such a clear-cut choice is rarely possible; it is usually necessary to take into account both the magnitude of the $\Delta Gr°$ and the type of reaction. Although some reactions with small positive $\Delta Gr°$'s may give appreciable yields of the desired product under feasible experimental constraints, reactions with small negative $\Delta Gr°$'s may give disappointingly low yields. Quantitative calculation of the composition of a system from Gibbs energy data is the subject of the next section. Qualitatively, however, reactions with negative standard Gibbs energy increments should be considered promising; reactions with zero or small positive increments may be worthy of further investigation and practicable; but those reactions with positive increments larger than 10 kcal are possible only under unusual conditions.

GIBBS ENERGY AND EQUILIBRIUM COMPOSITION

In Chapter 1 a system that is not measurably changing was defined as being at equilibrium. Two types of equilibrium were recognized: stable equilibrium, in which the system is undergoing no change at all, and a frozen or metastable equilibrium, in which the system is changing but at an imperceptible rate. A stable equilibrium is a state in which the Gibbs energy of the system is at a minimum with respect to all possible states of the system under constraints of constant pressure and temperature. An unstable equilibrium is a state in which the Gibbs energy is at a minimum with respect to *some* but *not all* other states of the system. An example of stable equilibrium is carbon dioxide gas in the vicinity of 300°K. There are no chemical species composed of carbon and oxygen with a lower Gibbs energy. An example of metastable equilibrium is a mixture of carbon monoxide and oxygen at room temperature. This system has a lower Gibbs energy than free carbon and oxygen and is thus at equilibrium with respect to these elements. However, further spontaneous change should take place, since reaction to form carbon dioxide would result in a decrease in the Gibbs energy. This reaction is so slow at room temperature that for all practical purposes no reaction takes place and the system can be regarded as at rest. It is not always possible

Gibbs Energy and Equilibrium Composition

to realize metastable equilibria experimentally, since the reaction may not stop at the intermediate stage. Kinetic considerations provide the determining factors.

For the quantitative application of Gibbs energies to chemical problems it is convenient to develop an expression for reversible processes involving only PV work by utilizing the first and second laws in differential form; that is $dE = \delta Q - P\,dV$ and $dS = \delta Q/T$. By substitution

$$dE = T\,dS - P\,dV. \qquad [5\text{-}4]$$

Since $H = E + PV$ (see Chapter 2), $dH = dE + d(PV)$. Hence adding $d(PV) - d(TS)$ to both members of [5-4] gives $dE + d(PV) - d(TS) = T\,dS - P\,dV + d(PV) - d(TS)$. By comparison with [5-1] in differential form, $dG = dH - d(TS)$, the left member is seen to be dG, and

$$dG = -S\,dT + V\,dP. \qquad [5\text{-}5]$$

Therefore at constant temperature ($dT = 0$)

$$dG = V\,dP. \qquad [5\text{-}6]$$

For an ideal gas $V = RT/P$; hence

$$dG = \frac{RT\,(dP)}{P} = RT\,d(\ln P)$$

and, upon integrating from P_1 to P,

$$G(p) = RT\ln\frac{P}{P_1} + G(p_1).$$

Or, if P_1 is arbitrarily selected as a unit pressure (conventionally in units of atmospheres), this equation becomes

$$G(p) = RT\ln P + G^\circ, \qquad [5\text{-}7]$$

relating the Gibbs energy of an ideal gas at a pressure P, $G(p)$, to that in its standard state, G°.

In treating real gases one might proceed in a similar manner by employing the proper equation of state. Considerable convenience, however, accrues from the use of a "fudge" function devised by G. N. Lewis, designated fugacity and symbolized by f. This permits the retention of the simple form of [5-7] even for real gases, provided that a corresponding state of unit fugacity ($f = 1$ atm) is selected. Thus

$$G(f) = RT\ln f + G^\circ. \qquad [5\text{-}8]$$

Obviously the fugacity and the pressure are identical for ideal gases. For real gases the fugacity may be calculated by methods described in a subsequent section of this chapter.

For some applications, especially those involving condensed phases, it is convenient to use a standard state other than unit pressure. Another function, designated *activity*, symbolized by a, and defined by $a = f/f°$, was also proposed by Lewis (859). Hence

$$G(a) = RT \ln a + G° \qquad [5\text{-}9]$$

and obviously the standard state corresponds to one of unit activity ($a = 1$). In this state the fugacity f is equal to the standard fugacity $f°$; however, the latter quantity is selected arbitrarily and need not be unity.

Noting once again that a chemical reaction may be generalized to

$$\sum \nu_i \mathscr{S}_i = 0,$$

we may write the expression for the Gibbs energy of reaction as

$$\Delta Gr° = \sum \nu_i G_i.$$

For a reaction involving only ideal gases (or, to a close approximation, real gases at low pressures) the individual G_i's are given by [5-7]; hence

$$\Delta Gr = \sum \nu_i (RT \ln P_i) + \sum \nu_i G_i°$$
$$= RT \sum \ln (P_i) \nu_i + \Delta G°$$
$$= RT \ln \Pi \, P_i \nu_i + \Delta G°. \qquad [5\text{-}10]$$

In this equation, often termed the reaction isotherm, the function $\Pi \, (P_i)^{\nu_i}$ has the form of a quotient of pressures, each raised to a power corresponding to the stoichiometric coefficient, and hence resembles the equilibrium constant expressed in pressures. Only under conditions of equilibrium, however, does it become identical with the equilibrium constant. At equilibrium $\Delta Gr = 0$ and

$$\Delta Gr° = -RT \ln \left[\Pi \, (P_i) \nu_i\right]_{eq} = -RT \ln Kp. \qquad [5\text{-}11]$$

Since the standard Gibbs energy increments for reactions are readily calculable from tables given in succeeding chapters, the equilibrium constant is calculated from [5-11]. The partial pressure of each reactant and product is calculated at equilibrium from the equilibrium constant using the initial pressures of the reactants and products and the total pressure of the system.

Gibbs Energy and Equilibrium Composition

It is often more convenient to express yields in moles rather than in partial pressures. In a mixture of ideal gases the partial pressure P_i of each constituent is equal to its mole fraction N_i times the total pressure

$$P_i = N_i P_{tot}. \qquad [5\text{-}12]$$

The mole fraction is the number of moles of each constituent, n_i, divided by the total number of moles, n_{tot}; thus $N_i = n_i/n_{tot}$, and Kp may be expressed as

$$Kp = Kn \left(\frac{P_{tot}}{n_{tot}}\right)^{\Delta \nu}, \qquad [5\text{-}13]$$

in which Kn is the equilibrium constant expressed in number of moles and $\Delta \nu$ is the increment in number of moles of gas in the reaction; that is,

$$\Delta \nu = \sum \nu_i.$$

Several examples will make the use of [5-11] and [5-13] evident. Consider the dissociation reaction

$$C_2H_5Cl(g) = C_2H_4(g) + HCl(g).$$

At 400°K the standard Gibbs energy increment from tables in later chapters is $\Delta Gr° = -23.00 + 17.69 - (-9.96) = 4.65$ kcal, from which the equilibrium constant is calculated as

$$Kp = 10^{-\Delta Gr°/(4.5758\,T)} = 0.00292 = \frac{(P_{HCl})(P_{C_2H_4})}{(P_{C_2H_5Cl})}.$$

The equilibrium constant expression in terms of numbers of moles at 1 atm total pressure is

$$Kp = Kn \left(\frac{P_{tot}}{n_{tot}}\right)^{\Delta \nu} = \frac{(n_{HCl})(n_{C_2H_4})}{(n_{C_2H_5Cl})} \left(\frac{1}{n_{tot}}\right)^1.$$

Assuming that only 1 mole of C_2H_5Cl is present initially and that x moles of HCl and C_2H_4 are formed, $n_{HCl} = n_{C_2H_4} = x$, $n_{C_2H_5Cl} = (1-x)$, $n_{tot} = (1+x)$, and

$$Kp = \frac{x^2}{(1-x)(1+x)} = 0.00292.$$

Solution by the quadratic formula or by successive approximation gives $x = 0.054$. Thus, even though $\Delta Gr°$ is positive by over 4 kcal, there is an appreciable concentration of ethylene and hydrogen chloride at equilibrium, since 5.4% of the ethyl chloride dissociates. Because there is an increase in volume as the reaction proceeds, higher pressures diminish the degree of dissociation. Under a total pressure of 2 atm

$$\frac{2x^2}{(1-x)(1+x)} = 0.00292 \quad \text{and} \quad x = 0.038.$$

If either or both products (hydrogen chloride or ethylene) are present initially, the dissociation is also diminished. Assuming that initially 1 mole of ethyl chloride and 1 mole of hydrogen chloride are present, at equilibrium there will be x moles of ethylene, $(1+x)$ moles of hydrogen chloride, $(1-x)$ moles of ethyl chloride, and $(2+x)$ total moles. Assuming 1 atm total pressure,

$$\frac{x(1+x)}{(1-x)(2+x)} = 0.00292 \quad \text{and} \quad x = 0.006.$$

Thus dissociation of ethyl chloride is considerably repressed by the presence of one of the products in an initial concentration equal to that of the reactants.

Finally, the effect of the presence of an inert gas, such as neon or helium, is considered. Assuming an initial mixture of 1 mole of ethyl chloride and 1 mole of inert gas, at equilibrium there will be x moles of ethylene, x moles of hydrogen chloride, $(1-x)$ moles of ethyl chloride, and $(2+x)$ total moles of mixture. At 1 atm total pressure

$$\frac{x^2}{(1-x)(2+x)} = 0.00292 \quad \text{and} \quad x = 0.076.$$

Thus addition of an inert gas at constant pressure favors the reaction essentially by increasing the volume. For reactions having an equal number of moles on each side, neither change of pressure nor addition of inert gas changes the equilibrium composition for ideal gases.

The type of reaction is also important in the relation between equilibrium composition and the standard Gibbs energy increment. Assuming that the $\Delta Gr°$'s for each of the following three reactions,

$$A + B = C$$
$$A + B = 2C$$
$$A + B = 3C,$$

are the same and such that the Kp's $= 0.05$, the equilibrium constants for

Gibbs Energy and Equilibrium Composition

these three reactions at 1 atm pressure are, respectively,

$$Kp = \frac{n_C \times n_{tot}}{n_A \times n_B} = 0.05 = \frac{x(2-x)}{(1-x)^2}$$

$$Kp = \frac{(n_C)^2}{n_A \times n_B} = 0.05 = \frac{(2y)^2}{(1-y)^2}$$

$$Kp = \frac{(n_C)^3}{n_A \times n_B \times n_{tot}} = 0.05 = \frac{(3z)^3}{(1-z)^2(2+z)}.$$

Assuming that 1 mole each of A and of B are taken as reactants and that equilibrium values of N_C are taken to be x, $2y$, and $3z$ in the Kp expressions, these equations may be solved and n_C evaluated for the respective reactions:

$x = 0.024 \qquad n_C = 0.024$ mole

$y = 0.101 \qquad n_C = 0.202$ mole

$z = 0.143 \qquad n_C = 0.429$ mole.

Hence, although the Gibbs energy increment is the same in each instance, the yield of C is much larger for the second and third types than for the first.

Calculation of Compositions in Gaseous Equilibria

The calculation of equilibrium compositions from the expression for the equilibrium constant often involves the solution of cubic, quartic, or even higher-degree equations. These equations are often difficult or tedious to solve, and shortcut methods are frequently employed. One method that can considerably decrease time and effort is a successive approximation method based on "sensitive factors." In the above reaction $A + B = 3C$, for example, inspection of the equilibrium constant expression

$$Kp = 0.05 = \frac{(3x)^3}{(1-x)^2(2+x)}$$

reveals that x will be small and hence that $(1-x)$ will be nearly equal to 1 and $(2+x)$ will be nearly equal to 2. The sensitive factor in this case is $(3x)^3$, since $(1-x)$ and $(2+x)$ are only slightly affected by a small x. As a first approximation, $0.05 \approx (3x)^3/2$ and $x \approx 0.155$. This value is used in the $(1-x)$ and $(2+x)$ factors to yield a second approximation, $0.05 \approx (3x)^3/[(0.845)^2(2.155)]$ and $x \approx 0.142$. The third approximation,

$0.05 = (3x)^3/[(0.858)^2(2.142)]$, yields $x = 0.143$. In three trials x has been obtained correct to three digits.

If the equilibrium constant for this reaction has a value of 100, then $100 = (3x)^3/[(1-x)^2(2+x)]$. Since x must be less than unity, the numerator cannot exceed 27. By inspection, therefore, x will be close to 1 and the sensitivity term now is $(1-x)$ in the denominator. As a first approximation, assuming $x = 1$ and solving for $(1-x)$, $(1-x)^2 = 27/(3 \times 100)$ and $x = 0.7$. Substitution of this value for x into the original expression, gives $(1-x)^2 = 9.26/(100 \times 2.7)$ and $x = 0.815$. Successive approximations give 0.772, 0.788, 0.782, and 0.784. Although convergence is not as rapid, a value correct to within 1% is obtained after only four successive trials. The method is most valuable for small or large values of the equilibrium constant. If the constant is near unity, it may be necessary to assume more than one sensitive term to achieve rapid convergence.

If it is desired to solve an equilibrium expression for several or many values of the equilibrium constant, it is convenient to calculate Kp or log Kp for appropriately spaced values of x, to plot the results as Kp or log Kp versus x, and to read intermediate values from the graph. Graphs for several important types of gaseous equilibria have been calculated and are given in Figures 5-1 through 5-6.

The simplest equilibrium, that of isomerization, may be represented by the equation $A = B$. Examples are equilibria between normal and isopropyl alcohol or between acetone and propionaldehyde. These reactions are unaffected by pressure and all have the same Gibbs energy-composition relation shown in Figure 5-1. To use this graph it is only necessary to ascertain from the tables the logarithm of the equilibrium constant of B for the temperature of interest, to subtract the corresponding value for A to get log Kp for the reaction, and to read the number of moles of B at equilibrium from the graph. This assumes that the starting material is 1 mole of A. As an example, the extent of the reaction

$$cis\text{-CHCl}{=}\text{CHCl}(g) = trans\text{-CHCl}{=}\text{CHCl}(g)$$

is calculated at 500°K. From the tables log Kp for the reaction is found as -0.214, and from the graph the number of moles of *trans*-1,2-dichloroethylene at equilibrium is 0.38, leaving 0.62 mole of *cis*-1,2-dichloroethylene.

Two identical molecules can react with each other to form a new molecule, giving an equation of the type $2A = B$. Since there is a change in the number of moles, this reaction is affected by pressure. Equilibrium yields of B at various pressures are given graphically in Figure 5-2. An

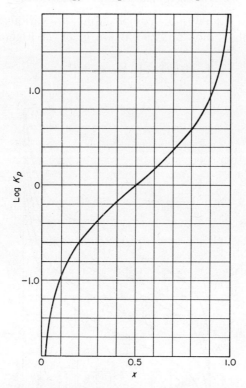

Figure 5-1 Equilibrium constant plot for isomerization reactions of the type $A = B$. At equilibrium $n_B/n_A = x/(1-x)$.

example of this type of reaction is the dimerization of tetrafluoroethylene to cyclooctafluorobutane,

$$2C_2F_4(g) = C_4F_8(g).$$

From the tables for these substances log Kp at 1000°K is found as 0.645, and from the graph the equilibrium composition at 1 atm pressure is 0.62 mole fraction of dimer.

Reactions of the type $A = B + C$ are very important, including such examples as dehydration, dehydrogenation, and dehydrohalogenation. This reaction depends on both the pressure and the initial composition of the system. One important case is a starting material of pure A, and therefore an equilibrium mixture containing equimolal amounts of B and C. Figure 5-3 gives equilibrium compositions for this case at various pressures. If it is desired to repress the dissociation, B or C may be added initially. Equilibrium compositions at various pressures for initially equimolal ratios of A to B are given in Figure 5-4. For the reaction

$$C_2H_6(g) = C_2H_4(g) + H_2(g),$$

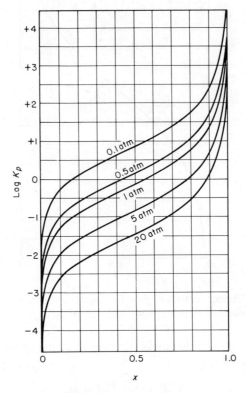

Figure 5-2 Equilibrium constant plot for reactions of the type $2A = B$. At equilibrium $n_B/n_A = x/[2(1-x)]$.

data in Chapter 9 give log Kp at 1000°K as -0.464. For 1 mole of pure ethane as starting material, Figure 5-3 indicates 0.50 mole of ethylene and 0.50 mole of ethane in equilibrium at 1 atm pressure. For an initial mixture of 1 mole of ethane and 1 mole of hydrogen, Figure 5-4 shows 0.38 mole of ethylene in equilibrium at 1 atm.

For some reactions 2 moles of starting material react to form two new molecular types, $2A = B + C$. This reaction type is independent of pressure since there is no change in the number of moles of material. Equilibrium does vary according to composition, however, and compositions for various initial ratios are given in Figure 5–5. An example of this type of reaction is

$$2CH_3OH(g) = (CH_3)_2O(g) + H_2O(g).$$

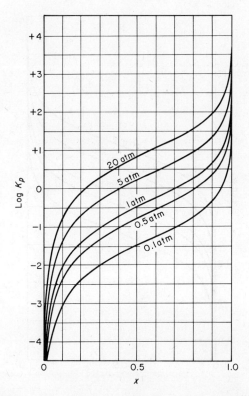

Figure 5-3 Equilibrium constant plot for reactions of the type $A = B + C$. Initial composition: pure A or equimolal amounts of B and C. At equilibrium $n_B/n_A = n_C/n_A = x/(1-x)$.

Data in Chapters 8 and 10 give log Kp at 800°K as 0.493. Reference to Figure 5–5 shows 0.39 mole of dimethyl ether formed at equilibrium from 1 mole of pure methanol. From an initial mixture of 1 mole of methanol and 1 mole of water, 0.32 mole of dimethyl ether is obtained at equilibrium irrespective of pressure.

Two different molecules may react to form two entirely new molecules in a metathesis reaction, $A + B = C + D$. This reaction does not depend on pressure, but yields vary with the initial ratio of A to B. Figure 5–6 gives equilibrium compositions for various starting mixtures. An example of this type of reaction is

$$CH_3NH_2(g) + CH_3Cl(g) = (CH_3)_2NH(g) + HCl(g).$$

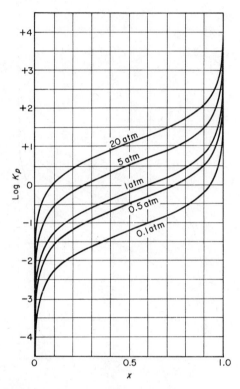

Figure 5-4 Equilibrium constant plot for reactions of the type $A = B + C$. Initial composition: equimolal amounts of A and B or $n_B/n_C = 2$. At equilibrium $n_C/n_A = x/(1-x)$, $n_B/n_A = (1+x)/(1-x)$.

Data from Chapters, 8, 11, and 12 give log Kp at 500°K as -0.671. From Figure 5-6 the yield of dimethylamine for 1 mole each of methylamine and methyl chloride is 0.31 mole, while 2 moles of methylamine and 1 mole of methyl chloride at equilibrium give 0.44 mole of dimethylamine.

There are, of course, many other types of reactions for which space does not permit consideration. Similar graphs, however, may be devised readily for equations of interest. Noddings and Mullet (1080) have presented an extensive *Handbook of Compositions at Thermodynamic Equilibrium*, prepared by digital electronic computer.

Gibbs Energy and Equilibrium Composition

Equilibria in Gases at High Pressure

For most organic vapors at pressures less than 2 or 3 atm the ideal gas law and the assumption that $f = P$ are adequate. At high pressures

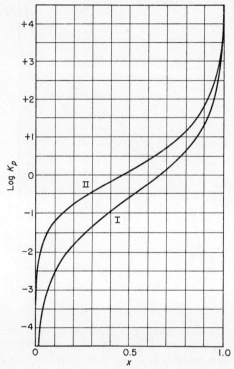

Figure 5-5 Equilibrium constant plot for reactions of the type $2A = B + C$. For curve I, initial composition: pure A or equimolal amounts of B and C. At equilibrium $n_B/n_A = n_C/n_A = x/[2(1-x)]$. For curve II, initial composition: equimolal amounts of A and B. At equilibrium $n_C/n_A = x/[2(1-x)]$, $n_B/n_A = (2+x)/[2(1-x)]$.

deviations from ideality become appreciable, and a correction factor, designated the fugacity coefficient γ, must be introduced:

$$f = \gamma P.$$

The fugacity coefficient may be evaluated by integrating the differential forms of [5-8] and [5-7] and invoking also [5-6] to obtain

$$\ln \gamma = \ln \left(\frac{f}{P}\right) = \int_0^P \left[V - \frac{RT}{P}\right] dP. \qquad [5\text{-}14]$$

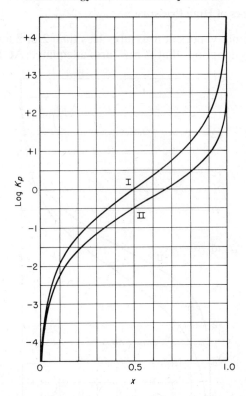

Figure 5-6 Equilibrium constant plot for (metathesis) reactions of the type $A+B=C+D$. For curve I, initial composition: equimolar amounts of A and B. At equilibrium $n_C/n_A = n_C/n_B = n_D/n_A = n_D/n_B = x/(1-x)$. For curve II, initial composition: $n_A/n_B = 2$. At equilibrium $n_C/n_A = n_D/n_A = x/(2-x)$, $n_C/n_B = n_D/n_B = x/(1-x)$.

Substituting the equation of state (if known) and performing the indicated integration give the coefficient. Graphical means can also be used. Since equation of state data are lacking for most gases, a generalized method must be employed. For example, [5-14] may be represented with fair accuracy by the approximate relationship

$$f = \frac{P^2}{P*},$$

in which $P*$ is the hypothetical pressure exerted by an ideal gas occupying the volume of the real gas; that is, $P* = RT/V$. Alternatively, the method of Gamson and Watson (458) is used for many engineering calculations. Gases at equal reduced temperatures and pressures are assumed to have

Gibbs Energy and Equilibrium Composition

the same fugacity coefficient, an assumption usually reliable to within about 10%. The reduced temperature is defined as $Tr = T/Tc$, and the reduced pressure as $Pr = P/Pc$. Here Tc is the critical temperature on the absolute scale and Pc is the critical pressure. From Figure 5-7 it is possible to read the fugacity coefficient over a considerable range of conditions.

For mixtures of gases it is customary to assume in the Lewis and Randall (861) approximation that the fugacity of each gas in the mixture is equal to its fugacity at the total pressure in the system times its mole fraction in the system, or

$$f_i = \gamma_i \left(\frac{n_i}{n_{tot}}\right) P_{tot},$$

and for real gases the thermodynamically exact equilibrium constant Kf may then be written

$$Kf = Kp K\gamma$$

in which $K\gamma = (\Pi \gamma_i)^{\nu_i}$ is a factor evaluated in terms of the fugacity coefficient at P_{tot}. Taking again the reaction

$$C_2H_5Cl(g) = C_2H_4(g) + HCl(g)$$

at 400°K and 20 atm as an example, the reduced pressures and temperatures may be calculated and the fugacity coefficients read from Figure 5-7, giving the data in Table 5-1. Then $K\gamma = (0.91 \times 0.96)/0.79 = 1.11$,

Table 5-1 Fugacity Coefficient for the Reaction $C_2H_5Cl(g) = C_2H_4(g) + HCl(g)$ at 20 atm Pressure and 400°K

Substance	Tc(°K)	Pc(atm)	Tr(°K)	Pr(atm)	γ
C_2H_5Cl	460	52.0	0.87	0.38	0.79
C_2H_4	283	50.5	1.41	0.40	0.96
HCl	355	51.4	1.13	0.39	0.91

and Kp (and Kn) is 11% less than Kf. Hence the dissociation is less than would be predicted from the ideal gas laws. The ethyl chloride has a considerably lower activity than that predicted by the ideal gas law, and high pressures therefore shift the equilibrium point to the left for the reaction as written.

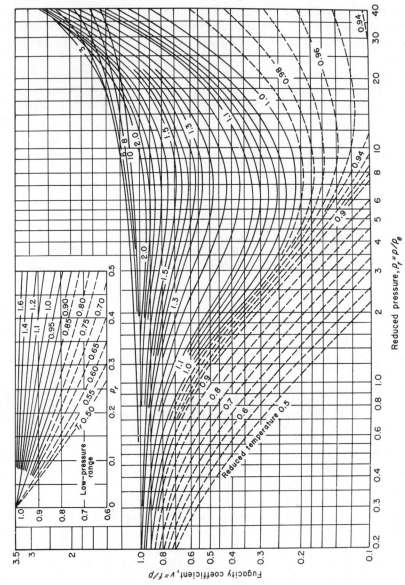

Figure 5-7 Fugacity coefficient plot of gases. Reproduced by permission from *Chemical Process Principles, Part II, Thermodynamics*, by O. A. Hougen and K. M. Watson, published by John Wiley & Sons, Inc., 1947; based on data taken from Gamson and Watson (458).

Equilibria in Systems Involving Condensed Phases

Thus far the systems discussed have been composed only of gases. Equilibria in such systems in which a single phase is involved are termed *homogeneous*. When more than one phase is present, situations are said to involve *heterogeneous* equilibria. The simplest case of such a system is the equilibrium between a pure liquid or solid and its own vapor. The pressure exerted by the vapor in equilibrium with the condensed phase at a given temperature is called the *vapor pressure* of the compound at that temperature. Since at equilibrium $\Delta Gv = 0$, the process of isothermal vaporization or sublimation at saturation pressure does not produce a Gibbs energy change. The fugacity of a liquid or solid is therefore equal to its fugacity as a gas at its vapor pressure at the given temperature. The activity of a pure liquid or solid with respect to a standard state of ideal gas at 1 atm is then

$$a = \frac{f}{f^\circ} \simeq P, \qquad [5\text{-}15]$$

in which P is now the vapor pressure of the condensed phase in atmospheres. For any system in which a pure solid or liquid is present, the activity in the equilibrium constant expression may be taken as the vapor pressure at the desired temperature. Tables of vapor pressure data are given in such sources as Rossini, Pitzer, Arnett, Braun, and Pimentel (1248), Stull (1432), Timmermans (1501), Dreisbach (350), and Jordan (703).

It is sometimes more convenient, however, to adopt a new standard state for condensed phases. If the standard state is taken as pure liquid or solid under 1 atm pressure, the activity in a system in which the pure liquid or solid is present in any amount is equal to unity and is relatively insensitive to pressure. The Gibbs energy of this standard state follows from [5-15] as the Gibbs energy of the gas state plus that of condensation,

$$\Delta Gf^\circ(l) = \Delta Gf^\circ(g) + RT \ln P. \qquad [5\text{-}16]$$

The standard Gibbs energy of a pure liquid or solid is thus readily calculable from tables for the ideal gas and vapor pressure data. If the pure liquid or solid is present in the system, its activity is equal to unity. It is assumed in [5-16] that the vapor pressures are less than 2 or 3 atm and that the compound is not associated in the vapor phase. At high pressures or for associated substances a fugacity coefficient must be introduced as outlined in the preceding section.

Consideration of Gibbs energy data for the substances in the reaction

$$CH_3Cl(g) + CO(g) = CH_3COCl(g)$$

permits calculation of the partial pressures of methyl chloride and carbon monoxide needed to produce liquid acetyl chloride at 298°K. The problem can be handled in the two ways described above. Jordan (703) gave the vapor pressure of acetyl chloride at 298°K as 300 mm (or 0.394 atm). The activity of liquid acetyl chloride *with respect to a standard state of ideal gas* is therefore constant at 0.394. From the appropriate tables for the ideal gas states of methyl chloride, carbon monoxide, and acetyl chloride, the equilibrium constant of the reaction at 298°K is obtained: $\log Kp = 37.084 + 10.960 - 24.046 = 2.078$. Hence

$$Kp = \frac{a_{CH_3COCl}}{P_{CH_3Cl} \times P_{CO}} = \frac{0.394}{P_{CH_3Cl} \times P_{CO}} = 120$$

and

$$P_{CH_3Cl} \times P_{CO} = \frac{0.394}{120} = 0.00328.$$

Therefore any combination of pressures of CH_3Cl and CO such that their product equals 0.00328 will give liquid acetyl chloride at equilibrium at 298°K.

Alternatively, the Gibbs energy of formation of liquid acetyl chloride can be calculated from [5-8]:

$$\Delta Gf(l) = \Delta Gf°(g) + RT \ln Kp = -50.59 + (4.5758 \times 298.15 \times \log P)$$

$$= -51.14 \text{ kcal/mole.}$$

The standard Gibbs energy and equilibrium constant for the reaction

$$CH_3Cl(g) + CO(g) = CH_3COCl(l)$$

is then $\Delta Gr° = -51.14 - (-14.95) - (-32.80) = -3.39$ kcal, and $\log Kp' = -(-3,390)/(4.5758 \times 298.15) = 2.485$. Since the activity of liquid acetyl chloride *with respect to a standard state of pure liquid* is unity,

$$Kp' = \frac{a_{CH_3COCl}}{P_{CH_3Cl} \times P_{CO}} = \frac{1}{P_{CH_3Cl} \times P_{CO}} = 305$$

and

$$P_{CH_3Cl} \times P_{CO} = \frac{1}{305} = 0.00328,$$

which is identical with the previous result.

For multicomponent systems in which an ideal liquid (or solid) solution is formed, the activity with respect to a standard state of pure liquid (or pure solid) of each species in the mixture is equal to its mole fraction,

N_i, in the liquid (or solid) phase. This follows from the definition of an ideal solution as one that follows Raoult's law,

$$P_i = N_i P_i^\circ, \qquad [5\text{-}17]$$

in which P_i is the partial pressure of a component and P_i° the vapor pressure of the pure substance. For a nonideal solution the activities may be determined experimentally by measurement of the pressure and composition of the vapor phase in equilibrium with the solution. An informative example of such measurements is the study by Halford and Brundage (564) of the reaction of acetic acid and ethanol in solution. Several examples of the treatment of nonideal solutions by Parks and Huffman (1105) and the review article by Waring (1571) on formic acid are also instructive.

Other methods for determining activity coefficients, such as freezing point measurements and solubility data, are discussed by Lewis and Randall (861) and in other standard texts (321, 492, 1129, 1245, 1410). In the absence of any activity data, the ideal solution law and judgment must be used to derive approximate values and to assess the reliability of such results. For mixtures of similar compounds, such as mixtures of alcohols and ketones, the ideal solution is a good approximation. For aqueous solutions appreciable deviations from ideality are to be expected.

Complex Equilibria

Processes involving organic compounds are often accompanied by side reactions, which may possibly be suppressed by selective catalysis. Occasionally the side reactions may also give useful products. For a process that involves more than one reaction, the equilibrium constants for all the reactions must be simultaneously satisfied. Calculation of the equilibrium composition becomes very difficult when more than two or three equations must be solved. A problem of particular interest is the calculation of the equilibrium composition of rocket exhausts, in which as many as 10 or more equilibrium constants are involved. On desk calculators such a task is laborious, but methods (170, 640, 1605) have been developed to solve these problems with high-speed electronic computers. Such programs are useful in the computation of equilibria among isomers and other complex mixtures common in organic chemistry.

A simple case of complex equilibrium is the system

[1] $CH_3Cl(g) + H_2O(g) = CH_3OH(g) + HCl(g)$

[2] $2CH_3OH(g) = (CH_3)_2O(g) + H_2O(g).$

The methanol formed in the first reaction may react further to form dimethyl ether. From the appropriate tables the equilibrium constants derived for 600°K are $Kp(1) = 0.00154$ and $Kp(2) = 10.6$. Since there is no change in the number of moles in either reaction, the equilibrium constants may be written

$$Kp(1) = \frac{n_{CH_3OH} \times n_{HCl}}{n_{CH_3Cl} \times n_{H_2O}} = Kn(1) \text{ and } Kp(2) = \frac{n_{CH_3OCH_3} \times n_{H_2O}}{n_{CH_3OH}^2} = Kn(2).$$

For 1 mole of methyl chloride and 1 mole of water as starting materials, the amount of hydrogen chloride formed may be designated as x, and the amount of dimethly ether formed as y; then $n_{CH_3OH} = x - 2y$, $n_{HCl} = x$, $n_{CH_3OCH_3} = y$, $n_{H_2O} = 1 - x + y$, $n_{CH_3Cl} = 1 - x$, and the equilibrium constants are

$$Kp(1) = \frac{x(x-2y)}{(1-x)(1-x+y)} = 0.00154$$

and

$$Kp(2) = \frac{(1-x+y)(y)}{(x-2y)^2} = 10.6.$$

These equations may be solved by successive approximations to find $x = 0.048$ and $y = 0.009$, from which the equilibrium composition can be calculated. It is sometimes advantageous to calculate values of y for various values of x for each expression. Plotting x versus y then gives two curves representing the two equations. The point of intersection gives values satisfying both equilibrium constants.

Temperature Dependence of Equilibrium Constants

Differentiation of [5-5] with respect to temperature gives

$$\left(\frac{\partial G}{\partial T}\right)_P = -S \qquad [5\text{-}18]$$

or, for a process or reaction at constant pressure,

$$\left(\frac{\partial \Delta Gr}{\partial T}\right)_P = -\Delta Sr. \qquad [5\text{-}19]$$

Alternatively, division of [5-2] by T and differentiation with respect to temperature yields

Gibbs Energy and Equilibrium Composition

$$\left[\frac{\partial(\Delta Gr/T)}{\partial T}\right]_P = \left[\frac{\partial}{\partial T}\left(\frac{\Delta Hr}{T} - \Delta Sr\right)\right]_P$$

$$= \left[T\left(\frac{\partial \Delta Hr}{\partial T}\right)_P - \Delta Hr - T^2\left(\frac{\partial \Delta Sr}{\partial T}\right)_P\right]T^{-2}$$

$$= \left[T\Delta Crp - \Delta Hr - T^2\frac{\Delta Crp}{T}\right]T^{-2}$$

$$= \frac{-\Delta Hr}{T^2} \qquad [5\text{-}20]$$

and, on introducing [5-9],

$$\left[\frac{\partial \ln Kp}{\partial T}\right]_P = \frac{\Delta Hr^\circ}{T^2}. \qquad [5\text{-}21]$$

This important equation was first derived by van't Hoff and still bears his name. It shows that the variation of the equilibrium constant with temperature depends only on the standard enthalpy increment of the reaction. For reactions that absorb heat the equilibrium constant increases with temperature; that is, the reaction is favored at higher temperatures. On the other hand, exothermic reactions are favored by keeping the temperature as low as possible. The most striking example of the importance of endothermic reactions at high temperatures is the calculation of the composition of flames such as rocket exhausts. Dissociation of water vapor, for example, although negligible at room temperature, occurs to a considerable extent at 4000°K. A more prosaic example is the ordinary dehydrogenation reaction in common use at petrochemical plants. The dissociation of hydrogen from butane is highly endothermic and does not take place at room temperature, but increasing the temperature shifts the equilibrium according to [5-21] until reasonable yields are obtained at 900°K.

The van't Hoff equation was integrated and used in Chapter 3 to derive enthalpies of reaction from measured equilibrium constants. The first two methods discussed there can be reversed to yield equilibrium constants from a known enthalpy of reaction, provided that a measured value of the equilibrium constant at some temperature is available. These two methods are usually termed *second-law* methods, since only changes in enthalpy and entropy are involved, and the absolute entropies of products and reactants are not necessarily known.

Gibbs Energy Functions

A third way of calculating equilibrium constants from enthalpies of reaction is by means of the Gibbs energy function related to 0°K, which is defined by the relation

$$\left(\frac{G_T^\circ - H_0^\circ}{T}\right) = \left(\frac{H_T^\circ - H_0^\circ}{T}\right) - S_T^\circ. \qquad [5\text{-}22]$$

The Gibbs energy function varies sufficiently slowly with temperature (in contrast to G°, for example) to permit tabular interpolation at 100°K intervals, and, moreover, may be evaluated directly for gases if the energy levels are known from spectroscopic data. Such functions are more conveniently applied than analytical expressions for G° or for ΔGr°. For crystalline solids that obey the third law of thermodynamics the Gibbs energy function may readily be calculated by graphical evaluation of either of two pairs of double integrals:

$$\left(\frac{G_T^\circ - H_0^\circ}{T}\right) = -\frac{1}{T}\int_0^T \left[\int_0^T \frac{Cp^\circ}{T}dT\right]dT = -\int_0^T \left[\frac{\int_0^T Cp^\circ dT}{T^2}\right]dT,$$

or by the difference of two integrals, since

$$\frac{G^\circ}{T} - \frac{G_0^\circ}{T} = \frac{(H^\circ - H_0^\circ)}{T} - S^\circ:$$

$$\left(\frac{G_T^\circ - H_0^\circ}{T}\right) = \frac{1}{T}\int_0^T Cp^\circ dT - \int_0^T \frac{Cp^\circ dT}{T}.$$

Alternatively, Gibbs energy functions may be related to a reference temperature of 298°K:

$$\left(\frac{G_T^\circ - H_{298}^\circ}{T}\right) = \left(\frac{H_T^\circ - H_{298}^\circ}{T}\right) - S_T^\circ. \qquad [5\text{-}23]$$

Since these two functions differ only in the reference temperature used for the enthalpies, they are related by

$$\left(\frac{G_T^\circ - H_{298}^\circ}{T}\right) = \left(\frac{G_T^\circ - H_0^\circ}{T}\right) - \left(\frac{H_{298}^\circ - H_0^\circ}{T}\right).$$

All Gibbs energy functions tabulated in succeeding chapters are based on 298°K. In the literature the function based on 0°K is frequently tabulated, since calculation by statistical methods or low-temperature heat capacity

Gibbs Energy and Equilibrium Composition

data gives values based on 0°K. For the present purposes, functions based on 298°K are more practical since most enthalpy of formation data are tabulated for that temperature. Combination of the Gibbs energy functions for reactants and products of chemical reactions yields

$$-R \ln Kp = \frac{\Delta Gr_T^\circ}{T} = \sum v_i \left(\frac{G_T^\circ - H_{298}^\circ}{T}\right)_i + \sum v_i \frac{[\Delta Hf_{298}^\circ]_i}{T}$$

$$= \Delta \left[\frac{Gr_T^\circ - Hr_{298}^\circ}{T}\right] + \frac{\Delta Hr_{298}^\circ}{T}. \qquad [5\text{-}24]$$

Thus if the Gibbs energy functions at temperature T and the enthalpy of reaction at 298°K are known, the equilibrium constant and hence the equilibrium composition are readily calculated. As mentioned in Chapter 3, it is often most convenient to calculate $[\Delta(G_T^\circ - H_{298}^\circ)/T]$ at the even temperatures given in the tables and then interpolate. At temperatures above 400°K a linear interpolation is frequently adequate. Tabulated Gibbs energy functions prove particularly useful for cases in which the enthalpy of formation is estimated or uncertain. When a more accurate enthalpy of formation becomes available, the Gibbs energies of formation will be in error at all temperatures, but the Gibbs energy functions will not be affected. The use of [5-24] is the most efficient way of incorporating new enthalpies of formation into the tabulated data that follow.

Since entropies—almost invariably obtained by methods involving the third law of thermodynamics—are involved in the free energy functions, use of these functions in calculating equilibrium constants from enthalpies of reaction and vice versa is termed a *third-law* method.

CHAPTER 6

METHODS FOR ESTIMATING THERMODYNAMIC QUANTITIES

INTRODUCTION

As noted in earlier chapters, the enthalpy of formation of a vast number of chemical substances is known, at least to moderate accuracy. Inasmuch as entropy and heat capacity data are also required in various typical thermochemical calculations, there is often a need to estimate these latter quantities in making chemical thermodynamic evaluations. Although our knowledge of entropy values of organic compounds has increased rapidly since the third law became firmly established and generally applied (1598), nevertheless there are many more substances for which enthalpies of formation are known but for which entropies have not been ascertained. Hence there is a particular need to estimate entropy values to an accuracy corresponding to that of the thermochemical data. Since an uncertainty of several tenths of a kilocalorie per mole is typical of older thermochemical data, an uncertainty of 1–2 cal/(mole °K) in the entropy has an equivalent effect at room temperature on the Gibbs energy increment in view of the equation

$$\Delta G° = \Delta H° - T \Delta S° \qquad [6\text{-}1]$$

or, of course, on the equilibrium constant, since it is related by

$$\Delta G° = -RT \ln Kp. \qquad [5\text{-}9]$$

In the evaluation of the Gibbs energy at other temperatures, heat capacities of both reactants and products are required, and it is frequently necessary and feasible to estimate these values if measured gaseous heat capacities are not available.

Discussions in subsequent sections of this chapter of means of estimating heat capacity are also applicable to the calculation of entropy.

The well-known relation

$$S_T - S_0 = \int_0^T Cpd\ln T$$

is indicative of the close relationship between these two quantities. It is obvious that the importance of the entropy increases at higher temperatures, since it is multiplied by the temperature in [6-1].

In other instances it is necessary to estimate the enthalpies of reaction together with the entropies and heat capacities. Here, too, meaningful estimates are possible.

It should be stated at the outset that in order to determine the importance of various compounds under particular conditions of interest, it may be adequate to use first a very simple and crude method for estimating bonding energies, to neglect ΔCp completely, and to use rough rules for estimating the entropy increments of reaction. After one has limited the possible compounds that need consideration, it is advantageous to employ more elaborate methods of estimating enthalpies of formation and entropies in order to define more accurately those species of importance under the specific conditions of interest. If these rough calculations have demonstrated that the major species involved are compounds that have been studied previously and for which adequate thermodynamic data are available, then it will be quite possible to obtain accurate results by applying standard thermodynamic calculations of the type indicated in previous chapters. Alternatively, if the preliminary calculations have indicated the importance of new species that have not been studied previously, one may be restricted to estimates, but at least one will not make the error of assuming that only well-known compounds need consideration. Very serious errors may thus be avoided by the above approach.

A number of monographs and treatments of estimation procedures already exist, and the aim of this chapter, therefore, is to survey rather than to treat the subject exhaustively. Although the extensive tabulations of estimation methods described by Janz (681) and by Reid and Sherwood (1220) should be consulted for details, several of these methods are briefly described here.

METHODS INVOLVING VALENCE BOND CONTRIBUTIONS

Because many organic compounds vaporize readily, the heat capacity of the gaseous state is of prime importance. Since gaseous heat capacity data are relatively scarce, the ability to estimate heat capacity values is obviously desirable. The behavior of the atomic groups in compounds of a homologous series is very similar. For example, the vibrations within a methyl group are much the same whether the methyl group is on a propane, butane, or octane molecule, and the heat capacity contribution of the methyl group will be very nearly the same in every case. This suggests that the total heat capacity of a gaseous molecule is the sum of its component parts, and to a considerable degree this is true. Mecke (980) observed that the vibrational frequencies of a molecule could be grouped with the valence bonds of which the molecule was constructed. Each valence bond has a certain stretching frequency and another bending frequency perpendicular to the bond. These two modes of vibration are designated as valence frequencies (ν) and deformation frequencies (δ). Mecke further observed that the frequencies associated with given bonds in organic molecules vary only by small amounts from one molecule to another. Bennewitz and Rossner (103) employed these generalized frequencies in an empirical equation to calculate gaseous heat capacities for molecules containing carbon, hydrogen, and oxygen:

$$Cp° = 4R + \sum q_i E_{\nu_i} + \left(3n - 6 - \sum q_i\right) \frac{\sum q_i E_{\delta_i}}{\sum q_i}. \qquad [6\text{-}2]$$

Here n is the number of carbon atoms and q is a valence bond, while E_{ν_i} and E_{δ_i} represent the Einstein functions for a given bond whose characteristic valence and deformation frequencies are ν_i and δ_i, respectively. Bennewitz and Rossner (103), Dobratz (331), and Stull and Mayfield (1436) referred to the valence vibration as ν. Since ν has been defined as frequency (per second), the valence (ν) and deformation frequencies (δ) are here defined in wave numbers (that is, per centimeter) to avoid ambiguity. Equation 6-2 reproduced the experimental heat capacity data of Bennewitz and Rossner in the vicinity of 400°K to within 5–15%.

Bennewitz and Rossner had treated the molecule as a rigid rotator, which led to poor agreement at lower temperatures. Dobratz (331) was able to improve the agreement by taking into account the bonds permitting free rotation (about C—C bonds and the like). He modified [6-2] as follows:

$$Cp° = 4R + \frac{aR}{2} + \sum q_i C_{\nu_i} + \left(3n - 6 - a - \sum q_i\right) \frac{\sum q_i C_{\delta_i}}{\sum q_i}, \qquad [6\text{-}3]$$

where a is the number of bonds capable of free rotation and C_{ν_i} and C_{δ_i} are quadratic expressions to fit the valence and deformational contributions to the heat capacity. These quadratic expressions deviate somewhat above 700°K and seriously above 800°K. Dobratz suggested the extension of this procedure to organic molecules containing sulfur, nitrogen, and the halogens by listing values of the frequencies associated with valence bonds involving these substances.

To extend this method above 700°K, Stull and Mayfield (1436) presented the contributions of the various bonds in tabular form based on the Einstein equation [2-10] and reassigned some of the frequencies to agree with later data. The agreement with experimental data was usually within 4% or better.

An empirical correlation of the enthalpy of combustion of an organic compound was put forward by Kharasch and Sher (745), who hypothesized that the enthalpy of combustion was due to an interdisplacement of electrons between atoms and molecules. As a standard of reference, they assumed that the combustion of an organic compound in gaseous oxygen displaced an electron from a "methane arrangement" to a "carbon dioxide arrangement," thereby evolving heat equal to 26.05 kcal/mole. They proposed an equation of the type

$$\Delta Hc = 26.05n + w, \qquad [6\text{-}4]$$

where ΔHc is the calculated enthalpy of combustion, n is the number of electrons displaced in the process, and w is an empirical constant correcting for the deviation of the electron distribution from the reference standard. For simple aromatics and saturated hydrocarbons w is zero, but for olefins $w = 13$ kcal. In 1929 Kharasch (744) compiled the available combustion data on some 1500 organic compounds of all types. He compared the observed values with those calculated from [6-4] and found in most cases that the calculated values agreed with the experimental values within 1–2%.

Many attempts have been made in recent decades to evaluate the enthalpies (and/or energies) of substances using quantum mechanical, quasiempirical, and frankly empirical methods. Although it is beyond the scope of this treatment to discuss these in detail, a brief summary of the nature of these methods is incorporated.

The earliest of all of the models, Fajans' (406), introduced in 1920, associates the energy of a molecule with the summation of that in the constituent bonds. Approximations based on Fajans' model may certainly be described as zero-order approximations, since bonds are not strictly additive nor are they constant from one molecule to another, as is shown by Table 6-1. Pauling (1130) presented a table of bond energies developed on the assumption that a given bond possessed identical energy in every molecule and that these bond energies were additive.

To avert the problem of nontransferability of the bond energy terms from one molecule to another, several authors sought to classify the types

Table 6-1 Comparison of Bond Energy Schemes

Compound	Method of Average Bond Energies	Δ	Observed[a]	Δ	Skinner-Modified[b] Allen Scheme
Hexane	−41.89	+1.93	−39.96	0.00	−39.96
2-Methylpentane	−41.89	+0.23	−41.66	0.00	−41.66
3-Methylpentane	−41.89	+0.87	−41.02	+0.01	−41.03
2,2-Dimethylbutane	−41.89	−2.46	−44.35	+0.01	−44.36
2,3-Dimethylbutane	−41.89	−0.60	−42.49	+0.03	−42.52

[a] See Chapter 14.
[b] Skinner (1361).

of bonds. In the method of Klages (765) bond contributions are summed according to the bonding types within the molecule. This method has been used for the estimation of resonance energies, and improved bond calculations have been devised for this purpose by Wheland [(1602), page 86]. Laidler (838) sought to explain the energies of the paraffins on the basis of three types of C—H bonds and one type of C—C bond. On the other hand Tatevskii, Benderskii and Yarovoi (1468) employed three types of C—H bonds and 10 types of C—C bonds. Bond length–bond energy relationships were used by Glockler (494–496), Feilchenfeld (414), Dewar and Schmeising (326), Mackle and O'Hare (905, 907), Cox (286), and Bernstein (117) to obtain bond energy values for alkene and alkyne bonds. Somayajulu (1394) followed Fox and Martin (437), using Sutherland's relationship (1448) involving bond energy, bond length, and force constants in order to obtain bond energy values applicable in alkenes and alkynes. Semiempirical methods were used by others to calculate bond energies. As long ago as 1934, Zahn (1641) recognized the inadequacy of the assumption of bond additivity and suggested association of part of the energy of the molecule with the bonds and part with pairs of bonds to a single atom. Zahn's model was supported by Dewar and Pettit's (325) study, and the model was the basis for the scheme devised by Allen (10). This scheme makes an adjustment for each pair of 1,4-C—H bonds *gauche* to each other. Skinner (1361) refined the method of estimation of the magnitude of steric repulsion on the basis of an "angle release mechanism". Other correlation procedures that explicitly consider steric effects have been devised by Tatevskii and Papulov (1470), Tatevskii, Benderskii, and Yarovoi (1468), Greenshields and Rossini (543), Platt (1169, 1170), Mulliken (1035), and Souders, Matthews, and Hurd (1396, 1397). A comparison of these methods has been presented by Skinner and Pilcher (1363) and by Somayajulu, Kudchadker, and Zwolinski (1395).

Methods Involving Valence Bond Contributions

Skinner (1361) applied the Allen scheme to many types of alkanes and also to substituted alkanes. The corresponding extension of Allen's scheme to ethers was reported by Pilcher, Pell, and Coleman (1145). McCullough and Good (957) applied the Allen formula to alkanethiols and other types of sulfur compounds. The treatment of electron correlation and bond properties in selected sulfur compounds by Bent (109) is of interest in this connection. Moreover, the monograph by Mortimer (1029) on the correlation of reaction heats with bond strengths is also relevant. Laidler's (838) scheme, involving both bond and group energies, has been extended and applied to alkanethiols, thiaalkanes, and dithiaalkanes, as well as to alcohols, ethers, and alkyl peroxides.

The scheme of Benson and Buss (108) for the evaluation of ΔHf°_{298}, S°_{298} and Cp° between 300° and 1500°K on the basis of compiled values of atomic, bond, and group contributions has been extended and updated by Golden, O'Neal, and Benson (504). An agreement of entropy and heat capacity to ± 0.5 cal/(mole °K) and in the enthalpies of formation to 0.5 kcal/mole was obtained up to 1500°K. The enthalpies of formation of fluorocarbons and fluorohalogenated hydrocarbons have been re-evaluated in view of recent revisions in the accepted standard enthalpies of formation of aqueous hydrofluoric acid and gaseous carbon tetrafluoride. These data, re-evaluated by Lacher and Skinner (833), have also been examined according to the bond additivity and bond interaction scheme of Allen. There appears to be a satisfactory correlation with respect to the halogen-substituted methanes and fluorocarbons, but a less successful correlation obtains for the halogenated ethanes and olefins. Lacher and Skinner suggested that steric repulsion effects between halogen atoms play a significant role.

Average C—C and C—H bond energies have been assigned by applying the method of least squares to the standard enthalpies of atomization of some 50 alkanes by Overmars and Blinder (1096). Their average error of ± 0.58 kcal/mole compares favorably with the deviations in more elaborate schemes.

It should be noted that in the use of bond energy schemes, as well as in the comparison with enthalpies of atomization, the values are usually referred to 25°C. They of course include the thermal energies of the atoms and molecules as well as the zero-point energies of the fundamental modes of nuclear vibration. It seems to be the consensus that the values may be referred either to 0°K or to 298°K in many instances. Some considerations pertaining to polycyclic molecules by Nelander and Sunner (1050) have heralded the importance of recognizing that although the term "energy" is used in connection with bond energies (and also with strain energies), enthalpies are dealt with in all of the usual schemes, and this must be taken into account in ring closure. Moreover, these authors pointed out the appropriateness of referring the data to 0°K when possible. Studies by Schleyer and his colleagues (1289) indicate that even a further refined scheme appears to give a surprising strain energy result in the case of a very symmetrical globular molecule. There is reason to believe that this difficulty might be resolved if the comparison were made at 0°K.

Bond dissociation energies may be evaluated by the use of Franklin's method (439), since values for group increments for the free radicals are provided. In the method of Voevodskii (1541), a correlation based on the stabilizing effect of methyl groups was used to estimate the C—H bond dissociation energies in paraffinic hydrocarbons. The method was generalized by Vedeneev (1530) to include bonds other than C—H.

METHODS INVOLVING GROUP CONTRIBUTIONS

As mentioned earlier, the various methods based on the properties of group contributions were developed on the studies of Parks and Huffman for the purpose of estimating numerical values of thermodynamic properties in as simple a manner with the help of as few data as possible.

The method of Anderson, Beyer, and Watson (18) is somewhat more difficult to apply than the other methods; its construction permits more quantitative consideration to be given to the arrangement of the carbon skeleton. In this method each compound is considered to be composed of a parent molecule that has been modified by substitution of some of its atoms by appropriate groups to achieve the final molecule. Paraffins therefore are derived from the parent molecule methane by replacing the hydrogen atoms by suitable groups. Similarly, the values for all ethers are obtained by starting with dimethyl ether as the parent compound and summing the increments and thermodynamic data corresponding to structural modifications by group substitutions. Deviations from the naïve principle of additivity are taken into account insofar as possible by specification of the related structural environment for each increment. Contributions are therefore included for primary methyl substitution, secondary methyl group substitutions, and multiple bond contributions (including correction for conjugated double bonds and double bonds conjugated with a ring). Thermodynamic increments are correlated with structure for the three properties $\Delta Hf°_{298}$, $S°_{298}$, and $Cp° = a + bT + cT^2$ (in terms of the three coefficients in the simple quadratic equation). The correlation of the temperature-dependent heat capacity together with the other two properties makes possible the calculation of Gibbs energies, enthalpies of formation, and entropies at temperatures other than 298°K for the gaseous molecules. Anderson, Beyer, and Watson (18) also presented contributions resulting from the substitution of various groups for one or two methyl groups. Replacement of a methyl group on a hydrocarbon molecule by a substituent atom or group leads to a relatively constant increment in the heat capacity. Thus the increment may be evaluated from pairs of molecules whose heat capacities are firmly established. This increment may then be used in conjunction with the appropriate hydrocarbon to estimate the heat capacity for an unmeasured compound. The appropriate parent hydrocarbon should be as structurally similar to the desired compound as possible. In this method the data for the hydrocarbons are the foundation on which the thermodynamic superstructure of organic chemistry rests.

The method of group substitution can readily be used to estimate the gaseous heat capacity of a substance from that of its parent hydrocarbon. If one compares the difference in heat capacity between the alkylthiols and their parent hydrocarbons, the increment is constant to about 0.5 cal/(mole °K) at 298.15°K. As an example, the gaseous heat capacity of pentanethiol is calculated from its parent hydrocarbon hexane and is presented in Table 6-2. The values estimated here are in reasonable agreement with those presented in Chapter 13.

The method of group substitution can also be applied successfully to entropy estimation, provided that appropriate adjustment is made for the symmetry number. Consider the two gaseous molecules propane ($CH_3CH_2CH_3$) and ethyl chloride (CH_3CH_2Cl). They may be regarded as ethyl groups to which a methyl group and a chlorine atom have been attached. The entropy of the ethyl group should be nearly the same in both molecules, and the entropy difference between the two will result from differences in such factors as symmetry (cf [4-19]), molecular weight, structure, and internal rotation. For convenience these can all be combined into two groups, (a) different symmetry and (b) all other factors. Rossini, Pitzer, Arnett, Braun, and Pimentel (1248) gave $S°_{298} = 64.51$ cal/(mole °K) for gaseous propane; Gordon and Giauque (518) reported $S°_{298} = 65.94$ cal/(mole °K) for gaseous ethyl chloride. Propane has $\sigma_t = 18$, while ethyl chloride has $\sigma_t = 3$.

For Propane	For Ethyl Chloride
$S°_{298} = 64.51$	$S°_{298} = 65.94$
$R \ln 18$ 5.74	$R \ln 3$ 2.18
$S^*_{298} = 70.25$	$S^*_{298} = 68.12$

Table 6-2 Estimate of Heat Capacity of —SH Grouping[a]
Average of six examples and estimation of pentanethiol.
Hydrocarbon data (Chapter 9) less thiol data (Chapter 13)

Compound	298°K	300°K	400°K	500°K	600°K	700°K	800°K	900°K	1000°K
H H —CH H H	12.59	12.65	15.68	18.66	21.35	23.72	25.83	27.69	29.33
H HC—SH H	12.01	12.05	14.04	15.91	17.57	19.03	20.32	21.46	22.48
Δ	−0.58	−0.60	−1.64	−2.75	−3.78	−4.69	−5.51	−6.23	−6.85
H H H HC—C—CH H H H	17.57	17.66	22.54	27.04	30.88	34.20	37.08	39.61	41.83
H H H—C—C—SH H H	17.37	17.44	21.08	24.36	27.21	29.68	31.83	33.71	35.38
Δ	−0.20	−0.22	−1.46	−2.68	−3.67	−4.52	−5.25	−5.90	−6.45
H H H H H—C—C—C—CH H H H H	23.29	23.40	29.60	35.34	40.30	44.55	48.23	51.44	54.22
H H H H—C—C—C—SH H H H	22.65	22.75	27.86	32.56	36.72	40.37	43.60	46.47	49.01
Δ	−0.64	−0.65	−1.74	−2.78	−3.58	−4.18	−4.63	−4.97	−5.21

[a] In calories per (mole degrees Kelvin).

Table 6-2 (*Continued*)

Compound	298°K	300°K	400°K	500°K	600°K	700°K	800°K	900°K	1000°K

`H H H H H`
`H—C—C—C—C—CH`
`H H H H H`
	28.73	28.87	36.53	43.58	49.64	54.83	59.30	63.18	66.55

`H H H H`
`HC—C—C—C—SH`
`H H H H`
	28.24	28.37	34.95	41.07	46.54	51.37	55.68	59.54	62.95

Δ	−0.49	−0.50	−1.58	−2.51	−3.10	−3.46	−3.62	−3.64	−3.60

`   H`
`   HCH    H`
`HC—C—C—CH`
`H  H  H  H`
	23.14	23.25	29.77	35.62	40.62	44.85	48.49	51.65	54.40

`H  H  H`
`HC—C—C—CH`
`H  H  SH`
	22.94	23.04	28.35	33.06	37.02	40.38	43.26	45.74	47.92

Δ	−0.20	−0.21	−1.42	−2.56	−3.60	−4.47	−5.23	−5.91	−6.48

`   H`
`   HCH   H`
`HC—C—CH`
`H  HCH`
`   H`
	29.07	29.21	37.55	45.00	51.21	56.40	60.78	64.55	67.80

Table 6-2 (*Continued*)

Compound	298°K	300°K	400°K	500°K	600°K	700°K	800°K	900°K	1000°K
H HCH H HC—C—CH H S H H	28.91	29.04	36.13	42.39	47.60	51.92	55.53	58.61	61.24
Δ	−0.16	−0.17	−1.42	−2.61	−3.61	−4.48	−5.25	−5.94	−6.56
Average Δ	−0.38	−0.39	−1.54	−2.65	−3.56	−4.30	−4.92	−5.43	−5.86
H H H H H HC—C—C—C—CH H H H H H	34.20	34.37	43.47	51.83	58.99	65.10	70.36	74.93	78.89
H H H H HC—C—C—C—SH H H H H (estimated)	(33.82)	(33.98)	(41.93)	(49.18)	(55.43)	(60.80)	(65.44)	(69.50)	(73.03)

Methods Involving Group Contributions

S^*_{298} is the entropy that would be calculated if the terms in [4-22] and [4-23] involving the symmetry were omitted. Thus the substitution of a methyl group by a chlorine atom gives the increment (*parent* hydrocarbon less the substituted hydrocarbon), or $\Delta S^*_{298} = 70.25 - 68.12 = 2.13$ cal/(mole °K). Note that S^*_{298} is larger for the parent hydrocarbon.

The entropy of chloroform ($CHCl_3$, $\sigma_t = 3$) is calculated from that of the parent hydrocarbon, 2-methylpropane [$(CH_3)_3CH$, $\sigma_t = 81$], listed as $S°_{298} = 70.42$ cal/(mole °K) by Rossini, Pitzer, Arnett, Braun, and Pimentel (1248). Thus in calories per (mole degrees Kelvin):

$$S°_{298} = 70.42$$
$$R \ln 81 \quad \underline{8.73}$$
$$S^*_{298} = 79.15$$
$$-3(2.13) \quad -6.39$$
$$-R \ln 3 \quad \underline{-2.18}$$
$$\text{and } S°_{298}, CHCl_3(g) = 70.58.$$

This value for the gaseous entropy of chloroform at 298.15°K compares quite well with the value of 70.66 cal/(mole °K) calculated from spectroscopic data by Gelles and Pitzer (465). The methyl group substitution increments to the entropy (omitting symmetry) indicated as ΔS^*_{298} are presented in various chapters.

The group substitution method (18) may be employed to estimate the enthalpy of formation of an organic compound from knowledge of the increment in enthalpy of formation between the compound and its parent hydrocarbon. As a detailed example, consider the two compounds, propane ($H_3C—CH_2—CH_3$) and ethanethiol ($H_3C—CH_2—SH$). Each has a methyl and methylene group composing an ethyl group that is common to each molecule. The difference in the enthalpies of formation of these two molecules resides in the fact that the enthalpies of formation of a methyl group and a thiol group are different. This difference is relatively constant, as is shown by Table 6-3. The substitution of a thiol group for a methyl group in a parent hydrocarbon makes the $\Delta Hf°_{298}$ of the parent hydrocarbon more positive by 13.93 kcal/mole.

A second example of the estimation of the enthalpy of formation of the alkyl sulfides is given in Table 6-4. The substitution of a sulfide group for a methylene group in a parent hydrocarbon makes the $\Delta Hf°_{298}$ of the parent hydrocarbon more positive by 15.53 kcal/mole.

Pitzer (1155) developed an approximate, statistical method for calculating the gaseous heat capacity of the normal paraffins and related compounds. Later Person and Pimentel (1140) revised the factors and presented gaseous heat capacity data for the normal paraffins from butane through heptane, and CH_2 increments, permitting calculations of values

Table 6-3 Enthalpies of Formation of Hydrocarbons and Thiols for the Process R—CH$_3$ = R—SH at 298°K

Parent Hydrocarbon[a]	$\Delta Hf°_{298}$ (kcal/mole)	Thiol[b]	$\Delta Hf°_{298}$ (kcal/mole)	$\Delta[\Delta Hf°_{298}]$ (kcal/mole)
n-Propane	−24.82	Ethanethiol	−11.02	13.80
n-Butane	−30.15	1-Propanethiol	−16.22	13.93
n-Pentane	−35.00	1-Butanethiol	−21.05	13.95
n-Hexane	−39.96	1-Pentanethiol	−25.91	14.05
			Average	13.93 ± 0.03

[a]Data of Rossini, Pitzer, Arnett, Braun, and Pimentel (1248).
[b]Data of Scott and McCullough (1316).

at least through eicosane (C$_{20}$H$_{42}$). Their equation for a normal paraffin of n carbon atoms is

$$Cp° = 4R + (n-1)[C-C_{str}] + (n-2)[C-C_{bend}] +$$
$$(n-3)[\text{internal rotation}] + [F_{steric}] + 2[CH_3] + (n-2)[CH_2],$$
$$\Delta Cp \text{ per } CH_2 \text{ group} = [C-C_{str}] + [C-C_{bend}]$$
$$+ [\text{internal rotation}] + [B] + [CH_2],$$

where [C—C$_{str}$] is the contribution arising from stretching the C—C bond. [C—C$_{bend}$] is the contribution of C—C bending modes, [internal

Table 6-4 Enthalpies of Formation of Hydrocarbons and Aliphatic Sulfides for the Process R—CH$_2$—R' = R—S—R' at 298°K

Parent Hydrocarbon[a]	$\Delta Hf°_{298}$ (kcal/mole)	Alkyl Sulfide[b]	$\Delta Hf°_{298}$ (kcal/mole)	$\Delta[\Delta Hf°_{298}]$ (kcal/mole)
n-Propane	−24.82	Methyl sulfide	−8.97	15.85
n-Butane	−30.15	Ethyl methyl sulfide	−14.25	15.90
n-Pentane	−35.00	Methyl propyl sulfide	−19.54	15.46
n-Pentane	−35.00	Ethyl sulfide	−19.95	15.05
Isopentane	−36.92	Isopropyl methyl sulfide	−21.61	15.31
			Average	15.51 ± 0.11

[a]Data of Rossini, Pitzer, Arnett, Braun, and Pimentel (1248).
[b]Data of Scott and McCullough (1316).

rotation] is the contribution of the internal rotations, $[F_{\text{steric}}]$ is the contribution of the steric interaction energy, $[CH_3]$ and $[CH_2]$ are the contributions from the methyl and methylene groups, and $[B]$ is the steric interaction coefficient in the expression $[F_{\text{steric}}] = A + (n-7)$ involving the steric constant A. These methods are applicable to other families of molecules. Bremner and Thomas (165, 166) extended this treatment to aromatic systems.

Rossini (1243) reviewed the enthalpy of formation data for organic compounds and emphasized the inadequacy of most of the work prior to 1937. Much of the work leads to errors of several kilocalories per mole in $\Delta Hf°$ values. Rossini recommended an empirical correlation of enthalpy of formation data for a homologous series of molecules. For normal paraffins he found that the enthalpy of formation may be represented by

$$\Delta Hf°_{288} = A + Bn + \Delta,$$

where A is a constant characteristic of the type of end group, B is a constant, n represents the number of carbon atoms in the molecule, and Δ is the deviation from linearity.

The method of Franklin (439) is simpler to use, and data are provided at several rounded temperatures. Additionally, increments for free radicals and gaseous phase carbonium ions are given. This method is based on an extension of the relations and principles developed by Pitzer (1150, 1153) for the long-chain alkanes. Pitzer showed from theoretical considerations that the enthalpy and Gibbs energy for gaseous normal alkanes can be expressed as additive functions of the number of carbon atoms and constants characteristic of temperature, bond stretching, bending, internal rotation, and the symmetry number of the molecule. In this method contributions for structural groups are taken to obtain $\Delta Hf°_T$ and $\Delta Gf°_T$ at the desired temperature. Additional terms are added to the Gibbs energy values to account for structural symmetry and optical isomers. The contribution of symmetry number has already been discussed, and for compounds that contain an asymmetrical carbon atom (that is, one set of optical isomers) an additional correction of $-RT \ln 2$ is made to the Gibbs energy summation. If two sets of optical isomers are possible, the contribution is $-RT \ln 4$. In practice, correction factors for branching and for substituents are placed on aromatic nuclei.

The correlation method of Souders, Matthews, and Hurd (1396, 1397) presents structural group contributions to the heat capacity, heat content, and entropy for the vibrational and characteristic internal rotational contributions. Inspection of the structural formula provides guidance to the selection of the increments most appropriate for the environment within the molecule, and these, summed together with the translational and external rotational contributions for the system as a whole, lead to the desired thermodynamic property. The correlation method, moreover, leads directly to a value for the entropy of formation as distinct from the entropy of the molecule under consideration. Contributions for non-hydrocarbon groups apparently have not been given for this method.

The original method of Van Krevelen and Chermin (1528) is quite similar in application to that of Franklin, except that the result of the calculation is given directly in the form of the convenient temperature-dependent equation for the Gibbs energy of the molecule in question. Because of the advent of better data, the method was completely revised by Chermin (220), who applied it to alkanes, alkenes, cycloalkanes, cycloalkenes, alkynes, and aromatic compounds. A larger number of increments for non-hydrocarbon groups are listed for this method than for the others. The increments for the various hydrocarbon groups are corrected for branching in paraffinic chains and in various ring sizes, as well as for ring formation in aromatic compounds.

CHAPTER 7

APPLICATION TO INDUSTRIAL PROBLEMS

INTRODUCTION

Chemical industry is concerned with improving the utility and value of raw materials by transforming them into chemical substances that may be either intermediates in a processing sequence or end product commodities. Since chemical reactions are characterized by a certain arrangement of atoms in the reactants and a different arrangement of the same atoms in the products, it is the task of the chemist to establish the conditions of temperature, pressure, and other constraints under which the reactant materials will preferentially rearrange to form the desired product molecules. A penetrating thermodynamic analysis of the system may often provide considerable information on the proper conditions for obtaining the desired products. Reactions that cannot occur may be definitely eliminated, directing attention toward the feasible reactions. The fruitfulness of such a thermodynamic analysis will depend on the extent to which the basic thermodynamic properties of the substances involved are known, but even in the absence of complete data chemical thermodynamics permits intelligent speculation concerning the reaction products. As appropriate data become more abundant, the utility of this tool in modern research will increase even further.

In planning chemical research, according to Berg, Carpenter, Daly, Dev, Herzel, Hippely, Kindschy, and Popovac (110), it is customary first to survey the literature and then, before attempting experimental work, to make a thermodynamic analysis of the system, to determine the feasibility of the reaction, and if possible, ascertain the maximum yield and the optimum conditions for achieving it. It is, however, necessary to keep clearly in mind the rules, generalizations, and limitations of the thermodynamic method to avoid false conclusions. For example, thermodynamics is concerned only with equilibrium conditions, and not with the time rate of approach of the reaction to the equilibrium state. Even reactions having sizeable negative Gibbs energy changes may actually proceed exceedingly slowly in the absence of a suitable catalyst. On the other hand, a reaction having a sizable positive Gibbs energy increment will not occur spontaneously; hence search for a catalyst would be entirely futile.

The presence or absence of a catalyst may favor the production of a particular product. When Fischer (427) began his study of fuels from water gas, he was concerned with the reactions

$$CO + H_2 = HCHO, \qquad [7\text{-}1]$$

$$CO + 2H_2 = CH_3OH, \qquad [7\text{-}2]$$

and

$$CO + 3H_2 = CH_4 + H_2O. \qquad [7\text{-}3]$$

While thermodynamic equilibrium predicts a large preponderance of methane with small amounts of methanol at 600°K and 150 atm, he found a large proportion of methanol under these conditions when zinc oxide was employed as a catalyst. Actually the catalytic suppression of reaction [7-3] is a matter of reaction kinetics that could not be predicted from the laws of thermodynamics.

These three reactions also demonstrate that the same reactants may react in a number of ways, which is the case with many organic substances. When one reaction is selected for thermodynamic study, the other reactions are assumed to be inoperative in that they will not disturb the equilibrium of the reaction chosen for study. By selection of a suitable catalyst it may be possible to accelerate the desired reaction or to suppress the undesired reactions, so that one reaction can be studied separately and employed to prepare a given product. Consequently it is necessary to assume that the rates of all side reactions are negligible when compared with the rate of the reaction selected for study. The additional assumption that all intermediate products at equilibrium have negligibly small concentrations compared to those of the main products is also required. It is important to realize that these three assumptions are taken for granted in all thermodynamic calculations, and that the reliability of the results hinges on the degree to which they are good assumptions.

Practical problems often contain complexities that are disposed of by making additional simplifying assumptions. The resultant solution to the problem may be only a rough approximation to the unsimplified facts. The real system may have such a large throughput that it will not be at the assumed equilibrium; the constant temperature that has been assumed may vary considerably; a clean system of three or four components may actually contain many more.

Introduction

It is often necessary to employ correlations of a generalized nature to approximate nonexistent data. Such estimated data are frequently several percent in error. Under some circumstances these errors can be seriously magnified. For example, the Gibbs energy change for the reaction

$$CH_3CHOHCH_3(g) = CH_3COCH_3(g) + H_2(g)$$

using the data on pages 424 and 444 is $\Delta Gf^\circ_{298} = 4.91$ kcal/mole and $\Delta Gf^\circ_{600} = -3.71$ kcal/mole. The enthalpy of combustion of liquid 2-propanol is -479.480 kcal/mole, but if this value is taken 1% too high the figures in Table 7-1 show how the Gibbs energy of dehydrogenation

Table 7-1 Effect of 1% Change in the Enthalpy of Combustion of 2-Propanol on Equilibrium

ΔHc of C_3H_8O (kcal/mole) at 298°K	ΔGr° of Dehydrogenation (kcal/mole) at		Equilibrium Constant Kr (atm) at	
	298°K	600°K	298°K	600°K
−479.480 (best)	4.91	−3.71	2.52×10^{-4}	22.5
−484.270 (1% lower)	0.12	−8.50	8.17×10^{-1}	12,500

and the equilibrium constant for the reaction are altered. The 1% change in the enthalpy of combustion of 2-propanol changes the equilibrium constant more than 3000-fold at 298°K and 500-fold at 600°K. The worth of accurate values is immediately clear.

Once the Gibbs energy change is established for a reaction, its feasibility may readily be calculated. At the temperature where $\Delta Gr^\circ = 0$, the equilibrium constant Kr equals unity, indicating that the reaction will progress to a considerable extent toward completion. As ΔGr° takes on more positive values, the reaction becomes less and less favored, until the yield of product shrinks to the level where the reaction is no longer of interest. However, since the point at which feasibility ceases is a function of numerous factors, there is no definite positive value of the Gibbs energy that bounds the feasibility of a reaction. Even though the ΔGr° for reaction [7-2], the synthesis of methanol at 600°K, is 10.84 kcal/mole, the reaction is certainly feasible at this temperature. High pressure is employed to overcome this unfavorable Gibbs energy change and displace the equilibrium. Favorable displacement of equilibria can also be achieved by changing the ratio of the reactants or by removing one of the reaction products. Each case must be carefully analyzed on its own merits. Table 7-2 is intended as a rough guide to the feasibility of a

Table 7-2   Use of $\Delta Gr°$ as an Index to Reaction Feasibility

$\Delta Gr° < 0$	Reaction is promising
$\Delta Gr° = 0\text{--}10$ kcal/mole	Reaction is of questionable feasibility; warrants further study
$\Delta Gr° > 10$ kcal/mole	Reaction is very unfavorable; unusual circumstances might make useful

reaction, and must be regarded as presenting approximate criteria for use in preliminary exploratory work.

The literature contains reviews of the applications of thermodynamics to problems of chemical engineering, among which should be mentioned those of Dodge (333) and Edmister (369).

THE PETROLEUM INDUSTRY

Because of the emphasis on hydrocarbon thermochemistry during the past three decades, it is not surprising that there are numerous examples of applications of this information to industrial problems. Following World War II the oil industry was called on to develop quantitative production of aromatic materials to meet the increasing demand for benzene as a basic raw material, and for other aromatic hydrocarbons to supply the needs of the plastics industry, the synthetic rubber program, and for aviation gasoline. Marshall (933) described the hydroforming operations at the Baytown Ordnance Works for making toluene during the war and for the production of high-purity aromatic solvents and aviation blending stocks after the war. Haensel and Berger (559) discussed the application of Universal Oil Products' platforming process to the production of aromatics from various feed stocks. They were able to show that the gross molal yield of aromatics from a commercial unit was 95% of the theoretical value.

Glasgow, Willingham, and Rossini (491) made a careful study of the hydrocarbon components of a gasoline produced by the catalytic cracking process and reported the composition of the C_8 aromatic fraction. It is of interest to compare the distribution of the compounds in the C_8 aromatic group produced by these several catalytic decomposition and reforming processes. The magnitudes of these compositions of C_8 isomers are very similar to the quantities predicted from thermodynamic equilibria at processing temperatures; as noted by Rossini (1246), the distribution is about the same as found in the C_8 aromatic fractions occurring naturally in petroleum. These values are compared in Table 7-3.

Table 7-3   Composition of C_8 Aromatic Fractions[a]

Compound	Hydroforming	Platforming	Catalytic Cracking	Virgin Petroleum	Thermodynamic Equilibrium
o-Xylene	20	23	20	20	23
m-Xylene	43	40	50	50	47
p-Xylene	17	21	20	20	21
Ethylbenzene	20	16	10	10	9
Total	100	100	100	100	100
Temperature (°K)	755–810		727		727
Source of data	(933)	(559)	(491)	(1246)	(1246)

[a] In volume percent (= mole or weight percent).

Thorne, Murphy, Ball, Stanfield and Horne (1497), in their study of oil shale and shale oil, presented an analysis of the character of Pumpherston shale oil naphtha. Their comparison (presented in Table 7-4)

Table 7-4   Equilibrium Attained by Shale Oil Naphtha[a]

| Compound | Thermodynamic Equilibria | | | | Pumpherston Naphtha |
	300°C	400°C	500°C	600°C	
C_8-Benzenes					
Ethylbenzene	5	8	10	13	9
p-Xylene	23	22	21	20	19
m-Xylene	51	48	46	44	46
o-Xylene	21	22	23	23	26
C_9-Benzenes					
Isopropylbenzene	1	1	1	2	0
n-Propylbenzene	1	2	3	4	0
1-Methyl-4-ethylbenzene	8	11	12	13	11
1-Methyl-3-ethylbenzene	12	17	20	23	22
1-Methyl-2-ethylbenzene	4	6	8	10	} 25
1,3,5-Trimethylbenzene	21	17	14	12	
1,2,4-Trimethylbenzene	53	46	42	26	42
Source of data		(1477)			(1497)

[a] All values in volume percent.

reveals a remarkable similarity between the composition of the C_8 and C_9 alkylbenzenes in this shale oil and that calculated from thermodynamic equilibria by Taylor, Wagman, Williams, Pitzer, and Rossini (1477).

Draeger, Gwin, Leesemann, and Morrow (347) reviewed thoroughly the production of high-octane gasoline components and studied their thermodynamic stability. They observed that reactions producing low molecular weight compounds through the decomposition of high molecular weight paraffins, naphthenes, and aromatics containing long alkyl side chains are the thermodynamically probable ones. The differences in stability between low and high molecular weight compounds increase with increasing temperature, indicating the desirability of conducting cracking operations at high temperatures. For paraffins of the same molecular weight, *normal* compounds are more stable at high temperatures (above about 590°K) but less stable at low temperatures than the corresponding branched paraffins, while the branched paraffins have higher octane ratings. These thermodynamic predictions apply to equilibrium reactions; equilibrium may not always be attained in actual refinery practice.

Heinemann, Schall, and Stevenson (583) reported on the application of Houdriforming to produce aromatics and high-octane motor gasoline. They presented plant operating data (Table 7-5) showing the conversion

Table 7-5 Houdriforming of Cyclohexane at 300 psig and a Hydrogen-to-Oil Molal Ratio of 4[a]

	Cyclohexane = Benzene (percent conversion)	
Severity	Found	Equilibrium Value
1	50	55
2	70	74
3	92	94

[a]Data of Heinemann, Schall, and Stevenson (583).

of cyclohexane to benzene over a typical Houdriforming catalyst at 300-psig pressure and a hydrogen-to-oil molal ratio of 4; the three different severities (conditions) used were obtained by changing temperature and space rate. Comparison with equilibrium conversion under the same conditions shows that equilibrium was essentially reached at each level of severity. These data were obtained by feeding a pure hydrocarbon to an experimental unit and analyzing the reaction products. The authors noted that the conclusions derived from work with pure compounds established trends, but were not entirely applicable to the mixed feed stock encountered in refinery practice.

Hastings and Nicholson (578) made a detailed study with a digital computer of the equilibria involving benzene and the 12 methylbenzenes based on the Gibbs energies of formation. Computations in complex systems can become quite involved but can be simplified, as Kandiner and Brinkley (712) showed, by a proper choice of mathematical relations within the system. The disproportionation reaction of any methylbenzene leads to the following array of species at equilibrium.

Compound	Mole Fraction
Benzene	N_3
Toluene	N_4
1,2-Dimethylbenzene 1,3-Dimethylbenzene 1,4-Dimethylbenzene	N_1
1,2,3-Trimethylbenzene 1,2,4-Trimethylbenzene 1,3,5-Trimethylbenzene	N_2
1,2,3,4-Tetramethylbenzene 1,2,3,5-Tetramethylbenzene 1,2,4,5-Tetramethylbenzene	N_5
Pentamethylbenzene	N_6
Hexamethylbenzene	N_7

Here N_i represents the mole fractions of the indicated compounds present at equilibrium. The concentrations of isomeric species can be related to that of a selected isomer. Since

$$1,3\text{-dimethylbenzene} \overset{K_x}{=} 1,2\text{-dimethylbenzene}$$

and

$$1,3\text{-dimethylbenzene} \overset{K_y}{=} 1,4\text{-dimethylbenzene},$$

in which K_x and K_y are the equilibrium constants for the reactions, the concentrations, in brackets, are

$$[1,3\text{-dimethylbenzene}] = N_1,$$
$$[1,2\text{-dimethylbenzene}] = K_x \quad [1,3\text{-dimethylbenzene}] = K_x N_1,$$
$$[1,4\text{-dimethylbenzene}] = K_y \quad [1,3\text{-dimethylbenzene}] = K_y N_1.$$

Hence the total concentration of the dimethylbenzenes is

$$N'_1 = (K_x + K_y + 1)N_1.$$

Treating the tri- and tetramethylbenzenes similarly reduces the number of species of immediate concern from 13 to 7 and thereby facilitates the calculation. Seven independent equations are required to evaluate the seven unknown concentrations. Two of these equations represent material balances for the phenyl and methyl radicals in the system. Then several independent components, equal in number to that of the different radicals involved, are selected. The remaining five equations, shown in Table 7-6, each contain the concentration of only one unknown species

Table 7-6 Chemical Equations for Use in the Kandiner and Brinkley Method[a]

---

3 (1,3-dimethylbenzene) $\overset{K_3}{=}$ benzene + 2 (1,2,4-trimethylbenzene)
$[N_1]$ $[N_3]$ $[N_2]$

2 (1,3-dimethylbenzene) $\overset{K_4}{=}$ methylbenzene + 1,2,4-trimethylbenzene
$[N_1]$ $[N_4]$ $[N_2]$

1,3-dimethylbenzene + 1,2,3,5-tetramethylbenzene $\overset{K_5}{=}$ 2 (1,2,4-trimethylbenzene)
$[N_1]$ $[N_5]$ $[N_2]$

2 (1,3-dimethylbenzene) + pentamethylbenzene $\overset{K_6}{=}$ 3 (1,2,4-trimethylbenzene)
$[N_1]$ $[N_6]$ $[N_2]$

3 (1,3-dimethylbenzene) + hexamethylbenzene $\overset{K_7}{=}$ 4 (1,2,4-trimethylbenzene)
$[N_1]$ $[N_7]$ $[N_2]$

---

[a] Kandiner and Brinkley (712).

in terms of the two specially selected independent components (1,3-dimethylbenzene and 1,2,4-trimethylbenzene). A typical equation (the last line of Table 7-6) reduces to

$$K_7 = \frac{(N_2)^4}{N_7(N_1)^3} \quad \text{or} \quad N_7 = \frac{(N_2)^4}{K_7(N_1)^3}.$$

By assuming tentative values for N_1 and N_2, the value of K_7 is calculated from the Gibbs energy data for the species involved. This permits calculation of N_7 from the above relation, which can then be checked against the material balances of the phenyl and methyl radicals. Lack of agreement requires the assumption of revised values for N_1 and N_2

and recalculation to improve the agreement. This method of successive approximations is facilitated by modern digital electronic computers. Similarly, the concentration of other constituents of the equilibrium mixture may be calculated, and those of the separate isomers finally obtained. The final calculated equilibrium concentrations of all the methylbenzenes are listed in Table 7-7 and are presented graphically in Figure 7-1.

Hastings and Nicholson (577) also compared the calculated and experimental isomerization equilibria among the tri- and tetramethylbenzenes at

Table 7-7   Theoretical Equilibrium Concentrations of Methylbenzenes[a]

$T(°K)$	Benzene	Toluene	1,2-Dimethylbenzene	1,3-Dimethylbenzene	1,4-Dimethylbenzene	1,2,3-Trimethylbenzene	1,2,4-Trimethylbenzene	1,3,5-Trimethylbenzene	1,2,3,4-Tetramethylbenzene	1,2,3,5-Tetramethylbenzene	1,2,4,5-Tetramethylbenzene	Pentamethylbenzene	Hexamethylbenzene
Toluene disproportionation equilibrium													
300	30.0	44.0	3.5	13.4	5.3	0.1	2.1	1.4	-----	0.2	-----	0.0	0.0
400	30.2	43.1	4.2	12.9	5.6	0.2	2.5	1.1	-----	0.2	-----	0.0	0.0
500	31.2	42.2	4.6	12.5	5.5	0.2	2.5	1.0	-----	0.3	-----	0.0	0.0
600	31.5	41.7	5.0	12.2	5.5	0.3	2.6	0.9	-----	0.3	-----	0.0	0.0
700	31.9	41.1	5.3	12.0	5.4	0.4	2.6	0.9	-----	0.4	-----	0.0	0.0
800	32.0	40.6	5.8	11.9	5.4	0.4	2.7	0.8	-----	0.4	-----	0.0	0.0
900	32.3	40.6	5.9	11.6	5.2	0.5	2.6	0.8	-----	0.5	-----	0.0	0.0
1000	32.4	40.3	6.1	11.5	5.2	0.5	2.7	0.8	-----	0.5	-----	0.0	0.0
Dimethylbenzene disproportionation equilibrium													
300	4.5	23.7	6.7	25.9	10.3	0.7	14.8	9.7	0.3	1.9	1.4	0.1	0.0
400	4.9	23.6	7.8	24.4	10.3	1.1	15.9	7.2	0.5	2.4	1.8	0.1	0.0
500	5.3	23.8	8.7	23.3	10.2	1.4	15.5	6.2	0.7	2.8	1.9	0.2	0.0
600	5.7	24.0	9.1	22.5	10.0	1.7	15.2	5.5	0.9	3.0	2.1	0.3	0.0
700	6.0	24.1	9.7	21.8	9.8	2.1	14.7	4.9	1.2	3.3	2.1	0.3	0.0
800	6.3	24.1	10.3	21.4	9.6	2.2	14.3	4.5	1.3	3.4	2.2	0.4	0.0
900	6.5	24.2	10.4	20.7	9.3	2.4	14.2	4.3	1.5	3.7	2.3	0.5	0.0
1000	6.7	24.4	10.8	20.4	9.1	2.6	13.8	4.0	1.6	3.8	2.3	0.5	0.0

Table 7-7 (Continued)

Trimethylbenzene disproportionation equilibrium

300	0.2	3.2	3.4	13.1	5.2	1.2	28.0	18.4	2.0	13.4	9.9	2.0	0.0
400	0.2	3.9	4.2	13.3	5.6	1.9	28.3	12.9	2.9	14.1	10.1	2.6	0.0
500	0.3	4.5	4.9	13.4	5.8	2.5	26.9	10.7	3.7	14.4	9.9	3.0	0.0
600	0.4	5.2	5.5	13.6	6.0	2.9	25.5	9.2	4.3	14.3	9.7	3.4	0.0
700	0.5	5.4	5.9	13.2	6.0	3.4	24.2	8.1	5.1	14.8	9.3	4.1	0.0
800	0.6	5.6	6.3	13.2	5.9	3.6	23.2	7.4	5.5	14.8	9.4	4.4	0.1
900	0.6	6.0	6.6	13.0	5.8	3.9	22.5	6.8	6.0	14.8	9.2	4.7	0.1
1000	0.7	6.4	6.9	13.1	5.9	4.1	21.7	6.3	6.3	14.7	8.9	4.9	0.1

Tetramethylbenzene disproportionation equilibrium

300	0.0	0.0	0.3	1.0	0.4	0.5	11.6	7.7	4.5	29.1	21.5	23.3	0.1
400	0.0	0.1	0.4	1.1	0.5	0.8	11.9	5.4	5.9	28.3	20.2	25.2	0.2
500	0.0	0.1	0.6	1.5	0.6	1.2	12.5	5.0	7.1	27.7	19.2	24.1	0.4
600	0.0	0.2	0.7	1.8	0.8	1.5	12.7	4.6	8.1	26.8	18.2	24.0	0.6
700	0.0	0.3	1.0	2.2	0.9	1.8	12.9	4.3	9.1	26.2	16.3	24.1	0.9
800	0.0	0.3	1.1	2.2	1.0	2.0	12.6	4.0	9.6	25.7	16.2	24.3	1.0
900	0.0	0.3	1.2	2.3	1.0	2.1	12.4	3.7	10.2	25.2	15.7	24.5	1.4
1000	0.0	0.4	1.3	2.5	1.1	2.3	12.2	3.6	10.7	24.7	15.0	24.6	1.6

Pentamethylbenzene disproportionation equilibrium

300	0.0	0.0	0.0	0.0	0.0	0.0	0.1	0.1	0.6	4.0	3.0	84.4	7.4
400	0.0	0.0	0.0	0.0	0.0	0.0	0.1	0.1	1.1	5.5	3.9	77.9	11.4
500	0.0	0.0	0.0	0.0	0.0	0.0	0.3	0.1	1.8	7.0	4.8	71.9	14.1
600	0.0	0.0	0.0	0.0	0.0	0.0	0.4	0.1	2.3	7.8	5.2	67.9	16.3
700	0.0	0.0	0.0	0.0	0.0	0.1	0.5	0.2	2.9	8.5	5.3	63.8	18.7
800	0.0	0.0	0.0	0.0	0.0	0.1	0.6	0.2	3.3	9.0	5.7	61.5	19.6
900	0.0	0.0	0.0	0.0	0.0	0.1	0.7	0.2	3.8	9.4	5.9	58.7	21.2
1000	0.0	0.0	0.0	0.0	0.0	0.1	0.8	0.2	4.1	9.6	5.8	57.5	21.9

[a]Data of Hastings and Nicholson (578), in mole percent.

300° and 700°K, and believed the agreement to be within the limits imposed by probable errors in product analyses and the assumptions in the statistical calculations.

It is difficult to place a direct monetary value on such information. A petroleum refiner's knowledge of these equilibria can be used repeatedly. Once the Gibbs energy is determined for a substance, it may be used to establish the equilibrium behavior in any reaction in which the Gibbs energy is available for the other substances involved. It is the continued use in successive problems of thermodynamic equilibrium that readily justifies the initial expenditure to obtain the information.

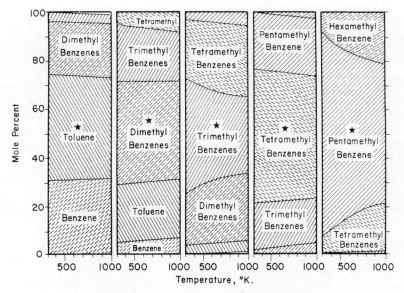

Figure 7-1 Disproportionation equilibria of the methyl benzenes. Compounds present initially are indicated with a star (★). From Hastings and Nicholson (578).

Teranishi and Benson (1478) studied the homogeneous gas phase dehydrogenation of isobutane by iodine from 522.5° to 582.9°K. The equilibrium constants shown in Table 7-8 are represented by the expression

$$\log Kp \text{ (atm)} = -\frac{25{,}900 \pm 300}{4.575T} + 8.130,$$

from which the mean deviation of the experimental points was ±1%. These data are shown in Figure 7-2 and are compared with equilibrium constant values calculated from thermodynamic data. The observed and calculated temperature dependencies are almost identical, although the observed values have a $\Delta Gr°$ about 0.10 kcal higher than those calculated. The authors believed that this difference was outside their own limit of error, but well within the limits of error of the thermodynamic data for the hydrocarbons involved.

CHEMICALS FROM METHANE

By virtue of geographical location each chemical manufacturer has a certain stock of raw materials that are economical for him to use. His

Table 7-8 Equilibrium Data for the Homogeneous Gas Reaction
$i\text{-}C_4H_{10} + I_2 = i\text{-}C_4H_8 + 2HI$[a]

T (°K)	Initial Pressure (mm)		Equilibrium Pressure (mm)				Kp (atm × 10^3)	
	I_2	$i\text{-}C_4H_{10}$	I_2	$i\text{-}C_4H_{10}$	$i\text{-}C_4H_8$	HI	Experimental	Calculated
525.1	9.67	200.3	3.24	193.9	6.35	12.7	2.15	2.02
522.5	14.6	119.4	7.82	112.6	6.90	13.8	1.96	1.79
525.7	3.89	275.4	0.33	271.8	3.44	6.88	2.34	2.08
551.9	7.50	206.5	0.88	199.9	6.35	12.7	7.66	6.77
552.5	14.8	121.2	5.25	111.6	9.55	19.1	7.83	6.95
552.5	4.32	158.7	0.26	154.6	3.91	7.82	7.83	6.95
582.7	8.37	131.9	0.72	124.3	7.60	15.2	25.8	23.7
582.9	3.56	136.7	0.06	133.2	3.42	6.83	26.3	23.9
582.7	5.95	168.8	0.26	163.1	5.92	11.8	25.5	23.7

[a] Data of Teranishi and Benson (1478).

problem is to convert these raw materials to profitably salable products. The market in his location will value certain products more highly than others, and he will seek to expend his manufacturing efforts on the most

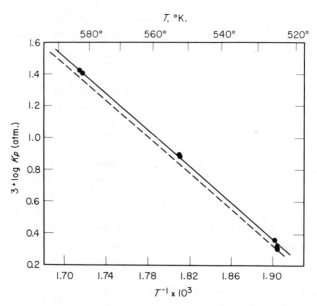

Figure 7-2 Plot of log Kp versus $1/T$ for the reaction isobutane + I_2 = isobute + 2HI. Dashed line shows the Kp values calculated from thermodynamic data. From Teranishi and Benson (1478).

valuable products. Thus he is faced with the need to explore the largest number of reactions and products that his raw material supplies will permit. The least expensive and most rapid way of sifting all the possible reactions is to apply the thermodynamic approach to such a problem. Let us suppose that the manufacturer owns some nearby natural gas wells that are very rich in methane. He desires to investigate quickly all of the possible reactions at his disposal to convert this methane into salable products.

He may immediately draw up a list of all of the reactions he can think of, and tabulate the Gibbs energy changes for these reactions at a series of temperatures. Thermodynamic studies of some of these technical methane reactions have been considered by Miller (1007), Fuchs, Andres, Plenz, and Veiser (451), and others. Table 7-9 contains a partial list of reactions for converting methane to other materials at 1 atm total pressure. The Gibbs energy changes for the reactions at 400°, 600°, 800°, and 1000°K are given. Only reactions for which data for all constituents are available in this book are considered; the list is not exhaustive by any means. Even so, the information concerning the reactivity of methane with other materials is startling and serves as a primary survey of chemicals from methane.

Table 7-9 Gibbs Energy Changes for Some Gas Phase Reactions Utilizing Gaseous Methane

Reaction Type	400°K	600°K	800°K	1000°K
	\multicolumn{4}{c}{$\Delta Gr°$ (kcal/mole of CH_4)}			
Decomposition				
[1] $CH_4 = C + 2H_2$	10.07	5.50	0.55	−4.61
[2] $2CH_4 = C_2H_2 + 3H_2$	34.36	28.42	22.14	15.69
[3] $2CH_4 = C_2H_4 + 2H_2$	18.92	15.96	12.79	9.51
[4] $2CH_4 = C_2H_6 + H_2$	8.34	8.48	8.51	8.46
Reactions with elements				
[5] $CH_4 + C = C_2H_4$	27.77	26.42	25.04	23.64
[6a] $CH_4 + Cl_2 = CH_3Cl + HCl$	−25.96	−26.57	−27.21	−27.86
[7a] $CH_4 + 2Cl_2 = CH_2Cl_2 + 2HCl$	−50.18	−50.92	−51.76	−52.66
[8a] $CH_4 + 3Cl_2 = CHCl_3 + 3HCl$	−72.64	−73.07	−73.67	−74.40
[9a] $CH_4 + 4Cl_2 = CCl_4 + 4HCl$	−92.48	−92.15	−92.12	−92.33
	\multicolumn{4}{c}{$\Delta Gr°$ (kcal/mole of HCl)}			
[6b] $CH_4 + Cl_2 = CH_3Cl + HCl$	−25.96	−26.57	−27.71	−27.86
[7b] $CH_4 + 2Cl_2 = CH_2Cl_2 + 2HCl$	−25.09	−25.46	−25.88	−26.33
[8b] $CH_4 + 3Cl_2 = CHCl_3 + 3HCl$	−24.21	−24.36	−24.56	−24.80
[9b] $CH_4 + 4Cl_2 = CCl_4 + 4HCl$	−23.12	−23.04	−23.03	−23.08

Table 7-9 (*Continued*)

Reaction Type	400°K	600°K	800°K	1000°K
	\multicolumn{4}{c}{$\Delta Gr°$ (kcal/mole of CH_4)}			
[10] $CH_4 + N_2 = HCN + NH_3$	36.56	35.59	34.49	33.33
[11] $CH_4 + \frac{1}{2}O_2 = CO + 2H_2$	−24.95	−33.87	−43.14	−52.56
[12] $CH_4 + \frac{1}{2}O_2 = CH_3OH$	−25.48	−23.03	−20.55	−18.07
[13] $CH_4 + O_2 = C + 2H_2O$	−96.99	−96.82	−96.82	−96.71
[14] $CH_4 + 2O_2 = C + 2H_2O_2$	−35.25	−29.44	−23.83	−18.35
[15] $2CH_4 + \frac{3}{2}O_2 = C_2H_2 + 3H_2O$	−91.87	−96.64	−101.70	−106.76
[16] $CH_4 + \frac{3}{2}O_2 = HCOOH + H_2O$	−125.03	−122.37	−119.77	−117.19
[17] $CH_4 + O_2 = HCHO + H_2O$	−69.19	−70.03	−70.88	−71.71
[18] $CH_4 + O_3 = HCHO + H_2O_2$	−78.92	−80.28	−81.67	−83.07
[19] $CH_4 + S = CH_3SH$	−6.47	−7.66	−9.85	−7.81
[20] $CH_4 + 4S = CS_2 + 2H_2S$	4.31	−9.64	−27.56	−19.17
Reactions with inorganic compounds				
[21] $CH_4 + CO = CH_3CHO$	16.05	21.90	27.70	33.43
[22] $CH_4 + CO_2 = 2CO + 2H_2$	34.36	21.21	7.72	−5.89
[23] $CH_4 + CO_2 = CH_3COOH$	19.25	24.88	30.44	35.90
[24] $CH_4 + CO_2 = H_2C{=}CO + H_2O$	36.56	34.82	32.94	31.03
[25] $CH_4 + CO = H_2C{=}CO + H_2$	30.78	30.90	30.74	30.41
[26] $CH_4 + CO = H_2C\!\!-\!\!CH_2\ (O\text{-ring})$	45.74	52.71	59.64	66.48
[27] $CH_4 + COS = C + H_2 + CO + H_2S$	7.96	1.63	−4.90	−11.51
[28] $CH_4 + COS = 2C + H_2O + H_2S$	−10.55	−10.16	−9.87	−9.61
[29] $CH_4 + COS = CH_3CHO + S$	22.81	28.12	34.38	36.34
[30] $CH_4 + COS = CH_3SH + CO$	0.29	−1.44	−3.17	−4.90
[31] $CH_4 + COS = H_2C{=}CO + H_2S$	28.67	27.03	25.29	23.51
[32] $CH_4 + COCl_2 = CH_3Cl + CO + HCl$	−12.51	−19.70	−26.87	−33.97
[33] $CH_4 + COCl_2 = CH_3COCl + HCl$	−10.60	−11.24	−11.94	−12.67
[34] $CH_4 + COCl_2 = H_2C{=}CO + 2HCl$	−1.79	−9.07	−16.46	−23.88
[35] $CH_4 + COCl_2 = HCHO + CH_2Cl_2$	18.58	17.79	17.04	16.31
[36] $CH_4 + COCl_2 = H_2C{=}CCl_2 + H_2O$	12.71	12.20	11.55	10.87
[37] $CH_4 + H_2O = CO + 3H_2$	28.58	17.29	5.52	−6.51
[38] $CH_4 + H_2O = CH_3OH + H_2$	28.05	28.13	28.11	27.98
[39] $CH_4 + NH_3 = HCN + 3H_2$	39.32	27.89	15.91	3.63
[40] $CH_4 + NH_3 = CH_3NH_2 + H_2$	23.84	23.85	23.72	23.00
[41] $CH_4 + N_2H_4 = CH_3NH_2 + NH_3$	−22.10	−22.41	−22.65	−23.30
[42] $CH_4 + NO_2 = HCOOH + H_2 + \frac{1}{2}N_2$	−85.43	−88.17	−91.11	−94.17
[43] $CH_4 + 2NO_2 = C + 2H_2O_2 + N_2$	−63.11	−63.36	−63.83	−64.37
[44] $CH_4 + NO + NO_2 = CO + 2H_2O + N_2$	−166.36	−172.97	−179.67	−186.30
[45] $CH_4 + NO + NO_2 = HCOOH + H_2O + N_2$	−159.38	−159.15	−158.98	−158.83
[46] $CH_4 + 2NO = HCHO + H_2O + N_2$	−110.03	−109.67	−109.30	−108.93
[47] $CH_4 + HNO_3 = CH_3NO_2 + H_2O$ (nitromethane)	−26.81	−27.34	−27.87	−28.34
[48] $CH_4 + HNO_3 = CH_3NO_2 + H_2O$ (methyl nitrite)	−25.16	−26.28	−27.50	−28.72

Table 7-9 (*Continued*)

	Reaction Type	400°K	600°K	800°K	1000°K
[49]	$CH_4 + HNO_3 = CH_3NO_3 + H_2$ (methyl nitrate)	22.99	23.57	23.89	24.07
[50]	$CH_4 + H_2S = CH_3SH + H_2$	2.40	2.43	2.28	2.00
[51]	$CH_4 + 2H_2S = CS_2 + 4H_2$	39.79	30.72	20.96	10.85
[52]	$CH_4 + SO_2 = CO_2 + H_2S + H_2$	−21.18	−27.24	−33.55	−39.96
[53]	$CH_4 + SO_2 = CO + H_2S + H_2O$	−15.40	−23.32	−31.35	−39.34
[54]	$CH_4 + SO_2 = HCOOH + H_2S$	−8.42	−9.50	−10.66	−11.87
[55]	$CH_4 + SO_2 = CH_3SH + O_2$	65.48	64.14	62.73	61.27
[56]	$CH_4 + SO_2 = HCHO + H_2O + S$	2.76	1.77	1.70	−2.63
[57]	$CH_4 + SO_3 = CH_3OH + SO_2$	−11.01	−13.09	−15.05	−16.91
[58]	$CH_4 + SO_3 = HCOOH + H_2O + S$	−38.68	−40.63	−41.69	−46.95
[59]	$CH_4 + SO_2Cl_2 = CH_3Cl + SO_2 + HCl$	−52.22	−55.63	−58.94	−63.88
[60]	$CH_4 + 2SO_2Cl_2 = CH_2Cl_2 + 2SO_2 + 2HCl$	−102.70	−109.04	−115.22	−121.34
[61]	$CH_4 + 3SO_2Cl_2 = CHCl_3 + 3SO_2 + 3HCl$	−151.42	−160.25	−168.86	−177.42
[62]	$CH_4 + 4SO_2Cl_2 = CCl_4 + 4SO_2 + 4HCl$	−197.52	−208.39	−219.04	−229.69
[63]	$CH_4 + ClCN = CH_3CN + HCl$[a]	−15.59	−16.04	−16.51	−17.00
[64]	$CH_4 + ClCN = CH_3Cl + HCN$[a]	−4.51	−5.17	−5.84	−6.50
[65]	$CH_4 + (CN)_2 = CH_3CN + HCN$	−5.33	−6.07	−6.81	−7.57
Simple organic compounds					
[66]	$CH_4 + C_2H_2 = CH_3CH{=}CH_2$	−19.88	−13.87	−7.81	−1.79
[67]	$CH_4 + C_2H_4 = C_3H_8$	−6.43	0.08	6.51	12.85
[68]	$CH_4 + HCHO = C_2H_4 + H_2O$	−0.03	−0.37	−0.85	−1.36
[69]	$CH_4 + HCHO = C_2H_5OH$	1.19	7.04	12.77	18.42
[70]	$CH_4 + CH_3OH = C_2H_6 + H_2O$	−11.37	−11.16	−11.09	−11.06
[71]	$CH_4 + CH_3OH = CH_3OCH_3 + H_2$	24.26	25.22	25.93	26.45
[72]	$CH_4 + HCOOH = CH_3OH + HCHO$	30.36	29.31	28.34	27.41
[73]	$CH_4 + HCOOH = CH_3OCH_3 + H_2O$	9.07	8.08	7.01	5.96
[74]	$CH_4 + HCOOH = HCOOCH_3 + H_2$	25.03	24.79	24.30	23.66
[75]	$CH_4 + CH_3COOH = CH_3COCH_3 + H_2O$	10.57	9.76	8.90	8.06
[76]	$CH_4 + CH_3COOH = CH_3OH + CH_3CHO$	44.82	48.79	52.95	57.19
[77]	$CH_4 + (CH_3CO)_2O$ $= CH_3COCH_3 + CH_3COOH$	−0.57	−0.74	−0.93	−1.12
[78]	$CH_4 + CH_3CHO = CH_3CH{=}CH_2 + H_2O$	4.21	3.78	3.25	2.68
[79]	$CH_4 + CH_3CHO = CH_3COCH_3 + H_2$	7.99	8.82	9.44	9.91
[80]	$CH_4 + CH_3COCH_3$ $= (CH_3)_2CH{=}CH_2 + H_2O$	7.89	7.54	7.03	6.46
[81]	$CH_4 + CH_3SH = CH_3SCH_3 + H_2$	16.72	17.48	17.97	18.28
[82]	$CH_4 + H_2C{=}CO = CH_3COCH_3$	−6.74	−0.18	6.40	12.93
[83]	$CH_4 + CH_3Cl = C_2H_6 + HCl$	−3.38	−3.30	−3.31	−3.40
[84]	$CH_4 + CH_2Cl_2 = C_2H_5Cl + HCl$	−8.67	−9.11	−9.62	−10.23
[85]	$CH_4 + CHCl_3 = H_2C{=}CHCl + 2HCl$	−8.54	−16.18	−23.88	−31.61
[86]	$CH_4 + CHCl_3 = C_2H_2 + 3HCl$	3.30	−10.61	−24.67	−38.75
[87]	$CH_4 + CHCl_3 = H_3C{-}CHCl_2 + HCl$	−11.98	−12.58	−13.26	−14.03
[88]	$CH_4 + CCl_4 = 2CH_2Cl_2$	−7.88	−9.69	−11.40	−12.99
[89]	$CH_4 + CCl_4 = H_2C{=}CCl_2 + 2HCl$	−17.74	−25.75	−33.77	−41.74
[90]	$CH_4 + C_2H_5Cl = C_3H_8 + HCl$	−1.78	−1.65	−1.60	−1.59

Table 7-9 (*Continued*)

Reaction Type	400°K	600°K	800°K	1000°K
[91] $CH_4 + C_3H_7Cl = C_4H_{10} + HCl$	−2.42	−2.19	−2.03	−1.95
[92] $CH_4 + H_2C=CHCl = C_3H_6 + HCl$	−8.04	−8.30	−8.60	−8.93
[93] $CH_4 + H_2C=CHCl = CH_3CHClCH_3$	−11.56	−4.64	2.14	8.81
[94] $CH_4 + H_2C=CHCl = CH_3CH_2CH_2Cl$	−9.08	−2.80	3.38	9.48
[95] $CH_4 + H_2C=CHCN = C_3H_6 + HCN$	8.97	8.52	8.03	7.50
[96] $CH_4 + H_2C=CCl_2 = H_3C-CCl_2-CH_3$	−10.26	−3.36	3.34	9.86
[97] $CH_4 + CH_3COCl = CH_3COCH_3 + HCl$	2.07	1.99	1.88	1.72
[98] $CH_4 + CH_3COCl = CH_3CHO + CH_3Cl$	14.14	13.44	12.77	12.13
[99] $CH_4 + ClCH_2CH_2Cl$ $= H_3C-CHCl-CH_3 + HCl$	−8.12	−8.24	−8.48	−8.77
[100] $CH_4 + ClCH_2CH_2Cl = C_3H_7Cl + HCl$	−5.37	−5.96	−6.63	−7.36
[101] $CH_4 + CH_3CHCl-CH_3$ $= (CH_3)_2CH-CH_3 + HCl$	−0.46	−0.14	0.13	0.34
[102] $CH_4 + CH_3CCl_2CH_3 = (CH_3)_3CCl + HCl$	−5.66	−5.57	−5.53	−5.53
[103] $CH_4 + CH_3NH_2 = (CH_3)_2NH + H_2$	21.25	22.10	22.66	23.04
[104] $CH_4 + (CH_3)_2NH = (CH_3)_3N + H_2$	20.44	21.90	23.07	24.03
[105] $CH_4 + C_5H_5N = C_6H_7N + H_2$ (pyridine) (2-picoline)	9.27	9.53	9.58	9.46
[106] $CH_4 + H_2C\!-\!CH_2 = CH_3CH_2CH_2NH_2$ $\backslash\,/$ N H	−18.16	−12.98	−7.92	−2.95
[107] $CH_4 + H_2C\!-\!CH_2 = C_2H_5OCH_3$ $\backslash\,/$ O	−10.68	−5.20	0.18	5.45
[108] $CH_4 + C_6H_5Cl = C_6H_6 + CH_3Cl$	4.34	4.37	4.49	4.66
[109] $CH_4 + C_6H_5Cl = C_6H_5 \cdot CH_3 + HCl$	−5.35	−5.58	−5.89	−6.21
[110] $CH_4 + C_6H_4Cl_2 = C_6H_4(CH_3)_2 + 2HCl$	−22.24	−18.42	−14.27	−9.98
[111] $CH_4 + C_6H_5OH = C_6H_5 \cdot CH_3 + H_2O$	−5.55	−5.81	−6.02	−6.17
[112] $CH_4 + C_6H_5OH = C_6H_4 \cdot CH_3 \cdot OH$	11·41	11·60	11.53	11.31
[113] $CH_4 + C_6H_5 \cdot C_2H_5 = C_6H_4 \cdot CH_3 \cdot C_2H_5 + H_2$	12.59	12.89	12.91	12.78
[114] $CH_4 + C_6H_5 \cdot CH=CH_2 = C_6H_5 \cdot C_3H_7$	−2.95	3.33	9.53	15.63

[a] Other possible products are chlorocyanomethanes.

Certain reactions are worthy of special notice. Reactions 2, 3, and 4 show the stability of methane with respect to the three C_2 hydrocarbons in the presence of hydrogen. Reactions 6a, 7a, 8a, and 9a indicate the feasibility of the replacement of methane hydrogens by chlorine, while the $\Delta Gr°$'s of Reactions 6b, 7b, 8b, and 9b emphasize that the replacement of methane hydrogens by chlorine becomes progressively more difficult. Reactions 11 and 37 show the suitability of methane as a "synthesis gas." Reactions 14, 18, and 43 indicate that H_2O_2 can be formed by oxidation of methane. Although Reaction 23 indicates that the reaction of carbon dioxide and methane to form acetic acid is not feasible, the same calculation proves the feasibility of decarboxylation of acetic acid to form methane and carbon dioxide, since the reversal of a reaction that is not feasible yields one that is. Reaction 32, a milder method of chlorination, can be used to replace each of the hydrogens of methane and yield any of the chlorine-substituted methanes. Comparison of the various agents for oxidation of methane involves a large group of related reactions. Reactions 59, 60, 61, and 62 demonstrate a more vigorous method of chlorination of methane. Reactions 63, 64, and 65 demonstrate the action of a bifunctional reactant and suggest the feasibility of producing chlorocyanomethanes. Reaction 67 is a good example of a reaction that produces a larger yield of product in a lower temperature range, while Reaction 86 is an example of a reaction that produces a larger yield of product in the higher temperature range. Reaction 82 is an example of a reaction that goes to the right at low temperatures, and to the left at high temperatures. In considering the yields of reactions over large ranges of temperature, it is important to note that both the Gibbs energy increment and T are factors related to the yields. Thus $\Delta Gr°/T = -R \ln K$. For Reaction 6a, for example,

$T°K$	400°K	600°K	800°K	1000°K
$\Delta Gr° =$	−25.96	−26.57	−27.21	−27.86
but $\Delta Gr°/T =$	−6.49	−4.43	−3.26	−2.79,

indicating a higher yield of product at the lowest temperature.

When the organic chemist is called on to investigate an area outside his own field of competence, he will first extrapolate his own experience as far as he dares. Beyond this point, however, chemical thermodynamics will provide knowledge drawn from Gibbs energy data that will enable him to extend his own experience into the areas of high pressure or high temperature consistent with his needs. It must be emphasized that the equilibrium predictions of thermodynamics rest solely on the basic heat and energy data for the compounds involved.

STYRENE MANUFACTURE

A comprehensive thermodynamic analysis of the engineering problems associated with styrene manufacture has been reported by Stull (1433). The heat absorbed by the dehydration of ethylbenzene to styrene,

Application to Industrial Problems

$$C_6H_5 \cdot C_2H_5(g) = C_6H_5 \cdot CH=CH_2(g) + H_2(g), \qquad [7\text{-}4]$$

is shown as a function of temperature in Figure 7-3. The equilibrium constant for this reaction has been studied by Ghosh, Guha, and Roy (472), Mitchell (1014), and Webb and Corson (1581); also, Guttman, Westrum, and Pitzer (554) calculated it by making certain assumptions. The theoretically calculated equilibrium constants are in good agreement with measurements based on laboratory experiments and industrial processes. These data are presented in Table 7-10, while the calculated equilibrium ratios for 1 atm pressure and for various partial pressures of hydrogen are presented in Figures 7-4 and 7-5. Further dehydrogenation of styrene to ethynylbenzene,

$$C_6H_5 \cdot CH=CH_2(g) = C_6H_5 \cdot C\equiv CH(g) + H_2(g), \qquad [7\text{-}5]$$

Table 7-10 Comparison of Calculated and Observed Values for the Equilibrium Constant Kp for the Reaction Ethylbenzene = Styrene + Hydrogen

	Stull (1433)		Other Sources		
T(°K)	$Gr°$(kcal/mole)	Kp(atm)	$Gr°$(kcal/mole)	Kp(atm)	Source
300	19.843	3.53×10^{-15}			
381			17.900	5.5×10^{-10}	(554)
400	16.999	5.14×10^{-10}	17.350	3.3×10^{-9}	(554)
500	14.066	7.12×10^{-7}	14.410	5.0×10^{-6}	(554)
600	11.024	9.65×10^{-5}	11.400	7.1×10^{-4}	(554)
633			9.636	4.7×10^{-4}	(472)
668			8.545	1.6×10^{-3}	(472)
700	7.949	3.30×10^{-3}			
703			7.414	4.95×10^{-3}	(472)
733			6.442	1.20×10^{-2}	(472)
768			5.299	3.10×10^{-2}	(472)
800	4.858	4.71×10^{-2}	5.160	3.89×10^{-1}	(554)
900	1.751	3.75×10^{-1}			
903			0.863	6.18×10^{-1}	(1014)
973			−1.157	2.6	(1581)
1000	−1.383	2.00	−1.230	1.86	(554)
1023			−1.825	3.9	(1581)
1073			−2.111	4.3	(1581)
1100	−4.510	7.87			

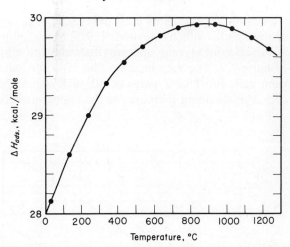

Figure 7-3 Enthalpy of reaction for the process ethylbenzene = styrene + hydrogen. From Stull (1433).

must also be considered, since it leads to branching and cross-linking during polymerization. Figure 7-6 shows the calculated equilibrium ratios for [7-5], which requires temperatures some 500° higher to reach the same

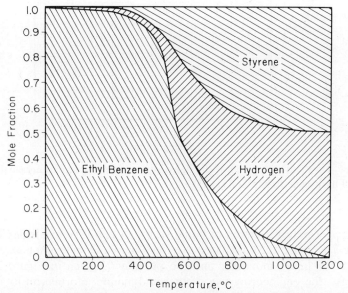

Figure 7-4 Calculated thermodynamic equilibria at 1 atm pressure for the reaction ethylbenzene = styrene + hydrogen. From Stull (1433).

degree of conversion as [7-4]. This cues the reaction operator to control [7-4] at temperatures for which the yield of [7-5] is negligible. A study of the disproportionation of styrene into ethylbenzene and ethynylbenzene is shown in Figure 7-7. The data indicate that even at room temperature the equilibrium calls for 1 or 2 parts of ethynylbenzene per million of styrene, and that this amount increases as the temperature is raised. The

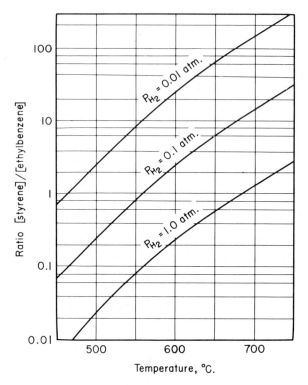

Figure 7-5 Yield of styrene in the dehydrogenation step at various partial pressures of hydrogen. From Stull (1433).

enthalpy of polymerization of liquid styrene to polystyrene is depicted in Figure 7-8, while the calculated thermodynamic equilibrium between the gaseous or liquid monomer and the polymer is given in Figure 7-9. In this case the engineering use of thermodynamic information in styrene technology minimized plant start-up delays and ensured the soundness of the process.

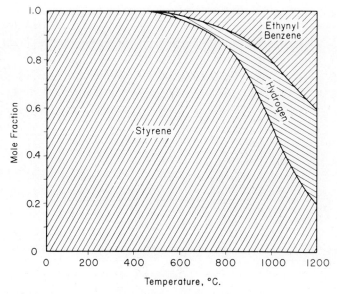

Figure 7-6 Calculated thermodynamic equilibria at 1 atm for the reaction styrene = ethynylbenzene + hydrogen. From Stull (1433).

ACRYLONITRILE SYNTHESIS

The synthesis of acrylonitrile on an industrial scale has been carried out for some years. During the early laboratory stages Stamm, Halverson, and Whalen (1404) developed thermodynamic functions for vinylacetylene and hydrogen cyanide and published equilibrium data for two reactions involved in the synthesis of acrylonitrile. Table 7-11 and Figure 7-10 present the Gibbs energies and calculated equilibrium constants for the two reactions

$$HCN(g) + HC\equiv CH(g) = H_2C=CHCN(g) \qquad [7\text{-}6]$$

and

$$2HC\equiv CH(g) = H_2C=CH-C\equiv CH(g). \qquad [7\text{-}7]$$

During the exploratory stages it is very helpful to know with reasonable certainty the equilibrium concentrations of products and reactants to be expected of the reaction under consideration. If the thermodynamic equilibria of the reactions are favorable (as both of these are), further laboratory work on the reactions is warranted. If the conditions of the

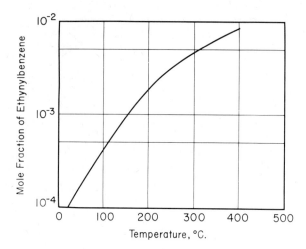

Figure 7-7 Disproportionation of styrene to ethynylbenzene and ethylbenzene. From Stull (1433).

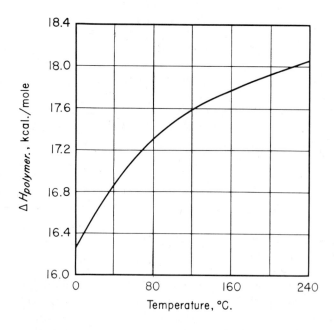

Figure 7-8 Enthalpy of polymerization of liquid styrene monomer to 100% solid polystyrene. From Stull (1433).

Acrylonitrile Synthesis

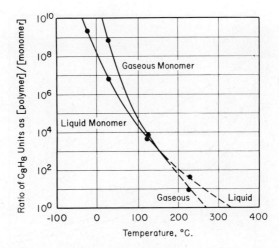

Figure 7-9 Calculated thermodynamic reaction equilibrium for styrene polymerization. From Stull (1433).

Table 7-11 Gibbs Energy Changes and Calculated Equilibrium Constants for the Reactions

$$HCN(g) + HC\equiv CH(g) = H_2C=CHCN(g) \quad [7\text{-}6]$$
and
$$2HC\equiv CH(g) = H_2C=CH-C\equiv CH(g)^a \quad [7\text{-}7]$$

	$\Delta Gr°$(kcal)		$\text{Log}_{10}Kp$(atm)	
T(°K)	[7-6]	[7-7]	[7-6]	[7-7]
298	−31.99	−26.90	23.445	19.72
300	−31.93	−26.90	23.262	19.59
400	−28.80	−23.90	15.735	13.06
500	−25.59	−20.80	11.184	9.09
600	−22.34	−17.80	8.135	6.48
700	−19.07	−14.70	5.952	4.59
800	−15.78	−11.60	4.316	3.17
900	−12.52	−8.50	3.040	2.06
1000	−9.26	−5.40	2.024	1.18

[a] Data of Stamm, Halverson, and Whalen (1404).

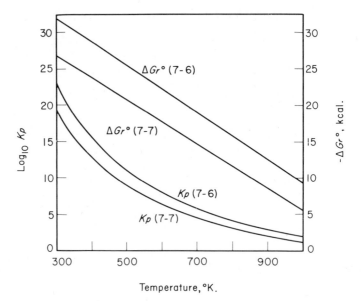

Figure 7-10 Temperature dependence of the Gibbs energy increment and the logarithm of the equilibrium constant for the two reactions

$$HCN(g) + C_2H_2(g) = H_2C\!\!=\!\!CHCN(g) \qquad [7\text{-}6]$$

and

$$2C_2H_2(g) = H_2C\!\!=\!\!CH-C\!\!\equiv\!\!CH(g). \qquad [7\text{-}7]$$

From the data of Stamm, Halverson, and Whalen (1404).

laboratory work are optimal the theoretical equilibria will be achieved. However, when the theoretical equilibria are not attained in practice, it is possible that the kinetics of the reaction are unfavorable, in which case a search for an appropriate catalyst to promote the reaction rate is indicated.

VINYL CHLORIDE SYNTHESIS

A thermodynamic analysis of the gas phase synthesis of vinyl chloride from acetylene and hydrogen chloride under atmospheric and elevated pressures was reported by Skupinski (1369). Also included was an economic study and comparison of unpressurized and pressurized processes utilizing concentrated and dilute acetylene. The primary reaction is

$$C_2H_2(g) + HCl(g) \stackrel{K_8}{=} CH_2\!\!=\!\!CHCl(g), \qquad [7\text{-}8]$$

accompanied by a side reaction,

$$CH_2{=}CHCl(g) + HCl(g) \stackrel{K_9}{=} CH_3{-}CHCl_2(g), \qquad [7\text{-}9]$$

in which 1,1-dichloroethane is formed. The industrial synthesis is carried out at 120–180°C and normal pressure in multitube reactors packed with the catalyst (mercuric chloride) deposited on activated carbon.

Our calculated equilibrium constants listed in Table 7-12 are in good agreement with those reported by Skupinski. The equilibrium composition of the reaction mixture for the synthesis from concentrated and dilute acetylene for various pressures is given in Tables 7-13 and 7-14.

The following conclusions result from the assembled information. The total conversion of acetylene decreases slightly with increasing temperature, and changes insignificantly with increasing pressure; however, the amount of 1,1-dichloroethane produced increases slightly with pressure. Hence the optimum (thermodynamic) conditions for the reaction are low

Table 7-12 Calculated Thermodynamic Equilibrium Constants for [7-8] and [7-9]

Equilibrium Constant	300°K	400°K	500°K
K_8	6.58×10^{10}	2.94×10^6	6.55×10^3
K_9	1.06×10^5	7.13×10^1	8.77×10^{-1}

Table 7-13 Calculated Equilibrium Composition of the Reaction for the Synthesis of Vinyl Chloride from Concentrated (Carbide) Acetylene[a]

		Equilibrium Composition			(mole percent)
Pressure (atm)	Temperature (°C)	C_2H_2	CH_3CHCl_2	HCl	CH_2CHCl
1	100	0.02	0.02	0.01	99.95
	200	0.49	0.05	0.43	99.03
5	100	0.02	0.02	0.00	99.96
	200	0.26	0.10	0.16	99.48
10	100	0.02	0.02	0.00	99.96
	200	0.22	0.12	0.10	99.56

[a] Data of Skupinski (1369).

Table 7-14 Calculated Equilibrium Composition of the Reaction Mixture for the Synthesis of Vinyl Chloride from Dilute (7.7 volume percent) Acetylene[a]

Pressure (atm)	Temperature (°C)	Equilibrium Composition (mole percent)					Conversion of C_2H_2 to CH_2CHCl
		Inert	C_2H_2	CH_3CHCl_2	HCl	CH_2CHCl	
5	100	92.31	0.00	0.00	0.00	7.69	100
	200	92.26	0.06	0.00	0.06	7.63	99.22
10	100	92.31	0.00	0.00	0.00	7.69	100
	200	92.27	0.04	0.00	0.04	7.64	99.35

[a] Data of Skupinski (1369).

pressures and temperatures. Although thermodynamic considerations do not suggest the use of pressure in the synthesis of vinyl chloride, the reduction in the required quantity of catalyst is found to be almost proportional to the increase of pressure for both concentrated and dilute acetylene. Costs of pressurization and catalyst will determine which conditions of operation are most economical.

THERMODYNAMICS IN METHANOL SYNTHESIS

Although at least a score of plants synthesize methanol from carbon monoxide and hydrogen,

$$CO(g) + 2H_2(g) = CH_3OH(g), \qquad [7\text{-}10]$$

very little information has been published on the best conditions for this reaction. This prompted Thomas and Portalski (1486) to employ thermodynamics to establish the optimum conditions by determining the pressure and temperature effects on the enthalpy of reaction and chemical equilibrium. Expressing the equation of state in the form

$$\left(\frac{\partial H}{\partial P}\right)_T = V - T\left(\frac{\partial V}{\partial T}\right)_P$$

indicates that, if ideal gas conditions apply, pressure will not affect the enthalpy of the reaction. At elevated pressures, however, the conditions will not be ideal, causing pressure to affect the enthalpy of the reaction.

Berthelot equations of state were applied to all components, summed algebraically, and then integrated, leading to

$$\Delta H = \Delta H_0 + \int_{P^\circ}^{P} \Delta V \, dP - T \int_{P^\circ}^{P} \left(\frac{\partial(\Delta V)}{\partial T}\right)_P dP,$$

the change in the enthalpy of reaction at elevated pressures. Substitution of this enthalpy of reaction in the van't Hoff equation leads to the expression

$$\frac{d \ln Ka}{dT} = \frac{\Delta H}{RT^2},$$

where Ka is the equilibrium constant in terms of activities. The activity of each component may be replaced by its fugacity f, which in turn is replaced by the product of the partial pressure p and the fugacity coefficient γ, since $f = p\gamma$. Although data for the fugacities of individual gases in mixtures are not usually available, the fugacity may be approximated by $f \simeq Nf'$, where f is the fugacity of the component in the mixture, f' is the fugacity of the pure component at total pressure P, $f' = P\gamma'$, and N is the mole fraction of the component in the mixture. Then

$$Kf' = \left(\frac{N_{CH_3OH}}{N_{CO} \times (N_{H_2})^2}\right)\left(\frac{\gamma'_{CH_3OH}}{\gamma'_{CO} \times (\gamma'_{H_2})^2}\right) P^{-2}, \qquad [7\text{-}11]$$

in which -2 is the change in the number of moles in the reaction.
Now

$$Kp' = \left(\frac{N_{CH_3OH}}{N_{CO} \times (N_{H_2})^2}\right) P^{-2}$$

and the γ' values in [7-11] may be obtained from the activity and fugacity charts of Newton and Dodge (1063), permitting the calculation of Kp', the equilibrium constants at higher pressures and temperatures. The calculated conversions for [7–10] are shown in Table 7-15 and Figure 7-11.

MANUFACTURE OF FORMALDEHYDE FROM METHANOL

Jones and Fowlie (699) treated the manufacture of formaldehyde from methanol as an exercise in chemical thermodynamics. They began by assembling the thermodynamic equilibrium constants for the pyrolysis and oxidation of methanol shown in Table 7-16. The first stage in the pyrolytic decomposition of methanol is

$$CH_3OH = HCHO + H_2.$$

Table 7-15 Calculated Equilibrium Conversion for the Reaction $CO(g) + 2H_2(g) = CH_3OH(g)$[a]

	Mole Fraction of CH_3OH in the Equilibrium Reaction Mixture at			
P (atm)	$t = 200°C$ ($\times 10^1$)	$t = 300°C$ ($\times 10^3$)	$t = 400°C$ ($\times 10^5$)	$t = 500°C$ ($\times 10^6$)
100	1.2	1.4	5.5	4.8
200	1.8	2.0	6.9	5.6
300	2.4	2.9	8.3	6.8
400	2.9	3.8	10.8	7.9
500	3.4	4.8	12.8	8.9
600	3.8	5.9	15.1	10.2
700	4.2	6.9	17.2	11.4
800	4.5	7.7	19.3	12.5
900	—	8.8	21.8	13.6
1000	—	9.8	24.2	15.6

[a] Data of Thomas and Portalski (1486).

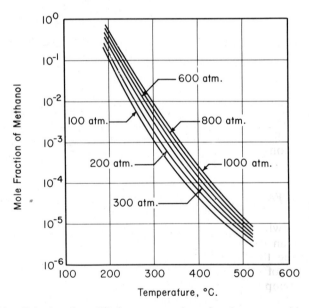

Figure 7-11 Calculated equilibrium conversion of carbon monoxide and hydrogen to methanol. Data of Thomas and Portalski (1486).

The mole fractions for this reaction, shown in Figure 7-12, may be calculated from the equilibrium values given in Table 7-16. It will be

Table 7-16   Thermodynamic Equilibrium Constants for Pyrolysis and Oxidation of Methanol[a]

	$\text{Log}_{10} K$ at				
Gaseous Chemical Reaction	600°K	700°K	800°K	900°K	1000°K
[1] $CH_3OH = HCHO + H_2$	−1.22	−0.14	0.69	1.33	1.86
[2] $HCHO = CO + H_2$	5.16	5.34	5.48	5.62	5.70
[3] $CH_3OH + \frac{1}{2}O_2 = HCHO + H_2O$	17.41	15.44	13.98	12.83	11.92
[4] $HCHO + \frac{1}{2}O_2 = CO + H_2O$	23.79	20.92	18.77	17.12	15.76
[5] $CO + \frac{1}{2}O_2 = CO_2$	20.06	16.54	13.90	11.84	10.20

[a] Data of Jones and Fowlie (699).

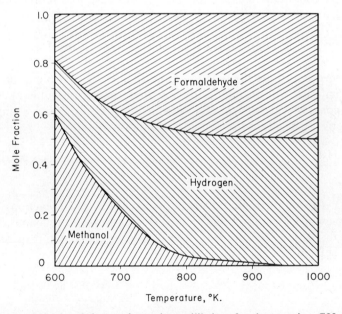

Figure 7-12   Calculated thermodynamic equilibrium for the reaction $CH_3OH(g) = HCHO(g) + H_2(g)$. Data of Jones and Fowlie (699).

observed that this process of dehydrogenation of methanol to yield formaldehyde improves with rise of temperature and is substantially complete at

1000°K or somewhat above. However, Newton and Dodge (1062) showed that formaldehyde is catalytically hydrogenated to methanol in the range 390–470°K, indicating that this is a reversible reaction.

The second stage of the pyrolytic decomposition of methanol is

$$HCHO = CO + H_2,$$

but according to Table 7-16 the equilibrium constant for this reaction ranges from 10^5 to 10^6, indicating that the reaction is in reality an irreversible process. Thus, as quickly as formaldehyde forms by Reaction 1 of Table 7-16, it is promptly destroyed by Reaction 2, making the dehydrogenation of methanol an unattractive process for the manufacture of formaldehyde.

The production of formaldehyde involves catalytic oxidation of methanol by Reaction 3 of Table 7-16 at temperatures sufficiently low that Reactions 4 and 5 have negligible rates. Since pure formaldehyde vapor begins to decompose at about 573°K, this is the upper temperature boundary for the oxidation process. Air is a desirable source of oxygen for such a process; hence the reaction may be written

$$CH_3OH + \tfrac{1}{2}\underbrace{(O_2 + 4N_2)}_{\text{(air)}} = HCHO + H_2O + 2N_2.$$

In these proportions the reaction mixture is explosive, but dilution with additional air will maintain the reaction temperature below that of decomposition and place the concentration out of the explosive range. The equation may be altered to include this dilution as follows:

$$CH_3OH + \tfrac{1}{2}x(O_2 + 4N_2) = HCHO + H_2O + \tfrac{1}{2}(x-1)O_2 + 2xN_2,$$

[7-12]

where x is twice the required amount of air. Enthalpy of reaction data and enthalpies of gases for application to this and related reactions are presented in Tables 7-17 and 7-18. The change in enthalpy for the reactants of [7-12] from 298° to 573°K is given by

$$3.665 + \tfrac{1}{2}x(1.995 + 4 \times 1.933) = 3.665 + 4.863x.$$

The enthalpy of the reaction at 573°K is given as 36.7 kcal, and equating these we get $x = 6.8$. If the assumptions are correct, [7-12] indicates the ideal reaction mixture to be $[CH_3OH + 3.4(O_2 + 4N_2)]$, or 5.6% methanol in air. If the operation is to be carried out at 673°K the value of x is found to be 4.6, and the methanol content of the reactant mixture is 8.0%.

Table 7-17 Enthalpies of Reaction at Constant Pressure for Reactions Related to the Manufacture of Formaldehyde[a]

Gaseous Chemical Reaction	Enthalpy of Reaction (kcal) at					
	500°K	600°K	700°K	800°K	900°K	1000°K
[1] $CH_3OH = HCHO + H_2$	21.5	21.8	22.0	22.1	22.2	22.3
[2] $HCHO = CO + H_2$	2.2	2.6	2.8	3.0	3.1	3.1
[3] $CH_3OH + \frac{1}{2}O_2 = HCHO + H_2O$	−36.8	−36.7	−36.7	−36.8	−36.9	−36.9
[4] $HCHO + \frac{1}{2}O_2 = CO + H_2O$	−56.1	−55.9	−55.9	−55.9	−56.0	−56.1
[5] $CO + \frac{1}{2}O_2 = CO_2$	−67.80	−67.79	−67.76	−67.70	−67.64	−67.55

[a] Data of Jones and Fowlie (699).

Table 7-18 Enthalpies of Gases Related to the Manufacture of Formaldehyde[a]

Gas	Enthalpy (kcal/mole) at							
	300°K	400°K	500°K	600°K	700°K	800°K	900°K	1000°K
Oxygen	0.013	0.722	1.454	2.209	2.988	3.786	4.600	5.427
Hydrogen	0.013	0.708	1.407	2.106	2.808	3.514	4.225	4.943
Nitrogen	0.013	0.710	1.413	2.126	2.853	3.597	4.356	5.131
Steam	0.016	0.828	1.659	2.516	3.405	4.324	5.269	6.242
Carbon monoxide	0.014	0.712	1.418	2.138	2.874	3.628	4.399	5.185
Carbon dioxide	0.017	0.958	1.986	3.086	4.244	5.453	6.702	7.985
Methanol	0.021	1.181	2.541	4.081	5.771	7.631	9.621	11.711
Formaldehyde	0.017	0.906	1.896	2.993	4.194	5.483	6.865	8.314

[a] Data of Jones and Fowlie (699).

Patents on this straight oxidation method issued to Craver (294) indicate working temperatures from 250° to 450°C, and the gas composition at 5–10% of methanol in air.

As seen in Table 7-17, the dehydrogenation reactions are endothermic whereas the oxidation reactions are exothermic. If these two main reaction types could be combined in the proper way, their enthalpies of reaction could be made to balance at the reaction temperature. A plant operating in this way should stabilize itself at the desired temperature and could be controlled by feeding the proper ratio of reactants. Table 7-16 shows that the equilibrium constant for the oxidation reaction 3 has a value of 10^{11}–10^{17} over the temperature range considered, indicating that the oxidation part of the composite process will be complete and irreversible at all temperatures considered. However, because the equilibrium constant for the dehydrogenation reaction 1 in Table 7-16 extends from 10^{-1} to 10^2, with excess reactant methanol the reaction products will be methanol, formaldehyde, steam, hydrogen, and nitrogen in thermodynamic equilibrium at the reaction temperature. Representing the ratio of formaldehyde to excess methanol in the products as $(1+x)/y$ permits the combined process to be represented as

$$CH_3OH + \tfrac{1}{2}(O_2 + 4N_2) + (x+y)CH_3OH = \\ HCHO + H_2O + 2N_2 + xHCHO + xH_2 + yCH_3OH,$$

which reduces to

$$(1+x+y)CH_3OH + \tfrac{1}{2}(O_2 + 4N_2) = (1+x)HCHO \\ + yCH_3OH + xH_2 + H_2O + 2N_2.$$

Table 7-16 gives the equilibrium constant as a function of temperature for the dehydrogenation reaction 1, which now becomes

$$K = \frac{N_{H_2} N_{HCHO}}{N_{CH_3OH}} P_{tot} = \frac{x(1+x)}{y(4+2x+y)} P_{tot}$$

and provides one relationship between x and y. Another relationship can be obtained as before from Table 7-18 by equating the enthalpy of the reactants at the reaction temperature to the enthalpy increment of the reaction at that temperature, as given in Table 7-17. Thus x and y may be evaluated at any temperature and permit calculation of the formaldehyde yield at various temperatures and gas compositions.

Operation of the combined processes will require higher mole fractions of methanol, a condition that can be achieved by preheating the reactant gases to 383°K. Assuming a reaction temperature of 1000°K, and 1 atm we have

$$K = \frac{x(1+x)}{y(4+2x+y)} = 10^{1.86} = 72.4. \qquad [7\text{-}13]$$

The enthalpy of the reactants between 383° and 1000°K equals the sum of the enthalpies of the oxidation and dehydrogenation reactions at 1000°K.

Using the enthalpy data from Table 7-18, this gives

$$36.9 - 22.3x = (1 + x + y)10.73 + \tfrac{1}{2}(4.825) + 2(4.539),$$

which reduces to

$$y = 1.369 - 3.078x. \qquad [7\text{-}14]$$

Simultaneous solution of [7-13] and [7-14] leads to $x = 0.443$ and $y = 0.004$. Consequently at 1000°K the composite reaction would be

$$1.447\ CH_3OH + \tfrac{1}{2}(O_2 + 4N_2) = 1.443\ HCHO + 0.004\ CH_3OH$$
$$+ 0.443\ H_2 + H_2O + 2N_2.$$

Similar calculations were performed for other temperatures and are presented in Table 7-19. These calculations show several interesting

Table 7-19 Effect of Reaction Temperature on Composition for the Gaseous Production of Formaldehyde by [7-12][a]

$T(°K)$	Values of		Reactants				Products					Percent
			Moles			Percent	Moles					
	x	y	CH_3OH	Air[b]	H_2O	CH_3OH	HCHO	CH_3OH	H_2	H_2O	N_2	HCHO
600	0.83	2.92	4.75	2.50		65.5	1.83	2.92	0.83	1.00	2.00	21.3
700	0.92	0.35	2.27	2.50		47.6	1.92	0.35	0.92	1.00	2.00	31.0
800	0.77	0.051	1.82	2.50		42.2	1.77	0.051	0.77	1.00	2.00	31.7
900	0.61	0.004	1.61	2.50		39.2	1.61	0.004	0.61	1.00	2.00	30.8
1000	0.44	0.004	1.45	2.50		36.7	1.44	0.004	0.44	1.00	2.00	29.5
600	0.78	2.73	4.51	2.50	1.00	56.3	1.78	2.73	0.78	2.00	2.00	19.2
700	0.83	0.285	2.11	2.50	1.00	37.6	1.83	0.285	0.83	2.00	2.00	26.3
800	0.65	0.020	1.67	2.50	1.00	32.3	1.65	0.020	0.65	2.00	2.00	26.1
900	0.46	0.008	1.46	2.50	1.00	29.5	1.46	0.008	0.46	2.00	2.00	24.6
1000	0.22	0.003	1.22	2.50	1.00	25.9	1.22	0.003	0.22	2.00	2.00	22.4

[a]Data of Jones and Fowlie (699).
[b]One mole of air $= 0.2(O_2 + 4N_2)$.

features. At higher temperatures a smaller amount of reactant methanol is required. According to White (1604) the upper explosive limit of methanol in air is 36.5%, indicating that the reaction mixtures are explosive above 1000°K, so that conditions above this temperature may be expected to degenerate. The last column of Table 7-19 shows that a maximum concentration of formaldehyde is achieved at about 750°K from a mixture of 45% methanol and 55% air, while the maximum conversion of methanol to formaldehyde takes place at about 1000°K from a mixture of 37% methanol and 63% air. Hence optimum conditions for this process occur in the range between these two limits.

Addition of steam is a convenient way of heating a reaction, even though in this case steam is a product of the reaction. Incorporation of steam into the reaction leads to new equations like [7-13] and [7-14], whose solutions result in the figures given in the lower half of Table 7-19. The steam added to the combined process acts to lower the over-all temperature of the reaction, an effect to be expected of a diluent. Although the use of steam is a means of controlling the reaction temperature, there seems to be no improvement in yield.

MANUFACTURE OF ACETIC ACID

During his historic synthesis of liquid fuels from water gas, Fischer (427) identified acetic acid and methanol among the products of his catalytic reactions. Later Hardy (570) studied the catalytic interaction of carbon monoxide and methanol and noted that at room temperature the formation of acetic acid,

$$CH_3OH(g) + CO(g) = CH_3COOH(g), \qquad [7\text{-}15]$$

is an exothermic process. Hardy employed thermodynamic reasoning and calculated that up to about 643°K at 1 atm pressure the equilibrium should be in favor of acetic acid, and that for an increase of pressure from 1 to 50 atm at 600°K the reversal of [7-15] would be decreased from 25% to 4%. Hardy studied the effects of temperature (up to 643°K), pressure (up to 200 atm), rate of circulation over the catalyst, and catalyst quantity; Figures 7-13 and 7-14 show the effects of temperature and pressure upon the yield of acetic acid. In addition to acetic acid, a number of other organic compounds were identified, showing that side reactions also occurred.

At about the same time Singh and Krase (1352) also studied the vapor phase, catalytic synthesis of acetic acid from methanol and carbon monoxide under pressure. These authors deplored the lack of accurate thermodynamic information on acetic acid and its mixtures, but from approximate data they decided that "the reaction is exothermic to the extent of 20 to 30 kcal and that the equilibrium at temperatures between 300° and 500°C and even at pressures below 300 atm is very

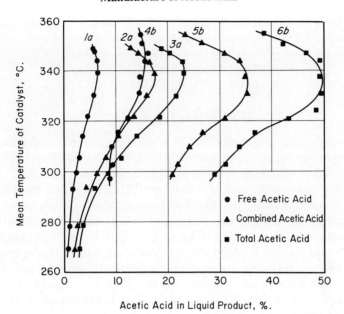

Figure 7-13 Effect of catalyst temperature on yield of acetic acid. From Hardy (570).

favorable to acetic acid formation." They studied the effects of temperature (to 500°C), pressure (to 4000 psig), space velocity over the catalyst, relative amounts of reactants, and recirculation of by-products. The

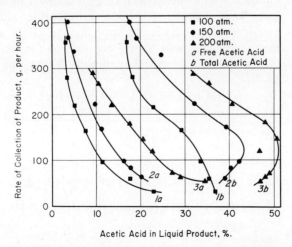

Figure 7-14 Effect of pressure on acetic acid yield. From Hardy (570).

by-products found experimentally were accounted for by the following reactions:
$$2CH_3OH = CH_3OCH_3 + H_2O,$$
$$CH_3OH + CO = HCOOCH_3,$$
$$CH_3OH + CO = CO_2 + CH_4,$$
or
$$CH_3COOH = CO_2 + CH_4.$$

The process could be carried out in a manner such that water-insoluble oils need not be formed and that the organic by-products were capable of ultimate and complete conversion to acetic acid.

Kutepow, Himmele, and Hohenschutz (818) published the technical details of a continuous 1000 ton/month acetic acid plant that operates in the liquid phase under about 100 atm at 300°C with a cobalt carbonyl-iodine catalyst. With every 100 kg of CH_3COOH (greater than 99.8% purity) about 4 kg of by-products are produced; half of this is propionic acid, while the rest consists of ethyl acetate, butyl acetate, and 2-ethylbutanal. The enthalpy of reaction for the plant is -29.03 kcal/mole of acetic acid produced. Both the synthesis plant and the product separation plant are completely automated.

THE GATTERMANN-KOCH REACTION

Dilke and Eley (328) studied the thermodynamics of the Gattermann-Koch reaction. This is a reaction in which the —CHO group is introduced into aromatic molecules with the assistance of a catalyst of the (HCl + $AlCl_3$) type. Some have postulated that formyl chloride, although unisolated, is a transient intermediate. By applying the Gibbs energy increment between acetic acid and acetyl chloride to formic acid, one estimates $\Delta Gf°_{298}$ ClCHO(g) $= -43.17$ kcal/mole. From this one may calculate

$$HCl(g) + CO(g) = ClCHO(g),$$
$\Delta Gf°_{298}$ -22.78 -32.82 -43.17

$\Delta Gr°_{298}$ = 12.43 kcal/mole.

It becomes immediately apparent why formyl chloride has never been isolated in a reaction of HCl and CO. Completing the reaction, we have

$$ClCHO(g) + C_6H_6(g) = C_6H_5CHO(g) + HCl(g),$$
$\Delta Gf°_{298}$ -43.17 30.99 -1.07 -22.78

$\Delta Gr°_{298}$ $= -11.67$ kcal/mole,

which is ample indication that the elusive formyl chloride will react with benzene to produce benzaldehyde and hydrogen chloride.

In view of the difficulty of preparing formyl chloride (although formyl fluoride is known and can be isolated), Dilke and Eley (328) investigated the thermodynamics of formation of the catalytic complex. For the reaction

$$C_6H_6(l) + CO(g) = C_6H_5CHO(l)$$

they reported the thermodynamic properties in Table 7-20, from which

Table 7-20 Thermodynamic Properties of Benzene, Carbon Monoxide, and Benzaldehyde[a]

Compound	ΔHc_{298}[b]	Source	$\Delta Hf°_{298}$[b]	$S°_{298}$[c]	Source	Cp_{298}[G]	Source
Benzene (l)	781.0	(1194)	11.737	41.5	(1477)	31.9	(1577)
Carbon monoxide (g)	67.6	(1244)	−26.452	47.3	(1244)	7.0	(1577)
Benzaldehyde (l)	841.3	(744)	−22.015	57.4[d]	(1105)	45.4	(1577)

[a] Data of Dilke and Eley (328).
[b] In kilocalories per mole.
[c] In calories per (mole degrees Kelvin).
[d] Estimated from data in this source.

may be calculated: $\Delta Hr°_{298} = -7.3$ kcal/mole, $\Delta Sr°_{298} = -31.4$ cal/(mole °K), and $\Delta Cp°_{298} = 6.5$ cal/(mole °K). The Gibbs energy increment and equilibrium constant as a function of temperature may then be calculated by

$$-RT \ln K = \Delta G°_T = \Delta H°_{298} - T \Delta S°_{298} + \Delta Cp (T - 298) - T \Delta Cp \ln \frac{T}{298}$$

as shown in Table 7-21. The yield of benzaldehyde at room temperature is about 2% (calculated on the benzene) and decreases as the temperature is raised. Since the calculated value of K is very sensitive to small changes in ΔHr, the uncertainty in the benzaldehyde enthalpy of formation data makes accurate comparison difficult.

In an endeavor to clarify the situation, Dilke and Eley (328) measured the Gibbs energy increments for mixing benzaldehyde with some of the halide salts that catalyze this reaction. They carried out two kinds of measurements: (a) the calorimetric enthalpy of mixing of the solid catalytic halide with liquid benzaldehyde, that is, the enthalpy of complex formation as in [7-16], and (b) the concentration of the species present at equilibrium, yielding the equilibrium constant K for each reaction in the enthalpy of mixing series

Table 7-21 Calculated Benzaldehyde Equilibria for the Reaction $C_6H_6(l) + CO(g) = C_6H_5CHO(l)$[a]

$T(°K)$	$\Delta G°$(kcal)	$K(\times 10^3)$
298	2.06	31.6
323	2.83	12.6
373	4.35	2.9

[a] Data of Dilke and Eley (328), from data in Table 7-20.

$$C_6H_5CHO(l) + \tfrac{1}{2}Al_2Cl_6(s) = C_6H_5CHO \cdot AlCl_3(s). \qquad [7\text{-}16]$$

These measurements are summarized in Table 7-22 and show that the increments in the enthalpy, Gibbs energy (or equilibrium constant), and entropy rank the catalysts in the same order, with the one exception of the

Table 7-22 Thermodynamic Order of Catalyst Activity in [7-16] at 298°K[a]

Complex	$\Delta Hr°$(kcal/mole)	K	$\Delta G°$(kcal/mole)	$\Delta S°$[cal/(mole°K)]
$(C_6H_5 \cdot CHO)_2 \cdot SnCl_4$	−28.0	3.6×10^5	−7.6	−68
$C_6H_5 \cdot CHO \cdot AlCl_3$	−24.7	60	−2.4	−75
$C_6H_5 \cdot CHO \cdot FeCl_3$	−12.8	30	−2.0	−36
$C_6H_5 \cdot CHO \cdot SbCl_3$	−3.9	9	−1.3	−9

[a] Data of Dilke and Eley (328).

entropy change for aluminum chloride. This, however, may be satisfactorily explained by the additional dissociation equilibrium $Al_2Cl_6 = 2AlCl_3$, associated with this case. It is clear from these data that complex formation occurs and will make the Gibbs energy change more favorable for the Gattermann-Koch synthesis.

Obviously this section points up the need for further thermochemical investigation to clarify the role of the catalyst in such problems.

UTILIZATION OF THERMODYNAMICS IN CATALYST SELECTION

Although the theory of heterogeneous catalysis has not advanced to the point where catalysts can be selected on a theoretical basis, the thermodynamic characteristics of catalytic substances and reactions are helpful in the selection of appropriate catalysts. Because chemical forces operate between the reactants and the catalyst, the catalyst must have a chemical affinity for at least one of the components of the reaction. Hence substances that are chemically inert for a given system are immediately eliminated as prospective catalysts. The change in Gibbs energy accompanying the interaction of a catalyst with the reactants must have a smaller negative value than that for the reaction catalyzed. This criterion eliminates as catalysts those materials that form stable compounds with the reactants. Thus it is necessary to search for some relationship between thermodynamic and catalytic properties that will indicate the usefulness of a given material as a catalyst for a given reaction. Among the efforts made to employ thermodynamic information should be mentioned the works of Schwab (1293), Frankenburg (438), Beeck (97), Kurin and Zakharov (815), Kul'kova and Temkin (813), and the review article by Golodets and Roiter (506).

An example of the application of the Gibbs energy change to catalyst selection is shown in Figure 7-15. Line a indicates the temperature dependence of $\frac{1}{2}\Delta Gr^\circ$ of some over-all reaction, while lines b and c represent the temperature dependence of ΔGr° of a proposed intermediate stage utilizing catalysts B and C, respectively. This situation indicates that

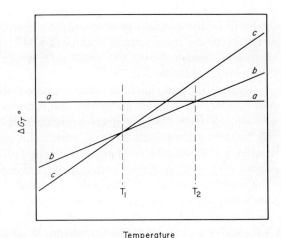

Figure 7-15 Temperature dependence of the change in the Gibbs energy for the formation of intermediates with catalysts. Here a represents the temperature dependence of $\frac{1}{2}\Delta Gr^\circ$ for over-all reaction; b and c represent the temperature dependencies of ΔGr° for formation of hypothetical intermediates utilizing catalysts B and C, respectively. Data of Golodets and Roiter (506).

catalyst B will be more active in the temperature ranges below T_1 and above T_2, while catalyst C will be more active in the temperature range between T_1 and T_2. If the structural relationships of the reaction intermediates yield an entropy increment near zero, then the relationship

$$\Delta Gr° = \Delta Hr° - T \Delta Sr°$$

indicates that the enthalpy change for such a reaction will also be a reliable index to catalyst effectiveness.

Makishima, Yoneda, and Saito (915) showed a series of examples substantiating the principle that the most suitable substances for catalysis are those for which the enthalpy increment on interaction with the components of the reaction catalyzed is about half the over-all enthalpy increment for the reaction. As an example, consider the oxidation

$$CO + \tfrac{1}{2}O_2 = CO_2$$

and assume that the catalyst K acts in two stages:

$$K + \tfrac{1}{2}O_2 = K' \qquad [7\text{-}17]$$

$$K' + CO = CO_2 + K. \qquad [7\text{-}18]$$

The best catalyst will be the one that most evenly distributes the Gibbs energy change between the intermediate reactions [7-17] and [7-18]. These requirements are satisfied by CuO and Cu_2O, which are indeed efficient catalysts for this oxidation.

Fahrenfort, van Reijen, and Sachtler (403) studied the catalytic decomposition of formic acid by various metals and found a correlation between the enthalpy of formation per gram-equivalent of the metallic formates in bulk with the temperature Tr at which the catalyzed decomposition reaches a definite rate r. Their data, represented in Figure 7-16, show at about 80 kcal an optimum enthalpy of formation per gram-equivalent for the metallic formate. At this point the catalytic activity is maximized, as indicated by the lowest temperature required to achieve a given rate.

One theory of catalysis postulates the formation of an intermediate compound on the surface of the catalyst. For a doublet reaction of the type

$$\begin{array}{c} A \quad B \\ | \quad | \\ C \quad D \end{array} + K \underset{\Delta H_I}{\overset{(I)}{\rightleftharpoons}} \begin{array}{c} A\text{-----}B \\ | \quad | \quad K \\ C\text{-----}D \end{array} \underset{\Delta H_{II}}{\overset{(II)}{\rightleftharpoons}} \begin{array}{c} A\text{---}B \\ \\ C\text{---}D \end{array} + K$$

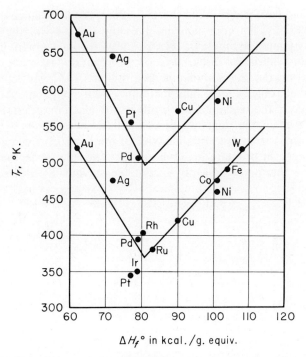

Figure 7-16 Correlation of the enthalpies of formation per gram-equivalent weight of metal formates and the catalytic activities of the metals in formic acid decomposition. The points of the upper curve involve values of $\ln r = 1.0$. Those on the lower curve have $\ln r = 0.8$. Data of Fahrenfort, van Reijen, and Sachtler (403).

Stage I represents the formation of an intermediate compound of the reactants on the surface of the catalyst K, while Stage II represents the decomposition of the intermediate compound into the products of the reaction. Balandin (66) theorized that in this case, for the optimum catalyst, the enthalpy of reaction ΔH_I for Stage I should be equal to the enthalpy of reaction ΔH_{II} for Stage II, and that the value of the over-all reaction [7-19] should be equal to the sum of the enthalpies of reaction in Stages I and II:

$$A\text{—}C + B\text{—}D. = A\text{—}B + C\text{—}D \qquad [7\text{-}19]$$

$$\Delta Hr = \Delta H_I + \Delta H_{II}.$$

This concept may be generalized to any catalytic process that appears to pass through a series of stages. It also indicates the most advantageous route to be that for which the thermal effects of all the stages are approximately equal. This approach involves the selection of the appropriate thermodynamic functions, in correlation with catalytic activity and other applicable criteria for each stage of the reaction. It is not adequate to consider only the thermodynamic requirements, however. Kinetic factors, type of electronic binding, and matching of atomic geometry of the catalyst to that of the reactants and products are also important. Such factors help to explain the dependence of catalytic activity on the method of preparation, variation in the behavior of the faces of single crystals and the presence of promoters and modifiers. These latter factors determine whether a thermodynamically acceptable substance will become a real catalyst.

SECTION II

THERMAL AND THERMOCHEMICAL PROPERTIES IN THE IDEAL GASEOUS STATE FROM 298° TO 1000°K

CHAPTER 8

THERMODYNAMIC SYMBOLS, CONSTANTS, METHODS OF CALCULATION, AND DATA FOR ELEMENTS AND SELECTED INORGANIC COMPOUNDS

The practical application of thermodynamics requires consistency in notation and numerical constants. It is especially important that tabulations of chemical thermodynamic properties be based on the same set of values for the elements. Hence the notation, constants, and reference state data for the elements used in the presentation of the remaining chapters are described here. The method of computation is discussed, and tables are given also for some inorganic compounds more or less frequently involved in organic chemistry.

DEFINITION OF THERMODYNAMIC SYMBOLS

The confusion in symbols for thermodynamic functions noted by Parks and Huffman (1105) persists to the present. Particularly unfortunate is the dichotomy existing between American and European chemical literature with respect to the function known in the United States as *free energy* and in Europe as *free enthalpy* and denoted by F and G, respectively. Pitzer and Brewer (861) proposed the adoption of the symbol G and the name *Gibbs energy* for this function. This course has also been recommended by the Commission on Symbols and Nomenclature in the Section on Physical Chemistry of the International Union of Pure and Applied Chemistry (1656). In the firm conviction that future compilations of thermodynamic data will follow the above recommendation, we have adopted this terminology. The remaining symbols, the most important of which are listed below, are less controversial. A more extensive list is given in Appendix 1.

E is the (intrinsic) energy.
H is the enthalpy ($= E + PV$).

G is the Gibbs energy ($= H - TS$).
Cp is the heat capacity at constant pressure.
Unless otherwise noted, molal values are given throughout.

The superscript ° when applied to any thermodynamic property, denotes that the value given is for the standard reference state. For our purposes the standard reference state for gases is the ideal gas state at 1 atm, while for condensed phases it is the pure liquid or solid at 1 atm. For example, ΔHv indicates the enthalpy of vaporization of the liquid phase to the *real* gas at saturation pressure, while $\Delta Hv°$ indicates the enthalpy of vaporization of pure liquid to the *ideal* gas at 1 atm.

Italicized letters *f, t, m, v, s, c,* and *r* used as modifiers immediately following thermodynamic symbols refer respectively to the following processes: formation from the elements, solid state transition, melting (that is, fusion), vaporization (liquid to vapor), sublimation (solid to vapor), combustion, and reaction.

Numerical subscripts, for example 298.15 or 0, denote the Kelvin temperature. In all tables and text the subscript 298 denotes 298.15°K.

$H_T° - H_{298}°$ is the enthalpy in the standard state at temperature T, less the enthalpy in the standard state at 298.15°K.

$(G_T° - H_{298}°)/T$ is the Gibbs energy function. It is defined in [5–22] as equal to $(H_T° - H_{298}°)/T - S_T°$. This function is tabulated in addition to the Gibbs energy itself, since it shows greater linearity and less variance with temperature, thus facilitating interpolation.

$S°$ is the "practical" entropy in the standard state at the indicated temperature, omitting contributions from isotopic mixing, and nuclear spins.

Δ denotes the increment of a given property for a given process or reaction. It represents the value for the final state less that for the initial state. For example, ΔHf is the enthalpy of formation, which is the enthalpy increment associated with the chemical reaction forming the given compound from its elements with each substance in a specified thermodynamic state at the given temperature. On the other hand, $\Delta Hf°$ is the standard enthalpy of formation, which is the enthalpy increment associated with the same reaction with each substance in its *standard* state at the given temperature.

Kp is the equilibrium constant for a given process or reaction expressed in terms of pressure. For the schematic chemical reaction $cC + dD \rightleftarrows mM + nN$, with each of its reactants and products at equilibrium, the equilibrium constant K is given by

$$K = \frac{(a_M)^m (a_N)^n}{(a_C)^c (a_D)^d},$$

in which a_i represents the activity of the ith substance. For a pure condensed phase the activity may be taken as unity; for a gas the activity may be equated with the fugacity. For an ideal gas the fugacity is equal to the pressure; for a real gas at relatively low pressures and at ambient atmospheric or higher temperatures the pressure is approximately equal to the fugacity; hence under these conditions

$$K \simeq Kp = \frac{(P_M)^m(P_N)^n}{(P_C)^c(P_D)^d}.$$

On the basis of [5-11], the equilibrium constant is related to the Gibbs energy by

$$\log_{10} Kp = -\frac{\Delta Gr°}{2.302585RT}.$$

$\log_{10} Kp$ as provided in the tabulations represents the logarithm (to the base 10) of the equilibrium constant in terms of pressure for the reaction forming 1 mole of substance from its elements with each substance in its standard state at the given temperature. It is similarly related to the standard Gibbs energy of formation by

$$\log_{10} Kp = -\frac{\Delta Gf°}{2.302585RT}.$$

ΔHm and ΔSm are the enthalpy and entropy increments of the melting (or fusion) process, with each substance in a specified thermodynamic state, usually saturation pressure. Since only condensed phases are involved, there are insignificant differences between the standard enthalpy and entropy increments $\Delta Hm°$ and $\Delta Sm°$ and the corresponding ΔHm and ΔSm, except at high pressures. For pure liquids and solids the effect of moderate pressure on the thermodynamic properties is nearly always negligible. The low-temperature heat capacity of a compound, for example, is commonly determined at the saturation vapor pressure plus a temperature-dependent pressure of helium gas to hasten attainment of thermal equilibrium, yet the results are within experimental precision of the heat capacity measured at a constant pressure of 1 atm. The temperature of melting is sufficiently pressure-dependent, however, to necessitate adoption of a special symbol, Ttp, for the frequently occurring case of triple points, that is, "melting points," measured at the saturation pressure of the compound.

ΔHv and ΔSv are the enthalpy and entropy increments for the vaporization process with liquid and gas at specified pressure, typically saturation values. Usually the difference between ΔHv (at the normal boiling point) and $\Delta Hv°$ is small, but for some associated vapors it is significant, as examples given in an earlier chapter show. The entropy of a gas is, of course, strongly dependent on pressure. Even at 1 atm significant differences between real and ideal gas reference state values often occur, especially when there is a tendency toward association or dissociation. Hence ΔSv and $\Delta Sv°$ may be expected to differ both because the saturation pressure may differ from 1 atm and because of gas imperfection.

ΔHs and ΔSs are enthalpy and entropy increments for the sublimation process (that is, solid to gas). They bear a relation to the standard quantities similar to that of the corresponding vaporization increments.

$\Delta Hc°$ is the enthalpy increment for the reaction involving the combustion of the substance in gaseous oxygen to form combustion products at constant temperature and pressure with all reactants and products in their standard states.

ΔHr is the enthalpy increment for a reaction, the nature of which is specified by the context. It is often loosely designated the "heat of reaction."

Temperatures given in degrees Celsius (centigrade, °C), based upon experimental determination, are referred to the international temperature scale (183, 1034, 1423) and have been converted to degrees Kelvin (°K) by the relation °K = 273.15 + °C. Differences between the international and thermodynamic scales have been discussed by Mueller (1034) and Stimson (1423).

CONSTANTS

To recalculate all the thermodynamic functions in the literature using modern constants is beyond the scope of this endeavor. Fortunately, recent changes in fundamental constants have been insignificant for the present purposes. For the cases in which thermodynamic functions have been calculated directly, the constants used were those selected by Cohen, Crowe, and Dumond (240). As noted elsewhere, the ice point defined as 273.15°K has been adopted. This change affects only the most accurately determined melting and boiling points. A selected list of constants used is given in Table 8-1.

The 1957 scale of atomic weights was used (1609) except for carbon, for which 12.010 was retained. The time-consuming task of converting all the enthalpies of combustion and thermodynamic functions to the latest value of 12.01115 is not justified by the slight increase in accuracy. Consistency is more important, and since the bulk of the tabular literature, including the extensive tables of Rossini, Wagman, Evans, Levine, and Jaffe (1249) and Rossini, Pitzer, Arnett, Braun, and Pimentel (1248),

Constants

Table 8-1 Physical Constants

Name	Value	Units
Basic constants		
Velocity of light	2.997930×10^{10}	cm/sec
Planck constant	6.62517×10^{-27}	erg-sec
Avogadro constant	6.02338×10^{23}	mole^{-1}
Absolute temperature of the ice point (0°C)	273.1500	°K
Molar volume of ideal gas	22,414.6	cm^3/mole
Pressure-volume product for 1 mole of gas at 0°C and zero pressure	2271.16	abs joule/mole
Derived constants		
Electronic charge	1.601886×10^{-19}	abs coulomb
Gas constant	8.31469	abs joule/(mole °K)
	1.98726	cal/(mole °K)
Boltzmann constant	1.380403×10^{-16}	erg/(molecule °K)
Defined constants		
Standard gravity	980.665	cm/sec^2
Standard atmosphere	1,013,250	dyne/cm^2
	760	mm Hg (at 298.15°K, sea level, 45° latitude)
Calorie (defined thermochemical)	4.1840	abs joule
	4.1833	int joule

plus the supplementary material (1653), are on the basis of 12.010; intercomparison is facilitated by retaining this value.

For the derivation of enthalpies of formation, the modern values for CO_2 and H_2O selected by Wagman, Kilpatrick, Taylor, Pitzer, and Rossini (1562) are used. In this connection it is interesting that a recent study of the enthalpy of combustion of graphite showed a significant dependence on the method used in preparation of the graphite (858). This finding may be of considerable importance in reactions that involve carbon as a product or reactant, for example, pyrolysis of hydrocarbons.

Enthalpies of combustion of compounds containing chlorine usually refer to a final state of aqueous hydrochloric acid. The enthalpy of formation of HCl (in 600 H_2O) is taken from Rossini, Wagman, Evans, Levine, and Jaffe (1249). Sulfur compounds are generally referred to a final state of aqueous sulfuric acid. The enthalpy of formation of H_2SO_4 (in 115 H_2O) was averaged from the determinations of the enthalpy of formation of aqueous sulfuric acid by the Petroleum Research Center in

Bartlesville, Oklahoma (509), and the Thermochemical Laboratory at Lund, Sweden (926). Enthalpies of dilution to other concentrations are from Rossini, Wagman, Evans, Levine, and Jaffe (1249). The limited combustion data on bromine compounds use liquid bromine as the final state. Conversion from aqueous hydrogen bromide to liquid bromine is also taken from the compilation of Rossini et al. (1249). The value for HF was taken from the recent revision of Lacher and Skinner (833). These enthalpy constants are summarized in Table 8-2.

Table 8-2 Enthalpy of Formation Constants

Substance	State	$\Delta H f^\circ_{298}$ (kcal/mole)
CO_2	(g)	−94.0519
H_2O	(l)	−68.3174
HF	(in 50 H_2O)	−77.03
HCl	(in 600 H_2O)	−39.881
HBr	(in 600 H_2O)	−28.775
H_2SO_4	(in 115 H_2O)	−212.20

Although these values are considered to be the most reliable available as this volume goes to press, it should be noted that additional studies are under way in several laboratories. This work may require future revision of such values.

MACHINE COMPUTATION OF THE TABLES

The practical utilization of chemical thermodynamic data frequently is more readily achieved from pertinent functions tabulated at selected temperatures than from equations that require the user to calculate his own values at each temperature. Hence a systematic digital computer method for compiling suitable tables is employed. The following information constitutes the basic data on each compound for the thermodynamic property table:

1. Smoothed values of the ideal gaseous heat capacity at 1 atm at nine temperatures (298.15°, 300°, 400°, 500°, 600°, 700°, 800°, 900°, and 1000°K*).

*Organic reactions operating above 1000°K are rare, and, although literature values have often been given as high as 1500°K, industrial organic chemistry (except, e.g., for the cracking of saturated hydrocarbons to produce ethylene and acetylene) is carried out below 1000°K.

2. Practical entropy of the ideal gas at 1 atm and 298.15°K.

3. The isothermal enthalpy of formation of the ideal gaseous compound at 1 atm and 298.15°K, from its elements in their reference states at 298.15°K.

4. The identity and number of each type of constituent atom in the compound.

5. Data on the reference elements (from this chapter) as follows: (a) gram atomic weight, (b) practical gram atomic entropy values at each of the nine temperatures, and (c) gram atomic enthalpy values at each of the nine temperatures with the value zero assigned to 298.15°K.

The elements considered in this study are C, H, Br, Cl, F, I, N, O, and S.

The appropriate values of Items 1 through 5 are punched in cards as input data for a Burroughs type 5500 electronic digital computer. For subsequent use in the computations Item 5 data (that is, for the elements) are stored by the computer. Data on Items 1 through 4 on one compound are given to the computer. From heat capacities at 300°, 400°, and 500°K the coefficients of the quadratic equation [2-13] are generated as shown on pages 48 and 49. Enthalpy increments, $H°_{300} - H°_{298}$ and $H°_{400} - H°_{300}$, are computed by integration of the heat capacity as shown in [2-17] to give enthalpy values at 300° and 400°K, respectively. The corresponding entropy increments are obtained by integration of the heat capacity data as shown in [4-25] and added to Item 2 (the practical entropy at 298.15°K) to give entropies at 300° and 400°K. The Gibbs energy functions are next calculated at 300° and 400°K from the relation

$$-\frac{(G°_T - H°_{298})}{T} = S°_T - 1000\frac{(H°_T - H°_{298})}{T},$$

in which T is the absolute temperature of the calculation and 1000 is a scaling factor to convert kilocalories to calories. The enthalpies of formation at 300 and 400°K are calculated from

$$\Delta Hf°_T = \Delta Hf°_{298} + [H°_T - H°_{298}] \text{ (compound)} - \sum [H°_T - H°_{298}] \text{ (elements)},$$

in which the summation is over the constituent elements in the compound. The standard Gibbs energy of formation is calculated from the relation

$$\Delta Gf°_T = \Delta Hf°_T - T \Delta Sf°_T.$$

Here $\Delta Sf°_T$, the standard entropy of formation for the compound from its elements at $T°K$, is calculated from

$$\Delta Sf°_T = S°_T \text{ (compound)} - \sum S°_T \text{ (elements)},$$

in which the summation includes all of the elements making up the compound.

The common logarithm of the standard equilibrium constant in terms of pressure is calculated from the equation

$$\log_{10}Kp = \frac{-\Delta Gf_T^\circ}{0.004575845T}.$$

Subsequently, using Item 1 data (heat capacities) at 400°, 500°, and 600°K, the coefficients of [2-13] are solved as above to give enthalpy and entropy increments relative to 400°K. These are added to the 400°K value to give the new values at 500°K, and the higher temperature values of the ΔHf°, ΔSf°, and $\log_{10}Kp$ are calculated successively for each temperature to 1000°K. The molecular weight is computed from data on Items 4 and 5(a).

This method of calculating thermodynamic functions gives tables with high internal consistency. When accurate heat capacities are available, the derived functions are also accurate. Table 8-3 compares thermody-

Table 8-3 Thermodynamic Functions of Benzene

	$H^\circ - H_{298}^\circ$ (kcal/mole)		S° [cal/(mole °K)]	
T(°K)	Rossini et al. (1248)	Present Method	Rossini et al. (1248)	Present Method
298	0.000	0.00	64.34	64.34
300	0.036	0.04	64.46	64.46
400	2.361	2.36	71.10	71.11
500	5.349	5.35	77.74	77.75
600	8.884	8.89	84.17	84.18
700	12.866	12.87	90.30	90.31
800	17.211	17.21	96.10	96.10
900	21.859	21.86	101.57	101.58
1000	26.762	26.76	106.73	106.74

namic functions for benzene calculated by the present method with those of Rossini, Pitzer, Arnett, Braun, and Pimentel (1248), which were computed directly from molecular and spectroscopic constants. The agreement is very satisfactory. On the other hand, errors in heat capacity data are propagated throughout a given table. The heat capacities given by Rossini et al. for the n-alkylcyclopentanes above the methyl derivative are not smooth and appear to be inaccurate. Table 8-4 shows first and second differences in Cp for methylcyclopentane and ethylcyclopentane. The value for the ethyl derivative at 400°K is obviously too high. Because of the lack of smoothness the derived functions would not be expected to agree as well as those for benzene. Enthalpies and entropies derived by the present program are compared in Table 8-5 with those of Rossini et al.,

Table 8-4 Heat Capacities of Two Cyclopentanes[a]

T(°K)	Methylcyclopentane			Ethylcyclopentane		
	Cp	ΔCp	$\Delta^2 Cp$	Cp	ΔCp	$\Delta^2 Cp$
300	26.46			31.75		
		9.65			12.14	
400	36.11		−0.82	43.89		−3.00
		8.83			9.14	
500	44.94		−1.34	53.03		−0.47
		7.49			8.67	
600	52.43		−1.24	61.70		−1.35
		6.25			7.32	
700	58.68		−0.93	69.02		−1.12
		5.32			6.20	
800	64.00		−0.79	75.22		−0.88
		4.53			5.32	
900	68.53		−0.62	80.54		−0.70
		3.91			4.62	
1000	72.44			85.16		

[a] Data of Rossini, Pitzer, Arnett, Braun, and Pimentel (1248).

and show the expected deviations. Other apparent discrepancies of a similar type have been found in the data of Rossini et al., but in most cases the resulting differences are not serious.

The authors have closely checked and verified the input data and punching to ensure the accuracy of the reproduced tables from the machine output. By frequently running a check problem the electronic computer can be monitored and machine errors readily detected. The authors have found more human errors than machine errors, and will appreciate, with thanks in advance, notices of any errors that may be found in this volume.

Table 8-5 Thermodynamic Functions of Ethylcyclopentane

	$H° - H°_{298}$ (kcal/mole)		$S°$ [cal/(mole °K)]	
$T(°K)$	Rossini et al. (1248)	Present Method	Rossini et al. (1248)	Present Method
298	0.000	0.00	90.42	90.42
300	0.063	0.06	90.64	90.62
400	3.791	3.87	101.40	101.48
500	8.592	8.72	112.18	112.27
600	14.337	14.46	122.69	122.72
700	20.898	21.01	132.85	132.80
800	28.103	28.23	142.53	142.43
900	35.913	36.02	151.71	151.60
1000	44.187	44.31	160.42	160.33

REFERENCE STATES OF THE ELEMENTS

The reference state corresponding to that form of the element stable at 1 atm has usually been selected. Six of the nine elements present obvious choices, with carbon as graphite, and hydrogen, oxygen, nitrogen, fluorine, and chlorine as ideal diatomic gases at 1 atm over the entire range of 298–1000°K. Bromine, iodine, and sulfur, however, have stable condensed phases at 1 atm from 298°K to their respective boiling points. We have therefore adopted these condensed phases as reference states up to the normal boiling points, and the ideal diatomic gases above these temperatures. It should be noted that this causes discontinuities at the temperatures of phase changes in the enthalpies of formation of all compounds involving these elements. In the case of sulfur there is also a slight discontinuity in the Gibbs energies of formation, because the vapor in equilibrium with liquid sulfur at the boiling point is a complex mixture of polyatomic molecules, S_x, with $x = 4, 6, 8$, and possibly other values. Vaporization to S_2 gas at 1 atm at 717.75°K is not an equilibrium process and involves a change in Gibbs energy.

These discontinuities could be eliminated by using ideal diatomic gas over the whole temperature range of 298–1000°K, but this would result in enthalpies and Gibbs energies of formation subject to misinterpretation when the reference state deviates significantly from that met in practice. These misleading implications would lead to much more confusion than the discontinuities of the selected reference states. A further advantage of the selected states is their identity at 298°K with those employed by Rossini, Wagman, Evans, Levine, and Jaffe (1249). Values from their comprehensive compilation can be directly used with those of the present work without the necessity of tedious and awkward conversions for differences in reference states.

No. 1 Carbon (Graphite) C (Solid) Atomic Wt 12.010

T°K	$Cp°$	$S°$	$-(G°-H°_{298})/T$	$H°-H°_{298}$	$\Delta Hf°$	$\Delta Gf°$	Log Kp
	cal/(mole °K)			kcal/mole			
298	2.066	1.361	1.361	0.000	0.000	0.000	0.000
300	2.083	1.374	1.361	0.004	0.000	0.000	0.000
400	2.851	2.081	1.453	0.251	0.000	0.000	0.000
500	3.496	2.788	1.649	0.569	0.000	0.000	0.000
600	4.030	3.474	1.896	0.947	0.000	0.000	0.000
700	4.430	4.127	2.169	1.370	0.000	0.000	0.000
800	4.750	4.740	2.453	1.830	0.000	0.000	0.000
900	4.980	5.314	2.739	2.318	0.000	0.000	0.000
1000	5.140	5.846	3.023	2.823	0.000	0.000	0.000

Recent reviews by Kelley (736, 738) and by the staff of the JANAF Thermochemical Tables (1435) indicate essentially no changes in the thermodynamic functions of graphite since the compilation of Rossini, Pitzer, Arnett, Braun, and Pimentel (1248). We have therefore adopted the values of Rossini et al. for the reference state. This has the advantage of making the present hydrocarbon tables consistent with the extensive tables of American Petroleum Institute Research Project 44. The atomic weight of carbon used by Rossini et al. is also retained in order to avoid the necessity of recalculating all the American Petroleum Institute tables without a significant increase in accuracy.

No. 2 Hydrogen H_2 (Ideal Diatomic Gas State) Mol Wt 2.0160

T°K	$Cp°$	$S°$	$-(G°-H°_{298})/T$	$H°-H°_{298}$	$\Delta Hf°$	$\Delta Gf°$	Log Kp
	cal/(mole °K)			kcal/mole			
298	6.892	31.211	31.211	0.000	0.000	0.000	0.000
300	6.895	31.253	31.211	0.013	0.000	0.000	0.000
400	6.974	33.250	31.482	0.707	0.000	0.000	0.000
500	6.993	34.809	31.997	1.406	0.000	0.000	0.000
600	7.008	36.085	32.575	2.106	0.000	0.000	0.000
700	7.035	37.167	33.156	2.808	0.000	0.000	0.000
800	7.078	38.108	33.707	3.514	0.000	0.000	0.000
900	7.139	38.946	34.253	4.224	0.000	0.000	0.000
1000	7.217	39.704	34.762	4.942	0.000	0.000	0.000

The thermodynamic functions for hydrogen are also taken from the review of Rossini, Pitzer, Arnett, Braun, and Pimentel (1248) to maintain consistency between the present hydrocarbon tables and their extensive tables. The values are based in large part, if not completely, on the thorough work of Woolley, Scott, and Brickwedde (1630). Rossini, Wagman, Evans, Levine, and Jaffe (1249) listed $Ttp = 13.96°K$, $\Delta Hm° = 0.028$ kcal/mole, $Tb = 20.39°K$, and $\Delta Hv = 0.216$ kcal/mole.

No. 3 Bromine Br_2 (Liquid from 298°K to 332.62°K; Ideal Diatomic Gas State from 332.62°K to 1000°K) Mol Wt 159.832

	cal/(mole °K)				kcal/mole		
T°K	$Cp°$	$S°$	$-(G°-H°_{298})/T$	$H°-H°_{298}$	$\Delta Hf°$	$\Delta Gf°$	Log Kp
298	18.090	36.384	36.384	0.000	0.000	0.000	0.000
300	18.077	36.496	36.384	0.033	0.000	0.000	0.000
400	8.775	61.203	40.520	8.273	0.000	0.000	0.000
500	8.857	63.172	44.860	9.156	0.000	0.000	0.000
600	8.908	64.791	48.051	10.044	0.000	0.000	0.000
700	8.944	66.167	50.543	10.937	0.000	0.000	0.000
800	8.970	67.363	52.573	11.833	0.000	0.000	0.000
900	8.992	68.421	54.276	12.731	0.000	0.000	0.000
1000	9.011	69.370	55.739	13.631	0.000	0.000	0.000

A recent review of the thermodynamic functions of elemental bromine was reported in the JANAF Thermochemical Tables (1435). The low-temperature heat capacity and the enthalpy of fusion were reported by Hildenbrand, Kramer, McDonald, and Stull (596). The thermodynamic functions of the ideal gas were calculated on the basis of spectroscopic data listed by Herzberg (590) and are in good agreement with those computed by Evans, Munson, and Wagman (398). Thermodynamic functions of liquid and gas were then combined with vapor pressure data to calculate the enthalpy of vaporization at 298°K by third-law methods. Data of Fischer and Bingle (428) and of Scheffer and Voogd (1280) show decided trends in the calculated $Hv°_{298}$ with temperature, while the data of Ramsay and Young (1212) show only a slight trend, with an average in good agreement with the calorimetric value reported by Hildenbrand et al. (596). The reference state values given here are derived from the calorimetric value of 7.39 kcal/mole. The calculated boiling point is 332.62°K.

No. 4 Chlorine Cl_2 (Ideal Diatomic Gas State) Mol Wt 70.914

We have adopted the thermodynamic functions recently calculated by Potter (1183) by direct summation over the energy levels of the molecule. A previous calculation by Evans, Munson, and Wagman (398) by less rigorous methods gave results not significantly different in the temperature range of interest here. Rossini, Wagman, Evans, Levine, and Jaffe

No. 4   Chlorine Cl_2 (Ideal Diatomic Gas State)   Mol Wt 70.914

T°K	Cp°	S°	$-(G°-H°_{298})/T$	$H°-H°_{298}$	ΔHf°	ΔGf°	Log Kp
	cal/(mole °K)				kcal/mole		
298	8.111	53.289	53.289	0.000	0.000	0.000	0.000
300	8.119	53.339	53.289	0.015	0.000	0.000	0.000
400	8.437	55.724	53.612	0.845	0.000	0.000	0.000
500	8.624	57.628	54.231	1.698	0.000	0.000	0.000
600	8.741	59.212	54.933	2.567	0.000	0.000	0.000
700	8.821	60.565	55.643	3.445	0.000	0.000	0.000
800	8.878	61.747	56.334	4.331	0.000	0.000	0.000
900	8.922	62.796	56.995	5.221	0.000	0.000	0.000
1000	8.956	63.737	57.623	6.115	0.000	0.000	0.000

(1249) listed $Tm = 172.16°K$, $\Delta Hm° = 1.531$ kcal/mole, $Tb = 239.10°K$, and $\Delta Hv = 4.878$ kcal/mole.

No. 5   Fluorine F_2 (Ideal Diatomic Gas State)   Mol Wt 38.00

T°K	Cp°	S°	$-(G°-H°_{298})/T$	$H°-H°_{298}$	ΔHf°	ΔGf°	Log Kp
	cal/(mole °K)				kcal/mole		
298	7.490	48.447	48.447	0.000	0.000	0.000	0.000
300	7.498	48.493	48.447	0.014	0.000	0.000	0.000
400	7.900	50.707	48.746	0.784	0.000	0.000	0.000
500	8.207	52.505	49.324	1.591	0.000	0.000	0.000
600	8.429	54.022	49.983	2.423	0.000	0.000	0.000
700	8.591	55.334	50.656	3.274	0.000	0.000	0.000
800	8.713	56.489	51.314	4.140	0.000	0.000	0.000
900	8.808	57.521	51.948	5.016	0.000	0.000	0.000
1000	8.885	58.453	52.553	5.901	0.000	0.000	0.000

Modern calculations of the thermodynamic functions of fluorine have been reported by Murphy and Vance (1038), Cole, Farber, and Elverum (243), Evans, Munson, and Wagman (398), Potter (1182), and in the JANAF Thermochemical Tables (1435). The differences are minor, and we have adopted the most recent work (1435), which is based on molecular constants derived by Rees (1219) from the Raman spectra of Andrychuk (23) and the continuous spectra of Steunenberg and Vogel (1417). Hu, White, and Johnston (626) measured $Tt = 45.55°K$, $\Delta Ht° = 0.174$ kcal/mole, $Ttp = 53.54°K$, $\Delta Hm° = 0.122$ kcal/mole, $Tb = 85.02°K$, and $\Delta Hv = 1.562$ kcal/mole.

No. 6   Iodine I_2 (Crystal from 298°K to 386.75°K; Liquid from 386.75°K to 458.39°K to 1000°K.)   Mol Wt 253.82

Thermodynamic data for iodine have been reported in the JANAF Thermochemical Tables (1435). The recent measurements of the low-

No. 6 Iodine I$_2$ (Crystal from 298°K to 386.75°K; Liquid from 386.75°K to 458.39°K to 1000°K.) Mol Wt 253.82

T°K	Cp°	cal/(mole °K)			kcal/mole		Log Kp
		S°	$-(G°-H°_{298})/T$	$H°-H°_{298}$	$\Delta Hf°$	$\Delta Gf°$	
298	13.011	27.758	27.758	0.000	0.000	0.000	0.000
300	13.028	27.839	27.758	0.024	0.000	0.000	0.000
400	19.281	41.602	28.612	5.196	0.000	0.000	0.000
500	8.948	66.877	33.438	16.719	0.000	0.000	0.000
600	8.980	68.511	39.152	17.616	0.000	0.000	0.000
700	9.005	69.897	43.447	18.515	0.000	0.000	0.000
800	9.025	71.101	46.830	19.416	0.000	0.000	0.000
900	9.044	72.165	49.587	20.320	0.000	0.000	0.000
1000	9.061	73.119	51.894	21.225	0.000	0.000	0.000

temperature heat capacity by Shirley and Giauque (1347) were used to derive the entropy of the crystal at 298°K. No weight was given to the earlier work of Lange (846), which deviates by 1% or more. The enthalpy of the solid and liquid above 298°K and the derived enthalpy of fusion were taken from Frederick and Hildebrand (444), in preference to the heat capacity data of Carpenter and Harle (193). The results of Frederick and Hildebrand were corrected for a change in the enthalpy of copper, which they used for calibration, by a factor that varied linearly from zero at 300°K to 0.995 at 400°K and 0.990 at 500°K. The corrected enthalpies and the heat capacity data of Shirley and Giauque were fitted by equations (in kilocalories per mole).

Crystal: $H°_T - H°_{298} = -0.0121048T + 2.9506 \times 10^{-5}T^2 - 668.6T^{-1} + 3.22865,$

Liquid: $H°_T - H°_{298} = 0.019281T - 2.5165.$

At the melting point, 386.75°K, $\Delta Hm° = 3.7087$ kcal/mole. Thermodynamic functions of the ideal gas were calculated using molecular constants given by Rank and Baldwin (1213) with the revised ω_e as given by Verma (1539). The results are in good agreement with Evans, Munson, and Wagman (398). The thermodynamic functions of crystal and gas were combined with the vapor pressure data of Baxter, Hickey, and Holmes (86) and Baxter and Grose (85) to derive $\Delta Hv°_{298} = 14.924$ kcal/mole. Other vapor pressure investigations were checked and either confirmed this result or appeared to be of too low precision to be given any weight. The calculated boiling point is 458.39°K.

No. 7 Nitrogen N_2 (Ideal Diatomic Gas State) Mol Wt 28.016

T°K	Cp°	S°	$-(G° - H°_{298})/T$	$H° - H°_{298}$	ΔHf°	ΔGf°	Log Kp
	cal/(mole °K)				kcal/mole		
298	6.960	45.767	45.767	0.000	0.000	0.000	0.000
300	6.961	45.809	45.767	0.013	0.000	0.000	0.000
400	6.991	47.818	46.043	0.710	0.000	0.000	0.000
500	7.070	49.385	46.559	1.413	0.000	0.000	0.000
600	7.197	50.685	47.142	2.126	0.000	0.000	0.000
700	7.351	51.805	47.729	2.853	0.000	0.000	0.000
800	7.512	52.797	48.302	3.596	0.000	0.000	0.000
900	7.671	53.692	48.852	4.356	0.000	0.000	0.000
1000	7.816	54.509	49.379	5.130	0.000	0.000	0.000

The values selected by Rossini, Pitzer, Arnett, Braun, and Pimentel (1248) have been adopted. A recent recalculation of the thermodynamic functions of nitrogen reported in the JANAF Thermochemical Tables (1435) using the latest spectroscopic and molecular constants did not give changes significant for our purposes. Rossini, Wagman, Evans, Levine, and Jaffe (1249) listed $Tt = 35.62°K$, $\Delta Ht° = 0.055$ kcal/mole, $Ttp = 63.15°K$, $\Delta Hm° = 0.172$ kcal/mole, $Tb = 77.34°K$, and $\Delta Hv = 1.333$ kcal/mole.

No. 8 Oxygen O_2 (Ideal Diatomic Gas State) Mol Wt 32.000

T°K	Cp°	S°	$-(G° - H°_{298})/T$	$H° - H°_{298}$	ΔHf°	ΔGf°	Log Kp
	cal/(mole °K)				kcal/mole		
298	7.017	49.003	49.003	0.000	0.000	0.000	0.000
300	7.019	49.048	49.004	0.013	0.000	0.000	0.000
400	7.194	51.093	49.285	0.723	0.000	0.000	0.000
500	7.429	52.723	49.815	1.454	0.000	0.000	0.000
600	7.670	54.100	50.418	2.209	0.000	0.000	0.000
700	7.885	55.296	51.027	2.988	0.000	0.000	0.000
800	8.064	56.364	51.631	3.786	0.000	0.000	0.000
900	8.212	57.322	52.211	4.600	0.000	0.000	0.000
1000	8.335	58.194	52.767	5.427	0.000	0.000	0.000

Thermodynamic functions for oxygen are those listed by Rossini, Pitzer, Arnett, Braun, and Pimentel (1248), which are based primarily, if not completely, on the thorough work of Woolley (1629). Rossini, Wagman, Evans, Levine, and Jaffe (1249) listed $Tt = 23.66°K$, $\Delta Ht° = 0.022$ kcal/mole, $Tt = 43.77°K$, $\Delta Ht° = 0.178$ kcal/mole, $Ttp = 54.40°K$, $\Delta Hm° = 0.106$ kcal/mole, $Tb = 90.19°K$, and $\Delta Hv = 1.630$ kcal/mole.

No. 9 Sulfur S_2 (Rhombic Crystal from 298°K to 368.46°K; Monoclinic Crystal from 368.46°K to 388.36°K; Liquid from 388.36°K to 717.75°K; Ideal Diatomic Gas State from 717.75°K to 1000°K.) Atomic Wt 32.066

T°K	cal/(mole °K)				kcal/mole		Log Kp
	$Cp°$	$S°$	$-(G°-H°_{298})/T$	$H°-H°_{298}$	$\Delta Hf°$	$\Delta Gf°$	
298	5.401	7.631	7.631	0.000	0.000	0.000	0.000
300	5.412	7.665	7.632	0.010	0.000	0.000	0.000
400	7.734	10.674	7.901	1.109	0.000	0.000	0.000
500	9.081	12.768	8.673	2.047	0.000	0.000	0.000
600	8.200	14.333	9.492	2.904	0.000	0.000	0.000
700	7.799	15.601	10.309	3.704	0.000	0.000	0.000
800	4.368	31.363	9.452	17.529	0.000	0.000	0.000
900	4.396	31.879	11.916	17.967	0.000	0.000	0.000
1000	4.418	32.344	13.936	18.408	0.000	0.000	0.000

Thermodynamic data for sulfur were reported in the JANAF Thermochemical Tables (1435). The low-temperature heat capacity and derived entropy of the rhombic form were reported by Eastman and McGavock (362). Above 298°K the heat capacity and enthalpies of transition are taken from the work of West (1590), with corrections pointed out by McCullough and Scott (961). At the transition temperature, 368.46°K, $\Delta Ht° = 0.0960$ kcal/mole, and at the melting point, 388.36°K, $\Delta Hm° = 0.4105$ kcal/mole. The boiling point, 717.75°K, is a reference temperature on the international temperature scale. Sulfur vapor is complex and consists of a mixture of polyatomic molecules with $n = 2$ to $n = 8$. As the best approximation consistent with a reasonably simple reference state, the ideal diatomic gas was adopted for the range 717.75–1000°K. For those needing a more accurate treatment, the tables of Stull and Sinke (1437) are recommended. The thermodynamic functions and enthalpy of formation of $S_2(g)$ used here are from the review of Evans and Wagman (399).

THERMAL PROPERTIES OF SOME INORGANIC COMPOUNDS

For the convenience of the reader we list thermal properties of a number of inorganic compounds frequently encountered in organic chemistry. In many cases critical evaluations of the existing data have already been made and may be consulted for a complete discussion. In some instances recent work made it necessary to review or revise previously compiled values, and more details are given. The tables are arranged according to the formula index system used in *Chemical Abstracts*.

Thermal Properties of Some Inorganic Compounds

No. 10   Hydrogen Bromide   BrH   Ideal Gas State   Mol Wt 80.924

$T°K$	cal/(mole °K)				kcal/mole		$\text{Log} K_p$
	$C_p°$	$S°$	$-(G°-H°_{298})/T$	$H°-H°_{298}$	$\Delta Hf°$	$\Delta Gf°$	
298	6.96	47.44	47.44	0.00	-8.66	-12.73	9.328
300	6.97	47.49	47.44	0.02	-8.67	-12.75	9.289
400	6.98	49.49	47.72	0.71	-12.44	-13.34	7.290
500	7.03	51.05	48.23	1.41	-12.53	-13.56	5.926
600	7.12	52.34	48.81	2.12	-12.62	-13.76	5.011
700	7.24	53.45	49.40	2.84	-12.70	-13.94	4.352
800	7.38	54.42	49.97	3.57	-12.77	-14.11	3.855
900	7.53	55.30	50.51	4.32	-12.83	-14.28	3.467
1000	7.68	56.10	51.03	5.08	-12.88	-14.43	3.154

Thermodynamic functions based on the molecular constants determined by Plyler (1171) have recently been adjusted to the natural isotopic mixture and reported in the JANAF Thermochemical Tables (1435). The enthalpy of formation taken from the compilation of Rossini, Wagman, Evans, Levine, and Jaffe (1249) was confirmed by a determination by Lacher, Kianpour, Oetting, and Park (827) within experimental error. Rossini et al. (1249) listed $Tm = 186.28°K$, $\Delta Hm° = 0.575$ kcal/mole, $Tb = 206.43°K$ and $\Delta Hv = 4.210$ kcal/mole.

No. 11   Cyanogen Bromide   CBrN   (Ideal Gas State)   Mol Wt 105.934

$T°K$	cal/(mole °K)				kcal/mole		$\text{Log} K_p$
	$C_p°$	$S°$	$-(G°-H°_{298})/T$	$H°-H°_{298}$	$\Delta Hf°$	$\Delta Gf°$	
298	11.12	59.07	59.07	0.00	43.35	38.39	-28.140
300	11.14	59.14	59.08	0.03	43.34	38.36	-27.944
400	11.85	62.45	59.52	1.18	39.78	37.44	-20.454
500	12.32	65.15	60.39	2.39	39.88	36.84	-16.102
600	12.68	67.43	61.37	3.64	39.95	36.22	-13.194
700	12.98	69.40	62.38	4.92	40.00	35.60	-11.114
800	13.24	71.16	63.37	6.23	40.03	34.97	-9.552
900	13.46	72.73	64.33	7.57	40.05	34.33	-8.337
1000	13.64	74.16	65.24	8.92	40.06	33.70	-7.364

The enthalpy of formation was determined by Lord and Woolf (883) as $\Delta Hf°_{298}(s) = 32.5$ kcal/mole from the enthalpy of alkaline hydrolysis to sodium cyanate, sodium bromide, and water. The enthalpy of formation of aqueous sodium cyanate was established in separate experiments. Lord and Woolf (883) also measured the vapor pressure from 273° to 308°K. We combine their data with those of Baxter, Bezzenberger, and Wilson (84) and by least squares derive an Antoine equation, $\log P$ (mm)$= 9.4559 - 2041.8/(t + 251.70)$, from which we derive $\Delta Hs°_{298} = 10.85$ kcal/mole. Lord and Woolf (883) calculated the entropy of the gas at

298°K, Kobe and Long (779) and Stevenson (1418) reported ideal gas heat capacities. The vapor pressure equation and calculated enthalpy of sublimation give $S°_{298}(s) = 26.37$ cal/(mole °K). Rossini, Wagman, Evans, Levine, and Jaffe (1249) listed $Tm = 324.5°K$ and $Tb = 334.5°K$.

No. 12 Cyanogen Chloride CClN (Ideal Gas State) Mol Wt 61.475

	cal/(mole °K)				kcal/mole		
$T°K$	$Cp°$	$S°$	$-(G°-H°_{298})/T$	$H°-H°_{298}$	$\Delta Hf°$	$\Delta Gf°$	Log Kp
298	10.69	56.28	56.28	0.00	31.60	29.99	-21.984
300	10.72	56.35	56.29	0.02	31.60	29.98	-21.841
400	11.54	59.55	56.72	1.14	31.71	29.43	-16.078
500	12.07	62.19	57.56	2.32	31.79	28.85	-12.609
600	12.48	64.43	58.52	3.55	31.85	28.25	-10.290
700	12.83	66.38	59.50	4.82	31.89	27.65	-8.632
800	13.12	68.11	60.47	6.11	31.92	27.04	-7.387
900	13.36	69.67	61.41	7.44	31.93	26.43	-6.418
1000	13.56	71.09	62.31	8.78	31.93	25.82	-5.643

The enthalpy of formation of gaseous cyanogen chloride was derived by Lord and Woolf (883) from their measurements of the enthalpy of alkaline hydrolysis. The enthalpy of formation of sodium cyanate was established in separate experiments using the enthalpy of reaction of iodine and sodium cyanide and the enthalpy of hydrolysis of cyanogen. Lord and Woolf also calculated $S°_{298}(g) = 56.28$ cal/(mole °K); Kobe and Long (779) and Stevenson (1418) computed heat capacity values for the ideal gas. Douglas and Winkler (337) reported $Tm = 266.25°K$, $\Delta Hm° = 2.720$ kcal/mole, and $Tb = 286.1°K$.

No. 13 Phosgene CCL$_2$O (Ideal Gas State) Mol. Wt 98.924

	cal/(mole °K)				kcal/mole		
$T°K$	$Cp°$	$S°$	$-(G°-H°_{298})/T$	$H°-H°_{298}$	$\Delta Hf°$	$\Delta Gf°$	Log Kp
298	13.79	67.82	67.82	0.00	-52.80	-49.42	36.225
300	13.82	67.91	67.83	0.03	-52.80	-49.40	35.986
400	15.28	72.10	68.39	1.49	-52.77	-48.27	26.372
500	16.27	75.62	69.49	3.07	-52.73	-47.15	20.607
600	16.98	78.65	70.77	4.73	-52.69	-46.04	16.768
700	17.52	81.31	72.09	6.46	-52.65	-44.93	14.027
800	17.92	83.68	73.39	8.23	-52.63	-43.83	11.973
900	18.24	85.81	74.66	10.04	-52.60	-42.73	10.376
1000	18.49	87.74	75.87	11.88	-52.58	-41.64	9.099

Thermodynamic functions have been reported in the JANAF Thermochemical Tables (1435) on the basis of the molecular dimensions reported by Robinson (1235) and the vibrational assignment of Catalano and Pitzer

(203). This new vibrational assignment, which postulated an accidental degeneracy, was confirmed by the low-temperature thermal data of Giauque and Ott (477), who found three forms of crystalline phosgene, with the following melting points and enthalpies of fusion: crystal I, $Tm = 145.37°K$, $\Delta Hm° = 1.3715$ kcal/mole; crystal II, $Tm = 142.09°K$, $\Delta Hm° = 1.3354$ kcal/mole; and crystal III, $Tm = 139.19°K$, $\Delta Hm° = 1.131$ kcal/mole. Earlier work by Giauque and Jones (476) gave $Tb = 280.7°K$, at which $\Delta Hv = 5.832$ kcal/mole. Thompson (1487) reviewed values for the enthalpy of formation derived from equilibrium data on the dissociation into carbon monoxide and chlorine. Recalculation using the present thermodynamic functions yields $\Delta Hf°_{298}(g) = -52.7$ kcal/mole. The thermochemical data of Thomsen (1495) give an average of -52.9 kcal/mole. The close agreement is somewhat fortuitous in view of the uncertainties. An average of -52.8 kcal/mole is adopted.

No. 14 Carbonyl Fluoride CF_2O (Ideal Gas State) Mol Wt 66.010

$T°K$	$Cp°$	$S°$	$-(G°-H°_{298})/T$	$H°-H°_{298}$	$\Delta Hf°$	$\Delta Gf°$	Log Kp
		cal/(mole °K)			kcal/mole		
298	11.29	61.84	61.84	0.00	-153.00	-149.28	109.421
300	11.33	61.91	61.85	0.03	-153.00	-149.26	108.730
400	13.09	65.42	62.31	1.25	-153.15	-147.99	80.851
500	14.46	68.50	63.25	2.63	-153.26	-146.68	64.111
600	15.51	71.23	64.36	4.13	-153.35	-145.36	52.944
700	16.31	73.69	65.52	5.72	-153.42	-144.02	44.963
800	16.93	75.90	66.68	7.39	-153.48	-142.67	38.974
900	17.41	77.93	67.82	9.10	-153.53	-141.32	34.315
1000	17.79	79.78	68.92	10.87	-153.58	-139.98	30.590

The enthalpy of hydrolysis was measured by Wartenberg (1572) and yields $\Delta Hf°_{298}(g) = -153.2$ kcal/mole. The equilibrium constants of the reaction $2CF_2O = CF_4 + CO_2$ at various temperatures, reported by Ruff and Li (1267), are combined with free energy functions to derive $\Delta Hf°_{298}(g) = -152.5$ kcal/mole. A value of -153 kcal/mole is adopted, giving the most weight to the calorimetric value. Thermodynamic functions have been calculated and reported in the JANAF Thermochemical Tables (1435), based on molecular data of Lovell, Stephenson, and Jones (888). Rossini, Wagman, Evans, Levine, and Jaffe (1249) selected $Ttp = 159.2°K$, $Tb = 189.9°K$, and $\Delta Hv = 3.86$ kcal/mole.

No. 15 Hydrogen Cyanide CHN (Ideal Gas State) Mol Wt 27.026

Thermodynamic functions of the ideal gas were computed by Bradley, Haar, and Friedman (157). Less extensive calculations are given by

No. 15   Hydrogen Cyanide   CHN   (Ideal Gas State)   Mol Wt 27.026

T°K	$Cp°$	$S°$	$-(G°-H°_{298})/T$	$H°-H°_{298}$	$\Delta Hf°$	$\Delta Gf°$	Log Kp
		cal/(mole °K)			kcal/mole		
298	8.59	48.21	48.21	0.00	31.20	28.71	-21.042
300	8.61	48.27	48.22	0.02	31.20	28.69	-20.901
400	9.42	50.86	48.56	0.92	31.16	27.86	-15.223
500	10.04	53.03	49.25	1.90	31.11	27.04	-11.820
600	10.56	54.91	50.04	2.93	31.06	26.23	-9.555
700	11.03	56.57	50.85	4.01	31.00	25.43	-7.941
800	11.45	58.07	51.66	5.13	30.94	24.64	-6.732
900	11.83	59.44	52.45	6.30	30.88	23.86	-5.793
1000	12.18	60.71	53.22	7.50	30.83	23.08	-5.044

Stamm, Halverson, and Whalen (1404) and by Kobe and Long (779). Giauque and Ruehrwein (478) reported low-temperature thermal data, which yield $Tm = 259.91°K$, $\Delta Hm° = 2.009$ kcal/mole, $Tb = 298.85°K$, $\Delta Hv = 6.027$ kcal/mole, and $Cp(l) = 16.9$ cal/(mole °K) (260–300°K). When corrected for the association of the gas state, the calorimetric entropy is in reasonable agreement with calculations based on spectroscopic methods. Gas heat capacity measurements by Felsing and Drake (415) also indicate polymerization in the gas phase. Rossini, Wagman, Evans, Levine, and Jaffe (1249) reviewed enthalpy of formation data and selected $\Delta Hf°_{298}(l) = 25.2$ kcal/mole and $\Delta Hf°_{298}(g) = 31.2$ kcal/mole. These values are in good agreement with the results of Nakamura (1045), based on equilibrium constants of the reaction of carbon monoxide and ammonia.

No. 16   Cyanogen Iodide   CIN   (Ideal Gas State)   Mol Wt 152.928

T°K	$Cp°$	$S°$	$-(G°-H°_{298})/T$	$H°-H°_{298}$	$\Delta Hf°$	$\Delta Gf°$	Log Kp
		cal/(mole °K)			kcal/mole		
298	11.55	61.33	61.33	0.00	53.80	46.88	-34.363
300	11.57	61.41	61.34	0.03	53.80	46.84	-34.120
400	12.15	64.82	61.80	1.21	51.80	44.60	-24.365
500	12.54	67.57	62.69	2.45	46.61	43.28	-18.919
600	12.85	69.89	63.70	3.72	46.70	42.61	-15.520
700	13.11	71.89	64.73	5.02	46.76	41.92	-13.089
800	13.34	73.65	65.74	6.34	46.80	41.23	-11.263
900	13.55	75.24	66.71	7.68	46.82	40.53	-9.842
1000	13.72	76.68	67.63	9.05	46.84	39.83	-8.705

Lord and Woolf (883) determined the enthalpy of alkaline hydrolysis of solid cyanogen iodide and derived $\Delta Hf°_{298}(s) = 38.3$ kcal/mole. The enthalpy of formation of sodium cyanate, one of the hydrolysis products, was measured in separate experiments on the hydrolysis of cyanogen and reaction of iodine and sodium cyanide. Lord and Woolf also calculated a statistical value for the entropy of the ideal gas. Kobe and Long (779) and Stevenson (1418) computed values for gaseous heat capacity, which are in good agreement. We derive an Antoine equation for the vapor pressure from the data of Yost and Stone (1639) and of Ketelaar and Kruyer (742) and use it to calculate the enthalpy of sublimation at 373°K, the midpoint of the temperature range covered. Assuming $\Delta Cp = 10$ cal/(mole °K), this result is corrected to 298°K, giving $\Delta Hs^\circ_{298} = 15.5$ kcal/mole, from which we calculate $\Delta Hf^\circ_{298}(g) = 53.8$ kcal/mole. Rossini, Wagman, Evans, Levine, and Jaffe (1249) listed $Tm = 419°K$, $Ts = 413.0°K$, and $\Delta Hs = 14.2$ kcal/mole.

No. 17 Carbon Monoxide CO (Ideal Gas State) Mol Wt 28.010

	cal/(mole °K)				kcal/mole		
T°K	$Cp°$	$S°$	$-(G°-H°_{298})/T$	$H°-H°_{298}$	$\Delta Hf°$	$\Delta Gf°$	Log Kp
298	6.97	47.30	47.30	0.00	-26.42	-32.81	24.050
300	6.97	47.35	47.31	0.02	-26.42	-32.85	23.931
400	7.01	49.36	47.58	0.72	-26.32	-35.01	19.128
500	7.12	50.93	48.10	1.42	-26.30	-37.19	16.254
600	7.28	52.24	48.68	2.14	-26.33	-39.36	14.337
700	7.45	53.38	49.27	2.88	-26.41	-41.53	12.966
800	7.62	54.39	49.85	3.63	-26.52	-43.68	11.933
900	7.79	55.29	50.41	4.40	-26.64	-45.82	11.126
1000	7.93	56.12	50.94	5.19	-26.77	-47.95	10.478

The thermodynamic functions and enthalpy of formation of carbon monoxide are taken from the review article of Wagman, Kilpatrick, Taylor, Pitzer, and Rossini (1562). Clayton and Giauque (233) reported $Tt = 61.53°K$, $\Delta Ht° = 0.151$ kcal/mole, $Ttp = 68.10°K$, $\Delta Hm° = 0.200$ kcal/mole, $Tb = 81.66°K$, and $\Delta Hv = 1.444$ kcal/mole. Comparison of the third-law and statistical entropies indicates random orientation in the solid.

No. 18 Carbonyl Sulfide COS (Ideal Gas State) Mol Wt 60.076

	cal/(mole °K)				kcal/mole		
T°K	$Cp°$	$S°$	$-(G°-H°_{298})/T$	$H°-H°_{298}$	$\Delta Hf°$	$\Delta Gf°$	Log Kp
298	9.92	55.32	55.32	0.00	-33.08	-39.59	29.017
300	9.94	55.39	55.33	0.02	-33.08	-39.63	28.867
400	10.96	58.39	55.73	1.07	-33.74	-41.77	22.821
500	11.69	60.92	56.52	2.20	-34.22	-43.72	19.110
600	12.25	63.10	57.44	3.40	-34.64	-45.58	16.602
700	12.70	65.03	58.39	4.65	-35.00	-47.35	14.784
800	13.07	66.75	59.33	5.94	-48.40	-50.36	13.758
900	13.37	68.30	60.24	7.26	-48.41	-50.61	12.289
1000	13.62	69.73	61.12	8.61	-48.42	-50.85	11.113

Thermodynamic functions have been computed by Gordon (519). The enthalpy of formation is based on the equilibrium constants reported by Terres and Wesemann (1479) for the reaction $CO_2(g) + H_2S(g) = COS(g) + H_2O(g)$, and the enthalpies of formation of CO_2, H_2S, and H_2O. Kemp and Giauque (739) measured low-temperature thermal data and reported $Tm = 134.34°K$, $\Delta Hm° = 1.130$ kcal/mole, $Tb = 222.92°K$, and $\Delta Hv = 4.423$ kcal/mole.

No. 19   Carbon Dioxide CO_2 (Ideal Gas State)   Mol Wt 44.010

	cal/(mole °K)				kcal/mole		
$T°K$	$Cp°$	$S°$	$-(G°-H°_{298})/T$	$H°-H°_{298}$	$\Delta Hf°$	$\Delta Gf°$	Log Kp
298	8.87	51.07	51.07	0.00	-94.05	-94.26	69.091
300	8.89	51.13	51.08	0.02	-94.05	-94.26	68.666
400	9.87	53.83	51.44	0.96	-94.07	-94.33	51.535
500	10.66	56.12	52.15	1.99	-94.09	-94.39	41.255
600	11.31	58.12	52.98	3.09	-94.12	-94.45	34.400
700	11.84	59.90	53.84	4.25	-94.17	-94.50	29.503
800	12.29	61.51	54.70	5.45	-94.22	-94.54	25.826
900	12.66	62.98	55.54	6.70	-94.27	-94.58	22.966
1000	12.97	64.33	56.36	7.98	-94.32	-94.61	20.676

Accurate thermodynamic functions recently reported by Gordon (519) have been adopted, although they differ little from those given earlier by Wagman, Kilpatrick, Taylor, Pitzer, and Rossini (1562). The enthalpy of formation is that given in the review by Wagman et al. (1562). Giauque and Egan (474) reported $Ts = 194.71°K$, $\Delta Hs = 6.030$ kcal/mole, $Tc = 304.1°K$, and $Pc = 72.8$ atm. Rossini, Wagman, Evans, Levine, and Jaffe (1249) selected $Tm = 217.0°K$ and $\Delta Hm = 1.99$ kcal/mole.

No. 20   Carbon Disulfide CS_2 (Ideal Gas State)   Mol Wt 76.142

	cal/(mole °K)				kcal/mole		
$T°K$	$Cp°$	$S°$	$-(G°-H°_{298})/T$	$H°-H°_{298}$	$\Delta Hf°$	$\Delta Gf°$	Log Kp
298	10.87	56.83	56.83	0.00	27.98	15.99	-11.722
300	10.90	56.90	56.84	0.03	27.98	15.92	-11.596
400	11.82	60.17	57.27	1.16	26.67	11.97	-6.542
500	12.48	62.88	58.13	2.38	25.69	8.42	-3.678
600	12.97	65.20	59.12	3.65	24.87	5.04	-1.835
700	13.35	67.23	60.14	4.97	24.17	1.84	-0.574
800	13.64	69.03	61.14	6.32	-2.59	-3.84	1.050
900	13.86	70.65	62.11	7.70	-2.58	-4.00	0.971
1000	14.03	72.12	63.04	9.09	-2.57	-4.16	0.909

Brown and Manov (175) reported low-temperature thermal data including $Ttp = 161.11°K$ and $\Delta Hm = 1.05$ kcal/mole, and derived $S°_{298}(l) = 36.10$ cal/(mole °K). Waddington, Smith, Williamson, and Scott (1550) measured the enthalpy of vaporization, vapor pressure, and vapor heat capacity over a range of temperatures. They calculated thermodynamic functions from the complete spectroscopic information, which allows inclusion of effects of anharmonicity, centrifugal stretching, vibration-rotation interaction, and isotopic composition. Calculated heat capacities were in excellent agreement with calorimetric values, and CS_2 was proposed as a standard substance for vapor flow calorimetry. The calculated entropy, 56.84 cal/(mole °K), is in good agreement with the third-law result of 56.75 cal/(mole °K). Calculated functions by Gordon (519) are in good agreement, but those of McBride and Gordon (944) deviate somewhat. Earlier, less accurate calculations to the rigid rotator, harmonic oscillator approximation were reported by Cross (295).

Liquid heat capacities in the range 270–320°K were reported by Brown and Manov (175), Staveley, Tupman, and Hart (1407), Zhdanov (1644), and Mazur (943). The data are discordant, and no equation is attempted. Good, Lacina, and McCullough (511) measured the enthalpy of combustion by rotating bomb calorimetry and derived $\Delta Hf^\circ_{298}(l) = 21.37$ kcal/mole and $\Delta Hf^\circ_{298}(g) = 27.98$ kcal/mole. Good et al. also discussed earlier work on the enthalpy of formation.

No. 21   Cyanogen C_2N_2 (Ideal Gas State)   Mol Wt 52.036

T°K	cal/(mole °K)				kcal/mole			Log Kp
	Cp°	S°	$-(G^\circ - H^\circ_{298})/T$	$H^\circ - H^\circ_{298}$	ΔHf°	ΔGf°		
298	13.60	57.90	57.90	0.00	73.84	71.03	-52.067
300	13.63	57.99	57.91	0.03	73.84	71.02	-51.733
400	14.80	62.08	58.45	1.45	74.08	70.04	-38.266
500	15.65	65.48	59.53	2.98	74.26	69.01	-30.162
600	16.33	68.39	60.77	4.58	74.39	67.94	-24.747
700	16.91	70.95	62.04	6.24	74.48	66.86	-20.874
800	17.43	73.24	63.30	7.96	74.54	65.77	-17.966
900	17.87	75.32	64.52	9.72	74.57	64.67	-15.703
1000	18.25	77.83	65.70	11.53	74.59	63.57	-13.898

Ruehrwein and Giauque (1263) measured low-temperature thermal data leading to $Ttp = 245.31°K$, $\Delta Hm° = 1.938$ kcal/mole, $Tb = 252.00°K$, and $\Delta Hv = 5.576$ kcal/mole. The calorimetric value for the entropy agreed within experimental error with that calculated by Stevenson (1418) from spectroscopic data. The heat capacity of the ideal gas calculated by Stevenson was extended over a larger temperature range by Thompson (1488) and by Kobe and Long (779). Calculated values are in agreement with experimental determinations of Burcik and Yost (182) and of Stitt (1425).

The definitive work on the enthalpy of combustion was done by Knowlton and Prosen (769). They found $\Delta Hc^\circ_{298}(g) = -261.94$ kcal/

mole and $\Delta Hf^\circ_{298}(g) = 73.84$ kcal/mole. The two most recent among previous values listed by Knowlton and Prosen are those of Wartenberg and Schütza (1576) (which are in good agreement) and of McMorris and Badger (978) (which are too low).

No. 22 Carbon Suboxide C_3O_2 (Ideal Gas State) Mol Wt 68.030

T°K	Cp°	S°	$-(G^\circ - H^\circ_{298})/T$	$H^\circ - H^\circ_{298}$	ΔHf°	ΔGf°	Log Kp
		cal/(mole °K)			kcal/mole		
298	16.01	66.05	66.05	0.00	-22.38	-26.25	19.237
300	16.05	66.15	66.06	0.03	-22.38	-26.27	19.136
400	17.92	71.04	66.71	1.74	-22.12	-27.60	15.081
500	19.31	75.19	68.00	3.60	-21.95	-29.00	12.673
600	20.43	78.81	69.51	5.59	-21.85	-30.42	11.079
700	21.37	82.04	71.07	7.68	-21.80	-31.85	9.944
800	22.15	84.94	72.63	9.86	-21.80	-33.29	9.093
900	22.97	87.60	74.14	12.12	-21.82	-34.72	8.431
1000	23.34	90.04	75.61	14.44	-21.85	-36.15	7.901

Kybett, Johnson, Barker, and Margrave (819) measured the enthalpy of combustion of the liquid and derived $\Delta Hf^\circ_{298}(l) = -28.03$ kcal/mole. McDougall and Kilpatrick (971) reported low-temperature thermal data, including $Ttp = 160.96°K$, $\Delta Hm = 1.291$ kcal/mole, $\Delta Hv^\circ_{230} = 6.421$ kcal/mole, and $S^\circ_{230}(g) = 62.12$ cal/(mole °K). Third-law entropy established the lowest vibrational frequency as 63 cm^{-1}. The remaining frequencies are taken from Long, Murfin, and Williams (881), and the moment of inertia from Lafferty, Maki, and Plyler (837). Miller, Lemmon, and Witkowski (1004) confirmed a frequency at 63 cm^{-1} by direct observation in the far infrared region.

Heat capacity values for liquid and vapor are combined with the measured enthalpy of vaporization of McDougall and Kilpatrick to derive $\Delta Hv^\circ_{298} = 5.65$ kcal/mole and $\Delta Hf^\circ_{298}(g) = -22.38$ kcal/mole.

No. 23 Acetylenedicarbonitrile C_4N_2 (Ideal Gas State) Mol Wt 76.056

T°K	Cp°	S°	$-(G^\circ - H^\circ_{298})/T$	$H^\circ - H^\circ_{298}$	ΔHf°	ΔGf°	Log Kp
		cal/(mole °K)			kcal/mole		
298	20.53	69.31	69.31	0.00	127.50	122.10	-89.500
300	20.58	69.44	69.32	0.04	127.51	122.07	-88.923
400	22.66	75.66	70.15	2.21	127.99	120.19	-65.663
500	24.16	80.88	71.79	4.55	128.36	118.19	-51.657
600	25.37	85.40	73.69	7.03	128.61	116.12	-42.296
700	26.39	89.39	75.65	9.62	128.78	114.03	-35.601
800	27.26	92.97	77.60	12.30	128.88	111.92	-30.573
900	28.00	96.23	79.49	15.07	128.94	109.79	-26.659
1000	28.62	99.21	81.31	17.90	128.97	107.66	-23.529

Armstrong and Marantz (26) measured the enthalpy of combustion and derived $\Delta H f^\circ_{298}(g) = 127.5$ kcal/mole. Thermodynamic functions have been reported in the JANAF Thermochemical Tables (1435) using structural parameters of Miller and Hannan (1002) and vibrational frequencies assigned by Miller, Hannan, and Cousins (1003).

No. 24 Perchloryl Fluoride ClFO$_3$ (Ideal Gas State) Mol Wt 102.457

	cal/(mole °K)				kcal/mole		
$T°K$	$Cp°$	$S°$	$-(G°-H°_{298})/T$	$H°-H°_{298}$	$\Delta Hf°$	$\Delta Gf°$	Log Kp
298	15.52	66.65	66.65	0.00	-6.49	10.72	-7.858
300	15.57	66.75	66.66	0.03	-6.50	10.83	-7.887
400	18.15	71.60	67.30	1.73	-6.67	16.64	-9.089
500	20.00	75.86	68.59	3.64	-6.68	22.47	-9.819
600	21.32	79.63	70.13	5.71	-6.60	28.29	-10.304
700	22.27	82.99	71.73	7.89	-6.45	34.09	-10.642
800	22.97	86.01	73.33	10.15	-6.26	39.87	-10.891
900	23.49	88.75	74.89	12.48	-6.04	45.62	-11.078
1000	23.88	91.24	76.40	14.85	-5.80	51.34	-11.220

The enthalpy of formation was determined by Neugebauer and Margrave (1060). Thermodynamic functions have been computed and reported in the JANAF Thermochemical Tables (1435) on the basis of the vibrational assignment of Lide and Mann (869) and estimated bond lengths and angles. Low-temperature thermal data reported by Koehler and Giauque (785) gave evidence of random orientation in the solid. They found $Tm = 125.41°K$, $\Delta Hm° = 0.916$ kcal/mole, $Tb = 226.48°K$, and $\Delta Hv = 4.619$ kcal/mole.

No. 25 Hydrogen Chloride ClH (Ideal Gas State) Mol Wt 36.465

	cal/(mole °K)				kcal/mole		
$T°K$	$Cp°$	$S°$	$-(G°-H°_{298})/T$	$H°-H°_{298}$	$\Delta Hf°$	$\Delta Gf°$	Log Kp
298	6.96	44.64	44.64	0.00	-22.06	-22.77	16.692
300	6.96	44.69	44.65	0.02	-22.06	-22.78	16.592
400	6.97	46.69	44.92	0.71	-22.13	-23.01	12.570
500	7.00	48.25	45.43	1.41	-22.20	-23.22	10.148
600	7.07	49.53	46.01	2.12	-22.29	-23.41	8.528
700	7.17	50.63	46.60	2.83	-22.36	-23.59	7.366
800	7.29	51.59	47.16	3.55	-22.44	-23.77	6.492
900	7.42	52.46	47.70	4.29	-22.50	-23.93	5.810
1000	7.56	53.25	48.22	5.03	-22.56	-24.08	5.263

Thermodynamic functions of hydrogen chloride were calculated by Potter (1183). The enthalpy of formation is that listed in the compilation of Rossini, Wagman, Evans, Levine, and Jaffe (1249). Rossini et al. also selected $Tt = 98.38°K$, $\Delta Ht° = 0.284$ kcal/mole, $Ttp = 158.94°K$, $\Delta Hm° = 0.476$ kcal/mole, $Tb = 188.11°K$, and $\Delta Hv = 3.86$ kcal/mole.

No. 26 Nitrosyl Chloride ClNO (Ideal Gas State) Mol Wt 65.465

T°K	cal/(mole °K)				kcal/mole		Log Kp
	$Cp°$	$S°$	$-(G°-H°_{298})/T$	$H°-H°_{298}$	$\Delta Hf°$	$\Delta Gf°$	
298	10.68	62.53	62.53	0.00	12.57	16.00	-11.727
300	10.69	62.60	62.54	0.02	12.57	16.02	-11.670
400	11.31	65.76	62.96	1.13	12.55	17.18	-9.384
500	11.79	68.34	63.79	2.28	12.56	18.33	-8.012
600	12.20	70.53	64.73	3.48	12.60	19.48	-7.096
700	12.54	72.43	65.70	4.72	12.64	20.62	-6.439
800	12.83	74.13	66.65	5.99	12.70	21.76	-5.944
900	13.07	75.65	67.57	7.28	12.76	22.89	-5.558
1000	13.27	77.04	68.45	8.60	12.83	24.01	-5.248

Gordon (520) calculated accurate thermodynamic functions for nitrosyl chloride. The enthalpy of formation is from the compilation of Rossini, Wagman, Evans, Levine, and Jaffe (1249), which also listed $Ttp = 211.7°K$, $Tb = 267.4°K$, and $\Delta Hv = 6.0$ kcal/mole.

No. 27 Nitryl Chloride ClNO$_2$ (Ideal Gas State) Mol Wt 81.465

T°K	cal/(mole °K)				kcal/mole		Log Kp
	$Cp°$	$S°$	$-(G°-H°_{298})/T$	$H°-H°_{298}$	$\Delta Hf°$	$\Delta Gf°$	
298	12.71	65.01	65.01	0.00	3.12	13.11	-9.613
300	12.74	65.09	65.02	0.03	3.12	13.18	-9.599
400	14.26	68.98	65.53	1.38	3.00	16.55	-9.044
500	15.41	72.29	66.56	2.87	2.97	19.95	-8.718
600	16.29	75.18	67.76	4.45	3.01	23.34	-8.501
700	16.97	77.74	69.01	6.12	3.10	26.72	-8.341
800	17.49	80.04	70.25	7.84	3.21	30.09	-8.219
900	17.89	82.13	71.45	9.61	3.34	33.44	-8.120
1000	18.20	84.03	72.62	11.42	3.48	36.78	-8.037

The enthalpy of formation was measured by Ray and Ogg (1217). Thermodynamic functions calculated by Larmann, Martire, and Pollara (848) and by Puranik and Rao (1206) are in good agreement. Rossini, Wagman, Evans, Levine, and Jaffe (1249) selected $Tm = 128°K$, $Tb = 257.9°K$, and $\Delta Hv = 5.0$ kcal/mole.

No. 28 Thionyl Chloride Cl$_2$OS (Ideal Gas State) Mol Wt 118.980

The enthalpy of formation of liquid thionyl chloride was measured by Neale and Williams (1048). Their value is combined with the enthalpy of vaporization at the boiling point listed by Rossini, Wagman, Evans,

No. 28   Thionyl Chloride Cl_2OS (Ideal Gas State)   Mol Wt 118.980

| T°K | $Cp°$ | cal/(mole °K) | | | kcal/mole | | Log Kp |
		$S°$	$-(G°-H°_{298})/T$	$H°-H°_{298}$	$\Delta Hf°$	$\Delta Gf°$	
298	15.88	73.23	73.23	0.00	-50.60	-46.97	34.425
300	15.91	73.33	73.24	0.03	-50.60	-46.94	34.196
400	17.03	78.07	73.87	1.68	-51.24	-45.69	24.960
500	17.75	81.95	75.11	3.43	-51.65	-44.25	19.340
600	18.25	85.24	76.53	5.23	-51.95	-42.74	15.566
700	18.60	88.08	77.99	7.07	-52.18	-41.16	12.850
800	18.85	90.58	79.41	8.94	-65.41	-40.84	11.157
900	19.04	92.81	80.77	10.84	-65.25	-37.78	9.173
1000	19.18	94.82	82.08	12.75	-65.09	-34.73	7.591

Levine, and Jaffe (1249) and an estimated correction to 298°K to derive the enthalpy of formation of the gas. The frequencies adopted by Kelley (738) to compute the entropy are used here to calculate the heat capacity of the ideal gas. Rossini et al. (1249) listed $Tm = 168.7°K$, $Tb = 348.9°K$, and $\Delta Hv = 7.41$ kcal/mole.

No. 29   Sulfuryl Chloride Cl_2O_2S (Ideal Gas State)   Mol Wt 134.980

| T°K | $Cp°$ | cal/(mole °K) | | | kcal/mole | | Log Kp |
		$S°$	$-(G°-H°_{298})/T$	$H°-H°_{298}$	$\Delta Hf°$	$\Delta Gf°$	
298	18.34	74.37	74.37	0.00	-85.40	-74.80	54.827
300	18.38	74.49	74.38	0.04	-85.40	-74.73	54.441
400	20.06	80.02	75.12	1.97	-86.12	-71.13	38.859
500	21.10	84.62	76.57	4.03	-86.58	-67.32	29.425
600	21.79	88.53	78.25	6.17	-86.91	-63.44	23.106
700	22.25	91.92	79.96	8.38	-87.17	-59.48	18.571
800	22.58	94.92	81.65	10.62	-100.43	-56.78	15.511
900	22.82	97.59	83.27	12.89	-100.30	-51.33	12.465
1000	23.00	100.00	84.83	15.18	-100.17	-45.90	10.031

The enthalpy of formation of the liquid, recently determined by Neale and Williams (1048), is combined with the enthalpy of vaporization at the boiling point listed by Rossini, Wagman, Evans, Levine, and Jaffe (1249) and an estimated correction to 298°K to derive the enthalpy of formation of the ideal gas. Kelley (738) calculated the entropy from molecular data; the frequencies adopted by him are used to calculate the heat capacities given here. Rossini et al. (1249) selected $Tm = 227°K$, $Tb = 342.4°K$, and $\Delta Hv = 7.50$ kcal/mole.

No. 30   Sulfur Monochloride Cl_2S_2 (Ideal Gas State)   Mol Wt 135.046

The enthalpy of formation is derived from values listed by Rossini, Wagman, Evans, Levine, and Jaffe (1249). Kelley (738) calculated the

No. 30 Sulfur Monochloride Cl_2S_2 (Ideal Gas State) Mol Wt 135.046

$T°K$	$Cp°$	$S°$	$-(G°-H°_{298})/T$	$H°-H°_{298}$	$\Delta Hf°$	$\Delta Gf°$	Log Kp
		cal/(mole °K)			kcal/mole		
298	17.41	76.35	76.35	0.00	-4.66	-6.99	5.120
300	17.43	76.46	76.36	0.04	-4.66	-7.00	5.099
400	18.37	81.62	77.05	1.83	-5.90	-7.71	4.214
500	18.87	85.77	78.39	3.69	-6.76	-8.06	3.525
600	19.16	89.24	79.92	5.60	-7.44	-8.26	3.008
700	19.34	92.21	81.47	7.52	-8.00	-8.30	2.592
800	19.46	94.80	82.98	9.46	-34.59	-10.85	2.963
900	19.54	97.10	84.42	11.41	-34.41	-7.89	1.916
1000	19.61	99.16	85.79	13.37	-34.23	-4.95	1.083

entropy from molecular data. The fundamental frequencies adopted by Kelley are used to derive heat capacity values for the ideal gas. There is some doubt about the torsional frequency, and Luft and Todhunter (893) computed somewhat different values based on restricted internal rotation. Rossini et al. (1249) selected $Tm = 193°K$, $Tb = 411.2°K$, and $\Delta Hv = 8.61$ kcal/mole.

No. 31 Hydrogen Fluoride FH (Ideal Gas State) Mol Wt 20.008

$T°K$	$Cp°$	$S°$	$-(G°-H°_{298})/T$	$H°-H°_{298}$	$\Delta Hf°$	$\Delta Gf°$	Log Kp
		cal/(mole °K)			kcal/mole		
298	6.96	41.51	41.51	0.00	-64.80	-65.30	47.865
300	6.96	41.56	41.52	0.02	-64.80	-65.30	47.572
400	6.97	43.56	41.79	0.71	-64.84	-65.47	35.768
500	6.97	45.12	42.30	1.41	-64.89	-65.62	28.681
600	6.99	46.39	42.88	2.11	-64.96	-65.76	23.951
700	7.02	47.47	43.46	2.81	-65.04	-65.89	20.569
800	7.06	48.41	44.02	3.51	-65.12	-66.00	18.030
900	7.13	49.24	44.56	4.22	-65.20	-66.11	16.052
1000	7.21	50.00	45.06	4.94	-65.29	-66.21	14.470

Thermochemical data for hydrogen fluoride gas were critically reviewed in the JANAF Thermochemical Tables (1435), and the selected $\Delta Hf°_{298}(g) = -64.8$ kcal/mole is adopted here. Potter (1182) calculated accurate thermodynamic functions. Hu, White, and Johnston (625) reported low-temperature thermal data, including $Ttp = 189.79°K$ and $\Delta Hm° = 0.939$ kcal/mole. Rossini, Wagman, Evans, Levine, and Jaffe (1249) selected $Tb = 293.1°K$ and $\Delta Hv = 1.8$ kcal/mole. This corresponds to an abnormally low entropy of vaporization, owing to association in the gas phase. In considering the thermodynamics of HF at low temperatures and/or high pressures, the effects of this association should be taken into account.

No. 32 Nitrosyl Fluoride FNO (Ideal Gas State) Mol Wt 49.008

T°K	$C_p°$	cal/(mole °K)			kcal/mole		Log K_p
		$S°$	$-(G°-H°_{298})/T$	$H°-H°_{298}$	$\Delta H_f°$	$\Delta G_f°$	
298	9.88	59.27	59.27	0.00	-15.70	-12.02	8.811
300	9.90	59.34	59.28	0.02	-15.70	-12.00	8.741
400	10.65	62.29	59.67	1.05	-15.76	-10.75	5.874
500	11.23	64.73	60.45	2.15	-15.79	-9.50	4.151
600	11.69	66.82	61.34	3.29	-15.79	-8.24	3.000
700	12.06	68.65	62.26	4.48	-15.78	-6.98	2.180
800	12.35	70.28	63.16	5.70	-15.76	-5.72	1.564
900	12.59	71.75	64.03	6.95	-15.74	-4.47	1.086
1000	12.78	73.09	64.87	8.22	-15.71	-3.23	0.706

The enthalpy of formation was determined by Johnston and Bertin (696). Thermodynamic functions have been calculated (1435) from molecular dimensions and fundamental frequencies reported by Stephenson and Jones (1412). Rossini, Wagman, Evans, Levine, and Jaffe (1249) listed $Tm = 140.7°K$, $Tb = 213.3°K$, and $\Delta Hv = 4.61$ kcal/mole.

No. 33 Nitryl Fluoride FNO_2 (Ideal Gas State) Mol Wt 65.008

T°K	$C_p°$	cal/(mole °K)			kcal/mole		Log K_p
		$S°$	$-(G°-H°_{298})/T$	$H°-H°_{298}$	$\Delta H_f°$	$\Delta G_f°$	
298	11.92	62.24	62.24	0.00	-19.00	-8.90	6.525
300	11.96	62.32	62.25	0.03	-19.00	-8.84	6.439
400	13.64	66.00	62.74	1.31	-19.16	-5.42	2.961
500	14.90	69.18	63.71	2.74	-19.22	-1.98	0.864
600	15.88	71.99	64.86	4.28	-19.21	1.47	-0.537
700	16.62	74.50	66.06	5.91	-19.15	4.91	-1.534
800	17.19	76.75	67.26	7.60	-19.06	8.35	-2.280
900	17.64	78.81	68.43	9.34	-18.95	11.77	-2.857
1000	17.98	80.68	69.56	11.12	-18.82	15.16	-3.314

Thermodynamic functions of nitryl fluoride were reported by Puranik and Rao (1206). Values in excellent agreement were calculated by Tschuikow-Roux (1520), who also reported $\Delta Hf°_{298}(g) = -19$ kcal/mole. Rossini, Wagman, Evans, Levine, and Jaffe (1249) selected $Tm = 107.2°K$, $Tb = 200.8°K$, and $\Delta Hv = 4.31$ kcal/mole.

No. 34 Sulfur Tetrafluoride F_4S (Ideal Gas State) Mol Wt 108.066

The enthalpy of formation of sulfur tetrafluoride was measured by Vaughn and Muetterties (1529). Thermodynamic functions were calculated on the basis of molecular constants and vibrational frequencies reported by Dodd, Woodward, and Roberts (332). Since some of the values are estimated, the entropy is uncertain by ±1 cal/(mole °K).

No. 34 Sulfur Tetrafluoride F₄S (Ideal Gas State) Mol Wt 108.066

$T°K$	$Cp°$	$S°$	$-(G°-H°_{298})/T$	$H°-H°_{298}$	$\Delta Hf°$	$\Delta Gf°$	Log Kp
	cal/(mole °K)				kcal/mole		
298	17.21	69.58	69.58	0.00	-174.10	-163.68	119.976
300	17.27	69.69	69.59	0.04	-174.11	-163.62	119.189
400	19.99	75.06	70.30	1.91	-174.87	-160.06	87.448
500	21.69	79.71	71.73	4.00	-175.34	-156.30	68.316
600	22.78	83.77	73.40	6.22	-175.63	-152.46	55.532
700	23.50	87.34	75.14	8.54	-175.82	-148.56	46.380
800	24.01	90.51	76.87	10.92	-189.00	-145.93	39.864
900	24.36	93.36	78.55	13.34	-188.77	-140.56	34.130
1000	24.63	95.94	80.16	15.79	-188.53	-135.25	29.557

No. 35 Hydrogen Iodide HI (Ideal Gas State) Mol Wt 127.918

$T°K$	$Cp°$	$S°$	$-(G°-H°_{298})/T$	$H°-H°_{298}$	$\Delta Hf°$	$\Delta Gf°$	Log Kp
	cal/(mole °K)				kcal/mole		
298	6.97	49.35	49.35	0.00	6.30	0.38	-0.276
300	6.97	49.40	49.36	0.02	6.29	0.34	-0.248
400	7.01	51.41	49.63	0.72	4.06	-1.53	0.836
500	7.11	52.98	50.15	1.42	-1.35	-2.41	1.054
600	7.25	54.29	50.73	2.14	-1.43	-2.62	0.954
700	7.42	55.42	51.32	2.87	-1.49	-2.81	0.878
800	7.60	56.42	51.90	3.62	-1.55	-3.00	0.818
900	7.77	57.33	52.45	4.39	-1.58	-3.17	0.771
1000	7.92	58.15	52.98	5.18	-1.61	-3.35	0.732

Thermodynamic functions based on the molecular constants of Cowan and Gordy (282) and of Boyd and Thompson (153) were reported in the JANAF Thermochemical Tables (1435). The enthalpy of formation was calculated from the free energy functions adopted in this text for I_2, H_2, and HI, and the equilibrium constants determined by Taylor and Crist (1473), which are the most concordant of the available data. Rossini, Wagman, Evans, Levine, and Jaffe (1249) selected $Tm = 222.36°K$, $\Delta Hm° = 0.686$ kcal/mole, $Tb = 237.80°K$, and $\Delta Hv = 4.724$ kcal/mole.

No. 36 Hydrogen Nitrate HNO₃ (Ideal Gas State) Mol Wt 63.016

Thermodynamic functions for hydrogen nitrate were calculated by Palm and Kilpatrick (1099). The enthalpy of formation listed by Rossini, Wagman, Evans, Levine, and Jaffe (1249) is in good agreement with a value calculated from the dissociation pressure of ammonium nitrate reported by Feick (413). Rossini et al. (1249) listed $Tm = 231.56°K$, $\Delta Hm° = 2.503$ kcal/mole, $Tb = 293°K$, and $\Delta Hv = 9.43$ kcal/mole.

No. 36 Hydrogen Nitrate HNO_3 (Ideal Gas State) Mol Wt 63.016

	cal/(mole °K)				kcal/mole		
$T°K$	$Cp°$	$S°$	$-(G°-H°_{298})/T$	$H°-H°_{298}$	$\Delta Hf°$	$\Delta Gf°$	Log Kp
298	12.82	63.68	63.68	0.00	-32.02	-17.62	12.912
300	12.87	63.76	63.69	0.03	-32.03	-17.53	12.767
400	15.21	67.80	64.22	1.44	-32.38	-12.63	6.900
500	17.09	71.40	65.30	3.05	-32.56	-7.67	3.352
600	18.57	74.65	66.59	4.84	-32.61	-2.68	0.977
700	19.71	77.60	67.96	6.76	-32.58	2.30	-0.718
800	20.64	80.30	69.34	8.78	-32.48	7.28	-1.989
900	21.37	82.77	70.69	10.88	-32.34	12.24	-2.973
1000	21.95	85.06	72.02	13.04	-32.16	17.19	-3.756

No. 37 Water H_2O (Ideal Gas State) Mol Wt 18.016

	cal/(mole °K)				kcal/mole		
$T°K$	$Cp°$	$S°$	$-(G°-H°_{298})/T$	$H°-H°_{298}$	$\Delta Hf°$	$\Delta Gf°$	Log Kp
298	8.03	45.11	45.11	0.00	-57.80	-54.64	40.049
300	8.03	45.16	45.12	0.02	-57.80	-54.62	39.788
400	8.19	47.49	45.43	0.83	-58.04	-53.52	29.241
500	8.42	49.35	46.04	1.66	-58.28	-52.36	22.887
600	8.68	50.90	46.72	2.52	-58.50	-51.16	18.634
700	8.95	52.26	47.42	3.40	-58.71	-49.92	15.585
800	9.25	53.48	48.10	4.31	-58.91	-48.65	13.290
900	9.55	54.58	48.76	5.25	-59.08	-47.36	11.499
1000	9.85	55.61	49.39	6.22	-59.24	-46.04	10.062

Thermodynamic functions for water were reported by Friedman and Haar (445). The enthalpy of formation is taken from the review of Wagman, Kilpatrick, Taylor, Pitzer, and Rossini (1562). The melting and boiling points of 273.15°K and 373.15°K are reference temperatures on the international scale. Rossini, Wagman, Evans, Levine, and Jaffe (1249) listed $\Delta Hm° = 1.4363$ kcal/mole and $\Delta Hv = 9.7171$ kcal/mole.

No. 38 Hydrogen Peroxide H_2O_2 (Ideal Gas State) Mol Wt 34.016

	cal/(mole °K)				kcal/mole		
$T°K$	$Cp°$	$S°$	$-(G°-H°_{298})/T$	$H°-H°_{298}$	$\Delta Hf°$	$\Delta Gf°$	Log Kp
298	10.31	55.66	55.66	0.00	-32.53	-25.21	18.478
300	10.33	55.73	55.67	0.02	-32.54	-25.16	18.331
400	11.58	58.88	56.09	1.12	-32.84	-22.66	12.378
500	12.56	61.57	56.92	2.33	-33.06	-20.08	8.777
600	13.31	63.93	57.90	3.63	-33.22	-17.47	6.363
700	13.86	66.03	58.91	4.99	-33.35	-14.84	4.632
800	14.30	67.91	59.92	6.39	-33.44	-12.18	3.328
900	14.69	69.61	60.90	7.84	-33.52	-9.52	2.312
1000	15.02	71.18	61.85	9.33	-33.57	-6.85	1.497

Thermal and thermochemical data for hydrogen peroxide were the subject of careful investigations by Giguère (480), Giguère and Liu (481), Giguère, Liu, Dugdale, and Morrison (482), and Giguère, Morissette, Olmos, and Knop (483). Results include $Tm = 272.75°K$, $\Delta Hm° = 2.987$ kcal/mole, and $\Delta Hv°_{298} = 12.34$ kcal/mole.

No. 39 Sulfuric Acid H_2O_4S (Anhydrous Liquid) Mol Wt 98.082

| $T°K$ | $Cp°$ | cal/(mole °K) | | | kcal/mole | | Log Kp |
		$S°$	$-(G°-H°_{298})/T$	$H°-H°_{298}$	$\Delta Hf°$	$\Delta Gf°$	
298	33.18	37.49	37.49	0.00	-194.45	-164.83	120.815
300	33.30	37.70	37.50	0.07	-194.44	-164.64	119.936
400	36.70	47.78	38.85	3.58	-194.14	-154.80	84.577
500	38.70	56.20	41.50	7.35	-193.46	-145.05	63.397
600	40.00	63.37	44.56	11.29	-192.59	-135.44	49.332
700	41.10	69.62	47.70	15.35	-191.60	-125.98	39.329

Rubin and Giauque (1260) reported low-temperature thermal data for liquid hydrogen sulfate as $Tm = 283.53°K$, $\Delta Hm° = 2.560$ kcal/mole, and $S°_{298}(l) = 37.49$ cal/(mole °K). Their heat capacity data near 300°K are extrapolated in order to estimate higher temperature values. Good, Lacina, and McCullough (509) determined the enthalpy of formation of aqueous hydrogen sulfate with high accuracy. The enthalpy of formation of anhydrous H_2SO_4 is calculated using enthalpy of solution data listed by Rossini, Wagman, Evans, Levine, and Jaffe (1249).

No. 40 Hydrogen Sulfide H_2S (Ideal Gas State) Mol Wt 34.082

| $T°K$ | $Cp°$ | cal/(mole °K) | | | kcal/mole | | Log Kp |
		$S°$	$-(G°-H°_{298})/T$	$H°-H°_{298}$	$\Delta Hf°$	$\Delta Gf°$	
298	8.17	49.18	49.18	0.00	-4.82	-7.90	5.792
300	8.18	49.24	49.19	0.02	-4.83	-7.92	5.771
400	8.51	51.63	49.51	0.85	-5.79	-8.87	4.845
500	8.90	53.57	50.13	1.72	-6.55	-9.55	4.174
600	9.32	55.23	50.85	2.64	-7.20	-10.09	3.673
700	9.75	56.70	51.58	3.59	-7.75	-10.50	3.278
800	10.18	58.03	52.31	4.59	-21.28	-12.13	3.313
900	10.59	59.25	53.01	5.62	-21.39	-10.97	2.665
1000	10.97	60.39	53.69	6.70	-21.47	-9.81	2.144

Thermodynamic functions for hydrogen sulfide have been computed by Haar, Bradley, and Friedman (557). The enthalpy of formation is that selected by Evans and Wagman (399) from a review of the literature. A recent determination by Kapustinskii and Kan'kovskii (715) is in accord within experimental error. Rossini, Wagman, Evans, Levine, and Jaffe (1249) selected $Tt = 103.54$ kcal/mole, $\Delta Ht° = 0.366$ kcal/mole, $Tt =$

126.24°K, $\Delta Ht° = 0.108$ kcal/mole, $Ttp = 187.63°K$, $\Delta Hm° = 0.568$ kcal/mole, $Tb = 212.82°K$, and $\Delta Hv = 4.463$ kcal/mole.

No. 41   Ammonia H_3N (Ideal Gas State)   Mol Wt 17.032

T°K	cal/(mole °K)				kcal/mole		Log Kp
	$Cp°$	$S°$	$-(G°-H°_{298})/T$	$H°-H°_{298}$	$\Delta Hf°$	$\Delta Gf°$	
298	8.51	46.03	46.03	0.00	-10.92	-3.86	2.831
300	8.53	46.09	46.04	0.02	-10.93	-3.82	2.783
400	9.24	48.64	46.38	0.91	-11.43	-1.37	0.749
500	10.04	50.79	47.05	1.87	-11.87	1.20	-0.522
600	10.81	52.68	47.83	2.92	-12.23	3.84	-1.400
700	11.54	54.41	48.65	4.03	-12.53	6.55	-2.044
800	12.23	55.99	49.47	5.22	-12.77	9.29	-2.537
900	12.87	57.47	50.28	6.48	-12.96	12.06	-2.928
1000	13.47	58.86	51.07	7.80	-13.11	14.85	-3.245

The chemical thermodynamic properties of ammonia were reviewed and revised by Harrison and Kobe (573). Their values have been adopted. Rossini, Wagman, Evans, Levine, and Jaffe (1249) selected $Ttp = 195.40°K$, $\Delta Hm° = 1.351$ kcal/mole, $Tb = 239.73°K$, and $\Delta Hv = 5.581$ kcal/mole.

No. 42   Hydrazine H_4N_2 (Ideal Gas State)   Mol Wt 32.048

T°K	cal/(mole °K)				kcal/mole		Log Kp
	$Cp°$	$S°$	$-(G°-H°_{298})/T$	$H°-H°_{298}$	$\Delta Hf°$	$\Delta Gf°$	
298	12.60	57.41	57.41	0.00	22.75	37.89	-27.773
300	12.61	57.49	57.42	0.03	22.73	37.98	-27.669
400	15.10	61.48	57.94	1.42	22.04	43.18	-23.590
500	16.90	65.05	59.01	3.02	21.54	48.52	-21.208
600	18.30	68.26	60.29	4.78	21.19	53.95	-19.652
700	19.50	71.17	61.64	6.68	20.95	59.43	-18.555
800	20.60	73.85	63.00	8.68	20.80	64.94	-17.740
900	21.50	76.33	64.34	10.79	20.73	70.47	-17.111
1000	22.30	78.63	65.66	12.98	20.71	76.00	-16.609

Low-temperature thermal data were measured by Scott, Oliver, Gross, Hubbard, and Huffman (1323). They reported $Tm = 274.69°K$, $\Delta Hm° = 3.025$ kcal/mole, and $S°_{298}(l) = 28.97$ cal/(mole °K). From vapor pressure measurements they derived the enthalpy of vaporization at 298°K as 10.70 kcal/mole, which was combined with the enthalpy of formation of the liquid reported by Hughes, Corruccini, and Gilbert (658) to derive $\Delta Hf°_{298}(g) = 22.75$ kcal/mole. Scott et al. calculated thermodynamic functions from a vibrational assignment and molecular data. The calculated entropy was 0.44 cal/(mole °K) higher than the measured value;

this discrepancy, however, was less than the combined experimental uncertainty.

No. 43   Nitric Oxide   NO   (Ideal Gas State)   Mol Wt 30.008

$T°K$	$Cp°$	cal/(mole °K)			kcal/mole		
		$S°$	$-(G°-H°_{298})/T$	$H°-H°_{298}$	$\Delta Hf°$	$\Delta Gf°$	$\text{Log } Kp$
298	7.13	50.35	50.35	0.00	21.60	20.72	-15.184
300	7.13	50.40	50.36	0.02	21.60	20.71	-15.087
400	7.16	52.45	50.63	0.73	21.61	20.41	-11.153
500	7.29	54.06	51.16	1.45	21.62	20.11	-8.791
600	7.47	55.41	51.76	2.19	21.62	19.81	-7.217
700	7.66	56.57	52.37	2.95	21.62	19.51	-6.091
800	7.83	57.61	52.96	3.72	21.63	19.21	-5.247
900	7.99	58.54	53.53	4.51	21.63	18.91	-4.591
1000	8.12	59.39	54.07	5.32	21.64	18.60	-4.066

The enthalpy of formation is taken from the compilation of Rossini, Wagman, Evans, Levine, and Jaffe (1249). Thermodynamic functions have been reported in the JANAF Thermochemical Tables (1435), based on spectroscopic data listed by Herzberg (590). Rossini et al. (1249) selected $Ttp = 109.51°K$, $\Delta Hm° = 0.550$ kcal/mole, $Tb = 121.39°K$, and $\Delta Hv = 3.293$ kcal/mole.

No. 44   Nitrogen Dioxide NO_2 (Ideal Gas State)   Mol Wt 46.008

$T°K$	$Cp°$	cal/(mole °K)			kcal/mole		
		$S°$	$-(G°-H°_{298})/T$	$H°-H°_{298}$	$\Delta Hf°$	$\Delta Gf°$	$\text{Log } Kp$
298	8.86	57.35	57.35	0.00	8.09	12.42	-9.107
300	8.87	57.41	57.36	0.02	8.09	12.45	-9.070
400	9.64	60.07	57.71	0.95	7.95	13.93	-7.611
500	10.38	62.30	58.41	1.95	7.87	15.43	-6.746
600	11.02	64.25	59.23	3.02	7.83	16.95	-6.174
700	11.56	65.99	60.07	4.15	7.82	18.47	-5.766
800	12.00	67.56	60.91	5.33	7.83	19.99	-5.461
900	12.35	69.00	61.73	6.55	7.85	21.51	-5.223
1000	12.63	70.32	62.52	7.80	7.89	23.03	-5.032

Thermodynamic functions were calculated by Gordon (519). The enthalpy of formation is that listed in the compilation of Rossini, Wagman, Evans, Levine, and Jaffe (1249). At low temperatures NO_2 dimerizes to N_2O_4 and vaporization is complex, involving dissociation as well as the usual enthalpy of vaporization.

No. 45   Nitrous Oxide N_2O (Ideal Gas State)   Mol Wt 44.016

Thermodynamic functions were calculated and reported in the JANAF Thermochemical Tables (1435), based on frequencies of Herzberg and

No. 45 Nitrous Oxide N_2O (Ideal Gas State) Mol Wt 44.016

$T°K$	cal/(mole °K)				kcal/mole		
	$Cp°$	$S°$	$-(G°-H°_{298})/T$	$H°-H°_{298}$	$\Delta Hf°$	$\Delta Gf°$	Log Kp
298	9.23	52.56	52.56	0.00	19.49	24.77	-18.156
300	9.25	52.62	52.57	0.02	19.49	24.80	-18.068
400	10.20	55.42	52.94	1.00	19.41	26.59	-14.528
500	10.95	57.78	53.68	2.05	19.40	28.39	-12.407
600	11.56	59.83	54.54	3.18	19.44	30.18	-10.993
700	12.06	61.65	55.42	4.36	19.50	31.97	-9.980
800	12.48	63.29	56.31	5.59	19.59	33.74	-9.218
900	12.82	64.78	57.17	6.86	19.69	35.51	-8.622
1000	13.11	66.14	58.00	8.15	19.79	37.26	-8.143

Herzberg (591) and rotational constants of Coles and Hughes (248). Rossini, Wagman, Evans, Levine, and Jaffe (1249) selected $\Delta Hf°_{298}(g) = 19.49$ kcal/mole, $Ttp = 182.30°K$, $\Delta Hm° = 1.563$ kcal/mole, $Tb = 184.68°K$, and $\Delta Hv = 3.956$ kcal/mole.

No. 46 Sulfur Dioxide O_2S (Ideal Gas State) Mol Wt 64.066

$T°K$	cal/(mole °K)				kcal/mole		
	$Cp°$	$S°$	$-(G°-H°_{298})/T$	$H°-H°_{298}$	$\Delta Hf°$	$\Delta Gf°$	Log Kp
298	9.53	59.30	59.30	0.00	-70.95	-71.74	52.588
300	9.55	59.36	59.31	0.02	-70.96	-71.75	52.267
400	10.39	62.23	59.69	1.02	-71.77	-71.95	39.309
500	11.13	64.63	60.44	2.10	-72.36	-71.93	31.437
600	11.72	66.71	61.32	3.24	-72.83	-71.79	26.149
700	12.18	68.56	62.22	4.44	-73.21	-71.57	22.343
800	12.53	70.21	63.12	5.67	-86.60	-72.58	19.826
900	12.81	71.70	63.99	6.94	-86.58	-70.82	17.198
1000	13.02	73.06	64.83	8.23	-86.56	-69.07	15.095

Thermodynamic functions calculated by Gordon (519) are used, although they differ only slightly from those given by Evans and Wagman (399). The enthalpy of formation is from the review of Evans and Wagman (399) and is based on the enthalpy of combustion of rhombic sulfur as measured by Eckman and Rossini (364). Rossini, Wagman, Evans, Levine, and Jaffe (1249) selected $Ttp = 197.68°K$, $\Delta Hm° = 1.769$ kcal/mole, $Tb = 263.14°K$, and $\Delta Hv = 5.955$ kcal/mole.

No. 47 Ozone O_3 (Ideal Gas State) Mol Wt 48.000

The enthalpy of formation is taken from the compilation of Rossini, Wagman, Evans, Levine, and Jaffe (1249). Thermodynamic functions were computed by Birdsall, Jenkins, DiPaolo, Beattie, and Apt (133) and

No. 47 Ozone O_3 (Ideal Gas State) Mol Wt 48.000

T°K	Cp°	S°	$-(G°-H°_{298})/T$	$H°-H°_{298}$	$\Delta Hf°$	$\Delta Gf°$	Log Kp
	cal/(mole °K)				kcal/mole		
298	9.37	57.05	57.05	0.00	34.00	38.91	-28.517
300	9.39	57.11	57.06	0.02	34.00	38.94	-28.364
400	10.43	59.96	57.44	1.02	33.93	40.60	-22.181
500	11.26	62.38	58.19	2.10	33.92	42.27	-18.475
600	11.87	64.49	59.07	3.26	33.94	43.94	-16.004
700	12.30	66.36	59.98	4.47	33.98	45.60	-14.235
800	12.62	68.02	60.88	5.72	34.03	47.26	-12.909
900	12.86	69.52	61.76	6.99	34.09	48.91	-11.875
1000	13.04	70.89	62.60	8.29	34.14	50.55	-11.047

by Klein, Cleveland, and Meister (766). Rossini, Wagman, Evans, Levine, and Jaffe (1249) selected $Tb = 162.65°K$, and $\Delta Hv = 2.59$ kcal/mole.

No. 48 Sulfur Trioxide O_3S (Ideal Gas State) Mol Wt 80.066

T°K	Cp°	S°	$-(G°-H°_{298})/T$	$H°-H°_{298}$	$\Delta Hf°$	$\Delta Gf°$	Log Kp
	cal/(mole °K)				kcal/mole		
298	12.10	61.19	61.19	0.00	-94.47	-88.52	64.886
300	12.13	61.27	61.20	0.03	-94.48	-88.49	64.458
400	14.06	65.03	61.70	1.34	-95.33	-86.41	47.212
500	15.66	68.35	62.70	2.83	-95.87	-84.12	36.767
600	16.90	71.32	63.89	4.46	-96.23	-81.73	29.769
700	17.86	74.00	65.15	6.20	-96.46	-79.27	24.749
800	18.61	76.43	66.41	8.02	-109.66	-78.07	21.328
900	19.23	78.66	67.65	9.92	-109.43	-74.14	18.003
1000	19.76	80.71	68.85	11.87	-109.16	-70.23	15.348

Evans and Wagman (399) reviewed enthalpy of formation data and calculated thermodynamic functions for sulfur trioxide. Rossini, Wagman, Evans, Levine, and Jaffe (1249) listed melting points and enthalpies of fusion of three crystalline forms: crystal III, $Ttp = 290.0°K$, $\Delta Hm° = 0.47$ kcal/mole; crystal II, $Ttp = 305.7°K$, $\Delta Hm° = 2.47$ kcal/mole; crystal I, $Ttp = 335.4°K$, $\Delta Hm° = 6.09$ kcal/mole. The liquid boils at 316.5°K, and $\Delta Hv = 9.99$ kcal/mole.

CHAPTER 9

CHEMICAL THERMODYNAMICS OF HYDROCARBON COMPOUNDS

HISTORICAL INTRODUCTION

Of the 154 compounds covered in the pioneering monograph entitled *The Free Energies of Some Organic Compounds* by George S. Parks and Hugh M. Huffman (1105), more than a third were hydrocarbons. Their systematic treatment of the thermodynamic properties was based mainly on data determined by the authors and their students, who were aided financially by the Universal Oil Products Company, the American Petroleum Institute, and some petroleum companies.

Early in World War II the need for more extensive and precise thermodynamic data on petroleum constituents became urgent. Consequently in 1942 the American Petroleum Institute underwrote A.P.I. Project 44, a thorough, definitive, painstaking study of the thermodynamic properties of hydrocarbons at the National Bureau of Standards under the direction of Frederick D. Rossini. This continuing study has been a tripartite, cooperative effort among experimentalists providing accurate data, theoreticians calculating thermodynamic properties statistically from first principles, and the petroleum industry financing the work and providing well-characterized, high-purity samples for the program. Both the project and the director transferred to Carnegie Institute of Technology in 1950 and remained there for the next decade. In September 1960 Professor Rossini resigned to devote his full attention to the field of education, at which time Bruno J. Zwolinski became Director. In April 1961 the new director and A.P.I. Project 44 were moved to the Chemical Thermodynamic Properties Center of the A & M College of Texas*. During its history, A.P.I. Project 44 has benefited from the endeavors of a sizable group of capable workers. The reader is referred to the official record

---

*B. J. Zwolinski, Department of Chemistry, Agricultural and Mechanical College of Texas, College Station.

(1248) and supplementary looseleaf sheets (1653) distributed semi-annually for the details of personnel, lists of original publications describing the work accomplished, and the complete bibliography of hydrocarbon thermodynamics, as well as the selected values of the physical and thermodynamic properties of hydrocarbons. Many of the values tabulated in this chapter are based on this compilation.

In the same era a Petroleum Thermodynamics Laboratory was established in 1943 at the Petroleum Research Center of the U.S. Bureau of Mines at Bartlesville, Okla. This program for the determination of basic chemical thermodynamic data on petroleum and related compounds has continued and been expanded to provide data for the burgeoning postwar economy. The work of this thermodynamics laboratory was initiated and directed by Hugh M. Huffman until his untimely death in January 1950. Since that time this productive laboratory has been directed successively by Guy Waddington, John P. McCullough, Donald W. Scott, and Donald R. Douslin. Because of the pre-eminence of its thermal measurements on hydrocarbon materials, one finds it instructive to review their present objectives and the plan of their systematic research program. Since neither the determination of reliable experimental thermodynamic data nor the calculation of such data by theoretical methods for the multiplicity of organic substances of technical importance is presently feasible, a systematic "ideal" program as outlined below is employed by the Petroleum Thermodynamics Laboratory:

"1. Select an important class of substances for which information is needed.

2. Select for individual study 'key' members of each homologous series of the class.

3. Determine as completely and accurately as possible experimental thermodynamic data for each 'key' substance in the solid, liquid, and vapor states and over wide ranges of temperature and pressure.

4. Correlate and interpret experimental results for all 'key' compounds.

5. By the methods of statistical mechanics, calculate thermodynamic functions over ranges of variables inaccessible to experimentation.

6. Apply semitheoretical methods to calculate thermodynamic properties for *all* important members of each homologous series" (736).

A complete index to the definitive work of this laboratory published in scientific journals since 1943 has recently been issued (947).

In 1940 Kenneth S. Pitzer (1150, 1153) developed semiempirical statistical methods for evaluating the thermodynamic properties of the straight-chain paraffins and related hydrocarbons. In the following 12 years much experimental and theoretical work was accumulated. Willis B. Person and George C. Pimentel (1140) in 1953 re-examined and revised the earlier calculations of Pitzer and refined the parameters. They presented the basic equations for the thermodynamic functions of the normal alkanes and tabulated the calculated values from butane through heptane, together with a table of the increments required to compute values for each member of the alkane series through eicosane. Many other workers have also contributed to our present knowledge of thermodynamic properties of the hydrocarbons.

IMPORTANCE OF THE HYDROCARBONS

It is of prime importance that the thermodynamic properties of the hydrocarbons be considered first, since they form the base on which the properties of the other organic compounds can be correlated. More research endeavor has gone into the hydrocarbons than into all other organic compounds; consequently the foundation of this structure (that is, "the thermodynamics of organic compounds") is fortunately the most extensive and best established section of the whole.

Moreover, in a generic sense also, the hydrocarbons are the root compounds from which all other organic compounds are derived. They are the largest group of related compounds containing a pair of covalently bonded elements. The chemical and physical characteristics resulting from these bonds change in a fairly regular and predictable manner from one member to the next in a homologous series of hydrocarbons. Usually the properties of the first three or four members of such a hydrocarbon series deviate the most from the systematic progression of values established by the higher members of the series. Even so, it is possible to predict with considerable assurance the properties expected of a wide variety of hydrocarbons.

Although available measurements on some of the other types of organic compounds are few, the resolution of a compound's structure into a carbon skeleton and various appendages permits prediction of unknown properties with considerable confidence. The measurement of the thermodynamic properties of a few key alcohols suffices to establish the incremental change in the thermodynamic properties resulting from the substitution of a hydroxyl group for a methyl group on the parent hydrocarbon. Once the incremental change is established, one may apply it to each of the hydrocarbons and, in principle, generate reliable information on a group of alcohols of magnitude commensurate with the number of hydrocarbons characterized.

In like fashion, establishment of the thermodynamic properties of a few key chlorohydrocarbons permits calculation of the incremental change attributable to the replacement of a methyl group on a parent hydrocarbon by a chlorine atom. Application of the appropriate incremental change to each of the hydrocarbons will generate useful thermodynamic information on a corresponding group of chlorohydrocarbons. It is thus possible to extend by many fold the information known for the hydrocarbons by consideration of the effect on thermodynamic properties brought about by substitution of the elements Br, Cl, F, I, N, O, and S to the hydrocarbons either separately or in combination.

REGULARITY OF PROPERTIES

Regularities in the enthalpy of formation are well demonstrated by the study of Prosen, Johnson, and Rossini (1196), which established the deviation from linearity in the enthalpies of formation for several homologous hydrocarbon series, with number of carbon atoms, m, in the normal alkyl radical. Their results are expressed by the relation

$$\Delta H f^\circ_{298}[g, Y-(CH_2)_m-H] = A + Bm + \delta \quad \text{kcal/mole} \qquad [9\text{-}1]$$

in which $-(CH_2)_m-H$ is a *normal* alkyl radical (methyl, ethyl, propyl, and so on) attached to any end group Y (methyl, vinyl, phenyl, cyclopentyl, cyclohexyl, and so on); A is a constant peculiar to the end group Y; B, the increment per $-CH_2-$ group, is a constant for all *normal* alkyl series, independent of Y; δ is a term that has a small finite value for the lower members, being largest for $m = 0$ and becoming zero for the higher members beyond $m = 4$, as shown graphically in Figure 9-1.

Table 9-1 gives numerical values of the constants A and B and for the term δ in [9-1] for the lower members of the five series. The uncertainty assigned to the value of δ in each case includes the uncertainty associated with the extrapolation of the linear part of [9-1] to lower values of m. Values of δ for $m = 0$ are characteristic of the end group in each series and can be seen to vary regularly with changes in the number of hydrogen atoms and the number and kind of carbon atoms bonded to the main or attaching carbon atom of the end group. The value of B represents the increment in the enthalpy of formation from one member to the next in each series. In other words, insertion of a $-CH_2-$ group into an alkyl radical (thereby lengthening the alkyl radical by 1 carbon atom) always makes the enthalpy of formation more negative by 4.926 kcal/mole.

Regularities in the gaseous thermodynamic functions for a homologous

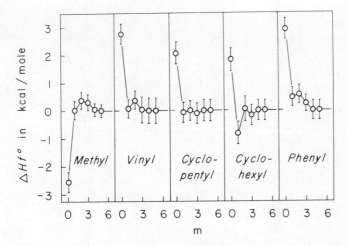

Figure 9-1 Plot of the deviations from linearity with number of carbon atoms, m, in the normal alkyl radical of the enthalpies of formation of the lower members of several homologous series of hydrocarbons, expressed as the value of δ in [9-1]

series were well established by Taylor, Wagman, Williams, Pitzer, and Rossini (1477) in their study of the alkylbenzene series. Rigorous calculations of the temperature-dependent thermodynamic functions (such as enthalpy function, Gibbs energy function, entropy, enthalpy, and heat capacity) were made for nine of the simpler alkylbenzene members: benzene, toluene, m-, o-, and p-xylene, ethylbenzene, 1,2,3- and 1,2,4-trimethylbenzene, and mesitylene. This computation included (a) the translational and free rotational contributions, (b) the vibrational contributions, and (c) restricted rotational contributions where applicable. In the instances of 1,2,3- and 1,2,4-trimethylbenzenes, however, the vibrational contributions were estimated empirically from the values calculated rigorously for mesitylene and for the three xylenes. On the assumption that positional effects were additive, the empirical relations for the vibrational contributions were

$$(1,2,3\text{-trimethylbenzene}) = (\text{mesitylene}) + 2[(o\text{-xylene}) - (m\text{-xylene})] \quad [9\text{-}2]$$

and

$$(1,2,4\text{-trimethylbenzene}) = (\text{mesitylene}) + [(o\text{-xylene}) + (p\text{-xylene}) - 2(m\text{-xylene})]. \quad [9\text{-}3]$$

Table 9-1 Values of the Deviations of the Increment δ from Linearity with Number of Carbon Atoms *m* in the *normal* Alkyl Radical[a]

Series	Structure	A	B per $-CH_2-$ group	δ(kcal/mole)				
				$m=0$	$m=1$	$m=2$	$m=3$	$m=4$	$m=5$
Alkanes	Methyl$-(CH_2)_m-$H	−15.334	−4.926	−2.55 ± 0.37	0.02 ± 0.33	0.37 ± 0.30	0.30 ± 0.28	0.04 ± 0.23	0.00 ± 0.23
(1-Alkenes)	Vinyl$-(CH_2)_m-$H	9.740	−4.926	2.76 ± 0.37	0.07 ± 0.33	0.39 ± 0.33	0.04 ± 0.48	0.00 ± 0.47	0.00 ± 0.47
Alkylbenzenes	Phenyl$-(CH_2)_m-$H	16.404	−4.926	3.42 ± 0.38	0.47 ± 0.34	0.57 ± 0.33	0.24 ± 0.30	0.00 ± 0.35	0.00 ± 0.35
Alkylcyclopentanes	Cyclopentyl$-(CH_2)_m-$H	−20.516	−4.926	2.05 ± 0.40	−0.06 ± 0.37	−0.01 ± 0.36	−0.10 ± 0.37	0.00 ± 0.39	0.00 ± 0.39
Alkylcyclohexanes	Cyclohexyl$-(CH_2)_m-$H	−31.246	−4.926	1.82 ± 0.41	−0.82 ± 0.40	0.05 ± 0.45	−0.18 ± 0.37	0.00 ± 0.37	0.00 ± 0.37

[a] These values are based on enthalpies of formation of the lower members of several homologous series of hydrocarbons in the gaseous state at 25°C and the relationship $\Delta H f° = A + Bm + δ$; Prosen, Johnson, and Rossini (1196).

Regularity of Properties

From the thermodynamic functions of the nine members of this series calculated as described above, thermodynamic functions for 18 additional members of this series were estimated by the method of increments as follows:

For the three methylethylbenzenes the functions were estimated from those for appropriate lower alkylbenzenes and alkanes by the formula

$$x\text{-ethyltoluene} = x\text{-xylene} + \text{ethylbenzene} - \text{toluene} + R \ln 2,$$
[9-4]

in which $x\text{-} = m\text{-}, o\text{-},$ and $p\text{-}$. For the two propylbenzenes,

$$\text{propylbenzene} = \text{ethylbenzene} + \text{butane} - \text{propane}$$
[9-5]

and

$$\text{isopropylbenzene} = \text{ethylbenzene} + 2\text{-methylpropane} - \text{propane} + R \ln(3/2).$$
[9-6]

The 13 alkylbenzenes from butylbenzene ($C_{10}H_{14}$) to hexadecylbenzene ($C_{22}H_{38}$) were calculated by

$$\text{alkylbenzene } (C_n H_{2n-6}) = \text{ethylbenzene} + \text{alkane } (C_{n-5}H_{2n-8}) - \text{propane} \,(n \geq 9).$$
[9-7]

When $n = 9$, this reduces to [9-5] for propylbenzene (C_9H_{12}).

The logarithmic terms correct for the discrepancies in total symmetry numbers for over-all and internal rotation in the formulas as written. They are to be included in the calculation of the entropy and the negative of the Gibbs energy function, $-(G_T^\circ - H_{298}^\circ)/T$, but omitted from the calculation of the enthalpy and the heat capacity.

Some idea of the accuracy of the values for propylbenzene may be gained by comparison of the directly calculated values for ethylbenzene with those calculated by

$$\text{ethylbenzene} = \text{toluene} + \text{propane} - \text{ethane}.$$
[9-8]

The values for the thermodynamic functions for these and other alkylbenzenes may be found elsewhere in this book. Table 9-2 lists the increments for each thermodynamic function that results from insertion of a $-CH_2-$ group into an alkylbenzene molecule beyond pentylbenzene ($C_{11}H_{16}$).

Allen (10) deduced one of the most accurate methods of correlating and predicting enthalpies of formation of hydrocarbons using thermo-

Table 9-2 Values of the $-CH_2-$ Increments Applicable for the Gaseous Alkylbenzenes beyond Pentylbenzene ($C_{11}H_{16}$)

T (°K)	$Cp°$ [a]	$S°_T$ [a]	$-(G°_T - H°_{298})/T$ [a]	$H°_T - H°_{298}$ [b]	$\Delta Hf°$ [b]	$\Delta Gf°$ [b]
298	5.466	9.309	−2.649	0.000	−4.926	2.009
300	5.494	9.343	−2.687	0.010	−4.932	2.053
400	6.941	11.124	−4.577	0.632	−5.252	4.431
500	8.246	12.817	−6.056	1.394	−5.507	6.882
600	9.342	14.409	−7.317	2.270	−5.708	9.382
700	10.276	15.918	−8.437	3.250	−5.854	11.910
800	11.065	17.348	−9.464	4.321	−5.948	14.451
900	11.746	18.692	−10.612	5.455	−6.002	17.009
1000	12.33	19.962	−11.306	6.670	−6.02	19.57

[a] In calories per (mole degrees Kelvin).
[b] In kilocalories per mole.

chemical bond energies. The enthalpies of atomization, $\Delta Ha°$, of alkanes, for example, are given by the expression

$$\Delta Ha°_{298} = N_{CH}E_{CH} + N_{CC}E_{CC} + X\alpha_{CCC} - T\beta_{CCC} - SA, \qquad [9\text{-}9]$$

in which N_{CH} and N_{CC} are the number of C−H and C−C bonds in the molecule; E_{CH} and E_{CC} are the C−H and C−C thermochemical bond energies; α_{CCC} is the interaction energy for a pair of next-nearest-neighbor carbon atoms joined to a carbon atom; X is the number of such pairs; β_{CCC} is a trigonal interaction energy involving three carbon atoms, each of which is a next-nearest-neighbor of the other two; T is the number of trigonal interactions; A is the *gauche* butane interaction energy; and S is the number of such interactions.

This method is based on the assumption that the enthalpy of formation of an alkane hydrocarbon may be computed by addition of the appropriate number of constant-valued methylene increments and corresponding interaction energy terms to the enthalpy of formation of methane. Thus, for gaseous C_nH_{2n+2} hydrocarbons,

$$\Delta Hf°_{298} = \Delta Hf°(CH_4, g) + (n-1)[\Delta Hf°(C_2H_6, g) - \Delta Hf°(CH_4, g)]$$
$$- X\alpha_{CCC} - T\beta_{CCC} + SA. \qquad [9\text{-}10]$$

Substitution of the appropriate values gives

$$\Delta Hf°(\text{alkane}, g) = -15.54 - 2.35n - 2.58X + 0.65T + 0.5S. \qquad [9\text{-}11]$$

Equivalent reasoning for gaseous cyclanes, C_nH_{2n}, gives

$$\Delta Hf°_{298} = -n[\Delta Hf°(C_2H_6, g) - \Delta Hf°(CH_4, g)] - X\alpha_{CCC} + SA$$
$$+ T\beta_{CCC} + E_s, \quad [9\text{-}12]$$

in which E_s is the strain energy of the cyclic molecule. Substitution of the appropriate values gives

$$\Delta Hf°_{298}(\text{cyclane}, g) = -2.35n - 2.58X + 0.5S + 0.65T + E_s. \quad [9\text{-}13]$$

ALKANE IDEAL GAS TABLES

No. 101 Methane CH_4 (Ideal Gas State) Mol Wt 16.042

$T°K$	$Cp°$	$S°$	$-(G°-H°_{298})/T$	$H°-H°_{298}$	$\Delta Hf°$	$\Delta Gf°$	Log Kp
	cal/(mole °K)			kcal/mole			
298	8.54	44.52	44.52	0.00	-17.89	-12.15	8.903
300	8.55	44.58	44.53	0.02	-17.90	-12.11	8.823
400	9.71	47.19	44.87	0.93	-18.63	-10.07	5.502
500	11.12	49.51	45.57	1.97	-19.30	-7.85	3.432
600	12.55	51.66	46.41	3.16	-19.90	-5.51	2.005
700	13.91	53.70	47.31	4.48	-20.40	-3.06	0.957
800	15.18	55.64	48.23	5.94	-20.82	-0.56	0.154
900	16.34	57.50	49.16	7.51	-21.15	1.99	-0.484
1000	17.40	59.28	50.08	9.20	-21.40	4.58	-1.001

Colwell, Gill, and Morrison (257) made a careful study of the low-temperature heat capacity and thermal transitions, finding $Ttp = 90.67°K$ and $\Delta Hm° = 0.222$ kcal/mole. They cited previous work in good agreement. Hestermans and White (593) measured the vapor pressure, enthalpy of vaporization, and liquid heat capacity from the boiling point to the critical temperature. They reported $Tb = 111.42°K$, where $\Delta Hv = 1.953$ kcal/mole, $Tc = 190.55°K$, and $Pc = 45.41$ atm. Earlier published results were compared with their values.

Thermodynamic functions adopted here are those calculated by McDowell and Kruse (972) and include corrections for anharmonicity. Pitzer (1155) computed thermodynamic functions for the rigid rotator, harmonic oscillator approximation. Colwell, Gill, and Morrison (257) derived a calorimetric entropy of $S°_{298}(g) = 45.09$ cal/(mole °K), which is 0.57 cal/(mole °K) higher than the "practical" spectroscopic value. They related the difference to the nuclear spin of the hydrogen atoms.

The enthalpy of combustion of methane was accurately determined by Rossini (1239). His value was corrected for changes in atomic weights and converted to zero pressure by Prosen and Rossini (1203), who derived $\Delta Hf^\circ_{298}(g) = -17.889$ kcal/mole.

Douslin, Harrison, Moore, and McCullough (340) reported *PVT* data for methane over a wide range of temperatures and pressures.

No. 102 Ethane C_2H_6 (Ideal Gas State) Mol Wt 30.068

$T^\circ K$	Cp°	S°	$-(G^\circ-H^\circ_{298})/T$	$H^\circ-H^\circ_{298}$	ΔHf°	ΔGf°	Log Kp
	cal/(mole °K)				kcal/mole		
298	12.58	54.85	54.85	0.00	-20.24	-7.87	5.765
300	12.64	54.93	54.86	0.03	-20.26	-7.79	5.675
400	15.68	58.99	55.39	1.44	-21.42	-3.45	1.886
500	18.66	62.81	56.49	3.16	-22.44	1.16	-0.508
600	21.35	66.46	57.85	5.17	-23.29	5.96	-2.171
700	23.72	69.93	59.33	7.42	-23.99	10.90	-3.402
800	25.83	73.24	60.86	9.90	-24.54	15.91	-4.347
900	27.69	76.39	62.42	12.58	-24.97	21.00	-5.100
1000	29.33	79.39	63.96	15.43	-25.28	26.13	-5.711

Prosen and Rossini (1203) applied small corrections to the accurate enthalpy of combustion reported by Rossini (1241) and calculated $\Delta Hf^\circ_{298}(g) = -20.236$ kcal/mole.

Pitzer (1157) reviewed published work on the third-law entropy and vapor heat capacity and derived a value of 2.875 ± 0.125 kcal/mole for the barrier to internal rotation. Lide (866) found 3.03 ± 0.30 kcal/mole for the barrier from spectroscopic information. Thermodynamic functions based on Pitzer's work are presented by Rossini, Pitzer, Arnett, Braun, and Pimentel (1248), who also selected $Tm = 89.88°$K., $\Delta Hm^\circ = 0.683$ kcal/mole, and $Tb = 184.52°$K, where $\Delta Hv = 3.517$ kcal/mole. Kobe and Lynn (780) reviewed work on the critical constants and selected as best the values of Beattie, Su, and Simard (88), $Tc = 305.4°$K and $Pc = 48.2$ atm.

No. 103 Propane C_3H_8 (Ideal Gas State) Mol Wt 44.094

$T^\circ K$	Cp°	S°	$-(G^\circ-H^\circ_{298})/T$	$H^\circ-H^\circ_{298}$	ΔHf°	ΔGf°	Log Kp
	cal/(mole °K)				kcal/mole		
298	17.57	64.51	64.51	0.00	-24.82	-5.61	4.115
300	17.66	64.62	64.52	0.04	-24.85	-5.50	4.004
400	22.54	70.38	65.27	2.05	-26.36	1.19	-0.651
500	27.02	75.90	66.85	4.53	-27.62	8.23	-3.596
600	30.88	81.18	68.80	7.43	-28.66	15.50	-5.644
700	34.20	86.20	70.93	10.69	-29.48	22.93	-7.157
800	37.08	90.95	73.14	14.26	-30.11	30.45	-8.318
900	39.61	95.47	75.37	18.10	-30.58	38.05	-9.240
1000	41.83	99.76	77.60	22.17	-30.89	45.71	-9.988

The values selected by Rossini, Pitzer, Arnett, Braun, and Pimentel (1248) are adopted. $Tm = 85.46°K$, $\Delta Hm° = 0.842$ kcal/mole, and $Tb = 231.08°K$, where $\Delta Hv = 4.487$ kcal/mole. Kobe and Lynn (780) chose $Tc = 379.9°K$, $Pc = 42.0$ atm, and $dc = 0.220$ g/cc.

No. 104 Butane C_4H_{10} (Ideal Gas State) Mol Wt 58.120

T°K	cal/(mole °K)				kcal/mole		Log Kp
	$Cp°$	$S°$	$-(G°-H°_{298})/T$	$H°-H°_{298}$	$\Delta Hf°$	$\Delta Gf°$	
298	23.29	74.12	74.12	0.00	-30.15	-4.10	3.004
300	23.40	74.27	74.13	0.05	-30.19	-3.94	2.869
400	29.60	81.86	75.12	2.70	-31.99	5.10	-2.784
500	35.34	89.10	77.20	5.96	-33.51	14.55	-6.358
600	40.30	95.99	79.76	9.74	-34.73	24.27	-8.841
700	44.55	102.53	82.55	13.99	-35.68	34.19	-10.674
800	48.23	108.72	85.44	18.63	-36.41	44.21	-12.078
900	51.44	114.59	88.35	23.62	-36.93	54.33	-13.193
1000	54.22	120.16	91.26	28.91	-37.25	64.50	-14.095

The values chosen by Rossini, Pitzer, Arnett, Braun, and Pimentel (1248) are adopted. $Tt = 107.55°K$, $\Delta Ht° = 0.494$ kcal/mole, $Tm = 134.81°K$, $\Delta Hm° = 1.114$ kcal/mole, and $Tb = 272.65°K$, where $\Delta Hv = 5.352$ kcal/mole. Kobe and Lynn (780) selected $Tc = 425.1°K$, $Pc = 37.5$ atm., and $dc = 0.228$ g/cc.

No. 105 2-Methylpropane (Isobutane) C_4H_{10} (Ideal Gas State) Mol Wt 58.120

T°K	cal/(mole °K)				kcal/mole		Log Kp
	$Cp°$	$S°$	$-(G°-H°_{298})/T$	$H°-H°_{298}$	$\Delta Hf°$	$\Delta Gf°$	
298	23.14	70.42	70.42	0.00	-32.15	-4.99	3.661
300	23.25	70.57	70.43	0.05	-32.19	-4.83	3.518
400	29.77	78.17	71.42	2.70	-33.99	4.58	-2.500
500	35.62	85.45	73.50	5.98	-35.48	14.39	-6.291
600	40.62	92.40	76.08	9.80	-36.67	24.48	-8.917
700	44.85	98.99	78.88	14.08	-37.60	34.75	-10.850
800	48.49	105.22	81.79	18.75	-38.30	45.13	-12.329
900	51.65	111.12	84.72	23.76	-38.79	55.60	-13.500
1000	54.40	116.71	87.65	29.07	-39.09	66.11	-14.448

The values adopted are those selected by Rossini, Pitzer, Arnett, Braun, and Pimentel (1248). $Tm = 113.56°K$, $\Delta Hm° = 1.085$ kcal/mole, and $Tb = 261.42°K$, where $\Delta Hv = 5.089$ kcal/mole. Kobe and Lynn (780) reported $Tc = 408.0°K$, $Pc = 36.0$ atm, and $dc = 0.228$ g/cc.

No. 106 Pentane C_5H_{12} (Ideal Gas State) Mol Wt 72.146

$T°K$	$Cp°$	$S°$	$-(G°-H°_{298})/T$	$H°-H°_{298}$	$\Delta Hf°$	$\Delta Gf°$	Log Kp
		cal/(mole °K)			kcal/mole		
298	28.73	83.40	83.40	0.00	-35.00	-2.00	1.468
300	28.87	83.58	83.41	0.06	-35.04	-1.80	1.313
400	36.53	92.95	84.63	3.33	-37.17	9.61	-5.253
500	43.58	101.88	87.20	7.35	-38.94	21.52	-9.406
600	49.64	110.37	90.36	12.02	-40.36	33.75	-12.291
700	54.83	118.42	93.80	17.24	-41.46	46.19	-14.422
800	59.30	126.04	97.36	22.96	-42.28	58.76	-16.053
900	63.18	133.26	100.95	29.08	-42.85	71.44	-17.347
1000	66.55	140.09	104.52	35.58	-43.20	84.17	-18.394

The values adopted are taken from the selected values of Rossini, Pitzer, Arnett, Braun, and Pimentel (1248) and the measurements of Pitzer (1154) and of Messerly and Kennedy (992). $Tm = 143.43°K$, $\Delta Hm° = 2.006$ kcal/mole, and $Tb = 309.22°K$, where $\Delta Hv = 6.160$ kcal/mole. Partington, Rowlinson, and Weston (1123) observed $Tc = 469.5°K$, and Kobe and Lynn (780) reported $Pc = 33.3$ atm and $dc = 0.232$ g/cc.

No. 107 2-Methylbutane (Isopentane) C_5H_{12} (Ideal Gas State) Mol Wt 72.146

$T°K$	$Cp°$	$S°$	$-(G°-H°_{298})/T$	$H°-H°_{298}$	$\Delta Hf°$	$\Delta Gf°$	Log Kp
		cal/(mole °K)			kcal/mole		
298	28.39	82.12	82.12	0.00	-36.92	-3.54	2.596
300	28.54	82.30	82.13	0.06	-36.97	-3.34	2.431
400	36.49	91.62	83.34	3.32	-39.11	8.21	-4.486
500	43.71	100.56	85.90	7.33	-40.87	20.25	-8.851
600	49.89	109.08	89.06	12.02	-42.27	32.61	-11.876
700	55.19	117.18	92.50	17.28	-43.34	45.18	-14.105
800	59.71	124.85	96.07	23.03	-44.13	57.87	-15.809
900	63.66	132.12	99.68	29.20	-44.65	70.66	-17.159
1000	67.12	139.01	103.27	35.75	-44.94	83.50	-18.249

The measurements of Guthrie and Huffman (550) and the selected values of Rossini, Pitzer, Arnett, Braun, and Pimentel (1248) are used. $Tm = 113.26°K$, $\Delta Hm° = 1.231$ kcal/mole, and $Tb = 301.00°K$, where $\Delta Hv = 5.901$ kcal/mole. Vohra and Kobe (1544) measured $Tc = 460.9°K$, and Kobe and Lynn (780) reported $Pc = 32.9$ atm and $dc = 0.234$ g/cc.

No. 108 2,2-Dimethylpropane (Neopentane) C_5H_{12} (Ideal Gas State) Mol Wt 72.146

The measurements of Aston and Messerly (39a) and the selected values of Rossini, Pitzer, Arnett, Braun, and Pimentel (1248) are adopted. $Tt = 140.01°K$, $\Delta Ht° = 0.616$ kcal/mole, $Tm = 256.60°K$, $\Delta Hm° = 0.778$

No. 108 2,2-Dimethylpropane (Neopentane) C_5H_{12} (Ideal Gas State) Mol Wt 72.146

$T°K$	$Cp°$	cal/(mole °K) $S°$	$-(G°-H°_{298})/T$	$H°-H°_{298}$	kcal/mole $\Delta Hf°$	$\Delta Gf°$	Log Kp
298	29.07	73.23	73.23	0.00	-39.67	-3.64	2.669
300	29.21	73.42	73.24	0.06	-39.71	-3.42	2.492
400	37.55	82.98	74.48	3.40	-41.77	9.00	-4.919
500	45.00	92.18	77.11	7.54	-43.41	21.89	-9.569
600	51.21	100.95	80.36	12.36	-44.68	35.07	-12.775
700	56.40	109.25	83.90	17.75	-45.62	48.45	-15.127
800	60.78	117.07	87.56	23.61	-46.30	61.93	-16.917
900	64.55	124.45	91.25	29.88	-46.73	75.49	-18.331
1000	67.80	131.43	94.93	36.51	-46.94	89.10	-19.471

kcal/mole, and $Tb = 282.65°K$, where $\Delta Hv = 5.438$ kcal/mole. Partington, Rowlinson, and Weston (1123) observed $Tc = 433.7°K$, and Kobe and Lynn (780) reported $Pc = 31.6$ atm and $dc = 0.238$ g/cc.

No. 109 Hexane C_6H_{14} (Ideal Gas State) Mol Wt 86.172

$T°K$	$Cp°$	cal/(mole °K) $S°$	$-(G°-H°_{298})/T$	$H°-H°_{298}$	kcal/mole $\Delta Hf°$	$\Delta Gf°$	Log Kp
298	34.20	92.83	92.83	0.00	-39.96	-0.06	0.047
300	34.37	93.05	92.84	0.07	-40.01	0.18	-0.131
400	43.47	104.20	94.29	3.97	-42.45	13.96	-7.629
500	51.83	114.82	97.35	8.74	-44.48	28.31	-12.373
600	58.99	124.92	101.11	14.29	-46.10	43.02	-15.669
700	65.10	134.48	105.20	20.50	-47.34	57.98	-18.101
800	70.36	143.52	109.43	27.28	-48.26	73.08	-19.963
900	74.93	152.08	113.70	34.55	-48.89	88.30	-21.440
1000	78.89	160.19	117.94	42.25	-49.25	103.57	-22.635

The values presented are those selected by Rossini, Pitzer, Arnett, Braun, and Pimentel (1248). Kobe and Lynn (780) reported $Tc = 507.8°K$, $Pc = 29.9$ atm, and $dc = 0.234$ g/cc. $Tm = 177.80°K$, $\Delta Hm° = 3.114$ kcal/mole, and $Tb = 341.89°K$, where $\Delta Hv = 6.896$ kcal/mole.

No. 110 2-Methylpentane C_6H_{14} (Ideal Gas State) Mol Wt 86.172

$T°K$	$Cp°$	cal/(mole °K) $S°$	$-(G°-H°_{298})/T$	$H°-H°_{298}$	kcal/mole $\Delta Hf°$	$\Delta Gf°$	Log Kp
298	34.46	90.95	90.95	0.00	-41.66	-1.20	0.882
300	34.63	91.17	90.96	0.07	-41.71	-0.96	0.696
400	44.00	102.44	92.43	4.01	-44.11	13.01	-7.108
500	52.50	113.19	95.51	8.84	-46.08	27.53	-12.031
600	59.60	123.41	99.32	14.46	-47.63	42.39	-15.440
700	65.70	133.07	103.46	20.73	-48.81	57.50	-17.951
800	70.80	142.18	107.73	27.56	-49.68	72.74	-19.870
900	75.30	150.78	112.04	34.87	-50.27	88.09	-21.389
1000	79.20	158.92	116.33	42.60	-50.60	103.49	-22.616

The values employed are those selected by Rossini, Pitzer, Arnett, Braun, and Pimentel (1248). Kobe and Lynn (780) listed $Tc = 497.8°K$, $Pc = 29.9$ atm, and $dc = 0.235$ g/cc. $Tm = 119.49°K$, $\Delta Hm° = 1.500$ kcal/mole, and $Tb = 333.42°K$, where $\Delta Hv = 6.643$ kcal/mole.

No. 111 3-Methylpentane C_6H_{14} (Ideal Gas State) Mol Wt 86.172

$T°K$	$Cp°$	$S°$	$-(G°-H°_{298})/T$	$H°-H°_{298}$	$\Delta Hf°$	$\Delta Gf°$	Log Kp
		cal/(mole °K)			kcal/mole		
298	34.20	90.77	90.77	0.00	-41.02	-0.51	0.373
300	34.37	90.99	90.78	0.07	-41.07	-0.26	0.191
400	43.47	102.14	92.23	3.97	-43.51	13.73	-7.500
500	51.83	112.76	95.29	8.74	-45.54	28.28	-12.360
600	59.00	122.86	99.05	14.29	-47.16	43.19	-15.733
700	65.10	132.42	103.14	20.50	-48.40	58.36	-18.221
800	70.40	141.47	107.37	27.29	-49.32	73.67	-20.124
900	74.90	150.03	111.64	34.55	-49.95	89.09	-21.633
1000	78.90	158.13	115.89	42.25	-50.31	104.57	-22.853

Rossini, Pitzer, Arnett, Braun, and Pimentel (1248) selected the values employed here. Kobe and Lynn (780) listed $Tc = 504.6°K$, $Pc = 30.8$ atm, and $dc = 0.235$ g/cc. $Tb = 336.43°K$, where $\Delta Hv = 6.711$ kcal/mole.

No. 112 2,2-Dimethylbutane C_6H_{14} (Ideal Gas State) Mol Wt 86.172

$T°K$	$Cp°$	$S°$	$-(G°-H°_{298})/T$	$H°-H°_{298}$	$\Delta Hf°$	$\Delta Gf°$	Log Kp
		cal/(mole °K)			kcal/mole		
298	33.91	85.62	85.62	0.00	-44.35	-2.30	1.689
300	34.09	85.84	85.63	0.07	-44.40	-2.05	1.491
400	43.70	96.98	87.08	3.96	-46.85	12.46	-6.807
500	52.50	107.70	90.14	8.78	-48.83	27.52	-12.030
600	60.00	117.95	93.93	14.42	-50.36	42.94	-15.639
700	66.10	127.67	98.06	20.73	-51.50	58.59	-18.291
800	71.40	136.86	102.34	27.62	-52.32	74.36	-20.313
900	75.60	145.51	106.66	34.97	-52.86	90.24	-21.912
1000	79.70	153.69	110.96	42.73	-53.15	106.17	-23.202

The low-temperature measurements of Kilpatrick and Pitzer (750) and the values selected by Rossini, Pitzer, Arnett, Braun, and Pimentel (1248) are adopted. $Tt = 125.81°K$, $\Delta Ht° = 1.289$ kcal/mole, $Tt = 140.87°K$, $\Delta Ht° = 0.068$ kcal/mole, $Tm = 173.28°K$, $\Delta Hm° = 0.138$ kcal/mole, and $Tb = 322.89°K$, where $\Delta Hv = 6.289$ kcal/mole. Kobe and Lynn (780) presented $Tc = 489.3°K$, $Pc = 30.7$ atm, and $dc = 0.240$ g/cc.

No. 113 2,3-Dimethylbutane C_6H_{14} (Ideal Gas State) Mol Wt 86.172

Rossini, Pitzer, Arnett, Braun, and Pimentel (1248) selected the values adopted here. $Tm = 144.61°K$, $\Delta Hm° = 0.194$ kcal/mole, and $Tb = $

No. 113 2,3-Dimethylbutane C_6H_{14} (Ideal Gas State) Mol Wt 86.172

$T°K$	$Cp°$	$S°$	$-(G°-H°_{298})/T$	$H°-H°_{298}$	$\Delta Hf°$	$\Delta Gf°$	Log Kp
		cal/(mole °K)			kcal/mole		
298	33.59	87.42	87.42	0.00	-42.49	-0.98	0.719
300	33.76	87.63	87.43	0.07	-42.54	-0.73	0.529
400	43.30	98.67	88.87	3.93	-45.02	13.60	-7.433
500	51.90	109.28	91.90	8.70	-47.05	28.50	-12.458
600	59.20	119.41	95.65	14.26	-48.66	43.77	-15.941
700	65.40	129.01	99.73	20.50	-49.87	59.28	-18.506
800	70.70	138.10	103.97	27.31	-50.76	74.92	-20.466
900	75.20	146.69	108.24	34.61	-51.36	90.68	-22.018
1000	79.10	154.82	112.50	42.33	-51.69	106.49	-23.273

331.14°K, where $\Delta Hv = 6.519$ kcal/mole. Kobe and Lynn (780) reported $Tc = 500.2°K$, $Pc = 30.9$ atm, and $dc = 0.241$ g/cc.

No. 114 Heptane C_7H_{16} (Ideal Gas State) Mol Wt 100.198

$T°K$	$Cp°$	$S°$	$-(G°-H°_{298})/T$	$H°-H°_{298}$	$\Delta Hf°$	$\Delta Gf°$	Log Kp
		cal/(mole °K)			kcal/mole		
298	39.67	102.27	102.27	0.00	-44.88	1.91	-1.402
300	39.86	102.52	102.28	0.08	-44.94	2.20	-1.602
400	50.42	115.46	103.97	4.60	-47.70	18.35	-10.025
500	60.07	127.77	107.51	10.14	-49.98	35.13	-15.355
600	68.33	139.47	111.87	16.57	-51.80	52.32	-19.058
700	75.38	150.55	116.61	23.76	-53.18	69.80	-21.792
800	81.43	161.01	121.51	31.61	-54.20	87.43	-23.883
900	86.68	170.92	126.46	40.02	-54.88	105.19	-25.541
1000	91.20	180.29	131.38	48.92	-55.26	123.01	-26.882

The measurements of Huffman, Gross, Scott, and McCullough (654) are included in the selections made by Rossini, Pitzer, Arnett, Braun, and Pimentel (1248) for the values used. $Tm = 182.54°K$, $\Delta Hm° = 3.354$ kcal/mole, and $Tb = 371.58°K$, where $\Delta Hv = 7.575$ kcal/mole. Kobe and Lynn (780) found $Tc = 540.1°K$, $Pc = 27.0$ atm, and $dc = 0.235$ g/cc.

No. 115 2-Methylhexane C_7H_{16} (Ideal Gas State) Mol Wt 100.198

$T°K$	$Cp°$	$S°$	$-(G°-H°_{298})/T$	$H°-H°_{298}$	$\Delta Hf°$	$\Delta Gf°$	Log Kp
		cal/(mole °K)			kcal/mole		
298	39.67	100.38	100.38	0.00	-46.59	0.77	-0.562
300	39.86	100.63	100.39	0.08	-46.65	1.06	-0.770
400	50.42	113.57	102.08	4.60	-49.41	17.39	-9.503
500	60.07	125.88	105.62	10.14	-51.69	34.37	-15.021
600	68.33	137.58	109.98	16.57	-53.51	51.75	-18.848
700	75.38	148.66	114.72	23.76	-54.89	69.41	-21.671
800	81.43	159.12	119.62	31.61	-55.91	87.23	-23.829
900	86.68	169.03	124.57	40.02	-56.59	105.18	-25.539
1000	91.20	178.40	129.49	48.92	-56.97	123.19	-26.921

250   Chemical Thermodynamics of Hydrocarbon Compounds

The selected values of Rossini, Pitzer, Arnett, Braun, and Pimentel (1248) are consistent with the measurements of Huffman, Gross, Scott, and McCullough (654). $Tm = 154.87°K$, $\Delta Hm° = 2.195$ kcal/mole, and $Tb = 363.20°K$, where $\Delta Hv = 7.329$ kcal/mole. Kobe and Lynn (780) gave $Tc = 531.0°K$, $Pc = 27.2$ atm, and $dc = 0.234$ g/cc.

No. 116   3-Methylhexane C_7H_{16} (Ideal Gas State)   Mol Wt 100.198

	cal/(mole °K)				kcal/mole		
$T°K$	$Cp°$	$S°$	$-(G°-H°_{298})/T$	$H°-H°_{298}$	$\Delta Hf°$	$\Delta Gf°$	Log Kp
298	39.67	101.37	101.37	0.00	-45.96	1.10	-0.807
300	39.86	101.62	101.38	0.08	-46.02	1.39	-1.012
400	50.42	114.56	103.07	4.60	-48.78	17.63	-9.631
500	60.07	126.87	106.61	10.14	-51.06	34.50	-15.080
600	68.33	138.57	110.97	16.57	-52.88	51.78	-18.861
700	75.38	149.65	115.71	23.76	-54.26	69.35	-21.651
800	81.43	160.11	120.61	31.61	-55.28	87.07	-23.785
900	86.68	170.02	125.56	40.02	-55.96	104.92	-25.476
1000	91.20	179.39	130.48	48.92	-56.34	122.83	-26.842

The values selected by Rossini, Pitzer, Arnett, Braun, and Pimentel (1248) are used. $Tb = 365.00°K$, where $\Delta Hv = 7.358$ kcal/mole. Kobe and Lynn (780) indicated $Tc = 535.5°K$, $Pc = 28.1$ atm, and $dc = 0.240$ g/cc.

No. 117   3-Ethylpentane C_7H_{16} (Ideal Gas State)   Mol Wt 100.198

	cal/(mole °K)				kcal/mole		
$T°K$	$Cp°$	$S°$	$-(G°-H°_{298})/T$	$H°-H°_{298}$	$\Delta Hf°$	$\Delta Gf°$	Log Kp
298	39.67	98.35	98.35	0.00	-45.33	2.63	-1.929
300	39.86	98.60	98.36	0.08	-45.39	2.93	-2.131
400	50.42	111.54	100.05	4.60	-48.15	19.47	-10.636
500	60.07	123.85	103.59	10.14	-50.43	36.64	-16.015
600	68.33	135.55	107.95	16.57	-52.25	54.23	-19.751
700	75.38	146.63	112.69	23.76	-53.63	72.09	-22.508
800	81.43	157.09	117.59	31.61	-54.65	90.11	-24.617
900	86.68	167.00	122.54	40.02	-55.33	108.26	-26.289
1000	91.20	176.37	127.46	48.92	-55.71	126.48	-27.640

The values selected by Rossini, Pitzer, Arnett, Braun, and Pimentel (1248) are based in part on the measurements of Huffman, Gross, Scott, and McCullough (654). $Tm = 154.55°K$, $\Delta Hm° = 2.282$ kcal/mole, and $Tb = 366.62°K$, where $\Delta Hv = 7.398$ kcal/mole. Kobe and Lynn (780) found $Tc = 540.7°K$, $Pc = 28.6$ atm, and $dc = 0.241$ g/cc.

Alkane Ideal Gas Tables

No. 118 2,2-Dimethylpentane C_7H_{16} (Ideal Gas State) Mol Wt 100.198

T°K	cal/(mole °K)			kcal/mole			
	$Cp°$	$S°$	$-(G°-H°_{298})/T$	$H°-H°_{298}$	$\Delta Hf°$	$\Delta Gf°$	Log Kp
298	39.67	93.90	93.90	0.00	-49.27	0.02	-0.014
300	39.86	94.15	93.91	0.08	-49.33	0.32	-0.233
400	50.42	107.09	95.60	4.60	-52.09	17.31	-9.455
500	60.07	119.40	99.14	10.14	-54.37	34.93	-15.266
600	68.33	131.10	103.50	16.57	-56.19	52.96	-19.288
700	75.38	142.18	108.24	23.76	-57.57	71.27	-22.250
800	81.43	152.64	113.14	31.61	-58.59	89.73	-24.513
900	86.68	162.55	118.09	40.02	-59.27	108.33	-26.304
1000	91.20	171.92	123.01	48.92	-59.65	126.99	-27.752

The values selected by Rossini, Pitzer, Arnett, Braun, and Pimentel (1248) are based in part on the measurements of Huffman, Gross, Scott, and McCullough (654). $Tm = 149.34°K$, $\Delta Hm° = 1.392$ kcal/mole, and $Tb = 352.35°K$, where $\Delta Hv = 6.969$ kcal/mole. Kobe and Lynn (780) found $Tc = 520.8°K$, $Pc = 28.4$ atm, and $dc = 0.248$ g/cc.

No. 119 2,3-Dimethylpentane C_7H_{16} (Ideal Gas State) Mol Wt 100.198

T°K	cal/(mole °K)			kcal/mole			
	$Cp°$	$S°$	$-(G°-H°_{298})/T$	$H°-H°_{298}$	$\Delta Hf°$	$\Delta Gf°$	Log Kp
298	39.67	98.96	98.96	0.00	-47.62	0.16	-0.117
300	39.86	99.21	98.97	0.08	-47.68	0.45	-0.330
400	50.42	112.15	100.66	4.60	-50.44	16.93	-9.251
500	60.07	124.46	104.20	10.14	-52.72	34.05	-14.881
600	68.33	136.16	108.56	16.57	-54.54	51.57	-18.783
700	75.38	147.24	113.30	23.76	-55.92	69.38	-21.660
800	81.43	157.70	118.20	31.61	-56.94	87.34	-23.858
900	86.68	167.61	123.15	40.02	-57.62	105.42	-25.599
1000	91.20	176.98	128.07	48.92	-58.00	123.58	-27.006

The values adopted are based on the selected values of Rossini, Pitzer, Arnett, Braun, and Pimentel (1248). $Tb = 362.93°K$, where $\Delta Hv = 7.262$ kcal/mole. Kobe and Lynn (780) reported $Tc = 537.7°K$, $Pc = 29.2$ atm, and $dc = 0.247$ g/cc.

No. 120 2,4-Dimethylpentane C_7H_{16} (Ideal Gas State) Mol Wt 100.198

The measurements of Huffman, Gross, Scott, and McCullough (654) are included in the selected values of Rossini, Pitzer, Arnett, Braun, and Pimentel (1248). $Tm = 153.91°K$, $\Delta Hm° = 1.635$ kcal/mole, and $Tb = 353.65°K$, where $\Delta Hv = 7.050$ kcal/mole. Kobe and Lynn (780) indicated $Tc = 520.2°K$, $Pc = 27.4$ atm, and $dc = 0.239$ g/cc.

No. 120 2,4-Dimethylpentane C_7H_{16} (Ideal Gas State) Mol Wt 100.198

$T°K$	$Cp°$	$S°$	$-(G°-H°_{298})/T$	$H°-H°_{298}$	$\Delta Hf°$	$\Delta Gf°$	Log Kp
		cal/(mole °K)			kcal/mole		
298	39.67	94.80	94.80	0.00	-48.28	0.74	-0.543
300	39.86	95.05	94.81	0.08	-48.34	1.04	-0.758
400	50.42	107.99	96.50	4.60	-51.10	17.94	-9.800
500	60.07	120.30	100.04	10.14	-53.38	35.47	-15.502
600	68.33	132.00	104.40	16.57	-55.20	53.41	-19.452
700	75.38	143.08	109.14	23.76	-56.58	71.63	-22.363
800	81.43	153.54	114.04	31.61	-57.60	90.00	-24.587
900	86.68	163.45	118.99	40.02	-58.28	108.51	-26.348
1000	91.20	172.82	123.91	48.92	-58.66	127.08	-27.771

No. 121 3,3-Dimethylpentane C_7H_{16} (Ideal Gas State) Mol Wt 100.198

$T°K$	$Cp°$	$S°$	$-(G°-H°_{298})/T$	$H°-H°_{298}$	$\Delta Hf°$	$\Delta Gf°$	Log Kp
		cal/(mole °K)			kcal/mole		
298	39.67	95.53	95.53	0.00	-48.17	0.63	-0.464
300	39.86	95.78	95.54	0.08	-48.23	0.93	-0.678
400	50.42	108.72	97.23	4.60	-50.99	17.75	-9.700
500	60.07	121.03	100.77	10.14	-53.27	35.21	-15.390
600	68.33	132.73	105.13	16.57	-55.09	53.08	-19.333
700	75.38	143.81	109.87	23.76	-56.47	71.23	-22.237
800	81.43	154.27	114.77	31.61	-57.49	89.53	-24.457
900	86.68	164.18	119.72	40.02	-58.17	107.96	-26.215
1000	91.20	173.55	124.64	48.92	-58.55	126.46	-27.636

The values of Rossini, Pitzer, Arnett, Braun, and Pimentel (1248) are adopted. $Tt = 137.79°K$, $Tt = 138.20°K$, $Tm = 138.69°K$, $\Delta Hm° = 1.689$ kcal/mole, and $Tb = 359.21°K$, where $\Delta Hv = 7.058$ kcal/mole. Kobe and Lynn (780) listed $Tc = 536°K$ and $Pc = 36$ atm.

No. 122 2,2,3-Trimethylbutane C_7H_{16} (Ideal Gas State) Mol Wt 100.198

$T°K$	$Cp°$	$S°$	$-(G°-H°_{298})/T$	$H°-H°_{298}$	$\Delta Hf°$	$\Delta Gf°$	Log Kp
		cal/(mole °K)			kcal/mole		
298	39.33	91.61	91.61	0.00	-48.95	1.02	-0.749
300	39.54	91.86	91.62	0.08	-49.01	1.33	-0.967
400	50.83	104.81	93.30	4.61	-51.76	18.54	-10.131
500	61.04	117.28	96.86	10.21	-53.97	36.39	-15.903
600	69.61	129.18	101.26	16.76	-55.67	54.62	-19.893
700	76.74	140.46	106.07	24.08	-56.92	73.11	-22.825
800	82.73	151.11	111.04	32.06	-57.81	91.74	-25.060
900	87.88	161.16	116.05	40.60	-58.37	110.48	-26.826
1000	92.32	170.66	121.04	49.62	-58.64	129.27	-28.250

Measurements by Huffman, Gross, Scott, and McCullough (654) reveal complex low-temperature behavior. As this material was warmed from 10°K, two nonisothermal transitions were found [$Cp = 39.00$ cal/(mole °K) at 86.8°K and $Cp = 50.13$ cal/(mole °K) at 108.0°K], followed by an isothermal transition, $Tt = 121.4$°K, $\Delta Ht° = 0.586$ kcal/mole, and finally melting at $Ttp = 248.56$°K, $\Delta Hm° = 0.540$ kcal/mole. $Tb = 354.02$°K, where $\Delta Hv = 6.918$ kcal/mole. Kobe and Lynn (780) reported $Tc = 531.4$°K, $Pc = 29.75$ atm, and $dc = 0.254$ g/cc.

No. 123 Octane C_8H_{18} (Ideal Gas State) Mol Wt 114.224

	cal/(mole °K)			kcal/mole			
$T°K$	$Cp°$	$S°$	$-(G°-H°_{298})/T$	$H°-H°_{298}$	$\Delta Hf°$	$\Delta Gf°$	Log Kp
298	45.14	111.55	111.55	0.00	-49.82	3.92	-2.872
300	45.35	111.83	111.56	0.09	-49.89	4.25	-3.093
400	57.36	126.55	113.48	5.23	-52.96	22.78	-12.445
500	68.32	140.56	117.51	11.53	-55.50	42.02	-18.364
600	77.67	153.86	122.47	18.84	-57.51	61.71	-22.475
700	85.66	166.45	127.86	27.02	-59.04	81.71	-25.511
800	92.50	178.34	133.43	35.93	-60.16	101.88	-27.832
900	98.43	189.59	139.05	45.49	-60.90	122.20	-29.672
1000	103.60	200.23	144.64	55.59	-61.29	142.58	-31.160

The values selected by Rossini, Pitzer, Arnett, Braun, and Pimentel (1248) have been adopted. $Tm = 216.36$°K, $\Delta Hm° = 4.957$ kcal/mole, and $Tb = 398.81$°K, where $\Delta Hv = 8.224$ kcal/mole. Ambrose, Cox, and Townsend (13) measured $Tc = 568.5$°K.

No. 124 2-Methylheptane C_8H_{18} (Ideal Gas State) Mol Wt 114.224

	cal/(mole °K)			kcal/mole			
$T°K$	$Cp°$	$S°$	$-(G°-H°_{298})/T$	$H°-H°_{298}$	$\Delta Hf°$	$\Delta Gf°$	Log Kp
298	45.14	108.81	108.81	0.00	-51.50	3.05	-2.239
300	45.35	109.09	108.82	0.09	-51.57	3.39	-2.468
400	57.36	123.81	110.74	5.23	-54.64	22.19	-12.125
500	68.32	137.82	114.77	11.53	-57.18	41.71	-18.228
600	77.67	151.12	119.73	18.84	-59.19	61.67	-22.462
700	85.66	163.71	125.12	27.02	-60.72	81.95	-25.585
800	92.50	175.60	130.69	35.93	-61.84	102.40	-27.972
900	98.43	186.85	136.31	45.49	-62.58	122.98	-29.863
1000	103.60	197.49	141.90	55.59	-62.97	143.64	-31.391

The values presented are those selected by Rossini, Pitzer, Arnett, Braun, and Pimentel (1248). $Tm = 164.11$°K, $\Delta Hm° = 2.45$ kcal/mole, and $Tb = 390.80$°K, where $\Delta Hv = 8.08$ kcal/mole. Ambrose, Cox, and Townsend (13) measured $Tc = 559.5$°K.

No. 125 3-Methylheptane C_8H_{18} (Ideal Gas State) Mol Wt 114.224

$T°K$	$Cp°$	$S°$	$-(G°-H°_{298})/T$	$H°-H°_{298}$	$\Delta Hf°$	$\Delta Gf°$	Log Kp
	cal/(mole °K)				kcal/mole		
298	45.14	110.32	110.32	0.00	-50.82	3.28	-2.407
300	45.35	110.60	110.33	0.09	-50.89	3.62	-2.634
400	57.36	125.32	112.25	5.23	-53.96	22.27	-12.167
500	68.32	139.33	116.28	11.53	-56.50	41.63	-18.196
600	77.67	152.63	121.24	18.84	-58.51	61.44	-22.380
700	85.66	165.22	126.63	27.02	-60.04	81.57	-25.467
800	92.50	177.11	132.20	35.93	-61.16	101.87	-27.828
900	98.43	188.36	137.82	45.49	-61.90	122.31	-29.698
1000	103.60	199.00	143.41	55.59	-62.29	142.81	-31.210

These values were selected by Rossini, Pitzer, Arnett, Braun, and Pimentel (1248). $Tm = 152.65°K$, $\Delta Hm° = 2.72$ kcal/mole, and $Tb = 392.07°K$, where $\Delta Hv = 8.10$ kcal/mole.

No. 126 4-Methylheptane C_8H_{18} (Ideal Gas State) Mol Wt 114.224

$T°K$	$Cp°$	$S°$	$-(G°-H°_{298})/T$	$H°-H°_{298}$	$\Delta Hf°$	$\Delta Gf°$	Log Kp
	cal/(mole °K)				kcal/mole		
298	45.14	108.35	108.35	0.00	-50.69	4.00	-2.933
300	45.35	108.63	108.36	0.09	-50.76	4.34	-3.159
400	57.36	123.35	110.28	5.23	-53.83	23.19	-12.669
500	68.32	137.36	114.31	11.53	-56.37	42.75	-18.683
600	77.67	150.66	119.27	18.84	-58.38	62.76	-22.858
700	85.66	163.25	124.66	27.02	-59.91	83.08	-25.938
800	92.50	175.14	130.23	35.93	-61.03	103.57	-28.294
900	98.43	186.39	135.85	45.49	-61.77	124.21	-30.161
1000	103.60	197.03	141.44	55.59	-62.16	144.91	-31.669

The values selected by Rossini, Pitzer, Arnett, Braun, and Pimentel (1248) are presented. $Tm = 152.20°K$, $\Delta Hm° = 2.59$ kcal/mole, and $Tb = 390.86°K$, where $\Delta Hv = 8.100$ kcal/mole.

No. 127 3-Ethylhexane C_8H_{18} (Ideal Gas State) Mol Wt 114.224

$T°K$	$Cp°$	$S°$	$-(G°-H°_{298})/T$	$H°-H°_{298}$	$\Delta Hf°$	$\Delta Gf°$	Log Kp
	cal/(mole °K)				kcal/mole		
298	45.14	109.51	109.51	0.00	-50.40	3.95	-2.892
300	45.35	109.79	109.52	0.09	-50.47	4.28	-3.117
400	57.36	124.51	111.44	5.23	-53.54	23.01	-12.573
500	68.32	138.52	115.47	11.53	-56.08	42.46	-18.556
600	77.67	151.82	120.43	18.84	-58.09	62.35	-22.710
700	85.66	164.41	125.82	27.02	-59.62	82.56	-25.775
800	92.50	176.30	131.39	35.93	-60.74	102.94	-28.119
900	98.43	187.55	137.01	45.49	-61.48	123.45	-29.977
1000	103.60	198.19	142.60	55.59	-61.87	144.04	-31.479

Alkane Ideal Gas Tables

The values selected by Rossini, Pitzer, Arnett, Braun, and Pimentel (1248) are adopted. $Tb = 391.68°K$, where $\Delta Hv = 8.032$ kcal/mole.

No. 128 2,2-Dimethylhexane C_8H_{18} (Ideal Gas State) Mol Wt 114.224

	cal/(mole °K)			kcal/mole			
$T°K$	$Cp°$	$S°$	$-(G°-H°_{298})/T$	$H°-H°_{298}$	$\Delta Hf°$	$\Delta Gf°$	Log Kp
298	45.14	103.06	103.06	0.00	-53.71	2.56	-1.876
300	45.35	103.34	103.07	0.09	-53.78	2.90	-2.115
400	57.36	118.06	104.99	5.23	-56.85	22.28	-12.175
500	68.32	132.07	109.02	11.53	-59.39	42.37	-18.519
600	77.67	145.37	113.98	18.84	-61.40	62.91	-22.914
700	85.66	157.96	119.37	27.02	-62.93	83.77	-26.152
800	92.50	169.85	124.94	35.93	-64.05	104.79	-28.625
900	98.43	181.10	130.56	45.49	-64.79	125.95	-30.583
1000	103.60	191.74	136.15	55.59	-65.18	147.18	-32.165

Rossini, Pitzer, Arnett, Braun, and Pimentel (1248) selected these values. $Tm = 151.97°K$, $\Delta Hm° = 1.62$ kcal/mole, and $Tb = 379.99°K$, where $\Delta Hv = 7.71$ kcal/mole.

No. 129 2,3-Dimethylhexane C_8H_{18} (Ideal Gas State) Mol Wt 114.224

	cal/(mole °K)			kcal/mole			
$T°K$	$Cp°$	$S°$	$-(G°-H°_{298})/T$	$H°-H°_{298}$	$\Delta Hf°$	$\Delta Gf°$	Log Kp
298	45.14	106.11	106.11	0.00	-51.13	4.23	-3.100
300	45.35	106.39	106.12	0.09	-51.20	4.57	-3.328
400	57.36	121.11	108.04	5.23	-54.27	23.64	-12.918
500	68.32	135.12	112.07	11.53	-56.81	43.43	-18.980
600	77.67	148.42	117.03	18.84	-58.82	63.66	-23.187
700	85.66	161.01	122.42	27.02	-60.35	84.21	-26.290
800	92.50	172.90	127.99	35.93	-61.47	104.93	-28.663
900	98.43	184.15	133.61	45.49	-62.21	125.78	-30.543
1000	103.60	194.79	139.20	55.59	-62.60	146.71	-32.062

The values selected by Rossini, Pitzer, Arnett, Braun, and Pimentel (1248) are employed. $Tb = 388.76°K$, where $\Delta Hv = 7.935$ kcal/mole.

No. 130 2,4-Dimethylhexane C_8H_{18} (Ideal Gas State) Mol Wt 114.224

	cal/(mole °K)			kcal/mole			
$T°K$	$Cp°$	$S°$	$-(G°-H°_{298})/T$	$H°-H°_{298}$	$\Delta Hf°$	$\Delta Gf°$	Log Kp
298	45.14	106.51	106.51	0.00	-52.44	2.80	-2.053
300	45.35	106.79	106.52	0.09	-52.51	3.14	-2.286
400	57.36	121.51	108.44	5.23	-55.58	22.17	-12.115
500	68.32	135.52	112.47	11.53	-58.12	41.92	-18.320
600	77.67	148.82	117.43	18.84	-60.13	62.11	-22.623
700	85.66	161.41	122.82	27.02	-61.66	82.62	-25.794
800	92.50	173.30	128.39	35.93	-62.78	103.30	-28.218
900	98.43	184.55	134.01	45.49	-63.52	124.11	-30.138
1000	103.60	195.19	139.60	55.59	-63.91	145.00	-31.689

Rossini, Pitzer, Arnett, Braun, and Pimentel's (1248) selected values are used. $Tb = 382.58°K$, where $\Delta Hv = 7.79$ kcal/mole.

No. 131 2,5-Dimethylhexane C_8H_{18} (Ideal Gas State) Mol Wt 114.224

	cal/(mole °K)				kcal/mole		
$T°K$	$Cp°$	$S°$	$-(G°-H°_{298})/T$	$H°-H°_{298}$	$\Delta Hf°$	$\Delta Gf°$	Log Kp
298	45.14	104.93	104.93	0.00	-53.21	2.50	-1.833
300	45.35	105.21	104.94	0.09	-53.28	2.84	-2.071
400	57.36	119.93	106.86	5.23	-56.35	22.04	-12.039
500	68.32	133.94	110.89	11.53	-58.89	41.94	-18.329
600	77.67	147.24	115.85	18.84	-60.90	62.29	-22.687
700	85.66	159.83	121.24	27.02	-62.43	82.96	-25.899
800	92.50	171.72	126.81	35.93	-63.55	103.79	-28.353
900	98.43	182.97	132.43	45.49	-64.29	124.77	-30.296
1000	103.60	193.61	138.02	55.59	-64.68	145.81	-31.866

The selected values of Rossini, Pitzer, Arnett, Braun, and Pimentel (1248) are adopted. $Tm = 181.95°K$, $\Delta Hm° = 3.07$ kcal/mole, and $Tb = 382.25°K$, where $\Delta Hv = 7.80$ kcal/mole.

No. 132 3,3-Dimethylhexane C_8H_{18} (Ideal Gas State) Mol Wt 114.224

	cal/(mole °K)				kcal/mole		
$T°K$	$Cp°$	$S°$	$-(G°-H°_{298})/T$	$H°-H°_{298}$	$\Delta Hf°$	$\Delta Gf°$	Log Kp
298	45.14	104.70	104.70	0.00	-52.61	3.17	-2.324
300	45.35	104.98	104.71	0.09	-52.68	3.51	-2.558
400	57.36	119.70	106.63	5.23	-55.75	22.73	-12.417
500	68.32	133.71	110.66	11.53	-58.29	42.65	-18.642
600	77.67	147.01	115.62	18.84	-60.30	63.03	-22.956
700	85.66	159.60	121.01	27.02	-61.83	83.72	-26.137
800	92.50	171.49	126.58	35.93	-62.95	104.57	-28.567
900	98.43	182.74	132.20	45.49	-63.69	125.57	-30.492
1000	103.60	193.38	137.79	55.59	-64.08	146.64	-32.047

The values selected by Rossini, Pitzer, Arnett, Braun, and Pimentel (1248) are presented. $Tm = 147.05°K$, $\Delta Hm° = 1.7$ kcal/mole, and $Tb = 385.12°K$, where $\Delta Hv = 7.76$ kcal/mole.

No. 133 3,4-Dimethylhexane C_8H_{18} (Ideal Gas State) Mol Wt 114.224

These values were selected by Rossini, Pitzer, Arnett, Braun, and Pimentel (1248). $Tb = 390.87°K$, where $\Delta Hv = 7.952$ kcal/mole.

No. 134 3-Ethyl-2-Methylpentane C_8H_{18} (Ideal Gas State) Mol Wt 114.224

The values listed here are those selected by Rossini, Pitzer, Arnett, Braun, and Pimentel (1248). $Tm = 258.19°K$, $\Delta Hm° = 2.71$ kcal/mole, and $Tb = 388.80°K$, where $\Delta Hv = 7.878$ kcal/mole.

No. 133 3,4-Dimethylhexane C_8H_{18} (Ideal Gas State) Mol Wt 114.224

$T°K$	$Cp°$	cal/(mole °K) $S°$	$-(G°-H°_{298})/T$	$H°-H°_{298}$	kcal/mole $\Delta Hf°$	$\Delta Gf°$	Log Kp
298	45.14	107.15	107.15	0.00	-50.91	4.14	-3.034
300	45.35	107.43	107.16	0.09	-50.98	4.48	-3.261
400	57.36	122.15	109.08	5.23	-54.05	23.45	-12.811
500	68.32	136.16	113.11	11.53	-56.59	43.13	-18.849
600	77.67	149.46	118.07	18.84	-58.60	63.26	-23.040
700	85.66	162.05	123.46	27.02	-60.13	83.70	-26.132
800	92.50	173.94	129.03	35.93	-61.25	104.31	-28.496
900	98.43	185.19	134.65	45.49	-61.99	125.07	-30.369
1000	103.60	195.83	140.24	55.59	-62.38	145.89	-31.883

No. 134 3-Ethyl-2-Methylpentane C_8H_{18} (Ideal Gas State) Mol Wt 114.224

$T°K$	$Cp°$	cal/(mole °K) $S°$	$-(G°-H°_{298})/T$	$H°-H°_{298}$	kcal/mole $\Delta Hf°$	$\Delta Gf°$	Log Kp
298	45.14	105.43	105.43	0.00	-50.48	5.08	-3.725
300	45.35	105.71	105.44	0.09	-50.55	5.42	-3.950
400	57.36	120.43	107.36	5.23	-53.62	24.57	-13.421
500	68.32	134.44	111.39	11.53	-56.16	44.42	-19.413
600	77.67	147.74	116.35	18.84	-58.17	64.72	-23.573
700	85.66	160.33	121.74	27.02	-59.70	85.34	-26.642
800	92.50	172.22	127.31	35.93	-60.82	106.12	-28.989
900	98.43	183.47	132.93	45.49	-61.56	127.05	-30.850
1000	103.60	194.11	138.52	55.59	-61.95	148.04	-32.353

No. 135 3-Ethyl-3-Methylpentane C_8H_{18} (Ideal Gas State) Mol Wt 114.224

$T°K$	$Cp°$	cal/(mole °K) $S°$	$-(G°-H°_{298})/T$	$H°-H°_{298}$	kcal/mole $\Delta Hf°$	$\Delta Gf°$	Log Kp
298	45.14	103.48	103.48	0.00	-51.38	4.76	-3.492
300	45.35	103.76	103.49	0.09	-51.45	5.11	-3.721
400	57.36	118.48	105.41	5.23	-54.52	24.45	-13.356
500	68.32	132.49	109.44	11.53	-57.06	44.49	-19.446
600	77.67	145.79	114.40	18.84	-59.07	64.99	-23.671
700	85.66	158.38	119.79	27.02	-60.60	85.80	-26.787
800	92.50	170.27	125.36	35.93	-61.72	106.78	-29.170
900	98.43	181.52	130.98	45.49	-62.46	127.90	-31.057
1000	103.60	192.16	136.57	55.59	-62.85	149.09	-32.582

The selected values of Rossini, Pitzer, Arnett, Braun, and Pimentel (1248) are given here. $Tm = 182.28°K$, $\Delta Hm° = 2.59$ kcal/mole, and $Tb = 391.41°K$, where $\Delta Hv = 7.837$ kcal/mole.

No. 136 2,2,3,-Trimethylpentane C_8H_{18} (Ideal Gas State) Mol Wt 114.224

$T°K$	$Cp°$	$S°$	$-(G°-H°_{298})/T$	$H°-H°_{298}$	$\Delta Hf°$	$\Delta Gf°$	Log Kp
		cal/(mole °K)			kcal/mole		
298	45.14	101.62	101.62	0.00	-52.61	4.09	-2.997
300	45.35	101.90	101.63	0.09	-52.68	4.44	-3.231
400	57.36	116.62	103.55	5.23	-55.75	23.96	-13.090
500	68.32	130.63	107.58	11.53	-58.29	44.19	-19.315
600	77.67	143.93	112.54	18.84	-60.30	64.87	-23.629
700	85.66	156.52	117.93	27.02	-61.83	85.87	-26.810
800	92.50	168.41	123.50	35.93	-62.95	107.04	-29.240
900	98.43	179.66	129.12	45.49	-63.69	128.35	-31.165
1000	103.60	190.30	134.71	55.59	-64.08	149.72	-32.720

The values selected by Rossini, Pitzer, Arnett, Braun, and Pimentel (1248) are employed. $Tm = 160.88°K$, $\Delta Hm° = 2.06$ kcal/mole, and $Tb = 382.99°K$, where $\Delta Hv = 7.65$ kcal/mole.

No. 137 2,2,4-Trimethylpentane C_8H_{18} (Ideal Gas State) Mol Wt 114.224

$T°K$	$Cp°$	$S°$	$-(G°-H°_{298})/T$	$H°-H°_{298}$	$\Delta Hf°$	$\Delta Gf°$	Log Kp
		cal/(mole °K)			kcal/mole		
298	45.14	101.15	101.15	0.00	-53.57	3.27	-2.396
300	45.35	101.43	101.16	0.09	-53.64	3.62	-2.634
400	57.36	116.15	103.08	5.23	-56.71	23.19	-12.669
500	68.32	130.16	107.11	11.53	-59.25	43.47	-18.998
600	77.67	143.46	112.07	18.84	-61.26	64.20	-23.382
700	85.66	156.05	117.46	27.02	-62.79	85.24	-26.613
800	92.50	167.94	123.03	35.93	-63.91	106.45	-29.081
900	98.43	179.19	128.65	45.49	-64.65	127.81	-31.035
1000	103.60	189.83	134.24	55.59	-65.04	149.23	-32.613

The low-temperature measurements of Pitzer (1152) and the selected values of Rossini, Pitzer, Arnett, Braun, and Pimentel (1248) are used. $Tm = 165.77°K$, $\Delta Hm° = 2.202$ kcal/mole, and $Tb = 372.39°K$, where $\Delta Hv = 7.410$ kcal/mole. Ambrose, Cox, and Townsend (13) measured $Tc = 543.6°K$.

No. 138 2,3,3-Trimethylpentane C_8H_{18} (Ideal Gas State) Mol Wt 114.224

The selected values of Rossini, Pitzer, Arnett, Braun, and Pimentel (1248) are adopted. $Tm = 172.45°K$, $\Delta Hm° = 0.37$ kcal/mole, and $Tb = 387.91°K$, where $\Delta Hv = 7.73$ kcal/mole.

No. 138 2,3,3-Trimethylpentane C_8H_{18} (Ideal Gas State) Mol Wt 114.224

$T°K$	$Cp°$	$S°$	$-(G°-H°_{298})/T$	$H°-H°_{298}$	$\Delta Hf°$	$\Delta Gf°$	Log Kp
	cal/(mole °K)			kcal/mole			
298	45.14	103.14	103.14	0.00	-51.73	4.52	-3.309
300	45.35	103.42	103.15	0.09	-51.80	4.86	-3.540
400	57.36	118.14	105.07	5.23	-54.87	24.23	-13.239
500	68.32	132.15	109.10	11.53	-57.41	44.31	-19.367
600	77.67	145.45	114.06	18.84	-59.42	64.84	-23.618
700	85.66	158.04	119.45	27.02	-60.95	85.69	-26.752
800	92.50	169.93	125.02	35.93	-62.07	106.70	-29.148
900	98.43	181.18	130.64	45.49	-62.81	127.86	-31.047
1000	103.60	191.82	136.23	55.59	-63.20	149.08	-32.580

No. 139 2,3,4-Trimethylpentane C_8H_{18} (Ideal Gas State) Mol Wt 114.224

$T°K$	$Cp°$	$S°$	$-(G°-H°_{298})/T$	$H°-H°_{298}$	$\Delta Hf°$	$\Delta Gf°$	Log Kp
	cal/(mole °K)			kcal/mole			
298	45.14	102.31	102.31	0.00	-51.97	4.52	-3.315
300	45.35	102.59	102.32	0.09	-52.04	4.87	-3.546
400	57.36	117.31	104.24	5.23	-55.11	24.32	-13.289
500	68.32	131.32	108.27	11.53	-57.65	44.49	-19.444
600	77.67	144.62	113.23	18.84	-59.66	65.10	-23.712
700	85.66	157.21	118.62	27.02	-61.19	86.03	-26.859
800	92.50	169.10	124.19	35.93	-62.31	107.13	-29.264
900	98.43	180.35	129.81	45.49	-63.05	128.36	-31.170
1000	103.60	190.99	135.40	55.59	-63.44	149.67	-32.709

The selected values of Rossini, Pitzer, Arnett, Braun, and Pimentel (1248) include the low-temperature measurements of Pitzer and Scott (1165), who also presented a simple semiempirical formula for calculation of the entropies of larger branched-chain paraffins. $Tm = 163.94°K$, $\Delta Hm° = 2.215$ kcal/mole, and $Tb = 386.62°K$, where $\Delta Hv = 7.822$ kcal/mole.

No. 140 2,2,3,3-Tetramethylbutane C_8H_{18} (Ideal Gas State) Mol Wt 114.224

$T°K$	$Cp°$	$S°$	$-(G°-H°_{298})/T$	$H°-H°_{298}$	$\Delta Hf°$	$\Delta Gf°$	Log Kp
	cal/(mole °K)			kcal/mole			
298	46.03	93.06	93.06	0.00	-53.99	5.26	-3.856
300	46.29	93.35	93.07	0.09	-54.05	5.62	-4.096
400	59.88	108.57	95.05	5.41	-56.95	25.98	-14.194
500	71.76	123.25	99.23	12.01	-59.19	46.98	-20.535
600	81.52	137.22	104.41	19.69	-60.83	68.37	-24.903
700	89.52	150.40	110.05	28.25	-61.97	90.01	-28.101
800	96.18	162.80	115.87	37.55	-62.71	111.76	-30.530
900	101.80	174.46	121.74	47.45	-63.10	133.61	-32.443
1000	106.60	185.45	127.57	57.88	-63.18	155.49	-33.980

Properties of this material were well established by the complete study of Scott, Douslin, Gross, Oliver, and Huffman (1300). Enthalpy of formation was well established by Prosen and Rossini (1202). $Tt = 152.5°K$, $\Delta Ht° = 0.478$ kcal/mole, $Ttp = 373.96°K$, $\Delta Hm° = 1.802$ kcal/mole, and $Tb = 379.62°K$, where $\Delta Hv = 7.51$ kcal/mole.

No. 141   Nonane C_9H_{20} (Ideal Gas State)   Mol Wt 128.250

	cal/(mole °K)				kcal/mole		
$T°K$	$Cp°$	$S°$	$-(G°-H°_{298})/T$	$H°-H°_{298}$	$\Delta Hf°$	$\Delta Gf°$	Log Kp
298	50.60	120.86	120.86	0.00	-54.74	5.93	-4.349
300	50.85	121.18	120.87	0.10	-54.81	6.30	-4.593
400	64.30	137.68	123.02	5.87	-58.21	27.21	-14.869
500	76.56	153.38	127.54	12.92	-61.00	48.90	-21.375
600	87.01	168.28	133.10	21.12	-63.21	71.09	-25.894
700	95.93	182.38	139.14	30.27	-64.88	93.62	-29.229
800	103.56	195.70	145.39	40.26	-66.10	116.34	-31.780
900	110.17	208.29	151.68	50.95	-66.90	139.21	-33.802
1000	115.90	220.20	157.94	62.26	-67.31	162.15	-35.436

The low-temperature measurements of Finke, Gross, Waddington, and Huffman (420) are combined with the values selected by Rossini, Pitzer, Arnett, Braun, and Pimentel (1248). $Tt = 217.2°K$, $\Delta Ht° = 1.501$ kcal/mole, $Tm = 219.63°K$, $\Delta Hm° = 3.697$ kcal/mole, and $Tb = 423.95°K$, where $\Delta Hv = 9.030$ kcal/mole. Ambrose, Cox, and Townsend (13) measured $Tc = 594.6°K$.

No. 142   2-Methyloctane C_9H_{20} (Ideal Gas State)   Mol Wt 128.250

	cal/(mole °K)				kcal/mole		
$T°K$	$Cp°$	$S°$	$-(G°-H°_{298})/T$	$H°-H°_{298}$	$\Delta Hf°$	$\Delta Gf°$	Log Kp
298	51.88	118.52	118.52	0.00	-56.45	4.92	-3.607
300	52.12	118.85	118.53	0.10	-56.52	5.30	-3.858
400	65.26	135.67	120.73	5.98	-59.80	26.42	-14.436
500	77.24	151.55	125.32	13.12	-62.52	48.30	-21.112
600	87.63	166.57	130.96	21.37	-64.66	70.67	-25.739
700	96.44	180.76	137.07	30.59	-66.28	93.37	-29.149
800	103.93	194.14	143.37	40.62	-67.45	116.24	-31.753
900	110.32	206.75	149.72	51.34	-68.22	139.26	-33.816
1000	116.04	218.68	156.03	62.66	-68.62	162.36	-35.481

The enthalpy of formation was calculated by Labbauf, Greenshields, and Rossini (820); the thermodynamic functions were calculated by the methods of Souders, Matthews, and Hurd (1396, 1397). Transitional temperatures and their enthalpies are the values selected by Rossini, Pitzer, Arnett, Braun, and Pimentel (1248). $Tm = 192.75°K$ and $Tb = 416.41°K$, where $\Delta Hv = 8.76$ kcal/mole.

No. 143 3-Methyloctane C_9H_{20} (Ideal Gas State) Mol Wt 128.250

	cal/(mole °K)			kcal/mole			
$T°K$	$Cp°$	$S°$	$-(G°-H°_{298})/T$	$H°-H°_{298}$	$\Delta Hf°$	$\Delta Gf°$	Log Kp
298	50.81	119.90	119.90	0.00	-55.77	5.19	-3.804
300	51.05	120.22	119.91	0.10	-55.84	5.56	-4.052
400	64.67	136.80	122.08	5.90	-59.21	26.56	-14.513
500	76.91	152.58	126.61	12.99	-61.97	48.34	-21.127
600	87.47	167.56	132.20	21.22	-64.14	70.60	-25.715
700	96.36	181.73	138.27	30.43	-65.76	93.20	-29.097
800	103.89	195.10	144.55	40.45	-66.94	115.98	-31.682
900	110.31	207.72	150.87	51.16	-67.71	138.90	-33.729
1000	116.04	219.64	157.16	62.49	-68.12	161.90	-35.382

Labbauf, Greenshields, and Rossini (820) calculated the enthalpy of formation. The thermodynamic functions were calculated by the methods of Souders, Matthews, and Hurd (1396, 1397). Transitional temperatures and their enthalpies are the values selected by Rossini, Pitzer, Arnett, Braun, and Pimentel (1248). $Tm = 165.55°K$ and $Tb = 417.36°K$, where $\Delta Hv = 8.79$ kcal/mole.

No. 144 4-Methyloctane C_9H_{20} (Ideal Gas State) Mol Wt 128.250

	cal/(mole °K)			kcal/mole			
$T°K$	$Cp°$	$S°$	$-(G°-H°_{298})/T$	$H°-H°_{298}$	$\Delta Hf°$	$\Delta Gf°$	Log Kp
298	50.81	119.90	119.90	0.00	-55.77	5.19	-3.804
300	51.05	120.22	119.91	0.10	-55.84	5.56	-4.052
400	64.67	136.80	122.08	5.90	-59.21	26.56	-14.513
500	76.91	152.58	126.61	12.99	-61.97	48.34	-21.127
600	87.47	167.56	132.20	21.22	-64.14	70.60	-25.715
700	96.36	181.73	138.27	30.43	-65.76	93.20	-29.097
800	103.89	195.10	144.55	40.45	-66.94	115.98	-31.682
900	110.31	207.72	150.87	51.16	-67.71	138.90	-33.729
1000	116.04	219.64	157.16	62.49	-68.12	161.90	-35.382

The enthalpy of formation was calculated by Labbauf, Greenshields, and Rossini (820); the thermodynamic functions were evaluated by the methods of Souders, Matthews, and Hurd (1396, 1397). Transition temperatures and enthalpy increments were selected by Rossini, Pitzer, Arnett, Braun, and Pimentel (1248). $Tm = 159.9°K$ and $Tb = 415.57°K$, where $\Delta Hv = 8.75$ kcal/mole.

No. 145 3-Ethylheptane C_9H_{20} (Ideal Gas State) Mol Wt 128.250

The enthalpy of formation was calculated by Labbauf, Greenshields, and Rossini (820); the thermodynamic functions were evaluated by the

262   Chemical Thermodynamics of Hydrocarbon Compounds

No. 145 3-Ethylheptane C_9H_{20} (Ideal Gas State) Mol Wt 128.250

T°K	$Cp°$	cal/(mole °K) $S°$	$-(G°-H°_{298})/T$	$H°-H°_{298}$	kcal/mole $\Delta Hf°$	$\Delta Gf°$	Log Kp
298	49.74	118.52	118.52	0.00	-55.08	6.29	-4.611
300	49.98	118.83	118.53	0.10	-55.15	6.67	-4.856
400	64.08	135.18	120.66	5.81	-58.60	27.82	-15.200
500	76.58	150.86	125.14	12.86	-61.40	49.76	-21.749
600	87.31	165.79	130.68	21.07	-63.60	72.20	-26.297
700	96.28	179.94	136.72	30.26	-65.23	94.98	-29.652
800	103.85	193.31	142.97	40.28	-66.42	117.93	-32.216
900	110.30	205.92	149.27	50.99	-67.20	141.04	-34.247
1000	116.04	217.84	155.53	62.31	-67.60	164.22	-35.888

methods of Souders, Matthews, and Hurd (1396, 1397). Transition temperatures and enthalpy increments were selected by Rossini, Pitzer, Arnett, Braun, and Pimentel (1248). $Tb = 416.1°K$, where $\Delta Hv = 8.78$ kcal/mole.

No. 146 4-Ethylheptane C_9H_{20} (Ideal Gas State) Mol Wt 128.250

T°K	$Cp°$	cal/(mole °K) $S°$	$-(G°-H°_{298})/T$	$H°-H°_{298}$	kcal/mole $\Delta Hf°$	$\Delta Gf°$	Log Kp
298	49.74	118.52	118.52	0.00	-55.08	6.29	-4.611
300	49.98	118.83	118.53	0.10	-55.15	6.67	-4.856
400	64.08	135.18	120.66	5.81	-58.60	27.82	-15.200
500	76.58	150.86	125.14	12.86	-61.40	49.76	-21.749
600	87.31	165.79	130.68	21.07	-63.60	72.20	-26.297
700	96.28	179.94	136.72	30.26	-65.23	94.98	-29.652
800	103.85	193.31	142.97	40.28	-66.42	117.93	-32.216
900	110.30	205.92	149.27	50.99	-67.20	141.04	-34.247
1000	116.04	217.84	155.53	62.31	-67.60	164.22	-35.888

The enthalpy of formation was calculated by Labbauf, Greenshields, and Rossini (820); the thermodynamic functions were evaluated by the methods of Souders, Matthews, and Hurd (1396, 1397). Transition temperatures and enthalpy increments were selected by Rossini, Pitzer, Arnett, Braun, and Pimentel (1248). $Tb = 414.3°K$, where $\Delta Hv = 8.76$ kcal/mole.

No. 147 2,2-Dimethylheptane C_9H_{20} (Ideal Gas State) Mol Wt 128.250

The enthalpy of formation was calculated by Labbauf, Greenshields, and Rossini (820); the thermodynamic functions were evaluated by the methods of Souders, Matthews, and Hurd (1396, 1397). Transition temperatures and enthalpy increments were selected by Rossini, Pitzer,

No. 147 2,2-Dimethylheptane C_9H_{20} (Ideal Gas State) Mol Wt 128.250

T°K	$Cp°$	$S°$	$-(G°-H°_{298})/T$	$H°-H°_{298}$	$\Delta Hf°$	$\Delta Gf°$	Log Kp
	cal/(mole °K)			kcal/mole			
298	52.01	113.07	113.07	0.00	-59.00	4.00	-2.929
300	52.25	113.40	113.08	0.10	-59.07	4.38	-3.192
400	66.16	130.37	115.30	6.03	-62.30	26.05	-14.230
500	78.66	146.51	119.94	13.29	-64.89	48.44	-21.174
600	89.17	161.81	125.65	21.70	-66.89	71.30	-25.969
700	97.94	176.23	131.86	31.06	-68.35	94.46	-29.491
800	105.36	189.80	138.26	41.24	-69.38	117.78	-32.173
900	111.66	202.58	144.71	52.09	-70.01	141.22	-34.292
1000	117.33	214.65	151.10	63.55	-70.28	164.73	-36.000

Arnett, Braun, and Pimentel (1248). $Tm = 160.15°K$, $\Delta Hm° = 2.14$ kcal/mole, and $Tb = 405.84°K$, where $\Delta Hv = 8.31$ kcal/mole.

No. 148 2,3-Dimethylheptane C_9H_{20} (Ideal Gas State) Mol Wt 128.250

T°K	$Cp°$	$S°$	$-(G°-H°_{298})/T$	$H°-H°_{298}$	$\Delta Hf°$	$\Delta Gf°$	Log Kp
	cal/(mole °K)			kcal/mole			
298	50.30	116.79	116.79	0.00	-56.32	5.57	-4.080
300	50.53	117.11	116.80	0.10	-56.39	5.95	-4.331
400	64.53	133.60	118.95	5.87	-59.79	27.26	-14.896
500	76.84	149.35	123.47	12.95	-62.56	49.36	-21.573
600	87.50	164.33	129.04	21.18	-64.73	71.94	-26.204
700	96.43	178.51	135.11	30.39	-66.35	94.87	-29.618
800	103.97	191.89	141.38	40.42	-67.52	117.97	-32.225
900	110.41	204.51	147.70	51.14	-68.29	141.21	-34.290
1000	116.09	216.45	153.98	62.47	-68.68	164.53	-35.957

The enthalpy of formation was calculated by Labbauf, Greenshields, and Rossini (820); the thermodynamic functions were evaluated by the methods of Souders, Matthews, and Hurd (1396, 1397). Transition temperatures and enthalpy increments were selected by Rossini, Pitzer, Arnett, Braun, and Pimentel (1248). $Tb = 413.6°K$, where $\Delta Hv = 8.63$ kcal/mole.

No. 149 2,4-Dimethylheptane C_9H_{20} (Ideal Gas State) Mol Wt 128.250

The enthalpy of formation was calculated by Labbauf, Greenshields, and Rossini (820); the thermodynamic functions were evaluated by the methods of Souders, Matthews, and Hurd (1396, 1397). Transition temperatures and enthalpy increments were selected by Rossini, Pitzer, Arnett, Braun, and Pimentel (1248). $Tb = 406.04°K$, where $\Delta Hv = 8.45$ kcal/mole.

No. 149 2,4-Dimethylheptane C_9H_{20} (Ideal Gas State) Mol Wt 128.250

$T°K$	$Cp°$	$S°$	$-(G°-H°_{298})/T$	$H°-H°_{298}$	$\Delta Hf°$	$\Delta Gf°$	Log Kp
		cal/(mole °K)			kcal/mole		
298	49.25	116.79	116.79	0.00	-57.48	4.41	-3.230
300	49.48	117.10	116.80	0.10	-57.55	4.79	-3.486
400	63.96	133.36	118.92	5.78	-61.03	26.12	-14.270
500	76.53	149.02	123.38	12.82	-63.84	48.24	-21.085
600	87.37	163.96	128.91	21.03	-66.03	70.86	-25.811
700	96.40	178.12	134.94	30.23	-67.66	93.83	-29.292
800	103.99	191.50	141.18	40.26	-68.83	116.96	-31.951
900	110.45	204.13	147.48	50.99	-69.60	140.25	-34.055
1000	116.12	216.07	153.75	62.33	-69.99	163.60	-35.754

No. 150 2,5-Dimethylheptane C_9H_{20} (Ideal Gas State) Mol Wt 128.250

$T°K$	$Cp°$	$S°$	$-(G°-H°_{298})/T$	$H°-H°_{298}$	$\Delta Hf°$	$\Delta Gf°$	Log Kp
		cal/(mole °K)			kcal/mole		
298	49.25	116.79	116.79	0.00	-57.48	4.41	-3.230
300	49.48	117.10	116.80	0.10	-57.55	4.79	-3.486
400	63.96	133.36	118.92	5.78	-61.03	26.12	-14.270
500	76.53	149.02	123.38	12.82	-63.84	48.24	-21.085
600	87.37	163.96	128.91	21.03	-66.03	70.86	-25.811
700	96.40	178.12	134.94	30.23	-67.66	93.83	-29.292
800	103.99	191.50	141.18	40.26	-68.83	116.96	-31.951
900	110.45	204.13	147.48	50.99	-69.60	140.25	-34.055
1000	116.12	216.07	153.75	62.33	-69.99	163.60	-35.754

The enthalpy of formation was calculated by Labbauf, Greenshields, and Rossini (820); the thermodynamic functions are evaluated by the methods of Souders, Matthews, and Hurd (1396, 1397). Transition temperatures and enthalpy increments were selected by Rossini, Pitzer, Arnett, Braun, and Pimentel (1248). $Tb = 409.1°K$, where $\Delta Hv = 8.51$ kcal/mole.

No. 151 2,6-Dimethylheptane C_9H_{20} (Ideal Gas State) Mol Wt 128.250

$T°K$	$Cp°$	$S°$	$-(G°-H°_{298})/T$	$H°-H°_{298}$	$\Delta Hf°$	$\Delta Gf°$	Log Kp
		cal/(mole °K)			kcal/mole		
298	50.32	114.03	114.03	0.00	-58.17	4.54	-3.327
300	50.55	114.35	114.04	0.10	-58.24	4.92	-3.587
400	64.55	130.85	116.19	5.87	-61.64	26.52	-14.488
500	76.86	146.60	120.71	12.95	-64.40	48.89	-21.367
600	87.53	161.59	126.29	21.19	-66.57	71.75	-26.133
700	96.48	175.77	132.35	30.40	-68.19	94.95	-29.642
800	104.03	189.16	138.62	40.43	-69.35	118.32	-32.321
900	110.46	201.79	144.95	51.16	-70.11	141.84	-34.441
1000	116.12	213.73	151.23	62.50	-70.50	165.43	-36.153

The enthalpy of formation was calculated by Labbauf, Greenshields, and Rossini (820); the thermodynamic functions were evaluated by the methods of Souders, Matthews, and Hurd (1396, 1397). Transition temperatures and enthalpy increments were selected by Rossini, Pitzer, Arnett, Braun, and Pimentel (1248). $Tm = 170.2°K$ and $Tb = 408.36°K$, where $\Delta Hv = 8.49$ kcal/mole.

No. 152   3,3-Dimethylheptane C_9H_{20} (Ideal Gas State)   Mol Wt 128.250

T°K	Cp°	S°	$-(G°-H°_{298})/T$	$H°-H°_{298}$	$\Delta Hf°$	$\Delta Gf°$	Log Kp
	cal/(mole °K)				kcal/mole		
298	50.94	115.25	115.25	0.00	-57.74	4.61	-3.376
300	51.18	115.57	115.26	0.10	-57.81	4.99	-3.633
400	65.57	132.30	117.44	5.95	-61.12	26.45	-14.450
500	78.33	148.35	122.03	13.16	-63.76	48.66	-21.267
600	89.01	163.60	127.70	21.55	-65.78	71.33	-25.981
700	97.86	178.00	133.87	30.90	-67.25	94.32	-29.446
800	105.32	191.57	140.24	41.07	-68.29	117.45	-32.086
900	111.65	204.35	146.66	51.92	-68.92	140.73	-34.171
1000	117.33	216.41	153.04	63.38	-69.20	164.06	-35.853

The enthalpy of formation was calculated by Labbauf, Greenshields, and Rossini (820); the thermodynamic functions were evaluated by the methods of Souders, Matthews, and Hurd (1396, 1397). Transition temperatures and enthalpy increments were selected by Rossini, Pitzer, Arnett, Braun, and Pimentel (1248). $Tb = 410.16°K$, where $\Delta Hv = 8.44$ kcal/mole.

No. 153   3,4-Dimethylheptane C_9H_{20} (Ideal Gas State)   Mol Wt 128.250

T°K	Cp°	S°	$-(G°-H°_{298})/T$	$H°-H°_{298}$	$\Delta Hf°$	$\Delta Gf°$	Log Kp
	cal/(mole °K)				kcal/mole		
298	49.23	117.48	117.48	0.00	-55.63	6.05	-4.435
300	49.46	117.79	117.49	0.10	-55.70	6.43	-4.683
400	63.94	134.05	119.60	5.78	-59.18	27.69	-15.130
500	76.51	149.70	124.07	12.82	-62.00	49.75	-21.744
600	87.34	164.63	129.60	21.03	-64.19	72.30	-26.335
700	96.35	178.79	135.62	30.22	-65.82	95.20	-29.720
800	103.93	192.16	141.86	40.25	-67.00	118.27	-32.307
900	110.40	204.79	148.16	50.97	-67.77	141.49	-34.356
1000	116.09	216.72	154.42	62.30	-68.16	164.78	-36.010

The enthalpy of formation was calculated by Labbauf, Greenshields, and Rossini (820); the thermodynamic functions were evaluated by the methods of Souders, Matthews, and Hurd (1396, 1397). Transition temperatures and enthalpy increments were selected by Rossini, Pitzer,

Arnett, Braun, and Pimentel (1248). $Tb = 413.7°K$, where $\Delta Hv = 8.69$ kcal/mole.

No. 154 3,5-Dimethylheptane C_9H_{20} (Ideal Gas State) Mol Wt 128.250

	cal/(mole °K)				kcal/mole		
$T°K$	$Cp°$	$S°$	$-(G°-H°_{298})/T$	$H°-H°_{298}$	$\Delta Hf°$	$\Delta Gf°$	Log Kp
298	48.18	116.10	116.10	0.00	-56.79	5.30	-3.887
300	48.41	116.40	116.11	0.09	-56.87	5.68	-4.140
400	63.37	132.43	118.19	5.70	-60.42	27.10	-14.805
500	76.20	147.98	122.60	12.69	-63.28	49.32	-21.557
600	87.21	162.88	128.08	20.88	-65.50	72.05	-26.243
700	96.32	177.02	134.07	30.07	-67.14	95.12	-29.696
800	103.95	190.40	140.29	40.09	-68.31	118.37	-32.335
900	110.44	203.02	146.56	50.82	-69.08	141.76	-34.423
1000	116.12	214.96	152.81	62.15	-69.47	165.23	-36.109

The enthalpy of formation was calculated by Labbauf, Greenshields, and Rossini (820); the thermodynamic functions were evaluated by the methods of Souders, Matthews, and Hurd (1396, 1397). Transition temperatures and enthalpy increments were selected by Rossini, Pitzer, Arnett, Braun, and Pimentel (1248). $Tb = 409.1°K$, where $\Delta Hv = 8.52$ kcal/mole.

No. 155 4,4-Dimethylheptane C_9H_{20} (Ideal Gas State) Mol Wt 128.250

	cal/(mole °K)				kcal/mole		
$T°K$	$Cp°$	$S°$	$-(G°-H°_{298})/T$	$H°-H°_{298}$	$\Delta Hf°$	$\Delta Gf°$	Log Kp
298	50.94	113.87	113.87	0.00	-57.74	5.02	-3.678
300	51.18	114.19	113.88	0.10	-57.81	5.40	-3.935
400	65.57	130.92	116.06	5.95	-61.12	27.00	-14.752
500	78.33	146.97	120.65	13.16	-63.76	49.35	-21.569
600	89.01	162.22	126.32	21.55	-65.78	72.16	-26.283
700	97.86	176.62	132.49	30.90	-67.25	95.28	-29.747
800	105.32	190.19	138.86	41.07	-68.29	118.56	-32.387
900	111.65	202.97	145.28	51.92	-68.92	141.97	-34.473
1000	117.33	215.03	151.66	63.38	-69.20	165.44	-36.154

The enthalpy of formation was calculated by Labbauf, Greenshields, and Rossini (820); the thermodynamic functions were evaluated by the methods of Souders, Matthews, and Hurd (1396, 1397). Transition temperatures and enthalpy increments were selected by Rossini, Pitzer, Arnett, Braun, and Pimentel (1248). $Tb = 408.3°K$, where $\Delta Hv = 8.45$ kcal/mole.

No. 156 3-Ethyl-2-Methylhexane C_9H_{20} (Ideal Gas State) Mol Wt 128.250

$T°K$	$Cp°$	cal/(mole °K)			kcal/mole		Log Kp
		$S°$	$-(G°-H°_{298})/T$	$H°-H°_{298}$	$\Delta Hf°$	$\Delta Gf°$	
298	49.23	116.79	116.79	0.00	-55.63	6.26	-4.586
300	49.46	117.10	116.80	0.10	-55.70	6.64	-4.834
400	63.94	133.36	118.91	5.78	-59.18	27.97	-15.281
500	76.51	149.01	123.38	12.82	-62.00	50.09	-21.894
600	87.34	163.94	128.91	21.03	-64.19	72.72	-26.485
700	96.35	178.10	134.93	30.22	-65.82	95.68	-29.871
800	103.93	191.47	141.17	40.25	-67.00	118.82	-32.458
900	110.40	204.10	147.47	50.97	-67.77	142.11	-34.507
1000	116.09	216.03	153.73	62.30	-68.16	165.47	-36.161

The enthalpy of formation was calculated by Labbauf, Greenshields, and Rossini (820); the thermodynamic functions were evaluated by the methods of Souders, Matthews, and Hurd (1396, 1397). Transition temperatures and enthalpy increments were selected by Rossini, Pitzer, Arnett, Braun, and Pimentel (1248). $Tb = 411.1°K$, where $\Delta Hv = 8.60$ kcal/mole.

No. 157 4-Ethyl-2-Methylhexane C_9H_{20} (Ideal Gas State) Mol Wt 128.250

$T°K$	$Cp°$	cal/(mole °K)			kcal/mole		Log Kp
		$S°$	$-(G°-H°_{298})/T$	$H°-H°_{298}$	$\Delta Hf°$	$\Delta Gf°$	
298	48.18	115.41	115.41	0.00	-56.79	5.51	-4.037
300	48.41	115.71	115.42	0.09	-56.87	5.89	-4.290
400	63.37	131.74	117.50	5.70	-60.42	27.37	-14.956
500	76.20	147.29	121.91	12.69	-63.28	49.67	-21.708
600	87.21	162.19	127.39	20.88	-65.50	72.46	-26.393
700	96.32	176.33	133.38	30.07	-67.14	95.60	-29.847
800	103.95	189.71	139.60	40.09	-68.31	118.92	-32.485
900	110.44	202.33	145.87	50.82	-69.08	142.38	-34.574
1000	116.12	214.27	152.12	62.15	-69.47	165.92	-36.260

The enthalpy of formation was calculated by Labbauf, Greenshields, and Rossini (820); the thermodynamic functions were evaluated by the methods of Souders, Matthews, and Hurd (1396, 1397). Transition temperatures and enthalpy increments were selected by Rossini, Pitzer, Arnett, Braun, and Pimentel (1248). $Tb = 406.9°K$, where $\Delta Hv = 8.52$ kcal/mole.

No. 158 3-Ethyl-3-Methylhexane C_9H_{20} (Ideal Gas State) Mol Wt 128.250

The enthalpy of formation was calculated by Labbauf, Greenshields, and Rossini (820); the thermodynamic functions were evaluated by the

No. 158 3-Ethyl-3-Methylhexane C_9H_{20} (Ideal Gas State) Mol Wt 128.250

$T°K$	$Cp°$	$S°$	$-(G°-H°_{298})/T$	$H°-H°_{298}$	$\Delta Hf°$	$\Delta Gf°$	Log Kp
		cal/(mole °K)			kcal/mole		
298	49.87	115.25	115.25	0.00	-56.48	5.87	-4.300
300	50.11	115.56	115.26	0.10	-56.55	6.25	-4.551
400	64.98	132.06	117.40	5.87	-59.95	27.72	-15.146
500	78.00	148.00	121.94	13.03	-62.63	49.96	-21.837
600	88.85	163.21	127.56	21.39	-64.68	72.67	-26.470
700	97.78	177.59	133.69	30.74	-66.16	95.70	-29.877
800	105.28	191.15	140.04	40.90	-67.20	118.88	-32.474
900	111.64	203.93	146.43	51.75	-67.84	142.19	-34.527
1000	117.33	215.99	152.79	63.20	-68.11	165.56	-36.182

methods of Souders, Matthews, and Hurd (1396, 1397). Transition temperatures and enthalpy increments were selected by Rossini, Pitzer, Arnett, Braun, and Pimentel (1248). $Tb = 413.7°K$, where $\Delta Hv = 8.54$ kcal/mole.

No. 159 3-Ethyl-4-Methylhexane C_9H_{20} (Ideal Gas State) Mol Wt 128.250

$T°K$	$Cp°$	$S°$	$-(G°-H°_{298})/T$	$H°-H°_{298}$	$\Delta Hf°$	$\Delta Gf°$	Log Kp
		cal/(mole °K)			kcal/mole		
298	48.16	116.79	116.79	0.00	-54.94	6.95	-5.092
300	48.39	117.09	116.80	0.09	-55.02	7.33	-5.336
400	63.35	133.11	118.88	5.70	-58.57	28.67	-15.665
500	76.18	148.66	123.29	12.69	-61.44	50.83	-22.215
600	87.18	163.55	128.77	20.87	-63.65	73.49	-26.766
700	96.27	177.69	134.76	30.06	-65.30	96.49	-30.125
800	103.89	191.06	140.97	40.08	-66.48	119.67	-32.691
900	110.39	203.68	147.24	50.80	-67.25	143.00	-34.724
1000	116.09	215.61	153.49	62.13	-67.64	166.40	-36.366

The enthalpy of formation was calculated by Labbauf, Greenshields, and Rossini (820); the thermodynamic functions were evaluated by the methods of Souders, Matthews, and Hurd (1396, 1397). Transition temperatures and enthalpy increments were selected by Rossini, Pitzer, Arnett, Braun, and Pimentel (1248). $Tb = 413.5°K$, where $\Delta Hv = 8.70$ kcal/mole.

No. 160 2,2,3-Trimethylhexane C_9H_{20} (Ideal Gas State) Mol Wt 128.250

The enthalpy of formation was calculated by Labbauf, Greenshields, and Rossini (820); the thermodynamic functions were evaluated by the methods of Souders, Matthews, and Hurd (1396, 1397). Transition temperatures and enthalpy increments were selected by Rossini, Pitzer,

Alkane Ideal Gas Tables

No. 160 2,2,3-Trimethylhexane C_9H_{20} (Ideal Gas State) Mol Wt 128.250

$T°K$	cal/(mole °K)			kcal/mole			
	$Cp°$	$S°$	$-(G°-H°_{298})/T$	$H°-H°_{298}$	$\Delta Hf°$	$\Delta Gf°$	Log Kp
298	50.43	111.34	111.34	0.00	-57.65	5.86	-4.296
300	50.66	111.66	111.35	0.10	-57.72	6.25	-4.553
400	65.43	128.30	113.52	5.92	-61.06	28.11	-15.357
500	78.26	144.32	118.09	13.12	-63.72	50.72	-22.168
600	89.04	159.57	123.74	21.50	-65.74	73.80	-26.879
700	97.93	173.98	129.90	30.86	-67.20	97.18	-30.341
800	105.40	187.56	136.26	41.04	-68.23	120.72	-32.979
900	111.75	200.34	142.68	51.90	-68.86	144.40	-35.063
1000	117.38	212.42	149.06	63.36	-69.12	168.13	-36.742

Arnett, Braun, and Pimentel (1248). $Tb = 406.75°K$, where $\Delta Hv = 8.31$ kcal/mole.

No. 161 2,2,4-Trimethylhexane C_9H_{20} (Ideal Gas State) Mol Wt 128.250

$T°K$	cal/(mole °K)			kcal/mole			
	$Cp°$	$S°$	$-(G°-H°_{298})/T$	$H°-H°_{298}$	$\Delta Hf°$	$\Delta Gf°$	Log Kp
298	49.38	111.34	111.34	0.00	-58.13	5.38	-3.945
300	49.61	111.65	111.35	0.10	-58.20	5.77	-4.204
400	64.86	128.06	113.48	5.84	-61.63	27.64	-15.102
500	77.95	143.98	118.00	13.00	-64.32	50.28	-21.977
600	88.91	159.19	123.61	21.36	-66.36	73.39	-26.733
700	97.90	173.59	129.73	30.71	-67.84	96.82	-30.228
800	105.42	187.17	136.07	40.88	-68.86	120.40	-32.890
900	111.79	199.96	142.47	51.75	-69.49	144.11	-34.993
1000	117.41	212.04	148.82	63.22	-69.74	167.88	-36.688

The enthalpy of formation was calculated by Labbauf, Greenshields, and Rossini (820); the thermodynamic functions were evaluated by the methods of Souders, Matthews, and Hurd (1396, 1397). Transition temperatures and enthalpy increments were selected by Rossini, Pitzer, Arnett, Braun, and Pimentel (1248). $Tm = 153.1°K$, $\Delta Hm° = 2.8$ kcal/mole, and $Tb = 399.69°K$, where $\Delta Hv = 8.13$ kcal/mole.

No. 162 2,2,5-Trimethylhexane C_9H_{20} (Ideal Gas State) Mol Wt 128.250

The enthalpy of formation was calculated by Labbauf, Greenshields, and Rossini (820); the thermodynamic functions were evaluated by the methods of Souders, Matthews, and Hurd (1396, 1397). Transition temperatures and enthalpy increments were selected by Rossini, Pitzer, Arnett, Braun, and Pimentel (1248). $Tm = 167.37°K$, $\Delta Hm° = 1.48$ kcal/mole, and $Tb = 397.23°K$, where $\Delta Hv = 8.07$ kcal/mole.

No. 162 2,2,5-Trimethylhexane C_9H_{20} (Ideal Gas State) Mol Wt 128.250

$T°K$	$Cp°$	$S°$	$-(G°-H°_{298})/T$	$H°-H°_{298}$	$\Delta Hf°$	$\Delta Gf°$	Log Kp
		cal/(mole °K)			kcal/mole		
298	50.45	109.96	109.96	0.00	-60.71	3.21	-2.355
300	50.68	110.28	109.97	0.10	-60.78	3.60	-2.626
400	65.45	126.93	112.14	5.92	-64.12	25.60	-13.987
500	78.28	142.95	116.71	13.12	-66.77	48.35	-21.132
600	89.07	158.20	122.36	21.51	-68.79	71.56	-26.065
700	97.98	172.62	128.52	30.87	-70.25	95.09	-29.686
800	105.46	186.20	134.89	41.05	-71.27	118.76	-32.443
900	111.80	199.00	141.31	51.92	-71.89	142.57	-34.619
1000	117.41	211.07	147.69	63.39	-72.15	166.43	-36.372

No. 163 2,3,3,-Trimethylhexane C_9H_{20} (Ideal Gas State) Mol Wt 128.250

$T°K$	$Cp°$	$S°$	$-(G°-H°_{298})/T$	$H°-H°_{298}$	$\Delta Hf°$	$\Delta Gf°$	Log Kp
		cal/(mole °K)			kcal/mole		
298	50.43	112.14	112.14	0.00	-57.08	6.19	-4.539
300	50.66	112.46	112.15	0.10	-57.15	6.58	-4.794
400	65.43	129.10	114.32	5.92	-60.49	28.36	-15.494
500	78.26	145.12	118.89	13.12	-63.15	50.89	-22.243
600	89.04	160.37	124.54	21.50	-65.17	73.89	-26.911
700	97.93	174.78	130.70	30.86	-66.63	97.19	-30.344
800	105.40	188.36	137.06	41.04	-67.66	120.65	-32.960
900	111.75	201.14	143.48	51.90	-68.29	144.25	-35.026
1000	117.38	213.22	149.86	63.36	-68.55	167.90	-36.692

The enthalpy of formation was calculated by Labbauf, Greenshields, and Rossini (820); the thermodynamic functions were evaluated by the methods of Souders, Matthews, and Hurd (1396, 1397). Transition temperatures and enthalpy increments were selected by Rossini, Pitzer, Arnett, Braun, and Pimentel (1248). $Tm = 156.35°K$, $\Delta Hm° = 2.17$ kcal/mole, and $Tb = 410.83°K$, where $\Delta Hv = 8.36$ kcal/mole.

No. 164 2,3,4-Trimethylhexane C_9H_{20} (Ideal Gas State) Mol Wt 128.250

$T°K$	$Cp°$	$S°$	$-(G°-H°_{298})/T$	$H°-H°_{298}$	$\Delta Hf°$	$\Delta Gf°$	Log Kp
		cal/(mole °K)			kcal/mole		
298	48.72	114.37	114.37	0.00	-56.18	6.43	-4.712
300	48.94	114.68	114.38	0.10	-56.26	6.81	-4.962
400	63.80	130.84	116.48	5.75	-59.76	28.39	-15.512
500	76.44	146.47	120.93	12.78	-62.59	50.77	-22.190
600	87.37	161.40	126.44	20.98	-64.78	73.65	-26.824
700	96.42	175.57	132.45	30.18	-66.41	96.86	-30.241
800	104.01	188.95	138.69	40.22	-67.58	120.26	-32.851
900	110.50	201.58	144.98	50.95	-68.34	143.80	-34.917
1000	116.14	213.53	151.24	62.29	-68.72	167.41	-36.585

Alkane Ideal Gas Tables

The enthalpy of formation was calculated by Labbauf, Greenshields, and Rossini (820); the thermodynamic functions were evaluated by the methods of Souders, Matthews, and Hurd (1396, 1397). Transition temperatures and enthalpy increments were selected by Rossini, Pitzer, Arnett, Braun, and Pimentel (1248). $Tb = 412.19°K$, where $\Delta Hv = 8.53$ kcal/mole.

No. 165 2,3,5-Trimethylhexane C_9H_{20} (Ideal Gas State) Mol Wt 128.250

$T°K$	$Cp°$	$S°$	$-(G°-H°_{298})/T$	$H°-H°_{298}$	$\Delta Hf°$	$\Delta Gf°$	Log Kp
	cal/(mole °K)			kcal/mole			
298	48.74	112.30	112.30	0.00	-58.03	5.20	-3.808
300	48.96	112.61	112.31	0.10	-58.11	5.58	-4.067
400	63.82	128.78	114.41	5.75	-61.61	27.37	-14.954
500	76.46	144.41	118.86	12.78	-64.43	49.95	-21.833
600	87.40	159.35	124.37	20.99	-66.63	73.04	-26.602
700	96.47	173.52	130.39	30.20	-68.25	96.46	-30.115
800	104.07	186.91	136.63	40.23	-69.41	120.06	-32.796
900	110.55	199.55	142.92	50.97	-70.17	143.80	-34.918
1000	116.17	211.50	149.19	62.31	-70.55	167.61	-36.630

The enthalpy of formation was calculated by Labbauf, Greenshields, and Rossini (820); the thermodynamic functions were evaluated by the methods of Souders, Matthews, and Hurd (1396, 1397). Transition temperatures and enthalpy increments were selected by Rossini, Pitzer, Arnett, Braun, and Pimentel (1248). $Tm = 145.3°K$ and $Tb = 404.49°K$, where $\Delta Hv = 8.32$ kcal/mole.

No. 166 2,4,4-Trimethylhexane C_9H_{20} (Ideal Gas State) Mol Wt 128.250

$T°K$	$Cp°$	$S°$	$-(G°-H°_{298})/T$	$H°-H°_{298}$	$\Delta Hf°$	$\Delta Gf°$	Log Kp
	cal/(mole °K)			kcal/mole			
298	49.38	112.14	112.14	0.00	-57.56	5.71	-4.188
300	49.61	112.45	112.15	0.10	-57.63	6.10	-4.444
400	64.86	128.86	114.28	5.84	-61.06	27.89	-15.239
500	77.95	144.78	118.80	13.00	-63.75	50.45	-22.052
600	88.91	159.99	124.41	21.36	-65.79	73.48	-26.765
700	97.90	174.39	130.53	30.71	-67.27	96.83	-30.231
800	105.42	187.97	136.87	40.88	-68.29	120.33	-32.871
900	111.79	200.76	143.27	51.75	-68.92	143.96	-34.957
1000	117.41	212.84	149.62	63.22	-69.17	167.65	-36.638

The enthalpy of formation was calculated by Labbauf, Greenshields, and Rossini (820); the thermodynamic functions were evaluated by the methods of Souders, Matthews, and Hurd (1396, 1397). Transition temperatures and enthalpy increments were selected by Rossini, Pitzer,

Arnett, Braun, and Pimentel (1248). $Tm = 159.77°K$, $\Delta Hm° = 2.71$ kcal/mole, and $Tb = 403.80°K$, where $\Delta Hv = 8.20$ kcal/mole.

No. 167 3,3,4-Trimethylhexane C_9H_{20} (Ideal Gas State) Mol Wt 128.250

	cal/(mole °K)				kcal/mole		
T°K	Cp°	S°	$-(G°-H°_{298})/T$	$H°-H°_{298}$	$\Delta Hf°$	$\Delta Gf°$	Log Kp
298	49.36	113.52	113.52	0.00	-56.39	6.47	-4.744
300	49.59	113.83	113.53	0.10	-56.46	6.86	-4.995
400	64.84	130.24	115.66	5.84	-59.89	28.51	-15.577
500	77.93	146.15	120.18	12.99	-62.58	50.93	-22.262
600	88.88	161.36	125.78	21.35	-64.63	73.83	-26.891
700	97.85	175.75	131.90	30.70	-66.11	97.04	-30.296
800	105.36	189.32	138.24	40.87	-67.14	120.40	-32.891
900	111.74	202.11	144.63	51.73	-67.77	143.90	-34.942
1000	117.38	214.18	150.99	63.19	-68.03	167.45	-36.595

The enthalpy of formation was calculated by Labbauf, Greenshields, and Rossini (820); the thermodynamic functions were evaluated by the methods of Souders, Matthews, and Hurd (1396, 1397). Transition temperatures and enthalpy increments were selected by Rossini, Pitzer, Arnett, Braun, and Pimentel (1248). $Tm = 171.95°K$, $\Delta Hm° = 1.94$ kcal/mole, and $Tb = 413.61°K$, where $\Delta Hv = 8.40$ kcal/mole.

No. 168 3,3-Diethylpentane C_9H_{20} (Ideal Gas State) Mol Wt 128.250

	cal/(mole °K)				kcal/mole		
T°K	Cp°	S°	$-(G°-H°_{298})/T$	$H°-H°_{298}$	$\Delta Hf°$	$\Delta Gf°$	Log Kp
298	48.80	110.31	110.31	0.00	-55.44	8.38	-6.141
300	49.04	110.62	110.32	0.10	-55.52	8.77	-6.388
400	64.39	126.88	112.43	5.78	-58.99	30.75	-16.802
500	77.67	142.71	116.91	12.91	-61.72	53.52	-23.391
600	88.69	157.88	122.49	21.24	-63.79	76.76	-27.958
700	97.70	172.25	128.58	30.57	-65.28	100.32	-31.319
800	105.24	185.80	134.89	40.73	-66.33	124.03	-33.882
900	111.63	198.57	141.27	51.58	-66.97	147.88	-35.909
1000	117.33	210.63	147.60	63.03	-67.24	171.79	-37.542

The low-temperature measurements of Staveley, Warren, Paget, and Dowrick (1408) indicated $Tt = 208.8°K$, $Tt = 210.1°K$, $\Delta Ht°$ (combined) $= 0.304$ kcal/mole, $Ttp = 240.12°K$, and $\Delta Hm° = 2.398$ kcal/mole. Rossini, Pitzer, Arnett, Braun, and Pimentel (1248) reported $Tb = 419.32°K$, where $\Delta Hv = 8.60$ kcal/mole. The enthalpy of formation was calculated by Labbauf, Greenshields, and Rossini (820); the thermodynamic functions were calculated by the methods of Souders, Matthews, and Hurd (1396, 1397).

No. 169   3-Ethyl-2,2-Dimethylpentane C_9H_{20} (Ideal Gas State)   Mol Wt 128.250

$T°K$	$Cp°$	$S°$	$-(G°-H°_{298})/T$	$H°-H°_{298}$	$\Delta Hf°$	$\Delta Gf°$	Log Kp
		cal/(mole °K)			kcal/mole		
298	49.49	109.96	109.96	0.00	-56.96	6.96	-5.104
300	49.72	110.27	109.97	0.10	-57.03	7.35	-5.357
400	64.85	126.70	112.10	5.84	-60.45	29.36	-16.042
500	77.80	142.60	116.62	12.99	-63.15	52.14	-22.790
600	88.70	157.78	122.23	21.33	-65.21	75.39	-27.460
700	97.65	172.14	128.34	30.66	-66.71	98.96	-30.896
800	105.15	185.68	134.67	40.81	-67.76	122.69	-33.515
900	111.53	198.44	141.06	51.65	-68.41	146.55	-35.585
1000	117.16	210.49	147.40	63.09	-68.70	170.47	-37.254

The enthalpy of formation was calculated by Labbauf, Greenshields, and Rossini (820); the thermodynamic functions were evaluated by the methods of Souders, Matthews, and Hurd (1396, 1397). Transition temperatures and enthalpy increments were selected by Rossini, Pitzer, Arnett, Braun, and Pimentel (1248). $Tm = 173.66°K$ and $Tb = 406.98°K$, where $\Delta Hv = 8.32$ kcal/mole.

No. 170   3-Ethyl-2,3-Dimethylpentane C_9H_{20} (Ideal Gas State)   Mol Wt 128.250

$T°K$	$Cp°$	$S°$	$-(G°-H°_{298})/T$	$H°-H°_{298}$	$\Delta Hf°$	$\Delta Gf°$	Log Kp
		cal/(mole °K)			kcal/mole		
298	49.36	112.14	112.14	0.00	-55.82	7.45	-5.463
300	49.59	112.45	112.15	0.10	-55.89	7.84	-5.712
400	64.84	128.86	114.28	5.84	-59.32	29.63	-16.190
500	77.93	144.77	118.80	12.99	-62.01	52.19	-22.813
600	88.88	159.98	124.40	21.35	-64.06	75.23	-27.400
700	97.85	174.37	130.52	30.70	-65.54	98.58	-30.775
800	105.36	187.94	136.86	40.87	-66.57	122.08	-33.348
900	111.74	200.73	143.25	51.73	-67.20	145.71	-35.382
1000	117.38	212.80	149.61	63.19	-67.46	169.40	-37.021

The enthalpy of formation was calculated by Labbauf, Greenshields, and Rossini (820); the thermodynamic functions were evaluated by the methods of Souders, Matthews, and Hurd (1396, 1397). Transition temperatures and enthalpy increments were selected by Rossini, Pitzer, Arnett, Braun, and Pimentel (1248). $Tb = 417.85°K$, where $\Delta Hv = 8.44$ kcal/mole.

No. 171   3-Ethyl-2,4-Dimethylpentane C_9H_{20} (Ideal Gas State)   Mol Wt 128.250

The enthalpy of formation was calculated by Labbauf, Greenshields, and Rossini (820); the thermodynamic functions were evaluated by the

No. 171 3-Ethyl-2,4-Dimethylpentane C_9H_{20} (Ideal Gas State) Mol Wt 128.250

$T°K$	$Cp°$	$S°$	$-(G°-H°_{298})/T$	$H°-H°_{298}$	$\Delta Hf°$	$\Delta Gf°$	Log Kp
		cal/(mole °K)			kcal/mole		
298	48.72	112.30	112.30	0.00	-56.18	7.05	-5.164
300	48.94	112.61	112.31	0.10	-56.26	7.43	-5.414
400	63.80	128.77	114.41	5.75	-59.76	29.22	-15.965
500	76.44	144.40	118.86	12.78	-62.59	51.80	-22.643
600	87.37	159.33	124.37	20.98	-64.78	74.89	-27.277
700	96.42	173.50	130.38	30.18	-66.41	98.31	-30.693
800	104.01	186.88	136.62	40.22	-67.58	121.91	-33.303
900	110.50	199.51	142.91	50.95	-68.34	145.66	-35.369
1000	116.14	211.46	149.17	62.29	-68.72	169.48	-37.037

methods of Souders, Matthews, and Hurd (1396, 1397). Transition temperatures and enthalpy increments were selected by Rossini, Pitzer, Arnett, Braun, and Pimentel (1248). $Tm = 150.79°K$ and $Tb = 409.84°K$, where $\Delta Hv = 8.46$ kcal/mole.

No. 172 2,2,3,3-Tetramethylpentane C_9H_{20} (Ideal Gas State) Mol Wt 128.250

$T°K$	$Cp°$	$S°$	$-(G°-H°_{298})/T$	$H°-H°_{298}$	$\Delta Hf°$	$\Delta Gf°$	Log Kp
		cal/(mole °K)			kcal/mole		
298	50.69	106.69	106.69	0.00	-56.70	8.20	-6.009
300	50.92	107.01	106.70	0.10	-56.77	8.60	-6.262
400	66.34	123.82	108.88	5.98	-60.05	30.91	-16.888
500	79.55	140.09	113.51	13.29	-62.59	53.96	-23.584
600	90.40	155.58	119.24	21.81	-64.48	77.44	-28.208
700	99.23	170.20	125.49	31.30	-65.81	101.22	-31.601
800	106.62	183.94	131.95	41.60	-66.71	125.13	-34.182
900	112.88	196.87	138.45	52.59	-67.22	149.16	-36.218
1000	118.45	209.06	144.91	64.16	-67.37	173.23	-37.857

The enthalpy of formation was calculated by Labbauf, Greenshields, and Rossini (820); the thermodynamic functions were evaluated by the methods of Souders, Matthews, and Hurd (1396, 1397). Transition temperatures and enthalpy increments were selected by Rossini, Pitzer, Arnett, Braun, and Pimentel (1248). $Tm = 263.25°K$, $\Delta Hm° = 0.555$ kcal/mole, and $Tb = 413.42°K$, where $\Delta Hv = 8.43$ kcal/mole.

No. 173 2,2,3,4-Tetramethylpentane C_9H_{20} (Ideal Gas State) Mol Wt 128.250

The enthalpy of formation was calculated by Labbauf, Greenshields, and Rossini (820); the thermodynamic functions were evaluated by the methods of Souders, Matthews, and Hurd (1396, 1397). Transition temperatures and enthalpy increments were selected by Rossini, Pitzer,

No. 173   2,2,3,4-Tetramethylpentane C_9H_{20} (Ideal Gas State)   Mol Wt 128.250

	cal/(mole °K)				kcal/mole		
T°K	$Cp°$	$S°$	$-(G°-H°_{298})/T$	$H°-H°_{298}$	$\Delta Hf°$	$\Delta Gf°$	Log Kp
298	49.92	108.23	108.23	0.00	-56.64	7.80	-5.716
300	50.14	108.54	108.24	0.10	-56.71	8.19	-5.969
400	65.29	125.10	110.39	5.89	-60.09	30.37	-16.592
500	78.19	141.09	114.94	13.08	-62.75	53.30	-23.297
600	89.07	156.34	120.58	21.46	-64.77	76.70	-27.937
700	98.00	170.76	126.73	30.82	-66.23	100.41	-31.348
800	105.48	184.34	133.09	41.01	-67.25	124.27	-33.948
900	111.88	197.14	139.50	51.88	-67.87	148.27	-36.002
1000	117.43	209.23	145.88	63.35	-68.12	172.31	-37.657

Arnett, Braun, and Pimentel (1248). $Tm = 152.06°K$, $\Delta Hm° = 0.124$ kcal/mole, and $Tb = 406.17°K$, where $\Delta Hv = 8.19$ kcal/mole.

No. 174   2,2,4,4-Tetramethylpentane C_9H_{20} (Ideal Gas State)   Mol Wt 128.250

	cal/(mole °K)				kcal/mole		
T°K	$Cp°$	$S°$	$-(G°-H°_{298})/T$	$H°-H°_{298}$	$\Delta Hf°$	$\Delta Gf°$	Log Kp
298	50.58	103.13	103.13	0.00	-57.83	8.13	-5.959
300	50.81	103.45	103.14	0.10	-57.90	8.53	-6.216
400	66.35	120.25	105.32	5.98	-61.19	31.21	-17.049
500	79.70	136.53	109.95	13.30	-63.72	54.61	-23.869
600	90.61	152.06	115.68	21.83	-65.59	78.45	-28.574
700	99.48	166.71	121.94	31.35	-66.90	102.58	-32.025
800	106.89	180.49	128.40	41.68	-67.77	126.83	-34.648
900	113.14	193.45	134.92	52.68	-68.25	151.20	-36.716
1000	118.70	205.66	141.39	64.28	-68.38	175.62	-38.379

The enthalpy of formation was calculated by Labbauf, Greenshields, and Rossini (820); the thermodynamic functions were evaluated by the methods of Souders, Matthews, and Hurd (1396, 1397). Transition temperatures and enthalpy increments were selected by Rossini, Pitzer, Arnett, Braun, and Pimentel (1248). $Tm = 206.61°K$, $\Delta Hm° = 2.32$ kcal/mole, and $Tb = 395.43°K$, where $\Delta Hv = 7.85$ kcal/mole.

No. 175   2,3,3,4-Tetramethylpentane C_9H_{20} (Ideal Gas State)   Mol Wt 128.250

The enthalpy of formation was calculated by Labbauf, Greenshields, and Rossini (820); the thermodynamic functions were evaluated by the methods of Souders, Matthews, and Hurd (1396, 1397). Transition temperatures and enthalpy increments were selected by Rossini, Pitzer, Arnett, Braun, and Pimentel (1248). $Tm = 171.03°K$, $\Delta Hm° = 2.14$ kcal/mole, and $Tb = 414.70°K$, where $\Delta Hv = 8.35$ kcal/mole.

No. 175 2,3,3,4-Tetramethylpentane C_9H_{20} (Ideal Gas State) Mol Wt 128.250

$T°K$	$Cp°$	$S°$	$-(G°-H°_{298})/T$	$H°-H°_{298}$	$\Delta Hf°$	$\Delta Gf°$	Log Kp
	cal/(mole °K)				kcal/mole		
298	49.92	107.65	107.65	0.00	-56.46	8.15	-5.975
300	50.14	107.96	107.66	0.10	-56.53	8.55	-6.227
400	65.29	124.52	109.81	5.89	-59.91	30.78	-16.817
500	78.19	140.51	114.36	13.08	-62.57	53.77	-23.502
600	89.07	155.76	120.00	21.46	-64.59	77.23	-28.129
700	98.00	170.18	126.15	30.82	-66.05	101.00	-31.531
800	105.48	183.76	132.51	41.01	-67.07	124.92	-34.124
900	111.85	196.56	138.92	51.88	-67.69	148.97	-36.173
1000	117.43	208.64	145.30	63.35	-67.94	173.07	-37.824

No. 176 Decane $C_{10}H_{22}$ (Ideal Gas State) Mol Wt 142.276

$T°K$	$Cp°$	$S°$	$-(G°-H°_{298})/T$	$H°-H°_{298}$	$\Delta Hf°$	$\Delta Gf°$	Log Kp
	cal/(mole °K)				kcal/mole		
298	56.07	130.17	130.17	0.00	-59.67	7.94	-5.819
300	56.34	130.52	130.18	0.11	-59.75	8.35	-6.085
400	71.24	148.80	132.57	6.50	-63.46	31.64	-17.287
500	84.81	166.20	137.57	14.32	-66.51	55.78	-24.381
600	96.36	182.70	143.73	23.39	-68.92	80.47	-29.308
700	106.21	198.32	150.42	33.53	-70.73	105.53	-32.945
800	114.63	213.06	157.34	44.58	-72.05	130.78	-35.725
900	121.92	226.99	164.31	56.42	-72.90	156.20	-37.929
1000	128.20	240.17	171.24	68.93	-73.33	181.70	-39.709

The low-temperature measurements of Finke, Gross, Waddington, and Huffman (420) are combined with the selected values of Rossini, Pitzer, Arnett, Braun, and Pimentel (1248). $Tm = 243.49°K$, $\Delta Hm° = 6.863$ kcal/mole, and $Tb = 447.27°K$, where $\Delta Hv = 9.387$ kcal/mole. Ambrose, Cox, and Townsend (13) measured $Tc = 617.5°K$.

No. 177 2-Methylnonane $C_{10}H_{22}$ (Ideal Gas State) Mol Wt 142.276

$T°K$	$Cp°$	$S°$	$-(G°-H°_{298})/T$	$H°-H°_{298}$	$\Delta Hf°$	$\Delta Gf°$	Log Kp
	cal/(mole °K)				kcal/mole		
298	57.86	127.74	127.74	0.00	-61.38	6.95	-5.097
300	58.13	128.10	127.75	0.11	-61.46	7.37	-5.370
400	72.51	146.82	130.20	6.65	-65.02	30.88	-16.870
500	85.69	164.45	135.31	14.58	-67.96	55.20	-24.128
600	97.11	181.11	141.56	23.73	-70.29	80.05	-29.158
700	106.81	196.83	148.35	33.94	-72.03	105.27	-32.864
800	115.06	211.64	155.34	45.05	-73.29	130.67	-35.694
900	122.09	225.61	162.38	56.91	-74.12	156.23	-37.935
1000	128.40	238.80	169.37	69.44	-74.54	181.87	-39.745

The low-temperature data of Parks, West, and Moore (1120) are coupled with the selected values of Rossini, Pitzer, Arnett, Braun, and Pimentel (1248); the thermodynamic functions were calculated by the methods of Souders, Matthews, and Hurd (1396, 1397). $Tm = 198.50°K$, $\Delta Hm° = 4.180$ kcal/mole, and $Tb = 440.15°K$.

No. 178 3-Methylnonane $C_{10}H_{22}$ (Ideal Gas State) Mol Wt 142.276

	cal/(mole °K)			kcal/mole			
$T°K$	$Cp°$	$S°$	$-(G°-H°_{298})/T$	$H°-H°_{298}$	$\Delta Hf°$	$\Delta Gf°$	Log Kp
298	56.79	129.12	129.12	0.00	-60.70	7.22	-5.293
300	57.06	129.48	129.13	0.11	-60.78	7.64	-5.564
400	71.92	147.96	131.55	6.57	-64.42	31.02	-16.948
500	85.36	165.49	136.60	14.45	-67.41	55.24	-24.143
600	96.95	182.10	142.81	23.58	-69.76	79.99	-29.134
700	106.73	197.80	149.55	33.78	-71.52	105.10	-32.813
800	115.02	212.61	156.52	44.88	-72.78	130.40	-35.623
900	122.08	226.57	163.54	56.74	-73.61	155.87	-37.849
1000	128.40	239.77	170.50	69.27	-74.03	181.41	-39.646

The gaseous thermodynamic functions were calculated by the methods of Souders, Matthews and Hurd (1396, 1397); the enthalpy of formation was calculated by Labbauf, Greenshields, and Rossini (820). Transitional temperatures were selected by Rossini, Pitzer, Arnett, Braun, and Pimentel (1248). $Tm = 188.35°K$ and $Tb = 440.9°K$.

No. 179 4-Methylnonane $C_{10}H_{22}$ (Ideal Gas State) Mol Wt 142.276

	cal/(mole °K)			kcal/mole			
$T°K$	$Cp°$	$S°$	$-(G°-H°_{298})/T$	$H°-H°_{298}$	$\Delta Hf°$	$\Delta Gf°$	Log Kp
298	56.79	129.12	129.12	0.00	-60.70	7.22	-5.293
300	57.06	129.48	129.13	0.11	-60.78	7.64	-5.564
400	71.92	147.96	131.55	6.57	-64.42	31.02	-16.948
500	85.36	165.49	136.60	14.45	-67.41	55.24	-24.143
600	96.95	182.10	142.81	23.58	-69.76	79.99	-29.134
700	106.73	197.80	149.55	33.78	-71.52	105.10	-32.813
800	115.02	212.61	156.52	44.88	-72.78	130.40	-35.623
900	122.08	226.57	163.54	56.74	-73.61	155.87	-37.849
1000	128.40	239.77	170.50	69.27	-74.03	181.41	-39.646

The gaseous thermodynamic functions were calculated by the methods of Souders, Matthews, and Hurd (1396, 1397); the enthalpy of formation was calculated by Labbauf, Greenshields, and Rossini (820). Transitional temperatures were selected by Rossini, Pitzer, Arnett, Braun, and Pimentel (1248). $Tm = 174.4°K$ and $Tb = 438.8°K$.

No. 180 5-Methylnonane $C_{10}H_{22}$ (Ideal Gas State) Mol Wt 142.276

T°K	$Cp°$	$S°$	$-(G°-H°_{298})/T$	$H°-H°_{298}$	$\Delta Hf°$	$\Delta Gf°$	Log Kp
		cal/(mole °K)			kcal/mole		
298	56.79	127.74	127.74	0.00	-60.70	7.63	-5.595
300	57.06	128.10	127.75	0.11	-60.78	8.05	-5.865
400	71.92	146.58	130.17	6.57	-64.42	31.57	-17.250
500	85.36	164.11	135.22	14.45	-67.41	55.93	-24.445
600	96.95	180.72	141.43	23.58	-69.76	80.81	-29.435
700	106.73	196.42	148.17	33.78	-71.52	106.07	-33.114
800	115.02	211.23	155.14	44.88	-72.78	131.51	-35.925
900	122.08	225.19	162.16	56.74	-73.61	157.11	-38.150
1000	128.40	238.39	169.12	69.27	-74.03	182.79	-39.947

The low-temperature measurements of Parks, West, and Moore (1120) were combined with the selected values of Rossini, Pitzer, Arnett, Braun, and Pimentel (1248) and with the calculated enthalpy of formation of Labbauf, Greenshields, and Rossini (820). The gaseous thermodynamic functions were calculated by the methods of Souders, Matthews, and Hurd (1396, 1397). $Tm = 185.45°K$, $\Delta Hm° = 3.977$ kcal/mole, and $Tb = 438.2°K$.

No. 181 3-Ethyloctane $C_{10}H_{22}$ (Ideal Gas State) Mol Wt 142.276

T°K	$Cp°$	$S°$	$-(G°-H°_{298})/T$	$H°-H°_{298}$	$\Delta Hf°$	$\Delta Gf°$	Log Kp
		cal/(mole °K)			kcal/mole		
298	55.72	127.74	127.74	0.00	-60.01	8.32	-6.101
300	55.99	128.09	127.75	0.11	-60.09	8.74	-6.368
400	71.33	146.34	130.13	6.49	-63.81	32.28	-17.634
500	85.03	163.76	135.13	14.32	-66.85	56.66	-24.766
600	96.79	180.33	141.29	23.43	-69.22	81.59	-29.716
700	106.65	196.01	148.00	33.61	-70.99	106.88	-33.368
800	114.98	210.81	154.94	44.71	-72.26	132.36	-36.157
900	122.07	224.77	161.93	56.56	-73.09	158.01	-38.367
1000	128.40	237.97	168.88	69.09	-73.51	183.73	-40.152

The gaseous thermodynamic functions were calculated by the methods of Souders, Matthews, and Hurd (1396, 1397); the enthalpy of formation was calculated by Labbauf, Greenshields, and Rossini (820). The normal boiling point was selected by Rossini, Pitzer, Arnett, Braun, and Pimentel (1248). $Tb = 439.6°K$.

No. 182 4-Ethyloctane $C_{10}H_{22}$ (Ideal Gas State) Mol Wt 142.276

The gaseous thermodynamic functions were calculated by the methods of Souders, Matthews, and Hurd (1396, 1397); the enthalpy of formation

No. 182 4-Ethyloctane $C_{10}H_{22}$ (Ideal Gas State) Mol Wt 142.276

T°K	$Cp°$	$S°$	$-(G°-H°_{298})/T$	$H°-H°_{298}$	$\Delta Hf°$	$\Delta Gf°$	Log Kp
	cal/(mole °K)			kcal/mole			
298	55.72	129.12	129.12	0.00	-60.01	7.91	-5.799
300	55.99	129.47	129.13	0.11	-60.09	8.33	-6.066
400	71.33	147.72	131.51	6.49	-63.81	31.72	-17.333
500	85.03	165.14	136.51	14.32	-66.85	55.97	-24.464
600	96.79	181.71	142.67	23.43	-69.22	80.76	-29.415
700	106.65	197.39	149.38	33.61	-70.99	105.91	-33.066
800	114.98	212.19	156.32	44.71	-72.26	131.26	-35.856
900	122.07	226.15	163.31	56.56	-73.09	156.76	-38.066
1000	128.40	239.35	170.26	69.09	-73.51	182.35	-39.850

was calculated by Labbauf, Greenshields, and Rossini (820). The normal boiling point was selected by Rossini, Pitzer, Arnett, Braun, and Pimentel (1248). $Tb = 436.79°K$.

No. 183 2,2-Dimethyloctane $C_{10}H_{22}$ (Ideal Gas State) Mol Wt 142.276

T°K	$Cp°$	$S°$	$-(G°-H°_{298})/T$	$H°-H°_{298}$	$\Delta Hf°$	$\Delta Gf°$	Log Kp
	cal/(mole °K)			kcal/mole			
298	57.99	122.29	122.29	0.00	-63.93	6.03	-4.419
300	58.26	122.65	122.30	0.11	-64.01	6.46	-4.703
400	73.41	141.52	124.77	6.71	-67.51	30.50	-16.665
500	87.11	159.42	129.92	14.75	-70.34	55.34	-24.190
600	98.65	176.35	136.26	24.06	-72.52	80.68	-29.388
700	108.31	192.30	143.14	34.42	-74.11	106.36	-33.206
800	116.49	207.31	150.23	45.67	-75.22	132.20	-36.115
900	123.43	221.44	157.37	57.67	-75.91	158.19	-38.412
1000	129.69	234.77	164.45	70.33	-76.20	184.24	-40.263

The gaseous thermodynamic functions were calculated by the methods of Souders, Matthews, and Hurd (1396, 1397); the enthalpy of formation was calculated by Labbauf, Greenshields, and Rossini (820). The normal boiling point was selected by Rossini, Pitzer, Arnett, Braun, and Pimentel (1248). $Tb = 430.0°K$.

No. 184 2,3-Dimethyloctane $C_{10}H_{22}$ (Ideal Gas State) Mol Wt 142.276

The gaseous thermodynamic functions were calculated by the methods of Souders, Matthews, and Hurd (1396, 1397); the enthalpy of formation was calculated by Labbauf, Greenshields, and Rossini (820). The normal boiling point was selected by Rossini, Pitzer, Arnett, Braun, and Pimentel (1248). $Tb = 437.46°K$.

No. 184 2,3-Dimethyloctane $C_{10}H_{22}$ (Ideal Gas State) Mol Wt 142.276

T°K	$Cp°$	$S°$	$-(G°-H°_{298})/T$	$H°-H°_{298}$	$\Delta Hf°$	$\Delta Gf°$	Log Kp
	cal/(mole °K)				kcal/mole		
298	55.23	126.01	126.01	0.00	-62.41	6.44	-4.720
300	55.49	126.36	126.02	0.11	-62.49	6.86	-4.998
400	71.21	144.52	128.39	6.46	-66.24	30.57	-16.704
500	84.98	161.93	133.37	14.28	-69.29	55.14	-24.102
600	96.85	178.50	139.52	23.39	-71.66	80.25	-29.230
700	106.77	194.19	146.22	33.59	-73.42	105.73	-33.008
800	115.12	209.01	153.15	44.69	-74.68	131.39	-35.892
900	122.22	222.99	160.14	56.56	-75.49	157.21	-38.175
1000	128.48	236.19	167.09	69.11	-75.90	183.11	-40.018

No. 185 2,4-Dimethyloctane $C_{10}H_{22}$ (Ideal Gas State) Mol Wt 142.276

T°K	$Cp°$	$S°$	$-(G°-H°_{298})/T$	$H°-H°_{298}$	$\Delta Hf°$	$\Delta Gf°$	Log Kp
	cal/(mole °K)				kcal/mole		
298	56.28	126.01	126.01	0.00	-61.25	7.60	-5.570
300	56.54	126.36	126.02	0.11	-61.33	8.02	-5.843
400	71.78	144.76	128.42	6.54	-65.00	31.72	-17.330
500	85.29	162.26	133.45	14.41	-68.00	56.26	-24.590
600	96.98	178.87	139.65	23.54	-70.35	81.33	-29.623
700	106.80	194.58	146.39	33.74	-72.10	106.77	-33.333
800	115.10	209.40	153.35	44.84	-73.36	132.39	-36.166
900	122.18	223.37	160.36	56.72	-74.18	158.18	-38.409
1000	128.45	236.57	167.33	69.25	-74.59	184.04	-40.220

The gaseous thermodynamic functions were calculated by the methods of Souders, Matthews, and Hurd (1396, 1397); the enthalpy of formation was calculated by Labbauf, Greenshields, and Rossini (820). The normal boiling point was selected by Rossini, Pitzer, Arnett, Braun, and Pimentel (1248). $Tb = 429.0°K$.

No. 186 2,5-Dimethyloctane $C_{10}H_{22}$ (Ideal Gas State) Mol Wt 142.276

T°K	$Cp°$	$S°$	$-(G°-H°_{298})/T$	$H°-H°_{298}$	$\Delta Hf°$	$\Delta Gf°$	Log Kp
	cal/(mole °K)				kcal/mole		
298	55.23	126.01	126.01	0.00	-62.41	6.44	-4.720
300	55.49	126.36	126.02	0.11	-62.49	6.86	-4.998
400	71.21	144.52	128.39	6.46	-66.24	30.57	-16.704
500	84.98	161.93	133.37	14.28	-69.29	55.14	-24.102
600	96.85	178.50	139.52	23.39	-71.66	80.25	-29.230
700	106.77	194.19	146.22	33.59	-73.42	105.73	-33.008
800	115.12	209.01	153.15	44.69	-74.68	131.39	-35.892
900	122.22	222.99	160.14	56.56	-75.49	157.21	-38.175
1000	128.48	236.19	167.09	69.11	-75.90	183.11	-40.018

Alkane Ideal Gas Tables

The gaseous thermodynamic functions were calculated by the methods of Souders, Matthews, and Hurd (1396, 1397); the enthalpy of formation was calculated by Labbauf, Greenshields, and Rossini (820). The normal boiling point was selected by Rossini, Pitzer, Arnett, Braun, and Pimentel (1248). $Tb = 431.6°K$.

No. 187 2,6-Dimethyloctane $C_{10}H_{22}$ (Ideal Gas State) Mol Wt 142.276

$T°K$	$Cp°$	cal/(mole °K)			kcal/mole		
		$S°$	$-(G°-H°_{298})/T$	$H°-H°_{298}$	$\Delta Hf°$	$\Delta Gf°$	Log Kp
298	55.23	126.01	126.01	0.00	-62.41	6.44	-4.720
300	55.49	126.36	126.02	0.11	-62.49	6.86	-4.998
400	71.21	144.52	128.39	6.46	-66.24	30.57	-16.704
500	84.98	161.93	133.37	14.28	-69.29	55.14	-24.102
600	96.85	178.50	139.52	23.39	-71.66	80.25	-29.230
700	106.77	194.19	146.22	33.59	-73.42	105.73	-33.008
800	115.12	209.01	153.15	44.69	-74.68	131.39	-35.892
900	122.22	222.99	160.14	56.56	-75.49	157.21	-38.175
1000	128.48	236.19	167.09	69.11	-75.90	183.11	-40.018

The gaseous thermodynamic functions were calculated by the methods of Souders, Matthews, and Hurd (1396, 1397); the enthalpy of formation was calculated by Labbauf, Greenshields, and Rossini (820). The normal boiling point was selected by Rossini, Pitzer, Arnett, Braun, and Pimentel (1248). $Tb = 431.69°K$.

No. 188 2,7-Dimethyloctane $C_{10}H_{22}$ (Ideal Gas State) Mol Wt 142.276

$T°K$	$Cp°$	cal/(mole °K)			kcal/mole		
		$S°$	$-(G°-H°_{298})/T$	$H°-H°_{298}$	$\Delta Hf°$	$\Delta Gf°$	Log Kp
298	56.30	123.25	123.25	0.00	-63.10	6.57	-4.817
300	56.56	123.60	123.26	0.11	-63.18	7.00	-5.098
400	71.80	142.00	125.66	6.54	-66.85	30.97	-16.923
500	85.31	159.51	130.70	14.41	-69.85	55.79	-24.384
600	97.01	176.13	136.90	23.54	-72.20	81.13	-29.552
700	106.85	191.84	143.63	33.75	-73.94	106.85	-33.358
800	115.16	206.66	150.59	44.86	-75.20	132.74	-36.262
900	122.23	220.64	157.61	56.74	-76.01	158.80	-38.561
1000	128.48	233.85	164.58	69.28	-76.42	184.94	-40.416

The gaseous thermodynamic functions were calculated by the methods of Souders, Matthews, and Hurd (1396, 1397); the enthalpy of formation was calculated by Labbauf, Greenshields, and Rossini (820). Transitional temperatures were selected by Rossini, Pitzer, Arnett, Braun, and Pimentel (1248). $Tm = 219°K$ and $Tb = 433.02°K$.

No. 189 3,3-Dimethyloctane $C_{10}H_{22}$ (Ideal Gas State) Mol Wt 142.276

$T°K$	$Cp°$	$S°$	$-(G°-H°_{298})/T$	$H°-H°_{298}$	$\Delta Hf°$	$\Delta Gf°$	Log Kp
	cal/(mole °K)			kcal/mole			
298	56.92	124.47	124.47	0.00	-62.67	6.64	-4.866
300	57.19	124.83	124.48	0.11	-62.75	7.06	-5.145
400	72.82	143.46	126.91	6.62	-66.34	30.90	-16.885
500	86.78	161.25	132.01	14.62	-69.21	55.56	-24.284
600	98.49	178.14	138.31	23.90	-71.41	80.72	-29.400
700	108.23	194.07	145.15	34.25	-73.01	106.22	-33.161
800	116.45	209.07	152.21	45.50	-74.13	131.88	-36.027
900	123.42	223.20	159.32	57.50	-74.82	157.69	-38.291
1000	129.69	236.53	166.38	70.16	-75.11	183.57	-40.116

The gaseous thermodynamic functions were calculated by the methods of Souders, Matthews, and Hurd (1396, 1397); the enthalpy of formation was calculated by Labbauf, Greenshields, and Rossini (820). The normal boiling point was selected by Rossini, Pitzer, Arnett, Braun, and Pimentel (1248). $Tb = 434.3°K$.

No. 190 3,4-Dimethyloctane $C_{10}H_{22}$ (Ideal Gas State) Mol Wt 142.276

$T°K$	$Cp°$	$S°$	$-(G°-H°_{298})/T$	$H°-H°_{298}$	$\Delta Hf°$	$\Delta Gf°$	Log Kp
	cal/(mole °K)			kcal/mole			
298	55.21	126.70	126.70	0.00	-60.56	8.08	-5.925
300	55.47	127.05	126.71	0.11	-60.64	8.50	-6.195
400	71.19	145.20	129.08	6.46	-64.40	32.15	-17.564
500	84.96	162.60	134.05	14.28	-67.44	56.65	-24.760
600	96.82	179.17	140.21	23.39	-69.82	81.69	-29.754
700	106.72	194.86	146.90	33.58	-71.58	107.10	-33.436
800	115.06	209.67	153.83	44.67	-72.84	132.69	-36.249
900	122.17	223.64	160.82	56.54	-73.67	158.45	-38.476
1000	128.45	236.84	167.77	69.08	-74.08	184.29	-40.274

The gaseous thermodynamic functions were calculated by the methods of Souders, Matthews, and Hurd (1396, 1397); the enthalpy of formation was calculated by Labbauf, Greenshields, and Rossini (820). The normal boiling point was selected by Rossini, Pitzer, Arnett, Braun, and Pimentel (1248). $Tb = 436.5°K$.

No. 191 3,5-Dimethyloctane $C_{10}H_{22}$ (Ideal Gas State) Mol Wt 142.276

The gaseous thermodynamic functions were calculated by the methods of Souders, Matthews, and Hurd (1396, 1397); the enthalpy of formation was calculated by Labbauf, Greenshields, and Rossini (820). The normal boiling point was selected by Rossini, Pitzer, Arnett, Braun, and Pimentel (1248). $Tb = 432.5°K$.

No. 191 3,5-Dimethyloctane $C_{10}H_{22}$ (Ideal Gas State) Mol Wt 142.276

T°K	$Cp°$	$S°$	$-(G°-H°_{298})/T$	$H°-H°_{298}$	$\Delta Hf°$	$\Delta Gf°$	Log Kp
	cal/(mole °K)			kcal/mole			
298	54.16	126.70	126.70	0.00	-61.72	6.92	-5.075
300	54.42	127.04	126.71	0.11	-61.80	7.34	-5.350
400	70.62	144.97	129.04	6.38	-65.64	31.00	-16.938
500	84.65	162.27	133.97	14.16	-68.73	55.53	-24.272
600	96.69	178.80	140.07	23.24	-71.12	80.61	-29.360
700	106.69	194.47	146.74	33.42	-72.89	106.06	-33.110
800	115.08	209.28	153.64	44.52	-74.16	131.69	-35.974
900	122.21	223.26	160.60	56.39	-74.98	157.49	-38.241
1000	128.48	236.47	167.54	68.93	-75.38	183.36	-40.071

No. 192 3,6-Dimethyloctane $C_{10}H_{22}$ (Ideal Gas State) Mol Wt 142.276

T°K	$Cp°$	$S°$	$-(G°-H°_{298})/T$	$H°-H°_{298}$	$\Delta Hf°$	$\Delta Gf°$	Log Kp
	cal/(mole °K)			kcal/mole			
298	54.16	125.32	125.32	0.00	-61.72	7.33	-5.376
300	54.42	125.66	125.33	0.11	-61.80	7.76	-5.651
400	70.62	143.59	127.66	6.38	-65.64	31.55	-17.240
500	84.65	160.89	132.59	14.16	-68.73	56.22	-24.573
600	96.69	177.42	138.69	23.24	-71.12	81.44	-29.661
700	106.69	193.09	145.36	33.42	-72.89	107.02	-33.412
800	115.08	207.90	152.26	44.52	-74.16	132.79	-36.276
900	122.21	221.88	159.22	56.39	-74.98	158.73	-38.543
1000	128.48	235.09	166.16	68.93	-75.38	184.74	-40.373

The gaseous thermodynamic functions were calculated by the methods of Souders, Matthews, and Hurd (1396, 1397); the enthalpy of formation was calculated by Labbauf, Greenshields, and Rossini (820). The normal boiling point was selected by Rossini, Pitzer, Arnett, Braun, and Pimentel (1248). $Tb = 433.9°K$.

No. 193 4,4-Dimethyloctane $C_{10}H_{22}$ (Ideal Gas State) Mol Wt 142.276

T°K	$Cp°$	$S°$	$-(G°-H°_{298})/T$	$H°-H°_{298}$	$\Delta Hf°$	$\Delta Gf°$	Log Kp
	cal/(mole °K)			kcal/mole			
298	56.92	124.47	124.47	0.00	-62.67	6.64	-4.866
300	57.19	124.83	124.48	0.11	-62.75	7.06	-5.145
400	72.82	143.46	126.91	6.62	-66.34	30.90	-16.885
500	86.78	161.25	132.01	14.62	-69.21	55.56	-24.284
600	98.49	178.14	138.31	23.90	-71.41	80.72	-29.400
700	108.23	194.07	145.15	34.25	-73.01	106.22	-33.161
800	116.45	209.07	152.21	45.50	-74.13	131.88	-36.027
900	123.42	223.20	159.32	57.50	-74.82	157.69	-38.291
1000	129.69	236.53	166.38	70.16	-75.11	183.57	-40.116

The gaseous thermodynamic functions were calculated by the methods of Souders, Matthews, and Hurd (1396, 1397); the enthalpy of formation was calculated by Labbauf, Greenshields, and Rossini (820). The normal boiling point was selected by Rossini, Pitzer, Arnett, Braun, and Pimentel (1248). $Tb = 430.6°K$.

No. 194 4,5-Dimethyloctane $C_{10}H_{22}$ (Ideal Gas State) Mol Wt 142.276

	cal/(mole °K)				kcal/mole		
$T°K$	$Cp°$	$S°$	$-(G°-H°_{298})/T$	$H°-H°_{298}$	$\Delta Hf°$	$\Delta Gf°$	Log Kp
298	55.21	125.32	125.32	0.00	−60.56	8.49	−6.227
300	55.47	125.67	125.33	0.11	−60.64	8.92	−6.496
400	71.19	143.82	127.70	6.46	−64.40	32.70	−17.866
500	84.96	161.22	132.67	14.28	−67.44	57.34	−25.061
600	96.82	177.79	138.83	23.39	−69.82	82.52	−30.055
700	106.72	193.48	145.52	33.58	−71.58	108.06	−33.737
800	115.06	208.29	152.45	44.67	−72.84	133.80	−36.550
900	122.17	222.26	159.44	56.54	−73.67	159.70	−38.777
1000	128.45	235.46	166.39	69.08	−74.08	185.67	−40.576

The gaseous thermodynamic functions were calculated by the methods of Souders, Matthews, and Hurd (1396, 1397); the enthalpy of formation was calculated by Labbauf, Greenshields, and Rossini (820). The normal boiling point was selected by Rossini, Pitzer, Arnett, Braun, and Pimentel (1248). $Tb = 435.28°K$.

No. 195 4-Propylheptane $C_{10}H_{22}$ (Ideal Gas State) Mol Wt 142.276

	cal/(mole °K)				kcal/mole		
$T°K$	$Cp°$	$S°$	$-(G°-H°_{298})/T$	$H°-H°_{298}$	$\Delta Hf°$	$\Delta Gf°$	Log Kp
298	55.72	125.56	125.56	0.00	−60.01	8.97	−6.577
300	55.99	125.91	125.57	0.11	−60.09	9.40	−6.844
400	71.33	144.16	127.95	6.49	−63.81	33.15	−18.111
500	85.03	161.58	132.95	14.32	−66.85	57.75	−25.242
600	96.79	178.15	139.11	23.43	−69.22	82.89	−30.193
700	106.65	193.83	145.82	33.61	−70.99	108.41	−33.844
800	114.98	208.63	152.76	44.71	−72.26	134.10	−36.634
900	122.07	222.59	159.75	56.56	−73.09	159.97	−38.844
1000	128.40	235.79	166.70	69.09	−73.51	185.91	−40.628

The gaseous thermodynamic functions were calculated by the methods of Souders, Matthews, and Hurd (1396, 1397); the enthalpy of formation was calculated by Labbauf, Greenshields, and Rossini (820). The normal boiling point was selected by Rossini, Pitzer, Arnett, Braun, and Pimentel (1248). $Tb = 430.6°K$.

No. 196   4-Isopropylheptane $C_{10}H_{22}$ (Ideal Gas State)   Mol Wt 142.276

$T°K$	$Cp°$	$S°$	$-(G°-H°_{298})/T$	$H°-H°_{298}$	$\Delta Hf°$	$\Delta Gf°$	Log Kp
		cal/(mole °K)			kcal/mole		
298	55.21	124.63	124.63	0.00	-60.02	9.24	-6.773
300	55.47	124.98	124.64	0.11	-60.10	9.66	-7.040
400	71.19	143.13	127.01	6.46	-63.86	33.52	-18.312
500	84.96	160.53	131.98	14.28	-66.90	58.22	-25.448
600	96.82	177.10	138.14	23.39	-69.28	83.47	-30.403
700	106.72	192.79	144.83	33.58	-71.04	109.09	-34.057
800	115.06	207.60	151.76	44.67	-72.30	134.89	-36.848
900	122.17	221.57	158.75	56.54	-73.13	160.86	-39.059
1000	128.45	234.77	165.70	69.08	-73.54	186.90	-40.844

The gaseous thermodynamic functions were calculated by the methods of Souders, Matthews, and Hurd (1396, 1397); the enthalpy of formation was calculated by Labbauf, Greenshields, and Rossini (820). The normal boiling point was selected by Rossini, Pitzer, Arnett, Braun, and Pimentel (1248). $Tb = 432.0°K$.

No. 197   3-Ethyl-2-Methylheptane $C_{10}H_{22}$ (Ideal Gas State)   Mol Wt 142.276

$T°K$	$Cp°$	$S°$	$-(G°-H°_{298})/T$	$H°-H°_{298}$	$\Delta Hf°$	$\Delta Gf°$	Log Kp
		cal/(mole °K)			kcal/mole		
298	55.21	126.01	126.01	0.00	-60.56	8.29	-6.076
300	55.47	126.36	126.02	0.11	-60.64	8.71	-6.345
400	71.19	144.51	128.39	6.46	-64.40	32.42	-17.715
500	84.96	161.91	133.36	14.28	-67.44	56.99	-24.911
600	96.82	178.48	139.52	23.39	-69.82	82.10	-29.904
700	106.72	194.17	146.21	33.58	-71.58	107.58	-33.587
800	115.06	208.98	153.14	44.67	-72.84	133.25	-36.399
900	122.17	222.95	160.13	56.54	-73.67	159.07	-38.627
1000	128.45	236.15	167.08	69.08	-74.08	184.98	-40.425

The gaseous thermodynamic functions were calculated by the methods of Souders, Matthews, and Hurd (1396, 1397); the enthalpy of formation was calculated by Labbauf, Greenshields, and Rossini (820). The normal boiling point was selected by Rossini, Pitzer, Arnett, Braun, and Pimentel (1248). $Tb = 434.3°K$.

No. 198   4-Ethyl-2-Methylheptane $C_{10}H_{22}$ (Ideal Gas State)   Mol Wt 142.276

The gaseous thermodynamic functions were calculated by the methods of Souders, Matthews, and Hurd (1396, 1397); the enthalpy of formation was calculated by Labbauf, Greenshields, and Rossini (820). The normal boiling point was selected by Rossini, Pitzer, Arnett, Braun, and Pimentel (1248). $Tb = 429.3°K$.

No. 198 4-Ethyl-2-Methylheptane $C_{10}H_{22}$ (Ideal Gas State) Mol Wt 142.276

$T°K$	$Cp°$	cal/(mole °K) $S°$	$-(G°-H°_{298})/T$	$H°-H°_{298}$	kcal/mole $\Delta Hf°$	$\Delta Gf°$	Log Kp
298	54.16	126.01	126.01	0.00	-61.72	7.13	-5.226
300	54.42	126.35	126.02	0.11	-61.80	7.55	-5.500
400	70.62	144.28	128.35	6.38	-65.64	31.28	-17.089
500	84.65	161.58	133.28	14.16	-68.73	55.88	-24.423
600	96.69	178.11	139.38	23.24	-71.12	81.02	-29.511
700	106.69	193.78	146.05	33.42	-72.89	106.54	-33.261
800	115.08	208.59	152.95	44.52	-74.16	132.24	-36.125
900	122.21	222.57	159.91	56.39	-74.98	158.11	-38.392
1000	128.48	235.78	166.85	68.93	-75.38	184.05	-40.222

No. 199 5-Ethyl-2-Methylheptane $C_{10}H_{22}$ (Ideal Gas State) Mol Wt 142.276

$T°K$	$Cp°$	cal/(mole °K) $S°$	$-(G°-H°_{298})/T$	$H°-H°_{298}$	kcal/mole $\Delta Hf°$	$\Delta Gf°$	Log Kp
298	54.16	124.63	124.63	0.00	-61.72	7.54	-5.527
300	54.42	124.97	124.64	0.11	-61.80	7.96	-5.802
400	70.62	142.90	126.97	6.38	-65.64	31.83	-17.391
500	84.65	160.20	131.90	14.16	-68.73	56.57	-24.724
600	96.69	176.73	138.00	23.24	-71.12	81.85	-29.812
700	106.69	192.40	144.67	33.42	-72.89	107.50	-33.563
800	115.08	207.21	151.57	44.52	-74.16	133.35	-36.427
900	122.21	221.19	158.53	56.39	-74.98	159.35	-38.694
1000	128.48	234.40	165.47	68.93	-75.38	185.43	-40.524

The gaseous thermodynamic functions were calculated by the methods of Souders, Matthews, and Hurd (1396, 1397); the enthalpy of formation was calculated by Labbauf, Greenshields, and Rossini (820). The normal boiling point was selected by Rossini, Pitzer, Arnett, Braun, and Pimentel (1248). $Tb = 432.8°K$.

No. 200 3-Ethyl-3-Methylheptane $C_{10}H_{22}$ (Ideal Gas State) Mol Wt 142.276

$T°K$	$Cp°$	cal/(mole °K) $S°$	$-(G°-H°_{298})/T$	$H°-H°_{298}$	kcal/mole $\Delta Hf°$	$\Delta Gf°$	Log Kp
298	55.85	124.47	124.47	0.00	-61.41	7.90	-5.789
300	56.12	124.82	124.48	0.11	-61.49	8.32	-6.063
400	72.23	143.22	126.88	6.54	-65.16	32.18	-17.581
500	86.45	160.91	131.93	14.50	-68.08	56.86	-24.854
600	98.33	177.75	138.17	23.75	-70.30	82.06	-29.888
700	108.15	193.66	144.97	34.09	-71.91	107.60	-33.592
800	116.41	208.66	152.01	45.33	-73.04	133.31	-36.415
900	123.41	222.78	159.09	57.32	-73.74	159.16	-38.647
1000	129.69	236.12	166.14	69.98	-74.02	185.07	-40.445

The gaseous thermodynamic functions were calculated by the methods of Souders, Matthews, and Hurd (1396, 1397); the enthalpy of formation was calculated by Labbauf, Greenshields, and Rossini (820). The normal boiling point was selected by Rossini, Pitzer, Arnett, Braun, and Pimentel (1248). $Tb = 436.9°K$.

No. 201 4-Ethyl-3-Methylheptane $C_{10}H_{22}$ (Ideal Gas State) Mol Wt 142.276

	cal/(mole °K)				kcal/mole		
$T°K$	$Cp°$	$S°$	$-(G°-H°_{298})/T$	$H°-H°_{298}$	$\Delta Hf°$	$\Delta Gf°$	Log Kp
298	54.14	126.70	126.70	0.00	-59.87	8.77	-6.431
300	54.40	127.04	126.71	0.11	-59.95	9.19	-6.697
400	70.60	144.96	129.04	6.37	-63.79	32.85	-17.949
500	84.63	162.26	133.97	14.15	-66.88	57.38	-25.081
600	96.66	178.78	140.07	23.23	-69.28	82.46	-30.035
700	106.64	194.45	146.73	33.41	-71.05	107.91	-33.689
800	115.02	209.25	153.63	44.50	-72.32	133.55	-36.481
900	122.16	223.22	160.59	56.37	-73.15	159.35	-38.693
1000	128.45	236.43	167.52	68.91	-73.56	185.22	-40.479

The gaseous thermodynamic functions were calculated by the methods of Souders, Matthews, and Hurd (1396, 1397); the enthalpy of formation was calculated by Labbauf, Greenshields, and Rossini (820). The normal boiling point was selected by Rossini, Pitzer, Arnett, Braun, and Pimentel (1248). $Tb = 435.3°K$.

No. 202 3-Ethyl-5-Methylheptane $C_{10}H_{22}$ (Ideal Gas State) Mol Wt 142.276

	cal/(mole °K)				kcal/mole		
$T°K$	$Cp°$	$S°$	$-(G°-H°_{298})/T$	$H°-H°_{298}$	$\Delta Hf°$	$\Delta Gf°$	Log Kp
298	53.09	126.01	126.01	0.00	-61.04	7.81	-5.724
300	53.35	126.34	126.02	0.10	-61.12	8.23	-5.996
400	70.03	144.03	128.32	6.29	-65.04	31.97	-17.468
500	84.32	161.23	133.19	14.03	-68.17	56.60	-24.739
600	96.53	177.72	139.25	23.09	-70.59	81.78	-29.788
700	106.61	193.38	145.87	33.26	-72.37	107.34	-33.511
800	115.04	208.18	152.74	44.35	-73.65	133.08	-36.355
900	122.20	222.15	159.69	56.22	-74.47	158.99	-38.607
1000	128.48	235.36	166.60	68.76	-74.88	184.98	-40.425

The gaseous thermodynamic functions were calculated by the methods of Souders, Matthews, and Hurd (1396, 1397); the enthalpy of formation was calculated by Labbauf, Greenshields, and Rossini (820). The normal boiling point was selected by Rossini, Pitzer, Arnett, Braun, and Pimentel (1248). $Tb = 431.3°K$.

No. 203 3-Ethyl-4-Methylheptane $C_{10}H_{22}$ (Ideal Gas State) Mol Wt 142.276

	cal/(mole °K)			kcal/mole			
$T°K$	$Cp°$	$S°$	$-(G°-H°_{298})/T$	$H°-H°_{298}$	$\Delta Hf°$	$\Delta Gf°$	Log Kp
298	54.14	126.01	126.01	0.00	-59.87	8.98	-6.582
300	54.40	126.35	126.02	0.11	-59.95	9.40	-6.848
400	70.60	144.27	128.35	6.37	-63.79	33.13	-18.100
500	84.63	161.57	133.28	14.15	-66.88	57.73	-25.232
600	96.66	178.09	139.38	23.23	-69.28	82.87	-30.185
700	106.64	193.76	146.04	33.41	-71.05	108.39	-33.840
800	115.02	208.56	152.94	44.50	-72.32	134.10	-36.632
900	122.16	222.53	159.90	56.37	-73.15	159.97	-38.844
1000	128.45	235.74	166.83	68.91	-73.56	185.91	-40.629

The gaseous thermodynamic functions were calculated by the methods of Souders, Matthews, and Hurd (1396, 1397); the enthalpy of formation was calculated by Labbauf, Greenshields, and Rossini (820). The normal boiling point was selected by Rossini, Pitzer, Arnett, Braun, and Pimentel (1248). $Tb = 436.1°K$.

No. 204 4-Ethyl-4-Methylheptane $C_{10}H_{22}$ (Ideal Gas State) Mol Wt 142.276

	cal/(mole °K)			kcal/mole			
$T°K$	$Cp°$	$S°$	$-(G°-H°_{298})/T$	$H°-H°_{298}$	$\Delta Hf°$	$\Delta Gf°$	Log Kp
298	55.85	124.47	124.47	0.00	-61.41	7.90	-5.789
300	56.12	124.82	124.48	0.11	-61.49	8.32	-6.063
400	72.23	143.22	126.88	6.54	-65.16	32.18	-17.581
500	86.45	160.91	131.93	14.50	-68.08	56.86	-24.854
600	98.33	177.75	138.17	23.75	-70.30	82.06	-29.888
700	108.15	193.66	144.97	34.09	-71.91	107.60	-33.592
800	116.41	208.66	152.01	45.33	-73.04	133.31	-36.415
900	123.41	222.78	159.09	57.32	-73.74	159.16	-38.647
1000	129.69	236.12	166.14	69.98	-74.02	185.07	-40.445

The gaseous thermodynamic functions were calculated by the methods of Souders, Matthews, and Hurd (1396, 1397); the enthalpy of formation was calculated by Labbauf, Greenshields, and Rossini (820). The normal boiling point was selected by Rossini, Pitzer, Arnett, Braun, and Pimentel (1248). $Tb = 433.9°K$.

No. 205 2,2,3-Trimethylheptane $C_{10}H_{22}$ (Ideal Gas State) Mol Wt 142.276

The gaseous thermodynamic functions were calculated by the methods of Souders, Matthews, and Hurd (1396, 1397); the enthalpy of formation was calculated by Labbauf, Greenshields, and Rossini (820). The normal boiling point was selected by Rossini, Pitzer, Arnett, Braun, and Pimentel (1248). $Tb = 430.7°K$.

No. 205 2,2,3-Trimethylheptane $C_{10}H_{22}$ (Ideal Gas State) Mol Wt 142.276

	cal/(mole °K)				kcal/mole		
$T°K$	$Cp°$	$S°$	$-(G°-H°_{298})/T$	$H°-H°_{298}$	$\Delta Hf°$	$\Delta Gf°$	Log Kp
298	56.41	120.56	120.56	0.00	-62.58	7.89	-5.786
300	56.67	120.91	120.57	0.11	-62.66	8.33	-6.065
400	72.68	139.46	122.99	6.59	-66.28	32.56	-17.792
500	86.71	157.22	128.07	14.58	-69.16	57.62	-25.185
600	98.52	174.11	134.35	23.86	-71.36	83.18	-30.298
700	108.30	190.05	141.18	34.21	-72.96	109.08	-34.056
800	116.53	205.06	148.24	45.47	-74.07	135.15	-36.920
900	123.52	219.20	155.34	57.47	-74.75	161.36	-39.182
1000	129.74	232.54	162.40	70.14	-75.03	187.64	-41.006

No. 206 2,2,4-Trimethylheptane $C_{10}H_{22}$ (Ideal Gas State) Mol Wt 142.276

	cal/(mole °K)				kcal/mole		
$T°K$	$Cp°$	$S°$	$-(G°-H°_{298})/T$	$H°-H°_{298}$	$\Delta Hf°$	$\Delta Gf°$	Log Kp
298	55.36	120.56	120.56	0.00	-63.06	7.41	-5.434
300	55.62	120.91	120.57	0.11	-63.14	7.85	-5.715
400	72.11	139.22	122.95	6.51	-66.84	32.10	-17.537
500	86.40	156.89	127.98	14.46	-69.76	57.18	-24.994
600	98.39	173.73	134.22	23.71	-71.99	82.78	-30.152
700	108.27	189.66	141.01	34.06	-73.59	108.72	-33.943
800	116.55	204.67	148.04	45.31	-74.71	134.83	-36.831
900	123.56	218.81	155.13	57.32	-75.38	161.08	-39.113
1000	129.77	232.16	162.17	70.00	-75.66	187.39	-40.952

The gaseous thermodynamic functions were calculated by the methods of Souders, Matthews, and Hurd (1396, 1397); the enthalpy of formation was calculated by Labbauf, Greenshields, and Rossini (820). The normal boiling point was selected by Rossini, Pitzer, Arnett, Braun, and Pimentel (1248). $Tb = 421.4°K$.

No. 207 2,2,5-Trimethylheptane $C_{10}H_{22}$ (Ideal Gas State) Mol Wt 142.276

	cal/(mole °K)				kcal/mole		
$T°K$	$Cp°$	$S°$	$-(G°-H°_{298})/T$	$H°-H°_{298}$	$\Delta Hf°$	$\Delta Gf°$	Log Kp
298	55.36	120.56	120.56	0.00	-64.95	5.52	-4.049
300	55.62	120.91	120.57	0.11	-65.03	5.96	-4.338
400	72.11	139.22	122.95	6.51	-68.73	30.21	-16.504
500	86.40	156.89	127.98	14.46	-71.65	55.29	-24.168
600	98.39	173.73	134.22	23.71	-73.88	80.89	-29.463
700	108.27	189.66	141.01	34.06	-75.48	106.83	-33.353
800	116.55	204.67	148.04	45.31	-76.60	132.94	-36.315
900	123.56	218.81	155.13	57.32	-77.27	159.19	-38.654
1000	129.77	232.16	162.17	70.00	-77.55	185.50	-40.538

The gaseous thermodynamic functions were calculated by the methods of Souders, Matthews, and Hurd (1396, 1397); the enthalpy of formation was calculated by Labbauf, Greenshields, and Rossini (820). The normal boiling point was selected by Rossini, Pitzer, Arnett, Braun, and Pimentel (1248). $Tb = 423.9°K$.

No. 208 2,2,6-Trimethylheptane $C_{10}H_{22}$ (Ideal Gas State) Mol Wt 142.276

	cal/(mole °K)			kcal/mole			
$T°K$	$Cp°$	$S°$	$-(G°-H°_{298})/T$	$H°-H°_{298}$	$\Delta Hf°$	$\Delta Gf°$	Log Kp
298	56.43	119.18	119.18	0.00	-65.64	5.25	-3.845
300	56.69	119.53	119.19	0.11	-65.72	5.68	-4.137
400	72.70	138.08	121.61	6.60	-69.34	30.06	-16.421
500	86.73	155.85	126.69	14.59	-72.22	55.25	-24.148
600	98.55	172.74	132.97	23.87	-74.41	80.95	-29.484
700	108.35	188.69	139.81	34.22	-76.01	106.99	-33.401
800	116.59	203.71	146.86	45.48	-77.12	133.19	-36.384
900	123.57	217.85	153.97	57.50	-77.79	159.54	-38.739
1000	129.77	231.20	161.04	70.17	-78.07	185.94	-40.635

The gaseous thermodynamic functions were calculated by the methods of Souders, Matthews, and Hurd (1396, 1397); the enthalpy of formation was calculated by Labbauf, Greenshields, and Rossini (820). The normal boiling point was selected by Rossini, Pitzer, Arnett, Braun, and Pimentel (1248). $Tb = 422.08°K$.

No. 209 2,3,3-Trimethylheptane $C_{10}H_{22}$ (Ideal Gas State) Mol Wt 142.276

	cal/(mole °K)			kcal/mole			
$T°K$	$Cp°$	$S°$	$-(G°-H°_{298})/T$	$H°-H°_{298}$	$\Delta Hf°$	$\Delta Gf°$	Log Kp
298	56.41	121.36	121.36	0.00	-62.01	8.23	-6.029
300	56.67	121.71	121.37	0.11	-62.09	8.66	-6.305
400	72.68	140.26	123.79	6.59	-65.71	32.81	-17.928
500	86.71	158.02	128.87	14.58	-68.59	57.79	-25.259
600	98.52	174.91	135.15	23.86	-70.79	83.27	-30.330
700	108.30	190.85	141.98	34.21	-72.39	109.09	-34.059
800	116.53	205.86	149.04	45.47	-73.50	135.08	-36.901
900	123.52	220.00	156.14	57.47	-74.18	161.21	-39.146
1000	129.74	233.34	163.20	70.14	-74.46	187.41	-40.955

The gaseous thermodynamic functions were calculated by the methods of Souders, Matthews, and Hurd (1396, 1397); the enthalpy of formation was calculated by Labbauf, Greenshields, and Rossini (820). The normal boiling point was selected by Rossini, Pitzer, Arnett, Braun, and Pimentel (1248). $Tb = 433.3°K$.

No. 210 2,3,4-Trimethylheptane $C_{10}H_{22}$ (Ideal Gas State) Mol Wt 142.276

	cal/(mole °K)				kcal/mole		
$T°K$	$Cp°$	$S°$	$-(G°-H°_{298})/T$	$H°-H°_{298}$	$\Delta Hf°$	$\Delta Gf°$	Log Kp
298	54.70	123.59	123.59	0.00	-61.11	8.46	-6.201
300	54.95	123.93	123.60	0.11	-61.19	8.89	-6.474
400	71.05	142.00	125.95	6.43	-64.98	32.85	-17.947
500	84.89	159.38	130.91	14.24	-68.03	57.67	-25.206
600	96.85	175.94	137.05	23.34	-70.41	83.03	-30.243
700	106.79	191.64	143.74	33.54	-72.17	108.77	-33.956
800	115.14	206.46	150.66	44.64	-73.43	134.68	-36.792
900	122.27	220.44	157.64	56.52	-74.24	160.76	-39.037
1000	128.50	233.65	164.59	69.07	-74.64	186.92	-40.849

The gaseous thermodynamic functions were calculated by the methods of Souders, Matthews, and Hurd (1396, 1397); the enthalpy of formation was calculated by Labbauf, Greenshields, and Rossini (820). The normal boiling point was selected by Rossini, Pitzer, Arnett, Braun, and Pimentel (1248). $Tb = 433.0°K$.

No. 211 2,3,5-Trimethylheptane $C_{10}H_{22}$ (Ideal Gas State) Mol Wt 142.276

	cal/(mole °K)				kcal/mole		
$T°K$	$Cp°$	$S°$	$-(G°-H°_{298})/T$	$H°-H°_{298}$	$\Delta Hf°$	$\Delta Gf°$	Log Kp
298	53.65	123.59	123.59	0.00	-62.27	7.30	-5.351
300	53.90	123.93	123.60	0.10	-62.35	7.73	-5.629
400	70.48	141.76	125.92	6.34	-66.22	31.70	-17.321
500	84.58	159.04	130.82	14.11	-69.32	56.55	-24.718
600	96.72	175.57	136.92	23.20	-71.72	81.95	-29.850
700	106.76	191.25	143.57	33.38	-73.48	107.72	-33.631
800	115.16	206.07	150.46	44.49	-74.74	133.68	-36.518
900	122.31	220.06	157.43	56.37	-75.55	159.80	-38.802
1000	128.53	233.27	164.36	68.92	-75.95	185.99	-40.646

The gaseous thermodynamic functions were calculated by the methods of Souders, Matthews, and Hurd (1396, 1397); the enthalpy of formation was calculated by Labbauf, Greenshields, and Rossini (820). The normal boiling point was selected by Rossini, Pitzer, Arnett, Braun, and Pimentel (1248). $Tb = 433.8°K$.

No. 212 2,3,6-Trimethylheptane $C_{10}H_{22}$ (Ideal Gas State) Mol Wt 142.276

The gaseous thermodynamic functions were calculated by the methods of Souders, Matthews, and Hurd (1396, 1397); the enthalpy of formation was calculated by Labbauf, Greenshields, and Rossini (820). The normal boiling point was selected by Rossini, Pitzer, Arnett, Braun, and Pimentel (1248). $Tb = 429.1°K$.

No. 212 2,3,6-Trimethylheptane $C_{10}H_{22}$ (Ideal Gas State) Mol Wt 142.276

	cal/(mole °K)				kcal/mole		
$T°K$	$Cp°$	$S°$	$-(G°-H°_{298})/T$	$H°-H°_{298}$	$\Delta Hf°$	$\Delta Gf°$	Log Kp
298	54.72	122.90	122.90	0.00	-62.96	6.82	-4.996
300	54.97	123.24	122.91	0.11	-63.04	7.24	-5.277
400	71.07	141.32	125.26	6.43	-66.82	31.27	-17.087
500	84.91	158.70	130.22	14.24	-69.88	56.16	-24.548
600	96.88	175.27	136.36	23.35	-72.25	81.59	-29.719
700	106.84	190.97	143.05	33.55	-74.01	107.39	-33.528
800	115.20	205.80	149.98	44.66	-75.26	133.38	-36.436
900	122.32	219.78	156.96	56.54	-76.07	159.52	-38.736
1000	128.53	233.00	163.91	69.09	-76.46	185.74	-40.592

No. 213 2,4,4-Trimethylheptane $C_{10}H_{22}$ (Ideal Gas State) Mol Wt 142.276

	cal/(mole °K)				kcal/mole		
$T°K$	$Cp°$	$S°$	$-(G°-H°_{298})/T$	$H°-H°_{298}$	$\Delta Hf°$	$\Delta Gf°$	Log Kp
298	55.36	121.36	121.36	0.00	-62.49	7.75	-5.677
300	55.62	121.71	121.37	0.11	-62.57	8.18	-5.956
400	72.11	140.02	123.75	6.51	-66.27	32.35	-17.673
500	86.40	157.69	128.78	14.46	-69.19	57.35	-25.068
600	98.39	174.53	135.02	23.71	-71.42	82.87	-30.184
700	108.27	190.46	141.81	34.06	-73.02	108.73	-33.946
800	116.55	205.47	148.84	45.31	-74.14	134.76	-36.812
900	123.56	219.61	155.93	57.32	-74.81	160.93	-39.077
1000	129.77	232.96	162.97	70.00	-75.09	187.16	-40.901

The gaseous thermodynamic functions were calculated by the methods of Souders, Matthews, and Hurd (1396, 1397); the enthalpy of formation was calculated by Labbauf, Greenshields, and Rossini (820). The normal boiling point was selected by Rossini, Pitzer, Arnett, Braun, and Pimentel (1248). $Tb = 424.1°K$.

No. 214 2,4,5-Trimethylheptane $C_{10}H_{22}$ (Ideal Gas State) Mol Wt 142.276

	cal/(mole °K)				kcal/mole		
$T°K$	$Cp°$	$S°$	$-(G°-H°_{298})/T$	$H°-H°_{298}$	$\Delta Hf°$	$\Delta Gf°$	Log Kp
298	53.65	123.59	123.59	0.00	-62.27	7.30	-5.351
300	53.90	123.93	123.60	0.10	-62.35	7.73	-5.629
400	70.48	141.76	125.92	6.34	-66.22	31.70	-17.321
500	84.58	159.04	130.82	14.11	-69.32	56.55	-24.718
600	96.72	175.57	136.92	23.20	-71.72	81.95	-29.850
700	106.76	191.25	143.57	33.38	-73.48	107.72	-33.631
800	115.16	206.07	150.46	44.49	-74.74	133.68	-36.518
900	122.31	220.06	157.43	56.37	-75.55	159.80	-38.802
1000	128.53	233.27	164.36	68.92	-75.95	185.99	-40.646

The gaseous thermodynamic functions were calculated by the methods of Souders, Matthews, and Hurd (1396, 1397); the enthalpy of formation was calculated by Labbauf, Greenshields, and Rossini (820). The normal boiling point was selected by Rossini, Pitzer, Arnett, Braun, and Pimentel (1248). $Tb = 429.6°K$.

No. 215 2,4,6-Trimethylheptane $C_{10}H_{22}$ (Ideal Gas State) Mol Wt 142.276

	cal/(mole °K)				kcal/mole		
$T°K$	$Cp°$	$S°$	$-(G°-H°_{298})/T$	$H°-H°_{298}$	$\Delta Hf°$	$\Delta Gf°$	Log Kp
298	53.67	121.52	121.52	0.00	-60.52	9.67	-7.086
300	53.92	121.86	121.53	0.10	-60.60	10.10	-7.356
400	70.50	139.70	123.85	6.35	-64.47	34.28	-18.729
500	84.60	156.98	128.76	14.12	-67.56	59.34	-25.935
600	96.75	173.51	134.85	23.20	-69.96	84.94	-30.939
700	106.81	189.20	141.50	33.39	-71.72	110.92	-34.628
800	115.22	204.03	148.40	44.51	-72.97	137.08	-37.446
900	122.36	218.02	155.37	56.39	-73.78	163.40	-39.677
1000	128.56	231.24	162.30	68.95	-74.17	189.80	-41.478

The gaseous thermodynamic functions were calculated by the methods of Souders, Matthews, and Hurd (1396, 1397); the enthalpy of formation was calculated by Labbauf, Greenshields, and Rossini (820). The normal boiling point was selected by Rossini, Pitzer, Arnett, Braun, and Pimentel (1248). $Tb = 420.7°K$.

No. 216 2,5,5-Trimethylheptane $C_{10}H_{22}$ (Ideal Gas State) Mol Wt 142.276

	cal/(mole °K)				kcal/mole		
$T°K$	$Cp°$	$S°$	$-(G°-H°_{298})/T$	$H°-H°_{298}$	$\Delta Hf°$	$\Delta Gf°$	Log Kp
298	55.36	121.36	121.36	0.00	-64.38	5.86	-4.292
300	55.62	121.71	121.37	0.11	-64.46	6.29	-4.579
400	72.11	140.02	123.75	6.51	-68.16	30.46	-16.641
500	86.40	157.69	128.78	14.46	-71.08	55.46	-24.242
600	98.39	174.53	135.02	23.71	-73.31	80.98	-29.496
700	108.27	190.46	141.81	34.06	-74.91	106.84	-33.356
800	116.55	205.47	148.84	45.31	-76.03	132.87	-36.296
900	123.56	219.61	155.93	57.32	-76.70	159.04	-38.618
1000	129.77	232.96	162.97	70.00	-76.98	185.27	-40.488

The gaseous thermodynamic functions were calculated by the methods of Souders, Matthews, and Hurd (1396, 1397); the enthalpy of formation was calculated by Labbauf, Greenshields, and Rossini (820). The normal boiling point was selected by Rossini, Pitzer, Arnett, Braun, and Pimentel (1248). $Tb = 425.95°K$.

No. 217 3,3,4-Trimethylheptane $C_{10}H_{22}$ (Ideal Gas State) Mol Wt 142.276

	cal/(mole °K)				kcal/mole		
$T°K$	$Cp°$	$S°$	$-(G°-H°_{298})/T$	$H°-H°_{298}$	$\Delta Hf°$	$\Delta Gf°$	Log Kp
298	55.34	122.74	122.74	0.00	-61.32	8.50	-6.233
300	55.60	123.09	122.75	0.11	-61.40	8.93	-6.506
400	72.09	141.39	125.13	6.51	-65.10	32.97	-18.011
500	86.38	159.06	130.16	14.45	-68.03	57.83	-25.278
600	98.36	175.90	136.39	23.71	-70.25	83.22	-30.310
700	108.22	191.82	143.19	34.05	-71.86	108.94	-34.011
800	116.49	206.83	150.21	45.30	-72.98	134.83	-36.832
900	123.51	220.96	157.30	57.30	-73.67	160.87	-39.062
1000	129.74	234.30	164.34	69.97	-73.95	186.96	-40.858

The gaseous thermodynamic functions were calculated by the methods of Souders, Matthews, and Hurd (1396, 1397); the enthalpy of formation was calculated by Labbauf, Greenshields, and Rossini (820). The normal boiling point was selected by Rossini, Pitzer, Arnett, Braun, and Pimentel (1248). $Tb = 435.0°K$.

No. 218 3,3,5-Trimethylheptane $C_{10}H_{22}$ (Ideal Gas State) Mol Wt 142.276

	cal/(mole °K)				kcal/mole		
$T°K$	$Cp°$	$S°$	$-(G°-H°_{298})/T$	$H°-H°_{298}$	$\Delta Hf°$	$\Delta Gf°$	Log Kp
298	54.29	122.74	122.74	0.00	-61.80	8.02	-5.881
300	54.55	123.08	122.75	0.11	-61.88	8.45	-6.157
400	71.52	141.16	125.10	6.43	-65.66	32.50	-17.757
500	86.07	158.72	130.08	14.33	-68.63	57.40	-25.087
600	98.23	175.52	136.26	23.56	-70.88	82.81	-30.164
700	108.19	191.44	143.02	33.90	-72.50	108.58	-33.898
800	116.51	206.44	150.02	45.14	-73.62	134.51	-36.744
900	123.55	220.58	157.08	57.15	-74.30	160.58	-38.992
1000	129.74	233.92	164.10	69.82	-74.57	186.71	-40.804

The gaseous thermodynamic functions were calculated by the methods of Souders, Matthews, and Hurd (1396, 1397); the enthalpy of formation was calculated by Labbauf, Greenshields, and Rossini (820). The normal boiling point was selected by Rossini, Pitzer, Arnett, Braun, and Pimentel (1248). $Tb = 428.83°K$.

No. 219 3,4,4-Trimethylheptane $C_{10}H_{22}$ (Ideal Gas State) Mol Wt 142.276

The gaseous thermodynamic functions were calculated by the methods of Souders, Matthews, and Hurd (1396, 1397); the enthalpy of formation was calculated by Labbauf, Greenshields, and Rossini (820). The normal boiling point was selected by Rossini, Pitzer, Arnett, Braun, and Pimentel (1248). $Tb = 434.3°K$.

No. 219 3,4,4-Trimethylheptane $C_{10}H_{22}$ (Ideal Gas State) Mol Wt 142.276

$T°K$	$Cp°$	$S°$	$-(G°-H°_{298})/T$	$H°-H°_{298}$	$\Delta Hf°$	$\Delta Gf°$	Log Kp
		cal/(mole °K)			kcal/mole		
298	55.34	122.74	122.74	0.00	-61.32	8.50	-6.233
300	55.60	123.09	122.75	0.11	-61.40	8.93	-6.506
400	72.09	141.39	125.13	6.51	-65.10	32.97	-18.011
500	86.38	159.06	130.16	14.45	-68.03	57.83	-25.278
600	98.36	175.90	136.39	23.71	-70.25	83.22	-30.310
700	108.22	191.82	143.19	34.05	-71.86	108.94	-34.011
800	116.49	206.83	150.21	45.30	-72.98	134.83	-36.832
900	123.51	220.96	157.30	57.30	-73.67	160.87	-39.062
1000	129.74	234.30	164.34	69.97	-73.95	186.96	-40.858

No. 220 3,4,5-Trimethylheptane $C_{10}H_{22}$ (Ideal Gas State) Mol Wt 142.276

$T°K$	$Cp°$	$S°$	$-(G°-H°_{298})/T$	$H°-H°_{298}$	$\Delta Hf°$	$\Delta Gf°$	Log Kp
		cal/(mole °K)			kcal/mole		
298	53.63	123.59	123.59	0.00	-60.43	9.14	-6.700
300	53.88	123.93	123.60	0.10	-60.51	9.57	-6.969
400	70.46	141.76	125.92	6.34	-64.38	33.54	-18.326
500	84.56	159.03	130.82	14.11	-67.48	58.39	-25.523
600	96.69	175.55	136.91	23.19	-69.88	83.79	-30.521
700	106.71	191.23	143.56	33.37	-71.65	109.57	-34.207
800	115.10	206.04	150.46	44.47	-72.92	135.53	-37.022
900	122.26	220.02	157.42	56.35	-73.73	161.65	-39.251
1000	128.50	233.23	164.34	68.90	-74.13	187.84	-41.051

The gaseous thermodynamic functions were calculated by the methods of Souders, Matthews, and Hurd (1396, 1397); the enthalpy of formation was calculated by Labbauf, Greenshields, and Rossini (820). The normal boiling point was selected by Rossini, Pitzer, Arnett, Braun, and Pimentel (1248). $Tb = 435.6°K$.

No. 221 3-Isopropyl-2-Methylhexane $C_{10}H_{22}$ (Ideal Gas State) Mol Wt 142.276

$T°K$	$Cp°$	$S°$	$-(G°-H°_{298})/T$	$H°-H°_{298}$	$\Delta Hf°$	$\Delta Gf°$	Log Kp
		cal/(mole °K)			kcal/mole		
298	54.70	121.52	121.52	0.00	-61.11	9.08	-6.654
300	54.95	121.86	121.53	0.11	-61.19	9.51	-6.926
400	71.05	139.93	123.88	6.43	-64.98	33.68	-18.399
500	84.89	157.31	128.84	14.24	-68.03	58.71	-25.659
600	96.85	173.87	134.98	23.34	-70.41	84.28	-30.696
700	106.79	189.57	141.67	33.54	-72.17	110.21	-34.409
800	115.14	204.39	148.59	44.64	-73.43	136.34	-37.244
900	122.27	218.37	155.57	56.52	-74.24	162.63	-39.489
1000	128.50	231.58	162.52	69.07	-74.64	188.99	-41.301

The gaseous thermodynamic functions were calculated by the methods of Souders, Matthews, and Hurd (1396, 1397); the enthalpy of formation was calculated by Labbauf, Greenshields, and Rossini (820). The normal boiling point was selected by Rossini, Pitzer, Arnett, Braun, and Pimentel (1248). $Tb = 439.8°K$.

No. 222 3,3-Diethylhexane $C_{10}H_{22}$ (Ideal Gas State)   Mol Wt 142.276

	cal/(mole °K)				kcal/mole		
$T°K$	$Cp°$	$S°$	$-(G°-H°_{298})/T$	$H°-H°_{298}$	$\Delta Hf°$	$\Delta Gf°$	Log Kp
298	54.78	122.29	122.29	0.00	-60.15	9.81	-7.189
300	55.05	122.63	122.30	0.11	-60.23	10.24	-7.457
400	71.64	140.79	124.66	6.46	-63.98	34.33	-18.753
500	86.12	158.38	129.66	14.37	-66.94	59.26	-25.900
600	98.17	175.18	135.86	23.60	-69.19	84.71	-30.853
700	108.07	191.08	142.62	33.92	-70.82	110.51	-34.500
800	116.37	206.06	149.62	45.16	-71.95	136.47	-37.280
900	123.40	220.18	156.69	57.15	-72.65	162.58	-39.479
1000	129.69	233.52	163.71	69.81	-72.93	188.76	-41.251

The gaseous thermodynamic functions were calculated by the methods of Souders, Matthews, and Hurd (1396, 1397); the enthalpy of formation was calculated by Labbauf, Greenshields, and Rossini (820). The normal boiling point was selected by Rossini, Pitzer, Arnett, Braun, and Pimentel (1248). $Tb = 439.4°K$.

No. 223 3,4-Diethylhexane $C_{10}H_{22}$ (Ideal Gas State)   Mol Wt 142.276

	cal/(mole °K)				kcal/mole		
$T°K$	$Cp°$	$S°$	$-(G°-H°_{298})/T$	$H°-H°_{298}$	$\Delta Hf°$	$\Delta Gf°$	Log Kp
298	53.07	123.25	123.25	0.00	-59.17	10.50	-7.698
300	53.33	123.58	123.26	0.10	-59.25	10.93	-7.961
400	70.01	141.27	125.55	6.29	-63.17	34.95	-19.093
500	84.30	158.46	130.43	14.02	-66.31	59.85	-26.160
600	96.50	174.94	136.48	23.08	-68.73	85.31	-31.073
700	106.56	190.60	143.11	33.25	-70.52	111.15	-34.699
800	114.98	205.39	149.98	44.34	-71.79	137.17	-37.471
900	122.15	219.35	156.92	56.20	-72.62	163.36	-39.666
1000	128.45	232.56	163.83	68.74	-73.03	189.62	-41.439

The gaseous thermodynamic functions were calculated by the methods of Souders, Matthews, and Hurd (1396, 1397); the enthalpy of formation was calculated by Labbauf, Greenshields, and Rossini (820). The normal boiling point was selected by Rossini, Pitzer, Arnett, Braun, and Pimentel (1248). $Tb = 437.0°K$.

Alkane Ideal Gas Tables

No. 224   3-Ethyl-2,2-Dimethylhexane $C_{10}H_{22}$ (Ideal Gas State)   Mol Wt 142.276

$T°K$	$Cp°$	$S°$	$-(G°-H°_{298})/T$	$H°-H°_{298}$	$\Delta Hf°$	$\Delta Gf°$	Log Kp
		cal/(mole °K)			kcal/mole		
298	55.34	120.56	120.56	0.00	-61.89	8.58	-6.292
300	55.60	120.91	120.57	0.11	-61.97	9.02	-6.568
400	72.09	139.21	122.95	6.51	-65.67	33.27	-18.176
500	86.38	156.88	127.98	14.45	-68.60	58.35	-25.506
600	98.36	173.72	134.21	23.71	-70.82	83.95	-30.579
700	108.22	189.64	141.01	34.05	-72.43	109.90	-34.309
800	116.49	204.65	148.03	45.30	-73.55	136.00	-37.153
900	123.51	218.78	155.12	57.30	-74.24	162.26	-39.400
1000	129.74	232.12	162.16	69.97	-74.52	188.57	-41.210

The gaseous thermodynamic functions were calculated by the methods of Souders, Matthews, and Hurd (1396, 1397); the enthalpy of formation was calculated by Labbauf, Greenshields, and Rossini (820). The normal boiling point was selected by Rossini, Pitzer, Arnett, Braun, and Pimentel (1248). $Tb = 429.2°K$.

No. 225   4-Ethyl-2,2-Dimethylhexane $C_{10}H_{22}$ (Ideal Gas State)   Mol Wt 142.276

$T°K$	$Cp°$	$S°$	$-(G°-H°_{298})/T$	$H°-H°_{298}$	$\Delta Hf°$	$\Delta Gf°$	Log Kp
		cal/(mole °K)			kcal/mole		
298	54.29	119.18	119.18	0.00	-62.37	8.52	-6.242
300	54.55	119.52	119.19	0.11	-62.45	8.95	-6.519
400	71.52	137.60	121.54	6.43	-66.23	33.35	-18.223
500	86.07	155.16	126.52	14.33	-69.20	58.61	-25.616
600	98.23	171.96	132.70	23.56	-71.45	84.38	-30.734
700	108.19	187.88	139.46	33.90	-73.07	110.50	-34.498
800	116.51	202.88	146.46	45.14	-74.19	136.78	-37.366
900	123.55	217.02	153.52	57.15	-74.87	163.21	-39.632
1000	129.77	230.36	160.54	69.83	-75.14	189.70	-41.458

The gaseous thermodynamic functions were calculated by the methods of Souders, Matthews, and Hurd (1396, 1397); the enthalpy of formation was calculated by Labbauf, Greenshields, and Rossini (820). The normal boiling point was selected by Rossini, Pitzer, Arnett, Braun, and Pimentel (1248). $Tb = 420°K$.

No. 226   3-Ethyl-2,3-Dimethylhexane $C_{10}H_{22}$ (Ideal Gas State)   Mol Wt 142.276

The gaseous thermodynamic functions were calculated by the methods of Souders, Matthews, and Hurd (1396, 1397); the enthalpy of formation was calculated by Labbauf, Greenshields, and Rossini (820). The normal boiling point was selected by Rossini, Pitzer, Arnett, Braun, and Pimentel (1248). $Tb = 436.8°K$.

No. 226 3-Ethyl-2,3-Dimethylhexane $C_{10}H_{22}$ (Ideal Gas State) Mol Wt 142.276

$T°K$	$Cp°$	$S°$	$-(G°-H°_{298})/T$	$H°-H°_{298}$	$\Delta Hf°$	$\Delta Gf°$	Log Kp
	cal/(mole °K)				kcal/mole		
298	55.34	122.74	122.74	0.00	-60.75	9.07	-6.651
300	55.60	123.09	122.75	0.11	-60.83	9.50	-6.922
400	72.09	141.39	125.13	6.51	-64.53	33.54	-18.323
500	86.38	159.06	130.16	14.45	-67.46	58.40	-25.527
600	98.36	175.90	136.39	23.71	-69.68	83.79	-30.517
700	108.22	191.82	143.19	34.05	-71.29	109.51	-34.189
800	116.49	206.83	150.21	45.30	-72.41	135.40	-36.988
900	123.51	220.96	157.30	57.30	-73.10	161.44	-39.200
1000	129.74	234.30	164.34	69.97	-73.38	187.53	-40.983

No. 227 4-Ethyl-2,3-Dimethylhexane $C_{10}H_{22}$ (Ideal Gas State) Mol Wt 142.276

$T°K$	$Cp°$	$S°$	$-(G°-H°_{298})/T$	$H°-H°_{298}$	$\Delta Hf°$	$\Delta Gf°$	Log Kp
	cal/(mole °K)				kcal/mole		
298	53.63	122.90	122.90	0.00	-60.43	9.35	-6.851
300	53.88	123.24	122.91	0.10	-60.51	9.77	-7.120
400	70.46	141.07	125.23	6.34	-64.38	33.82	-18.477
500	84.56	158.34	130.13	14.11	-67.48	58.74	-25.674
600	96.69	174.86	136.22	23.19	-69.88	84.21	-30.671
700	106.71	190.54	142.87	33.37	-71.65	110.05	-34.357
800	115.10	205.35	149.77	44.47	-72.92	136.08	-37.173
900	122.26	219.33	156.73	56.35	-73.73	162.27	-39.402
1000	128.50	232.54	163.65	68.90	-74.13	188.53	-41.202

The gaseous thermodynamic functions were calculated by the methods of Souders, Matthews, and Hurd (1396, 1397); the enthalpy of formation was calculated by Labbauf, Greenshields, and Rossini (820). The normal boiling point was selected by Rossini, Pitzer, Arnett, Braun, and Pimentel (1248). $Tb = 434.0°K$.

No. 228 3-Ethyl-2,4-Dimethylhexane $C_{10}H_{22}$ (Ideal Gas State) Mol Wt 142.276

$T°K$	$Cp°$	$S°$	$-(G°-H°_{298})/T$	$H°-H°_{298}$	$\Delta Hf°$	$\Delta Gf°$	Log Kp
	cal/(mole °K)				kcal/mole		
298	53.63	123.59	123.59	0.00	-60.43	9.14	-6.700
300	53.88	123.93	123.60	0.10	-60.51	9.57	-6.969
400	70.46	141.76	125.92	6.34	-64.38	33.54	-18.326
500	84.56	159.03	130.82	14.11	-67.48	58.39	-25.523
600	96.69	175.55	136.91	23.19	-69.88	83.79	-30.521
700	106.71	191.23	143.56	33.37	-71.65	109.57	-34.207
800	115.10	206.04	150.46	44.47	-72.92	135.53	-37.022
900	122.26	220.02	157.42	56.35	-73.73	161.65	-39.251
1000	128.50	233.23	164.34	68.90	-74.13	187.84	-41.051

The gaseous thermodynamic functions were calculated by the methods of Souders, Matthews, and Hurd (1396, 1397); the enthalpy of formation was calculated by Labbauf, Greenshields, and Rossini (820). The normal boiling point was selected by Rossini, Pitzer, Arnett, Braun, and Pimentel (1248). $Tb = 433.2°K$.

No. 229 4-Ethyl-2,4-Dimethylhexane $C_{10}H_{22}$ (Ideal Gas State) Mol Wt 142.276

$T°K$	$Cp°$	cal/(mole °K) $S°$	$-(G° - H°_{298})/T$	$H° - H°_{298}$	kcal/mole $\Delta Hf°$	$\Delta Gf°$	Log Kp
298	54.29	121.36	121.36	0.00	-61.23	9.01	-6.601
300	54.55	121.70	121.37	0.11	-61.31	9.44	-6.873
400	71.52	139.78	123.72	6.43	-65.09	33.62	-18.370
500	86.07	157.34	128.70	14.33	-68.06	58.66	-25.638
600	98.23	174.14	134.88	23.56	-70.31	84.21	-30.673
700	108.19	190.06	141.64	33.90	-71.93	110.11	-34.377
800	116.51	205.06	148.64	45.14	-73.05	136.18	-37.201
900	123.55	219.20	155.70	57.15	-73.73	162.39	-39.432
1000	129.77	232.54	162.72	69.83	-74.00	188.66	-41.230

The gaseous thermodynamic functions were calculated by the methods of Souders, Matthews, and Hurd (1396, 1397); the enthalpy of formation was calculated by Labbauf, Greenshields, and Rossini (820). The normal boiling point was selected by Rossini, Pitzer, Arnett, Braun, and Pimentel (1248). $Tb = 434.2°K$.

No. 230 3-Ethyl-2,5-Dimethylhexane $C_{10}H_{22}$ (Ideal Gas State) Mol Wt 142.276

$T°K$	$Cp°$	cal/(mole °K) $S°$	$-(G° - H°_{298})/T$	$H° - H°_{298}$	kcal/mole $\Delta Hf°$	$\Delta Gf°$	Log Kp
298	53.65	122.90	122.90	0.00	-62.27	7.51	-5.502
300	53.90	123.24	122.91	0.10	-62.35	7.93	-5.779
400	70.48	141.07	125.23	6.34	-66.22	31.98	-17.471
500	84.58	158.35	130.13	14.11	-69.32	56.90	-24.869
600	96.72	174.88	136.23	23.20	-71.72	82.37	-30.000
700	106.76	190.56	142.88	33.38	-73.48	108.21	-33.782
800	115.16	205.38	149.77	44.49	-74.74	134.23	-36.668
900	122.31	219.37	156.74	56.37	-75.55	160.42	-38.953
1000	128.53	232.58	163.67	68.92	-75.95	186.68	-40.797

The gaseous thermodynamic functions were calculated by the methods of Souders, Matthews, and Hurd (1396, 1397); the enthalpy of formation was calculated by Labbauf, Greenshields, and Rossini (820). The normal boiling point was selected by Rossini, Pitzer, Arnett, Braun, and Pimentel (1248). $Tb = 427.2°K$.

No. 231 4-Ethyl-3,3-Dimethylhexane $C_{10}H_{22}$ (Ideal Gas State) Mol Wt 142.276

T°K	cal/(mole °K)				kcal/mole		Log Kp
	$Cp°$	$S°$	$-(G°-H°_{298})/T$	$H°-H°_{298}$	$\Delta Hf°$	$\Delta Gf°$	
298	54.27	121.36	121.36	0.00	-60.63	9.61	-7.041
300	54.53	121.70	121.37	0.11	-60.71	10.04	-7.311
400	71.50	139.77	123.72	6.43	-64.49	34.22	-18.698
500	86.05	157.33	128.69	14.32	-67.47	59.26	-25.901
600	98.20	174.13	134.88	23.56	-69.72	84.82	-30.892
700	108.14	190.03	141.63	33.89	-71.34	110.72	-34.566
800	116.45	205.03	148.63	45.13	-72.46	136.79	-37.367
900	123.50	219.16	155.69	57.13	-73.15	163.00	-39.580
1000	129.74	232.50	162.71	69.80	-73.43	189.28	-41.364

The gaseous thermodynamic functions were calculated by the methods of Souders, Matthews, and Hurd (1396, 1397); the enthalpy of formation was calculated by Labbauf, Greenshields, and Rossini (820). The normal boiling point was selected by Rossini, Pitzer, Arnett, Braun, and Pimentel (1248). $Tb = 436.0°K$.

No. 232 3-Ethyl-3,4-Dimethylhexane $C_{10}H_{22}$ (Ideal Gas State) Mol Wt 142.276

T°K	cal/(mole °K)				kcal/mole		Log Kp
	$Cp°$	$S°$	$-(G°-H°_{298})/T$	$H°-H°_{298}$	$\Delta Hf°$	$\Delta Gf°$	
298	54.27	122.74	122.74	0.00	-60.06	9.76	-7.157
300	54.53	123.08	122.75	0.11	-60.14	10.19	-7.424
400	71.50	141.15	125.10	6.43	-63.92	34.24	-18.707
500	86.05	158.71	130.07	14.32	-66.90	59.14	-25.848
600	98.20	175.51	136.26	23.56	-69.15	84.56	-30.798
700	108.14	191.41	143.01	33.89	-70.77	110.32	-34.442
800	116.45	206.41	150.01	45.13	-71.89	136.25	-37.221
900	123.50	220.54	157.07	57.13	-72.58	162.33	-39.417
1000	129.74	233.88	164.09	69.80	-72.86	188.47	-41.187

The gaseous thermodynamic functions were calculated by the methods of Souders, Matthews, and Hurd (1396, 1397); the enthalpy of formation was calculated by Labbauf, Greenshields, and Rossini (820). The normal boiling point was selected by Rossini, Pitzer, Arnett, Braun, and Pimentel (1248). $Tb = 435.2°K$.

No. 233 2,2,3,3-Tetramethylhexane $C_{10}H_{22}$ (Ideal Gas State) Mol Wt 142.276

The gaseous thermodynamic functions were calculated by the methods of Souders, Matthews, and Hurd (1396, 1397); the enthalpy of formation was calculated by Labbauf, Greenshields, and Rossini (820). Transitional temperatures were selected by Rossini, Pitzer, Arnett, Braun, and Pimentel (1248). $Tm = 219.15°K$ and $Tb = 433.46°K$.

No. 233   2,2,3,3-Tetramethylhexane $C_{10}H_{22}$ (Ideal Gas State)   Mol Wt 142.276

T°K	Cp°	S°	$-(G°-H°_{298})/T$	$H°-H°_{298}$	$\Delta Hf°$	$\Delta Gf°$	Log Kp
		cal/(mole °K)			kcal/mole		
298	56.67	115.91	115.91	0.00	-61.63	10.23	-7.499
300	56.93	116.27	115.92	0.11	-61.71	10.67	-7.773
400	73.59	134.98	118.35	6.65	-65.27	35.37	-19.323
500	88.00	153.00	123.49	14.76	-68.04	60.86	-26.600
600	99.88	170.12	129.85	24.17	-70.10	86.83	-31.626
700	109.60	186.27	136.77	34.65	-71.57	113.12	-35.316
800	117.75	201.45	143.92	46.03	-72.56	139.56	-38.123
900	124.65	215.73	151.11	58.16	-73.12	166.12	-40.338
1000	130.81	229.18	158.25	70.94	-73.29	192.74	-42.120

No. 234   2,2,3,4-Tetramethylhexane $C_{10}H_{22}$ (Ideal Gas State)   Mol Wt 142.276

T°K	Cp°	S°	$-(G°-H°_{298})/T$	$H°-H°_{298}$	$\Delta Hf°$	$\Delta Gf°$	Log Kp
		cal/(mole °K)			kcal/mole		
298	54.83	118.14	118.14	0.00	-60.55	10.65	-7.803
300	55.08	118.48	118.15	0.11	-60.63	11.08	-8.073
400	71.95	136.70	120.52	6.48	-64.36	35.58	-19.441
500	86.31	154.34	125.53	14.41	-67.30	60.92	-26.627
600	98.39	171.18	131.75	23.66	-69.53	86.77	-31.606
700	108.29	187.11	138.53	34.01	-71.13	112.97	-35.269
800	116.57	202.12	145.55	45.27	-72.24	139.33	-38.062
900	123.61	216.27	152.63	57.28	-72.92	165.84	-40.269
1000	129.79	229.62	159.67	69.96	-73.19	192.40	-42.047

The gaseous thermodynamic functions were calculated by the methods of Souders, Matthews, and Hurd (1396, 1397); the enthalpy of formation was calculated by Labbauf, Greenshields, and Rossini (820). The normal boiling point was selected by Rossini, Pitzer, Arnett, Braun, and Pimentel (1248). $Tb = 431.9°K$.

No. 235   2,2,3,5-Tetramethylhexane $C_{10}H_{22}$ (Ideal Gas State)   Mol Wt 142.276

T°K	Cp°	S°	$-(G°-H°_{298})/T$	$H°-H°_{298}$	$\Delta Hf°$	$\Delta Gf°$	Log Kp
		cal/(mole °K)			kcal/mole		
298	54.85	117.45	117.45	0.00	-64.29	7.11	-5.212
300	55.10	117.80	117.46	0.11	-64.37	7.55	-5.499
400	71.97	136.02	119.83	6.48	-68.10	32.12	-17.548
500	86.33	153.66	124.84	14.42	-71.04	57.53	-25.143
600	98.42	170.50	131.06	23.67	-73.26	83.45	-30.394
700	108.34	186.44	137.84	34.02	-74.86	109.71	-34.251
800	116.63	201.46	144.87	45.28	-75.97	136.14	-37.189
900	123.66	215.61	151.95	57.30	-76.64	162.71	-39.509
1000	129.82	228.97	158.99	69.98	-76.90	189.34	-41.378

The gaseous thermodynamic functions were calculated by the methods of Souders, Matthews, and Hurd (1396, 1397); the enthalpy of formation was calculated by Labbauf, Greenshields, and Rossini (820). The normal boiling point was selected by Rossini, Pitzer, Arnett, Braun, and Pimentel (1248). $Tb = 421.5°K$.

No. 236 2,2,4,4-Tetramethylhexane $C_{10}H_{22}$ (Ideal Gas State) Mol Wt 142.276

T°K	Cp°	S°	$-(G°-H°_{298})/T$	$H°-H°_{298}$	$\Delta Hf°$	$\Delta Gf°$	Log Kp
		cal/(mole °K)			kcal/mole		
298	55.49	115.91	115.91	0.00	-61.50	10.36	-7.594
300	55.75	116.26	115.92	0.11	-61.58	10.80	-7.868
400	73.01	134.72	118.32	6.57	-65.23	35.51	-19.402
500	87.82	152.65	123.40	14.63	-68.03	61.04	-26.677
600	99.93	169.77	129.72	24.04	-70.10	87.04	-31.704
700	109.77	185.93	136.60	34.54	-71.56	113.37	-35.393
800	117.98	201.14	143.73	45.93	-72.53	139.84	-38.200
900	124.90	215.44	150.91	58.08	-73.07	166.43	-40.413
1000	131.06	228.93	158.05	70.89	-73.21	193.07	-42.194

The gaseous thermodynamic functions were calculated by the methods of Souders, Matthews, and Hurd (1396, 1397); the enthalpy of formation was calculated by Labbauf, Greenshields, and Rossini (820). The normal boiling point was selected by Rossini, Pitzer, Arnett, Braun, and Pimentel (1248). $Tb = 426.9°K$.

No. 237 2,2,4,5-Tetramethylhexane $C_{10}H_{22}$ (Ideal Gas State) Mol Wt 142.276

T°K	Cp°	S°	$-(G°-H°_{298})/T$	$H°-H°_{298}$	$\Delta Hf°$	$\Delta Gf°$	Log Kp
		cal/(mole °K)			kcal/mole		
298	54.85	117.45	117.45	0.00	-63.61	7.79	-5.711
300	55.10	117.80	117.46	0.11	-63.69	8.23	-5.994
400	71.97	136.02	119.83	6.48	-67.42	32.80	-17.919
500	86.33	153.66	124.84	14.42	-70.36	58.21	-25.440
600	98.42	170.50	131.06	23.67	-72.58	84.13	-30.641
700	108.34	186.44	137.84	34.02	-74.18	110.39	-34.463
800	116.63	201.46	144.87	45.28	-75.29	136.82	-37.375
900	123.66	215.61	151.95	57.30	-75.96	163.39	-39.674
1000	129.82	228.97	158.99	69.98	-76.22	190.02	-41.526

The gaseous thermodynamic functions were calculated by the methods of Souders, Matthews, and Hurd (1396, 1397); the enthalpy of formation was calculated by Labbauf, Greenshields, and Rossini (820). The normal boiling point was selected by Rossini, Pitzer, Arnett, Braun, and Pimentel (1248). $Tb = 421.03°K$.

No. 238 2,2,5,5-Tetramethylhexane $C_{10}H_{22}$ (Ideal Gas State) Mol Wt 142.276

$T°K$	cal/(mole °K)				kcal/mole		
	$Cp°$	$S°$	$-(G°-H°_{298})/T$	$H°-H°_{298}$	$\Delta Hf°$	$\Delta Gf°$	Log Kp
298	56.56	112.35	112.35	0.00	-68.18	4.74	-3.476
300	56.82	112.71	112.36	0.11	-68.26	5.19	-3.780
400	73.60	131.40	114.79	6.65	-71.82	30.24	-16.523
500	88.15	149.44	119.93	14.76	-74.58	56.09	-24.516
600	100.09	166.60	126.29	24.19	-76.63	82.42	-30.019
700	109.85	182.78	133.22	34.70	-78.07	109.06	-34.048
800	118.02	198.00	140.37	46.10	-79.04	135.84	-37.108
900	124.91	212.30	147.58	58.26	-79.57	162.75	-39.519
1000	131.06	225.79	154.73	71.06	-79.72	189.70	-41.458

The gaseous thermodynamic functions were calculated by the methods of Souders, Matthews, and Hurd (1396, 1397); the enthalpy of formation was calculated by Labbauf, Greenshields, and Rossini (820). Transitional temperatures were selected by Rossini, Pitzer, Arnett, Braun, and Pimentel (1248). $Tm = 260.55°K$ and $Tb = 410.61°K$.

No. 239 2,3,3,4-Tetramethylhexane $C_{10}H_{22}$ (Ideal Gas State) Mol Wt 142.276

$T°K$	cal/(mole °K)				kcal/mole		
	$Cp°$	$S°$	$-(G°-H°_{298})/T$	$H°-H°_{298}$	$\Delta Hf°$	$\Delta Gf°$	Log Kp
298	54.83	119.63	119.63	0.00	-60.66	10.09	-7.397
300	55.08	119.97	119.64	0.11	-60.74	10.52	-7.667
400	71.95	138.19	122.01	6.48	-64.47	34.88	-19.055
500	86.31	155.83	127.02	14.41	-67.41	60.07	-26.254
600	98.39	172.67	133.24	23.66	-69.64	85.77	-31.240
700	108.29	188.60	140.02	34.01	-71.24	111.82	-34.909
800	116.57	203.61	147.04	45.27	-72.35	138.03	-37.706
900	123.61	217.76	154.12	57.28	-73.03	164.39	-39.916
1000	129.79	231.11	161.16	69.96	-73.30	190.80	-41.697

The gaseous thermodynamic functions were calculated by the methods of Souders, Matthews and Hurd (1396, 1397); the enthalpy of formation was calculated by Labbauf, Greenshields, and Rossini (820). The normal boiling point was selected by Rossini, Pitzer, Arnett, Braun, and Pimentel (1248). $Tb = 437.74°K$.

No. 240 2,3,3,5-Tetramethylhexane $C_{10}H_{22}$ (Ideal Gas State) Mol Wt 142.276

The gaseous thermodynamic functions were calculated by the methods of Souders, Matthews, and Hurd (1396, 1397); the enthalpy of formation was calculated by Labbauf, Greenshields, and Rossini (820). The normal boiling point was selected by Rossini, Pitzer, Arnett, Braun, and Pimentel (1248). $Tb = 426.3°K$.

No. 240 2,3,3,5-Tetramethylhexane $C_{10}H_{22}$ (Ideal Gas State) Mol Wt 142.276

T°K	cal/(mole °K)				kcal/mole		
	$Cp°$	$S°$	$-(G°-H°_{298})/T$	$H°-H°_{298}$	$\Delta Hf°$	$\Delta Gf°$	Log Kp
298	54.85	118.25	118.25	0.00	-61.83	9.33	-6.841
300	55.10	118.60	118.26	0.11	-61.91	9.77	-7.116
400	71.97	136.82	120.63	6.48	-65.64	34.26	-18.717
500	86.33	154.46	125.64	14.42	-68.58	59.59	-26.043
600	98.42	171.30	131.86	23.67	-70.80	85.43	-31.115
700	108.34	187.24	138.64	34.02	-72.40	111.61	-34.844
800	116.63	202.26	145.67	45.28	-73.51	137.96	-37.686
900	123.66	216.41	152.75	57.30	-74.18	164.45	-39.931
1000	129.82	229.77	159.79	69.98	-74.44	191.00	-41.740

No. 241 2,3,4,4-Tetramethylhexane $C_{10}H_{22}$ (Ideal Gas State) Mol Wt 142.276

T°K	cal/(mole °K)				kcal/mole		
	$Cp°$	$S°$	$-(G°-H°_{298})/T$	$H°-H°_{298}$	$\Delta Hf°$	$\Delta Gf°$	Log Kp
298	54.83	119.63	119.63	0.00	-59.98	10.77	-7.895
300	55.08	119.97	119.64	0.11	-60.06	11.20	-8.162
400	71.95	138.19	122.01	6.48	-63.79	35.56	-19.426
500	86.31	155.83	127.02	14.41	-66.73	60.75	-26.551
600	98.39	172.67	133.24	23.66	-68.96	86.45	-31.488
700	108.29	188.60	140.02	34.01	-70.56	112.50	-35.122
800	116.57	203.61	147.04	45.27	-71.67	138.71	-37.892
900	123.61	217.76	154.12	57.28	-72.35	165.07	-40.081
1000	129.79	231.11	161.16	69.96	-72.62	191.48	-41.846

The gaseous thermodynamic functions were calculated by the methods of Souders, Matthews, and Hurd (1396, 1397); the enthalpy of formation was calculated by Labbauf, Greenshields, and Rossini (820). The normal boiling point was selected by Rossini, Pitzer, Arnett, Braun, and Pimentel (1248). $Tb = 434.7°K$.

No. 242 2,3,4,5-Tetramethylhexane $C_{10}H_{22}$ (Ideal Gas State) Mol Wt 142.276

T°K	cal/(mole °K)				kcal/mole		
	$Cp°$	$S°$	$-(G°-H°_{298})/T$	$H°-H°_{298}$	$\Delta Hf°$	$\Delta Gf°$	Log Kp
298	54.19	119.10	119.10	0.00	-61.67	9.24	-6.772
300	54.43	119.44	119.11	0.11	-61.75	9.67	-7.047
400	70.91	137.42	121.45	6.39	-65.57	34.09	-18.625
500	84.82	154.77	126.39	14.20	-68.64	59.37	-25.950
600	96.88	171.33	132.51	23.30	-71.01	85.20	-31.031
700	106.86	187.04	139.19	33.50	-72.76	111.39	-34.775
800	115.22	201.86	146.10	44.61	-74.02	137.77	-37.634
900	122.37	215.86	153.09	56.50	-74.82	164.31	-39.897
1000	128.55	229.08	160.03	69.05	-75.21	190.92	-41.723

The gaseous thermodynamic functions were calculated by the methods of Souders, Matthews, and Hurd (1396, 1397); the enthalpy of formation was calculated by Labbauf, Greenshields, and Rossini (820). The normal boiling point was selected by Rossini, Pitzer, Arnett, Braun, and Pimentel (1248). $Tb = 429.3°K$.

No. 243 3,3,4,4-Tetramethylhexane $C_{10}H_{22}$ (Ideal Gas State) Mol Wt 142.276

	cal/(mole °K)				kcal/mole		
$T°K$	$Cp°$	$S°$	$-(G°-H°_{298})/T$	$H°-H°_{298}$	$\Delta Hf°$	$\Delta Gf°$	Log Kp
298	55.60	116.71	116.71	0.00	-60.37	11.25	-8.247
300	55.86	117.06	116.72	0.11	-60.45	11.69	-8.516
400	73.00	135.54	119.12	6.57	-64.09	36.32	-19.844
500	87.67	153.45	124.21	14.63	-66.90	61.76	-26.995
600	99.72	170.53	130.52	24.02	-69.00	87.69	-31.940
700	109.52	186.66	137.40	34.49	-70.47	113.94	-35.573
800	117.71	201.84	144.51	45.86	-71.47	140.34	-38.337
900	124.64	216.11	151.68	57.99	-72.03	166.87	-40.519
1000	130.81	229.57	158.80	70.77	-72.20	193.44	-42.275

The gaseous thermodynamic functions were calculated by the methods of Souders, Matthews, and Hurd (1396, 1397); the enthalpy of formation was calculated by Labbauf, Greenshields, and Rossini (820). The normal boiling point was selected by Rossini, Pitzer, Arnett, Braun, and Pimentel (1248). $Tb = 443.1°K$.

No. 244 2,4-Dimethyl-3-Isopropylpentane $C_{10}H_{22}$ (Ideal Gas State) Mol Wt 142.276

	cal/(mole °K)				kcal/mole		
$T°K$	$Cp°$	$S°$	$-(G°-H°_{298})/T$	$H°-H°_{298}$	$\Delta Hf°$	$\Delta Gf°$	Log Kp
298	54.83	116.23	116.23	0.00	-61.67	10.10	-7.399
300	55.08	116.57	116.24	0.11	-61.75	10.53	-7.674
400	71.95	134.79	118.61	6.48	-65.48	35.23	-19.246
500	86.31	152.43	123.62	14.41	-68.42	60.76	-26.555
600	98.39	169.27	129.84	23.66	-70.65	86.80	-31.615
700	108.29	185.20	136.62	34.01	-72.25	113.19	-35.337
800	116.57	200.21	143.64	45.27	-73.36	139.74	-38.173
900	123.61	214.36	150.72	57.28	-74.04	166.44	-40.414
1000	129.79	227.71	157.76	69.96	-74.31	193.19	-42.220

The gaseous thermodynamic functions were calculated by the methods of Souders, Matthews, and Hurd (1396, 1397); the enthalpy of formation was calculated by Labbauf, Greenshields, and Rossini (820). Transitional temperatures were selected by Rossini, Pitzer, Arnett, Braun, and Pimentel (1248). $Tm = 191.45°K$ and $Tb = 430.19°K$.

No. 245 3,3-Diethyl-2-Methylpentane $C_{10}H_{22}$ (Ideal Gas State) Mol Wt 142.276

	cal/(mole °K)				kcal/mole		
T°K	$Cp°$	$S°$	$-(G°-H°_{298})/T$	$H°-H°_{298}$	$\Delta Hf°$	$\Delta Gf°$	Log Kp
298	54.27	119.18	119.18	0.00	-59.49	11.40	-8.353
300	54.53	119.52	119.19	0.11	-59.57	11.83	-8.617
400	71.50	137.59	121.54	6.43	-63.35	36.23	-19.797
500	86.05	155.15	126.51	14.32	-66.33	61.49	-26.876
600	98.20	171.95	132.70	23.56	-68.58	87.26	-31.784
700	108.14	187.85	139.45	33.89	-70.20	113.38	-35.398
800	116.45	202.85	146.45	45.13	-71.32	139.67	-38.154
900	123.50	216.98	153.51	57.13	-72.01	166.10	-40.333
1000	129.74	230.32	160.53	69.80	-72.29	192.60	-42.090

The gaseous thermodynamic functions were calculated by the methods of Souders, Matthews, and Hurd (1396, 1397); the enthalpy of formation was calculated by Labbauf, Greenshields, and Rossini (820). The normal boiling point was selected by Rossini, Pitzer, Arnett, Braun, and Pimentel (1248). $Tb = 442.8°K$.

No. 246 3-Ethyl-2,2,3-Trimethylpentane $C_{10}H_{22}$ (Ideal Gas State) Mol Wt 142.276

	cal/(mole °K)				kcal/mole		
T°K	$Cp°$	$S°$	$-(G°-H°_{298})/T$	$H°-H°_{298}$	$\Delta Hf°$	$\Delta Gf°$	Log Kp
298	55.60	117.29	117.29	0.00	-60.37	11.08	-8.121
300	55.86	117.64	117.30	0.11	-60.45	11.52	-8.389
400	73.00	136.12	119.70	6.57	-64.09	36.09	-19.717
500	87.67	154.03	124.79	14.63	-66.90	61.47	-26.868
600	99.72	171.11	131.10	24.02	-69.00	87.34	-31.813
700	109.52	187.24	137.98	34.49	-70.47	113.54	-35.446
800	117.71	202.42	145.09	45.86	-71.47	139.88	-38.210
900	124.64	216.69	152.26	57.99	-72.03	166.35	-40.392
1000	130.81	230.15	159.38	70.77	-72.20	192.86	-42.148

The gaseous thermodynamic functions were calculated by the methods of Souders, Matthews, and Hurd (1396, 1397); the enthalpy of formation was calculated by Labbauf, Greenshields, and Rossini (820). The normal boiling point was selected by Rossini, Pitzer, Arnett, Braun, and Pimentel (1248). $Tb = 442.6°K$.

No. 247 3-Ethyl-2,2,4-Trimethylpentane $C_{10}H_{22}$ (Ideal Gas State) Mol Wt 142.276

The gaseous thermodynamic functions were calculated by the methods of Souders, Matthews, and Hurd (1396, 1397); the enthalpy of formation was calculated by Labbauf, Greenshields, and Rossini (820). The normal boiling point was selected by Rossini, Pitzer, Arnett, Braun, and Pimentel (1248). $Tb = 428.4°K$.

No. 247 3-Ethyl-2,2,4-Trimethylpentane $C_{10}H_{22}$ (Ideal Gas State) Mol Wt 142.276

$T°K$	$Cp°$	cal/(mole °K) $S°$	$-(G°-H°_{298})/T$	$H°-H°_{298}$	kcal/mole $\Delta Hf°$	$\Delta Gf°$	Log Kp
298	54.83	115.91	115.91	0.00	-60.55	11.31	-8.290
300	55.08	116.25	115.92	0.11	-60.63	11.75	-8.560
400	71.95	134.47	118.29	6.48	-64.36	36.47	-19.928
500	86.31	152.11	123.30	14.41	-67.30	62.04	-27.115
600	98.39	168.95	129.52	23.66	-69.53	88.11	-32.093
700	108.29	184.88	136.30	34.01	-71.13	114.53	-35.757
800	116.57	199.89	143.32	45.27	-72.24	141.12	-38.549
900	123.61	214.04	150.40	57.28	-72.92	167.84	-40.756
1000	129.79	227.39	157.44	69.96	-73.19	194.63	-42.534

No. 248 3-Ethyl-2,3,4-Trimethylpentane $C_{10}H_{22}$ (Ideal Gas State) Mol Wt 142.276

$T°K$	$Cp°$	cal/(mole °K) $S°$	$-(G°-H°_{298})/T$	$H°-H°_{298}$	kcal/mole $\Delta Hf°$	$\Delta Gf°$	Log Kp
298	54.83	118.25	118.25	0.00	-60.09	11.07	-8.116
300	55.08	118.59	118.26	0.11	-60.17	11.51	-8.384
400	71.95	136.81	120.63	6.48	-63.90	36.00	-19.668
500	86.31	154.45	125.64	14.41	-66.84	61.33	-26.804
600	98.39	171.29	131.86	23.66	-69.07	87.17	-31.749
700	108.29	187.22	138.64	34.01	-70.67	113.35	-35.389
800	116.57	202.23	145.66	45.27	-71.78	139.70	-38.163
900	123.61	216.38	152.74	57.28	-72.46	166.20	-40.356
1000	129.79	229.73	159.78	69.96	-72.73	192.75	-42.124

The gaseous thermodynamic functions were calculated by the methods of Souders, Matthews, and Hurd (1396, 1397); the enthalpy of formation was calculated by Labbauf, Greenshields, and Rossini (820). The normal boiling point was selected by Rossini, Pitzer, Arnett, Braun, and Pimentel (1248). $Tb = 442.59°K$.

No. 249 2,2,3,3,4-Pentamethylpentane $C_{10}H_{22}$ (Ideal Gas State) Mol Wt 142.276

$T°K$	$Cp°$	cal/(mole °K) $S°$	$-(G°-H°_{298})/T$	$H°-H°_{298}$	kcal/mole $\Delta Hf°$	$\Delta Gf°$	Log Kp
298	56.42	112.80	112.80	0.00	-59.08	13.71	-10.047
300	56.67	113.15	112.81	0.11	-59.16	14.15	-10.310
400	73.47	131.82	115.24	6.64	-62.73	39.16	-21.397
500	87.67	149.78	120.36	14.71	-65.53	64.97	-28.399
600	99.55	166.85	126.70	24.09	-67.63	91.27	-33.243
700	109.27	182.95	133.60	34.55	-69.13	117.89	-36.805
800	117.41	198.08	140.72	45.89	-70.15	144.66	-39.518
900	124.33	212.32	147.90	57.99	-70.74	171.57	-41.660
1000	130.42	225.74	155.02	70.73	-70.95	198.52	-43.385

The gaseous thermodynamic functions were calculated by the methods of Souders, Matthews, and Hurd (1396, 1397); the enthalpy of formation was calculated by Labbauf, Greenshields, and Rossini (820). Transitional temperatures were selected by Rossini, Pitzer, Arnett, Braun, and Pimentel (1248). $Tm = 236.70°K$ and $Tb = 439.20°K$.

No. 250 2,2,3,4,4-Pentamethylpentane $C_{10}H_{22}$ (Ideal Gas State) Mol Wt 142.276

	cal/(mole °K)				kcal/mole		
$T°K$	$Cp°$	$S°$	$-(G°-H°_{298})/T$	$H°-H°_{298}$	$\Delta Hf°$	$\Delta Gf°$	Log Kp
298	56.03	110.62	110.62	0.00	-59.04	14.40	-10.553
300	56.28	110.97	110.63	0.11	-59.12	14.85	-10.816
400	73.44	129.57	113.05	6.62	-62.72	40.08	-21.898
500	88.06	147.58	118.16	14.71	-65.49	66.11	-28.897
600	100.09	164.73	124.51	24.14	-67.54	92.63	-33.737
700	109.87	180.92	131.42	34.65	-68.98	119.45	-37.293
800	118.04	196.14	138.57	46.06	-69.94	146.42	-39.999
900	124.96	210.45	145.77	58.21	-70.48	173.52	-42.134
1000	131.08	223.94	152.92	71.02	-70.62	200.66	-43.852

The gaseous thermodynamic functions were calculated by the methods of Souders, Matthews, and Hurd (1396, 1397); the enthalpy of formation was calculated by Labbauf, Greenshields, and Rossini (820). Transitional temperatures were selected by Rossini, Pitzer, Arnett, Braun, and Pimentel (1248). $Tm = 234.40°K$ and $Tb = 432.44°K$.

No. 251 Undecane $C_{11}H_{24}$ (Ideal Gas State) Mol Wt 156.302

	cal/(mole °K)				kcal/mole		
$T°K$	$Cp°$	$S°$	$-(G°-H°_{298})/T$	$H°-H°_{298}$	$\Delta Hf°$	$\Delta Gf°$	Log Kp
298	61.53	139.48	139.48	0.00	-64.60	9.94	-7.289
300	61.84	139.87	139.49	0.12	-64.69	10.40	-7.576
400	78.18	159.93	142.11	7.13	-68.72	36.07	-19.706
500	93.05	179.01	147.60	15.71	-72.02	62.66	-27.387
600	105.70	197.13	154.36	25.67	-74.63	89.84	-32.722
700	116.48	214.25	161.70	36.79	-76.58	117.43	-36.660
800	125.69	230.42	169.29	48.91	-78.00	145.22	-39.671
900	133.66	245.69	176.94	61.88	-78.91	173.20	-42.056
1000	140.60	260.14	184.54	75.60	-79.36	201.26	-43.983

The low-temperature measurements of Finke, Gross, Waddington, and Huffman (420) were merged with the selected values of Rossini, Pitzer, Arnett, Braun, and Pimentel (1248). $Tt = 236.6°K$, $\Delta Ht° = 1.639$ kcal/mole, $Tm = 247.56°K$, $\Delta Hm° = 5.301$ kcal/mole, and $Tb = 470.04°K$, where $\Delta Hv = 9.92$ kcal/mole. Ambrose, Cox, and Townsend (13) measured $Tc = 638.7°K$.

No. 252 Dodecane C$_{12}$H$_{26}$ (Ideal Gas State) Mol Wt 170.328

T°K	Cp°	S°	$-(G°-H°_{298})/T$	$H°-H°_{298}$	$\Delta Hf°$	$\Delta Gf°$	Log Kp
	cal/(mole °K)				kcal/mole		
298	67.00	148.78	148.78	0.00	-69.52	11.96	-8.769
300	67.33	149.20	148.79	0.13	-69.61	12.46	-9.078
400	85.13	171.05	151.64	7.77	-73.96	40.51	-22.132
500	101.30	191.83	157.62	17.11	-77.52	69.55	-30.400
600	115.04	211.54	164.98	27.94	-80.33	99.23	-36.142
700	126.76	230.18	172.97	40.05	-82.42	129.34	-40.381
800	136.76	247.77	181.23	53.23	-83.93	159.68	-43.621
900	145.41	264.39	189.56	67.35	-84.90	190.21	-46.188
1000	152.90	280.11	197.84	82.28	-85.37	220.83	-48.260

The low-temperature measurements of Finke, Gross, Waddington, and Huffman (420) are combined with the selected values of Rossini, Pitzer, Arnett, Braun, and Pimentel (1248). $Tm = 263.56°K$, $\Delta Hm° = 8.803$ kcal/mole, and $Tb = 489.43°K$, where $\Delta Hv = 10.43$ kcal/mole. Ambrose, Cox, and Townsend (13) measured $Tc = 658.2°K$.

No. 253 Tridecane C$_{13}$H$_{28}$ (Ideal Gas State) Mol Wt 184.354

T°K	Cp°	S°	$-(G°-H°_{298})/T$	$H°-H°_{298}$	$\Delta Hf°$	$\Delta Gf°$	Log Kp
	cal/(mole °K)				kcal/mole		
298	72.47	158.09	158.09	0.00	-74.45	13.97	-10.239
300	72.82	158.54	158.10	0.14	-74.55	14.51	-10.570
400	92.07	182.17	161.19	8.40	-79.22	44.94	-24.551
500	109.55	204.64	167.65	18.50	-83.03	76.43	-33.406
600	124.38	225.96	175.61	30.22	-86.03	108.60	-39.557
700	137.03	246.11	184.25	43.30	-88.27	141.25	-44.097
800	147.82	265.13	193.19	57.56	-89.88	174.12	-47.566
900	157.16	283.09	202.19	72.82	-90.91	207.21	-50.315
1000	165.20	300.08	211.13	88.95	-91.40	240.39	-52.534

The low-temperature measurements of Finke, Gross, Waddington, and Huffman (420) are combined with the selected values of Rossini, Pitzer, Arnett, Braun, and Pimentel (1248). $Tt = 255.0°K$, $\Delta Ht° = 1.831$ kcal/mole, $Tm = 267.78°K$, $\Delta Hm° = 6.812$ kcal/mole, and $Tb = 508.58°K$, where $\Delta Hv = 10.91$ kcal/mole.

No. 254 Tetradecane C$_{14}$H$_{30}$ (Ideal Gas State) Mol Wt 198.380

The low-temperature measurements of Finke, Gross, Waddington, and Huffman (420) are merged with the selected values of Rossini, Pitzer, Arnett, Braun, and Pimentel (1248). $Tm = 279.02°K$, $\Delta Hm° = 10.772$ kcal/mole, and $Tb = 526.66°K$, where $\Delta Hv = 11.38$ kcal/mole.

No. 254 Tetradecane $C_{14}H_{30}$ (Ideal Gas State) Mol Wt 198.380

T°K	Cp°	S°	$-(G°-H°_{298})/T$	$H°-H°_{298}$	$\Delta Hf°$	$\Delta Gf°$	Log Kp
	cal/(mole °K)				kcal/mole		
298	77.93	167.40	167.40	0.00	-79.38	15.97	-11.709
300	78.32	167.89	167.41	0.15	-79.49	16.56	-12.062
400	99.01	193.30	170.73	9.03	-84.47	49.36	-26.970
500	117.79	217.46	177.68	19.90	-88.55	83.31	-36.413
600	133.72	240.38	186.24	32.49	-91.74	117.98	-42.971
700	147.31	262.04	195.54	46.56	-94.13	153.15	-47.812
800	158.89	282.49	205.14	61.88	-95.83	188.57	-51.512
900	168.90	301.79	214.82	78.28	-96.91	224.21	-54.442
1000	177.60	320.05	224.43	95.62	-97.42	259.94	-56.808

No. 255 Pentadecane $C_{15}H_{32}$ (Ideal Gas State) Mol Wt 212.406

T°K	Cp°	S°	$-(G°-H°_{298})/T$	$H°-H°_{298}$	$\Delta Hf°$	$\Delta Gf°$	Log Kp
	cal/(mole °K)				kcal/mole		
298	83.40	176.71	176.71	0.00	-84.31	17.98	-13.179
300	83.81	177.23	176.72	0.16	-84.42	18.61	-13.554
400	105.95	204.42	180.27	9.66	-89.73	53.79	-29.389
500	126.04	230.28	187.71	21.29	-94.06	90.19	-39.419
600	143.07	254.80	196.87	34.77	-97.45	127.35	-46.386
700	157.59	277.98	206.82	49.82	-99.98	165.05	-51.528
800	169.95	299.85	217.09	66.21	-101.78	203.01	-55.457
900	180.65	320.49	227.44	83.75	-102.92	241.20	-58.569
1000	189.90	340.02	237.73	102.29	-103.44	279.50	-61.081

The low-temperature measurements of Finke, Gross, Waddington, and Huffman (420) are coupled with the selected values of Rossini, Pitzer, Arnett, Braun, and Pimentel (1248). $Tt = 270.9°K$, $\Delta Ht° = 2.191$ kcal/mole, $Ttp = 283.10°K$, $\Delta Hm° = 8.268$ kcal/mole, and $Tb = 543.76°K$, where $\Delta Hv = 11.82$ kcal/mole.

No. 256 Hexadecane $C_{16}H_{34}$ (Ideal Gas State) Mol Wt 226.432

T°K	Cp°	S°	$-(G°-H°_{298})/T$	$H°-H°_{298}$	$\Delta Hf°$	$\Delta Gf°$	Log Kp
	cal/(mole °K)				kcal/mole		
298	88.86	186.02	186.02	0.00	-89.23	20.00	-14.656
300	89.31	186.58	186.03	0.17	-89.35	20.66	-15.053
400	112.89	215.55	189.82	10.30	-94.97	58.23	-31.813
500	134.28	243.10	197.74	22.68	-99.56	97.08	-42.430
600	152.41	269.23	207.50	37.04	-103.15	136.74	-49.804
700	167.86	293.91	218.10	53.07	-105.82	176.96	-55.246
800	181.02	317.20	229.04	70.53	-107.72	217.46	-59.405
900	192.39	339.20	240.07	89.22	-108.92	258.21	-62.699
1000	202.20	359.99	251.03	108.96	-109.46	299.06	-65.357

The low-temperature measurements of Finke, Gross, Waddington, and Huffman (420) are merged with the selected values of Rossini, Pitzer, Arnett, Braun, and Pimentel (1248). $Ttp = 291.33°K$, $\Delta Hm° = 12.753$ kcal/mole, and $Tb = 559.94°K$, where $\Delta Hv = 12.24$ kcal/mole.

No. 257   Heptadecane $C_{17}H_{36}$ (Ideal Gas State)   Mol Wt 240.458

	cal/(mole °K)				kcal/mole		
$T°K$	$Cp°$	$S°$	$-(G°-H°_{298})/T$	$H°-H°_{298}$	$\Delta Hf°$	$\Delta Gf°$	Log Kp
298	94.33	195.33	195.33	0.00	-94.15	22.01	-16.133
300	94.80	195.92	195.34	0.18	-94.28	22.72	-16.552
400	119.83	226.68	199.36	10.93	-100.22	62.67	-34.237
500	142.53	255.92	207.77	24.08	-105.06	103.96	-45.440
600	161.75	283.65	218.13	39.32	-108.85	146.12	-53.222
700	178.14	309.84	229.38	56.33	-111.66	188.87	-58.965
800	192.08	334.56	241.00	74.86	-113.66	231.91	-63.353
900	204.14	357.90	252.70	94.68	-114.91	275.22	-66.828
1000	214.60	379.96	264.33	115.63	-115.47	318.63	-69.633

These values were selected by Rossini, Pitzer, Arnett, Braun, and Pimentel (1248). $Tm = 295.14°K$, $\Delta Hm° = 9.676$ kcal/mole, and $Tb = 575.30°K$, where $\Delta Hv = 12.64$ kcal/mole.

No. 258   Octadecane $C_{18}H_{38}$ (Ideal Gas State)   Mol Wt 254.484

	cal/(mole °K)				kcal/mole		
$T°K$	$Cp°$	$S°$	$-(G°-H°_{298})/T$	$H°-H°_{298}$	$\Delta Hf°$	$\Delta Gf°$	Log Kp
298	99.80	204.64	204.64	0.00	-99.08	24.02	-17.604
300	100.29	205.26	204.65	0.19	-99.21	24.77	-18.044
400	126.77	237.80	208.90	11.56	-105.47	67.09	-36.656
500	150.78	268.74	217.80	25.47	-110.57	110.84	-48.447
600	171.09	298.07	228.76	41.59	-114.55	155.50	-56.636
700	188.42	325.78	240.66	59.59	-117.51	200.77	-62.681
800	203.15	351.92	252.95	79.18	-119.61	246.36	-67.298
900	215.89	376.60	265.33	100.15	-120.92	292.21	-70.955
1000	226.90	399.93	277.63	122.30	-121.49	338.18	-73.906

The values selected by Rossini, Pitzer, Arnett, Braun, and Pimentel (1248) are used. $Tm = 301.33°K$, $\Delta Hm° = 14.815$ kcal/mole, and $Tb = 589.86°K$, where $\Delta Hv = 13.02$ kcal/mole.

No. 259   Nonadecane $C_{19}H_{40}$ (Ideal Gas State)   Mol Wt 268.510

These values are taken from the selected values of Rossini, Pitzer, Arnett, Braun, and Pimentel (1248). $Tm = 305.05°K$, $\Delta Hm° = 12.0$ kcal/mole, and $Tb = 603.75°K$, where $\Delta Hv = 13.39$ kcal/mole.

No. 259 Nonadecane $C_{19}H_{40}$ (Ideal Gas State) Mol Wt 268.510

	cal/(mole °K)				kcal/mole		
T°K	$Cp°$	$S°$	$-(G°-H°_{298})/T$	$H°-H°_{298}$	$\Delta Hf°$	$\Delta Gf°$	Log Kp
298	105.26	213.95	213.95	0.00	-104.00	26.03	-19.081
300	105.79	214.61	213.96	0.20	-104.14	26.83	-19.543
400	133.71	248.93	218.45	12.20	-110.72	71.53	-39.080
500	159.02	281.56	227.83	26.87	-116.07	117.73	-51.457
600	180.43	312.49	239.39	43.86	-120.25	164.88	-60.055
700	198.69	341.71	251.94	62.84	-123.35	212.68	-66.399
800	214.21	369.28	264.90	83.51	-125.55	260.81	-71.246
900	227.63	395.30	277.96	105.61	-126.91	309.22	-75.085
1000	239.20	419.90	290.93	128.97	-127.51	357.75	-78.182

No. 260 Eicosane $C_{20}H_{42}$ (Ideal Gas State) Mol Wt 282.536

	cal/(mole °K)				kcal/mole		
T°K	$Cp°$	$S°$	$-(G°-H°_{298})/T$	$H°-H°_{298}$	$\Delta Hf°$	$\Delta Gf°$	Log Kp
298	110.73	223.26	223.26	0.00	-108.93	28.04	-20.551
300	111.28	223.95	223.27	0.21	-109.08	28.88	-21.035
400	140.65	260.05	227.99	12.83	-115.97	75.96	-41.499
500	167.27	294.37	237.86	28.26	-121.58	124.61	-54.464
600	189.78	326.91	250.02	46.14	-125.96	174.25	-63.469
700	208.97	357.64	263.22	66.10	-129.20	224.58	-70.115
800	225.28	386.64	276.85	87.83	-131.50	275.25	-75.192
900	239.38	414.00	290.59	111.08	-132.92	326.21	-79.212
1000	251.60	439.87	304.23	135.65	-133.53	377.30	-82.456

The values selected by Rossini, Pitzer, Arnett, Braun, and Pimentel (1248) are adopted. $Tm = 309.59°K$, $\Delta Hm° = 16.8$ kcal/mole, and $Tb = 616.95°K$, where $\Delta Hv = 13.74$ kcal/mole.

ALKENE IDEAL GAS TABLES

No. 261 Ethylene C_2H_4 (Ideal Gas State) Mol Wt 28.052

	cal/(mole °K)				kcal/mole		
T°K	$Cp°$	$S°$	$-(G°-H°_{298})/T$	$H°-H°_{298}$	$\Delta Hf°$	$\Delta Gf°$	Log Kp
298	10.41	52.45	52.45	0.00	12.50	16.28	-11.936
300	10.45	52.52	52.46	0.02	12.49	16.31	-11.879
400	12.90	55.87	52.89	1.19	11.77	17.69	-9.666
500	15.16	58.99	53.81	2.60	11.14	19.25	-8.412
600	17.10	61.93	54.92	4.21	10.60	20.92	-7.619
700	18.76	64.70	56.12	6.01	10.15	22.68	-7.079
800	20.20	67.30	57.36	7.96	9.77	24.49	-6.690
900	21.46	69.75	58.60	10.04	9.45	26.35	-6.399
1000	22.57	72.07	59.83	12.25	9.21	28.25	-6.173

Alkene Ideal Gas Tables

Egan and Kemp (372) determined low-temperature thermal data, including $Ttp = 103.95°K$ and $\Delta Hm° = 0.800$ kcal/mole. Thermodynamic functions and enthalpy of formation are the selections of Rossini, Pitzer, Arnett, Braun, and Pimentel (1248), who also listed $Tb = 169.44°K$, where $\Delta Hv = 3.237$ kcal/mole. Kobe and Lynn (780) selected $Tc = 282.6°K$, $Pc = 50.5$ atm, and $dc = 0.227$ g/cc.

No. 262 Propene C_3H_6 (Ideal Gas State) Mol Wt 42.078

$T°K$	$Cp°$	$S°$	$-(G°-H°_{298})/T$	$H°-H°_{298}$	$\Delta Hf°$	$\Delta Gf°$	Log Kp
	cal/(mole °K)				kcal/mole		
298	15.27	63.80	63.80	0.00	4.88	14.99	-10.989
300	15.34	63.90	63.81	0.03	4.86	15.05	-10.966
400	19.10	68.83	64.45	1.76	3.76	18.62	-10.175
500	22.62	73.48	65.80	3.85	2.80	22.45	-9.814
600	25.70	77.88	67.45	6.27	1.98	26.46	-9.638
700	28.37	82.05	69.24	8.97	1.31	30.60	-9.553
800	30.68	85.99	71.09	11.93	0.77	34.82	-9.511
900	32.70	89.73	72.96	15.10	0.35	39.10	-9.495
1000	34.46	93.26	74.81	18.46	0.04	43.43	-9.491

Thermodynamic functions and enthalpy of formation were selected by Rossini, Pitzer, Arnett, Braun, and Pimentel (1248), who also presented $Tm = 87.91°K$, $\Delta Hm° = 0.718$ kcal/mole, and $Tb = 225.45°K$, where $\Delta Hv = 4.402$ kcal/mole. Nangia and Benson (1046) determined equilibrium constants for the dehydrogenation of propane and found excellent agreement with values calculated from thermal data. Kobe and Lynn (780) selected $Tc = 364.9°K$, $Pc = 45.6$ atm, and $dc = 0.233$ g/cc.

No. 263 1-Butene C_4H_8 (Ideal Gas State) Mol Wt 56.104

$T°K$	$Cp°$	$S°$	$-(G°-H°_{298})/T$	$H°-H°_{298}$	$\Delta Hf°$	$\Delta Gf°$	Log Kp
	cal/(mole °K)				kcal/mole		
298	20.47	73.04	73.04	0.00	-0.03	17.04	-12.489
300	20.57	73.17	73.05	0.04	-0.06	17.14	-12.488
400	26.04	79.85	73.92	2.38	-1.49	23.10	-12.622
500	30.93	86.20	75.75	5.23	-2.70	29.39	-12.847
600	35.14	92.22	78.00	8.54	-3.71	35.91	-13.078
700	38.71	97.91	80.44	12.24	-4.51	42.58	-13.293
800	41.80	103.29	82.96	16.27	-5.14	49.34	-13.479
900	44.49	108.37	85.50	20.58	-5.62	56.19	-13.644
1000	46.82	113.18	88.03	25.15	-5.94	63.08	-13.786

Thermodynamic functions are those selected by Rossini, Pitzer, Arnett, Braun, and Pimentel (1248), who also listed $Tm = 87.80°K$, $\Delta Hm° = 0.920$ kcal/mole, and $Tb = 266.89°K$, where $\Delta Hv = 5.240$ kcal/mole. These selections are based in part on the low-temperature thermal data of Aston, Fink, Bestul, Pace, and Szasz (33). Golden, Egger, and Benson (503) made careful determinations of the equilibrium constants for dehydrogenation of butane and the positional and geometrical isomerization of the butenes. Agreement with values calculated from thermal data is good, except that the entropy of 1-butene from thermal data appears about 1 cal/(mole °K) too low. Kobe and Lynn (780) selected $Tc = 419.5°K$, $Pc = 39.7$ atm, and $dc = 0.234$ g/cc.

No. 264 2-Butene, cis C_4H_8 (Ideal Gas State) Mol Wt 56.104

T°K	cal/(mole °K)				kcal/mole		
	$Cp°$	$S°$	$-(G°-H°_{298})/T$	$H°-H°_{298}$	$\Delta Hf°$	$\Delta Gf°$	Log Kp
298	18.86	71.90	71.90	0.00	-1.67	15.74	-11.536
300	18.96	72.02	71.91	0.04	-1.70	15.84	-11.542
400	24.33	78.22	72.71	2.21	-3.30	21.94	-11.989
500	29.39	84.20	74.42	4.90	-4.68	28.42	-12.421
600	33.80	89.96	76.53	8.06	-5.82	35.14	-12.800
700	37.60	95.46	78.85	11.64	-6.75	42.05	-13.129
800	40.87	100.70	81.25	15.56	-7.49	49.07	-13.404
900	43.70	105.68	83.69	19.80	-8.05	56.18	-13.641
1000	46.15	110.42	86.13	24.29	-8.44	63.34	-13.843

The values adopted are from the selections of Rossini, Pitzer, Arnett, Braun, and Pimentel (1248) and the measurements of Scott, Ferguson, and Brickwedde (1326). $Tm = 134.24°K$, $\Delta Hm° = 1.747$ kcal/mole, and $Tb = 276.87°K$, where $\Delta Hv = 5.553$ kcal/mole. Ambrose, Cox, and Townsend (13) measured $Tc = 435.55°K$, and Kobe and Lynn (780) reported $Pc = 41$ atm and $dc = 0.238$ g/cc for a mixture of the cis and trans isomers of 2-butene.

No. 265 2-Butene, trans C_4H_8 (Ideal Gas State) Mol Wt 56.104

T°K	cal/(mole °K)				kcal/mole		
	$Cp°$	$S°$	$-(G°-H°_{298})/T$	$H°-H°_{298}$	$\Delta Hf°$	$\Delta Gf°$	Log Kp
298	20.99	70.86	70.86	0.00	-2.67	15.05	-11.030
300	21.08	71.00	70.87	0.04	-2.70	15.16	-11.041
400	26.02	77.74	71.75	2.40	-4.11	21.33	-11.653
500	30.68	84.06	73.59	5.24	-5.33	27.83	-12.165
600	34.80	90.03	75.84	8.52	-6.37	34.56	-12.588
700	38.38	95.66	78.27	12.18	-7.20	41.46	-12.943
800	41.50	101.00	80.78	16.18	-7.87	48.45	-13.235
900	44.20	106.05	83.31	20.47	-8.38	55.52	-13.482
1000	46.58	110.83	85.82	25.01	-8.73	62.65	-13.692

The measurements of Guttman and Pitzer (553) and the review by Rossini, Pitzer, Arnett, Braun, and Pimentel (1248) are the sources of the adopted data. $Tm = 167.60°K$, $\Delta Hm° = 2.332$ kcal/mole, and $Tb = 274.03°K$, where $\Delta Hv = 5.439$ kcal/mole. Ambrose, Cox, and Townsend (13) measured $Tc = 428.6°K$, and Kobe and Lynn (780) reported $Pc = 41$ atm and $dc = 0.238$ g/cc for a mixture of the *cis* and *trans* isomers of 2-butene.

No. 266 2-Methylpropene C_4H_8 (Ideal Gas State) Mol Wt 56.104

$T°K$	$Cp°$	$S°$	$-(G°-H°_{298})/T$	$H°-H°_{298}$	$\Delta Hf°$	$\Delta Gf°$	Log Kp
		cal/(mole °K)			kcal/mole		
298	21.30	70.17	70.17	0.00	-4.04	13.88	-10.177
300	21.39	70.31	70.18	0.04	-4.07	13.99	-10.194
400	26.57	77.18	71.08	2.45	-5.43	20.23	-11.052
500	31.24	83.62	72.95	5.34	-6.60	26.78	-11.706
600	35.30	89.69	75.24	8.67	-7.58	33.55	-12.220
700	38.81	95.40	77.71	12.38	-8.37	40.47	-12.636
800	41.86	100.78	80.27	16.42	-9.00	47.49	-12.973
900	44.53	105.87	82.83	20.74	-9.47	54.59	-13.254
1000	46.85	110.69	85.38	25.31	-9.79	61.73	-13.490

The measurements of Todd and Parks (1512) and the selected values of Rossini, Pitzer, Arnett, Braun, and Pimentel (1248) are adopted. $Tm = 132.80°K$, $\Delta Hm° = 1.415$ kcal/mole, and $Tb = 266.25°K$, where we calculate $\Delta Hv = 5.210$ kcal/mole. Kobe and Lynn (780) selected $Tc = 417.8°K$, $Pc = 39.5$ atm, and $dc = 0.235$ g/cc. Teranishi and Benson (1478) determined the equilibrium constant for dehydrogenation of isobutane over a range of temperatures and found excellent agreement with values calculated from thermal data.

No. 267 1-Pentene C_5H_{10} (Ideal Gas State) Mol Wt 70.130

$T°K$	$Cp°$	$S°$	$-(G°-H°_{298})/T$	$H°-H°_{298}$	$\Delta Hf°$	$\Delta Gf°$	Log Kp
		cal/(mole °K)			kcal/mole		
298	26.19	82.65	82.65	0.00	-5.00	18.91	-13.864
300	26.31	82.82	82.66	0.05	-5.04	19.06	-13.885
400	33.10	91.33	83.77	3.03	-6.77	27.37	-14.951
500	39.23	99.39	86.10	6.65	-8.23	36.07	-15.766
600	44.56	107.03	88.95	10.85	-9.42	45.04	-16.406
700	49.06	114.24	92.06	15.54	-10.36	54.20	-16.922
800	52.95	121.05	95.26	20.64	-11.08	63.47	-17.338
900	56.32	127.49	98.49	26.11	-11.61	72.83	-17.684
1000	59.21	133.58	101.69	31.89	-11.94	82.24	-17.972

The measurements of Todd, Oliver, and Huffman (1511) and the selected values of Rossini, Pitzer, Arnett, Braun, and Pimentel (1248) are adopted. $Tm = 107.94°K$, $\Delta Hm° = 1.388$ kcal/mole, and $Tb = 303.12°K$, where $\Delta Hv = 6.022$ kcal/mole. Ambrose, Cox, and Townsend (13) measured $Tc = 464.7°K$. Kobe and Lynn (780) reported $Pc = 40$ atm.

No. 268 2-Pentene, cis C_5H_{10} (Ideal Gas State) Mol Wt 70.130

	cal/(mole °K)				kcal/mole		
$T°K$	$Cp°$	$S°$	$-(G°-H°_{298})/T$	$H°-H°_{298}$	$\Delta Hf°$	$\Delta Gf°$	Log Kp
298	24.32	82.76	82.76	0.00	-6.71	17.17	-12.587
300	24.45	82.92	82.77	0.05	-6.75	17.32	-12.615
400	31.57	90.94	83.81	2.86	-8.65	25.64	-14.008
500	38.05	98.70	86.02	6.35	-10.24	34.40	-15.036
600	43.62	106.14	88.76	10.44	-11.54	43.45	-15.826
700	48.25	113.22	91.75	15.03	-12.57	52.71	-16.455
800	52.29	119.93	94.86	20.07	-13.37	62.08	-16.959
900	55.76	126.30	98.00	25.47	-13.95	71.55	-17.375
1000	58.78	132.33	101.13	31.20	-14.34	81.09	-17.720

The selected values of Rossini, Pitzer, Arnett, Braun, and Pimentel (1248) and the measured data of Todd, Oliver, and Huffman (1511) are adopted. $Tm = 121.77°K$, $\Delta Hm° = 1.700$ kcal/mole, and $Tb = 310.08°K$, where $\Delta Hv = 6.240$ kcal/mole. Kobe and Lynn (780) reported $Tc = 475.5°K$ and $Pc = 40.4$ atm, but the cis:trans isomer ratio was not mentioned.

No. 269 2-Pentene, trans C_5H_{10} (Ideal Gas State) Mol Wt 70.130

	cal/(mole °K)				kcal/mole		
$T°K$	$Cp°$	$S°$	$-(G°-H°_{298})/T$	$H°-H°_{298}$	$\Delta Hf°$	$\Delta Gf°$	Log Kp
298	25.92	81.36	81.36	0.00	-7.59	16.71	-12.248
300	26.04	81.53	81.37	0.05	-7.63	16.86	-12.280
400	32.67	89.94	82.47	2.99	-9.39	25.30	-13.821
500	38.75	97.90	84.76	6.57	-10.90	34.15	-14.925
600	44.02	105.44	87.59	10.72	-12.14	43.27	-15.761
700	48.54	112.57	90.65	15.35	-13.14	52.60	-16.420
800	52.45	119.31	93.82	20.40	-13.91	62.03	-16.946
900	55.85	125.69	97.01	25.82	-14.48	71.57	-17.378
1000	58.81	131.73	100.18	31.56	-14.86	81.16	-17.736

The measured values of Todd, Oliver, and Huffman (1511) and the values selected by Rossini, Pitzer, Arnett, Braun, and Pimentel (1248) are employed. $Tm = 132.91°K$, $\Delta Hm° = 1.996$ kcal/mole, and $Tb = 309.50°K$, where $\Delta Hv = 6.230$ kcal/mole. Kobe and Lynn (780) reported $Tc = 475.5°K$ and $Pc = 40.4$ atm, but the cis:trans isomer ratio was not given.

No. 270 2-Methyl-1-Butene C_5H_{10} (Ideal Gas State) Mol Wt 70.130

The measured values of Scott, Waddington, Smith, and Huffman (1324), the statistical calculations of McCullough and Scott (960), and the selected values of Rossini, Pitzer, Arnett, Braun, and Pimentel (1248) are

Alkene Ideal Gas Tables

No. 270 2-Methyl-1-Butene C_5H_{10} (Ideal Gas State) Mol Wt 70.130

T°K	$Cp°$	$S°$	$-(G°-H°_{298})/T$	$H°-H°_{298}$	$\Delta Hf°$	$\Delta Gf°$	Log Kp
	cal/(mole °K)			kcal/mole			
298	26.28	81.15	81.15	0.00	-8.68	15.68	-11.495
300	26.41	81.32	81.16	0.05	-8.72	15.83	-11.532
400	33.20	89.86	82.27	3.04	-10.44	24.28	-13.268
500	39.40	97.95	84.61	6.68	-11.88	33.14	-14.483
600	44.72	105.62	87.48	10.89	-13.06	42.25	-15.389
700	49.26	112.86	90.59	15.59	-13.98	51.55	-16.094
800	53.15	119.70	93.80	20.72	-14.69	60.95	-16.651
900	56.52	126.16	97.04	26.21	-15.19	70.45	-17.106
1000	59.43	132.27	100.26	32.01	-15.50	79.99	-17.480

used. $Tm = 135.60°K$, $\Delta Hm° = 1.891$ kcal/mole, and $Tb = 304.31°K$, where $\Delta Hv = 6.094$ kcal/mole.

No. 271 3-Methyl-1-Butene C_5H_{10} (Ideal Gas State) Mol Wt 70.130

T°K	$Cp°$	$S°$	$-(G°-H°_{298})/T$	$H°-H°_{298}$	$\Delta Hf°$	$\Delta Gf°$	Log Kp
	cal/(mole °K)			kcal/mole			
298	28.35	79.70	79.70	0.00	-6.92	17.87	-13.101
300	28.47	79.88	79.71	0.06	-6.95	18.03	-13.131
400	35.26	89.03	80.91	3.25	-8.46	26.59	-14.528
500	40.97	97.52	83.39	7.07	-9.73	35.50	-15.518
600	45.90	105.44	86.41	11.42	-10.77	44.65	-16.262
700	50.15	112.84	89.67	16.23	-11.59	53.96	-16.845
800	53.85	119.78	93.00	21.43	-12.21	63.36	-17.307
900	57.03	126.31	96.34	26.98	-12.66	72.84	-17.686
1000	59.83	132.47	99.65	32.82	-12.93	82.36	-17.999

The measurements of Todd, Oliver, and Huffman (1511) and the selected values of Rossini, Pitzer, Arnett, Braun, and Pimentel (1248) are used. $Tm = 104.67°K$, $\Delta Hm° = 1.281$ kcal/mole, and $Tb = 293.21°K$, where $\Delta Hv = 5.750$ kcal/mole. Kobe and Lynn (780) reported $Tc = 464.7°K$ and $Pc = 33.9$ atm.

No. 272 2-Methyl-2-Butene C_5H_{10} (Ideal Gas State) Mol Wt 70.130

T°K	$Cp°$	$S°$	$-(G°-H°_{298})/T$	$H°-H°_{298}$	$\Delta Hf°$	$\Delta Gf°$	Log Kp
	cal/(mole °K)			kcal/mole			
298	25.10	80.92	80.92	0.00	-10.17	14.26	-10.453
300	25.22	81.08	80.93	0.05	-10.21	14.41	-10.497
400	31.93	89.27	82.00	2.91	-12.05	22.91	-12.514
500	38.07	97.07	84.24	6.42	-13.63	31.83	-13.913
600	43.42	104.49	87.00	10.50	-14.94	41.05	-14.950
700	48.04	111.54	90.01	15.08	-15.99	50.47	-15.756
800	52.05	118.22	93.12	20.09	-16.81	60.01	-16.393
900	55.52	124.56	96.26	25.47	-17.41	69.66	-16.914
1000	58.55	130.57	99.40	31.18	-17.82	79.36	-17.344

The measured values of Scott, Waddington, Smith, and Huffman (1324), of Todd, Oliver, and Huffman (1511) and the values selected by Rossini, Pitzer, Arnett, Braun, and Pimentel (1248) are adopted. $Tm = 139.38°K$, $\Delta Hm° = 1.816$ kcal/mole, and $Tb = 311.72°K$, where $\Delta Hv = 6.287$ kcal/mole. Kobe and Lynn (780) reported $Tc = 470°K$ and $Pc = 34$ atm.

No. 273 1-Hexene C_6H_{12} (Ideal Gas State) Mol Wt 84.156

$T°K$	$Cp°$	$S°$	$-(G°-H°_{298})/T$	$H°-H°_{298}$	$\Delta Hf°$	$\Delta Gf°$	Log Kp
		cal/(mole °K)			kcal/mole		
298	31.63	91.93	91.93	0.00	-9.96	20.90	-15.319
300	31.78	92.13	91.94	0.06	-10.00	21.09	-15.361
400	40.03	102.42	93.28	3.66	-12.05	31.78	-17.360
500	47.47	112.17	96.09	8.04	-13.77	42.94	-18.766
600	53.90	121.41	99.55	13.12	-15.16	54.41	-19.817
700	59.34	130.14	103.30	18.79	-16.24	66.10	-20.636
800	64.02	138.37	107.18	24.96	-17.07	77.91	-21.282
900	68.06	146.15	111.08	31.57	-17.65	89.82	-21.811
1000	71.54	153.51	114.96	38.56	-18.00	101.80	-22.246

The selected values of Rossini, Pitzer, Arnett, Braun, and Pimentel (1248) are combined with the measured combustion values of Prosen and Rossini (1204) and the low-temperature values of McCullough, Finke, Gross, Messerly, and Waddington (950). $Ttp = 133.39°K$, $\Delta Hm° = 2.234$ kcal/mole, and $Tb = 336.63°K$.

No. 274 2-Hexene, cis C_6H_{12} (Ideal Gas State) Mol Wt 84.156

$T°K$	$Cp°$	$S°$	$-(G°-H°_{298})/T$	$H°-H°_{298}$	$\Delta Hf°$	$\Delta Gf°$	Log Kp
		cal/(mole °K)			kcal/mole		
298	30.04	92.37	92.37	0.00	-12.51	18.22	-13.353
300	30.19	92.56	92.38	0.06	-12.56	18.41	-13.408
400	38.60	102.41	93.66	3.51	-14.76	29.07	-15.884
500	46.40	111.89	96.37	7.77	-16.60	40.25	-17.593
600	53.00	120.94	99.71	12.74	-18.09	51.76	-18.852
700	58.60	129.55	103.37	18.33	-19.25	63.50	-19.826
800	63.40	137.69	107.15	24.44	-20.14	75.38	-20.591
900	67.60	145.41	110.98	30.99	-20.78	87.37	-21.214
1000	71.20	152.72	114.79	37.94	-21.17	99.42	-21.726

The measured enthalpy of combustion values of Bartolo and Rossini (80) and the selected values of Rossini, Pitzer, Arnett, Braun, and Pimentel (1248) are used. $Tm = 132.03°K$ and $Tb = 342.04°K$.

Alkene Ideal Gas Tables

No. 275 2-Hexene, trans C_6H_{12} (Ideal Gas State) Mol Wt 84.156

	cal/(mole °K)			kcal/mole			
$T°K$	$Cp°$	$S°$	$-(G°-H°_{298})/T$	$H°-H°_{298}$	$\Delta Hf°$	$\Delta Gf°$	Log Kp
298	31.64	90.97	90.97	0.00	-12.88	18.27	-13.388
300	31.78	91.17	90.98	0.06	-12.92	18.46	-13.444
400	39.70	101.41	92.32	3.64	-14.99	29.24	-15.976
500	47.10	111.08	95.11	7.99	-16.74	40.51	-17.705
600	53.40	120.24	98.54	13.02	-18.18	52.09	-18.973
700	58.90	128.90	102.27	18.64	-19.31	63.90	-19.950
800	63.60	137.08	106.11	24.78	-20.17	75.84	-20.717
900	67.60	144.80	109.99	31.34	-20.80	87.89	-21.341
1000	71.20	152.11	113.84	38.28	-21.19	100.00	-21.853

The values selected by Rossini, Pitzer, Arnett, Braun, and Pimentel (1248) and the measured enthalpy of combustion of Bartolo and Rossini (80) are adopted. $Tm = 140.18°K$ and $Tb = 341.03°K$.

No. 276 3-Hexene, cis C_6H_{12} (Ideal Gas State) Mol Wt 84.156

	cal/(mole °K)			kcal/mole			
$T°K$	$Cp°$	$S°$	$-(G°-H°_{298})/T$	$H°-H°_{298}$	$\Delta Hf°$	$\Delta Gf°$	Log Kp
298	29.55	90.73	90.73	0.00	-11.38	19.84	-14.540
300	29.71	90.92	90.74	0.06	-11.43	20.03	-14.589
400	38.50	100.69	92.01	3.48	-13.66	30.86	-16.863
500	46.40	110.15	94.70	7.73	-15.50	42.22	-18.451
600	53.20	119.23	98.03	12.72	-16.98	53.90	-19.631
700	58.70	127.85	101.68	18.32	-18.13	65.81	-20.546
800	63.50	136.01	105.47	24.44	-19.01	77.85	-21.268
900	67.60	143.73	109.30	31.00	-19.64	90.01	-21.856
1000	71.20	151.05	113.11	37.94	-20.03	102.23	-22.340

The experimental enthalpy of combustion of Bartolo and Rossini (80) and the values selected by Rossini, Pitzer, Arnett, Braun, and Pimentel (1248) are employed. $Tm = 135.33°K$ and $Tb = 339.60°K$.

No. 277 3-Hexene, trans C_6H_{12} (Ideal Gas State) Mol Wt 84.156

	cal/(mole °K)			kcal/mole			
$T°K$	$Cp°$	$S°$	$-(G°-H°_{298})/T$	$H°-H°_{298}$	$\Delta Hf°$	$\Delta Gf°$	Log Kp
298	31.75	89.59	89.59	0.00	-13.01	18.55	-13.594
300	31.90	89.79	89.60	0.06	-13.05	18.74	-13.651
400	40.20	100.13	90.95	3.68	-15.09	29.66	-16.204
500	47.60	109.91	93.77	8.08	-16.79	41.05	-17.941
600	53.90	119.16	97.24	13.16	-18.17	52.74	-19.211
700	59.20	127.88	101.00	18.82	-19.26	64.66	-20.187
800	63.90	136.10	104.88	24.98	-20.10	76.70	-20.952
900	67.80	143.85	108.78	31.57	-20.70	88.84	-21.573
1000	71.40	151.19	112.66	38.53	-21.07	101.05	-22.082

The selected values of Rossini, Pitzer, Arnett, Braun, and Pimentel (1248) are combined with the enthalpy of combustion measured by Bartolo and Rossini (80). $Tm = 159.72°K$ and $Tb = 340.24°K$.

No. 278 2-Methyl-1-Pentene C_6H_{12} (Ideal Gas State) Mol Wt 84.156

T°K	cal/(mole °K)			kcal/mole			
	$Cp°$	$S°$	$-(G°-H°_{298})/T$	$H°-H°_{298}$	$\Delta Hf°$	$\Delta Gf°$	Log Kp
298	32.41	91.34	91.34	0.00	-12.49	18.55	-13.593
300	32.56	91.55	91.35	0.07	-12.53	18.73	-13.647
400	40.80	102.06	92.72	3.74	-14.50	29.47	-16.100
500	48.10	111.97	95.59	8.19	-16.15	40.66	-17.770
600	54.40	121.31	99.11	13.33	-17.49	52.14	-18.992
700	59.80	130.11	102.91	19.04	-18.52	63.84	-19.931
800	64.40	138.40	106.84	25.26	-19.30	75.65	-20.666
900	68.40	146.23	110.78	31.90	-19.84	87.56	-21.262
1000	71.80	153.61	114.70	38.92	-20.17	99.52	-21.750

The selected values of Rossini, Pitzer, Arnett, Braun, and Pimentel (1248) are employed with the measured enthalpy of combustion of Bartolo and Rossini (80). $Tm = 137.43°K$ and $Tb = 335.26°K$.

No. 279 3-Methyl-1-Pentene C_6H_{12} (Ideal Gas State) Mol Wt 84.156

T°K	cal/(mole °K)			kcal/mole			
	$Cp°$	$S°$	$-(G°-H°_{298})/T$	$H°-H°_{298}$	$\Delta Hf°$	$\Delta Gf°$	Log Kp
298	34.04	90.06	90.06	0.00	-10.76	20.66	-15.141
300	34.19	90.28	90.07	0.07	-10.80	20.85	-15.187
400	42.50	101.28	91.51	3.91	-12.60	31.68	-17.311
500	49.60	111.55	94.50	8.53	-14.09	42.93	-18.765
600	55.60	121.13	98.15	13.79	-15.29	54.45	-19.831
700	60.70	130.09	102.08	19.61	-16.22	66.15	-20.653
800	65.20	138.50	106.11	25.92	-16.91	77.96	-21.297
900	69.00	146.40	110.16	32.63	-17.39	89.86	-21.819
1000	72.30	153.85	114.16	39.70	-17.66	101.80	-22.247

The experimental enthalpy of combustion of Bartolo and Rossini (80) is combined with the values selected by Rossini, Pitzer, Arnett, Braun, and Pimentel (1248). $Tm = 120.1°K$ and $Tb = 327.33°K$.

No. 280 4-Methyl-1-Pentene C_6H_{12} (Ideal Gas State) Mol Wt 84.156

The measured enthalpy of combustion of Bartolo and Rossini (80) is combined with the values selected by Rossini, Pitzer, Arnett, Braun, and Pimentel (1248). $Tm = 119.52°K$ and $Tb = 327.01°K$.

Alkene Ideal Gas Tables

No. 280   4-Methyl-1-Pentene C_6H_{12} (Ideal Gas State)   Mol Wt 84.156

T°K	$Cp°$	$S°$	$-(G°-H°_{298})/T$	$H°-H°_{298}$	$\Delta Hf°$	$\Delta Gf°$	Log Kp
		cal/(mole °K)			kcal/mole		
298	30.23	87.89	87.89	0.00	-10.54	21.52	-15.776
300	30.44	88.08	87.90	0.06	-10.59	21.72	-15.822
400	38.90	98.02	89.19	3.54	-12.76	32.83	-17.937
500	46.40	107.53	91.92	7.81	-14.59	44.45	-19.426
600	52.90	116.58	95.28	12.78	-16.08	56.39	-20.539
700	58.40	125.15	98.94	18.35	-17.26	68.57	-21.408
800	63.10	133.27	102.73	24.44	-18.17	80.89	-22.096
900	67.10	140.93	106.55	30.95	-18.85	93.32	-22.660
1000	70.70	148.19	110.36	37.84	-19.29	105.82	-23.126

No. 281   2-Methyl-2-Pentene C_6H_{12} (Ideal Gas State)   Mol Wt 84.156

T°K	$Cp°$	$S°$	$-(G°-H°_{298})/T$	$H°-H°_{298}$	$\Delta Hf°$	$\Delta Gf°$	Log Kp
		cal/(mole °K)			kcal/mole		
298	30.26	90.45	90.45	0.00	-14.28	17.02	-12.476
300	30.42	90.64	90.46	0.06	-14.33	17.21	-12.538
400	39.00	100.59	91.75	3.54	-16.49	28.07	-15.334
500	46.60	110.13	94.48	7.83	-18.31	39.42	-17.231
600	53.20	119.23	97.86	12.83	-19.77	51.10	-18.614
700	58.60	127.84	101.53	18.42	-20.93	63.02	-19.675
800	63.40	135.99	105.33	24.53	-21.82	75.06	-20.506
900	67.50	143.70	109.17	31.08	-22.46	87.22	-21.180
1000	71.10	151.00	112.99	38.01	-22.86	99.44	-21.732

The selected values of Rossini, Pitzer, Arnett, Braun, and Pimentel (1248) are combined with the measured enthalpy of combustion of Bartolo and Rossini (80). $Tm = 138.08°K$ and $Tb = 340,46°K$.

No. 282   3-Methyl-2-Pentene, cis C_6H_{12} (Ideal Gas State)   Mol Wt 84.156

T°K	$Cp°$	$S°$	$-(G°-H°_{298})/T$	$H°-H°_{298}$	$\Delta Hf°$	$\Delta Gf°$	Log Kp
		cal/(mole °K)			kcal/mole		
298	30.26	90.45	90.45	0.00	-13.80	17.50	-12.827
300	30.42	90.64	90.46	0.06	-13.85	17.69	-12.888
400	39.00	100.59	91.75	3.54	-16.01	28.55	-15.596
500	46.60	110.13	94.48	7.83	-17.83	39.90	-17.440
600	53.20	119.23	97.86	12.83	-19.29	51.58	-18.789
700	58.60	127.84	101.53	18.42	-20.45	63.50	-19.824
800	63.40	135.99	105.33	24.53	-21.34	75.54	-20.637
900	67.50	143.70	109.17	31.08	-21.98	87.70	-21.296
1000	71.10	151.00	112.99	38.01	-22.38	99.92	-21.837

The enthalpy of combustion measured by Bartolo and Rossini (80) is joined with the selected values of Rossini, Pitzer, Arnett, Braun, and Pimentel (1248). $Tm = 134.72°K$ and $Tb = 343.59°K$.

No. 283 3-Methyl-2-Pentene, *trans* C_6H_{12} (Ideal Gas State) Mol Wt 84.156

$T°K$	$Cp°$	$S°$	$-(G°-H°_{298})/T$	$H°-H°_{298}$	$\Delta Hf°$	$\Delta Gf°$	Log Kp
		cal/(mole °K)			kcal/mole		
298	30.26	91.26	91.26	0.00	-14.02	17.04	-12.489
300	30.42	91.45	91.27	0.06	-14.07	17.23	-12.550
400	39.00	101.40	92.56	3.54	-16.23	28.00	-15.299
500	46.60	110.94	95.29	7.83	-18.05	39.28	-17.167
600	53.20	120.04	98.67	12.83	-19.51	50.88	-18.531
700	58.60	128.65	102.34	18.42	-20.67	62.71	-19.579
800	63.40	136.80	106.14	24.53	-21.56	74.68	-20.400
900	67.50	144.51	109.98	31.08	-22.20	86.75	-21.066
1000	71.10	151.81	113.80	38.01	-22.60	98.89	-21.612

The values selected by Rossini, Pitzer, Arnett, Braun, and Pimentel (1248) are used with the measured enthalpy of combustion of Bartolo and Rossini (80). $Tm = 138.31°K$ and $Tb = 340.85°K$.

No. 284 4-Methyl-2-Pentene, *cis* C_6H_{12} (Ideal Gas State) Mol Wt 84.156

$T°K$	$Cp°$	$S°$	$-(G°-H°_{298})/T$	$H°-H°_{298}$	$\Delta Hf°$	$\Delta Gf°$	Log Kp
		cal/(mole °K)			kcal/mole		
298	31.92	89.23	89.23	0.00	-12.03	19.63	-14.391
300	32.07	89.43	89.24	0.06	-12.07	19.83	-14.444
400	40.05	99.76	90.59	3.67	-14.11	30.78	-16.818
500	47.80	109.55	93.41	8.08	-15.81	42.21	-18.448
600	54.10	118.84	96.88	13.18	-17.17	53.94	-19.646
700	59.40	127.59	100.65	18.86	-18.24	65.89	-20.569
800	64.00	135.83	104.54	25.04	-19.06	77.95	-21.294
900	68.00	143.60	108.45	31.64	-19.65	90.12	-21.883
1000	71.50	150.95	112.34	38.62	-20.01	102.35	-22.367

The enthalpy of combustion measured by Bartolo and Rossini (80) is linked with the selected values presented by Rossini, Pitzer, Arnett, Braun, and Pimentel (1248). $Tm = 138.72°K$ and $Tb = 329.53°K$.

No. 285 4-Methyl-2-Pentene, *trans* C_6H_{12} (Ideal Gas State) Mol Wt 84.156

$T°K$	$Cp°$	$S°$	$-(G°-H°_{298})/T$	$H°-H°_{298}$	$\Delta Hf°$	$\Delta Gf°$	Log Kp
		cal/(mole °K)			kcal/mole		
298	33.80	88.02	88.02	0.00	-12.99	19.03	-13.952
300	33.94	88.23	88.03	0.07	-13.03	19.23	-14.009
400	41.90	99.11	89.46	3.87	-14.87	30.28	-16.541
500	48.80	109.22	92.41	8.41	-16.43	41.75	-18.247
600	54.80	118.66	96.01	13.60	-17.72	53.50	-19.488
700	60.00	127.51	99.88	19.34	-18.72	65.46	-20.437
800	64.50	135.82	103.86	25.57	-19.49	77.53	-21.180
900	68.40	143.65	107.85	32.22	-20.02	89.70	-21.781
1000	71.80	151.03	111.80	39.24	-20.35	101.92	-22.274

The selected values presented by Rossini, Pitzer, Arnett, Braun, and Pimentel (1248) are employed with the measured enthalpy of combustion of Bartolo and Rossini (80). $Tm = 132.35°K$ and $Tb = 331.76°K$.

No. 286 2-Ethyl-1-Butene C_6H_{12} (Ideal Gas State) Mol Wt 84.156

$T°K$	$Cp°$	$S°$	$-(G°-H°_{298})/T$	$H°-H°_{298}$	$\Delta Hf°$	$\Delta Gf°$	Log Kp
	cal/(mole °K)				kcal/mole		
298	31.92	90.01	90.01	0.00	-12.32	19.11	-14.008
300	32.08	90.21	90.02	0.06	-12.36	19.30	-14.062
400	40.70	100.65	91.38	3.71	-14.36	30.18	-16.487
500	48.20	110.56	94.23	8.17	-16.01	41.51	-18.142
600	54.50	119.92	97.74	13.31	-17.33	53.13	-19.352
700	59.80	128.73	101.55	19.03	-18.36	64.97	-20.283
800	64.40	137.02	105.47	25.25	-19.14	76.92	-21.012
900	68.40	144.84	109.41	31.89	-19.69	88.97	-21.603
1000	71.90	152.23	113.33	38.91	-20.00	101.07	-22.087

The enthalpy of combustion measured by Bartolo and Rossini (80) and the values selected by Rossini, Pitzer, Arnett, Braun, and Pimentel (1248) are employed. $Tm = 141.62°K$ and $Tb = 337.83°K$.

No. 287 2,3-Dimethyl-1-Butene C_6H_{12} (Ideal Gas State) Mol Wt 84.156

$T°K$	$Cp°$	$S°$	$-(G°-H°_{298})/T$	$H°-H°_{298}$	$\Delta Hf°$	$\Delta Gf°$	Log Kp
	cal/(mole °K)				kcal/mole		
298	34.29	87.39	87.39	0.00	-13.32	18.89	-13.848
300	34.44	87.61	87.40	0.07	-13.36	19.09	-13.906
400	42.60	98.66	88.85	3.93	-15.14	30.19	-16.494
500	49.50	108.93	91.85	8.54	-16.63	41.70	-18.226
600	55.40	118.49	95.50	13.80	-17.85	53.48	-19.478
700	60.50	127.42	99.43	19.60	-18.80	65.45	-20.433
800	65.00	135.80	103.46	25.88	-19.51	77.52	-21.177
900	68.80	143.68	107.49	32.57	-20.01	89.69	-21.779
1000	72.20	151.11	111.49	39.62	-20.29	101.91	-22.271

The values selected by Rossini, Pitzer, Arnett, Braun, and Pimentel (1248) and the measured enthalpy of combustion of Bartolo and Rossini (80) are chosen. $Tm = 115.89°K$ and $Tb = 328.77°K$.

No. 288 3,3-Dimethyl-1-Butene C_6H_{12} (Ideal Gas State) Mol Wt 84.156

The measurements of Kennedy, Shomate, and Parks (741) and of Bartolo and Rossini (80) are employed with the selected values of Rossini, Pitzer, Arnett, Braun, and Pimentel (1248). $Tt = 124.9°K$, $\Delta Ht° = 1.037$ kcal/mole. $Tm = 157.95°K$, $\Delta Hm° = 0.261$ kcal/mole, and $Tb = 314.40°K$.

No. 288   3,3-Dimethyl-1-Butene C_6H_{12} (Ideal Gas State)   Mol Wt 84.156

T°K	Cp°	S°	$-(G°-H°_{298})/T$	$H°-H°_{298}$	$\Delta Hf°$	$\Delta Gf°$	Log Kp
	cal/(mole °K)				kcal/mole		
298	30.23	82.16	82.16	0.00	-10.31	23.46	-17.197
300	30.39	82.35	82.17	0.06	-10.36	23.67	-17.242
400	38.90	92.28	83.46	3.53	-12.53	35.35	-19.315
500	46.70	101.82	86.18	7.82	-14.34	47.54	-20.779
600	53.40	110.94	89.56	12.84	-15.80	60.05	-21.873
700	58.90	119.60	93.24	18.46	-16.93	72.80	-22.727
800	63.60	127.78	97.05	24.59	-17.79	85.66	-23.401
900	67.30	135.48	100.89	31.14	-18.43	98.64	-23.952
1000	71.00	142.77	104.72	38.05	-18.85	111.68	-24.407

No. 289   2,3-Dimethyl-2-Butene C_6H_{12} (Ideal Gas State)   Mol Wt 84.156

T°K	Cp°	S°	$-(G°-H°_{298})/T$	$H°-H°_{298}$	$\Delta Hf°$	$\Delta Gf°$	Log Kp
	cal/(mole °K)				kcal/mole		
298	29.54	87.15	87.15	0.00	-14.15	18.13	-13.292
300	29.68	87.34	87.16	0.06	-14.20	18.33	-13.354
400	37.48	96.95	88.41	3.42	-16.48	29.53	-16.135
500	45.04	106.14	91.05	7.55	-18.45	41.27	-18.038
600	51.78	114.96	94.30	12.40	-20.07	53.37	-19.438
700	57.67	123.40	97.86	17.88	-21.34	65.72	-20.517
800	62.78	131.44	101.56	23.91	-22.31	78.21	-21.366
900	67.25	139.10	105.31	30.41	-22.99	90.83	-22.055
1000	71.14	146.39	109.05	37.34	-23.41	103.51	-22.621

The low-temperature measurements of Scott, Finke, McCullough, Gross, Messerly, Pennington, and Waddington (1306) and the experimental enthalpy of combustion of Bartolo and Rossini (80) are adopted. $Tt = 196.81°K$, $\Delta Ht° = 0.844$ kcal/mole, $Tm = 198.88°K$, $\Delta Hm° = 1.542$ kcal/mole, and $Tb = 346.35°K$, where $\Delta Hv = 7.083$ kcal/mole.

No. 290   1-Heptene C_7H_{14} (Ideal Gas State)   Mol Wt 98.182

T°K	Cp°	S°	$-(G°-H°_{298})/T$	$H°-H°_{298}$	$\Delta Hf°$	$\Delta Gf°$	Log Kp
	cal/(mole °K)				kcal/mole		
298	37.10	101.24	101.24	0.00	-14.89	22.90	-16.789
300	37.27	101.48	101.25	0.07	-14.94	23.14	-16.853
400	46.97	113.55	102.83	4.29	-17.31	36.20	-19.779
500	55.72	124.99	106.13	9.44	-19.28	49.81	-21.773
600	63.24	135.83	110.18	15.40	-20.87	63.78	-23.231
700	69.62	146.07	114.58	22.05	-22.09	78.00	-24.351
800	75.08	155.73	119.13	29.29	-23.02	92.35	-25.228
900	79.81	164.86	123.71	37.04	-23.65	106.82	-25.938
1000	83.90	173.48	128.26	45.23	-24.02	121.35	-26.520

The measurements of McCullough, Finke, Gross, Messerly, and Waddington (950) are used to revise the values selected by Rossini, Pitzer, Arnett, Braun, and Pimentel (1248). The former workers found two crystalline forms of 1-heptene, and studied each from 11°K to its triple point. Initial crystallization always gave the lower-melting polymorph (designated crystal II; $Ttp = 153.88°K$, $\Delta Hm° = 3.021$ kcal/mole), which after being more than half melted and warmed to the melting point of the higher-melting form (designated crystal I; $Ttp = 154.29°K$, $\Delta Hm° = 2.964$ kcal/mole) transformed into the higher-melting polymorph and could then be recrystallized in the higher-melting form. Thus the enthalpy of melting of crystal I is 0.057 kcal/mole less than that of crystal II; this figure corresponds to the enthalpy increment for the transformation of crystal II → crystal I at the melting point. $Tb = 366.79°K$, where $\Delta Hv = 7.440$ kcal/mole. Ambrose, Cox, and Townsend (13) measured $Tc = 537.2°K$.

No. 291   1-Octene C_8H_{16} (Ideal Gas State)   Mol Wt 112.208

	cal/(mole °K)				kcal/mole		
$T°K$	$Cp°$	$S°$	$-(G°-H°_{298})/T$	$H°-H°_{298}$	$\Delta Hf°$	$\Delta Gf°$	Log Kp
298	42.56	110.55	110.55	0.00	-19.82	24.91	-18.259
300	42.77	110.82	110.56	0.08	-19.88	25.18	-18.345
400	53.91	124.67	112.37	4.93	-22.56	40.63	-22.198
500	63.96	137.81	116.16	10.83	-24.79	56.69	-24.779
600	72.58	150.25	120.81	17.67	-26.58	73.16	-26.645
700	79.89	162.00	125.87	25.30	-27.95	89.90	-28.067
800	86.15	173.09	131.08	33.61	-28.97	106.79	-29.173
900	91.55	183.56	136.34	42.50	-29.66	123.82	-30.065
1000	96.20	193.45	141.56	51.90	-30.05	140.91	-30.794

The measurements of McCullough, Finke, Gross, Messerly, and Waddington (950) and the selected values of Rossini, Pitzer, Arnett, Braun, and Pimentel (1248) are presented. $Ttp = 171.45°K$, $\Delta Hm° = 3.660$ kcal/mole, and $Tb = 394.43°K$; Rockenfeller and Rossini (1236) reported $\Delta Hv_{298} = 9.70$ kcal/mole. Ambrose, Cox, and Townsend (13) measured $Tc = 566.5°K$.

No. 292   1-Nonene C_9H_{18} (Ideal Gas State)   Mol Wt 126.234

The selected values of Rossini, Pitzer, Arnett, Braun, and Pimentel (1248) are adopted. $Tm = 191.78°K$ and $Tb = 420.02°K$.

No. 292   1-Nonene C_9H_{18} (Ideal Gas State)   Mol Wt 126.234

	cal/(mole °K)				kcal/mole		
$T°K$	$Cp°$	$S°$	$-(G°-H°_{298})/T$	$H°-H°_{298}$	$\Delta Hf°$	$\Delta Gf°$	Log Kp
298	48.03	119.86	119.86	0.00	−24.74	26.93	−19.736
300	48.26	120.16	119.87	0.09	−24.80	27.24	−19.845
400	60.85	135.80	121.91	5.56	−27.81	45.07	−24.622
500	72.21	150.63	126.19	12.23	−30.29	63.58	−27.790
600	81.93	164.68	131.44	19.95	−32.28	82.54	−30.064
700	90.17	177.94	137.15	28.56	−33.79	101.81	−31.785
800	97.22	190.45	143.03	37.94	−34.90	121.25	−33.121
900	103.30	202.26	148.97	47.97	−35.65	140.82	−34.195
1000	108.50	213.42	154.86	58.57	−36.06	160.47	−35.069

No. 293   1-Decene $C_{10}H_{20}$ (Ideal Gas State)   Mol Wt 140.260

	cal/(mole °K)				kcal/mole		
$T°K$	$Cp°$	$S°$	$-(G°-H°_{298})/T$	$H°-H°_{298}$	$\Delta Hf°$	$\Delta Gf°$	Log Kp
298	53.49	129.17	129.17	0.00	−29.67	28.93	−21.206
300	53.76	129.51	129.18	0.10	−29.74	29.29	−21.336
400	67.79	146.93	131.46	6.19	−33.06	49.49	−27.041
500	80.45	163.45	136.22	13.62	−35.80	70.46	−30.796
600	91.27	179.10	142.07	22.22	−37.98	91.91	−33.478
700	100.44	193.87	148.43	31.82	−39.64	113.71	−35.501
800	108.28	207.81	154.99	42.26	−40.85	135.69	−37.067
900	115.04	220.96	161.59	53.44	−41.66	157.82	−38.322
1000	120.90	233.39	168.16	65.24	−42.09	180.03	−39.343

The low-temperature measurements of McCullough, Finke, Gross, Messerly, and Waddington (950) are combined with the selected values of Rossini, Pitzer, Arnett, Braun, and Pimentel (1248). Tt = 198.3°K, $\Delta Ht°$ = 1.900 kcal/mole, Tm = 206.84°K, $\Delta Hm°$ = 3.300 kcal/mole, and Tb = 443.72°K.

No. 294   1-Undecene $C_{11}H_{22}$ (Ideal Gas State)   Mol Wt 154.286

	cal/(mole °K)				kcal/mole		
$T°K$	$Cp°$	$S°$	$-(G°-H°_{298})/T$	$H°-H°_{298}$	$\Delta Hf°$	$\Delta Gf°$	Log Kp
298	58.96	138.48	138.48	0.00	−34.60	30.94	−22.676
300	59.25	138.85	138.49	0.11	−34.68	31.34	−22.828
400	74.74	158.05	141.00	6.83	−38.32	53.92	−29.459
500	88.70	176.27	146.25	15.02	−41.31	77.34	−33.802
600	100.61	193.52	152.70	24.50	−43.69	101.29	−36.892
700	110.72	209.81	159.71	35.08	−45.49	125.61	−39.216
800	119.34	225.17	166.94	46.59	−46.80	150.13	−41.012
900	126.79	239.67	174.22	58.90	−47.66	174.82	−42.449
1000	133.20	253.37	181.46	71.91	−48.11	199.58	−43.616

Alkene Ideal Gas Tables

The low-temperature measurements of McCullough, Finke, Gross, Messerly, and Waddington (950) are combined with the selected values of Rossini, Pitzer, Arnett, Braun, and Pimentel (1248). $Tt = 217.3°K$, $\Delta Ht° = 2.202$ kcal/mole, $Tm = 223.96°K$, $\Delta Hm° = 4.061$ kcal/mole, and $Tb = 465.82°K$.

No. 295 1-Dodecene $C_{12}H_{24}$ (Ideal Gas State) Mol Wt 168.312

$T°K$	$Cp°$	$S°$	$-(G°-H°_{298})/T$	$H°-H°_{298}$	$\Delta Hf°$	$\Delta Gf°$	Log Kp
		cal/(mole °K)			kcal/mole		
298	64.43	147.78	147.78	0.00	-39.52	32.96	-24.156
300	64.74	148.18	147.79	0.12	-39.60	33.40	-24.330
400	81.68	169.17	150.53	7.46	-43.56	58.36	-31.886
500	96.95	189.08	156.27	16.41	-46.82	84.23	-36.815
600	109.95	207.93	163.32	26.77	-49.39	110.68	-40.313
700	121.00	225.73	170.98	38.33	-51.33	137.53	-42.937
800	130.41	242.52	178.88	50.91	-52.74	164.59	-44.962
900	138.54	258.36	186.84	64.37	-53.66	191.83	-46.581
1000	145.50	273.33	194.75	78.58	-54.12	219.16	-47.894

The low-temperature measurements of McCullough, Finke, Gross, Messerly, and Waddington (950) are joined with the selected values of Rossini, Pitzer, Arnett, Braun, and Pimentel (1248). $Tt = 212.9°K$, $\Delta Ht° = 1.088$ kcal/mole, $Tm = 237.92°K$, $\Delta Hm° = 4.758$ kcal/mole, and $Tb = 486.51°K$.

No. 296 1-Tridecene $C_{13}H_{26}$ (Ideal Gas State) Mol Wt 182.338

$T°K$	$Cp°$	$S°$	$-(G°-H°_{298})/T$	$H°-H°_{298}$	$\Delta Hf°$	$\Delta Gf°$	Log Kp
		cal/(mole °K)			kcal/mole		
298	69.89	157.09	157.09	0.00	-44.45	34.96	-25.626
300	70.24	157.53	157.10	0.13	-44.54	35.45	-25.822
400	88.62	180.30	160.08	8.09	-48.82	62.79	-34.304
500	105.19	201.90	166.30	17.80	-52.33	91.11	-39.822
600	119.29	222.35	173.95	29.05	-55.10	120.05	-43.727
700	131.27	241.66	182.26	41.59	-57.18	149.43	-46.653
800	141.48	259.88	190.83	55.24	-58.69	179.03	-48.907
900	150.28	277.06	199.47	69.84	-59.66	208.83	-50.708
1000	157.80	293.29	208.05	85.25	-60.15	238.71	-52.168

The values used are those selected by Rossini, Pitzer, Arnett, Braun, and Pimentel (1248). $Tm = 250.08°K$ and $Tb = 505.93°K$.

No. 297 1-Tetradecene $C_{14}H_{28}$ (Ideal Gas State) Mol Wt 196.364

The values of Rossini, Pitzer, Arnett, Braun, and Pimentel (1248) are used. $Tm = 260.30°K$ and $Tb = 524.25°K$.

No. 297 1-Tetradecene $C_{14}H_{28}$ (Ideal Gas State) Mol Wt 196.364

$T°K$	$Cp°$	$S°$	$-(G°-H°_{298})/T$	$H°-H°_{298}$	$\Delta Hf°$	$\Delta Gf°$	Log Kp
	cal/(mole °K)				kcal/mole		
298	75.36	166.40	166.40	0.00	-49.36	36.99	-27.111
300	75.73	166.87	166.41	0.14	-49.46	37.51	-27.328
400	95.56	191.42	169.62	8.73	-54.05	67.24	-36.734
500	113.44	214.72	176.33	19.20	-57.82	98.01	-42.837
600	128.64	236.78	184.58	31.32	-60.79	129.45	-47.149
700	141.55	257.60	193.54	44.85	-63.01	161.35	-50.375
800	152.54	277.24	202.79	59.56	-64.62	193.50	-52.858
900	162.03	295.76	212.10	75.30	-65.65	225.84	-54.840
1000	170.20	313.27	221.35	91.93	-66.15	258.29	-56.446

No. 298 1-Pentadecene $C_{15}H_{30}$ (Ideal Gas State) Mol Wt 210.390

$T°K$	$Cp°$	$S°$	$-(G°-H°_{298})/T$	$H°-H°_{298}$	$\Delta Hf°$	$\Delta Gf°$	Log Kp
	cal/(mole °K)				kcal/mole		
298	80.82	175.71	175.71	0.00	-54.31	38.97	-28.566
300	81.23	176.22	175.72	0.15	-54.42	39.54	-28.806
400	102.50	202.55	179.16	9.36	-59.33	71.64	-39.142
500	121.68	227.53	186.36	20.59	-63.35	104.86	-45.834
600	137.98	251.20	195.21	33.60	-66.51	138.80	-50.556
700	151.82	273.53	204.82	48.10	-68.88	173.24	-54.084
800	163.60	294.59	214.74	63.89	-70.59	207.92	-56.798
900	173.77	314.46	224.73	80.77	-71.68	242.82	-58.962
1000	182.50	333.24	234.65	98.59	-72.20	277.82	-60.715

The values of Rossini, Pitzer, Arnett, Braun, and Pimentel (1248) are used. $Tm = 269.42°K$ and $Tb = 541.54°K$.

No. 299 1-Hexadecene $C_{16}H_{32}$ (Ideal Gas State) Mol Wt 224.416

$T°K$	$Cp°$	$S°$	$-(G°-H°_{298})/T$	$H°-H°_{298}$	$\Delta Hf°$	$\Delta Gf°$	Log Kp
	cal/(mole °K)				kcal/mole		
298	86.29	185.02	185.02	0.00	-59.23	40.99	-30.043
300	86.72	185.56	185.03	0.17	-59.34	41.60	-30.305
400	109.44	213.67	188.71	9.99	-64.57	76.08	-41.566
500	129.93	240.35	196.39	21.99	-68.85	111.75	-48.845
600	147.32	265.62	205.84	35.87	-72.21	148.19	-53.974
700	162.10	289.47	216.10	51.36	-74.72	185.15	-57.803
800	174.67	311.95	226.69	68.21	-76.53	222.37	-60.746
900	185.52	333.17	237.35	86.24	-77.67	259.83	-63.091
1000	194.80	353.21	247.95	105.26	-78.21	297.39	-64.991

The values of Rossini, Pitzer, Arnett, Braun, and Pimentel (1248) are adopted. $Tm = 277.27°K$ and $Tb = 558.02°K$.

Alkene Ideal Gas Tables

No. 300   1-Heptadecene $C_{17}H_{34}$ (Ideal Gas State)   Mol Wt 238.442

$T°K$	$Cp°$	$S°$	$-(G°-H°_{298})/T$	$H°-H°_{298}$	$\Delta Hf°$	$\Delta Gf°$	Log Kp
		cal/(mole °K)			kcal/mole		
298	91.76	194.33	194.33	0.00	-64.15	43.00	-31.521
300	92.21	194.90	194.34	0.18	-64.27	43.66	-31.804
400	116.38	224.80	198.25	10.62	-69.82	80.52	-43.990
500	138.18	253.17	206.42	23.38	-74.35	118.64	-51.856
600	156.66	280.04	216.47	38.15	-77.91	157.57	-57.392
700	172.38	305.40	227.38	54.62	-80.56	197.06	-61.521
800	185.74	329.31	238.64	72.54	-82.46	236.82	-64.694
900	197.27	351.87	249.98	91.70	-83.67	276.83	-67.221
1000	207.20	373.18	261.25	111.94	-84.22	316.95	-69.266

The selected values of Rossini, Pitzer, Arnett, Braun, and Pimentel (1248) are used. $Tm = 284.4°K$ and $Tb = 573.48°K$.

No. 301   1-Octadecene $C_{18}H_{36}$ (Ideal Gas State)   Mol Wt 252.468

$T°K$	$Cp°$	$S°$	$-(G°-H°_{298})/T$	$H°-H°_{298}$	$\Delta Hf°$	$\Delta Gf°$	Log Kp
		cal/(mole °K)			kcal/mole		
298	97.22	203.64	203.64	0.00	-69.08	45.01	-32.991
300	97.71	204.25	203.65	0.19	-69.21	45.71	-33.296
400	123.32	235.92	207.79	11.26	-75.07	84.94	-46.409
500	146.42	265.99	216.45	24.77	-79.86	125.52	-54.862
600	166.00	294.46	227.10	40.42	-83.62	166.94	-60.806
700	182.65	321.33	238.66	57.87	-86.42	208.96	-65.237
800	196.80	346.67	250.60	76.86	-88.42	251.27	-68.639
900	209.01	370.57	262.61	97.17	-89.67	293.83	-71.348
1000	219.50	393.15	274.55	118.61	-90.25	336.51	-73.540

Rossini, Pitzer, Arnett, Braun, and Pimentel (1248) selected the values used. $Tm = 290.8°K$ and $Tb = 587.97°K$.

No. 302   1-Nonadecene $C_{19}H_{38}$ (Ideal Gas State)   Mol Wt 266.494

$T°K$	$Cp°$	$S°$	$-(G°-H°_{298})/T$	$H°-H°_{298}$	$\Delta Hf°$	$\Delta Gf°$	Log Kp
		cal/(mole °K)			kcal/mole		
298	102.69	212.95	212.95	0.00	-74.00	47.02	-34.468
300	103.20	213.59	212.96	0.20	-74.13	47.77	-34.795
400	130.26	247.05	217.34	11.89	-80.32	89.38	-48.833
500	154.67	278.81	226.48	26.17	-85.36	132.41	-57.873
600	175.35	308.88	237.73	42.69	-89.32	176.33	-64.225
700	192.93	337.27	249.94	61.13	-92.26	220.87	-68.955
800	207.86	364.03	262.55	81.19	-94.35	265.72	-72.587
900	220.76	389.27	275.24	102.63	-95.67	310.84	-75.478
1000	231.80	413.12	287.85	125.28	-96.26	356.07	-77.816

The values employed are those selected by Rossini, Pitzer, Arnett, Braun, and Pimentel (1248). $Tm = 296.6°K$ and $Tb = 602.3°K$.

No. 303 1-Eicosene $C_{20}H_{40}$ (Ideal Gas State) Mol Wt 280.520

$T°K$	$Cp°$	$S°$	$-(G°-H°_{298})/T$	$H°-H°_{298}$	$\Delta Hf°$	$\Delta Gf°$	Log Kp
		cal/(mole °K)			kcal/mole		
298	108.15	222.26	222.26	0.00	-78.93	49.03	-35.938
300	108.70	222.94	222.27	0.21	-79.07	49.81	-36.287
400	137.20	258.18	226.88	12.52	-85.57	93.81	-51.252
500	162.91	291.63	236.51	27.56	-90.87	139.29	-60.879
600	184.69	323.30	248.36	44.97	-95.03	185.70	-67.639
700	203.20	353.20	261.23	64.39	-98.11	232.77	-72.671
800	218.93	381.38	274.50	85.51	-100.30	280.16	-76.533
900	232.50	407.97	287.87	108.10	-101.68	327.83	-79.605
1000	244.20	433.09	301.15	131.95	-102.29	375.63	-82.090

The values presented are those selected by Rossini, Pitzer, Arnett, Braun, and Pimentel (1248). $Tm = 301.8°K$ and $Tb = 615.6°K$.

ALKADIENE IDEAL GAS TABLES

No. 304 Allene (Propadiene) C_3H_4 (Ideal Gas State) Mol Wt 40.062

$T°K$	$Cp°$	$S°$	$-(G°-H°_{298})/T$	$H°-H°_{298}$	$\Delta Hf°$	$\Delta Gf°$	Log Kp
		cal/(mole °K)			kcal/mole		
298	14.10	58.30	58.30	0.00	45.92	48.37	-35.452
300	14.16	58.39	58.31	0.03	45.91	48.38	-35.243
400	17.21	62.89	58.90	1.60	45.35	49.29	-26.931
500	19.82	67.02	60.12	3.46	44.85	50.34	-22.001
600	22.00	70.83	61.59	5.55	44.41	51.47	-18.748
700	23.84	74.37	63.17	7.85	44.04	52.68	-16.447
800	25.42	77.66	64.77	10.31	43.71	53.94	-14.734
900	26.80	80.73	66.38	12.92	43.44	55.23	-13.412
1000	28.00	83.62	67.96	15.67	43.23	56.56	-12.360

The values selected by Rossini, Pitzer, Arnett, Braun, and Pimentel (1248) are adopted. $Tm = 136.85°K$ and $Tb = 238.6°K$. Kobe and Lynn (780) reported $Tc = 393°K$.

No. 305 1,2-Butadiene C_4H_6 (Ideal Gas State) Mol Wt 54.088

The values adopted are from the measurements of Aston and Szasz (44) and from the selections of Rossini, Pitzer, Arnett, Braun, and Pimentel (1248). $Tm = 136.96°K$, $\Delta Hm° = 1.664$ kcal/mole, and $Tb = 284.00°K$, where $\Delta Hv = 5.77$ kcal/mole.

Alkadiene Ideal Gas Tables

No. 305 1,2-Butadiene C_4H_6 (Ideal Gas State) Mol Wt 54.088

$T°K$	$Cp°$	$S°$	$-(G°-H°_{298})/T$	$H°-H°_{298}$	$\Delta Hf°$	$\Delta Gf°$	Log Kp
		cal/(mole °K)			kcal/mole		
298	19.15	70.03	70.03	0.00	38.77	47.43	-34.766
300	19.23	70.15	70.04	0.04	38.75	47.48	-34.589
400	23.54	76.29	70.84	2.18	37.82	50.54	-27.612
500	27.39	81.96	72.51	4.73	37.00	53.82	-23.521
600	30.72	87.26	74.53	7.64	36.30	57.24	-20.849
700	33.54	92.21	76.70	10.86	35.72	60.78	-18.976
800	36.01	96.85	78.94	14.34	35.24	64.39	-17.590
900	38.16	101.22	81.17	18.05	34.87	68.06	-16.526
1000	40.02	105.34	83.39	21.96	34.61	71.77	-15.684

No. 306 1,3-Butadiene C_4H_6 (Ideal Gas State) Mol Wt 54.088

$T°K$	$Cp°$	$S°$	$-(G°-H°_{298})/T$	$H°-H°_{298}$	$\Delta Hf°$	$\Delta Gf°$	Log Kp
		cal/(mole °K)			kcal/mole		
298	19.01	66.62	66.62	0.00	26.33	36.01	-26.393
300	19.11	66.74	66.63	0.04	26.31	36.07	-26.272
400	24.29	72.97	67.44	2.22	25.42	39.46	-21.559
500	28.52	78.86	69.14	4.87	24.70	43.06	-18.819
600	31.84	84.37	71.23	7.89	24.11	46.78	-17.039
700	34.55	89.48	73.47	11.21	23.63	50.60	-15.799
800	36.84	94.25	75.78	14.78	23.25	54.48	-14.882
900	38.81	98.71	78.08	18.57	22.95	58.40	-14.182
1000	40.52	102.89	80.35	22.54	22.74	62.36	-13.628

The measurements of Scott, Meyers, Rands, Brickwedde, and Bekkedahl (1327) and the selections of Rossini, Pitzer, Arnett, Braun, and Pimentel (1248) are the sources of the values adopted. $Tm = 164.24°K$, $\Delta Hm° = 1.908$ kcal/mole, and $Tb = 268.74°K$, where $\Delta Hv = 5.572$ kcal/mole. Kobe and Lynn (780) selected $Tc = 425°K$, $Pc = 42.7$ atm, and $dc = 0.245$ g/cc.

No. 307 1,2-Pentadiene C_5H_8 (Ideal Gas State) Mol Wt 68.114

$T°K$	$Cp°$	$S°$	$-(G°-H°_{298})/T$	$H°-H°_{298}$	$\Delta Hf°$	$\Delta Gf°$	Log Kp
		cal/(mole °K)			kcal/mole		
298	25.20	79.70	79.70	0.00	34.80	50.29	-36.861
300	25.30	79.86	79.71	0.05	34.77	50.38	-36.702
400	31.40	88.00	80.77	2.90	33.61	55.77	-30.471
500	36.50	95.57	82.98	6.30	32.62	61.43	-26.849
600	40.80	102.61	85.68	10.17	31.80	67.26	-24.500
700	44.50	109.19	88.57	14.44	31.15	73.23	-22.864
800	47.70	115.34	91.54	19.05	30.64	79.27	-21.656
900	50.40	121.12	94.51	23.96	30.27	85.38	-20.733
1000	52.80	126.56	97.44	29.12	30.03	91.53	-20.002

The values adopted are those selected by Rossini, Pitzer, Arnett, Braun, and Pimentel (1248). $Tm = 135.89°K$ and $Tb = 318.01°K$.

No. 308   1,3-Pentadiene, cis C_5H_8 (Ideal Gas State)   Mol Wt 68.114

T°K	Cp°	S°	$-(G°-H°_{298})/T$	$H°-H°_{298}$	ΔHf°	ΔGf°	Log Kp
		cal/(mole °K)			kcal/mole		
298	22.60	77.50	77.50	0.00	18.70	34.84	-25.540
300	22.70	77.65	77.51	0.05	18.67	34.94	-25.454
400	29.50	85.13	78.48	2.67	17.28	40.59	-22.176
500	35.30	92.36	80.54	5.92	16.14	46.55	-20.347
600	39.90	99.21	83.09	9.68	15.22	52.72	-19.202
700	43.80	105.66	85.85	13.87	14.48	59.04	-18.431
800	47.00	111.73	88.71	18.42	13.90	65.43	-17.874
900	49.80	117.43	91.59	23.26	13.47	71.91	-17.460
1000	52.20	122.80	94.45	28.36	13.17	78.42	-17.138

The selected values of Rossini, Pitzer, Arnett, Braun, and Pimentel (1248) are adopted. $Tm = 132.33°K$ and $Tb = 317.22°K$.

No. 309   1,3-Pentadiene, trans C_5H_8 (Ideal Gas State)   Mol Wt 68.114

T°K	Cp°	S°	$-(G°-H°_{298})/T$	$H°-H°_{298}$	ΔHf°	ΔGf°	Log Kp
		cal/(mole °K)			kcal/mole		
298	24.70	76.40	76.40	0.00	18.60	35.07	-25.708
300	24.90	76.56	76.41	0.05	18.57	35.17	-25.622
400	31.20	84.61	77.46	2.86	17.38	40.90	-22.344
500	36.60	92.17	79.65	6.26	16.39	46.89	-20.497
600	40.90	99.23	82.33	10.14	15.58	53.07	-19.329
700	44.60	105.82	85.22	14.42	14.94	59.38	-18.537
800	47.70	111.98	88.19	19.04	14.43	65.75	-17.962
900	50.30	117.76	91.16	23.94	14.05	72.20	-17.531
1000	52.60	123.18	94.09	29.09	13.80	78.68	-17.194

Rossini, Pitzer, Arnett, Braun, and Pimentel (1248) selected the values adopted. $Tm = 185.68°K$ and $Tb = 315.18°K$.

No. 310   1,4-Pentadiene C_5H_8 (Ideal Gas State)   Mol Wt 68.114

The measurements of Parks, Todd, and Shomate (1119) and the review by Rossini, Pitzer, Arnett, Braun, and Pimentel (1248) are adopted. $Tm = 124.88°K$, $\Delta Hm° = 1.468$ kcal/mole, and $Tb = 299.12°K$.

No. 311   2,3-Pentadiene C_5H_8 (Ideal Gas State)   Mol Wt 68.114

The values adopted are those selected by Rossini, Pitzer, Arnett, Braun, and Pimentel (1248). $Tm = 147.50°K$ and $Tb = 321.41°K$.

Alkadiene Ideal Gas Tables

No. 310 1,4-Pentadiene C_5H_8 (Ideal Gas State) Mol Wt 68.114

$T°K$	$Cp°$	$S°$	$-(G°-H°_{298})/T$	$H°-H°_{298}$	$\Delta Hf°$	$\Delta Gf°$	Log Kp
		cal/(mole °K)			kcal/mole		
298	25.10	79.70	79.70	0.00	25.20	40.69	-29.824
300	25.20	79.86	79.71	0.05	25.17	40.78	-29.709
400	31.30	87.97	80.77	2.88	24.00	46.17	-25.227
500	36.50	95.53	82.97	6.28	23.01	51.83	-22.656
600	40.80	102.57	85.66	10.15	22.19	57.67	-21.007
700	44.40	109.14	88.55	14.42	21.53	63.65	-19.871
800	47.60	115.28	91.51	19.02	21.01	69.69	-19.039
900	50.30	121.05	94.48	23.92	20.63	75.81	-18.408
1000	52.70	126.47	97.41	29.07	20.38	81.96	-17.912

No. 311 2,3-Pentadiene C_5H_8 (Ideal Gas State) Mol Wt 68.114

$T°K$	$Cp°$	$S°$	$-(G°-H°_{298})/T$	$H°-H°_{298}$	$\Delta Hf°$	$\Delta Gf°$	Log Kp
		cal/(mole °K)			kcal/mole		
298	24.20	77.60	77.60	0.00	33.10	49.21	-36.074
300	24.30	77.76	77.61	0.05	33.07	49.31	-35.922
400	29.90	85.52	78.63	2.76	31.78	54.93	-30.011
500	35.00	92.76	80.74	6.01	30.64	60.85	-26.597
600	39.40	99.54	83.31	9.74	29.68	66.98	-24.397
700	43.20	105.90	86.09	13.87	28.89	73.27	-22.875
800	46.60	111.90	88.94	18.37	28.26	79.65	-21.758
900	49.50	117.56	91.81	23.18	27.78	86.11	-20.908
1000	52.00	122.91	94.66	28.25	27.47	92.61	-20.239

No. 312 3-Methyl-1,2-Butadiene C_5H_8 (Ideal Gas State) Mol Wt 68.114

$T°K$	$Cp°$	$S°$	$-(G°-H°_{298})/T$	$H°-H°_{298}$	$\Delta Hf°$	$\Delta Gf°$	Log Kp
		cal/(mole °K)			kcal/mole		
298	25.20	76.40	76.40	0.00	31.00	47.47	-34.797
300	25.30	76.56	76.41	0.05	30.97	47.57	-34.655
400	31.00	84.64	77.47	2.87	29.78	53.29	-29.117
500	36.00	92.10	79.66	6.23	28.75	59.29	-25.915
600	40.30	99.06	82.32	10.05	27.88	65.48	-23.850
700	44.00	105.55	85.18	14.27	27.18	71.81	-22.418
800	47.20	111.64	88.11	18.83	26.62	78.22	-21.366
900	50.00	117.37	91.05	23.69	26.20	84.70	-20.566
1000	52.40	122.76	93.95	28.82	25.93	91.22	-19.935

Values from the compilation by Rossini, Pitzer, Arnett, Braun, and Pimentel (1248) are adopted. $Tm = 159.53°K$ and $Tb = 314.00°K$.

No. 313 2-Methyl-1,3-Butadiene (Isoprene) C_5H_8 (Ideal Gas State) Mol Wt 68.114

The measurements of Bekkedahl, Wood, and Wojciechowski (99, 100) on isoprene and the values selected by Rossini, Pitzer, Arnett, Braun,

No. 313 2-Methyl-1,3-Butadiene (Isoprene) C_5H_8 (Ideal Gas State) Mol Wt 68.114

T°K	Cp°	S°	$-(G°-H°_{298})/T$	$H°-H°_{298}$	$\Delta Hf°$	$\Delta Gf°$	Log Kp
	cal/(mole °K)				kcal/mole		
298	25.00	75.44	75.44	0.00	18.10	34.86	−25.551
300	25.20	75.60	75.45	0.05	18.07	34.96	−25.467
400	31.80	83.78	76.52	2.91	16.92	40.77	−22.277
500	37.10	91.47	78.75	6.37	15.99	46.85	−20.476
600	41.40	98.62	81.47	10.30	15.23	53.09	−19.336
700	45.00	105.28	84.40	14.62	14.63	59.45	−18.561
800	48.00	111.49	87.41	19.27	14.16	65.88	−17.996
900	50.60	117.30	90.41	24.21	13.82	72.37	−17.573
1000	52.90	122.75	93.37	29.38	13.60	78.90	−17.242

and Pimentel (1248) are adopted. $Tm = 127.20°K$, $\Delta Hm° = 1.155$ kcal/mole, and $Tb = 307.22°K$, where $\Delta Hv = 6.191$ kcal/mole.

ALKYNE IDEAL GAS TABLES

No. 314 Acetylene (Ethyne) C_2H_2 (Ideal Gas State) Mol Wt 26.036

T°K	Cp°	S°	$-(G°-H°_{298})/T$	$H°-H°_{298}$	$\Delta Hf°$	$\Delta Gf°$	Log Kp
	cal/(mole °K)				kcal/mole		
298	10.50	48.00	48.00	0.00	54.19	50.00	−36.646
300	10.53	48.07	48.01	0.02	54.19	49.97	−36.401
400	11.97	51.31	48.44	1.15	54.13	48.57	−26.538
500	12.97	54.09	49.30	2.40	54.04	47.19	−20.627
600	13.73	56.52	50.30	3.74	53.92	45.83	−16.693
700	14.37	58.69	51.35	5.14	53.78	44.50	−13.892
800	14.93	60.65	52.39	6.61	53.62	43.18	−11.795
900	15.45	62.43	53.41	8.13	53.45	41.88	−10.170
1000	15.92	64.09	54.39	9.70	53.29	40.61	−8.875

The values selected by Rossini, Pitzer, Arnett, Braun, and Pimentel (1248) and by Rossini, Wagman, Evans, Levine, and Jaffe (1249) are adopted. They reported $Tm = 191.7°K$ at 1.18 atm, $\Delta Hm = 0.900$ kcal/mole, and $Ts = 189.2°K$ where $\Delta Hs = 5.1$ kcal/mole. $Tc = 309°K$, $Pc = 61.6$ atm, and $dc = 0.231$ g/cc were reported by Kobe and Lynn (780).

No. 315 Propyne (Methylacetylene) C_3H_4 (Ideal Gas State) Mol Wt 40.062

Rossini, Pitzer, Arnett, Braun, and Pimentel (1248) selected the values adopted here. $Tm = 170.4°K$ and $Tb = 249.93°K$. Kobe and Lynn (780) selected $Tc = 401°K$ and $Pc = 52.8$ atm.

No. 315 Propyne (Methylacetylene) C₃H₄ (Ideal Gas State) Mol Wt 40.062

T°K	Cp°	S°	$-(G°-H°_{298})/T$	$H°-H°_{298}$	$\Delta Hf°$	$\Delta Gf°$	Log Kp
	cal/(mole °K)				kcal/mole		
298	14.50	59.30	59.30	0.00	44.32	46.47	-34.060
300	14.55	59.39	59.31	0.03	44.31	46.48	-33.859
400	17.33	63.97	59.91	1.63	43.78	47.29	-25.836
500	19.74	68.10	61.14	3.49	43.28	48.22	-21.077
600	21.80	71.89	62.62	5.56	42.83	49.25	-17.939
700	23.58	75.39	64.20	7.84	42.42	50.36	-15.722
800	25.14	78.64	65.80	10.27	42.07	51.51	-14.072
900	26.51	81.68	67.40	12.86	41.77	52.71	-12.800
1000	27.71	84.54	68.97	15.57	41.53	53.95	-11.789

No. 316 Butadiyne (Biacetylene) C₄H₂ (Ideal Gas State) Mol Wt 50.056

T°K	Cp°	S°	$-(G°-H°_{298})/T$	$H°-H°_{298}$	$\Delta Hf°$	$\Delta Gf°$	Log Kp
	cal/(mole °K)				kcal/mole		
298	17.60	59.76	59.76	0.00	113.00	106.11	-77.778
300	17.65	59.87	59.77	0.04	113.00	106.07	-77.266
400	20.17	65.32	60.49	1.94	113.22	103.72	-56.670
500	21.86	70.01	61.94	4.04	113.35	101.33	-44.290
600	23.14	74.11	63.63	6.29	113.39	98.92	-36.030
700	24.20	77.76	65.39	8.66	113.37	96.51	-30.131
800	25.11	81.05	67.15	11.13	113.29	94.11	-25.708
900	25.90	84.05	68.86	13.68	113.18	91.72	-22.270
1000	26.61	86.82	70.52	16.30	113.07	89.34	-19.524

Ferigle and Weber (417) used the vibrational assignment of Jones (698) and reported the thermodynamic functions listed here. Kraus (810) reported the enthalpy of dissociation (presumably to elements at 298°K) of biacetylene as −113 kcal/mole.

No. 317 1-Buten-3-yne (Vinylacetylene) C₄H₄ (Ideal Gas State) Mol Wt 52.072

T°K	Cp°	S°	$-(G°-H°_{298})/T$	$H°-H°_{298}$	$\Delta Hf°$	$\Delta Gf°$	Log Kp
	cal/(mole °K)				kcal/mole		
298	17.49	66.77	66.77	0.00	72.80	73.13	-53.601
300	17.57	66.88	66.78	0.04	72.79	73.13	-53.271
400	21.26	72.46	67.51	1.98	72.36	73.31	-40.052
500	24.25	77.54	69.02	4.27	71.97	73.59	-32.165
600	26.67	82.18	70.83	6.81	71.61	73.95	-26.933
700	28.68	86.44	72.76	9.58	71.28	74.37	-23.217
800	30.40	90.39	74.72	12.54	70.99	74.82	-20.439
900	31.87	94.06	76.67	15.66	70.73	75.32	-18.289
1000	33.16	97.48	78.58	18.91	70.53	75.84	-16.575

Stamm, Halverson, and Whalen (1404) studied the enthalpy of the reaction forming vinylacetylene from 2 moles of acetylene and reported

$\Delta Hr° = -35.6$ kcal/mole at 291°K, from which we calculated $\Delta Hf°_{298}(g)$ = 72.8 kcal/mole. These authors also made a frequency assignment and calculated the gaseous heat capacities and entropy used here.

No. 318   1-Butyne (Ethylacetylene) C_4H_6 (Ideal Gas State)   Mol Wt 54.088

	cal/(mole °K)				kcal/mole		
T°K	$Cp°$	$S°$	$-(G°-H°_{298})/T$	$H°-H°_{298}$	$\Delta Hf°$	$\Delta Gf°$	Log Kp
298	19.46	69.51	69.51	0.00	39.48	48.30	-35.400
300	19.54	69.64	69.52	0.04	39.46	48.35	-35.220
400	23.87	75.86	70.34	2.22	38.57	51.45	-28.111
500	27.63	81.60	72.02	4.80	37.78	54.77	-23.937
600	30.83	86.93	74.07	7.72	37.09	58.23	-21.208
700	33.57	91.89	76.26	10.95	36.52	61.80	-19.294
800	35.95	96.54	78.51	14.42	36.04	65.44	-17.876
900	38.02	100.89	80.76	18.12	35.66	69.14	-16.789
1000	39.84	104.99	82.98	22.02	35.38	72.88	-15.928

The values adopted are from the measurements of Aston, Mastrangelo, and Moessen (39) and from the selections of Rossini, Wagman, Evans, Levine, and Jaffe (1249). $Tm = 147.43°K$, $\Delta Hm° = 1.441$ kcal/mole, and $Tb = 281.22°K$, where $\Delta Hv = 5.861$ kcal/mole.

No. 319   2-Butyne (Dimethylacetylene) C_4H_6 (Ideal Gas State)   Mol Wt 54.088

	cal/(mole °K)				kcal/mole		
T°K	$Cp°$	$S°$	$-(G°-H°_{298})/T$	$H°-H°_{298}$	$\Delta Hf°$	$\Delta Gf°$	Log Kp
298	18.63	67.71	67.71	0.00	34.97	44.32	-32.487
300	18.70	67.83	67.72	0.04	34.95	44.38	-32.328
400	22.62	73.75	68.50	2.11	33.95	47.68	-26.049
500	26.36	79.21	70.10	4.56	33.03	51.22	-22.386
600	29.68	84.31	72.05	7.36	32.22	54.93	-20.007
700	32.59	89.11	74.15	10.48	31.54	58.77	-18.349
800	35.14	93.63	76.30	13.87	30.97	62.70	-17.127
900	37.36	97.90	78.47	17.50	30.52	66.69	-16.195
1000	39.29	101.94	80.61	21.33	30.18	70.74	-15.459

The values adopted are from the measurements of Yost, Osborne, and Garner (1638) and the selections of Rossini, Wagman, Evans, Levine, and Jaffe (1249). $Tm = 240.89°K$, $\Delta Hm° = 2.207$ kcal/mole, and $Tb = 300.14°K$, where $\Delta Hv = 6.340$ kcal/mole.

No. 320   1-Pentyne C_5H_8 (Ideal Gas State)   Mol Wt 68.114

Rossini, Pitzer, Arnett, Braun, and Pimentel (1248) selected the values presented. $Tm = 167.45°K$ and $Tb = 313.33°K$.

Alkyne Ideal Gas Tables

No. 320   1-Pentyne C_5H_8 (Ideal Gas State)   Mol Wt 68.114

$T°K$	$Cp°$	$S°$	$-(G°-H°_{298})/T$	$H°-H°_{298}$	$\Delta Hf°$	$\Delta Gf°$	Log Kp
		cal/(mole °K)			kcal/mole		
298	25.50	78.82	78.82	0.00	34.50	50.25	-36.833
300	25.65	78.98	78.83	0.05	34.48	50.35	-36.676
400	31.10	87.12	79.90	2.89	33.31	55.82	-30.498
500	36.10	94.61	82.10	6.26	32.29	61.57	-26.911
600	40.40	101.58	84.77	10.09	31.43	67.51	-24.588
700	44.00	108.09	87.64	14.31	30.73	73.58	-22.972
800	47.10	114.17	90.58	18.87	30.16	79.74	-21.782
900	49.80	119.87	93.52	23.72	29.73	85.96	-20.874
1000	52.20	125.25	96.43	28.82	29.43	92.24	-20.157

No. 321   2-Pentyne C_5H_8 (Ideal Gas State)   Mol Wt 68.114

$T°K$	$Cp°$	$S°$	$-(G°-H°_{298})/T$	$H°-H°_{298}$	$\Delta Hf°$	$\Delta Gf°$	Log Kp
		cal/(mole °K)			kcal/mole		
298	23.59	79.30	79.30	0.00	30.80	46.41	-34.016
300	23.69	79.45	79.31	0.05	30.77	46.50	-33.875
400	29.20	87.03	80.30	2.70	29.41	51.96	-28.388
500	34.30	94.11	82.36	5.88	28.20	57.74	-25.237
600	38.70	100.76	84.88	9.53	27.17	63.74	-23.217
700	42.60	107.02	87.60	13.60	26.31	69.91	-21.827
800	45.90	112.93	90.40	18.03	25.62	76.18	-20.811
900	48.90	118.52	93.22	22.77	25.08	82.54	-20.043
1000	51.40	123.80	96.01	27.79	24.70	88.95	-19.440

The values used were selected by Rossini, Pitzer, Arnett, Braun, and Pimentel (1248). $Tm = 163.85°K$ and $Tb = 329.22°K$.

No. 322   3-Methyl-1-Butyne C_5H_8 (Ideal Gas State)   Mol Wt 68.114

$T°K$	$Cp°$	$S°$	$-(G°-H°_{298})/T$	$H°-H°_{298}$	$\Delta Hf°$	$\Delta Gf°$	Log Kp
		cal/(mole °K)			kcal/mole		
298	25.02	76.23	76.23	0.00	32.60	49.12	-36.006
300	25.13	76.39	76.24	0.05	32.57	49.22	-35.857
400	31.10	84.46	77.30	2.87	31.38	54.96	-30.029
500	36.20	91.96	79.49	6.24	30.37	60.98	-26.652
600	40.60	98.96	82.16	10.09	29.52	67.18	-24.468
700	44.20	105.49	85.03	14.33	28.84	73.51	-22.951
800	47.40	111.61	87.97	18.92	28.30	79.93	-21.834
900	50.10	117.35	90.92	23.79	27.90	86.41	-20.982
1000	52.40	122.75	93.84	28.92	27.63	92.93	-20.309

The selected values of Rossini, Pitzer, Arnett, Braun, and Pimentel (1248) are employed. $Tm = 183.4°K$ and $Tb = 299.50°K$.

No. 323 1-Hexyne C_6H_{10} (Ideal Gas State) Mol Wt 82.140

T°K	$Cp°$	$S°$	$-(G°-H°_{298})/T$	$H°-H°_{298}$	$\Delta Hf°$	$\Delta Gf°$	Log Kp
	cal/(mole °K)				kcal/mole		
298	30.65	88.13	88.13	0.00	29.55	52.24	-38.288
300	30.77	88.32	88.14	0.06	29.52	52.37	-38.153
400	37.87	98.17	89.43	3.50	28.00	60.23	-32.908
500	44.18	107.31	92.10	7.61	26.71	68.44	-29.915
600	49.59	115.86	95.36	12.31	25.64	76.89	-28.005
700	54.20	123.86	98.86	17.50	24.79	85.51	-26.695
800	58.16	131.36	102.46	23.12	24.12	94.22	-25.738
900	61.60	138.41	106.07	29.11	23.63	103.02	-25.015
1000	64.56	145.06	109.64	35.43	23.32	111.86	-24.447

The values chosen by Rossini, Pitzer, Arnett, Braun, and Pimentel (1248) are adopted. $Tm = 141.3°K$ and $Tb = 344.48°K$.

No. 324 1-Heptyne C_7H_{12} (Ideal Gas State) Mol Wt 96.166

T°K	$Cp°$	$S°$	$-(G°-H°_{298})/T$	$H°-H°_{298}$	$\Delta Hf°$	$\Delta Gf°$	Log Kp
	cal/(mole °K)				kcal/mole		
298	36.11	97.44	97.44	0.00	24.62	54.24	-39.759
300	36.27	97.67	97.45	0.07	24.58	54.42	-39.645
400	44.81	109.29	98.97	4.13	22.75	64.66	-35.327
500	52.42	120.13	102.13	9.00	21.20	75.32	-32.922
600	58.93	130.28	105.99	14.58	19.93	86.26	-31.419
700	64.47	139.79	110.14	20.76	18.93	97.41	-30.411
800	69.23	148.72	114.41	27.45	18.17	108.66	-29.683
900	73.34	157.11	118.70	34.58	17.62	120.01	-29.142
1000	76.90	165.03	122.94	42.10	17.30	131.42	-28.720

The values selected by Rossini, Pitzer, Arnett, Braun, and Pimentel (1248) are employed. $Tm = 192.3°K$ and $Tb = 372.89°K$.

No. 325 1-Octyne C_8H_{14} (Ideal Gas State) Mol Wt 110.192

T°K	$Cp°$	$S°$	$-(G°-H°_{298})/T$	$H°-H°_{298}$	$\Delta Hf°$	$\Delta Gf°$	Log Kp
	cal/(mole °K)				kcal/mole		
298	41.58	106.75	106.75	0.00	19.70	56.26	-41.236
300	41.76	107.01	106.76	0.08	19.65	56.48	-41.144
400	51.75	120.42	108.52	4.77	17.50	69.10	-37.751
500	60.67	132.95	112.16	10.40	15.70	82.21	-35.932
600	68.28	144.70	116.62	16.86	14.23	95.65	-34.837
700	74.75	155.72	121.43	24.01	13.09	109.32	-34.129
800	80.30	166.08	126.37	31.77	12.23	123.11	-33.631
900	85.09	175.82	131.33	40.05	11.63	137.02	-33.271
1000	89.20	185.00	136.24	48.77	11.29	150.98	-32.996

The values selected by Rossini, Pitzer, Arnett, Braun, and Pimentel (1248) are employed. $Tm = 193.9°K$ and $Tb = 399.35°K$.

Alkyne Ideal Gas Tables

No. 326   1-Nonyne C_9H_{16} (Ideal Gas State)   Mol Wt 124.218

$T°K$	$Cp°$	$S°$	$-(G°-H°_{298})/T$	$H°-H°_{298}$	$\Delta Hf°$	$\Delta Gf°$	Log Kp
		cal/(mole °K)			kcal/mole		
298	47.04	116.06	116.06	0.00	14.77	58.26	-42.706
300	47.26	116.36	116.07	0.09	14.72	58.53	-42.636
400	58.69	131.55	118.06	5.40	12.25	73.52	-40.170
500	68.91	145.77	122.19	11.79	10.19	89.09	-38.939
600	77.62	159.12	127.25	19.13	8.52	105.02	-38.252
700	85.02	171.66	132.71	27.27	7.24	121.22	-37.845
800	91.36	183.43	138.32	36.10	6.28	137.56	-37.577
900	96.83	194.52	143.95	45.51	5.62	154.02	-37.399
1000	101.60	204.97	149.54	55.44	5.26	170.54	-37.270

The values selected by Rossini, Pitzer, Arnett, Braun, and Pimentel (1248) are employed. $Tm = 223.2°K$ and $Tb = 423.9°K$.

No. 327   1-Decyne $C_{10}H_{18}$ (Ideal Gas State)   Mol Wt 138.244

$T°K$	$Cp°$	$S°$	$-(G°-H°_{298})/T$	$H°-H°_{298}$	$\Delta Hf°$	$\Delta Gf°$	Log Kp
		cal/(mole °K)			kcal/mole		
298	52.51	125.36	125.36	0.00	9.85	60.28	-44.186
300	52.75	125.69	125.37	0.10	9.79	60.59	-44.137
400	65.64	142.66	127.59	6.03	7.01	77.97	-42.596
500	77.16	158.58	132.22	13.19	4.69	95.98	-41.951
600	86.96	173.54	137.87	21.41	2.83	114.41	-41.672
700	95.30	187.58	143.98	30.53	1.40	133.14	-41.565
800	102.42	200.78	150.26	40.42	0.34	152.02	-41.527
900	108.58	213.21	156.57	50.98	-0.37	171.03	-41.530
1000	113.90	224.93	162.83	62.11	-0.75	190.11	-41.547

The values selected by Rossini, Pitzer, Arnett, Braun, and Pimentel (1248) are employed. $Tm = 229.2°K$ and $Tb = 447°K$.

No. 328   1-Undecyne $C_{11}H_{20}$ (Ideal Gas State)   Mol Wt 152.270

$T°K$	$Cp°$	$S°$	$-(G°-H°_{298})/T$	$H°-H°_{298}$	$\Delta Hf°$	$\Delta Gf°$	Log Kp
		cal/(mole °K)			kcal/mole		
298	57.98	134.67	134.67	0.00	4.92	62.29	-45.656
300	58.24	135.03	134.68	0.11	4.85	62.64	-45.629
400	72.58	153.79	137.14	6.67	1.75	82.39	-45.015
500	85.41	171.40	142.25	14.58	-0.82	102.86	-44.958
600	96.30	187.96	148.50	23.68	-2.88	123.79	-45.086
700	105.58	203.52	155.26	33.79	-4.45	145.04	-45.281
800	113.49	218.14	162.21	44.75	-5.61	166.46	-45.472
900	120.33	231.92	169.20	56.45	-6.38	188.03	-45.657
1000	126.20	244.91	176.13	68.78	-6.78	209.67	-45.821

The values selected by Rossini, Pitzer, Arnett, Braun, and Pimentel (1248) are employed. $Tm = 248.2°K$ and $Tb = 468°K$.

No. 329 1-Dodecyne $C_{12}H_{22}$ (Ideal Gas State) Mol Wt 166.296

$T°K$	$Cp°$	$S°$	$-(G°-H°_{298})/T$	$H°-H°_{298}$	$\Delta Hf°$	$\Delta Gf°$	Log Kp
		cal/(mole °K)			kcal/mole		
298	63.44	143.98	143.98	0.00	-0.01	64.29	-47.126
300	63.74	144.38	143.99	0.12	-0.08	64.69	-47.121
400	79.52	164.92	146.68	7.30	-3.50	86.82	-47.434
500	93.65	184.22	152.28	15.98	-6.33	109.74	-47.964
600	105.64	202.38	159.13	25.96	-8.59	133.16	-48.501
700	115.85	219.45	166.54	37.04	-10.30	156.94	-48.997
800	124.56	235.50	174.17	49.07	-11.56	180.90	-49.417
900	132.07	250.62	181.83	61.91	-12.38	205.02	-49.784
1000	138.60	264.88	189.43	75.45	-12.80	229.23	-50.095

The values selected by Rossini, Pitzer, Arnett, Braun, and Pimentel (1248) are employed. $Tm = 254.2°K$ and $Tb = 488°K$.

No. 330 1-Tridecyne $C_{13}H_{24}$ (Ideal Gas State) Mol Wt 180.322

$T°K$	$Cp°$	$S°$	$-(G°-H°_{298})/T$	$H°-H°_{298}$	$\Delta Hf°$	$\Delta Gf°$	Log Kp
		cal/(mole °K)			kcal/mole		
298	68.91	153.29	153.29	0.00	-4.93	66.31	-48.603
300	69.27	153.72	153.30	0.13	-5.01	66.74	-48.621
400	86.46	176.05	156.22	7.93	-8.75	91.26	-49.850
500	101.90	197.04	162.31	17.37	-11.83	116.63	-50.974
600	114.99	216.81	169.76	28.23	-14.29	142.54	-51.918
700	126.13	235.39	177.82	40.30	-16.14	168.85	-52.715
800	135.62	252.87	186.12	53.40	-17.49	195.35	-53.365
900	143.82	269.32	194.46	67.38	-18.38	222.03	-53.913
1000	150.90	284.85	202.73	82.13	-18.81	248.79	-54.369

The values selected by Rossini, Pitzer, Arnett, Braun, and Pimentel (1248) are employed. $Tm = 268.2°K$ and $Tb = 507°K$.

No. 331 1-Tetradecyne $C_{14}H_{26}$ (Ideal Gas State) Mol Wt 194.348

$T°K$	$Cp°$	$S°$	$-(G°-H°_{298})/T$	$H°-H°_{298}$	$\Delta Hf°$	$\Delta Gf°$	Log Kp
		cal/(mole °K)			kcal/mole		
298	74.37	162.60	162.60	0.00	-9.86	68.31	-50.073
300	74.73	163.07	162.61	0.14	-9.95	68.79	-50.112
400	93.40	187.17	165.77	8.57	-14.00	95.68	-52.277
500	110.14	209.85	172.34	18.76	-17.35	123.50	-53.981
600	124.33	231.22	180.39	30.50	-20.00	151.92	-55.333
700	136.40	251.32	189.10	43.56	-21.99	180.75	-56.431
800	146.68	270.22	198.07	57.72	-23.44	209.80	-57.311
900	155.56	288.02	207.09	72.84	-24.38	239.03	-58.041
1000	163.20	304.82	216.03	88.79	-24.84	268.35	-58.644

The values selected by Rossini, Pitzer, Arnett, Braun, and Pimentel (1248) are employed. $Tm = 273.2°K$ and $Tb = 525°K$.

Alkyne Ideal Gas Tables

No. 332 1-Pentadecyne $C_{15}H_{28}$ (Ideal Gas State) Mol Wt 208.374

T°K	cal/(mole °K)				kcal/mole		
	$Cp°$	$S°$	$-(G°-H°_{298})/T$	$H°-H°_{298}$	$\Delta Hf°$	$\Delta Gf°$	Log Kp
298	79.84	171.91	171.91	0.00	-14.78	70.33	-51.550
300	80.22	172.41	171.92	0.15	-14.87	70.85	-51.612
400	100.34	198.29	175.31	9.20	-19.25	100.12	-54.701
500	118.39	222.67	182.37	20.16	-22.85	130.39	-56.992
600	133.67	245.64	191.02	32.78	-25.69	161.30	-58.751
700	146.68	267.25	200.38	46.81	-27.83	192.66	-60.150
800	157.75	287.58	210.02	62.05	-29.38	224.25	-61.259
900	167.31	306.72	219.71	78.31	-30.38	256.03	-62.170
1000	175.60	324.79	229.33	95.47	-30.85	287.91	-62.920

The values selected by Rossini, Pitzer, Arnett, Braun, and Pimentel (1248) are employed. $Tm = 283.2°K$ and $Tb = 542°K$.

No. 333 1-Hexadecyne $C_{16}H_{30}$ (Ideal Gas State) Mol Wt 222.400

T°K	cal/(mole °K)				kcal/mole		
	$Cp°$	$S°$	$-(G°-H°_{298})/T$	$H°-H°_{298}$	$\Delta Hf°$	$\Delta Gf°$	Log Kp
298	85.31	181.22	181.22	0.00	-19.71	72.34	-53.020
300	85.71	181.75	181.23	0.16	-19.81	72.90	-53.104
400	107.28	209.42	184.85	9.83	-24.50	104.55	-57.120
500	126.64	235.49	192.40	21.55	-28.36	137.27	-59.998
600	143.01	260.07	201.65	35.05	-31.40	170.68	-62.166
700	156.96	283.18	211.66	50.07	-33.68	204.57	-63.865
800	168.82	304.94	221.98	66.37	-35.33	238.69	-65.204
900	179.06	325.43	232.34	83.78	-36.38	273.03	-66.297
1000	187.90	344.76	242.63	102.14	-36.87	307.47	-67.193

The values selected by Rossini, Pitzer, Arnett, Braun, and Pimentel (1248) are employed. $Tm = 288.2°K$ and $Tb = 558°K$.

No. 334 1-Heptadecyne $C_{17}H_{32}$ (Ideal Gas State) Mol Wt 236.426

T°K	cal/(mole °K)				kcal/mole		
	$Cp°$	$S°$	$-(G°-H°_{298})/T$	$H°-H°_{298}$	$\Delta Hf°$	$\Delta Gf°$	Log Kp
298	90.77	190.53	190.53	0.00	-24.64	74.34	-54.490
300	91.21	191.10	190.54	0.17	-24.75	74.95	-54.596
400	114.22	220.54	194.40	10.46	-29.76	108.98	-59.538
500	134.88	248.31	202.43	22.95	-33.87	144.15	-63.004
600	152.35	274.49	212.28	37.33	-37.11	180.05	-65.580
700	167.23	299.12	222.94	53.33	-39.54	216.47	-67.581
800	179.88	322.29	233.93	70.70	-41.28	253.13	-69.150
900	190.80	344.12	244.97	89.24	-42.39	290.03	-70.425
1000	200.20	364.73	255.93	108.81	-42.90	327.02	-71.467

The values selected by Rossini, Pitzer, Arnett, Braun, and Pimentel (1248) are employed. $Tm = 295.2°K$ and $Tb = 573°K$.

No. 335　1-Octadecyne　$C_{18}H_{34}$　(Ideal Gas State)　Mol Wt 250.452

$T°K$	$Cp°$	$S°$	$-(G°-H°_{298})/T$	$H°-H°_{298}$	$\Delta Hf°$	$\Delta Gf°$	Log Kp
	cal/(mole °K)			kcal/mole			
298	96.24	199.84	199.84	0.00	-29.56	76.36	-55.968
300	96.70	200.44	199.85	0.18	-29.67	77.00	-56.095
400	121.16	231.67	203.94	11.10	-35.00	113.41	-61.963
500	143.13	261.13	212.46	24.34	-39.37	151.04	-66.015
600	161.70	288.91	222.91	39.60	-42.81	189.44	-68.998
700	177.51	315.05	234.22	56.58	-45.38	228.38	-71.299
800	190.94	339.65	245.88	75.02	-47.22	267.59	-73.098
900	202.55	362.83	257.60	94.71	-48.39	307.03	-74.554
1000	212.60	384.70	269.23	115.48	-48.91	346.59	-75.743

The values selected by Rossini, Pitzer, Arnett, Braun, and Pimentel (1248) are employed. $Tm = 300.2°K$ and $Tb = 588°K$.

No. 336　1-Nonadecyne　$C_{19}H_{36}$　(Ideal Gas State)　Mol Wt 264.478

$T°K$	$Cp°$	$S°$	$-(G°-H°_{298})/T$	$H°-H°_{298}$	$\Delta Hf°$	$\Delta Gf°$	Log Kp
	cal/(mole °K)			kcal/mole			
298	101.70	209.15	209.15	0.00	-34.49	78.36	-57.438
300	102.20	209.79	209.16	0.19	-34.61	79.05	-57.587
400	128.10	242.80	213.48	11.73	-40.26	117.84	-64.381
500	151.37	273.95	222.49	25.73	-44.88	157.92	-69.021
600	171.04	303.33	233.54	41.88	-48.52	198.81	-72.413
700	187.78	330.98	245.50	59.84	-51.23	240.28	-75.015
800	202.01	357.01	257.83	79.35	-53.17	282.03	-77.043
900	214.29	381.53	270.23	100.18	-54.39	324.03	-78.681
1000	224.90	404.67	282.53	122.15	-54.94	366.14	-80.016

The values selected by Rossini, Pitzer, Arnett, Braun, and Pimentel (1248) are employed. $Tm = 306.2°K$ and $Tb = 602°K$.

No. 337　1-Eicosyne　$C_{20}H_{38}$　(Ideal Gas State)　Mol Wt 278.504

$T°K$	$Cp°$	$S°$	$-(G°-H°_{298})/T$	$H°-H°_{298}$	$\Delta Hf°$	$\Delta Gf°$	Log Kp
	cal/(mole °K)			kcal/mole			
298	107.17	218.46	218.46	0.00	-39.41	80.38	-58.915
300	107.69	219.13	218.47	0.20	-39.54	81.11	-59.086
400	135.04	253.92	223.03	12.36	-45.50	122.28	-66.805
500	159.62	286.77	232.52	27.13	-50.38	164.80	-72.032
600	180.38	317.75	244.17	44.15	-54.22	208.19	-75.831
700	198.06	346.92	256.79	63.10	-57.07	252.19	-78.734
800	213.08	374.37	269.78	83.67	-59.11	296.48	-80.991
900	226.04	400.23	282.86	105.64	-60.39	341.04	-82.811
1000	237.20	424.64	295.83	128.82	-60.95	385.71	-84.292

The values selected by Rossini, Pitzer, Arnett, Braun, and Pimentel (1248) are employed. $Tm = 309.2°K$ and $Tb = 615°K$.

CYCLOALKANE IDEAL GAS TABLES

No. 338 Cyclopropane C_3H_6 (Ideal Gas State) Mol Wt 42.078

$T°K$	$Cp°$	$S°$	$-(G°-H°_{298})/T$	$H°-H°_{298}$	$\Delta Hf°$	$\Delta Gf°$	$Log\,Kp$
		cal/(mole °K)			kcal/mole		
298	13.37	56.75	56.75	0.00	12.74	24.95	-18.291
300	13.44	56.84	56.76	0.03	12.71	25.03	-18.232
400	18.31	61.38	57.34	1.62	11.48	29.33	-16.023
500	22.65	65.95	58.61	3.68	10.49	33.91	-14.821
600	26.15	70.40	60.21	6.12	9.70	38.67	-14.084
700	29.02	74.65	61.97	8.88	9.09	43.55	-13.596
800	31.45	78.69	63.81	11.91	8.61	48.50	-13.249
900	33.57	82.52	65.68	15.16	8.27	53.51	-12.994
1000	35.39	86.15	67.54	18.61	8.05	58.56	-12.797

Ruehrwein and Powell (1266) made a low-temperature third-law study of cylcopropane. They reported $Ttp = 145.59°K$, $\Delta Hm° = 1.301$ kcal/mole, and $Tb = 240.35°K$, where $\Delta Hv = 4.793$ kcal/mole. Their calorimetric entropy of the ideal gas at 298.15°K of 56.75 cal/(mole °K) is in good agreement with the spectroscopic value 56.84 cal/(mole °K) calculated by Kistiakowsky and Rice (757), who also listed the vapor heat capacities used here. More recently Baker and Lord (63) repeated the spectroscopic measurements, and with a modified assignment calculated $S°_{298}(g) = 56.79$ cal/(mole °K). Kobe and Pennington (782) and Linnett (876) also reported thermodynamic functions for cyclopropane. Knowlton and Rossini (771) measured the enthalpy of combustion and reported $\Delta Hf°_{298}(g) = 12.74$ kcal/mole.

No. 339 Cyclobutane C_4H_8 (Ideal Gas State) Mol Wt 56.104

$T°K$	$Cp°$	$S°$	$-(G°-H°_{298})/T$	$H°-H°_{298}$	$\Delta Hf°$	$\Delta Gf°$	$Log\,Kp$
		cal/(mole °K)			kcal/mole		
298	17.26	63.43	63.43	0.00	6.37	26.30	-19.280
300	17.37	63.54	63.44	0.04	6.33	26.43	-19.250
400	23.89	69.45	64.20	2.10	4.64	33.39	-18.243
500	29.86	75.44	65.85	4.80	3.27	40.74	-17.808
600	34.76	81.33	67.94	8.04	2.19	48.34	-17.607
700	38.89	87.00	70.26	11.73	1.38	56.10	-17.516
800	42.42	92.43	72.69	15.80	0.78	63.96	-17.471
900	45.41	97.61	75.18	20.19	0.39	71.88	-17.455
1000	47.96	102.53	77.67	24.86	0.17	79.85	-17.450

Rathjens, Freeman, Gwinn, and Pitzer (1214) studied the spectrum of cyclobutane, assigned the fundamental frequencies, and calculated the thermodynamic functions reported here. Rathjens and Gwinn (1215) studied the low-temperature properties and reported $Tt = 145.74°K$, $\Delta Ht° = 1.413$ kcal/mole. $Ttp = 182.48°K$, $\Delta Hm° = 0.260$ kcal/mole, and $Tb = 285.72°K$, at which $\Delta Hv = 5.781$ kcal/mole, while the entropy of the ideal gas at Tb was 62.72 cal/(mole °K). From the calorimetric data we calculate $S°_{298}(g) = 63.44$ cal/(mole °K) compared to 63.43 cal/(mole °K) calculated from the spectrum. Coops and Kaarsemaker (267, 707) measured the enthalpy of combustion, which leads to $\Delta Hf°_{298}(g) = 6.37$ kcal/mole.

No. 340　Cyclopentane　C_5H_{10}　(Ideal Gas State)　Mol Wt 70.130

$T°K$	$Cp°$	$S°$	$-(G°-H°_{298})/T$	$H°-H°_{298}$	$\Delta Hf°$	$\Delta Gf°$	Log Kp
		cal/(mole °K)			kcal/mole		
298	19.84	70.00	70.00	0.00	-18.46	9.23	-6.763
300	19.99	70.13	70.01	0.04	-18.51	9.40	-6.844
400	28.38	77.04	70.89	2.47	-20.79	19.06	-10.412
500	36.07	84.23	72.84	5.70	-22.64	29.24	-12.780
600	42.57	91.39	75.34	9.64	-24.09	39.75	-14.480
700	48.01	98.37	78.13	14.18	-25.18	50.49	-15.763
800	52.60	105.09	81.08	19.21	-25.97	61.35	-16.759
900	56.50	111.52	84.11	24.67	-26.50	72.30	-17.557
1000	59.84	117.65	87.16	30.49	-26.80	83.31	-18.206

The measurements of Douslin and Huffman (341) and the statistical calculations of McCullough, Pennington, Smith, Hossenlopp, and Waddington (959) are adopted. $Tt = 122.39°K$, $\Delta Ht° = 1.167$ kcal/mole, $Tt = 138.07°K$, $\Delta Ht° = 0.082$ kcal/mole, $Ttp = 179.28°K$, $\Delta Hm° = 0.145$ kcal/mole, and $Tb = 322.41°K$, at which $\Delta Hv = 6.524$ kcal/mole. Kobe and Lynn (780) reported $Tc = 511.7°K$ and $Pc = 44.55$ atm. The enthalpy of formation is that selected by Rossini, Pitzer, Arnett, Braun, and Pimentel (1248).

No. 341　Cyclohexane　C_6H_{12}　(Ideal Gas State)　Mol Wt 84.156

$T°K$	$Cp°$	$S°$	$-(G°-H°_{298})/T$	$H°-H°_{298}$	$\Delta Hf°$	$\Delta Gf°$	Log Kp
		cal/(mole °K)			kcal/mole		
298	25.40	71.28	71.28	0.00	-29.43	7.59	-5.560
300	25.58	71.44	71.29	0.05	-29.48	7.81	-5.691
400	35.82	80.22	72.41	3.13	-32.06	20.65	-11.284
500	45.47	89.27	74.88	7.20	-34.08	34.07	-14.893
600	53.83	98.32	78.03	12.18	-35.57	47.85	-17.428
700	60.87	107.16	81.57	17.92	-36.58	61.84	-19.308
800	66.76	115.69	85.30	24.31	-37.19	75.94	-20.745
900	71.68	123.84	89.13	31.24	-37.45	90.11	-21.880
1000	75.80	131.61	93.00	38.62	-37.40	104.29	-22.791

The measurements of Ruehrwein and Huffman (1264) and the selections of Rossini, Pitzer, Arnett, Braun, and Pimentel (1248) are adopted.

$Tt = 186.10°K$, $\Delta Ht° = 1.611$ kcal/mole, $Tm = 279.69°K$, $\Delta Hm° = 0.640$ kcal/mole, and $Tb = 353.88°K$, where $\Delta Hv = 7.190$ kcal/mole. Kobe and Lynn (780) selected $Tc = 553°K$, $Pc = 40.0$ atm, and $dc = 0.273$ g/cc.

No. 342 Cycloheptane C_7H_{14} (Ideal Gas State) Mol Wt 98.182

	cal/(mole °K)				kcal/mole		
$T°K$	$Cp°$	$S°$	$-(G°-H°_{298})/T$	$H°-H°_{298}$	$\Delta Hf°$	$\Delta Gf°$	Log Kp
298	29.42	81.82	81.82	0.00	-28.52	15.06	-11.042
300	29.65	82.01	81.83	0.06	-28.58	15.33	-11.168
400	41.82	92.23	83.14	3.64	-31.59	30.45	-16.635
500	52.94	102.79	86.01	8.39	-33.96	46.24	-20.212
600	62.42	113.30	89.68	14.17	-35.72	62.45	-22.746
700	70.36	123.53	93.79	20.82	-36.95	78.92	-24.640
800	77.03	133.38	98.13	28.20	-37.73	95.52	-26.094
900	82.65	142.78	102.57	36.19	-38.13	112.21	-27.248
1000	87.40	151.74	107.05	44.70	-38.18	128.93	-28.177

Finke, Scott, Gross, Messerly, and Waddington (424) measured the low-temperature properties and have reported three solid-state transitions. At $Tt = 134.8°K$, $\Delta Ht° = 1.187$ kcal/mole; at $Tt = 198.2°K$, $\Delta Ht° = 0.069$ kcal/mole; and at $Tt = 212.4°K$, $\Delta Ht° = 0.107$ kcal/mole; $Ttp = 265.10°K$, $\Delta Hm° = 0.576$ kcal/mole, and $Tb = 391.94°K$, where $\Delta Hv = 9.21$ kcal/mole. The third-law entropy from these measurements is $S°_{298}$ (ideal gas) = 81.82 cal/(mole °K). Kaarsemaker and Coops (707) burned this material, and the measured enthalpy of combustion leads to $\Delta Hf°_{298}(g) = -28.52$ kcal/mole.

No. 343 Cyclooctane C_8H_{16} (Ideal Gas State) Mol Wt 112.208

	cal/(mole °K)				kcal/mole		
$T°K$	$Cp°$	$S°$	$-(G°-H°_{298})/T$	$H°-H°_{298}$	$\Delta Hf°$	$\Delta Gf°$	Log Kp
298	33.45	87.66	87.66	0.00	-30.06	21.49	-15.755
300	33.71	87.87	87.67	0.07	-30.13	21.81	-15.888
400	47.82	99.54	89.16	4.16	-33.57	39.67	-21.675
500	60.41	111.60	92.44	9.58	-36.28	58.31	-25.486
600	71.00	123.58	96.64	16.17	-38.32	77.42	-28.199
700	79.85	135.20	101.32	23.72	-39.77	96.84	-30.234
800	87.30	146.36	106.26	32.09	-40.73	116.41	-31.801
900	93.62	157.02	111.31	41.14	-41.26	136.10	-33.048
1000	99.01	167.17	116.39	50.78	-41.40	155.83	-34.055

Finke, Scott, Gross, Messerly, and Waddington (424) made a complete low-temperature study and reported $Tt = 166.5°K$, $\Delta Ht° = 1.507$ kcal/mole, $Tt = 183.8°K$, $\Delta Ht° = 0.114$ kcal/mole, $Ttp = 287.97°K$, $\Delta Hm° = 0.576$ kcal/mole, $Tb = 423.84°K$, $\Delta Hv_{298} = 10.360$ kcal/mole, $S°_{298}(l) =$

62.62 cal/(mole °K), and $S°_{298}(g) = 87.66$ cal/(mole °K). They used the enthalpy of combustion given by Kaarsemaker and Coops (707) to calculate $\Delta Hf°_{298}(l) = -40.42$ kcal/mole and $\Delta Hf°_{298}(g) = -30.06$ kcal/mole. Bellis and Slowinski (101) assigned most of the fundamental frequencies. The remaining ones are estimated for calculation of the ideal gas heat capacity.

CYCLOALKENE IDEAL GAS TABLES

No. 344 Cyclobutene C_4H_6 (Ideal Gas State) Mol Wt 54.088

	cal/(mole °K)				kcal/mole		
$T°K$	$Cp°$	$S°$	$-(G°-H°_{298})/T$	$H°-H°_{298}$	$\Delta Hf°$	$\Delta Gf°$	Log Kp
298	16.03	62.98	62.98	0.00	31.00	41.76	-30.611
300	16.13	63.08	62.99	0.03	30.97	41.83	-30.470
400	21.59	68.49	63.69	1.93	29.80	45.63	-24.931
500	26.37	73.84	65.18	4.33	28.83	49.71	-21.726
600	30.30	79.00	67.06	7.17	28.06	53.95	-19.651
700	33.55	83.92	69.12	10.37	27.46	58.32	-18.208
800	36.26	88.58	71.26	13.86	26.99	62.76	-17.144
900	38.57	92.99	73.44	17.61	26.66	67.25	-16.330
1000	40.53	97.16	75.60	21.56	26.44	71.78	-15.687

From other members of the cycloalkanes and cycloalkenes a value of $\Delta Hf°_{298}(g) = 31 \pm 5$ kcal/mole is estimated. The thermodynamic functions were calculated statistically by Danti (306) from the fundamental frequencies assigned by Lord and Rea (886).

No. 345 Cyclopentene C_5H_8 (Ideal Gas State) Mol Wt 68.114

	cal/(mole °K)				kcal/mole		
$T°K$	$Cp°$	$S°$	$-(G°-H°_{298})/T$	$H°-H°_{298}$	$\Delta Hf°$	$\Delta Gf°$	Log Kp
298	17.95	69.23	69.23	0.00	7.87	26.48	-19.410
300	18.08	69.35	69.24	0.04	7.83	26.59	-19.372
400	25.08	75.52	70.03	2.20	5.98	33.14	-18.106
500	31.62	81.83	71.76	5.04	4.44	40.11	-17.532
600	37.19	88.11	73.96	8.49	3.20	47.36	-17.251
700	41.86	94.20	76.42	12.45	2.23	54.81	-17.111
800	45.78	100.05	79.01	16.84	1.50	62.36	-17.036
900	49.11	105.64	81.66	21.59	0.96	70.01	-17.000
1000	51.94	110.97	84.33	26.64	0.62	77.71	-16.983

The measurements of Huffman, Eaton, and Oliver (645) and the selections of Rossini, Pitzer, Arnett, Braun, and Pimentel (1248) are adopted. $Tt = 87.07°K$, $\Delta Ht° = 0.115$ kcal/mole, $Tm = 138.07°K$, $\Delta Hm° =$

0.804 kcal/mole, and $Tb = 317.39°K$. Ambrose, Cox, and Townsend (13) calculated $Tc = 506.0°K$.

No. 346 Cyclohexene C_6H_{10} (Ideal Gas State) Mol Wt 82.140

T°K	cal/(mole °K)			kcal/mole			
	$Cp°$	$S°$	$-(G°-H°_{298})/T$	$H°-H°_{298}$	$\Delta Hf°$	$\Delta Gf°$	Log Kp
298	25.10	74.27	74.27	0.00	-1.28	25.54	-18.720
300	25.28	74.43	74.28	0.05	-1.32	25.70	-18.723
400	34.64	83.01	75.38	3.06	-3.27	35.02	-19.134
500	42.78	91.65	77.77	6.94	-4.79	44.78	-19.571
600	49.45	100.05	80.79	11.56	-5.93	54.80	-19.959
700	54.92	108.10	84.12	16.79	-6.76	65.00	-20.292
800	59.49	115.74	87.60	22.52	-7.32	75.28	-20.564
900	63.34	122.97	91.13	28.66	-7.65	85.63	-20.793
1000	66.62	129.82	94.66	35.16	-7.77	96.01	-20.982

Labbauf and Rossini (821) measured the enthalpy of combustion and reported $\Delta Hf°_{298}(l) = -9.28$ kcal/mole. Epstein, Pitzer, and Rossini (380) calculated $\Delta Hv_{298} = 8.00$ kcal/mole; hence $\Delta Hf°_{298}(g) = -1.28$ kcal/mole. Huffman, Eaton, and Oliver (645) measured $Tt = 138.8°K$, $\Delta Ht° = 1.016$ kcal/mole, $Tm = 169.63°K$, and $\Delta Hm° = 0.787$ kcal/mole. Rossini, Pitzer, Arnett, Braun, and Pimentel (1248) selected $Tb = 356.12$ °K, where $\Delta Hv = 7.285$ kcal/mole, according to Mathews (940). Beckett, Freeman, and Pitzer (92) calculated the ideal gaseous thermodynamic properties from spectroscopic data. Ambrose, Cox, and Townsend (13) measured $Tc = 560.4°K$.

ALKYLCYCLOPENTANE IDEAL GAS TABLES

No. 347 Methylcyclopentane C_6H_{12} (Ideal Gas State) Mol Wt 84.156

T°K	cal/(mole °K)			kcal/mole			
	$Cp°$	$S°$	$-(G°-H°_{298})/T$	$H°-H°_{298}$	$\Delta Hf°$	$\Delta Gf°$	Log Kp
298	26.24	81.24	81.24	0.00	-25.50	8.55	-6.264
300	26.46	81.41	81.25	0.05	-25.55	8.75	-6.377
400	36.11	90.36	82.40	3.19	-28.06	20.59	-11.248
500	44.94	99.39	84.90	7.25	-30.10	33.00	-14.421
600	52.43	108.27	88.06	12.13	-31.69	45.76	-16.669
700	58.68	116.83	91.56	17.69	-32.88	58.78	-18.351
800	64.00	125.02	95.23	23.84	-33.73	71.92	-19.648
900	68.53	132.83	98.98	30.47	-34.29	85.17	-20.682
1000	72.44	140.26	102.74	37.52	-34.57	98.47	-21.521

The measurements of Douslin and Huffman (341) and the selected values of Rossini, Pitzer, Arnett, Braun, and Pimentel (1248) are used.

$Tm = 130.70°K$, $\Delta Hm° = 1.656$ kcal/mole, and $Tb = 344.96°K$, where $\Delta Hv = 6.916$ kcal/mole. Kobe and Lynn (780) reported $Tc = 532.8°K$, $Pc = 37.4$ atm, and $dc = 0.264$ g/cc.

No. 348 Ethylcyclopentane C_7H_{14} (Ideal Gas State) Mol Wt 98.182

	cal/(mole °K)				kcal/mole		
T°K	$Cp°$	$S°$	$-(G°-H°_{298})/T$	$H°-H°_{298}$	$\Delta Hf°$	$\Delta Gf°$	Log Kp
298	31.49	90.42	90.42	0.00	-30.37	10.65	-7.807
300	31.75	90.62	90.43	0.06	-30.43	10.90	-7.941
400	43.89	101.49	91.83	3.87	-33.21	25.12	-13.726
500	53.03	112.27	94.84	8.72	-35.48	39.98	-17.473
600	61.70	122.73	98.62	14.47	-37.28	55.24	-20.119
700	69.02	132.80	102.79	21.01	-38.61	70.77	-22.096
800	75.22	142.43	107.15	28.23	-39.55	86.45	-23.617
900	80.54	151.61	111.58	36.03	-40.14	102.25	-24.829
1000	85.16	160.34	116.03	44.32	-40.41	118.10	-25.810

Gross, Oliver, and Huffman (544) studied the low-temperature properties and reported two solid crystalline forms. The higher-melting form (designated crystal I; $Ttp = 134.71°K$, $\Delta Hm° = 1.642$ kcal/mole) and the lower-melting form (designated crystal II; $Ttp = 134.02°K$, $\Delta Hm° = 1.889$ kcal/mole) were studied from 12°K to the triple point. Either form could be obtained at will, since the rate of cooling at the melting point determined which form would crystallize; rapid cooling produced crystal I, while slow cooling produced crystal II. The crystal II form thus has 247 cal/mole more enthalpy at the melting point than crystal I. The remaining values are those selected by Rossini, Pitzer, Arnett, Braun, and Pimentel (1248). $Tb = 376.62°K$, where $\Delta Hv = 7.715$ kcal/mole.

No. 349 1,1-Dimethylcyclopentane C_7H_{14} (Ideal Gas State) Mol Wt 98.182

	cal/(mole °K)				kcal/mole		
T°K	$Cp°$	$S°$	$-(G°-H°_{298})/T$	$H°-H°_{298}$	$\Delta Hf°$	$\Delta Gf°$	Log Kp
298	31.86	85.87	85.87	0.00	-33.05	9.33	-6.837
300	32.16	86.07	85.88	0.06	-33.11	9.59	-6.983
400	43.55	96.91	87.28	3.86	-35.90	24.26	-13.256
500	54.01	107.78	90.29	8.75	-38.13	39.57	-17.296
600	62.78	118.42	94.10	14.60	-39.82	55.27	-20.132
700	70.08	128.67	98.31	21.25	-41.05	71.23	-22.238
800	76.18	138.43	102.72	28.57	-41.89	87.32	-23.853
900	81.38	147.71	107.21	36.46	-42.39	103.51	-25.135
1000	85.83	156.52	111.70	44.83	-42.58	119.75	-26.169

The measurements of Gross, Oliver, and Huffman (544) are used to extend the values selected by Rossini, Pitzer, Arnett, Braun, and Pimentel

Alkylcyclopentane Ideal Gas Tables

(1248). $Tt = 146.79°K$, $\Delta Ht° = 1.551$ kcal/mole, $Tm = 203.36°K$, $\Delta Hm° = 0.258$ kcal/mole, and $Tb = 361.00°K$, where $\Delta Hv = 7.239$ kcal/mole.

No. 350 1,2-Dimethylcyclopentane, *cis* C_7H_{14} (Ideal Gas State) Mol Wt 98.182

	cal/(mole °K)				kcal/mole		
$T°K$	$Cp°$	$S°$	$-(G°-H°_{298})/T$	$H°-H°_{298}$	$\Delta Hf°$	$\Delta Gf°$	Log Kp
298	32.06	87.51	87.51	0.00	-30.96	10.93	-8.010
300	32.34	87.71	87.52	0.06	-31.02	11.18	-8.148
400	43.67	98.59	88.92	3.87	-33.80	25.69	-14.038
500	54.03	109.48	91.95	8.77	-36.02	40.83	-17.847
600	62.72	120.12	95.76	14.62	-37.71	56.36	-20.529
700	69.92	130.35	99.98	21.26	-38.95	72.15	-22.526
800	75.98	140.09	104.39	28.57	-39.81	88.07	-24.060
900	81.14	149.34	108.87	36.43	-40.33	104.10	-25.278
1000	85.57	158.13	113.36	44.77	-40.55	120.18	-26.263

The measurements of Gross, Oliver, and Huffman (544) extend the values selected by Rossini, Pitzer, Arnett, Braun, and Pimentel (1248). $Tt = 141.49°K$, $\Delta Ht° = 1.594$ kcal/mole, $Tm = 219.25°K$, $\Delta Hm° = 0.396$ kcal/mole, and $Tb = 372.68°K$, where $\Delta Hv = 7.576$ kcal/mole.

No. 351 1,2-Dimethylcyclopentane, *trans* C_7H_{14} (Ideal Gas State) Mol Wt 98.182

	cal/(mole °K)				kcal/mole		
$T°K$	$Cp°$	$S°$	$-(G°-H°_{298})/T$	$H°-H°_{298}$	$\Delta Hf°$	$\Delta Gf°$	Log Kp
298	32.14	87.67	87.67	0.00	-32.67	9.17	-6.722
300	32.44	87.87	87.68	0.06	-32.73	9.43	-6.867
400	43.71	98.77	89.09	3.88	-35.50	23.92	-13.068
500	54.03	109.67	92.11	8.78	-37.72	39.04	-17.063
600	62.66	120.30	95.93	14.63	-39.42	54.55	-19.869
700	69.88	130.52	100.15	21.26	-40.66	70.32	-21.955
800	75.84	140.25	104.56	28.56	-41.53	86.23	-23.555
900	80.98	149.49	109.04	36.40	-42.06	102.24	-24.826
1000	85.43	158.25	113.53	44.73	-42.30	118.30	-25.854

The values adopted are those selected by Rossini, Pitzer, Arnett, Braun, and Pimentel (1248). $Tm = 155.57°K$, $\Delta Hm° = 1.540$ kcal/mole, and $Tb = 365.02°K$, where $\Delta Hv = 7.375$ kcal/mole.

No. 352 1,3-Dimethylcyclopentane, *cis* C_7H_{14} (Ideal Gas State) Mol Wt 98.182

The values selected by Rossini, Pitzer, Arnett, Braun, and Pimentel (1248) are adopted, $Tm = 139.45°K$, $\Delta Hm° = 1.768$ kcal/mole, and $Tb = 363.92°K$, where McCullough, Pennington, Smith, Hossenlopp, and Waddington (959) measured $\Delta Hv = 7.265$ kcal/mole.

No. 352 1,3-Dimethylcyclopentane, cis C_7H_{14} (Ideal Gas State) Mol Wt 98.182

T°K	$C_p°$	$S°$	$-(G°-H°_{298})/T$	$H°-H°_{298}$	$\Delta H f°$	$\Delta G f°$	Log K_p
	cal/(mole °K)				kcal/mole		
298	32.14	87.67	87.67	0.00	-32.47	9.37	-6.868
300	32.44	87.87	87.68	0.06	-32.53	9.63	-7.013
400	43.71	98.77	89.09	3.88	-35.30	24.12	-13.177
500	54.03	109.67	92.11	8.78	-37.52	39.24	-17.151
600	62.66	120.30	95.93	14.63	-39.22	54.75	-19.942
700	69.88	130.52	100.15	21.26	-40.46	70.52	-22.018
800	75.84	140.25	104.56	28.56	-41.33	86.43	-23.610
900	80.98	149.49	109.04	36.40	-41.86	102.44	-24.875
1000	85.43	158.25	113.53	44.73	-42.10	118.50	-25.897

No. 353 1,3-Dimethylcyclopentane, trans C_7H_{14} (Ideal Gas State) Mol Wt 98.182

T°K	$C_p°$	$S°$	$-(G°-H°_{298})/T$	$H°-H°_{298}$	$\Delta H f°$	$\Delta G f°$	Log K_p
	cal/(mole °K)				kcal/mole		
298	32.14	87.67	87.67	0.00	-31.93	9.91	-7.264
300	32.44	87.87	87.68	0.06	-31.99	10.17	-7.406
400	43.71	98.77	89.09	3.88	-34.76	24.66	-13.472
500	54.03	109.67	92.11	8.78	-36.98	39.78	-17.387
600	62.66	120.30	95.93	14.63	-38.68	55.29	-20.139
700	69.88	130.52	100.15	21.26	-39.92	71.06	-22.186
800	75.84	140.25	104.56	28.56	-40.79	86.97	-23.758
900	80.98	149.49	109.04	36.40	-41.32	102.98	-25.006
1000	85.43	158.25	113.53	44.73	-41.56	119.04	-26.015

The measurements of Gross, Oliver, and Huffman (544) supplement the selected values of Rossini, Pitzer, Arnett, Braun, and Pimentel (1248). $Tm = 139.18°K$, $\Delta Hm° = 1.738$ kcal/mole, and $Tb = 364.87°K$, where $\Delta Hv = 7.361$ kcal/mole.

No. 354 Propylcyclopentane C_8H_{16} (Ideal Gas State) Mol Wt 112.208

T°K	$C_p°$	$S°$	$-(G°-H°_{298})/T$	$H°-H°_{298}$	$\Delta H f°$	$\Delta G f°$	Log K_p
	cal/(mole °K)				kcal/mole		
298	36.96	99.73	99.73	0.00	-35.39	12.57	-9.211
300	37.24	99.96	99.74	0.07	-35.46	12.86	-9.368
400	50.83	112.61	101.37	4.50	-38.56	29.46	-16.095
500	61.28	125.09	104.87	10.11	-41.08	46.76	-20.440
600	71.04	137.15	109.25	16.74	-43.08	64.52	-23.500
700	79.30	148.74	114.07	24.27	-44.55	82.59	-25.783
800	86.28	159.79	119.10	32.56	-45.59	100.81	-27.538
900	92.29	170.31	124.21	41.49	-46.24	119.16	-28.934
1000	97.50	180.31	129.33	50.99	-46.53	137.57	-30.064

Messerly, Todd, and Finke (993) measured low-temperature thermal properties, including $Ttp = 155.79°K$, $\Delta Hm° = 2.398$ kcal/mole, $S°_{298}(l) =$

74.29 cal/(mole °K), $S°_{298}(g) = 99.06$ cal/(mole °K), and $Tb = 404.10°$K. For the sake of consistency the correlated thermodynamic functions and enthalpy of formation presented by Rossini, Pitzer, Arnett, Braun, and Pimentel (1248) are adopted. The correlated entropy is in reasonable agreement with the calorimetric result.

No. 355   Butylcyclopentane C_9H_{18} (Ideal Gas State)   Mol Wt 126.234

T°K	cal/(mole °K)				kcal/mole			
	$Cp°$	$S°$	$-(G°-H°_{298})/T$	$H°-H°_{298}$	$\Delta Hf°$	$\Delta Gf°$	Log Kp
298	42.42	109.04	109.04	0.00	-40.22	14.67	-10.754
300	42.74	109.31	109.05	0.08	-40.29	15.01	-10.933
400	57.77	123.74	110.91	5.14	-43.71	33.99	-18.569
500	69.52	137.91	114.90	11.51	-46.49	53.74	-23.490
600	80.38	151.57	119.88	19.02	-48.68	73.99	-26.951
700	89.57	164.67	125.35	27.53	-50.30	94.59	-29.530
800	97.35	177.15	131.05	36.88	-51.44	115.35	-31.511
900	104.03	188.96	136.84	46.92	-52.19	136.26	-33.086
1000	114.80	200.44	142.62	57.82	-52.29	157.23	-34.360

Low-temperature thermal data were reported by Messerly, Todd, and Finke (993), including $Ttp = 165.18°$K, $\Delta Hm° = 2.704$ kcal/mole, $S°_{298}(l) = 82.18$ cal/(mole °K), $S°_{298}(g) = 108.46$ cal/(mole °K), and $Tb = 429.75°$K. The calorimetric entropy is in reasonable agreement with the correlated functions given by Rossini, Pitzer, Arnett, Braun, and Pimentel (1248) and, for the sake of consistency, the correlated functions and enthalpy of formation are adopted.

No. 356   1-Cyclopentylpentane $C_{10}H_{20}$ (Ideal Gas State)   Mol Wt 140.260

T°K	cal/(mole °K)				kcal/mole			
	$Cp°$	$S°$	$-(G°-H°_{298})/T$	$H°-H°_{298}$	$\Delta Hf°$	$\Delta Gf°$	Log Kp
298	47.89	118.35	118.35	0.00	-45.15	16.68	-12.224
300	48.23	118.65	118.36	0.09	-45.23	17.06	-12.424
400	64.71	134.87	120.45	5.77	-48.97	38.41	-20.987
500	77.77	150.73	124.93	12.90	-52.00	60.62	-26.496
600	89.73	165.99	130.51	21.29	-54.39	83.37	-30.366
700	99.85	180.61	136.63	30.78	-56.15	106.49	-33.245
800	108.42	194.51	143.01	41.21	-57.39	129.79	-35.456
900	115.78	207.72	149.47	52.43	-58.15	153.25	-37.213
1000	122.20	220.25	155.93	64.33	-58.47	176.78	-38.633

The values selected by Rossini, Pitzer, Arnett, Braun, and Pimentel (1248) are adopted. $Tm = 190°$K and $Tb = 453.6°$K.

No. 357 1-Cyclopentylhexane $C_{11}H_{22}$ (Ideal Gas State) Mol Wt 154.286

$T°K$	$Cp°$	$S°$	$-(G°-H°_{298})/T$	$H°-H°_{298}$	$\Delta Hf°$	$\Delta Gf°$	Log Kp
		cal/(mole °K)			kcal/mole		
298	53.35	127.66	127.66	0.00	-50.07	18.69	-13.702
300	53.73	128.00	127.67	0.10	-50.16	19.11	-13.924
400	71.65	145.99	130.00	6.40	-54.21	42.85	-23.412
500	86.01	163.55	134.96	14.30	-57.50	67.51	-29.507
600	99.07	180.41	141.14	23.57	-60.09	92.75	-33.784
700	110.12	196.54	147.91	34.04	-61.99	118.40	-36.964
800	119.48	211.87	154.96	45.53	-63.33	144.25	-39.404
900	127.52	226.41	162.10	57.89	-64.15	170.26	-41.342
1000	134.50	240.22	169.23	71.00	-64.49	196.34	-42.909

The values selected by Rossini, Pitzer, Arnett, Braun, and Pimentel (1248) are adopted. $Tm = 200°K$ and $Tb = 476.1°K$.

No. 358 1-Cyclopentylheptane $C_{12}H_{24}$ (Ideal Gas State) Mol Wt 168.312

$T°K$	$Cp°$	$S°$	$-(G°-H°_{298})/T$	$H°-H°_{298}$	$\Delta Hf°$	$\Delta Gf°$	Log Kp
		cal/(mole °K)			kcal/mole		
298	58.82	136.96	136.96	0.00	-55.00	20.70	-15.174
300	59.22	137.33	136.97	0.11	-55.09	21.16	-15.418
400	78.60	157.11	139.53	7.04	-59.46	47.28	-25.832
500	94.26	176.36	144.99	15.69	-63.01	74.39	-32.515
600	108.41	194.83	151.76	25.84	-65.80	102.13	-37.200
700	120.40	212.46	159.19	37.30	-67.84	130.31	-40.681
800	130.54	229.22	166.90	49.86	-69.28	158.69	-43.351
900	139.27	245.11	174.72	63.36	-70.15	187.26	-45.471
1000	146.80	260.18	182.52	77.67	-70.51	215.91	-47.184

The values selected by Rossini, Pitzer, Arnett, Braun, and Pimentel (1248) are adopted. $Tm = 220°K$ and $Tb = 497.1°K$.

No. 359 1-Cyclopentyloctane $C_{13}H_{26}$ (Ideal Gas State) Mol Wt 182.338

$T°K$	$Cp°$	$S°$	$-(G°-H°_{298})/T$	$H°-H°_{298}$	$\Delta Hf°$	$\Delta Gf°$	Log Kp
		cal/(mole °K)			kcal/mole		
298	64.29	146.27	146.27	0.00	-59.92	22.72	-16.651
300	64.71	146.67	146.28	0.12	-60.02	23.22	-16.917
400	85.54	168.23	149.07	7.67	-64.71	51.72	-28.257
500	102.51	189.18	155.02	17.09	-68.51	81.28	-35.526
600	117.75	209.25	162.39	28.12	-71.50	111.52	-40.618
700	130.68	228.40	170.47	40.56	-73.68	142.22	-44.400
800	141.61	246.58	178.85	54.18	-75.21	173.15	-47.299
900	151.02	263.81	187.35	68.83	-76.15	204.27	-49.601
1000	159.20	280.16	195.82	84.35	-76.52	235.47	-51.460

The values selected by Rossini, Pitzer, Arnett, Braun, and Pimentel (1248) are adopted. $Tm = 229°K$ and $Tb = 516.6°K$.

No. 360 1-Cyclopentylnonane $C_{14}H_{28}$ (Ideal Gas State) Mol Wt 196.364

	cal/(mole °K)				kcal/mole		
$T°K$	$Cp°$	$S°$	$-(G°-H°_{298})/T$	$H°-H°_{298}$	$\Delta Hf°$	$\Delta Gf°$	Log Kp
298	69.75	155.58	155.58	0.00	−64.85	24.72	−18.121
300	70.21	156.02	155.59	0.13	−64.96	25.27	−18.409
400	92.48	179.36	158.62	8.30	−69.96	56.15	−30.675
500	110.75	202.00	165.05	18.48	−74.03	88.16	−38.532
600	127.09	223.67	173.02	30.39	−77.20	120.89	−44.033
700	140.95	244.33	181.75	43.81	−79.53	154.12	−48.116
800	152.68	263.94	190.81	58.51	−81.16	187.59	−51.245
900	162.76	282.51	199.97	74.29	−82.15	221.27	−53.728
1000	171.50	300.13	209.12	91.01	−82.55	255.03	−55.734

The values selected by Rossini, Pitzer, Arnett, Braun, and Pimentel (1248) are adopted. $Tm = 244°K$ and $Tb = 535.2°K$.

No. 361 1-Cyclopentyldecane $C_{15}H_{30}$ (Ideal Gas State) Mol Wt 210.390

	cal/(mole °K)				kcal/mole		
$T°K$	$Cp°$	$S°$	$-(G°-H°_{298})/T$	$H°-H°_{298}$	$\Delta Hf°$	$\Delta Gf°$	Log Kp
298	75.22	164.89	164.89	0.00	−69.78	26.73	−19.591
300	75.70	165.36	164.90	0.14	−69.90	27.32	−19.901
400	99.42	190.49	168.16	8.94	−75.22	60.57	−33.094
500	119.00	214.81	175.08	19.87	−79.54	95.04	−41.538
600	136.44	238.09	183.65	32.67	−82.91	130.27	−47.447
700	151.23	260.27	193.03	47.07	−85.38	166.02	−51.831
800	163.74	281.30	202.76	62.83	−87.11	202.03	−55.190
900	174.51	301.22	212.60	79.76	−88.16	238.26	−57.855
1000	183.80	320.10	222.42	97.69	−88.57	274.58	−60.007

Messerly, Todd, and Finke (993) determined low-temperature thermal properties and reported $Ttp = 251.02°K$, $\Delta Hm° = 7.917$ kcal/mole $S°_{298}(l) = 128.71$ cal/(mole °K), $S°_{298}(g) = 164.45$ cal/(mole °K), and $Tb = 552.53°K$. The calorimetric entropy is in good agreement with the correlated selection of Rossini, Pitzer, Arnett, Braun, and Pimentel (1248). To retain consistency with related compounds, the correlated thermodynamic functions of Rossini et al. are adopted. Loeffler and Rossini (880) measured the enthalpy of combustion and derived $\Delta Hf°_{298}(l) = -87.52$ kcal/mole and $\Delta Hf°_{298}(g) = -69.78$ kcal/mole.

No. 362 1-Cyclopentylundecane $C_{16}H_{32}$ (Ideal Gas State) Mol Wt 224.416

The values selected by Rossini, Pitzer, Arnett, Braun, and Pimentel (1248) are adopted. $Tm = 263°K$ and $Tb = 569.0°K$.

No. 362 1-Cyclopentylundecane $C_{16}H_{32}$ (Ideal Gas State) Mol Wt 224.416

$T°K$	$Cp°$	$S°$	$-(G°-H°_{298})/T$	$H°-H°_{298}$	$\Delta Hf°$	$\Delta Gf°$	Log Kp
		cal/(mole °K)			kcal/mole		
298	80.68	174.20	174.20	0.00	-74.70	28.74	-21.069
300	81.20	174.71	174.21	0.15	-74.82	29.38	-21.400
400	106.36	201.61	177.71	9.57	-80.46	65.01	-35.518
500	127.24	227.63	185.11	21.27	-85.04	101.92	-44.549
600	145.78	252.51	194.28	34.94	-88.61	139.65	-50.865
700	161.50	276.20	204.31	50.33	-91.23	177.93	-55.550
800	174.80	298.65	214.71	67.16	-93.05	216.49	-59.138
900	186.25	319.92	225.23	85.22	-94.16	255.27	-61.985
1000	196.20	340.07	235.72	104.36	-94.59	294.15	-64.283

No. 363 1-Cyclopentyldodecane $C_{17}H_{34}$ (Ideal Gas State) Mol Wt 238.442

$T°K$	$Cp°$	$S°$	$-(G°-H°_{298})/T$	$H°-H°_{298}$	$\Delta Hf°$	$\Delta Gf°$	Log Kp
		cal/(mole °K)			kcal/mole		
298	86.15	183.51	183.51	0.00	-80.28	30.10	-22.062
300	86.69	184.05	183.52	0.16	-80.41	30.78	-22.419
400	113.30	212.74	187.25	10.20	-86.37	68.79	-37.582
500	135.49	240.45	195.14	22.66	-91.20	108.15	-47.271
600	155.12	266.94	204.91	37.22	-94.97	148.37	-54.043
700	171.78	292.13	215.59	53.58	-97.73	189.18	-59.063
800	185.87	316.01	226.66	71.48	-99.65	230.28	-62.906
900	198.00	338.62	237.86	90.69	-100.81	271.62	-65.954
1000	208.50	360.04	249.02	111.03	-101.26	313.05	-68.415

The values selected by Rossini, Pitzer, Arnett, Braun, and Pimentel (1248) are adopted. $Tm = 268°K$ and $Tb = 584.4°K$.

No. 364 1-Cyclopentyltridecane $C_{18}H_{36}$ (Ideal Gas State) Mol Wt 252.468

$T°K$	$Cp°$	$S°$	$-(G°-H°_{298})/T$	$H°-H°_{298}$	$\Delta Hf°$	$\Delta Gf°$	Log Kp
		cal/(mole °K)			kcal/mole		
298	91.62	192.89	192.89	0.00	-84.55	32.74	-24.001
300	92.18	193.46	192.90	0.18	-84.69	33.46	-24.376
400	120.24	223.93	196.86	10.83	-90.96	73.85	-40.346
500	143.74	253.34	205.24	24.06	-96.05	115.66	-50.551
600	164.46	281.43	215.61	39.49	-100.02	158.37	-57.682
700	182.06	308.13	226.94	56.84	-102.92	201.70	-62.969
800	196.94	333.44	238.69	75.81	-104.94	245.32	-67.016
900	209.75	357.39	250.56	96.16	-106.15	289.21	-70.226
1000	220.80	380.08	262.39	117.70	-106.63	333.20	-72.817

The values selected by Rossini, Pitzer, Arnett, Braun, and Pimentel (1248) are adopted. $Tm = 278°K$ and $Tb = 599.1°K$.

Alkylcyclopentane Ideal Gas Tables

No. 365 1-Cyclopentyltetradecane $C_{19}H_{38}$ (Ideal Gas State) Mol Wt 266.494

$T°K$	$Cp°$	$S°$	$-(G°-H°_{298})/T$	$H°-H°_{298}$	$\Delta Hf°$	$\Delta Gf°$	Log Kp
	cal/(mole °K)				kcal/mole		
298	97.08	202.13	202.13	0.00	-89.48	34.77	-25.486
300	97.68	202.74	202.14	0.19	-89.62	35.53	-25.883
400	127.18	234.99	206.33	11.47	-96.22	78.30	-42.780
500	151.98	266.09	215.20	25.45	-101.56	122.57	-53.572
600	173.80	295.78	226.17	41.77	-105.73	167.78	-61.112
700	192.33	324.00	238.15	60.10	-108.77	213.65	-66.700
800	208.00	350.73	250.57	80.13	-110.89	259.82	-70.977
900	221.49	376.02	263.11	101.62	-112.16	306.27	-74.369
1000	233.20	399.98	275.61	124.37	-112.65	352.83	-77.106

The values selected by Rossini, Pitzer, Arnett, Braun, and Pimentel (1248) are adopted. $Tm = 282°K$ and $Tb = 614°K$.

No. 366 1-Cyclopentylpentadecane $C_{20}H_{40}$ (Ideal Gas State) Mol Wt 280.520

$T°K$	$Cp°$	$S°$	$-(G°-H°_{298})/T$	$H°-H°_{298}$	$\Delta Hf°$	$\Delta Gf°$	Log Kp
	cal/(mole °K)				kcal/mole		
298	102.55	211.44	211.44	0.00	-94.41	36.78	-26.956
300	103.17	212.08	211.45	0.20	-94.56	37.58	-27.375
400	134.12	246.12	215.88	12.10	-101.47	82.73	-45.199
500	160.23	278.91	225.23	26.84	-107.07	129.45	-56.579
600	183.15	310.20	236.80	44.04	-111.43	177.16	-64.527
700	202.61	339.93	249.43	63.36	-114.62	225.55	-70.415
800	219.06	368.09	262.52	84.46	-116.84	274.27	-74.922
900	233.24	394.73	275.74	107.09	-118.17	323.27	-78.496
1000	245.50	419.95	288.91	131.04	-118.67	372.38	-81.380

The values selected by Rossini, Pitzer, Arnett, Braun, and Pimentel (1248) are adopted. $Tm = 290°K$ and $Tb = 627°K$.

No. 367 1-Cyclopentylhexadecane $C_{21}H_{42}$ (Ideal Gas State) Mol Wt 294.546

$T°K$	$Cp°$	$S°$	$-(G°-H°_{298})/T$	$H°-H°_{298}$	$\Delta Hf°$	$\Delta Gf°$	Log Kp
	cal/(mole °K)				kcal/mole		
298	108.01	220.75	220.75	0.00	-99.33	38.79	-28.434
300	108.67	221.43	220.76	0.21	-99.49	39.64	-28.874
400	141.06	257.24	225.42	12.73	-106.72	87.17	-47.623
500	168.47	291.73	235.26	28.24	-112.57	136.34	-59.589
600	192.49	324.62	247.43	46.32	-117.13	186.54	-67.945
700	212.88	355.86	260.71	66.61	-120.46	237.46	-74.134
800	230.13	385.44	274.47	88.78	-122.78	288.72	-78.870
900	244.98	413.43	288.37	112.55	-124.16	340.27	-82.625
1000	257.80	439.92	302.21	137.71	-124.69	391.95	-85.655

The values selected by Rossini, Pitzer, Arnett, Braun, and Pimentel (1248) are adopted. $Tm = 294°K$ and $Tb = 639°K$.

ALKYLCYCLOPENTENE IDEAL GAS TABLES

No. 368 1-Methylcyclopentene C_6H_{10} (Ideal Gas State) Mol Wt 82.140

$T°K$	cal/(mole °K)				kcal/mole		
	$Cp°$	$S°$	$-(G°-H°_{298})/T$	$H°-H°_{298}$	$\Delta Hf°$	$\Delta Gf°$	Log Kp
298	24.10	78.00	78.00	0.00	-1.30	24.41	-17.890
300	24.30	78.15	78.01	0.05	-1.34	24.56	-17.894
400	32.50	86.28	79.06	2.89	-3.45	33.53	-18.320
500	40.20	94.38	81.31	6.54	-5.21	42.99	-18.789
600	46.80	102.29	84.15	10.89	-6.63	52.76	-19.217
700	53.30	110.03	87.30	15.92	-7.65	62.75	-19.591
800	57.00	117.39	90.61	21.43	-8.42	72.85	-19.901
900	60.90	124.33	93.97	27.33	-9.00	83.05	-20.167
1000	64.30	130.93	97.34	33.59	-9.36	93.31	-20.392

Labbauf and Rossini (821) measured the enthalpy of combustion and reported $\Delta Hf°_{298}(l) = -9.05$ kcal/mole. Estimating $\Delta Hv_{298} = 7.75$ kcal/mole, we calculate $\Delta Hf°_{298}(g) = -1.30$ kcal/mole. Rossini, Pitzer, Arnett, Braun, and Pimentel (1248) selected $Tm = 146°K$ and $Tb = 348.9°K$. Hrostowski and Pimentel (622) assigned the fundamental vibrational frequencies and calculated the ideal gaseous thermodynamic functions.

No. 369 3-Methylcyclopentene C_6H_{10} (Ideal Gas State) Mol Wt 82.140

$T°K$	cal/(mole °K)				kcal/mole		
	$Cp°$	$S°$	$-(G°-H°_{298})/T$	$H°-H°_{298}$	$\Delta Hf°$	$\Delta Gf°$	Log Kp
298	23.90	79.00	79.00	0.00	2.07	27.48	-20.141
300	24.10	79.15	79.01	0.05	2.03	27.63	-20.130
400	32.60	87.27	80.05	2.89	-0.09	36.50	-19.944
500	40.50	95.41	82.31	6.56	-1.82	45.86	-20.044
600	47.10	103.40	85.16	10.94	-3.20	55.52	-20.224
700	52.60	111.08	88.32	15.94	-4.26	65.41	-20.420
800	57.20	118.41	91.63	21.43	-5.05	75.41	-20.599
900	61.10	125.38	95.00	27.35	-5.61	85.50	-20.762
1000	64.50	132.00	98.37	33.64	-5.95	95.66	-20.904

Labbauf and Rossini (821) measured the enthalpy of combustion and reported $\Delta Hf°_{298}(l) = -5.68$ kcal/mole. Estimating $\Delta Hv_{298} = 7.75$ kcal/mole, we find $\Delta Hf°_{298}(g) = 2.07$ kcal/mole. Rossini, Pitzer, Arnett, Braun, and Pimentel (1248) selected $Tb = 338.1°K$. Hrostowski and Pimentel (622) assigned the fundamental vibrational frequencies and calculated the ideal gaseous thermodynamic functions.

No. 370 4-Methylcyclopentene C_6H_{10} (Ideal Gas State) Mol Wt 82.140

Labbauf and Rossini (821) measured the enthalpy of combustion and reported $\Delta Hf°_{298}(l) = -4.22$ kcal/mole. Estimating $\Delta Hv_{298} = 7.75$ kcal/

Alkylcyclohexane Ideal Gas Tables

No. 370   4-Methylcyclopentene C_6H_{10} (Ideal Gas State)   Mol Wt 82.140

$T°K$	$Cp°$	cal/(mole °K) $S°$	$-(G°-H°_{298})/T$	$H°-H°_{298}$	kcal/mole $\Delta Hf°$	$\Delta Gf°$	Log Kp
298	23.90	78.60	78.60	0.00	3.53	29.06	-21.299
300	24.10	78.75	78.61	0.05	3.49	29.21	-21.281
400	32.60	86.87	79.65	2.89	1.37	38.12	-20.829
500	40.40	95.00	81.91	6.55	-0.37	47.52	-20.770
600	47.00	102.97	84.76	10.93	-1.76	57.23	-20.844
700	52.50	110.64	87.91	15.91	-2.82	67.15	-20.965
800	57.10	117.96	91.22	21.40	-3.63	77.20	-21.088
900	61.10	124.92	94.58	27.31	-4.19	87.34	-21.208
1000	64.40	131.53	97.94	33.59	-4.53	97.54	-21.316

mole, we calculate $\Delta Hf°_{298} = 3.53$ kcal/mole. Rossini, Pitzer, Arnett, Braun, and Pimentel (1248) selected $Tb = 348.3°K$. Hrostowski and Pimentel (622) assigned the fundamental vibrational frequencies and calculated the ideal gaseous thermodynamic functions.

ALKYLCYCLOHEXANE IDEAL GAS TABLES

No. 371   Methylcyclohexane C_7H_{14} (Ideal Gas State)   Mol Wt 98.182

$T°K$	$Cp°$	cal/(mole °K) $S°$	$-(G°-H°_{298})/T$	$H°-H°_{298}$	kcal/mole $\Delta Hf°$	$\Delta Gf°$	Log Kp
298	32.27	82.06	82.06	0.00	-36.99	6.52	-4.781
300	32.51	82.27	82.07	0.06	-37.05	6.79	-4.946
400	44.35	93.26	83.49	3.92	-39.78	21.84	-11.932
500	55.21	104.36	86.55	8.91	-41.91	37.50	-16.391
600	64.46	115.26	90.43	14.90	-43.46	53.53	-19.498
700	72.23	125.80	94.74	21.75	-44.49	69.79	-21.789
800	78.74	135.88	99.25	29.31	-45.10	86.15	-23.534
900	84.20	145.48	103.86	37.46	-45.33	102.58	-24.909
1000	88.79	154.59	108.48	46.12	-45.23	119.03	-26.012

The values selected by Rossini, Pitzer, Arnett, Braun, and Pimentel (1248) are employed. $Tm = 146.56°K$, $\Delta Hm° = 1.613$ kcal/mole, and $Tb = 374.08°K$, where $\Delta Hv = 7.580$ kcal/mole. Kobe and Lynn (780) reported $Tc = 572.1°K$, $Pc = 34.32$ atm. and $dc = 0.285$ g/cc.

No. 372   Ethylcyclohexane C_8H_{16} (Ideal Gas State)   Mol Wt 112.208

Huffman, Todd, and Oliver (657) measured the low-temperature properties; their values are merged with those selected by Rossini, Pitzer, Arnett, Braun, and Pimentel (1248). $Tm = 161.83°K$, $\Delta Hm° = 1.992$ kcal/mole, and $Tb = 404.93°K$, where $\Delta Hv = 8.29$ kcal/mole.

No. 372 Ethylcyclohexane C_8H_{16} (Ideal Gas State) Mol Wt 112.208

$T°K$	$Cp°$	$S°$	$-(G°-H°_{298})/T$	$H°-H°_{298}$	$\Delta Hf°$	$\Delta Gf°$	Log Kp
	cal/(mole °K)				kcal/mole		
298	37.96	91.44	91.44	0.00	-41.05	9.38	-6.874
300	38.23	91.68	91.45	0.08	-41.12	9.69	-7.056
400	51.60	104.54	93.11	4.58	-44.14	27.10	-14.808
500	63.80	117.40	96.68	10.36	-46.49	45.20	-19.756
600	74.10	129.96	101.19	17.27	-48.21	63.70	-23.201
700	82.80	142.06	106.17	25.13	-49.35	82.46	-25.743
800	90.10	153.60	111.38	33.78	-50.02	101.32	-27.679
900	96.20	164.58	116.69	43.11	-50.29	120.27	-29.204
1000	101.30	174.99	122.00	52.99	-50.19	139.23	-30.428

No. 373 1,1-Dimethylcyclohexane C_8H_{16} (Ideal Gas State) Mol Wt 112.208

$T°K$	$Cp°$	$S°$	$-(G°-H°_{298})/T$	$H°-H°_{298}$	$\Delta Hf°$	$\Delta Gf°$	Log Kp
	cal/(mole °K)				kcal/mole		
298	36.90	87.24	87.24	0.00	-43.26	8.42	-6.172
300	37.20	87.47	87.25	0.07	-43.33	8.74	-6.364
400	50.70	100.05	88.87	4.48	-46.45	26.59	-14.527
500	63.30	112.75	92.37	10.19	-48.87	45.14	-19.731
600	74.10	125.27	96.82	17.08	-50.61	64.11	-23.351
700	83.20	137.39	101.75	24.95	-51.74	83.34	-26.018
800	90.70	149.01	106.94	33.66	-52.36	102.67	-28.046
900	97.00	160.06	112.23	43.05	-52.55	122.07	-29.641
1000	102.20	170.56	117.55	53.02	-52.36	141.48	-30.919

The selected values of Rossini, Pitzer, Arnett, Braun, and Pimentel (1248) are combined with the measured values of Huffman, Todd, and Oliver (657). $Tt = 153.14°K$, $\Delta Ht° = 1.430$ kcal/mole, $Tm = 239.66°K$, $\Delta Hm° = 0.483$ kcal/mole, and $Tb = 392.69°K$, where $\Delta Hv = 7.880$ kcal/mole.

No. 374 1,2-Dimethylcyclohexane, cis C_8H_{16} (Ideal Gas State) Mol Wt 112.208

$T°K$	$Cp°$	$S°$	$-(G°-H°_{298})/T$	$H°-H°_{298}$	$\Delta Hf°$	$\Delta Gf°$	Log Kp
	cal/(mole °K)				kcal/mole		
298	37.40	89.51	89.51	0.00	-41.15	9.85	-7.222
300	37.70	89.75	89.52	0.07	-41.22	10.17	-7.405
400	51.10	102.45	91.16	4.52	-44.30	27.78	-15.180
500	63.50	115.22	94.69	10.27	-46.69	46.09	-20.146
600	74.00	127.75	99.17	17.16	-48.42	64.81	-23.607
700	82.80	139.84	104.12	25.01	-49.57	83.79	-26.160
800	90.10	151.39	109.31	33.66	-50.24	102.88	-28.104
900	96.30	162.37	114.60	42.99	-50.50	122.05	-29.636
1000	101.40	172.79	119.90	52.89	-50.39	141.23	-30.865

The measurements of Huffman, Todd, and Oliver (657) are combined with the selected values of Rossini, Pitzer, Arnett, Braun, and Pimentel (1248). $Tt = 172.49°K$, $\Delta Ht° = 1.973$ kcal/mole, $Ttp = 223.27°K$, $\Delta Hm° = 0.393$ kcal/mole, and $Tb = 402.88°K$, where $\Delta Hv = 8.18$ kcal/mole.

No. 375 1,2-Dimethylcyclohexane, trans C_8H_{16} (Ideal Gas State) Mol Wt 112.208

	cal/(mole °K)				kcal/mole		
T°K	$Cp°$	$S°$	$-(G°-H°_{298})/T$	$H°-H°_{298}$	$\Delta Hf°$	$\Delta Gf°$	Log Kp
298	38.00	88.65	88.65	0.00	-43.02	8.24	-6.040
300	38.30	88.89	88.66	0.08	-43.09	8.55	-6.231
400	51.90	101.80	90.32	4.60	-46.09	26.25	-14.340
500	64.20	114.74	93.92	10.42	-48.41	44.61	-19.499
600	74.60	127.39	98.45	17.37	-50.08	63.37	-23.083
700	83.30	139.56	103.46	25.28	-51.17	82.39	-25.721
800	90.50	151.17	108.70	33.98	-51.80	101.50	-27.726
900	96.60	162.19	114.04	43.34	-52.02	120.69	-29.305
1000	101.70	172.64	119.38	53.26	-51.88	139.88	-30.570

The selected values of Rossini, Pitzer, Arnett, Braun, and Pimentel (1248) are combined with the measured values of Huffman, Todd, and Oliver (657). $Ttp = 184.99°K$, $\Delta Hm° = 2.508$ kcal/mole, and $Tb = 396.57°K$, where $\Delta Hv = 7.98$ kcal/mole.

No. 376 1,3-Dimethylcyclohexane, cis C_8H_{16} (Ideal Gas State) Mol Wt 112.208

	cal/(mole °K)				kcal/mole		
T°K	$Cp°$	$S°$	$-(G°-H°_{298})/T$	$H°-H°_{298}$	$\Delta Hf°$	$\Delta Gf°$	Log Kp
298	37.60	88.54	88.54	0.00	-44.16	7.13	-5.228
300	37.90	88.78	88.55	0.07	-44.23	7.45	-5.425
400	51.20	101.52	90.19	4.54	-47.29	25.16	-13.746
500	63.60	114.32	93.74	10.29	-49.67	43.56	-19.039
600	74.20	126.87	98.22	17.20	-51.39	62.37	-22.717
700	83.10	139.00	103.19	25.07	-52.51	81.44	-25.424
800	90.50	150.59	108.39	33.76	-53.15	100.60	-27.483
900	96.70	161.62	113.70	43.13	-53.37	119.85	-29.102
1000	102.00	172.09	119.02	53.07	-53.21	139.11	-30.400

The measurements of Huffman, Todd, and Oliver (657) are included with the selected values of Rossini, Pitzer, Arnett, Braun, and Pimentel (1248). $Ttp = 197.58°K$, $\Delta Hm° = 2.586$ kcal/mole, and $Tb = 393.24°K$, where $\Delta Hv = 7.96$ kcal/mole.

No. 377 1,3-Dimethylcyclohexane, trans C_8H_{16} (Ideal Gas State) Mol Wt 112.208

The selected values of Rossini, Pitzer, Arnett, Braun, and Pimentel (1248) and the measurements of Huffman, Todd, and Oliver (657) are

No. 377   1,3-Dimethylcyclohexane, trans C_8H_{16} (Ideal Gas State)   Mol Wt 112.208

	cal/(mole °K)				kcal/mole		
$T°K$	$Cp°$	$S°$	$-(G°-H°_{298})/T$	$H°-H°_{298}$	$\Delta Hf°$	$\Delta Gf°$	Log Kp
298	37.60	89.92	89.92	0.00	-42.20	8.68	-6.363
300	37.90	90.16	89.93	0.07	-42.27	8.99	-6.551
400	51.10	102.89	91.57	4.53	-45.34	26.57	-14.515
500	63.40	115.65	95.11	10.27	-47.73	44.83	-19.596
600	73.80	128.15	99.58	17.15	-49.48	63.51	-23.133
700	82.50	140.20	104.53	24.97	-50.65	82.45	-25.742
800	89.80	151.71	109.72	33.60	-51.36	101.51	-27.729
900	95.90	162.65	114.99	42.89	-51.65	120.65	-29.295
1000	101.10	173.03	120.28	52.75	-51.58	139.80	-30.552

employed. $Ttp = 183.05°K$, $\Delta Hm° = 2.358$ kcal/mole, and $Tb = 397.60°K$, where $\Delta Hv = 8.09$ kcal/mole.

No. 378   1,4-Dimethylcyclohexane, cis C_8H_{16} (Ideal Gas State)   Mol Wt 112.208

	cal/(mole °K)				kcal/mole		
$T°K$	$Cp°$	$S°$	$-(G°-H°_{298})/T$	$H°-H°_{298}$	$\Delta Hf°$	$\Delta Gf°$	Log Kp
298	37.60	88.54	88.54	0.00	-42.22	9.07	-6.650
300	37.90	88.78	88.55	0.07	-42.29	9.39	-6.838
400	51.10	101.51	90.19	4.53	-45.36	27.10	-14.806
500	63.40	114.27	93.73	10.27	-47.75	45.50	-19.889
600	73.80	126.77	98.20	17.15	-49.50	64.32	-23.427
700	82.50	138.82	103.15	24.97	-50.67	83.40	-26.037
800	89.80	150.33	108.34	33.60	-51.38	102.59	-28.025
900	95.90	161.27	113.61	42.89	-51.67	121.87	-29.592
1000	101.10	171.65	118.90	52.75	-51.60	141.16	-30.849

The determinations of Huffman, Todd, and Oliver (657) are used with the selected values of Rossini, Pitzer, Arnett, Braun, and Pimentel (1248). $Ttp = 185.72°K$, $\Delta Hm° = 2.244$ kcal/mole, and $Tb = 397.47°K$, where $\Delta Hv = 8.07$ kcal/mole.

No. 379   1,4-Dimethylcyclohexane, trans C_8H_{16} (Ideal Gas State)   Mol Wt 112.208

	cal/(mole °K)				kcal/mole		
$T°K$	$Cp°$	$S°$	$-(G°-H°_{298})/T$	$H°-H°_{298}$	$\Delta Hf°$	$\Delta Gf°$	Log Kp
298	37.70	87.19	87.19	0.00	-44.12	7.58	-5.552
300	38.00	87.43	87.20	0.08	-44.19	7.89	-5.749
400	51.60	100.25	88.85	4.57	-47.22	25.74	-14.061
500	64.00	113.13	92.42	10.36	-49.56	44.26	-19.345
600	74.60	125.77	96.93	17.31	-51.24	63.18	-23.013
700	83.30	137.94	101.93	25.21	-52.34	82.36	-25.712
800	90.60	149.55	107.16	33.92	-52.96	101.63	-27.763
900	96.80	160.59	112.49	43.30	-53.17	120.98	-29.377
1000	101.90	171.06	117.83	53.24	-53.00	140.34	-30.669

The measurements of Huffman, Todd, and Oliver (657) are combined with the selected values of Rossini, Pitzer, Arnett, Braun, and Pimentel (1248). $Ttp = 236.21°K$, $\Delta Hm° = 2.947$ kcal/mole, and $Tb = 392.50°K$, at which $\Delta Hv = 7.90$ kcal/mole.

No. 380 Propylcyclohexane C_9H_{18} (Ideal Gas State) Mol Wt 126.234

	cal/(mole °K)				kcal/mole		
$T°K$	$Cp°$	$S°$	$-(G°-H°_{298})/T$	$H°-H°_{298}$	$\Delta Hf°$	$\Delta Gf°$	Log Kp
298	44.03	100.27	100.27	0.00	-46.20	11.31	-8.288
300	44.32	100.55	100.28	0.09	-46.27	11.66	-8.493
400	59.10	115.35	102.19	5.27	-49.56	31.49	-17.207
500	72.60	130.03	106.30	11.87	-52.11	52.06	-22.756
600	83.80	144.29	111.45	19.71	-53.98	73.07	-26.616
700	93.30	157.94	117.12	28.58	-55.23	94.37	-29.461
800	101.20	170.93	123.04	38.31	-55.99	115.78	-31.628
900	107.80	183.24	129.05	48.77	-56.31	137.28	-33.335
1000	113.40	194.89	135.06	59.84	-56.25	158.81	-34.706

Finke, Messerly, and Todd (423) measured low-temperature thermal properties, including $Ttp = 178.25°K$, $\Delta Hm° = 2.479$ kcal/mole, $S°_{298}(l) = 74.54$ cal/(mole °K), and $S°_{298}(g) = 100.35$ cal/(mole °K). Thermodynamic functions and enthalpy of formation are the selections of Rossini, Pitzer, Arnett, Braun, and Pimentel (1248), who also listed $Tb = 429.87°K$. The calorimetric entropy is in excellent agreement with the correlated selection of Rossini et al.

No. 381 1,3,5-Trimethylcyclohexane, cis, cis C_9H_{18} (Ideal Gas State) Mol Wt 126.234

	cal/(mole °K)				kcal/mole		
$T°K$	$Cp°$	$S°$	$-(G°-H°_{298})/T$	$H°-H°_{298}$	$\Delta Hf°$	$\Delta Gf°$	Log Kp
298	42.93	93.30	93.30	0.00	-51.48	8.10	-5.941
300	43.29	93.57	93.31	0.08	-51.55	8.47	-6.170
400	58.05	108.07	95.18	5.16	-54.95	29.02	-15.854
500	71.99	122.55	99.21	11.68	-57.58	50.33	-21.998
600	83.94	136.77	104.29	19.49	-59.47	72.09	-26.258
700	93.97	150.48	109.92	28.40	-60.69	94.13	-29.388
800	102.26	163.58	115.81	38.22	-61.36	116.28	-31.766
900	109.20	176.04	121.82	48.80	-61.56	138.52	-33.635
1000	115.21	187.86	127.84	60.03	-61.34	160.75	-35.131

Egan and Buss (371) measured the hydrogenation equilibria of mesitylene (1,3,5-trimethylbenzene) from 473° to 573°K and presented thermodynamic functions for the cis and trans isomers of 1,3,5-trimethylcyclohexane.

No. 382 1,3,5-Trimethylcyclohexane, cis, trans C_9H_{18} (Ideal Gas State) Mol Wt 126.234

T°K	cal/(mole °K)				kcal/mole		Log Kp
	$Cp°$	$S°$	$-(G°-H°_{298})/T$	$H°-H°_{298}$	$\Delta Hf°$	$\Delta Gf°$	
298	42.93	95.60	95.60	0.00	-49.37	9.53	-6.985
300	43.29	95.87	95.61	0.08	-49.44	9.89	-7.204
400	57.85	110.34	97.48	5.15	-52.85	30.21	-16.505
500	71.59	124.76	101.50	11.64	-55.51	51.30	-22.420
600	83.14	138.87	106.56	19.39	-57.46	72.84	-26.531
700	92.77	152.43	112.15	28.20	-58.78	94.68	-29.559
800	100.86	165.36	118.00	37.89	-59.58	116.65	-31.864
900	107.60	177.63	123.95	48.32	-59.93	138.71	-33.682
1000	113.41	189.28	129.90	59.38	-59.88	160.80	-35.140

Egan and Buss (371) measured the hydrogenation equilibria of mesitylene (1,3,5-trimethylbenzene) from 473° to 573°K and presented thermodynamic functions for the *cis* and *trans* isomers of 1,3,5-trimethylcyclohexane.

No. 383 Butylcyclohexane $C_{10}H_{20}$ (Ideal Gas State) Mol Wt 140.260

T°K	cal/(mole °K)				kcal/mole		Log Kp
	$Cp°$	$S°$	$-(G°-H°_{298})/T$	$H°-H°_{298}$	$\Delta Hf°$	$\Delta Gf°$	
298	49.50	109.58	109.58	0.00	-50.95	13.49	-9.890
300	49.81	109.89	109.59	0.10	-51.03	13.89	-10.116
400	66.00	126.47	111.74	5.90	-54.64	36.10	-19.724
500	80.80	142.84	116.33	13.26	-57.45	59.12	-25.842
600	93.10	158.68	122.07	21.97	-59.52	82.63	-30.097
700	103.60	173.85	128.40	31.82	-60.92	106.45	-33.235
800	112.30	188.26	134.99	42.63	-61.77	130.41	-35.624
900	119.60	201.92	141.67	54.23	-62.14	154.47	-37.509
1000	125.70	214.85	148.35	66.51	-62.10	178.56	-39.022

Low-temperature thermal data were reported by Finke, Messerly, and Todd (423), including $Ttp = 198.42°K$, $\Delta Hm° = 3.384$ kcal/mole, $S°_{298}(l) = 82.45$ cal/(mole °K), and $S°_{298}(g) = 109.89$ cal/(mole °K). Thermodynamic functions and enthalpy of formation are the selection of Rossini, Pitzer, Arnett, Braun, and Pimentel (1248), who also listed $Tb = 454.10°K$. The calorimetric entropy is in good agreement with the correlated selection of Rossini et al.

No. 384 Pentylcyclohexane $C_{11}H_{22}$ (Ideal Gas State) Mol Wt 154.286

The values selected by Rossini, Pitzer, Arnett, Braun, and Pimentel (1248) are employed. $Tm = 215.7°K$ and $Tb = 476.82°K$.

Alkylcyclohexane Ideal Gas Tables

No. 384 Pentylcyclohexane $C_{11}H_{22}$ (Ideal Gas State) Mol Wt 154.286

$T°K$	$Cp°$	$S°$	$-(G°-H°_{298})/T$	$H°-H°_{298}$	$\Delta Hf°$	$\Delta Gf°$	Log Kp
	cal/(mole °K)			kcal/mole			
298	54.96	118.89	118.89	0.00	-55.88	15.50	-11.360
300	55.31	119.24	118.90	0.11	-55.97	15.93	-11.608
400	73.00	137.61	121.28	6.54	-59.89	40.53	-22.142
500	89.10	155.68	126.36	14.66	-62.95	66.00	-28.847
600	102.50	173.14	132.71	24.26	-65.21	92.00	-33.510
700	113.90	189.82	139.69	35.10	-66.75	118.35	-36.948
800	123.30	205.66	146.95	46.97	-67.70	144.84	-39.566
900	131.30	220.65	154.32	59.71	-68.14	171.45	-41.632
1000	138.10	234.85	161.67	73.19	-68.11	198.09	-43.291

No. 385 1-Cyclohexylhexane $C_{12}H_{24}$ (Ideal Gas State) Mol Wt 168.312

$T°K$	$Cp°$	$S°$	$-(G°-H°_{298})/T$	$H°-H°_{298}$	$\Delta Hf°$	$\Delta Gf°$	Log Kp
	cal/(mole °K)			kcal/mole			
298	60.43	128.20	128.20	0.00	-60.80	17.51	-12.837
300	60.80	128.58	128.21	0.12	-60.89	17.99	-13.107
400	79.90	148.73	130.82	7.17	-65.14	44.97	-24.567
500	97.40	168.49	136.39	16.06	-68.45	72.89	-31.859
600	111.80	187.55	143.34	26.53	-70.91	101.39	-36.929
700	124.10	205.74	150.97	38.35	-72.60	130.26	-40.667
800	134.40	223.00	158.90	51.28	-73.65	159.30	-43.515
900	143.00	239.34	166.94	65.16	-74.14	188.46	-45.763
1000	150.40	254.80	174.96	79.84	-74.14	217.67	-47.568

The values chosen by Rossini, Pitzer, Arnett, Braun, and Pimentel (1248) are used. $Tm = 230.2°K$ and $Tb = 497.85°K$.

No. 386 1-Cyclohexylheptane $C_{13}H_{26}$ (Ideal Gas State) Mol Wt 182.338

$T°K$	$Cp°$	$S°$	$-(G°-H°_{298})/T$	$H°-H°_{298}$	$\Delta Hf°$	$\Delta Gf°$	Log Kp
	cal/(mole °K)			kcal/mole			
298	65.89	137.51	137.51	0.00	-65.73	19.52	-14.307
300	66.30	137.92	137.52	0.13	-65.83	20.04	-14.599
400	86.90	159.86	140.37	7.80	-70.39	49.39	-26.985
500	105.60	181.32	146.42	17.45	-73.96	79.77	-34.864
600	121.20	201.99	153.98	28.81	-76.61	110.76	-40.342
700	134.40	221.69	162.25	41.61	-78.44	142.16	-44.381
800	145.40	240.37	170.86	55.62	-79.59	173.73	-47.459
900	154.80	258.06	179.58	70.64	-80.14	205.45	-49.888
1000	162.70	274.79	188.27	86.53	-80.15	237.21	-51.840

The selected values of Rossini, Pitzer, Arnett, Braun, and Pimentel (1248) are adopted. $Tm = 242.7°K$ and $Tb = 518.05°K$.

No. 387 1-Cyclohexyloctane $C_{14}H_{28}$ (Ideal Gas State) Mol Wt 196.364

$T°K$	$Cp°$	$S°$	$-(G°-H°_{298})/T$	$H°-H°_{298}$	$\Delta Hf°$	$\Delta Gf°$	Log Kp
		cal/(mole °K)			kcal/mole		
298	71.36	146.82	146.82	0.00	-70.65	21.53	-15.784
300	71.79	147.27	146.83	0.14	-70.76	22.10	-16.098
400	93.80	170.98	149.91	8.43	-75.63	53.83	-29.409
500	113.80	194.12	156.45	18.84	-79.46	86.66	-37.875
600	130.50	216.39	164.60	31.08	-82.32	120.15	-43.761
700	144.70	237.60	173.53	44.86	-84.29	154.07	-48.102
800	156.50	257.71	182.80	59.93	-85.54	188.19	-51.409
900	166.60	276.74	192.19	76.10	-86.14	222.47	-54.020
1000	175.00	294.75	201.56	93.19	-86.17	256.79	-56.118

The selected values of Rossini, Pitzer, Arnett, Braun, and Pimentel (1248) are used. $Tm = 253.5°K$ and $Tb = 536.75°K$.

No. 388 1-Cyclohexylnonane $C_{15}H_{30}$ (Ideal Gas State) Mol Wt 210.390

$T°K$	$Cp°$	$S°$	$-(G°-H°_{298})/T$	$H°-H°_{298}$	$\Delta Hf°$	$\Delta Gf°$	Log Kp
		cal/(mole °K)			kcal/mole		
298	76.83	156.12	156.12	0.00	-75.58	23.54	-17.257
300	77.28	156.60	156.13	0.15	-75.69	24.15	-17.592
400	100.70	182.09	159.44	9.06	-80.89	58.26	-31.831
500	122.10	206.93	166.47	20.23	-84.98	93.54	-40.884
600	139.80	230.79	175.22	33.35	-88.03	129.53	-47.179
700	155.00	253.52	184.79	48.11	-90.14	165.99	-51.820
800	167.60	275.06	194.74	64.26	-91.49	202.65	-55.358
900	178.30	295.43	204.81	81.57	-92.15	239.48	-58.150
1000	187.40	314.70	214.84	99.86	-92.20	276.36	-60.395

The selected values of Rossini, Pitzer, Arnett, Braun, and Pimentel (1248) are used. $Tm = 263.0°K$ and $Tb = 554.65°K$.

No. 389 1-Cyclohexyldecane $C_{16}H_{32}$ (Ideal Gas State) Mol Wt 224.416

$T°K$	$Cp°$	$S°$	$-(G°-H°_{298})/T$	$H°-H°_{298}$	$\Delta Hf°$	$\Delta Gf°$	Log Kp
		cal/(mole °K)			kcal/mole		
298	82.29	165.43	165.43	0.00	-80.51	25.55	-18.727
300	82.78	165.95	165.44	0.16	-80.63	26.20	-19.084
400	107.70	193.23	168.99	9.70	-86.14	62.69	-34.249
500	130.40	219.77	176.50	21.64	-90.48	100.42	-43.890
600	149.20	245.25	185.85	35.64	-93.72	138.90	-50.591
700	165.30	269.49	196.09	51.39	-95.97	177.88	-55.533
800	178.60	292.45	206.71	68.60	-97.42	217.08	-59.299
900	190.00	314.16	217.45	87.04	-98.14	256.46	-62.273
1000	199.70	334.70	228.16	106.54	-98.21	295.89	-64.664

Finke, Messerly, and Todd (423) measured low-temperature thermal data, including $Ttp = 271.43°K$, $\Delta Hm° = 9.225$ kcal/mole, $S°_{298}(l) = 129.10$ cal/(mole °K), and $S°_{298}(g) = 165.97$ cal/(mole °K). Thermodynamic functions are the selections of Rossini, Pitzer, Arnett, Braun, and Pimentel (1248), who also listed $Tb = 570.74°K$. The calorimetric entropy is in good agreement with the correlated selection of Rossini et al. Loeffler and Rossini (880) determined the enthalpy of combustion and derived $\Delta Hf°_{298}(l) = -100.03$ kcal/mole and $\Delta Hf°_{298}(g) = -80.51$ kcal/kcal/mole.

No. 390 1-Cylohexylundecane $C_{17}H_{34}$ (Ideal Gas State) Mol Wt 238.442

T°K	Cp°	S°	$-(G°-H°_{298})/T$	$H°-H°_{298}$	$\Delta Hf°$	$\Delta Gf°$	Log Kp
	cal/(mole °K)				kcal/mole		
298	87.76	174.74	174.74	0.00	-85.43	27.56	-20.204
300	88.27	175.29	174.75	0.17	-85.56	28.26	-20.584
400	114.60	204.34	178.53	10.33	-91.39	67.12	-36.673
500	138.60	232.57	186.53	23.02	-95.99	107.31	-46.901
600	158.50	259.64	196.48	37.90	-99.43	148.29	-54.011
700	175.50	285.39	207.36	54.63	-101.83	189.80	-59.254
800	189.70	309.77	218.65	72.90	-103.38	231.54	-63.250
900	201.80	332.84	230.07	92.50	-104.15	273.48	-66.406
1000	212.00	354.64	241.45	113.20	-104.24	315.47	-68.944

The selected values of Rossini, Pitzer, Arnett, Braun, and Pimentel (1248) are used. $Tm = 278.95°K$ and $Tb = 586.35°K$.

No. 391 1-Cyclohexyldodecane $C_{18}H_{36}$ (Ideal Gas State) Mol Wt 252.468

T°K	Cp°	S°	$-(G°-H°_{298})/T$	$H°-H°_{298}$	$\Delta Hf°$	$\Delta Gf°$	Log Kp
	cal/(mole °K)				kcal/mole		
298	93.22	184.05	184.05	0.00	-90.36	29.57	-21.674
300	93.77	184.63	184.06	0.18	-90.49	30.30	-22.076
400	121.60	215.48	188.08	10.97	-96.64	71.55	-39.092
500	146.80	245.40	196.57	24.42	-101.49	114.18	-49.907
600	167.90	274.08	207.11	40.18	-105.13	157.66	-57.424
700	185.80	301.34	218.65	57.89	-107.67	201.69	-62.968
800	200.70	327.15	230.61	77.24	-109.32	245.97	-67.194
900	213.60	351.55	242.71	97.97	-110.15	290.47	-70.531
1000	224.40	374.64	254.76	119.89	-110.25	335.02	-73.215

The selected values of Rossini, Pitzer, Arnett, Braun, and Pimentel (1248) are used. $Tm = 285.65°K$ and $Tb = 601.1°K$.

No. 392 1-Cyclohexyltridecane $C_{19}H_{38}$ (Ideal Gas State) Mol Wt 266.494

The selected values of Rossini, Pitzer, Arnett, Braun, and Pimentel (1248) are used. $Tm = 291.65°K$ and $Tb = 615.1°K$.

No. 392 1-Cyclohexyltridecane $C_{19}H_{38}$ (Ideal Gas State) Mol Wt 266.494

$T°K$	$Cp°$	cal/(mole °K)			kcal/mole		Log Kp
		$S°$	$-(G°-H°_{298})/T$	$H°-H°_{298}$	$\Delta Hf°$	$\Delta Gf°$	
298	98.69	193.36	193.36	0.00	-95.28	31.59	-23.151
300	99.26	193.98	193.37	0.19	-95.42	32.36	-23.575
400	128.50	226.60	197.62	11.60	-101.89	75.99	-41.516
500	155.10	258.22	206.59	25.82	-106.99	121.07	-52.918
600	177.20	288.50	217.74	42.46	-110.83	167.04	-60.843
700	196.10	317.27	229.92	61.15	-113.52	213.61	-66.687
800	211.80	344.51	242.56	81.56	-115.26	260.43	-71.143
900	225.30	370.25	255.33	103.43	-116.15	307.47	-74.661
1000	236.70	394.60	268.05	126.55	-116.27	354.59	-77.491

No. 393 1-Cyclohexyltetradecane $C_{20}H_{40}$ (Ideal Gas State) Mol Wt 280.520

$T°K$	$Cp°$	cal/(mole °K)			kcal/mole		Log Kp
		$S°$	$-(G°-H°_{298})/T$	$H°-H°_{298}$	$\Delta Hf°$	$\Delta Gf°$	
298	104.16	202.67	202.67	0.00	-100.21	33.59	-24.621
300	104.75	203.32	202.68	0.20	-100.36	34.41	-25.067
400	135.40	237.72	207.16	12.23	-107.15	80.42	-43.935
500	163.40	271.03	216.62	27.21	-112.51	127.95	-55.925
600	186.50	302.91	228.37	44.73	-116.54	176.42	-64.258
700	206.40	333.20	241.20	64.40	-119.37	225.51	-70.404
800	222.90	361.86	254.51	85.89	-121.21	274.88	-75.089
900	237.00	388.95	267.96	108.90	-122.15	324.47	-78.789
1000	249.00	414.56	281.35	133.22	-122.30	374.15	-81.766

The selected values of Rossini, Pitzer, Arnett, Braun, and Pimentel (1248) are used. $Tm = 297.15°K$ and $Tb = 628°K$.

No. 394 1-Cyclohexylpentadecane $C_{21}H_{42}$ (Ideal Gas State) Mol Wt 294.546

$T°K$	$Cp°$	cal/(mole °K)			kcal/mole		Log Kp
		$S°$	$-(G°-H°_{298})/T$	$H°-H°_{298}$	$\Delta Hf°$	$\Delta Gf°$	
298	109.62	211.98	211.98	0.00	-105.14	35.60	-26.092
300	110.25	212.67	211.99	0.21	-105.29	36.46	-26.559
400	142.40	248.86	216.70	12.87	-112.40	84.84	-46.354
500	171.60	283.86	226.66	28.61	-118.01	134.83	-58.930
600	195.90	317.35	239.00	47.01	-122.25	185.79	-67.671
700	216.70	349.15	252.49	67.67	-125.21	237.41	-74.118
800	233.90	379.24	266.47	90.22	-127.15	289.31	-79.032
900	248.80	407.67	280.59	114.37	-128.15	341.46	-82.914
1000	261.40	434.56	294.66	139.90	-128.31	393.69	-86.037

The selected values of Rossini, Pitzer, Arnett, Braun, and Pimentel (1248) are used. $Tm = 302.15°K$ and $Tb = 641°K$.

No. 395 1-Cyclohexylhexadecane $C_{22}H_{44}$ (Ideal Gas State) Mol Wt 308.572

$T°K$	$Cp°$	$S°$	$-(G°-H°_{298})/T$	$H°-H°_{298}$	$\Delta Hf°$	$\Delta Gf°$	$\text{Log } Kp$
		cal/(mole °K)			kcal/mole		
298	115.09	221.29	221.29	0.00	-110.06	37.61	-27.569
300	115.74	222.01	221.30	0.22	-110.22	38.52	-28.058
400	149.30	259.97	226.25	13.50	-117.64	89.28	-48.778
500	179.80	296.66	236.68	29.99	-123.52	141.72	-61.942
600	205.20	331.74	249.63	49.27	-127.96	195.18	-71.090
700	226.90	365.05	263.76	70.91	-131.07	249.32	-77.838
800	245.00	396.56	278.41	94.53	-133.11	303.77	-82.983
900	260.60	426.34	293.21	119.83	-134.16	358.48	-87.047
1000	273.70	454.50	307.94	146.56	-134.33	413.27	-90.316

The selected values of Rossini, Pitzer, Arnett, Braun, and Pimentel (1248) are used. $Tm = 306.75°K$ and $Tb = 653°K$.

ALKYLBENZENE IDEAL GAS TABLES

No. 396 Benzene C_6H_6 (Ideal Gas State) Mol Wt 78.108

$T°K$	$Cp°$	$S°$	$-(G°-H°_{298})/T$	$H°-H°_{298}$	$\Delta Hf°$	$\Delta Gf°$	$\text{Log } Kp$
		cal/(mole °K)			kcal/mole		
298	19.52	64.34	64.34	0.00	19.82	30.99	-22.714
300	19.65	64.47	64.35	0.04	19.79	31.06	-22.623
400	26.74	71.11	65.20	2.37	18.56	35.01	-19.126
500	32.80	77.75	67.05	5.36	17.54	39.24	-17.152
600	37.74	84.18	69.38	8.89	16.71	43.66	-15.901
700	41.75	90.31	71.93	12.87	16.04	48.21	-15.051
800	45.06	96.11	74.60	17.22	15.51	52.84	-14.434
900	47.83	101.58	77.29	21.86	15.10	57.53	-13.970
1000	50.16	106.74	79.98	26.77	14.82	62.27	-13.608

The measurements of Oliver, Eaton, and Huffman (1085) and of Pitzer and Scott (1166), with the selected values of Rossini, Pitzer, Arnett, Braun, and Pimentel (1248), are adopted. $Tm = 278.68°K$, $\Delta Hm° = 2.351$ kcal/mole, and $Tb = 353.25°K$, where $\Delta Hv = 7.353$ kcal/mole. Ambrose, Cox, and Townsend (13) measured $Tc = 562.0°K$, while Kobe and Lynn (780) reported $Pc = 48.6$ atm and $dc = 0.300$ g/cc.

No. 397 Toluene C_7H_8 (Ideal Gas State) Mol Wt 92.134

Scott, Guthrie, Messerly, Todd, Berg, Hossenlopp, and McCullough (1314) measured low-temperature thermal data, including $Ttp = 178.15°K$, $\Delta Hm° = 1.586$ kcal/mole, and $Tb = 383.77°K$, where $\Delta Hv = 7.933$ kcal/

No. 397 Toluene C_7H_8 (Ideal Gas State) Mol Wt 92.134

$T°K$	$Cp°$	$S°$	$-(G°-H°_{298})/T$	$H°-H°_{298}$	$\Delta Hf°$	$\Delta Gf°$	$\text{Log } Kp$
		cal/(mole °K)			kcal/mole		
298	24.77	76.64	76.64	0.00	11.95	29.16	-21.376
300	24.94	76.80	76.65	0.05	11.92	29.27	-21.320
400	33.48	85.17	77.73	2.98	10.34	35.30	-19.287
500	40.98	93.47	80.05	6.71	9.05	41.70	-18.225
600	47.20	101.51	82.96	11.13	8.02	48.32	-17.599
700	52.33	109.18	86.16	16.12	7.24	55.11	-17.205
800	56.61	116.45	89.50	21.57	6.65	61.98	-16.931
900	60.23	123.33	92.88	27.41	6.24	68.93	-16.736
1000	63.32	129.85	96.25	33.60	6.01	75.91	-16.589

mole. Scott et al. also determined the vapor heat capacity and reported new values for the two lowest vibrational frequencies. Assumption of free rotation for the methyl group then gave calculated functions (adopted here) in good agreement with experimental values. Scott et al. give references to earlier experimental work on toluene.

The enthalpy of formation is that selected by Rossini, Pitzer, Arnett, Braun, and Pimentel (1248). Ambrose, Cox, and Townsend (13) measured $Tc = 591.7°K$; Kobe and Lynn (780) selected $Pc = 41.6$ atm and $dc = 0.29$ g/cc.

No. 398 Ethylbenzene C_8H_{10} (Ideal Gas State) Mol Wt 106.160

$T°K$	$Cp°$	$S°$	$-(G°-H°_{298})/T$	$H°-H°_{298}$	$\Delta Hf°$	$\Delta Gf°$	$\text{Log } Kp$
		cal/(mole °K)			kcal/mole		
298	30.69	86.15	86.15	0.00	7.12	31.21	-22.875
300	30.88	86.35	86.16	0.06	7.08	31.35	-22.841
400	40.76	96.61	87.49	3.65	5.23	39.74	-21.713
500	49.35	106.66	90.32	8.17	3.71	48.55	-21.222
600	56.44	116.30	93.86	13.47	2.48	57.63	-20.992
700	62.28	125.45	97.72	19.42	1.53	66.91	-20.890
800	67.15	134.10	101.73	25.89	0.80	76.29	-20.842
900	71.27	142.25	105.79	32.82	0.27	85.77	-20.827
1000	74.77	149.94	109.82	40.13	-0.05	95.30	-20.826

The selected values of Rossini, Pitzer, Arnett, Braun, and Pimentel (1248) are used. $Tm = 178.18°K$, $\Delta Hm° = 2.190$ kcal/mole, and $Tb = 409.34°K$, where $\Delta Hv = 8.600$ kcal/mole. Kobe and Lynn (780) indicated $Tc = 619.5°K$, $Pc = 38$ atm, and $dc = 0.29$ g/cc.

No. 399 m-Xylene C_8H_{10} (Ideal Gas State) Mol Wt 106.160

The values presented by Rossini, Pitzer, Arnett, Braun, and Pimentel (1248) are selected. $Tm = 225.28°K$, $\Delta Hm° = 2.765$ kcal/mole, and $Tb =$

No. 399 m-Xylene C$_8$H$_{10}$ (Ideal Gas State) Mol Wt 106.160

T°K	Cp°	S°	$-(G°-H°_{298})/T$	$H°-H°_{298}$	$\Delta Hf°$	$\Delta Gf°$	Log Kp
	cal/(mole °K)				kcal/mole		
298	30.49	85.49	85.49	0.00	4.12	28.41	-20.821
300	30.66	85.68	85.50	0.06	4.08	28.55	-20.800
400	40.03	95.81	86.81	3.60	2.18	37.01	-20.222
500	48.43	105.67	89.60	8.04	0.57	45.91	-20.068
600	55.51	115.14	93.08	13.24	-0.75	55.10	-20.069
700	61.43	124.16	96.88	19.10	-1.79	64.50	-20.138
800	66.41	132.69	100.83	25.50	-2.60	74.02	-20.220
900	70.63	140.77	104.82	32.36	-3.19	83.64	-20.310
1000	74.23	148.40	108.80	39.60	-3.58	93.32	-20.394

412.25°K, where $\Delta Hv = 8.70$ kcal/mole. Kobe and Lynn (780) reported $Tc = 619°K, Pc = 36$ atm, and $dc = 0.27$ g/cc.

No. 400 o-Xylene C$_8$H$_{10}$ (Ideal Gas State) Mol Wt 106.160

T°K	Cp°	S°	$-(G°-H°_{298})/T$	$H°-H°_{298}$	$\Delta Hf°$	$\Delta Gf°$	Log Kp
	cal/(mole °K)				kcal/mole		
298	31.85	84.31	84.31	0.00	4.54	29.18	-21.386
300	32.02	84.51	84.32	0.06	4.50	29.33	-21.364
400	41.03	94.98	85.68	3.72	2.72	37.89	-20.699
500	49.11	105.03	88.55	8.24	1.19	46.86	-20.481
600	55.98	114.60	92.10	13.50	-0.07	56.10	-20.435
700	61.76	123.68	95.97	19.40	-1.07	65.56	-20.467
800	66.64	132.25	99.98	25.83	-1.85	75.12	-20.521
900	70.80	140.35	104.02	32.70	-2.43	84.78	-20.587
1000	74.35	148.00	108.04	39.97	-2.79	94.50	-20.653

The values selected by Rossini, Pitzer, Arnett, Braun, and Pimentel (1248) are employed. $Tm = 247.97°K$, $\Delta Hm° = 3.250$ kcal/mole, and $Tb = 417.56°K$, where $\Delta Hv = 8.800$ kcal/mole. Kobe and Lynn (780) indicated $Tc = 631.5°K, Pc = 36.9$ atm, and $dc = 0.28$ g/cc.

No. 401 p-Xylene C$_8$H$_{10}$ (Ideal Gas State) Mol Wt 106.160

T°K	Cp°	S°	$-(G°-H°_{298})/T$	$H°-H°_{298}$	$\Delta Hf°$	$\Delta Gf°$	Log Kp
	cal/(mole °K)				kcal/mole		
298	30.32	84.23	84.23	0.00	4.29	28.95	-21.221
300	30.49	84.42	84.24	0.06	4.25	29.10	-21.199
400	39.70	94.48	85.54	3.58	2.32	37.69	-20.592
500	48.06	104.26	88.31	7.98	0.68	46.73	-20.424
600	55.16	113.66	91.76	13.15	-0.67	56.06	-20.419
700	61.12	122.63	95.54	18.97	-1.75	65.61	-20.485
800	66.14	131.12	99.46	25.34	-2.59	75.29	-20.566
900	70.39	139.17	103.43	32.17	-3.21	85.06	-20.655
1000	74.02	146.78	107.38	39.40	-3.61	94.90	-20.740

The values selected by Rossini, Pitzer, Arnett, Braun, and Pimentel (1248) are presented. $Tm = 286.41°K$, $\Delta Hm° = 4.090$ kcal/mole, and $Tb = 411.50°K$, where $\Delta Hv = 8.62$ kcal/mole. Kobe and Lynn (780) listed $Tc = 618°K, Pc = 35$ atm, and $dc = 0.29$ g/cc.

No. 402 Propylbenzene C_9H_{12} (Ideal Gas State) Mol Wt 120.186

	cal/(mole °K)				kcal/mole		
$T°K$	$Cp°$	$S°$	$-(G°-H°_{298})/T$	$H°-H°_{298}$	$\Delta Hf°$	$\Delta Gf°$	Log Kp
298	36.41	95.76	95.76	0.00	1.87	32.80	-24.045
300	36.62	95.99	95.77	0.07	1.82	32.99	-24.034
400	47.82	108.09	97.34	4.31	-0.33	43.73	-23.890
500	57.65	119.85	100.67	9.59	-2.10	54.95	-24.018
600	65.86	131.11	104.82	15.78	-3.51	66.49	-24.218
700	72.63	141.78	109.34	22.71	-4.60	78.26	-24.432
800	78.30	151.86	114.03	30.27	-5.42	90.14	-24.624
900	83.10	161.37	118.77	38.34	-6.00	102.13	-24.799
1000	87.16	170.34	123.48	46.86	-6.33	114.17	-24.951

Messerly, Todd, and Finke (994) measured low-temperature thermal properties and reported $Ttp = 173.59°K$, $\Delta Hm° = 2.215$ kcal/mole, $S°_{298}(l) = 68.78$ cal/(mole °K), and $S°_{298}(g) = 95.09$ cal/(mole °K). Thermodynamic functions and enthalpy of formation are the selections of Rossini, Pitzer, Arnett, Braun, and Pimentel (1248), who also listed $Tb = 432.37°K$, at which $\Delta Hv = 9.14$ kcal/mole. The calorimetric entropy is in reasonable agreement with the correlated selection of Rossini et al., which is retained here for the sake of consistency.

No. 403 Cumene C_9H_{12} (Ideal Gas State) Mol Wt 120.186

	cal/(mole °K)				kcal/mole		
$T°K$	$Cp°$	$S°$	$-(G°-H°_{298})/T$	$H°-H°_{298}$	$\Delta Hf°$	$\Delta Gf°$	Log Kp
298	36.26	92.87	92.87	0.00	0.94	32.74	-23.995
300	36.47	93.10	92.88	0.07	0.89	32.93	-23.988
400	48.00	105.21	94.45	4.31	-1.26	43.95	-24.014
500	57.90	117.01	97.79	9.62	-3.00	55.46	-24.242
600	66.20	128.33	101.94	15.84	-4.39	67.28	-24.507
700	72.90	139.05	106.49	22.80	-5.44	79.33	-24.766
800	78.60	149.17	111.19	30.38	-6.24	91.48	-24.990
900	83.30	158.70	115.95	38.48	-6.79	103.74	-25.189
1000	87.30	167.69	120.68	47.02	-7.10	116.05	-25.361

The selected values of Rossini, Pitzer, Arnett, Braun, and Pimentel (1248) are used. $Tm = 177.12°K$, $\Delta Hm° = 1.7$ kcal/mole, and $Tb = 425.54°K$, where $\Delta Hv = 8.97$ kcal/mole. Kobe and Lynn (780) indicated $Tc = 635.8°K$ and $Pc = 32$ atm.

No. 404 m-Ethyltoluene C_9H_{12} (Ideal Gas State) Mol Wt 120.186

T°K	$Cp°$	$S°$	$-(G°-H°_{298})/T$	$H°-H°_{298}$	$\Delta Hf°$	$\Delta Gf°$	Log Kp
	cal/(mole °K)				kcal/mole		
298	36.38	96.60	96.60	0.00	-0.46	30.22	-22.154
300	36.59	96.83	96.61	0.07	-0.51	30.41	-22.153
400	47.50	108.88	98.17	4.29	-2.68	41.06	-22.434
500	57.20	120.55	101.49	9.53	-4.49	52.21	-22.821
600	65.40	131.73	105.61	15.68	-5.95	63.69	-23.197
700	72.10	142.32	110.10	22.56	-7.08	75.40	-23.539
800	77.80	152.33	114.76	30.06	-7.96	87.23	-23.828
900	82.70	161.79	119.47	38.09	-8.58	99.17	-24.081
1000	86.80	170.72	124.15	46.58	-8.95	111.18	-24.296

These values are those selected by Rossini, Pitzer, Arnett, Braun, and Pimentel (1248). $Tt = 176.18°K$, $\Delta Ht° = 1.79$ kcal/mole, $Tm = 177.58°K$, $\Delta Hm° = 1.82$ kcal/mole, and $Tb = 434.45°K$, where $\Delta Hv = 9.21$ kcal/mole.

No. 405 o-Ethyltoluene C_9H_{12} (Ideal Gas State) Mol Wt 120.186

T°K	$Cp°$	$S°$	$-(G°-H°_{298})/T$	$H°-H°_{298}$	$\Delta Hf°$	$\Delta Gf°$	Log Kp
	cal/(mole °K)				kcal/mole		
298	37.74	95.42	95.42	0.00	0.29	31.33	-22.961
300	37.94	95.66	95.43	0.08	0.25	31.51	-22.958
400	48.50	108.05	97.04	4.41	-1.81	42.27	-23.091
500	57.90	119.91	100.44	9.74	-3.53	53.49	-23.379
600	65.80	131.18	104.63	15.93	-4.94	65.02	-23.682
700	72.50	141.84	109.19	22.86	-6.03	76.78	-23.971
800	78.10	151.90	113.91	30.40	-6.87	88.66	-24.219
900	82.80	161.37	118.66	38.45	-7.48	100.64	-24.439
1000	86.90	170.32	123.38	46.94	-7.84	112.69	-24.627

The values selected by Rossini, Pitzer, Arnett, Braun, and Pimentel (1248) are employed. $Tt = 186.59°K$, $\Delta Ht° = 2.28$ kcal/mole, $Tm = 192.32°K$, $\Delta Hm° = 2.54$ kcal/mole, and $Tb = 438.30°K$, where $\Delta Hv = 9.29$ kcal/mole.

No. 406 p-Ethyltoluene C_9H_{12} (Ideal Gas State) Mol Wt 120.186

T°K	$Cp°$	$S°$	$-(G°-H°_{298})/T$	$H°-H°_{298}$	$\Delta Hf°$	$\Delta Gf°$	Log Kp
	cal/(mole °K)				kcal/mole		
298	36.22	95.34	95.34	0.00	-0.78	30.28	-22.195
300	36.42	95.57	95.35	0.07	-0.83	30.47	-22.196
400	47.20	107.55	96.91	4.26	-3.02	41.25	-22.537
500	56.90	119.15	100.20	9.48	-4.86	52.54	-22.963
600	65.00	130.26	104.29	15.59	-6.36	64.15	-23.367
700	71.80	140.81	108.76	22.43	-7.53	76.01	-23.731
800	77.60	150.78	113.40	29.91	-8.43	88.00	-24.039
900	82.40	160.21	118.08	37.92	-9.07	100.10	-24.306
1000	86.60	169.11	122.74	46.37	-9.47	112.26	-24.533

The values selected by Rossini, Pitzer, Arnett, Braun, and Pimentel (1248) are used. $Tm = 210.80°K$, $\Delta Hm° = 3.04$ kcal/mole, and $Tb = 435.14°K$, where $\Delta Hv = 9.18$ kcal/mole.

No. 407 1,2,3-Trimethylbenzene C_9H_{12} (Ideal Gas State) Mol Wt 120.186

	cal/(mole °K)				kcal/mole		
$T°K$	$Cp°$	$S°$	$-(G°-H°_{298})/T$	$H°-H°_{298}$	$\Delta Hf°$	$\Delta Gf°$	Log Kp
298	36.85	91.98	91.98	0.00	-2.29	29.77	-21.822
300	37.04	92.21	91.99	0.07	-2.34	29.97	-21.830
400	46.90	104.24	93.56	4.28	-4.52	41.08	-22.444
500	56.10	115.71	96.85	9.44	-6.42	52.70	-23.036
600	64.00	126.66	100.91	15.45	-8.00	64.67	-23.555
700	70.90	137.05	105.34	22.20	-9.27	76.90	-24.008
800	76.70	146.91	109.93	29.59	-10.26	89.27	-24.385
900	81.60	156.23	114.56	37.51	-10.99	101.76	-24.709
1000	85.90	165.06	119.17	45.89	-11.46	114.32	-24.984

The measurements of Taylor, Johnson, and Kilpatrick (1475) are merged with the selected values of Rossini, Pitzer, Arnett, Braun, and Pimentel (1248). $Tt = 218.68°K$, $\Delta Ht° = 0.157$ kcal/mole, $Tt = 230.25°K$, $\Delta Ht° = 0.320$ kcal/mole, $Tm = 247.78°K$, $\Delta Hm° = 1.953$ kcal/mole, and $Tb = 449.23°K$, where $\Delta Hv = 9.57$ kcal/mole. Kobe and Lynn (780) reported $Tc = 668°K$, $Pc = 31$ atm, and $dc = 0.28$ g/cc.

No. 408 1,2,4-Trimethylbenzene C_9H_{12} (Ideal Gas State) Mol Wt 120.186

	cal/(mole °K)				kcal/mole		
$T°K$	$Cp°$	$S°$	$-(G°-H°_{298})/T$	$H°-H°_{298}$	$\Delta Hf°$	$\Delta Gf°$	Log Kp
298	36.81	94.59	94.59	0.00	-3.33	27.95	-20.489
300	36.99	94.82	94.60	0.07	-3.38	28.14	-20.502
400	46.96	106.85	96.17	4.28	-5.56	39.00	-21.305
500	56.26	118.35	99.46	9.45	-7.44	50.36	-22.010
600	64.29	129.33	103.53	15.49	-9.01	62.06	-22.604
700	71.12	139.77	107.97	22.26	-10.25	74.02	-23.108
800	76.93	149.65	112.57	29.67	-11.21	86.11	-23.523
900	81.87	159.01	117.21	37.62	-11.92	98.33	-23.876
1000	86.10	167.86	121.84	46.02	-12.37	110.61	-24.174

The low-temperature measurements of Putnam and Kilpatrick (1208) indicated two melting points at 229.32°K and at about 224.2°K, but the only solid phase giving reproducible heat capacities was the one with the higher melting point. Hastings and Nicholson (577) restudied the spectrum and calculated the ideal gaseous thermodynamic functions. The remainder of the properties are those selected by Rossini, Pitzer, Arnett, Braun, and Pimentel (1248). $Tm = 229.35°K$, $\Delta Hm° = 3.152$ kcal/mole, and $Tb = 442.50°K$, where $\Delta Hv = 9.38$ kcal/mole. Kobe and Lynn (780) listed $Tc = 654.3°K$ and $Pc = 33$ atm.

No. 409 Mesitylene C_9H_{12} (Ideal Gas State) Mol Wt 120.186

$T°K$	$Cp°$	cal/(mole °K)			kcal/mole		Log Kp
		$S°$	$-(G°-H°_{298})/T$	$H°-H°_{298}$	$\Delta Hf°$	$\Delta Gf°$	
298	35.91	92.09	92.09	0.00	-3.84	28.19	-20.662
300	36.10	92.32	92.10	0.07	-3.89	28.38	-20.677
400	46.41	104.13	93.64	4.20	-6.14	39.50	-21.579
500	55.92	115.54	96.88	9.33	-8.07	51.14	-22.351
600	64.08	126.47	100.91	15.34	-9.66	63.12	-22.991
700	70.99	136.88	105.31	22.10	-10.92	75.37	-23.530
800	76.84	146.75	109.88	29.50	-11.90	87.75	-23.971
900	81.81	156.10	114.50	37.44	-12.61	100.26	-24.345
1000	86.07	164.94	119.11	45.84	-13.06	112.84	-24.659

The low-temperature measurements of Taylor and Kilpatrick (1476) are joined with the selected values of Rossini, Pitzer, Arnett, Braun, and Pimentel (1248). $Tt = 221.44°K$, $\Delta Ht° = 1.892$ kcal/mole, $Tt = 223.32°K$, $\Delta Ht° = 1.932$ kcal/mole, $Tm = 228.43°K$, $\Delta Hm° = 2.274$ kcal/mole, and $Tb = 437.87°K$, at which $\Delta Hv = 9.33$ kcal/mole. Ambrose, Cox, and Townsend (13) measured $Tc = 637.3°K$, while Kobe and Lynn (780) gave $Pc = 33$ atm and $dc = 0.28$ g/cc.

No. 410 Butylbenzene $C_{10}H_{14}$ (Ideal Gas State) Mol Wt 134.212

$T°K$	$Cp°$	cal/(mole °K)			kcal/mole		Log Kp
		$S°$	$-(G°-H°_{298})/T$	$H°-H°_{298}$	$\Delta Hf°$	$\Delta Gf°$	
298	41.85	105.04	105.04	0.00	-3.30	34.58	-25.346
300	42.09	105.30	105.05	0.08	-3.35	34.81	-25.358
400	54.75	119.18	106.85	4.94	-5.83	47.93	-26.184
500	65.89	132.63	110.67	10.98	-7.85	61.61	-26.927
600	75.20	145.49	115.41	18.05	-9.46	75.64	-27.552
700	82.91	157.68	120.59	25.97	-10.69	89.94	-28.080
800	89.37	169.18	125.95	34.59	-11.61	104.37	-28.511
900	94.84	180.03	131.36	43.81	-12.25	118.92	-28.875
1000	99.49	190.27	136.75	53.53	-12.60	133.52	-29.180

Low-temperature thermal data were determined by Messerly, Todd, and Finke (994), including $Ttp = 185.30°K$, $\Delta Hm° = 2.682$ kcal/mole, $S°_{298}(l) = 76.77$ cal/(mole °K), and $S°_{298}(g) = 104.65$ cal/(mole °K). Thermodynamic functions and enthalpy of formation were selected by Rossini, Pitzer, Arnett, Braun, and Pimentel (1248), who also listed $Tb = 456.42°K$. The calorimetric entropy is in reasonable agreement with the correlated selection of Rossini et al., which is retained here for the sake of consistency. Ambrose, Cox, and Townsend (13) measured $Tc = 660.4°K$.

No. 411 m-Diethylbenzene $C_{10}H_{14}$ (Ideal Gas State) Mol Wt 134.212

T°K	$Cp°$	$S°$	$-(G°-H°_{298})/T$	$H°-H°_{298}$	$\Delta Hf°$	$\Delta Gf°$	Log Kp
	cal/(mole °K)				kcal/mole		
298	42.27	104.99	104.99	0.00	-5.22	32.67	-23.949
300	42.58	105.26	105.00	0.08	-5.27	32.91	-23.970
400	55.01	119.24	106.82	4.97	-7.71	46.02	-25.143
500	66.03	132.74	110.66	11.04	-9.71	59.69	-26.090
600	75.19	145.61	115.42	18.12	-11.32	73.72	-26.850
700	82.89	157.79	120.61	26.03	-12.55	88.01	-27.475
800	89.31	169.29	125.99	34.65	-13.47	102.42	-27.979
900	94.69	180.13	131.41	43.85	-14.12	116.96	-28.400
1000	99.37	190.35	136.79	53.56	-14.49	131.55	-28.750

By a correlation method Prosen, Johnson, and Rossini (1195) calculated $\Delta Hf°_{298}(g) = -5.22$ kcal/mole. The 9.75 cal/(mole °K) entropy increase for conversion of an alkyl methyl group to an alkyl ethyl group was applied to the entropy of gaseous m-xylene, to give $S°_{298}(g) = 104.99$ cal/(mole °K). The gaseous heat capacities were calculated by adding the nine increments for conversion of alkyl methyl groups to alkyl ethyl groups to the gaseous heat capacities of m-xylene.

No. 412 o-Diethylbenzene $C_{10}H_{14}$ (Ideal Gas State) Mol Wt 134.212

T°K	$Cp°$	$S°$	$-(G°-H°_{298})/T$	$H°-H°_{298}$	$\Delta Hf°$	$\Delta Gf°$	Log Kp
	cal/(mole °K)				kcal/mole		
298	43.63	103.81	103.81	0.00	-4.53	33.72	-24.713
300	43.94	104.09	103.82	0.09	-4.58	33.95	-24.731
400	56.01	118.41	105.69	5.09	-6.90	47.16	-25.767
500	66.71	132.09	109.61	11.25	-8.82	60.91	-26.621
600	75.66	145.07	114.45	18.38	-10.37	74.99	-27.314
700	83.22	157.31	119.71	26.33	-11.56	89.33	-27.889
800	89.54	168.85	125.14	34.98	-12.46	103.79	-28.353
900	94.86	179.71	130.60	44.20	-13.08	118.37	-28.743
1000	99.49	189.95	136.03	53.93	-13.43	133.01	-29.068

Prosen, Johnson, and Rossini (1195) calculated $\Delta Hf°_{298}(g) = -4.53$ kcal/mole by a correlation method. The standard entropy increase for conversion of an alkyl methyl group to an alkyl ethyl group is 9.75 cal/(mole °K) and was applied to the entropy of gaseous o-xylene, giving $S°_{298}(g) = 103.81$ cal/(mole °K). Heat capacity increments for the conversion of alkyl methyl groups to alkyl ethyl groups were applied to the nine heat capacities of o-xylene to give the gaseous heat capacity.

No. 413 p-Diethylbenzene $C_{10}H_{14}$ (Ideal Gas State) Mol Wt 134.212

$T°K$	$Cp°$	$S°$	$-(G°-H°_{298})/T$	$H°-H°_{298}$	$\Delta Hf°$	$\Delta Gf°$	Log Kp
		cal/(mole °K)			kcal/mole		
298	42.10	103.73	103.73	0.00	-5.32	32.95	-24.152
300	42.41	104.00	103.74	0.08	-5.37	33.18	-24.173
400	54.68	117.91	105.55	4.95	-7.84	46.43	-25.365
500	65.66	131.32	109.37	10.98	-9.88	60.24	-26.328
600	74.84	144.13	114.11	18.02	-11.52	74.41	-27.102
700	82.58	156.26	119.27	25.90	-12.78	88.85	-27.738
800	89.04	167.72	124.62	34.49	-13.73	103.42	-28.251
900	94.45	178.53	130.01	43.67	-14.40	118.11	-28.680
1000	99.16	188.73	135.38	53.36	-14.79	132.87	-29.037

A value of $\Delta Hf°_{298}(g) = -5.32$ kcal/mole was calculated by Prosen, Johnson, and Rossini (1195) by a correlation method. To the entropy of gaseous p-xylene was added the 9.75 cal/(mole °K) entropy increase for conversion of an alkyl methyl group to an alkyl ethyl group, to give $S°_{298}(g) = 103.73$ cal/(mole °K). To the gaseous heat capacities of p-xylene were added the nine increments for conversion of alkyl methyl groups to alkyl ethyl groups, giving the gaseous heat capacities of p-diethylbenzene.

No. 414 1,2,3,4-Tetramethylbenzene $C_{10}H_{14}$ (Ideal Gas State) Mol Wt 134.212

$T°K$	$Cp°$	$S°$	$-(G°-H°_{298})/T$	$H°-H°_{298}$	$\Delta Hf°$	$\Delta Gf°$	Log Kp
		cal/(mole °K)			kcal/mole		
298	45.31	99.55	99.55	0.00	-10.02	29.50	-21.620
300	45.50	99.84	99.56	0.09	-10.07	29.74	-21.662
400	56.81	114.50	101.48	5.21	-12.27	43.36	-23.687
500	67.01	128.30	105.48	11.42	-14.14	57.48	-25.125
600	75.68	141.30	110.38	18.56	-15.67	71.95	-26.205
700	83.13	153.54	115.68	26.51	-16.87	86.66	-27.055
800	89.42	165.07	121.14	35.15	-17.78	101.50	-27.727
900	94.82	175.92	126.63	44.37	-18.41	116.46	-28.278
1000	99.47	186.15	132.07	54.09	-18.76	131.48	-28.733

Banse and Parks (71) measured the enthalpy of combustion and reported $\Delta Hf°_{298}(l) = -26.2$ kcal/mole or $\Delta Hf°_{298}(g) = -10.02$ kcal/mole. Huffman, Parks, and Barmore (655) measured the heat capacity from 90° to 298°K and reported $S°_{298}(l) = 69.45$ cal/(mole °K) and $\Delta Hm° = 2.68$ kcal/mole, while Rossini, Pitzer, Arnett, Braun, and Pimentel (1248) selected $Tm = 266.90°K$ and $Tb = 478.19°K$. Hastings and Nicholson (577) calculated the gaseous thermodynamic functions from fundamental frequencies.

No. 415 1,2,3,5-Tetramethylbenzene $C_{10}H_{14}$ (Ideal Gas State)   Mol Wt 134.212

	cal/(mole °K)				kcal/mole		
$T°K$	$Cp°$	$S°$	$-(G°-H°_{298})/T$	$H°-H°_{298}$	$\Delta Hf°$	$\Delta Gf°$	Log Kp
298	44.39	100.99	100.99	0.00	-10.71	28.38	-20.800
300	44.57	101.27	101.00	0.09	-10.76	28.62	-20.845
400	55.76	115.65	102.88	5.11	-13.06	42.10	-23.004
500	66.03	129.22	106.80	11.21	-15.03	56.13	-24.533
600	74.81	142.05	111.62	18.27	-16.66	70.51	-25.682
700	82.39	154.17	116.84	26.14	-17.94	85.16	-26.586
800	88.79	165.60	122.23	34.70	-18.91	99.94	-27.301
900	94.27	176.38	127.65	43.86	-19.60	114.85	-27.887
1000	99.01	186.56	133.04	53.53	-20.01	129.82	-28.371

Banse and Parks (71) measured the enthalpy of combustion, leading to $\Delta Hf°_{298}(g) = -10.71$ kcal/mole. The low-temperature measurements of Huffman, Parks, and Barmore (655), the selected values of Rossini, Pitzer, Arnett, Braun, and Pimentel (1248), and the calculated gaseous thermodynamic functions of Hastings and Nicholson (577) are used. $Tm = 249.46°K$, $\Delta Hm° = 2.8$ kcal/mole, and $Tb = 471.15°K$.

No. 416 1,2,4,5-Tetramethylbenzene $C_{10}H_{14}$ (Ideal Gas State)   Mol Wt 134.212

	cal/(mole °K)				kcal/mole		
$T°K$	$Cp°$	$S°$	$-(G°-H°_{298})/T$	$H°-H°_{298}$	$\Delta Hf°$	$\Delta Gf°$	Log Kp
298	44.58	100.03	100.03	0.00	-10.82	28.55	-20.929
300	44.77	100.31	100.04	0.09	-10.87	28.79	-20.975
400	55.50	114.67	101.92	5.11	-13.18	42.38	-23.153
500	65.62	128.17	105.83	11.17	-15.18	56.50	-24.697
600	74.38	140.93	110.63	18.18	-16.85	70.99	-25.858
700	81.97	152.98	115.83	26.01	-18.17	85.76	-26.773
800	88.41	164.35	121.19	34.54	-19.19	100.66	-27.497
900	93.94	175.09	126.59	43.66	-19.91	115.69	-28.093
1000	98.71	185.25	131.95	53.30	-20.35	130.80	-28.584

Banse and Parks (71) measured the enthalpy of combustion, leading to $\Delta Hf°_{298}(g) = -10.82$ kcal/mole. The low-temperature measurements of Huffman, Parks, and Barmore (655), the selected values of Rossini, Pitzer, Arnett, Braun, and Pimentel (1248), and the calculated gaseous thermodynamic functions of Hastings and Nicholson (577) are employed. $Tm = 352.39°K$, $\Delta Hm° = 5.020$ kcal/mole, and $Tb = 469.95°K$.

No. 417 Pentylbenzene $C_{11}H_{16}$ (Ideal Gas State)   Mol Wt 148.238

The values selected by Rossini, Pitzer, Arnett, Braun, and Pimentel (1248) are adopted. $Tm = 198°K$ and $Tb = 478.61°K$.

No. 417 Pentylbenzene $C_{11}H_{16}$ (Ideal Gas State) Mol Wt 148.238

$T°K$	$C_p°$	$S°$	$-(G°-H°_{298})/T$	$H°-H°_{298}$	$\Delta Hf°$	$\Delta Gf°$	$\text{Log } K_p$
	cal/(mole °K)				kcal/mole		
298	47.32	114.47	114.47	0.00	-8.23	36.55	-26.790
300	47.59	114.77	114.48	0.09	-8.29	36.82	-26.824
400	61.69	130.43	116.52	5.57	-11.08	52.31	-28.577
500	74.14	145.57	120.82	12.38	-13.36	68.42	-29.907
600	84.55	160.04	126.16	20.33	-15.17	84.95	-30.940
700	93.18	173.73	131.99	29.23	-16.54	101.76	-31.769
800	100.43	186.66	138.02	38.92	-17.56	118.72	-32.430
900	106.59	198.86	144.11	49.27	-18.25	135.80	-32.976
1000	111.83	210.37	150.17	60.20	-18.62	152.96	-33.427

No. 418 Pentamethylbenzene $C_{11}H_{16}$ (Ideal Gas State) Mol Wt 148.238

$T°K$	$C_p°$	$S°$	$-(G°-H°_{298})/T$	$H°-H°_{298}$	$\Delta Hf°$	$\Delta Gf°$	$\text{Log } K_p$
	cal/(mole °K)				kcal/mole		
298	51.74	106.09	106.09	0.00	-17.80	29.48	-21.606
300	51.99	106.42	106.10	0.10	-17.85	29.77	-21.684
400	65.00	123.19	108.30	5.96	-20.26	46.02	-25.145
500	76.43	138.96	112.87	13.05	-22.26	62.83	-27.462
600	86.08	153.77	118.46	21.19	-23.88	80.00	-29.137
700	94.27	167.67	124.51	30.21	-25.12	97.43	-30.416
800	101.29	180.73	130.73	40.00	-26.04	114.98	-31.410
900	107.20	193.00	136.97	50.43	-26.66	132.66	-32.212
1000	112.33	204.57	143.16	61.42	-26.98	150.39	-32.867

Huffman, Parks, and Barmore (655) investigated the low-temperature properties and found a solid-state transition at 296.8°K, $\Delta Ht° = 0.473$ kcal/mole, and $S°_{298}(s) = 70.22$ cal/(mole °K). Ferry and Thomas (418) reported $Tm = 327.5°K$ and $\Delta Hm° = 2.950$ kcal/mole. Parks, West, Naylor, Fujii, and McClaine (1121) measured the enthalpy of combustion and reported $\Delta Hf°_{298}(s) = -32.33$ kcal/mole, in agreement with Banse and Parks (71). Hastings and Nicholson (577) tabulated the thermodynamic properties of the ideal gas.

No. 419 Hexylbenzene $C_{12}H_{18}$ (Ideal Gas State) Mol Wt 162.264

$T°K$	$C_p°$	$S°$	$-(G°-H°_{298})/T$	$H°-H°_{298}$	$\Delta Hf°$	$\Delta Gf°$	$\text{Log } K_p$
	cal/(mole °K)				kcal/mole		
298	52.79	123.78	123.78	0.00	-13.15	38.56	-28.267
300	53.08	124.11	123.79	0.10	-13.22	38.88	-28.323
400	68.63	141.56	126.06	6.20	-16.33	56.74	-31.001
500	82.39	158.39	130.85	13.77	-18.86	75.31	-32.917
600	93.89	174.46	136.79	22.60	-20.87	94.33	-34.358
700	103.46	189.67	143.27	32.48	-22.38	113.67	-35.488
800	111.50	204.02	149.98	43.24	-23.50	133.17	-36.378
900	118.34	217.56	156.74	54.74	-24.25	152.81	-37.105
1000	124.20	230.34	163.47	66.88	-24.63	172.52	-37.703

The values selected by Rossini, Pitzer, Arnett, Braun, and Pimentel (1248) are adopted $Tm = 212°K$ and $Tb = 499.25°K$.

No. 420 1,2,3-Triethylbenzene $C_{12}H_{18}$ (Ideal Gas State) Mol Wt 162.264

$T°K$	cal/(mole °K)				kcal/mole		
	$Cp°$	$S°$	$-(G°-H°_{298})/T$	$H°-H°_{298}$	$\Delta Hf°$	$\Delta Gf°$	Log Kp
298	54.52	121.23	121.23	0.00	-16.25	36.22	-26.552
300	54.92	121.57	121.24	0.11	-16.31	36.55	-26.622
400	69.37	139.38	123.57	6.33	-19.30	54.64	-29.852
500	82.50	156.31	128.44	13.94	-21.79	73.42	-32.090
600	93.52	172.35	134.43	22.76	-23.82	92.65	-33.745
700	103.09	187.51	140.94	32.60	-25.37	112.20	-35.029
800	111.05	201.81	147.66	43.32	-26.52	131.92	-36.036
900	117.69	215.28	154.44	54.76	-27.33	151.78	-36.856
1000	123.61	227.99	161.16	66.83	-27.78	171.73	-37.529

The correlation method developed by Prosen, Johnson, and Rossini (1195) was used to calculate $\Delta Hf°_{298}(g) = -16.25$ kcal/mole. The standard entropy increase for conversion of an alkyl methyl group to an alkyl ethyl group was added to the gaseous entropy of 1,2,3-trimethylbenzene, giving $S°_{298}(g) = 121.23$ cal/(mole °K). The heat capacity increase increments for the conversion of alkyl methyl groups to alkyl ethyl groups were added to the nine gaseous heat capacities of 1,2,3-trimethylbenzene to give the gaseous heat capacity.

No. 421 1,2,4-Triethylbenzene $C_{12}H_{18}$ (Ideal Gas State) Mol Wt 162.264

$T°K$	cal/(mole °K)				kcal/mole		
	$Cp°$	$S°$	$-(G°-H°_{298})/T$	$H°-H°_{298}$	$\Delta Hf°$	$\Delta Gf°$	Log Kp
298	54.48	123.84	123.84	0.00	-16.99	34.71	-25.439
300	54.87	124.18	123.85	0.11	-17.05	35.02	-25.512
400	69.43	141.99	126.18	6.33	-20.04	52.86	-28.877
500	82.66	158.95	131.05	13.95	-22.52	71.37	-31.196
600	93.81	175.03	137.05	22.79	-24.52	90.34	-32.903
700	103.31	190.22	143.57	32.66	-26.05	109.62	-34.223
800	111.28	204.55	150.31	43.40	-27.18	129.06	-35.257
900	117.96	218.05	157.09	54.87	-27.96	148.65	-36.096
1000	123.81	230.79	163.83	66.96	-28.38	168.32	-36.784

A value of $\Delta Hf°_{298}(g) = -16.99$ kcal/mole was calculated by the correlation method of Prosen, Johnson, and Rossini (1195). The 9.75 cal/(mole °K) entropy increase for conversion of an alkyl methyl group to an alkyl ethyl group was added to the entropy of gaseous 1,2,4-trimethylbenzene, to give $S°_{298}(g) = 123.84$ cal/(mole °K). The gaseous heat capacities were calculated by adding the nine increments for conversion of alkyl methyl groups to alkyl ethyl groups to the gaseous heat capacities of 1,2,4-trimethylbenzene.

No. 422   1,3,5-Triethylbenzene $C_{12}H_{18}$ (Ideal Gas State)   Mol Wt 162.264

$T°K$	cal/(mole °K)				kcal/mole		
	$Cp°$	$S°$	$-(G°-H°_{298})/T$	$H°-H°_{298}$	$\Delta Hf°$	$\Delta Gf°$	Log Kp
298	53.58	121.34	121.34	0.00	-17.86	34.58	-25.348
300	53.98	121.68	121.35	0.10	-17.93	34.90	-25.425
400	68.88	139.28	123.65	6.26	-20.98	53.00	-28.955
500	82.32	156.14	128.47	13.84	-23.51	71.79	-31.379
600	93.60	172.17	134.43	22.65	-25.54	91.04	-33.159
700	103.18	187.34	140.91	32.50	-27.08	110.61	-34.532
800	111.19	201.65	147.62	43.23	-28.22	130.34	-35.606
900	117.90	215.14	154.38	54.69	-29.01	150.22	-36.477
1000	123.78	227.88	161.10	66.78	-29.44	170.18	-37.191

The correlation developed by Prosen, Johnson, and Rossini (1195) was used to calculate $\Delta Hf°_{298}(g) = -17.86$ kcal/mole. To the entropy of gaseous 1,3,5-trimethylbenzene was added the 9.75 cal/(mole °K) entropy increase for conversion of an alkyl methyl group to an alkyl ethyl group, to give $S°_{298}(g) = 121.34$ cal/(mole °K). To the gaseous heat capacities of 1,3,5-trimethylbenzene were added the nine increments for conversion of alkyl methyl groups to alkyl ethyl groups, giving the gaseous heat capacities of 1,3,5-triethylbenzene.

No. 423   Hexamethylbenzene $C_{12}H_{18}$ (Ideal Gas State)   Mol Wt 162.264

$T°K$	cal/(mole °K)				kcal/mole		
	$Cp°$	$S°$	$-(G°-H°_{298})/T$	$H°-H°_{298}$	$\Delta Hf°$	$\Delta Gf°$	Log Kp
298	59.42	108.12	108.12	0.00	-25.26	31.12	-22.813
300	59.73	108.49	108.13	0.12	-25.31	31.47	-22.923
400	74.18	127.70	110.65	6.83	-27.81	50.80	-27.753
500	86.65	145.63	115.87	14.89	-29.86	70.69	-30.898
600	97.13	162.38	122.24	24.09	-31.50	90.95	-33.127
700	105.97	178.04	129.11	34.25	-32.72	111.47	-34.802
800	113.51	192.69	136.15	45.24	-33.61	132.12	-36.092
900	119.99	206.44	143.20	56.92	-34.18	152.88	-37.123
1000	125.55	219.38	150.18	69.20	-34.42	173.70	-37.960

Frankosky and Aston (441) investigated the low-temperature properties and found $Tt = 116.48°K$, $\Delta Ht = 0.269$ kcal/mole, and $S°_{298}(s) = 71.66$ cal/(mole °K). They compared previous work in the temperature range 14–340°K. Spaght, Thomas, and Parks (1400) found $Tt = 383.8°K$, $\Delta Ht = 0.422$ kcal/mole, $Tm = 438.7°K$, and $\Delta Hm° = 4.93$ kcal/mole. Parks, West, Naylor, Fujii, and McClaine (1121) reviewed the combustion literature when they measured the enthalpy of combustion, and on the basis of their measurements reported $\Delta Hf°_{298}(s) = -39.19$ kcal/mole. Hastings and Nicholson (577) tabulated the thermodynamic data for the ideal gas. Momotani, Suga, Seki, and Nitta (1020) repeated the measurements on temperatures and enthalpies of transition, and their findings agree with those of Parks and his students.

No. 424 1-Phenylheptane $C_{13}H_{20}$ (Ideal Gas State)   Mol Wt 176.290

	cal/(mole °K)				kcal/mole		
$T°K$	$Cp°$	$S°$	$-(G°-H°_{298})/T$	$H°-H°_{298}$	$\Delta Hf°$	$\Delta Gf°$	Log Kp
298	58.25	133.09	133.09	0.00	-18.08	40.57	-29.737
300	58.58	133.46	133.10	0.11	-18.15	40.93	-29.815
400	75.57	152.68	135.60	6.84	-21.58	61.17	-33.420
500	90.63	171.21	140.88	15.17	-24.37	82.19	-35.924
600	103.23	188.88	147.42	24.88	-26.58	103.71	-37.773
700	113.73	205.60	154.55	35.74	-28.24	125.57	-39.203
800	122.56	221.38	161.93	47.56	-29.45	147.61	-40.324
900	130.08	236.26	169.37	60.21	-30.25	169.81	-41.233
1000	136.50	250.31	176.77	73.54	-30.66	192.08	-41.976

The values selected by Rossini, Pitzer, Arnett, Braun, and Pimentel (1248) are adopted. $Tm = 225°K$ and $Tb = 519.25°K$.

No. 425 1-Phenyloctane $C_{14}H_{22}$ (Ideal Gas State)   Mol Wt 190.316

	cal/(mole °K)				kcal/mole		
$T°K$	$Cp°$	$S°$	$-(G°-H°_{298})/T$	$H°-H°_{298}$	$\Delta Hf°$	$\Delta Gf°$	Log Kp
298	63.72	142.40	142.40	0.00	-23.00	42.59	-31.214
300	64.07	142.80	142.41	0.12	-23.08	42.99	-31.314
400	82.51	163.81	145.15	7.47	-26.83	65.61	-35.844
500	98.88	184.03	150.91	16.56	-29.88	89.08	-38.934
600	112.58	203.30	158.05	27.15	-32.28	113.09	-41.191
700	124.01	221.54	165.83	39.00	-34.08	137.48	-42.922
800	133.62	238.74	173.88	51.89	-35.39	162.06	-44.272
900	141.83	254.96	182.00	65.67	-36.25	186.81	-45.362
1000	148.80	270.28	190.07	80.22	-36.67	211.64	-46.252

The values selected by Rossini, Pitzer, Arnett, Braun, and Pimentel (1248) are adopted. $Tm = 237°K$ and $Tb = 537.55°K$.

No. 426 1,2,3,4-Tetraethylbenzene $C_{14}H_{22}$ (Ideal Gas State)   Mol Wt 190.316

The correlation method developed by Prosen, Johnson, and Rossini (1195) was used to calculate $\Delta Hf°_{298}(g) = -29.46$ kcal/mole. To the entropy of gaseous 1,2,3,4-tetramethylbenzene was added the 9.75 cal/

No. 426 1,2,3,4-Tetraethylbenzene $C_{14}H_{22}$ (Ideal Gas State) Mol Wt 190.316

T°K	Cp°	S°	$-(G°-H°_{298})/T$	$H°-H°_{298}$	ΔHf°	ΔGf°	Log Kp
	cal/(mole °K)				kcal/mole		
298	68.87	138.55	138.55	0.00	-29.46	37.27	-27.321
300	69.34	138.98	138.56	0.13	-29.53	37.68	-27.449
400	86.77	161.37	141.49	7.95	-32.80	60.61	-33.113
500	102.21	182.43	147.59	17.43	-35.47	84.28	-36.837
600	115.04	202.23	155.07	28.30	-37.59	108.42	-39.490
700	126.05	220.81	163.15	40.37	-39.16	132.90	-41.492
800	135.22	238.26	171.46	53.45	-40.29	157.54	-43.037
900	142.94	254.64	179.80	67.36	-41.02	182.33	-44.274
1000	149.75	270.06	188.06	82.01	-41.34	207.19	-45.278

(mole °K) entropy increase for conversion of an alkyl methyl group to an alkyl ethyl group, to give $S°_{298}(g) = 138.55$ cal/(mole °K). To the gaseous heat capacities of 1,2,3,4-tetramethylbenzene were added the nine increments for conversion of alkyl methyl groups to alkyl ethyl groups, giving the gaseous heat capacities of 1,2,3,4-tetraethylbenzene.

No. 427 1,2,3,5-Tetraethylbenzene $C_{14}H_{22}$ (Ideal Gas State) Mol Wt 190.316

T°K	Cp°	S°	$-(G°-H°_{298})/T$	$H°-H°_{298}$	ΔHf°	ΔGf°	Log Kp
	cal/(mole °K)				kcal/mole		
298	67.95	139.99	139.99	0.00	-29.36	36.94	-27.079
300	68.41	140.42	140.00	0.13	-29.43	37.35	-27.208
400	85.72	162.51	142.89	7.85	-32.80	60.15	-32.862
500	101.23	183.35	148.92	17.22	-35.58	83.72	-36.591
600	114.17	202.98	156.31	28.01	-37.78	107.78	-39.255
700	125.31	221.44	164.31	40.00	-39.44	132.19	-41.269
800	134.59	238.79	172.55	53.00	-40.64	156.77	-42.826
900	142.39	255.11	180.82	66.86	-41.42	181.51	-44.074
1000	149.29	270.47	189.03	81.45	-41.80	206.32	-45.089

A value of $\Delta Hf°_{298}(g) = -29.36$ kcal/mole was calculated by the correlation method of Prosen, Johnson, and Rossini (1195). The 9.75 cal/(mole °K) entropy increase for conversion of an alkyl methyl group to an alkyl ethyl group was added to the entropy of gaseous 1,2,3,5-tetramethylbenzene, to give $S°_{298}(g) = 139.99$ cal/(mole °K). The gaseous heat capacities were calculated by adding the nine increments for conversion of alkyl methyl groups to alkyl ethyl groups to the gaseous heat capacities of 1,2,3,5-tetramethylbenzene.

No. 428 1,2,4,5-Tetraethylbenzene $C_{14}H_{22}$ (Ideal Gas State) Mol Wt 190.316

The correlation method developed by Prosen, Johnson, and Rossini (1195) was used to calculate $\Delta Hf°_{298}(g) = -29.46$ kcal/mole. The standard entropy increase for conversion of an alkyl methyl group to an alkyl

No. 428 1,2,4,5-Tetraethylbenzene $C_{14}H_{22}$ (Ideal Gas State) Mol Wt 190.316

$T°K$	$Cp°$	$S°$	$-(G°-H°_{298})/T$	$H°-H°_{298}$	$\Delta Hf°$	$\Delta Gf°$	Log Kp
		cal/(mole °K)			kcal/mole		
298	69.14	139.03	139.03	0.00	-29.46	37.13	-27.216
300	68.61	139.46	139.04	0.13	-29.53	37.54	-27.345
400	85.46	161.54	141.93	7.85	-32.91	60.43	-33.016
500	100.82	182.31	147.95	17.18	-35.71	84.10	-36.758
600	113.74	201.86	155.32	27.93	-37.96	108.27	-39.434
700	124.89	220.25	163.30	39.87	-39.66	132.80	-41.459
800	134.21	237.55	171.51	52.84	-40.90	157.50	-43.025
900	142.06	253.82	179.76	66.66	-41.72	182.36	-44.282
1000	148.99	269.16	187.94	81.22	-42.13	207.31	-45.305

ethyl group was added to the gaseous entropy of 1,2,4,5-tetramethylbenzene, giving $S°_{298}(g) = 139.03$ cal/(mole °K). The heat capacity increments for the conversion of alkyl methyl groups to alkyl ethyl groups were added to the nine gaseous heat capacities of 1,2,4,5-tetramethylbenzene, to give the gaseous heat capacity.

No. 429 1-Phenylnonane $C_{15}H_{24}$ (Ideal Gas State) Mol Wt 204.342

$T°K$	$Cp°$	$S°$	$-(G°-H°_{298})/T$	$H°-H°_{298}$	$\Delta Hf°$	$\Delta Gf°$	Log Kp
		cal/(mole °K)			kcal/mole		
298	69.18	151.71	151.71	0.00	-27.93	44.59	-32.685
300	69.57	152.14	151.72	0.13	-28.02	45.03	-32.806
400	89.45	174.93	154.69	8.10	-32.08	70.03	-38.262
500	107.12	196.85	160.94	17.96	-35.39	95.96	-41.941
600	121.92	217.72	168.68	29.43	-37.98	122.46	-44.605
700	134.28	237.47	177.11	42.25	-39.93	149.38	-46.638
800	144.69	256.10	185.83	56.22	-41.34	176.51	-48.217
900	153.57	273.66	194.63	71.14	-42.25	203.81	-49.489
1000	161.20	290.25	203.37	86.89	-42.70	231.20	-50.526

The values selected by Rossini, Pitzer, Arnett, Braun, and Pimentel (1248) are adopted. $Tm = 249°K$ and $Tb = 555.15°K$.

No. 430 1-Phenyldecane $C_{16}H_{26}$ (Ideal Gas State) Mol Wt 218.368

Loeffler and Rossini (880) measured the enthalpy of combustion and derived $\Delta Hf°_{298}(l) = -51.58$ kcal/mole and $\Delta Hf°_{298}(g) = -32.86$ kcal/mole. Thermodynamic functions were presented by Rossini, Pitzer, Arnett, Braun, and Pimentel (1248), who also selected $Tm = 258.77°K$ and $Tb = 571.04°K$.

No. 430 1-Phenyldecane $C_{16}H_{26}$ (Ideal Gas State) Mol Wt 218.368

$T°K$	$Cp°$	$S°$	$-(G°-H°_{298})/T$	$H°-H°_{298}$	$\Delta Hf°$	$\Delta Gf°$	Log Kp
	cal/(mole °K)				kcal/mole		
298	74.65	161.02	161.02	0.00	-32.86	46.60	-34.155
300	75.06	161.49	161.03	0.14	-32.95	47.08	-34.298
400	96.40	186.06	164.23	8.74	-37.34	74.46	-40.681
500	115.37	209.67	170.98	19.35	-40.90	102.84	-44.947
600	131.26	232.15	179.31	31.70	-43.69	131.84	-48.019
700	144.56	253.40	188.39	45.51	-45.78	161.29	-50.353
800	155.76	273.46	197.79	60.54	-47.29	190.95	-52.162
900	165.32	292.37	207.26	76.61	-48.26	220.80	-53.616
1000	173.50	310.22	216.67	93.56	-48.72	250.75	-54.799

No. 431 Pentaethylbenzene $C_{16}H_{26}$ (Ideal Gas State) Mol Wt 218.368

$T°K$	$Cp°$	$S°$	$-(G°-H°_{298})/T$	$H°-H°_{298}$	$\Delta Hf°$	$\Delta Gf°$	Log Kp
	cal/(mole °K)				kcal/mole		
298	81.19	154.84	154.84	0.00	-41.87	39.43	-28.901
300	81.79	155.35	154.85	0.16	-41.95	39.93	-29.085
400	102.45	181.77	158.31	9.39	-45.69	67.82	-37.053
500	120.43	206.62	165.51	20.56	-48.70	96.56	-42.203
600	135.28	229.93	174.33	33.36	-51.04	125.82	-45.828
700	147.92	251.75	183.85	47.54	-52.76	155.46	-48.534
800	158.54	272.22	193.63	62.88	-53.96	185.27	-50.610
900	167.35	291.41	203.44	79.18	-54.69	215.23	-52.262
1000	175.18	309.46	213.15	96.32	-54.97	245.26	-53.599

A value of $\Delta Hf°_{298}(g) = -41.87$ kcal/mole was calculated by the correlation method of Prosen, Johnson, and Rossini (1195). The 9.75 cal/(mole °K) entropy increase for conversion of an alkyl methyl group to an alkyl ethyl group was added to the entropy of gaseous pentamethylbenzene, to give $S°_{298}(g) = 154.84$ cal/(mole °K). The gaseous heat capacities were calculated by adding the nine increments for conversion of alkyl methyl groups to alkyl ethyl groups to the gaseous heat capacities of pentamethylbenzene.

No. 432 1-Phenylundecane $C_{17}H_{28}$ (Ideal Gas State) Mol Wt 232.394

$T°K$	$Cp°$	$S°$	$-(G°-H°_{298})/T$	$H°-H°_{298}$	$\Delta Hf°$	$\Delta Gf°$	Log Kp
	cal/(mole °K)				kcal/mole		
298	80.12	170.32	170.32	0.00	-37.78	48.62	-35.634
300	80.55	170.82	170.33	0.15	-37.88	49.14	-35.799
400	103.34	197.18	173.77	9.37	-42.58	78.90	-43.108
500	123.62	222.48	181.00	20.74	-46.40	109.73	-47.960
600	140.60	246.56	189.93	33.98	-49.39	141.23	-51.440
700	154.84	269.33	199.67	48.77	-51.62	173.20	-54.074
800	166.82	290.81	209.73	64.87	-53.22	205.41	-56.112
900	177.07	311.06	219.87	82.07	-54.25	237.82	-57.748
1000	185.80	330.18	229.96	100.23	-54.73	270.33	-59.077

The values selected by Rossini, Pitzer, Arnett, Braun, and Pimentel (1248) are adopted. $Tm = 268°K$ and $Tb = 586.35°K$.

No. 433 1-Phenyldodecane $C_{18}H_{30}$ (Ideal Gas State) Mol Wt 246.420

	cal/(mole °K)				kcal/mole		
$T°K$	$Cp°$	$S°$	$-(G°-H°_{298})/T$	$H°-H°_{298}$	$\Delta Hf°$	$\Delta Gf°$	Log Kp
298	85.58	179.63	179.63	0.00	-42.71	50.62	-37.104
300	86.05	180.17	179.64	0.16	-42.82	51.19	-37.291
400	110.28	208.30	183.31	10.00	-47.84	83.33	-45.526
500	131.86	235.29	191.03	22.14	-51.91	116.61	-50.966
600	149.94	260.98	200.56	36.25	-55.10	150.60	-54.854
700	165.11	285.26	210.95	52.03	-57.47	185.10	-57.789
800	177.88	308.16	221.68	69.19	-59.17	219.85	-60.058
900	188.81	329.76	232.50	87.54	-60.26	254.82	-61.875
1000	198.20	350.15	243.26	106.90	-60.76	289.88	-63.351

The values selected by Rossini, Pitzer, Arnett, Braun, and Pimentel (1248) are adopted. $Tm = 276°K$ and $Tb = 600.75°K$.

No. 434 Hexaethylbenzene $C_{18}H_{30}$ (Ideal Gas State) Mol Wt 246.420

	cal/(mole °K)				kcal/mole		
$T°K$	$Cp°$	$S°$	$-(G°-H°_{298})/T$	$H°-H°_{298}$	$\Delta Hf°$	$\Delta Gf°$	Log Kp
298	94.76	166.62	166.62	0.00	-53.60	43.61	-31.965
300	95.49	167.21	166.63	0.18	-53.69	44.20	-32.201
400	119.12	198.00	170.66	10.94	-57.79	77.50	-42.340
500	139.45	226.83	179.05	23.90	-61.04	111.71	-48.825
600	156.17	253.77	189.28	38.70	-63.54	146.48	-53.354
700	170.35	278.94	200.31	55.05	-65.34	181.66	-56.714
800	182.21	302.48	211.63	72.69	-66.57	217.00	-59.280
900	192.17	324.53	222.96	91.42	-67.27	252.51	-61.315
1000	200.97	345.24	234.16	111.08	-67.46	288.08	-62.958

The correlation method developed by Prosen, Johnson, and Rossini (1195) was used to calculate $\Delta Hf°_{298}(g) = -53.60$ kcal/mole. To the entropy of gaseous hexamethylbenzene was added the 9.75 cal/(mole °K) entropy increase for the conversion of an alkyl methyl group to an alkyl ethyl group, to give $S°_{298}(g) = 166.62$ cal/(mole °K). To the gaseous heat capacities of hexamethylbenzene were added the nine increments for conversion of alkyl methyl groups to alkyl ethyl groups, giving the gaseous heat capacity of hexaethylbenzene.

No. 435 1-Phenyltridecane $C_{19}H_{32}$ (Ideal Gas State) Mol Wt 260.446

The values selected by Rossini, Pitzer, Arnett, Braun, and Pimentel (1248) are adopted. $Tm = 283°K$ and $Tb = 614.45°K$.

Alkylbenzene Ideal Gas Tables

No. 435 1-Phenyltridecane $C_{19}H_{32}$ (Ideal Gas State) Mol Wt 260.446

$T°K$	$Cp°$	$S°$	$-(G° - H°_{298})/T$	$H° - H°_{298}$	$\Delta Hf°$	$\Delta Gf°$	Log Kp
	cal/(mole °K)			kcal/mole			
298	91.05	188.94	188.94	0.00	-47.63	52.64	-38.582
300	91.54	189.51	188.95	0.17	-47.75	53.25	-38.790
400	117.22	219.43	192.85	10.64	-53.08	87.77	-47.950
500	140.11	248.11	201.06	23.53	-57.41	123.49	-53.977
600	159.29	275.40	211.19	38.53	-60.80	159.99	-58.272
700	175.39	301.20	222.23	55.28	-63.31	197.02	-61.508
800	188.95	325.52	233.63	73.52	-65.11	234.30	-64.006
900	200.56	348.46	245.13	93.01	-66.25	271.82	-66.004
1000	210.50	370.13	256.56	113.57	-66.77	309.45	-67.626

No. 436 1-Phenyltetradecane $C_{20}H_{34}$ (Ideal Gas State) Mol Wt 274.472

$T°K$	$Cp°$	$S°$	$-(G° - H°_{298})/T$	$H° - H°_{298}$	$\Delta Hf°$	$\Delta Gf°$	Log Kp
	cal/(mole °K)			kcal/mole			
298	96.51	198.25	198.25	0.00	-52.56	54.64	-40.052
300	97.04	198.85	198.26	0.18	-52.68	55.30	-40.282
400	124.16	230.55	202.40	11.27	-58.34	92.19	-50.369
500	148.35	260.93	211.09	24.93	-62.92	130.37	-56.983
600	168.63	289.82	221.82	40.80	-66.50	169.36	-61.687
700	185.66	317.13	233.51	58.54	-69.16	208.92	-65.224
800	200.02	342.88	245.59	77.84	-71.06	248.75	-67.951
900	212.30	367.17	257.76	98.47	-72.26	288.82	-70.132
1000	222.80	390.09	269.86	120.24	-72.80	329.00	-71.900

The values selected by Rossini, Pitzer, Arnett, Braun, and Pimentel (1248) are adopted. $Tm = 289°K$ and $Tb = 627°K$.

No. 437 1-Phenylpentadecane $C_{21}H_{36}$ (Ideal Gas State) Mol Wt 288.498

$T°K$	$Cp°$	$S°$	$-(G° - H°_{298})/T$	$H° - H°_{298}$	$\Delta Hf°$	$\Delta Gf°$	Log Kp
	cal/(mole °K)			kcal/mole			
298	101.98	207.56	207.56	0.00	-57.49	56.65	-41.522
300	102.53	208.20	207.57	0.19	-57.62	57.35	-41.774
400	131.10	241.68	211.94	11.90	-63.59	96.62	-52.788
500	156.60	273.75	221.12	26.32	-68.43	137.25	-59.989
600	177.97	304.24	232.45	43.08	-72.21	178.74	-65.101
700	195.94	333.06	244.79	61.80	-75.01	220.82	-68.939
800	211.08	360.24	257.54	82.17	-77.01	263.19	-71.896
900	224.05	385.87	270.39	103.94	-78.27	305.82	-74.259
1000	235.20	410.07	283.16	126.92	-78.82	348.56	-76.173

The values selected by Rossini, Pitzer, Arnett, Braun, and Pimentel (1248) are adopted. $Tm = 295°K$ and $Tb = 639°K$.

No. 438 1-Phenylhexadecane $C_{22}H_{38}$ (Ideal Gas State) Mol Wt 302.524

$T°K$	$Cp°$	$S°$	$-(G°-H°_{298})/T$	$H°-H°_{298}$	$\Delta Hf°$	$\Delta Gf°$	Log Kp
	cal/(mole °K)				kcal/mole		
298	107.45	216.87	216.87	0.00	-62.41	58.66	-42.999
300	108.02	217.54	216.88	0.20	-62.55	59.40	-43.274
400	138.04	252.80	221.48	12.53	-68.84	101.06	-55.212
500	164.85	286.57	231.15	27.71	-73.93	144.14	-63.000
600	187.31	318.66	243.08	45.35	-77.91	188.12	-68.519
700	206.22	349.00	256.07	65.05	-80.85	232.73	-72.658
800	222.14	377.60	269.49	86.49	-82.95	277.64	-75.844
900	235.80	404.57	283.01	109.40	-84.26	322.82	-78.388
1000	247.50	430.04	296.46	133.59	-84.83	368.12	-80.449

The values selected by Rossini, Pitzer, Arnett, Braun, and Pimentel (1248) are adopted. $Tm = 300°K$ and $Tb = 651°K$.

STYRENE IDEAL GAS TABLES

No. 439 Styrene C_8H_8 (Ideal Gas State) Mol Wt 104.144

$T°K$	$Cp°$	$S°$	$-(G°-H°_{298})/T$	$H°-H°_{298}$	$\Delta Hf°$	$\Delta Gf°$	Log Kp
	cal/(mole °K)				kcal/mole		
298	29.18	82.48	82.48	0.00	35.22	51.10	-37.453
300	29.35	82.67	82.49	0.06	35.19	51.19	-37.292
400	38.32	92.37	83.75	3.45	33.83	56.75	-31.003
500	45.94	101.77	86.42	7.68	32.72	62.61	-27.364
600	52.14	110.71	89.73	12.59	31.81	68.67	-25.010
700	57.21	119.14	93.33	18.07	31.09	74.88	-23.376
800	61.40	127.06	97.06	24.00	30.52	81.16	-22.171
900	64.93	134.50	100.81	30.32	30.10	87.52	-21.252
1000	67.92	141.50	104.53	36.97	29.83	93.92	-20.526

The values selected by Rossini, Pitzer, Arnett, Braun, and Pimentel (1248) are adopted. [See also the styrene monograph edited by Boundy and Boyer (151).] $Tm = 242.52°K$, $\Delta Hm° = 2.617$ kcal/mole, and $Tb = 418.29°K$, at which $\Delta Hv = 8.850$ kcal/mole. $Tc = 546°K$, $Pc = 40$ atm, and $dc = 0.30$ g/cc.

No. 440 α-Methylstyrene C_9H_{10} (Ideal Gas State) Mol Wt 118.170

The values used here are those selected by Rossini, Pitzer, Arnett, Braun, and Pimentel (1248). $Tm = 249.9°K$ and $Tb = 438.6°K$.

Styrene Ideal Gas Tables

No. 440 α-Methylstyrene C_9H_{10} (Ideal Gas State) Mol Wt 118.170

T°K	cal/(mole °K)				kcal/mole		
	$Cp°$	$S°$	$-(G°-H°_{298})/T$	$H°-H°_{298}$	$\Delta Hf°$	$\Delta Gf°$	Log Kp
298	34.70	91.70	91.70	0.00	27.00	49.84	-36.531
300	34.90	91.92	91.71	0.07	26.96	49.98	-36.407
400	44.80	103.34	93.20	4.06	25.27	57.92	-31.645
500	53.50	114.30	96.33	8.99	23.84	66.25	-28.958
600	60.70	124.71	100.20	14.71	22.65	74.84	-27.260
700	66.80	134.54	104.41	21.10	21.72	83.63	-26.109
800	71.80	143.79	108.76	28.03	20.99	92.51	-25.272
900	76.10	152.51	113.14	35.43	20.44	101.49	-24.645
1000	79.80	160.72	117.49	43.23	20.11	110.53	-24.155

No. 441 Propenylbenzene (β-Methylstyrene), cis C_9H_{10} (Ideal Gas State) Mol Wt 118.170

T°K	cal/(mole °K)				kcal/mole		
	$Cp°$	$S°$	$-(G°-H°_{298})/T$	$H°-H°_{298}$	$\Delta Hf°$	$\Delta Gf°$	Log Kp
298	34.70	91.70	91.70	0.00	29.00	51.84	-37.997
300	34.90	91.92	91.71	0.07	28.96	51.98	-37.864
400	44.80	103.34	93.20	4.06	27.27	59.92	-32.738
500	53.50	114.30	96.33	8.99	25.84	68.25	-29.833
600	60.70	124.71	100.20	14.71	24.65	76.84	-27.989
700	66.80	134.54	104.41	21.10	23.72	85.63	-26.734
800	71.80	143.79	108.76	28.03	22.99	94.51	-25.819
900	76.10	152.51	113.14	35.43	22.44	103.49	-25.130
1000	79.80	160.72	117.49	43.23	22.11	112.53	-24.592

The values selected by Rossini, Pitzer, Arnett, Braun, and Pimentel (1248) are adopted. $Tm = 211.47°K$ and $Tb = 440.58°K$.

No. 442 Propenylbenzene (β-Methylstyrene), trans C_9H_{10} (Ideal Gas State) Mol Wt 118.170

T°K	cal/(mole °K)				kcal/mole		
	$Cp°$	$S°$	$-(G°-H°_{298})/T$	$H°-H°_{298}$	$\Delta Hf°$	$\Delta Gf°$	Log Kp
298	34.90	90.90	90.90	0.00	28.00	51.08	-37.439
300	35.10	91.12	90.91	0.07	27.96	51.22	-37.310
400	45.20	102.63	92.41	4.10	26.30	59.24	-32.364
500	54.00	113.69	95.57	9.07	24.91	67.64	-29.562
600	61.20	124.19	99.47	14.84	23.78	76.28	-27.784
700	67.20	134.09	103.72	21.27	22.89	85.12	-26.573
800	72.20	143.40	108.10	28.24	22.20	94.04	-25.690
900	76.40	152.15	112.52	35.68	21.69	103.06	-25.025
1000	80.00	160.39	116.90	43.50	21.38	112.13	-24.504

The selected values of Rossini, Pitzer, Arnett, Braun, and Pimentel (1248) are employed. $Tm = 243.82°K$ and $Tb = 451.41°K$.

No. 443 m-Methylstyrene C_9H_{10} (Ideal Gas State) Mol Wt 118.170

T°K	$Cp°$	$S°$	$-(G°-H°_{298})/T$	$H°-H°_{298}$	$\Delta Hf°$	$\Delta Gf°$	Log Kp
	cal/(mole °K)				kcal/mole		
298	34.70	93.10	93.10	0.00	27.60	50.02	-36.665
300	34.90	93.32	93.11	0.07	27.56	50.16	-36.538
400	44.80	104.74	94.60	4.06	25.87	57.96	-31.667
500	53.50	115.70	97.73	8.99	24.44	66.15	-28.915
600	60.70	126.11	101.60	14.71	23.25	74.60	-27.173
700	66.80	135.94	105.81	21.10	22.32	83.25	-25.990
800	71.80	145.19	110.16	28.03	21.59	91.99	-25.130
900	76.10	153.91	114.54	35.43	21.04	100.83	-24.485
1000	79.80	162.12	118.89	43.23	20.71	109.73	-23.980

The data presented here were selected by Rossini, Pitzer, Arnett, Braun, and Pimentel (1248). $Tm = 186.81°K$ and $Tb = 444.7°K$, at which $\Delta Hv = 9.27$ kcal/mole according to Clements, Wise, and Johnsen (234).

No. 444 o-Methylstyrene C_9H_{10} (Ideal Gas State) Mol Wt 118.170

T°K	$Cp°$	$S°$	$-(G°-H°_{298})/T$	$H°-H°_{298}$	$\Delta Hf°$	$\Delta Gf°$	Log Kp
	cal/(mole °K)				kcal/mole		
298	34.70	91.70	91.70	0.00	28.30	51.14	-37.484
300	34.90	91.92	91.71	0.07	28.26	51.28	-37.354
400	44.80	103.34	93.20	4.06	26.57	59.22	-32.355
500	53.50	114.30	96.33	8.99	25.14	67.55	-29.527
600	60.70	124.71	100.20	14.71	23.95	76.14	-27.734
700	66.80	134.54	104.41	21.10	23.02	84.93	-26.515
800	71.80	143.79	108.76	28.03	22.29	93.81	-25.628
900	76.10	152.51	113.14	35.43	21.74	102.79	-24.960
1000	79.80	160.72	117.49	43.23	21.41	111.83	-24.439

Rossini, Pitzer, Arnett, Braun, and Pimentel (1248) selected the values presented here. $Tm = 204.58°K$ and $Tb = 442.96°K$, at which $\Delta Hv = 9.46$ kcal/mole according to Clements, Wise, and Johnsen (234).

No. 445 p-Methylstyrene C_9H_{10} (Ideal Gas State) Mol Wt 118.170

T°K	$Cp°$	$S°$	$-(G°-H°_{298})/T$	$H°-H°_{298}$	$\Delta Hf°$	$\Delta Gf°$	Log Kp
	cal/(mole °K)				kcal/mole		
298	34.70	91.70	91.70	0.00	27.40	50.24	-36.825
300	34.90	91.92	91.71	0.07	27.36	50.38	-36.699
400	44.80	103.34	93.20	4.06	25.67	58.32	-31.864
500	53.50	114.30	96.33	8.99	24.24	66.65	-29.133
600	60.70	124.71	100.20	14.71	23.05	75.24	-27.406
700	66.80	134.54	104.41	21.10	22.12	84.03	-26.234
800	71.80	143.79	108.76	28.03	21.39	92.91	-25.382
900	76.10	152.51	113.14	35.43	20.84	101.89	-24.742
1000	79.80	160.72	117.49	43.23	20.51	110.93	-24.242

The values listed are those selected by Rossini, Pitzer, Arnett, Braun, and Pimentel (1248). $Tm = 239.00°K$, and $Tb = 445.9°K$, at which $\Delta Hv = 9.31$ kcal/mole according to Clements, Wise, and Johnsen (234).

ALKYLNAPHTHALENE IDEAL GAS TABLES

No. 446 Naphthalene $C_{10}H_8$ (Ideal Gas State) Mol Wt 128.164

	cal/(mole °K)				kcal/mole		
$T°K$	$Cp°$	$S°$	$-(G° - H°_{298})/T$	$H° - H°_{298}$	$\Delta Hf°$	$\Delta Gf°$	Log Kp
298	31.68	80.22	80.22	0.00	36.08	53.44	-39.172
300	31.89	80.42	80.23	0.06	36.05	53.55	-39.007
400	42.83	91.13	81.61	3.81	34.55	59.62	-32.575
500	52.13	101.72	84.58	8.58	33.34	66.04	-28.863
600	59.67	111.92	88.29	14.18	32.36	72.66	-26.465
700	65.77	121.59	92.36	20.46	31.60	79.45	-24.805
800	70.77	130.70	96.59	27.29	31.01	86.32	-23.580
900	74.91	139.28	100.86	34.58	30.58	93.26	-22.646
1000	78.38	147.36	105.11	42.25	30.33	100.25	-21.908

Coleman and Pilcher (264) and Speros and Rossini (1401) measured the enthalpy of combustion of carefully purified naphthalene and are in good agreement at $\Delta Hf°_{298}(s) = 18.66$ kcal/mole. Coleman and Pilcher listed references to earlier work.

Low-temperature thermal data from 10° to 370°K were determined by McCullough, Finke, Messerly, Todd, Kincheloe, and Waddington (954), including $Ttp = 353.43°K$ and $\Delta Hm° = 4.536$ kcal/mole. Barrow and McClellan (77) measured the vapor heat capacity at two temperatures, 451.0° and 522.7°K, and the enthalpy of vaporization at 440.9°K. Miller (1005) measured the vapor pressure in the submicron region. He showed that the vibrational assignment presented by Mitra and Bernstein (1015) was consistent with his own measurements, with the calorimetric entropy, with the enthalpy of vaporization, and with the vapor pressures in the millimeter range reported by Camin and Rossini (189). The assignment of Mitra and Bernstein (1015) was therefore used to calculate the thermodynamic functions presented here. The enthalpy of sublimation at 298°K was calculated by Miller (1005) as 17.42 kcal/mole, which yields $\Delta Hf°_{298}(g) = 36.08$ kcal/mole.

Camin and Rossini (189) reported $Tb = 491.1°K$, and Ambrose, Cox, and Townsend (13) measured $Tc = 748.3°K$.

No. 447 1-Methylnaphthalene $C_{11}H_{10}$ (Ideal Gas State) Mol Wt 142.190

T°K	$Cp°$	$S°$	$-(G°-H°_{298})/T$	$H°-H°_{298}$	$\Delta Hf°$	$\Delta Gf°$	Log Kp
		cal/(mole °K)			kcal/mole		
298	38.13	90.21	90.21	0.00	27.93	52.03	-38.134
300	38.37	90.45	90.22	0.08	27.89	52.17	-38.005
400	50.74	103.23	91.87	4.55	26.18	60.54	-33.077
500	61.25	115.72	95.40	10.16	24.80	69.30	-30.289
600	69.79	127.66	99.79	16.73	23.71	78.29	-28.517
700	76.76	138.96	104.59	24.07	22.88	87.47	-27.309
800	82.48	149.59	109.56	32.04	22.26	96.73	-26.425
900	87.19	159.59	114.57	40.53	21.83	106.07	-25.757
1000	91.21	168.99	119.54	49.45	21.61	115.46	-25.232

The combustion measurements of Speros and Rossini (1401) are coupled with the correlated functions of Milligan, Becker, and Pitzer (1009). The condensed state was studied by McCullough, Finke, Messerly, Todd, Kincheloe, and Waddington (954), who reported $Tt = 240.78°K$, $\Delta Ht° = 1.190$ kcal/mole, $Ttp = 242.67°K$, $\Delta Hm° = 1.160$ kcal/mole, and $Tb = 517.83°K$.

No. 448 2-Methylnaphthalene $C_{11}H_{10}$ (Ideal Gas State) Mol Wt 142.190

T°K	$Cp°$	$S°$	$-(G°-H°_{298})/T$	$H°-H°_{298}$	$\Delta Hf°$	$\Delta Gf°$	Log Kp
		cal/(mole °K)			kcal/mole		
298	38.19	90.83	90.83	0.00	27.75	51.66	-37.866
300	38.42	91.07	90.84	0.08	27.71	51.81	-37.738
400	50.50	103.82	92.49	4.54	25.99	60.12	-32.844
500	60.87	116.24	96.01	10.12	24.58	68.82	-30.078
600	69.31	128.10	100.38	16.64	23.44	77.76	-28.324
700	76.28	139.33	105.15	23.93	22.57	86.90	-27.131
800	82.03	149.90	110.09	31.86	21.90	96.13	-26.260
900	86.81	159.84	115.07	40.30	21.43	105.44	-25.603
1000	90.86	169.21	120.02	49.19	21.17	114.80	-25.088

The combustion measurements of Speros and Rossini (1401) are combined with the correlated functions of Milligan, Becker, and Pitzer (1009). A study of the condensed states was reported by McCullough, Finke, Messerly, Todd, Kincheloe, and Waddington (954), who found $Tt = 288.5°K$, $\Delta Ht° = 1.341$ kcal/mole, $Ttp = 307.73°K$, $\Delta Hm° = 2.808$ kcal/mole, and $Tb = 514.20°K$.

No. 449 1-Ethylnaphthalene $C_{12}H_{12}$ (Ideal Gas State) Mol Wt 156.216

Since the difference in $\Delta Hf°$ of gaseous benzene and toluene is nearly the same as the difference between naphthalene and 1-methylnaph-

No. 449 1-Ethylnaphthalene $C_{12}H_{12}$ (Ideal Gas State) Mol Wt 156.216

T°K	$Cp°$	$S°$	$-(G°-H°_{298})/T$	$H°-H°_{298}$	$\Delta Hf°$	$\Delta Gf°$	Log Kp
		cal/(mole °K)			kcal/mole		
298	44.02	99.94	99.94	0.00	23.10	54.01	-39.585
300	44.30	100.22	99.95	0.09	23.06	54.19	-39.478
400	58.25	114.92	101.86	5.23	21.07	64.89	-35.455
500	70.06	129.23	105.91	11.67	19.50	76.04	-33.235
600	79.65	142.88	110.94	19.17	18.26	87.45	-31.854
700	87.47	155.76	116.44	27.53	17.34	99.08	-30.932
800	93.91	167.88	122.12	36.61	16.66	110.79	-30.265
900	99.24	179.25	127.84	46.28	16.21	122.59	-29.767
1000	103.79	189.95	133.52	56.43	16.00	134.43	-29.378

thalene, the difference between benzene and ethylbenzene is applied to naphthalene to estimate the enthalpy of formation of gaseous 1-ethylnaphthalene. The remainder of the functions were correlated by Milligan, Becker, and Pitzer (1009). Rossini, Pitzer, Arnett, Braun, and Pimentel (1248) found $Tm = 259.27°K$ and $Tb = 531.82°K$.

No. 450 2-Ethylnaphthalene $C_{12}H_{12}$ (Ideal Gas State) Mol Wt 156.216

T°K	$Cp°$	$S°$	$-(G°-H°_{298})/T$	$H°-H°_{298}$	$\Delta Hf°$	$\Delta Gf°$	Log Kp
		cal/(mole °K)			kcal/mole		
298	44.08	100.56	100.56	0.00	22.92	53.64	-39.318
300	44.35	100.84	100.57	0.09	22.88	53.83	-39.211
400	58.01	115.52	102.47	5.22	20.88	64.47	-35.221
500	69.68	129.75	106.52	11.62	19.28	75.56	-33.024
600	79.17	143.32	111.53	19.08	18.00	86.92	-31.660
700	86.99	156.13	117.00	27.40	17.03	98.51	-30.754
800	93.46	168.18	122.65	36.43	16.30	110.18	-30.099
900	98.86	179.51	128.34	46.05	15.81	121.95	-29.613
1000	103.44	190.17	134.00	56.18	15.56	133.77	-29.235

The enthalpy of formation is estimated as described for 1-ethylnaphthalene. The correlated functions of Milligan, Becker, and Pitzer (1009) are used. Rossini, Pitzer, Arnett, Braun, and Pimentel (1248) selected $Tm = 265.7°K$ and $Tb = 531.0°K$.

No. 451 1,2-Dimethylnaphthalene $C_{12}H_{12}$ (Ideal Gas State) Mol Wt 156.216

The enthalpy of formation is estimated by correlation of similarities between alkyl groups attached to naphthalene and identical groups attached to benzene. Milligan, Becker, and Pitzer (1009) calculated the gaseous functions by correlative methods. Rossini, Pitzer, Arnett, Braun, and Pimentel (1248) selected the melting and boiling points. $Tm = 272.1°K$ and $Tb = 541°K$.

No. 451 1,2-Dimethylnaphthalene $C_{12}H_{12}$ (Ideal Gas State) Mol Wt 156.216

$T°K$	$Cp°$	$S°$	$-(G°-H°_{298})/T$	$H°-H°_{298}$	$\Delta Hf°$	$\Delta Gf°$	$\log Kp$
		cal/(mole °K)			kcal/mole		
298	44.18	97.23	97.23	0.00	19.97	51.68	-37.883
300	44.46	97.51	97.24	0.09	19.93	51.88	-37.790
400	58.28	112.24	99.15	5.24	17.95	62.85	-34.336
500	70.06	126.55	103.21	11.68	16.38	74.26	-32.457
600	79.64	140.20	108.25	19.18	15.14	85.94	-31.303
700	87.51	153.08	113.74	27.54	14.22	97.83	-30.544
800	93.96	165.20	119.42	36.63	13.55	109.81	-29.998
900	99.25	176.58	125.15	46.29	13.10	121.88	-29.595
1000	103.83	187.28	130.83	56.45	12.89	133.99	-29.282

No. 452 1,3-Dimethylnaphthalene $C_{12}H_{12}$ (Ideal Gas State) Mol Wt 156.216

$T°K$	$Cp°$	$S°$	$-(G°-H°_{298})/T$	$H°-H°_{298}$	$\Delta Hf°$	$\Delta Gf°$	$\log Kp$
		cal/(mole °K)			kcal/mole		
298	44.24	97.86	97.86	0.00	19.55	51.08	-37.438
300	44.51	98.14	97.87	0.09	19.51	51.27	-37.347
400	58.04	112.84	99.78	5.23	17.52	62.18	-33.969
500	69.68	127.08	103.83	11.63	15.92	73.53	-32.139
600	79.16	140.65	108.84	19.09	14.63	85.17	-31.020
700	87.03	153.46	114.31	27.41	13.67	97.02	-30.288
800	93.51	165.52	119.97	36.45	12.95	108.96	-29.765
900	98.87	176.85	125.66	46.07	12.46	121.00	-29.381
1000	103.48	187.51	131.32	56.20	12.21	133.08	-29.084

The enthalpy of formation is estimated by correlation of similarities between alkyl groups attached to naphthalene and identical groups attached to benzene. Milligan, Becker, and Pitzer (1009) calculated the gaseous functions by correlative methods. Rossini, Pitzer, Arnett, Braun, and Pimentel (1248) selected the melting and boiling points: $Tm = 269.1°K$ and $Tb = 538°K$.

No. 453 1,4-Dimethylnaphthalene $C_{12}H_{12}$ (Ideal Gas State) Mol Wt 156.216

$T°K$	$Cp°$	$S°$	$-(G°-H°_{298})/T$	$H°-H°_{298}$	$\Delta Hf°$	$\Delta Gf°$	$\log Kp$
		cal/(mole °K)			kcal/mole		
298	44.18	95.86	95.86	0.00	19.72	51.84	-37.999
300	44.46	96.14	95.87	0.09	19.68	52.04	-37.907
400	58.28	110.87	97.78	5.24	17.70	63.14	-34.499
500	70.06	125.18	101.84	11.68	16.13	74.69	-32.647
600	79.64	138.83	106.88	19.18	14.89	86.52	-31.512
700	87.51	151.71	112.37	27.54	13.97	98.54	-30.765
800	93.96	163.83	118.05	36.63	13.30	110.66	-30.229
900	99.25	175.21	123.78	46.29	12.85	122.86	-29.834
1000	103.83	185.91	129.46	56.45	12.64	135.11	-29.527

The enthalpy of formation is estimated by correlation of similarities between alkyl groups attached to naphthalene and identical groups attached to benzene. Milligan, Becker, and Pitzer (1009) calculated the gaseous functions by correlative methods. Rossini, Pitzer, Arnett, Braun, and Pimentel (1248) selected the melting and boiling points: $Tm = 280.81°K$ and $Tb = 541.6°K$.

No. 454 1,5-Dimethylnaphthalene $C_{12}H_{12}$ (Ideal Gas State) Mol Wt 156.216

	cal/(mole °K)			kcal/mole			
$T°K$	$Cp°$	$S°$	$-(G°-H°_{298})/T$	$H°-H°_{298}$	$\Delta Hf°$	$\Delta Gf°$	Log Kp
298	44.18	95.86	95.86	0.00	19.55	51.67	-37.875
300	44.46	96.14	95.87	0.09	19.51	51.87	-37.784
400	58.28	110.87	97.78	5.24	17.53	62.97	-34.406
500	70.06	125.18	101.84	11.68	15.96	74.52	-32.573
600	79.64	138.83	106.88	19.18	14.72	86.35	-31.450
700	87.51	151.71	112.37	27.54	13.80	98.37	-30.712
800	93.96	163.83	118.05	36.63	13.13	110.49	-30.183
900	99.25	175.21	123.78	46.29	12.68	122.69	-29.792
1000	103.83	185.91	129.46	56.45	12.47	134.94	-29.490

The enthalpy of formation is estimated by correlation of similarities between alkyl groups attached to naphthalene and identical groups attached to benzene. Milligan, Becker, and Pitzer (1009) calculated the gaseous functions by correlative methods. Rossini, Pitzer, Arnett, Braun, and Pimentel (1248) selected the melting and boiling points: $Tm = 355.1°K$ and $Tb = 538°K$.

No. 455 1,6-Dimethylnaphthalene $C_{12}H_{12}$ (Ideal Gas State) Mol Wt 156.216

	cal/(mole °K)			kcal/mole			
$T°K$	$Cp°$	$S°$	$-(G°-H°_{298})/T$	$H°-H°_{298}$	$\Delta Hf°$	$\Delta Gf°$	Log Kp
298	44.24	97.86	97.86	0.00	19.72	51.25	-37.562
300	44.51	98.14	97.87	0.09	19.68	51.44	-37.470
400	58.04	112.84	99.78	5.23	17.69	62.35	-34.062
500	69.68	127.08	103.83	11.63	16.09	73.70	-32.213
600	79.16	140.65	108.84	19.09	14.80	85.34	-31.082
700	87.03	153.46	114.31	27.41	13.84	97.19	-30.341
800	93.51	165.52	119.97	36.45	13.12	109.13	-29.811
900	98.87	176.85	125.66	46.07	12.63	121.17	-29.422
1000	103.48	187.51	131.32	56.20	12.38	133.25	-29.121

The enthalpy of formation is estimated by correlation of similarities between alkyl groups attached to naphthalene and identical groups attached to benzene. Milligan, Becker, and Pitzer (1009) calculated the gaseous functions by correlative methods. Rossini, Pitzer, Arnett, Braun, and Pimentel (1248) selected the melting and boiling points: $Tm = 257°K$ and $Tb = 536°K$.

No. 456 1,7-Dimethylnaphthalene $C_{12}H_{12}$ (Ideal Gas State) Mol Wt 156.216

$T°K$	$Cp°$	$S°$	$-(G°-H°_{298})/T$	$H°-H°_{298}$	$\Delta Hf°$	$\Delta Gf°$	Log Kp
		cal/(mole °K)			kcal/mole		
298	44.24	97.86	97.86	0.00	19.55	51.08	-37.438
300	44.51	98.14	97.87	0.09	19.51	51.27	-37.347
400	58.04	112.84	99.78	5.23	17.52	62.18	-33.969
500	69.68	127.08	103.83	11.63	15.92	73.53	-32.139
600	79.16	140.65	108.84	19.09	14.63	85.17	-31.020
700	87.03	153.46	114.31	27.41	13.67	97.02	-30.288
800	93.51	165.52	119.97	36.45	12.95	108.96	-29.765
900	98.87	176.85	125.66	46.07	12.46	121.00	-29.381
1000	103.48	187.51	131.32	56.20	12.21	133.08	-29.084

The enthalpy of formation is estimated by correlation of similarities between alkyl groups attached to naphthalene and identical groups attached to benzene. Milligan, Becker, and Pitzer (1009) calculated the gaseous functions by correlative methods. Rossini, Pitzer, Arnett, Braun, and Pimentel (1248) selected the melting and boiling points: $Tm = 259.3°K$ and $Tb = 536°K$.

No. 457 2,3-Dimethylnaphthalene $C_{12}H_{12}$ (Ideal Gas State) Mol Wt 156.216

$T°K$	$Cp°$	$S°$	$-(G°-H°_{298})/T$	$H°-H°_{298}$	$\Delta Hf°$	$\Delta Gf°$	Log Kp
		cal/(mole °K)			kcal/mole		
298	44.41	98.22	98.22	0.00	19.97	51.39	-37.667
300	44.65	98.50	98.23	0.09	19.93	51.58	-37.574
400	57.49	113.14	100.14	5.21	17.92	62.45	-34.121
500	68.75	127.21	104.15	11.53	16.24	73.79	-32.250
600	78.18	140.61	109.12	18.89	14.86	85.42	-31.112
700	86.02	153.26	114.53	27.12	13.79	97.28	-30.371
800	92.58	165.19	120.13	36.05	12.97	109.25	-29.844
900	98.10	176.42	125.76	45.59	12.40	121.33	-29.460
1000	102.78	187.01	131.36	55.65	12.08	133.46	-29.166

The enthalpy of formation is estimated by correlation of similarities between alkyl groups attached to naphthalene and identical groups attached to benzene. Milligan, Becker, and Pitzer (1009) calculated the gaseous functions by correlative methods. Rossini, Pitzer, Arnett, Braun, and Pimentel (1248) selected the melting and boiling points: $Tm = 378.1°K$ and $Tb = 542°K$.

Alkylnaphthalene Ideal Gas Tables

No. 458 2,6-Dimethylnaphthalene $C_{12}H_{12}$ (Ideal Gas State) Mol Wt 156.216

$T°K$	$Cp°$	$S°$	$-(G°-H°_{298})/T$	$H°-H°_{298}$	$\Delta Hf°$	$\Delta Gf°$	Log Kp
		cal/(mole °K)			kcal/mole		
298	44.71	97.68	97.68	0.00	19.72	51.30	-37.602
300	44.91	97.96	97.69	0.09	19.68	51.49	-37.510
400	58.07	112.73	99.61	5.25	17.71	62.41	-34.099
500	69.45	126.95	103.67	11.64	16.10	73.78	-32.248
600	78.77	140.46	108.68	19.07	14.78	85.43	-31.117
700	86.49	153.19	114.14	27.34	13.77	97.31	-30.379
800	93.09	165.19	119.78	36.33	13.00	109.28	-29.852
900	98.51	176.47	125.46	45.92	12.47	121.35	-29.467
1000	103.14	187.10	131.10	56.01	12.19	133.48	-29.170

The enthalpy of formation is estimated by correlation of similarities between alkyl groups attached to naphthalene and identical groups attached to benzene. Milligan, Becker, and Pitzer (1009) calculated the gaseous functions by correlative methods. Rossini, Pitzer, Arnett, Braun, and Pimentel (1248) selected the melting and boiling points: $Tm = 385.1°K$ and $Tb = 535°K$.

No. 459 2,7-Dimethylnaphthalene $C_{12}H_{12}$ (Ideal Gas State) Mol Wt 156.216

$T°K$	$Cp°$	$S°$	$-(G°-H°_{298})/T$	$H°-H°_{298}$	$\Delta Hf°$	$\Delta Gf°$	Log Kp
		cal/(mole °K)			kcal/mole		
298	44.71	97.68	97.68	0.00	19.72	51.30	-37.602
300	44.97	97.96	97.69	0.09	19.68	51.49	-37.510
400	58.07	112.73	99.61	5.25	17.72	62.41	-34.099
500	69.45	126.95	103.67	11.65	16.10	73.78	-32.247
600	78.77	140.46	108.69	19.07	14.79	85.43	-31.116
700	86.49	153.20	114.14	27.34	13.77	97.30	-30.378
800	93.09	165.19	119.78	36.33	13.00	109.28	-29.851
900	98.51	176.48	125.46	45.92	12.47	121.35	-29.466
1000	103.14	187.10	131.10	56.01	12.20	133.47	-29.169

The enthalpy of formation is estimated by correlation of similarities between alkyl groups attached to naphthalene and identical groups attached to benzene. Milligan, Becker, and Pitzer (1009) calculated the gaseous functions by correlative methods. Rossini, Pitzer, Arnett, Braun, and Pimentel (1248) selected the melting and boiling points: $Tm = 371.1°K$ and $Tb = 535°K$.

No. 460 1-Propylnaphthalene $C_{13}H_{14}$ (Ideal Gas State) Mol Wt 170.242

The enthalpy of formation is estimated by correlation of similarities between alkyl groups attached to naphthalene and identical groups

No. 460 1-Propylnaphthalene $C_{13}H_{14}$ (Ideal Gas State) Mol Wt 170.242

T°K	Cp°	S°	$-(G°-H°_{298})/T$	$H°-H°_{298}$	$\Delta Hf°$	$\Delta Gf°$	Log Kp
	cal/(mole °K)				kcal/mole		
298	49.74	109.55	109.55	0.00	17.85	55.60	-40.755
300	50.04	109.86	109.56	0.10	17.80	55.83	-40.671
400	65.30	126.40	111.71	5.88	15.52	68.88	-37.631
500	78.40	142.42	116.26	13.09	13.69	82.44	-36.032
600	89.10	157.69	121.90	21.48	12.27	96.31	-35.080
700	97.80	172.10	128.06	30.84	11.21	110.42	-34.474
800	105.10	185.65	134.42	40.99	10.45	124.63	-34.046
900	111.10	198.38	140.82	51.81	9.95	138.94	-33.738
1000	116.20	210.36	147.19	63.18	9.73	153.30	-33.503

attached to benzene. Milligan, Becker, and Pitzer (1009) calculated the gaseous functions by correlative methods. Rossini, Pitzer, Arnett, Braun, and Pimentel (1248) selected the melting and boiling points: $Tm = 264.68°K$ and $Tb = 545.93°K$.

No. 461 2-Propylnaphthalene $C_{13}H_{14}$ (Ideal Gas State) Mol Wt 170.242

T°K	Cp°	S°	$-(G°-H°_{298})/T$	$H°-H°_{298}$	$\Delta Hf°$	$\Delta Gf°$	Log Kp
	cal/(mole °K)				kcal/mole		
298	49.80	110.18	110.18	0.00	17.65	55.21	-40.471
300	50.04	110.49	110.19	0.10	17.60	55.44	-40.388
400	65.10	127.00	112.34	5.87	15.31	68.43	-37.385
500	78.00	142.96	116.88	13.05	13.45	81.93	-35.809
600	88.60	158.14	122.50	21.39	11.98	95.75	-34.877
700	97.40	172.48	128.63	30.70	10.88	109.82	-34.287
800	104.60	185.97	134.96	40.81	10.07	124.00	-33.873
900	110.70	198.65	141.34	51.59	9.53	138.28	-33.577
1000	115.80	210.59	147.67	62.92	9.27	152.61	-33.352

The enthalpy of formation is estimated by correlation of similarities between alkyl groups attached to naphthalene and identical groups attached to benzene. Milligan, Becker, and Pitzer (1009) calculated the gaseous functions by correlative methods. Rossini, Pitzer, Arnett, Braun, and Pimentel (1248) selected the melting and boiling points: $Tm = 270°K$ and $Tb = 546.6°K$.

No. 462 2-Ethyl-3-Methylnaphthalene $C_{13}H_{14}$ (Ideal Gas State) Mol Wt 170.242

The enthalpy of formation is estimated by correlation of similarities between alkyl groups attached to naphthalene and identical groups attached to benzene. Milligan, Becker, and Pitzer (1009) calculated the gaseous functions by correlative methods. Rossini, Pitzer, Arnett, Braun, and Pimentel (1248) selected the boiling point: $Tb = 550°K$.

Alkylnaphthalene Ideal Gas Tables

No. 462 2-Ethyl-3-Methylnaphthalene $C_{13}H_{14}$ (Ideal Gas State) Mol Wt 170.242

T°K	Cp°	S°	$-(G°-H°_{298})/T$	$H°-H°_{298}$	$\Delta Hf°$	$\Delta Gf°$	Log Kp
		cal/(mole °K)			kcal/mole		
298	50.30	109.33	109.33	0.00	15.72	53.54	-39.242
300	50.57	109.65	109.34	0.10	15.67	53.77	-39.168
400	65.00	126.22	111.50	5.89	13.40	66.83	-36.514
500	77.50	142.10	116.04	13.03	11.51	80.42	-35.148
600	88.00	157.19	121.65	21.32	9.99	94.33	-34.359
700	96.80	171.43	127.76	30.58	8.83	108.50	-33.874
800	104.00	184.84	134.06	40.63	7.95	122.79	-33.542
900	110.10	197.45	140.41	51.34	7.35	137.18	-33.311
1000	115.30	209.33	146.72	62.62	7.04	151.64	-33.140

No. 463 2-Ethyl-6-Methylnaphthalene $C_{13}H_{14}$ (Ideal Gas State) Mol Wt 170.242

T°K	Cp°	S°	$-(G°-H°_{298})/T$	$H°-H°_{298}$	$\Delta Hf°$	$\Delta Gf°$	Log Kp
		cal/(mole °K)			kcal/mole		
298	50.60	108.79	108.79	0.00	14.65	52.63	-38.576
300	50.90	109.11	108.80	0.10	14.60	52.86	-38.506
400	65.60	125.82	110.97	5.94	12.37	65.97	-36.043
500	78.20	141.85	115.56	13.15	10.55	79.59	-34.786
600	88.70	157.06	121.22	21.51	9.10	93.52	-34.064
700	97.30	171.44	127.38	30.85	8.03	107.70	-33.623
800	101.40	184.64	133.72	40.74	7.00	121.99	-33.324
900	110.60	197.16	140.08	51.37	6.32	136.42	-33.124
1000	115.70	209.11	146.39	62.72	6.08	150.90	-32.977

The enthalpy of formation is estimated by correlation of similarities between alkyl groups attached to naphthalene and identical groups attached to benzene. Milligan, Becker, and Pitzer (1009) calculated the gaseous functions by correlative methods. Rossini, Pitzer, Arnett, Braun, and Pimentel (1248) selected the melting and boiling points: $Tm = 318°K$ and $Tb = 543°K$.

No. 464 2-Ethyl-7-Methylnaphthalene $C_{13}H_{14}$ (Ideal Gas State) Mol Wt 170.242

T°K	Cp°	S°	$-(G°-H°_{298})/T$	$H°-H°_{298}$	$\Delta Hf°$	$\Delta Gf°$	Log Kp
		cal/(mole °K)			kcal/mole		
298	50.60	108.79	108.79	0.00	14.65	52.63	-38.576
300	50.90	109.11	108.80	0.10	14.60	52.86	-38.506
400	65.60	125.82	110.97	5.94	12.37	65.97	-36.043
500	78.20	141.85	115.56	13.15	10.55	79.59	-34.786
600	88.70	157.06	121.22	21.51	9.10	93.52	-34.064
700	97.30	171.44	127.38	30.85	8.03	107.70	-33.623
800	101.40	184.64	133.72	40.74	7.00	121.99	-33.324
900	110.60	197.16	140.08	51.37	6.32	136.42	-33.124
1000	115.70	209.11	146.39	62.72	6.08	150.90	-32.977

The enthalpy of formation is estimated by correlation of similarities between alkyl groups attached to naphthalene and identical groups attached to benzene. Milligan, Becker, and Pitzer (1009) calculated the gaseous functions by correlative methods. Rossini, Pitzer, Arnett, Braun, and Pimentel (1248) selected the boiling point: $Tb = 543°K$.

No. 465 1-Butylnaphthalene $C_{14}H_{16}$ (Ideal Gas State) Mol Wt 184.268

$T°K$	$Cp°$	$S°$	$-(G°-H°_{298})/T$	$H°-H°_{298}$	$\Delta Hf°$	$\Delta Gf°$	Log Kp
		cal/(mole °K)			kcal/mole		
298	55.18	118.83	118.83	0.00	12.68	57.38	-42.056
300	55.51	119.18	118.84	0.11	12.62	57.65	-41.995
400	72.20	137.49	121.22	6.51	10.02	73.08	-39.926
500	86.60	155.19	126.26	14.47	7.93	89.09	-38.940
600	98.40	172.06	132.50	23.74	6.31	105.47	-38.415
700	108.10	187.98	139.30	34.08	5.11	122.11	-38.123
800	116.10	202.95	146.33	45.30	4.24	138.87	-37.935
900	122.80	217.02	153.41	57.25	3.69	155.74	-37.817
1000	128.50	230.26	160.44	69.83	3.44	172.67	-37.734

The enthalpy of formation is estimated by correlation of similarities between alkyl groups attached to naphthalene and identical groups attached to benzene. Milligan, Becker, and Pitzer (1009) calculated the gaseous functions by correlative methods. Rossini, Pitzer, Arnett, Braun, and Pimentel (1248) selected the melting and boiling points: $Tm = 253.39°K$ and $Tb = 562.49°K$.

No. 466 2-Butylnaphthalene $C_{14}H_{16}$ (Ideal Gas State) Mol Wt 184.268

$T°K$	$Cp°$	$S°$	$-(G°-H°_{298})/T$	$H°-H°_{298}$	$\Delta Hf°$	$\Delta Gf°$	Log Kp
		cal/(mole °K)			kcal/mole		
298	55.24	119.46	119.46	0.00	12.50	57.01	-41.786
300	55.56	119.81	119.47	0.11	12.44	57.28	-41.726
400	72.00	138.10	121.85	6.50	9.83	72.65	-39.690
500	86.20	155.73	126.87	14.43	7.72	88.60	-38.726
600	97.90	172.51	133.10	23.66	6.04	104.93	-38.219
700	107.70	188.36	139.87	33.95	4.80	121.53	-37.942
800	115.70	203.28	146.87	45.13	3.89	138.25	-37.767
900	122.40	217.31	153.93	57.04	3.30	155.09	-37.660
1000	128.10	230.51	160.93	69.58	3.02	171.99	-37.587

The enthalpy of formation is estimated by correlation of similarities between alkyl groups attached to naphthalene and identical groups attached to benzene. Milligan, Becker, and Pitzer (1009) calculated the gaseous functions by correlative methods. Rossini, Pitzer, Arnett, Braun, and Pimentel (1248) selected the melting and boiling points: $Tm = 268°K$ and $Tb = 565°K$.

No. 467 1-Pentylnaphthalene $C_{15}H_{18}$ (Ideal Gas State) Mol Wt 198.294

T°K	Cp°	S°	$-(G°-H°_{298})/T$	$H°-H°_{298}$	$\Delta Hf°$	$\Delta Gf°$	Log Kp
	cal/(mole °K)				kcal/mole		
298	60.65	128.26	128.26	0.00	7.75	59.35	-43.500
300	61.01	128.64	128.27	0.12	7.69	59.66	-43.461
400	79.20	148.74	130.88	7.15	4.77	77.46	-42.318
500	94.90	168.16	136.41	15.88	2.43	95.91	-41.920
600	107.80	186.63	143.26	26.03	0.62	114.77	-41.801
700	118.40	204.07	150.71	37.36	-0.72	133.92	-41.810
800	127.20	220.47	158.42	49.65	-1.68	153.20	-41.851
900	134.60	235.89	166.18	62.75	-2.29	172.61	-41.914
1000	140.90	250.41	173.88	76.53	-2.55	192.08	-41.976

The enthalpy of formation is estimated by correlation of similarities between alkyl groups attached to naphthalene and identical groups attached to benzene. Milligan, Becker, and Pitzer (1009) calculated the gaseous functions by correlative methods. Rossini, Pitzer, Arnett, Braun, and Pimentel (1248) selected the melting and boiling points: $Tm = 251°K$ and $Tb = 580°K$.

No. 468 2-Pentylnaphthalene $C_{15}H_{18}$ (Ideal Gas State) Mol Wt 198.294

T°K	Cp°	S°	$-(G°-H°_{298})/T$	$H°-H°_{298}$	$\Delta Hf°$	$\Delta Gf°$	Log Kp
	cal/(mole °K)				kcal/mole		
298	60.71	128.89	128.89	0.00	7.57	58.98	-43.230
300	61.06	129.27	128.90	0.12	7.51	59.29	-43.192
400	79.00	149.35	131.51	7.14	4.58	77.02	-42.082
500	94.50	168.70	137.03	15.84	2.21	95.42	-41.706
600	107.30	187.09	143.85	25.95	0.35	114.23	-41.605
700	117.90	204.45	151.28	37.22	-1.04	133.34	-41.628
800	126.80	220.79	158.96	49.47	-2.04	152.59	-41.683
900	134.20	236.16	166.69	62.53	-2.69	171.97	-41.757
1000	140.50	250.64	174.37	76.27	-2.99	191.41	-41.830

The enthalpy of formation is estimated by correlation of similarities between alkyl groups attached to naphthalene and identical groups attached to benzene. Milligan, Becker, and Pitzer (1009) calculated the gaseous functions by correlative methods. Rossini, Pitzer, Arnett, Braun, and Pimentel (1248) selected the melting and boiling points: $Tm = 269°K$ and $Tb = 583°K$.

MISCELLANEOUS HYDROCARBON IDEAL GAS TABLES

No. 469 Spiropentane C_5H_8 (Ideal Gas State) Mol Wt 68.114

Scott, Finke, Hubbard, McCullough, Gross, Williamson, Waddington, and Huffman (1303) made a thorough study of the low-temperature

No. 469   Spiropentane C_5H_8 (Ideal Gas State)   Mol Wt 68.114

$T°K$	$Cp°$	$S°$	$-(G°-H°_{298})/T$	$H°-H°_{298}$	$\Delta Hf°$	$\Delta Gf°$	Log Kp
		cal/(mole °K)			kcal/mole		
298	21.06	67.45	67.45	0.00	44.27	63.41	-46.479
300	21.19	67.59	67.46	0.04	44.24	63.53	-46.277
400	28.55	74.71	68.38	2.54	42.72	70.20	-38.354
500	34.91	81.79	70.35	5.72	41.52	77.21	-33.749
600	40.10	88.63	72.83	9.48	40.59	84.44	-30.755
700	44.36	95.14	75.56	13.71	39.89	91.81	-28.663
800	47.91	101.30	78.40	18.33	39.39	99.26	-27.114
900	50.93	107.12	81.27	23.27	39.05	106.77	-25.925
1000	53.51	112.62	84.13	28.50	38.88	114.31	-24.981

thermal and spectroscopic properties of spiropentane. They reported $Ttp = 166.13°K$, $\Delta Hm° = 1.538$ kcal/mole, and $Tb = 312.12°K$, at which $\Delta Hv = 6.393$ kcal/mole, and third-law entropies at 298°K of 46.29 cal/(mole °K) for the liquid and 67.45 cal/(mole °K) for the ideal gas, in excellent agreement with their spectroscopic assignment. Fraser and Prosen (442) measured the enthalpy of combustion of spiropentane and reported $\Delta Hf°_{298}(l) = 37.67$ kcal/mole and $\Delta Hf°_{298}(g) = 44.27$ kcal/mole.

No. 470   1,3,5-Cycloheptatriene C_7H_8 (Ideal Gas State)   Mol Wt 92.134

$T°K$	$Cp°$	$S°$	$-(G°-H°_{298})/T$	$H°-H°_{298}$	$\Delta Hf°$	$\Delta Gf°$	Log Kp
		cal/(mole °K)			kcal/mole		
298	28.15	75.44	75.44	0.00	43.47	61.04	-44.741
300	28.33	75.62	75.45	0.06	43.44	61.15	-44.543
400	37.13	85.01	76.67	3.34	42.22	67.25	-36.740
500	44.32	94.10	79.25	7.43	41.29	73.62	-32.176
600	50.07	102.70	82.45	12.16	40.57	80.14	-29.191
700	54.73	110.78	85.93	17.40	40.05	86.79	-27.097
800	58.58	118.35	89.51	23.07	39.67	93.49	-25.539
900	61.82	125.44	93.11	29.10	39.44	100.23	-24.339
1000	64.58	132.10	96.68	35.42	39.36	107.00	-23.384

Finke, Scott, Gross, Messerly, and Waddington (424) measured the low-temperature properties and found a solid-state transition at $Tt = 153.96°K$, $\Delta Ht° = 0.561$ kcal/mole, $Ttp = 197.90°K$, $\Delta Hm° = 0.277$ kcal/mole. $Tb = 388.64°K$, at which $\Delta Hv = 9.250$ kcal/mole. The third-law entropy from these measurements is $S°_{298}$(ideal gas) $= 75.44$ cal/(mole °K). Conn, Kistiakowsky, and Smith (261) measured the catalyzed enthalpy of hydrogenation of cycloheptatriene to cycloheptane at 355°K. Finke et al. (424) calculated from these data that $\Delta Hf°_{298}(g) = 43.47$ kcal/mole of cycloheptatriene. Evans and Lord (397) studied the spectra, assigned fundamental vibrational frequencies, and calculated $S°_{298}$ (ideal gas) $= 76.0$ cal/(mole °K).

No. 471 Ethynylbenzene (Phenylacetylene) C_8H_6 (Ideal Gas State) Mol Wt 102.128

$T°K$	$Cp°$	$S°$	$-(G°-H°_{298})/T$	$H°-H°_{298}$	$\Delta Hf°$	$\Delta Gf°$	$\text{Log } Kp$
		cal/(mole °K)			kcal/mole		
298	27.46	76.88	76.88	0.00	78.22	86.46	-63.375
300	27.63	77.06	76.89	0.06	78.20	86.51	-63.019
400	35.95	86.18	78.07	3.25	77.33	89.42	-48.856
500	42.67	94.95	80.58	7.19	76.64	92.53	-40.442
600	48.01	103.22	83.67	11.73	76.05	95.75	-34.877
700	52.29	110.95	87.02	16.76	75.59	99.09	-30.934
800	55.79	118.17	90.47	22.16	75.20	102.46	-27.990
900	58.71	124.91	93.93	27.89	74.89	105.89	-25.712
1000	61.17	131.23	97.34	33.89	74.70	109.35	-23.897

Flitcroft and Skinner (432) measured the enthalpy of hydrogenation of ethynylbenzene to ethylbenzene, from which we calculate $\Delta Hf°_{298}(l) = 67.72$ kcal/mole and $\Delta Hf°_{298}(g) = 78.22$ kcal/mole. Evans and Nyquist (395) presented a complete vibrational assignment and calculated the thermodynamic functions. Smith and Andrews (1388) studied the low-temperature properties from liquid air temperatures upward and reported $Tm = 228.3°K$ and $Tb = 416°K$ on a sample of 99% purity.

No. 472 1,3,5,7-Cyclooctatetraene C_8H_8 (Ideal Gas State) Mol Wt 104.144

$T°K$	$Cp°$	$S°$	$-(G°-H°_{298})/T$	$H°-H°_{298}$	$\Delta Hf°$	$\Delta Gf°$	$\text{Log } Kp$
		cal/(mole °K)			kcal/mole		
298	29.16	78.10	78.10	0.00	71.23	88.41	-64.805
300	29.32	78.29	78.11	0.06	71.20	88.52	-64.481
400	38.45	88.00	79.37	3.46	69.85	94.51	-51.634
500	46.38	97.46	82.05	7.71	68.76	100.80	-44.059
600	52.77	106.50	85.38	12.68	67.90	107.29	-39.077
700	57.90	115.03	89.01	18.22	67.25	113.91	-35.564
800	62.23	123.05	92.77	24.23	66.76	120.61	-32.946
900	65.75	130.59	96.56	30.64	66.42	127.36	-30.926
1000	68.88	137.68	100.32	37.37	66.24	134.15	-29.317

Prosen, Johnson, and Rossini (1197) burned a high-purity sample of this material. From these measurements we calculate $\Delta Hf°_{298}(l) = 60.93$ kcal/mole and $\Delta Hf°_{298}(g) = 71.23$ kcal/mole. Scott, Gross, Oliver, and Huffman (1312) measured the low-temperature properties on the same sample as above from liquid hydrogen temperatures upward. They found $Ttp = 268.47°K$, $\Delta Hm° = 2.695$ kcal/mole, and $Tb = 413.7°K$, at which $\Delta Hv = 8.700$ kcal/mole, $S°_{298}(l) = 52.65$ cal/(mole °K), $S°_{298}(g) = 78.10$ cal/(mole °K), and $\Delta Hv_{298} = 10.300$ kcal/mole. Lippincott and Lord (877) made a complete spectral assignment and calculated the thermodynamic functions.

No. 473 Azulene $C_{10}H_8$ (Ideal Gas State) Mol Wt 128.164

T°K	$Cp°$	$S°$	$-(G°-H°_{298})/T$	$H°-H°_{298}$	$\Delta Hf°$	$\Delta Gf°$	Log Kp
		cal/(mole °K)			kcal/mole		
298	30.69	80.75	80.75	0.00	66.90	84.10	-61.647
300	30.93	80.95	80.76	0.06	66.86	84.21	-61.343
400	42.15	91.42	82.11	3.73	65.29	90.24	-49.305
500	51.69	101.89	85.02	8.44	64.02	96.64	-42.238
600	59.32	112.01	88.68	14.00	63.00	103.25	-37.606
700	65.56	121.63	92.71	20.25	62.22	110.03	-34.352
800	70.59	130.73	96.90	27.07	61.61	116.90	-31.933
900	74.75	139.29	101.14	34.34	61.16	123.84	-30.070
1000	78.24	147.35	105.36	42.00	60.89	130.82	-28.590

Kováts, Günthard, and Plattner (802, 803) made a complete thermodynamic study of azulene (cyclopentacycloheptene, isomeric with naphthalene), and reported, from their measured $\Delta Hc°$, $\Delta Hf°_{298}(g) = 66.90$ kcal/mole. From their own frequency assignment they calculated the thermodynamic functions and found $S°_{298}(g) = 80.75$ cal/(mole °K). They listed $Tm = 373.64°K$, $\Delta Hm° = 2.89$ kcal/mole, and $Tb = 523°K$ (calculated), at which $\Delta Hv = 13.26$ kcal/mole, and included the transitional enthalpies given by Heilbronner and Wieland (581).

No. 474 Decahydronaphthalene, cis $C_{10}H_{18}$ (Ideal Gas State) Mol Wt 138.244

T°K	$Cp°$	$S°$	$-(G°-H°_{298})/T$	$H°-H°_{298}$	$\Delta Hf°$	$\Delta Gf°$	Log Kp
		cal/(mole °K)			kcal/mole		
298	39.84	90.28	90.28	0.00	-40.38	20.51	-15.034
300	40.14	90.53	90.29	0.08	-40.46	20.88	-15.213
400	56.64	104.38	92.06	4.93	-44.33	41.95	-22.918
500	71.64	118.67	95.95	11.37	-47.36	63.88	-27.922
600	84.14	132.87	100.93	19.17	-49.64	86.35	-31.450
700	94.71	146.66	106.48	28.13	-51.23	109.16	-34.079
800	103.36	159.88	112.33	38.04	-52.27	132.13	-36.094
900	110.65	172.49	118.32	48.75	-52.83	155.23	-37.692
1000	116.91	184.48	124.34	60.14	-52.95	178.37	-38.980

Parks and Hatton (1102), as well as McCullough, Finke, Messerly, Todd, Kincheloe, and Waddington (954), measured the low-temperature properties of this material. The data reported by Parks and Hatton were obtained on supercooled material and were superseded by those of the latter group, who found unusually complex behavior. A sluggish, solid-state transition was observed at 216.1°K, with $\Delta Ht° = 0.511$ kcal/mole. They reported $Ttp = 230.17°K$, $\Delta Hm° = 2.268$ kcal/mole, and $S°_{298}(l) = 63.34$ cal/(mole °K). Camin and Rossini (189) measured the vapor pressure and reported $Tb = 468.91°K$, where Seyer and Mann (1339) calculated

$\Delta Hv = 9.940$ kcal/mole. Miyazawa and Pitzer (1016) used the calorimetric data as a guide in calculating thermodynamic functions for the ideal gas by spectroscopic methods.

Speros and Rossini (1401) determined the enthalpy of combustion and derived $\Delta Hf°_{298}(l) = -52.45$ kcal/mole and $\Delta Hf°_{298}(g) = -40.38$ kcal/mole. Frye (450) measured equilibrium constants for the hydrogenation of naphthalene to decalin, *trans*, and found excellent agreement with the values calculated from thermodynamic data. The ratio of the *trans* to *cis* isomers of decalin was slightly lower than calculated from thermodynamic constants.

No. 475 Decahydronaphthalene, *trans* $C_{10}H_{18}$ (Ideal Gas State)   Mol Wt 138.244

	cal/(mole °K)				kcal/mole		
T°K	$Cp°$	$S°$	$-(G°-H°_{298})/T$	$H°-H°_{298}$	$\Delta Hf°$	$\Delta Gf°$	Log Kp
298	40.04	89.52	89.52	0.00	-43.57	17.55	-12.862
300	40.36	89.77	89.53	0.08	-43.65	17.92	-13.055
400	56.78	103.68	91.31	4.95	-47.49	39.06	-21.340
500	71.14	117.92	95.21	11.36	-50.56	61.06	-26.690
600	84.20	132.08	100.18	19.15	-52.85	83.61	-30.452
700	94.77	145.88	105.73	28.11	-54.43	106.49	-33.247
800	103.40	159.11	111.58	38.03	-55.47	129.54	-35.388
900	110.67	171.72	117.57	48.74	-56.03	152.72	-37.083
1000	116.93	183.71	123.59	60.13	-56.15	175.94	-38.449

Parks and Hatton (1102), as well as McCullough, Finke, Messerly, Todd, Kincheloe, and Waddington (954), investigated the low-temperature properties of this material. Although the findings of the two groups of workers are in substantial agreement, the work of the latter group is cited here: $Ttp = 242.76°K$, $\Delta Hm° = 3.455$ kcal/mole, and $S°_{298}(l) = 63.32$ cal/(mole °K). Camin and Rossini (189) gave $Tb = 460.41°K$, where Seyer and Mann (1339) calculated an average $\Delta Hv = 9.260$ kcal/mole. Miyazawa and Pitzer (1016) used the calorimetric data as a guide in calculating thermodynamic functions from spectroscopic data.

Speros and Rossini (1401) determined the enthalpy of combustion and derived $\Delta Hf°_{298}(l) = -55.14$ kcal/mole and $\Delta Hf°_{298}(g) = -43.57$ kcal/mole. Frye (450) measured equilibrium constants for the hydrogenation of naphthalene to decalin, *trans*, in excellent agreement with values calculated from the above thermodynamic data.

No. 476 Biphenyl $C_{12}H_{10}$ (Ideal Gas State)   Mol Wt 154.200

Coleman and Pilcher (246) made a careful determination of the enthalpy of combustion and derived $\Delta Hf°_{298}(s) = 24.02$ kcal/mole. They also listed

No. 476 Biphenyl $C_{12}H_{10}$ (Ideal Gas State) Mol Wt 154.200

T°K	Cp°	S°	$-(G°-H°_{298})/T$	$H°-H°_{298}$	$\Delta Hf°$	$\Delta Gf°$	Log Kp
		cal/(mole °K)			kcal/mole		
298	38.80	93.85	93.85	0.00	43.52	66.94	-49.063
300	39.05	94.10	93.86	0.08	43.48	67.08	-48.864
400	52.83	107.27	95.56	4.69	41.66	75.24	-41.107
500	64.38	120.34	99.21	10.57	40.23	83.81	-36.631
600	73.54	132.92	103.79	17.48	39.10	92.62	-33.736
700	80.93	144.83	108.81	25.22	38.25	101.63	-31.728
800	86.92	156.04	114.02	33.62	37.60	110.71	-30.244
900	91.86	166.57	119.28	42.56	37.14	119.88	-29.110
1000	96.00	176.47	124.51	51.96	36.89	129.10	-28.214

previous work on this compound. From available vapor pressure data on the solid, they selected as most accurate the values of Bradley and Cleasby (158) and calculated $\Delta Hf°_{298}(g) = 43.52$ kcal/mole.

Huffman, Parks, and Daniels (656) measured low-temperature heat capacity from 90° to 300°K and derived $S°_{298}(s) = 49.2$ cal/(mole °K). The vapor pressure data of Bradley and Cleasby (158) yield $S°_{298}(g) = 92.2$ cal/(mole °K). Katon and Lippincott (725) assigned vibrational frequencies and assumed free rotation in calculating $S°_{298}(g) = 93.9$ cal/(mole °K), in agreement within experimental error. Thermodynamic functions given by Katon and Lippincott (725) are adopted. They compared earlier work on spectroscopic calculations.

Stull (1432) selected $Tm = 342.7°K$ and $Tb = 528.1°K$. Spaght, Thomas, and Parks (1400) measured $\Delta Hm° = 4.44$ kcal/mole. Heat capacity of the liquid in the range 350–600°K was determined by Walker, Brooks, Ewing, and Miller (1567) and can be represented by Cp[cal/(mole °K)] = 33.46 + 0.094T. Walker et al. listed earlier references giving liquid heat capacities or enthalpies. Combination of the selected values for ΔHs_{298}, ΔHm, and heat capacities yields $\Delta Hv = 10.9$ kcal/mole at the normal boiling point.

CHAPTER 10

CHEMICAL THERMODYNAMICS OF COMPOUNDS OF CARBON, HYDROGEN, AND OXYGEN

INTRODUCTION

To the impressive number of combinations of carbon and hydrogen as hydrocarbons, a great proliferation develops upon the addition of a third kind of atom—oxygen. The thermochemical behavior of these compounds with oxygen-carbon and oxygen-hydrogen covalent bonds exhibits the order and regularity characteristic of the properties of hydrocarbons, showing only occasional departures; these are expected for polar compounds such as the alcohols, aldehydes, and acids, where terminal hydrogen atoms tend to form "hydrogen bonds" with electronegative atoms of other molecules.

The thermodynamic properties of the hydrocarbons and their oxidation products are of particular interest because the fuel value of a hydrocarbon depends on the difference between its internal energy and that of its combustion products. However, since not all combustion reactions proceed completely to yield carbon dioxide and water, it is also necessary to know the thermodynamic properties of the many stable and transitory intermediate compounds of carbon, hydrogen, and oxygen that are formed. Animals also provide their heat and energy needs by this process of oxidation, thereby producing numerous transient and stable intermediate compounds. Plants complete other portions of the cycle. With sunlight as the prime source of energy for the photosynthetic process, plants avidly sequester carbon dioxide from the atmosphere, uniting it with water to synthesize energetically less degraded compounds. After this upgrading process has gone through numerous stages, the intermediate compound (for instance, a sugar) may represent a useful fuel to the animal degradation cycle. Thus the energy-degrading and energy-upgrading processes are constantly meshing through the many common intermediate compounds containing carbon, hydrogen, and oxygen.

Chemical Thermodynamics of Compounds

These compounds are important because they pervade all living matter and have been synthesized and used by man to satisfy his needs. The thermochemistry of these compounds has not been as extensively studied as that of the hydrocarbons and presents a vast area that is ripe for the experimentalist and theoretician alike. The energy of a C—C bond is less than the energy contained in a C—H bond, so that the mildest way in which an oxygen atom can enter a hydrocarbon molecule is to form an ether type of compound. This causes the least degradation of the hydrocarbon molecule, and is arbitrarily discussed next.

The Ethers

Because the energy of a C—C bond is less than that of a C—H bond, the most favorable way in which an oxygen atom may combine with a hydrocarbon is to enter the molecule between two carbon atoms, thereby forming an ether. Ether formation may also be thought of artificially as the replacement of a —CH_2— group in a hydrocarbon by an oxygen atom, as has been done in this chapter.

The available thermochemical data concerning the effect of replacement by an oxygen atom of a —CH_2— group in a normal parent hydro-

Table 10-1 Enthalpy of Formation of Ideal Gaseous Ethers at 298.15°K

	Measured $\Delta Hf°_{298} (g)$[a]			Source of
Ether	Ether	Parent Hydrocarbon	$\Delta[\Delta Hf°_{298}]$	$\Delta Hf°$, Ether
Methyl	−43.99	−24.82	−19.17	(1145)
Ethyl methyl	−51.73	−30.15	−21.58	(1145)
Ethyl	−60.28	−35.00	−25.28	(1146)
Methyl propyl	−56.82	−35.00	−21.82	(1145)
Propyl	−70.00	−44.89	−25.11	(252)
Butyl	−79.80	−54.75	−25.05	See text
Methyl isopropyl	−60.24	−36.92	−23.32	(1145)
Methyl *tert*-butyl	−70.00	−44.35	−25.65	(1390)
Isopropyl	−76.20	−48.28	−27.92	See text
Isopropyl *tert*-butyl	−85.60	−53.57	−32.03	(1390)
sec-Butyl	−86.20	−56.79	−29.41	(252)
tert-Butyl	−87.20	−57.83	−29.37	(1390)
Average			−25.5	

[a] In kilocalories per mole.

Introduction

Table 10-2 Entropy of Formation of Ideal Gaseous Ethers at 298.15°K

	Observed $S^\circ_{298}(g)$[a]		
Ether	Ether	Parent Hydrocarbon	ΔS^*_{298}[a,b]
Methyl	63.75	64.51	−0.76
Ethyl	81.90	83.40	−1.50
Isopropyl	93.27	94.80	−1.53
Average			−1.26

[a] In calories per (mole degrees Kelvin).
[b] S^*_{298} = symmetryless entropy = $S^\circ_{298} + R \ln \sigma_t$.

carbon are collected in Tables 10-1 and 10-2. Large variations in the enthalpy of formation increment ΔHf°_{298} are apparent. These variations are related to the degree of branching and chain length of the alkyl groups. In these examples the $\Delta Hf^\circ_{298}(g)$ of the ether is more negative than that of the corresponding parent hydrocarbon by an average of 25.5 ± 7.0 kcal/mole, while the $S^*_{298}(g)$ for the ether is smaller than for the parent hydrocarbon by 1.26 ± 0.5 cal/(mole °K). A more sophisticated approach, such as described in Chapter 6, is needed to account adequately for the influences of branching and chain length.

Averaged ΔCp increments are provisionally summarized in Table 10-8.

The Alcohols

Addition of an atom of oxygen to a hydrocarbon at a C—H linkage may produce a C—O—H linkage, characteristic of an alcohol. The properties of normal alcohols provide an example of how a limited amount of data can be fitted into a realistic pattern and expanded to predict the properties of unmeasured compounds. The values of ΔHf°_{298} were measured on the compounds in the liquid state, but the values of $\Delta Hf^\circ_{298}(g)$ are more regular and have thus been selected for tabulation. The required values for enthalpies of vaporization at 298.15°K which have been calculated from the difference between $\Delta Hf(g)$ and $\Delta Hf(l)$ are listed in Table 10-3. These are in good agreement with the recent measurements of McCurdy and Laidler (966). Beyond propyl alcohol, the increment for the addition of a —CH$_2$— group increases the enthalpy of vaporization about 1.00

Table 10-3 Enthalpy of Vaporization of Normal Alcohols at 298.15°K

Alcohol	$\Delta H v_{298}$(kcal/mole)	Source
CH_3OH	8.96	(1249)
C_2H_5OH	10.08	(1553)
C_3H_7OH	11.24	(1553)
C_4H_9OH	12.28	(1553)
$C_5H_{11}OH$	13.28	(1553)

kcal/mole; hence for the normal alcohols with more than 3 carbon atoms ($n > 3$) above propyl,

$$\Delta H v_{298} = 8.28 + 1.00\, n \text{ (kcal/mole)}, \qquad [10\text{-}1]$$

in which n indicates the number of carbon atoms in the normal alcohol.

Green (530) published a review and comparison of the enthalpies of formation of the normal alcohols. Since that time the measurements of Skinner and Snelson (1365) and of Chao (214) were reported. Water is a difficult impurity to remove from alcohols and, when present, it causes the ΔHc to be too small and ΔHf°_{298} to have too negative a value. The values of ΔHf° reported by Chao (214) are more positive than the values from other sources (except for methanol), and therefore are the basis for the present study. Table 10-4 lists the gaseous enthalpy of formation values for each of the normal alcohols and its parent hydrocarbon, the increment between these two quantities, and the value for the alcohol derived by applying the average value of $\Delta[\Delta Hf^\circ_{298}]$ to the enthalpy of formation for the parent hydrocarbon.

Table 10-5 presents the measured gaseous entropy values for the alcohols, the increments in the symmetryless entropies, the gaseous entropies for the parent hydrocarbons, and the gaseous entropies of the alcohols derived from the parent hydrocarbons using the increment in the symmetryless entropy. These increments and the averaged ΔCp increments are given in Table 10-8 and have been used along with the corresponding values of the parent hydrocarbons to calculate the thermodynamic functions from tridecyl alcohol through eicosyl alcohol.

The Aldehydes

Addition of 1 atom of oxygen to any noncyclic hydrocarbon may produce at least one, and sometimes more than one, (isomeric) aldehyde.

Table 10-4 Enthalpy of Formation of Ideal Gaseous Normal Alcohols at 298.15°K

Alcohol	Measured $\Delta Hf°_{298}(g)$[a] Alcohol	Parent Hydrocarbon	$\Delta[\Delta Hf°_{298}]$[a]	$\Delta Hf°_{298}(g)$,[a] Alcohol, Derived	Source of ΔHf, Alcohol
CH_3OH	−48.08	−20.24	−27.84[b]	−51.63	(530)
C_2H_5OH	−56.12	−24.82	−31.30	−56.21	(214)
C_3H_7OH	−61.55	−30.15	−31.40	−61.54	(214)
C_4H_9OH	−65.59	−35.00	−30.59	−66.39	(214)
$C_5H_{11}OH$	−72.27	−39.96	−32.31	−71.35	(214)
$C_6H_{13}OH$	−76.39	−44.89	−31.50	−76.28	(214)
$C_7H_{15}OH$	−80.03	−49.82	−30.21	−81.21	(214)
$C_8H_{17}OH$	−85.34	−54.74	−30.60	−86.13	(214)
$C_9H_{19}OH$	−92.47	−59.67	−32.80	−91.06	(214)
$C_{10}H_{21}OH$	−96.38	−64.60	−31.78	−95.99	(214)
Average			−31.39		

[a]In kilocalories per mole.
[b]Not included in average.

Table 10-5 Entropy of Formation of Ideal Gaseous Normal Alcohols at 298.15°K

Alcohol	Measured $S°_{298}(g)$[a] Alcohol	Parent Hydrocarbon	$\Delta S^*_{298}(g)$[a]	$S°_{298}(g)$,[a] Alcohol, Derived	Source of $S°$, Alcohol
CH_3OH	57.29	54.85	−1.12[b]	57.82	(532)
C_2H_5OH	67.54	64.51	−0.53	67.48	(536)
C_3H_7OH	77.63	74.12	−0.05	77.09	(939)
C_4H_9OH	87.30	83.40	−0.66	86.37	(532)
$C_5H_{11}OH$	95.80	92.83	−0.59	95.80	(532)
$C_6H_{13}OH$	105.10	102.27	−0.72	105.24	(532)
$C_7H_{15}OH$	116.10	111.55	−0.99	114.52	(532)
Average			−0.59		

[a]In calories per (mole degrees Kelvin).
[b]Not included in average.

There is, unfortunately, less information on the aldehydes, and so the predicted properties are somewhat less certain. Chermin (221) reviewed the information on aldehydes and employed a —CH_2— incremental method to develop complete thermodynamic properties for the series of normal aldehydes from formaldehyde through decanal. More recently, additional measurements have been reported, and the current information is summarized in Tables 10-6 and 10-7. This information has been used to calculate the increments shown in Table 10-8 between values of the heat of formation, $\Delta[\Delta Hf°_{298}]$, the symmetryless entropy, S^*_{298}, and the heat capacity, ΔCp, for the normal aldehyde and its parent hydrocarbon. These

Table 10-6 Enthalpy of Formation of Ideal Gaseous Normal Aldehydes at 298.15°K

Aldehyde	Measured $\Delta Hf°_{298}(g)$[a] Aldehyde	Parent Hydrocarbon	$\Delta[\Delta Hf°_{298}]$[a]	$\Delta Hf°_{298}(g)$,[a] Aldehyde, Derived	Source of $\Delta Hf°$, Aldehyde
Acetaldehyde	−39.76	−24.82	−14.94	−39.31	(1249)
Propionaldehyde	−45.9	−30.15	−15.75	−44.64	(1508)
Butyraldehyde	−49.0	−35.00	−14.00	−49.49	(1065)
Heptaldehyde	−63.1	−49.82	−13.28	−64.31	(1065)
Average			−14.49		

[a] In kcal/mole.

Table 10-7 Entropy of Ideal Gaseous Normal Aldehydes at 298.15°K

Aldehyde	Observed $S°_{298}(g)$[a] Aldehyde	Parent Hydrocarbon	$\Delta S^*_{298}(g)$[a]	$S°_{298}(g)$,[a] Aldehyde, Derived	Source of $S°$, Aldehyde
Acetaldehyde	63.15[b]	64.51	−4.92	63.31	(1167)
Propionaldehyde	72.83[b]	74.12	−4.85	72.92	(221)
Butyraldehyde	82.44[c]	83.40	−4.52	82.20	(221)
Heptaldehyde	110.34[c]	111.55	−4.77	110.35	(221)
Average			−4.76		

[a] In cal/(mole °K).
[b] Calculated from spectra.
[c] $S°_{298}(l)$ converted with vapor pressure data.

Table 10-8 Increments for Calculating the Thermodynamic Properties of Ethers, Alcohols, and Aldehydes from the Parent Hydrocarbon

For the Ideal Gas State Process	$\Delta[\Delta H_f^\circ{}_{298}]$ (kcal/mole)	ΔS^*_{298} [cal/(mole °K)]	$\Delta C p$ [cal/(mole °K)] at								
			298°K	300°K	400°K	500°K	600°K	700°K	800°K	900°K	1000°K
n-R—CH$_2$—n-R′ → n-R—O—n-R′	−25.5	−1.26	−1.84	−1.87	−3.52	−4.81	−5.72	−6.44	−7.04	−7.57	−8.04
n-R—CH$_3$ → n-R—OH	−31.39	−0.59	−2.44	−2.45	−3.74	−4.81	−5.75	−6.51	−7.18	−7.78	−8.30
n-R—CH$_2$—CH$_3$ → n-R—CH=O	−14.49	−4.76	−4.32	−4.35	−6.47	−8.44	−10.09	−11.51	−12.74	−13.84	−14.88

increments have been used to revise the thermodynamic functions of Chermin (221).

The remaining oxygen-bearing organic compounds for which thermodynamic information is available contain features too widely variant to constitute a single class. This fact emphasizes the desirability of additional work.

OXYGEN COMPOUND TABLES

Tables of thermodynamic properties have been compiled from available data and are presented on the following pages. These tables are grouped for systematic consideration.

ALIPHATIC ETHER IDEAL GAS TABLES

No. 501 Methyl Ether C_2H_6O (Ideal Gas State) Mol Wt 46.068

	cal/(mole °K)				kcal/mole		
$T°K$	$Cp°$	$S°$	$-(G°-H°_{298})/T$	$H°-H°_{298}$	$\Delta Hf°$	$\Delta Gf°$	Log Kp
298	15.73	63.83	63.83	0.00	-43.99	-26.99	19.781
300	15.79	63.93	63.84	0.03	-44.01	-26.88	19.583
400	19.02	68.92	64.49	1.77	-45.20	-20.99	11.466
500	22.23	73.51	65.84	3.84	-46.24	-14.81	6.473
600	25.16	77.83	67.48	6.21	-47.10	-8.44	3.075
700	27.76	81.91	69.26	8.86	-47.79	-1.94	0.606
800	30.04	85.77	71.08	11.75	-48.34	4.64	-1.268
900	32.04	89.42	72.92	14.86	-48.74	11.30	-2.743
1000	33.79	92.89	74.74	18.15	-49.03	17.99	-3.931

Kennedy, Sagenkahn, and Aston (740) determined low-temperature thermal data, including $Ttp = 131.65°K$, $\Delta Hm = 1.180$ kcal/mole, and $Tb = 248.33°K$, where $\Delta Hv = 5.141$ kcal/mole, and derived $S°_{298}(g) = 63.72$ cal/(mole °K). Kistiakowsky and Rice (758) and Eucken and Franck (383) measured the vapor heat capacity at 300.76°K and at 200° and 280°K, respectively. These experimental thermal data are in good agreement with values calculated from the following spectroscopic information: the vibrational assignment of Kanazawa and Nukada (711), the moments of inertia reported by Kasai and Myers (722), and a barrier to internal rotation of 2.7 kcal/mole selected from 2.72 kcal/mole found by Kasai and Myers (722) and 2.625 kcal/mole determined by Fateley and Miller (408). Other calculations by Seha (1329), Hadni (558), and Banerjee and Doraiswamy (70) are in reasonable agreement, except that the heat capacity values of Banerjee and Doraiswamy deviate in a random fashion.

Pilcher, Pell, and Coleman (1145) used a flame calorimeter to determine $\Delta H f°_{298}(g) = -43.99$ kcal/mole, which is adopted. An old combustion result of Thomsen (1495) as calculated by Rossini, Wagman, Evans, Levine, and Jaffe (1249) is in good agreement at -44.3 kcal/mole. Given (489) summarized equilibrium constant data for the reaction $2CH_3OH(g) \rightleftarrows (CH_3)_2O(g) + H_2O(g)$, which yield -44.4 kcal/mole. Kobe and Lynn (780) listed $Tc = 400.1°K$ and $Pc = 53$ atm.

No. 502   Ethyl Methyl Ether C_3H_8O (Ideal Gas State)   Mol Wt 60.094

	cal/(mole °K)				kcal/mole		
$T°K$	$Cp°$	$S°$	$-(G°-H°_{298})/T$	$H°-H°_{298}$	$\Delta Hf°$	$\Delta Gf°$	Log Kp
298	21.45	74.24	74.24	0.00	-51.73	-28.12	20.611
300	21.53	74.38	74.25	0.04	-51.76	-27.98	20.379
400	26.08	81.20	75.14	2.43	-53.25	-19.81	10.825
500	30.53	87.50	76.99	5.26	-54.53	-11.30	4.939
600	34.58	93.43	79.24	8.52	-55.58	-2.56	0.931
700	38.11	99.04	81.67	12.16	-56.41	6.35	-1.983
800	41.19	104.33	84.18	16.13	-57.05	15.36	-4.196
900	43.87	109.34	86.70	20.38	-57.50	24.44	-5.936
1000	46.18	114.09	89.20	24.89	-57.80	33.57	-7.337

Pilcher, Pell, and Coleman (1145) used a flame calorimeter to determine $\Delta H f°_{298}(g) = -51.73$ kcal/mole, which is adopted. An old combustion value of Thomsen (1495) as listed by Kharasch (744) is in good agreement at -52.0 kcal/mole. Thermodynamic functions of the ideal gas are estimated using the constants of Table 10-8. Eucken and Franck (383) measured the vapor heat capacity at 200° and 280°K. Extrapolation of the estimated values shows reasonable agreement. Timmermans (1501) listed $Tb = 280.8°K$ and $Tc = 437.9°K$.

No. 503   Ethyl Ether $C_4H_{10}O$ (Ideal Gas State)   Mol Wt 74.120

	cal/(mole °K)				kcal/mole		
$T°K$	$Cp°$	$S°$	$-(G°-H°_{298})/T$	$H°-H°_{298}$	$\Delta Hf°$	$\Delta Gf°$	Log Kp
298	26.89	81.90	81.90	0.00	-60.28	-29.24	21.434
300	27.00	82.07	81.91	0.05	-60.32	-29.05	21.163
400	33.01	90.67	83.04	3.06	-62.13	-18.35	10.023
500	38.77	98.66	85.37	6.65	-63.67	-7.22	3.154
600	43.92	106.20	88.22	10.79	-64.92	4.19	-1.526
700	48.39	113.31	91.30	15.41	-65.89	15.79	-4.930
800	52.26	120.03	94.48	20.45	-66.62	27.50	-7.514
900	55.61	126.39	97.67	25.85	-67.13	39.31	-9.545
1000	58.51	132.40	100.85	31.56	-67.44	51.16	-11.181

Pilcher, Skinner, Pell, and Pope (1146) determined $\Delta Hf°_{298}(g) = -60.28$ kcal/mole by flame calorimetry. The only previous work was listed by Kharasch (744) and is seriously in error. Parks and Huffman (1103) measured the heat capacity from 90° to 298°K as well as $Tm = 156.9°K$ and $\Delta Hm = 1.745$ kcal/mole. Parks and Huffman (1105) used their extrapolation procedure to derive $S°_{298}(l) = 60.5$ cal/(mole °K). Timmermans (1501) listed a dozen determinations of the boiling point, from which we select $Tb = 307.7°K$. Mathews (940) measured $\Delta Hv = 6.38$ kcal/mole at the boiling point. Employing a ΔCp of vaporization of -14.3 cal/(mole °K), we calculate $\Delta Hv_{298} = 6.516$ kcal/mole. The vapor pressure at 298.15°K is 531 mm by interpolation of the values listed by Stull (1432). A gas imperfection correction is calculated using the Berthelot equation and critical constants of Kay and Donham (727), $Tc = 466.56°K$ and $Pc = 36.0$ atm. The ideal gas entropy is then calculated as follows.

$S°_{298}(l)$	60.50 cal/(mole °K)
ΔSv Vaporization, 6516/298.15	21.85
ΔS Compression, $-R \ln 760/531$	-0.71
ΔS Gas imperfection	0.18
$S°_{298}(g)$	81.82 cal/(mole °K)

An independent value of the entropy can be calculated from the equilibrium constant for the reaction $2C_2H_5OH(g) \rightleftarrows (C_2H_5)_2O(g) + H_2O(g)$ reported by Kabel and Johanson (708). They found $Kp = 24.7$ at 394.0°K, from which $\Delta G°_{394} = -2.511$ kcal/mole of ether. Using the enthalpy of formation of ethyl ether adopted above, with heat capacity data, yields $\Delta H°_{394} = -5.303$ kcal/mole of ether and, by difference, $\Delta S°_{394} = 7.09$ cal/(mole °K). Adjustment to 298.15°K yields $S°_{298}(g) = 82.0$ cal/(mole °K), in excellent agreement with the third-law result. An average of 81.9 cal/(mole °K) is adopted. Other equilibrium constant data listed by Kabel and Johanson appear to be in error, except for that of Atherton (49).

Vapor heat capacities are estimated using the constants of Table 10-8. An experimental value by Eucken and Franck (383) appears to be too high.

No. 504 Methyl Propyl Ether $C_4H_{10}O$ (Ideal Gas State) Mol Wt 74.120

Pilcher, Pell, and Coleman (1145) measured $\Delta Hf°_{298}(g) = -56.82$ kcal/mole by flame calorimetry. Thermodynamic functions of the ideal gas are estimated using values for pentane and the constants of Table 10-8.

No. 504 Methyl Propyl Ether $C_4H_{10}O$ (Ideal Gas State) Mol Wt 74.120

T°K	Cp°	S°	$-(G°-H°_{298})/T$	$H°-H°_{298}$	$\Delta Hf°$	$\Delta Gf°$	Log Kp
	cal/(mole °K)				kcal/mole		
298	26.89	83.52	83.52	0.00	-56.82	-26.27	19.252
300	27.00	83.69	83.53	0.05	-56.86	-26.08	18.997
400	33.01	92.29	84.66	3.06	-58.67	-15.53	8.486
500	38.77	100.28	86.99	6.65	-60.21	-4.57	1.996
600	43.92	107.82	89.84	10.79	-61.46	6.68	-2.433
700	48.39	114.93	92.92	15.41	-62.43	18.12	-5.656
800	52.26	121.65	96.10	20.45	-63.16	29.67	-8.105
900	55.61	128.01	99.29	25.85	-63.67	41.31	-10.031
1000	58.51	134.02	102.47	31.56	-63.98	53.00	-11.583

Timmermans (1501) listed four boiling point determinations, from which we select $Tb = 312.2°K$.

No. 505 Methyl Isopropyl Ether $C_4H_{10}O$ (Ideal Gas State) Mol Wt 74.120

T°K	Cp°	S°	$-(G°-H°_{298})/T$	$H°-H°_{298}$	$\Delta Hf°$	$\Delta Gf°$	Log Kp
	cal/(mole °K)				kcal/mole		
298	26.55	80.86	80.86	0.00	-60.24	-28.89	21.178
300	26.67	81.03	80.87	0.05	-60.28	-28.70	20.907
400	32.97	89.57	81.99	3.04	-62.11	-17.89	9.772
500	38.90	97.58	84.31	6.64	-63.64	-6.65	2.906
600	44.17	105.15	87.16	10.80	-64.87	4.87	-1.773
700	48.75	112.31	90.25	15.45	-65.81	16.57	-5.173
800	52.67	119.08	93.43	20.52	-66.50	28.38	-7.753
900	56.09	125.49	96.64	25.97	-66.97	40.28	-9.780
1000	59.08	131.55	99.83	31.73	-67.23	52.22	-11.412

Pilcher, Pell, and Coleman (1145) measured $\Delta Hf°_{298}(g) = -60.24$ kcal/mole by flame calorimetry. Thermodynamic functions of the ideal gas are estimated using values for 2-methylbutane and the constants of Table 10-8.

No. 506 Methyl tert-Butyl Ether $C_5H_{12}O$ (Ideal Gas State) Mol Wt 88.146

T°K	Cp°	S°	$-(G°-H°_{298})/T$	$H°-H°_{298}$	$\Delta Hf°$	$\Delta Gf°$	Log Kp
	cal/(mole °K)				kcal/mole		
298	32.07	84.36	84.36	0.00	-70.00	-29.98	21.978
300	32.22	84.56	84.37	0.06	-70.05	-29.74	21.664
400	40.18	94.93	85.73	3.69	-72.18	-15.97	8.723
500	47.70	104.72	88.55	8.09	-73.92	-1.71	0.745
600	54.28	114.02	92.03	13.20	-75.28	12.87	-4.686
700	59.66	122.80	95.81	18.90	-76.30	27.65	-8.631
800	64.36	131.09	99.70	25.11	-77.02	42.54	-11.620
900	68.03	138.88	103.63	31.73	-77.51	57.52	-13.967
1000	71.66	146.23	107.52	38.72	-77.77	72.55	-15.855

Smutny and Bondi (1390) measured the enthalpy of combustion and enthalpy of vaporization and derived $\Delta Hf°_{298}(g) = -70.0$ kcal/mole. Thermodynamic functions of the ideal gas are estimated using values for 2,2-dimethylbutane and the constants of Table 10-8.

No. 507 Propyl Ether $C_6H_{14}O$ (Ideal Gas State)   Mol Wt 102.172

	cal/(mole °K)				kcal/mole		
$T°K$	$Cp°$	$S°$	$-(G°-H°_{298})/T$	$H°-H°_{298}$	$\Delta Hf°$	$\Delta Gf°$	Log Kp
298	37.83	100.98	100.98	0.00	-70.00	-25.23	18.492
300	37.99	101.22	100.99	0.08	-70.05	-24.95	18.178
400	46.90	113.38	102.58	4.32	-72.50	-9.54	5.209
500	55.26	124.76	105.89	9.44	-74.55	6.45	-2.819
600	62.61	135.50	109.94	15.34	-76.19	22.80	-8.306
700	68.94	145.64	114.32	21.93	-77.45	39.41	-12.304
800	74.39	155.21	118.84	29.10	-78.38	56.16	-15.342
900	79.11	164.25	123.39	36.78	-79.00	73.03	-17.732
1000	83.17	172.80	127.91	44.90	-79.35	89.95	-19.658

Colomina, Pell, Skinner, and Coleman (252) measured the enthalpy of combustion and derived $\Delta Hf°_{298}(l) = -78.59$ kcal/mole. They used vapor pressure data to calculate $\Delta Hv°_{298} = 8.6$ kcal/mole and $\Delta Hf°_{298}(g) = -70.0$ kcal/mole. Thermodynamic functions are estimated using the values for heptane and the constants of Table 10-8. Timmermans (1501) listed two boiling point determinations, in good agreement at $Tb = 363.2°K$.

No. 508 Isopropyl Ether $C_6H_{14}O$ (Ideal Gas State)   Mol Wt 102.172

	cal/(mole °K)				kcal/mole		
$T°K$	$Cp°$	$S°$	$-(G°-H°_{298})/T$	$H°-H°_{298}$	$\Delta Hf°$	$\Delta Gf°$	Log Kp
298	37.83	93.27	93.27	0.00	-76.20	-29.13	21.352
300	37.99	93.51	93.28	0.08	-76.25	-28.84	21.010
400	46.90	105.67	94.87	4.32	-78.70	-12.65	6.912
500	55.26	117.05	98.18	9.44	-80.75	4.10	-1.794
600	62.61	127.79	102.23	15.34	-82.39	21.23	-7.732
700	68.94	137.93	106.61	21.93	-83.65	38.61	-12.054
800	74.39	147.50	111.13	29.10	-84.58	56.13	-15.333
900	79.11	156.54	115.68	36.78	-85.20	73.77	-17.912
1000	83.17	165.09	120.20	44.90	-85.55	91.46	-19.988

Colomina, Pell, Skinner, and Coleman (252) and Parks and Manchester (1111) reported the enthalpy of combustion as -958.51 and -958.64 kcal/mole, respectively, from which is derived $\Delta Hf°_{298}(l) = -83.95$ kcal/mole. Nicolini (1069) calculated $\Delta Hv°_{298} = 7.6$ kcal/mole from vapor pressure and vapor density data, while Smutny and Bondi (1390) quoted 7.9 kcal/mole. An average of 7.75 kcal/mole is adopted to calculate $\Delta Hf°_{298}(g) = -76.2$ kcal/mole. Parks, Huffman, and Barmore (1106) measured the heat capacity from 90° to 298°K and reported $Tm = 186.3°K$, $\Delta Hm = 2.635$ kcal/mole, and $S°_{298}(l) = 70.4$ cal/(mole °K). Nicolini (1069) gave the vapor pressure at 298.15°K as 150.7 mm. The following values are calculated using the above data and an estimated gas imperfection correction of 0.1 cal/(mole °K).

$S°_{298}(l)$	70.40 cal/(mole °K)
ΔSv Vaporization, 7750/298.15	25.99
ΔS Compression, $-R \ln 760/150.7$	-3.22
ΔS Gas imperfection	0.10
$S°_{298}(g)$	93.27 cal/(mole °K)

Vapor heat capacities are calculated from the values for 2,4-dimethylpentane and the constants of Table 10-8. Kobe, Ravicz, and Vohra (783) determined $Tc = 500.1°K$ and $Pc = 28.4$ atm. Stull (1432) listed $Tb = 340.7°K$.

No. 509   Isopropyl *tert*-Butyl Ether $C_7H_{16}O$ (Ideal Gas State)   Mol Wt 116.198

	cal/(mole °K)				kcal/mole		
T°K	$Cp°$	$S°$	$-(G° - H°_{298})/T$	$H° - H°_{298}$	$\Delta Hf°$	$\Delta Gf°$	Log Kp
298	43.30	99.89	99.89	0.00	-85.60	-30.79	22.570
300	43.69	100.16	99.90	0.09	-85.66	-30.46	22.186
400	53.84	114.13	101.73	4.97	-88.41	-11.62	6.348
500	63.52	127.21	105.53	10.84	-90.72	7.85	-3.432
600	71.95	139.56	110.18	17.63	-92.56	27.74	-10.104
700	79.22	151.21	115.22	25.19	-93.96	47.91	-14.958
800	85.46	162.20	120.41	33.44	-94.98	68.24	-18.641
900	90.86	172.58	125.64	42.26	-95.67	88.70	-21.537
1000	95.56	182.41	130.83	51.58	-96.03	109.22	-23.868

Smutny and Bondi (1390) reported the enthalpy of combustion and enthalpy of vaporization and derived $\Delta Hf°_{298}(l) = -94.0$ kcal/mole and $\Delta Hf°_{298}(g) = -85.6$ kcal/mole. Thermodynamic functions of the ideal gas are estimated using the values for 2,2,4-trimethylpentane and the constants of Table 10-8.

No. 510   Butyl Ether $C_8H_{18}O$ (Ideal Gas State)   Mol Wt 130.224

Colomina, Pell, Skinner, and Coleman (252) and Skuratov and Kozina (1370) measured the enthalpy of combustion of the liquid as -1276.9 and -1277.1 kcal/mole, respectively, from which is derived $\Delta Hf°_{298}(l) = -90.3$ kcal/mole. Mathews and Fehlandt (941) measured $\Delta Hv = 8.83$ kcal/mole at 414.5°K. An estimate of ΔCp of vaporization as -14

No. 510 Butyl Ether $C_8H_{18}O$ (Ideal Gas State) Mol Wt 130.224

$T°K$	$Cp°$	$S°$	$-(G°-H°_{298})/T$	$H°-H°_{298}$	$\Delta Hf°$	$\Delta Gf°$	Log Kp
	cal/(mole °K)				kcal/mole		
298	48.76	119.60	119.60	0.00	-79.80	-21.16	15.508
300	48.98	119.91	119.61	0.10	-79.87	-20.80	15.150
400	60.78	135.63	121.67	5.59	-82.95	-0.62	0.339
500	71.75	150.40	125.95	12.23	-85.51	20.27	-8.858
600	81.29	164.35	131.20	19.89	-87.55	41.61	-15.156
700	89.49	177.51	136.89	28.44	-89.09	63.27	-19.754
800	96.52	189.93	142.75	37.75	-90.21	85.11	-23.249
900	102.60	201.66	148.65	47.71	-90.95	107.08	-26.001
1000	107.86	212.74	154.51	58.24	-91.34	129.12	-28.218

cal/(mole °K) yields $\Delta Hv°_{298} = 10.5$ kcal/mole and $\Delta Hf°_{298}(g) = -79.8$ kcal/mole. Timmermans (1501) listed $Tm = 175.3°K$ and $Tb = 415.6°K$.

No. 511 sec-Butyl Ether $C_8H_{18}O$ (Ideal Gas State) Mol Wt 130.224

$T°K$	$Cp°$	$S°$	$-(G°-H°_{298})/T$	$H°-H°_{298}$	$\Delta Hf°$	$\Delta Gf°$	Log Kp
	cal/(mole °K)				kcal/mole		
298	48.76	110.57	110.57	0.00	-86.20	-24.87	18.226
300	48.98	110.88	110.58	0.10	-86.27	-24.49	17.839
400	60.78	126.60	112.64	5.59	-89.35	-3.41	1.862
500	71.75	141.37	116.92	12.23	-91.91	18.38	-8.034
600	81.29	155.32	122.17	19.89	-93.95	40.63	-14.799
700	89.49	168.48	127.86	28.44	-95.49	63.20	-19.729
800	96.52	180.90	133.72	37.75	-96.61	85.93	-23.474
900	102.60	192.63	139.62	47.71	-97.35	108.81	-26.421
1000	107.86	203.71	145.48	58.24	-97.74	131.75	-28.793

Colomina, Pell, Skinner, and Coleman (252) measured the enthalpy of combustion and derived $\Delta Hf°_{298}(l) = -95.98$ kcal/mole. They employed vapor pressure data to calculate $\Delta Hv°_{298} = 9.8$ kcal/mole and $\Delta Hf°_{298}(g) = -86.2$ kcal/mole. Thermodynamic functions of the ideal gas are estimated using values for 3,5-dimethylheptane and the constants of Table 10-8.

No. 512 tert-Butyl Ether $C_8H_{18}O$ (Ideal Gas State) Mol Wt 130.224

Smutny and Bondi (1390) measured the enthalpy of combustion and derived $\Delta Hf°_{298}(l) = -96.1$ kcal/mole. They also measured the vapor pressure over the range 4–109°C and calculated $\Delta Hv°_{298} = 9.0$ kcal/mole and $\Delta Hf°_{298}(g) = -87.2$ kcal/mole. Thermodynamic functions of the ideal gas are estimated using the values for 2,2,4,4-tetramethylpentane and the constants of Table 10-8. Smutny and Bondi (1390) calculated the strain energy in this molecule as 7.6 kcal/mole.

Cyclic Ether Ideal Gas Tables

No. 512 tert-Butyl Ether C$_8$H$_{18}$O   (Ideal Gas State)   Mol Wt 130.224

T°K	Cp°	S°	$-(G°-H°_{298})/T$	$H°-H°_{298}$	$\Delta Hf°$	$\Delta Gf°$	Log Kp
	cal/(mole °K)				kcal/mole		
298	48.76	102.12	102.12	0.00	-87.20	-23.35	17.112
300	48.98	102.43	102.13	0.10	-87.27	-22.95	16.721
400	60.78	118.15	104.19	5.59	-90.35	-1.03	0.562
500	71.75	132.92	108.47	12.23	-92.91	21.61	-9.443
600	81.29	146.87	113.72	19.89	-94.95	44.70	-16.281
700	89.49	160.03	119.41	28.44	-96.49	68.11	-21.264
800	96.52	172.45	125.27	37.75	-97.61	91.69	-25.047
900	102.60	184.18	131.17	47.71	-98.35	115.41	-28.024
1000	107.86	195.26	137.03	58.24	-98.74	139.20	-30.421

CYCLIC ETHER IDEAL GAS TABLES

No. 513 Ethylene Oxide C$_2$H$_4$O   (Ideal Gas State)   Mol Wt 44.052

T°K	Cp°	S°	$-(G°-H°_{298})/T$	$H°-H°_{298}$	$\Delta Hf°$	$\Delta Gf°$	Log Kp
	cal/(mole °K)				kcal/mole		
298	11.54	57.94	57.94	0.00	-12.58	-3.13	2.292
300	11.60	58.02	57.95	0.03	-12.60	-3.07	2.236
400	14.95	61.82	58.44	1.36	-13.51	0.25	-0.137
500	18.03	65.49	59.49	3.01	-14.25	3.78	-1.652
600	20.62	69.02	60.78	4.95	-14.85	7.44	-2.711
700	22.78	72.36	62.20	7.12	-15.32	11.20	-3.496
800	24.60	75.53	63.67	9.49	-15.68	15.01	-4.100
900	26.15	78.52	65.16	12.03	-15.94	18.86	-4.580
1000	27.47	81.34	66.63	14.71	-16.12	22.74	-4.970

Pell and Pilcher (1133) measured the enthalpy of combustion in a flame calorimeter and derived $\Delta Hf°_{298}(g) = -12.58$ kcal/mole. They also compared previous work on this compound.

Giauque and Gordon (475) made a complete low-temperature study and reported $Ttp = 160.71°K$, $\Delta Hm = 1.236$ kcal/mole, and $Tb = 283.71°K$, where $\Delta Hv = 6.101 \pm 0.006$ kcal/mole and $S°_{298}(g) = 57.94$ cal/(mole °K). Günthard and Heilbronner (548) assigned the vibrational frequencies and calculated the thermodynamic functions. Their calculated vapor heat capacities agree with the measured values of Kistiakowsky and Rice (758) within 0.1 cal/(mole °K), while their calculated gaseous entropy is 0.30 cal/(mole °K) above that reported by Giauque and Gordon (475). Kobe and Pennington (781) also repeated the calculation of the thermodynamic functions, while more recently Lord and Nolin (885) restudied the spectrum with no significant change in the calculated thermodynamic properties. Kobe and Lynn (780) reported $Tc = 468°K$ and $Pc = 71.0$ atm.

No. 514 Propylene Oxide C_3H_6O (Ideal Gas State) Mol Wt 58.078

$T°K$	$Cp°$	$S°$	$-(G°-H°_{298})/T$	$H°-H°_{298}$	$\Delta Hf°$	$\Delta Gf°$	$Log\ Kp$
		cal/(mole °K)			kcal/mole		
298	17.29	68.53	68.53	0.00	-22.17	-6.16	4.517
300	17.38	68.64	68.54	0.04	-22.20	-6.07	4.418
400	22.16	74.31	69.28	2.02	-23.39	-0.50	0.272
500	26.46	79.73	70.83	4.45	-24.37	5.34	-2.335
600	30.07	84.88	72.74	7.29	-25.15	11.36	-4.137
700	33.11	89.75	74.83	10.45	-25.75	17.50	-5.462
800	35.68	94.34	76.98	13.89	-26.21	23.70	-6.475
900	37.89	98.68	79.16	17.57	-26.53	29.96	-7.276
1000	39.79	102.77	81.31	21.46	-26.72	36.26	-7.924

Sinke and Hildenbrand (1356) measured the heat of combustion and the enthalpy of vaporization of the liquid and reported $\Delta Hf°_{298}(l) = -28.84$ kcal/mole and $\Delta Hf°_{298}(g) = -22.17$ kcal/mole. Oetting (1083) measured the heat capacity from 11° to 300°K and reported $Tm = 161.22°K$, $\Delta Hm = 1.561$ kcal/mole, and $S°_{298}(l) = 46.91$ cal/(mole °K). A residual entropy of $R \ln 2$ was included for the entropy of mixing of the optical d and l isomers. The entropy summary follows [in cal/(mole °K.)].

S(mixing), $R \ln 2$	1.377
$S_{11} - S_0$ (Debye-Einstein)	0.119
$S_{161.22} - S_{11°K}$ crystal (graphical)	19.297
ΔSm, 1561.4/161.22	9.686
$S_{298.15} - S_{161.22°K}$ liquid (graphical)	16.431
$S°_{298.15}(l)$	46.91 ± 0.15
ΔSv, 6667/298.15	22.361
ΔS (gas imperfection)	0.108
ΔS (compression), $-R \ln(760/538)$	-0.686
$S°_{298.15}$ (ideal gas)	68.69 ± 0.20

The moment of inertia $= 824.7 \times 10^{-117}$ g^3-cm^6 from Swalen and Herschbach (1457), a liquid phase vibrational assignment substantially like that of Tobin (1510), and a barrier to internal rotation of 2.56 kcal/mole given by Herschbach and Swalen (588) and substantiated by Fateley and Miller (409) are used to calculate the thermodynamic functions, including $S°_{298}(g) = 68.53$ cal/(mole °K), in good agreement with the third-law measurement and the spectroscopic value calculated by Green (531).

No. 515 Furan C_4H_4O (Ideal Gas State) Mol Wt 68.072

Guthrie, Scott, Hubbard, Katz, McCullough, Gross, Williamson, and Waddington (551) made a very thorough study of this material and

No. 515 Furan C_4H_4O (Ideal Gas State) Mol Wt 68.072

	cal/(mole °K)				kcal/mole		
T°K	$Cp°$	$S°$	$-(G°-H°_{298})/T$	$H°-H°_{298}$	$\Delta Hf°$	$\Delta Gf°$	Log Kp
298	15.64	63.86	63.86	0.00	-8.29	0.21	-0.154
300	15.75	63.96	63.87	0.03	-8.31	0.26	-0.190
400	21.20	69.26	64.55	1.89	-9.19	3.26	-1.781
500	25.73	74.50	66.02	4.24	-9.87	6.45	-2.821
600	29.31	79.52	67.85	7.00	-10.40	9.77	-3.557
700	32.13	84.25	69.86	10.08	-10.81	13.16	-4.110
800	34.41	88.70	71.94	13.41	-11.13	16.61	-4.536
900	36.30	92.86	74.04	16.95	-11.37	20.09	-4.878
1000	37.89	96.77	76.12	20.66	-11.53	23.60	-5.157

reported $\Delta Hf°_{298}(g) = -8.293$ kcal/mole, $Tt = 150.0°$K, $\Delta Ht° = 0.489$ kcal/mole, $Tm = 187.54°$K, $\Delta Hm° = 0.909$ kcal/mole, $\Delta Hv = 6.474$ kcal/mole at $Tb = 304.5°$K, and $S°_{298}(g) = 63.86$ cal/(mole °K), as well as complete thermodynamic functions. Bak, Brodersen, and Hansen (62) recently completed a new vibrational assignment and calculated thermodynamic functions inconsistent with the observed values. The experimentally based values of the former workers are adopted here.

No. 516 p-Dioxane $C_4H_8O_2$ (Ideal Gas State) Mol Wt 88.104

	cal/(mole °K)				kcal/mole		
T°K	$Cp°$	$S°$	$-(G°-H°_{298})/T$	$H°-H°_{298}$	$\Delta Hf°$	$\Delta Gf°$	Log Kp
298	22.48	71.65	71.65	0.00	-75.30	-43.21	31.670
300	22.60	71.79	71.66	0.05	-75.34	-43.01	31.331
400	30.23	79.35	72.63	2.69	-77.17	-31.94	17.450
500	37.49	86.90	74.73	6.09	-78.57	-20.46	8.944
600	43.44	94.28	77.38	10.14	-79.58	-8.74	3.185
700	48.25	101.35	80.30	14.74	-80.27	3.12	-0.975
800	52.15	108.05	83.36	19.76	-80.70	15.06	-4.114
900	55.36	114.39	86.46	25.14	-80.93	27.05	-6.569
1000	58.05	120.36	89.55	30.82	-80.98	39.06	-8.537

Snelson and Skinner (1391) measured the heat of combustion and reported $\Delta Hf°_{298}(l) = -84.50$ kcal/mole, which yields $\Delta Hf°_{298}(g) = -75.30$ kcal/mole with the value $\Delta Hv_{298} = 9.20$ kcal/mole of McDonald (968). Jacobs and Parks (677) measured the third-law entropy and reported $\Delta Ht° = 0.562$ kcal/mole at $Tt = 272.9°$K, $\Delta Hm° = 3.07$ kcal/mole at $Tm = 284.1°$K, and $S°_{298}(l) = 46.67$ cal/(mole °K). The vapor pressure data of Stull (1432) indicated $Tb = 374.3°$K and permit calculation of $\Delta Hv = 8.30$ kcal/mole at Tb. The vibrational frequency assignment of Malherbe and Bernstein (916) is utilized to calculate $S°_{298}(g) = 71.65$ cal/(mole °K) and the other thermodynamic functions. The consistency of these values is bolstered by the following calculations (neglecting gas imperfection).

$S°_{298}(l)$	46.67 cal/(mole °K)
ΔSv_{298}, 9200/298.15	30.86
ΔS, compression, $-R \ln 760/40$	-5.85
$S°_{298}$ (ideal gas state)	71.68 cal/(mole °K)

ALKANOL IDEAL GAS TABLES

No. 517 Methanol CH_4O (Ideal Gas State) Mol Wt 32.042

	cal/(mole °K)				kcal/mole		
T°K	$Cp°$	$S°$	$-(G°-H°_{298})/T$	$H°-H°_{298}$	$\Delta Hf°$	$\Delta Gf°$	Log Kp
298	10.49	57.29	57.29	0.00	-48.08	-38.84	28.468
300	10.52	57.36	57.30	0.02	-48.10	-38.78	28.252
400	12.29	60.62	57.73	1.16	-48.95	-35.54	19.420
500	14.22	63.58	58.61	2.49	-49.70	-32.11	14.032
600	16.02	66.33	59.67	4.00	-50.34	-28.52	10.389
700	17.62	68.92	60.81	5.69	-50.88	-24.84	7.756
800	19.04	71.37	61.98	7.52	-51.31	-21.10	5.763
900	20.29	73.69	63.15	9.49	-51.66	-17.30	4.200
1000	21.38	75.88	64.31	11.57	-51.93	-13.46	2.941

Chao and Rossini (215) burned liquid methanol and reported $\Delta Hf°_{298}(l) = -57.24$ kcal/mole, compared to the value $\Delta Hf°_{298}(l) = -57.02$ kcal/mole reported by Rossini (1240). For reasons stated on page 408, the value of Rossini is selected, yielding $\Delta Hf°_{298}(g) = -48.08$ kcal/mole by employing $\Delta Hv_{298} = 8.96$ kcal/mole as listed by Rossini, Wagman, Evans, Levine, and Jaffe (1249). Kelley (732) measured the low-temperature heat capacity and reported $Tt = 157.4°K$, $\Delta Ht° = 0.154$ kcal/mole, $Tm = 175.25°K$, $\Delta Hm° = 0.757$ kcal/mole, and $S°_{298}(l) = 30.3$ cal/(mole °K) Rossini et al. (1249) selected $Tb = 337.9°K$. ΔHv is selected as 8.43 kcal/mole from the following work.

$\Delta Hv_{337.9}$ (kcal/mole)	Source	
8.429	Fiock, Ginnings, and Holton	(426)
8.43	Mathews	(940)
8.442	Weltner and Pitzer	(1588)

Ivash, Li, and Pitzer (674) reviewed the spectroscopic data, made a rigorous calculation of the internal rotational contribution [the tables of Pitzer and Gwinn (1162) do not apply to this small molecule], and calculated complete thermodynamic properties, reporting $S°_{298}(g) = 57.29$ cal/(mole °K). These calculated properties are in good agreement with the measured vapor heat capacities of Eucken and Franck (383), De Vries and Collins (323), Rowlinson (1256), and Weltner and Pitzer (1588), and with the third-law entropy of Kelley (732) when corrected using the gas imperfection data of Weltner and Pitzer (1588). The latter investigators were the first to treat methanol vapor as a mixture of dimers and tetramers. The more recent infrared absorption measurements of Inskeep, Kelliher, McMahon, and Somers (671) support the model that methanol vapor is composed of monomer, dimer, and tetramer, with little or no trimer present. Kobe and Lynn (780) reported $Tc = 513°K$, $Pc = 78.5$ atm, and $dc = 0.272$ g/cc.

No. 518 Ethyl Alcohol C_2H_6O (Ideal Gas State) Mol Wt 46.068

	cal/(mole °K)				kcal/mole		
$T°K$	$Cp°$	$S°$	$-(G°-H°_{298})/T$	$H°-H°_{298}$	$\Delta Hf°$	$\Delta Gf°$	Log Kp
298	15.64	67.54	67.54	0.00	-56.12	-40.22	29.483
300	15.71	67.64	67.55	0.03	-56.14	-40.13	29.231
400	19.36	72.67	68.21	1.79	-57.32	-34.60	18.904
500	22.77	77.36	69.57	3.90	-58.31	-28.80	12.590
600	25.69	81.78	71.24	6.33	-59.11	-22.83	8.314
700	28.19	85.93	73.05	9.02	-59.76	-16.73	5.222
800	30.33	89.84	74.90	11.95	-60.27	-10.55	2.881
900	32.19	93.52	76.77	15.08	-60.65	-4.30	1.045
1000	33.83	97.00	78.62	18.38	-60.93	1.98	-0.433

Rossini, Wagman, Evans, Levine, and Jaffe (1249) selected $\Delta Hf°_{298}(g) = -56.240$ kcal/mole, but more recently Chao and Rossini (215) measured $\Delta Hf°_{298}(l) = -66.20$ kcal/mole, which is adopted for the reasons stated on page 408. Wadsö and Wadsö (1553) measured $\Delta Hv_{298} = 10.08$ kcal/mole; hence $\Delta Hf°_{298}(g) = -56.12$ kcal/mole. Kelley (733) measured the heat capacity from 16° to 298°K and reported $Tm = 158.5°K$, $\Delta Hm° = 1.200$ kcal/mole, and $S°_{298}(l) = 38.4$ cal/(mole °K). Fiock, Ginnings, and Holton (426) measured and reported $Tb = 351.7°K$, at which $\Delta Hv = 9.220$ kcal/mole. Sinke and De Vries (1355), Halford and Miller (565), and Barrow (76) measured the gaseous heat capacity. Barrow assigned the fundamental vibrational frequencies, reviewed the equilibrium data, and achieved a good fit of all the experimental data. Nearly a decade later Green (536) reviewed the data, revised the spectroscopic assignment slightly, calculated the thermodynamic functions, and reported $S°_{298}(g) = 67.54$ cal/(mole °K).

Kobe and Lynn (780) found $Tc = 516°K$, $Pc = 63.0$ atm, and $dc = 0.276$ g/cc.

No. 519 Propyl Alcohol C_3H_8O (Ideal Gas State) Mol Wt 60.094

$T°K$	$Cp°$	$S°$	$-(G°-H°_{298})/T$	$H°-H°_{298}$	$\Delta Hf°$	$\Delta Gf°$	Log Kp
		cal/(mole °K)			kcal/mole		
298	20.82	77.63	77.63	0.00	-61.55	-38.95	28.550
300	20.91	77.76	77.64	0.04	-61.58	-38.81	28.273
400	25.86	84.46	78.52	2.38	-63.11	-30.98	16.926
500	30.51	90.75	80.34	5.21	-64.40	-22.79	9.963
600	34.56	96.68	82.57	8.47	-65.46	-14.37	5.235
700	38.03	102.27	84.99	12.10	-66.29	-5.79	1.807
800	41.04	107.55	87.48	16.06	-66.94	2.90	-0.791
900	43.65	112.54	89.99	20.29	-67.41	11.66	-2.831
1000	45.93	117.26	92.49	24.78	-67.73	20.47	-4.474

Subsequent to the review of combustion literature by Green (530), the measurements of Chao and Rossini (215) became available and are adopted. Chao and Rossini reported $\Delta Hf°_{298}(l) = -72.79$ kcal/mole. The measured $\Delta Hv_{298} = 11.24$ kcal/mole of Wadsö and Wadsö (1553) gives $\Delta Hf°_{298}(g) = -61.55$ kcal/mole. Parks, Kelley, and Huffman (1108) measured the thermal properties from 90° to 298°K and reported $S°_{298}(l) = 46.1$ cal/(mole °K). Parks and Huffman (1103) reported $Tm = 147°K$ and $\Delta Hm° = 1.240$ kcal/mole. Mathews and McKetta (939) measured the vapor pressure and reported $\Delta Hv = 9.852$ kcal/mole at $Tb = 370.35°K$. They also measured the vapor heat capacity, assigned the fundamental vibrational frequencies and barrier to internal rotation, reported $S°_{298}(g) = 77.63$ cal/(mole °K), and calculated the thermodynamic properties of the ideal gas adopted here. Kobe and Lynn (780) reported $Tc = 537°K$, $Pc = 50.2$ atm, and $dc = 0.273$ g/cc.

No. 520 Isopropyl Alcohol C_3H_8O (Ideal Gas State) Mol Wt 60.094

$T°K$	$Cp°$	$S°$	$-(G°-H°_{298})/T$	$H°-H°_{298}$	$\Delta Hf°$	$\Delta Gf°$	Log Kp
		cal/(mole °K)			kcal/mole		
298	21.21	74.07	74.07	0.00	-65.15	-41.49	30.411
300	21.31	74.21	74.08	0.04	-65.18	-41.34	30.118
400	26.78	81.09	74.98	2.45	-66.65	-33.17	18.120
500	31.89	87.64	76.86	5.40	-67.82	-24.66	10.776
600	35.76	93.81	79.18	8.78	-68.74	-15.94	5.804
700	39.21	99.58	81.68	12.53	-69.46	-7.07	2.208
800	42.13	105.01	84.26	16.61	-69.99	1.87	-0.511
900	44.63	110.12	86.86	20.95	-70.36	10.88	-2.642
1000	46.82	114.94	89.43	25.52	-70.58	19.93	-4.355

Parks, Mosley, and Peterson (1116), Parks and Manchester (1111), Chao and Rossini (215), and Snelson and Skinner (1391) measured the heat of combustion. The last two workers also reviewed the hydrogenation of acetone by Dolliver, Gresham, Kistiakowsky, Smith, and Vaughan (334) and found good agreement with their value of $\Delta Hf°_{298}(l) = -76.04$ kcal/mole. Wadsö and Wadsö (1553), Berman, Larkam, and McKetta (113), and Hales, Cox, and Lees (561) are in good agreement with the measured value $\Delta Hv_{298} = 10.89$ kcal/mole, leading to $\Delta Hf°_{298}(g) = -65.15$ kcal/mole. The latter two groups of workers also measured vapor heat capacities and treated the complex relationship between Cp and temperature, arising from molecular association, as an equilibrium mixture of monomers, dimers, and tetramers. Kelley (734) measured the heat capacity from 20° to 298.15°K. More recently Andon, Counsell, and Martin (20) measured the heat capacity from 12° to 327°K and reported $\Delta Hm° = 1.293$ kcal/mole at $Ttp = 185.20°K$ and $S°_{298}(l) = 43.16$ cal/(mole °K). Green (538) recently reviewed the spectral data, assigned the fundamental vibrational frequencies, calculated $S°_{298}(g) = 74.07$ cal/(mole °K), and determined the complete thermodynamic functions. He obtained the best over-all agreement between the experimental and calculated values, with symmetrical three-fold barriers to rotation of the CH_3 and OH groups of 4.00 and 0.80 kcal/mole, respectively. Hales, Cox, and Lees (561) reported $\Delta Hv = 9.512$ kcal/mole at $Tb = 355.39°K$. Kobe and Lynn (780) reported $Tc = 508.8°K$, $Pc = 53$ atm, and $dc = 0.274$ g/cc.

No. 521 Butyl Alcohol $C_4H_{10}O$ (Ideal Gas State)   Mol Wt 74.120

$T°K$	$Cp°$	$S°$	$-(G°-H°_{298})/T$	$H°-H°_{298}$	$\Delta Hf°$	$\Delta Gf°$	Log Kp
		cal/(mole °K)			kcal/mole		
298	26.29	86.80	86.80	0.00	-65.59	-36.01	26.397
300	26.41	86.97	86.81	0.05	-65.63	-35.83	26.102
400	32.80	95.45	87.92	3.02	-67.48	-25.61	13.991
500	38.76	103.43	90.23	6.60	-69.03	-14.96	6.537
600	43.90	110.96	93.06	10.74	-70.28	-4.02	1.466
700	48.31	118.06	96.13	15.36	-71.25	7.10	-2.217
800	52.11	124.77	99.30	20.38	-72.00	18.34	-5.010
900	55.40	131.10	102.48	25.76	-72.53	29.67	-7.205
1000	58.26	137.09	105.65	31.45	-72.86	41.05	-8.972

Tjebbes (1506), Skinner and Snelson (1365), Gundry, Head, and Lewis (547), and Chao and Rossini (215) all measured the enthalpy of combustion. Except for the work of Tjebbes, the results are in good agreement, and the value of Chao and Rossini, $\Delta Hf°_{298}(l) = -77.87$ kcal/mole, is selected. Wadsö and Wadsö (1553) measured $\Delta Hv°_{298} = 12.28$ kcal/mole, making $\Delta Hf°_{298}(g) = -65.59$ kcal/mole. Counsell, Hales, and Martin (280) determined low-temperature thermal data as well as vapor heat capacity and enthalpy of vaporization, finding $Ttp = 184.51°K$, $\Delta Hm° = 2.240$ kcal/mole, $Tb = 390.88°K$, where $\Delta Hv = 10.31$ kcal/mole, $S°_{298}(l) = 53.95$ cal/(mole °K), and $S°_{298}(g) = 86.8$ cal/(mole °K). Green (532) used a method of increments to derive $S°_{298}(g) = 86.90$ cal/(mole °K), as well as the remaining thermodynamic functions adopted here. Ambrose and Townsend (14) measured $Tc = 563.0°K$ and $Pc = 43.55$ atm.

No. 522 sec-Butyl Alcohol $C_4H_{10}O$ (Ideal Gas State) Mol Wt 74.120

	cal/(mole °K)				kcal/mole		
$T°K$	$Cp°$	$S°$	$-(G°-H°_{298})/T$	$H°-H°_{298}$	$\Delta Hf°$	$\Delta Gf°$	Log Kp
298	27.08	85.81	85.81	0.00	-69.86	-39.99	29.311
300	27.20	85.98	85.82	0.06	-69.90	-39.81	28.997
400	33.70	94.71	86.96	3.10	-71.66	-29.49	16.114
500	39.70	102.89	89.34	6.78	-73.12	-18.78	8.208
600	44.72	110.59	92.24	11.01	-74.28	-7.80	2.842
700	49.02	117.81	95.39	15.70	-75.18	3.35	-1.047
800	52.68	124.60	98.62	20.79	-75.86	14.61	-3.992
900	55.88	130.99	101.86	26.22	-76.33	25.96	-6.303
1000	58.62	137.03	105.08	31.95	-76.63	37.35	-8.162

Combustion measurements of Skinner and Snelson (1365) and Chao and Rossini (215) are in excellent agreement and lead to $\Delta Hf°_{298}(l) = -81.90$ kcal/mole. Berman and McKetta (114) measured $\Delta Hv_{298} = 12.038$ kcal/mole, leading to $\Delta Hf°_{298}(g) = -69.86$ kcal/mole. The latter workers measured the vapor heat capacity from 365° to 455°K. They selected an equilibrium mixture of monomers, dimers, and tetramers as a model for an equation of state. From the molecular structure, spectra, and calorimetric data, barriers were evaluated and tables of thermodynamic functions computed. They reported $S°_{298}(g) = 85.81$ cal/(mole °K). Biddiscombe, Collerson, Handley, Herington, Martin, and Sprake (126) measured the vapor pressure, found $Tb = 372.66°K$, and calculated $\Delta Hv_{373} = 9.80$ kcal/mole, in good agreement with Berman and McKetta (114). Ambrose and Townsend (14) reported $Tc = 535.95°K$ and $Pc = 41.39$ atm.

No. 523 tert-Butyl Alcohol $C_4H_{10}O$ (Ideal Gas State) Mol Wt 74.120

Skinner and Snelson (1365) measured the heat of combustion and reported $\Delta Hf°_{298}(l) = -85.87$ kcal/mole, which becomes $\Delta Hf°_{298}(g) = -77.87$ kcal/mole when coupled with the measured heat of vaporization of Beynon and McKetta (124). Oetting (1082) studied the thermal behavior from 15° to 330°K and reported $\Delta Ht° = 0.198$ kcal/mole to transform crystal II phase to crystal I phase at $Tt = 286.14°K$, $\Delta Ht° = 0.155$

Alkanol Ideal Gas Tables

No. 523 tert-Butyl Alcohol $C_4H_{10}O$ (Ideal Gas State) Mol Wt 74.120

$T°K$	$Cp°$	$S°$	$-(G°-H°_{298})/T$	$H°-H°_{298}$	$\Delta Hf°$	$\Delta Gf°$	Log Kp
	cal/(mole °K)			kcal/mole			
298	27.10	77.98	77.98	0.00	-77.87	-45.66	33.471
300	27.23	78.15	77.99	0.06	-77.91	-45.47	33.120
400	34.16	86.96	79.14	3.13	-79.64	-34.38	18.781
500	40.27	95.25	81.54	6.86	-81.05	-22.89	10.005
600	45.37	103.06	84.48	11.15	-82.15	-11.16	4.064
700	49.64	110.38	87.67	15.91	-82.98	0.75	-0.233
800	53.32	117.26	90.94	21.06	-83.60	12.75	-3.482
900	56.42	123.72	94.23	26.55	-84.02	24.82	-6.027
1000	59.16	129.81	97.48	32.33	-84.26	36.94	-8.072

kcal/mole to transform crystal II phase to crystal III phase at $Tt = 281.54°K$, and $\Delta Ht° = 0.117$ kcal/mole to transform crystal III phase to crystal I phase at $Tt = 294.47°K$, $\Delta Hm = 1.602$ kcal/mole at $Ttp = 298.97°K$, $S°_{298}(c) = 40.84$ cal/(mole °K), and $S°_{298.97}(l) = 46.30$ cal/(mole °K). Beynon and McKetta (124) measured the vapor pressure, heat of vaporization, and the vapor heat capacity. They reported that the vapor heat capacities agree within experimental error with those measured by Reynolds and De Vries (1221) and by Sinke and De Vries (1355). Beynon and McKetta (124) correlated the vapor heat capacities with a model assuming an equilibrium mixture of monomers, dimers, and tetramers. They were able to reconcile the third-law entropy with the spectroscopic data and molecular structure, and calculated the thermodynamic functions. They reported $\Delta Hv = 7.86$ kcal/mole at $Tb = 355.48°K$, $\Delta Hv_{298} = 8.00$ kcal/mole, and $S°_{298}(g) = 77.98$ cal/(mole °K). Ambrose and Townsend (14) measured $Tc = 506.2°K$, $Pc = 39.2$ atm, and $dc = 0.270$ g/cc.

Eberz and Lucas (363) and Taft and Riesz (1464) studied the equilibrium hydration of isobutene to tert-butyl alcohol and found $\Delta G°_{298} = -1.300$ kcal/mole for the following reaction.

$$C_4H_8(g) + H_2O(l) = C_4H_{10}O(l) \qquad \Delta$$

$S°_{298}$	70.17	16.72	46.30	-40.59
$\Delta Hf°_{298}$	-4.04	-68.32	-85.87	-13.51

Below the compounds in this reaction are the entropy and enthalpy of formation and reaction information assembled from thermal data tabulated in this book. From these values $\Delta Gr = -13.51 - [298.15(-40.59)] = -1.408$ kcal/mole, which is in good agreement with the equilibrium values.

No. 524 Pentyl Alcohol $C_5H_{12}O$ (Ideal Gas State) Mol Wt 88.146

	cal/(mole °K)			kcal/mole			
$T°K$	$Cp°$	$S°$	$-(G°-H°_{298})/T$	$H°-H°_{298}$	$\Delta Hf°$	$\Delta Gf°$	Log Kp
298	31.76	96.21	96.21	0.00	-72.27	-35.79	26.232
300	31.91	96.41	96.22	0.06	-72.32	-35.56	25.907
400	39.74	106.68	97.56	3.65	-74.48	-22.97	12.550
500	47.01	116.35	100.36	8.00	-76.29	-9.88	4.318
600	53.24	125.48	103.79	13.02	-77.73	3.54	-1.289
700	58.59	134.10	107.51	18.61	-78.85	17.18	-5.364
800	63.18	142.23	111.35	24.71	-79.69	30.95	-8.455
900	67.15	149.91	115.21	31.23	-80.28	44.83	-10.885
1000	70.59	157.16	119.05	38.12	-80.63	58.76	-12.841

Chao and Rossini (215) measured the heat of combustion and reported $\Delta Hf°_{298}(l) = -85.55$ kcal/mole, in good agreement with the value proposed by Rossini (1242). Wadsö and Wadsö (1553) measured $\Delta Hv_{298} = 13.28$ kcal/mole, which leads to $\Delta Hf°_{298}(g) = -72.27$ kcal/mole. Parks, Huffman, and Barmore (1106) studied the thermal properties from 90° to 298°K and reported $Tm = 195.0°K$, $\Delta Hm° = 2.350$ kcal/mole, and $Tb = 411.0°K$. Thence $\Delta Hv_{411} = 10.20$ kcal/mole is calculated. Parks, Huffman, and Barmore (1106) reported $S°_{298}(l) = 60.9$ cal/(mole °K). Sinke and De Vries (1355) measured the vapor heat capacity and are in agreement with the values calculated by Green (532), who also calculated complete thermodynamic properties and reported $S°_{298}(g) = 96.21$ cal/(mole °K).

No. 525 *tert*-Pentyl Alcohol $C_5H_{12}O$ (Ideal Gas State) Mol Wt 88.146

	cal/(mole °K)			kcal/mole			
$T°K$	$Cp°$	$S°$	$-(G°-H°_{298})/T$	$H°-H°_{298}$	$\Delta Hf°$	$\Delta Gf°$	Log Kp
298	31.47	87.68	87.68	0.00	-78.65	-39.62	29.044
300	31.63	87.88	87.69	0.06	-78.70	-39.39	28.691
400	39.97	98.14	89.03	3.65	-80.86	-25.94	14.171
500	47.68	107.91	91.83	8.04	-82.62	-11.99	5.243
600	54.25	117.20	95.29	13.15	-83.98	2.26	-0.823
700	59.59	125.97	99.06	18.85	-85.00	16.72	-5.221
800	64.22	134.24	102.94	25.04	-85.74	31.30	-8.550
900	67.82	142.01	106.86	31.65	-86.24	45.97	-11.162
1000	71.40	149.34	110.74	38.61	-86.53	60.68	-13.262

Chao and Rossini (215) burned this material and reported $\Delta Hf°_{298}(l) = -90.71$ kcal/mole. Adjusting the measurements of McCurdy and Laidler (966) to achieve harmony with those of Wadsö and Wadsö (1553), $\Delta Hv_{298} = 12.06$ kcal/mole is found, leading to $\Delta Hf°_{298}(g) = -78.65$ kcal/mole. Parks, Huffman, and Barmore (1106) studied the low-temperature properties from 90° to 298°K and reported $Tt = 146.0°K$, $\Delta Ht° = 0.469$

kcal/mole, $Tm = 264°K$, $\Delta Hm° = 1.065$ kcal/mole, $Tb = 375.5°K$, and $S°_{298}(l) = 54.8$ cal/(mole °K). $\Delta Hv_{375} = 9.40$ kcal/mole according to Butler, Ramchandani, and Thomson (187). The calculated third-law entropy is

$S°_{298}(l)$	54.80 cal/(mole °K)
$\Delta Sv = \Delta Hv/T = 12,060/298.15$	40.45
ΔS Compression, $-R \ln (760/16.8)$	-7.57
$S°_{298}(g)$	87.68 cal/(mole °K).

Taking total symmetry into consideration, one may estimate the entropy of *tert*-amyl alcohol as follows [units are cal/(mole °K)]:

$S°_{298}$ neopentane (g)	73.23	$S°_{298}$ *tert*-butyl alcohol (g)	78.00
$R \ln 972$	13.66	$R \ln 27$	6.54
	86.89		84.54

Thus conversion of a methyl group on neopentane to an —OH group decreases the entropy by $(86.89 - 84.54) = 2.35$ cal/(mole °K).

$S°_{298}$ 2,2-dimethylbutane (g)	85.62	$S°_{298}$ *tert*-amyl alcohol (g)	x
$R \ln 243$	10.90	$R \ln 27$	6.54
	96.52	(less 2.35)	$= 94.17$

Now $(96.52 - 2.35 - 6.54) = x = 87.63$ cal/(mole °K). This degree of agreement is fortuitous, but the estimated value should be correct within 0.5 cal/(mole °K). The gaseous heat capacity is calculated by applying the difference in heat capacity between pentyl alcohol and hexane to 2,2-dimethylbutane.

No. 526 Hexyl Alcohol $C_6H_{14}O$ (Ideal Gas State) Mol Wt 102.172

	cal/(mole °K)				kcal/mole		
$T°K$	$Cp°$	$S°$	$-(G°-H°_{298})/T$	$H°-H°_{298}$	$\Delta Hf°$	$\Delta Gf°$	Log Kp
298	37.23	105.52	105.52	0.00	-76.39	-32.97	24.168
300	37.41	105.76	105.53	0.07	-76.44	-32.71	23.825
400	46.68	117.80	107.11	4.28	-78.93	-17.73	9.689
500	55.26	129.17	110.39	9.39	-80.99	-2.19	0.957
600	62.58	139.90	114.42	15.29	-82.63	13.72	-4.998
700	68.87	150.03	118.79	21.87	-83.89	29.89	-9.332
800	74.25	159.59	123.30	29.03	-84.83	46.20	-12.621
900	78.90	168.61	127.84	36.70	-85.47	62.63	-15.208
1000	82.92	177.14	132.35	44.79	-85.85	79.12	-17.291

Chao and Rossini (215) measured the heat of combustion and reported

$\Delta Hf°_{298}(l) = -90.67$ kcal/mole. Equation 10-1 leads to $\Delta Hv_{298} = 14.28$ kcal/mole, making $\Delta Hf°_{298}(g) = -76.39$ kcal/mole. Kelley (733) studied the low-temperature properties from 18° to 298°K and reported $Ttp = 225.8°K$, $\Delta Hm° = 3.676$ kcal/mole, and $S°_{298}(l) = 68.6$ cal/(mole °K). Green (532) converted the third-law entropy to $S°_{298}(g) = 105.1$ cal/(mole °K), and by a method of increments calculated $S°_{298}(g) = 105.52$ cal/(mole °K) as well as complete thermodynamic properties, which are adopted.

No. 527   Heptyl Alcohol $C_7H_{16}O$ (Ideal Gas State)   Mol Wt 116.198

	cal/(mole °K)				kcal/mole		
T°K	$Cp°$	$S°$	$-(G°-H°_{298})/T$	$H°-H°_{298}$	$\Delta Hf°$	$\Delta Gf°$	Log Kp
298	42.70	114.83	114.83	0.00	-80.03	-29.68	21.752
300	42.91	115.10	114.84	0.08	-80.09	-29.37	21.393
400	53.62	128.93	116.65	4.92	-82.89	-12.02	6.565
500	63.51	141.99	120.42	10.79	-85.21	5.98	-2.613
600	71.92	154.33	125.05	17.57	-87.05	24.39	-8.883
700	79.15	165.97	130.08	25.13	-88.45	43.08	-13.451
800	85.32	176.95	135.26	33.36	-89.49	61.94	-16.919
900	90.65	187.31	140.47	42.17	-90.19	80.92	-19.649
1000	95.25	197.11	145.65	51.47	-90.58	99.97	-21.847

Chao and Rossini (215) measured the heat of combustion and reported $\Delta Hf°_{298}(l) = -95.31$ kcal/mole. Equation [10-1] leads to $\Delta Hv_{298} = 15.28$ kcal/mole, making $\Delta Hf°_{298}(g) = -80.03$ kcal/mole. Parks, Kennedy, Gates, Mosley, Moore, and Renquist (1109) studied the thermal properties from 80° to 300°K and reported $Tm = 240.4°K$, $\Delta Hm° = 4.344$ kcal/mole, and $S°_{298}(l) = 77.9$ cal/(mole °K). Rossini, Pitzer, Arnett, Braun and Pimentel (1248) indicated $Tb = 449.4°K$. Green (532) converted the third-law entropy to $S°_{298}(g) = 116.1$ cal/(mole °K), and by a method of increments calculated $S°_{298}(g) = 114.83$ cal/(mole °K) as well as complete properties, which are adopted.

No. 528   Octyl Alcohol $C_8H_{18}O$ (Ideal Gas State)   Mol Wt 130.224

	cal/(mole °K)				kcal/mole		
T°K	$Cp°$	$S°$	$-(G°-H°_{298})/T$	$H°-H°_{298}$	$\Delta Hf°$	$\Delta Gf°$	Log Kp
298	48.17	124.14	124.14	0.00	-85.34	-28.05	20.561
300	48.41	124.44	124.15	0.09	-85.41	-27.70	20.178
400	60.56	140.06	126.19	5.55	-88.53	-7.97	4.354
500	71.76	154.81	130.45	12.18	-91.10	12.48	-5.453
600	81.26	168.75	135.69	19.84	-93.14	33.38	-12.159
700	89.43	181.90	141.36	28.39	-94.68	54.60	-17.047
800	96.39	194.31	147.21	37.69	-95.82	76.00	-20.760
900	102.40	206.02	153.10	47.63	-96.57	97.53	-23.683
1000	107.58	217.08	158.95	58.14	-96.98	119.14	-26.037

Chao and Rossini (215) measured the heat of combustion and reported $\Delta Hf°_{298}(l) = -101.62$ kcal/mole, which becomes $\Delta Hf°_{298}(g) = -85.34$ kcal/mole when coupled with $\Delta Hv_{298} = 16.28$ kcal/mole from [10-1]. Green (532) calculated the thermodynamic functions, including $S°_{298}(g) = 124.14$ cal/(mole °K). Cline and Andrews (237) observed $Tm = 231.7°K$, while Rossini, Pitzer, Arnett, Braun, and Pimentel (1248) listed $Tb = 468.4°K$.

No. 529 Nonyl Alcohol $C_9H_{20}O$ (Ideal Gas State) Mol Wt 144.250

	cal/(mole °K)				kcal/mole		
T°K	Cp°	S°	$-(G°-H°_{298})/T$	$H°-H°_{298}$	$\Delta Hf°$	$\Delta Gf°$	Log Kp
298	53.64	133.45	133.45	0.00	-92.47	-28.25	20.703
300	53.91	133.79	133.46	0.10	-92.54	-27.85	20.289
400	67.50	151.18	135.74	6.18	-95.98	-5.74	3.138
500	80.01	167.63	140.48	13.58	-98.81	17.15	-7.498
600	90.60	183.17	146.32	22.12	-101.04	40.56	-14.772
700	99.71	197.84	152.64	31.65	-102.73	64.31	-20.076
800	107.46	211.67	159.16	42.01	-103.97	88.24	-24.105
900	114.15	224.72	165.73	53.10	-104.78	112.33	-27.276
1000	119.91	237.06	172.25	64.81	-105.20	136.50	-29.830

Chao and Rossini (215) measured the heat of combustion and reported $\Delta Hf°_{298}(l) = -109.75$ kcal/mole, which converts to $\Delta Hf°_{298}(g) = -92.47$ kcal/mole with $\Delta Hv_{298} = 17.28$ kcal/mole from [10-1]. Green (532) calculated $S°_{298}(g) = 133.45$ cal/(mole °K) and the complete thermodynamic functions. Rossini, Pitzer, Arnett, Braun, and Pimentel (1248) selected $Tb = 486.7°K$.

No. 530 Decyl Alcohol $C_{10}H_{22}O$ (Ideal Gas State) Mol Wt 158.276

	cal/(mole °K)				kcal/mole		
T°K	Cp°	S°	$-(G°-H°_{298})/T$	$H°-H°_{298}$	$\Delta Hf°$	$\Delta Gf°$	Log Kp
298	59.11	142.76	142.76	0.00	-96.38	-25.22	18.486
300	59.41	143.13	142.77	0.11	-96.46	-24.78	18.054
400	74.44	162.31	145.28	6.82	-100.22	-0.30	0.162
500	88.26	180.45	150.52	14.97	-103.30	25.05	-10.950
600	99.94	197.59	156.95	24.39	-105.73	50.95	-18.557
700	109.99	213.77	163.92	34.90	-107.56	77.23	-24.110
800	118.53	229.03	171.11	46.34	-108.89	103.70	-28.328
900	125.90	243.43	178.36	58.57	-109.76	130.35	-31.651
1000	132.24	257.03	185.55	71.48	-110.21	157.07	-34.326

Chao and Rossini (215) measured the heat of combustion and reported $\Delta Hf°_{298}(l) = -114.66$ kcal/mole, which becomes $\Delta Hf°_{298}(g) = -96.38$ kcal/mole when coupled with $\Delta Hv_{298} = 18.28$ kcal/mole from [10-1].

Green (532) calculated the thermodynamic functions, including $S°_{298}$ (g) = 142.76 cal/(mole °K). Rossini, Pitzer, Arnett, Braun, and Pimentel (1248) selected Tb = 503.2°K.

No. 531 Undecyl Alcohol $C_{11}H_{24}O$ (Ideal Gas State) Mol Wt 172.302

T°K	$Cp°$	$S°$	$-(G°-H°_{298})/T$	$H°-H°_{298}$	$\Delta Hf°$	$\Delta Gf°$	Log Kp
	cal/(mole °K)				kcal/mole		
298	64.58	152.07	152.07	0.00	-100.91	-22.81	16.722
300	64.91	152.48	152.08	0.12	-101.00	-22.34	16.271
400	81.38	173.44	154.82	7.45	-105.07	4.53	-2.476
500	96.51	193.27	160.55	16.36	-108.41	32.33	-14.131
600	109.28	212.02	167.58	26.67	-111.04	60.72	-22.117
700	120.27	229.71	175.20	38.16	-113.01	89.53	-27.950
800	129.60	246.39	183.07	50.66	-114.44	118.54	-32.383
900	137.65	262.13	190.99	64.04	-115.36	147.74	-35.875
1000	144.57	277.00	198.85	78.16	-115.83	177.02	-38.687

$\Delta Hf°_{298}(g) = -100.91$ kcal/mole is found from Table 10-8 and the value for the parent hydrocarbon. Green (532) calculated $S°_{298}(g) = 152.07$ cal/(mole °K) and the complete thermodynamic functions. Rossini, Pitzer, Arnett, Braun, and Pimentel (1248) selected $Tb = 519$°K.

No. 532 Dodecyl Alcohol $C_{12}H_{26}O$ (Ideal Gas State) Mol Wt 186.328

T°K	$Cp°$	$S°$	$-(G°-H°_{298})/T$	$H°-H°_{298}$	$\Delta Hf°$	$\Delta Gf°$	Log Kp
	cal/(mole °K)				kcal/mole		
298	70.05	161.38	161.38	0.00	-105.84	-20.81	15.252
300	70.41	161.82	161.39	0.13	-105.93	-20.29	14.779
400	88.32	184.56	164.37	8.08	-110.33	8.96	-4.894
500	104.76	206.09	170.58	17.76	-113.92	39.21	-17.137
600	118.62	226.44	178.21	28.94	-116.75	70.10	-25.532
700	130.55	245.64	186.48	41.42	-118.87	101.43	-31.666
800	140.67	263.75	195.02	54.99	-120.39	132.99	-36.328
900	149.40	280.84	203.62	69.50	-121.37	164.74	-40.001
1000	156.90	296.98	212.15	84.83	-121.85	196.58	-42.960

$\Delta Hf°_{298}(g) = -105.84$ kcal/mole is calculated from the parent hydrocarbon and the increment in Table 10-8. Green (532) calculated $S°_{298}(g) = 161.38$ cal/(mole °K) and the complete thermodynamic functions. Rossini, Pitzer, Arnett, Braun, and Pimentel (1248) selected $Tb = 534$°K.

No. 533 1-Tridecanol $C_{13}H_{28}O$ (Ideal Gas State) Mol Wt 200.354

$\Delta Hf°_{298}(g) = -110.77$ kcal/mole, $S°_{298}(g) = 170.37$ cal/(mole °K), and the gaseous heat capacity are calculated from the parent hydrocarbon

Alkanol Ideal Gas Tables

No. 533 1-Tridecanol $C_{13}H_{28}O$ (Ideal Gas State) Mol Wt 200.354

	cal/(mole °K)				kcal/mole		
$T°K$	$Cp°$	$S°$	$-(G°-H°_{298})/T$	$H°-H°_{298}$	$\Delta Hf°$	$\Delta Gf°$	Log Kp
298	75.49	170.37	170.37	0.00	-110.77	-18.71	13.712
300	75.87	170.84	170.38	0.15	-110.87	-18.14	13.217
400	95.27	195.37	173.59	8.72	-115.58	13.51	-7.383
500	112.98	218.58	180.29	19.15	-119.43	46.25	-20.214
600	127.97	240.54	188.52	31.22	-122.46	79.67	-29.017
700	140.80	261.25	197.44	44.67	-124.72	113.55	-35.452
800	151.71	280.78	206.65	59.31	-126.34	147.69	-40.344
900	161.12	299.20	215.92	74.96	-127.38	182.02	-44.199
1000	169.30	316.61	225.13	91.49	-127.88	216.46	-47.305

with the increments in Table 10-8. Rossini, Pitzer, Arnett, Braun, and Pimentel (1248) selected $Tb = 549°K$.

No. 534 1-Tetradecanol $C_{14}H_{30}O$ (Ideal Gas State) Mol Wt 214.380

	cal/(mole °K)				kcal/mole		
$T°K$	$Cp°$	$S°$	$-(G°-H°_{298})/T$	$H°-H°_{298}$	$\Delta Hf°$	$\Delta Gf°$	Log Kp
298	80.96	179.68	179.68	0.00	-115.70	-16.70	12.242
300	81.36	180.19	179.69	0.16	-115.81	-16.10	11.725
400	102.21	206.49	183.13	9.35	-120.84	17.94	-9.802
500	121.23	231.40	190.32	20.55	-124.94	53.13	-23.220
600	137.32	254.96	199.15	33.49	-128.17	89.04	-32.431
700	151.08	277.19	208.72	47.93	-130.57	125.46	-39.167
800	162.77	298.14	218.60	63.64	-132.29	162.13	-44.289
900	172.87	317.91	228.55	80.43	-133.39	199.02	-48.326
1000	181.60	336.59	238.43	98.16	-133.91	236.01	-51.578

The gaseous heat capacity, $S°_{298}(g) = 179.68$ cal/(mole °K), and $\Delta Hf°_{298}(g) = -115.70$ kcal/mole are calculated from the parent hydrocarbon with increments from Table 10-8. Rossini, Pitzer, Arnett, Braun, and Pimentel (1248) selected $Tb = 564°K$.

No. 535 1-Pentadecanol $C_{15}H_{32}O$ (Ideal Gas State) Mol Wt 228.406

	cal/(mole °K)				kcal/mole		
$T°K$	$Cp°$	$S°$	$-(G°-H°_{298})/T$	$H°-H°_{298}$	$\Delta Hf°$	$\Delta Gf°$	Log Kp
298	86.42	188.99	188.99	0.00	-120.62	-14.69	10.765
300	86.86	189.53	189.00	0.17	-120.73	-14.04	10.226
400	109.15	217.62	192.68	9.98	-126.08	22.38	-12.226
500	129.47	244.22	200.35	21.94	-130.44	60.01	-26.231
600	146.66	269.38	209.78	35.77	-133.86	98.42	-35.849
700	161.35	293.12	220.00	51.19	-136.41	137.37	-42.886
800	173.84	315.50	230.55	67.96	-138.23	176.58	-48.237
900	184.61	336.61	241.18	85.89	-139.38	216.03	-52.456
1000	193.90	356.55	251.73	104.83	-139.92	255.58	-55.854

The increments from Table 10-8 are applied to the parent hydrocarbon to calculate $\Delta Hf^{\circ}_{298}(g) = -120.62$ kcal/mole, $S^{\circ}_{298}(g) = 188.99$ cal/(mole °K), and the vapor heat capacity. Rossini, Pitzer, Arnett, Braun, and Pimentel (1248) selected $Tb = 578°$K.

No. 536 1-Hexadecanol $C_{16}H_{34}O$ (Ideal Gas State) Mol Wt 242.432

$T°K$	$Cp°$	$S°$	$-(G°-H^{\circ}_{298})/T$	$H°-H^{\circ}_{298}$	$\Delta Hf°$	$\Delta Gf°$	Log Kp
		cal/(mole °K)			kcal/mole		
298	91.89	198.30	198.30	0.00	-125.54	-12.67	9.288
300	92.35	198.87	198.31	0.18	-125.66	-11.98	8.726
400	116.09	228.74	202.22	10.62	-131.33	26.81	-14.650
500	137.72	257.03	210.38	23.33	-135.94	66.90	-29.241
600	156.00	283.80	220.41	38.04	-139.56	107.81	-39.267
700	171.63	309.05	231.28	54.44	-142.25	149.28	-46.604
800	184.90	332.86	242.51	72.28	-144.17	191.03	-52.185
900	196.36	355.31	253.81	91.36	-145.38	233.03	-56.586
1000	206.30	376.53	265.03	111.51	-145.93	275.14	-60.130

Parks, Kennedy, Gates, Mosley, Moore, and Renquist (1109) measured the thermal properties from 80° to 300°K and reported $Tt = 307°$K, $\Delta Ht° = 3.976$ kcal/mole, $Tm = 321.8°$K, $\Delta Hm° = 8.461$ kcal/mole, $S^{\circ}_{298}(s) = 108.0$ cal/(mole °K), and $S^{\circ}_{298}(l) = 145.0$ cal/(mole °K). The gaseous heat capacity, $S^{\circ}_{298}(g) = 198.30$ cal/(mole °K), and $\Delta Hf^{\circ}_{298}(g) = -125.54$ kcal/mole are calculated from the parent hydrocarbon with increments from Table 10-8. Rossini, Pitzer, Arnett, Braun, and Pimentel (1248) selected $Tb = 592°$K.

No. 537 1-Heptadecanol $C_{17}H_{36}O$ (Ideal Gas State) Mol Wt 256.458

$T°K$	$Cp°$	$S°$	$-(G°-H^{\circ}_{298})/T$	$H°-H^{\circ}_{298}$	$\Delta Hf°$	$\Delta Gf°$	Log Kp
		cal/(mole °K)			kcal/mole		
298	97.36	207.61	207.61	0.00	-130.47	-10.67	7.818
300	97.84	208.22	207.62	0.19	-130.60	-9.93	7.235
400	123.03	239.87	211.76	11.25	-136.58	31.24	-17.069
500	145.97	269.85	220.41	24.73	-141.46	73.78	-32.248
600	165.34	298.22	231.04	40.32	-145.27	117.18	-42.682
700	181.91	324.99	242.56	57.70	-148.10	161.18	-50.320
800	195.97	350.22	254.46	76.61	-150.12	205.48	-56.131
900	208.11	374.01	266.43	96.83	-151.39	250.03	-60.713
1000	218.60	396.50	278.33	118.18	-151.96	294.70	-64.403

Values from the parent hydrocarbon are used with the increments from Table 10-8 to calculate $\Delta Hf^{\circ}_{298}(g) = -130.47$ kcal/mole, $S^{\circ}_{298}(g) = 207.61$ cal/(mole °K), and the vapor heat capacity. Rossini, Pitzer, Arnett, Braun, and Pimentel (1248) selected $Tb = 605°$K.

Alkanol Ideal Gas Tables

No. 538   1-Octadecanol $C_{18}H_{38}O$ (Ideal Gas State)   Mol Wt 270.484

T°K	Cp°	S°	$-(G°-H°_{298})/T$	$H°-H°_{298}$	$\Delta Hf°$	$\Delta Gf°$	Log Kp
		cal/(mole °K)			kcal/mole		
298	102.82	216.92	216.92	0.00	-135.39	-8.65	6.340
300	103.34	217.56	216.93	0.20	-135.52	-7.87	5.735
400	129.97	251.00	221.31	11.88	-141.83	35.68	-19.493
500	154.21	282.67	230.44	26.12	-146.96	80.67	-35.258
600	174.68	312.64	241.67	42.59	-150.97	126.57	-46.100
700	192.18	340.92	253.84	60.96	-153.95	173.09	-54.039
800	207.03	367.57	266.41	80.93	-156.06	219.93	-60.079
900	219.85	392.71	279.06	102.29	-157.38	267.04	-64.842
1000	230.90	416.46	291.63	124.84	-157.98	314.27	-68.679

The gaseous heat capacity, $S°_{298}(g) = 216.92$ cal/(mole °K), and $\Delta Hf°_{298}(g) = -135.39$ kcal/mole are calculated from the parent hydrocarbon with increments from Table 10-8. Rossini, Pitzer, Arnett, Braun, and Pimentel (1248) selected $Tb = 618°K$.

No. 539   1-Nonadecanol $C_{19}H_{40}O$ (Ideal Gas State)   Mol Wt 284.510

T°K	Cp°	S°	$-(G°-H°_{298})/T$	$H°-H°_{298}$	$\Delta Hf°$	$\Delta Gf°$	Log Kp
		cal/(mole °K)			kcal/mole		
298	108.29	226.23	226.23	0.00	-140.32	-6.64	4.870
300	108.83	226.91	226.24	0.21	-140.46	-5.83	4.243
400	136.91	262.12	230.85	12.51	-147.08	40.11	-21.912
500	162.46	295.49	240.47	27.52	-152.47	87.55	-38.265
600	184.03	327.06	252.30	44.87	-156.68	135.94	-49.514
700	202.46	356.85	265.13	64.21	-159.79	184.99	-57.754
800	218.10	384.93	278.36	85.26	-162.01	234.37	-64.024
900	231.60	411.42	291.69	107.76	-163.39	284.03	-68.969
1000	243.30	436.44	304.93	131.52	-164.00	333.82	-72.953

The increments from Table 10-8 are applied to the parent hydrocarbon to calculate $\Delta Hf°_{298}(g) = -140.32$ kcal/mole, $S°_{298}(g) = 226.23$ cal/(mole °K), and the vapor heat capacity. Rossini, Pitzer, Arnett, Braun, and Pimentel (1248) selected $Tb = 631°K$.

No. 540   1-Eicosanol $C_{20}H_{42}O$ (Ideal Gas State)   Mol Wt 298.536

Values from the parent hydrocarbon are used with the increments from Table 10-8 to calculate $\Delta Hf°_{298}(g) = -145.25$ kcal/mole, $S°_{298}(g) = 235.54$ cal/(mole °K), and the vapor heat capacity. Rossini, Pitzer, Arnett, Braun, and Pimentel (1248) selected $Tb = 643°K$.

No. 540 1-Eicosanol $C_{20}H_{42}O$ (Ideal Gas State) Mol Wt 298.536

T°K	$Cp°$	$S°$	$-(G°-H°_{298})/T$	$H°-H°_{298}$	$\Delta Hf°$	$\Delta Gf°$	Log Kp
	cal/(mole °K)			kcal/mole			
298	113.76	235.54	235.54	0.00	-145.25	-4.64	3.400
300	114.32	236.25	235.55	0.22	-145.40	-3.78	2.751
400	143.85	273.25	240.39	13.15	-152.34	44.53	-24.331
500	170.71	308.31	250.50	28.91	-157.98	94.43	-41.271
600	193.38	341.49	262.93	47.14	-162.38	145.32	-52.929
700	212.74	372.79	276.41	67.47	-165.64	196.89	-61.470
800	229.17	402.29	290.32	89.59	-167.96	248.81	-67.969
900	243.35	430.12	304.32	113.23	-169.39	301.03	-73.096
1000	255.70	456.42	318.23	138.20	-170.02	353.37	-77.226

ALKENOL IDEAL GAS TABLE

No. 541 Allyl Alcohol C_3H_6O (Ideal Gas State) Mol Wt 58.078

T°K	$Cp°$	$S°$	$-(G°-H°_{298})/T$	$H°-H°_{298}$	$\Delta Hf°$	$\Delta Gf°$	Log Kp
	cal/(mole °K)			kcal/mole			
298	18.17	73.51	73.51	0.00	-31.55	-17.03	12.481
300	18.25	73.63	73.52	0.04	-31.57	-16.94	12.340
400	22.81	79.52	74.29	2.10	-32.69	-11.88	6.492
500	26.79	85.05	75.89	4.58	-33.62	-6.57	2.871
600	30.11	90.23	77.85	7.43	-34.39	-1.09	0.396
700	32.91	95.09	79.97	10.59	-35.00	4.52	-1.410
800	35.28	99.64	82.15	14.00	-35.48	10.19	-2.784
900	37.35	103.92	84.33	17.63	-35.85	15.92	-3.867
1000	39.06	107.95	86.50	21.46	-36.11	21.70	-4.742

Dolliver, Gresham, Kistiakowsky, Smith, and Vaughan (334) measured the heat of hydrogenation of allyl alcohol to propyl alcohol, a value that permits calculation of $\Delta Hf°_{298}(g) = -31.55$ kcal/mole for allyl alcohol. Kobe, Harrison, and Pennington (777) assigned the fundamental vibrational frequencies and barrier heights and calculated the thermodynamic functions. They reported $S°_{298}(g) = 73.51$ cal/(mole °K).

ALKANEDIOL IDEAL GAS TABLE

No. 542 Ethylene Glycol $C_2H_6O_2$ (Ideal Gas State) Mol Wt 62.068

Parks, West, Naylor, Fujii, and McClaine (1121) measured the heat of combustion and found $\Delta Hf°_{298}(l) = -108.73$ kcal/mole. From vapor pressure data selected by Stull (1432), $\Delta Hv_{298} = 15.68$ kcal/mole is calculated,

Cycloalkanol Ideal Gas Tables

No. 542   Ethylene Glycol $C_2H_6O_2$ (Ideal Gas State)   Mol Wt 62.068

$T°K$	$Cp°$	$S°$	$-(G°-H°_{298})/T$	$H°-H°_{298}$	$\Delta Hf°$	$\Delta Gf°$	Log Kp
	cal/(mole °K)			kcal/mole			
298	23.20	77.33	77.33	0.00	-93.05	-72.77	53.337
300	23.28	77.48	77.34	0.05	-93.07	-72.64	52.917
400	27.06	84.71	78.30	2.57	-93.83	-65.71	35.901
500	30.10	91.08	80.23	5.43	-94.43	-58.61	25.617
600	32.72	96.81	82.52	8.58	-94.90	-51.40	18.721
700	35.00	102.03	84.94	11.97	-95.24	-44.12	13.775
800	36.90	106.84	87.38	15.57	-95.48	-36.81	10.054
900	38.00	111.24	89.79	19.31	-95.66	-29.46	7.153
1000	39.88	115.33	92.14	23.19	-95.76	-22.09	4.827

which leads to $\Delta Hf°_{298}(g) = -93.05$ kcal/mole. Parks, Kelley, and Huffman (1108) measured the heat capacity from 90° to 298°K and reported the following.

$S°_{298}(l)$	39.9 cal/(mole °K)
ΔSv, 15,680/298	52.59
ΔS(compression), $-R \ln 760/0.098$	-17.80
ΔS, Gas imperfection (negligible)	0.00
$S°_{298}$(ideal gas)	74.69 cal/(mole °K)

Dyatkina (361) assigned the fundamental vibrational frequencies, calculated the thermodynamic functions, and reported $S°_{298}(g) = 77.33$ cal/(mole °K). The value of $S°_{298}(g) = 77.30$ cal/(mole °K), calculated from the parent hydrocarbon and $S^*_{298} = -0.59$ cal/(mole °K), is in good agreement. Uncertainties in the heat capacity below 90°K, and the vapor pressure extrapolation at 298°K, recommend adoption of Dyatkina's value.

CYCLOALKANOL IDEAL GAS TABLE

No. 543   Cyclohexanol $C_6H_{12}O$ (Ideal Gas State)   Mol Wt 100.156

Parks, Mosley, and Peterson (1116) and Sellers and Sunner (1331) measured the heat of combustion and are in agreement within experimental error. The value of the latter workers yields $\Delta Hf°_{298}(l) = -83.22$ kcal/mole, from which we derive $\Delta Hf°_{298}(g) = -70.40$ kcal/mole. Kelley (735) studied the low-temperature properties and listed $Tt = 263.5°K$, $\Delta Ht = 1.960$ kcal/mole, $\Delta Hm = 0.406$ kcal/mole, and $S°_{298}(l) = 47.7$ cal/(mole °K). Timmermans (1501) gave three values for the melting point; $Tm = 296.6°K$ is adopted. Stull (1432) reviewed the vapor pressure data

No. 543 Cyclohexanol $C_6H_{12}O$ (Ideal Gas State) Mol Wt 100.156

$T°K$	$Cp°$	$S°$	$-(G°-H°_{298})/T$	$H°-H°_{298}$	$\Delta Hf°$	$\Delta Gf°$	Log Kp
	cal/(mole °K)				kcal/mole		
298	30.41	78.32	78.32	0.00	-70.40	-28.18	20.654
300	30.59	78.51	78.33	0.06	-70.45	-27.92	20.338
400	41.14	88.78	79.65	3.65	-72.86	-13.36	7.297
500	50.90	99.03	82.51	8.27	-74.71	1.74	-0.762
600	59.29	109.07	86.10	13.79	-76.04	17.16	-6.251
700	66.33	118.76	90.08	20.08	-76.89	32.77	-10.232
800	72.18	128.01	94.25	27.01	-77.35	48.47	-13.239
900	77.06	136.80	98.49	34.48	-77.48	64.21	-15.591
1000	81.13	145.14	102.74	42.40	-77.31	79.96	-17.474

and gave $Tb = 434°K$; $\Delta Hv_{434} = 10.875$ kcal/mole according to Mathews and Fehlandt (941). Estimation of an average $\Delta Cp = 14.5$ cal/(mole °K) from 434° to 298°K yields $\Delta Hv_{298} = 12.820$ kcal/mole. These data indicate the following.

$S°_{298}(l)$	47.70 cal/(mole °K)
ΔSv, 12,820/298.15	43.00
ΔS (compression), $-R \ln (760/1.5)$	-12.38
$S°_{298}(g)$	78.32 cal/(mole °K)

The heat capacity of gaseous cyclohexanol is calculated as the sum of the values for cyclohexane plus ethyl alcohol, less that of ethane.

ALKANAL IDEAL GAS TABLES

No. 544 Formaldehyde CH_2O (Ideal Gas State) Mol Wt 30.026

$T°K$	$Cp°$	$S°$	$-(G°-H°_{298})/T$	$H°-H°_{298}$	$\Delta Hf°$	$\Delta Gf°$	Log Kp
	cal/(mole °K)				kcal/mole		
298	8.46	52.29	52.29	0.00	-27.70	-26.27	19.258
300	8.47	52.35	52.30	0.02	-27.71	-26.27	19.133
400	9.38	54.90	52.64	0.91	-28.11	-25.72	14.053
500	10.46	57.11	53.32	1.90	-28.50	-25.08	10.961
600	11.52	59.12	54.12	3.00	-28.86	-24.36	8.873
700	12.49	60.97	54.97	4.20	-29.17	-23.59	7.363
800	13.37	62.69	55.83	5.50	-29.44	-22.77	6.220
900	14.16	64.31	56.68	6.88	-29.67	-21.92	5.323
1000	14.81	65.84	57.52	8.33	-29.86	-21.05	4.599

Wartenberg and Lerner-Steinberg (1573) found $\Delta Hf°_{298}(g) = -27.7$

kcal/mole by combustion measurements. Newton and Dodge (1062) measured equilibrium data for the reaction

$$HCHO(g) + H_2(g) = CH_3OH(g)$$

and reported $Kp = 2090$ at 197°C, leading to $\Delta G^\circ_{470}/470 = -15.19$ cal/(mole °K). Using Gibbs energy functions as follows

Species	$-\dfrac{(G_{470}-H_{298})}{470}$ [cal/(mole °K)]	Source
HCHO(g)	53.09	Pillai and Cleveland (1149)
H$_2$(g)	31.84	Rossini et al. (1248)
CH$_3$OH(g)	58.34	Ivash, Li, and Pitzer (674)
$\Delta(G_{470}-H_{298})/470$	-26.59	

we calculate $\Delta Hr_{298} = 470(-15.19 - 26.59) = -19{,}637$ cal. From $\Delta Hf^\circ_{298}CH_3OH(g) = -48.080$ kcal/mole, we calculate for formaldehyde $\Delta Hf^\circ_{298}(g) = -28.44$ kcal/mole. From consideration of this information we select $\Delta Hf^\circ_{298}HCHO(g) = -28.0 \pm 1.5$ kcal/mole. Rossini, Wagman, Evans, Levine, and Jaffe (1249) reviewed the literature and selected $Tm = 154.9$°K and $Tb = 253.9$°K, where $\Delta Hv = 5.85$ kcal/mole. Pillai and Cleveland (1149) reviewed the spectroscopic information for some planar XYZ$_2$ molecules and carried out a normal coordinate treatment for five such molecules, including HCHO and DCDO. They employed the infrared frequencies measured by Blau and Nielsen (143). Structural parameters taken from the microwave measurements of Lawrance and Strandberg (852) are in agreement with the microwave measurements of Erlandsson (38?). Pillai and Cleveland (1149) also calculated complete gaseous thermodynamic properties, from which we select $S^\circ_{298}(g) = 52.29$ cal/(mole °K). This entropy value is only 0.03 cal/(mole °K) greater than the value calculated a score of years ago by Thompson (1487).

No. 545 Acetaldehyde C$_2$H$_4$O   (Ideal Gas State)   Mol Wt 44.052

Rossini, Wagman, Evans, Levine, and Jaffe (1249) reviewed the old combustion literature and the more recent studies of Dolliver, Gresham,

No. 545 Acetaldehyde C_2H_4O (Ideal Gas State) Mol Wt 44.052

$T°K$	$Cp°$	$S°$	$-(G°-H°_{298})/T$	$H°-H°_{298}$	$\Delta Hf°$	$\Delta Gf°$	Log Kp
		cal/(mole °K)			kcal/mole		
298	13.06	63.15	63.15	0.00	-39.76	-31.86	23.353
300	13.11	63.24	63.16	0.03	-39.78	-31.81	23.174
400	15.73	67.37	63.70	1.47	-40.57	-29.03	15.862
500	18.27	71.16	64.82	3.17	-41.27	-26.07	11.393
600	20.52	74.69	66.17	5.12	-41.86	-22.97	8.366
700	22.50	78.01	67.63	7.27	-42.35	-19.78	6.176
800	24.20	81.12	69.12	9.61	-42.74	-16.53	4.516
900	25.68	84.06	70.62	12.10	-43.05	-13.24	3.214
1000	26.96	86.84	72.10	14.74	-43.27	-9.91	2.165

Kistiakowsky, Smith, and Vaughan (334) on the heat of hydrogenation to ethanol, and reported $\Delta Hf°_{298}(g) = -39.76$ kcal/mole, $Tm = 155°K$, $\Delta Hm = 0.770$ kcal/mole, and $Tb = 293.3°K$. Coleman and De Vries (245) measured $\Delta Hv_{293.3} = 6.145$ kcal/mole. The vibrational frequency assignment originally used by Smith (1381) was revised by Pitzer and Weltner (1167) to obtain better agreement with the hydrogenation equilibria of Rideal (1227) as well as with the measured gaseous heat capacities of Coleman and De Vries (245). Pitzer and Weltner (1167) obtained best agreement with a 1.00 kcal/mole potential barrier to internal rotation, calculated thermodynamic functions, and listed $S°_{298}(g) = 63.15$ cal/(mole °K). The more recent microwave studies of Lin and Kilb (874) indicated a potential barrier to internal rotation of 1.14 ± 0.06 kcal/mole, while Fateley and Miller (407) found a potential barrier of 1.18 ± 0.15 kcal/mole from far infrared data. Evans and Bernstein (394) studied the spectra more intensively and improved the vibrational assignment, without significant alteration of the thermodynamic functions.

No. 546 Propionaldehyde C_3H_6O (Ideal Gas State) Mol Wt 58.078

$T°K$	$Cp°$	$S°$	$-(G°-H°_{298})/T$	$H°-H°_{298}$	$\Delta Hf°$	$\Delta Gf°$	Log Kp
		cal/(mole °K)			kcal/mole		
298	18.80	72.83	72.83	0.00	-45.90	-31.18	22.851
300	18.87	72.95	72.84	0.04	-45.92	-31.09	22.644
400	23.09	78.97	73.63	2.14	-47.00	-25.97	14.188
500	26.89	84.54	75.26	4.64	-47.91	-20.60	9.005
600	30.22	89.74	77.24	7.50	-48.66	-15.07	5.489
700	33.03	94.62	79.38	10.67	-49.26	-9.42	2.941
800	35.45	99.19	81.57	14.10	-49.73	-3.70	1.011
900	37.55	103.49	83.77	17.75	-50.08	2.08	-0.505
1000	39.27	107.54	85.95	21.59	-50.32	7.89	-1.725

Tjebbes (1508) measured the heat of combustion and reported $\Delta Hf°_{298}$ $(l) = -52.95$ kcal/mole and $\Delta Hf°_{298}(g) = -45.9$ kcal/mole. Chermin (221) summarized structural and spectroscopic information and reported $S°_{298}(g) = 72.83$ cal/(mole °K) in addition to the gaseous heat capacity data, which are adopted here. Möller (1019) also studied the spectrum and calculated the heat capacity a few percent lower than that of Chermin Tjebbes (1508) measured $Tm = 170°$K and $Tb = 321°$K.

No. 547 Butyraldehyde C_4H_8O (Ideal Gas State) Mol Wt 72.104

	cal/(mole °K)				kcal/mole		
$T°K$	$Cp°$	$S°$	$-(G°-H°_{298})/T$	$H°-H°_{298}$	$\Delta Hf°$	$\Delta Gf°$	Log Kp
298	24.52	82.44	82.44	0.00	-49.00	-27.43	20.105
300	24.61	82.60	82.45	0.05	-49.03	-27.30	19.885
400	30.20	90.46	83.48	2.80	-50.40	-19.84	10.837
500	35.20	97.74	85.61	6.07	-51.56	-12.06	5.269
600	39.60	104.56	88.21	9.82	-52.51	-4.07	1.481
700	43.40	110.96	91.01	13.97	-53.24	4.07	-1.271
800	46.60	116.97	93.88	18.48	-53.80	12.29	-3.358
900	49.30	122.61	96.76	23.27	-54.20	20.58	-4.998
1000	51.70	127.94	99.62	28.33	-54.45	28.91	-6.319

Nicholson (1065) burned this substance and reported a value in close agreement with $\Delta Hf°_{298}(l) = -57.06$ kcal/mole and $\Delta Hf°_{298}(g) = -49.0$ kcal/mole, measured by Tjebbes (1508). Parks, Kennedy, Gates, Mosley, Moore, and Renquist (1109) measured the heat capacity from 80° to 300°K and reported $Tm = 176.8°$K, $\Delta Hm = 2.654$ kcal/mole, and $S°_{298}(l) = 59.0$ cal/(mole °K). Chermin (221) converted this value to $S°_{298}(g) = 82.44$ cal/(mole °K) and calculated the other thermodynamic functions for the ideal gas. Rossini, Pitzer, Arnett, Braun, and Pimentel (1248) selected $Tb = 348.02°$K.

No. 548 Valeraldehyde $C_5H_{10}O$ (Ideal Gas State) Mol Wt 86.130

	cal/(mole °K)				kcal/mole		
$T°K$	$Cp°$	$S°$	$-(G°-H°_{298})/T$	$H°-H°_{298}$	$\Delta Hf°$	$\Delta Gf°$	Log Kp
298	29.96	91.53	91.53	0.00	-54.45	-25.88	18.968
300	30.08	91.72	91.54	0.06	-54.49	-25.70	18.724
400	37.10	101.35	92.80	3.43	-56.18	-15.84	8.654
500	43.40	110.32	95.42	7.46	-57.60	-5.59	2.441
600	49.00	118.74	98.61	12.08	-58.74	4.92	-1.793
700	53.70	126.66	102.06	17.22	-59.61	15.61	-4.874
800	57.70	134.10	105.60	22.80	-60.27	26.40	-7.211
900	61.10	141.09	109.16	28.74	-60.72	37.26	-9.048
1000	64.00	147.69	112.69	35.00	-60.99	48.18	-10.528

Increments from Table 10-8 are used to derive $\Delta Hf°_{298}(g) = -54.45$ kcal/mole and $S°_{298}(g) = 91.53$ cal/(mole °K). Chermin (221) presented the heat capacity of the ideal gas. Rossini, Pitzer, Arnett, Braun, and Pimentel (1248) selected $Tm = 182°K$ and $Tb = 375.7°K$.

No. 549 Hexanal $C_6H_{12}O$ (Ideal Gas State) Mol Wt 100.156

	cal/(mole °K)				kcal/mole		
$T°K$	$Cp°$	$S°$	$-(G°-H°_{298})/T$	$H°-H°_{298}$	$\Delta Hf°$	$\Delta Gf°$	Log Kp
298	35.43	101.07	101.07	0.00	-59.37	-23.93	17.541
300	35.57	101.29	101.08	0.07	-59.41	-23.71	17.275
400	44.00	112.70	102.57	4.06	-61.43	-11.49	6.280
500	51.70	123.37	105.68	8.85	-63.10	1.19	-0.520
600	58.30	133.39	109.47	14.36	-64.44	14.17	-5.162
700	63.90	142.81	113.57	20.47	-65.46	27.36	-8.543
800	68.70	151.66	117.78	27.11	-66.22	40.67	-11.110
900	72.90	160.00	122.01	34.20	-66.73	54.07	-13.129
1000	76.40	167.87	126.21	41.67	-67.01	67.52	-14.756

Increments from Table 10-8 are used to derive $\Delta Hf°_{298}(g) = -59.37$ kcal/mole and $S°_{298}(g) = 101.07$ cal/(mole °K). Chermin (221) presented the heat capacity of the ideal gas. Rossini, Pitzer, Arnett, Braun, and Pimentel (1248) selected $Tm = 217°K$ and $Tb = 401.9°K$.

No. 550 Heptanal $C_7H_{14}O$ (Ideal Gas State) Mol Wt 114.182

	cal/(mole °K)				kcal/mole		
$T°K$	$Cp°$	$S°$	$-(G°-H°_{298})/T$	$H°-H°_{298}$	$\Delta Hf°$	$\Delta Gf°$	Log Kp
298	40.89	110.34	110.34	0.00	-63.10	-20.71	15.183
300	41.07	110.60	110.35	0.08	-63.15	-20.45	14.900
400	51.00	123.80	112.08	4.69	-65.48	-5.85	3.197
500	59.90	136.16	115.67	10.25	-67.41	9.29	-4.058
600	67.70	147.79	120.06	16.64	-68.94	24.77	-9.021
700	74.20	158.72	124.81	23.74	-70.10	40.49	-12.640
800	79.80	169.00	129.70	31.45	-70.96	56.34	-15.390
900	84.60	178.69	134.61	39.67	-71.53	72.29	-17.554
1000	88.70	187.82	139.48	48.34	-71.83	88.30	-19.298

Nicholson (1065) measured the heat of combustion and reported $\Delta Hf°_{298}(l) = -74.5$ kcal/mole, which leads to $\Delta Hf°_{298}(g) = -63.1$ kcal/mole. Parks, Kennedy, Gates, Mosley, Moore, and Renquist (1109) measured the heat capacity from 80° to 300°K and reported $Tm = 229.8°K$, $\Delta Hm = 5.637$ kcal/mole, and $S°_{298}(l) = 83.3$ cal/(mole °K). Vapor pressure data of Stull (1432) permit calculation of $\Delta Hv_{298} = 11.40$ kcal/mole. Chermin (221) assembled pertinent data and calculated complete thermodynamic functions, including $S°_{298}(g) = 110.34$ cal/(mole °K), in good agreement with the third-law entropy. Rossini, Pitzer, Arnett, Braun, and Pimentel (1248) selected $Tb = 425.8°K$.

No. 551 Octanal $C_8H_{16}O$ (Ideal Gas State)   Mol Wt 128.208

$T°K$	$Cp°$	$S°$	$-(G°-H°_{298})/T$	$H°-H°_{298}$	$\Delta Hf°$	$\Delta Gf°$	Log Kp
	cal/(mole °K)				kcal/mole		
298	46.36	119.66	119.66	0.00	-69.23	-19.91	14.594
300	46.56	119.95	119.67	0.09	-69.29	-19.61	14.284
400	57.90	134.92	121.63	5.32	-71.94	-2.63	1.436
500	68.20	148.98	125.71	11.64	-74.12	14.96	-6.539
600	77.00	162.21	130.70	18.91	-75.85	32.94	-11.997
700	84.50	174.66	136.10	27.00	-77.16	51.19	-15.980
800	90.90	186.37	141.66	35.77	-78.11	69.57	-19.006
900	96.40	197.40	147.25	45.15	-78.73	88.08	-21.389
1000	101.00	207.80	152.79	55.02	-79.05	106.65	-23.307

Chermin (221) presented heat capacity functions. Increments from Table 10-8 were used to derive $\Delta Hf°_{298}(g) = -69.23$ kcal/mole and $S°_{298}(g) = 119.66$ cal/(mole °K). Rossini, Pitzer, Arnett, Braun, and Pimentel (1248) selected $Tm = 246°K$ and $Tb = 446°K$.

No. 552 Nonanal $C_9H_{18}O$ (Ideal Gas State)   Mol Wt 142.234

$T°K$	$Cp°$	$S°$	$-(G°-H°_{298})/T$	$H°-H°_{298}$	$\Delta Hf°$	$\Delta Gf°$	Log Kp
	cal/(mole °K)				kcal/mole		
298	51.82	128.97	128.97	0.00	-74.16	-17.91	13.124
300	52.06	129.30	128.98	0.10	-74.22	-17.56	12.792
400	64.80	146.05	131.17	5.95	-77.19	1.80	-0.983
500	76.40	161.78	135.74	13.03	-79.64	21.84	-9.546
600	86.40	176.62	141.33	21.18	-81.57	42.31	-15.412
700	94.80	190.58	147.38	30.25	-83.01	63.09	-19.697
800	101.90	203.72	153.61	40.09	-84.06	84.02	-22.952
900	108.10	216.09	159.87	50.60	-84.74	105.09	-25.517
1000	113.40	227.76	166.08	61.68	-85.08	126.21	-27.583

Heat capacity functions were calculated by Chermin (221). Increments from Table 10-8 are used to derive $\Delta Hf°_{298}(g) = -74.16$ kcal/mole and $S°_{298}(g) = 128.97$ cal/(mole °K). $Tm = 255°K$ and $Tb = 467°K$ were selected by Rossini, Pitzer, Arnett, Braun, and Pimentel (1248).

No. 553 Decanal $C_{10}H_{20}O$ (Ideal Gas State)   Mol Wt 156.260

Chermin (221) calculated heat capacity functions. Increments from Table 10-8 are used to derive $\Delta Hf°_{298}(g) = -79.09$ kcal/mole and

No. 553 Decanal $C_{10}H_{20}O$ (Ideal Gas State) Mol Wt 156.260

T°K	Cp°	S°	$-(G°-H°_{298})/T$	$H°-H°_{298}$	$\Delta Hf°$	$\Delta Gf°$	Log Kp
	cal/(mole °K)				kcal/mole		
298	57.29	138.28	138.28	0.00	-79.09	-15.90	11.654
300	57.55	138.64	138.29	0.11	-79.16	-15.51	11.300
400	71.80	157.18	140.72	6.59	-82.45	6.23	-3.402
500	84.70	174.62	145.77	14.43	-85.14	28.72	-12.551
600	95.70	191.06	151.96	23.46	-87.26	51.68	-18.825
700	105.00	206.53	158.67	33.51	-88.86	74.99	-23.410
800	113.00	221.09	165.57	44.42	-90.01	98.46	-26.896
900	119.90	234.80	172.51	56.07	-90.74	122.07	-29.642
1000	125.70	247.75	179.39	68.36	-91.10	145.76	-31.854

$S°_{298}(g) = 138.28$ cal/(mole °K). Rossini, Pitzer, Arnett, Braun, and Pimentel (1248) selected $Tm = 267°K$ and $Tb = 486°K$.

ALKANONE IDEAL GAS TABLES

No. 554 Acetone C_3H_6O (Ideal Gas State) Mol Wt 58.078

T°K	Cp°	S°	$-(G°-H°_{298})/T$	$H°-H°_{298}$	$\Delta Hf°$	$\Delta Gf°$	Log Kp
	cal/(mole °K)				kcal/mole		
298	17.90	70.49	70.49	0.00	-52.00	-36.58	26.811
300	17.97	70.61	70.50	0.04	-52.02	-36.48	26.577
400	22.00	76.33	71.25	2.04	-53.20	-31.12	17.001
500	25.89	81.67	72.80	4.44	-54.22	-25.48	11.135
600	29.34	86.70	74.70	7.20	-55.07	-19.65	7.156
700	32.34	91.45	76.76	10.29	-55.74	-13.69	4.273
800	34.93	95.94	78.88	13.65	-56.28	-7.64	2.088
900	37.19	100.19	81.01	17.26	-56.67	-1.54	0.373
1000	39.15	104.21	83.13	21.08	-56.93	4.61	-1.007

From measurements of the heat of vaporization and vapor heat capacity, Pennington and Kobe (1135) assigned the fundamental vibrational frequencies and derived the thermodynamic properties of the ideal gas, including $S°_{298}(g) = 70.49$ cal/(mole °K). They reported $\Delta Hv = 6.952$ kcal/mole at $Tb = 329.28°K$, and their calculated entropy, $S°_{298}(g) = 70.49$ cal/(mole °K), is in good agreement with the third-law value measured by Kelley (734), who reported $S°_{298}(l) = 47.9$ cal/(mole °K), $Tm = 176.61°K$, and $\Delta Hm = 1.366$ kcal/mole. Miles and Hunt (1000) used a flame calorimeter to measure the enthalpy of combustion of the vapor. Small adjustments for changes in atomic weights and in the enthalpy of combustion of hydrogen yield $\Delta Hf°_{298}(g) = -51.74$ kcal/mole. Dolliver, Gresham, Kistiakowsky, Smith, and Vaughan (334) measured the heat of hydrogenation of acetone vapor at 355°K. Their measurements can be combined with the accurately known values for isopropyl alcohol to yield $\Delta Hf°_{298}(g) = -52.07$ kcal/mole. A third source of information is the equilibrium study of Kolb and Burwell (791) on the acetone–hydrogen–isopropyl alcohol system. These data yield $\Delta Hf°_{298}(g) = -52.03$ kcal/mole, as described on page 78. A rounded average of -52.00 kcal/mole is adopted. Additional data on the same system were presented by Buckley and Herington (180). The new data are in good agreement with our selected values. Kobe and Lynn (780) listed $Tc = 508.7$°K and $Pc = 46.6$ atm.

No. 555   2-Butanone C_4H_8O (Ideal Gas State)   Mol Wt 72.104

$T°K$	cal/(mole °K)				kcal/mole		Log Kp
	$Cp°$	$S°$	$-(G°-H°_{298})/T$	$H°-H°_{298}$	$\Delta Hf°$	$\Delta Gf°$	
298	24.59	80.81	80.81	0.00	-56.97	-34.91	25.591
300	24.68	80.97	80.82	0.05	-57.00	-34.78	25.335
400	29.81	88.77	81.85	2.78	-58.39	-27.15	14.834
500	34.76	95.97	83.96	6.01	-59.59	-19.20	8.392
600	39.09	102.70	86.53	9.71	-60.58	-11.03	4.017
700	42.83	109.01	89.29	13.81	-61.37	-2.70	0.843
800	46.08	114.95	92.13	18.26	-61.99	5.72	-1.562
900	48.90	120.54	94.98	23.01	-62.43	14.21	-3.451
1000	51.33	125.82	97.80	28.02	-62.72	22.75	-4.973

Sinke and Oetting (1358) measured the enthalpy of combustion and low-temperature thermal data, finding $Ttp = 186.48$°K, $\Delta Hm° = 2.017$ kcal/mole, $S°_{298}(l) = 57.08$ cal/(mole °K), and $\Delta Hf°_{298}(l) = -65.31$ kcal/mole. Sinke and Oetting used their own data and the vapor heat capacities and enthalpies of vaporization reported by Nickerson, Kobe, and McKetta (1068) to derive barriers to internal rotation by the method of Scott and McCullough (1317). These barriers were employed to calculate thermodynamic functions, while the $\Delta Hv°_{298} = 8.34$ kcal/mole reported by Nickerson et al. (1068) yields $\Delta Hf°_{298}(g) = -56.97$ kcal/mole. Sinke and Oetting (1358) listed references to previous work on this compound. Buckley and Herington (180) presented equilibrium constants for hydrogenation of 2-butanone to 2-butanol. Their results are in excellent agreement with our selected values. Buckley and Herington listed references to earlier work on the hydrogenation equilibrium. Nickerson et al. (1068) reported $Tb = 352.61$°K, where $\Delta Hv = 7.475$ kcal/mole. Kobe, Crawford, and Stephenson (773) measured $Tc = 535.7$°K and $Pc = 41$ atm.

No. 556 2-Pentanone $C_5H_{10}O$ (Ideal Gas State) Mol Wt 86.130

T°K	$Cp°$	$S°$	$-(G°-H°_{298})/T$	$H°-H°_{298}$	$\Delta Hf°$	$\Delta Gf°$	Log Kp
		cal/(mole °K)			kcal/mole		
298	28.91	89.91	89.91	0.00	-61.82	-32.76	24.016
300	29.06	90.09	89.92	0.06	-61.86	-32.59	23.738
400	36.42	99.49	91.15	3.34	-63.64	-22.55	12.319
500	42.80	108.31	93.71	7.31	-65.12	-12.10	5.288
600	48.32	116.62	96.84	11.87	-66.32	-1.38	0.504
700	53.06	124.43	100.23	16.95	-67.26	9.52	-2.973
800	57.13	131.79	103.72	22.46	-67.98	20.53	-5.609
900	60.62	138.72	107.23	28.35	-68.48	31.64	-7.682
1000	63.61	145.27	110.71	34.57	-68.80	42.79	-9.351

Nickerson, Kobe, and McKetta (1068) measured the vapor heat capacity and enthalpy of vaporization of 2-pentanone and of 2-butanone. They studied the molecular structure, spectroscopic data, and barriers to internal rotation, and computed tables of thermodynamic functions for 2-butanone, which were later modified by Sinke and Oetting (1358). The former workers employed their measured information to establish the values for the —CH_2— increment for the thermodynamic functions of the higher members of methyl ketone series. The 2-pentanone values presented here employ these increments with the Sinke and Oetting (1358) values of 2-butanone, yielding $S°_{298}(g) = 89.91$ cal/(mole °K) for 2-pentanone. The vapor heat capacity values of Nickerson, Kobe, and McKetta (1068) do not require modification. The enthalpy of formation for gaseous pentane less that of gaseous butane yields the —CH_2— increment $\Delta Hf°_{298}(g) = (-35.00) - (-30.15) = -4.85$ kcal/mole. Application of this increment to the 2-butanone values of Sinke and Oetting(1358) give, for 2-pentanone, the values $\Delta Hf°_{298}(l) = -71.70$ kcal/mole and $\Delta Hf°_{298}(g) = -61.82$ kcal/mole, while $\Delta Hv_{298} = 9.885$ kcal/mole and $\Delta Hv = 7.980$ kcal/mole at $Tb = 375.42°K$ from the measurements of Nickerson, Kobe, and McKetta (1068).

ALKENONE IDEAL GAS TABLE

No. 557 Ketene C_2H_2O (Ideal Gas State) Mol Wt 42.036

Rice and Greenberg (1223) measured the heat of reaction of ketene with dilute aqueous sodium hydroxide and with several aliphatic alcohols. From their data Rossini, Wagman, Evans, Levine, and Jaffe (1249) derived $\Delta Hf°_{298}(g) = -14.6$ kcal/mole.

Recent microwave and infrared spectroscopic measurements by Cox

Cycloalkanone Ideal Gas Tables

No. 557 Ketene C_2H_2O (Ideal Gas State) Mol Wt 42.036

T°K	$Cp°$	$S°$	$-(G°-H°_{298})/T$	$H°-H°_{298}$	$\Delta Hf°$	$\Delta Gf°$	Log Kp
		cal/(mole °K)			kcal/mole		
298	12.37	57.79	57.79	0.00	-14.60	-14.41	10.561
300	12.41	57.87	57.80	0.03	-14.60	-14.41	10.495
400	14.22	61.70	58.30	1.36	-14.81	-14.31	7.817
500	15.68	65.03	59.32	2.86	-15.02	-14.16	6.188
600	16.89	68.00	60.53	4.49	-15.22	-13.97	5.088
700	17.91	70.68	61.79	6.23	-15.42	-13.74	4.290
800	18.80	73.13	63.06	8.07	-15.60	-13.49	3.686
900	19.57	75.39	64.30	9.99	-15.78	-13.22	3.209
1000	20.25	77.49	65.52	11.98	-15.93	-12.92	2.824

and Esbitt (283) and by Moore and Pimentel (1023) yield significantly improved thermodynamic functions compared to those calculated from the earlier spectroscopic data listed by Drayton and Thompson (349), Bak and Andersen (61), and Arendale and Fletcher (25). The thermodynamic properties calculated by Moore and Pimentel (1023) are selected for use here and include $S°_{298}(g) = 57.79$ cal/(mole °K). Rossini, Wagman, Evans, Levine, and Jaffe (1249) listed $Tm = 122°K$ and $Tb = 217°K$. No ΔHm or ΔHv values have been reported.

CYCLOALKANONE IDEAL GAS TABLE

No. 558 Cyclohexanone $C_6H_{10}O$ (Ideal Gas State) Mol Wt 98.140

T°K	$Cp°$	$S°$	$-(G°-H°_{298})/T$	$H°-H°_{298}$	$\Delta Hf°$	$\Delta Gf°$	Log Kp
		cal/(mole °K)			kcal/mole		
298	26.21	77.00	77.00	0.00	-55.00	-21.69	15.898
300	26.39	77.17	77.01	0.05	-55.05	-21.49	15.652
400	36.00	86.09	78.16	3.18	-57.23	-9.95	5.436
500	45.00	95.11	80.65	7.24	-58.94	2.08	-0.907
600	52.90	104.03	83.80	12.14	-60.18	14.40	-5.244
700	59.50	112.70	87.32	17.77	-60.99	26.90	-8.398
800	65.00	121.01	91.01	24.01	-61.44	39.48	-10.785
900	69.50	128.94	94.79	30.74	-61.59	52.11	-12.654
1000	73.00	136.45	98.58	37.87	-61.49	64.75	-14.151

Conn, Kistiakowsky, and Smith (261) measured the enthalpy of hydrogenation of gaseous cyclohexanone to gaseous cyclohexanol at 355°K and reported $\Delta Hr°_{355} = -15.424$ kcal/mole, which is corrected to $\Delta Hr°_{298} = -15.30$ kcal/mole. Employing for cyclohexanol $\Delta Hf°_{298}(g) = -70.40$ kcal/mole yields, for cyclohexanone, $\Delta Hf°_{298}(g) = -55.10$ kcal/mole. Sellers and Sunner (1331) measured the enthalpy of combustion and reported $\Delta Hc°(l) = -841.04$ kcal/mole, from which $\Delta Hf°_{298}(l) = -64.86$ kcal/mole and $\Delta Hf°_{298}(g) = -54.75$ kcal/mole, since Kolossovsky and Alimov (799) gave $\Delta Hv_{302} = 10.11$ kcal/mole. The value of $\Delta Hf°_{298}(g) = -55.00$ kcal/mole is adopted. The gaseous entropy at 298°K is estimated by comparing entropy differences of ethyl alcohol and acetaldehyde, ethane and acetaldehyde, propane and acetone, and isopropyl alcohol and acetone, then applying the average to cyclohexanol and cyclohexanone. From this comparison $S°_{298}(g) = 77 \pm 1$ cal/(mole °K). The gaseous heat capacity is obtained by adding the values for cyclohexane and acetone and subtracting that for propane. Timmermans (1501) listed $Tm = 242.0$°K and $Tb = 428.8$°K. At the boiling point $\Delta Hv = 9.00$ kcal/mole may be estimated from the 293°K value given by Kolossovsky and Alimov (799).

ALKANOIC ACID AND DERIVATIVE IDEAL GAS TABLES

No. 559   Formic Acid CH_2O_2 (Ideal Gas State)   Mol Wt 46.026

	cal/(mole °K)				kcal/mole		
T°K	$Cp°$	$S°$	$-(G°-H°_{298})/T$	$H°-H°_{298}$	$\Delta Hf°$	$\Delta Gf°$	Log Kp
298	10.81	59.45	59.45	0.00	-90.49	-83.89	61.492
300	10.84	59.52	59.46	0.03	-90.50	-83.85	61.084
400	12.85	62.92	59.91	1.21	-90.96	-81.56	44.561
500	14.62	65.99	60.82	2.59	-91.34	-79.17	34.602
600	16.02	68.78	61.92	4.12	-91.64	-76.70	27.938
700	17.32	71.35	63.08	5.79	-91.87	-74.20	23.164
800	18.35	73.73	64.27	7.58	-92.05	-71.66	19.576
900	19.21	75.94	65.44	9.45	-92.18	-69.10	16.779
1000	19.95	78.00	66.60	11.41	-92.27	-66.53	14.539

Sinke (1353) burned a high-purity sample and reported $\Delta Hf°_{298}(l) = -101.52$ kcal/mole, which may be coupled with $\Delta Hv_{298} = 4.78$ kcal/mole and the enthalpy change from saturated vapor to ideal gas monomer of 6.25 kcal/mole reported by Waring (1571) to give $\Delta Hf°_{298}(g) = -90.49$ kcal/mole. Stout and Fisher (1428) measured the low-temperature properties and reported $Ttp = 281.45$°K, $\Delta Hm = 3.035$ kcal/mole, and $S°_{298}(l) = 30.82 \pm 0.1$ cal/(mole °K). Millikan and Pitzer (1010) remeasured the infrared spectrum, reviewed the previous spectra, revised the fundamental vibrational frequency assignment, and calculated $S°_{298}(g) = 59.43$ cal/(mole °K). Green (534) more recently reviewed all the data, including microwave information, and made a refined calculation of all the thermodynamic functions, from which $S°_{298}(g) = 59.45$ cal/(mole °K) is adopted here. Halford (562) and Waring (1571) computed the entropy of vaporization to the equilibrium vapor (monomer and dimer) and the transformation of equilibrium vapor to ideal gas at 298.15°K. They believed that dimers existed in the crystalline state, with the possibility of random orientation of the H atoms in the hydrogen bonds. Later X-ray structure determinations showed that the crystal has an infinite chain type of structure. Consequently the term for random orientation in the crystal has been subtracted from the original calculations, giving the following values for the entropy of ideal gaseous formic acid at 298.15°K.

	$S°_{298}$ [cal/(mole °K)]	Source
Spectroscopic	59.427	Millikan and Pitzer (1010)
Thermal	59.30 ± 0.3	Halford (562)
Thermal	59.44 ± 0.1	Waring (1571)

This agreement is well within the experimental error. Waring (1571) reported $\Delta Hv = 5.235$ kcal/mole at $Tb = 373.7°K$.

No. 560   Acetic acid $C_2H_4O_2$ (Ideal Gas State)   Mol Wt 60.052

	cal/(mole °K)				kcal/mole		
T°K	$Cp°$	$S°$	$-(G°-H°_{298})/T$	$H°-H°_{298}$	$\Delta Hf°$	$\Delta Gf°$	Log Kp
298	15.90	67.52	67.52	0.00	-103.93	-90.03	65.989
300	15.97	67.62	67.53	0.03	-103.95	-89.94	65.520
400	19.52	72.72	68.20	1.81	-104.76	-85.14	46.518
500	22.60	77.41	69.58	3.92	-105.42	-80.16	35.036
600	25.15	81.76	71.25	6.31	-105.94	-75.06	27.339
700	27.35	85.81	73.04	8.94	-106.34	-69.88	21.817
800	29.08	89.58	74.88	11.77	-106.64	-64.65	17.662
900	30.60	93.09	76.71	14.75	-106.87	-59.39	14.421
1000	31.99	96.39	78.51	17.88	-107.01	-54.10	11.823

Evans and Skinner (389) reviewed earlier combustion work, measured the heat of combustion, and reported $\Delta Hf°_{298}(l) = -115.72$ kcal/mole. From the measurements of Brown (174), MacDougall (895), and Ritter and Simons (1233) we calculate an enthalpy increment of 11.75 kcal/mole for the conversion of the real liquid to the ideal monomeric gas at 298.15°K, leading to $\Delta Hf°_{298}(g) = -103.93$ kcal/mole. Parks, Kelley, and Huffman (1108) measured the heat capacity from 90° to 298°K and reported $S°_{298}(l) = 38.20$ cal/(mole °K). More recently Weltner (1587) measured the vapor heat capacity and heat of vaporization and derived the monomer-dimer equilibrium constant. He made a frequency assignment, calculated the thermodynamic functions, and reported $S°_{298}(g) = 67.52$ cal/(mole °K). Rossini, Wagman, Evans, Levine, and Jaffe (1249) selected $Tm = 298.76°K$, $\Delta Hm = 2.80$ kcal/mole, and $Tb = 391.25°K$, at which temperature Weltner (1587) measured $\Delta Hv = 5.663$ kcal/mole. The values $Tc = 594.8°K$ and $Pc = 56.0$ atm are taken from the measurements of Young as quoted by Timmermans (1501).

No. 561 Methyl Formate $C_2H_4O_2$ (Ideal Gas State) Mol Wt 60.052

T°K	Cp°	S°	$-(G°-H°_{298})/T$	$H°-H°_{298}$	$\Delta Hf°$	$\Delta Gf°$	Log Kp
	cal/(mole °K)				kcal/mole		
298	15.90	72.00	72.00	0.00	-83.60	-71.03	52.067
300	16.00	72.10	72.01	0.03	-83.62	-70.96	51.689
400	19.50	77.20	72.68	1.81	-84.43	-66.61	36.390
500	22.60	81.89	74.06	3.92	-85.09	-62.07	27.130
600	25.20	86.25	75.73	6.32	-85.60	-57.42	20.913
700	27.40	90.30	77.52	8.95	-86.00	-52.69	16.449
800	29.10	94.07	79.36	11.78	-86.30	-47.91	13.088
900	30.60	97.59	81.19	14.76	-86.53	-43.09	10.464
1000	32.00	100.89	83.00	17.89	-86.67	-38.26	8.360

Rossini, Wagman, Evans, Levine, and Jaffe (1249) reviewed the combustion literature and selected $\Delta Hf°_{298}(g) = -83.6$ kcal/mole. An entropy value of $S°_{298}(g) = 72.0$ cal/(mole °K) is estimated from the equilibrium measurements of Lacy, Dunning, and Storch (836). Rossini et al. (1249) selected $\Delta Hv = 6.750$ kcal/mole at $Tb = 304.7°K$. Timmermans (1500) reported $Tm = 173°K$ and $\Delta Hm = 1.800$ kcal/mole. Curl (299) observed the microwave spectrum and fixed the potential barrier of the methyl group to internal rotation at 1.190 kcal/mole. The heat capacity of the gas is assumed to be the same as that of acetic acid gas.

No. 562 Acetic Anhydride $C_4H_6O_3$ (Ideal Gas State) Mol Wt 102.088

T°K	Cp°	S°	$-(G°-H°_{298})/T$	$H°-H°_{298}$	$\Delta Hf°$	$\Delta Gf°$	Log Kp
	cal/(mole °K)				kcal/mole		
298	23.78	93.20	93.20	0.00	-137.60	-113.93	83.511
300	23.91	93.35	93.21	0.05	-137.63	-113.79	82.889
400	30.86	101.21	94.23	2.80	-139.02	-105.61	57.702
500	36.78	108.75	96.39	6.19	-140.09	-97.13	42.455
600	41.62	115.89	99.05	10.11	-140.91	-88.46	32.221
700	45.74	122.63	101.94	14.49	-141.50	-79.67	24.874
800	48.91	128.95	104.93	19.23	-141.92	-70.81	19.344
900	51.64	134.87	107.93	24.25	-142.19	-61.91	15.032
1000	54.11	140.44	110.90	29.54	-142.32	-52.97	11.576

Conn, Kistiakowsky, Roberts, and Smith (258) and Wadsö (1558) are in good agreement on the enthalpy of hydrolysis of liquid acetic anhydride to liquid acetic acid, finding -13.96 and -14.00 kcal/mole, respectively.

An average of -13.98 kcal/mole is used with the enthalpies of formation of acetic acid and water adopted for this work to derive for acetic anhydride $\Delta Hf^\circ_{298}(l) = -149.14$ kcal/mole. Measured low-temperature thermal data (968) include $Tm = 199.11°K$, $\Delta Hm° = 2.51$ kcal/mole, and $S^\circ_{298}(l) = 64.2$ cal/(mole °K). These data lead to a Gibbs energy of hydrolysis of 12.66 kcal/(mole °K), in reasonable agreement with the 12.2 kcal/(mole °K) reported by Knopp, Linnell, and Child (767) on the basis of equilibrium data. Vapor pressure measurements by McDonald, Shrader, and Stull (969) are used to calculate $\Delta H v^\circ_{298} = 11.54$ kcal/mole, $\Delta Hf^\circ_{298}(g) = -137.6$ kcal/mole, $S^\circ_{298}(g) = 93.2$ cal/(mole °K), and $\Delta Hv = 9.85$ kcal/mole at $Tb = 411.8°K$. The vapor heat capacity is approximated by subtracting values for water vapor from twice those of acetic acid. Kobe and Lynn (780) selected $Tc = 569°K$ and $Pc = 46.2$ atm.

No. 563   Ethyl Acetate $C_4H_8O_2$ (Ideal Gas State)   Mol Wt 88.104

T°K	Cp°	S°	$-(G°-H°_{298})/T$	$H°-H°_{298}$	$\Delta Hf°$	$\Delta Gf°$	Log Kp
	cal/(mole °K)				kcal/mole		
298	27.16	86.70	86.70	0.00	-105.86	-78.25	57.359
300	27.24	86.87	86.71	0.06	-105.89	-78.08	56.882
400	32.84	95.47	87.84	3.06	-107.36	-68.58	37.470
500	38.70	103.44	90.17	6.64	-108.58	-58.74	25.675
600	43.65	110.95	93.02	10.77	-109.52	-48.69	17.733
700	47.69	117.99	96.09	15.34	-110.23	-38.49	12.015
800	51.01	124.58	99.24	20.28	-110.75	-28.21	7.705
900	53.75	130.76	102.40	25.52	-111.11	-17.86	4.337
1000	56.05	136.54	105.53	31.01	-111.34	-7.48	1.635

Wadsö (1555) measured the enthalpy of hydrolysis, $\Delta H^\circ_{298} = 0.89$ kcal/mole, for the reaction

$$CH_3COOC_2H_5(l) + H_2O(l) = C_2H_5OH(l) + CH_3COOH(l).$$

With previously adopted values for ethyl alcohol, acetic acid, and water, the value for ethyl acetate becomes $\Delta Hf^\circ_{298}(l) = -114.49$ kcal/mole. The vapor pressure values selected by Stull (1432) are 71 mm of mercury at 20°C, 90 mm at 25°C, and 115 mm at 30°C. From the Clausius-Clapeyron equation we deduce

$$\Delta Hv_{298} = R\frac{\Delta P}{P}\frac{T^2}{\Delta T} = 1.9872 \cdot \frac{(115-71)}{90 \times 1000} \cdot \frac{(298.15)^2}{(30-20)} = 8.63 \text{ kcal/mole,}$$

leading to $\Delta Hf^\circ_{298}(g) = -105.86$ kcal/mole, which is adopted. This value agrees reasonably well with the heat of combustion of the liquid measured by Schjanberg (1285), who reported $\Delta Hf^\circ_{298}(g) = -106.58$ kcal/mole.

Parks, Huffman, and Barmore (1106) measured the heat capacity from 90° to 298°K and reported $Tm = 189.3°K$, $\Delta Hm° = 2.505$ kcal/mole, and

$S°_{298}(l)$	62.0 cal/(mole °K)
ΔSv, 8630/298.15	28.94
ΔS(compression), $- R \ln 760/90$	-4.24
$S°_{298}$(ideal gas)	86.7 cal/(mole °K).

Mathews (940) measured $\Delta Hv = 7.720$ kcal/mole at $Tb = 349.15°K$. The gaseous heat capacity was measured by Jatkar (683) and by Bennewitz and Rossner (103), whose values accord well and are extrapolated to higher temperatures subject to the assumption that 80% of the classical value is reached at 1500°K.

ALKENOIC ACID IDEAL GAS TABLES

No. 564   Acrylic Acid $C_3H_4O_2$ (Ideal Gas State)   Mol Wt 72.062

	cal/(mole °K)				kcal/mole		
T°K	Cp°	S°	$-(G°-H°_{298})/T$	$H°-H°_{298}$	$\Delta Hf°$	$\Delta Gf°$	Log Kp
298	18.59	75.29	75.29	0.00	-80.36	-68.37	50.113
300	18.67	75.41	75.30	0.04	-80.38	-68.30	49.751
400	22.94	81.38	76.08	2.13	-81.13	-64.15	35.046
500	26.56	86.90	77.70	4.61	-81.73	-59.83	26.149
600	29.50	92.01	79.66	7.41	-82.21	-55.40	20.179
700	32.00	96.75	81.77	10.49	-82.59	-50.90	15.892
800	33.93	101.16	83.92	13.79	-82.88	-46.36	12.664
900	35.61	105.25	86.07	17.27	-83.10	-41.78	10.145
1000	37.12	109.08	88.18	20.91	-83.24	-37.17	8.124

As an example of a combination of calculational procedures, the following is presented. Schjanberg (1285) measured the heat of combustion of liquid propionic acid, leading to $\Delta Hf°_{298}(l) = -121.92$ kcal/mole. Skinner and Snelson (1364) measured the enthalpy of hydrogenation of liquid acrylic acid to liquid propionic acid as -30.35 kcal/mole, giving for acrylic acid $\Delta Hf°_{298}(l) = -91.75$ kcal/mole. Stull (1432) selected vapor pressure values for acrylic acid of 3.10 mm of mercury pressure at 20°C, 4.28 mm at 25°C, and 5.85 mm at 30°C. From the Clausius-Clapeyron equation we deduce $\Delta Hv = 11.21$ kcal/mole. Hence for acrylic acid $\Delta Hf°_{298}(g) = -80.54$ kcal/mole.

Alkenoic Acid Ideal Gas Tables

Parks and Huffman (1105) listed $S°_{298}(l)$ for butyric acid = 54.1 cal/(mole °K). From vapor pressure values selected by Stull (1432) we interpolate pressures of 0.68, 0.98, and 1.40 mm for 20°, 25°, and 30°C, respectively, from which we calculate ΔHv_{298} for butyric acid = 12.98 kcal/mole. We can now calculate the entropy of butyric acid as follows.

$S°_{298}(l)$	54.10 cal/(mole °K)
ΔSv, 12,980/298.15	43.53
ΔS (compression), $-R \ln (760/0.98)$	-13.22
Gas imperfection correction (estimated)	0.02
$S°_{298}$(ideal gas)	84.43 cal/(mole °K)

We may now estimate the entropy of gaseous acrylic acid from the following gaseous entropies at 298.15°K.

Replacement of —CH_3 by —COOH	
H_3CCH_3 (54.85), H_3CCOOH (67.52)	12.67
$H_3CCH_2CH_2CH_3$ (74.12), $H_3CCH_2CH_2COOH$ (84.43)	10.31
Average	11.49

For H_2CCHCH_3, $S°_{298}(g) = 63.80$ cal/(mole °K). Thus, $63.80 + 11.49 = 75.29$ cal/(mole °K) for $H_2CCHCOOH(g)$. Finally, the gaseous heat capacity of acrylic acid may be estimated by the following combination of heat capacities.

T (°K)	H_3CCOOH [1]	H_3CCH_3 [2]	[1]−[2]	H_2CCHCH_3 [3]	$H_2CCHCOOH$ [3]+[1]−[2]
298	15.90	12.58	3.32	15.27	18.59
300	15.97	12.64	3.33	15.34	18.67
400	19.52	15.68	3.84	19.10	22.94
500	22.60	18.66	3.94	22.62	26.56
600	25.15	21.35	3.80	25.70	29.50
700	27.35	23.72	3.63	28.37	32.00
800	29.08	25.83	3.25	30.68	33.93
900	30.60	27.69	2.91	32.70	35.61
1000	31.99	29.33	2.66	34.46	37.12

Dimerization or polymerization of these species has been neglected.

PHENOL IDEAL GAS TABLES

No. 565   Phenol C_6H_6O (Ideal Gas State)   Mol Wt 94.108

$T°K$	$Cp°$	$S°$	$-(G°-H°_{298})/T$	$H°-H°_{298}$	$\Delta Hf°$	$\Delta Gf°$	Log Kp
		cal/(mole °K)			kcal/mole		
298	24.75	75.43	75.43	0.00	-23.03	-7.86	5.763
300	24.90	75.59	75.44	0.05	-23.05	-7.77	5.661
400	32.45	83.82	76.51	2.93	-24.09	-2.51	1.369
500	38.64	91.75	78.77	6.50	-24.90	2.99	-1.306
600	43.54	99.24	81.56	10.61	-25.53	8.62	-3.140
700	47.44	106.26	84.59	15.17	-26.01	14.36	-4.482
800	50.62	112.80	87.72	20.07	-26.38	20.14	-5.502
900	53.26	118.92	90.85	25.27	-26.64	25.98	-6.308
1000	55.49	124.65	93.94	30.71	-26.80	31.84	-6.958

Parks, Manchester, and Vaughan (1112) and Cox (285) measured the heat of combustion; the latter worker reported $\Delta Hf°_{298}(c) = -39.44$ kcal/mole, which is adopted. Biddiscombe and Martin (129) carefully studied the vapor pressure and reported $\Delta Hv = 9.73$ kcal/mole at $Tb = 454.99°K$ and $\Delta Hs_{298} = 16.41$ kcal/mole for the solid, from which is derived $\Delta Hf°_{298}(g) = -23.03$ kcal/mole. Parks, Huffman, and Barmore (1106) measured thermal properties from 90° to 298°K and reported $S°_{298}(c) = 34.10$ cal/(mole °K); more recently Andon, Counsell, Herington, and Martin (19) measured the thermal properties from 12° to 336°K and reported $S°_{298}(c) = 34.42$ cal/(mole °K) and $\Delta Hm = 2.752$ kcal/mole at $Ttp = 314.06°K$. The latter workers noted that ΔHs and vapor pressure values at 298.15°K were not sufficiently accurate to achieve agreement of observed and spectroscopically derived $S°_{298}(g) = 75.43$ cal/(mole °K), but their extrapolation of Cp/T values provided good agreement between observed and spectroscopically derived $S°_{400}(g)$. Evans (392) and later Green (533) deduced essentially the same vibrational assignment and calculated virtually the same thermodynamic functions. Calculations (968) indicate $Tc = 692°K$ and $Pc = 60.5$ atm.

No. 566 m-Cresol C_7H_8O (Ideal Gas State)   Mol Wt 108.134

$T°K$	$Cp°$	$S°$	$-(G°-H°_{298})/T$	$H°-H°_{298}$	$\Delta Hf°$	$\Delta Gf°$	Log Kp
		cal/(mole °K)			kcal/mole		
298	29.27	85.27	85.27	0.00	-31.63	-9.69	7.099
300	29.91	85.46	85.28	0.06	-31.66	-9.55	6.958
400	38.74	95.30	86.56	3.50	-33.08	-1.95	1.067
500	46.20	104.78	89.26	7.76	-34.21	5.96	-2.607
600	52.26	113.75	92.60	12.69	-35.10	14.08	-5.128
700	57.19	122.19	96.23	18.17	-35.78	22.34	-6.974
800	61.27	130.10	99.98	24.10	-36.29	30.67	-8.378
900	64.93	137.53	103.74	30.41	-36.65	39.06	-9.486
1000	68.50	144.55	107.47	37.08	-36.79	47.49	-10.379

Phenol Ideal Gas Tables

Green (537) calculated vapor heat capacities and $S°_{298}(g) = 85.27$ cal/(mole °K) from spectra and reported $\Delta Hf°_{298}(g) = -31.63$ kcal/mole. Pardee and Weinrich (1101) reported $Tm = 284.7°K$ and $\Delta Hv = 10.32$ kcal/mole at $Tb = 475.4°K$.

No. 567 o-Cresol C₇H₈O (Ideal Gas State) Mol Wt 108.134

T°K	Cp°	S°	$-(G°-H°_{298})/T$	$H°-H°_{298}$	$\Delta Hf°$	$\Delta Gf°$	Log Kp
	cal/(mole °K)				kcal/mole		
298	31.15	85.47	85.47	0.00	-30.74	-8.86	6.491
300	31.31	85.67	85.48	0.06	-30.77	-8.72	6.353
400	39.74	95.86	86.81	3.63	-32.07	-1.16	0.635
500	46.91	105.52	89.59	7.97	-33.11	6.69	-2.923
600	52.77	114.61	93.01	12.96	-33.94	14.72	-5.362
700	57.56	123.11	96.71	18.49	-34.58	22.89	-7.147
800	61.55	131.06	100.52	24.44	-35.06	31.13	-8.503
900	65.25	138.53	104.33	30.78	-35.38	39.43	-9.573
1000	68.82	145.59	108.10	37.49	-35.50	47.75	-10.436

Green (537) presented thermodynamic functions, $\Delta Hf°_{298}(g) = -30.74$ kcal/mole and $S°_{298}(g) = 85.47$ cal/(mole °K), and vapor heat capacities calculated from spectra. Pardee and Weinrich (1101) reported $Tm = 303.2°K$ and $\Delta Hv = 10.20$ kcal/mole at $Tb = 464.0°K$.

No. 568 p-Cresol C₇H₈O (Ideal Gas State) Mol Wt 108.134

T°K	Cp°	S°	$-(G°-H°_{298})/T$	$H°-H°_{298}$	$\Delta Hf°$	$\Delta Gf°$	Log Kp
	cal/(mole °K)				kcal/mole		
298	29.75	83.09	83.09	0.00	-29.97	-7.38	5.406
300	29.91	83.28	83.10	0.06	-30.00	-7.24	5.272
400	38.65	93.11	84.38	3.50	-31.42	0.58	-0.317
500	46.07	102.56	87.08	7.75	-32.56	8.72	-3.810
600	52.10	111.51	90.41	12.66	-33.47	17.05	-6.212
700	57.03	119.92	94.03	18.13	-34.16	25.54	-7.973
800	61.11	127.81	97.77	24.04	-34.70	34.10	-9.315
900	64.85	135.22	101.52	30.34	-35.06	42.72	-10.374
1000	68.40	142.24	105.24	37.00	-35.22	51.38	-11.229

Green (537) calculated $S°_{298}(g) = 83.09$ cal/(mole °K) and the vapor heat capacity from the spectrum, and reported $\Delta Hf°_{298}(g) = -29.97$ kcal/mole. Pardee and Weinrich (1101) reported $Tm = 307.9°K$ and $\Delta Hv = 10.32$ kcal/mole at $Tb = 475.3°K$.

BENZENECARBOXYLIC ACID IDEAL GAS TABLE

No. 569   Benzoic Acid $C_7H_6O_2$ (Ideal Gas State)   Mol Wt 122.118

	cal/(mole °K)				kcal/mole		
T°K	$Cp°$	$S°$	$-(G°-H°_{298})/T$	$H°-H°_{298}$	$\Delta Hf°$	$\Delta Gf°$	Log Kp
298	24.73	88.19	88.19	0.00	-69.36	-50.29	36.859
300	24.86	88.35	88.20	0.05	-69.39	-50.17	36.547
400	33.07	96.64	89.27	2.95	-71.01	-43.50	23.768
500	40.76	104.87	91.57	6.66	-72.36	-36.46	15.938
600	47.02	112.87	94.46	11.05	-73.47	-29.18	10.629
700	52.06	120.51	97.64	16.02	-74.35	-21.72	6.782
800	56.14	127.74	100.95	21.43	-75.07	-14.16	3.869
900	59.50	134.55	104.31	27.22	-75.64	-6.51	1.581
1000	62.30	140.97	107.66	33.31	-76.07	1.20	-0.262

Jessup (687, 688) and Prosen and Rossini (1201) are in substantial agreement on the heat of combustion, leading to $\Delta Hf°_{298}(c) = -92.062$ kcal/mole. Davies and Jones (310) gave $\Delta Hs°_{357} = 21.85$ kcal/mole, which yielded $\Delta Hs°_{298} = 22.70$ kcal/mole, making $\Delta Hf°_{298}(g) = -69.36$ kcal/mole. Furukawa, McCoskey, and King (454) carefully measured the low-temperature heat capacities and reported $S°_{298}(c) = 40.05$ cal/(mole °K), $Tm = 395.52°K$, and $\Delta Hm° = 4.32$ kcal/mole. Stull (1432) reported $Tb = 522.4°K$, at which $\Delta Hv = 12.10$ kcal/mole according to Davies and Jones (310). We calculate the thermodynamic functions of the ideal gas state from a provisional assignment of the fundamental vibrational frequencies and find $S°_{298}(g) = 88.19$ cal/(mole °K).

CHAPTER 11

CHEMICAL THERMODYNAMICS OF NITROGEN COMPOUNDS

INTRODUCTION

Compounds containing nitrogen in addition to carbon, hydrogen, and occasionally oxygen are considered in this chapter. Those containing halogen or sulfur as well as nitrogen are found in the following two chapters.

Many organic nitrogen compounds are commercially important, with annual production in the millions of pounds. Astle (28) summarized the chemistry and uses of these industrial organic nitrogen compounds. Since many of the processes he described appear to be limited by thermodynamic equilibria, the application of thermodynamic considerations to such reactions should prove fruitful. However, lack of accurate free-energy data for organic nitrogen compounds severely limits the effectiveness of this approach, in contrast to the situation in hydrocarbon technology where thermodynamics is used routinely. Parks and Huffman (1105) discussed free-energy values for 24 organic nitrogen compounds, but nearly all the quoted values were of low reliability and such important classes as aliphatic amines, nitriles, and nitro compounds were not represented. Accurate values on these and other classes are still scant; hence the many estimates incorporated in the tables of this chapter.

Although complete and highly accurate thermodynamic data are available for relatively few nitrogen compounds, many enthalpies of formation of moderate accuracy have been reported. This is in part a consequence of the importance of organic nitrogen compounds as explosives and high-energy fuels. As noted by Pauling (1130), the bond strength in molecular nitrogen is unexpectedly large. Molecules containing C—N, O—N, and H—N bonds are therefore highly unstable with respect to rearrangement to molecular nitrogen and other products, such as carbon monoxide and water. Well-known examples are nitroglycerin, trinitrotoluene, and cyclonite. A recent application of this property of nitrogen compounds is in rocket propellants; typical systems are aniline–nitric acid and dimethylhydrazine–liquid oxygen. The theoretical maximum performance of an explosive or propellant can be determined from its enthalpy of formation; hence one finds such extensive investigations as those of Williams, McEwan, and Henry (1616), Holcomb and Dorsey (613), Murrin and Goldhagen (1039), and Médard and Thomas (981–986).

Organic nitrogen compounds are highly important in biochemistry. Interest in the thermodynamics of biochemical processes led Huffman and his collaborators to measure the entropies and enthalpies of combustion of several amino acids. Borsook and Huffman (148) summarized the results and calculated equilibria for the syntheses of amino acids under physiological conditions; for example, the formation of alanine from pyruvic acid and ammonia in the liver. Interest in this field has recently been revived; thermal data for some of the compounds have been redetermined with higher accuracy, and properties of other amino acids not studied previously have also been measured (241, 666).

Organic nitrogen compounds provide interesting studies in structure, resonance, and internal rotation. An accurate and comprehensive study of pyrrolidine by McCullough (946) reinforced the concept of pseudorotation (rotation of the angle of maximum puckering around the ring) in saturated five-membered rings. Pseudorotation may be free, as in cyclopentane (753), or slightly hindered, as found in pyrrolidine (948).

In unsaturated six-membered rings, nitrogen can substitute for carbon to form heteroaromatic compounds. Resonance stabilization in these compounds is comparable to that in benzene, although exact comparison is not possible because of uncertainty in the "normal" C=N bond strength. Bedford, Beezer, and Mortimer (93) reported the resonance energy of pyridine and pyrazine as 32 and 24 kcal/mole, respectively, compared to 36 kcal/mole in benzene. Resonance is also important in the explanation of barriers to internal rotation. The nitro group rotates freely in aliphatic compounds, but a nitro group substituted on a benzene ring can undergo resonance involving the structure (1130)

$$\overset{+}{\underset{}{\bigcirc}} = \overset{+}{N} \underset{\diagdown O^-}{\diagup O^-}$$

giving the C—N link some double-bond character. The barrier to rotation in nitrobenzene is found experimentally to be about 6 kcal/mole. Similarly, amino groups in aliphatic compounds have a barrier to rotation of about

2 kcal/mole, but this is increased for aniline to 3.4 kcal/mole because of a resonance structure involving the ring (1130):

$$:\overline{\bigcirc} = \overset{+}{N}H_2.$$

On the other hand, methyl groups substituted on aromatic rings rotate freely, as evidenced by studies on the picolines (methylpyridines). An important development in the picoline studies was the unexpectedly large shift of low vibrational frequencies in going from liquid to vapor phase, an effect first observed in toluene. The use of liquid frequencies for the vapor phase caused an error in the calculated functions nearly as large as the usual calorimetric uncertainty, and precluded a fit with calorimetric data even if free rotation was assumed. Other noteworthy studies of internal rotation barriers include the methyl nitrate molecule, for which thermal data indicate free rotation of the nitro and methyl groups but microwave spectra indicate large barriers of 9 kcal/mole and 2.3 kcal/mole for the nitro and methyl groups, respectively. Further work is needed to resolve the discrepancy.

As noted earlier, many of the ideal gas tables of this chapter are based partly on estimates, particularly with respect to heat capacities and entropies. Unfortunately, so few experimental data on the nitrogen compounds exist that the estimates are not of high reliability. Table 11-1 gives various "methyl" substitution constants used for these estimates, and the data on which they are based. Some of the discrepancies in this table may be due to experimental uncertainties, but it appears that the methyl substitution method itself may be inadequate to make highly accurate estimates for some of the series.

PRIMARY ALKYLAMINE IDEAL GAS TABLES

No. 601 Methylamine CH_5N (Ideal Gas State) Mol Wt 31.058

Low-temperature thermal data for methylamine were obtained by Aston, Siller, and Messerly (43) as follows: $Ttp = 179.69°K$, $\Delta Hm° = 1.466$ kcal/mole, and $\Delta Hv = 6.169$ kcal/mole at $Tb = 266.82°K$. The entropy of the ideal gas at the boiling point was calculated to be 56.42 ± 0.3 cal/(mole °K). The relatively large uncertainty was due to thermal hysteresis effects in the solid state. The heat capacity was found to be strongly dependent on the thermal history of the sample.

Table 11-1 Substitution Constants for Gaseous Aliphatic Nitrogen Compounds

Type of Compound	$\Delta[\Delta H_{f\,298}^{\circ}]$ (kcal/mole)	ΔS_{298}° [cal/(mole °K)]	ΔC_p [cal/(mole °K)] at								
			298°K	300°K	400°K	500°K	600°K	700°K	800°K	900°K	1000°K
Primary amines											
$CH_3CH_3 \rightarrow CH_3NH_2$	14.74	−0.41	−0.59	−0.62	−1.29	−1.92	−2.48	−2.96	−3.38	−3.75	−4.07
$CH_3CH_2CH_3 \rightarrow CH_3CH_2NH_2$	13.82	0.01	−0.21	−0.22	−0.89	−1.58	−2.20	−2.73	−3.19	−3.59	−3.95
$CH_3CH_2CH_2CH_3 \rightarrow CH_3CH_2CH_2NH_2$	12.85										
$CH_3CH_2CH_2CH_2CH_3 \rightarrow CH_3CH_2CH_2CH_2NH_2$	13.00										
$R-CH_3 \rightarrow R-NH_2$	13.60	−0.20	−0.40	−0.42	−1.09	−1.75	−2.34	−2.85	−3.29	−3.67	−4.01
Secondary amines											
$CH_3CH_2CH_3 \rightarrow CH_3NHCH_3$	20.32	−0.60	−1.07	−1.08	−1.65	−2.09	−2.47	−2.78	−3.14	−3.37	−3.64
$CH_3CH_2CH_2CH_3 \rightarrow CH_3CH_2NHCH_2CH_3$	17.70										
$R-CH_2-R' \rightarrow R-NH-R'$	19.0	−0.60	−1.07	−1.08	−1.65	−2.09	−2.47	−2.78	−3.14	−3.37	−3.64
Tertiary amines											
$CH_3-\overset{\overset{H}{\mid}}{\underset{\underset{CH_3}{\mid}}{C}}-CH_3 \rightarrow CH_3-\underset{\underset{CH_3}{\mid}}{N}-CH_3$	26.45	−1.40	−1.21	−1.21	−1.69	−1.99	−2.28	−2.56	−2.87	−3.15	−3.42
$CH_3CH_2-\overset{\overset{H}{\mid}}{\underset{\underset{CH_2CH_3}{\mid}}{C}}-CH_2CH_3 \rightarrow CH_3CH_2-\underset{\underset{CH_2CH_3}{\mid}}{N}-CH_2CH_3$	21.54										
$R-\overset{\overset{H}{\mid}}{\underset{\underset{R''}{\mid}}{C}}-R' \rightarrow R-\underset{\underset{R''}{\mid}}{N}-R'$	24.0	−1.40	−1.21	−1.21	−1.69	−1.99	−2.28	−2.56	−2.87	−3.15	−3.42

Nitriles

CH₃CH₃ → CH₃CN	41.24	−0.22	−0.10	−0.12	−1.06	−2.07	−3.00	−3.82	−4.57	−5.24	−5.83
CH₃CH₂CH₃ → CH₃CH₂CN	35.92	0.43	−0.11	−0.13	−1.36	−2.50	−3.46	−4.26	−4.94	−5.56	−6.13
CH₃CH₂CH₂CH₃ → CH₃CH₂CH₂CN	38.29									
R—CH₃ → R—CN	**38.50**	**0.10**	**−0.10**	**−0.12**	**−1.21**	**−2.29**	**−3.23**	**−4.04**	**−4.75**	**−5.40**	**−6.00**

Nitro compounds

CH₃CH₃ → CH₃NO₂	2.38	8.70	1.12	1.12	1.12	0.90	0.57	0.18	−0.27	−0.72	−1.16
CH₃CH₂CH₃ → CH₃CH₂NO₂	0.62									
CH₃CH₂CH₂CH₃ → CH₃CH₂CH₂NO₂	0.35									
CH₃CH₂CH₂CH₂CH₃ → CH₃CH₂CH₂CH₂NO₂	0.60									
R—CH₃ → R—NO₂	**0.52**	**8.70**	**1.12**	**1.12**	**1.12**	**0.90**	**0.57**	**0.18**	**−0.27**	**−0.72**	**−1.16**

Nitrates

CH₃CH₃ → CH₃ONO₂	−8.86	15.12	5.70	6.19	6.29	6.19	5.97	5.64	5.27	4.86
CH₃CH₂CH₃ → CH₃CH₂ONO₂	−11.98	16.56								
CH₃CH₂CH₂CH₃ → CH₃CH₂CH₂ONO₂	−11.45									
R—CH₃ → R—ONO₂	**−10.8**	**15.8**	**5.70**	**6.19**	**6.29**	**6.19**	**5.97**	**5.64**	**5.27**	**4.86**

No. 601   Methylamine   CH$_5$N   (Ideal Gas State)   Mol Wt 31.058

	cal/(mole °K)				kcal/mole		
$T°K$	$Cp°$	$S°$	$-(G°-H°_{298})/T$	$H°-H°_{298}$	$\Delta Hf°$	$\Delta Gf°$	Log Kp
298	11.97	57.98	57.98	0.00	-5.50	7.71	-5.648
300	12.01	58.06	57.99	0.03	-5.52	7.79	-5.672
400	14.38	61.84	58.48	1.35	-6.53	12.38	-6.764
500	16.73	65.30	59.51	2.90	-7.39	17.21	-7.522
600	18.86	68.55	60.75	4.69	-8.09	22.20	-8.085
700	20.76	71.60	62.08	6.67	-8.65	27.29	-8.521
800	22.44	74.48	63.45	8.83	-9.09	32.46	-8.866
900	23.94	77.21	64.83	11.15	-9.41	37.67	-9.148
1000	25.26	79.81	66.20	13.61	-9.64	42.92	-9.380

Gray and Lord (523) made a vibrational assignment based on a thorough spectroscopic study. Wu, Zerbi, Califano, and Crawford (1631) revised a single frequency of the assignment on the basis of a normal coordinate calculation. Identical structural data and barrier to internal rotation have been reported by Itoh (672), Nishikawa (1077) and Nishikawa, Itoh, and Shimoda (1078). The revised vibrational assignment and the structural data yield an entropy of the ideal gas at the boiling point of 56.69 cal/(mole °K), in agreement, within experimental error, with the third-law value. Thermodynamic functions are calculated on the basis of the molecular data, which seem well established. The vapor heat capacity results of Felsing and Jessen (416) appear to be too high, possibly because of association in the gas phase. Kobe and Harrison (775) gave thermodynamic functions based on less accurate molecular data and the third-law entropy.

Jaffe (678) measured the enthalpy of combustion of liquid methylamine and calculated $\Delta Hf°_{298}(l) = -11.31$ kcal/mole. From the enthalpy of vaporization measurement of Aston et al. (43) and heat capacity data, Jaffe derived $\Delta Hv_{298} = 5.8$ kcal/mole, from which $\Delta Hf°_{298}(g) = -5.5$ kcal/mole. Cottrell and Gill (277) derived $\Delta Hf°_{298}(g) = -2.8$ kcal/mole from the enthalpy of combustion of methylamine nitrate and enthalpies of solution of methylamine and methylamine nitrate. Rossini, Wagman, Evans, Levine, and Jaffe (1249) listed -6.7 kcal/mole, based on very old combustion data. The value of Jaffe is adopted. Kobe and Lynn (780) selected $Tc = 430.1°K$ and $Pc = 73.6$ atm.

No. 602   Ethylamine   C$_2$H$_7$N   (Ideal Gas State)   Mol Wt 45.084

Wagner (1564) compared frequencies for several molecules having a C—C—X skeleton. From his scheme and the spectra determined

No. 602 Ethylamine C_2H_7N (Ideal Gas State) Mol Wt 45.084

T°K	Cp°	S°	$-(G°-H°_{298})/T$	$H°-H°_{298}$	$\Delta Hf°$	$\Delta Gf°$	Log Kp
	cal/(mole °K)				kcal/mole		
298	17.36	68.08	68.08	0.00	-11.00	8.91	-6.528
300	17.44	68.19	68.09	0.04	-11.03	9.03	-6.576
400	21.65	73.80	68.82	2.00	-12.34	15.92	-8.698
500	25.44	79.04	70.35	4.35	-13.42	23.11	-10.103
600	28.68	83.98	72.21	7.06	-14.27	30.50	-11.109
700	31.47	88.61	74.23	10.07	-14.92	38.02	-11.870
800	33.89	92.98	76.30	13.35	-15.42	45.61	-12.460
900	36.02	97.09	78.38	16.84	-15.76	53.27	-12.935
1000	37.88	100.99	80.45	20.54	-15.97	60.96	-13.321

by Stewart (1420) and Kohlrausch (787), fundamentals (in cm^{-1}) are assigned as follows: 414, 773, 780, 889, 997, 1046, 1082, 1122, 1223, 1300, 1376, 1450[3], 1620, 2950[5], 3320[2]. The over-all and reduced moments of inertia are calculated using standard bond distances and angles, and the barriers to internal rotation are estimated as 3.30 and 1.99 kcal/mole for the methyl and amino tops, respectively. Thermodynamic functions are calculated from these molecular values. Jaffe (678) measured the enthalpy of combustion of the liquid and derived $\Delta Hf°_{298}(l) = -17.71$ kcal/mole. From vapor pressures reported by Mehl (987) Jaffe calculated $\Delta Hv_{298} = 6.7$ kcal/mole, which yields $\Delta Hf°_{298}(g) = -11.0$ kcal/mole. Rossini, Wagman, Evans, Levine, and Jaffe (1249) selected $\Delta Hf°_{298}(g) = -11.6$ kcal/mole, based on old combustion data, which are not given any weight. Rossini et al. (1249) also listed $Tm = 192.2°K$ and $\Delta Hv = 6.7$ kcal/mole at $Tb = 289.7°K$.

Pohland and Mehl (1173) reported $Cp_{298}(l) = 31.1$ cal/(mole °K), as well as vapor pressures and liquid densities. Mehl (987) calculated tables of properties of ethylamine for use as a refrigerant. Kobe and Lynn (780) selected $Tc = 456°K$ and $Pc = 55.5$ atm.

No. 603 Propylamine C_3H_9N (Ideal Gas State) Mol Wt 59.110

T°K	Cp°	S°	$-(G°-H°_{298})/T$	$H°-H°_{298}$	$\Delta Hf°$	$\Delta Gf°$	Log Kp
	cal/(mole °K)				kcal/mole		
298	22.89	77.48	77.48	0.00	-17.30	9.51	-6.974
300	22.98	77.63	77.49	0.05	-17.33	9.68	-7.050
400	28.51	85.01	78.45	2.63	-18.97	18.94	-10.349
500	33.59	91.93	80.46	5.74	-20.31	28.58	-12.491
600	37.96	98.45	82.92	9.32	-21.37	38.46	-14.007
700	41.70	104.59	85.58	13.31	-22.17	48.50	-15.141
800	44.94	110.37	88.32	17.64	-22.76	58.63	-16.016
900	47.77	115.83	91.08	22.28	-23.16	68.83	-16.714
1000	50.21	120.99	93.82	27.18	-23.40	79.08	-17.281

Thermodynamic functions are estimated, using Table 11-1 and values for n-butane. Kharasch (744) listed two values for the enthalpy of combustion, based on the work of Thomsen and of Lemoult, respectively. The work of Lemoult is known to be unreliable [cf. (387)], and in this case his result appears to be too low. The determination of Thomsen leads to $\Delta Hf^\circ_{298}(g) = -17.3$ kcal/mole. Rossini and co-workers (1250) listed $Tm = 190.2°K$, $Tb = 321.0°K$, and constants for an Antoine vapor pressure equation. The Antoine constants and the critical constants selected by Kobe and Lynn (780), $Tc = 507°K$ and $Pc = 46.8$ atm, are used with the Haggenmacher equation (560) to derive $\Delta Hv_{298} = 7.46$ kcal/mole, from which $\Delta Hf^\circ_{298}(l) = -24.8$ kcal/mole.

No. 604   Butylamine $C_4H_{11}N$ (Ideal Gas State)   Mol Wt 73.136

		cal/(mole °K)			kcal/mole		
T°K	$Cp°$	$S°$	$-(G°-H°_{298})/T$	$H°-H°_{298}$	$\Delta Hf°$	$\Delta Gf°$	Log Kp
298	28.33	86.76	86.76	0.00	-22.00	11.76	-8.619
300	28.45	86.94	86.77	0.06	-22.04	11.97	-8.716
400	35.44	96.10	87.97	3.26	-24.00	23.61	-12.900
500	41.83	104.71	90.46	7.13	-25.59	35.70	-15.605
600	47.30	112.83	93.52	11.59	-26.85	48.08	-17.512
700	51.98	120.48	96.83	16.56	-27.79	60.65	-18.935
800	56.01	127.69	100.24	21.96	-28.49	73.33	-20.031
900	59.51	134.50	103.67	27.74	-28.94	86.09	-20.905
1000	62.54	140.93	107.08	33.85	-29.19	98.90	-21.613

Evans, Fairbrother, and Skinner (387) determined the enthalpy of combustion and reported $\Delta Hf^\circ_{298}(l) = -30.53$ kcal/mole. This is nearly 11 kcal less negative than the old value listed by Kharasch (744), which was based on the work of Lemoult. Rossini and co-workers (1250) selected $Tm = 224.1°K$ and $Tb = 351.0°K$, as well as constants for an Antoine vapor pressure equation. These constants are used with estimates of $Tc = 525°K$ and $Pc = 40$ atm and the Haggenmacher equation (560) to derive $\Delta Hv_{298} = 8.50$ kcal/mole and $\Delta Hf^\circ_{298}(g) = -22.10$ kcal/mole. Thermodynamic functions are estimated by means of Table 11-1 and values for n-pentane.

No. 605   sec-Butylamine $C_4H_{11}N$ (Ideal Gas State)   Mol Wt 73.136

Evans, Fairbrother, and Skinner (387) measured the enthalpy of combustion of 2-aminobutane and derived $\Delta Hf^\circ_{298}(l) = -32.88$ kcal/mole. The old work of Lemoult as listed by Kharasch (744) gave -39.0 kcal/mole. The enthalpy of vaporization of 298°K is estimated as 8.0

No. 605 *sec*-Butylamine $C_4H_{11}N$ (Ideal Gas State) Mol Wt 73.136

$T°K$	$Cp°$	$S°$	$-(G°-H°_{298})/T$	$H°-H°_{298}$	$\Delta Hf°$	$\Delta Gf°$	Log Kp
	cal/(mole °K)				kcal/mole		
298	27.99	83.90	83.90	0.00	-24.90	9.71	-7.118
300	28.12	84.08	83.91	0.06	-24.94	9.92	-7.229
400	35.40	93.18	85.10	3.24	-26.91	21.86	-11.942
500	41.96	101.80	87.58	7.11	-28.51	34.24	-14.966
600	47.55	109.96	90.64	11.60	-29.74	46.91	-17.085
700	52.34	117.66	93.96	16.60	-30.66	59.76	-18.658
800	56.42	124.92	97.38	22.04	-31.31	72.72	-19.865
900	59.99	131.78	100.82	27.87	-31.72	85.76	-20.824
1000	62.54	138.24	104.24	34.00	-31.94	98.83	-21.599

kcal/mole to calculate $\Delta Hf°_{298}(g) = -24.9$ kcal/mole. Rossini and co-workers (1250) listed $Tm = 168.7°K$ and $Tb = 336°K$. Thermodynamic functions of the ideal gas are estimated using the constants of Table 11-1 and the values of 2-methylbutane.

No. 606 *tert*-Butylamine $C_4H_{11}N$ (Ideal Gas State) Mol Wt 73.136

$T°K$	$Cp°$	$S°$	$-(G°-H°_{298})/T$	$H°-H°_{298}$	$\Delta Hf°$	$\Delta Gf°$	Log Kp
	cal/(mole °K)				kcal/mole		
298	28.67	80.76	80.76	0.00	-28.65	6.90	-5.056
300	28.79	80.94	80.77	0.06	-28.69	7.12	-5.183
400	36.46	90.30	81.99	3.33	-30.57	19.35	-10.573
500	43.25	99.18	84.55	7.32	-32.05	32.01	-13.991
600	48.87	107.58	87.69	11.94	-33.15	44.93	-16.364
700	53.55	115.47	91.10	17.06	-33.94	58.01	-18.111
800	57.49	122.89	94.62	22.62	-34.48	71.18	-19.443
900	60.88	129.86	98.15	28.54	-34.79	84.41	-20.497
1000	63.79	136.43	101.65	34.78	-34.91	97.67	-21.346

Evans, Fairbrother, and Skinner (387) derived $\Delta Hf°_{298}(l) = -35.95$ kcal/mole from their measurements of the enthalpy of combustion. Kharasch (744) listed a value in excellent agreement, based on the work of Lemoult. The agreement must be regarded as fortuitous, since the data of Lemoult for other isomers are known to be in error by several kilocalories per mole. The enthalpy of vaporization at 298°K was estimated by Evans et al. (387) as 7.3 kcal/mole, which appears reasonable and yields $\Delta Hf°_{298}(g) = -28.65$ kcal/mole. Rossini and co-workers (1250) selected $Tm = 205.7°K$ and $Tb = 317.6°K$. Thermodynamic functions are estimated using the constants of Table 11-1 and values for neopentane.

SECONDARY ALKYLAMINE IDEAL GAS TABLES

No. 607   Dimethylamine C_2H_7N (Ideal Gas State)   Mol Wt 45.084

T°K	$Cp°$	$S°$	$-(G°-H°_{298})/T$	$H°-H°_{298}$	$\Delta Hf°$	$\Delta Gf°$	Log Kp
	cal/(mole °K)				kcal/mole		
298	16.50	65.24	65.24	0.00	-4.50	16.25	-11.913
300	16.58	65.35	65.25	0.04	-4.53	16.38	-11.932
400	20.89	70.71	65.95	1.91	-5.93	23.57	-12.877
500	24.93	75.82	67.42	4.21	-7.06	31.08	-13.584
600	28.41	80.68	69.22	6.88	-7.96	38.79	-14.130
700	31.42	85.29	71.19	9.87	-8.63	46.64	-14.562
800	33.94	89.65	73.23	13.14	-9.12	54.57	-14.906
900	36.24	93.79	75.29	16.66	-9.45	62.55	-15.189
1000	38.19	97.71	77.34	20.38	-9.63	70.57	-15.423

Aston, Eidinoff, and Forster (32) reported low-temperature data, including $Ttp = 180.96°K$, $\Delta Hm° = 1.420$ kcal/mole, and $Tb = 280.04°K$, at which $\Delta Hv = 6.330$ kcal/mole, as well as $S°_{298}(g) = 65.24$ cal/(mole °K). Kobe and Harrison (775) used the vibrational assignment and barrier to internal rotation derived by Aston et al. to calculate thermodynamic functions to 1000°K. Vapor heat capacities measured by Felsing and Jessen (416) are in good agreement with the calculated values. Jaffe (678) measured the enthalpy of combustion of the liquid and derived $\Delta Hf°_{298}(l) = -10.50$ kcal/mole. Jaffe employed the enthalpy of vaporization result and heat capacity data of Aston et al (32) to derive $\Delta Hv_{298} = 6.0$ kcal/mole and $\Delta Hf°_{298}(g) = -4.5$ kcal/mole. Rossini, Wagman, Evans, Levine, and Jaffe (1249) listed $\Delta Hf°_{298}(g) = -6.6$ kcal/mole, based on old, less reliable data. Kobe and Lynn (780) selected $Tc = 437.7°K$ and $Pc = 52.4$ atm.

No. 608   Diethylamine $C_4H_{11}N$ (Ideal Gas State)   Mol Wt 73.136

T°K	$Cp°$	$S°$	$-(G°-H°_{298})/T$	$H°-H°_{298}$	$\Delta Hf°$	$\Delta Gf°$	Log Kp
	cal/(mole °K)				kcal/mole		
298	27.66	84.18	84.18	0.00	-17.30	17.23	-12.628
300	27.79	84.36	84.19	0.06	-17.34	17.44	-12.704
400	34.88	93.33	85.36	3.19	-19.36	29.35	-16.037
500	41.49	101.84	87.81	7.02	-21.00	41.73	-18.238
600	47.14	109.92	90.83	11.46	-22.28	54.39	-19.812
700	52.05	117.57	94.11	16.42	-23.23	67.26	-20.997
800	56.16	124.79	97.50	21.84	-23.91	80.22	-21.915
900	59.81	131.62	100.91	27.64	-24.35	93.28	-22.649
1000	62.91	138.09	104.31	33.78	-24.56	106.37	-23.245

Jaffe (678) measured the enthalpy of combustion of the liquid and found $\Delta H f°_{298}(l) = -24.79$ kcal/mole. An average of the enthalpies of vaporization at 298°K reported by Kolossovsky and Alimov (799) and Pohland and Mehl (1173) yields $\Delta H f°_{298}(g) = -17.3$ kcal/mole. Kharasch (744) listed old literature data leading to $\Delta H f°_{298}(l) = -29.2$ and $\Delta H f°_{298}(g) = -21.3$ kcal/mole. These values are not given any weight. Thermodynamic functions of the ideal gas were estimated using values for pentane and the constants of Table 11-1. Pohland and Mehl (1173) determined $Tm = 223.2°$K and $Tb = 327.3°$K. Kobe and Lynn (780) selected $Tc = 496°$K and $Pc = 36.6$ atm.

TERTIARY ALKYLAMINE IDEAL GAS TABLES

No. 609   Trimethylamine C_3H_9N (Ideal Gas State)   Mol Wt 59.110

	cal/(mole °K)				kcal/mole		
$T°K$	$Cp°$	$S°$	$-(G°-H°_{298})/T$	$H°-H°_{298}$	$\Delta Hf°$	$\Delta Gf°$	Log Kp
298	21.93	69.02	69.02	0.00	-5.70	23.64	-17.325
300	22.05	69.16	69.03	0.05	-5.74	23.82	-17.349
400	28.08	76.34	69.96	2.56	-7.44	33.94	-18.542
500	33.63	83.22	71.93	5.65	-8.80	44.44	-19.426
600	38.34	89.78	74.36	9.25	-9.83	55.19	-20.102
700	42.29	95.99	77.01	13.29	-10.59	66.10	-20.635
800	45.62	101.86	79.76	17.69	-11.12	77.08	-21.056
900	48.50	107.41	82.52	22.40	-11.45	88.13	-21.401
1000	50.98	112.65	85.28	27.38	-11.60	99.22	-21.682

Low-temperature thermal data for trimethylamine were measured by Aston, Sagenkahn, Szasz, Moessen, and Zuhr (42) as $Ttp = 156.07°$K, $\Delta Hm° = 1.564$ kcal/mole, $\Delta Hv = 5.482$ kcal/mole at $Tb = 276.02°$K, and $S°_{276.02}(g) = 67.31$ cal/(mole °K). They calculated a barrier to internal rotation based on a vibrational assignment, standard bond angles and distances, and the calorimetric entropy. The derived barrier was confirmed by the microwave studies of Lide and Mann (872). Kobe and Harrison (775) reported thermodynamic functions of the ideal gas, based on the above values. Jaffe (678) determined the enthalpy of combustion of the liquid and derived $\Delta H f°_{298}(l) = -10.94$ kcal/mole. The enthalpy of vaporization of Aston et al. (42) is adjusted to 298°K, to derive $\Delta H f°_{298}(g) = -5.70$ kcal/mole. Kharasch (744) listed three older, less reliable enthalpy of combustion results. Two differ widely from Jaffe's value, and one is in excellent agreement.

No. 610 Triethylamine $C_6H_{15}N$ (Ideal Gas State) Mol Wt 101.188

	cal/(mole °K)				kcal/mole		
$T°K$	$Cp°$	$S°$	$-(G°-H°_{298})/T$	$H°-H°_{298}$	$\Delta Hf°$	$\Delta Gf°$	Log Kp
298	38.46	96.90	96.90	0.00	-23.80	26.36	-19.320
300	38.66	97.14	96.91	0.08	-23.86	26.67	-19.425
400	48.70	109.66	98.54	4.45	-26.52	43.93	-24.000
500	58.10	121.56	101.97	9.80	-28.67	61.80	-27.011
600	66.10	132.88	106.18	16.02	-30.32	80.05	-29.156
700	72.80	143.58	110.77	22.98	-31.54	98.55	-30.767
800	78.56	153.69	115.51	30.55	-32.39	117.18	-32.011
900	83.50	163.23	120.29	38.66	-32.91	135.92	-33.005
1000	87.80	172.26	125.04	47.23	-33.14	154.71	-33.811

Jaffe (678) and Lautsch, Erzberger, and Tröber (851) measured the enthalpy of combustion, leading to $\Delta Hf°_{298}(l) = -32.07$ and -29.6 kcal/mole, respectively. The work of Jaffe appears to be the more careful investigation of the two and is accepted. Copp and Findlay (273) measured the vapor pressure and calculated Antoine constants. Their equation fits the older data of Joukovsky (705) very well, but is in poor agreement with Pohland and Mehl (1173). The equation of Copp and Findlay is judged most accurate and is used with the critical constants selected by Kobe and Lynn (780), $Tc = 532°K$ and $Pc = 30$ atm, and the Haggenmacher equation (560) to calculate $\Delta Hv_{298} = 8.29$ kcal/mole. From this we derive $\Delta Hf°_{298}(g) = -23.8$ kcal/mole. Kharasch (744) listed a value determined over 50 years ago, which deviates by 8 kcal from Jaffe's result.

Pohland and Mehl (1173) gave $Tm = 158.5°K$, and Copp and Findlay (273) gave $Tb = 362.7°K$. Thermodynamic functions are estimated using the constants of Table 11-1 and the values for 2-ethylpentane.

CYCLIC AMINE IDEAL GAS TABLES

No. 611 Ethylenimine C_2H_5N (Ideal Gas State) Mol Wt 43.068

The enthalpy of combustion of liquid ethylenimine was measured by Nelson and Jessup (1052) in a careful investigation, which yielded $\Delta Hf°_{298}(l) = 21.96$ kcal/mole. Wenker (1589) reported $Tb = 329°K$. An estimate of the entropy of vaporization of 22 cal/(mole °K) gives $\Delta Hv = 7.24$ kcal/mole. An estimate of ΔCp of vaporization of -10 cal/(mole °K) then yields $\Delta Hv_{298} = 7.55$ kcal/mole, from which $\Delta Hf°_{298}(g) = 29.5$ kcal/mole. Timmermans (1502) reported $Tm = 195.2°K$.

Thompson and Cave (1489) and Hoffman, Evans, and Glockler (612) derived fundamental frequencies from studies of the infrared and Raman

No. 611   Ethylenimine C_2H_5N (Ideal Gas State)   Mol Wt 43.068

	cal/(mole °K)			kcal/mole			
$T°K$	$Cp°$	$S°$	$-(G°-H°_{298})/T$	$H°-H°_{298}$	$\Delta Hf°$	$\Delta Gf°$	Log Kp
298	12.55	59.90	59.90	0.00	29.50	42.54	-31.180
300	12.63	59.98	59.91	0.03	29.48	42.62	-31.046
400	16.83	64.20	60.45	1.51	28.38	47.17	-25.774
500	20.52	68.37	61.62	3.38	27.51	51.98	-22.718
600	23.56	72.39	63.08	5.59	26.86	56.93	-20.737
700	26.05	76.21	64.69	8.07	26.38	61.99	-19.352
800	28.14	79.83	66.36	10.78	26.03	67.10	-18.329
900	29.92	83.25	68.04	13.69	25.81	72.25	-17.543
1000	31.45	86.48	69.73	16.76	25.69	77.42	-16.919

spectra. Although they differ somewhat in the assignment of fundamentals to the various vibrational modes, they agree reasonably well on numerical values for 16 of the 18 frequencies. Thompson and Cave did not give a value for one mode and listed another as doubtful. Hoffman, Evans, and Glockler proposed 1654 cm^{-1} and 1209 cm^{-1} for these two frequencies. The 1654 cm^{-1} does not appear reasonable for the NH bending motion to which it is assigned. A value of 800 cm^{-1} is arbitrarily adopted for this mode by comparison with other molecules. Turner, Fiora, and Kendrick (1523) derived structural constants from microwave spectroscopy and gave references to previous work. Results of Igarashi (670) by electron diffraction are in agreement within experimental error. Thermodynamic functions are calculated using the modified frequency assignment of Hoffman, Evans, and Glockler and the molecular constants of Turner, Fiora, and Kendrick.

No. 612   Pyrrolidine C_4H_9N (Ideal Gas State)   Mol Wt 71.120

	cal/(mole °K)			kcal/mole			
$T°K$	$Cp°$	$S°$	$-(G°-H°_{298})/T$	$H°-H°_{298}$	$\Delta Hf°$	$\Delta Gf°$	Log Kp
298	19.39	73.97	73.97	0.00	-0.86	27.41	-20.089
300	19.53	74.10	73.98	0.04	-0.90	27.58	-20.091
400	27.33	80.80	74.84	2.39	-3.02	37.41	-20.439
500	34.39	87.68	76.72	5.49	-4.69	47.72	-20.856
600	40.31	94.49	79.11	9.23	-5.96	58.32	-21.242
700	45.22	101.08	81.78	13.51	-6.90	69.12	-21.578
800	49.35	107.39	84.59	18.25	-7.55	80.01	-21.858
900	52.85	113.41	87.46	23.36	-7.96	90.99	-22.095
1000	55.84	119.14	90.35	28.80	-8.16	102.01	-22.293

Low-temperature thermal data for pyrrolidine were reported by Hildenbrand, Sinke, McDonald, Kramer, and Stull (600) and by McCullough, Douslin, Hubbard, Todd, Messerly, Hossenlopp, Frow, Dawson, and Waddington (948). The results are in excellent agreement. The slightly more precise results of McCullough et al. are given here: $Tt = 207.14°K$, $\Delta Ht° = 0.129$ kcal/mole, $Ttp = 215.31°K$, $\Delta Hm° = 2.050$ kcal/mole, $Tb = 359.72°K$, at which $\Delta Hv = 7.89$ kcal/mole, and $S°_{298}(l) = 48.76$ cal/(mole °K). Measurements of the enthalpy of combustion reported by the two groups are also in excellent agreement and yield $\Delta Hf°_{298}(l) = -9.84$ kcal/mole. McCullough et al. also determined the vapor heat capacity and the enthalpy of vaporization as a function of temperature and calculated $\Delta Hv°_{298} = 8.98$ kcal/mole and $\Delta Hf°_{298}(g) = -0.86$ kcal/mole. Using a vibrational assignment and adjusting the parameters of a slightly restricted "pseudorotation," they calculated thermodynamic functions that fit the experimental data.

Helm, Lanum, Cook, and Ball (584) measured the density, viscosity, surface tension, and refractive index of pure pyrrolidine. Briegleb (169) compared base strengths and resonance energies of several organic nitrogen bases, including pyrrolidine.

No. 613 Pyridine C_5H_5N (Ideal Gas State) Mol Wt 79.098

T°K	Cp°	S°	$-(G°-H°_{298})/T$	$H°-H°_{298}$	$\Delta Hf°$	$\Delta Gf°$	Log Kp
		cal/(mole °K)			kcal/mole		
298	18.67	67.59	67.59	0.00	33.50	45.46	-33.324
300	18.80	67.71	67.60	0.04	33.48	45.54	-33.171
400	25.42	74.05	68.41	2.26	32.38	49.73	-27.172
500	31.11	80.35	70.17	5.09	31.52	54.18	-23.679
600	35.72	86.44	72.38	8.44	30.87	58.77	-21.404
700	39.44	92.24	74.81	12.21	30.40	63.46	-19.812
800	42.49	97.71	77.33	16.31	30.07	68.20	-18.630
900	45.04	102.87	79.89	20.69	29.85	72.98	-17.722
1000	47.17	107.73	82.43	25.30	29.76	77.79	-16.999

The definitive work on the thermodynamic properties of pyridine was reported by McCullough, Douslin, Messerly, Hossenlopp, Kincheloe, and Waddington (949), including $Ttp = 231.48°K$, $\Delta Hm° = 1.979$ kcal/mole, $Tb = 388.38°K$, at which $\Delta Hv = 8.392$ kcal/mole, $S°_{298}(l) = 42.52$ cal/(mole °K), $S°_{298}(g) = 67.59$ cal/(mole °K), $\Delta Hf°_{298}(l) = 23.89$ kcal/mole, and $\Delta Hf°_{298}(g) = 33.50$ kcal/mole. Earlier work was reviewed by Li(864) who derived a consistent set of values of slightly lower accuracy. The data of Herington and Martin (585) on the vapor pressure and of Cox, Challoner, and Meetham (288) on the enthalpy of formation are in excellent agreement with those of McCullough et al.

No. 614 2-Picoline C_6H_7N (Ideal Gas State) Mol Wt 93.124

$T°K$	$Cp°$	$S°$	$-(G°-H°_{298})/T$	$H°-H°_{298}$	$\Delta Hf°$	$\Delta Gf°$	Log Kp
		cal/(mole °K)			kcal/mole		
298	23.90	77.68	77.68	0.00	23.65	42.32	-31.017
300	24.05	77.83	77.69	0.05	23.62	42.43	-30.909
400	31.92	85.85	78.73	2.86	22.17	48.93	-26.735
500	38.82	93.74	80.94	6.40	21.01	55.77	-24.374
600	44.55	101.34	83.71	10.58	20.11	62.80	-22.873
700	49.27	108.57	86.75	15.28	19.45	69.98	-21.846
800	53.21	115.41	89.91	20.41	18.97	77.22	-21.095
900	56.52	121.88	93.11	25.90	18.67	84.52	-20.524
1000	59.34	127.98	96.29	31.69	18.54	91.86	-20.074

Complete thermodynamic data for 2-picoline are taken from the work of Scott, Hubbard, Messerly, Todd, Hossenlopp, Good, Douslin, and McCullough (1315). Results include $Ttp = 206.45°K$, $\Delta Hm° = 2.324$ kcal/mole, $\Delta Hv = 8.654$ kcal/mole at $Tb = 402.54°K$, $\Delta Hv°_{298} = 10.15$ kcal/mole, $S°_{298}(l) = 52.07$ cal/(mole °K), $S°_{298}(g) = 77.68$ cal/(mole °K), $\Delta Hf°_{298}(l) = 13.50$ kcal/mole, and $\Delta Hf°_{298}(g) = 23.65$ kcal/mole. The measured entropy corresponds to free rotation of the methyl group. Vapor pressures measured by Herington and Martin (585) and the melting point and cryoscopic enthalpy of melting determined by Biddiscombe, Coulson, Handley, and Herington (127) are in excellent agreement. Second virial coefficients and calculated enthalpies of vaporization reported by Andon, Cox, Herington, and Martin (21) and by Cox and Andon (287) are also in reasonable agreement with results of Scott et al. The enthalpy of combustion determined by Cox, Challoner, and Meetham (288) leads to $\Delta Hf°_{298}(l) = 14.10$ kcal/mole. There is no obvious reason for the disagreement, which is slightly more than the sum of the experimental errors.

No. 615 3-Picoline C_6H_7N (Ideal Gas State) Mol Wt 93.124

$T°K$	$Cp°$	$S°$	$-(G°-H°_{298})/T$	$H°-H°_{298}$	$\Delta Hf°$	$\Delta Gf°$	Log Kp
		cal/(mole °K)			kcal/mole		
298	23.80	77.67	77.67	0.00	25.37	44.04	-32.280
300	23.94	77.82	77.68	0.05	25.34	44.15	-32.164
400	31.82	85.81	78.71	2.85	23.87	50.66	-27.678
500	38.74	93.68	80.92	6.38	22.71	57.50	-25.130
600	44.47	101.26	83.68	10.55	21.80	64.54	-23.506
700	49.19	108.48	86.72	15.24	21.13	71.72	-22.392
800	53.12	115.31	89.87	20.36	20.65	78.98	-21.574
900	56.43	121.77	93.06	25.84	20.34	86.29	-20.953
1000	59.23	127.86	96.24	31.63	20.20	93.63	-20.463

Complete thermodynamic data for 3-picoline are taken from the work of Scott, Good, Guthrie, Todd, Hossenlopp, Osborn, and McCullough (1310). Their results include $Tm = 255.01°K$, $\Delta Hm° = 3.389$ kcal/mole, $\Delta Hv = 8.932$ kcal/mole at $Tb = 417.29°K$, $\Delta Hv°_{298} = 10.62$ kcal/mole, $S°_{298}(l) = 51.70$ cal/(mole °K), $S°_{298}(g) = 77.67$ cal/(mole °K), $\Delta Hf°_{298}(l) = 14.75$ kcal/mole, and $\Delta Hf°_{298}(g) = 25.37$ kcal/mole. The measured entropy indicates free rotation of the methyl group. Melting point and cryoscopic enthalpy of melting were reported by Biddiscombe, Coulson, Handley, and Herington (127). Vapor pressures measured by Herington and Martin (585) and second virial coefficients and calculated enthalpies of vaporization reported by Andon, Cox, Herington, and Martin (21) and by Cox and Andon (287) are in agreement with Scott et al. The enthalpy of combustion determined by Cox, Challoner, and Meetham (288), however, leads to $\Delta Hf°_{298}(l) = 16.32$ kcal/mole. No obvious reason was found for the discrepancy with the value of Scott et al., which is over five times the sum of the experimental errors.

ARYLAMINE IDEAL GAS TABLE

No. 616 Aniline C_6H_7N (Ideal Gas State) Mol Wt 93.124

$T°K$	cal/(mole °K)				kcal/mole			
	$Cp°$	$S°$	$-(G°-H°_{298})/T$	$H°-H°_{298}$	$\Delta Hf°$	$\Delta Gf°$	Log Kp
298	25.91	76.28	76.28	0.00	20.76	39.84	-29.205
300	26.07	76.45	76.29	0.05	20.73	39.96	-29.109
400	34.17	85.09	77.41	3.08	19.50	46.57	-25.444
500	40.81	93.45	79.79	6.84	18.55	53.45	-23.363
600	46.09	101.38	82.73	11.19	17.83	60.50	-22.035
700	50.32	108.81	85.93	16.02	17.30	67.66	-21.123
800	53.79	115.76	89.23	21.23	16.91	74.87	-20.454
900	56.71	122.27	92.54	26.76	16.64	82.14	-19.946
1000	59.18	128.38	95.83	32.56	16.51	89.43	-19.545

Hatton, Hildenbrand, Sinke, and Stull (579) combined their own measurements with literature values to derive the chemical thermodynamic properties of aniline. They reported low-temperature thermal data, finding $Ttp = 267.13°K$, $\Delta Hm° = 2.519$ kcal/mole, and $S°_{298}(l) = 45.72$ cal/(mole °K), all in good agreement with previous, less extensive work of Parks, Huffman, and Barmore (1106). The liquid heat capacity results near 298°K were combined with measurements of Hough, Mason, and Sage (616, 617) at higher temperatures, to give $Cp(l) = 33.71 + 0.0409\ T$[cal/(mole °K)]. Hatton et al. also reported enthalpy of combustion measurements in good agreement with those of Anderson and Gilbert

(17) as revised by Cole and Gilbert (244). Corrected to carbon = 12.010 g, the average gives $\Delta Hf°_{298}(l) = 7.43$ kcal/mole. The vibrational assignment of Evans (391), except for a slight revision of the barrier to internal rotation, was used to compute thermodynamic functions of the ideal gas. Free energy functions of liquid and gas were then combined with vapor pressure values from the literature to derive $\Delta Hv_{298} = 13.325$ kcal/mole and $\Delta Hf°_{298}(g) = 20.76$ kcal/mole. The calculated boiling point is 457.55°K, where $\Delta Hv = 10.643$ kcal/mole.

NITRILE IDEAL GAS TABLES

No. 617 Acetonitrile C_2H_3N (Ideal Gas State) Mol Wt 41.052

	cal/(mole °K)				kcal/mole		
$T°K$	$Cp°$	$S°$	$-(G°-H°_{298})/T$	$H°-H°_{298}$	$\Delta Hf°$	$\Delta Gf°$	Log Kp
298	12.48	58.19	58.19	0.00	21.00	25.24	-18.503
300	12.52	58.27	58.20	0.03	20.99	25.27	-18.407
400	14.62	62.16	58.71	1.39	20.46	26.78	-14.630
500	16.59	65.64	59.76	2.95	19.99	28.41	-12.418
600	18.35	68.83	61.01	4.70	19.58	30.13	-10.976
700	19.90	71.77	62.34	6.61	19.23	31.92	-9.967
800	21.26	74.52	63.69	8.67	18.94	33.76	-9.221
900	22.45	77.10	65.04	10.86	18.70	35.62	-8.650
1000	23.50	79.52	66.37	13.16	18.53	37.52	-8.199

Thermodynamic functions of acetonitrile were recently calculated by Pillai and Cleveland (1148) using the accepted vibrational assignment and the molecular structure constants of Thomas, Sherrard, and Sheridan (1485). Work by Costain (275) and Venkateswarlu, Baker, and Gordy (1533) confirmed the structural constants. Earlier calculations by Ewell and Bourland (400), Thompson (1488), and Günthard and Kováts (549) are in good agreement with those of Pillai and Cleveland. Iwanciow (675) measured the vapor heat capacity and heat of vaporization in a modern flow calorimeter. The vapor heat capacities calculated to zero pressure are in excellent agreement with spectroscopic values. Iwanciow determined $Tb = 354.8°K$, at which $\Delta Hv = 7.3$ kcal/mole.

Putnam, McEachern, and Kilpatrick (1209) measured low-temperature thermal properties and found $Tt = 216.9°K$, $\Delta Ht° = 0.215$ kcal/mole, $Ttp = 229.32°K$, $\Delta Hm° = 1.952$ kcal/mole, and $\Delta Hv_{298} = 7.94$ kcal/mole, and derived $S°_{298}(l) = 35.76$ cal/(mole °K) and $S°_{298}(g) = 58.67$ cal/(mole °K). The third-law entropy is 0.48 cal/(mole °K) higher than the spectroscopic value. The discrepancy may be associated with the measured ΔHv_{298}, since an extrapolation of the results of Iwanciow (675) yields $\Delta Hv_{298} = 7.86$ kcal/mole and $S°_{298}(g) = 58.40$ cal/(mole °K).

The only available enthalpy of formation data were listed by Rossini, Wagman, Evans, Levine, and Jaffe (1249). Their value for the gas, $\Delta Hf°_{298}(g) = 21.0$ kcal/mole, based on the work of Thomsen (1495), is taken as most reliable. Rossini et al. also selected $Tm = 228.3°K$ and $\Delta Hm° = 2.13$ kcal/mole. Prausnitz and Carter (1185) obtained values for the second virial coefficient at low pressures in the range 40–100°C and derived equilibrium constants of dimerization.

No. 618 Acrylonitrile C_3H_3N (Ideal Gas State) Mol Wt 53.062

	cal/(mole °K)				kcal/mole		
$T°K$	$Cp°$	$S°$	$-(G°-H°_{298})/T$	$H°-H°_{298}$	$\Delta Hf°$	$\Delta Gf°$	Log Kp
298	15.24	65.47	65.47	0.00	44.20	46.68	-34.215
300	15.30	65.57	65.48	0.03	44.19	46.69	-34.014
400	18.36	70.40	66.11	1.72	43.75	47.60	-26.006
500	20.95	74.78	67.42	3.69	43.36	48.61	-21.245
600	23.11	78.80	68.98	5.90	43.03	49.69	-18.097
700	24.90	82.50	70.65	8.30	42.74	50.82	-15.866
800	26.43	85.93	72.35	10.87	42.50	51.99	-14.202
900	27.75	89.12	74.04	13.58	42.30	53.19	-12.915
1000	28.88	92.10	75.70	16.41	42.16	54.41	-11.890

Thermodynamic functions of the ideal gas were calculated from spectroscopic data by Halverson, Stamm, and Whalen (568). The enthalpy of combustion of the liquid was measured by Davis and Wiedeman (314) with the result $\Delta Hf°_{298}(l) = 35.9$ kcal/mole. These workers also reported $Tm = 189.5°K$ and $Tb = 350.5°K$, at which $\Delta Hv = 7.8$ kcal/mole. Estimating the enthalpy of vaporization at 298°K yields $\Delta Hf°_{298}(g) = 44.2$ kcal/mole.

No. 619 Propionitrile C_3H_5N (Ideal Gas State) Mol Wt 55.078

	cal/(mole °K)				kcal/mole		
$T°K$	$Cp°$	$S°$	$-(G°-H°_{298})/T$	$H°-H°_{298}$	$\Delta Hf°$	$\Delta Gf°$	Log Kp
298	17.46	68.50	68.50	0.00	12.10	22.98	-16.844
300	17.53	68.61	68.51	0.04	12.08	23.05	-16.789
400	21.18	74.16	69.24	1.98	11.19	26.84	-14.665
500	24.52	79.26	70.74	4.26	10.43	30.84	-13.481
600	27.42	83.99	72.56	6.86	9.79	34.99	-12.743
700	29.94	88.41	74.51	9.73	9.27	39.23	-12.248
800	32.14	92.55	76.51	12.84	8.86	43.53	-11.893
900	34.05	96.45	78.51	16.15	8.56	47.89	-11.629
1000	35.70	100.13	80.49	19.64	8.35	52.28	-11.425

Weber and Kilpatrick (1583) reported low-temperature thermal data for propionitrile, including $Tt = 176.96°K$, $\Delta Ht° = 0.408$ kcal/mole,

$Tm = 180.37°K$, $\Delta Hm° = 1.202$ kcal/mole, $S°_{298}(l) = 45.25$ cal/(mole °K), $\Delta Hv_{298} = 8.632$ kcal/mole, and $S°_{298}(g) = 68.75$ cal/(mole °K). Duncan and Janz (357) assigned fundamental frequencies and calculated thermodynamic functions using a potential barrier to internal rotation of 5.2 kcal/mole. Weber and Kilpatrick (1583) revised the $S°_{298}(g) = 67.81$ cal/(mole °K) given by Duncan and Janz to $S°_{298}(g) = 68.58$ cal/(mole °K), using the moments of inertia and barrier of 3.05 kcal/mole determined by Lerner and Dailey (856) and by Laurie (849). A check on these calculations yields $S°_{298}(g) = 67.61$ cal/(mole °K) with the parameters used by Duncan and Janz and $S°_{298}(g) = 68.24$ cal/(mole °K) with the parameters adopted by Weber and Kilpatrick. The cause for these discrepancies is not apparent. For the present purpose we adopt $S°_{298}(g) = 68.50$ cal/(mole °K), which requires a potential barrier of 2500 cal/mole combined with the assignment of Duncan and Janz. The heat capacity is also calculated using this barrier and assignment.

Iwanciow (675) measured the enthalpy of vaporization at several temperatures, including $\Delta Hv = 7.353$ kcal/mole at $Tb = 370.32°K$. The only available enthalpy of formation is $\Delta Hf°_{298}(l) = 3.45$ kcal/mole derived from the enthalpy of combustion listed by Kharasch (744), which is based on the work of Lemoult. Since Lemoult's work on butyronitrile is in good agreement with a modern determination, this value is felt to be reasonably reliable. The enthalpy of vaporization of Weber and Kilpatrick (1583) yields $\Delta Hf°_{298}(g) = 12.1$ kcal/mole.

No. 620   Butyronitrile C_4H_7N (Ideal Gas State)   Mol Wt 69.104

	cal/(mole °K)				kcal/mole		
T°K	$Cp°$	$S°$	$-(G°-H°_{298})/T$	$H°-H°_{298}$	$\Delta Hf°$	$\Delta Gf°$	Log Kp
298	23.19	77.78	77.78	0.00	8.14	25.97	-19.032
300	23.28	77.93	77.79	0.05	8.11	26.07	-18.994
400	28.39	85.34	78.76	2.64	6.94	32.25	-17.618
500	33.05	92.18	80.77	5.71	5.94	38.69	-16.912
600	37.07	98.57	83.21	9.22	5.14	45.32	-16.506
700	40.51	104.55	85.84	13.11	4.51	52.07	-16.256
800	43.48	110.16	88.53	17.31	4.03	58.89	-16.088
900	46.04	115.43	91.23	21.79	3.69	65.78	-15.972
1000	48.22	120.40	93.90	26.50	3.49	72.69	-15.886

Evans and Skinner (388) measured the enthalpy of combustion and derived $\Delta Hf°_{298}(l) = -1.39$ kcal/mole. A 50-year-old determination by Lemoult as listed by Kharasch (744) gave $\Delta Hf°_{298}(l) = -2.0$ kcal/mole, in remarkably good agreement. Iwanciow (675) determined the enthalpy of vaporization in the range 85–116°C and found $Tb = 390.47°K$, at which $\Delta Hv = 8.13$ kcal/mole. From an extrapolation of Iwanciow's results there is derived $\Delta Hv_{298} = 9.53$ kcal/mole and $\Delta Hf°_{298}(g) = 8.14$ kcal/mole. Thermodynamic functions are estimated using the constants of Table 11-1 and values for n-butane.

Heim (582) reported vapor pressures in excellent agreement with Iwanciow. Timmermans (1500) measured $Tm = 161°K$ and $\Delta Hm° = 1.2$ kcal/mole.

No. 621 Isobutyronitrile C_4H_7N (Ideal Gas State) Mol Wt 69.104

$T°K$	$Cp°$	$S°$	$-(G°-H°_{298})/T$	$H°-H°_{298}$	$\Delta Hf°$	$\Delta Gf°$	$\text{Log } Kp$
	cal/(mole °K)				kcal/mole		
298	23.04	74.88	74.88	0.00	6.07	24.76	-18.149
300	23.13	75.03	74.89	0.05	6.04	24.87	-18.120
400	28.56	82.44	75.86	2.64	4.87	31.34	-17.121
500	33.33	89.34	77.87	5.74	3.90	38.07	-16.640
600	37.39	95.79	80.33	9.28	3.12	44.98	-16.382
700	40.81	101.81	82.97	13.19	2.52	52.01	-16.236
800	43.74	107.46	85.68	17.42	2.07	59.10	-16.145
900	46.25	112.76	88.40	21.93	1.76	66.25	-16.087
1000	48.40	117.75	91.09	26.66	1.57	73.44	-16.049

Evans and Skinner (388) measured the enthalpy of combustion and derived $\Delta Hf°_{298}(l) = -2.92$ kcal/mole. There are no previous values. Iwanciow (675) determined the enthalpy of vaporization in the range 70–102°C and derived $Tb = 376.76°K$, at which $\Delta Hv = 7.754$ kcal/mole. An extrapolation of Iwanciow's results yields $\Delta Hv_{298} = 8.99$ kcal/mole and $\Delta Hf°_{298}(g) = 6.07$ kcal/mole. Thermodynamic functions of the ideal gas are estimated using the constants of Table 11-1 and values for isobutane. Timmermans (1502) reported $Tm = 201.7°K$.

No. 622 Benzonitrile C_7H_5N (Ideal Gas State) Mol Wt 103.118

$T°K$	$Cp°$	$S°$	$-(G°-H°_{298})/T$	$H°-H°_{298}$	$\Delta Hf°$	$\Delta Gf°$	$\text{Log } Kp$
	cal/(mole °K)				kcal/mole		
298	26.07	76.73	76.73	0.00	52.30	62.35	-45.702
300	26.21	76.90	76.74	0.05	52.28	62.41	-45.464
400	33.65	85.49	77.86	3.06	51.47	65.92	-36.015
500	39.85	93.68	80.21	6.74	50.83	69.61	-30.424
600	44.80	101.40	83.11	10.98	50.32	73.41	-26.737
700	48.79	108.62	86.24	15.67	49.93	77.30	-24.131
800	52.08	115.35	89.46	20.72	49.62	81.22	-22.187
900	54.80	121.65	92.69	26.06	49.39	85.18	-20.685
1000	57.08	127.54	95.89	31.66	49.28	89.17	-19.488

Evans and Skinner (388) reported enthalpy of combustion measurements and derived $\Delta Hf°_{298}(l) = 39.0$ kcal/mole. An old, less reliable enthalpy of combustion listed by Kharasch (744) gave 36.3 kcal/mole. From vapor pressure data listed by Jordan (703) Evans and Skinner calculated $\Delta H v_{298} = 13.26$ kcal/mole and $\Delta H f°_{298}(g) = 52.3$ kcal/mole. Green (535) assigned vibrational frequencies and calculated thermodynamic functions to 1000°K. Witschonke (1623) measured $Tm = 260.40$ °K and $\Delta H m° = 2.60$ kcal/mole. Timmermans (1501) reported $Tm = 259.35$ °K, which may indicate a second crystalline form, since Stull (1432) agrees with Witschonke. An average $Tb = 464.15$°K is selected from Timmermans (1501), Jordan (703), and Stull (1432), Philip and Waterton (1141) measured $\Delta H v = 11.0$ kcal/mole at Tb.

NITROALKANE IDEAL GAS TABLES

No. 623   Nitromethane CH$_3$NO$_2$ (Ideal Gas State)   Mol Wt 61.042

$T°K$	cal/(mole °K)				kcal/mole		
	$Cp°$	$S°$	$-(G°-H°_{298})/T$	$H°-H°_{298}$	$\Delta H f°$	$\Delta G f°$	Log Kp
298	13.70	65.73	65.73	0.00	-17.86	-1.66	1.217
300	13.76	65.82	65.74	0.03	-17.88	-1.56	1.137
400	16.80	70.20	66.31	1.56	-18.69	4.01	-2.191
500	19.56	74.25	67.50	3.38	-19.32	9.76	-4.267
600	21.92	78.03	68.94	5.46	-19.78	15.62	-5.691
700	23.90	81.57	70.50	7.75	-20.11	21.55	-6.728
800	25.56	84.87	72.09	10.23	-20.32	27.52	-7.517
900	26.97	87.96	73.68	12.86	-20.44	33.51	-8.136
1000	28.17	90.87	75.26	15.61	-20.48	39.51	-8.634

Jones and Giauque (702) measured low-temperature thermal data, including $Ttp = 244.73$°K and $\Delta H m° = 2.319$ kcal/mole, and calculated $S°_{298}(l) = 41.05$ cal/(mole °K). In a careful investigation of the freezing point, Toops (1515) reported $Tm = 244.60$°K, more than 0.1° lower than the triple point found by Jones and Giauque. The reason for this discrepancy is not obvious. McCullough, Scott, Pennington, Hossenlopp, and Waddington (964) measured the vapor heat capacity, vapor pressure, and enthalpy of vaporization over a range of temperatures. Experimental results included $Tb = 374.34$°K, at which $\Delta H v = 8.120$ kcal/mole, $\Delta H v_{298} = 9.17$ kcal/mole, and $S°_{298}(g) = 65.73$ cal/(mole °K). Agreement between calorimetrically and spectroscopically derived functions was obtained by assuming a reasonable contribution from anharmonicity and free internal rotation. Thermodynamic functions to 1000°K were calculated on this basis. McCullough et al. gave references to previous work, including vapor heat capacities of Pitzer and Gwinn (1161) and of De Vries and Collins (324), and enthalpy of vaporization of Mathews (940). The vibrational assignment was taken from Smith, Pan, and Nielsen (1379). Free internal rotation was confirmed by microwave spectroscopy by Tannenbaum, Johnson, Myers, and Gwinn (1466).

McCullough et al. (964) cited unpublished combustion calorimetry at the National Bureau of Standards, leading to $\Delta Hf^\circ_{298}(l) = -27.03$ kcal/mole. The enthalpies of combustion of Holcomb and Dorsey (613) and of Cass, Fletcher, Mortimer, Quincey, and Springall (200) are in fair agreement with each other, but differ by 5–6 kcal/mole from the unpublished value. When the nitroalkane series is considered as a whole, the value cited by McCullough et al. appears more likely to be correct and is adopted. Older data listed by Bichowsky and Rossini (125) are in agreement with the selected result.

Hough, Mason, and Sage (616) measured the liquid heat capacity from 310° to 360°K. Their results can be fitted by the expression $Cp(l) = 18.70 + 0.023T$ [cal/(mole °K)]. Extrapolation to lower temperatures indicates that their results may be too high, since the data of Jones and Giauque show a lower rate of increase with temperature.

No. 624 Nitroethane $C_2H_5NO_2$ (Ideal Gas State) Mol Wt 75.068

	cal/(mole °K)				kcal/mole		
T°K	Cp°	S°	$-(G^\circ - H^\circ_{298})/T$	$H^\circ - H^\circ_{298}$	ΔHf°	ΔGf°	Log Kp
298	18.69	75.39	75.39	0.00	-24.20	-1.17	0.857
300	18.78	75.51	75.40	0.04	-24.23	-1.03	0.748
400	23.66	81.60	76.19	2.17	-25.39	6.89	-3.766
500	27.92	87.35	77.85	4.75	-26.27	15.07	-6.586
600	31.45	92.76	79.89	7.73	-26.91	23.40	-8.523
700	34.38	97.83	82.10	11.02	-27.36	31.82	-9.934
800	36.81	102.59	84.36	14.58	-27.65	40.29	-11.007
900	38.89	107.05	86.64	18.37	-27.81	48.80	-11.850
1000	40.67	111.24	88.89	22.35	-27.85	57.32	-12.527

Cass, Fletcher, Mortimer, Quincey, and Springall (200) measured the enthalpy of combustion and quoted an earlier result from Holcomb and Dorsey (613) and an unpublished value from the National Bureau of Standards. The three sources are in only fair agreement, and the Bureau of Standards figure is given the most weight in adopting $\Delta Hf^\circ_{298}(l) = -33.9$ kcal/mole. A careful investigation by Toops (1515) gave $Tm = 233.63°K$, $Tb = 387.22°K$, and an equation for the vapor pressure. The equation is used with the Haggenmacher equation (560) and estimates of $Tc = 582°K$ and $Pc = 37$ atm to calculate $\Delta Hv = 8.4$ kcal/mole at Tb, $\Delta Hv_{298} = 9.7$ kcal/mole, and $\Delta Hf^\circ_{298}(g) = -24.2$ kcal/mole. Thermodynamic functions of the ideal gas are estimated using the constants of Table 11-1 and values for propane. Beard (87) measured the heat capacity of nitroethane and found $Cp(l) = 32.9$ cal/(mole °K) in the range 298–323°K.

No. 625 1-Nitropropane $C_3H_7NO_2$ (Ideal Gas State) Mol Wt 89.094

T°K	Cp°	S°	$-(G°-H°_{298})/T$	$H°-H°_{298}$	$\Delta Hf°$	$\Delta Gf°$	Log Kp
		cal/(mole °K)			kcal/mole		
298	24.41	85.00	85.00	0.00	-29.80	0.08	-0.056
300	24.52	85.16	85.01	0.05	-29.83	0.26	-0.190
400	30.72	93.08	86.04	2.82	-31.29	10.53	-5.751
500	36.24	100.54	88.20	6.17	-32.42	21.12	-9.230
600	40.87	107.57	90.85	10.04	-33.25	31.91	-11.621
700	44.73	114.17	93.72	14.32	-33.84	42.81	-13.367
800	47.96	120.36	96.66	18.96	-34.22	53.79	-14.694
900	50.72	126.17	99.62	23.90	-34.42	64.81	-15.737
1000	53.06	131.64	102.55	29.09	-34.47	75.84	-16.575

Cass, Fletcher, Mortimer, Quincey, and Springall (200) reported enthalpy of combustion measurements and quoted a previous value of Holcomb and Dorsey (613) and an unpublished result of the National Bureau of Standards. The three are in reasonable agreement, and the Bureau of Standards value is given the most weight in selecting $\Delta Hf°_{298}(l) = -40.15$ kcal/mole. A careful investigation by Toops (1515) yielded $Tm = 169.16°K$ and $Tb = 404.33°K$, as well as an expression for the vapor pressure. The Haggenmacher equation (560) and estimates of $Tc = 595°K$ and $Pc = 36$ atm are used to derive $\Delta Hv = 8.8$ kcal/mole at Tb and $\Delta Hv_{298} = 10.3$ kcal/mole. Holcomb and Dorsey calculated $\Delta Hv_{298} = 10.4$ kcal/mole from their own vapor pressure measurements, covering a wider range than Toops. Averaging yields $\Delta Hf°_{298}(g) = -29.8$ kcal/mole.

Thermodynamic functions of the ideal gas are estimated using the constants of Table 11-1 and values for n-butane. Beard (87) measured the heat capacity of 1-nitropropane from 298° to 323°K. The data can be fitted within ±2% by $Cp(l) = 8.11 + 0.107T$ [cal/(mole °K)].

No. 626 2-Nitropropane $C_3H_7NO_2$ (Ideal Gas State) Mol Wt 89.094

The enthalpy of combustion was measured by Holcomb and Dorsey (613) and by Cass, Fletcher, Mortimer, Quincey, and Springall (200). The latter group also quoted an unpublished value from the National Bureau of Standards. The three are in good agreement, and the Bureau of Standards value is given the most weight in selecting $\Delta Hf°_{298}(l) = -43.3$ kcal/mole. Toops (1515) reported accurate measurements of $Tm = 181.83°K$ and $Tb = 393.40°K$, as well as an equation for the vapor pressure. This is used with the Haggenmacher equation (560) and

No. 626 2-Nitropropane $C_3H_7NO_2$ (Ideal Gas State) Mol Wt 89.094

$T°K$	$Cp°$	$S°$	$-(G°-H°_{298})/T$	$H°-H°_{298}$	$\Delta Hf°$	$\Delta Gf°$	Log Kp
		cal/(mole °K)			kcal/mole		
298	24.26	83.10	83.10	0.00	-33.50	-3.06	2.240
300	24.37	83.26	83.11	0.05	-33.53	-2.87	2.090
400	30.89	91.18	84.14	2.82	-34.99	7.59	-4.146
500	36.52	98.70	86.31	6.20	-36.09	18.37	-8.027
600	41.19	105.78	88.97	10.09	-36.90	29.33	-10.685
700	45.03	112.43	91.85	14.41	-37.45	40.42	-12.619
800	48.22	118.65	94.82	19.07	-37.80	51.57	-14.087
900	50.93	124.49	97.79	24.04	-37.99	62.75	-15.238
1000	53.24	129.98	100.74	29.25	-38.02	73.96	-16.163

estimates of $Tc = 595°K$ and $Pc = 36$ atm to calculate $\Delta Hv = 8.4$ kcal/mole at Tb and $\Delta Hv_{298} = 9.76$ kcal/mole. Holcomb and Dorsey (613) calculated 9.9 kcal/mole from their own vapor pressure data, which is in good agreement with Toops. An average is selected to derive $\Delta Hf°_{298}(g) = -33.5$ kcal/mole.

Thermodynamic functions of the ideal gas are estimated using the constants of Table 11-1 and values for isobutane.

No. 627 1-Nitrobutane $C_4H_9NO_2$ (Ideal Gas State) Mol Wt 103.120

$T°K$	$Cp°$	$S°$	$-(G°-H°_{298})/T$	$H°-H°_{298}$	$\Delta Hf°$	$\Delta Gf°$	Log Kp
		cal/(mole °K)			kcal/mole		
298	29.85	94.28	94.28	0.00	-34.40	2.42	-1.775
300	29.99	94.47	94.29	0.06	-34.44	2.65	-1.929
400	37.65	104.17	95.56	3.45	-36.22	15.30	-8.357
500	44.48	113.32	98.20	7.56	-37.60	28.34	-12.388
600	50.21	121.95	101.45	12.31	-38.63	41.63	-15.163
700	55.01	130.06	104.96	17.57	-39.36	55.07	-17.193
800	59.03	137.68	108.58	23.28	-39.84	68.59	-18.737
900	62.46	144.83	112.22	29.36	-40.10	82.17	-19.951
1000	65.39	151.57	115.82	35.76	-40.17	95.76	-20.928

The enthalpy of combustion was measured by Holcomb and Dorsey (613), with the result $\Delta Hf°_{298}(l) = -46.03$ kcal/mole. They also measured the vapor pressure in order to derive $\Delta Hv_{298} = 11.6$ kcal/mole and $\Delta Hf°_{298}(g) = -34.4$ kcal/mole. Careful work by Toops (1515) yielded $Tm = 191.82°K$ and $Tb = 425.92°K$. Vapor pressures reported by Toops agree well with those of Holcomb and Dorsey in the range in which they overlap, and are used with the Haggenmacher equation (560) and estimates of $Tc = 600°K$ and $Pc = 35$ atm to calculate $\Delta Hv = 9.3$ kcal/mole at Tb. Thermodynamic functions of the ideal gas are estimated using the constants of Table 11-1 and values for n-pentane.

Alkyl Nitrite and Alkyl Nitrate Ideal Gas Tables

No. 628 2-Nitrobutane $C_4H_9NO_2$ (Ideal Gas State) Mol Wt 103.120

$T°K$	$Cp°$	$S°$	$-(G°-H°_{298})/T$	$H°-H°_{298}$	$\Delta Hf°$	$\Delta Gf°$	Log Kp
		cal/(mole °K)			kcal/mole		
298	29.51	91.62	91.62	0.00	-39.10	-1.49	1.089
300	29.66	91.81	91.63	0.06	-39.14	-1.25	0.914
400	37.61	101.45	92.89	3.43	-40.94	11.66	-6.373
500	44.61	110.62	95.52	7.55	-42.32	24.98	-10.919
600	50.46	119.28	98.77	12.31	-43.33	38.54	-14.036
700	55.37	127.44	102.29	17.61	-44.02	52.24	-16.310
800	59.44	135.11	105.92	23.36	-44.47	66.02	-18.035
900	62.94	142.31	109.56	29.48	-44.68	79.85	-19.390
1000	65.96	149.11	113.18	35.93	-44.70	93.70	-20.477

The enthalpy of combustion was measured by Holcomb and Dorsey (613), with the result $\Delta Hf°_{298}(l) = -49.61$ kcal/mole. They also measured the vapor pressure and calculated $\Delta Hv_{298} = 10.48$ kcal/mole, from which $\Delta Hf°_{298}(g) = -39.1$ kcal/mole. Toops (1515) could not induce crystallization, but did measure vapor pressures in good agreement with those of Holcomb and Dorsey, and found $Tb = 412.65°K$. From his vapor pressures and estimates of $Tc = 600°K$ and $Pc = 35$ atm, $\Delta Hv = 8.8$ kcal/mole at Tb is calculated on the basis of the Haggenmacher equation (560). Thermodynamic functions of the ideal gas are estimated using the constants of Table 11-1 and values for 2-methylbutane.

ALKYL NITRITE AND ALKYL NITRATE IDEAL GAS TABLES

No. 629 Methyl Nitrite CH_3NO_2 (Ideal Gas State) Mol Wt 61.042

$T°K$	$Cp°$	$S°$	$-(G°-H°_{298})/T$	$H°-H°_{298}$	$\Delta Hf°$	$\Delta Gf°$	Log Kp
		cal/(mole °K)			kcal/mole		
298	15.11	67.95	67.95	0.00	-15.30	0.24	-0.174
300	15.17	68.05	67.96	0.03	-15.31	0.33	-0.243
400	18.24	72.84	68.59	1.71	-15.99	5.66	-3.093
500	21.01	77.22	69.88	3.67	-16.47	11.13	-4.865
600	23.35	81.26	71.45	5.89	-16.79	16.68	-6.077
700	25.32	85.01	73.12	8.33	-16.97	22.28	-6.955
800	26.97	88.50	74.83	10.95	-17.04	27.89	-7.619
900	28.36	91.76	76.53	13.71	-17.02	33.51	-8.136
1000	29.52	94.81	78.21	16.61	-16.92	39.12	-8.550

Leermakers and Ramsperger (854) measured equilibrium compositions in the gas phase system

$$CH_3OH(g) + NOCl(g) = CH_3ONO(g) + HCl(g)$$

at 25° and 50°C. These data yield $\Delta Hf°_{298}(g) = -16.28$ kcal/mole by a second-law calculation. Geiseler and Thierfelder (464) burned methyl nitrite in a flame calorimeter and derived $\Delta Hf°_{298}(g) = -16.8 \pm 0.8$ kcal/mole. Wheeler, Whittaker, and Pike (1601) determined the enthalpy of explosion of methyl nitrate gas, which, when combined with the enthalpy of reaction of nitrogen pentoxide and methyl nitrite reported by Ray and Ogg (1218), gives $\Delta Hf°_{298}(g) = -14.46 \pm 0.9$ kcal/mole. Gray and Pratt (525) cited an unpublished value of $\Delta Hf°_{298}(g) = -14.93 \pm 0.26$ kcal/mole, based on an enthalpy of hydrolysis communicated by Baldrey, Lotzgesell, and Style. Ray and Gershon (1216) measured the enthalpy of reaction of methanol and nitrosyl chloride and found $\Delta Hf°_{298}(g) = -15.64 \pm 0.20$ kcal/mole. An average of -15.3 kcal/mole is adopted, giving the most weight to values reported by Gray and Pratt and by Ray and Gershon.

Cox and Ray (293) measured the equilibrium constant in the reaction of methanol and nitrosyl chloride at 298°K and obtained a value considerably lower than Leermakers and Ramsperger. The result of Cox and Ray is selected and used to calculate $S°_{298}(g) = 67.95$ cal/(mole °K). Gray and Pratt (525) calculated the entropy from structural and spectroscopic data for the *cis* and *trans* isomers. They cited evidence that the energy difference is small, and it is assumed for the present purpose that equal amounts of each isomer exist in the vapor. The calculated entropy of the mixture is 74.17 cal/(mole °K), assuming free internal rotation for both CH_3 and NO groups. Gray and Reeves (527) found a barrier of 10.5 kcal/mole for the $-NO$ group, which lowers the entropy to 71.17 cal/(mole °K). The difference between this and the experimental value of 67.95 cal/(mole °K) implies a very high barrier to internal rotation of the methyl group, and a torsional frequency of 705 cm^{-1} is assigned to this degree of freedom. The heat capacity of the ideal gas is calculated using this frequency and the vibrational assignment of Gray and Pratt.

Thompson and Dainton (1490) measured $Tb = 255.8°K$, at which $\Delta Hv = 5.0$ kcal/mole. Gray and Pratt reported a higher value of $\Delta Hv_{298} = 5.4$ kcal/mole. Geiseler and Rätzsch (462) calculated thermodynamic functions using a barrier of 9.00 kcal/mole for the nitroso group and assuming free internal rotation for the methyl group.

No. 630 Methyl Nitrate CH_3NO_3 (Ideal Gas State) Mol Wt 77.042

Ray and Ogg (1218) measured the enthalpy of reaction of nitrogen pentoxide and methyl nitrite. Their result is combined with our selected value for methyl nitrite to give $\Delta Hf°_{298}(g) = -28.2$ kcal/mole. Ray and Ogg combined the enthalpy of explosion of liquid methyl nitrate reported by Wheeler, Whittaker, and Pike (1601) with an enthalpy of vaporization

No. 630 Methyl Nitrate CH_3NO_3 (Ideal Gas State) Mol Wt 77.042

T°K	Cp°	S°	$-(G°-H°_{298})/T$	$H°-H°_{298}$	$\Delta Hf°$	$\Delta Gf°$	Log Kp
	cal/(mole °K)				kcal/mole		
298	18.28	72.15	72.15	0.00	-28.80	-7.21	5.284
300	18.34	72.27	72.16	0.04	-28.82	-7.08	5.154
400	21.87	78.04	72.92	2.05	-29.50	0.29	-0.156
500	24.95	83.26	74.47	4.40	-29.97	7.79	-3.405
600	27.54	88.04	76.34	7.03	-30.26	15.37	-5.599
700	29.69	92.45	78.33	9.89	-30.40	22.99	-7.177
800	31.47	96.54	80.36	12.95	-30.43	30.62	-8.364
900	32.96	100.33	82.37	16.18	-30.36	38.25	-9.287
1000	34.19	103.87	84.34	19.53	-30.21	45.87	-10.024

calculated from the vapor pressures of McKinley-McKee and Moelwyn-Hughes (977) and derived $\Delta Hf°_{298}(g) = -29.4$ kcal/mole. An average of -28.8 kcal/mole is adopted. Gray and Smith (528) measured low-temperature thermal data, including $Tm = 190.2°K$, $\Delta Hm° = 1.97$ kcal/mole, and $S°_{298}(l) = 51.86$ cal/(mole °K). Gray and Pratt (524) used the vapor pressures of McKinley-McKee and Moelwyn-Hughes (977) to derive $\Delta Hv = 7.54$ kcal/mole at $Tb = 337.8°K$, $\Delta Hv_{298} = 8.1$ kcal/mole, and $S°_{298}(g) = 76.05$ cal/(mole °K). Brand and Cawthon (159) assigned fundamental frequencies and calculated $S°_{298}(g) = 75.97$ cal/(mole °K), assuming two free internal rotations. The agreement is impressive but apparently fortuitous, since the microwave spectrum reported by Dixon and Wilson (330) can only be interpreted by assuming barriers of 9.10 and 2.32 kcal/mole for the nitro and methyl group rotations, respectively. The calculated entropy using these barriers is 72.15 cal/(mole °K). The third-law data are difficult to evaluate. No standard substance was run for comparison, and even the empty calorimeter was not checked. Gray and Smith (529) measured ethyl nitrate and derived $S°_{298}(g) = 83.25$ cal/(mole °K). The increment from the methyl to the ethyl compound is only 7.2 cal/(mole °K), compared to an average of about 10 cal/(mole °K) for other classes of compounds. In this case the spectroscopic value is judged most accurate, and the vibrational assignment of Brand and Cawthon, together with the rotational barriers and moments of inertia of Dixon and Wilson, are used to calculate thermodynamic functions.

No. 631 Ethyl Nitrate $C_2H_5NO_3$ (Ideal Gas State) Mol Wt 91.068

Low-temperature thermal data for ethyl nitrate were determined by Gray and Smith (529) as $Tm = 178.6°K$, $\Delta Hm = 2.04$ kcal/mole, and $S°_{298}(l) = 59.08$ cal/(mole °K). The data are not considered to be of high

No. 631 Ethyl Nitrate $C_2H_5NO_3$ (Ideal Gas State) Mol Wt 91.068

$T°K$	$Cp°$	$S°$	$-(G°-H°_{298})/T$	$H°-H°_{298}$	$\Delta Hf°$	$\Delta Gf°$	$\log K_p$
	cal/(mole °K)				kcal/mole		
298	23.27	83.25	83.25	0.00	-36.80	-8.81	6.456
300	23.36	83.40	83.26	0.05	-36.82	-8.63	6.290
400	28.73	90.87	84.24	2.66	-37.85	0.93	-0.509
500	33.31	97.79	86.27	5.77	-38.58	10.72	-4.684
600	37.07	104.21	88.73	9.29	-39.05	20.62	-7.511
700	40.17	110.16	91.37	13.16	-39.31	30.59	-9.550
800	42.72	115.70	94.07	17.31	-39.42	40.58	-11.086
900	44.88	120.86	96.76	21.69	-39.39	50.58	-12.283
1000	46.69	125.68	99.42	26.27	-39.24	60.58	-13.239

accuracy (cf. methyl nitrate), but are judged acceptable in this case. The enthalpy of combustion was measured by Fairbrother, Skinner, and Evans (405) and yields $\Delta Hf°_{298}(l) = -45.5$ kcal/mole. Gray and Pratt (524) used the vapor pressure data of Gray, Pratt, and Larkin (526) to derive $\Delta Hv = 7.92$ kcal/mole at $Tb = 360.8°K$ and $\Delta Hv_{298} = 8.67$ kcal/mole. From these data are calculated $\Delta Hf°_{298}(g) = -36.8$ kcal/mole and $S°_{298}(g) = 83.25$ cal/(mole °K). The heat capacity of the ideal gas is estimated. Gray, Pratt, and Larkin (526) reviewed early work on the enthalpy of formation.

No. 632 Propyl Nitrate $C_3H_7NO_3$ (Ideal Gas State) Mol Wt 105.094

$T°K$	$Cp°$	$S°$	$-(G°-H°_{298})/T$	$H°-H°_{298}$	$\Delta Hf°$	$\Delta Gf°$	$\log K_p$
	cal/(mole °K)				kcal/mole		
298	28.99	92.10	92.10	0.00	-41.60	-6.53	4.790
300	29.10	92.28	92.11	0.06	-41.63	-6.32	4.603
400	35.79	101.59	93.33	3.31	-42.96	5.67	-3.098
500	41.63	110.22	95.86	7.19	-43.93	17.95	-7.844
600	46.49	118.26	98.93	11.60	-44.59	30.38	-11.067
700	50.52	125.73	102.23	16.46	-44.99	42.91	-13.398
800	53.87	132.71	105.61	21.68	-45.19	55.49	-15.157
900	56.71	139.22	108.99	27.22	-45.21	68.08	-16.530
1000	59.08	145.32	112.32	33.01	-45.07	80.66	-17.628

Fairbrother, Skinner, and Evans (405) measured the enthalpy of combustion and derived $\Delta Hf°_{298}(l) = -51.27$ kcal/mole. There are no previous values with which to compare this work. Gray and Pratt (524) determined the vapor pressure and calculated $\Delta Hv = 8.58$ kcal/mole at $Tb = 383.2°K$ and $\Delta Hv_{298} = 9.70$ kcal/mole. Duff (355) also reported vapor pressures, which agree with Gray and Pratt near the normal boiling point but are higher at 298°K and yield $\Delta Hv = 7.6$ kcal/mole. The results of Gray and Pratt are accepted and give $\Delta Hf°_{298}(g) = -41.6$ kcal/mole.

Thermodynamic functions of the ideal gas are estimated using the constants of Table 11-1 and values for n-butane.

No. 633   Isopropyl Nitrate $C_3H_7NO_3$ (Ideal Gas State)   Mol Wt 105.094

	cal/(mole °K)				kcal/mole		
$T°K$	$Cp°$	$S°$	$-(G°-H°_{298})/T$	$H°-H°_{298}$	$\Delta Hf°$	$\Delta Gf°$	Log Kp
298	28.84	89.20	89.20	0.00	-45.65	-9.72	7.125
300	28.95	89.38	89.21	0.06	-45.68	-9.50	6.919
400	35.96	98.70	90.43	3.31	-47.01	2.78	-1.519
500	41.91	107.38	92.96	7.22	-47.96	15.34	-6.706
600	46.81	115.47	96.05	11.66	-48.59	28.06	-10.222
700	50.82	122.99	99.36	16.55	-48.96	40.87	-12.760
800	54.13	130.00	102.76	21.80	-49.12	53.71	-14.673
900	56.92	136.54	106.16	27.35	-49.12	66.57	-16.165
1000	59.26	142.67	109.50	33.17	-48.96	79.43	-17.358

Fairbrother, Skinner, and Evans (405) measured the enthalpy of combustion and derived $\Delta Hf°_{298}(l) = -54.92$ kcal/mole. There are no previous values. Gray and Pratt (524) determined the vapor pressure and calculated $\Delta Hv = 8.35$ kcal/mole at $Tb = 375.2°K$ and $\Delta Hv_{298} = 9.27$ kcal/mole. Duff (355) also reported vapor pressures, which agree with Gray and Pratt near the normal boiling point but are more than 10 mm higher at 298°K and give $\Delta Hv = 7.3$ kcal/mole. The results of Gray and Pratt are accepted and give $\Delta Hf°_{298}(g) = -45.65$ kcal/mole. Thermodynamic functions of the ideal gas are estimated using the constants of Table 11-1 and values for isobutene.

CHAPTER 12

CHEMICAL THERMODYNAMICS OF HALOGEN COMPOUNDS

INTRODUCTION

The importance of organic halogen compounds is emphasized by a recent survey (1655) indicating that by 1967 the United States' capacity for chlorine production would total 7.5 million tons a year, 70% of which would be used for synthesis of organic chemicals. It was predicted that a large share would involve the manufacture of polyvinyl chloride, a widely used plastic. The high thermal stability of fluorocarbons and polyfluorocarbons makes these compounds useful despite their relatively high cost. Bromine and iodine compounds serve many specialized functions on a smaller scale.

The thermochemistry of organic halogen compounds is still in an unsatisfactory state. Patrick (1126, 1127) reviewed the thermochemistry of organic fluorine compounds; his survey shows the serious lack of good values for enthalpy of formation for these compounds. Techniques for accurate measurements are now worked out (513) but have thus far been applied to relatively few compounds. Lacher and Skinner (833) revised the enthalpies of formation of the organic fluorine compounds on the basis of a critical analysis of the extant data on $HF(aq)$ and $CF_4(g)$. All fluorine compounds in this chapter and in Chapter 14 have been calculated on this basis (cf. Table 8-2 and Ideal Gas Table 704). Our knowledge of many of

Introduction

the simple organic chlorine compounds rests on work at the University of Lund, Sweden, by a stationary bomb method. A valiant effort was made by Smith, Bjellerup, Krook, and Westermark (1384) to salvage this work by applying a number of significant corrections to the original data, but the results must still bear a relatively large uncertainty. Enthalpy of reaction measurements by Lacher, Park, and co-workers at the University of Colorado, as well as earlier work by Kistiakowsky and his students, serve to fix the enthalpies of formation of a number of fluorine and chlorine compounds. Bjellerup (139, 140) gave an interesting history of the combustion calorimetry of organic bromine compounds. As in the case of fluorine compounds, good techniques have been developed but have been applied to relatively few compounds.

Organic halogen compound thermochemistry and thermodynamics hold interest for theoreticians as well as the engineer. As Patrick (1126) pointed out in his review, the $C-F$ bond energy appears to increase with increasing substitution of fluorine on a carbon atom. The extent of and reasons for this increased stability are not well defined. An exuberant literature on the enthalpy of formation of the CF_2 radical and the related $C{=}C$ bond dissociation energy of C_2F_4 has not yet resulted in accurate values for these quantities. Barriers to internal rotation in these compounds warrant further study. Despite the wide variation in electronegativity and atomic size of the four halogens, barriers to internal rotation of methyl groups appear to vary only slightly in such series as the ethyl halides. In the C_2F_5X series, the barrier rises from 2.0 to 2.5 kcal, from $X = Cl$ to $X = I$ (1232). Low-temperature heat capacities on the condensed phases and flow calorimetry on the vapor should serve to define these trends better.

Because of the paucity of accurate data, methyl substitution constants are given in Tables 12-1, 12-2, 12-3, and 12-4 only for the monosubstituted compounds. At present it does not appear possible to estimate reasonably accurate enthalpies of formation of compounds with a high degree of halogen substitution. An exception to this may be the perfluoroalkanes, for which Bryant (178) and Good, Douslin, Scott, George, Lacina, Dawson, and Waddington (508) have given methods of estimation.

Table 12-1 Substitution Constants for Gaseous Fluorine Compounds

Type of Compound	$\Delta[\Delta H f°_{298}]$ (kcal/mole)	$\Delta S°_{298}$ [cal/(mole °K)]	$\Delta C p$ [cal/(mole °K)] at								
			298°K	300°K	400°K	500°K	600°K	700°K	800°K	900°K	1000°K
Monofluorides											
$CH_3CH_3 \rightarrow CH_3F$	−37.8	−5.16	−3.62	−3.65	−5.12	−6.40	−7.52	−8.49	−9.38	−10.18	−10.89
$CH_3CH_2CH_3 \rightarrow CH_3CH_2F$	−36.5	−4.77	−3.46	−3.49	−4.97	−6.30	−7.44	−8.43	−9.32	−10.12	−10.85
$CH_3CH_2CH_2CH_3 \rightarrow CH_3CH_2CH_2F$											
$CH_3-\overset{\overset{H}{|}}{\underset{\underset{CH_3}{|}}{C}}-CH_3 \rightarrow CH_3-\overset{\overset{H}{|}}{\underset{\underset{CH_3}{|}}{C}}-CH_3$	−36.3										
$R-CH_3 \rightarrow R-F$	**−36.8**	**−4.97**	**−3.54**	**−3.57**	**−5.05**	**−6.35**	**−7.48**	**−8.46**	**−9.35**	**−10.15**	**−10.87**
Difluorides											
$CH_3CH_2CH_3 \rightarrow FCH_2F$	−80.7	−9.98	−7.32	−7.38	−10.32	−12.92	−15.16	−17.12	−18.86	−20.44	−21.85
$CH_3CH(CH_3)_2 \rightarrow CH_3CHF_2$	−83.0	−9.44	−6.90	−6.96	−9.84	−12.55	−14.94	−16.99	−18.80	−20.41	−21.84
$RCH(CH_3)_2 \rightarrow RCHF_2$	**−81.9**	**−9.71**	**−7.11**	**−7.17**	**−10.08**	**−12.74**	**−15.05**	**−17.06**	**−18.83**	**−20.43**	**−21.85**
Trifluorides											
$HC(CH_3)_3 \rightarrow HCF_3$	−130.5	−14.92	−10.94	−11.01	−14.89	−19.07	−22.49	−25.49	−28.16	−30.66	−32.66
$CH_3C(CH_3)_3 \rightarrow CH_3CF_3$	−134.4	−13.87	−10.32	−10.38	−14.80	−19.10	−22.83	−26.03	−28.80	−31.23	−33.35
$RC(CH_3)_3 \rightarrow RCF_3$	**−132.5**	**−14.40**	**−10.63**	**−10.70**	**−14.85**	**−19.09**	**−22.66**	**−25.66**	**−28.48**	**−30.95**	**−33.01**
Aromatic fluorides											
$C_6H_5CH_3 \rightarrow C_6H_5F$	−38.43	−6.27	−2.23	−2.23	−3.26	−4.50	−5.72	−6.87	−7.89	−8.80	−9.61
$1,4\text{-}CH_3C_6H_4F \rightarrow 1,4\text{-}C_6H_4F_2$	−38.30	−6.64	−2.56	−2.56	−3.27	−4.33	−5.46	−6.55	−7.54	−8.41	−9.18
$R-CH_3 \rightarrow R-F$	**−38.37**	**−6.46**	**−2.40**	**−2.40**	**−3.27**	**−4.42**	**−5.59**	**−6.71**	**−7.72**	**−8.60**	**−9.40**

Table 12-2 Substitution Constants for Gaseous Chlorine Compounds

Type of Compound	$\Delta[\Delta H f^\circ_{298}]$ (kcal/mole)	ΔS°_{298} [cal/(mole °K)]	ΔC_p [cal/(mole °K)] at 298°K	300°K	400°K	500°K	600°K	700°K	800°K	900°K	1000°K
Aliphatic chlorides											
$CH_3CH_3 \rightarrow CH_3Cl$	−0.39	−2.37	−2.84	−2.87	−4.16	−5.47	−6.69	−7.79	−8.79	−9.68	−10.47
$CH_3CH_2CH_3 \rightarrow CH_3CH_2Cl$	−1.88	−2.15	−2.58	−2.61	−3.98	−5.34	−6.57	−7.67	−8.66	−9.55	−10.35
$CH_3CH_2CH_2CH_3 \rightarrow CH_3CH_2CH_2Cl$	−0.95	1.39	−3.05	−3.06	−4.24	−5.61	−6.87	−7.98	−8.99	−9.87	−10.63
$CH_3{-}\underset{CH_3}{\overset{H}{C}}{-}CH_3 \rightarrow CH_3{-}\underset{CH_3}{\overset{Cl}{C}}{-}CH_3$	−2.85	−2.09	−2.27	−2.28	−3.78	−5.14	−6.42	−7.55	−8.55	−9.45	−10.24
$CH_3{-}\underset{CH_3}{\overset{CH_3}{C}}{-}CH_3 \rightarrow CH_3{-}\underset{CH_3}{\overset{CH_3}{C}}{-}Cl$	−4.13		−1.17	−1.77	−1.81	−3.55	−5.40	−7.01	−8.30	−9.28	−10.05	−10.80
$\tfrac{1}{2}CH_3CH_2CH_2CH_3 \rightarrow \tfrac{1}{2}ClCH_2CH_2Cl$	−0.43	−2.41	−2.25	−2.20	−3.80	−5.32	−6.70	−7.88	−8.92	−9.82	−10.61
$CH_3(CH_2)_3CH_3 \rightarrow CH_3(CH_2)_3Cl$	−0.20										
$CH_3CH_2{-}\underset{CH_3}{\overset{H}{C}}{-}CH_3 \rightarrow CH_3CH_2{-}\underset{CH_3}{\overset{Cl}{C}}{-}CH_3$	−1.68										
$CH_3CH_2{-}\underset{CH_3}{\overset{H}{C}}{-}CH_3 \rightarrow ClCH_2{-}\underset{CH_3}{\overset{H}{C}}{-}CH_3$	−1.18										

Table 12-2 Substitution Constants for Gaseous Chlorine Compounds (contd.)

Type of Compound	$\Delta[\Delta H f^\circ_{298}]$ (kcal/mole)	ΔS°_{298} [cal/(mole °K)]	$\Delta C p$ [cal/(mole °K)] at 298°K	300°K	400°K	500°K	600°K	700°K	800°K	900°K	1000°K
Aliphatic Chlorides											
$CH_3CH_2CH_2\text{—}\overset{H}{\underset{CH_3}{C}}\text{—}CH_3 \to ClCH_2CH_2\text{—}\overset{H}{\underset{CH_3}{C}}\text{—}CH_3$	-1.44										
$CH_3CH_2\text{—}\overset{CH_3}{\underset{CH_3}{C}}\text{—}CH_3 \to CH_3CH_2\text{—}\overset{CH_3}{\underset{CH_3}{C}}\text{—}Cl$	-4.05										
$CH_3CH_2\text{—}\overset{H}{\underset{CH_3}{C}}\text{—}CH_3 \to ClCH_2\text{—}\overset{H}{\underset{CH_3}{C}}\text{—}Cl$	-1.34										
$\tfrac{1}{2}CH_3(CH_2)_3CH_3 \to \tfrac{1}{2}Cl(CH_2)_3Cl$	-1.80										
$\tfrac{1}{3}CH_3CH_2\text{—}\overset{H}{\underset{CH_3}{C}}\text{—}CH_2CH_3 \to \tfrac{1}{3}ClCH_2\text{—}\overset{H}{\underset{Cl}{C}}\text{—}CH_2Cl$	-1.07										
$R\text{—}CH_3 \to R\text{—}Cl$	-1.67	-1.93	-2.46	-2.47	-3.92	-5.38	-6.71	-7.86	-8.87	-9.74	-10.52
Aromatic chlorides											
$C_6H_5CH_3 \to C_6H_5Cl$	0.44	-3.38	-1.37	-1.38	-2.63	-4.05	-5.42	-6.71	-7.83	-8.82	-9.71

Table 12-3 Substitution Constants for Gaseous Bromine Compounds

Type of Compound	$\Delta[\Delta H_f^{\circ}{}_{298}]$ (kcal/mole)	ΔS°_{298} [cal/(mole °K)]	ΔC_p [cal/(mole °K)] at								
			298°K	300°K	400°K	500°K	600°K	700°K	800°K	900°K	1000°K

Aliphatic bromides

Type of Compound	$\Delta[\Delta H_f^{\circ}{}_{298}]$	ΔS°_{298}	298°K	300°K	400°K	500°K	600°K	700°K	800°K	900°K	1000°K
$CH_3CH_3 \rightarrow CH_3Br$	11.24	0.34	−2.44	−2.47	−3.75	−5.10	−6.38	−7.53	−8.58	−9.50	−10.33
$CH_3CH_2CH_3 \rightarrow CH_3CH_2Br$	9.52	0.64	−2.09	−2.18	−3.61	−5.03	−6.32	−7.47	−8.50	−9.42	−10.24
$CH_3 \underset{\underset{CH_3}{\|}}{\overset{\overset{H}{\|}}{C}} CH_3 \rightarrow CH_3 \underset{\underset{CH_3}{\|}}{\overset{\overset{Br}{\|}}{C}} CH_3$	8.95	0.74	−1.87	−1.94	−3.43	−4.86	−6.20	−7.36	−8.40	−9.33	−10.14
$CH_3 \underset{\underset{CH_3}{\|}}{\overset{\overset{CH_3}{\|}}{C}} CH_3 \rightarrow CH_3 \underset{\underset{CH_3}{\|}}{\overset{\overset{CH_3}{\|}}{C}} Br$	7.67	1.17	−1.22	−1.23	−2.62	−4.16	−5.63	−6.97	−8.13	−9.16	−10.06
$CH_3CH_2CH_2CH_3 \rightarrow CH_3CH_2CH_2Br$	9.15										
$CH_3(CH_2)_3CH_3 \rightarrow CH_3CH_2(CH_2)_3Br$	9.35										
$CH_3(CH_2)_4CH_3 \rightarrow CH_3(CH_2)_4Br$	9.09										
$CH_3CH_2 \underset{\underset{CH_3}{\|}}{\overset{\overset{H}{\|}}{C}} CH_3 \rightarrow CH_3CH_2 \underset{\underset{CH_3}{\|}}{\overset{\overset{Br}{\|}}{C}} CH_3$	8.22										
$\tfrac{1}{2}CH_3CH_2CH_2CH_3 \rightarrow \tfrac{1}{2}BrCH_2CH_2Br$	10.43	0.17									
$\tfrac{1}{2}CH_3CH_2 \underset{\underset{CH_3}{\|}}{\overset{\overset{H}{\|}}{C}} CH_3 \rightarrow \tfrac{1}{2}BrCH_2 \underset{\underset{CH_3}{\|}}{\overset{\overset{H}{\|}}{C}} CH_3$ Br	9.76										

Table 12-3 Substitution Constants for Gaseous Bromine Compounds (contd.)

Type of Compound	$\Delta[\Delta H f^{\circ}_{298}]$ (kcal/mole)	ΔS°_{298} [cal/(mole °K)]	ΔC_p [cal/(mole °K)] at							
			298°K	300°K	400°K	500°K	600°K	700°K	800°K	900°K	1000°K

Aliphatic bromides

Type of Compound	$\Delta[\Delta H f^{\circ}_{298}]$	ΔS°_{298}	298°K	300°K	400°K	500°K	600°K	700°K	800°K	900°K	1000°K
$\tfrac{1}{2}$CH$_3$CH$_2$—C(H)(CH$_3$)—CH$_2$CH$_3$ → $\tfrac{1}{2}$BrCH$_2$—C(Br)(CH$_3$)—CH$_2$CH$_3$	8.75										
$\tfrac{1}{2}$CH$_3$—C(H)(CH$_3$CH$_3$)—C(H)(H)—CH$_3$ → $\tfrac{1}{2}$CH$_3$—C(Br)(CH$_3$CH$_3$)—C(Br)(H)—CH$_3$	9.05										
$\tfrac{1}{2}$CH$_3$—C(CH$_3$)(H)—C(H)(CH$_3$)—CH$_3$ → $\tfrac{1}{2}$CH$_3$—C(CH$_3$)(Br)—C(Br)(CH$_3$)—CH$_3$	7.88										
R—CH$_3$ → R—Br	9.20	0.61				−1.91	−1.96	−3.35	−4.79	−6.13	−7.33	−8.40	−9.37	−10.19

Aromatic bromides

| C$_6$H$_5$CH$_3$ → C$_6$H$_5$Br | 13.15 | −1.07 | | | −1.45 | −1.46 | −2.81 | −4.24 | −5.79 | −6.83 | −7.94 | −8.92 | −9.79 |

Table 12-4 Substitution Constants for Gaseous Iodine Compounds

Type of Compound	$\Delta[\Delta H f^\circ_{298}]$ (kcal/mole)	ΔS°_{298} [cal/(mole °K)]	ΔCp[cal/(mole °K)] at								
			298°K	300°K	400°K	500°K	600°K	700°K	800°K	900°K	1000°K
Aliphatic iodides											
$CH_3CH_3 \to CH_3I$	23.6	2.30	−2.03	−2.06	−3.32	−4.71	−6.04	−7.24	−8.32	−9.29	−10.13
$CH_3CH_2CH_3 \to CH_3CH_2I$	22.8	2.75	−1.81	−1.84	−3.36	−4.89	−6.24	−7.40	−8.43	−9.36	−10.18
$CH_3 \to CH_3\!-\!\underset{CH_3}{\overset{H}{C}}\!-\!CH_3 \to CH_3\!-\!\underset{I}{\overset{H}{C}}\!-\!CH_3$	22.2	2.76	−1.61	−1.64	−3.18	−4.66	−6.04	−7.23	−8.28	−9.23	−10.06
$CH_3CH_2CH_2CH_3 \to CH_3CH_2CH_2CH_2I$	22.9	2.41									
$\tfrac{1}{2}CH_3CH_2CH_2CH_3 \to \tfrac{1}{2}ICH_2CH_2I$	23.0	2.07									
$\tfrac{1}{2}CH_3CH_2\!-\!\underset{CH_3}{\overset{H}{C}}\!-\!CH_3 \to \tfrac{1}{2}ICH_2\!-\!\underset{I}{\overset{H}{C}}\!-\!CH_3$	22.8	3.62	−0.80	−0.81	−2.28	−3.88	−5.39	−6.79	−7.83	−8.97	−9.89
$CH_3\!-\!\underset{CH_3}{\overset{CH_3}{C}}\!-\!CH_3 \to CH_3\!-\!\underset{CH_3}{\overset{CH_3}{C}}\!-\!I$	22.1										
R—CH₃ → R—I	**22.8**	**2.65**	**−1.56**	**−1.56**	**−3.05**	**−4.54**	**−5.93**	**−7.17**	**−8.22**	**−9.21**	**−10.07**
Aromatic iodides											
$C_6H_5CH_3 \to C_6H_5I$	22.8	1.24	−0.72	−0.73	−2.15	−3.68	−5.15	−6.49	−7.65	−8.67	−9.59

493

ALIPHATIC FLUORINE COMPOUND IDEAL GAS TABLES

No. 701　Fluoromethane　CH_3F　(Ideal Gas State)　Mol Wt 34.034

	cal/(mole °K)				kcal/mole		
$T°K$	$Cp°$	$S°$	$-(G°-H°_{298})/T$	$H°-H°_{298}$	$\Delta Hf°$	$\Delta Gf°$	Log Kp
298	8.96	53.25	53.25	0.00	-55.90	-50.19	36.789
300	8.99	53.31	53.26	0.02	-55.91	-50.16	36.536
400	10.56	56.11	53.63	1.00	-56.61	-48.13	26.295
500	12.26	58.65	54.38	2.14	-57.24	-45.93	20.077
600	13.83	61.03	55.29	3.45	-57.78	-43.62	15.889
700	15.23	63.27	56.27	4.90	-58.22	-41.23	12.870
800	16.45	65.38	57.28	6.49	-58.59	-38.77	10.592
900	17.51	67.38	58.29	8.18	-58.88	-36.28	8.809
1000	18.44	69.28	59.30	9.98	-59.11	-33.76	7.378

Lacher and Skinner (833) estimated $\Delta Hf°_{298}(g) = -55.9$ kcal/mole. Thermodynamic functions are calculated on the basis of fundamental frequencies selected by Smith and Mills (1389) and the moments of inertia determined from microwave spectroscopy by Gilliam, Edwards, and Gordy (485). Gelles and Pitzer (465) calculated functions in good agreement, using frequencies from Plyler and Benedict (1172) and an over-all moment of inertia of 5.92×10^{-117} g^3-cm^6.

Rossini, Wagman, Evans, Levine, and Jaffe (1249) listed $Tm = 131.4°K$ and $Tb = 194.8°K$. Michels, Visser, Lunbeck, and Wolkers (998) reported PVT measurements for this fluid.

No. 702　Difluoromethane　CH_2F_2　(Ideal Gas State)　Mol Wt 52.026

	cal/(mole °K)				kcal/mole		
$T°K$	$Cp°$	$S°$	$-(G°-H°_{298})/T$	$H°-H°_{298}$	$\Delta Hf°$	$\Delta Gf°$	Log Kp
298	10.25	58.94	58.94	0.00	-108.24	-101.66	74.513
300	10.28	59.01	58.95	0.02	-108.25	-101.62	74.024
400	12.22	62.23	59.37	1.15	-108.84	-99.31	54.260
500	14.10	65.17	60.24	2.47	-109.34	-96.87	42.341
600	15.72	67.88	61.29	3.96	-109.76	-94.34	34.361
700	17.08	70.41	62.42	5.60	-110.09	-91.74	28.641
800	18.22	72.77	63.56	7.37	-110.36	-89.10	24.340
900	19.17	74.97	64.71	9.24	-110.56	-86.43	20.987
1000	19.98	77.04	65.84	11.20	-110.71	-83.76	18.304

Neugebauer and Margrave (1061) measured the enthalpy of combustion of difluoromethane vapor. Their data give $\Delta Hf°_{298}(g) = -108.24$ kcal/mole. Thermodynamic functions are calculated on the basis of the fundamental frequencies of Plyler and Benedict (1172) and the moments of inertia derived by Lide (865) from microwave spectroscopy. Recent study of fine structure of infrared bands by Porto (1181) gave a structure in excellent agreement with that of Lide. Gelles and Pitzer (465), Meister, Dowling, and Bielecki (990), Sverdlin and Godnev (1453), and Glockler and Edgell (497, 498) calculated thermodynamic functions using less accurate molecular data. Rossini, Wagman, Evans, Levine, and Jaffe (1249) listed $Tb = 221°K$.

No. 703 Trifluoromethane CHF_3 (Ideal Gas State) Mol Wt 70.018

	cal/(mole °K)				kcal/mole		
$T°K$	$Cp°$	$S°$	$-(G°-H°_{298})/T$	$H°-H°_{298}$	$\Delta Hf°$	$\Delta Gf°$	Log Kp
298	12.20	62.04	62.04	0.00	-166.71	-158.48	116.165
300	12.24	62.12	62.05	0.03	-166.72	-158.43	115.412
400	14.88	66.03	62.56	1.39	-167.10	-155.61	85.015
500	16.55	69.53	63.61	2.96	-167.41	-152.70	66.740
600	18.13	72.69	64.86	4.70	-167.65	-149.73	54.536
700	19.36	75.58	66.19	6.58	-167.82	-146.73	45.808
800	20.33	78.23	67.53	8.56	-167.95	-143.71	39.257
900	20.99	80.66	68.86	10.63	-168.04	-140.67	34.158
1000	21.74	82.91	70.15	12.77	-168.09	-137.65	30.083

Neugebauer and Margrave (1061) measured the enthalpy of combustion of the vapor. Their work yields $\Delta Hf°_{298}(g) = -166.71$ kcal/mole. Gelles and Pitzer (465) calculated thermodynamic functions adopted here, based on the vibrational assignment of Plyler and Benedict (1172) and the moments of inertia derived from optical spectra by Bernstein and Herzberg (118). Recent microwave spectroscopy by Ghosh, Trambarulo, and Gordy (473) gave a structure in excellent agreement with that of Bernstein and Herzberg; Decker, Meister, and Cleveland (316) confirmed the vibrational assignment and calculated functions nearly identical with those of Gelles and Pitzer. Sverdlin and Godnev (1453) also reported thermodynamic functions, and Glockler and Edgell (497, 498) calculated heat capacities. Vanderkooi and De Vries (1527) measured the heat capacity at 300°K by a wire-ribbon method, in good agreement with the calculated values. Valentine, Brodale, and Giauque (1526) measured low-temperature thermal data, including $Ttp = 117.97°K$, $\Delta Hm = 0.970$ kcal/mole, and $Tb = 190.97°K$, where $\Delta Hv = 3.994$ kcal/mole. The third-law entropy of the vapor is in excellent agreement with that calculated from the spectroscopic data adopted here. Hou and Martin (615) measured PVT data and gave tables of thermodynamic properties in engineering units. They derived $Tc = 299°K$ and $Pc = 47.7$ atm.

No. 704   Carbon Tetrafluoride   CF$_4$ (Ideal Gas State)   Mol Wt 88.010

	cal/(mole °K)				kcal/mole		
T°K	Cp°	S°	$-(G°-H°_{298})/T$	$H°-H°_{298}$	$\Delta Hf°$	$\Delta Gf°$	Log Kp
298	14.68	62.50	62.50	0.00	-223.00	-212.34	155.641
300	14.73	62.60	62.51	0.03	-223.00	-212.27	154.634
400	17.41	67.22	63.12	1.64	-223.18	-208.67	114.004
500	19.43	71.33	64.36	3.49	-223.26	-205.03	89.613
600	20.91	75.01	65.83	5.51	-223.29	-201.38	73.348
700	21.99	78.32	67.38	7.66	-223.26	-197.73	61.729
800	22.84	81.31	68.94	9.90	-223.21	-194.08	53.018
900	23.42	84.03	70.47	12.22	-223.14	-190.45	46.244
1000	23.91	86.53	71.95	14.58	-223.05	-186.85	40.835

Lacher and Skinner (833) reviewed thermal data on CF$_4$ and selected $\Delta Hf°_{298}(g) = -223.0$ kcal/mole. Recent precise work by Walker (1568) on the enthalpy of reaction of cyanogen and nitrogen trifluoride and by Greenberg and Hubbard (542) on the combustion of graphite in fluorine confirms the above value within experimental error. Several investigations not considered by Lacher and Skinner, because of low accuracy, are also in agreement within experimental error. Kirkbride and Davidson (755) and Vorob'ev and Skuratov (1546) measured the enthalpy of reaction of CF$_4$ and alkali metals. Duus (360) and Neugebauer and Margrave (1059) reported data on the enthalpy of decomposition of tetrafluoroethylene to CF$_4$ and amorphous carbon. Jessup, McCoskey, and Nelson (691) determined the enthalpy of combustion of methane in fluorine, but their data cannot be accurately corrected for deviation of HF vapor from ideal behavior. Baibuz (59) used an "explosion" method to obtain $\Delta Hf°_{298}(g) = -220.1$ kcal/mole. Of historical interest is the determination of the enthalpy of direct combination of the elements by Wartenberg and Schütte (1575), which gave a value much too low.

Albright, Galegar, and Innes (8) calculated thermodynamic functions, using frequencies listed by Claassen (227), an over-all moment of inertia of 3.24×10^{-115} g^3-cm^6, and estimated anharmonicity corrections based on vapor heat capacity data for dichlorodifluoromethane by Masi (935). The functions of Albright et al. are adopted, except for a small revision in the entropy for an over-all moment of inertia of 3.15×10^{-115} g^3-cm^6 from electron diffraction data by Thornton (1498) and Hoffman and Livingston (611). Other calculations to the harmonic oscillator approximation were made by Gelles and Pitzer (465), Nagarajan (1043), Sverdlin and Godnev (1453), Decker, Meister, and Cleveland (316), and Kobe and Kobe (778). Hwang and Martin (669) measured the vapor heat capacity at constant volume, in good agreement with spectroscopic values. Eucken and Schröder (384) measured low-temperature thermal data, including $Tt = 76.23°K$, $\Delta Ht = 0.353$ kcal/mole, $Ttp = 89.47°K$, and $\Delta Hm = 0.167$ kcal/mole. The experimental entropy is in good agreement with the calculated result of Albright et al. Kostryukov, Samorukov, and Strelkov (801) also reported low-temperature thermal data in reasonable agreement except for ΔHt, which they gave as 0.415 kcal/mole.

Douslin, Harrison, Moore, and McCullough (339) measured PVT data and derived virial coefficients. Chari (217) measured vapor pressures and densities and calculated engineering tables as well as $Tc = 227.5°K$ and $Pc = 37$ atm.

No. 705   Fluoroethane C_2H_5F (Ideal Gas State)   Mol Wt 48.060

	cal/(mole °K)				kcal/mole		
$T°K$	$Cp°$	$S°$	$-(G°-H°_{298})/T$	$H°-H°_{298}$	$\Delta Hf°$	$\Delta Gf°$	Log Kp
298	14.11	63.32	63.32	0.00	-62.50	-50.08	36.709
300	14.17	63.41	63.33	0.03	-62.52	-50.01	36.427
400	17.57	67.96	63.92	1.62	-63.55	-45.67	24.953
500	20.72	72.23	65.16	3.54	-64.42	-41.10	17.964
600	23.44	76.25	66.68	5.75	-65.13	-36.37	13.247
700	25.77	80.04	68.32	8.21	-65.69	-31.53	9.843
800	27.76	83.62	70.01	10.89	-66.13	-26.62	7.273
900	29.49	86.99	71.71	13.76	-66.45	-21.66	5.260
1000	30.98	90.18	73.40	16.78	-66.67	-16.68	3.645

Thermodynamic functions are calculated on the basis of the frequency assignment of Smith, Saunders, Nielsen, and Ferguson (1380), the moments of inertia reported by Kraitchman and Dailey (808), and the barrier to internal rotation determined by Catalano and Pitzer (205). Calculations by Catalano and Pitzer at three temperatures agree with the present work. The enthalpy of formation was estimated by Lacher and Skinner (833) as -62.5 kcal/mole. Li and Rossini (863) reviewed vapor pressure data and selected constants for an Antoine equation that yields $Tb = 235.5°K$.

No. 706   1,1-Difluoroethane $C_2H_4F_2$ (Ideal Gas State)   Mol Wt 66.052

Thermodynamic functions are calculated on the basis of the vibrational assignment of Smith, Saunders, Nielsen, and Ferguson (1380), the moments of inertia reported by Solimene and Dailey (1392), and the barrier to internal rotation determined by Herschbach (586) and by Fateley and Miller (407). Mears, Stahl, Orfeo, Shair, Kells, Thompson, and McCann (979) calculated functions in good agreement and also reported liquid and vapor densities and vapor pressures, from which they derived $Tb = 248.15°K$, where $\Delta Hv = 5.1$ kcal/mole, $Tc = 386.7°K$,

No. 706 1,1-Difluoroethane $C_2H_4F_2$ (Ideal Gas State) Mol Wt 66.052

T°K	$Cp°$	$S°$	$-(G°-H°_{298})/T$	$H°-H°_{298}$	$\Delta Hf°$	$\Delta Gf°$	Log Kp
		cal/(mole °K)			kcal/mole		
298	16.24	67.52	67.52	0.00	-118.00	-104.26	76.424
300	16.31	67.63	67.53	0.04	-118.02	-104.18	75.891
400	19.93	72.83	68.21	1.85	-118.85	-99.43	54.326
500	23.07	77.62	69.62	4.01	-119.54	-94.50	41.303
600	25.68	82.06	71.33	6.45	-120.09	-89.44	32.576
700	27.86	86.19	73.16	9.13	-120.51	-84.29	26.316
800	29.69	90.03	75.03	12.01	-120.83	-79.10	21.608
900	31.24	93.62	76.90	15.05	-121.05	-73.87	17.937
1000	32.56	96.98	78.74	18.25	-121.19	-68.63	14.999

and $Pc = 44.4$ atm. Rossini, Wagman, Evans, Levine, and Jaffe (1249) listed $Tm = 156°K$.

Lacher and Skinner (833) reported $\Delta Hf°_{298}(g) = -118.0$ kcal/mole, based on an unpublished enthalpy of hydrogenation of 1,1-difluoroethylene. Their result is adopted.

Lacher, Kianpour, Oetting, and Park (827) measured the enthalpy of hydrogenation of 1,1-difluoro-2,2-dichloroethylene to 1,1-difluoroethane. The enthalpy of formation of 1,1-difluoro-2,2-dichloroethylene was determined by Wartenberg and Schiefer (1574) by reaction with potassium. Combination of these two yields $\Delta Hf°_{298}(g) = -117.9$ kcal/mole. The result is not of high accuracy, but is in good agreement with the adopted value.

No. 707 1,1,1-Trifluoroethane $C_2H_3F_3$ (Ideal Gas State) Mol Wt 84.044

T°K	$Cp°$	$S°$	$-(G°-H°_{298})/T$	$H°-H°_{298}$	$\Delta Hf°$	$\Delta Gf°$	Log Kp
		cal/(mole °K)			kcal/mole		
298	18.75	68.66	68.66	0.00	-178.20	-162.23	118.915
300	18.83	68.78	68.67	0.04	-178.21	-162.14	118.110
400	22.75	74.75	69.45	2.13	-178.82	-156.68	85.601
500	25.90	80.18	71.06	4.56	-179.28	-151.09	66.038
600	28.38	85.13	73.00	7.28	-179.61	-145.42	52.967
700	30.37	89.66	75.06	10.22	-179.85	-139.70	43.614
800	31.98	93.82	77.15	13.34	-180.00	-133.96	36.594
900	33.32	97.67	79.22	16.61	-180.09	-128.19	31.128
1000	34.45	101.24	81.25	20.00	-180.12	-122.45	26.760

Kolesov, Martynov, and Skuratov (793) measured the enthalpy of combustion in oxygen, leading to $\Delta Hf°_{298}(g) = -178.2$ kcal/mole. Russell, Golding, and Yost (1270) reported low-temperature thermal data: $Ttp = 161.82°K$, $\Delta Hm = 1.480$ kcal/mole, $Tb = 225.86°K$, and $\Delta Hv = 4.583$ kcal/mole at $224.40°K$. Edgell, Miller, and Amy (367) and Thomas,

Heeks, and Sheridan (1484) made microwave spectroscopy studies leading to practically identical over-all moments of inertia, although they differ in interpretation. Nielsen, Claassen, and Smith (1070) reported a generally accepted vibrational assignment. These data and the third-law entropy yield a barrier to internal rotation of 3200 cal/mole. Catalano and Pitzer (204) deduced $V = 3040$ cal/mole from infrared combination bands, while Vanderkooi and De Vries (1527) measured the vapor heat capacity at a single temperature and derived $V = 3370$ cal/mole. An average value of 3200 cal/mole is used to calculate thermodynamic functions.

Mears, Stahl, Orfeo, Shair, Kells, Thompson, and McCann (979) measured liquid and vapor densities as well as vapor pressures and derived $Tc = 346.3°K$ and $Pc = 37.1$ atm.

No. 708   Hexafluoroethane C_2F_6 (Ideal Gas State)   Mol Wt 138.020

	cal/(mole °K)				kcal/mole		
$T°K$	$Cp°$	$S°$	$-(G°-H°_{298})/T$	$H°-H°_{298}$	$\Delta Hf°$	$\Delta Gf°$	Log Kp
298	25.43	79.37	79.37	0.00	-321.00	-300.52	220.275
300	25.53	79.53	79.38	0.05	-321.00	-300.39	218.825
400	29.99	87.52	80.43	2.84	-321.02	-293.51	160.360
500	33.26	94.58	82.57	6.01	-320.91	-286.65	125.288
600	35.54	100.85	85.10	9.45	-320.71	-279.81	101.917
700	37.15	106.46	87.76	13.09	-320.47	-273.01	85.234
800	38.32	111.50	90.42	16.87	-320.22	-266.25	72.734
900	39.17	116.06	93.02	20.75	-319.94	-259.53	63.018
1000	39.78	120.23	95.54	24.69	-319.66	-252.88	55.265

Kirkbride and Davidson (755) measured the enthalpy of reaction of sodium and hexafluoroethane, which yields $\Delta Hf°_{298}(g) = -311$ kcal/mole. Sinke (1354) measured the enthalpy of reaction of nitrogen trifluoride and hexafluoroethane, giving $\Delta Hf°_{298}(g) = -321.0$ kcal/mole. Tschuikow-Roux (1521) reviewed the thermochemistry of C_2F_6 and CF_3 radicals and concluded that -311 kcal/mole was probably too positive. The result of Sinke is adopted.

Pace and Aston (1097) reported low-temperature thermal data: $Tt = 103.98°K$, $\Delta Ht = 0.893$ kcal/mole, $Ttp = 173.10°K$, $\Delta Hm = 0.642$ kcal/mole, and $Tb = 194.87°K$, at which $\Delta Hv = 3.860$ kcal/mole. The entropy of the ideal gas at $194.87°K$ was found to be 69.88 cal/(mole °K). Combination of this result with the electron diffraction data of Swick and Karle (1459) and the vibrational assignment of Carney, Piotrowski, Meister, Braun, and Cleveland (192) yields a barrier to internal rotation of 3.96 kcal/mole. Thermodynamic functions are calculated using this value for the barrier.

No. 709 1-Fluoropropane C_3H_7F (Ideal Gas State) Mol Wt 62.086

T°K	$Cp°$	$S°$	$-(G°-H°_{298})/T$	$H°-H°_{298}$	$\Delta Hf°$	$\Delta Gf°$	Log Kp
	cal/(mole °K)				kcal/mole		
298	19.75	72.71	72.71	0.00	-67.20	-47.87	35.087
300	19.83	72.84	72.72	0.04	-67.23	-47.75	34.785
400	24.55	79.20	73.55	2.26	-68.56	-41.05	22.427
500	28.99	85.16	75.28	4.95	-69.68	-34.04	14.877
600	32.82	90.79	77.40	8.04	-70.59	-26.82	9.770
700	36.09	96.11	79.70	11.49	-71.29	-19.47	6.078
800	38.88	101.11	82.07	15.24	-71.82	-12.03	3.287
900	41.29	105.83	84.45	19.25	-72.20	-4.53	1.100
1000	43.37	110.29	86.81	23.49	-72.43	3.00	-0.655

Thermodynamic functions of the ideal gas are estimated. Lacher, Kianpour, and Park (828) measured the enthalpy of hydrogenation to propane and hydrogen fluoride at 248°C. Adjustment to 298°K and combination with enthalpies of formation of propane and hydrogen fluoride adopted elsewhere in this volume give $\Delta Hf°_{298}(g) = -67.2$ kcal/mole. Li and Rossini (863) selected vapor pressure data leading to $Tb = 270.7°K$. Hirota (607) reported moments of inertia and barriers to internal rotation based on microwave spectroscopy.

No. 710 2-Fluoropropane C_3H_7F (Ideal Gas State) Mol Wt 62.086

T°K	$Cp°$	$S°$	$-(G°-H°_{298})/T$	$H°-H°_{298}$	$\Delta Hf°$	$\Delta Gf°$	Log Kp
	cal/(mole °K)				kcal/mole		
298	19.60	69.82	69.82	0.00	-69.00	-48.81	35.775
300	19.68	69.95	69.83	0.04	-69.03	-48.68	35.465
400	24.72	76.31	70.66	2.27	-70.36	-41.69	22.779
500	29.27	82.33	72.40	4.97	-71.46	-34.40	15.033
600	33.14	88.02	74.53	8.10	-72.33	-26.90	9.798
700	36.39	93.37	76.84	11.58	-73.00	-19.27	6.016
800	39.14	98.42	79.23	15.36	-73.51	-11.56	3.159
900	41.50	103.17	81.63	19.39	-73.86	-3.79	0.921
1000	43.55	107.65	84.01	23.65	-74.08	4.00	-0.874

Thermodynamic functions of the ideal gas are estimated. Lacher, Kianpour, and Park (828) measured the enthalpy of hydrogenation to propane and hydrogen fluoride at 248°C. Adjustment to 298°K and combination with enthalpies of formation of propane and hydrogen fluoride adopted elsewhere in this volume give $\Delta Hf°_{298}(g) = -69.0$ kcal/mole. Rossini (1250) listed $Tm = 139.8°K$ and $Tb = 263.8°K$.

No. 711 Octafluorocyclobutane C_4F_8 (Ideal Gas State) Mol Wt 200.040

Duus (360) measured the enthalpy of combustion of both C_2F_4 and C_4F_8, the products being a mixture of CF_4, CO_2, and CF_2O. The assump-

No. 711 Octafluorocyclobutane C_4F_8 (Ideal Gas State) Mol Wt 200.040

T°K	Cp°	S°	$-(G°-H°_{298})/T$	$H°-H°_{298}$	$\Delta Hf°$	$\Delta Gf°$	Log Kp
	cal/(mole °K)			kcal/mole			
298	37.32	95.69	95.69	0.00	-365.20	-334.33	245.057
300	37.45	95.93	95.70	0.07	-365.20	-334.14	243.408
400	44.50	107.70	97.25	4.18	-365.16	-323.78	176.895
500	50.10	118.27	100.42	8.93	-364.92	-313.46	137.007
600	53.85	127.75	104.20	14.14	-364.55	-303.20	110.437
700	56.55	136.26	108.18	19.66	-364.12	-293.01	91.477
800	58.65	143.95	112.18	25.42	-363.66	-282.89	77.277
900	60.25	150.96	116.10	31.37	-363.17	-272.82	66.246
1000	61.50	157.37	119.91	37.46	-362.64	-262.88	57.450

tion that any systematic errors are largely canceled out gives the enthalpy of dimerization of tetrafluoroethylene as -49.4 kcal/mole. Atkinson and Trenwith (51) derived the equilibrium constant for the dimerization of C_2F_4 at 500°C. Their data yield an enthalpy of dimerization of -50.4 kcal/mole. Employing the enthalpy of formation of C_2F_4 adopted elsewhere in this book allows the enthalpy of formation of C_4F_8 to be calculated as -364.2 and -365.2 kcal/mole, respectively. Kolesov, Talakin, and Skuratov (795) measured the enthalpy of reaction of sodium and C_4F_8 to form sodium fluoride and amorphous carbon as -714.5 kcal/mole. They found the enthalpy of formation of the amorphous carbon to be 0.700 kcal/g atom by combustion calorimetry. The enthalpy of formation of sodium fluoride listed by Rossini, Wagman, Evans, Levine, and Jaffe (1249) is adjusted to the value for HF(aq) adopted here in order to calculate for C_4F_8 an enthalpy of formation of -356.7 kcal/mole. The value selected as most accurate is that derived from Atkinson and Trenwith, $\Delta Hf°_{298}(g) = -365.2$ kcal/mole.

Furukawa, McCoskey, and Reilly (457) reported low-temperature thermal data as follows: $Ttp = 232.96°K$, $\Delta Hm = 0.662$ kcal/mole, and $Tb = 267.17°K$, where $\Delta Hv = 5.58$ kcal/mole. Four solid-state transitions were observed, but the exact characterization of the low-temperature solid state was difficult because of the tendency to glass formation. Masi (936) measured the vapor heat capacity at three temperatures. A vibrational assignment by Edgell (365) gives a calculated heat capacity lower than that measured by Masi, while an assignment by Claassen (226) gives too high a calculated heat capacity. A heat capacity curve was therefore constructed by joining smoothly the experimental data of Masi and calculated values at 800°, 900°, and 1000°K. The third-law entropy reported by Furukawa, McCoskey, and Reilly (457) may be uncertain because of glass formation, but is probably good to an entropy unit. The entropy calculated to 298°K is 95.69 cal/(mole °K).

Douslin, Moore, and Waddington (344), Bambach (68), and Martin (934) reported PVT data. Martin (934) reviewed the data and derived equations. Critical constants from the three investigations are in good agreement at $Tc = 388.5°K$ and $Pc = 27.5$ atm. De Nevers and Martin (322) determined Cv between 26 and 90% of the critical density.

No. 712 1,1-Difluoroethylene $C_2H_2F_2$ (Ideal Gas State) Mol Wt 64.036

$T°K$	$Cp°$	$S°$	$-(G°-H°_{298})/T$	$H°-H°_{298}$	$\Delta Hf°$	$\Delta Gf°$	Log Kp
		cal/(mole °K)			kcal/mole		
298	14.14	63.38	63.38	0.00	-82.50	-76.84	56.319
300	14.18	63.47	63.39	0.03	-82.51	-76.80	55.947
400	17.16	67.98	63.98	1.60	-82.89	-74.83	40.886
500	19.51	72.07	65.19	3.44	-83.20	-72.78	31.812
600	21.32	75.79	66.65	5.49	-83.44	-70.68	25.744
700	22.77	79.19	68.21	7.69	-83.63	-68.53	21.396
800	23.95	82.31	69.78	10.03	-83.79	-66.37	18.130
900	24.93	85.19	71.33	12.48	-83.91	-64.18	15.585
1000	25.74	87.86	72.85	15.01	-83.98	-62.00	13.550

Lacher and Skinner (833) reviewed work by Neugebauer and Margrave (1059) and Kolesov, Martynov, Shtekher, and Skuratov (792) on the enthalpy of combustion and selected $\Delta Hf°_{298}(g) = -82.5$ kcal/mole. Thermodynamic functions are based on the moments of inertia determined by Edgell, Kinsey, and Amy (366) by microwave spectroscopy and the fundamental frequencies assigned by Edgell and Ultee (368). Mears, Stahl, Orfeo, Shair, Kells, Thompson, and McCann (979) measured liquid and vapor densities and vapor pressures and derived $Tb = 187.5°K$, $Tc = 303.2°K$, and $Pc = 43.7$ atm. Mears et al. also calculated thermodynamic functions based on the electron diffraction results of Karle and Karle (717) and the vibrational assignment of Smith, Nielsen, and Claassen (1378). Their values are slightly higher than the present calculations, chiefly because of a lower value for the torsional frequency.

No. 713 Trifluoroethylene C_2HF_3 (Ideal Gas State) Mol Wt 82.028

$T°K$	$Cp°$	$S°$	$-(G°-H°_{298})/T$	$H°-H°_{298}$	$\Delta Hf°$	$\Delta Gf°$	Log Kp
		cal/(mole °K)			kcal/mole		
298	16.54	69.94	69.94	0.00	-118.50	-112.22	82.257
300	16.60	70.05	69.95	0.04	-118.50	-112.18	81.721
400	19.39	75.22	70.63	1.84	-118.70	-110.04	60.122
500	21.59	79.79	72.02	3.89	-118.84	-107.86	47.145
600	23.30	83.88	73.66	6.14	-118.95	-105.66	38.484
700	24.63	87.58	75.39	8.54	-119.02	-103.44	32.292
800	25.69	90.94	77.12	11.06	-119.07	-101.21	27.648
900	26.54	94.02	78.83	13.67	-119.11	-98.97	24.033
1000	27.23	96.85	80.50	16.36	-119.11	-96.76	21.146

Kolesov, Martynov, Shtekher, and Skuratov (792) measured the enthalpy of combustion. Their result yields $\Delta Hf°_{298}(g) = -118.5$ kcal/mole. Although not of high accuracy, this result is reasonable and is adopted. Mann, Acquista, and Plyler (919) assigned fundamental frequencies, estimated structural constants, and calculated thermodynamic functions. Rossini, Wagman, Evans, Levine, and Jaffe (1249) listed $Tb = 221°K$.

No. 714 Tetrafluoroethylene C_2F_4 (Ideal Gas State) Mol Wt 100.020

	cal/(mole °K)				kcal/mole		
$T°K$	$Cp°$	$S°$	$-(G°-H°_{298})/T$	$H°-H°_{298}$	$\Delta Hf°$	$\Delta Gf°$	Log Kp
298	19.24	71.69	71.69	0.00	-157.40	-149.07	109.269
300	19.29	71.81	71.70	0.04	-157.40	-149.02	108.558
400	21.97	77.75	72.49	2.11	-157.37	-146.23	79.893
500	23.99	82.87	74.06	4.41	-157.31	-143.46	62.701
600	25.53	87.39	75.91	6.89	-157.25	-140.69	51.244
700	26.71	91.42	77.85	9.50	-157.19	-137.93	43.062
800	27.61	95.05	79.77	12.22	-157.12	-135.19	36.930
900	28.31	98.34	81.66	15.02	-157.05	-132.45	32.162
1000	28.86	101.35	83.48	17.88	-156.97	-129.76	28.357

Furukawa, McCoskey, and Reilly (456) measured low-temperature thermal data, finding $Ttp = 142.00°K$, $\Delta Hm = 1.844$ kcal/mole, and $Tb = 197.53°K$, where $\Delta Hv = 4.020$ kcal/mole. The calorimetric entropy is in excellent agreement with that calculated by Mann, Acquista, and Plyler (920) from their vibrational assignment and the electron diffraction results of Karle and Karle (717). Since Mann, Meal, and Plyler (923) and Mann, Fano, Meal, and Shimanouchi (922) confirmed the vibrational assignment. the thermodynamic functions given by Mann, Acquista, and Plyler are adopted.

Lacher and Skinner (833) selected $\Delta Hf°_{298}(g) = -157.4$ kcal/mole, based on the enthalpy of reaction with hydrogen and the enthalpy of explosion to CF_4 and carbon, both reported by Neugebauer and Margrave (1059). Duus (360) made similar determinations in good agreement but of lower precision. Kirkbride and Davidson (755), Wartenberg and Schiefer (1574), and Kolesov, Zenkov, and Skuratov (796) all measured the enthalpy of reaction of tetrafluoroethylene and alkali metals, with results of low precision and doubtful accuracy. Kirkbride and Davidson (755) also determined the enthalpy of formation of 1,1,2,2-tetrafluoro-1,2-dichloroethane by reaction with potassium. Combination of this value with the enthalpy of addition of chlorine to tetrafluoroethylene reported by Lacher, McKinley, Snow, Michel, Nelson, and Park (831) yields $\Delta Hf°_{298}(g) = -162$ kcal/mole, in poor agreement with the selected value.

Wood, Lagow, and Margrave (1625) measured the enthalpy of combustion of polytetrafluoroethylene in fluorine and cited previous work in good agreement. Their result yields $\Delta Hf^\circ_{298}(s) = -199.2$ kcal/mole of monomer. Low-temperature heat capacity data for polytetrafluoroethylene were determined by Furukawa, McCoskey, and King (455).

AROMATIC FLUORINE COMPOUND IDEAL GAS TABLES

No. 715 Fluorobenzene C_6H_5F (Ideal Gas State) Mol Wt 96.100

	cal/(mole °K)				kcal/mole		
$T°K$	$Cp°$	$S°$	$-(G°-H°_{298})/T$	$H°-H°_{298}$	$\Delta Hf°$	$\Delta Gf°$	Log Kp
298	22.57	72.33	72.33	0.00	-27.86	-16.50	12.097
300	22.72	72.48	72.34	0.05	-27.88	-16.44	11.973
400	29.99	80.04	73.32	2.69	-28.84	-12.46	6.810
500	36.04	87.40	75.40	6.00	-29.59	-8.28	3.620
600	40.86	94.41	77.99	9.86	-30.17	-3.97	1.446
700	44.70	101.01	80.81	14.14	-30.60	0.44	-0.137
800	47.83	107.19	83.73	18.77	-30.93	4.89	-1.336
900	50.42	112.98	86.66	23.69	-31.15	9.38	-2.278
1000	52.58	118.40	89.57	28.84	-31.27	13.89	-3.035

Scott, McCullough, Good, Messerly, Pennington, Kincheloe, Hossenlopp, Douslin, and Waddington (1318) reported complete thermodynamic properties for fluorobenzene, including $Ttp = 230.94°K$, $\Delta Hm = 2.702$ kcal/mole, $Tb = 357.89°K$, at which $\Delta Hv° = 7.457$ kcal/mole, $S°_{298}(l) = 49.22$ cal/(mole °K), $S°_{298}(g) = 72.33$ cal/(mole °K), $\Delta Hv°_{298} = 8.27$ kcal/mole, and $Cp(l) = 16.39 + 0.0623T$ [cal/(mole °K), 290–350°K]. Earlier studies of the low-temperature heat capacity by Stull (1431) and of the vapor heat capacity by Montgomery and De Vries (1021) are in fair agreement. Scott et al. employed the microwave spectroscopy structure studies of McCulloh and Pollnow (945) and slightly revised the vibrational assignment of Smith, Ferguson, Hudson, and Nielsen (1377) to calculate thermodynamic functions in good agreement with the experimental values. The revised assignment is in good agreement with an independent assignment by Whiffen (1603).

Good, Scott, and Waddington (514) determined the enthalpy of combustion of fluorobenzene, which leads to $\Delta Hf^\circ_{298}(l) = -36.13$ kcal/mole and $\Delta Hf^\circ_{298}(g) = -27.86$ kcal/mole. The only previous value is that of Swarts (1458), which is now of only historical interest.

No. 716 m-Difluorobenzene $C_6H_4F_2$ (Ideal Gas State) Mol Wt 114.092

Green, Kynaston, and Paisley (541) made a vibrational assignment, estimated structural constants, and calculated thermodynamic functions.

Aromatic Fluorine Compound Ideal Gas Tables

No. 716   m-Difluorobenzene $C_6H_4F_2$ (Ideal Gas State)   Mol Wt 114.092

$T°K$	$Cp°$	$S°$	$-(G°-H°_{298})/T$	$H°-H°_{298}$	$\Delta Hf°$	$\Delta Gf°$	Log Kp
	cal/(mole °K)			kcal/mole			
298	25.40	76.57	76.57	0.00	-74.09	-61.43	45.026
300	25.54	76.73	76.58	0.05	-74.11	-61.35	44.693
400	32.72	85.10	77.67	2.98	-74.82	-56.98	31.132
500	38.55	93.05	79.96	6.55	-75.36	-52.46	22.928
600	43.13	100.49	82.77	10.64	-75.77	-47.84	17.425
700	46.74	107.42	85.80	15.14	-76.06	-43.15	13.473
800	49.67	113.86	88.91	19.97	-76.28	-38.45	10.502
900	51.99	119.85	92.02	25.05	-76.41	-33.71	8.185
1000	53.94	125.43	95.08	30.35	-76.47	-28.97	6.332

Electron diffraction data by Oosaka, Sekine, and Saito (1088) confirm the estimated structure within experimental error. Good, Lacina, Scott, and McCullough (512) measured the enthalpy of combustion by rotating bomb calorimetry, which yields $\Delta Hf°_{298}(l) = -82.38$ kcal/mole and $\Delta Hf°_{298}(g) = -74.09$ kcal/mole. No information concerning transitions was found in the literature.

No. 717   o-Difluorobenzene $C_6H_4F_2$ (Ideal Gas State)   Mol Wt 114.092

$T°K$	$Cp°$	$S°$	$-(G°-H°_{298})/T$	$H°-H°_{298}$	$\Delta Hf°$	$\Delta Gf°$	Log Kp
	cal/(mole °K)			kcal/mole			
298	25.46	76.94	76.94	0.00	-70.39	-57.84	42.395
300	25.60	77.10	76.95	0.05	-70.41	-57.76	42.078
400	32.76	85.48	78.04	2.98	-71.12	-53.43	29.192
500	38.65	93.44	80.33	6.56	-71.65	-48.94	21.393
600	43.33	100.92	83.15	10.67	-72.04	-44.37	16.161
700	47.08	107.89	86.19	15.19	-72.31	-39.73	12.403
800	50.12	114.38	89.31	20.06	-72.48	-35.07	9.580
900	52.62	120.43	92.44	25.20	-72.57	-30.38	7.378
1000	54.72	126.09	95.52	30.57	-72.55	-25.71	5.619

Scott, Messerly, Todd, Hossenlopp, Osborn, and McCullough (1322) made a complete thermal study, finding $Ttp = 226.01°K$, $\Delta Hm = 2.640$ kcal/mole, $Tb = 367.07°K$, at which temperature $\Delta Hv = 7.699$ kcal/mole, $\Delta Hv°_{298} = 8.65$ kcal/mole, $S°_{298}(g) = 76.94$ cal/(mole °K), and $Cp(l) = 19.52 + 0.062T$[cal/(mole °K), 280–360°K]. They estimated structural constants on the basis of data for fluorobenzene, since the electron diffraction values of Oosaka (1087) appeared less plausible. A vibrational assignment was then made, which gave calculated values in good agreement with experimental work. An independent assignment by Green, Kynaston, and Paisley (541) gave an entropy at 298°K more than 0.3 cal/(mole °K) higher, chiefly because of the use of 196 cm^{-1} in place of 240 cm^{-1} for one fundamental.

Good, Lacina, Scott, and McCullough (512) measured the enthalpy of combustion by rotating bomb combustion calorimetry, which gives $\Delta Hf^\circ_{298}(l) = -79.04$ kcal/mole and $\Delta Hf^\circ_{298}(g) = -70.39$ kcal/mole. Of historical interest is the 50-year-old work of Swarts (1458).

No. 718 p-Difluorobenzene $C_6H_4F_2$ (Ideal Gas State) Mol Wt 114.092

	cal/(mole °K)				kcal/mole		
$T°K$	$Cp°$	$S°$	$-(G°-H°_{298})/T$	$H°-H°_{298}$	$\Delta Hf°$	$\Delta Gf°$	Log Kp
298	25.55	75.43	75.43	0.00	-73.43	-60.43	44.294
300	25.70	75.59	75.44	0.05	-73.45	-60.35	43.963
400	32.84	84.00	76.53	2.99	-74.15	-55.87	30.523
500	38.64	91.97	78.83	6.57	-74.68	-51.23	22.393
600	43.20	99.43	81.65	10.67	-75.08	-46.51	16.941
700	46.80	106.37	84.69	15.18	-75.36	-41.72	13.025
800	49.68	112.82	87.81	20.01	-75.57	-36.91	10.082
900	52.04	118.81	90.93	25.10	-75.71	-32.06	7.786
1000	53.99	124.39	94.00	30.40	-75.75	-27.23	5.950

Thermodynamic functions are calculated on the basis of the vibrational assignment of Stojiljkovic and Whiffen (1426) as extended by Green, Kynaston, and Paisley (541) and by Scott, Messerly, Todd, Hossenlopp, Douslin, and McCullough (1321). The results differ only slightly from the values given by Green, Kynaston, and Paisley (541). The molecular structure determined by Oosaka, Sekine, and Saito (1088) confirms estimated values used by Green et al.

Good, Lacina, Scott, and McCullough (512) measured the enthalpy of combustion by rotating bomb calorimetry, which leads to $\Delta Hf^\circ_{298}(l) = -81.98$ kcal/mole and $\Delta Hf^\circ_{298}(g) = -73.43$ kcal/mole. Timmermans (1501) listed $Tm = 249.5°K$ and $Tb = 362.0°K$.

No. 719 Hexafluorobenzene C_6F_6 (Ideal Gas State) Mol Wt 186.060

	cal/(mole °K)				kcal/mole		
$T°K$	$Cp°$	$S°$	$-(G°-H°_{298})/T$	$H°-H°_{298}$	$\Delta Hf°$	$\Delta Gf°$	Log Kp
298	37.43	91.59	91.59	0.00	-228.64	-210.18	154.058
300	37.56	91.83	91.60	0.07	-228.64	-210.07	153.026
400	43.88	103.53	93.15	4.16	-228.34	-203.91	111.407
500	48.78	113.87	96.28	8.80	-228.03	-197.84	86.473
600	52.55	123.11	100.00	13.87	-227.72	-191.84	69.874
700	55.43	131.44	103.90	19.28	-227.41	-185.88	58.030
800	57.62	138.99	107.82	24.94	-227.11	-179.97	49.163
900	59.32	145.88	111.67	30.79	-226.81	-174.10	42.274
1000	60.63	152.20	115.41	36.79	-226.50	-168.31	36.782

Counsell, Green, Hales, and Martin (279) measured complete thermal data, including $Ttp = 278.25°K$, $\Delta Hm = 2.770$ kcal/mole, $Tb = 353.41°K$, at which temperature $\Delta Hv = 7.571$ kcal/mole, $\Delta Hv°_{298} = 8.61$ kcal/mole, $S°_{298}(l) = 66.90$ cal/(mole °K), and $Cp(l) = 39.54 + 0.045T$ [cal/(mole °K), 280–310°K]. The vibrational assignment of Steele and Whiffen (1409) was changed with respect to one low frequency in order to fit the experimental entropy and vapor heat capacity. Cox, Gundry, and Head (290) measured the enthalpy of combustion, which yields $\Delta Hf°_{298}(l) = -237.25$ kcal/mole and $\Delta Hf°_{298}(g) = -228.64$ kcal/mole. Counsell et al. also derived $Tc = 516.72°K$ and $Pc = 32.61$ atm. Cheng and McCoubrey (219) and Patrick and Prosser (1128) reported critical temperatures in reasonable agreement.

No. 720 α,α,α-Trifluorotoluene $C_7H_5F_3$ (Ideal Gas State)   Mol Wt 146.110

	cal/(mole °K)				kcal/mole		
T°K	Cp°	S°	$-(G°-H°_{298})/T$	$H°-H°_{298}$	$\Delta Hf°$	$\Delta Gf°$	Log Kp
298	31.17	89.05	89.05	0.00	-143.42	-122.20	89.570
300	31.35	89.25	89.06	0.06	-143.44	-122.07	88.923
400	40.59	99.57	90.40	3.67	-144.45	-114.78	62.708
500	48.20	109.48	93.23	8.13	-145.18	-107.27	46.886
600	54.20	118.81	96.73	13.26	-145.70	-99.64	36.293
700	58.94	127.54	100.51	18.92	-146.02	-91.93	28.701
800	62.75	135.66	104.40	25.01	-146.22	-84.20	23.001
900	65.86	143.24	108.30	31.45	-146.29	-76.44	18.561
1000	68.45	150.32	112.16	38.17	-146.23	-68.70	15.014

Scott, Douslin, Messerly, Todd, Hossenlopp, Kincheloe, and McCullough (1301) obtained complete thermal data, including $Ttp = 244.14°K$, $\Delta Hm = 3.294$ kcal/mole, $Tb = 375.20°K$, at which temperature $\Delta Hv° = 7.800$ kcal/mole, $\Delta Hv°_{298} = 8.98$ kcal/mole, and $S°_{298}(l) = 64.89$ cal/(mole °K). Assuming free rotation for the CF_3 group and a reasonable completion of the vibrational assignment of Narasimham, Nielsen, and Theimer (1047), functions were calculated in good agreement with the experimental values.

Good, Scott, and Waddington (514) measured the enthalpy of combustion by rotating bomb calorimetry, which gives $\Delta Hf°_{298}(l) = -152.40$ kcal/mole and $\Delta Hf°_{298}(g) = -143.42$ kcal/mole.

No. 721   p-Fluorotoluene C_7H_7F (Ideal Gas State)   Mol Wt 110.126

Scott, Messerly, Todd, Hossenlopp, Douslin, and McCullough (1321) reported complete thermal data for p-fluorotoluene, including $Ttp =$

No. 721 *p*-Fluorotoluene C_7H_7F (Ideal Gas State) Mol Wt 110.126

$T°K$	$Cp°$	$S°$	$-(G°-H°_{298})/T$	$H°-H°_{298}$	$\Delta Hf°$	$\Delta Gf°$	Log Kp
		cal/(mole °K)			kcal/mole		
298	27.76	81.15	81.15	0.00	-35.38	-16.94	12.419
300	27.93	81.33	81.16	0.06	-35.41	-16.83	12.261
400	36.43	90.55	82.36	3.28	-36.72	-10.43	5.696
500	43.73	99.49	84.90	7.30	-37.78	-3.72	1.628
600	49.70	108.01	88.04	11.98	-38.61	3.16	-1.151
700	54.57	116.05	91.48	17.20	-39.24	10.18	-3.179
800	58.60	123.61	95.03	22.87	-39.70	17.27	-4.717
900	61.98	130.71	98.60	28.90	-40.00	24.41	-5.927
1000	64.84	137.39	102.15	35.25	-40.15	31.57	-6.900

216.49°K, $\Delta Hm = 2.235$ kcal/mole, $Tb = 389.79°K$, at which $\Delta Hv = 8.144$ kcal/mole, $S°_{298}(l) = 56.67$ cal/(mole °K), $Cp(l) = 20.53 + 0.0685T$ [cal/(mole °K), 300–360°K], and $\Delta Hv°_{298} = 9.42$ kcal/mole. Estimated structural constants, a vibrational assignment made by comparison with *p*-difluorobenzene and *p*-xylene, and free rotation for the methyl group gave calculated functions in excellent agreement with the experimental values.

Good, Lacina, Scott, and McCullough (512) measured the enthalpy of combustion by rotating bomb calorimetry, from which $\Delta Hf°_{298}(l) = -44.80$ kcal/mole and $\Delta Hf°_{298}(g) = -35.38$ kcal/mole. The only previous work, reported by Swarts 50 years ago, is of low accuracy.

ALIPHATIC CHLORINE COMPOUND IDEAL GAS TABLES

No. 722 Chloromethane CH_3Cl (Ideal Gas State) Mol Wt 50.491

$T°K$	$Cp°$	$S°$	$-(G°-H°_{298})/T$	$H°-H°_{298}$	$\Delta Hf°$	$\Delta Gf°$	Log Kp
		cal/(mole °K)			kcal/mole		
298	9.74	56.04	56.04	0.00	-20.63	-15.03	11.017
300	9.77	56.11	56.05	0.02	-20.64	-15.00	10.924
400	11.52	59.16	56.45	1.09	-21.28	-13.01	7.110
500	13.19	61.91	57.27	2.33	-21.84	-10.88	4.756
600	14.66	64.45	58.26	3.72	-22.30	-8.65	3.149
700	15.93	66.81	59.31	5.25	-22.69	-6.34	1.979
800	17.04	69.01	60.39	6.90	-23.00	-3.98	1.088
900	18.01	71.07	61.46	8.65	-23.25	-1.59	0.386
1000	18.86	73.01	62.52	10.50	-23.43	0.83	-0.182

Lacher, Emery, Bohmfalk, and Park (824) measured the enthalpy of hydrogenation of chloromethane to methane and hydrogen chloride and derived $\Delta Hf°_{298}(g) = -20.63$ kcal/mole, which is adopted here. Rossini,

Wagman, Evans, Levine, and Jaffe (1249) listed $\Delta H f°_{298}(g) = -19.6$ kcal/mole, based on the old results of Thomsen (1495). Margrave (929) calculated $\Delta H f°_{298}(g) = -18.0$ kcal/mole from mass spectrometer appearance potentials. Messerly and Aston (991) measured low-temperature thermal data, including $Ttp = 175.43°$K, $\Delta Hm = 1.537$ kcal/mole, $Tb = 248.93°$K, for which $\Delta Hv = 5.147$ kcal/mole, and $S°_{298}(g) = 55.94$ cal/(mole °K). Cerny and Erdos (207) calculated the thermodynamic functions given here, using microwave structural constants of Miller, Aamodt, Dousmanis, Townes, and Kraitchman (1008) and frequencies listed by Herzberg (589). The calculated $S°_{298}(g) = 56.04$ cal/(mole °K) is in excellent agreement with the third-law result. Bartell and Brockway (79) determined bond lengths and angles by electron diffraction, in good agreement with the microwave values.

Other recent calculations of thermodynamic functions have been made by Gelles and Pitzer (465), Sverdlin (1451), and Kobe and Crawford (772). Whytlaw-Gray, Reeves, and Bottomley (1606) measured vapor densities and calculated virial coefficients. Hsu and McKetta (624) gave tables of thermodynamic properties in engineering units.

No. 723   Dichloromethane CH_2Cl_2 (Ideal Gas State)   Mol Wt 84.940

	cal/(mole °K)				kcal/mole		
T°K	$Cp°$	$S°$	$-(G°-H°_{298})/T$	$H°-H°_{298}$	$\Delta H f°$	$\Delta G f°$	Log Kp
298	12.22	64.59	64.59	0.00	-22.80	-16.46	12.063
300	12.26	64.67	64.60	0.03	-22.81	-16.42	11.961
400	14.69	68.56	65.11	1.39	-23.22	-14.22	7.770
500	15.87	71.96	66.15	2.91	-23.57	-11.93	5.215
600	17.36	74.99	67.37	4.58	-23.85	-9.58	3.489
700	18.47	77.75	68.66	6.37	-24.06	-7.18	2.242
800	19.38	80.28	69.96	8.26	-24.22	-4.76	1.301
900	20.15	82.61	71.24	10.24	-24.33	-2.32	0.563
1000	20.80	84.77	72.48	12.29	-24.40	0.13	-0.028

Gelles and Pitzer (465) calculated thermodynamic functions using the vibrational assignment of Plyler and Benedict (1172) and moments of inertia from the microwave study of Myers and Gwinn (1042). Gelles and Pitzer gave references to earlier calculations, while Sverdlin (1451) has since published similar values. Perlick (1138) and Riedel (1228, 1229) are in reasonable agreement on the liquid heat capacity as $Cp(l) = 17.16 + 0.023T$ [cal/(mole °K), 270–320°K]. Kurbatov (814) also measured liquid heat capacities, but his values are far too high. Mathews (940) measured $\Delta Hv = 6.69$ kcal/mole at 313.7°K. Using heat capacities of liquid and gas discussed above $\Delta Hv°_{298}$ is computed as 6.87 kcal/mole.

Dreisbach (352) reported $Tm = 178.01°K$, while Rossini, Wagman, Evans, Levine, and Jaffe (1249) listed $\Delta Hm = 1.1$ kcal/mole. Perez Masiá and Diaz Peña (1137) measured molar vapor volumes in the range 323–423°K. Dreisbach (352) reported $Tb = 312.90°K$ and constants for the Antoine vapor pressure equation, which, with the vapor volumes of Perez Masiá and Diaz Peña, yield $\Delta Hv°_{298} = 6.89$ kcal/mole by use of the exact Clapeyron equation. This is in excellent agreement with the calorimetric value discussed above.

The energy of combustion reported by Smith, Bjellerup, Krook, and Westermark (1384), corrected for a new enthalpy of oxidation for arsenious oxide (142) and converted to ΔHc_{298}, yields $\Delta Hf°_{298}(l) = -29.2$ kcal/mole. The $\Delta Hv°_{298}$ calculated above then gives $\Delta Hf°_{298}(g) = -22.3$ kcal/mole. Amador, Lacher, and Park (11) reported the enthalpy of hydrogenation of dichloromethane to methane and hydrogen chloride at 250°C as -40.07 kcal/mole. Correction to 298°K using a ΔCp of reaction of -5 cal/(mole °K) yields $\Delta Hf°_{298}(g) = -23.05$ kcal/mole. A weighted average of -22.8 kcal/mole is adopted.

No. 724 Chloroform CHCl$_3$ (Ideal Gas State) Mol Wt 119.389

	cal/(mole °K)				kcal/mole		
$T°K$	$Cp°$	$S°$	$-(G°-H°_{298})/T$	$H°-H°_{298}$	$\Delta Hf°$	$\Delta Gf°$	$\log Kp$
298	15.71	70.66	70.66	0.00	-24.20	-16.38	12.004
300	15.76	70.76	70.67	0.03	-24.20	-16.33	11.895
400	17.83	75.59	71.31	1.72	-24.36	-13.68	7.472
500	19.34	79.74	72.59	3.58	-24.44	-10.99	4.806
600	20.44	83.37	74.09	5.57	-24.48	-8.30	3.024
700	21.27	86.58	75.65	7.66	-24.49	-5.60	1.749
800	21.91	89.47	77.20	9.82	-24.47	-2.91	0.795
900	22.43	92.08	78.71	12.04	-24.43	-0.21	0.052
1000	22.86	94.47	80.17	14.30	-24.37	2.47	-0.540

Gelles and Pitzer (465) calculated the thermodynamic functions given here, using the vibrational assignment of Plyler and Benedict (1172) and moments of inertia approximating those recently determined by Jen and Lide (684) by microwave techniques. Functions nearly identical with those of Gelles and Pitzer were reported by Černý and Erdos (207), Madigan and Cleveland (912), and Sverdlin (1451). Staveley, Tupman, and Hart (1407), Harrison and Moelwyn-Hughes (572), and Richards and Wallace (1225) measured liquid heat capacities in reasonable agreement. Their results in the range 240–330°K. can be expressed as $Cp(l) = 21.86 + 0.018T$ [cal/(mole °K)]. Kurbatov (814) also measured liquid heat capacities, but his results are known to be of low accuracy. Dreisbach (352) reported $Tm = 209.56°K$ and $Tb = 334.88°K$, while Mathews (940) measured $\Delta Hv_{334.4} = 7.02$ kcal/mole. The heat capacity data previously adopted yield $\Delta Hv°_{298} = 7.44$ kcal/mole. Kolossovsky and Alimov (799) and Kusano (816) also reported enthalpy of vaporization data, but both appear too high.

Smith, Bjellerup, Krook, and Westermark (1384) reported energy of combustion data, which are corrected for a new enthalpy of oxidation of arsenious oxide (142) and adjusted to ΔHc_{298} to give $\Delta Hf°_{298}(l) = -32.3$ kcal/mole. Kirkbride (754) measured the enthalpy of reaction of chloroform and chlorine to give carbon tetrachloride and hydrogen chloride. Combination of his result with other data for carbon tetrachloride yields $\Delta Hf°_{298}(l) = -30.5$ kcal/mole. A weighted average of $\Delta Hf°_{298}(l) = -31.6$ kcal/mole is adopted. This and the enthalpy of vaporization of Mathews discussed above then yield $\Delta Hf°_{298}(g) = -24.2$ kcal/mole.

No. 725 Carbon Tetrachloride CCl$_4$ (Ideal Gas State) Mol Wt 153.838

	cal/(mole °K)				kcal/mole		
$T°K$	$Cp°$	$S°$	$-(G°-H°_{298})/T$	$H°-H°_{298}$	$\Delta Hf°$	$\Delta Gf°$	Log Kp
298	20.02	74.12	74.12	0.00	-24.00	-13.92	10.201
300	20.08	74.25	74.13	0.04	-24.00	-13.85	10.093
400	22.04	80.31	74.94	2.15	-23.79	-10.50	5.738
500	23.28	85.37	76.53	4.42	-23.55	-7.21	3.150
600	24.07	89.69	78.37	6.79	-23.29	-3.96	1.444
700	24.61	93.44	80.27	9.23	-23.04	-0.76	0.238
800	25.00	96.75	82.12	11.71	-22.79	2.40	-0.656
900	25.29	99.72	83.92	14.23	-22.54	5.53	-1.344
1000	25.52	102.39	85.63	16.77	-22.29	8.64	-1.888

Smith, Bjellerup, Krook, and Westermark (1384) reported an energy of combustion that, when corrected for a new enthalpy of oxidation of arsenious oxide (142) and adjusted to ΔHc_{298}, gives $\Delta Hf°_{298}(l) = -29.9$ kcal/mole. Only 4% of the measured enthalpy in each experiment resulted from combustion of the tetrachloride; the rest was due to auxiliary substances. Smith et al. also gave data for chloroform, which, when similarly corrected, give $\Delta Hf°_{298}(l) = -32.3$ kcal/mole; Kirkbride measured the enthalpy of reaction of chlorine and chloroform producing the tetrachloride and hydrogen chloride as -22.3 kcal/mole; combination of these data yields, for tetrachloromethane, $\Delta Hf°_{298}(l) = -32.5$ kcal/mole. Enthalpy of vaporization at 298°K, discussed later, gives $\Delta Hf°_{298}(g) = -22.2$ and -24.8 kcal/mole, respectively.

Bodenstein, Günther, and Hoffmeister (144) measured the enthalpy of explosion of mixtures of tetrachloromethane vapor, chlorine, and hydrogen, deriving $\Delta Hr = -62.6$ kcal/mole for the reaction of the tetrachloride and hydrogen to produce carbon and hydrogen chloride. The assumption that the carbon product has an enthalpy of formation of 1-2 kcal/mole (1059) yields $\Delta Hf^\circ_{298}(g) = -23.9$ to -24.9 kcal/mole for CCl_4. A rounded average of these values, $\Delta Hf^\circ_{298}(g) = -24.0$ kcal/mole, is adopted.

Hicks, Hooley, and Stephenson (594) measured low-temperature thermal data, including $Tt = 225.3°K$, $\Delta Ht = 1.095$ kcal/mole, $Ttp = 250.3°K$, and $\Delta Hm = 0.601$ kcal/mole. Earlier work by Lord and Blanchard (884) is in fair agreement with that of Stull (1431). Hildenbrand and McDonald (598) revised the calculations of Hicks et al. for extrapolation from 15° to 0°K and measured $\Delta Hv^\circ_{298} = 7.75$ kcal/mole, deriving $S^\circ_{298}(g) = 73.92$ cal/(mole °K), in agreement within experimental error with values based on spectroscopic data. Dunlop (359) carefully determined $Tm = 250.4°K$, while Timmermans (1501) listed several references in excellent agreement at $Tb = 349.9°K$. Pitzer (1154) measured $\Delta Hv = 7.17$ kcal/mole at 349.8°K, as well as vapor heat capacities. Mathews (940) determined $\Delta Hv = 7.16$ kcal/mole at 349.5°K, in good agreement. Kusano (816) measured $\Delta Hv_{303} = 7.72$ kcal/mole, in good agreement with Hildenbrand and McDonald. The enthalpies of vaporization are consistent with heat capacities of liquid and vapor. Staveley, Tupman, and Hart (1407), Harrison and Moelwyn-Hughes (572), and Richards and Wallace (1225) all reported liquid heat capacities in reasonable agreement with each other and with Hicks, Hooley, and Stephenson (594). Experimental points from 250° to 340°K can be fitted to within 1% by the equation $Cp(l) = 27.65 + 0.013T$ [cal/(mole °K)].

Numerous calculations of thermodynamic functions from spectroscopic values include those of Stevenson and Beach (1419), Vold (1545), Černý and Erdos (207), Sverdlin (1451), Gelles and Pitzer (465), and Madigan and Cleveland (912). Values adopted here are taken from Albright, Galegar, and Innes (8), who applied approximate anharmonicity corrections on the basis of vapor heat capacities of dichlorodifluoromethane reported by Masi (935).

No. 726 Chloroethane C_2H_5Cl (Ideal Gas State) Mol Wt 64.517

$T°K$	$Cp°$	$S°$	$-(G°-H°_{298})/T$	$H°-H°_{298}$	$\Delta Hf°$	$\Delta Gf°$	$\text{Log } Kp$
	cal/(mole °K)				kcal/mole		
298	14.99	65.93	65.93	0.00	-26.70	-14.34	10.509
300	15.05	66.03	65.94	0.03	-26.72	-14.26	10.389
400	18.56	70.85	66.57	1.72	-27.68	-9.96	5.440
500	21.68	75.33	67.88	3.73	-28.47	-5.43	2.374
600	24.31	79.52	69.47	6.04	-29.11	-0.76	0.278
700	26.53	83.44	71.19	8.58	-29.61	4.00	-1.250
800	28.42	87.11	72.96	11.33	-29.99	8.83	-2.411
900	30.06	90.56	74.72	14.26	-30.26	13.70	-3.327
1000	31.48	93.80	76.47	17.33	-30.43	18.60	-4.064

Gordon and Giauque (518) measured low-temperature thermal properties, including $Ttp = 134.8°K$, $\Delta Hm = 1.064$ kcal/mole, and $Tb = 285.4°K$, at which $\Delta Hv = 5.892$ kcal/mole. Liquid heat capacities were fitted in the range 260–290°K by $Cp(l) = 16.57 + 0.28T$ [cal/(mole °K)]. The experimental entropy at 285.4°K was $S°(g) = 65.31$ cal/(mole °K). Schwendeman and Jacobs (1296) derived moments of inertia by microwave spectroscopy and calculated the barrier to internal rotation as 3685 cal/mole from rotational splitting measurements of Lide (867). Using far infrared spectroscopy, Fateley and Miller (410) derived a barrier that agrees well with the result of Schwendeman and Jacobs. These values and the vibrational assignment of Daasch, Liang, and Nielsen (302) are used to calculate thermodynamic functions, including $S°(g) = 65.25$ cal/(mole °K) at 285.4°K, in excellent agreement with the third-law value. Green and Holden (540) calculated functions only slightly different, using earlier microwave data by Wagner and Dailey (1565) and the same vibrational assignment.

Lacher, Emery, Bohmfalk, and Park (824) measured the enthalpy of reaction of chloroethane and hydrogen to form ethane and hydrogen chloride, obtaining $\Delta Hf°_{298}(g) = -25.8$ kcal/mole. Casey and Fordham (199) made enthalpy of combustion measurements leading to $\Delta Hf°_{298}(g) = -23.6$ kcal/mole. Equilibrium constants in the reaction of ethylene and hydrogen chloride, reported by Lane, Linnett, and Oswin (845), Howlett (620), and Rudkovskii, Trifel, and Frost (1262), yielded $\Delta Hf°_{298}(g) = -26.7, -26.9,$ and -27.0 kcal/mole, respectively. The value selected from the equilibrium data is -26.7 kcal/mole.

Eucken and Franck (383) measured vapor heat capacities in good agreement with calculated values. Kobe and Harrison (776) and Kobe and Crawford (772) calculated thermodynamic properties using the vibrational assignment and barrier of 4700 cal/mole proposed by Daasch, Liang, and Nielsen (302).

No. 727 1,1-Dichloroethane $C_2H_4Cl_2$ (Ideal Gas State) Mol Wt 98.966

	cal/(mole °K)				kcal/mole		
T°K	$Cp°$	$S°$	$-(G°-H°_{298})/T$	$H°-H°_{298}$	$\Delta Hf°$	$\Delta Gf°$	Log Kp
298	18.22	72.89	72.89	0.00	-31.05	-17.47	12.806
300	18.29	73.01	72.90	0.04	-31.07	-17.39	12.667
400	21.85	78.77	73.66	2.05	-31.77	-12.72	6.948
500	24.82	83.98	75.21	4.39	-32.31	-7.89	3.448
600	27.24	88.72	77.07	7.00	-32.73	-2.96	1.080
700	29.21	93.07	79.05	9.82	-33.04	2.02	-0.632
800	30.85	97.08	81.06	12.83	-33.25	7.04	-1.924
900	32.25	100.80	83.05	15.98	-33.38	12.09	-2.936
1000	33.45	104.26	85.00	19.27	-33.43	17.15	-3.748

Li and Pitzer (862) reported low-temperature thermal data, including $Ttp = 176.18°K$, $\Delta Hm = 1.881$ kcal/mole, and $\Delta Hv_{293} = 7.409$ kcal/mole. Their smoothed heat capacities in the range 270–300°K can be expressed as $Cp(l) = 24.22 + 0.020T$ [cal/(mole °K)]. The derived third-law entropy, $S°_{298}(g) = 72.89$ cal/(mole °K), is in exact agreement with our calculated value based on the vibrational assignment of Daasch, Liang, and Nielsen (302) and the molecular structure from the electron diffraction work of Danford and Livingston (305). The torsional frequency selected by Daasch et al. implies a barrier to internal rotation of 3750 cal/mole.

Rossini, Wagman, Evans, Levine, and Jaffe (1249) listed $\Delta Hf°_{298}(l) = -36.4$ kcal/mole, based on very old combustion data. Smith, Bjellerup, Krook, and Westermark (1384) reported the enthalpy of combustion, which, when corrected for a new value of the enthalpy of oxidation of arsenious oxide (142) and adjusted to 1 atm pressure at 298°K, yields $\Delta Hf°_{298}(l) = -38.37$ kcal/mole. The enthalpy of vaporization measured by Li and Pitzer corrected to 298°K gives $\Delta Hf°_{298}(g) = -31.0$ kcal/mole.

Amador, Lacher, and Park (11) measured the enthalpy of hydrogenation of dichloroethane to ethane and hydrogen chloride at 250°C as -34.65 kcal/mole. Adjustment to 298°K yields $\Delta Hf°_{298}(g) = -31.09$ kcal/mole, in excellent agreement with the combustion result. An average of -31.05 kcal/mole is adopted. Timmermans (1501) listed boiling point data that average $Tb = 330.4°K$. The enthalpy of vaporization data of Li and Pitzer (862) are extrapolated to yield $\Delta Hv = 6.97$ kcal/mole at the boiling point.

No. 728   1,2-Dichloroethane $C_2H_4Cl_2$ (Ideal Gas State)   Mol Wt 98.966

	cal/(mole °K)				kcal/mole		
T°K	$Cp°$	$S°$	$-(G°-H°_{298})/T$	$H°-H°_{298}$	$\Delta Hf°$	$\Delta Gf°$	Log Kp
298	18.80	73.66	73.66	0.00	-31.00	-17.65	12.938
300	19.00	73.78	73.67	0.04	-31.01	-17.57	12.799
400	22.00	79.67	74.45	2.09	-31.67	-12.98	7.093
500	24.70	84.87	76.02	4.43	-32.22	-8.25	3.604
600	26.90	89.58	77.89	7.01	-32.66	-3.41	1.242
700	28.80	93.87	79.87	9.80	-33.00	1.50	-0.467
800	30.40	97.82	81.87	12.76	-33.26	6.44	-1.759
900	31.80	101.49	83.85	15.87	-33.44	11.42	-2.772
1000	33.00	104.90	85.79	19.12	-33.53	16.41	-3.586

Conn, Kistiakowsky, and Smith (259) measured the enthalpy of addition of chlorine to ethylene at 355°K at -43.65 kcal/mole. ΔCp for the reaction is 0.5 cal/(mole °K), and at 298°K $\Delta Hr = -43.68$ kcal/mole. Combination of this value with the enthalpy of formation of ethylene listed by Rossini, Wagman, Evans, Levine, and Jaffe (1249) yields

$\Delta Hf^\circ_{298}(g) = -31.18$ kcal/mole. Sinke and Stull (1359), reporting the enthalpy of combustion of the liquid, derived $\Delta Hf^\circ_{298}(l) = -39.57$ kcal/mole and also corrected an earlier ΔEc value of Smith, Bjellerup, Krook, and Westermark (1384) to give $\Delta Hf^\circ_{298}(l) = -38.52$ kcal/mole. The enthalpy of vaporization at 298°K of 8.44 kcal/mole quoted by Gwinn and Pitzer (556) is close to the 8.47 kcal/mole given by Sinke and Stull and is used here for consistency. Enthalpies of formation of the ideal gas are then -30.08 kcal/mole from Smith et al. and -31.13 kcal/mole from Sinke and Stull. Amador, Lacher, and Park (11) measured the enthalpy of reaction with hydrogen to form ethane and hydrogen chloride at 250°C as -35.32 kcal/mole. After adjustment to 298°K, $\Delta Hf^\circ_{298}(g) = -30.47$ kcal/mole may be deduced. A weighted average of $\Delta Hf^\circ_{298}(g) = -31.0$ kcal/mole is adopted.

Pitzer (1151) measured low-temperature thermal data, finding $Ttp = 237.2°K$, $\Delta Hm = 2.112$ kcal/mole, and $S^\circ_{298}(l) = 49.84$ cal/(mole °K). Pitzer also cited the less accurate values of Railing (1210) for comparison. Gwinn and Pitzer (556) calculated $S^\circ_{298}(g) = 73.66$ cal/(mole °K) from low-temperature thermal data and also determined vapor heat capacities. The heat capacity of the ideal gas was calculated at several temperatures from spectroscopic data, with a potential function for internal rotation selected to fit the experimental results. A smooth curve drawn through both experimental and calculated points is used to derive the values given here.

Heat capacity of the liquid was measured by Pitzer (1151), Staveley, Tupman, and Hart (1407), and Ruiter (1268), and can be represented by $Cp(l)$ (280–350°K) $= 21.05 + 0.033T$ [cal/(mole °K)]. Dreisbach (352) listed $Tb = 356.62°K$, while Mathews (940) measured $\Delta Hv = 7.65$ kcal/mole.

No. 729 1,1,2-Trichloroethane $C_2H_3Cl_3$ (Ideal Gas State) Mol Wt 133.415

T°K	cal/(mole °K)				kcal/mole		Log Kp
	$Cp°$	$S°$	$-(G°-H°_{298})/T$	$H°-H°_{298}$	$\Delta Hf°$	$\Delta Gf°$	
298	21.27	80.57	80.57	0.00	-33.10	-18.52	13.575
300	21.34	80.71	80.58	0.04	-33.11	-18.43	13.426
400	25.03	87.37	81.46	2.37	-33.57	-13.46	7.355
500	27.92	93.28	83.24	5.02	-33.88	-8.40	3.670
600	30.13	98.57	85.36	7.93	-34.08	-3.28	1.196
700	31.87	103.35	87.60	11.03	-34.19	1.86	-0.581
800	33.28	107.70	89.84	14.29	-34.24	7.01	-1.915
900	34.44	111.69	92.05	17.68	-34.23	12.17	-2.955
1000	35.42	115.37	94.20	21.17	-34.16	17.32	-3.786

Sinke (1354) measured the enthalpy of combustion using the techniques described by Sinke and Stull (1359), with the result $\Delta Hf^\circ_{298}(l) = -42.7$

kcal/mole. Kirkbride (754) reported the enthalpy of chlorination of 1,2-dichloroethane to 1,1,2-trichloroethane and hydrogen chloride as -27.8 kcal/mole, which yields $\Delta H f^\circ_{298}(l) = -45.3$ kcal/mole when combined with the well-established values for 1,2-dichloroethane and hydrogen chloride. As pointed out by Kirkbride, his result appears to be too high, since it gives too large a heat of addition of chlorine to vinyl chloride compared to other similar reactions. The combustion measurement is therefore accepted. Williamson and Harrison (1617) reported enthalpy of vaporization data in the range 57–85°C, which are extrapolated to give $\Delta H v^\circ_{298} = 9.57$ kcal/mole and $\Delta H f^\circ_{298}(g) = -33.1$ kcal/mole. Williamson and Harrison also reported the vapor heat capacity data employed by Harrison and Kobe (574) to derive a satisfactory vibrational assignment and barrier to internal rotation for the calculation of thermodynamic functions. Timmermans and Roland (1505) determined $Tm = 236.5°K$ and $Tb = 387.2°K$, while Crowe and Smyth (296) measured $\Delta Hm = 2.7$ kcal/mole. An extrapolation of the data of Williamson and Harrison gives $\Delta H v = 8.3$ kcal/mole at the boiling point.

No. 730   1,1,2,2-Tetrachloroethane $C_2H_2Cl_4$ (Ideal Gas State)   Mol Wt 167.864

T°K	$Cp°$	$S°$	$-(G°-H°_{298})/T$	$H°-H°_{298}$	$\Delta Hf°$	$\Delta Gf°$	Log Kp
		cal/(mole °K)			kcal/mole		
298	24.09	86.69	86.69	0.00	-36.50	-20.45	14.992
300	24.16	86.84	86.70	0.05	-36.51	-20.35	14.827
400	27.90	94.33	87.69	2.66	-36.74	-14.93	8.157
500	30.77	100.88	89.69	5.60	-36.85	-9.46	4.135
600	32.91	106.68	92.04	8.79	-36.85	-3.98	1.451
700	34.55	111.88	94.51	12.16	-36.78	1.49	-0.466
800	35.85	116.58	96.98	15.69	-36.66	6.95	-1.898
900	36.90	120.87	99.40	19.32	-36.48	12.39	-3.008
1000	37.76	124.80	101.75	23.06	-36.26	17.61	-3.892

Smith, Bjellerup, Krook, and Westermark (1384) reported an energy of combustion, which, after correction for the new enthalpy of oxidation of arsenious oxide (142) and conversion to $\Delta H c_{298}$, yields $\Delta H f^\circ_{298}(l) = -47.2$ kcal/mole. Dreisbach (353) listed $\Delta H v^\circ_{298} = 10.7$ kcal/mole, from which $\Delta H f^\circ_{298}(g) = -36.5$ kcal/mole. Dreisbach also listed $Tm = 229.4°K$ and $Tb = 419.4°K$, while Mathews (940) measured $\Delta H v = 9.24$ kcal/mole at 419°K. Kobe and Harrison (776) calculated thermodynamic functions, assuming the rotational isomers to be of equal energy. The functions are relatively uncertain, since the rotational isomers actually differ in energy, as shown by spectroscopic studies of Naito, Nakagawa, Kuratani, Ichishima, and Mizushima (1044), Shimanouchi (1345), and Zietlow, Cleveland, and Meister (1645).

No. 731 Pentachloroethane C₂HCl₅ (Ideal Gas State) Mol Wt 202.313

T°K	Cp°	S°	−(G°−H°₂₉₈)/T	H°−H°₂₉₈	ΔHf°	ΔGf°	Log Kp
	cal/(mole °K)				kcal/mole		
298	28.14	90.95	90.95	0.00	−34.00	−15.93	11.678
300	28.23	91.13	90.96	0.06	−34.00	−15.82	11.525
400	31.77	99.76	92.11	3.06	−33.91	−9.77	5.338
500	34.42	107.15	94.39	6.38	−33.71	−3.76	1.641
600	36.22	113.59	97.07	9.92	−33.45	2.21	−0.805
700	37.53	119.27	99.84	13.61	−33.15	8.13	−2.539
800	38.52	124.35	102.59	17.41	−32.84	14.01	−3.826
900	39.26	128.93	105.27	21.30	−32.50	19.84	−4.818
1000	39.88	133.10	107.85	25.26	−32.15	25.64	−5.603

Nielsen, Liang, and Daasch (1071) studied the spectrum, made a vibrational assignment, and calculated thermodynamic functions at selected temperatures. Kobe and Harrison (776) made more complete calculations using the same assignment. Smith, Bjellerup, Krook, and Westermark (1384) reported energy of combustion data, which are corrected for the new enthalpy of oxidation of arsenious oxide (142) and adjusted to ΔHc_{298} to yield $\Delta Hf°_{298}(l) = −45.3$ kcal/mole. Rossini, Wagman, Evans, Levine, and Jaffe (1249) listed $Tm = 244.2°K$, $\Delta Hm = 2.7$ kcal/mole, $Tb = 432.5°K$, and $\Delta Hv = 9.7$ kcal/mole at the boiling point. Estimating ΔCp of vaporization as $−12$ cal/(mole °K) yields $\Delta Hv°_{298} = 11.3$ kcal/mole and $\Delta Hf°_{298}(g) = −34.0$ kcal/mole. Kirkbride (754) measured the enthalpy of addition of chlorine to trichloroethylene. His data are consistent with the combustion results within experimental error. Kurbatov (814) reported liquid heat capacity data that appear to be far too high.

No. 732 Hexachloroethane C₂Cl₆ (Ideal Gas State) Mol Wt 236.762

T°K	Cp°	S°	−(G°−H°₂₉₈)/T	H°−H°₂₉₈	ΔHf°	ΔGf°	Log Kp
	cal/(mole °K)				kcal/mole		
298	32.59	94.77	94.77	0.00	−33.80	−13.58	9.954
300	32.67	94.98	94.78	0.07	−33.79	−13.45	9.801
400	36.11	104.88	96.10	3.51	−33.33	−6.74	3.684
500	38.29	113.18	98.71	7.24	−32.80	−0.15	0.068
600	39.69	120.29	101.73	11.14	−32.26	6.32	−2.302
700	40.63	126.49	104.83	15.16	−31.72	12.71	−3.968
800	41.29	131.96	107.89	19.26	−31.20	19.02	−5.195
900	41.76	136.85	110.84	23.41	−30.69	25.26	−6.135
1000	42.11	141.27	113.67	27.61	−30.19	31.45	−6.873

Carney, Piotrowski, Meister, Braun, and Cleveland (192) used their vibrational assignment and the electron diffraction results of Swick,

Karle, and Karle (1460) in calculating thermodynamic functions. Kobe and Harrison (776) published calculations in good agreement. Smith, Bjellerup, Krook, and Westermark (1384) reported the energy of combustion, which is corrected for a new value of the enthalpy of oxidation of arsenious oxide (142) and converted to ΔHc_{298} to give $\Delta Hf°_{298}(s) = -49.3$ kcal/mole. Rossini, Wagman, Evans, Levine, and Jaffe (1249) listed $Tt = 344.4°K$, $\Delta Ht = 1.9$ kcal/mole, $Tm = 459.9°K$, $\Delta Hm = 2.33$ kcal/mole, and $Ts = 457.6°K$, at which temperature $\Delta Hs = 12.2$ kcal/mole. If ΔCp of sublimation is taken as -9 cal/(mole °K), $\Delta Hs°_{298}$ is calculated as 15.5 kcal/mole and $\Delta Hf_{298}(g)$ as -33.8 kcal/mole. Kirkbride (754) measured the enthalpy of addition of chlorine to tetrachloroethylene. His data are consistent with enthalpy of combustion data on these compounds.

No. 733 1-Chloropropane C_3H_7Cl (Ideal Gas State) Mol Wt 78.543

	cal/(mole °K)				kcal/mole		
$T°K$	$Cp°$	$S°$	$-(G°-H°_{298})/T$	$H°-H°_{298}$	$\Delta Hf°$	$\Delta Gf°$	Log Kp
298	20.24	76.27	76.27	0.00	-31.10	-12.11	8.876
300	20.34	76.40	76.28	0.04	-31.13	-11.99	8.736
400	25.36	82.96	77.14	2.33	-32.42	-5.41	2.956
500	29.73	89.10	78.92	5.09	-33.49	1.47	-0.643
600	33.43	94.85	81.10	8.26	-34.34	8.54	-3.111
700	36.57	100.25	83.45	11.76	-35.01	15.75	-4.917
800	39.24	105.31	85.87	15.55	-35.51	23.03	-6.291
900	41.57	110.07	88.30	19.60	-35.86	30.37	-7.375
1000	43.59	114.55	90.70	23.86	-36.07	37.75	-8.250

Green and Holden (540) calculated thermodynamic properties using a reasonable vibrational assignment, conventional bond lengths and angles, methyl and skeletal barriers to internal rotation of 3100 and 2900 cal/mole, respectively, and an energy difference of 50 cal/mole between *trans* and *gauche* rotational isomers. Microwave spectroscopy by Sarachman (1276) and electron diffraction data of Morino and Kuchitsu (1026) largely confirm the above estimates; hence the values of Green and Holden are accepted.

Smith, Bjellerup, Krook, and Westermark (1384) misquoted an enthalpy of combustion value of Eftring (370). The correct result, adjusted for a new enthalpy of oxidation of arsenious oxide (142) and converted to ΔHc_{298}, yields $\Delta Hf_{298}(l) = -37.5$ kcal/mole. Mathews and Fehlandt (941) measured $\Delta Hv = 6.62$ kcal/mole at 319°K. If ΔCp of vaporization is estimated as -12 cal/(mole °K), $\Delta Hv°_{298}$ is found to be 6.9 kcal/mole and $\Delta Hf_{298}(g) = -30.6$ kcal/mole. Davies, Lacher, and Park (308) measured the enthalpy of reaction of 1-chloropropane and hydrogen to form propane and hydrogen chloride and derived $\Delta Hf^{\circ}_{298}(g) = -31.54$ kcal/mole. An average of -31.1 kcal/mole is adopted.

Transition temperature data listed by Timmermans (1501) are not very concordant. Values determined in Timmermans' laboratory are $Tm = 150.35°K$ and $Tb = 319.75°K$. Perez Masiá and Diaz Peña (1137) reported second virial coefficients for the vapor in the range 323–398°K.

No. 734 2-Chloropropane C_3H_7Cl (Ideal Gas State) Mol Wt 78.543

	cal/(mole °K)				kcal/mole		
$T°K$	$Cp°$	$S°$	$-(G°-H°_{298})/T$	$H°-H°_{298}$	$\Delta Hf°$	$\Delta Gf°$	Log Kp
298	20.87	72.70	72.70	0.00	-35.00	-14.94	10.954
300	20.97	72.83	72.71	0.04	-35.03	-14.82	10.797
400	25.99	79.57	73.59	2.40	-36.26	-7.89	4.312
500	30.48	85.86	75.42	5.23	-37.26	-0.68	0.297
600	34.20	91.76	77.66	8.47	-38.04	6.71	-2.443
700	37.30	97.27	80.07	12.04	-38.62	14.22	-4.439
800	39.94	102.43	82.55	15.91	-39.05	21.79	-5.952
900	42.20	107.26	85.03	20.02	-39.33	29.42	-7.143
1000	44.16	111.81	87.48	24.34	-39.49	37.07	-8.102

Thermodynamic functions are calculated using the vibrational assignment of Sheppard (1344), the average bond distances recommended by Sutton (1449), and an estimate of 4.0 kcal/mole for the barrier to internal rotation. Electron diffraction results of Yamaha (1637) agree with the bond distances of Sutton within experimental error. The calculated functions are used with equilibrium constants of dehydrochlorination measured by Howlett (620) to derive $\Delta Hf^{\circ}_{298}(g) = -35.2$ kcal/mole. Less precise equilibrium data of Entelis and Chirkov (379) yield $\Delta Hf^{\circ}_{298}(g) = -34.6$ kcal/mole. Smith, Bjellerup, Krook, and Westermark (1384) reported energy of combustion data that yield $\Delta Hf^{\circ}_{298}(l) = -42.3$ kcal/mole when corrected for a new enthalpy of oxidation of arsenious oxide (142) and adjusted to ΔHc_{298}. Mathews and Fehlandt (941) determined $\Delta Hv = 6.34$ kcal/mole at 309°K. An estimate for ΔCp of vaporization of -12 cal/(mole °K) gives $\Delta Hv^{\circ}_{298} = 6.47$ kcal/mole. Combination of this value with the result of Smith et al. yields $\Delta Hf^{\circ}_{298}(g) = -35.8$ kcal/mole, in fair agreement with the equilibrium results. Davies, Lacher, and Park (308) determined the enthalpy of reaction with hydrogen to form propane and hydrogen chloride and derived $\Delta Hf^{\circ}_{298}(g) = -32.93$ kcal/mole, which appears too positive. A weighted average of -35.0 kcal/mole is adopted.

Timmermans (1501) listed $Tm = 156.2°K$ and $Tb = 308.0°K$. Perez Masiá and Diaz Peña (1137) measured molar vapor volumes in the range 313–398°K.

No. 735 1,2-Dichloropropane $C_3H_6Cl_2$ (Ideal Gas State) Mol Wt 112.992

	cal/(mole °K)				kcal/mole		
$T°K$	$Cp°$	$S°$	$-(G°-H°_{298})/T$	$H°-H°_{298}$	$\Delta Hf°$	$\Delta Gf°$	Log Kp
298	23.47	84.80	84.80	0.00	-39.60	-19.86	14.558
300	23.60	84.95	84.81	0.05	-39.62	-19.74	14.380
400	28.60	92.44	85.79	2.66	-40.66	-12.95	7.074
500	32.95	99.30	87.82	5.75	-41.48	-5.92	2.588
600	36.47	105.63	90.27	9.22	-42.11	1.25	-0.455
700	39.47	111.48	92.89	13.02	-42.56	8.52	-2.659
800	41.97	116.92	95.55	17.10	-42.87	15.83	-4.325
900	44.18	122.00	98.21	21.41	-43.04	23.18	-5.630
1000	46.08	126.75	100.83	25.92	-43.09	30.55	-6.676

Amador, Lacher, and Park (11) reported the enthalpy of reaction with hydrogen to form propane and hydrogen chloride at 250°C as -31.215 kcal/mole; revision to 298°K gives $\Delta Hf°_{298}(g) = -39.08$ kcal/mole. Smith, Bjellerup, Krook, and Westermark (1384) reported energy of combustion measurements, which are corrected for a new enthalpy of oxidation of arsenious oxide (142), adjusted to ΔHc_{298}, and yield $\Delta Hf°_{298}(l) = -48.7$ kcal/mole. Dreisbach (353) listed $\Delta Hv°_{298} = 8.68$ kcal/mole, which gives $\Delta Hf°_{298}(g) = -40.0$ kcal/mole. An average of -39.6 kcal/mole is adopted.

Thermodynamic functions of the ideal gas are estimated using constants from Table 12-2. Dreisbach listed $Tm = 172.71°K$, $Tb = 369.52°K$, and $\Delta Hv = 7.59$ kcal/mole at the boiling point. Kurbatov (814) reported rough liquid heat capacity data. Studies of the structure and barriers to internal rotation have been made using determinations of dipole moment (1090), Raman spectrum (1027), low-temperature heat capacity (296), and electron diffraction (1627).

No. 736 1,3-Dichloropropane $C_3H_6Cl_2$ (Ideal Gas State) Mol Wt 112.992

	cal/(mole °K)				kcal/mole		
$T°K$	$Cp°$	$S°$	$-(G°-H°_{298})/T$	$H°-H°_{298}$	$\Delta Hf°$	$\Delta Gf°$	Log Kp
298	23.81	87.76	87.76	0.00	-38.60	-19.74	14.472
300	23.93	87.91	87.77	0.05	-38.62	-19.63	14.298
400	28.69	95.46	88.76	2.69	-39.64	-13.14	7.176
500	32.82	102.32	90.80	5.77	-40.46	-6.41	2.802
600	36.22	108.62	93.25	9.22	-41.11	0.46	-0.167
700	39.11	114.42	95.87	12.99	-41.59	7.43	-2.320
800	41.56	119.81	98.53	17.03	-41.94	14.45	-3.948
900	43.70	124.83	101.17	21.30	-42.16	21.52	-5.226
1000	45.50	129.53	103.78	25.76	-42.26	28.61	-6.251

Adjustment of the energy of combustion given by Smith, Bjellerup, Krook, and Westermark (1384) for the new enthalpy of oxidation of arsenious oxide (142) and adjustment to ΔHc_{298} yields $\Delta Hf°_{298}(l) = -48.3$ kcal/mole. Dreisbach (353) listed $\Delta Hv°_{298} = 9.66$ kcal/mole, which gives $\Delta Hf°_{298}(g) = -38.6$ kcal/mole. Thermodynamic functions of the ideal gas are estimated using Table 12-2. Dreisbach (353) also listed $Tm = 173.7°K$ and $Tb = 393.6°K$, at which temperature $\Delta Hv = 8.10$ kcal/mole.

No. 737 2,2-Dichloropropane $C_3H_6Cl_2$ (Ideal Gas State) Mol Wt 112.992

	cal/(mole °K)				kcal/mole		
$T°K$	$Cp°$	$S°$	$-(G°-H°_{298})/T$	$H°-H°_{298}$	$\Delta Hf°$	$\Delta Gf°$	Log Kp
298	25.30	77.92	77.92	0.00	-42.00	-20.21	14.813
300	25.40	78.08	77.93	0.05	-42.02	-20.08	14.625
400	30.56	86.12	78.99	2.86	-42.87	-12.62	6.897
500	34.75	93.40	81.15	6.13	-43.50	-4.99	2.180
600	38.06	100.04	83.76	9.78	-43.95	2.76	-1.004
700	40.75	106.12	86.52	13.72	-44.26	10.57	-3.300
800	43.00	111.71	89.33	17.91	-44.46	18.41	-5.030
900	44.91	116.89	92.10	22.31	-44.54	26.28	-6.382
1000	46.56	121.71	94.83	26.88	-44.53	34.16	-7.464

The energy of combustion reported by Smith, Bjellerup, Krook, and Westermark (1384) is corrected for a new enthalpy of oxidation of arsenious oxide (142) and converted to ΔHc_{298}, giving $\Delta Hf°_{298}(l) = -49.65$ kcal/mole. Dreisbach (353) listed $\Delta Hv°_{298} = 7.64$ kcal/mole, which yields $\Delta Hf°_{298}(g) = -42.0$ kcal/mole. Tobin (1509) assigned the vibrational frequencies, while Coutts and Livingston (281) determined the structural data. These are used with an estimated barrier to internal rotation of 4.0 kcal/mole to calculate the thermodynamic properties of the ideal gas. Dreisbach (353) also listed $Tm = 239.4°K$, $Tb = 343.7°K$, and $\Delta Hv = 7.0$ kcal/mole at the boiling point.

No. 738 1,2,3-Trichloropropane $C_3H_5Cl_3$ (Ideal Gas State) Mol Wt 147.441

	cal/(mole °K)				kcal/mole		
$T°K$	$Cp°$	$S°$	$-(G°-H°_{298})/T$	$H°-H°_{298}$	$\Delta Hf°$	$\Delta Gf°$	Log Kp
298	26.82	91.52	91.52	0.00	-44.40	-23.37	17.132
300	26.96	91.69	91.53	0.05	-44.42	-23.24	16.933
400	31.71	100.12	92.64	2.99	-45.20	-16.06	8.775
500	35.69	107.63	94.90	6.37	-45.80	-8.70	3.804
600	38.87	114.43	97.60	10.10	-46.26	-1.24	0.452
700	41.52	120.63	100.45	14.13	-46.58	6.29	-1.964
800	43.79	126.32	103.34	18.40	-46.78	13.85	-3.784
900	45.68	131.59	106.19	22.87	-46.88	21.44	-5.207
1000	47.34	136.49	108.97	27.52	-46.88	29.04	-6.346

Bjellerup and Smith (141) measured the energy of combustion in a rotating bomb calorimeter, while Smith, Bjellerup, Krook, and Westermark (1384) reported data measured by a stationary bomb method. The rotating bomb technique is regarded as more accurate and is corrected for the new value of the enthalpy of oxidation of arsenious oxide (142) and adjusted to ΔHc_{298} to give $\Delta Hf^{\circ}_{298}(l) = -55.65$ kcal/mole. The enthalpy of vaporization is listed by Dreisbach (353) as $\Delta Hv^{\circ}_{298} = 11.22$ kcal/mole, producing $\Delta Hf^{\circ}_{298}(g) = -44.4$ kcal/mole. Thermodynamic functions of the ideal gas are estimated using the constants of Table 12-2. Dreisbach (353) listed $Tm = 258.5°K$ and $Tb = 430.0°K$, at which temperature $\Delta Hv = 8.87$ kcal/mole. Nelson and Newton (1051) measured $Cp(l) = 34.3 + 0.032T$[cal/(mole °K), 270–350°K]. Kurbatov (814) also reported liquid heat capacities, but his data for other compounds are known to be of low accuracy.

No. 739 1-Chlorobutane C_4H_9Cl (Ideal Gas State) Mol Wt 92.569

	cal/(mole °K)				kcal/mole		
$T°K$	$Cp°$	$S°$	$-(G°-H^{\circ}_{298})/T$	$H°-H^{\circ}_{298}$	$\Delta Hf°$	$\Delta Gf°$	Log Kp
298	25.71	85.58	85.58	0.00	-35.20	-9.27	6.797
300	25.84	85.74	85.59	0.05	-35.23	-9.11	6.640
400	32.30	94.08	86.68	2.97	-36.85	-0.15	0.084
500	37.98	101.92	88.95	6.49	-38.17	9.18	-4.012
600	42.77	109.27	91.73	10.53	-39.22	18.75	-6.828
700	46.85	116.18	94.74	15.02	-40.03	28.48	-8.891
800	50.31	122.67	97.83	19.88	-40.62	38.30	-10.463
900	53.32	128.77	100.93	25.06	-41.03	48.20	-11.703
1000	55.92	134.53	104.00	30.53	-41.26	58.13	-12.704

Green and Holden (540) calculated thermodynamic properties using their values for 1-chloropropane and the methylene increments of Person and Pimentel (1140). Smith, Bjellerup, Krook, and Westermark (1384) listed energy of combustion data, which are corrected for a new enthalpy of oxidation of arsenious oxide (142) and adjusted to ΔHc_{298}, yielding $\Delta Hf^{\circ}_{298}(l) = -43.2$ kcal/mole. Mathews and Fehlandt (941) measured $\Delta Hv = 7.38$ kcal/mole at 349.7°K. An estimated ΔCp of vaporization of -13 cal/(mole °K) gives $\Delta Hv^{\circ}_{298} = 8.0$ kcal/mole and $\Delta Hf^{\circ}_{298}(g) = -35.2$ kcal/mole. Dreisbach (352) and Timmermans (1501) are in good agreement at $Tm = 150.0°K$ and $Tb = 351.6°K$. Perez Masiá and Diaz Peña (1137) determined the second virial coefficient in the range 358–423°K. Ukaji and Bonham (1525) derived the skeletal structure and some conclusions about the relative stabilities of rotational isomers from electron diffraction studies.

No. 740 2-Chlorobutane C_4H_9Cl (Ideal Gas State) Mol Wt 92.569

$T°K$	$Cp°$	$S°$	$-(G°-H°_{298})/T$	$H°-H°_{298}$	$\Delta Hf°$	$\Delta Gf°$	Log Kp
		cal/(mole °K)			kcal/mole		
298	25.93	85.94	85.94	0.00	-38.60	-12.78	9.368
300	26.07	86.11	85.95	0.05	-38.63	-12.62	9.195
400	32.52	94.50	87.05	2.99	-40.23	-3.70	2.022
500	38.33	102.40	89.34	6.54	-41.52	5.59	-2.441
600	43.18	109.83	92.14	10.62	-42.53	15.10	-5.500
700	47.33	116.81	95.17	15.15	-43.29	24.78	-7.735
800	50.84	123.36	98.29	20.06	-43.84	34.53	-9.433
900	53.92	129.53	101.42	25.30	-44.19	44.36	-10.770
1000	56.60	135.35	104.53	30.83	-44.36	54.21	-11.847

The energy of combustion reported by Smith, Bjellerup, Krook, and Westermark (1384), adjusted to ΔHc_{298}, is corrected for the new enthalpy of oxidation of arsenious oxide (142) to give $\Delta Hf°_{298}(l) = -46.2$ kcal/mole. Dreisbach (353) listed $\Delta Hv = 6.98$ kcal/mole at $Tm = 141.9°K$, and $\Delta Hv°_{298} = 7.60$ kcal/mole, from which is derived $\Delta Hf°_{298}(g) = -38.6$ kcal/mole. Thermodynamic functions of the ideal gas were estimated. Ukaji and Bonham (1525) derived the structure and some conclusions about the stabilities of rotational isomers from electron diffraction studies.

No. 741 1-Chloro-2-Methylpropane C_4H_9Cl (Ideal Gas State) Mol Wt 92.569

$T°K$	$Cp°$	$S°$	$-(G°-H°_{298})/T$	$H°-H°_{298}$	$\Delta Hf°$	$\Delta Gf°$	Log Kp
		cal/(mole °K)			kcal/mole		
298	25.93	84.56	84.56	0.00	-38.10	-11.87	8.700
300	26.07	84.73	84.57	0.05	-38.13	-11.71	8.530
400	32.52	93.12	85.67	2.99	-39.73	-2.65	1.447
500	38.33	101.02	87.96	6.54	-41.02	6.78	-2.962
600	43.18	108.45	90.76	10.62	-42.03	16.43	-5.984
700	47.33	115.43	93.79	15.15	-42.79	26.24	-8.192
800	50.84	121.98	96.91	20.06	-43.34	36.13	-9.871
900	53.92	128.15	100.04	25.30	-43.69	46.10	-11.193
1000	56.60	133.97	103.15	30.83	-43.86	56.09	-12.258

The energy of combustion reported by Smith, Bjellerup, Krook, and Westermark (1384) is corrected for the new value of the enthalpy of oxidation of arsenious oxide (142) and adjusted to ΔHc_{298} to yield $\Delta Hf°_{298}(l) = -45.8$ kcal/mole. Timmermans (1501) listed $Tm = 142.0°K$ and $Tb = 342.0°K$. The enthalpy of vaporization at 298°K is estimated by comparison with similar compounds to be 7.7 kcal/mole, giving $\Delta Hf°_{298}(g) = -38.1$ kcal/mole. Thermodynamic functions of the ideal gas are estimated using the constants of Table 12-2.

No. 742 2-Chloro-2-Methylpropane C_4H_9Cl (Ideal Gas State) Mol Wt 92.569

	cal/(mole °K)				kcal/mole		
$T°K$	$Cp°$	$S°$	$-(G°-H°_{298})/T$	$H°-H°_{298}$	$\Delta Hf°$	$\Delta Gf°$	Log Kp
298	27.30	77.00	77.00	0.00	-43.80	-15.32	11.226
300	27.40	77.17	77.01	0.06	-43.83	-15.14	11.030
400	34.00	85.98	78.16	3.13	-45.28	-5.35	2.921
500	39.60	94.19	80.56	6.82	-46.43	4.78	-2.087
600	44.20	101.83	83.47	11.02	-47.34	15.10	-5.500
700	48.10	108.94	86.61	15.64	-48.01	25.57	-7.982
800	51.50	115.59	89.82	20.62	-48.48	36.10	-9.863
900	54.50	121.83	93.03	25.92	-48.77	46.70	-11.340
1000	57.00	127.71	96.21	31.50	-48.89	57.33	-12.528

The ΔEc reported by Smith, Bjellerup, Krook, and Westermark (1384) is corrected for the new value of the enthalpy of oxidation of arsenious oxide (142) and converted to ΔHc_{298} to give $\Delta Hf°_{298}(l) = -50.1$ kcal/mole. Dreisbach (353) listed $\Delta Hv°_{298} = 7.0$ kcal/mole, which yields $\Delta Hf°_{298}(g) = -43.1$ kcal/mole. Howlett (619, 620) calculated thermodynamic functions using a vibrational assignment nearly identical with those of Sheppard (1343) and Tobin (1509), as well as the moments of inertia reported by Williams and Gordy (1615) and a barrier to internal rotation of 3600 cal/mole. Howlett (619), Kistiakowsky and Stauffer (763), and Levanova and Andreevskii (857) reported equilibrium constants for the reaction of 2-methylpropene and hydrogen chloride. The data are all in good agreement and yield an average value of $\Delta Hf°_{298}(g) = -43.8$ kcal/mole. The equilibrium data are judged more accurate than the combustion result, and -43.8 kcal/mole is adopted.

Kushner, Crowe, and Smyth (817) measured $Tt = 183.1°K$ with $\Delta Ht = 0.41$ kcal/mole, $Tt = 219.6°K$ with $\Delta Ht = 1.39$ kcal/mole, and $\Delta Hm = 0.48$ kcal/mole. Dreisbach (353) listed $Tm = 247.8°K$, $Tb = 323.9°K$, and $\Delta Hv = 6.6$ kcal/mole at the boiling point.

Recent studies on the structure confirm the work of Williams and Gordy: microwave spectroscopy by Lide and Jen (868) and Zeil, Winnewisser, and Mueller (1642), and electron diffraction work by Bowen, Gilchrist, and Sutton (152) and Bastiansen and Smedvik (81).

No. 743 1-Chloropentane $C_5H_{11}Cl$ (Ideal Gas State) Mol Wt 106.595

Green and Holden (540) calculated thermodynamic properties using their values for 1-chloropropane and the methylene increments of Person and Pimentel (1140). Bjellerup and Smith (141) used a rotating bomb calorimeter to measure the enthalpy of combustion. Their value is corrected for a new enthalpy of oxidation of arsenious oxide (142) and

Aliphatic Chlorine Compound Ideal Gas Tables

No. 743 1-Chloropentane $C_5H_{11}Cl$ (Ideal Gas State) Mol Wt 106.595

T°K	Cp°	S°	$-(G°-H°_{298})/T$	$H°-H°_{298}$	$\Delta Hf°$	$\Delta Gf°$	Log Kp
	cal/(mole °K)				kcal/mole		
298	31.18	94.89	94.89	0.00	-41.80	-8.94	6.551
300	31.34	95.09	94.90	0.06	-41.84	-8.74	6.364
400	39.24	105.21	96.22	3.60	-43.77	2.60	-1.423
500	46.23	114.74	98.98	7.88	-45.35	14.39	-6.288
600	52.11	123.70	102.36	12.81	-46.60	26.45	-9.634
700	57.13	132.12	106.02	18.27	-47.55	38.71	-12.085
800	61.38	140.03	109.78	24.20	-48.24	51.07	-13.952
900	65.07	147.48	113.56	30.53	-48.71	63.52	-15.425
1000	68.25	154.50	117.30	37.20	-48.96	76.02	-16.613

adjusted to 1 atm pressure at 298°K to derive $\Delta Hf°_{298}(l) = -50.8$ kcal/mole. Earlier static bomb work, discussed by Smith, Bjellerup, Krook, and Westermark (1384) and similarly corrected, yields $\Delta Hf°_{298}(l) = -51.1$ kcal/mole, in good agreement. Mathews and Fehlandt (941) measured $\Delta Hv = 7.93$ kcal/mole at 381°K. An estimated ΔCp of vaporization of -14 cal/(mole °K) gives $\Delta Hv°_{298} = 9.1$ kcal/mole. As most weight is given to the rotating bomb result, $\Delta Hf°_{298}(g)$ is selected as -41.8 kcal/mole. Dreisbach (353) listed $Tm = 174.2$°K and $Tb = 380.9$°K.

No. 744 1-Chloro-3-Methylbutane $C_5H_{11}Cl$ (Ideal Gas State) Mol Wt 106.595

T°K	Cp°	S°	$-(G°-H°_{298})/T$	$H°-H°_{298}$	$\Delta Hf°$	$\Delta Gf°$	Log Kp
	cal/(mole °K)				kcal/mole		
298	32.00	95.56	95.56	0.00	-43.10	-10.44	7.651
300	32.16	95.76	95.57	0.06	-43.14	-10.24	7.458
400	40.10	106.13	96.92	3.69	-44.99	1.02	-0.559
500	47.10	115.85	99.75	8.06	-46.48	12.70	-5.552
600	52.90	124.96	103.20	13.06	-47.64	24.65	-8.978
700	57.80	133.49	106.92	18.60	-48.52	36.78	-11.481
800	61.90	141.48	110.75	24.59	-49.15	49.00	-13.384
900	65.60	148.99	114.59	30.97	-49.56	61.30	-14.885
1000	68.70	156.07	118.38	37.69	-49.77	73.64	-16.093

The energy of combustion reported by Smith, Bjellerup, Krook, and Westermark (1384) is corrected for the new value of the enthalpy of oxidation of arsenious oxide (142) and adjusted to ΔHc_{298} to give $\Delta Hf°_{298}(l) = -51.86$ kcal/mole. Rossini and co-workers (1250) listed $Tm = 168.8$°K and $Tb = 371.4$°K. The enthalpy of vaporization at 298°K is estimated as 8.72 kcal/mole by comparison with similar compounds, to give $\Delta Hf°_{298}(g) = -43.1$ kcal/mole. Thermodynamic functions of the ideal gas are estimated using the constants of Table 12-2. Kurbatov (814) gave rough heat capacity data for the liquid.

No. 745 2-Chloro-2-Methylbutane $C_5H_{11}Cl$ (Ideal Gas State) Mol Wt 106.595

	cal/(mole °K)				kcal/mole		
$T°K$	$Cp°$	$S°$	$-(G°-H°_{298})/T$	$H°-H°_{298}$	$\Delta Hf°$	$\Delta Gf°$	Log Kp
298	31.45	88.06	88.06	0.00	-48.40	-13.50	9.896
300	31.62	88.26	88.07	0.06	-48.44	-13.29	9.680
400	39.80	98.50	89.41	3.64	-50.33	-1.27	0.694
500	47.10	108.18	92.20	8.00	-51.84	11.18	-4.885
600	53.30	117.34	95.64	13.03	-52.98	23.89	-8.700
700	58.20	125.93	99.36	18.61	-53.81	36.77	-11.480
800	62.50	133.99	103.19	24.65	-54.40	49.75	-13.589
900	65.90	141.55	107.03	31.07	-54.77	62.79	-15.248
1000	69.20	148.67	110.85	37.83	-54.93	75.88	-16.582

The energy of combustion reported by Smith, Bjellerup, Krook, and Westermark (1384) is adjusted for a new value of the enthalpy of oxidation of arsenious oxide (142) and converted to ΔHc_{298} to give $\Delta Hf°_{298}(l) = -56.6$ kcal/mole. Mathews and Fehlandt (941) measured $\Delta Hv = 7.34$ kcal/mole at 356.4°K. An estimate of -14 cal/(mole °K) for ΔCp of vaporization yields $\Delta Hf°_{298}(g) = -48.4$ kcal/mole. Dreisbach (353) listed $Tm = 199.7°K$ and $Tb = 358.8°K$. Thermodynamic functions of the ideal gas are estimated using the constants of Table 12-2.

No. 746 Chloroethylene C_2H_3Cl (Ideal Gas State) Mol Wt 62.501

	cal/(mole °K)				kcal/mole		
$T°K$	$Cp°$	$S°$	$-(G°-H°_{298})/T$	$H°-H°_{298}$	$\Delta Hf°$	$\Delta Gf°$	Log Kp
298	12.84	63.08	63.08	0.00	8.40	12.31	-9.021
300	12.89	63.16	63.09	0.03	8.39	12.33	-8.982
400	15.56	67.25	63.62	1.45	7.86	13.73	-7.500
500	17.80	70.97	64.73	3.13	7.43	15.24	-6.663
600	19.61	74.38	66.06	5.00	7.06	16.84	-6.134
700	21.10	77.52	67.47	7.04	6.76	18.50	-5.776
800	22.35	80.42	68.91	9.21	6.51	20.19	-5.516
900	23.43	83.11	70.34	11.50	6.31	21.92	-5.322
1000	24.35	85.63	71.75	13.89	6.17	23.66	-5.170

Thermodynamic functions given here are calculated using fundamental frequencies given by Gullikson and Nielsen (546) and structural constants from the microwave study of Kivelson, Wilson, and Lide (764). Recent electron diffraction results of Kaplan (714) and of Akishin, Vilkov, and Vesnin (7) are in reasonable agreement with the microwave results. Richards (1224) computed nearly identical thermodynamic functions based on the assignment of Thompson and Torkington (1494) and the early electron diffraction results of Brockway, Beach, and Pauling (171). Lacher, Gottlieb, and Park (825) derived $\Delta Hf°_{298}(g) = 8.072$ kcal/mole from the enthalpy of addition of hydrogen chloride to acetylene. Earlier work by Lacher, Emery, Bohmfalk, and Park (824) on the enthalpy of hydrogenation of chloroethylene to ethane and hydrogen chloride yielded $\Delta H f°_{298}(g) = 8.889$ kcal/mole. A value of 8.40 kcal/mole is selected. Ghosh and Guha (471) measured equilibrium composition in the pyrolysis of 1,2-dichloroethane. Their equilibrium constants give $\Delta H f°_{298}(g) = 13.54$ kcal/mole, which is almost surely too positive. Kobe and Crawford (772) reported thermodynamic properties based on the selected $\Delta H f°_{298}(g) = 7.5$ kcal/mole of Rossini, Wagman, Evans, Levine, and Jaffe (1249), which is based on the early work of Thomsen (1495). McDonald, Shrader, and Stull (969) measured $Tm = 94.43°K$ and $Tb = 259.35°K$. Sinke and Stull (1359) reported the enthalpy of combustion of polyvinyl chloride and derived $\Delta H f°_{298}(s) = -22.6$ kcal/(mole of monomer). The enthalpy of polymerization of the gas is calculated as -31.0 kcal/(mole of monomer).

No. 747 1,1-Dichloroethylene, $C_2H_2Cl_2$ (Ideal Gas State) Mol Wt 96.950

$T°K$	$Cp°$	$S°$	$-(G°-H°_{298})/T$	$H°-H°_{298}$	$\Delta H f°$	$\Delta G f°$	Log Kp
		cal/(mole °K)			kcal/mole		
298	16.02	68.85	68.85	0.00	0.30	5.78	-4.235
300	16.08	68.95	68.86	0.03	0.29	5.81	-4.233
400	18.80	73.97	69.52	1.78	0.03	7.69	-4.204
500	20.86	78.39	70.86	3.77	-0.18	9.64	-4.212
600	22.44	82.34	72.45	5.94	-0.33	11.61	-4.229
700	23.69	85.90	74.12	8.25	-0.45	13.61	-4.250
800	24.71	89.13	75.80	10.67	-0.54	15.63	-4.269
900	25.57	92.09	77.45	13.18	-0.60	17.65	-4.287
1000	26.29	94.82	79.05	15.78	-0.63	19.68	-4.302

Hildenbrand, McDonald, Kramer, and Stull (599) reported low-temperature thermal data, including $Ttp = 150.59°K$, $\Delta Hm = 1.557$ kcal/mole, $\Delta Hv°_{298} = 6.328$ kcal/mole, $S°_{298}(l) = 48.17$ cal/(mole °K), and $S°_{298}(g) = 69.04$ cal/(mole °K). The experimental entropy agreed within experimental error with $S°_{298}(g) = 68.85$ cal/(mole °K), based on the vibrational assignment recommended by Evans (390) and structural constants from the microwave study of Sekino and Nishikawa (1330). Hildenbrand et al. calculated thermodynamic functions to 1500°K, which are adopted here. Electron diffraction work by Livingston, Rao, Kaplan, and Rocks (879) is in good agreement with the data of Sekino and Nishikawa.

Sinke and Stull (1359) measured the enthalpy of combustion and reported $\Delta H f°_{298}(l) = -6.0$ kcal/mole. Combination of this value with the enthalpy of vaporization from Hildenbrand et al. yields $\Delta H f°_{298}(g) = 0.3$ kcal/mole. Sinke and Stull also measured the enthalpy of combustion of the polymer and derived $\Delta Hp = -18.0$ kcal/(mole of monomer). Joshi

(704) reported $\Delta Hp = -17.5$ kcal/mole of monomer, and Tong and Kenyon (1514) found $\Delta Hp = -14.4$ kcal/mole of monomer.

Hildenbrand et al. (599) measured the vapor pressure and derived $Tb = 304.71°K$ and $\Delta Hv = 6.26$ kcal/mole. Their liquid heat capacities can be fitted by the expression $Cp(l) = 15.55 + 0.037T$[cal/(mole °K), 270–305°K]. Tatevskii and Frost (1469) and Kobe and Harrison (774) gave thermodynamic properties based on estimates and less accurate spectroscopic data than now available.

No. 748 1,2-Dichloroethylene, *cis* $C_2H_2Cl_2$ (Ideal Gas State) Mol Wt 96.950

	cal/(mole °K)				kcal/mole		
$T°K$	$Cp°$	$S°$	$-(G°-H°_{298})/T$	$H°-H°_{298}$	$\Delta Hf°$	$\Delta Gf°$	Log Kp
298	15.55	69.20	69.20	0.00	0.45	5.82	-4.268
300	15.61	69.30	69.21	0.03	0.44	5.86	-4.266
400	18.41	74.19	69.85	1.74	0.13	7.71	-4.213
500	20.57	78.54	71.16	3.69	-0.10	9.64	-4.211
600	22.23	82.44	72.72	5.84	-0.29	11.60	-4.225
700	23.54	85.97	74.37	8.13	-0.42	13.59	-4.244
800	24.60	89.19	76.02	10.54	-0.52	15.60	-4.261
900	25.48	92.14	77.65	13.04	-0.60	17.62	-4.279
1000	26.23	94.86	79.24	15.63	-0.63	19.65	-4.294

Pitzer and Hollenberg (1163) studied the infrared spectrum and confirmed the vibrational assignment of Bernstein and Ramsay (119). Bond lengths and angles determined by Flygare and Howe (434) with microwave techniques and by Hoffman (610) with electron diffraction are in good agreement. These data are used to calculate thermodynamic functions. Broers, Ketelaar, and Van Velden (172) measured $Tm = 193.1°K$ and $\Delta Hm = 1.72$ kcal/mole. Ketelaar, Van Velden, and Zalm (743) measured the vapor pressure from 240° to 372°K and found $Tb = 333.8°K$. Their data from 273° to 328°K are fitted to an equation using the procedure of Prosen (1192) and an estimated ΔCp of vaporization of -12 cal/(mole °K), based on the single liquid heat capacity value of Mehl (988). This vapor pressure equation and estimates of $Tc = 500°K$ and $Pc = 50$ atm are employed to calculate $\Delta Hv°_{298} = 7.43$ kcal/mole, using the Haggenmacher equation (560). The estimated ΔCp of vaporization then gives $\Delta Hv_{333.8} = 7.0$ kcal/mole.

Smith, Bjellerup, Krook, and Westermark (1384) reported energy of combustion measurements, which are corrected to ΔHc_{298} and for a new value of the enthalpy of oxidation of arsenious oxide (142) to yield $\Delta Hf°_{298}(l) = -6.7$ kcal/mole. Smith et al. also reported the energy of combustion of 1,1,2,2-tetrachloroethane, which is similarly corrected and combined with the enthalpy of addition of chlorine to 1,2-dichloroethylene, *cis*, measured by Kirkbride (754) to yield $\Delta Hf°_{298}(l) = -6.8$

kcal/mole. The enthalpy of isomerization at 298°K of 1,2-dichloroethylene, cis, to 1,2-dichloroethylene, trans, is calculated as 0.55 kcal/mole from the equilibrium data of Maroney (932) and of Wood and Stevenson (1626). If data are considered for both cis and trans isomers, $\Delta Hf°_{298}(g) = 0.45$ kcal/mole is selected for 1,2-dichloroethylene, cis.

No. 749 1,2-Dichloroethylene, trans $C_2H_2Cl_2$ (Ideal Gas State) Mol Wt 96.950

T°K	Cp°	S°	$-(G°-H°_{298})/T$	$H°-H°_{298}$	$\Delta Hf°$	$\Delta Gf°$	Log Kp
		cal/(mole °K)			kcal/mole		
298	15.93	69.29	69.29	0.00	1.00	6.35	−4.652
300	15.99	69.39	69.30	0.03	0.99	6.38	−4.647
400	18.58	74.36	69.96	1.77	0.71	8.22	−4.491
500	20.65	78.74	71.28	3.73	0.49	10.13	−4.426
600	22.28	82.65	72.86	5.88	0.31	12.07	−4.396
700	23.57	86.19	74.51	8.18	0.18	14.04	−4.384
800	24.62	89.40	76.18	10.59	0.08	16.03	−4.378
900	25.50	92.36	77.81	13.09	0.01	18.03	−4.377
1000	26.24	95.08	79.40	15.68	−0.03	20.03	−4.377

Pitzer and Hollenberg (1163) studied the far infrared spectrum and revised two low-lying frequencies of the assignment of Bernstein and Ramsay (119). Pitzer and Hollenberg calculated the thermodynamic functions adopted here, using structural constants of Brockway, Beach, and Pauling (171). Electron diffraction measurements by Kaplan (714) gave different bond lengths, but very nearly the same over-all moment of inertia. Broers, Ketelaar, and Van Velden (172) measured $Tm = 193.1°K$ and $\Delta Hm = 1.72$ kcal/mole. Ketelaar, Van Velden, and Zalm (743) measured the vapor pressure from 273° to 358° and found $Tb = 320.8°K$. Their data from 273° to 358°K are fitted to an equation using the procedure of Prosen (1192) and an estimated ΔCp of vaporization of -12 cal/(mole °K), based on the single liquid heat capacity value of Mehl (988). The vapor pressure equation and estimates of $Tc = 500°K$ and $Pc = 50$ atm are employed to calculate $\Delta Hv°_{298} = 6.92$ kcal/mole, using the Haggenmacher equation (560). The estimated ΔCp of vaporization then gives $\Delta Hv_{320.8} = 6.65$ kcal/mole.

Smith, Bjellerup, Krook, and Westermark (1384) reported energy of combustion measurements, which are adjusted to ΔHc_{298} and corrected for a new value for the enthalpy of oxidation of arsenious oxide (142) to yield $\Delta Hf°_{298}(l) = -6.1$ kcal/mole. The enthalpy of isomerization of 1,2-dichloroethylene, cis, to 1,2-dichloroethylene, trans, is calculated as 0.55 kcal/mole from the equilibrium data of Maroney (932) and of Wood and Stevenson (1626). If data for both cis and trans isomers are considered, $\Delta Hf°_{298}(g)$ is selected as 1.00 kcal/mole for 1,2-dichloroethylene, trans.

No. 750 Trichloroethylene C_2HCl_3 (Ideal Gas State) Mol Wt 131.399

$T°K$	$Cp°$	$S°$	$-(G°-H°_{298})/T$	$H°-H°_{298}$	$\Delta Hf°$	$\Delta Gf°$	Log Kp
		cal/(mole °K)			kcal/mole		
298	19.17	77.63	77.63	0.00	-1.40	4.75	-3.482
300	19.25	77.75	77.64	0.04	-1.40	4.79	-3.488
400	21.80	83.66	78.42	2.10	-1.43	6.86	-3.747
500	23.67	88.73	79.99	4.38	-1.42	8.93	-3.904
600	25.06	93.17	81.82	6.82	-1.39	11.00	-4.005
700	26.12	97.12	83.73	9.38	-1.34	13.06	-4.077
800	26.94	100.66	85.63	12.03	-1.29	15.11	-4.127
900	27.61	103.88	87.48	14.76	-1.23	17.16	-4.166
1000	28.15	106.81	89.27	17.55	-1.15	19.19	-4.195

The energy of combustion reported by Smith, Bjellerup, Krook, and Westermark (1384), after correction for a change in the enthalpy of oxidation of arsenious oxide (142) and adjustment to ΔHc_{298}, yields $\Delta Hf°_{298}(l) = -9.6$ kcal/mole. Kirkbride (754) measured the enthalpy of addition of chlorine to trichloroethylene and the enthalpy of dehydrochlorination of 1,1,2,2-tetrachloroethane. Combining this value with energy of combustion data from Smith et al. for pentachloroethane and for 1,1,2,2-tetrachloroethane, we derive $\Delta Hf°_{298}(l)$ as -9.3 and -10.8 kcal/mole, respectively, in agreement within the experimental precision. Mathews (940) determined $\Delta Hv = 7.52$ kcal/mole at 358.8°K. An estimated ΔCp of vaporization of -11 cal/(mole °K) gives $\Delta Hv°_{298} = 8.2$ kcal/mole and $\Delta Hf°_{298}(g) = -1.4$ kcal/mole. McDonald (967) measured vapor pressures and liquid and vapor densities and calculated enthalpies of vaporization that appear too high compared to those of Mathews. Kurbatov (814) reported liquid heat capacity values, but results for other compounds indicate that his work is of low accuracy. Timmermans and Roland (1505) measured $Tm = 188.4°K$ and $Tb = 360.1°K$.

Allen and Bernstein (9) studied the spectrum, assigned fundamentals, estimated reasonable bond lengths and angles, and calculated thermodynamic functions, which are adopted here. Kobe and Harrison (774) also made a vibrational assignment, but their derived heat capacities appear too high when compared with the values for other chlorinated ethylenes.

No. 751 Tetrachloroethylene C_2Cl_4 (Ideal Gas State) Mol Wt 165.848

The energy of combustion reported by Smith, Bjellerup, Krook, and Westermark (1384) yields $\Delta Hf°_{298}(l) = -12.8$ kcal/mole when corrected for a new value of the enthalpy of oxidation of arsenious oxide (142)

No. 751 Tetrachloroethylene C_2Cl_4 (Ideal Gas State) Mol Wt 165.848

	cal/(mole °K)				kcal/mole		
$T°K$	$Cp°$	$S°$	$-(G°-H°_{298})/T$	$H°-H°_{298}$	$\Delta Hf°$	$\Delta Gf°$	Log Kp
298	22.69	81.46	81.46	0.00	-3.40	4.90	-3.592
300	22.73	81.61	81.47	0.05	-3.40	4.95	-3.607
400	25.10	88.49	82.39	2.44	-3.15	7.70	-4.206
500	26.73	94.27	84.20	5.04	-2.90	10.38	-4.539
600	27.86	99.25	86.30	7.77	-2.66	13.02	-4.741
700	28.68	103.61	88.47	10.60	-2.43	15.61	-4.874
800	29.29	107.48	90.61	13.50	-2.23	18.17	-4.964
900	29.73	110.96	92.68	16.45	-2.03	20.71	-5.029
1000	30.07	114.11	94.67	19.44	-1.84	23.23	-5.076

and adjusted to ΔHc_{298}. Smith et al. also gave a value for hexachloroethane, which is similarly corrected and combined with the result of Kirkbride (754) for the enthalpy of addition of chlorine to tetrachloroethylene to derive $\Delta Hf°_{298}(l) = -12.3$ kcal/mole, in reasonable agreement. Rossini, Wagman, Evans, Levine, and Jaffe (1249) selected $Tm = 250.8°K$, $\Delta Hm = 2.5$ kcal/mole, and $Tb = 394.2°K$, where $\Delta Hv = 8.3$ kcal/mole. If ΔCp of vaporization is estimated as -11 cal/(mole °K), $\Delta Hv°_{298} = 9.4$ kcal/mole and $\Delta Hf°_{298}(g) = -3.4$ kcal/mole. Kurbatov (814) made liquid heat capacity measurements, but his results for other compounds indicate his work to be of very low accuracy. Rossini, Wagman, Evans, Levine, and Jaffe (1249) listed $\Delta Hf°_{298}(l) = -3$ kcal/mole, based on old combustion data obtained by Thomsen (1495).

Mann, Meal, and Plyler (923) revised the vibrational assignment of Mann, Acquista, and Plyler (920) and used the structural constants of Karle and Karle (718) to calculate the thermodynamic functions given here. Kobe and Harrison (774) calculated heat capacities using the assignment of Torkington (1516). Their values are considerably higher than the later ones of Mann, Meal, and Plyler.

No. 752 3-Chloro-1-Propene C_3H_5Cl (Ideal Gas State) Mol Wt 76.527

	cal/(mole °K)				kcal/mole		
$T°K$	$Cp°$	$S°$	$-(G°-H°_{298})/T$	$H°-H°_{298}$	$\Delta Hf°$	$\Delta Gf°$	Log Kp
298	18.01	73.29	73.29	0.00	-0.15	10.42	-7.641
300	18.10	73.41	73.30	0.04	-0.17	10.49	-7.640
400	22.12	79.18	74.06	2.05	-1.04	14.18	-7.747
500	25.55	84.49	75.62	4.44	-1.78	18.07	-7.899
600	28.43	89.41	77.51	7.15	-2.40	22.10	-8.050
700	30.85	93.98	79.54	10.11	-2.90	26.23	-8.188
800	32.93	98.24	81.62	13.30	-3.29	30.41	-8.307
900	34.75	102.23	83.69	16.69	-3.59	34.65	-8.413
1000	36.30	105.97	85.73	20.24	-3.79	38.91	-8.503

Electron impact studies by Lossing, Ingold, and Henderson (887) related the enthalpies of formation of allyl iodide and allyl chloride. The results of Gellner and Skinner (466) for allyl iodide yield $\Delta Hf°_{298}(g) = 0.0$ for allyl chloride. Enthalpy of combustion measurements by Thomsen as corrected by Kharasch (744) yield $\Delta Hf°_{298}(g) = -0.3$ kcal/mole. An average of -0.15 kcal/mole is adopted. Thermodynamic functions are estimated using the constants of Table 12-2 with 1-butene as the parent hydrocarbon. Bowen, Gilchrist, and Sutton (152) reported structural constants determined by electron diffraction. Timmermans (1501) listed $Tm = 138.7°K$ and $Tb = 318.3°K$.

AROMATIC CHLORINE COMPOUND IDEAL GAS TABLES

No. 753 Chlorobenzene C_6H_5Cl (Ideal Gas State) Mol Wt 112.557

	cal/(mole °K)				kcal/mole		
$T°K$	$Cp°$	$S°$	$-(G°-H°_{298})/T$	$H°-H°_{298}$	$\Delta Hf°$	$\Delta Gf°$	Log Kp
298	23.43	74.92	74.92	0.00	12.39	23.70	-17.368
300	23.57	75.07	74.93	0.05	12.37	23.76	-17.311
400	30.62	82.84	75.94	2.77	11.46	27.71	-15.139
500	36.49	90.33	78.07	6.13	10.74	31.86	-13.925
600	41.16	97.41	80.71	10.02	10.18	36.13	-13.161
700	44.86	104.04	83.58	14.33	9.75	40.50	-12.644
800	47.89	110.24	86.53	18.97	9.43	44.91	-12.268
900	50.40	116.03	89.49	23.89	9.20	49.36	-11.986
1000	52.48	121.45	92.41	29.04	9.07	53.83	-11.765

Smith, Bjellerup, Krook, and Westermark (1384) applied several corrections to the original ΔEc measured by Karlsson (719). Their final value is corrected for a change in the enthalpy of oxidation of arsenious oxide (142) and adjusted to ΔHc_{298} to yield $\Delta Hf°_{298}(l) = 2.74$ kcal/mole. Hubbard, Knowlton, and Huffman (635) also measured the enthalpy of combustion. Their result yields $\Delta Hf°_{298}(l) = 2.43$ kcal/mole. An average of 2.58 kcal/mole is adopted. Hildenbrand (595) determined $\Delta Hv°_{298} = 9.81$ kcal/mole, from which $\Delta Hf°_{298}(g) = 12.39$ kcal/mole. Kirkbride (754) measured the enthalpy of chlorination of benzene. His result gives $\Delta Hf°_{298}(l) = 1.8$ kcal/mole, in agreement with the combustion results within his experimental error of about one kcal. Hubbard, et al. (635) gave references to early combustion work of historical interest.

Whiffen (1603) assigned fundamental frequencies and used the moments of inertia determined by Erlandsson (381) to calculate the thermodynamic functions given here. Scherer and Evans (1281) confirmed the vibrational frequencies, and Poynter (1184) confirmed the microwave structure studies. Godnev, Sverdlin, and Savogina (502) also assigned frequencies and calculated functions in good agreement with Whiffen.

Hildenbrand (595) measured $Tm = 227.81°K$ and $\Delta Hm = 2.43$ kcal/mole. Dreisbach (351) listed $Tm = 227.57°K$, Timmermans (1501) gave 228.0°K, and Stull (1431) found $Tm = 227.9°K$ and $\Delta Hm = 2.28$ kcal/mole. Stull (1431) measured the heat capacity from 90° to 320°K, but an accurate entropy cannot be derived because of long extrapolation. Dreisbach (351) gave $Tb = 404.85°K$, in good agreement with several determinations listed by Timmermans (1501). Mathews (940) measured $\Delta Hv = 8.73$ kcal/mole at 405°K. Liquid heat capacities given by Stull (1431) and Hildenbrand (595) can be fitted by $Cp(l) = 19.50 + 0.055T$ [cal/(mole °K), 290–350°K].

No. 754 o-Dichlorobenzene $C_6H_4Cl_2$ (Ideal Gas State) Mol Wt 147.006

	cal/(mole °K)				kcal/mole		
T°K	$Cp°$	$S°$	$-(G° - H°_{298})/T$	$H° - H°_{298}$	$\Delta Hf°$	$\Delta Gf°$	Log Kp
298	27.12	81.61	81.61	0.00	7.16	19.76	-14.485
300	27.26	81.78	81.62	0.06	7.15	19.84	-14.452
400	34.12	90.60	82.77	3.14	6.53	24.17	-13.207
500	39.69	98.83	85.17	6.84	6.07	28.64	-12.518
600	44.07	106.47	88.09	11.03	5.73	33.18	-12.086
700	47.51	113.53	91.23	15.62	5.49	37.79	-11.797
800	50.28	120.06	94.43	20.51	5.33	42.41	-11.584
900	52.54	126.11	97.62	25.65	5.23	47.05	-11.424
1000	54.42	131.75	100.75	31.00	5.22	51.70	-11.298

Sinke and Stull (1359) measured the enthalpy of combustion in a static bomb lined with glass cloth saturated with aqueous arsenious oxide and derived $\Delta Hf°_{298}(l) = -4.43$ kcal/mole. Hubbard, Knowlton, and Huffman (635) made similar measurements, except for the use of aqueous hydrazine dihydrochloride as a reducing agent; from their result there is calculated $\Delta Hf°_{298}(l) = -4.41$ kcal/mole. Bjellerup and Smith (141) employed aqueous arsenious oxide and a rotating bomb calorimeter. Smith, Bjellerup, Krook, and Westermark (1384) corrected enthalpies of combustion based on the "quartz-spiral" technique from Karlsson (719) and from Bjellerup (134). These values are further corrected for the new enthalpy of oxidation of arsenious oxide (142) to yield $\Delta Hf°_{298}(l) = -4.19, -5.49$, and -4.28 kcal/mole, respectively. All the values are in good agreement, except for that of Karlsson, made in a cellophane-covered crucible. The other investigations used glass ampoules. A weighted average of -4.40 kcal/mole is adopted. McDonald, Shrader, and Stull (969) reported $Tm = 256.15°K$, $Tb = 453.57°K$, and vapor pressures in the range 400–450°K. From their vapor pressure equation and an estimate of 0.98 for the compressibility factor, ΔHv is calculated as 10.0 kcal/mole at the average temperature of 428°K. An estimated

ΔCp of vaporization of -12 cal/(mole °K) then yields $\Delta Hv = 9.7$ kcal/mole at the boiling point and $\Delta Hv^{\circ}_{298} = 11.56$ kcal/mole, from which $\Delta Hf^{\circ}_{298}(g) = 7.16$ kcal/mole. Dreisbach (351) listed ΔHm as 3.19 kcal/mole.

The fundamental frequencies assigned by Scherer and Evans (1281) are employed with interatomic distances listed by Sutton (1449) to calculate thermodynamic functions. Godnev and Sverdlin (501) also assigned fundamentals and calculated functions in excellent agreement.

No. 755 m-Dichlorobenzene $C_6H_4Cl_2$ (Ideal Gas State) Mol Wt 147.006

	cal/(mole °K)				kcal/mole		
T°K	$Cp°$	$S°$	$-(G°-H°_{298})/T$	$H°-H°_{298}$	$\Delta Hf°$	$\Delta Gf°$	Log Kp
298	27.20	82.09	82.09	0.00	6.32	18.78	-13.765
300	27.34	82.26	82.10	0.06	6.31	18.85	-13.735
400	34.18	91.10	83.25	3.14	5.69	23.14	-12.642
500	39.74	99.34	85.66	6.85	5.24	27.56	-12.045
600	44.09	106.99	88.58	11.05	4.90	32.05	-11.673
700	47.53	114.05	91.72	15.63	4.67	36.60	-11.426
800	50.29	120.58	94.93	20.53	4.51	41.17	-11.245
900	52.55	126.64	98.12	25.67	4.41	45.76	-11.110
1000	54.42	132.28	101.26	31.03	4.40	50.35	-11.004

Hubbard, Knowlton, and Huffman (635) measured the enthalpy of combustion using a static bomb lined with glass cloth saturated with aqueous hydrazine hydrochloride. From their result ΔHf°_{298} (l) is calculated as -5.12 kcal/mole. Smith, Bjellerup, Krook, and Westermark (1384) recalculated the energy of combustion measured by Karlsson (719), who used a static bomb lined with quartz fibers saturated with aqueous arsenious oxide. The derived enthalpy of formation is about 1 kcal/mole more negative than that of Hubbard et al., and a redetermination of the enthalpy of oxidation of arsenious oxide (142) increases the discrepancy. As pointed up by results for o-dichlorobenzene, Karlsson's technique of enclosing the sample with cellophane gives systematically low results for volatile compounds. Hubbard et al. used glass ampoules to enclose the sample, and their result is accepted.

Dreisbach (351) listed $Tm = 248.39°K$ and $Tb = 446.23°K$, as well as constants for the Antoine vapor pressure equation. This equation and an estimated compressibility factor were used to calculate $\Delta Hv^{\circ}_{298} = 11.44$ kcal/mole, from which $\Delta Hf^{\circ}_{298}(g) = 6.32$ kcal/mole.

The fundamental frequencies assigned by Scherer and Evans (1281) and molecular structure constants listed by Sutton (1449) are used to calculate thermodynamic functions, which are in good agreement with those of Godnev and Sverdlin (501).

No. 756 p-Dichlorobenzene $C_6H_4Cl_2$ (Ideal Gas State) Mol Wt 147.006

$T°K$	$Cp°$	$S°$	$-(G°-H°_{298})/T$	$H°-H°_{298}$	$\Delta Hf°$	$\Delta Gf°$	Log Kp
		cal/(mole °K)			kcal/mole		
298	27.22	80.47	80.47	0.00	5.50	18.44	-13.518
300	27.36	80.64	80.48	0.06	5.49	18.52	-13.492
400	34.24	89.49	81.63	3.15	4.88	22.97	-12.548
500	39.80	97.75	84.04	6.86	4.43	27.54	-12.039
600	44.16	105.40	86.97	11.06	4.10	32.19	-11.726
700	47.60	112.48	90.12	15.66	3.87	36.90	-11.521
800	50.35	119.02	93.33	20.56	3.72	41.63	-11.371
900	52.60	125.08	96.52	25.71	3.63	46.37	-11.260
1000	54.46	130.73	99.66	31.07	3.62	51.12	-11.173

Hubbard, Knowlton, and Huffman (635) measured the enthalpy of combustion using a static bomb lined with glass cloth saturated with aqueous hydrazine hydrochloride. From their work $\Delta Hf°_{298}(s)$ is derived as -10.35 kcal/mole. Hubbard et al. also listed three previous determinations at the University of Lund in Sweden, in which arsenious oxide was used as a reducing agent. The result of Karlsson (719) from a static bomb lined with quartz fibers was recalculated by Smith, Bjellerup, Krook, and Westermark (1384). Smith and Sunner (1385) reported results using a "shaking bomb" and a rotating bomb. The three Lund values, when corrected for a new enthalpy of oxidation of arsenious oxide (142), yield $\Delta Hf°_{298}(s) = -9.72$, -10.16, and -9.87 kcal/mole, respectively. These three values were measured using cellophane to enclose the sample. Although for liquids this technique gave systematically low results (cf. o-dichlorobenzene), data for the crystalline solid in this case appear to be unaffected, and an overall average of $\Delta Hf°_{298}(s) = -10.0$ kcal/mole is adopted.

McDonald, Shrader, and Stull (969) reported $Tm = 326.14°K$, $Tb = 447.21°K$, and constants for an Antoine vapor pressure equation. The vapor pressure equation with an estimated compressibility factor is used to derive $\Delta Hv = 9.5$ kcal/mole at $447°K$. Walsh and Smith (1569) measured the vapor pressure of the solid and calculated $\Delta Hs_{298} = 15.5$ kcal/mole, which yields $\Delta Hf°_{298}(g) = 5.5$ kcal/mole. Walsh and Smith also reviewed literature data on the transition that takes place at $313° \pm 5°K$. Hrynakowski and Smoczkiewiczowa (623) reported $\Delta Hm = 4.4$ kcal/mole, while Ueberreiter and Orthmann (1524) found $\Delta Hm = 4.34$ kcal/mole. Heat capacity data of Ueberreiter and Orthmann (1524) and of Hildenbrand (595) can be fitted by $Cp(s) = -3.9 + 0.13T$[cal/(mole °K), 290–310°K].

Scherer and Evans (1281) assigned fundamental frequencies, while Sutton (1449) listed molecular structure parameters. Thermodynamic functions from these data are in good agreement with similar calculations by Godnev and Sverdlin (501).

No. 757 Hexachlorobenzene C_6Cl_6 (Ideal Gas State) Mol Wt 284.802

	cal/(mole °K)				kcal/mole		
T°K	$Cp°$	$S°$	$-(G°-H°_{298})/T$	$H°-H°_{298}$	$\Delta Hf°$	$\Delta Gf°$	Log Kp
298	41.90	105.45	105.45	0.00	-8.10	10.56	-7.740
300	42.03	105.71	105.46	0.08	-8.09	10.67	-7.776
400	48.08	118.67	107.18	4.60	-7.54	16.85	-9.207
500	52.52	129.90	110.63	9.64	-6.97	22.89	-10.003
600	55.78	139.78	114.68	15.06	-6.42	28.80	-10.490
700	58.18	148.56	118.90	20.77	-5.89	34.64	-10.813
800	59.96	156.45	123.11	26.68	-5.40	40.39	-11.032
900	61.31	163.60	127.22	32.74	-4.93	46.08	-11.189
1000	62.34	170.11	131.19	38.93	-4.46	51.72	-11.303

Sinke and Stull (1359) measured the enthalpy of combustion and derived $\Delta Hf°_{298}(s) = -31.3$ kcal/mole. Hildenbrand, Kramer, and Stull (597) determined the low-temperature heat capacity and calculated $S°_{298}(s) = 62.20$ cal/(mole °K). Scherer and Evans (1281) made a vibrational assignment that is used to compute thermodynamic functions of the ideal gas, assuming a planar hexagonal structure and C—Cl and C—C distances of 1.70 Å and 1.397 Å, respectively. The sublimation pressure in the range 370–400°K was measured by Sears and Hopke (1328). Their results are extrapolated to 298°K and combined with entropies of solid and vapor to calculate an enthalpy of sublimation of 23.2 kcal/mole, giving $\Delta Hf°_{298}(g)$ as -8.1 kcal/mole. Hildenbrand (595) found $Tm = 501.7°$K and $\Delta Hm = 6.1$ kcal/mole.

ACYL CHLORIDE IDEAL GAS TABLE

No. 758 Acetyl Chloride C_2H_3ClO (Ideal Gas State) Mol Wt 78.501

	cal/(mole °K)				kcal/mole		
T°K	$Cp°$	$S°$	$-(G°-H°_{298})/T$	$H°-H°_{298}$	$\Delta Hf°$	$\Delta Gf°$	Log Kp
298	16.21	70.47	70.47	0.00	-58.30	-49.29	36.130
300	16.26	70.58	70.48	0.04	-58.31	-49.24	35.867
400	18.86	75.62	71.14	1.79	-58.86	-46.12	25.200
500	21.19	80.08	72.49	3.80	-59.33	-42.88	18.744
600	23.18	84.12	74.10	6.02	-59.73	-39.56	14.409
700	24.86	87.83	75.80	8.42	-60.05	-36.17	11.292
800	26.30	91.24	77.52	10.98	-60.31	-32.74	8.945
900	27.54	94.41	79.22	13.68	-60.51	-29.28	7.111
1000	28.60	97.37	80.89	16.49	-60.65	-25.80	5.639

Overend, Nyquist, Evans, and Potts (1095) assigned fundamental frequencies and calculated thermodynamic functions, using preliminary structural constants and the potential barrier to internal rotation determined by Sinnott, who used microwave techniques. Final values reported by Sinnott (1360) are negligibly different. Carson and Skinner (198) measured the enthalpy of hydrolysis of acetyl chloride. Pritchard and Skinner (1187) repeated the measurements and recalculated the results of Carson and Skinner. The two values agree within experimental error and, when combined with the recent results of Evans and Skinner (389) for the enthalpy of formation of acetic acid, yield for acetyl chloride $\Delta Hf_{298}(l) = -65.5$ kcal/mole. Mathews and Fehlandt (941) measured the enthalpy of vaporization at the boiling point as 6.93 kcal/mole. An estimated ΔCp of vaporization of -11 cal/(mole °K) gives $\Delta Hv_{298}^\circ = 7.2$ kcal/mole. McDonald, Shrader, and Stull (969) measured the vapor pressure and calculated constants for the Antoine equation. Combination with estimated critical constants of 500°K and 50 atm and the Haggenmacher equation (560) also gives $\Delta Hv_{298}^\circ = 7.2$ kcal/mole, in good agreement. The enthalpy of formation of the ideal gas is then $\Delta Hf_{298}^\circ(g) = -58.3$ kcal/mole. McDonald, Shrader, and Stull (969) gave $Tm = 160.3°$K and $Tb = 323.9°$K.

ALIPHATIC BROMINE COMPOUND IDEAL GAS TABLES

No. 759   Bromomethane CH_3Br (Ideal Gas State)   Mol Wt 94.950

	cal/(mole °K)				kcal/mole		
$T°K$	$Cp°$	$S°$	$-(G°-H_{298}°)/T$	$H°-H_{298}°$	$\Delta Hf°$	$\Delta Gf°$	Log Kp
298	10.14	58.75	58.75	0.00	-9.00	-6.73	4.932
300	10.17	58.82	58.76	0.02	-9.02	-6.71	4.891
400	11.93	61.99	59.18	1.13	-13.32	-5.09	2.783
500	13.56	64.83	60.03	2.41	-13.85	-2.97	1.299
600	14.97	67.43	61.05	3.83	-14.30	-0.75	0.275
700	16.19	69.83	62.13	5.39	-14.66	1.53	-0.479
800	17.25	72.06	63.24	7.07	-14.96	3.86	-1.056
900	18.19	74.15	64.33	8.84	-15.18	6.23	-1.514
1000	19.00	76.11	65.41	10.70	-15.36	8.63	-1.885

Hartley, Pritchard, and Skinner (575) measured the enthalpy of reaction of dimethyl mercury with bromine and with mercuric bromide. Carson, Carson, and Wilmshurst (194) combined these results with their enthalpy of combustion of dimethyl mercury to derive for bromomethane $\Delta Hf_{298}^\circ(g) = -8.6$ kcal/mole. The equilibrium constant of the reaction of chloromethane and hydrogen bromide to form bromomethane and hydrogen chloride was reported by Bak (60). Using Gibbs energy functions and enthalpies of formation adopted in the present work, $\Delta Hf^°_{298}(g)$ for bromomethane is calculated as -9.66 kcal/mole. Rossini, Wagman, Evans, Levine, and Jaffe (1249) listed $\Delta Hf^°_{298}(g) = -8.5$ kcal/mole, based on combustion data over 50 years old. Margrave (929) reported mass spectrometer appearance potentials leading to $\Delta Hf^°_{298}(g) = -7.6$ kcal/mole. Fowell, Lacher, and Park (435) determined the enthalpy of hydrogenation to methane and hydrogen bromide and derived $\Delta Hf^°_{298}(g) = -8.97$ kcal/mole. A weighted average of $\Delta Hf^°_{298}(g) = -9.0$ kcal/mole is adopted.

Low-temperature thermal data were reported by Egan and Kemp (373) as follows: $Tt = 173.78°K$, $\Delta Ht = 0.113$ kcal/mole, $Ttp = 179.47°K$, $\Delta Hm = 1.429$ kcal/mole, $Tb = 276.71°K$, where $\Delta Hv = 5.715$ kcal/mole, and $S^°_{298}(g) = 58.61$ cal/(mole °K). The experimental entropy is in good agreement with the $S^°_{298}(g) = 58.75$ cal/(mole °K) calculated by Weissman, Bernstein, Rosser, Meister, and Cleveland (1584), using their own vibrational assignment and moments of inertia close to the microwave results of Miller, Aamodt, Dousmanis, Townes, and Kraitchman (1008). Thermodynamic functions given by Weissman et al. are adopted here. Previous calculations by Gelles and Pitzer (465) and Sverdlin and Godnev (1453) are in reasonable agreement.

No. 760   Bromoethane C_2H_5Br (Ideal Gas State)   Mol Wt 108.976

$T°K$	$Cp°$	$S°$	$-(G°-H^°_{298})/T$	$H°-H^°_{298}$	$\Delta Hf°$	$\Delta Gf°$	Log Kp
		cal/(mole °K)			kcal/mole		
298	15.45	68.71	68.71	0.00	-15.30	-6.29	4.608
300	15.48	68.81	68.72	0.03	-15.33	-6.23	4.539
400	18.93	73.75	69.36	1.76	-19.95	-2.29	1.254
500	21.99	78.31	70.70	3.81	-20.73	2.21	-0.967
600	24.56	82.55	72.33	6.14	-21.35	6.86	-2.498
700	26.73	86.50	74.07	8.71	-21.83	11.60	-3.622
800	28.58	90.20	75.86	11.47	-22.19	16.40	-4.480
900	30.19	93.66	77.65	14.41	-22.45	21.24	-5.158
1000	31.59	96.91	79.41	17.50	-22.62	26.11	-5.707

Thermodynamic functions are calculated using the vibrational assignment of Sheppard (1341) as revised by Green and Holden (540), and the moments of inertia from the microwave spectroscopy of Flanagan and Pierce (430). Flanagan and Pierce derived the barrier to internal rotation as 3684 cal/mole from microwave measurements by Lide (867). Green and Holden calculated functions in reasonable agreement with the present work, using earlier microwave work of Wagner, Dailey, and Solimene (1566) and a barrier of 3567 cal/mole given by Lide (867). Li and Rossini

(863) reviewed vapor pressure data and calculated Antoine constants, which yield $Tb = 311.5°K$. With $Tc = 503.9°K$ and $Pc = 61.5$ atm, selected by Kobe and Lynn (780), the Antoine constants give $\Delta Hv_{298}° = 6.57$ kcal/mole and $\Delta Hv_{311.5} = 6.41$ kcal/mole. Kolossovsky and Alimov (799) measured the enthalpy of vaporization in fair agreement. Timmermans (1500, 1502) reported $Tm = 154.6°K$ and $\Delta Hm = 1.4$ kcal/mole. Kurbatov (814) measured liquid heat capacities, but his data are known to be of low accuracy.

Lane, Linnett, and Oswin (845) measured equilibrium constants for the reaction of ethylene and hydrogen bromide, from which is derived for bromoethane $\Delta Hf_{298}°(g) = -15.3$ kcal/mole. Carson, Carson, and Wilmshurst (194) found $\Delta Hf_{298}°(l) = -22.1$ kcal/mole by combining their measurements of the enthalpy of combustion of diethyl mercury with data of Hartley, Pritchard, and Skinner (576) on the reaction of diethyl mercury with bromine and mercuric bromide. Correction to the vapor state yields $\Delta Hf_{298}°(g) = -15.5$ kcal/mole. Ashcroft, Carson, Carter, and Laye (27) measured the enthalpy of reaction of ethyl bromide and lithium hydride and derived $\Delta Hf_{298}°(l) = -23.0$ kcal/mole. Converting to the gas state yields $\Delta Hf_{298}°(g) = -16.4$ kcal/mole. Fowell, Lacher, and Park (435) reported the enthalpy of hydrogenation to ethane and hydrogen bromide and calculated $\Delta Hf_{298}°(g) = -14.79$ kcal/mole. A weighted average of -15.3 kcal/mole is adopted. Rossini, Wagman, Evans, Levine, and Jaffe (1249) listed $\Delta Hf_{298}°(g) = -13.0$ kcal/mole, based on old data of historical interest.

No. 761 1,2-Dibromoethane $C_2H_4Br_2$ (Ideal Gas State) Mol Wt 187.884

	cal/(mole °K)				kcal/mole		
$T°K$	$Cp°$	$S°$	$-(G°-H_{298}°)/T$	$H°-H_{298}°$	$\Delta Hf°$	$\Delta Gf°$	Log Kp
298	20.40	78.81	78.81	0.00	-9.30	-2.53	1.852
300	20.45	78.94	78.82	0.04	-9.33	-2.49	1.810
400	23.83	85.30	79.66	2.26	-17.23	1.40	-0.762
500	26.80	90.94	81.36	4.80	-17.61	6.10	-2.666
600	29.24	96.05	83.39	7.60	-17.85	10.86	-3.957
700	31.20	100.71	85.54	10.63	-17.97	15.66	-4.890
800	32.94	104.99	87.71	13.83	-17.99	20.47	-5.591
900	34.48	108.96	89.85	17.21	-17.91	25.27	-6.137
1000	35.80	112.67	91.95	20.72	-17.74	30.07	-6.571

Pitzer (1151) measured low-temperature thermal data, including $Ttp = 283.1°K$ and $\Delta Hm = 2.62$ kcal/mole, and derived $S_{298}°(l) = 53.37$ cal/(mole °K). Gwinn and Pitzer (556) used the enthalpy of vaporization of 8.69 kcal/mole at 405°K reported by Mathews (940) to calculate $S_{298}°(g) = 78.81$ cal/(mole °K). Gwinn and Pitzer (556) also measured vapor heat capacities, but did not succeed in finding parameters for the internal rotation that gave statistical calculations in agreement with both entropy and vapor heat capacities. The experimental values are therefore adopted, and estimates are used to extrapolate to higher temperatures. Timmermans (1501) listed eight melting point values in good agreement with that reported by Pitzer (1151), and three boiling point values in good agreement at 404.8°K.

Conn, Kistiakowsky, and Smith (259) measured the enthalpy of addition of bromine to ethylene at 355°K. Reduction of these values to 298°K and use of enthalpies of formation of gaseous ethylene and bromine adopted for the present work yields $\Delta Hf^{\circ}_{298}(g) = -9.3$ kcal/mole. Several attempts to measure the enthalpy of combustion, (709, 1180, 1383, 1385) are all considerably at variance with the above data, reflecting the inadequacy of the early combustion techniques for bromine compounds.

No. 762 1-Bromopropane C_3H_7Br (Ideal Gas State) Mol Wt 123.002

$T°K$	$Cp°$	$S°$	$-(G°-H^{\circ}_{298})/T$	$H°-H^{\circ}_{298}$	$\Delta Hf°$	$\Delta Gf°$	Log Kp
	cal/(mole °K)				kcal/mole		
298	20.66	79.08	79.08	0.00	-21.00	-5.37	3.934
300	20.73	79.21	79.09	0.04	-21.04	-5.27	3.840
400	25.70	85.87	79.96	2.37	-26.00	0.94	-0.515
500	30.02	92.08	81.77	5.16	-27.05	7.80	-3.410
600	33.66	97.89	83.98	8.35	-27.89	14.85	-5.409
700	36.76	103.31	86.36	11.88	-28.54	22.03	-6.878
800	39.41	108.40	88.80	15.69	-29.02	29.28	-7.999
900	41.74	113.18	91.24	19.75	-29.36	36.60	-8.887
1000	43.70	117.68	93.67	24.02	-29.56	43.94	-9.604

Green and Holden (540) calculated thermodynamic functions using literature data and reasonable estimates for molecular parameters. Their results are adopted. Bjellerup (137) measured the enthalpy of combustion and derived $\Delta Hf^{\circ}_{298}(l) = -29.73$ kcal/mole and $\Delta Hf^{\circ}_{298}(g) = -21.98$ kcal/mole. Bjellerup also calculated $\Delta Hf^{\circ}_{298}(g) = -21.77$ kcal/mole from the enthalpy of reaction of cyclopropane and hydrogen bromide reported by Lacher, Kianpour, and Park (829), and $\Delta Hf^{\circ}_{298}(g) = -21.15$ kcal/mole from the enthalpy of reaction of 1-bromopropane and hydrogen reported by Lacher in a private communication. Earlier work on the enthalpy of reaction of cyclopropane and hydrogen bromide by Lacher, Walden, Lea, and Park (834) yields $\Delta Hf^{\circ}_{298}(g) = -18.6$ kcal/mole. Davies, Lacher, and Park (308) derived $\Delta Hf^{\circ}_{298}(g) = -19.91$ kcal/mole from the enthalpy of reaction of 1-bromopropane and hydrogen. Rozhnov and Andreevskii (1259) and Brouwer and Wibaut (173) measured equilibrium constants of isomerization to 2-bromopropane. With the entropy values and enthalpy of formation of 2-bromopropane adopted for this compilation, their results yield $\Delta Hf^\circ_{298}(g) = -20.2$ and -20.5 kcal/mole, respectively. A weighted average of $\Delta Hf^\circ_{298}(g) = -21.0$ kcal/mole is adopted.

Timmermans (1499) measured $Tm = 163.1°K$ and $\Delta Hm = 1.56$ kcal/mole. He (1501) listed several determinations of the boiling point in good agreement at $Tb = 344.15°K$, while Mathews and Fehlandt (941) found $\Delta Hv = 7.14$ kcal/mole at the boiling point.

No. 763 2-Bromopropane C_3H_7Br (Ideal Gas State) Mol Wt 123.002

	cal/(mole °K)				kcal/mole		
$T°K$	$Cp°$	$S°$	$-(G°-H^\circ_{298})/T$	$H°-H^\circ_{298}$	$\Delta Hf°$	$\Delta Gf°$	Log Kp
298	21.27	75.53	75.53	0.00	-23.20	-6.51	4.771
300	21.31	75.67	75.54	0.04	-23.23	-6.41	4.667
400	26.34	82.50	76.43	2.43	-28.14	0.15	-0.083
500	30.76	88.87	78.29	5.29	-29.12	7.34	-3.209
600	34.42	94.81	80.55	8.56	-29.88	14.71	-5.356
700	37.49	100.35	82.99	12.16	-30.45	22.19	-6.927
800	40.09	105.53	85.49	16.04	-30.87	29.73	-8.122
900	42.32	110.38	87.99	20.16	-31.15	37.33	-9.064
1000	44.26	114.95	90.46	24.49	-31.29	44.95	-9.824

Thermodynamic functions are calculated using the vibrational assignment of Sheppard (1344), the average bond distances recommended by Sutton (1449), and an estimate of 4.0 kcal/mole for the barrier to internal rotation. The lowest skeletal frequency, not assigned by Sheppard, is estimated as 280 cm^{-1}. Microwave spectroscopy by Schwendeman and Tobiason (1297) yielded moments of inertia whose product is only slightly different from that adopted here. The calculated functions are used with equilibrium constants for the reaction of propylene and hydrogen bromide reported by Rozhnov and Andreevskii (1258) to derive $\Delta Hf^\circ_{298}(g) = -23.2$ kcal/mole. Bjellerup (137) measured the enthalpy of combustion and calculated $\Delta Hf^\circ_{298}(l) = -30.85$ and $\Delta Hf^\circ_{298}(g) = -23.55$ kcal/mole. Davies, Lacher, and Park (308) measured the enthalpy of reaction with hydrogen to form hydrogen bromide and propane and derived $\Delta Hf^\circ_{298}(g) = -22.7$ kcal/mole. Earlier work on the enthalpy of addition of hydrogen bromide to propylene by Lacher, Walden, Lea, and Park (834) and by Lacher, Kianpour, and Park (829) gave somewhat more negative enthalpies of formation, possibly as a result of polymer formation. The equilibrium value is close to an average of the calorimetric work and is adopted.

Timmermans (1501) listed three references in good agreement at $Tb = 332.6°K$, but two references on the melting point differed markedly at 184.2° and 183.2°K. Mathews and Fehlandt (941) measured $\Delta Hv = 6.79$ kcal/mole at the boiling point.

No. 764 1,2-Dibromopropane $C_3H_6Br_2$ (Ideal Gas State) Mol Wt 201.910

$T°K$	$Cp°$	$S°$	$-(G°-H°_{298})/T$	$H°-H°_{298}$	$\Delta Hf°$	$\Delta Gf°$	$Log\,Kp$
		cal/(mole °K)			kcal/mole		
298	24.57	89.90	89.90	0.00	-17.40	-4.22	3.094
300	24.82	90.06	89.91	0.05	-17.44	-4.14	3.017
400	29.74	97.88	90.94	2.78	-25.77	1.96	-1.070
500	34.13	105.01	93.05	5.98	-26.50	8.98	-3.924
600	37.63	111.55	95.60	9.58	-27.03	16.12	-5.872
700	40.53	117.58	98.31	13.49	-27.39	23.35	-7.289
800	42.91	123.15	101.07	17.66	-27.61	30.61	-8.361
900	44.92	128.32	103.82	22.06	-27.70	37.89	-9.201
1000	46.74	133.15	106.51	26.64	-27.69	45.19	-9.876

Conn, Kistiakowsky, and Smith (259) reacted propylene and bromine to form 1,2-dibromopropane at 355°K and found $\Delta Hr = -29.41$ kcal/mole. Correction to 298°K, using $\Delta Cp = 5.9$ cal/(mole °K) and enthalpies of formation of gaseous propylene and bromine adopted for the present work, yields $\Delta Hf°_{298}(g) = -17.4$ kcal/mole. Thermodynamic functions of the ideal gas are estimated using the constants of Table 12-3. Timmermans (1501) and Stull (1432) are in reasonable agreement at $Tm = 328.6°$K, but listed boiling point determinations vary from 413.7° to 414.8°K. Kurbatov (814) reported liquid heat capacity data, but his results are known to be of low accuracy.

No. 765 1-Bromobutane C_4H_9Br (Ideal Gas State) Mol Wt 137.028

$T°K$	$Cp°$	$S°$	$-(G°-H°_{298})/T$	$H°-H°_{298}$	$\Delta Hf°$	$\Delta Gf°$	$Log\,Kp$
		cal/(mole °K)			kcal/mole		
298	26.13	88.39	88.39	0.00	-25.65	-3.08	2.259
300	26.23	88.56	88.40	0.05	-25.69	-2.94	2.144
400	32.64	97.00	89.50	3.00	-30.97	5.65	-3.086
500	38.27	104.90	91.80	6.56	-32.28	14.96	-6.539
600	43.00	112.31	94.61	10.63	-33.32	24.50	-8.925
700	47.04	119.25	97.64	15.13	-34.11	34.21	-10.681
800	50.48	125.76	100.75	20.01	-34.69	44.01	-12.021
900	53.49	131.88	103.87	25.21	-35.09	53.87	-13.082
1000	56.03	137.66	106.97	30.69	-35.31	63.78	-13.938

Green and Holden (540) calculated thermodynamic functions by adding increments to their values for 1-bromopropane. Their values are adopted. They also calculated $\Delta Hf°_{298}(g) = -25.65$ kcal/mole from the enthalpy of combustion determined by Bjellerup (138), and $\Delta Hv°_{298} = 8.82$ kcal/mole from vapor pressure data. Timmermans (1499, 1500) determined $Tm = 160.8°$K and $\Delta Hm = 1.6$ kcal/mole. Boiling point determinations listed by Timmermans (1501) are in good agreement at 374.7°K, while Mathews and Fehlandt (941) reported $\Delta Hv = 7.78$ kcal/mole at the boiling point. Deese (318) measured the heat capacity from 93° to 298°K. In the range 250–290°K his results can be fitted by $Cp\,(l) = 23.0 + 0.046T$ [cal/(mole °K)]. Deese also determined $Tm = 160.4$°K and $\Delta Hm = 2.21$ kcal/mole; these values are in poor agreement with those of Timmermans (1499, 1500).

No. 766 2-Bromobutane C_4H_9Br (Ideal Gas State) Mol Wt 137.028

	cal/(mole °K)				kcal/mole		
T°K	$Cp°$	$S°$	$-(G°-H°_{298})/T$	$H°-H°_{298}$	$\Delta Hf°$	$\Delta Gf°$	Log Kp
298	26.48	88.50	88.50	0.00	-28.70	-6.16	4.518
300	26.58	88.67	88.51	0.05	-28.74	-6.03	4.390
400	33.09	97.22	89.63	3.04	-33.98	2.55	-1.393
500	38.92	105.25	91.96	6.65	-35.23	11.83	-5.171
600	43.76	112.79	94.81	10.79	-36.20	21.34	-7.771
700	47.86	119.85	97.89	15.38	-36.91	30.99	-9.675
800	51.31	126.47	101.05	20.34	-37.41	40.72	-11.123
900	54.29	132.69	104.22	25.62	-37.73	50.51	-12.265
1000	56.93	138.55	107.37	31.19	-37.86	60.33	-13.184

Bjellerup (137) measured the enthalpy of combustion and derived $\Delta Hf°_{298}(l) = -37.12$ kcal/mole. From vapor pressure data he calculated $\Delta Hv°_{298} = 8.45$ kcal/mole and $\Delta Hf°_{298}(g) = -28.66$ kcal/mole. Bjellerup also reviewed results of Lacher, Billings, Campion, Lea, and Park (822) on the enthalpy of addition of hydrogen bromide to 1-butene, 2-butene, *cis*, and 2-butene, *trans*, and calculated $\Delta Hf°_{298}(g) = -28.75, -28.76$, and -28.61 kcal/mole, respectively, all in excellent agreement with the combustion result. An average of -28.70 kcal/mole is adopted. Thermodynamic functions are estimated using constants from Table 12-3. Timmermans (1501) listed three references in good agreement for $Tm = 161.2$°K, and two references in fair agreement for $Tb = 364.4$°K.

No. 767 2-Bromo-2-Methylpropane C_4H_9Br (Ideal Gas State) Mol Wt 137.028

	cal/(mole °K)				kcal/mole		
T°K	$Cp°$	$S°$	$-(G°-H°_{298})/T$	$H°-H°_{298}$	$\Delta Hf°$	$\Delta Gf°$	Log Kp
298	27.85	79.34	79.34	0.00	-32.00	-6.73	4.935
300	27.98	79.52	79.35	0.06	-32.04	-6.58	4.792
400	34.93	88.54	80.53	3.21	-37.12	2.89	-1.578
500	40.84	96.99	82.99	7.01	-38.18	13.02	-5.690
600	45.58	104.87	85.99	11.34	-38.95	23.33	-8.497
700	49.43	112.20	89.21	16.09	-39.50	33.76	-10.540
800	52.65	119.01	92.52	21.20	-39.85	44.24	-12.086
900	55.39	125.38	95.82	26.61	-40.04	54.77	-13.300
1000	57.74	131.34	99.08	32.27	-40.09	65.32	-14.275

Tobin (1509) assigned vibrational frequencies, Williams and Gordy (1615) determined two moments of inertia by microwave techniques, and Bowen, Gilchrist, and Sutton (152) made electron diffraction studies. These data and an estimated barrier to internal rotation of 4400 cal/mole are used to calculate thermodynamic functions. Functions for isobutene and hydrogen bromide adopted elsewhere in this work are then combined with the equilibrium constants for dehydrobromination given by Howlett (621) and Kistiakowsky and Stauffer (763) to derive $\Delta Hf°_{298}(g) = -32.0$ kcal/mole.

Bryce-Smith and Howlett (179) and Timmermans and Delcourt (1504) are in agreement at $Tm = 257.0°K$ and $Tb = 346.2°K$. Kushner, Crowe, and Smyth (817) reported $\Delta Hm = 0.47$ kcal/mole, as well as two solid-state transitions at 208.7° and 231.6°K, with enthalpies of transition of 1.35 and 0.25 kcal/mole, respectively. Kushner et al. also reported heat capacities from 120° to 265°K. The value for the liquid from 257° to 265°K is 36 cal/(mole °K).

No. 768   1,2-Dibromobutane $C_4H_8Br_2$ (Ideal Gas State)   Mol Wt 215.936

	cal/(mole °K)				kcal/mole		
$T°K$	$Cp°$	$S°$	$-(G°-H°_{298})/T$	$H°-H°_{298}$	$\Delta Hf°$	$\Delta Gf°$	Log Kp
298	30.38	97.70	97.70	0.00	-23.70	-3.14	2.299
300	30.45	97.89	97.71	0.06	-23.74	-3.01	2.193
400	36.77	107.54	98.98	3.43	-32.38	5.62	-3.069
500	42.25	116.35	101.58	7.39	-33.37	15.23	-6.659
600	46.70	124.46	104.73	11.84	-34.12	25.03	-9.115
700	50.40	131.94	108.09	16.70	-34.65	34.93	-10.906
800	53.60	138.89	111.51	21.91	-35.01	44.89	-12.263
900	56.20	145.35	114.92	27.40	-35.20	54.90	-13.330
1000	58.50	151.40	118.27	33.14	-35.26	64.92	-14.187

Bjellerup (137) measured the enthalpy of combustion and derived $\Delta Hf°_{298}(l) = -35.23$ kcal/mole. He used $\Delta Hv°_{298} = 10.80$ kcal/mole, based on vapor pressure data, to derive $\Delta Hf°_{298}(g) = -24.43$ kcal/mole. Bjellerup also derived $\Delta Hf°_{298}(g) = -22.24$ kcal/mole from the enthalpy of bromination of 1-butene at 355°K reported by Conn, Kistiakowsky, and Smith (259) and -21.46 kcal/mole from the enthalpy of bromination of 1-butene in carbon tetrachloride at 300°K reported by Lister (878). Since combustion results for other compounds investigated by Bjellerup are in good agreement with enthalpy of reaction data, the most weight is given to the combustion value in the present instance, and $\Delta Hf°_{298}(g)$ is adopted as -23.7 kcal/mole. Thermodynamic functions of the ideal gas are estimated using the constants of Table 12-3. Stull (1432) listed $Tm = 208.7°K$ and $Tb = 439.5°K$.

Aliphatic Bromine Compound Ideal Gas Tables

No. 769 2,3-Dibromobutane $C_4H_8Br_2$ (Ideal Gas State) Mol Wt 215.936

T°K	$Cp°$	$S°$	$-(G°-H°_{298})/T$	$H°-H°_{298}$	$\Delta Hf°$	$\Delta Gf°$	Log Kp
		cal/(mole °K)			kcal/mole		
298	29.77	94.40	94.40	0.00	-24.40	-2.85	2.091
300	29.84	94.59	94.41	0.06	-24.45	-2.72	1.981
400	36.60	104.12	95.66	3.39	-33.12	6.24	-3.411
500	42.30	112.92	98.24	7.35	-34.12	16.20	-7.083
600	46.90	121.06	101.38	11.81	-34.85	26.34	-9.593
700	50.70	128.58	104.73	16.70	-35.36	36.58	-11.421
800	53.90	135.56	108.16	21.93	-35.68	46.87	-12.805
900	56.50	142.07	111.57	27.46	-35.85	57.21	-13.892
1000	58.70	148.14	114.92	33.22	-35.88	67.56	-14.765

Conn, Kistiakowsky, and Smith (259) measured the enthalpy of addition of bromine to both isomers of 2-butene at 355°K. The product of addition to 2-butene, *cis*, gives the internally compensated *meso* form of 2,3-dibromobutane, while addition to 2-butene, *trans*, yields a racemic (or *dl*) mixture. Thermodynamic functions of the *meso* form and *dl* product are equal and are estimated using the constants of Table 12-3. The enthalpies of addition of bromine, when adjusted to 298°K, yield $\Delta Hf°_{298}$ (g) as -24.5 kcal/mole for the *dl* product and $\Delta Hf°_{298}(g) = -24.3$ kcal/mole for the *meso* form. These are the same within experimental error, and an average of -24.4 kcal/mole is used. Stull (1432) listed for the *dl* mixture $Tb = 433.7°K$, and for the *meso* form $Tm = 238.7°K$ and $Tb = 430.5°K$.

No. 770 2,3-Dibromo-2-Methylbutane $C_5H_{10}Br_2$ (Ideal Gas State) Mol Wt 229.962

T°K	$Cp°$	$S°$	$-(G°-H°_{298})/T$	$H°-H°_{298}$	$\Delta Hf°$	$\Delta Gf°$	Log Kp
		cal/(mole °K)			kcal/mole		
298	35.51	98.60	98.60	0.00	-33.20	-3.19	2.340
300	35.62	98.82	98.61	0.07	-33.25	-3.01	2.192
400	44.10	110.26	100.11	4.07	-42.20	8.84	-4.829
500	51.50	120.92	103.22	8.86	-43.38	21.74	-9.503
600	57.40	130.85	107.00	14.31	-44.20	34.84	-12.691
700	62.10	140.06	111.08	20.30	-44.74	48.07	-15.007
800	65.90	148.61	115.24	26.70	-45.06	61.34	-16.756
900	69.10	156.56	119.40	33.45	-45.19	74.66	-18.128
1000	71.90	163.99	123.49	40.51	-45.15	87.98	-19.227

Conn, Kistiakowsky, and Smith (259) measured the enthalpy of addition of bromine to 2-methyl-2-butene at 355°K as -30.39 kcal/mole. Adjustment to 298°K and a standard state of liquid bromine yields $\Delta Hf°_{298}(g) = -33.2$ kcal/mole. Thermodynamic properties of the ideal gas are estimated using the constants of Table 12-3. Dreisbach (353) listed $Tm = 288°K$ and $Tb = 444°K$.

No. 771 1-Bromopentane $C_5H_{11}Br$ (Ideal Gas State) Mol Wt 151.054

T°K	Cp°	S°	$-(G°-H°_{298})/T$	$H°-H°_{298}$	$\Delta Hf°$	$\Delta Gf°$	Log Kp
	cal/(mole °K)				kcal/mole		
298	31.60	97.70	97.70	0.00	-30.87	-1.37	1.001
300	31.73	97.90	97.71	0.06	-30.92	-1.19	0.863
400	39.58	108.13	99.05	3.64	-36.52	9.79	-5.347
500	46.52	117.72	101.83	7.95	-38.08	21.55	-9.418
600	52.34	126.73	105.24	12.90	-39.31	33.59	-12.234
700	57.32	135.18	108.92	18.39	-40.25	45.82	-14.306
800	61.55	143.12	112.70	24.34	-40.93	58.16	-15.887
900	65.24	150.59	116.50	30.68	-41.38	70.58	-17.138
1000	68.36	157.63	120.27	37.37	-41.62	83.04	-18.148

Green and Holden (540) calculated thermodynamic functions by adding increments to their values for 1-bromopropane. They also used the enthalpy of formation of the liquid reported by Bjellerup (138), $\Delta Hf°_{298}(l) = -40.72$ kcal/mole, and $\Delta Hv°_{298} = 9.85$ kcal/mole calculated from vapor pressure data, to derive $\Delta Hf°_{298}(g) = -30.87$ kcal/mole. Kushner, Crowe, and Smyth (817) measured $Tm = 185.2°K$ and $\Delta Hm = 2.74$ kcal/mole, as well as the heat capacity from 122° to 207°K. Deese (318) reported $Tm = 185.1°K$, $\Delta Hm = 3.44$ kcal/mole, and heat capacity data from 96° to 290°K. The heat capacity data from the two sources are in poor agreement, and an average of 40.5 ± 1.5 cal/(mole °K) fits the scattered data for the liquid. The poor agreement on ΔHm may reflect the polymorphism of the solid and the slow rates of transition between phases. Dreisbach (353) listed $Tb = 402.73°K$, at which temperature $\Delta Hv = 8.24$ kcal/mole.

No. 772 Bromoethylene C_2H_3Br (Ideal Gas State) Mol Wt 106.960

T°K	Cp°	S°	$-(G°-H°_{298})/T$	$H°-H°_{298}$	$\Delta Hf°$	$\Delta Gf°$	Log Kp
	cal/(mole °K)				kcal/mole		
298	13.26	65.83	65.83	0.00	18.73	19.30	-14.144
300	13.31	65.92	65.84	0.03	18.71	19.30	-14.059
400	15.91	70.11	66.39	1.49	14.52	20.33	-11.109
500	18.07	73.90	67.52	3.20	14.10	21.84	-9.544
600	19.83	77.36	68.88	5.09	13.74	23.42	-8.529
700	21.28	80.53	70.32	7.15	13.46	25.05	-7.822
800	22.50	83.45	71.78	9.34	13.22	26.72	-7.300
900	23.55	86.16	73.23	11.65	13.03	28.42	-6.902
1000	24.46	88.69	74.65	14.05	12.90	30.15	-6.588

Gullikson and Nielsen (546) studied the spectrum, assigned fundamental vibrational frequencies, and calculated thermodynamic properties using structural constants of Hugill, Coop, and Sutton (659). Their results, which are here extended to 1000°K, differ only slightly from earlier calculations by Richards (1224), based on the assignment of Thompson and

Torkington (1494) and the same structural constants. Lacher, Kianpour, Montgomery, Knedler, and Park (826) measured the enthalpy of hydrogenation of vinyl bromide to ethane and hydrogen bromide at 401°K. Adjustment to 298°K and combination with enthalpies of formation of ethane and hydrogen bromide yield $\Delta Hf°_{298}(g) = 18.73$ kcal/mole. Guyer, Schütze, and Weidenmann (555) and Mehl (989) gave vapor pressure data that are not in good agreement; further work is needed to establish the enthalpy of vaporization. Timmermans (1501) reported $Tb = 288.95°K$ and $Tm = 135.35°K$. Dreisbach's (352) value, $Tm = 133.61°K$, is in poor agreement. Physical properties of this compound are obviously not well established. Mehl (988) reported $Cp(l) = 26.15$ cal/(mole °K) at 288°K.

No. 773 3-Bromo-1-Propene C_3H_5Br (Ideal Gas State) Mol Wt 120.986

		cal/(mole °K)			kcal/mole		
$T°K$	$Cp°$	$S°$	$-(G°-H°_{298})/T$	$H°-H°_{298}$	$\Delta Hf°$	$\Delta Gf°$	Log Kp
298	18.56	75.80	75.80	0.00	11.80	19.11	-14.004
300	18.61	75.92	75.81	0.04	11.77	19.15	-13.950
400	22.69	81.85	76.59	2.11	7.25	22.50	-12.292
500	26.14	87.29	78.19	4.56	6.55	26.39	-11.536
600	29.01	92.32	80.13	7.32	5.98	30.41	-11.078
700	31.38	96.97	82.21	10.34	5.54	34.53	-10.779
800	33.40	101.30	84.33	13.58	5.19	38.69	-10.568
900	35.12	105.33	86.44	17.01	4.92	42.89	-10.416
1000	36.63	109.11	88.52	20.60	4.75	47.13	-10.299

Gellner and Skinner (466) measured the enthalpy of hydrolysis and derived $\Delta Hf°_{298}(l) = 4.1$ kcal/mole. The enthalpy of vaporization at 298°K is estimated here as 7.8 kcal/mole, to give $\Delta Hf°_{298}(g) = 11.9$ kcal/mole. Electron impact studies by Lossing, Ingold, and Henderson (887) related the enthalpies of formation of allyl iodide and allyl bromide. Acceptance of Gellner and Skinner's (466) value for allyl iodide yields $\Delta Hf°_{298}(g) = 10.6$ kcal/mole for allyl bromide. Thomsen (1495) reported an enthalpy of combustion that yields $\Delta Hf°_{298}(g) = 12.9$ kcal/mole. An average of 11.8 kcal/mole is adopted. Thermodynamic functions of the ideal gas are estimated from the constants of Table 12-3, using 1-butene as the parent hydrocarbon. Bowen, Gilchrist, and Sutton (152) determined structural constants by electron diffraction techniques.

AROMATIC BROMINE COMPOUND IDEAL GAS TABLE

No. 774 Bromobenzene C_6H_5Br (Ideal Gas State) Mol Wt 157.016

Thermodynamic functions are calculated using the moments of inertia determined by microwave spectroscopy by Rosenthal and Dailey (1237) and the vibrational assignment of Whiffen (1603). Stull (1431) measured the low-temperature heat capacity from 90° to 320°K and estimated values

No. 774   Bromobenzene C_6H_5Br (Ideal Gas State)   Mol Wt 157.016

$T°K$	cal/(mole °K)				kcal/mole		Log K_p
	$C_p°$	$S°$	$-(G°-H°_{298})/T$	$H°-H°_{298}$	$\Delta H f°$	$\Delta G f°$	
298	23.35	77.53	77.53	0.00	25.10	33.11	-24.267
300	23.49	77.68	77.54	0.05	25.07	33.16	-24.152
400	30.44	85.42	78.54	2.75	20.44	36.76	-20.084
500	36.30	92.86	80.67	6.10	19.69	40.93	-17.890
600	40.99	99.91	83.29	9.97	19.10	45.23	-16.474
700	44.74	106.52	86.14	14.27	18.65	49.63	-15.494
800	47.78	112.69	89.08	18.90	18.31	54.07	-14.771
900	50.30	118.47	92.03	23.80	18.06	58.56	-14.219
1000	52.40	123.88	94.95	28.94	17.93	63.07	-13.783

from 0° to 90°K to calculate a third-law entropy. Mathews' (940) measured value of $\Delta Hv = 9.05$ kcal/mole at 429°K, together with an estimated -12 cal/(mole °K) for ΔCp of vaporization, yields $\Delta Hv°_{298} = 10.62$ kcal/mole. If data listed by Stull (1432) are used to estimate the vapor pressure at 25°C, the entropy of the ideal gas is calculated as 75.2 cal/(mole °K). This is more than 2 cal/(mole °K) lower than the spectroscopic value. Since the third-law value involves a long extrapolation and consequent uncertainty, the spectroscopic result is adopted.

Smith and Sunner (1385) and Smith and Bjellerup (1383) reported enthalpy of combustion data. Experimental details are lacking, and considerable uncertainty is involved in deriving $\Delta Hf°_{298}(l) = 14.8$ and 15.7 kcal/mole, respectively. Chernick, Skinner, and Wadsö (225) revised slightly the results of Hartley, Pritchard, and Skinner (576) for the enthalpy of reaction of diphenyl mercury and bromine to form mercuric bromide and bromobenzene. Rossini, Wagman, Evans, Levine, and Jaffe (1249) selected $\Delta Hf°_{298}(s) = -40.5$ kcal/mole for mercuric bromide. For diphenyl mercury, three values are available: Chernick, Skinner, and Wadsö (225) found 65.4 kcal/mole by reaction with hydrogen chloride and mercuric chloride; Fairbrother and Skinner (404) found 66.9 kcal/mole by enthalpy of combustion work; and Carson, Carson, and Wilmshurst (194) found 71.3 kcal/mole also by enthalpy of combustion. These data yield for bromobenzene $\Delta Hf°_{298}(l) = 13.7$, 14.4, and 16.7 kcal/mole, respectively. A weighted average of 14.5 kcal/mole is adopted. The enthalpy of vaporization previously derived yields $\Delta Hf°_{298}(g) = 25.1$ kcal/mole.

Stull (1431) determined Tm as 242.43°K and ΔHm as 2.54 kcal/mole. The heat capacity of the liquid in the range 270° to 320°K given by Stull (1431) can be expressed by $Cp = 30.1 + 0.024T$ [cal/(mole °K)]. Timmermans (1501) listed several determinations of the boiling point, in good agreement at 429.3°K.

ALIPHATIC IODINE COMPOUND IDEAL GAS TABLES

No. 775   Iodomethane CH_3I (Ideal Gas State)   Mol Wt 141.944

	cal/(mole °K)				kcal/mole		
$T°K$	$Cp°$	$S°$	$-(G°-H°_{298})/T$	$H°-H°_{298}$	$\Delta Hf°$	$\Delta Gf°$	Log Kp
298	10.55	60.71	60.71	0.00	3.34	3.74	-2.742
300	10.58	60.78	60.72	0.02	3.32	3.74	-2.727
400	12.36	64.07	61.15	1.17	0.60	4.07	-2.226
500	13.95	67.01	62.03	2.49	-5.21	5.51	-2.407
600	15.31	69.67	63.09	3.96	-5.62	7.69	-2.801
700	16.48	72.12	64.21	5.55	-5.96	9.94	-3.103
800	17.51	74.39	65.34	7.25	-6.23	12.23	-3.340
900	18.40	76.51	66.46	9.04	-6.44	14.55	-3.532
1000	19.20	78.49	67.57	10.92	-6.59	16.89	-3.691

Two recent determinations of the equilibrium constant of the reaction of methane and iodine to form iodomethane and hydrogen iodide are in excellent agreement. Goy and Pritchard (521) derived $\Delta Hf°_{298}(g) = 3.40$ kcal/mole, while Golden, Walsh, and Benson (505) calculated $\Delta Hf°_{298}(g)$ as 3.28 kcal/mole. Carson, Carter, and Pedley (195) made a careful study of the enthalpy of reaction of lithium hydride with iodomethane and with elemental iodine and derived $\Delta Hf°_{298}(l)$ as -2.9 kcal/mole. Li and Rossini (863) reviewed vapor pressure data and derived $Tb = 315.5°K$ and constants for the Antoine equation. With $Tc = 528°K$ listed by Kobe and Lynn (780) and an estimated $Pc = 65$ atm, the Haggenmacher equation (560) yields $\Delta Hv°_{298} = 6.63$ kcal/mole. From these results $\Delta Hf°_{298}(g) = 3.7$ kcal/mole is calculated. Cottrell (276) gave an excellent review of earlier calorimetric work on iodomethane and selected $\Delta Hf°_{298}(l) = -2.6$ kcal/mole. With the present $\Delta Hv°_{298}$ this yields $\Delta Hf°_{298}(g) = 4.0$ kcal/mole. Rossini, Wagman, Evans, Levine, and Jaffe (1249) listed $\Delta Hf°_{298}(g) = 4.9$ kcal/mole and $\Delta Hf°_{298}(l) = -2.0$ kcal/mole, based on data obtained over 50 years ago. The equilibrium values are adopted here, giving $\Delta Hf°_{298}(g) = 3.34$ kcal/mole. The calorimetric work is in agreement within experimental error.

Gelles and Pitzer (465) calculated thermodynamic functions based on the vibrational assignment of Pitzer and Gelles (1159). The over-all moment of inertia they used appears to be in error, and their entropy values are corrected using moments of inertia from the microwave study of Miller, Aamodt, Dousmanis, Townes, and Kraitchman (1008). Liquid heat capacity data of Harrison and Moelwyn-Hughes (572) can be fitted by $Cp(l) = 12.35 + 0.025T$ [cal/(mole °K), 270–310°K]. Kurbatov (814) also measured liquid heat capacity, but his values appear to be much too high. Similarly, enthalpy of vaporization measurements of Kolossovsky and Alimov (799) seem to be several percent too high. Timmermans (1501) selected $Tm = 315.65°K$.

No. 776 Diiodomethane CH_2I_2 (Ideal Gas State) Mol Wt 267.846

T°K	Cp°	S°	$-(G°-H°_{298})/T$	$H°-H°_{298}$	$\Delta Hf°$	$\Delta Gf°$	Log Kp
		cal/(mole °K)			kcal/mole		
298	13.83	73.88	73.88	0.00	28.20	24.16	-17.709
300	13.86	73.97	73.89	0.03	28.18	24.13	-17.581
400	15.76	78.23	74.45	1.52	23.56	23.04	-12.588
500	17.25	81.91	75.58	3.17	12.67	23.95	-10.469
600	18.38	85.16	76.91	4.95	12.48	26.23	-9.553
700	19.30	88.06	78.30	6.84	12.34	28.53	-8.908
800	20.08	90.69	79.69	8.81	12.24	30.85	-8.428
900	20.73	93.10	81.05	10.85	12.18	33.18	-8.057
1000	21.30	95.31	82.37	12.95	12.16	35.52	-7.762

Berthelot (122) measured the enthalpy of combustion listed by Kharasch (744), which yields $\Delta Hf°_{298}(l) = 16.0$ kcal/mole. Dreisbach (353) calculated $\Delta Hv°_{298} = 12.2$ kcal/mole, which gives $\Delta Hf°_{298}(g) = 28.2$ kcal/mole. Gelles and Pitzer (465) calculated the thermodynamic functions given here. Entropies calculated by Voelz, Cleveland, Meister, and Bernstein (1540) are too high by about 3 cal/(mole °K). Rossini, Wagman, Evans, Levine, and Jaffe (1249) listed melting points for two crystalline forms at 279.3° and 278.76°K, with enthalpies of melting of 2.88 and 3.02 kcal/mole, respectively. Dreisbach (353) selected $Tb = 455°K$; Kurbatov (814) reported liquid heat capacity data that appear to be too high.

No. 777 Triiodomethane CHI_3 (Ideal Gas State) Mol Wt 393.748

T°K	Cp°	S°	$-(G°-H°_{298})/T$	$H°-H°_{298}$	$\Delta Hf°$	$\Delta Gf°$	Log Kp
		cal/(mole °K)			kcal/mole		
298	17.91	85.00	85.00	0.00	50.40	42.53	-31.174
300	17.94	85.12	85.01	0.04	50.39	42.48	-30.946
400	19.57	90.51	85.73	1.92	43.91	40.16	-21.939
500	20.71	95.01	87.15	3.93	27.98	40.73	-17.803
600	21.52	98.86	88.79	6.05	28.02	43.28	-15.763
700	22.13	102.22	90.47	8.23	28.08	45.82	-14.304
800	22.61	105.21	92.13	10.47	28.15	48.35	-13.207
900	23.01	107.90	93.74	12.75	28.24	50.86	-12.351
1000	23.34	110.34	95.28	15.07	28.33	53.37	-11.664

Berthelot (122) measured the enthalpy of combustion listed by Kharasch (744), from which is derived $\Delta Hf°_{298}(s) = 33.7$ kcal/mole. Jones (697) listed $\Delta Hs°_{298} = 16.7$ kcal/mole, based on vapor pressure data, yielding $\Delta Hf°_{298}(g) = 50.4$ kcal/mole. Plyler and Benedict (1172) reported the infrared spectrum, and Stammreich and Forneris (1405) determined the Raman spectrum. Sverdlin (1452) calculated functions in reasonable agreement with those given here, which are calculated on the basis of the Raman spectrum and an estimated C−I distance of 2.1 Å. Rossini, Wagman, Evans, Levine, and Jaffe (1249) listed $Tm = 398°K$ and $\Delta Hm = 3.9$ kcal/mole.

No. 778   Iodoethane C_2H_5I (Ideal Gas State)   Mol Wt 155.970

$T°K$	$Cp°$	$S°$	$-(G°-H°_{298})/T$	$H°-H°_{298}$	$\Delta Hf°$	$\Delta Gf°$	Log Kp
		cal/(mole °K)			kcal/mole		
298	15.76	70.82	70.82	0.00	-2.00	5.10	-3.737
300	15.82	70.92	70.83	0.03	-2.02	5.14	-3.745
400	19.18	75.94	71.49	1.79	-5.08	7.77	-4.248
500	22.13	80.55	72.85	3.86	-11.16	11.59	-5.064
600	24.64	84.81	74.49	6.20	-11.77	16.19	-5.898
700	26.80	88.78	76.25	8.77	-12.25	20.90	-6.523
800	28.65	92.48	78.05	11.55	-12.61	25.65	-7.007
900	30.25	95.95	79.85	14.49	-12.87	30.45	-7.395
1000	31.65	99.21	81.62	17.59	-13.03	35.28	-7.710

Sheppard (1341) assigned fundamental frequencies of the ethyl halides. Green and Holden (540) indicated that one frequency of the chloroethane and bromoethane assignments should be revised; accordingly one Sheppard frequency for iodoethane at 985 cm^{-1} is estimated as 1245 cm^{-1}. Kasuya and Oka (724) derived moments of inertia from microwave spectroscopy, and Kasuya (723) found a barrier to internal rotation of 3220 cal/mole from rotational splitting. Thermodynamic functions are calculated employing these values.

Springall and White (1402) measured the enthalpy of combustion and found $\Delta Hf°_{298}(l) = -8.4$ kcal/mole. Carson, Hartley, and Skinner (196) measured the enthalpy of reaction of diethyl zinc with water and with iodine and derived $\Delta Hf°_{298}(l) = -9.1$ kcal/mole. The enthalpy of formation they found for diethyl zinc from the hydrolysis reaction was in good agreement with the enthalpy of combustion results of Long and Norrish (882). Hartley, Pritchard, and Skinner (576) determined the enthalpy of reaction of diethyl mercury with iodine and mercuric iodide. Carson, Carson, and Wilmshurst (194) combined the data of Hartley et al. and their own measurements of the enthalpy of combustion of diethyl mercury to derive $\Delta Hf°_{298}(l) = -10.13$ kcal/mole. Mathews (940) measured $\Delta Hv = 7.115$ kcal/mole at 345°K, whereas Kolossovsky and Alimov (799) measured $\Delta Hv = 7.66$ kcal/mole at 303°K. Adjustment of these results to 298°K with an estimated -12 cal/(mole °K) for ΔCp of vaporization gives good agreement for $\Delta Hv°_{298} = 7.7$ kcal/mole. Benson and Bose

(106) measured the equilibrium constant at 630°K for the pyrolysis of ethyl iodide to ethylene and hydrogen iodide, from which is calculated $\Delta Hf^\circ_{298}(g) = -3.0$ kcal/mole. An average value of $\Delta Hf^\circ_{298}(g) = -2.0$ kcal/mole is selected. Rossini, Wagman, Evans, Levine, and Jaffe (1249) listed $\Delta Hf^\circ_{298}(l) = -7.4$ kcal/mole, based on old, uncertain combustion data.

Timmermans (1501) listed $Tm = 162.3°K$, and Li and Rossini (863), reviewing vapor pressure data, selected $Tb = 345.5°K$. Kurbatov (814) measured liquid heat capacity, but his values are known to be of low accuracy.

No. 779 1,2-Diiodoethane $C_2H_4I_2$ (Ideal Gas State) Mol Wt 281.872

T°K	$Cp°$	$S°$	$-(G°-H°_{298})/T$	$H°-H°_{298}$	$\Delta Hf°$	$\Delta Gf°$	Log Kp
	cal/(mole °K)				kcal/mole		
298	19.67	83.30	83.30	0.00	15.90	18.76	-13.753
300	19.72	83.43	83.31	0.04	15.88	18.78	-13.680
400	22.94	89.55	84.12	2.18	10.96	20.05	-10.953
500	25.70	94.98	85.76	4.61	-0.16	23.39	-10.223
600	27.92	99.86	87.71	7.30	-0.53	28.13	-10.247
700	29.79	104.31	89.77	10.19	-0.79	32.93	-10.282
800	31.37	108.40	91.84	13.25	-0.96	37.76	-10.316
900	32.70	112.17	93.90	16.45	-1.06	42.61	-10.347
1000	33.84	115.68	95.90	19.78	-1.08	47.47	-10.374

Benson and Amano (105) measured equilibrium compositions in the system ethylene–iodine–diiodoethane and combined them with previous measurements by Abrams and Davis (1) and Cuthbertson and Kistiakowsky (301) to derive for diiodoethane $\Delta Hf^\circ_{298}(g) = 15.9$ kcal/mole and $S^\circ_{298}(g) = 83.3$ cal/(mole °K). Older data by Mooney and Ludlam (1022) are not in agreement.

Heat capacity values are estimated using the constants of Table 12-4. Dreisbach (353) listed $Tm = 354°K$ and $Tb = 473°K$.

Berthelot (122) measured the enthalpy of combustion, from which is derived $\Delta Hf^\circ_{298}(s) = -0.1$ kcal/mole. Abrams and Davis (1) derived $\Delta Hs^\circ_{298}(g) = 15.7$ kcal/mole from vapor pressure data reported by Mooney and Ludlam (1022). The calculated $\Delta Hf^\circ_{298}(g) = 15.6$ kcal/mole is in good agreement with the equilibrium value adopted above.

No. 780 1-Iodopropane C_3H_7I (Ideal Gas State) Mol Wt 169.996

Mortimer, Pritchard, and Skinner (1030) measured the enthalpies of reaction of dipropyl mercury with bromine and with iodine. These data relate the enthalpies of formation of 1-bromopropane, 1-iodopropane,

No. 780 1-Iodopropane C_3H_7I (Ideal Gas State) Mol Wt 169.996

	cal/(mole °K)			kcal/mole			
$T°K$	$Cp°$	$S°$	$-(G°-H°_{298})/T$	$H°-H°_{298}$	$\Delta Hf°$	$\Delta Gf°$	Log Kp
298	21.48	80.32	80.32	0.00	-7.30	6.68	-4.894
300	21.56	80.46	80.33	0.04	-7.33	6.76	-4.926
400	26.27	87.32	81.23	2.44	-10.69	11.75	-6.421
500	30.52	93.65	83.09	5.29	-17.01	17.99	-7.862
600	34.11	99.54	85.34	8.52	-17.80	25.06	-9.128
700	37.17	105.03	87.77	12.09	-18.41	32.26	-10.072
800	39.80	110.17	90.25	15.94	-18.86	39.53	-10.797
900	42.07	114.99	92.73	20.04	-19.17	46.85	-11.376
1000	44.03	119.53	95.19	24.34	-19.34	54.20	-11.845

mercuric bromide, and mercuric iodide. Employing enthalpies of formation of mercuric bromide and mercuric iodide listed by Rossini, Wagman, Evans, Levine, and Jaffe (1249) and the enthalpy of formation of 1-bromopropane evaluated elsewhere in this work yields for 1-iodopropane $\Delta Hf°_{298}(l) = -16.3$ kcal/mole. Dreisbach (353) selected $\Delta Hv°_{298} = 8.6$ kcal/mole, which gives $\Delta Hf°_{298}(g) = -7.7$ kcal/mole. Andreevskii and Rozhnov (22) measured the equilibrium between 1-iodopropane and 2-iodopropane. Their results, when combined with the present selected enthalpy of formation of 2-iodopropane, yield for 1-iodopropane $\Delta Hf°_{298}(g) = -6.9$ kcal/mole. An average of -7.3 kcal/mole is adopted.

Thermodynamic functions of the ideal gas are estimated using the constants of Table 12-4. Timmermans (1501) listed four determinations of the boiling point in good agreement at 375.6°K, but his melting point of 175.5°K is at variance with that of 171.9°K listed by Dreisbach (353).

No. 781 2-Iodopropane C_3H_7I (Ideal Gas State) Mol Wt 169.996

	cal/(mole °K)			kcal/mole			
$T°K$	$Cp°$	$S°$	$-(G°-H°_{298})/T$	$H°-H°_{298}$	$\Delta Hf°$	$\Delta Gf°$	Log Kp
298	21.53	77.55	77.55	0.00	-10.00	4.80	-3.521
300	21.61	77.69	77.56	0.04	-10.03	4.89	-3.565
400	26.59	84.60	78.46	2.46	-13.37	10.16	-5.550
500	30.96	91.02	80.34	5.34	-19.65	16.66	-7.282
600	34.58	96.99	82.62	8.63	-20.40	23.99	-8.739
700	37.62	102.56	85.08	12.24	-20.96	31.44	-9.816
800	40.21	107.75	87.59	16.13	-21.37	38.95	-10.641
900	42.42	112.62	90.11	20.27	-21.64	46.51	-11.294
1000	44.34	117.19	92.59	24.61	-21.78	54.10	-11.823

Thermodynamic functions are calculated using the vibrational assignment of Sheppard (1344), the average bond distances recommended by Sutton (1449), and an estimate of 4.0 kcal/mole for the barrier to internal rotation. The lowest fundamental frequency, not given by Sheppard, was estimated as 250 cm^{-1}. These functions, with the equilibrium constant of dissociation to propylene and hydrogen iodide at 511.7°K reported by Bose and Benson (149), yield $\Delta Hf°_{298}(g) = -10.5$ kcal/mole. Mortimer, Pritchard, and Skinner (1030) measured the enthalpy of reaction of diisopropyl mercury with bromine and iodine. If the enthalpies of formation of mercuric bromide and mercuric iodide selected by Rossini, Wagman, Evans, Levine, and Jaffe (1249) and of 2-bromopropane selected elsewhere in this chapter are used, $\Delta Hf°_{298}(l)$ is found to be -18.1 kcal/mole. Mortimer et al. used published vapor pressure data to calculate $\Delta Hv°_{298} = 8.7$ kcal/mole, which yields $\Delta Hf°_{298}(g) = -9.4$ kcal/mole. An average $\Delta Hf°_{298}(g) = -10.0$ kcal/mole is adopted.

Timmermans (1501) listed $Tb = 362.6°K$ and $Tm = 183.2°K$.

No. 782   1,2-Diiodopropane $C_3H_6I_2$ (Ideal Gas State)   Mol Wt 295.898

	cal/(mole °K)				kcal/mole		
T°K	Cp°	S°	$-(G°-H°_{298})/T$	$H°-H°_{298}$	$\Delta Hf°$	$\Delta Gf°$	Log Kp
298	24.77	94.60	94.60	0.00	8.60	17.81	-13.051
300	24.86	94.76	94.61	0.05	8.57	17.86	-13.011
400	29.83	102.61	95.64	2.79	3.32	21.31	-11.644
500	34.07	109.74	97.76	5.99	-8.06	26.91	-11.763
600	37.51	116.26	100.31	9.58	-8.60	33.96	-12.368
700	40.43	122.27	103.02	13.48	-8.98	41.09	-12.827
800	42.85	127.83	105.78	17.64	-9.21	48.25	-13.181
900	44.92	133.00	108.52	22.04	-9.32	55.44	-13.463
1000	46.74	137.83	111.21	26.62	-9.30	62.64	-13.690

Benson and Amano (105) measured equilibrium compositions in the system propylene-iodine-1,2-diiodopropane and derived $\Delta Hf°_{298}(g) = 8.6$ kcal/mole and $S°_{298}(g) = 94.6$ cal/(mole °K). Heat capacities above 298°K are estimated using the constants of Table 12-4.

No. 783   2-Iodo-2-Methylpropane C_4H_9I (Ideal Gas State   Mol Wt 184.022

	cal/(mole °K)				kcal/mole		
T°K	Cp°	S°	$-(G°-H°_{298})/T$	$H°-H°_{298}$	$\Delta Hf°$	$\Delta Gf°$	Log Kp
298	28.27	81.79	81.79	0.00	-17.60	5.65	-4.142
300	28.40	81.97	81.80	0.06	-17.63	5.79	-4.220
400	35.27	91.11	82.99	3.25	-21.14	13.92	-7.606
500	41.12	99.63	85.48	7.08	-27.49	23.32	-10.191
600	45.82	107.55	88.51	11.43	-28.24	33.55	-12.219
700	49.61	114.91	91.76	16.21	-28.77	43.89	-13.704
800	52.85	121.75	95.08	21.34	-29.11	54.29	-14.831
900	55.58	128.13	98.41	26.76	-29.28	64.74	-15.719
1000	57.91	134.12	101.68	32.44	-29.31	75.19	-16.432

Thermodynamic functions are calculated on the basis of the vibrational assignment of Tobin (1509), the structural parameters of Williams and Gordy (1615) and Bowen, Gilchrist, and Sutton (152), and an estimated barrier to internal rotation of 4.4 kcal/mole. Jones and Ogg (700) reported equilibrium data for the dehydrohalogenation to isobutene and hydrogen iodide. Using Gibbs energy functions adopted for the present work, we derive an average enthalpy of reaction at 298°K of 19.74 kcal/mole. Combined with enthalpies of formation of hydrogen iodide and isobutene, this yields $\Delta Hf°_{298}(g) = -17.6$ kcal/mole for 2-iodo-2-methylpropane.

Milazzo (999) measured the vapor pressure of the solid and liquid and found $\Delta Hm = 3.47$ kcal/mole at the melting point, $Tb = 235.0°$K.

No. 784 1,2-Diiodobutane $C_4H_8I_2$ (Ideal Gas State) Mol Wt 309.924

	cal/(mole °K)				kcal/mole		
T°K	$Cp°$	$S°$	$-(G°-H°_{298})/T$	$H°-H°_{298}$	$\Delta Hf°$	$\Delta Gf°$	Log Kp
298	30.58	101.80	101.80	0.00	2.85	19.62	-14.381
300	30.69	101.99	101.81	0.06	2.81	19.72	-14.367
400	36.81	111.68	103.08	3.44	-2.74	25.76	-14.074
500	42.19	120.49	105.70	7.40	-14.37	34.02	-14.868
600	46.60	128.58	108.85	11.85	-15.14	43.77	-15.941
700	50.30	136.05	112.21	16.70	-15.69	53.63	-16.744
800	53.50	142.98	115.62	21.89	-16.06	63.56	-17.362
900	56.20	149.44	119.03	27.38	-16.26	73.53	-17.854
1000	58.50	155.49	122.37	33.12	-16.32	83.51	-18.251

Cline and Kistiakowsky (236) measured equilibrium compositions in the system 1-butene–iodine–diiodobutane and derived $\Delta Hr = -12.0$ kcal/mole, from which is derived $\Delta Hf°_{298}(g) = 2.85$ kcal/mole. Thermodynamic functions of the ideal gas are estimated using the constants of Table 12-4.

No. 785 3-Iodo-1-Propene C_3H_5I (Ideal Gas State) Mol Wt 167.980

	cal/(mole °K)				kcal/mole		
T°K	$Cp°$	$S°$	$-(G°-H°_{298})/T$	$H°-H°_{298}$	$\Delta Hf°$	$\Delta Gf°$	Log Kp
298	19.75	76.46	76.46	0.00	22.90	28.72	-21.053
300	19.84	76.59	76.47	0.04	22.88	28.76	-20.949
400	23.89	82.87	77.29	2.23	20.01	30.93	-16.900
500	27.25	88.57	78.99	4.80	14.11	34.24	-14.965
600	30.00	93.79	81.03	7.66	13.64	38.31	-13.953
700	32.20	98.58	83.20	10.77	13.28	42.45	-13.253
800	34.20	103.02	85.40	14.10	13.01	46.63	-12.739
900	35.80	107.14	87.59	17.60	12.82	50.85	-12.348
1000	37.20	110.98	89.74	21.25	12.71	55.09	-12.039

Gellner and Skinner (466) measured the enthalpy of hydrolysis of 2-iodo-1-propene and derived $\Delta Hf°_{298}(l) = 14.0$ kcal/mole. Dreisbach (353) listed $Tm = 173.85°K$, $Tb = 375°K$, and $\Delta Hv°_{298} = 8.9$ kcal/mole, which gives $\Delta Hf°_{298}(g) = 22.9$ kcal/mole. Thermodynamic functions of the ideal gas are estimated by using the constants of Table 12-4, with 1-butene as the parent hydrocarbon. Bowen, Gilchrist, and Sutton (152) determined the structure by electron diffraction. Of historical interest is an old enthalpy of combustion reported by Berthelot and listed by Kharasch (744), which appears to be several kilocalories too high.

AROMATIC IODINE COMPOUND IDEAL GAS TABLE

No. 786   Iodobenzene C_6H_5I (Ideal Gas State)   Mol Wt 204.010

		cal/(mole °K)			kcal/mole		
$T°K$	$Cp°$	$S°$	$-(G°-H°_{298})/T$	$H°-H°_{298}$	$\Delta Hf°$	$\Delta Gf°$	Log Kp
298	24.08	79.84	79.84	0.00	38.85	44.88	−32.898
300	24.22	79.99	79.85	0.05	38.83	44.92	−32.721
400	31.10	87.93	80.88	2.83	35.80	47.19	−25.784
500	36.86	95.51	83.06	6.23	29.79	50.63	−22.130
600	41.43	102.65	85.73	10.16	29.25	54.85	−19.976
700	45.08	109.32	88.63	14.49	28.83	59.15	−18.468
800	48.07	115.54	91.61	19.15	28.52	63.50	−17.347
900	50.55	121.35	94.60	24.08	28.30	67.89	−16.485
1000	52.60	126.78	97.55	29.24	28.18	72.30	−15.800

Hartley, Pritchard, and Skinner (576) measured the enthalpy of reaction of diphenyl mercury with iodine to form mercuric iodide and iodobenzene. Their values were revised slightly by Chernick, Skinner, and Wadsö (225), who also measured the enthalpy of reaction of diphenyl mercury with hydrogen chloride. Combination of these data with enthalpies of formation of mercuric chloride and mercuric iodide (−53.5 and −25.2 kcal/mole, respectively) yields for iodobenzene $\Delta Hf°_{298}(l) = 26.7$ kcal/mole. Smith (1382) recalculated the enthalpy of combustion data of Karlsson (719), leading to $\Delta Hf°_{298}(l) = 28.0$ kcal/mole. Graham, Nichol, and Ubbelohde (522) measured the enthalpy of reaction of phenyl magnesium bromide with hydrogen iodide and iodine, giving $\Delta Hf°_{298}(l) = 21.9$ kcal/mole for iodobenzene. A weighted average of 27.0 kcal/mole is adopted.

Stull (1431) measured the low-temperature heat capacity from 90° to 290°K, as well as $Tm = 241.83°K$ and $\Delta Hm = 2.33$ kcal/mole. Dreisbach (351) listed $\Delta Hv°_{298} = 11.85$ kcal/mole and $Tb = 461.48°K$, at which temperature $\Delta Hv = 9.44$ kcal/mole. Whiffen (1603) assigned vibrational frequencies and calculated thermodynamic properties of the ideal gas. The entropy from the low-temperature thermal data is not in good agreement with that calculated by Whiffen; because of the uncertainty inherent in the long extrapolation from 90° to 0°K, the calculated value is adopted.

CHAPTER 13

CHEMICAL THERMODYNAMICS OF ORGANIC SULFUR COMPOUNDS

INTRODUCTION

Although thermochemical studies of several simple alkanethiols and sulfides were reported by Thomsen (1495) as early as 1886, the chemical thermodynamics of organic sulfur compounds have been generally neglected until very recently. At the time Parks and Huffman wrote their monograph (1105), chemical thermodynamic values for only two organic sulfur compounds (carbon disulfide and carbonyl sulfide) were available. As late as 1949, when Barrow and Pitzer (78) reviewed that subject, they could find data enough to merit compilation for only seven compounds. Utilizing theoretical calculations, they were able to provide relatively complete data for three alkanethiols, two alkane sulfides, one sulfone, and one sulfoxide. However, these calculations were based on inadequate experimental data, which, with a few notable exceptions, were of low accuracy. Today, largely because of the efforts of scientists at two laboratories—the Bartlesville Petroleum Research Center in Bartlesville, Okla., and the Thermochemical Laboratory at the University of Lund, Sweden—accurate Gibbs energy values have been measured for more than 20 organic sulfur compounds. The current program of studies on organic sulfur compounds in the Bartlesville laboratory anticipates detailed investigation of more than 60 individual sulfur compounds, each a "key" member of one or more homologous series. The information obtained in studies of these individual compounds will permit accurate and extensive calculation of tables of thermodynamic properties for about 2000 other sulfur compounds. These calculations are of the type that can be done rapidly and economically by automatic computers. However, currently available data for carbonyl sulfide are only slightly better than the 2 kcal/mole uncertainty estimated as pertinent by Parks and Huffman, indicating that our knowledge of the sulfur field is by no means complete. Such important classes of compounds as the thianes and the thiols, however, are now on a firm footing, and values for these compounds are of high accuracy.

There are several reasons for the paucity of chemical thermodynamic data for sulfur compounds until the last decade. First, thermodynamicists who studied organic compounds had rather naturally devoted most of their efforts to hydrocarbons, a class of greater economic importance than the sulfur compounds. The large and growing mass of reliable thermodynamic data for hydrocarbons has been accumulated at the consequent neglect of many other classes of organic compounds. Moreover, sulfur compounds have been avoided by physical chemists in general because some of the properties of such compounds are rather unpleasant. Finally, the study of sulfur compounds posed difficult problems, such as those discussed in the next paragraph, which had to be solved before much progress could be made. The situation at present is changing rapidly, and the rate of accumulation of thermodynamic data for organic sulfur compounds now exceeds that for any comparably large class of substances at any time. In addition to those laboratories already enumerated, that of the University of Belfast has recently produced much thermodynamic data on sulfur compounds as a consequence of the endeavors of Henry Mackle and his co-workers.

The heats of combustion of sulfur compounds had been determined by such well-known thermochemists as Berthelot, Andre, and Matignon (123), Roth and Rist-Schumacher (1254), and Huffman and Ellis (646, 647), but for reasons discussed by Waddington, Sunner, and Hubbard (1551) the results were not as accurate as normally expected from the well-established technique of combustion calorimetry. Hence the values listed in the compilation of Kharasch (744) are quite unreliable. The first combustion data of moderate accuracy were obtained by Huffman and Ellis (646) using an ordinary steel bomb to which no water was added. Air was not swept out of the bomb because nitrogen was needed to catalyze the oxidation of sulfur to sulfuric acid. The concentrated acid formed was deposited as a mist on the interior of the bomb and was assumed to be of uniform concentration. Results obtained by this "Huffman and Ellis method" can be demonstrated to be reasonably accurate by comparison of values for thiophene found by this method with those of the more accurate rotating bomb technique. Measurements on thiophene using the Huffman and Ellis method were made by Waddington, Knowlton, Scott, Oliver, Todd, Hubbard, Smith, and Huffman (1548), Moore, Renquist, and Parks (1025), and Franklin and Lumpkin (440). All agreed with the presently accepted figure within 0.5 kcal/mole.

Because of the nonuniform distribution of sulfuric acid, the Huffman and Ellis method is less reliable than the rotating bomb technique, in which a relatively large volume of water is added to the bomb and rotation thus assures homogeneous and definite end products. The rotating bomb development was perfected at the University of Lund by Sunner and at the U.S. Bureau of Mines at Bartlesville by Waddington and co-workers. A further refinement is the use of the principle of comparative measurements, in which the heat of combustion of a mixture of paraffin oil and rhombic sulfur in appropriate concentration is compared with that of the compound whose heat of formation is desired. Detailed descriptions of the bomb calorimetry of sulfur compounds and the methods of correction to standard states are given by Hubbard, Scott, and Waddington (637) and by Waddington, Sunner, and Hubbard (1551). Thianthrene has been suggested as a standard substance for combustion calorimetry on sulfur compounds.

The standard reference states adopted for sulfur are those given by Stull and Sinke (1437): the rhombic solid from 298° to 368.6°K, the monoclinic solid from 368.6° to 392°K, the equilibrium liquid from 392° to 717.75°K, and the ideal diatomic gas from 717.75° to 1000°K. This last reference state produces a small apparent anomaly in the tabulated Gibbs energies of sulfur compounds at 800°K, because sulfur vapor at this temperature is actually composed of species higher than diatomic; adequate data, however, are not yet available to define sulfur vapor completely in the region just above the boiling point. In spite of this, the tabulated Gibbs energies generally give a reasonable idea of the stability of a compound with respect to the elements. Although diatomic sulfur gas could have been used over the entire range from 298° to 1000°K, it was felt that this might give misleading indications of the stability of the compound at low temperatures. Conversion from the one standard state to the other, if desired, is readily made with the aid of the tables of Stull and Sinke (1437).

In calculating the enthalpy of formation from the reported heats of combustion, data were adjusted to the average value of two recent determinations of the enthalpy of formation of aqueous sulfuric acid by the Petroleum Research Center in Bartlesville, Okla. (509), and the Thermochemical Laboratory at Lund, Sweden, by Månsson and Sunner (926). Their values for ΔH°_{298} of the reaction

$$S(c, \text{rhombic}) + H_2(g) + 2O_2(g) + 115H_2O(l) = [H_2SO_4, 115H_2O](l),$$

-212.17 and -212.24 kcal/mole, respectively, are averaged as -212.20 kcal/mole. The value for the heat of formation of $S_2(g)$,

$$2S(c, \text{rhombic}) = S_2(g),$$

Introduction

$\Delta H f^\circ_{298.16} = 30.84 \pm 0.15$ kcal/mole, selected by Evans and Wagman (399), was used.

We thank Dr. Geoffrey Pilcher for making available his re-evaluation of data on organic sulfur compounds (for purposes of comparison) prior to formal publication.

The sulfur compounds are a good example of the extrapolation of data for selected "key" members of specific classes of compounds, usually the first few members of appropriate homologous series. From these data on the individual lower members and the application of the incremental method, thermodynamic properties are extrapolated to all members of such a homologous series within the molecular size range of practical interest. Methylene increments to the entropy and enthalpy for the lower normal thiaalkanes and the normal alkanes are shown in Figures 13-1 and 13-2 (952). Higher members of the series have not been studied experimentally, and some may not be known yet. However, it is expected that further development in this field will continue at a rapid pace over the next few years.

Replacement of a —CH_2— group in a hydrocarbon by sulfur raises the Gibbs energy by about 4 kcal/mole if a thiol is formed, and by about 6 kcal/mole if a sulfide is formed. These two classes of compounds are

Table 13-1 Comparison of Gibbs Energies of Thiaalkanes with Alkanes at 298.15° K

Parent Hydrocarbon[a]	$\Delta G f^\circ_{298}$ (kcal/mole)	Alkyl Sulfide[b]	$\Delta G f^\circ_{298}$ (kcal/mole)	$\Delta[\Delta G f^\circ_{298}]$ (kcal/mole)
Propane	−5.61	Methyl sulfide	1.66	7.27
Butane	−4.10	Ethyl methyl sulfide	2.73	6.83
2-Methylbutane	−3.54	Isopropyl methyl sulfide	3.21	6.75
Pentane	−2.00	Methyl propyl sulfide	4.40	6.40
Hexane	−0.06	Butyl methyl sulfide	6.37	6.43
Pentane	−2.00	Ethyl sulfide	4.25	6.25
Hexane	−0.06	Ethyl propyl sulfide	5.63	5.69
Heptane	1.91	Butyl ethyl sulfide	7.65	5.74
Butane	−4.10	Methyl disulfide	3.52	7.62
Hexane	−0.06	Ethyl disulfide	5.32	5.38
Octane	3.92	Propyl disulfide	8.84	4.92
Decane	7.94	Butyl disulfide	12.87	4.93

[a] From Chapter 9.
[b] From this chapter.

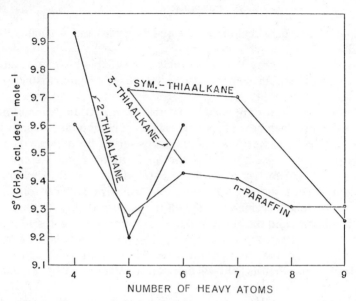

Figure 13-1 The methylene increments to the entropy, $\Delta S°_{298}(CH_2)$, for the lower normal thiaalkanes and normal alkanes (952).

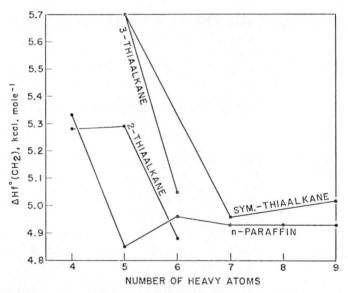

Figure 13-2 The methylene increments to the enthalpy, $\Delta H°_{298}(CH_2)$, for the lower normal thiaalkanes and normal alkanes (952).

therefore thermodynamically unstable with respect to sulfur and a hydrocarbon. A comparison of the Gibbs energies of the sulfur compounds and the hydrocarbons is given in Table 13-1.

Tables 6-2 and 6-3 also contain interesting comparisons involving the $\Delta Hf°$ values of parent hydrocarbons and those of thiols and alkyl sulfides.

Finally, brief mention should be made of other, less extensive tabulations and treatments on thermal properties of organic sulfur compounds: for example, that of Girelli and Burlamacchi (488) on mercaptans and aliphatic sulfides, those of Chermin (222–224), vapor pressures of 36 petroleum sulfur compounds by Osborn and Douslin (1091), formulas for estimating heat capacities by Maslov (937), and extension of a system of molecular thermochemistry to compounds containing sulfur and oxygen by Lovering and Laidler (889).

MONOTHIAALKANE IDEAL GAS TABLES

No. 800 Methyl Sulfide C_2H_6S (Ideal Gas State) Mol Wt 62.134

	cal/(mole °K)				kcal/mole		
$T°K$	$Cp°$	$S°$	$-(G°-H°_{298})/T$	$H°-H°_{298}$	$\Delta Hf°$	$\Delta Gf°$	Log Kp
298	17.71	68.32	68.32	0.00	-8.97	1.66	-1.220
300	17.77	68.43	68.33	0.04	-8.99	1.73	-1.259
400	21.12	74.01	69.06	1.98	-10.72	5.51	-3.010
500	24.24	79.07	70.57	4.26	-12.12	9.73	-4.254
600	27.01	83.74	72.38	6.82	-13.27	14.21	-5.177
700	29.44	88.08	74.31	9.65	-14.20	18.90	-5.899
800	31.59	92.16	76.29	12.70	-28.01	22.40	-6.120
900	33.50	95.99	78.27	15.96	-28.29	28.73	-6.976
1000	35.17	99.61	80.22	19.39	-28.46	35.08	-7.666

Osborne, Doescher, and Yost (1092) measured the low-temperature thermal data, finding $Tm = 174.85°K$, $\Delta Hm = 1.908$ kcal/mole, $S°_{298}(l) = 46.94$ cal/(mole °K), and $Cp(l, 270-290°K) = 21.35 + 0.023T$ [cal/(mole °K)]. These data were used by McCullough, Hubbard, Frow, Hossenlopp, and Waddington (958) to supplement their own measurements of vapor heat capacity, vapor pressure, enthalpy of vaporization, and enthalpy of combustion. Final results include $Tb = 350.33°K$, $\Delta Hv = 6.454$ kcal/mole at Tb, $\Delta Hf°_{298}(l) = -15.63$ kcal/mole, $\Delta Hf°_{298}(g) = -8.97$ kcal/mole when corrected to the new value of $\Delta Hf°_{298}$ for $H_2SO_4(aq)$ (cf. page 560), and $S°_{298}(g) = 68.32$ cal/(mole °K). McCullough et al (958) give references to previous work.

No. 801 Ethyl Methyl Sulfide C_3H_8S (Ideal Gas State) Mol Wt 76.160

$T°K$	$Cp°$	$S°$	$-(G°-H°_{298})/T$	$H°-H°_{298}$	$\Delta Hf°$	$\Delta Gf°$	Log Kp
		cal/(mole °K)			kcal/mole		
298	22.73	79.62	79.62	0.00	-14.25	2.73	-1.998
300	22.82	79.77	79.63	0.05	-14.28	2.83	-2.061
400	27.81	87.02	80.58	2.58	-16.36	8.80	-4.806
500	32.40	93.73	82.55	5.60	-18.04	15.28	-6.680
600	36.41	100.00	84.94	9.04	-19.38	22.08	-8.041
700	39.89	105.88	87.52	12.86	-20.44	29.10	-9.085
800	42.93	111.41	90.16	17.00	-34.33	34.96	-9.551
900	45.60	116.63	92.82	21.43	-34.64	43.65	-10.599
1000	47.94	121.55	95.44	26.11	-34.79	52.36	-11.443

Low-temperature thermal data, vapor pressure, vapor heat capacity, and enthalpy of vaporization were measured by Scott, Finke, McCullough, Gross, Williamson, Waddington, and Huffman (1308). Values found included $Tm = 167.22°K$, $\Delta Hm = 2.333$ kcal/mole, $\Delta Hv = 7.055$ kcal/mole at $Tb = 339.80°K$, $S°_{298}(l) = 57.14$ cal/(mole °K), $S°_{298}(g) = 79.62$ cal/(mole °K), and $Cp(l, 270-300°K) = 23.8 + 0.036T$ [cal/(mole °K)]. Spectroscopic studies were used with the above results to calculate thermodynamic functions of the ideal gas. The enthalpy of combustion reported by Hubbard and Waddington (638) led to the following enthalpies of formation: $\Delta Hf°_{298}(l) = -21.89$ kcal/mole and $\Delta Hf°_{298}(g) = -14.25$ kcal/mole, when adjusted to the new $\Delta Hf°_{298}$ for H_2SO_4 (aq) (cf. page 560).

No. 802 Ethyl Sulfide $C_4H_{10}S$ (Ideal Gas State) Mol Wt 90.186

$T°K$	$Cp°$	$S°$	$-(G°-H°_{298})/T$	$H°-H°_{298}$	$\Delta Hf°$	$\Delta Gf°$	Log Kp
		cal/(mole °K)			kcal/mole		
298	27.97	87.96	87.96	0.00	-19.95	4.25	-3.116
300	28.09	88.14	87.97	0.06	-19.99	4.40	-3.204
400	34.65	97.13	89.15	3.20	-22.41	12.84	-7.017
500	40.75	105.53	91.59	6.97	-24.33	21.89	-9.566
600	46.11	113.44	94.58	11.32	-25.85	31.27	-11.391
700	50.81	120.91	97.81	16.17	-27.01	40.92	-12.775
800	54.91	127.97	101.15	21.46	-40.91	49.41	-13.497
900	58.56	134.65	104.50	27.14	-41.17	60.72	-14.745
1000	61.79	140.99	107.84	33.16	-41.20	72.06	-15.747

Thermal data for ethyl sulfide used in the derivation of thermodynamic functions of the ideal gas were measured by Scott, Finke, Hubbard, McCullough, Oliver, Gross, Katz, Williamson, Waddington, and Huffman

(1305). These workers determined $Tm = 169.20°K$, $\Delta Hm = 2.845$ kcal/mole, $\Delta Hv = 7.591$ kcal/mole at $Tb = 365.25°K$, $S°_{298}(l) = 64.36$ cal/(mole °K), $S°_{298}(g) = 87.96$ cal/(mole °K), and $Cp(l, 270-320°K) = 26.7 + 0.048T$ [cal/(mole °K)]. Isomerization equilibria were reported by Scott, Guthrie, McCullough, and Waddington (1313). Hubbard, Good, and Waddington (631) measured the enthalpy of combustion by a rotating bomb technique, finding $\Delta Hf°_{298}(l) = -28.52$ kcal/mole and $\Delta Hf°_{298}(g) = -19.95$ kcal/mole when adjusted to the new $\Delta Hf°_{298}$ value for H_2SO_4 (aq) (cf. page 560).

No. 803 Isopropyl Methyl Sulfide $C_4H_{10}S$ (Ideal Gas State) Mol Wt 90.186

	cal/(mole °K)				kcal/mole		
$T°K$	$Cp°$	$S°$	$-(G°-H°_{298})/T$	$H°-H°_{298}$	$\Delta Hf°$	$\Delta Gf°$	Log Kp
298	28.00	85.87	85.87	0.00	-21.61	3.21	-2.356
300	28.12	86.05	85.88	0.06	-21.65	3.37	-2.452
400	34.69	95.05	87.06	3.20	-24.06	12.02	-6.567
500	40.72	103.45	89.51	6.98	-25.99	21.27	-9.296
600	46.01	111.35	92.50	11.32	-27.52	30.87	-11.242
700	50.65	118.80	95.73	16.16	-28.68	40.72	-12.713
800	54.95	125.85	99.06	21.44	-42.59	49.42	-13.501
900	58.82	132.55	102.41	27.13	-42.84	60.95	-14.799
1000	62.29	138.93	105.74	33.19	-42.83	72.49	-15.842

Calorimetric studies and thermodynamic properties of isopropyl methyl sulfide were reported by McCullough, Finke, Messerly, Pennington, Hossenlopp, and Waddington (953). Results included $Tm = 171.64°K$, $\Delta Hm = 2.236$ kcal/mole, $Tb = 357.97°K$, at which temperature $\Delta Hv = 7.338$ kcal/mole, $S°_{298}(l) = 62.88$ cal/(mole °K), $S°_{298}(g) = 85.87$ cal/(mole °K), and $Cp(l, 290-360°K) = 24.80 + 0.055T$ [cal/(mole °K)]. Combustion calorimetry by Hubbard, Good, and Waddington (631) was used to derive $\Delta Hf°_{298}(l) = -29.79$ kcal/mole and $\Delta Hf°_{298}(g) = -21.61$ kcal/mole when adjusted to the new value of $\Delta Hf°_{298}$ for H_2SO_4 (aq) (cf. page 560).

No. 804 Methyl Propyl Sulfide $C_4H_{10}S$ (Ideal Gas State) Mol Wt 90.186

Complete measurements of the thermodynamic properties were reported by Scott, Finke, McCullough, Messerly, Pennington, Hossenlopp, and Waddington (1309). Among the data are $Tm = 160.16°K$, $\Delta Hm = 2.369$ kcal/mole, $Tb = 368.69°K$, at which temperature $\Delta Hv = 7.667$ kcal/mole, $S_{298}(l) = 65.14$ cal/(mole °K), $S°_{298}(g) = 88.84$ cal/(mole °K), and $Cp(l, 270-320°K) = 27.35 + 0.046T$ [cal/(mole °K)]. $\Delta Hf°_{298}(l) = -28.21$ kcal/mole and $\Delta Hf°_{298}(g) = -19.54$ kcal/mole from the data of Hubbard,

No. 804 Methyl Propyl Sulfide $C_4H_{10}S$ (Ideal Gas State) Mol Wt 90.186

T°K	cal/(mole °K)				kcal/mole		Log Kp
	$Cp°$	$S°$	$-(G°-H°_{298})/T$	$H°-H°_{298}$	$\Delta Hf°$	$\Delta Gf°$	
298	28.05	88.84	88.84	0.00	-19.54	4.40	-3.224
300	28.17	89.02	88.85	0.06	-19.58	4.54	-3.311
400	34.64	98.02	90.03	3.20	-21.99	12.90	-7.049
500	40.61	106.41	92.47	6.97	-23.93	21.85	-9.552
600	45.86	114.28	95.46	11.30	-25.47	31.16	-11.349
700	50.43	121.70	98.68	16.12	-26.65	40.72	-12.713
800	54.45	128.70	102.00	21.37	-40.60	49.13	-13.422
900	58.02	135.33	105.34	26.99	-40.91	60.38	-14.661
1000	61.14	141.61	108.66	32.96	-41.00	71.65	-15.657

Good, and Waddington (631) when adjusted to the new value of $\Delta Hf°_{298}$ for H_2SO_4 (*aq*) (cf. page 560). Isomerization equilibria were reported by Scott, Guthrie, McCullough, and Waddington (1313). Thermodynamic properties were calculated by a refined method of increments using experimental vapor heat capacities and data for lower members of the series.

No. 805 Butyl Methyl Sulfide $C_5H_{12}S$ (Ideal Gas State) Mol Wt 104.212

T°K	cal/(mole °K)				kcal/mole		Log Kp
	$Cp°$	$S°$	$-(G°-H°_{298})/T$	$H°-H°_{298}$	$\Delta Hf°$	$\Delta Gf°$	
298	33.64	98.43	98.43	0.00	-24.42	6.37	-4.670
300	33.79	98.64	98.44	0.07	-24.47	6.56	-4.778
400	41.73	109.46	99.86	3.85	-27.18	17.26	-9.432
500	49.12	119.59	102.80	8.40	-29.36	28.63	-12.516
600	55.68	129.13	106.40	13.64	-31.06	40.39	-14.713
700	61.43	138.16	110.30	19.51	-32.32	52.44	-16.371
800	66.53	146.70	114.32	25.91	-46.28	63.33	-17.301
900	71.09	154.80	118.37	32.79	-46.53	77.06	-18.712
1000	75.08	162.51	122.40	40.11	-46.49	90.80	-19.844

Determinations of the thermodynamic properties of butyl methyl sulfide were made by McCullough, Finke, Hubbard, Todd, Messerly, Douslin, and Waddington (952), who found $Tm = 175.30°K$, $\Delta Hm = 2.976$ kcal/mole, $\Delta Hv_{298} = 9.730$ kcal/mole, $Tb = 396.57°K$, $S°_{298}(l) = 73.49$ cal/(mole °K), $S°_{298}(g) = 98.43$ cal/(mole °K), $Cs(l, 190-360°K) = 58.773 - 0.18710T + 6.914 \times 10^{-4}T^2 - 6.2000 \times 10^{-7}T^3$[cal/(mole °K)], $\Delta Hf°_{298}(l) = -34.15$ kcal/mole, and $\Delta Hf°_{298}(g) = -24.42$ kcal/mole when adjusted to the new $\Delta Hf°_{298}$ value for H_2SO_4 (*aq*) (cf. page 560). Gaseous thermodynamic functions were computed by Scott and McCullough (1316), who used a semirefined method of increments in which calculated values of the entropy—but not the heat capacity—were adjusted empirically to accord with the above entropy value.

Mackle and Mayrick (899) also determined ΔHc and ΔHv_{298} (9.9 ± 0.3 kcal/mole) and calculated $\Delta Hf°_{298}(g) = -24.2 \pm 0.6$ kcal/mole. Mackle and McClean (903) reported ΔHv_{298} by gas-liquid chromatography as 9.8 ± 0.2 kcal/mole and discussed accord with four empirical methods as well as with a twin ebulliometric method used by Mackle and Mayrick (898).

No. 806 Ethyl Propyl Sulfide $C_5H_{12}S$ (Ideal Gas State) Mol Wt 104.212

	cal/(mole °K)				kcal/mole		
$T°K$	$Cp°$	$S°$	$-(G°-H°_{298})/T$	$H°-H°_{298}$	$\Delta Hf°$	$\Delta Gf°$	Log Kp
298	33.25	98.97	98.97	0.00	-25.00	5.63	-4.126
300	33.40	99.18	98.98	0.07	-25.05	5.82	-4.237
400	41.43	109.90	100.38	3.81	-27.80	16.47	-9.001
500	48.95	119.97	103.30	8.34	-29.99	27.80	-12.152
600	55.62	129.50	106.88	13.57	-31.71	39.53	-14.397
700	61.50	138.52	110.76	19.44	-32.97	51.53	-16.089
800	66.68	147.08	114.77	25.85	-46.92	62.39	-17.044
900	71.32	155.20	118.82	32.75	-47.15	76.08	-18.474
1000	75.42	162.94	122.85	40.09	-47.09	89.78	-19.621

McCullough, Finke, Hubbard, Todd, Messerly, Douslin, and Waddington (952) determined the thermal and thermochemical constants for this compound as $Tm = 156.10°K$, $\Delta Hm = 2.529$ kcal/mole, $S°_{298}(l) = 73.98$ cal/(mole °K), $Tb = 391.65°K$, $\Delta Hv_{298} = 9.580$ kcal/mole, $\Delta Hf°_{298}(l) = -34.58$ kcal/mole, $\Delta Hf°_{298}(g) = -25.00$ kcal/mole when adjusted to the new value of $\Delta Hf°_{298}$ for H_2SO_4 (aq) (cf. page 560), and $S°_{298}(g) = 98.82$ cal/(mole °K). From 190° to 390°K, $Cs(l) = 60.865 - 0.22365T + 8.4708 \times 10^{-4}T^2 - 8.3246 \times 10^{-7}T^3$[cal/(mole °K)]. Thermodynamic functions for this compound were computed by an ordinary method of increments by Scott and McCullough (1316); that is, the calculated values were not adjusted empirically. The pair 1-propanethiol and ethanethiol was selected for obtaining the appropriate methylene increments instead of the pair 2-thiapentane and 2-thiabutane, in order to use only compounds for which thermodynamic functions had been calculated by detailed statistico-methanical methods.

No. 807 Butyl Ethyl Sulfide $C_6H_{14}S$ (Ideal Gas State) Mol Wt 118.238

Mackle and Mayrick (899) determined ΔHc and reported $\Delta Hf°_{298}(g)$ as -30.3 ± 0.6 kcal/mole and ΔHv_{298} as 10.8 ± 0.3 kcal/mole. Mackle and

No. 807 Butyl Ethyl Sulfide $C_6H_{14}S$ (Ideal Gas State) Mol Wt 118.238

$T°K$	$Cp°$	$S°$	$-(G°-H°_{298})/T$	$H°-H°_{298}$	$\Delta Hf°$	$\Delta Gf°$	Log Kp
		cal/(mole °K)			kcal/mole		
298	38.71	108.27	108.27	0.00	-29.92	7.65	-5.606
300	38.89	108.52	108.28	0.08	-29.97	7.88	-5.739
400	48.37	121.02	109.92	4.45	-33.04	20.92	-11.427
500	57.19	132.78	113.32	9.73	-35.50	34.70	-15.165
600	64.96	143.91	117.50	15.85	-37.41	48.92	-17.817
700	71.77	154.44	122.03	22.69	-38.81	63.45	-19.810
800	77.74	164.42	126.71	30.17	-52.86	76.85	-20.994
900	83.06	173.89	131.43	38.22	-53.15	93.10	-22.606
1000	87.75	182.89	136.13	46.76	-53.10	109.36	-23.899

McClean (903) determined ΔHv_{298} as 10.6 ± 0.2 kcal/mole by gas-liquid chromatography and compared this value with other determinations and estimates. Scott and McCullough (1316) calculated thermodynamic properties by adding to the values for ethyl propyl sulfide one methylene increment, recommended by Person and Pimentel (1140) for the thermodynamic functions, and the increment deduced by Prosen, Johnson, and Rossini (1196) for the enthalpy of formation. Their result, $\Delta Hf°_{298}(g) = -29.92$ kcal/mole, accords within the uncertainty of the experimental value cited and is adopted. Using the first cited value for ΔHv_{298}, $\Delta Hf°_{298}(l)$ is -40.7 kcal/mole.

No. 808 Isopropyl Sulfide $C_6H_{14}S$ (Ideal Gas State) Mol Wt 118.238

$T°K$	$Cp°$	$S°$	$-(G°-H°_{298})/T$	$H°-H°_{298}$	$\Delta Hf°$	$\Delta Gf°$	Log Kp
		cal/(mole °K)			kcal/mole		
298	40.45	99.30	99.30	0.00	-33.76	6.48	-4.752
300	40.65	99.56	99.31	0.08	-33.81	6.73	-4.902
400	50.64	112.65	101.02	4.66	-36.67	20.63	-11.272
500	59.17	124.90	104.59	10.16	-38.91	35.23	-15.396
600	66.22	136.33	108.94	16.44	-40.65	50.22	-18.290
700	72.10	146.99	113.62	23.36	-41.98	65.50	-20.450
800	77.12	156.95	118.42	30.83	-56.04	79.65	-21.758
900	81.46	166.29	123.23	38.76	-56.45	96.65	-23.468
1000	85.24	175.07	127.97	47.10	-56.60	113.68	-24.843

The availability of unpublished accurate entropy and vapor phase heat capacity values to about 500°K for isopropyl sulfide, together with statistical thermodynamic investigations of the conformational energetics by Scott and Crowder (1298), permitted evaluation of its chemical thermodynamic properties over the entire temperature range of likely interest.

The vibrational assignment was obtained by interpreting the molecular spectra for the crystal, liquid, and gaseous states with the guidance of force constant calculations. Barriers hindering internal rotation of the isopropyl group and the methyl groups were taken to be of the threefold cosine form. The barrier heights chosen to give agreement with the observed values of entropy and heat capacity were methyl, 3.75 kcal/mole and isopropyl, 2.80 kcal/mole. The thermodynamic functions of isopropyl sulfide were calculated without specific provision for conformational equilibrium, anharmonicity, and other small effects. How nearly the net contributions of these effects remain negligible at temperatures above the range of calorimetric measurements determines the accuracy of the thermodynamic functions for those temperatures. Scott and Crowder (1298) used $S^\circ_{298}(g) = 99.30$ cal/(mole °K) and $\Delta Hf^\circ_{298}(g) = -33.76$ kcal/mole when adjusted to the revised value of ΔHf°_{298} for H_2SO_4 (aq) (cf. page 560). A $\Delta Hf^\circ(g)$ based on the same experimental data was cited by McCullough and Good (957). Mackle and McClean (903) evaluated $\Delta Hv_{298} = 9.46 \pm 0.2$ kcal/mole by gas-liquid chromatography.

No. 809   Methyl Pentyl Sulfide $C_6H_{14}S$ (Ideal Gas State)   Mol Wt 118.238

	cal/(mole °K)				kcal/mole		
$T°K$	$Cp°$	$S°$	$-(G°-H°_{298})/T$	$H°-H°_{298}$	$\Delta Hf°$	$\Delta Gf°$	Log Kp
298	39.10	107.73	107.73	0.00	-29.34	8.39	-6.149
300	39.28	107.98	107.74	0.08	-29.39	8.62	-6.279
400	48.67	120.58	109.39	4.48	-32.43	21.71	-11.859
500	57.36	132.39	112.82	9.79	-34.86	35.53	-15.529
600	65.02	143.54	117.02	15.92	-36.75	49.79	-18.133
700	71.70	154.08	121.57	22.76	-38.16	64.36	-20.092
800	77.59	164.04	126.26	30.23	-52.22	77.79	-21.251
900	82.83	173.49	130.99	38.26	-52.53	94.08	-22.844
1000	87.41	182.46	135.69	46.77	-52.51	110.38	-24.123

Mackle and Mayrick (899) measured ΔHc and $\Delta Hv_{298} = 10.8 \pm 0.3$ kcal/mole. They reported $\Delta Hf^\circ_{298}(g) = -29.1 \pm 0.6$ kcal/mole. The value of ΔHv determined by Mackle and McClean (903) by gas-liquid chromatography, 11.0 ± 0.2 kcal/mole, accords within the combined experimental precision indices. $Tm = 179°K$ and $Tb = 418°K$ were reported by Rossini, Pitzer, Arnett, Braun, and Pimentel (1248). Thermodynamic properties were calculated by Scott and McCullough (1316) by adding one methylene increment, proposed by Person and Pimentel (1140) for each thermodynamic function and by Prosen, Johnson, and Rossini (1196) for $\Delta Hf°$, to the values for butyl methyl sulfide. $\Delta Hf^\circ_{298}(g) =$

−29.34 kcal/mole. This value is adopted and is within the precision index of the experimental value cited. Using the $\Delta Hv°$ of Mackle and Mayrick leads to $\Delta Hf°_{298}(l) = -40.14$ kcal/mole.

No. 810 Propyl Sulfide $C_6H_{14}S$ (Ideal Gas State) Mol Wt 118.238

T°K	Cp°	S°	$-(G°-H°_{298})/T$	$H°-H°_{298}$	$\Delta Hf°$	$\Delta Gf°$	Log Kp
		cal/(mole °K)			kcal/mole		
298	38.53	107.16	107.16	0.00	-29.96	7.94	-5.819
300	38.71	107.40	107.17	0.08	-30.01	8.17	-5.952
400	48.21	119.85	108.80	4.43	-33.10	21.32	-11.649
500	57.15	131.59	112.19	9.70	-35.56	35.22	-15.395
600	65.13	142.73	116.36	15.82	-37.47	49.56	-18.051
700	72.19	153.31	120.89	22.70	-38.85	64.21	-20.047
800	78.45	163.37	125.58	30.23	-52.84	77.72	-21.232
900	84.08	172.94	130.31	38.37	-53.04	94.07	-22.841
1000	89.05	182.06	135.03	47.03	-52.88	110.42	-24.131

Low-temperature thermal data, ΔHv, vapor pressure, and ΔHc were measured by McCullough, Finke, Hubbard, Todd, Messerly, Douslin, and Waddington (952). Values reported included $Tm = 170.44°K$, $\Delta Hm = 2.902$ kcal/mole, $\Delta Hv°_{298} = 10.66$ kcal/mole, $S°_{298}(l) = 80.85$ cal/(mole °K), $S°_{298}(g) = 107.16$ cal/(mole °K), and $Cs(l, 185-315°K) = 69.861 - 0.26760T + 10.219 \times 10^{-4}T^2 - 10.208 \times 10^{-7}T^3$[cal/(mole °K)]. From ΔHc measurements they give $\Delta Hf°_{298}(l) = -40.62$ kcal/mole and $\Delta Hf°_{298}(g) = -29.96$ kcal/mole when adjusted to the new $\Delta Hf°_{298}$ for H_2SO_4 (aq) (cf. page 560). Thermodynamic functions were calculated by Scott and McCullough (1316), who used an ordinary method of increments. Mackle and McClean (903) reported $\Delta Hv_{298} = 10.7 \pm 0.2$ kcal/mole by gas-liquid chromatography and 10.6 and 10.7 by ebulliometry.

No. 811 Butyl Propyl Sulfide $C_7H_{16}S$ (Ideal Gas State) Mol Wt 132.264

T°K	Cp°	S°	$-(G°-H°_{298})/T$	$H°-H°_{298}$	$\Delta Hf°$	$\Delta Gf°$	Log Kp
		cal/(mole °K)			kcal/mole		
298	43.99	117.90	117.90	0.00	-34.88	9.53	-6.984
300	44.20	118.18	117.91	0.09	-34.94	9.80	-7.139
400	55.15	132.41	119.77	5.06	-38.35	25.19	-13.761
500	65.39	145.84	123.65	11.10	-41.07	41.40	-18.093
600	74.47	158.58	128.42	18.10	-43.17	58.09	-21.157
700	82.46	170.67	133.60	25.95	-44.69	75.12	-23.453
800	89.51	182.15	138.96	34.56	-58.78	91.03	-24.867
900	95.82	193.06	144.37	43.83	-59.04	109.79	-26.659
1000	101.38	203.45	149.76	53.70	-58.89	128.56	-28.094

Thermodynamic functions for this compound were computed by Scott and McCullough (1316) by adding the appropriate methylene increment of Person and Pimentel (1140) to each of the functions for dipropyl sulfide and by adding the appropriate symmetry term, $R \ln 2$, to $S°$ and $-(G° - H_0°)/T$. The $\Delta Hf°_{298}(g) = -34.88$ kcal/mole value was likewise calculated by adding a methylene increment, -4.926 kcal/mole, of Prosen, Johnson, and Rossini (1196) to the value for dipropyl sulfide. Tb was reported as 426.7°K by Braun and Engelbertz (162).

No. 812 Ethyl Pentyl Sulfide $C_7H_{16}S$ (Ideal Gas State)   Mol Wt 132.264

	cal/(mole °K)			kcal/mole			
$T°K$	$Cp°$	$S°$	$-(G°-H°_{298})/T$	$H°-H°_{298}$	$\Delta Hf°$	$\Delta Gf°$	Log Kp
298	44.18	117.58	117.58	0.00	-34.85	9.65	-7.076
300	44.39	117.86	117.59	0.09	-34.91	9.93	-7.231
400	55.31	132.14	119.46	5.08	-38.30	25.34	-13.846
500	65.44	145.60	123.35	11.13	-41.01	41.58	-18.172
600	74.30	158.33	128.13	18.12	-43.11	58.29	-21.232
700	82.05	170.38	133.31	25.95	-44.66	75.35	-23.525
800	88.81	181.78	138.67	34.50	-58.81	91.30	-24.939
900	94.81	192.60	144.06	43.69	-59.15	110.09	-26.733
1000	100.08	202.86	149.43	53.44	-59.12	128.91	-28.173

Scott and McCullough (1316) calculated thermodynamic properties by adding to the values for ethyl propyl sulfide two methylene incremental values. That recommended by Person and Pimentel (1140) was used for the thermodynamic functions, and that of Prosen, Johnson, and Rossini (1196) for the enthalpy of formation $\Delta Hf°_{298}(g) = -34.85$ kcal/mole.

No. 813 Hexyl Methyl Sulfide $C_7H_{16}S$ (Ideal Gas State)   Mol Wt 132.264

	cal/(mole °K)			kcal/mole			
$T°K$	$Cp°$	$S°$	$-(G°-H°_{298})/T$	$H°-H°_{298}$	$\Delta Hf°$	$\Delta Gf°$	Log Kp
298	44.57	117.04	117.04	0.00	-34.27	10.39	-7.619
300	44.78	117.32	117.05	0.09	-34.33	10.67	-7.771
400	55.61	131.71	118.93	5.11	-37.68	26.13	-14.278
500	65.61	145.21	122.85	11.19	-40.37	42.41	-18.535
600	74.36	157.97	127.65	18.19	-42.46	59.16	-21.548
700	81.98	170.01	132.85	26.02	-44.01	76.26	-23.808
800	88.66	181.40	138.21	34.56	-58.17	92.24	-25.197
900	94.58	192.20	143.62	43.72	-58.54	111.07	-26.971
1000	99.74	202.43	148.99	53.45	-58.53	129.94	-28.396

Thermodynamic properties were calculated by Scott and McCullough (1316) by adding to the values for butyl methyl sulfide two methylene increments, proposed by Person and Pimentel (1140), for each thermodynamic function, and by Prosen, Johnson, and Rossini (1196), for $\Delta Hf°$. $\Delta Hf°_{298}(g)$ is -34.27 kcal/mole. A Tm of 206.6°K and a Tb of 444°K were reported by Rossini, Pitzer, Arnett, Braun, and Pimentel (1248).

No. 814 Butyl Sulfide $C_8H_{18}S$ (Ideal Gas State) Mol Wt 146.290

	cal/(mole °K)				kcal/mole		
$T°K$	$Cp°$	$S°$	$-(G°-H°_{298})/T$	$H°-H°_{298}$	$\Delta Hf°$	$\Delta Gf°$	Log Kp
298	49.46	125.84	125.84	0.00	-39.99	11.76	-8.622
300	49.70	126.15	125.85	0.10	-40.06	12.08	-8.799
400	62.09	142.17	127.95	5.69	-43.78	29.98	-16.381
500	73.64	157.29	132.32	12.49	-46.76	48.78	-21.320
600	83.81	171.63	137.68	20.37	-49.05	68.10	-24.805
700	92.74	185.24	143.51	29.21	-50.72	87.80	-27.412
800	100.58	198.14	149.54	38.88	-64.91	106.39	-29.063
900	107.57	210.40	155.63	49.30	-65.22	127.84	-31.041
1000	113.71	222.06	161.69	60.37	-65.10	149.30	-32.628

A determination of ΔHc by Sunner (1443) yielded a value of $\Delta Hf°_{298}(l)$, which was subsequently adjusted by Sunner (1447) to the new value for $\Delta Hf°_{298}$ for H_2SO_4 (aq) (cf. page 560) to yield -52.76 ± 0.45 kcal/mole. Low-temperature thermal and thermochemical values were reported by McCullough, Finke, Hubbard, Todd, Messerly, Douslin, and Waddington (952). They indicated $Tm = 198.13°K$, $\Delta Hm = 4.643$ kcal/mole, $Tb = 462.06°K$, $\Delta Hv_{298} = 12.75$ kcal/mole, $S°_{298}(l) = 96.82$ cal/(mole °K), $S°_{298}(g) = 125.84$ cal/(mole °K), $Cs(l, 205-355°K) = 85.483 - 0.30972T + 11.687 \times 10^{-4}T^2 - 10.974 \times 10^{-7}T^3$ [cal/(mole °K)], $\Delta Hf°_{298}(l) = -52.74$ kcal/mole, and $\Delta Hf°_{298}(g) = -39.99$ kcal/mole when adjusted to the new value for $\Delta Hf°_{298}$ for H_2SO_4 (aq) (cf. page 560). Thermodynamic functions for this compound were computed by adding two methylene increments of the value proposed by Person and Pimentel (1140) to the function for propyl sulfide, which accords with the data cited. Mackle and McClean (903) reported $\Delta Hv_{298} = 13.0 \pm 0.2$ kcal/mole by gas-liquid chromatography and 12.8 by a twin ebulliometric method.

No. 815 Ethyl Hexyl Sulfide $C_8H_{18}S$ (Ideal Gas State) Mol Wt 146.290

Scott and McCullough (1316) calculated thermodynamic properties by adding three methylene increments, recommended by Person and Pimentel (1140), for the thermodynamic functions, and that by Prosen, Johnson, and Rossini (1196), for the enthalpy of formation, to the ethyl propyl sulfide values. $\Delta Hf°_{298}(g)$ is -39.77 kcal/mole.

No. 815 Ethyl Hexyl Sulfide $C_8H_{18}S$ (Ideal Gas State) Mol Wt 146.290

		cal/(mole °K)			kcal/mole		
T°K	$Cp°$	$S°$	$-(G°-H°_{298})/T$	$H°-H°_{298}$	$\Delta Hf°$	$\Delta Gf°$	Log Kp
298	49.64	126.89	126.89	0.00	-39.77	11.67	-8.553
300	49.88	127.20	126.90	0.10	-39.84	11.98	-8.730
400	62.25	143.27	129.00	5.71	-43.54	29.78	-16.270
500	73.68	158.41	133.38	12.52	-46.51	48.46	-21.182
600	83.64	172.75	138.76	20.40	-48.81	67.68	-24.650
700	92.32	186.31	144.59	29.20	-50.51	87.27	-27.244
800	99.87	199.14	150.62	38.82	-64.75	105.75	-28.888
900	106.55	211.29	156.69	49.15	-65.15	127.10	-30.863
1000	112.41	222.83	162.73	60.10	-65.14	148.48	-32.449

No. 816 Heptyl Methyl Sulfide $C_8H_{18}S$ (Ideal Gas State) Mol Wt 146.290

		cal/(mole °K)			kcal/mole		
T°K	$Cp°$	$S°$	$-(G°-H°_{298})/T$	$H°-H°_{298}$	$\Delta Hf°$	$\Delta Gf°$	Log Kp
298	50.03	126.35	126.35	0.00	-39.19	12.41	-9.096
300	50.27	126.67	126.36	0.10	-39.26	12.73	-9.270
400	62.55	142.83	128.48	5.75	-42.93	30.57	-16.702
500	73.85	158.03	132.88	12.58	-45.87	49.29	-21.546
600	83.70	172.39	138.28	20.47	-48.16	68.54	-24.966
700	92.25	185.94	144.13	29.27	-49.86	88.17	-27.526
800	99.72	198.76	150.17	38.88	-64.11	106.69	-29.145
900	106.32	210.89	156.25	49.19	-64.53	128.08	-31.101
1000	112.07	222.40	162.29	60.12	-64.55	149.50	-32.672

Thermodynamic properties were calculated by Scott and McCullough (1316) by adding three methylene increments, proposed by Person and Pimentel (1140), for each thermodynamic function, and by Prosen, Johnson, and Rossini (1196), for $\Delta Hf°$, to the butyl methyl sulfide values. $\Delta Hf°_{298}(g)$ is -39.19 kcal/mole. Rossini, Pitzer, Arnett, Braun, and Pimentel (1248) selected values of $Tb = 468°K$ and $Tm = 209.9°K$.

No. 817 Pentyl Propyl Sulfide $C_8H_{18}S$ (Ideal Gas State) Mol Wt 146.290

One methylene increment, formulated by Person and Pimentel (1140), for thermodynamic functions, and that by Prosen, Johnson, and Rossini (1196), for the enthalpy of formation, were added to the appropriate butyl propyl sulfide values by Scott and McCullough (1316) to obtain the thermodynamic properties of the gaseous state. $\Delta Hf°_{298}(g)$ is -39.81 kcal/mole.

No. 817 Pentyl Propyl Sulfide $C_8H_{18}S$ (Ideal Gas State)   Mol Wt 146.290

	cal/(mole °K)				kcal/mole		
$T°K$	$Cp°$	$S°$	$-(G°-H°_{298})/T$	$H°-H°_{298}$	$\Delta Hf°$	$\Delta Gf°$	Log Kp
298	49.46	127.21	127.21	0.00	-39.81	11.53	-8.454
300	49.70	127.52	127.22	0.10	-39.88	11.85	-8.631
400	62.09	143.54	129.32	5.69	-43.60	29.61	-16.180
500	73.64	158.66	133.69	12.49	-46.58	48.27	-21.099
600	83.81	173.00	139.05	20.37	-48.87	67.46	-24.571
700	92.74	186.61	144.88	29.21	-50.54	87.02	-27.168
800	100.58	199.51	150.91	38.88	-64.73	105.47	-28.812
900	107.57	211.77	157.00	49.30	-65.04	126.78	-30.786
1000	113.71	223.43	163.06	60.37	-64.92	148.11	-32.368

No. 818 Butyl Pentyl Sulfide $C_9H_{20}S$ (Ideal Gas State)   Mol Wt 160.316

	cal/(mole °K)				kcal/mole		
$T°K$	$Cp°$	$S°$	$-(G°-H°_{298})/T$	$H°-H°_{298}$	$\Delta Hf°$	$\Delta Gf°$	Log Kp
298	54.92	136.52	136.52	0.00	-44.92	13.36	-9.792
300	55.19	136.87	136.53	0.11	-44.99	13.72	-9.992
400	69.03	154.66	138.86	6.33	-49.04	33.86	-18.500
500	81.88	171.48	143.72	13.88	-52.27	54.97	-24.027
600	93.15	187.42	149.68	22.65	-54.76	76.66	-27.920
700	103.01	202.54	156.16	32.47	-56.57	98.74	-30.828
800	111.64	216.87	162.86	43.21	-70.86	119.74	-32.709
900	119.31	230.47	169.63	54.76	-71.23	143.60	-34.869
1000	126.04	243.39	176.36	67.04	-71.12	167.49	-36.602

Thermodynamic functions for butyl pentyl sulfide were computed by Scott and McCullough (1316) by adding three methylene increments of Person and Pimentel (1140) to the functions for dipropyl sulfide and also adding the symmetry term, $R \ln 2$, to $S°$ and $-(G°-H°_0)/T$. $\Delta Hf°_{298}(g) = -44.92$ kcal/mole was calculated by adding one methylene increment to the corresponding value for dibutyl sulfide.

No. 819 Ethyl Heptyl Sulfide $C_9H_{20}S$ (Ideal Gas State)   Mol Wt 160.316

Scott and McCullough (1316) calculated thermodynamic properties by adding four methylene increments, recommended by Person and Pimentel (1140), for the thermodynamic functions, and by Prosen, Johnson and Rossini (1196), for the enthalpy of formation, to the values for ethyl propyl sulfide. $\Delta Hf°_{298}(g)$ is -44.70 kcal/mole. Tb of about 463°K at 732 torr was reported by Adams, Bramlet, and Tendick (2).

No. 819 Ethyl Heptyl Sulfide C₉H₂₀S (Ideal Gas State) Mol Wt 160.316

T°K	Cp°	S°	$-(G°-H°_{298})/T$	$H°-H°_{298}$	ΔHf°	ΔGf°	Log Kp
		cal/(mole °K)			kcal/mole		
298	55.11	136.20	136.20	0.00	-44.70	13.67	-10.023
300	55.38	136.55	136.21	0.11	-44.77	14.03	-10.222
400	69.19	154.40	138.55	6.34	-48.80	34.21	-18.689
500	81.93	171.23	143.42	13.91	-52.02	55.34	-24.189
600	92.98	187.17	149.39	22.67	-54.52	77.05	-28.064
700	102.60	202.24	155.88	32.46	-56.36	99.17	-30.960
800	110.94	216.50	162.57	43.15	-70.70	120.19	-32.833
900	118.30	230.00	169.32	54.62	-71.16	144.10	-34.990
1000	124.75	242.80	176.03	66.78	-71.16	168.03	-36.722

No. 820 Hexyl Propyl Sulfide C₉H₂₀S (Ideal Gas State) Mol Wt 160.316

T°K	Cp°	S°	$-(G°-H°_{298})/T$	$H°-H°_{298}$	ΔHf°	ΔGf°	Log Kp
		cal/(mole °K)			kcal/mole		
298	54.92	136.52	136.52	0.00	-44.73	13.55	-9.931
300	55.19	136.87	136.53	0.11	-44.80	13.91	-10.130
400	69.03	154.66	138.86	6.33	-48.85	34.05	-18.604
500	81.88	171.48	143.72	13.88	-52.08	55.16	-24.110
600	93.15	187.42	149.68	22.65	-54.57	76.85	-27.990
700	103.01	202.54	156.16	32.47	-56.38	98.93	-30.887
800	111.64	216.87	162.86	43.21	-70.67	119.93	-32.761
900	119.31	230.47	169.63	54.76	-71.04	143.79	-34.915
1000	126.04	243.39	176.36	67.04	-70.93	167.68	-36.644

Two methylene increments, formulated by Person and Pimentel (1140), for thermodynamic functions, and by Prosen, Johnson, and Rossini (1196), for the enthalpy of formation, were added to the appropriate values for butyl propyl sulfide by Scott and McCullough (1316) to obtain the thermodynamic properties of the gaseous state. $\Delta Hf°_{298}(g)$ is -44.73 kcal/mole.

No. 821 Methyl Octyl Sulfide C₉H₂₀S (Ideal Gas State) Mol Wt 160.316

Thermodynamic properties were calculated by Scott and McCullough (1316) by adding four methylene increments, proposed by Person and Pimentel (1140), for each thermodynamic function, and by Prosen, Johnson, and Rossini (1196), for $\Delta Hf°$, to the values for butyl methyl sulfide. $\Delta Hf°_{298}(g)$ is -44.12 kcal/mole. Rossini, Pitzer, Arnett, Braun, and Pimentel (1248) selected $Tm = 231°K$ and $Tb = 491°K$.

No. 821 Methyl Octyl Sulfide $C_9H_{20}S$ (Ideal Gas State)   Mol Wt 160.316

	cal/(mole °K)				kcal/mole		
$T°K$	$Cp°$	$S°$	$-(G°-H°_{298})/T$	$H°-H°_{298}$	$\Delta Hf°$	$\Delta Gf°$	$\log Kp$
298	55.50	135.66	135.66	0.00	-44.12	14.42	-10.567
300	55.77	136.01	135.67	0.11	-44.19	14.77	-10.762
400	69.49	153.96	138.02	6.38	-48.18	35.00	-19.120
500	82.10	170.85	142.91	13.97	-51.38	56.17	-24.552
600	93.04	186.81	148.91	22.74	-53.87	77.92	-28.380
700	102.53	201.88	155.41	32.53	-55.71	100.07	-31.242
800	110.79	216.12	162.12	43.21	-70.06	121.13	-33.090
900	118.07	229.60	168.87	54.66	-70.54	145.08	-35.228
1000	124.41	242.37	175.59	66.79	-70.57	169.06	-36.945

No. 822 Butyl Hexyl Sulfide $C_{10}H_{22}S$ (Ideal Gas State)   Mol Wt 174.342

	cal/(mole °K)				kcal/mole		
$T°K$	$Cp°$	$S°$	$-(G°-H°_{298})/T$	$H°-H°_{298}$	$\Delta Hf°$	$\Delta Gf°$	$\log Kp$
298	60.39	145.83	145.83	0.00	-49.84	15.37	-11.270
300	60.69	146.21	145.84	0.12	-49.92	15.77	-11.491
400	75.97	165.79	148.40	6.96	-54.28	38.30	-20.924
500	90.13	184.30	153.75	15.28	-57.77	61.86	-27.038
600	102.49	201.84	160.31	24.92	-60.46	86.04	-31.338
700	113.29	218.47	167.44	35.72	-62.41	110.66	-34.547
800	122.71	234.23	174.82	47.53	-76.80	134.19	-36.657
900	131.06	249.17	182.26	60.23	-77.23	160.61	-38.999
1000	138.38	263.37	189.66	73.71	-77.13	187.05	-40.878

Calculations based on butyl pentyl sulfide were made by Scott and McCullough (1316), using one methylene increment, devised by Person and Pimentel (1140), for the thermal properties, and by Prosen, Johnson, and Rossini (1196), for the enthalpy of formation. $\Delta Hf°_{298}(g)$ is -49.84 kcal/mole.

No. 823 Ethyl Octyl Sulfide $C_{10}H_{22}S$ (Ideal Gas State)   Mol Wt 174.342

Scott and McCullough (1316) calculated thermodynamic properties by adding five methylene increments, recommended by Person and Pimentel (1140), for the thermodynamic functions, and by Prosen, Johnson, and Rossini (1196), for the enthalpy of formation, to the values for ethyl propyl sulfide. $\Delta Hf°_{298}(g) = -49.63$ kcal/mole. Jones and Reid (701) reported Tb of about 428°K at 100 torr.

No. 823 Ethyl Octyl Sulfide $C_{10}H_{22}S$ (Ideal Gas State) Mol Wt 174.342

	cal/(mole °K)				kcal/mole		
$T°K$	$Cp°$	$S°$	$-(G°-H°_{298})/T$	$H°-H°_{298}$	$\Delta Hf°$	$\Delta Gf°$	Log Kp
298	60.58	145.51	145.51	0.00	-49.63	15.68	-11.493
300	60.87	145.89	145.52	0.12	-49.71	16.08	-11.714
400	76.13	165.52	148.09	6.98	-54.05	38.63	-21.107
500	90.18	184.05	153.45	15.31	-57.53	62.22	-27.195
600	102.33	201.59	160.02	24.95	-60.23	86.42	-31.479
700	112.88	218.18	167.16	35.72	-62.21	111.07	-34.675
800	122.00	233.86	174.52	47.47	-76.65	134.63	-36.778
900	130.05	248.70	181.95	60.08	-77.16	161.09	-39.117
1000	137.08	262.78	189.33	73.45	-77.19	187.59	-40.996

No. 824 Heptyl Propyl Sulfide $C_{10}H_{22}S$ (Ideal Gas State) Mol Wt 174.342

	cal/(mole °K)				kcal/mole		
$T°K$	$Cp°$	$S°$	$-(G°-H°_{298})/T$	$H°-H°_{298}$	$\Delta Hf°$	$\Delta Gf°$	Log Kp
298	60.39	145.83	145.83	0.00	-49.66	15.55	-11.402
300	60.69	146.21	145.84	0.12	-49.74	15.95	-11.622
400	75.97	165.79	148.40	6.96	-54.10	38.48	-21.023
500	90.13	184.30	153.75	15.28	-57.59	62.04	-27.116
600	102.49	201.84	160.31	24.92	-60.28	86.22	-31.404
700	113.29	218.47	167.44	35.72	-62.23	110.84	-34.603
800	122.71	234.23	174.82	47.53	-76.62	134.37	-36.706
900	131.06	249.17	182.26	60.23	-77.05	160.79	-39.042
1000	138.38	263.37	189.66	73.71	-76.95	187.23	-40.917

Three methylene increments, formulated by Person and Pimentel (1140), for thermodynamic functions, and by Prosen, Johnson, and Rossini (1196), for the enthalpy of formation, were added to the appropriate butyl propyl sulfide values by Scott and McCullough (1316) to obtain the thermodynamic properties of the gaseous state. $\Delta Hf°_{298}(g)$ is -49.66 kcal/mole.

No. 825 Methyl Nonyl Sulfide $C_{10}H_{22}S$ (Ideal Gas State) Mol Wt 174.342

Thermodynamic properties were calculated by Scott and McCullough (1316) by adding five methylene increments, proposed by Person and Pimentel (1140), for each thermodynamic function, and by Prosen, Johnson, and Rossini (1196), for $\Delta Hf°$, to the butyl methyl sulfide values. $\Delta Hf°_{298}(g)$ is -49.05 kcal/mole. Rossini, Pitzer, Arnett, Braun, and Pimentel (1248) selected $Tm = 238°K$ and $Tb = 513°K$.

No. 825 Methyl Nonyl Sulfide $C_{10}H_{22}S$ (Ideal Gas State) Mol Wt 174.342

T°K	Cp°	S°	$-(G°-H°_{298})/T$	$H°-H°_{298}$	$\Delta Hf°$	$\Delta Gf°$	Log Kp
	cal/(mole °K)				kcal/mole		
298	60.97	144.97	144.97	0.00	-49.05	16.42	-12.037
300	61.26	145.35	144.98	0.12	-49.13	16.82	-12.254
400	76.43	165.08	147.56	7.01	-53.44	39.42	-21.539
500	90.35	183.67	152.94	15.37	-56.89	63.05	-27.558
600	102.39	201.23	159.54	25.02	-59.58	87.29	-31.795
700	112.81	217.81	166.69	35.79	-61.56	111.97	-34.957
800	121.85	233.48	174.07	47.53	-76.01	135.57	-37.035
900	129.82	248.30	181.50	60.12	-76.54	162.07	-39.355
1000	136.74	262.35	188.89	73.46	-76.59	188.61	-41.219

No. 826 Pentyl Sulfide $C_{10}H_{22}S$ (Ideal Gas State) Mol Wt 174.342

T°K	Cp°	S°	$-(G°-H°_{298})/T$	$H°-H°_{298}$	$\Delta Hf°$	$\Delta Gf°$	Log Kp
	cal/(mole °K)				kcal/mole		
298	60.39	144.45	144.45	0.00	-49.84	15.79	-11.571
300	60.69	144.83	144.46	0.12	-49.92	16.19	-11.792
400	75.97	164.41	147.02	6.96	-54.28	38.85	-21.226
500	90.13	182.92	152.37	15.28	-57.77	62.55	-27.339
600	102.49	200.46	158.93	24.92	-60.46	86.87	-31.640
700	113.29	217.09	166.06	35.72	-62.41	111.62	-34.848
800	122.71	232.85	173.44	47.53	-76.80	135.29	-36.958
900	131.06	247.79	180.88	60.23	-77.23	161.85	-39.300
1000	138.38	261.99	188.28	73.71	-77.13	188.43	-41.179

From their ΔHc and ΔHv (14.7±0.3 kcal/mole) for pentyl sulfide, Mackle and Mayrick (899) computed $\Delta Hf°_{298}(l) = -63.7\pm0.4$ kcal/mole and $\Delta Hf°_{298}(g) = -49.0\pm0.6$ kcal/mole when adjusted to the revised value of $\Delta Hf°_{298}$ for H_2SO_4 (aq) (cf. page 560). Ayers and Agruss (53) reported $Tb = 500.5°K$. For this symmetrical thiaalkane Scott and McCullough (1316) calculated the gaseous thermal properties by adding four methylene increments of Person and Pimentel (1140) to dipropyl sulfide values. The enthalpy of formation increment of Prosen, Johnson, and Rossini (1196), similarly applied, gave $\Delta Hf°_{298}(g) = -49.84$ kcal/mole. For consistency, the latter figures are adopted.

No. 827 Butyl Heptyl Sulfide $C_{11}H_{24}S$ (Ideal Gas State) Mol Wt 188.368

Calculations based on butyl pentyl sulfide values were made by Scott and McCullough (1316), using two methylene increments, devised by Person and Pimentel (1140), for the thermal properties, and by Prosen,

No. 827 Butyl Heptyl Sulfide $C_{11}H_{24}S$ (Ideal Gas State) Mol Wt 188.368

	cal/(mole °K)				kcal/mole		
$T°K$	$Cp°$	$S°$	$-(G°-H°_{298})/T$	$H°-H°_{298}$	$\Delta Hf°$	$\Delta Gf°$	Log Kp
298	65.86	155.14	155.14	0.00	-54.77	17.38	-12.740
300	66.18	155.55	155.15	0.13	-54.86	17.82	-12.983
400	82.91	176.91	157.95	7.59	-59.54	42.73	-23.343
500	98.38	197.11	163.78	16.67	-63.28	68.74	-30.044
600	111.84	216.27	170.94	27.20	-66.17	95.41	-34.753
700	123.57	234.41	178.73	38.98	-68.26	122.56	-38.262
800	133.77	251.59	186.77	51.86	-82.74	148.63	-40.602
900	142.81	267.88	194.88	65.70	-83.23	177.60	-43.126
1000	150.71	283.34	202.96	80.38	-83.16	206.61	-45.151

Johnson, and Rossini (1196), for the enthalpy of formation. $\Delta Hf°_{298}(g)$ is −54.77 kcal/mole. Signaigo (1348) reported a boiling point of 408°K at 27 torr.

No. 828 Decyl Methyl Sulfide $C_{11}H_{24}S$ (Ideal Gas State) Mol Wt 188.368

	cal/(mole °K)				kcal/mole		
$T°K$	$Cp°$	$S°$	$-(G°-H°_{298})/T$	$H°-H°_{298}$	$\Delta Hf°$	$\Delta Gf°$	Log Kp
298	66.43	154.28	154.28	0.00	-53.97	18.44	-13.514
300	66.76	154.70	154.29	0.13	-54.06	18.88	-13.753
400	83.37	176.21	157.11	7.65	-58.68	43.86	-23.963
500	98.59	196.49	162.97	16.76	-62.39	69.94	-30.569
600	111.73	215.65	170.17	27.29	-65.28	96.68	-35.213
700	123.08	233.75	177.97	39.05	-67.40	123.88	-38.676
800	132.92	250.84	186.02	51.86	-81.95	150.03	-40.983
900	141.56	267.00	194.13	65.59	-82.54	179.08	-43.484
1000	149.07	282.32	202.19	80.13	-82.61	208.18	-45.495

Thermodynamic properties were calculated by Scott and McCullough (1316) by adding six methylene increments, proposed by Person and Pimentel (1140), for each thermodynamic function, and by Prosen, Johnson, and Rossini (1196), for $\Delta Hf°$, to the butyl methyl sulfide values. $\Delta Hf°_{298}(g)$ is −53.97 kcal/mole. Rossini, Pitzer, Arnett, Braun, and Pimentel (1248) selected $Tm = 254.7°K$ and $Tb = 533.1°K$.

No. 829 Ethyl Nonyl Sulfide $C_{11}H_{24}S$ (Ideal Gas State) Mol Wt 188.368

Scott and McCullough (1316) calculated thermodynamic properties by adding six methylene increments, recommended by Person and

No. 829 Ethyl Nonyl Sulfide $C_{11}H_{24}S$ (Ideal Gas State) Mol Wt 188.368

T°K	cal/(mole °K)				kcal/mole		Log Kp
	$Cp°$	$S°$	$-(G°-H°_{298})/T$	$H°-H°_{298}$	$\Delta Hf°$	$\Delta Gf°$	
298	66.04	154.82	154.82	0.00	-54.55	17.70	-12.971
300	66.37	155.23	154.83	0.13	-54.64	18.14	-13.213
400	83.07	176.65	157.63	7.61	-59.30	43.07	-23.532
500	98.42	196.87	163.48	16.70	-63.03	69.11	-30.206
600	111.67	216.01	170.65	27.22	-65.93	95.81	-34.897
700	123.15	234.11	178.44	38.98	-68.05	122.98	-38.394
800	133.07	251.22	186.48	51.80	-82.59	149.09	-40.726
900	141.79	267.40	194.58	65.55	-83.16	178.10	-43.246
1000	149.41	282.75	202.63	80.12	-83.20	207.16	-45.271

Pimentel (1140), for the thermodynamic functions, and by Prosen, Johnson, and Rossini (1196), for the enthalpy of formation, to the ethyl propyl sulfide values. $\Delta Hf°_{298}(g) = -54.55$ kcal/mole.

No. 830 Octyl Propyl Sulfide $C_{11}H_{24}S$ (Ideal Gas State) Mol Wt 188.368

T°K	cal/(mole °K)				kcal/mole		Log Kp
	$Cp°$	$S°$	$-(G°-H°_{298})/T$	$H°-H°_{298}$	$\Delta Hf°$	$\Delta Gf°$	
298	65.86	155.14	155.14	0.00	-54.56	17.59	-12.894
300	66.18	155.55	155.15	0.13	-54.65	18.03	-13.136
400	82.91	176.91	157.95	7.59	-59.33	42.94	-23.458
500	98.38	197.11	163.78	16.67	-63.07	68.95	-30.136
600	111.84	216.27	170.94	27.20	-65.96	95.62	-34.829
700	123.57	234.41	178.73	38.98	-68.05	122.77	-38.328
800	133.77	251.59	186.77	51.86	-82.53	148.84	-40.659
900	142.81	267.88	194.88	65.70	-83.02	177.81	-43.177
1000	150.71	283.34	202.96	80.38	-82.95	206.82	-45.197

Four methylene increments, formulated by Person and Pimentel (1140), for thermodynamic functions, and by Prosen, Johnson, and Rossini (1196), for the enthalpy of formation, were added to the appropriate butyl propyl sulfide values by Scott and McCullough (1316) to obtain the thermodynamic properties of the gaseous state. $\Delta Hf°_{298}(g)$ is -54.56 kcal/mole.

No. 831 Butyl Octyl Sulfide $C_{12}H_{26}S$ (Ideal Gas State) Mol Wt 202.394

Calculations based on butyl pentyl sulfide values were made by Scott and McCullough (1316), using three methylene increments, devised by Person and Pimentel (1140), for the thermal properties, and by Prosen,

No. 831 Butyl Octyl Sulfide $C_{12}H_{26}S$ (Ideal Gas State) Mol Wt 202.394

T°K	Cp°	S°	$-(G°-H°_{298})/T$	$H°-H°_{298}$	$\Delta Hf°$	$\Delta Gf°$	Log Kp
	cal/(mole °K)			kcal/mole			
298	71.32	164.45	164.45	0.00	-59.69	19.40	-14.217
300	71.68	164.90	164.46	0.14	-59.78	19.88	-14.482
400	89.85	188.04	167.49	8.23	-64.78	47.16	-25.767
500	106.62	209.93	173.81	18.07	-68.78	75.63	-33.055
600	121.18	230.69	181.57	29.47	-71.87	104.80	-38.171
700	133.84	250.34	190.01	42.24	-74.10	134.47	-41.981
800	144.84	268.94	198.72	56.18	-88.68	163.08	-44.550
900	154.55	286.58	207.51	71.16	-89.23	194.61	-47.255
1000	163.04	303.31	216.26	87.05	-89.17	226.17	-49.427

Johnson, and Rossini (1196), for the enthalpy of formation. $\Delta Hf°_{298}(g)$ is -59.69 kcal/mole.

No. 832 Decyl Ethyl Sulfide $C_{12}H_{26}S$ (Ideal Gas State) Mol Wt 202.394

T°K	Cp°	S°	$-(G°-H°_{298})/T$	$H°-H°_{298}$	$\Delta Hf°$	$\Delta Gf°$	Log Kp
	cal/(mole °K)			kcal/mole			
298	71.51	164.13	164.13	0.00	-59.84	19.34	-14.177
300	71.86	164.58	164.14	0.14	-59.93	19.83	-14.443
400	90.01	187.77	167.18	8.24	-64.91	47.14	-25.754
500	106.67	209.69	173.51	18.10	-68.90	75.63	-33.055
600	121.01	230.44	181.28	29.50	-71.99	104.82	-38.180
700	133.43	250.04	189.72	42.23	-74.26	134.52	-41.997
800	144.13	268.57	198.43	56.12	-88.89	163.17	-44.573
900	153.54	286.11	207.20	71.02	-89.52	194.74	-47.286
1000	161.75	302.72	215.93	86.79	-89.59	226.35	-49.466

Scott and McCullough (1316) calculated thermodynamic properties by adding seven methylene increments, recommended by Person and Pimentel (1140), for the thermodynamic functions, and by Prosen, Johnson, and Rossini (1196), for the enthalpy of formation, to the values for ethyl propyl sulfide. $\Delta Hf°_{298}(g)$ is -59.84 kcal/mole.

No. 833 Hexyl Sulfide $C_{12}H_{26}S$ (Ideal Gas State) Mol Wt 202.394

For this symmetrical thiaalkane Scott and McCullough (1316) calculated the gaseous thermal properties by adding six methylene increments of Person and Pimentel (1140) to dipropyl sulfide values. The enthalpy of formation increment of Prosen, Johnson, and Rossini (1196),

No. 833 Hexyl Sulfide $C_{12}H_{26}S$ (Ideal Gas State) Mol Wt 202.394

$T°K$	$Cp°$	$S°$	$-(G°-H°_{298})/T$	$H°-H°_{298}$	$\Delta Hf°$	$\Delta Gf°$	$Log\ Kp$
		cal/(mole °K)			kcal/mole		
298	71.32	163.07	163.07	0.00	-59.69	19.81	-14.519
300	71.68	163.52	163.08	0.14	-59.78	20.29	-14.784
400	89.85	186.66	166.11	8.23	-64.78	47.71	-26.069
500	106.62	208.55	172.43	18.07	-68.78	76.32	-33.356
600	121.18	229.31	180.19	29.47	-71.87	105.63	-38.472
700	133.84	248.96	188.63	42.24	-74.10	135.43	-42.282
800	144.84	267.56	197.34	56.18	-88.68	164.19	-44.852
900	154.55	285.20	206.13	71.16	-89.23	195.85	-47.557
1000	163.04	301.93	214.88	87.05	-89.17	227.55	-49.729

similarly applied, gave $\Delta Hf°_{298}(g) = -59.69$ kcal/mole. Vogel and Cowan (1542) reported $Tb = 386.5°K$ at 4 torr.

No. 834 Methyl Undecyl Sulfide $C_{12}H_{26}S$ (Ideal Gas State) Mol Wt 202.394

$T°K$	$Cp°$	$S°$	$-(G°-H°_{298})/T$	$H°-H°_{298}$	$\Delta Hf°$	$\Delta Gf°$	$Log\ Kp$
		cal/(mole °K)			kcal/mole		
298	71.90	163.59	163.59	0.00	-58.90	20.44	-14.984
300	72.25	164.04	163.60	0.14	-58.99	20.93	-15.245
400	90.31	187.34	166.65	8.28	-63.94	48.29	-26.382
500	106.84	209.31	173.00	18.16	-67.90	76.82	-33.575
600	121.07	230.07	180.80	29.57	-70.98	106.05	-38.627
700	133.36	249.68	189.25	42.30	-73.25	135.78	-42.392
800	143.98	268.20	197.98	56.18	-87.90	164.47	-44.929
900	153.31	285.70	206.76	71.06	-88.54	196.08	-47.611
1000	161.41	302.29	215.49	86.80	-88.63	227.73	-49.768

Thermodynamic properties were calculated by Scott and McCullough (1316) by adding seven methylene increments, proposed by Person and Pimentel (1140), for each thermodynamic function, and by Prosen, Johnson, and Rossini (1196), for $\Delta Hf°$, to the butyl methyl sulfide values. $\Delta Hf°_{298}(g)$ is -58.90 kcal/mole. Rossini, Pitzer, Arnett, Braun, and Pimentel (1248) selected $Tm = 259°K$ and $Tb = 552°K$.

No. 835 Nonyl Propyl Sulfide $C_{12}H_{26}S$ (Ideal Gas State) Mol Wt 202.394

Five methylene increments, formulated by Person and Pimentel (1140), for thermodynamic functions, and by Prosen, Johnson, and Rossini (1196), for the enthalpy of formation, were added to the appropriate values for butyl propyl sulfide by Scott and McCullough (1316) to obtain

No. 835 Nonyl Propyl Sulfide $C_{12}H_{26}S$ (Ideal Gas State) Mol Wt 202.394

$T°K$	$Cp°$	$S°$	$-(G°-H°_{298})/T$	$H°-H°_{298}$	$\Delta Hf°$	$\Delta Gf°$	$\text{Log } Kp$
		cal/(mole °K)			kcal/mole		
298	71.32	164.45	164.45	0.00	-59.51	19.58	-14.349
300	71.68	164.90	164.46	0.14	-59.60	20.06	-14.613
400	89.85	188.04	167.49	8.23	-64.60	47.34	-25.866
500	106.62	209.93	173.81	18.07	-68.60	75.81	-33.133
600	121.18	230.69	181.57	29.47	-71.69	104.98	-38.236
700	133.84	250.34	190.01	42.24	-73.92	134.65	-42.037
800	144.84	268.94	198.72	56.18	-88.50	163.26	-44.599
900	154.55	286.58	207.51	71.16	-89.05	194.79	-47.299
1000	163.04	303.31	216.26	87.05	-88.99	226.35	-49.466

the thermodynamic properties of the gaseous state. $\Delta Hf°_{298}(g)$ is -59.51 kcal/mole. Aibazov, Petrov, Khairullina, and Yaprintseva (5) selected $Tm = 243.3°K$ and $Tb = 405.4°K$ at 5 torr.

No. 836 Butyl Nonyl Sulfide $C_{13}H_{28}S$ (Ideal Gas State) Mol Wt 216.420

$T°K$	$Cp°$	$S°$	$-(G°-H°_{298})/T$	$H°-H°_{298}$	$\Delta Hf°$	$\Delta Gf°$	$\text{Log } Kp$
		cal/(mole °K)			kcal/mole		
298	76.79	173.76	173.76	0.00	-64.62	21.40	-15.687
300	77.17	174.24	173.77	0.15	-64.72	21.93	-15.974
400	96.79	199.17	177.03	8.86	-70.04	51.59	-28.186
500	114.87	222.75	183.84	19.46	-74.29	82.50	-36.061
600	130.52	245.11	192.20	31.75	-77.58	114.17	-41.585
700	144.12	266.27	201.29	45.50	-79.96	146.37	-45.696
800	155.90	286.30	210.67	60.51	-94.63	177.53	-48.495
900	166.30	305.28	220.14	76.63	-95.23	211.61	-51.382
1000	175.38	323.28	229.56	93.72	-95.20	245.73	-53.701

Calculations based on butyl pentyl sulfide values were made by Scott and McCullough (1316), using four methylene increments, devised by Person and Pimentel (1140), for the thermal properties, and by Prosen, Johnson, and Rossini (1196), for the enthalpy of formation. $\Delta Hf°_{298}(g)$ is -64.62 kcal/mole.

No. 837 Decyl Propyl Sulfide $C_{13}H_{28}S$ (Ideal Gas State) Mol Wt 216.420

Six methylene increments, formulated by Person and Pimentel (1140), for thermodynamic functions, and by Prosen, Johnson, and Rossini (1196), for the enthalpy of formation, were added to the appropriate butyl propyl sulfide values by Scott and McCullough (1316) to obtain the

No. 837 Decyl Propyl Sulfide $C_{13}H_{28}S$ (Ideal Gas State) Mol Wt 216.420

T°K	Cp°	S°	$-(G°-H°_{298})/T$	$H°-H°_{298}$	$\Delta Hf°$	$\Delta Gf°$	Log Kp
	cal/(mole °K)				kcal/mole		
298	76.79	173.76	173.76	0.00	-64.44	21.58	-15.819
300	77.17	174.24	173.77	0.15	-64.54	22.11	-16.105
400	96.79	199.17	177.03	8.86	-69.86	51.77	-28.284
500	114.87	222.75	183.84	19.46	-74.11	82.68	-36.140
600	130.52	245.11	192.20	31.75	-77.40	114.35	-41.651
700	144.12	266.27	201.29	45.50	-79.78	146.55	-45.753
800	155.90	286.30	210.67	60.51	-94.45	177.71	-48.545
900	166.30	305.28	220.14	76.63	-95.05	211.79	-51.426
1000	175.38	323.28	229.56	93.72	-95.02	245.91	-53.740

thermodynamic properties of the gaseous state. $\Delta Hf°_{298}(g)$ is -64.44 kcal/mole.

No. 838 Dodecyl Methyl Sulfide $C_{13}H_{28}S$ (Ideal Gas State) Mol Wt 216.420

T°K	Cp°	S°	$-(G°-H°_{298})/T$	$H°-H°_{298}$	$\Delta Hf°$	$\Delta Gf°$	Log Kp
	cal/(mole °K)				kcal/mole		
298	77.36	172.90	172.90	0.00	-63.82	22.46	-16.461
300	77.75	173.38	172.91	0.15	-63.92	22.99	-16.745
400	97.25	198.46	176.19	8.91	-69.18	52.73	-28.806
500	115.08	222.13	183.04	19.55	-73.40	83.71	-36.586
600	130.41	244.49	191.43	31.84	-76.68	115.44	-42.045
700	143.63	265.61	200.54	45.56	-79.09	147.70	-46.110
800	155.05	285.55	209.93	60.50	-93.84	178.92	-48.877
900	165.05	304.41	219.39	76.52	-94.54	213.08	-51.741
1000	173.74	322.26	228.79	93.47	-94.65	247.30	-54.044

Thermodynamic properties were calculated by Scott and McCullough (1316) by adding seven methylene increments, proposed by Person and Pimentel (1140), for each thermodynamic function, and by Prosen, Johnson, and Rossini (1196), for $\Delta Hf°$, to the values for butyl methyl sulfide. $\Delta Hf°_{298}(g)$ is -63.82 kcal/mole. Rossini, Pitzer, Arnett, Braun, and Pimentel (1248) selected $Tm = 271°K$ and $Tb = 570°K$.

No. 839 Ethyl Undecyl Sulfide $C_{13}H_{28}S$ (Ideal Gas State) Mol Wt 216.420

Scott and McCullough (1316) calculated thermodynamic properties by adding eight methylene increments, recommended by Person and Pimentel (1140), for the thermodynamic functions, and by Prosen,

No. 839 Ethyl Undecyl Sulfide $C_{13}H_{28}S$ (Ideal Gas State) Mol Wt 216.420

$T°K$	$Cp°$	$S°$	$-(G°-H°_{298})/T$	$H°-H°_{298}$	$\Delta Hf°$	$\Delta Gf°$	Log Kp
		cal/(mole °K)			kcal/mole		
298	76.97	173.44	173.44	0.00	-64.40	21.72	-15.918
300	77.36	173.92	173.45	0.15	-64.50	22.24	-16.204
400	96.95	198.90	176.72	8.88	-69.80	51.93	-28.374
500	114.91	222.51	183.54	19.49	-74.04	82.87	-36.223
600	130.35	244.86	191.91	31.77	-77.33	114.57	-41.729
700	143.70	265.98	201.00	45.49	-79.74	146.79	-45.828
800	155.20	285.93	210.38	60.45	-94.47	177.98	-48.620
900	165.28	304.81	219.83	76.48	-95.16	212.10	-51.503
1000	174.08	322.69	229.23	93.46	-95.24	246.28	-53.821

Johnson, and Rossini (1196), for the enthalpy of formation, to the ethyl propyl sulfide values. $\Delta Hf°_{298}(g)$ is -64.40 kcal/mole.

No. 840 Butyl Decyl Sulfide $C_{14}H_{30}S$ (Ideal Gas State) Mol Wt 230.446

$T°K$	$Cp°$	$S°$	$-(G°-H°_{298})/T$	$H°-H°_{298}$	$\Delta Hf°$	$\Delta Gf°$	Log Kp
		cal/(mole °K)			kcal/mole		
298	82.25	183.07	183.07	0.00	-69.55	23.41	-17.157
300	82.67	183.59	183.08	0.16	-69.66	23.98	-17.466
400	103.73	210.29	186.58	9.49	-75.29	56.02	-30.605
500	123.11	235.57	193.87	20.86	-79.80	89.38	-39.067
600	139.86	259.53	202.83	34.02	-83.28	123.55	-45.000
700	154.39	282.21	212.57	48.75	-85.81	158.27	-49.412
800	166.97	303.66	222.63	64.83	-100.58	191.97	-52.441
900	178.04	323.98	232.77	82.09	-101.24	228.60	-55.510
1000	187.71	343.25	242.86	100.39	-101.22	265.28	-57.974

Calculations based on butyl pentyl sulfide values were made by Scott and McCullough (1316), using five methylene increments, devised by Person and Pimentel (1140), for the thermal properties, and by Prosen, Johnson, and Rossini (1196), for the enthalpy of formation. $\Delta Hf°_{298}(g)$ is -69.55 kcal/mole. Aibazov, Petrov, Khairullina, and Yaprintseva (5) selected $Tb = 419°K$ at 9 torr.

No. 841 Dodecyl Ethyl Sulfide $C_{14}H_{30}S$ (Ideal Gas State) Mol Wt 230.446

Scott and McCullough (1316) calculated thermodynamic properties by adding nine methylene increments, recommended by Person and Pimentel (1140), for the thermodynamic functions, and by Prosen, Johnson, and

No. 841 Dodecyl Ethyl Sulfide $C_{14}H_{30}S$ (Ideal Gas State) Mol Wt 230.446

$T°K$	$Cp°$	$S°$	$-(G°-H°_{298})/T$	$H°-H°_{298}$	$\Delta Hf°$	$\Delta Gf°$	Log Kp
		cal/(mole °K)			kcal/mole		
298	82.44	182.75	182.75	0.00	-69.33	23.72	-17.388
300	82.85	183.27	182.76	0.16	-69.44	24.29	-17.696
400	103.89	210.02	186.26	9.51	-75.05	56.36	-30.793
500	123.16	235.33	193.57	20.88	-79.55	89.75	-39.229
600	139.69	259.28	202.54	34.05	-83.04	123.94	-45.144
700	153.98	281.91	212.28	48.75	-85.59	158.69	-49.544
800	166.26	303.29	222.33	64.77	-100.42	192.42	-52.565
900	177.03	323.51	232.46	81.95	-101.17	229.10	-55.630
1000	186.41	342.66	242.53	100.13	-101.26	265.83	-58.095

Rossini (1196), for the enthalpy of formation, to the values for ethyl propyl sulfide. $\Delta Hf°_{298}(g) = -69.33$ kcal/mole. Bost and Everett (150) reported $Tm = 267°K$ and $Tb = 430$ to $434°K$ at 18 torr.

No. 842 Heptyl Sulfide $C_{14}H_{30}S$ (Ideal Gas State) Mol Wt 230.446

$T°K$	$Cp°$	$S°$	$-(G°-H°_{298})/T$	$H°-H°_{298}$	$\Delta Hf°$	$\Delta Gf°$	Log Kp
		cal/(mole °K)			kcal/mole		
298	82.25	181.69	181.69	0.00	-69.54	23.83	-17.466
300	82.67	182.21	181.70	0.16	-69.65	24.40	-17.775
400	103.73	208.91	185.20	9.49	-75.28	56.58	-30.912
500	123.11	234.19	192.49	20.86	-79.79	90.08	-39.373
600	139.86	258.15	201.45	34.02	-83.27	124.39	-45.305
700	154.39	280.83	211.19	48.75	-85.80	159.25	-49.717
800	166.97	302.28	221.25	64.83	-100.57	193.08	-52.745
900	178.04	322.60	231.39	82.09	-101.23	229.85	-55.814
1000	187.71	341.87	241.48	100.39	-101.21	266.67	-58.278

For this symmetrical thiaalkane, Scott and McCullough (1316) calculated the gaseous thermal properties by adding eight methylene increments of Person and Pimentel (1140) to dipropyl sulfide values. The enthalpy of formation increment of Prosen, Johnson, and Rossini (1196), similarly applied, gave $\Delta Hf°_{298}(g) = -69.54$ kcal/mole. Rossini, Pitzer, Arnett, Braun, and Pimentel (1248) selected $Tb = 414.5°K$.

No. 843 Methyl Tridecyl Sulfide $C_{14}H_{30}S$ (Ideal Gas State) Mol Wt 230.446

Thermodynamic properties were calculated by Scott and McCullough (1316) by adding eight methylene increments, proposed by Person and Pimentel (1140), for each thermodynamic function, and by Prosen, Johnson, and Rossini (1196), for $\Delta Hf°$, to the butyl methyl sulfide values.

No. 843   Methyl Tridecyl Sulfide $C_{14}H_{30}S$ (Ideal Gas State)   Mol Wt 230.446

	cal/(mole °K)				kcal/mole		
$T°K$	$Cp°$	$S°$	$-(G°-H°_{298})/T$	$H°-H°_{298}$	$\Delta Hf°$	$\Delta Gf°$	Log Kp
298	82.83	182.21	182.21	0.00	-68.75	24.46	-17.931
300	83.24	182.73	182.22	0.16	-68.86	25.03	-18.237
400	104.19	209.59	185.74	9.55	-74.44	57.15	-31.225
500	123.33	234.94	193.07	20.94	-78.91	90.58	-39.592
600	139.75	258.92	202.06	34.12	-82.39	124.81	-45.460
700	153.91	281.55	211.82	48.82	-84.94	159.60	-49.826
800	166.11	302.91	221.88	64.83	-99.78	193.37	-52.822
900	176.80	323.11	232.02	81.99	-100.55	230.08	-55.868
1000	186.07	342.23	242.09	100.14	-100.67	266.85	-58.318

$\Delta Hf°_{298}(g)$ is -68.75 kcal/mole. Rossini, Pitzer, Arnett, Braun, and Pimentel (1248) selected $Tm = 276°K$ and $Tb = 587°K$.

No. 844   Propyl Undecyl Sulfide $C_{14}H_{30}S$ (Ideal Gas State)   Mol Wt 230.446

	cal/(mole °K)				kcal/mole		
$T°K$	$Cp°$	$S°$	$-(G°-H°_{298})/T$	$H°-H°_{298}$	$\Delta Hf°$	$\Delta Gf°$	Log Kp
298	82.25	183.07	183.07	0.00	-69.36	23.60	-17.296
300	82.67	183.59	183.08	0.16	-69.47	24.17	-17.604
400	103.73	210.29	186.58	9.49	-75.10	56.21	-30.708
500	123.11	235.57	193.87	20.86	-79.61	89.57	-39.150
600	139.86	259.53	202.83	34.02	-83.09	123.74	-45.069
700	154.39	282.21	212.57	48.75	-85.62	158.46	-49.471
800	166.97	303.66	222.63	64.83	-100.39	192.16	-52.493
900	178.04	323.98	232.77	82.09	-101.05	228.79	-55.556
1000	187.71	343.25	242.86	100.39	-101.03	265.47	-58.016

Seven methylene increments, formulated by Person and Pimentel (1140), for thermodynamic functions, and by Prosen, Johnson, and Rossini (1196), for the enthalpy of formation, were added to the appropriate values for butyl propyl sulfide by Scott and McCullough (1316) to obtain the thermodynamic properties of the gaseous state. $\Delta Hf°_{298}(g) = -69.36$ kcal/mole.

No. 845   Butyl Undecyl Sulfide $C_{15}H_{32}S$ (Ideal Gas State)   Mol Wt 244.472

Calculations based on butyl pentyl sulfide values were made by Scott and McCullough (1316), using six methylene increments, devised by Person and Pimentel (1140), for the thermal properties, and by Prosen, Johnson, and Rossini (1196), for the enthalpy of formation. $\Delta Hf°_{298}(g)$ is -74.47 kcal/mole.

No. 845 Butyl Undecyl Sulfide $C_{15}H_{32}S$ (Ideal Gas State) Mol Wt 244.472

T°K	Cp°	S°	$-(G°-H°_{298})/T$	$H°-H°_{298}$	ΔHf°	ΔGf°	Log Kp
		cal/(mole °K)			kcal/mole		
298	87.72	192.38	192.38	0.00	-74.47	25.42	-18.634
300	88.16	192.93	192.39	0.17	-74.59	26.03	-18.965
400	110.67	221.42	196.12	10.12	-80.54	60.45	-33.029
500	131.36	248.39	203.90	22.25	-85.30	96.27	-42.078
600	149.20	273.95	213.46	36.30	-88.98	132.93	-48.418
700	164.67	298.14	223.85	52.01	-91.65	170.18	-53.131
800	178.03	321.02	234.58	69.16	-106.52	206.42	-56.389
900	189.79	342.68	245.40	87.56	-107.24	245.61	-59.639
1000	200.04	363.22	256.16	107.06	-107.24	284.85	-62.250

No. 846 Dodecyl Propyl Sulfide $C_{15}H_{32}S$ (Ideal Gas State) Mol Wt 244.472

T°K	Cp°	S°	$-(G°-H°_{298})/T$	$H°-H°_{298}$	ΔHf°	ΔGf°	Log Kp
		cal/(mole °K)			kcal/mole		
298	87.72	192.38	192.38	0.00	-74.29	25.60	-18.766
300	88.16	192.93	192.39	0.17	-74.41	26.21	-19.096
400	110.67	221.42	196.12	10.12	-80.36	60.63	-33.127
500	131.36	248.39	203.90	22.25	-85.12	96.45	-42.157
600	149.20	273.95	213.46	36.30	-88.80	133.11	-48.483
700	164.67	298.14	223.85	52.01	-91.47	170.36	-53.187
800	178.03	321.02	234.58	69.16	-106.34	206.60	-56.438
900	189.79	342.68	245.40	87.56	-107.06	245.79	-59.683
1000	200.04	363.22	256.16	107.06	-107.06	285.03	-62.290

Eight methylene increments, formulated by Person and Pimentel (1140), for thermodynamic functions, and by Prosen, Johnson, and Rossini (1196), for the enthalpy of formation, were added to the appropriate butyl propyl values by Scott and McCullough (1316) to obtain the thermodynamic properties of the gaseous state. $\Delta Hf°_{298}(g)$ is -74.29 kcal/mole. Aibazov, Petrov, Khairullina, and Yaprintseva (5) selected $Tm = 288°K$ and $Tb = 618°K$.

No. 847 Ethyl Tridecyl Sulfide $C_{15}H_{32}S$ (Ideal Gas State) Mol Wt 244.472

Scott and McCullough (1316) calculated thermodynamic properties by adding 10 methylene increments, recommended by Person and Pimentel (1140), for the thermodynamic functions, and by Prosen, Johnson, and Rossini (1196), for the enthalpy of formation, to the values for ethyl propyl sulfide. $\Delta Hf°_{298}(g) = -74.26$ kcal/mole.

Monothiaalkane Ideal Gas Tables

No. 847 Ethyl Tridecyl Sulfide $C_{15}H_{32}S$ (Ideal Gas State) Mol Wt 244.472

$T°K$	$Cp°$	$S°$	$-(G°-H°_{298})/T$	$H°-H°_{298}$	$\Delta Hf°$	$\Delta Gf°$	$\mathrm{Log}\,Kp$
		cal/(mole °K)			kcal/mole		
298	87.91	192.06	192.06	0.00	-74.26	25.73	-18.858
300	88.35	192.61	192.07	0.17	-74.37	26.34	-19.188
400	110.84	221.15	195.81	10.14	-80.31	60.79	-33.212
500	131.41	248.15	203.60	22.28	-85.06	96.63	-42.235
600	149.04	273.70	213.17	36.32	-88.75	133.32	-48.558
700	164.26	297.85	223.56	52.01	-91.44	170.59	-53.259
800	177.33	320.65	234.29	69.10	-106.37	206.86	-56.510
900	188.78	342.22	245.09	87.42	-107.17	246.09	-59.757
1000	198.75	362.64	255.83	106.81	-107.28	285.38	-62.367

No. 848 Methyl Tetradecyl Sulfide $C_{15}H_{32}S$ (Ideal Gas State) Mol Wt 244.472

$T°K$	$Cp°$	$S°$	$-(G°-H°_{298})/T$	$H°-H°_{298}$	$\Delta Hf°$	$\Delta Gf°$	$\mathrm{Log}\,Kp$
		cal/(mole °K)			kcal/mole		
298	88.30	191.52	191.52	0.00	-73.68	26.47	-19.401
300	88.74	192.07	191.53	0.17	-73.79	27.08	-19.729
400	111.14	220.72	195.28	10.18	-79.69	61.58	-33.644
500	131.58	247.77	203.10	22.34	-84.42	97.46	-42.598
600	149.10	273.34	212.69	36.39	-88.10	134.18	-48.874
700	164.19	297.49	223.10	52.08	-90.79	171.50	-53.541
800	177.18	320.28	233.84	69.16	-105.73	207.81	-56.767
900	188.55	341.82	244.65	87.46	-106.55	247.07	-59.995
1000	198.41	362.21	255.39	106.82	-106.69	286.40	-62.591

Thermodynamic properties were calculated by Scott and McCullough (1316) by adding nine methylene increments, proposed by Person and Pimentel (1140), for each thermodynamic function, and by Prosen, Johnson, and Rossini (1196), for $\Delta Hf°$, to the values for butyl methyl sulfide. $\Delta Hf°_{298}(g)$ is -73.68 kcal/mole. Rossini, Pitzer, Arnett, Braun, and Pimentel (1248) selected $Tm = 284°K$ and $Tb = 603°K$.

No. 849 Butyl Dodecyl Sulfide $C_{16}H_{34}S$ (Ideal Gas State) Mol Wt 258.498

Calculations based on butyl pentyl sulfide values were made by Scott and McCullough (1316), using seven methylene increments, devised by Person and Pimentel (1140), for the thermal properties, and by Prosen, Johnson, and Rossini (1196), for the enthalpy of formation. $\Delta Hf°_{298}(g)$ is -79.40 kcal/mole. Drahowzal and Klamann (348) reported a Tb of about 343°K at 9 torr.

No. 849 Butyl Dodecyl Sulfide $C_{16}H_{34}S$ (Ideal Gas State) Mol Wt 258.498

	cal/(mole °K)				kcal/mole		
T°K	$Cp°$	$S°$	$-(G° - H°_{298})/T$	$H° - H°_{298}$	$\Delta Hf°$	$\Delta Gf°$	Log Kp
298	93.19	201.69	201.69	0.00	-79.40	27.43	-20.105
300	93.66	202.27	201.70	0.18	-79.52	28.08	-20.457
400	117.62	232.55	205.66	10.76	-85.79	64.88	-35.448
500	139.61	261.21	213.93	23.65	-90.81	103.15	-45.084
600	158.55	288.38	224.10	38.57	-94.69	142.31	-51.832
700	174.95	314.08	235.13	55.27	-97.50	182.08	-56.846
800	189.10	338.38	246.53	73.48	-112.47	220.86	-60.334
900	201.54	361.39	258.03	93.03	-113.24	262.60	-63.765
1000	212.38	383.20	269.46	113.74	-113.26	304.40	-66.523

No. 850 Ethyl Tetradecyl Sulfide $C_{16}H_{34}S$ (Ideal Gas State) Mol Wt 258.498

	cal/(mole °K)				kcal/mole		
T°K	$Cp°$	$S°$	$-(G° - H°_{298})/T$	$H° - H°_{298}$	$\Delta Hf°$	$\Delta Gf°$	Log Kp
298	93.37	201.36	201.36	0.00	-79.18	27.75	-20.338
300	93.84	201.94	201.37	0.18	-79.30	28.40	-20.689
400	117.78	232.27	205.34	10.78	-85.55	65.23	-35.638
500	139.65	260.96	213.62	23.67	-90.56	103.52	-45.248
600	158.38	288.11	223.79	38.60	-94.45	142.71	-51.978
700	174.53	313.77	234.83	55.26	-97.28	182.51	-56.980
800	188.39	338.00	246.23	73.42	-112.31	221.33	-60.460
900	200.52	360.90	257.71	92.88	-113.17	263.11	-63.889
1000	211.08	382.59	269.12	113.47	-113.30	304.96	-66.646

Scott and McCullough (1316) calculated thermodynamic properties by adding 11 methylene increments, recommended by Person and Pimentel (1140), for the thermodynamic functions, and by Prosen, Johnson, and Rossini (1196), for the enthalpy of formation, to the ethyl propyl sulfide values. $\Delta Hf°_{298}(g)$ is -79.18 kcal/mole.

No. 851 Methyl Pentadecyl Sulfide $C_{16}H_{34}S$ (Ideal Gas State) Mol Wt 258.498

	cal/(mole °K)				kcal/mole		
T°K	$Cp°$	$S°$	$-(G° - H°_{298})/T$	$H° - H°_{298}$	$\Delta Hf°$	$\Delta Gf°$	Log Kp
298	93.76	200.82	200.82	0.00	-78.60	28.49	-20.881
300	94.23	201.41	200.83	0.18	-78.72	29.14	-21.230
400	118.08	231.83	204.81	10.81	-84.94	66.02	-36.070
500	139.82	260.57	213.12	23.73	-89.93	104.36	-45.611
600	158.44	287.75	223.31	38.67	-93.80	143.57	-52.294
700	174.46	313.41	234.37	55.33	-96.63	183.42	-57.262
800	188.24	337.62	245.78	73.48	-111.67	222.27	-60.717
900	200.29	360.50	257.26	92.92	-112.55	264.09	-64.127
1000	210.74	382.16	268.68	113.49	-112.71	305.98	-66.869

Thermodynamic properties were calculated by Scott and McCullough (1316) by adding 10 methylene increments, proposed by Person and Pimentel (1140), for each thermodynamic function, and by Prosen, Johnson, and Rossini (1196), for $\Delta Hf°$, to the values for butyl methyl sulfide. $\Delta Hf°_{298}(g)$ is -78.60 kcal/mole. Rossini, Pitzer, Arnett, Braun, and Pimentel (1248) selected $Tm = 288°K$ and $Tb = 618°K$.

No. 852   Octyl Sulfide $C_{16}H_{34}S$ (Ideal Gas State)   Mol Wt 258.498

$T°K$	$Cp°$	$S°$	$-(G°-H°_{298})/T$	$H°-H°_{298}$	$\Delta Hf°$	$\Delta Gf°$	Log Kp
	cal/(mole °K)				kcal/mole		
298	93.19	200.31	200.31	0.00	-79.39	27.85	-20.413
300	93.66	200.89	200.32	0.18	-79.51	28.51	-20.766
400	117.62	231.17	204.28	10.76	-85.78	65.44	-35.755
500	139.61	259.83	212.55	23.65	-90.80	103.85	-45.390
600	158.55	287.00	222.72	38.57	-94.68	143.14	-52.137
700	174.95	312.70	233.75	55.27	-97.49	183.06	-57.150
800	189.10	337.00	245.15	73.48	-112.46	221.98	-60.638
900	201.54	360.01	256.65	93.03	-113.23	263.85	-64.070
1000	212.38	381.82	268.08	113.74	-113.25	305.79	-66.827

For this symmetrical thiaalkane Scott and McCullough (1316) calculated the gaseous thermal properties by adding 10 methylene increments of Person and Pimentel (1140) to dipropyl sulfide values. The enthalpy of formation increment of Prosen, Johnson, and Rossini (1196), similarly applied, gave $\Delta Hf°_{298}(g) = -79.39$ kcal/mole. Vogel and Cowan (1542) reported $Tb = 434.7°K$ at 4 torr.

No. 853   Propyl Tridecyl Sulfide $C_{16}H_{34}S$ (Ideal Gas State)   Mol Wt 258.498

$T°K$	$Cp°$	$S°$	$-(G°-H°_{298})/T$	$H°-H°_{298}$	$\Delta Hf°$	$\Delta Gf°$	Log Kp
	cal/(mole °K)				kcal/mole		
298	93.19	201.69	201.69	0.00	-79.22	27.61	-20.236
300	93.66	202.27	201.70	0.18	-79.34	28.26	-20.588
400	117.62	232.55	205.66	10.76	-85.61	65.06	-35.546
500	139.61	261.21	213.93	23.65	-90.63	103.33	-45.163
600	158.55	288.38	224.10	38.57	-94.51	142.49	-51.898
700	174.95	314.08	235.13	55.27	-97.32	182.26	-56.902
800	189.10	338.38	246.53	73.48	-112.29	221.04	-60.383
900	201.54	361.39	258.03	93.03	-113.06	262.78	-63.809
1000	212.38	383.20	269.46	113.74	-113.08	304.58	-66.562

Nine methylene increments, formulated by Person and Pimentel (1140), for thermodynamic functions, and by Prosen, Johnson, and Rossini

(1196), for the enthalpy of formation, were added to the appropriate values for butyl propyl sulfide by Scott and McCullough (1316) to obtain the thermodynamic properties of the gaseous state. $\Delta Hf°_{298}(g) = -79.22$ kcal/mole.

No. 854 Butyl Tridecyl Sulfide $C_{17}H_{36}S$ (Ideal Gas State) Mol Wt 272.524

$T°K$	$Cp°$	$S°$	$-(G°-H°_{298})/T$	$H°-H°_{298}$	$\Delta Hf°$	$\Delta Gf°$	Log Kp
		cal/(mole °K)			kcal/mole		
298	98.65	210.99	210.99	0.00	-84.32	29.45	-21.584
300	99.15	211.61	211.00	0.19	-84.45	30.14	-21.958
400	124.56	243.66	215.20	11.39	-91.04	69.32	-37.874
500	147.85	274.02	223.95	25.04	-96.31	110.04	-48.097
600	167.89	302.79	234.71	40.85	-100.39	151.70	-55.252
700	185.22	330.00	246.40	58.52	-103.34	194.00	-60.567
800	200.16	355.73	258.47	77.81	-118.41	235.32	-64.284
900	213.28	380.08	270.64	98.49	-119.24	279.62	-67.897
1000	224.71	403.15	282.75	120.41	-119.27	323.98	-70.801

Calculations based on butyl pentyl sulfide values were made by Scott and McCullough (1316), using eight methylene increments, devised by Person and Pimentel (1140), for the thermal properties, and by Prosen, Johnson, and Rossini (1196), for the enthalpy of formation. $\Delta Hf°_{298}(g)$ is -84.32 kcal/mole.

No. 855 Ethyl Pentadecyl Sulfide $C_{17}H_{36}S$ (Ideal Gas State) Mol Wt 272.524

$T°K$	$Cp°$	$S°$	$-(G°-H°_{298})/T$	$H°-H°_{298}$	$\Delta Hf°$	$\Delta Gf°$	Log Kp
		cal/(mole °K)			kcal/mole		
298	98.84	210.67	210.67	0.00	-84.11	29.75	-21.808
300	99.34	211.29	210.68	0.19	-84.24	30.45	-22.181
400	124.72	243.39	214.88	11.41	-90.81	69.66	-38.057
500	147.90	273.78	223.65	25.07	-96.07	110.40	-48.254
600	167.72	302.54	234.42	40.87	-100.15	152.08	-55.392
700	184.81	329.71	246.11	58.52	-103.13	194.41	-60.695
800	199.46	355.36	258.18	77.75	-118.26	235.77	-64.405
900	212.27	379.61	270.34	98.35	-119.17	280.11	-68.015
1000	223.41	402.57	282.42	120.15	-119.32	324.51	-70.919

Scott and McCullough (1316) calculated thermodynamic properties by adding 12 methylene increments, recommended by Person and Pimentel (1140), for the thermodynamic functions, and by Prosen, Johnson, and Rossini (1196), for the enthalpy of formation, to the ethyl propyl sulfide values. $\Delta Hf°_{298}(g) = -84.11$ kcal/mole.

No. 856 Hexadecyl Methyl Sulfide $C_{17}H_{36}S$ (Ideal Gas State) Mol Wt 272.524

$T°K$	cal/(mole °K)				kcal/mole		Log Kp
	$Cp°$	$S°$	$-(G°-H°_{298})/T$	$H°-H°_{298}$	$\Delta Hf°$	$\Delta Gf°$	
298	99.23	210.13	210.13	0.00	-83.53	30.49	-22.351
300	99.73	210.75	210.14	0.19	-83.66	31.19	-22.722
400	125.02	242.96	214.36	11.45	-90.19	70.45	-38.489
500	148.07	273.39	223.15	25.13	-95.44	111.23	-48.618
600	167.78	302.17	233.94	40.94	-99.50	152.95	-55.709
700	184.74	329.34	245.65	58.59	-102.48	195.32	-60.977
800	199.31	354.98	257.73	77.81	-117.62	236.71	-64.663
900	212.04	379.21	269.89	98.39	-118.55	281.09	-68.253
1000	223.07	402.14	281.98	120.16	-118.73	325.54	-71.142

Thermodynamic properties were calculated by Scott and McCullough (1316) by adding 11 methylene increments, proposed by Person and Pimentel (1140), for each thermodynamic function, and by Prosen, Johnson, and Rossini (1196), for $\Delta Hf°$, to the butyl methyl sulfide values. $\Delta Hf°_{298}(g) = -83.53$ kcal/mole. Rossini, Pitzer, Arnett, Braun, and Pimentel (1248) selected $Tm = 294°K$ and $Tb = 632°K$.

No. 857 Propyl Tetradecyl Sulfide $C_{17}H_{36}S$ (Ideal Gas State) Mol Wt 272.524

$T°K$	cal/(mole °K)				kcal/mole		Log Kp
	$Cp°$	$S°$	$-(G°-H°_{298})/T$	$H°-H°_{298}$	$\Delta Hf°$	$\Delta Gf°$	
298	98.65	210.99	210.99	0.00	-84.14	29.63	-21.716
300	99.15	211.61	211.00	0.19	-84.27	30.32	-22.090
400	124.56	243.66	215.20	11.39	-90.86	69.50	-37.972
500	147.85	274.02	223.95	25.04	-96.13	110.22	-48.176
600	167.89	302.79	234.71	40.85	-100.21	151.88	-55.318
700	185.22	330.00	246.40	58.52	-103.16	194.18	-60.623
800	200.16	355.73	258.47	77.81	-118.23	235.50	-64.333
900	213.28	380.08	270.64	98.49	-119.06	279.80	-67.941
1000	224.71	403.15	282.75	120.41	-119.09	324.16	-70.841

Ten methylene increments, formulated by Person and Pimentel (1140), for thermodynamic functions, and by Prosen, Johnson, and Rossini (1196), for the enthalpy of formation, were added to the appropriate butyl propyl sulfide values by Scott and McCullough (1316) to obtain the thermodynamic properties of the gaseous state. $\Delta Hf°_{298}(g)$ is -84.14 kcal/mole.

No. 858 Butyl Tetradecyl Sulfide $C_{18}H_{38}S$ (Ideal Gas State) Mol Wt 286.550

T°K	Cp°	S°	$-(G°-H°_{298})/T$	$H°-H°_{298}$	$\Delta Hf°$	$\Delta Gf°$	Log Kp
	cal/(mole °K)				kcal/mole		
298	104.12	220.30	220.30	0.00	-89.25	31.45	-23.054
300	104.65	220.95	220.31	0.20	-89.39	32.19	-23.450
400	131.50	254.79	224.74	12.02	-96.29	73.75	-40.293
500	156.10	286.84	233.98	26.43	-101.82	116.92	-51.103
600	177.23	317.21	245.35	43.12	-106.10	161.07	-58.667
700	195.50	345.93	257.68	61.78	-109.19	205.90	-64.282
800	211.23	373.09	270.43	82.13	-124.36	249.77	-68.229
900	225.03	398.78	283.27	103.96	-125.24	296.62	-72.024
1000	237.04	423.13	296.05	127.08	-125.29	343.53	-75.075

Calculations based on butyl pentyl sulfide values were made by Scott and McCullough (1316), using nine methylene increments, devised by Person and Pimentel (1140), for the thermal properties, and by Prosen, Johnson, and Rossini (1196), for the enthalpy of formation. $\Delta Hf°_{298}(g)$ is -89.25 kcal/mole.

No. 859 Ethyl Hexadecyl Sulfide $C_{18}H_{38}S$ (Ideal Gas State) Mol Wt 286.550

T°K	Cp°	S°	$-(G°-H°_{298})/T$	$H°-H°_{298}$	$\Delta Hf°$	$\Delta Gf°$	Log Kp
	cal/(mole °K)				kcal/mole		
298	104.30	219.98	219.98	0.00	-89.03	31.77	-23.285
300	104.83	220.63	219.99	0.20	-89.17	32.51	-23.681
400	131.66	254.52	224.43	12.04	-96.05	74.09	-40.481
500	156.14	286.60	233.68	26.46	-101.58	117.29	-51.265
600	177.06	316.96	245.05	43.15	-105.85	161.46	-58.811
700	195.08	345.64	257.39	61.77	-108.98	206.32	-64.414
800	210.52	372.72	270.13	82.07	-124.20	250.22	-68.354
900	224.01	398.31	282.97	103.81	-125.17	297.11	-72.145
1000	235.74	422.53	295.72	126.81	-125.34	344.08	-75.195

Scott and McCullough (1316) calculated thermodynamic properties by adding 13 methylene increments, recommended by Person and Pimentel (1140), for the thermodynamic functions, and by Prosen, Johnson, and Rossini (1196), for the enthalpy of formation, to the values for ethyl propyl sulfide. $\Delta Hf°_{298}(g)$ is -89.03 kcal/mole. Bost and Everett (150) reported $Tm = 292°K$ and a Tb of about $476°K$ at 12 torr.

No. 860 Heptadecyl Methyl Sulfide $C_{18}H_{38}S$ (Ideal Gas State) Mol Wt 286.550

Thermodynamic properties were calculated by Scott and McCullough (1316) by adding 12 methylene increments, proposed by Person and

No. 860 Heptadecyl Methyl Sulfide $C_{18}H_{38}S$ (Ideal Gas State) Mol Wt 286.550

T°K	$Cp°$	$S°$	$-(G°-H°_{298})/T$	$H°-H°_{298}$	$\Delta Hf°$	$\Delta Gf°$	Log Kp
	cal/(mole °K)			kcal/mole			
298	104.69	219.44	219.44	0.00	-88.45	32.51	-23.828
300	105.22	220.09	219.45	0.20	-88.58	33.25	-24.221
400	131.96	254.08	223.90	12.08	-95.44	74.88	-40.913
500	156.31	286.21	233.18	26.52	-100.94	118.12	-51.628
600	177.12	316.59	244.57	43.22	-105.20	162.33	-59.127
700	195.01	345.27	256.93	61.84	-108.33	207.23	-64.696
800	210.37	372.34	269.68	82.13	-123.56	251.16	-68.611
900	223.78	397.91	282.52	103.85	-124.55	298.09	-72.383
1000	235.40	422.10	295.28	126.83	-124.75	345.10	-75.418

Pimentel (1140), for each thermodynamic function, and by Prosen, Johnson, and Rossini (1196), for $\Delta Hf°$, to the values for butyl methyl sulfide. $\Delta Hf°_{298}(g) = -88.45$ kcal/mole. Rossini, Pitzer, Arnett, Braun, and Pimentel (1248) selected $Tm = 298°K$ and $Tb = 646°K$.

No. 861 Nonyl Sulfide $C_{18}H_{38}S$ (Ideal Gas State) Mol Wt 286.550

T°K	$Cp°$	$S°$	$-(G°-H°_{298})/T$	$H°-H°_{298}$	$\Delta Hf°$	$\Delta Gf°$	Log Kp
	cal/(mole °K)			kcal/mole			
298	104.12	218.92	218.92	0.00	-89.25	31.86	-23.356
300	104.65	219.57	218.93	0.20	-89.39	32.61	-23.752
400	131.50	253.41	223.36	12.02	-96.29	74.30	-40.594
500	156.10	285.46	232.60	26.43	-101.82	117.61	-51.405
600	177.23	315.83	243.97	43.12	-106.10	161.90	-58.968
700	195.50	344.55	256.30	61.78	-109.19	206.87	-64.584
800	211.23	371.71	269.05	82.13	-124.36	250.87	-68.531
900	225.03	397.40	281.89	103.96	-125.24	297.86	-72.326
1000	237.04	421.75	294.67	127.08	-125.29	344.91	-75.376

For this symmetrical thiaalkane Scott and McCullough (1316) calculated the gaseous thermal properties by adding 12 methylene increments of Person and Pimentel (1140) to dipropyl sulfide values. The enthalpy of formation increment of Prosen, Johnson, and Rossini (1196), similarly applied, gave $\Delta Hf°_{298}(g) = -89.25$ kcal/mole. Aibazov, Petrov, Khairullina, and Yaprintseva (5) selected $Tb = 459°K$.

No. 862 Pentadecyl Propyl Sulfide $C_{18}H_{38}S$ (Ideal Gas State) Mol Wt 286.550

Eleven methylene increments, formulated by Person and Pimentel (1140), for thermodynamic functions, and by Prosen, Johnson, and Ros-

No. 862 Pentadecyl Propyl Sulfide $C_{18}H_{38}S$ (Ideal Gas State)   Mol Wt 286.550

$T°K$	$Cp°$	$S°$	$-(G°-H°_{298})/T$	$H°-H°_{298}$	$\Delta Hf°$	$\Delta Gf°$	Log Kp
		cal/(mole °K)			kcal/mole		
298	104.12	220.30	220.30	0.00	−89.07	31.63	−23.186
300	104.65	220.95	220.31	0.20	−89.21	32.37	−23.582
400	131.50	254.79	224.74	12.02	−96.11	73.93	−40.391
500	156.10	286.84	233.98	26.43	−101.64	117.10	−51.182
600	177.23	317.21	245.35	43.12	−105.92	161.25	−58.732
700	195.50	345.93	257.68	61.78	−109.01	206.08	−64.338
800	211.23	373.09	270.43	82.13	−124.18	249.95	−68.278
900	225.03	398.78	283.27	103.96	−125.06	296.80	−72.068
1000	237.04	423.13	296.05	127.08	−125.11	343.71	−75.114

sini (1196), for the enthalpy of formation, were added to the appropriate butyl propyl sulfide values by Scott and McCullough (1316) to obtain the thermodynamic properties of the gaseous state. $\Delta Hf°_{298}(g)$ is −89.07 kcal/mole.

No. 863 Butyl Pentadecyl Sulfide $C_{19}H_{40}S$ (Ideal Gas State)   Mol Wt 300.576

$T°K$	$Cp°$	$S°$	$-(G°-H°_{298})/T$	$H°-H°_{298}$	$\Delta Hf°$	$\Delta Gf°$	Log Kp
		cal/(mole °K)			kcal/mole		
298	109.58	229.61	229.61	0.00	−94.18	33.46	−24.524
300	110.14	230.29	229.62	0.21	−94.32	34.24	−24.942
400	138.44	265.91	234.28	12.66	−101.55	78.18	−42.711
500	164.34	299.99	244.05	27.98	−107.19	123.78	−54.102
600	168.57	329.83	255.92	44.35	−112.85	170.48	−62.093
700	205.77	358.87	268.54	63.24	−116.84	218.10	−68.091
800	222.29	387.45	281.63	84.66	−132.11	264.80	−72.338
900	236.77	414.48	294.91	107.63	−133.05	314.51	−76.369
1000	249.37	440.10	308.16	131.95	−133.12	364.28	−79.610

Calculations based on butyl pentyl sulfide values were made by Scott and McCullough (1316), using 10 methylene increments devised by Person and Pimentel (1140), for the thermal properties, and by Prosen, Johnson, and Rossini (1196), for the enthalpy of formation. $\Delta Hf°_{298}(g)$ is −94.18 kcal/mole.

No. 864 Ethyl Heptadecyl Sulfide $C_{19}H_{40}S$ (Ideal Gas State)   Mol Wt 300.576

Scott and McCullough (1316) calculated thermodynamic properties by adding 14 methylene increments, recommended by Person and Pimentel (1140), for the thermodynamic functions, and by Prosen,

No. 864 Ethyl Heptadecyl Sulfide $C_{19}H_{40}S$ (Ideal Gas State) Mol Wt 300.576

T°K	$C_p°$	cal/(mole °K) $S°$	$-(G°-H°_{298})/T$	$H°-H°_{298}$	kcal/mole $\Delta Hf°$	$\Delta Gf°$	Log Kp
298	109.77	229.29	229.29	0.00	-93.96	33.77	-24.755
300	110.33	229.98	229.30	0.21	-94.10	34.56	-25.173
400	138.60	265.65	233.97	12.68	-101.31	78.52	-42.900
500	164.39	299.42	243.71	27.86	-107.09	124.17	-54.271
600	186.40	331.38	255.68	45.42	-111.56	170.84	-62.225
700	205.36	361.57	268.67	65.03	-114.83	218.22	-68.129
800	221.59	390.08	282.09	86.40	-130.15	264.66	-72.299
900	235.76	417.01	295.59	109.28	-131.17	314.11	-76.272
1000	248.08	442.51	309.02	133.49	-131.36	363.64	-79.468

Johnson, and Rossini (1196), for the enthalpy of formation, to the ethyl propyl sulfide values. $\Delta Hf°_{298}(g)$ is -93.96 kcal/mole.

No. 865 Hexadecyl Propyl Sulfide $C_{19}H_{40}S$ (Ideal Gas State) Mol Wt 300.576

T°K	$C_p°$	cal/(mole °K) $S°$	$-(G°-H°_{298})/T$	$H°-H°_{298}$	kcal/mole $\Delta Hf°$	$\Delta Gf°$	Log Kp
298	109.58	229.61	229.61	0.00	-93.99	33.65	-24.663
300	110.14	230.29	229.62	0.21	-94.13	34.43	-25.081
400	138.44	265.91	234.28	12.66	-101.36	78.37	-42.815
500	164.34	299.66	244.01	27.83	-107.15	123.99	-54.193
600	186.57	331.63	255.97	45.40	-111.61	170.63	-62.150
700	205.77	361.87	268.96	65.04	-114.85	217.99	-68.057
800	222.29	390.44	282.38	86.46	-130.12	264.40	-72.227
900	236.77	417.48	295.90	109.43	-131.06	313.80	-76.198
1000	249.37	443.09	309.35	133.75	-131.13	363.28	-79.390

Twelve methylene increments, formulated by Person and Pimentel (1140), for thermodynamic functions, and by Prosen, Johnson, and Rossini (1196), for the enthalpy of formation, were added to the appropriate values for butyl propyl sulfide by Scott and McCullough (1316) to obtain the thermodynamic properties of the gaseous state. $\Delta Hf°_{298}(g) = -93.99$ kcal/mole.

No. 866 Methyl Octadecyl Sulfide $C_{19}H_{40}S$ (Ideal Gas State) Mol Wt 300.576

Thermodynamic properties were calculated by Scott and McCullough (1316) by adding 13 methylene increments, proposed by Person and Pimentel (1140), for each thermodynamic function, and by Prosen, Johnson, and Rossini (1196), for $\Delta Hf°$, to the values for butyl methyl sul-

No. 866 Methyl Octadecyl Sulfide $C_{19}H_{40}S$ (Ideal Gas State) Mol Wt 300.576

$T°K$	$Cp°$	$S°$	$-(G°-H°_{298})/T$	$H°-H°_{298}$	$\Delta Hf°$	$\Delta Gf°$	Log Kp
		cal/(mole °K)			kcal/mole		
298	110.16	228.75	228.75	0.00	-93.38	34.51	-25.298
300	110.72	229.44	228.76	0.21	-93.52	35.30	-25.713
400	138.90	265.21	233.44	12.71	-100.69	79.31	-43.332
500	164.56	299.03	243.21	27.92	-106.45	125.00	-54.635
600	186.46	331.02	255.20	45.49	-110.91	171.71	-62.541
700	205.29	361.21	268.21	65.10	-114.18	219.13	-68.412
800	221.44	389.70	281.63	86.46	-129.51	265.60	-72.556
900	235.53	416.61	295.15	109.32	-130.55	315.09	-76.510
1000	247.74	442.08	308.58	133.50	-130.77	364.66	-79.692

fide. $\Delta Hf°_{298}(g) = -93.38$ kcal/mole. Rossini, Pitzer, Arnett, Braun, and Pimentel (1248) selected $Tm = 303°K$ and $Tb = 659°K$.

No. 867 Butyl Hexadecyl Sulfide $C_{20}H_{42}S$ (Ideal Gas State) Mol Wt 314.602

$T°K$	$Cp°$	$S°$	$-(G°-H°_{298})/T$	$H°-H°_{298}$	$\Delta Hf°$	$\Delta Gf°$	Log Kp
		cal/(mole °K)			kcal/mole		
298	115.05	238.92	238.92	0.00	-99.10	35.47	-26.002
300	115.64	239.64	238.93	0.22	-99.25	36.30	-26.442
400	145.38	277.04	243.83	13.29	-106.79	82.61	-45.136
500	172.59	312.48	254.04	29.22	-112.84	130.69	-57.120
600	195.91	346.05	266.61	47.67	-117.50	179.83	-65.499
700	216.05	377.80	280.24	68.29	-120.88	229.71	-71.716
800	233.36	407.80	294.33	90.78	-136.25	278.66	-76.123
900	248.52	436.18	308.53	114.89	-137.24	330.62	-80.281
1000	261.71	463.07	322.65	140.42	-137.33	382.65	-83.624

Calculations based on butyl pentyl sulfide values were made by Scott and McCullough (1316), using 11 methylene increments, devised by Person and Pimentel (1140), for the thermal properties, and by Prosen, Johnson, and Rossini (1196), for the enthalpy of formation. $\Delta Hf°_{298}(g)$ is -99.10 kcal/mole.

No. 868 Decyl Sulfide $C_{20}H_{42}S$ (Ideal Gas State) Mol Wt 314.602

For this symmetrical thiaalkane Scott and McCullough (1316) calculated the gaseous thermal properties by adding 14 methylene increments of Person and Pimentel (1140) to dipropyl sulfide values. The enthalpy of formation increment of Prosen, Johnson, and Rossini (1196), similarly applied, gave $\Delta Hf°_{298}(g) = -99.10$ kcal/mole. Kharasch and Zavist

No. 868 Decyl Sulfide $C_{20}H_{42}S$ (Ideal Gas State) Mol Wt 314.602

T°K	Cp°	S°	$-(G°-H°_{298})/T$	$H°-H°_{298}$	ΔHf°	ΔGf°	Log Kp
		cal/(mole °K)			kcal/mole		
298	115.05	237.54	237.54	0.00	-99.10	35.89	-26.303
300	115.64	238.26	237.55	0.22	-99.25	36.71	-26.743
400	145.38	275.66	242.45	13.29	-106.79	83.17	-45.437
500	172.59	311.10	252.66	29.22	-112.84	131.38	-57.422
600	195.91	344.67	265.23	47.67	-117.50	180.66	-65.801
700	216.05	376.42	278.86	68.29	-120.88	230.68	-72.018
800	233.36	406.42	292.95	90.78	-136.25	279.76	-76.424
900	248.52	434.80	307.15	114.89	-137.24	331.86	-80.583
1000	261.71	461.69	321.27	140.42	-137.33	384.03	-83.926

(746) reported $Tm = 300.5°K$, and Halpern and Glasser (566) found $Tb = 452°K$ at 1–2 torr.

No. 869 Ethyl Octadecyl Sulfide $C_{20}H_{42}S$ (Ideal Gas State) Mol Wt 314.602

T°K	Cp°	S°	$-(G°-H°_{298})/T$	$H°-H°_{298}$	ΔHf°	ΔGf°	Log Kp
		cal/(mole °K)			kcal/mole		
298	115.24	238.60	238.60	0.00	-98.89	35.78	-26.225
300	115.82	239.32	238.61	0.22	-99.04	36.60	-26.664
400	145.54	276.77	243.51	13.31	-106.56	82.95	-45.319
500	172.64	312.23	253.74	29.25	-112.60	131.05	-57.278
600	195.75	345.80	266.31	47.70	-117.27	180.21	-65.639
700	215.64	377.51	279.96	68.29	-120.68	230.13	-71.845
800	232.65	407.44	294.04	90.72	-136.09	279.11	-76.244
900	247.51	435.72	308.22	114.75	-137.18	331.10	-80.399
1000	260.41	462.48	322.32	140.16	-137.39	383.19	-83.742

Scott and McCullough (1316) calculated thermodynamic properties by adding 15 methylene increments, recommended by Person and Pimentel (1140), for the thermodynamic functions, and by Prosen, Johnson, and Rossini (1196), for the enthalpy of formation, to the ethyl propyl sulfide values. $\Delta Hf°_{298}(g)$ is -98.89 kcal/mole.

No. 870 Heptadecyl Propyl Sulfide $C_{20}H_{42}S$ (Ideal Gas State) Mol Wt 314.602

Thirteen methylene increments, formulated by Person and Pimentel (1140), for thermodynamic functions, and by Prosen, Johnson, and Rossini (1196), for the enthalpy of formation, were added to the appropriate butyl propyl sulfide values by Scott and McCullough (1316) to obtain the thermodynamic properties of the gaseous state. $\Delta Hf°_{298}(g)$ is -98.92 kcal/mole.

No. 870 Heptadecyl Propyl Sulfide C$_{20}$H$_{42}$S (Ideal Gas State) Mol Wt 314.602

T°K	Cp°	S°	$-(G°-H°_{298})/T$	$H°-H°_{298}$	ΔHf°	ΔGf°	Log Kp
	cal/(mole °K)				kcal/mole		
298	115.05	238.92	238.92	0.00	-98.92	35.65	-26.133
300	115.64	239.64	238.93	0.22	-99.07	36.48	-26.573
400	145.38	277.04	243.83	13.29	-106.61	82.79	-45.234
500	172.59	312.48	254.04	29.22	-112.66	130.87	-57.199
600	195.91	346.05	266.61	47.67	-117.32	180.01	-65.565
700	216.05	377.80	280.24	68.29	-120.70	229.89	-71.773
800	233.36	407.80	294.33	90.78	-136.07	278.84	-76.172
900	248.52	436.18	308.53	114.89	-137.06	330.80	-80.325
1000	261.71	463.07	322.65	140.42	-137.15	382.83	-83.663

No. 871 Methyl Nonadecyl Sulfide C$_{20}$H$_{42}$S (Ideal Gas State) Mol Wt 314.602

T°K	Cp°	S°	$-(G°-H°_{298})/T$	$H°-H°_{298}$	ΔHf°	ΔGf°	Log Kp
	cal/(mole °K)				kcal/mole		
298	115.63	238.06	238.06	0.00	-98.31	36.52	-26.769
300	116.21	238.78	238.07	0.22	-98.46	37.35	-27.205
400	145.84	276.34	242.99	13.34	-105.95	83.74	-45.750
500	172.81	311.85	253.24	29.31	-111.96	131.88	-57.641
600	195.81	345.44	265.83	47.77	-116.62	181.08	-65.956
700	215.57	377.14	279.49	68.36	-120.03	231.03	-72.127
800	232.50	407.06	293.59	90.78	-135.46	280.05	-76.501
900	247.28	435.31	307.78	114.79	-136.56	332.08	-80.637
1000	260.07	462.05	321.88	140.17	-136.79	384.21	-83.965

Thermodynamic properties were calculated by Scott and McCullough (1316) by adding 14 methylene increments, proposed by Person and Pimentel (1140), for each thermodynamic function, and by Prosen, Johnson, and Rossini (1196), for $\Delta Hf°$, to the butyl methyl sulfide values. $\Delta Hf°_{298}(g) = -98.31$ kcal/mole. Rossini, Pitzer, Arnett, Braun, and Pimentel (1248) selected $Tm = 308°K$ and $Tb = 671°K$.

DITHIAALKANE IDEAL GAS TABLES

No. 872 Methyl Disulfide C$_2$H$_6$S$_2$ (Ideal Gas State) Mol Wt 94.200

Scott, Finke, Gross, Guthrie, and Huffman (1302) measured the low-temperature thermal properties of methyl disulfide, reporting $Tm = 188.43°K$, $\Delta Hm = 2.197$ kcal/mole, $S°_{298}(l) = 56.26$ cal/(mole °K), and $Cp(l, 300-350°K) = 26.8 + 0.027T$ [cal/(mole °K)]. Further work by Hubbard, Douslin, McCullough, Scott, Todd, Messerly, Hossenlopp,

No. 872 Methyl Disulfide $C_2H_6S_2$ (Ideal Gas State) Mol Wt 94.200

T°K	Cp°	S°	$-(G°-H°_{298})/T$	$H°-H°_{298}$	$\Delta Hf°$	$\Delta Gf°$	Log Kp
		cal/(mole °K)			kcal/mole		
298	22.54	80.46	80.46	0.00	-5.77	3.52	-2.580
300	22.61	80.60	80.47	0.05	-5.80	3.58	-2.605
400	26.36	87.63	81.40	2.50	-8.12	6.94	-3.789
500	29.80	93.89	83.28	5.31	-9.92	10.91	-4.769
600	32.82	99.60	85.53	8.44	-11.35	15.22	-5.542
700	35.41	104.85	87.92	11.86	-12.49	19.79	-6.177
800	37.66	109.73	90.35	15.51	-39.52	21.92	-5.988
900	39.60	114.28	92.76	19.38	-39.64	29.61	-7.191
1000	41.31	118.55	95.12	23.43	-39.64	37.31	-8.155

George, and Waddington (627) yielded $\Delta Hv = 8.050$ kcal/mole at $Tb = 382.90°K$ and $S°_{298}(g) = 80.46$ cal/(mole °K). They calculated thermodynamic functions based on vapor heat capacity data. Combustion calorimetry was used to determine $\Delta Hf°_{298}(l) = -14.96$ kcal/mole and $\Delta Hf°_{298}(g) = -5.78$ kcal/mole when adjusted to the new value for $\Delta Hf°_{298}$ for H_2SO_4 (aq) (cf. page 560). Less accurate values were given by Franklin and Lumpkin (440).

No. 873 Ethyl Disulfide $C_4H_{10}S_2$ (Ideal Gas State) Mol Wt 122.252

T°K	Cp°	S°	$-(G°-H°_{298})/T$	$H°-H°_{298}$	$\Delta Hf°$	$\Delta Gf°$	Log Kp
		cal/(mole °K)			kcal/mole		
298	33.78	99.07	99.07	0.00	-17.84	5.32	-3.902
300	33.91	99.28	99.08	0.07	-17.88	5.47	-3.981
400	40.90	110.02	100.49	3.81	-20.79	13.58	-7.418
500	47.08	119.83	103.39	8.22	-23.02	22.43	-9.805
600	52.24	128.88	106.89	13.20	-24.78	31.69	-11.543
700	56.57	137.27	110.64	18.64	-26.13	41.27	-12.884
800	60.19	145.06	114.46	24.48	-53.31	48.43	-13.229
900	63.30	152.33	118.27	30.66	-53.51	61.16	-14.852
1000	65.97	159.15	122.02	37.13	-53.53	73.92	-16.154

Complete thermal data were reported by Scott, Finke, McCullough, Gross, Pennington, and Waddington (1307) as $Tm = 171.63°K$, $\Delta Hm = 2.248$ kcal/mole, $Tb = 427.13°K$, $\Delta Hv = 8.97$ kcal/mole at the boiling point, $S°_{298}(l) = 72.90$ cal/(mole °K), and $Cp(l, 270-300°K) = 36.47 + 0.0412T$[cal/(mole °K)]. Thermodynamic functions were calculated from these data by using the barrier to internal rotation about the S—S bond for methyl disulfide found by Hubbard, Douslin, McCullough, Scott,

Todd, Messerly, Hossenlopp, George, and Waddington (627). Combustion calorimetry by the latter group yielded $\Delta Hf°_{298}(l) = -28.69$ kcal/mole and $\Delta Hf°_{298}(g) = -17.84$ kcal/mole when adjusted to the new value for $\Delta Hf°_{298}$ for H_2SO_4 (aq) (cf. page 560). Franklin and Lumpkin (440) reported a less accurate value, which is about 4 kcal/mole more negative.

No. 874 Propyl Disulfide $C_6H_{14}S_2$ (Ideal Gas State) Mol Wt 150.304

	cal/(mole °K)				kcal/mole		
T°K	Cp°	S°	$-(G°-H°_{298})/T$	$H°-H°_{298}$	$\Delta Hf°$	$\Delta Gf°$	Log Kp
298	44.30	118.30	118.30	0.00	-28.01	8.84	-6.482
300	44.50	118.58	118.31	0.09	-28.06	9.07	-6.606
400	54.50	132.77	120.17	5.05	-31.64	21.88	-11.956
500	63.50	145.93	124.02	10.96	-34.41	35.59	-15.557
600	71.30	158.21	128.71	17.70	-36.54	49.80	-18.138
700	78.00	169.72	133.75	25.18	-38.12	64.37	-20.098
800	83.70	180.51	138.93	33.27	-65.38	76.55	-20.912
900	88.80	190.67	144.12	41.90	-65.53	94.31	-22.901
1000	93.20	200.26	149.26	51.00	-65.36	112.08	-24.494

Complete thermal data were reported by Hubbard, Douslin, McCullough, Scott, Todd, Messerly, Hossenlopp, George, and Waddington (627) as $Tm = 187.66°K$, $\Delta Hm = 3.300$ kcal/mole, $Tb = 469.01°K$, $S°_{298}(l) = 89.28$ cal/(mole °K), $S°_{298}(g) = 118.30$ cal/(mole °K), and $Cp°(l, 290-360°K) = 42.43 + 0.068T$ [cal/(mole °K)]. Thermodynamic functions of the ideal gas were estimated by a method of increments, and combustion calorimetry was used to determine $\Delta Hf°_{298}(l) = -40.95$ kcal/mole and $\Delta Hf°_{298}(g) = -28.01$ kcal/mole when adjusted to the new value of $\Delta Hf°_{298}$ for H_2SO_4 (aq) (cf. page 560).

No. 875 Butyl Disulfide $C_8H_{18}S_2$ (Ideal Gas State) Mol Wt 178.356

	cal/(mole °K)				kcal/mole		
T°K	Cp°	S°	$-(G°-H°_{298})/T$	$H°-H°_{298}$	$\Delta Hf°$	$\Delta Gf°$	Log Kp
298	55.23	136.91	136.91	0.00	-37.86	12.87	-9.431
300	55.49	137.26	136.92	0.11	-37.93	13.18	-9.599
400	68.38	155.02	139.25	6.31	-42.14	30.75	-16.801
500	79.99	171.55	144.08	13.74	-45.42	49.37	-21.576
600	89.98	187.04	149.96	22.25	-47.95	68.56	-24.973
700	98.55	201.57	156.31	31.69	-49.81	88.19	-27.534
800	105.83	215.22	162.83	41.92	-77.27	105.45	-28.807
900	112.29	228.06	169.37	52.83	-77.53	128.33	-31.160
1000	117.86	240.19	175.85	64.35	-77.40	151.21	-33.045

The thermodynamic functions were extrapolated by Scott and McCullough (1316). Their estimated $\Delta Hf°_{298}(g) = -37.86$ kcal/mole was in good accord with the value determined by Mackle and McClean (902). $\Delta Hf°_{298}(g) = -37.7 \pm 0.6$ kcal/mole, which they obtained by applying $\Delta Hv_{298} = 15.4 \pm 0.4$ kcal/mole to their $\Delta Hf°_{298}(l) = -53.1 \pm 0.5$ kcal/mole, obtained from ΔHc. Aibazov, Petrov, Khairullina, and Yaprintseva (5) selected $Tb = 375°K$ at 11 torr.

No. 876 Pentyl Disulfide $C_{10}H_{22}S_2$ (Ideal Gas State) Mol Wt 206.408

		cal/(mole °K)			kcal/mole		
$T°K$	$Cp°$	$S°$	$-(G°-H°_{298})/T$	$H°-H°_{298}$	$\Delta Hf°$	$\Delta Gf°$	Log Kp
298	66.16	155.53	155.53	0.00	-47.71	16.89	-12.379
300	66.48	155.95	155.54	0.13	-47.79	17.28	-12.590
400	82.26	177.27	158.34	7.58	-52.64	39.62	-21.644
500	96.48	197.19	164.14	16.53	-56.43	63.13	-27.593
600	108.66	215.88	171.22	26.80	-59.36	87.32	-31.805
700	119.10	233.44	178.87	38.20	-61.51	112.01	-34.968
800	127.96	249.93	186.73	50.57	-89.16	134.35	-36.701
900	135.78	265.46	194.62	63.76	-89.53	162.33	-39.417
1000	142.53	280.13	202.45	77.69	-89.44	190.33	-41.595

Scott and McCullough (1316) computed thermodynamic functions by adding methylene increments to butyl disulfide values. Aibazov, Petrov, Khairullina, and Yaprintseva (5) selected 401°K as Tb at 12 torr.

No. 877 Hexyl Disulfide $C_{12}H_{26}S_2$ (Ideal Gas State) Mol Wt 234.460

		cal/(mole °K)			kcal/mole		
$T°K$	$Cp°$	$S°$	$-(G°-H°_{298})/T$	$H°-H°_{298}$	$\Delta Hf°$	$\Delta Gf°$	Log Kp
298	77.09	174.15	174.15	0.00	-57.56	20.91	-15.326
300	77.47	174.63	174.16	0.15	-57.65	21.39	-15.582
400	96.14	199.52	177.42	8.84	-63.14	48.48	-26.487
500	112.97	222.83	184.20	19.32	-67.45	76.90	-33.610
600	127.35	244.73	192.48	31.35	-70.76	106.08	-38.638
700	139.65	265.31	201.43	44.72	-73.20	135.82	-42.403
800	150.09	284.65	210.63	59.22	-101.05	163.24	-44.594
900	159.27	302.87	219.88	74.69	-101.53	196.33	-47.673
1000	167.19	320.07	229.05	91.03	-101.47	229.45	-50.144

The calculated values of Scott and McCullough (1316) are adopted. Aibazov, Petrov, Khairullina, and Yaprintseva (5) selected $Tb = 455°K$ at 21 torr.

No. 878   Heptyl Disulfide $C_{14}H_{30}S_2$ (Ideal Gas State)   Mol Wt 262.512

T°K	Cp°	S°	$-(G°-H°_{298})/T$	$H°-H°_{298}$	$\Delta Hf°$	$\Delta Gf°$	Log Kp
	cal/(mole °K)				kcal/mole		
298	88.02	192.77	192.77	0.00	-67.41	24.93	-18.274
300	88.46	193.32	192.78	0.17	-67.52	25.50	-18.573
400	110.02	221.77	196.51	10.11	-73.64	57.34	-31.330
500	129.46	248.46	204.26	22.11	-78.46	90.66	-39.627
600	146.03	273.57	213.74	35.90	-82.17	124.84	-45.470
700	160.20	297.17	223.99	51.23	-84.89	159.63	-49.837
800	172.22	319.36	234.54	67.87	-112.94	192.14	-52.487
900	182.76	340.27	245.14	85.63	-113.53	230.33	-55.930
1000	191.86	360.01	255.65	104.37	-113.51	268.57	-58.694

Thermodynamic functions were computed from methylene increments by Scott and McCullough (1316). Aibazov, Petrov, Khairullina, and Yaprintseva (5) selected $Tb = 433°K$ at 10 torr.

No. 879   Octyl Disulfide $C_{16}H_{34}S_2$ (Ideal Gas State)   Mol Wt 290.564

T°K	Cp°	S°	$-(G°-H°_{298})/T$	$H°-H°_{298}$	$\Delta Hf°$	$\Delta Gf°$	Log Kp
	cal/(mole °K)				kcal/mole		
298	98.96	211.39	211.39	0.00	-77.27	28.94	-21.214
300	99.45	212.01	211.40	0.19	-77.39	29.59	-21.557
400	123.91	244.03	215.60	11.38	-84.15	66.20	-36.167
500	145.96	274.11	224.32	24.90	-89.48	104.42	-45.640
600	164.72	302.42	235.00	40.45	-93.58	143.59	-52.299
700	180.76	329.04	246.55	57.75	-96.59	183.43	-57.267
800	194.35	354.09	258.45	76.52	-124.83	221.02	-60.378
900	206.26	377.68	270.40	96.56	-125.54	264.33	-64.184
1000	216.53	399.96	282.25	117.72	-125.56	307.68	-67.240

Values calculated by Scott and McCullough (1316) are adopted. Aibazov, Petrov, Khairullina, and Yaprintseva (5) selected $Tb = 473°K$ at 10 torr.

No. 880   Nonyl Disulfide $C_{18}H_{38}S_2$ (Ideal Gas State)   Mol Wt 318.616

The thermodynamic functions that Scott and McCullough (1316) extrapolated by use of methylene increments are employed. Aibazov, Petrov, Khairullina, and Yaprintseva (5) selected $Tb = 485°K$ at 7 torr.

Thiacycloalkane Ideal Gas Tables

No. 880   Nonyl Disulfide $C_{18}H_{38}S_2$ (Ideal Gas State)   Mol Wt 318.616

	cal/(mole °K)			kcal/mole			
$T°K$	$Cp°$	$S°$	$-(G°-H°_{298})/T$	$H°-H°_{298}$	$\Delta Hf°$	$\Delta Gf°$	Log Kp
298	109.89	230.00	230.00	0.00	-87.12	32.97	-24.163
300	110.44	230.69	230.01	0.21	-87.26	33.70	-24.550
400	137.79	266.27	234.67	12.64	-94.65	75.07	-41.012
500	162.45	299.73	244.37	27.69	-100.49	118.19	-51.659
600	183.40	331.25	256.25	45.00	-104.99	162.35	-59.134
700	201.31	360.90	269.11	64.26	-108.28	207.25	-64.704
800	216.48	388.79	282.34	85.17	-136.72	249.93	-68.273
900	229.75	415.07	295.64	107.49	-137.55	298.34	-72.443
1000	241.19	439.89	308.84	131.06	-137.60	346.81	-75.792

No. 881   Decyl Disulfide $C_{20}H_{42}S_2$ (Ideal Gas State)   Mol Wt 346.668

	cal/(mole °K)			kcal/mole			
$T°K$	$Cp°$	$S°$	$-(G°-H°_{298})/T$	$H°-H°_{298}$	$\Delta Hf°$	$\Delta Gf°$	Log Kp
298	120.82	248.62	248.62	0.00	-96.97	36.99	-27.111
300	121.43	249.37	248.63	0.23	-97.12	37.81	-27.541
400	151.67	288.52	253.76	13.91	-105.15	83.93	-45.855
500	178.94	325.37	264.43	30.47	-111.50	131.96	-57.676
600	202.08	360.09	277.51	49.55	-116.40	181.11	-65.966
700	221.86	392.77	291.67	70.77	-119.98	231.07	-72.138
800	238.61	423.51	306.24	93.82	-148.61	278.82	-76.167
900	253.24	452.48	320.90	118.43	-149.55	332.34	-80.699
1000	265.86	479.83	335.44	144.40	-149.64	385.93	-84.341

Thermodynamic functions were computed by Scott and McCullough (1316) by adding methylene increments to butyl disulfide values. $Tm = 246°K$ and $Tb = 484°K$ at 7 torr were selected by Aibazov, Petrov, Khairullina, and Yaprintseva (5).

THIACYCLOALKANE IDEAL GAS TABLES

No. 882   Thiacyclopropane C_2H_4S (Ideal Gas State)   Mol Wt 60.118

Guthrie, Scott, and Waddington (552) combined their measurements of the infrared spectrum and vapor pressure with the enthalpy of combustion determined by Sunner (1442) to derive the thermodynamic properties of the ideal gas. Results included $\Delta Hv = 6.9$ kcal/mole at $Tb = 328.08°K$, $S°_{298}(g) = 61.01$ cal/(mole °K), and $S°_{298}(l) = 38.84$ cal/(mole °K). Sunner (1447) also determined ΔHc, leading to $\Delta Hf°_{298}(l) = 12.41$ kcal/mole and $\Delta Hf°_{298}(g) = 19.65 \pm 0.31$ kcal/mole when adjusted to the new $H_2SO_4(aq)$

No. 882 Thiacyclopropane C_2H_4S (Ideal Gas State) Mol Wt 60.118

$T°K$	$Cp°$	$S°$	$-(G°-H°_{298})/T$	$H°-H°_{298}$	$\Delta Hf°$	$\Delta Gf°$	Log Kp
		cal/(mole °K)			kcal/mole		
298	12.83	61.01	61.01	0.00	19.65	23.16	-16.974
300	12.90	61.09	61.02	0.03	19.63	23.18	-16.885
400	16.53	65.32	61.57	1.51	18.13	24.53	-13.405
500	19.56	69.34	62.72	3.31	16.96	26.28	-11.484
600	21.99	73.13	64.15	5.40	16.03	28.23	-10.281
700	23.96	76.67	65.68	7.70	15.28	30.35	-9.474
800	25.61	79.98	67.27	10.18	1.60	31.27	-8.542
900	27.03	83.08	68.85	12.81	1.40	34.99	-8.497
1000	28.21	85.99	70.42	15.57	1.28	38.74	-8.466

value (cf. page 560). Rossini, Pitzer, Arnett, Braun, and Pimentel (1248) selected $Tm = 164°K$. Aibazov, Petrov, Khairullina, and Yaprintseva (5) chose 328.08°K for Tb.

No. 883 Thiacyclobutane C_3H_6S (Ideal Gas State) Mol Wt 74.144

$T°K$	$Cp°$	$S°$	$-(G°-H°_{298})/T$	$H°-H°_{298}$	$\Delta Hf°$	$\Delta Gf°$	Log Kp
		cal/(mole °K)			kcal/mole		
298	16.57	68.17	68.17	0.00	14.61	25.69	-18.834
300	16.66	68.28	68.18	0.04	14.58	25.76	-18.766
400	21.89	73.80	68.89	1.97	12.59	29.74	-16.248
500	26.56	79.20	70.42	4.40	11.03	34.21	-14.953
600	30.45	84.40	72.32	7.25	9.79	38.97	-14.192
700	33.69	89.34	74.40	10.46	8.83	43.93	-13.716
800	36.40	94.02	76.56	13.97	-4.98	47.73	-13.038
900	38.70	98.44	78.75	17.73	-5.26	54.34	-13.195
1000	40.67	102.63	80.93	21.70	-5.40	60.98	-13.325

Complete thermal data were measured and all thermodynamic properties of the ideal gas were calculated for thiacyclobutane by Scott, Finke, Hubbard, McCullough, Katz, Gross, Messerly, Pennington, and Waddington (1304). Reported data include $Tt = 176.7°K$, $\Delta Ht = 0.160$ kcal/mole, $Tm = 199.90°K$, $\Delta Hm = 1.971$ kcal/mole, $Tb = 368.12°K$, $\Delta Hv = 7.82$ kcal/mole at the boiling point, $S°_{298}(l) = 44.72$ cal/(mole °K), $S°_{298}(g) = 68.17$ cal/(mole °K), and $Cp(l, 270-330°K) = 14.9 + 0.041T$[cal/(mole °K)]. Hubbard, Katz, and Waddington (633) determined the enthalpy of combustion and quoted an unpublished value of Sunner at the University of Lund, which is in good agreement. From the enthalpy of combustion, they derived $\Delta Hf°_{298}(l) = 6.04$ kcal/mole and $\Delta Hf°_{298}(g) = 14.61$ kcal/mole when adjusted to the revised value of $\Delta Hf°_{298}$ for $H_2SO_4(aq)$ (cf. page

560). Sunner (1447) derived $\Delta Hf°_{298}(l) = 5.80$ kcal/mole and $\Delta Hf°_{298}(g) = 14.38$ kcal/mole from ΔHc data when similarly adjusted. The former value of $\Delta Hf°_{298}(g)$ is adopted. Aibazov, Petrov, Khairullina, and Yaprintseva (5) selected $Tb = 368.11°K$.

No. 884 Thiacyclopentane C_4H_8S (Ideal Gas State) Mol Wt 88.170

		cal/(mole °K)			kcal/mole		
T°K	Cp°	S°	$-(G°-H°_{298})/T$	$H°-H°_{298}$	$\Delta Hf°$	$\Delta Gf°$	Log Kp
298	21.72	73.94	73.94	0.00	-8.08	11.00	-8.059
300	21.85	74.08	73.95	0.05	-8.12	11.11	-8.095
400	28.95	81.36	74.89	2.59	-10.43	17.82	-9.738
500	35.04	88.50	76.90	5.80	-12.23	25.10	-10.972
600	40.05	95.34	79.41	9.56	-13.64	32.70	-11.911
700	44.20	101.84	82.15	13.78	-14.72	40.54	-12.658
800	47.66	107.97	85.00	18.38	-28.61	47.22	-12.900
900	50.60	113.76	87.88	23.30	-28.92	56.73	-13.774
1000	53.14	119.22	90.74	28.49	-29.07	66.26	-14.480

Complete thermal data were measured by Hubbard, Finke, Scott, McCullough, Katz, Gross, Messerly, Pennington, and Waddington (628). Results included $Tm = 176.97°K$, $\Delta Hm = 1.757$ kcal/mole, $Tb = 394.27°K$, at which temperature $\Delta Hv = 8.279$ kcal/mole, $S°_{298}(l) = 49.67$ cal/(mole °K), $S°_{298}(g) = 73.94$ cal/(mole °K), and $Cp(l, 270-340°K) = 16.85 + 0.056T$[cal/(mole °K)]. Enthalpy of combustion measurements were reported by Hubbard, Katz, and Waddington (633), who derived $\Delta Hf°_{298}(l) = -17.31$ kcal/mole and $\Delta Hf°_{298}(g) = -8.08$ kcal/mole when adjusted to the revised value of $\Delta Hf°_{298}$ for H_2SO_4 (aq) (cf. page 560). Davies and Sunner (309) also measured ΔHc and obtained $\Delta Hf°_{298}(l) = -17.34$ kcal/mole and $\Delta Hf°_{298}(g) = -7.96$ kcal/mole, in excellent accord with the adopted former values despite the difference in the ΔHv used. Rossini, Pitzer, Arnett, Braun, and Pimentel (1248) selected $Tm = 176.99°K$ and $Tb = 394.27°K$.

No. 885 Thiacyclohexane $C_5H_{10}S$ (Ideal Gas State) Mol Wt 102.196

The chemical thermodynamic properties of thiacyclohexane were reported by McCullough, Finke, Hubbard, Good, Pennington, Messerly, and Waddington (951). Among the measured values were $Tt = 201.4°K$, $\Delta Ht = 0.2624$ kcal/mole, Tt (λ-type transition) = 207°K, $Tt = 240.01°K$, $\Delta Ht = 1.858$ kcal/mole, $Tm = 292.24°K$, $\Delta Hm = 0.585$ kcal/mole, $Tb = 414.90°K$, $\Delta Hv = 8.599$ kcal/mole at the boiling point, $S_{298}(l) = 52.16$ cal/(mole °K), $S°_{298}(g) = 77.26$ cal/(mole °K), and $Cp(l, 292-340°K) = 15.47 + 0.079T$ [cal/(mole °K)]. Combustion calorimetry was used to

No. 885 Thiacyclohexane $C_5H_{10}S$ (Ideal Gas State) Mol Wt 102.196

	cal/(mole °K)				kcal/mole		
$T°K$	$Cp°$	$S°$	$-(G°-H°_{298})/T$	$H°-H°_{298}$	$\Delta Hf°$	$\Delta Gf°$	Log Kp
298	25.86	77.26	77.26	0.00	-15.12	12.68	-9.292
300	26.03	77.43	77.27	0.05	-15.17	12.85	-9.358
400	35.71	86.26	78.40	3.15	-17.88	22.55	-12.322
500	44.73	95.22	80.87	7.18	-19.87	32.90	-14.381
600	52.37	104.07	84.00	12.04	-21.25	43.59	-15.877
700	58.71	112.63	87.48	17.61	-22.11	54.50	-17.015
800	64.00	120.83	91.14	23.75	-35.62	64.20	-17.538
900	68.49	128.63	94.88	30.38	-35.42	76.68	-18.619
1000	72.34	136.05	98.63	37.43	-34.93	89.12	-19.476

derive $\Delta Hf°_{298}(l) = -25.32$ kcal/mole and $\Delta Hf°_{298}(g) = -15.12$ kcal/mole when adjusted to the revised $\Delta Hf°_{298}$ for $H_2SO_4(aq)$ (cf. page 560). McCullough et al. corrected the combustion result of Sunner (1442) to 25°C to $\Delta Hf_{298}(l) = -25.57$ kcal/mole, in agreement within experimental error [cf. also Sunner (1447)]. Aibazov, Petrov, Khairullina, and Yaprintseva (5) selected $Tm = 292.14°K$ and $Tb = 414.90°K$.

No. 886 Thiacycloheptane $C_6H_{12}S$ (Ideal Gas State) Mol Wt 116.222

	cal/(mole °K)				kcal/mole		
$T°K$	$Cp°$	$S°$	$-(G°-H°_{298})/T$	$H°-H°_{298}$	$\Delta Hf°$	$\Delta Gf°$	Log Kp
298	29.78	86.50	86.50	0.00	-14.66	20.09	-14.728
300	30.00	86.69	86.51	0.06	-14.72	20.31	-14.792
400	42.00	96.96	87.83	3.66	-17.86	32.42	-17.712
500	54.00	107.64	90.72	8.47	-20.09	45.26	-19.783
600	65.00	118.51	94.44	14.44	-21.44	58.47	-21.296
700	73.00	129.15	98.65	21.36	-22.08	71.88	-22.439
800	79.00	139.30	103.10	28.97	-35.29	84.03	-22.955
900	84.00	148.91	107.66	37.13	-34.76	98.93	-24.021
1000	88.00	157.97	112.24	45.74	-33.93	113.75	-24.858

Sunner (1442) determined the enthalpy of combustion at 293°K by rotating bomb calorimetry. Correcting his results to 298°K yields $\Delta Hf_{298}(l) = -26.96$ kcal/mole and $\Delta Hf°_{298}(g) = -14.66$ kcal/mole after adjustment to the recently revised value of $\Delta Hf°_{298}$ for $H_2SO_4(aq)$ (cf. page 560). Sunner also reported $Tm = 273.65°K$ and $Tb = 446°K$. Thermodynamic functions of the ideal gas are estimated by comparing the known thiacycloalkanes with cycloalkanes.

THIACYCLOALKENE IDEAL GAS TABLES

No. 887   Thiophene C_4H_4S (Ideal Gas State)   Mol Wt 84.138

	cal/(mole °K)				kcal/mole		
T°K	$Cp°$	$S°$	$-(G°-H°_{298})/T$	$H°-H°_{298}$	$\Delta Hf°$	$\Delta Gf°$	Log Kp
298	17.42	66.65	66.65	0.00	27.66	30.30	-22.208
300	17.52	66.76	66.66	0.04	27.64	30.31	-22.082
400	23.02	72.58	67.41	2.07	26.20	31.37	-17.139
500	27.47	78.21	69.01	4.61	25.13	32.79	-14.332
600	30.95	83.54	71.00	7.53	24.28	34.40	-12.530
700	33.74	88.53	73.15	10.77	23.63	36.17	-11.292
800	36.01	93.19	75.36	14.26	10.04	36.73	-10.033
900	37.91	97.54	77.59	17.96	9.93	40.07	-9.730
1000	39.54	101.62	79.79	21.83	9.91	43.43	-9.490

Low-temperature thermal data, as well as vapor pressure, vapor heat capacity, and enthalpy of vaporization, were measured by Waddington, Knowlton, Scott, Oliver, Todd, Hubbard, Smith, and Huffman (1548), who found two λ-type transitions at 112° and 138°K and an isothermal transition at $Tt = 171.6°K$, with $\Delta Ht = 0.152$ kcal/mole, $Tm = 234.94°K$, $\Delta Hm = 1.2155$ kcal/mole, $Tb = 357.31°K$, at which temperature $\Delta Hv = 7.522$ kcal/mole, $S_{298}(l) = 43.30$ cal/(mole °K), $S°_{298}(g) = 66.62$ cal/(mole °K), and $Cp(l, 270-340°K) = 17.7 + 0.040T$ [cal/(mole °K)]. Complete thermodynamic functions calculated from spectroscopic data were revised by Hubbard, Scott, Frow, and Waddington (636) to include corrections for anharmonicity. Jacobs and Parks (677) reported results of heat capacity measurements from 93° to 293°K, leading to an estimated entropy 1.1 cal/(mole °K) less than the later and more accurate value given above. Sunner (1444) and Hubbard, Scott, Frow, and Waddington (636) determined the enthalpy of combustion by rotating bomb methods. Results are in excellent agreement and lead to $\Delta Hf°_{298}(l) = 19.37$ kcal/mole and $\Delta Hf°_{298}(g) = 27.66$ kcal/mole when adjusted to the revised value of $\Delta Hf°_{298}$ for $H_2SO_4(aq)$ (cf. page 560). Earlier measurements by Waddington et al. (1548), by Moore, Renquist, and Parks (1025), and by Franklin and Lumpkin (440) were all made using the Huffman and Ellis method (646). All, however, agreed to within 0.5 kcal/mole with the more accurate rotating bomb method. Critical properties of thiophene were reported by Kobe, Ravicz, and Vohra (783).

No. 888 2-Methylthiophene C_5H_6S (Ideal Gas State) Mol Wt 98.164

$T°K$	$Cp°$	$S°$	$-(G°-H°_{298})/T$	$H°-H°_{298}$	$\Delta Hf°$	$\Delta Gf°$	Log Kp
		cal/(mole °K)			kcal/mole		
298	22.80	76.62	76.62	0.00	20.00	29.38	-21.532
300	22.92	76.77	76.63	0.05	19.97	29.43	-21.441
400	29.43	84.27	77.61	2.67	18.18	32.81	-17.924
500	35.01	91.46	79.66	5.90	16.79	36.63	-16.009
600	39.57	98.26	82.20	9.64	15.68	40.70	-14.824
700	43.32	104.65	84.96	13.79	14.81	44.97	-14.040
800	46.43	110.64	87.80	18.28	1.05	48.06	-13.127
900	49.06	116.26	90.65	23.06	0.82	53.95	-13.100
1000	51.30	121.55	93.48	28.08	0.73	59.86	-13.082

Complete thermodynamic properties of 2-methylthiophene were measured and computed by Pennington, Finke, Hubbard, Messerly, Frow, Hossenlopp, and Waddington (1134). Results included $Tm = 209.78°K$, $\Delta Hm = 2.263$ kcal/mole, $\Delta Hv = 8.103$ kcal/mole at $Tb = 385.71°K$, $S_{298}(l) = 52.22$ cal/(mole °K), $S°_{298}(g) = 76.62$ cal/(mole °K), and $Cp(l, 280-340°K) = 20.90 + 0.05T$[cal/(mole °K)]. They determined ΔHc by the method of Hubbard, Katz, and Waddington (633) and computed $\Delta Hf°_{298}(l) = 10.70$ kcal/mole and $\Delta Hf°_{298}(g) = 20.00$ kcal/mole when adjusted for the revised value of $\Delta Hf°_{298}$ for $H_2SO_4(aq)$ (cf. page 560).

No. 889 3-Methylthiophene C_5H_6S (Ideal Gas State) Mol Wt 98.164

$T°K$	$Cp°$	$S°$	$-(G°-H°_{298})/T$	$H°-H°_{298}$	$\Delta Hf°$	$\Delta Gf°$	Log Kp
		cal/(mole °K)			kcal/mole		
298	22.67	76.79	76.79	0.00	19.79	29.12	-21.341
300	22.80	76.94	76.80	0.05	19.76	29.17	-21.251
400	29.38	84.42	77.77	2.66	17.96	32.53	-17.773
500	34.89	91.59	79.82	5.89	16.56	36.34	-15.882
600	39.34	98.36	82.36	9.61	15.43	40.40	-14.714
700	42.95	104.70	85.10	13.73	14.53	44.66	-13.943
800	45.95	110.64	87.92	18.17	0.74	47.74	-13.042
900	48.47	116.20	90.76	22.90	0.45	53.64	-13.025
1000	50.59	121.42	93.57	27.86	0.29	59.56	-13.017

Complete thermal data were obtained by McCullough, Sunner, Finke, Hubbard, Gross, Pennington, Messerly, Good, and Waddington (965). Results of interest included $Tm = 204.18°K$, $\Delta Hm = 2.518$ kcal/mole, $Tb = 388.59°K$, $\Delta Hv = 8.186$ kcal/mole at the boiling point, $S°_{298}(l) = 52.185$ cal/(mole °K), $S°_{298}(g) = 76.79$ cal/(mole °K), and $Cp(l, 270-340°K) = 21.40 + 0.049T$[cal/(mole °K)]. The enthalpy of combustion was determined by the method of Hubbard, Katz, and Waddington (633) and yielded the result of $\Delta Hf_{298}^\circ(l) = 10.34$ kcal/mole and $\Delta Hf_{298}^\circ(g) = 19.79$ kcal/mole when adjusted to the revised value of ΔHf_{298}° for $H_2SO_4(aq)$ (cf. page 560). Fawcett (411) reported data for several other physical properties of 3-methylthiophene.

ALKANETHIOL IDEAL GAS TABLES

No. 890   Methanethiol   C H_4S   (Ideal Gas State)   Mol Wt 48.108

	cal/(mole °K)				kcal/mole		
$T°K$	$Cp°$	$S°$	$-(G°-H_{298}°)/T$	$H°-H_{298}°$	$\Delta Hf°$	$\Delta Gf°$	Log Kp
298	12.01	60.96	60.96	0.00	-5.49	-2.37	1.739
300	12.05	61.04	60.97	0.03	-5.51	-2.35	1.715
400	14.04	64.78	61.46	1.33	-6.94	-1.14	0.625
500	15.91	68.12	62.47	2.83	-8.09	0.44	-0.192
600	17.57	71.17	63.67	4.51	-9.05	2.24	-0.815
700	19.03	73.99	64.94	6.34	-9.85	4.21	-1.314
800	20.32	76.62	66.24	8.31	-23.57	4.99	-1.364
900	21.46	79.08	67.53	10.40	-23.83	8.58	-2.084
1000	22.48	81.39	68.80	12.60	-24.01	12.20	-2.665

Low-temperature thermal data and vapor pressures were determined by Russell, Osborne, and Yost (1271), who found $Tm = 150.15°K$, $\Delta Hm = 1.411$ kcal/mole, $Tb = 279.11°K$, $\Delta Hv = 5.872$ kcal/mole at Tb, and $S_{279.11}°(g) = 60.16$ cal/(mole °K). Binder (131, 132) calculated thermodynamic functions from spectroscopic data and the barrier to internal rotation, based on the experimental entropy. The corresponding values by Barrow and Pitzer (78) are in good agreement. Raman spectral data of Venkateswaran (1531, 1532), Vogel-Högler (1543), and Wagner (1563), together with infrared spectral data of Thompson and Miller (1491), Thompson and Skerrett (1492), and Trotter and Thompson (1517), were used by Scott and McCullough (1316) to obtain the fundamental vibration frequencies. Structural data and barriers based on microwave studies have been published by Solimene and Dailey (1393), Kilb (749), Kojima (789), and Kojima and Nishikawa (790). They also used the data of the last mentioned authors to deduce the product of the principal moments of inertia (35.95×10^{-117} g^3-cm^6), the reduced moment for internal rotation (1.853×10^{-40} g-cm^2), and the height of the barrier as 1.270 ± 0.003 kcal/mole. Remeasurement of the vibrational spectra by Thompson and Miller (1491) provided confirmation for the previously made assignments.

Effects of vibrational anharmonicity were neglected to give a calculated value of the entropy in the ideal gaseous state of 60.17 cal/(mole °K) at 279.12°K. Values for the thermodynamic functions were taken from Scott and McCullough (1316). Good, Lacina, and McCullough (511) determined ΔHc and ΔHv, from which they derived $\Delta Hf°_{298}(l) = -11.18 \pm 0.13$ kcal/mole and $\Delta Hf°_{298}(g) = -5.49 \pm 0.14$ kcal/mole, adjusted to the new $\Delta Hf°_{298}$ value for $H_2SO_4(aq)$ (cf. page 560). This supersedes the early value of Thomsen (1495), which cannot be considered reliable.

No. 891 Ethanethiol C_2H_6S (Ideal Gas State) Mol Wt 62.134

	cal/(mole °K)				kcal/mole		
T°K	$Cp°$	$S°$	$-(G°-H°_{298})/T$	$H°-H°_{298}$	$\Delta Hf°$	$\Delta Gf°$	Log Kp
298	17.37	70.77	70.77	0.00	-11.02	-1.12	0.818
300	17.44	70.88	70.78	0.04	-11.04	-1.06	0.770
400	21.08	76.41	71.50	1.97	-12.79	2.48	-1.356
500	24.36	81.47	73.00	4.24	-14.19	6.47	-2.826
600	27.21	86.17	74.81	6.82	-15.32	10.70	-3.899
700	29.68	90.55	76.75	9.67	-16.22	15.14	-4.727
800	31.83	94.66	78.73	12.75	-30.01	18.40	-5.027
900	33.71	98.52	80.72	16.03	-30.27	24.47	-5.943
1000	35.38	102.16	82.68	19.48	-30.42	30.57	-6.681

Thermal properties of ethanethiol were measured accurately by McCullough, Scott, Finke, Gross, Williamson, Pennington, Waddington, and Huffman (962) and by McCullough, Hubbard, Frow, Hossenlopp, and Waddington (958). Results included $Tm = 125.26$°K, $\Delta Hm = 1.189$ kcal/mole, $Tb = 308.15$°K, $\Delta Hv = 6.401$ kcal/mole at Tb, $S°_{298}(l) = 49.48$ cal/(mole °K), $S°_{298}(g) = 70.77$ cal/(mole °K), $\Delta Hf°_{298}(l) = -17.60$ kcal/mole, $\Delta Hf°_{298}(g) = -11.02$ kcal/mole, and $Cp(l, 290-315°K) = 19.22 + 0.030T$[cal/(mole °K)]. The enthalpies of formation have been corrected to the new $\Delta Hf°_{298}$ value for $H_2SO_4(aq)$ (cf. page 560). The above papers also give references to previous, less accurate work.

No. 892 1-Propanethiol C_3H_8S (Ideal Gas State) Mol Wt 76.160

Complete and accurate thermodynamic properties of 1-propanethiol were reported by Pennington, Scott, Finke, McCullough, Messerly, Hossenlopp, and Waddington (1136). Data of interest are $Tt = 142.09$°K, $\Delta Ht = 0.9491$ kcal/mole, $Tm = 159.99$°K, $\Delta Hm = 1.309$ kcal/mole, $\Delta Hv = 7.059$ kcal/mole at $Tb = 340.87$°K, $S°_{298}(l) = 57.96$ cal/(mole °K), $S°_{298}(g) = 80.40$ cal/(mole °K), and $Cp(l, 270-320°K) = 23.55 + 0.037T$[cal/(mole °K)]. Hubbard and Waddington (638) gave enthalpy of combustion re-

No. 892 1-Propanethiol C_3H_8S (Ideal Gas State) Mol Wt 76.160

$T°K$	$Cp°$	$S°$	$-(G°-H°_{298})/T$	$H°-H°_{298}$	$\Delta Hf°$	$\Delta Gf°$	Log Kp
		cal/(mole °K)			kcal/mole		
298	22.65	80.40	80.40	0.00	-16.22	0.52	-0.384
300	22.75	80.55	80.41	0.05	-16.25	0.63	-0.456
400	27.86	87.80	81.36	2.58	-18.33	6.51	-3.559
500	32.56	94.53	83.33	5.61	-20.00	12.92	-5.649
600	36.72	100.84	85.73	9.07	-21.32	19.63	-7.152
700	40.37	106.78	88.32	12.93	-22.34	26.57	-8.296
800	43.60	112.39	90.98	17.13	-36.17	32.34	-8.834
900	46.47	117.69	93.65	21.64	-36.40	40.92	-9.937
1000	49.01	122.72	96.31	26.42	-36.45	49.53	-10.823

sults, which are used to derive $\Delta Hf°_{298}(l) = -23.87$ kcal/mole and $\Delta Hf°_{298}(g) = -16.22$ kcal/mole from the values listed by Scott and McCullough (1316), adjusted to the new $\Delta Hf°_{298}$ value for $H_2SO_4(aq)$ (cf. page 560). References to previous work on this compound have been given by Pennington et al. (1136).

No. 893 2-Propanethiol C_3H_8S (Ideal Gas State) Mol Wt 76.160

$T°K$	$Cp°$	$S°$	$-(G°-H°_{298})/T$	$H°-H°_{298}$	$\Delta Hf°$	$\Delta Gf°$	Log Kp
		cal/(mole °K)			kcal/mole		
298	22.94	77.51	77.51	0.00	-18.22	-0.61	0.451
300	23.04	77.66	77.52	0.05	-18.25	-0.51	0.370
400	28.35	85.03	78.48	2.62	-20.29	5.67	-3.095
500	33.06	91.87	80.49	5.70	-21.90	12.35	-5.396
600	37.02	98.26	82.92	9.21	-23.19	19.32	-7.036
700	40.38	104.22	85.54	13.08	-24.19	26.51	-8.277
800	43.26	109.81	88.23	17.27	-38.03	32.54	-8.888
900	45.74	115.05	90.92	21.72	-38.32	41.38	-10.048
1000	47.92	119.99	93.59	26.41	-38.46	50.25	-10.982

A complete and accurate study of 2-propanethiol was made by McCullough, Finke, Scott, Gross, Messerly, Pennington, and Waddington (955). The results obtained included $Tm = 142.63°K$, $\Delta Hm = 1.371$ kcal/mole, $Tb = 325.71°K$, at which $\Delta Hv = 6.670$ kcal/mole, $S°_{298}(l) = 55.82$ cal/(mole °K), $S°_{298}(g) = 77.51$ cal/(mole °K), and $Cp(l, 270-330°K) = 22.85 + 0.40T$[cal/(mole °K)]. Vapor heat capacities were used in conjunction with the measured entropy to derive thermodynamic functions of the ideal gas from spectroscopic data. Details of the determination of the enthalpy of combustion were given by Hubbard and Waddington (638). Enthalpies of formation were calculated as $\Delta Hf°_{298}(l) = -25.30$ kcal/mole and

$\Delta Hf^\circ_{298}(g) = -18.22$ kcal/mole when adjusted to the new ΔHf°_{298} value for $H_2SO_4(aq)$ (cf. page 560).

No. 894 1-Butanethiol $C_4H_{10}S$ (Ideal Gas State) Mol Wt 90.186

	cal/(mole °K)				kcal/mole		
T°K	$Cp°$	$S°$	$-(G°-H°_{298})/T$	$H°-H°_{298}$	$\Delta Hf°$	$\Delta Gf°$	Log Kp
298	28.24	89.68	89.68	0.00	-21.05	2.64	-1.934
300	28.37	89.86	89.69	0.06	-21.09	2.78	-2.027
400	34.95	98.93	90.88	3.23	-23.48	11.05	-6.038
500	41.07	107.40	93.34	7.03	-25.37	19.91	-8.702
600	46.54	115.38	96.36	11.42	-26.86	29.11	-10.602
700	51.37	122.93	99.62	16.32	-27.96	38.56	-12.037
800	55.68	130.07	102.98	21.68	-41.80	46.84	-12.795
900	59.54	136.86	106.37	27.44	-41.97	57.94	-14.068
1000	62.95	143.31	109.75	33.57	-41.90	69.04	-15.089

Complete thermodynamic measurements on 1-butanethiol were reported by Scott, Finke, McCullough, Messerly, Pennington, Hossenlopp, and Waddington (1309). Results include $Tm = 157.46°K$, $\Delta Hm = 2.500$ kcal/mole, $\Delta Hv = 7.702$ kcal/mole at $Tb = 371.61°K$, $S°_{298}(l) = 65.96$ cal/(mole °K), $S°_{298}(g) = 89.68$ cal/(mole °K), $Cp(l, 270-320°K) = 27.5 + 0.046T$[cal/(mole °K)], $\Delta Hf^\circ_{298}(l) = -29.79$ kcal/mole, and $\Delta Hf^\circ_{298}(g) = -21.05$ kcal/mole when adjusted to the new ΔHf°_{298} value for $H_2SO_4(aq)$ (cf. page 560). These heat of formation results were confirmed by Hubbard, Good, and Waddington (631) [$\Delta Hf^\circ_{298}(l) = -29.77$ kcal/mole and $\Delta Hf^\circ_{298}(g) = -21.03$ kcal/mole, when adjusted in the same manner]. Thermodynamic properties were calculated by a refined method of increments, based on experimental vapor heat capacities and data for low molecular weight sulfur compounds.

No. 895 2-Butanethiol $C_4H_{10}S$ (Ideal Gas State) Mol Wt 90.186

	cal/(mole °K)				kcal/mole		
T°K	$Cp°$	$S°$	$-(G°-H°_{298})/T$	$H°-H°_{298}$	$\Delta Hf°$	$\Delta Gf°$	Log Kp
298	28.51	87.65	87.65	0.00	-23.00	1.29	-0.948
300	28.64	87.83	87.66	0.06	-23.04	1.44	-1.050
400	35.38	97.01	88.86	3.27	-25.39	9.91	-5.413
500	41.37	105.57	91.36	7.11	-27.25	18.95	-8.284
600	46.42	113.57	94.40	11.51	-28.72	28.33	-10.320
700	50.68	121.05	97.68	16.37	-29.86	37.97	-11.853
800	54.29	128.06	101.04	21.62	-43.80	46.44	-12.686
900	57.37	134.64	104.41	27.21	-44.16	57.75	-14.023
1000	60.02	140.82	107.75	33.08	-44.34	69.09	-15.100

Complete thermal studies by McCullough, Finke, Scott, Pennington, Gross, Messerly, and Waddington (956) gave $Tm = 133.02°K$, $\Delta Hm = 1.548$ kcal/mole, $Tb = 358.13°K$, at which temperature $\Delta Hv = 7.312$ kcal/mole, $S°_{298}(l) = 64.87$ cal/(mole °K), $S°_{298}(g) = 87.65$ cal/(mole °K), $Cp(l, 270-310°K) = 27.51 + 0.045T$[cal/(mole °K)], $\Delta Hf°_{298}(l) = -31.13$ kcal/mole, and $\Delta Hf°_{298}(g) = -23.00$ kcal/mole when adjusted to the new $\Delta Hf°_{298}$ value for $H_2SO_4(aq)$ (cf. page 560). Hubbard, Good, and Waddington (631) confirmed these results [$\Delta Hf°_{298}(l) = -31.27$ kcal/mole and $\Delta Hf°_{298}(g) = -23.14$ kcal/mole when similarly adjusted]. Thermodynamic functions of the ideal gas were calculated from spectroscopic data, with adjustable parameters chosen to fit the experimental data.

No. 896 2-Methyl-1-Propanethiol $C_4H_{10}S$ (Ideal Gas State) Mol Wt 90.186

		cal/(mole °K)			kcal/mole		
T°K	Cp°	S°	$-(G°-H°_{298})/T$	$H°-H°_{298}$	$\Delta Hf°$	$\Delta Gf°$	Log Kp
298	28.28	86.73	86.73	0.00	-23.24	1.33	-0.973
300	28.41	86.91	86.74	0.06	-23.28	1.48	-1.076
400	35.31	96.05	87.94	3.25	-25.64	10.04	-5.485
500	41.30	104.59	90.42	7.09	-27.51	19.18	-8.384
600	46.26	112.57	93.46	11.48	-28.99	28.66	-10.439
700	50.35	120.02	96.73	16.31	-30.16	38.39	-11.986
800	53.77	126.97	100.08	21.52	-44.14	46.97	-12.832
900	56.68	133.48	103.43	27.05	-44.56	58.40	-14.180
1000	59.17	139.58	106.74	32.84	-44.81	69.86	-15.267

Complete thermodynamic data for 2-methyl-1-propanethiol were determined and reported by Scott, McCullough, Messerly, Pennington, Hossenlopp, Finke, and Waddington (1320). Results include $Tm = 128.31°K$, $\Delta Hm = 1.191$ kcal/mole, $Tb = 361.64°K$, at which temperature $\Delta Hv = 7.412$ kcal/mole, $S°_{298}(l) = 63.66$ cal/(mole °K), $S°_{298}(g) = 86.73$ cal/(mole °K), and $Cp(l, 270-350°K) = 26.26 + 0.050T$[cal/(mole °K)]. The enthalpy of combustion was measured by rotating bomb techniques, leading to $\Delta Hf°_{298}(l) = -31.54$ kcal/mole and $\Delta Hf°_{298}(g) = -23.24$ kcal/mole when adjusted to the new $\Delta Hf°_{298}$ value for $H_2SO_4(aq)$ (cf. page 560). Hubbard, Good, and Waddington (631) confirmed these, with $\Delta Hf°_{298}(l) = -31.52$ kcal/mole and $\Delta Hf°_{298}(g) = -23.22$ kcal/mole, when similarly adjusted.

No. 897 2-Methyl-2-Propanethiol $C_4H_{10}S$ (Ideal Gas State) Mol Wt 90.186

Complete thermal data for 2-methyl-2-propanethiol were obtained by McCullough, Scott, Finke, Hubbard, Gross, Katz, Pennington, Messerly,

No. 897 2-Methyl-2-Propanethiol $C_4H_{10}S$ (Ideal Gas State) Mol Wt 90.186

	cal/(mole °K)				kcal/mole		
$T°K$	$Cp°$	$S°$	$-(G°-H°_{298})/T$	$H°-H°_{298}$	$\Delta Hf°$	$\Delta Gf°$	Log Kp
298	28.91	80.79	80.79	0.00	-26.17	0.17	-0.124
300	29.04	80.97	80.80	0.06	-26.21	0.33	-0.240
400	36.13	90.32	82.02	3.32	-28.50	9.47	-5.176
500	42.39	99.07	84.57	7.26	-30.27	19.18	-8.383
600	47.60	107.28	87.68	11.77	-31.63	29.20	-10.635
700	51.92	114.95	91.03	16.75	-32.65	39.45	-12.316
800	55.53	122.12	94.47	22.12	-46.47	48.53	-13.256
900	58.61	128.84	97.92	27.84	-46.70	60.42	-14.672
1000	61.24	135.16	101.33	33.83	-46.75	72.34	-15.809

and Waddington (963) and used to calculate thermodynamic properties of the ideal gas. Three solid-state transistions were observed in addition to melting, with $Tm = 274.41°K$, $\Delta Hm = 0.593$ kcal/mole, $Tb = 337.37°K$, at which temperature $\Delta Hv = 6.797$ kcal/mole, $S°_{298}(l) = 58.90$ cal/(mole °K), $S°_{298}(g) = 80.79$ cal/(mole °K), and $Cp(l, 280-330°K) = 27.24 + 0.049T$ [cal/(mole °K)]. Hubbard, Good, and Waddington (631) reported the enthalpy of combustion, deriving $\Delta Hf°_{298}(l) = -33.56$ kcal/mole and $\Delta Hf°_{298}(g) = -26.17$ kcal/mole when adjusted to the new $\Delta Hf°_{298}$ value for $H_2SO_4(aq)$ (cf. page 560). A less accurate value of Franklin and Lumpkin (440) was in agreement within experimental error.

No. 898 2-Methyl-2-Butanethiol $C_5H_{12}S$ (Ideal Gas State) Mol Wt 104.212

	cal/(mole °K)				kcal/mole		
$T°K$	$Cp°$	$S°$	$-(G°-H°_{298})/T$	$H°-H°_{298}$	$\Delta Hf°$	$\Delta Gf°$	Log Kp
298	34.30	92.48	92.48	0.00	-30.36	2.20	-1.616
300	34.46	92.70	92.49	0.07	-30.40	2.40	-1.751
400	42.79	103.77	93.94	3.94	-33.03	13.69	-7.481
500	50.28	114.15	96.95	8.60	-35.09	25.62	-11.197
600	56.58	123.89	100.64	13.95	-36.69	37.91	-13.809
700	61.84	133.01	104.62	19.88	-37.89	50.47	-15.758
800	66.28	141.57	108.71	26.29	-51.84	61.88	-16.905
900	70.05	149.60	112.81	33.11	-52.15	76.13	-18.485
1000	73.30	157.15	116.87	40.29	-52.25	90.40	-19.755

The chemical thermodynamic properties were determined by Scott, Douslin, Finke, Hubbard, Messerly, Hossenlopp, and McCullough (1299). They reported $Tt = 159.1°K$, $\Delta Ht = 1907.1$ cal/mole, $Tm = 169.36°K$, $\Delta Hm = 145.4$ cal/mole, $Tb = 372.28°K$, at which temperature $\Delta Hv = 7.497$ kcal/mole, $S°_{298}(l) = 69.34$ cal/(mole °K), $S°_{298}(g) = 92.48$

cal/(mole °K), and $Cp(l, 200-350°K) = 53.373 - 0.14093T + 5.425 \times 10^{-4}T^2 - 4.6094 \times 10^{-7}T^3$[cal/(mole °K)]. From combustion calorimetry they determined $\Delta Hf°_{298}(l) = -38.90$ kcal/mole and $\Delta Hf°_{298}(g) = -30.36$ kcal/mole when adjusted to the new value of $\Delta Hf°_{298}$ for $H_2SO_4(aq)$ (cf. page 560). Thermodynamic functions for the gas phase were also evaluated from infrared spectral data on crystalline, solid, and gas phases, together with equation of state data. Values of barrier heights were taken as 1.5 kcal/mole for the thiol group, 3.4 for the methyl group of the ethyl group, and 5.0 for the other two methyl groups.

No. 899   1-Pentanethiol $C_5H_{12}S$ (Ideal Gas State)   Mol Wt 104.212

$T°K$	cal/(mole °K)				kcal/mole		Log Kp
	$Cp°$	$S°$	$-(G°-H°_{298})/T$	$H°-H°_{298}$	$\Delta Hf°$	$\Delta Gf°$	
298	33.75	99.28	99.28	0.00	-25.91	4.63	-3.392
300	33.91	99.49	99.29	0.07	-25.96	4.81	-3.507
400	41.93	110.36	100.71	3.86	-28.66	15.43	-8.431
500	49.36	120.53	103.67	8.44	-30.81	26.71	-11.675
600	55.92	130.12	107.29	13.71	-32.48	38.37	-13.977
700	61.69	139.19	111.20	19.59	-33.72	50.32	-15.709
800	66.78	147.76	115.24	26.02	-47.66	61.11	-16.693
900	71.33	155.90	119.31	32.93	-47.88	74.73	-18.145
1000	75.32	163.62	123.36	40.27	-47.82	88.36	-19.310

The low-temperature thermal data and the vapor pressure of 1-pentanethiol were determined by Finke, Scott, Gross, Waddington, and Huffman (425). Values found were $Tm = 197.45°K$, $\Delta Hm = 4.190$ kcal/mole, $\Delta Hv = 8.4$ kcal/mole at $Tb = 399.79°K$, $S°_{298}(l) = 74.18$ cal/(mole °K), $S°_{298}(g) = 99.28$ cal/(mole °K), and $Cp(l, 270-320°K) = 31.2 + 0.057T$[cal/(mole °K)]. Enthalpy of combustion measurements by Sunner (1443) and by Hubbard, Katz, and Waddington (633) [cf. McCullough and Good (957)] were in good agreement and led to $\Delta Hf°_{298}(l) = -35.74$ kcal/mole and $\Delta Hf°_{298}(g) = -25.91$ kcal/mole when adjusted to the new $\Delta Hf°_{298}$ value for $H_2SO_4(aq)$ (cf. page 560). Thermodynamic functions for the ideal gas state were computed by Scott and McCullough (1316), using a refined method of increments, in which calculated values of Cp and S were adjusted empirically to accord with the vapor phase calorimetric data of Finke, Hossenlopp, and Berg (422). They reported $\Delta Hv = 9.014$ kcal/mole at 356.08°K. Aibazov, Petrov, Khairullina, and Yaprintseva (5) selected $Tm = 197.45°K$ and $Tb = 399.79°K$.

No. 900 1-Hexanethiol $C_6H_{14}S$ (Ideal Gas State) Mol Wt 118.238

T°K	Cp°	S°	$-(G°-H°_{298})/T$	$H°-H°_{298}$	$\Delta Hf°$	$\Delta Gf°$	Log Kp
	cal/(mole °K)				kcal/mole		
298	39.21	108.58	108.58	0.00	-30.83	6.65	-4.871
300	39.40	108.83	108.59	0.08	-30.88	6.87	-5.008
400	48.87	121.48	110.25	4.50	-33.90	19.87	-10.858
500	57.60	133.34	113.69	9.83	-36.31	33.60	-14.687
600	65.26	144.53	117.91	15.98	-38.18	47.76	-17.397
700	71.96	155.11	122.47	22.85	-39.57	62.23	-19.430
800	77.84	165.11	127.18	30.34	-53.60	75.57	-20.643
900	83.07	174.58	131.93	38.39	-53.88	91.74	-22.277
1000	87.65	183.58	136.65	46.94	-53.84	107.94	-23.588

Thermodynamic functions were computed by Scott and McCullough (1316) using methylene increments. Unpublished measurements of $Cs(12-370°K)$ and ΔHm (at approximately 193°K) were made by Finke, Guthrie, Todd, Messerly, and McCullough (421). Morris, Lanum, Helm, Haines, Cook, and Ball (1028) reported $Tm = 192.66°K$ and $Tb = 425.75°K$. Good and DePrater (507) reported $\Delta Hf°_{298}(l) = -41.97$ kcal/mole and $\Delta Hf°_{298}(g) = -30.83$ kcal/mole when adjusted to the new $\Delta Hf°_{298}$ value for $H_2SO_4(aq)$ (cf. page 560).

No. 901 1-Heptanethiol $C_7H_{16}S$ (Ideal Gas State) Mol Wt 132.264

T°K	Cp°	S°	$-(G°-H°_{298})/T$	$H°-H°_{298}$	$\Delta Hf°$	$\Delta Gf°$	Log Kp
	cal/(mole °K)				kcal/mole		
298	44.68	117.89	117.89	0.00	-35.76	8.65	-6.341
300	44.90	118.17	117.90	0.09	-35.82	8.92	-6.500
400	55.81	132.60	119.79	5.13	-39.16	24.30	-13.276
500	65.85	146.16	123.72	11.22	-41.82	40.48	-17.694
600	74.60	158.96	128.54	18.26	-43.89	57.14	-20.811
700	82.24	171.04	133.75	26.11	-45.42	74.14	-23.145
800	88.91	182.47	139.14	34.67	-59.55	90.01	-24.588
900	94.82	193.29	144.56	43.86	-59.89	108.74	-26.404
1000	99.98	203.55	149.95	53.61	-59.86	127.49	-27.861

Scott and McCullough (1316), using a method of increments, calculated values of the thermodynamic fynctions for 1-heptanethiol. $Cs(13-359°K)$ and ΔHm (at about 230°K) have been measured but not yet published by Finke, Guthrie, Todd, Messerly, and McCullough (421). The values $Tm = 230.1°K$ and $Tb = 449.4°K$ were selected by Rossini, Pitzer, Arnett, Braun, and Pimentel (1248). Morris, Lanum, Helm, Haines, Cook, and Ball (1028) reported 229.88 and 450.0°K, respectively. Good and

DePrater (507) found $\Delta Hf^\circ_{298}(l) = -47.85$ kcal/mole and $\Delta Hf^\circ_{298}(g) = -35.76$ kcal/mole when adjusted to the new ΔHf°_{298} value for $H_2SO_4(aq)$ (cf. page 560).

No. 902 1-Octanethiol $C_8H_{18}S$ (Ideal Gas State) Mol Wt 146.290

	cal/(mole °K)				kcal/mole		
$T°K$	$Cp°$	$S°$	$-(G°-H°_{298})/T$	$H°-H°_{298}$	$\Delta Hf°$	$\Delta Gf°$	Log Kp
298	50.14	127.20	127.20	0.00	-40.68	10.67	-7.819
300	50.39	127.52	127.21	0.10	-40.75	10.98	-7.999
400	62.75	143.73	129.33	5.76	-44.40	28.74	-15.701
500	74.09	158.98	133.75	12.62	-47.32	47.37	-20.704
600	83.94	173.38	139.17	20.53	-49.59	66.52	-24.230
700	92.51	186.97	145.03	29.36	-51.26	86.05	-26.864
800	99.97	199.82	151.09	38.99	-65.49	104.46	-28.537
900	106.56	211.99	157.18	49.33	-65.89	125.75	-30.534
1000	112.31	223.52	163.25	60.28	-65.88	147.06	-32.138

A method of increments was used by Scott and McCullough (1316) to generate thermodynamic functions for 1-octanethiol. Rossini, Pitzer, Arnett, Braun, and Pimentel (1248) selected $Tm = 224.0°K$ and $Tb = 472.2°K$.

No. 903 1-Nonanethiol $C_9H_{20}S$ (Ideal Gas State) Mol Wt 160.316

	cal/(mole °K)				kcal/mole		
$T°K$	$Cp°$	$S°$	$-(G°-H°_{298})/T$	$H°-H°_{298}$	$\Delta Hf°$	$\Delta Gf°$	Log Kp
298	55.61	136.51	136.51	0.00	-45.61	12.67	-9.289
300	55.89	136.86	136.52	0.11	-45.68	13.03	-9.491
400	69.69	154.86	138.88	6.40	-49.66	33.16	-18.119
500	82.34	171.80	143.78	14.01	-52.83	54.25	-23.711
600	93.28	187.80	149.80	22.81	-55.30	75.90	-27.644
700	102.79	202.91	156.32	32.62	-57.11	97.95	-30.579
800	111.04	217.18	163.04	43.32	-71.43	118.90	-32.482
900	118.31	230.69	169.81	54.79	-71.89	142.74	-34.661
1000	124.65	243.49	176.55	66.95	-71.90	166.61	-36.411

Thermodynamic functions were computed for 1-nonanethiol by Scott and McCullough (1316) by means of methylene increments. The values of $Tm = 253.0°K$ and $Tb = 493.4°K$ were selected by Rossini, Pitzer, Arnett, Braun, and Pimentel (1248).

No. 904 1-Decanethiol $C_{10}H_{22}S$ (Ideal Gas State) Mol Wt 174.342

	cal/(mole °K)				kcal/mole		
$T°K$	$Cp°$	$S°$	$-(G°-H°_{298})/T$	$H°-H°_{298}$	$\Delta Hf°$	$\Delta Gf°$	Log Kp
298	61.08	145.82	145.82	0.00	-50.54	14.68	-10.759
300	61.38	146.20	145.83	0.12	-50.62	15.08	-10.983
400	76.63	165.98	148.42	7.03	-54.91	37.59	-20.538
500	90.59	184.62	153.81	15.41	-58.34	61.13	-26.717
600	102.63	202.22	160.43	25.08	-61.00	85.27	-31.058
700	113.07	218.84	167.60	35.88	-62.96	109.85	-34.295
800	122.10	234.54	174.99	47.65	-77.38	133.35	-36.427
900	130.06	249.39	182.44	60.26	-77.89	159.74	-38.788
1000	136.98	263.46	189.85	73.62	-77.92	186.17	-40.684

Scott and McCullough (1316) used a method of increments to calculate values of the thermodynamic functions for 1-decanethiol. Unpublished measurements of Cs (12–363°K) and ΔHm (at about 248°K) were made by Finke, Guthrie, Todd, Messerly, and McCullough (421). Rossini, Pitzer, Arnett, Braun, and Pimentel (1248) selected 247°K and 513.8°K for Tm and Tb, respectively. Good and DePrater (507) determined $\Delta Hc(l)$ and reported $\Delta Hf°_{298}(l) = -66.10$ kcal/mole and $\Delta Hf°_{298}(g) = -50.54$ kcal/mole when adjusted to the new $\Delta Hf°_{298}$ value for H_2SO_4 (aq) (cf. page 560).

No. 905 1-Undecanethiol $C_{11}H_{24}S$ (Ideal Gas State) Mol Wt 188.368

	cal/(mole °K)				kcal/mole		
$T°K$	$Cp°$	$S°$	$-(G°-H°_{298})/T$	$H°-H°_{298}$	$\Delta Hf°$	$\Delta Gf°$	Log Kp
298	66.54	155.13	155.13	0.00	-55.46	16.69	-12.236
300	66.88	155.55	155.14	0.13	-55.55	17.14	-12.482
400	83.57	177.11	157.96	7.66	-60.16	42.03	-22.962
500	98.83	197.43	163.84	16.80	-63.84	68.01	-29.728
600	111.97	216.64	171.06	27.36	-66.70	94.66	-34.476
700	123.34	234.78	178.88	39.13	-68.80	121.76	-38.013
800	133.17	251.90	186.95	51.97	-83.32	147.80	-40.375
900	141.80	268.09	195.07	65.73	-83.89	176.74	-42.917
1000	149.31	283.43	203.15	80.29	-83.94	205.73	-44.960

A method of increments was used by Scott and McCullough (1316) to generate thermodynamic functions for 1-undecanethiol. Values of $Tm = 270°K$ and $Tb = 532.6°K$ were selected by Rossini, Pitzer, Arnett, Braun, and Pimentel (1248).

Alkanethiol Ideal Gas Tables

No. 906 1-Dodecanethiol $C_{12}H_{26}S$ (Ideal Gas State) Mol Wt 202.394

$T°K$	$Cp°$	$S°$	$-(G°-H°_{298})/T$	$H°-H°_{298}$	$\Delta Hf°$	$\Delta Gf°$	Log Kp
		cal/(mole °K)			kcal/mole		
298	72.01	164.44	164.44	0.00	-60.39	18.70	-13.706
300	72.37	164.89	164.45	0.14	-60.48	19.18	-13.974
400	90.51	188.23	167.51	8.30	-65.41	46.46	-25.381
500	107.08	210.25	173.87	18.19	-69.35	74.89	-32.734
600	121.31	231.06	181.69	29.63	-72.41	104.03	-37.891
700	133.62	250.71	190.16	42.39	-74.65	133.66	-41.729
800	144.23	269.26	198.90	56.29	-89.27	162.24	-44.320
900	153.55	286.80	207.70	71.19	-89.90	193.74	-47.044
1000	161.65	303.40	216.45	86.96	-89.96	225.29	-49.234

Values of the thermodynamic functions for 1-dodecanethiol were calculated by Scott and McCullough (1316), using a method of methylene increments. $Tm = 266°K$ and $Tb = 550.4°K$ were selected by Rossini, Pitzer, Arnett, Braun, and Pimentel (1248).

No. 907 1-Tridecanethiol $C_{13}H_{28}S$ (Ideal Gas State) Mol Wt 216.420

$T°K$	$Cp°$	$S°$	$-(G°-H°_{298})/T$	$H°-H°_{298}$	$\Delta Hf°$	$\Delta Gf°$	Log Kp
		cal/(mole °K)			kcal/mole		
298	77.47	173.75	173.75	0.00	-65.31	20.71	-15.183
300	77.87	174.24	173.76	0.15	-65.41	21.24	-15.473
400	97.45	199.36	177.05	8.93	-70.66	50.89	-27.805
500	115.32	223.07	183.90	19.59	-74.85	81.78	-35.745
600	130.65	245.48	192.32	31.90	-78.11	113.41	-41.309
700	143.89	266.64	201.44	45.65	-80.49	145.57	-45.448
800	155.30	286.62	210.85	60.62	-95.21	176.70	-48.268
900	165.29	305.50	220.33	76.66	-95.89	210.75	-51.174
1000	173.98	323.37	229.75	93.63	-95.98	244.85	-53.510

Thermodynamic functions for 1-tridecanethiol were computed by Scott and McCullough (1316), using methylene increments. The values of $Tm = 282°K$ and $Tb = 567.2°K$ were selected by Rossini, Pitzer, Arnett, Braun, and Pimentel (1248).

No. 908 1-Tetradecanethiol $C_{14}H_{30}S$ (Ideal Gas State) Mol Wt 230.446

Scott and McCullough (1316), using a method of increments, calculated values of the thermodynamic functions for 1-tetradecanethiol. Rossini, Pitzer, Arnett, Braun, and Pimentel (1248) selected $Tm = 279°K$ and $Tb = 583°K$.

No. 908 1-Tetradecanethiol $C_{14}H_{30}S$ (Ideal Gas State) Mol Wt 230.446

T°K	Cp°	S°	$-(G°-H°_{298})/T$	$H°-H°_{298}$	$\Delta Hf°$	$\Delta Gf°$	Log Kp
	cal/(mole °K)				kcal/mole		
298	82.94	183.06	183.06	0.00	-70.24	22.72	-16.654
300	83.36	183.58	183.07	0.16	-70.35	23.29	-16.965
400	104.39	210.48	186.59	9.56	-75.91	55.32	-30.224
500	123.57	235.89	193.94	20.98	-80.37	88.66	-38.751
600	139.99	259.91	202.95	34.18	-83.82	122.79	-44.723
700	154.17	282.58	212.72	48.90	-86.35	157.47	-49.163
800	166.36	303.97	222.80	64.94	-101.16	191.14	-52.214
900	177.04	324.20	232.96	82.12	-101.90	227.74	-55.301
1000	186.31	343.34	243.05	100.30	-102.00	264.41	-57.783

No. 909 1-Pentadecanethiol $C_{15}H_{32}S$ (Ideal Gas State) Mol Wt 244.472

T°K	Cp°	S°	$-(G°-H°_{298})/T$	$H°-H°_{298}$	$\Delta Hf°$	$\Delta Gf°$	Log Kp
	cal/(mole °K)				kcal/mole		
298	88.41	192.37	192.37	0.00	-75.17	24.73	-18.124
300	88.86	192.92	192.38	0.17	-75.28	25.34	-18.457
400	111.34	221.61	196.14	10.20	-81.17	59.75	-32.643
500	131.82	248.71	203.97	22.38	-85.87	95.54	-41.757
600	149.34	274.33	213.58	36.46	-89.52	132.16	-48.138
700	164.45	298.51	224.00	52.16	-92.19	169.37	-52.878
800	177.43	321.34	234.76	69.27	-107.11	205.58	-56.159
900	188.79	342.91	245.59	87.59	-107.90	244.74	-59.428
1000	198.65	363.32	256.35	106.98	-108.02	283.96	-62.056

A method of increments was used by Scott and McCullough (1316) to generate thermodynamic functions. Values of $Tm = 291°K$ and $Tb = 598°K$ were selected by Rossini, Pitzer, Arnett, Braun, and Pimentel (1248).

No. 910 1-Hexadecanethiol $C_{16}H_{34}S$ (Ideal Gas State) Mol Wt 258.498

T°K	Cp°	S°	$-(G°-H°_{298})/T$	$H°-H°_{298}$	$\Delta Hf°$	$\Delta Gf°$	Log Kp
	cal/(mole °K)				kcal/mole		
298	93.87	201.67	201.67	0.00	-80.09	26.74	-19.603
300	94.35	202.26	201.68	0.18	-80.21	27.40	-19.959
400	118.28	232.73	205.67	10.83	-86.41	64.19	-35.069
500	140.06	261.52	213.99	23.77	-91.38	102.43	-44.770
600	158.68	288.74	224.20	38.73	-95.22	141.55	-51.558
700	174.72	314.44	235.27	55.42	-98.04	181.29	-56.599
800	188.49	338.68	246.70	73.59	-113.05	220.04	-60.109
900	200.53	361.60	258.20	93.06	-113.90	261.75	-63.560
1000	210.98	383.28	269.64	113.65	-114.04	303.54	-66.334

Values of the thermodynamic functions for 1-hexadecanethiol were calculated by Scott and McCullough (1316), using a method of methylene increments. Rossini, Pitzer, Arnett, Braun, and Pimentel (1248) selected 291°K and 612°K for Tm and Tb, respectively.

No. 911   1-Heptadecanethiol $C_{17}H_{36}S$ (Ideal Gas State)   Mol Wt 272.524

$T°K$	$Cp°$	$S°$	$-(G°-H°_{298})/T$	$H°-H°_{298}$	$\Delta Hf°$	$\Delta Gf°$	Log Kp
		cal/(mole °K)			kcal/mole		
298	99.34	210.98	210.98	0.00	-85.02	28.75	-21.073
300	99.85	211.60	210.99	0.19	-85.15	29.45	-21.451
400	125.22	243.86	215.21	11.46	-91.67	68.62	-37.488
500	148.31	274.34	224.02	25.17	-96.89	109.31	-47.776
600	168.02	303.16	234.83	41.01	-100.93	150.93	-54.972
700	185.00	330.37	246.55	58.68	-103.89	193.19	-60.315
800	199.56	356.04	258.65	77.92	-119.00	234.48	-64.054
900	212.28	380.30	270.83	98.53	-119.90	278.75	-67.686
1000	223.31	403.25	282.94	120.32	-120.06	323.09	-70.608

Thermodynamic functions were computed by Scott and McCullough (1316), using methylene increments. The values of $Tm = 300°K$ and $Tb = 626°K$ were selected by Rossini, Pitzer, Arnett, Braun, and Pimentel (1248).

No. 912   1-Octadecanethiol $C_{18}H_{38}S$ (Ideal Gas State)   Mol Wt 286.550

$T°K$	$Cp°$	$S°$	$-(G°-H°_{298})/T$	$H°-H°_{298}$	$\Delta Hf°$	$\Delta Gf°$	Log Kp
		cal/(mole °K)			kcal/mole		
298	104.80	220.29	220.29	0.00	-89.94	30.77	-22.551
300	105.34	220.94	220.30	0.20	-90.07	31.50	-22.950
400	132.16	254.98	224.76	12.09	-96.91	73.05	-39.912
500	156.55	287.16	234.05	26.56	-102.39	116.20	-50.787
600	177.36	317.58	245.46	43.28	-106.63	160.31	-58.390
700	195.27	346.30	257.83	61.93	-109.73	205.11	-64.034
800	210.62	373.40	270.60	82.24	-124.94	248.94	-68.003
900	224.02	399.00	283.46	103.99	-125.90	295.76	-71.816
1000	235.64	423.22	296.24	126.99	-126.08	342.66	-74.884

Scott and McCullough (1316), using a method of increments, calculated values of the thermodynamic functions for 1-octadecanethiol. $Tm = 301°K$ and $Tb = 639°K$ were selected by Rossini, Pitzer, Arnett, Braun, and Pimentel (1248).

No. 913 1-Nonadecanethiol $C_{19}H_{40}S$ (Ideal Gas State) Mol Wt 300.576

	cal/(mole °K)			kcal/mole			
$T°K$	$Cp°$	$S°$	$-(G°-H°_{298})/T$	$H°-H°_{298}$	$\Delta Hf°$	$\Delta Gf°$	Log Kp
298	110.37	229.60	229.60	0.00	-94.87	32.77	-24.021
300	110.84	230.29	229.61	0.21	-95.01	33.55	-24.442
400	139.10	266.11	234.30	12.73	-102.17	77.48	-42.331
500	164.80	299.98	244.08	27.95	-107.90	123.08	-53.793
600	186.70	332.01	256.09	45.56	-112.34	169.69	-61.805
700	205.55	362.24	269.12	65.19	-115.58	217.01	-67.749
800	221.69	390.76	282.56	86.57	-130.88	263.38	-71.948
900	235.77	417.70	296.09	109.46	-131.91	312.75	-75.943
1000	247.98	443.19	309.54	133.66	-132.10	362.21	-79.157

A method of increments was used by Scott and McCullough (1316) to generate the thermodynamic functions for 1-nonadecanethiol. Rossini, Pitzer, Arnett, Braun, and Pimentel (1248) selected the values of $Tm = 307°K$ and $Tb = 651°K$.

No. 914 1-Eicosanethiol $C_{20}H_{42}S$ (Ideal Gas State) Mol Wt 314.602

	cal/(mole °K)			kcal/mole			
$T°K$	$Cp°$	$S°$	$-(G°-H°_{298})/T$	$H°-H°_{298}$	$\Delta Hf°$	$\Delta Gf°$	Log Kp
298	115.74	238.91	238.91	0.00	-99.80	34.78	-25.491
300	116.33	239.63	238.92	0.22	-99.95	35.60	-25.934
400	146.04	277.23	243.84	13.36	-107.42	81.91	-44.749
500	173.05	312.80	254.11	29.35	-113.41	129.95	-56.800
600	196.05	346.43	266.72	47.83	-118.04	179.06	-65.219
700	215.83	378.17	280.40	68.45	-121.43	228.91	-71.465
800	232.75	408.12	294.51	90.90	-136.83	277.82	-75.893
900	247.52	436.41	308.72	114.93	-137.91	329.75	-80.070
1000	260.31	463.17	322.84	140.33	-138.12	381.77	-83.431

Values of the thermodynamic functions for 1-eicosanethiol were calculated by Scott and McCullough (1316), using a method of methylene increments. The values of $Tm = 310°K$ and $Tb = 662°K$ were selected by Rossini, Pitzer, Arnett, Braun, and Pimentel (1248).

CYCLOALKANETHIOL IDEAL GAS TABLE

No. 915 Cyclopentanethiol $C_5H_{10}S$ (Ideal Gas State) Mol Wt 102.196

Berg, Scott, Hubbard, Todd, Messerly, Hossenlopp, Osborn, Douslin, and McCullough (111) measured the thermal and thermochemical values of cyclopentanethiol. Their results included $Tm = 155.39°K$, $\Delta Hm = 1.872$

Cycloalkanethiol Ideal Gas Table

No. 915   Cyclopentanethiol $C_5H_{10}S$ (Ideal Gas State)   Mol Wt 102.196

T°K	cal/(mole °K)			kcal/mole			
	$Cp°$	$S°$	$-(G°-H°_{298})/T$	$H°-H°_{298}$	$\Delta Hf°$	$\Delta Gf°$	Log Kp
298	25.79	86.38	86.38	0.00	-11.45	13.63	-9.989
300	25.94	86.54	86.39	0.05	-11.50	13.78	-10.039
400	34.53	95.21	87.51	3.08	-14.27	22.58	-12.337
500	42.20	103.76	89.91	6.93	-16.45	32.05	-14.010
600	48.65	112.04	92.91	11.48	-18.14	41.92	-15.267
700	54.05	119.95	96.21	16.62	-19.43	52.06	-16.253
800	58.61	127.48	99.65	22.26	-33.44	61.06	-16.681
900	62.51	134.61	103.15	28.32	-33.81	72.91	-17.704
1000	65.84	141.37	106.63	34.75	-33.94	84.78	-18.528

kcal/mole, $Tb = 132.16°K$, at which temperature $\Delta Hv = 8.443$ kcal/mole, $S°_{298}(l) = 61.39$ cal/(mole °K), $S°_{298}(g) = 86.38$ cal/(mole °K), and $Cp(l, 162-366°K) = 55.249 - 0.24225T + 8.8181 \times 10^{-4}T^2 - 8.2702 \times 10^{-7}T^3$ [cal/(mole °K)]. The combustion calorimetric results of these authors gave $\Delta Hf°_{298}(l) = -21.35$ kcal/mole and $\Delta Hf°_{298}(g) = -11.45$ kcal/mole when adjusted to the new value of $\Delta Hf°_{298}$ for $H_2SO_4(aq)$ (cf. page 560). Thermodynamic functions were also evaluated from all available Raman and infrared spectral data and molecular structure values. The height of the potential barrier restricting internal rotation of the thiol group (1.2 kcal/mole), that of the potential barrier restricting pseudorotation of the five-membered ring (0, i.e., free pseudorotation), the effective moment of inertia for this pseudorotation (18.9×10^{-40} g-cm²), and the parameters of an empirical anharmonicity equation were selected to fit the experital values of the heat capacity and the entropy.

AROMATIC THIOL IDEAL GAS TABLE

No. 916   Benzenethiol C_6H_6S (Ideal Gas State)   Mol Wt 110.174

T°K	cal/(mole °K)			kcal/mole			
	$Cp°$	$S°$	$-(G°-H°_{298})/T$	$H°-H°_{298}$	$\Delta Hf°$	$\Delta Gf°$	Log Kp
298	25.07	80.51	80.51	0.00	26.66	35.28	-25.861
300	25.22	80.67	80.52	0.05	26.63	35.33	-25.740
400	32.76	88.99	81.60	2.96	24.88	38.45	-21.007
500	39.07	97.00	83.88	6.56	23.54	42.00	-18.359
600	44.13	104.58	86.71	10.73	22.48	45.79	-16.679
700	48.23	111.70	89.78	15.35	21.66	49.78	-15.540
800	51.59	118.37	92.94	20.35	7.95	52.56	-14.359
900	54.41	124.61	96.11	25.65	7.76	58.16	-14.121
1000	56.79	130.47	99.26	31.22	7.70	63.76	-13.935

Complete thermodynamic properties of benzenethiol were reported by Scott, McCullough, Hubbard, Messerly, Hossenlopp, Frow, and Waddington (1319). Results included $Tm = 258.26°K$, $\Delta Hm = 2.736$ kcal/mole, $Tb = 442.29°K$, at which temperature $\Delta Hv = 9.53$ kcal/mole, $S°_{298}(l) = 53.25$ cal/(mole °K), $S°_{298}(g) = 80.51$ cal/(mole °K), and $Cp(l, 300-370°K) = 27.65 + 0.46T$[cal/(mole °K)]. The enthalpy of formation was determined by rotating bomb calorimetry as $\Delta Hf°_{298}(l) = 15.27$ kcal/mole and $\Delta Hf°_{298}(g) = 26.66$ kcal/mole when adjusted to the $\Delta Hf°_{298}$ value for $H_2SO_4(aq)$ (cf. page 560). The low-temperature heat capacity and calculated entropy of Parks, Todd, and Moore (1118) are in reasonable agreement with the later data.

MISCELLANEOUS SULFUR COMPOUND IDEAL GAS TABLES

No. 917 Isothiocyanic Acid CHNS (Ideal Gas State) Mol Wt 59.092

	cal/(mole °K)				kcal/mole		
$T°K$	$Cp°$	$S°$	$-(G°-H°_{298})/T$	$H°-H°_{298}$	$\Delta Hf°$	$\Delta Gf°$	Log Kp
298	11.09	59.28	59.28	0.00	30.50	26.98	-19.777
300	11.11	59.35	59.29	0.03	30.49	26.96	-19.639
400	12.71	62.78	59.74	1.22	29.65	25.85	-14.124
500	13.78	65.74	60.65	2.55	29.02	24.98	-10.917
600	14.57	68.32	61.72	3.97	28.49	24.22	-8.821
700	15.21	70.62	62.83	5.46	28.05	23.57	-7.358
800	15.74	72.68	63.93	7.00	14.59	21.69	-5.924
900	16.18	74.56	65.01	8.60	14.52	22.58	-5.483
1000	16.57	76.29	66.06	10.24	14.47	23.48	-5.131

Ideal gas state thermal properties were computed by Mackle and O'Hare (911) from molecular parameters tabulated by Sutton (1449), and the vibrational assignment of Barakat, Legge, and Pullin (72). The gas phase enthalpy of formation of 30.5 kcal/mole was selected by Wagman, Evans, Halow, Parker, Bailey, and Schumm (1560) and is presumably based upon the work of Brandenburg (160).

No. 918 Thioacetic Acid C_2H_4OS (Ideal Gas State) Mol Wt 76.118

Mackle and O'Hare (909) made a statistico-mechanical computation of the ideal gas state thermodynamic properties from molecular parameters tabulated by Sutton (1449) and the assumption that $\angle HSC = 100°$ by analogy with methanethiol and the fundamental frequencies assigned by Sheppard (1342). A value of 2.5 kcal/mole was selected for the barrier to

Sulfur Compound Ideal Gas Tables

No. 918 Thioacetic Acid C_2H_4OS (Ideal Gas State) Mol Wt 76.118

	cal/(mole °K)				kcal/mole		
$T°K$	$Cp°$	$S°$	$-(G°-H°_{298})/T$	$H°-H°_{298}$	$\Delta Hf°$	$\Delta Gf°$	Log Kp
298	19.33	74.86	74.86	0.00	−43.49	−36.81	26.979
300	19.41	74.98	74.87	0.04	−43.50	−36.77	26.782
400	22.25	80.97	75.66	2.13	−44.75	−34.39	18.787
500	24.61	86.19	77.26	4.47	−45.75	−31.68	13.846
600	26.72	90.87	79.14	7.04	−46.57	−28.79	10.485
700	28.68	95.14	81.13	9.81	−47.24	−25.74	8.037
800	30.41	99.09	83.13	12.77	−60.83	−23.91	6.530
900	31.39	102.72	85.11	15.86	−60.98	−19.28	4.681
1000	32.62	106.09	87.04	19.06	−61.09	−14.63	3.198

rotation about the C—C bond, to be the same as that selected by Weltner (1586) for acetic acid; that for rotation about the C—S axis was assumed to be the same as in the thiols (1.5 ± 0.2 kcal/mole). Sunner (1447) derived the gas phase $\Delta Hf°_{298}(g) = -43.49$ kcal/mole and $\Delta Hf°_{298}(l) = -52.39$ kcal/mole when adjusted to the $\Delta Hf°_{298}$ value for $H_2SO_4(aq)$ (cf. page 560).

SECTION III
THERMAL AND THERMOCHEMICAL DATA AT 298°K

CHAPTER 14

SELECTED VALUES OF ENTHALPY OF FORMATION AND ENTROPY OF ORGANIC COMPOUNDS AT 298°K

INTRODUCTION

A serious effort was made to collect and evaluate the large bulk of thermodynamic information on organic compounds. Before about 1930 we depended on the tabulations of Parks and Huffman (1105) and Kharasch (744). For the years 1930 through 1965 *Chemical Abstracts* was systematically searched and pertinent references noted. These references were then examined, and the thermodynamic data for individual compounds were noted and filed. Valuable cross-checks for completeness and accuracy were available for hydrocarbons from the American Petroleum Institute Project 44 collection of Rossini, Pitzer, Arnett, Braun, and Pimentel (1248), for sulfur compounds from the tabulation of Scott and McCullough (1316), and for general coverage from the revised Landolt-Börnstein tables edited by Auer (52).

For a limited number of compounds, enough information was available to construct complete ideal gas tables of thermodynamic data, such as are presented in Chapters 8 through 13. Much more often, however, only fragmentary results were found, usually enthalpies of formation and/or entropies for the condensed state at 298°K. These data are tabulated here. Also repeated here are values at 298°K taken from the ideal gas tables of Chapters 8 through 13. This final chapter thus serves as an index to all compounds covered in this work.

Compounds are arranged according to the Hill indexing system (601) as used by *Chemical Abstracts*. The arrangement of atomic symbols in chemical formulas is alphabetical, except for carbon and hydrogen. The C always comes first, followed immediately by H if hydrogen is present, and the H is followed by any other elements in their normal alphabetical order. The number of atoms of an element also determines the order of indexing. All formulas with 1 carbon precede those with 2 carbon atoms; within the series of 1-carbon formulas all those with 1 hydrogen appear before those with 2 hydrogens. The following series illustrates the application of these rules: $CBrCl_3$, CCl_2O, CCl_4, $CHBrCl_2$, $CHCl_3$, CH_2Cl_2, CH_3Cl, CH_4, CI_4, C_2Cl_4, C_2Cl_6, $C_2H_2Cl_4$, C_2H_4, C_2H_6. "Inorganic" compounds (such as those discussed in Chapter 8) appear at the beginning of the summary table.

Following the empirical formula is the compound name, which in nearly all cases corresponds to *Chemical Abstracts* usage. Next is given the physical state of the compound: (*g*) for gas state, (*l*) for liquid state, and (*s*) for solid crystalline state. Molecular weights are calculated on the basis of the 1957 atomic weights, except for carbon, for which 12.010 is retained. Selected values for the enthalpy of formation and the absolute entropy are given under the column headings "Enthalpy" and "Entropy", respectively.

Immediately following the enthalpy of formation or entropy is the reference number in parentheses. The reference number is preceded by letters that serve to classify the quality or methodology. Enthalpy of formation references preceded by an "A" indicate that the work was carefully done and sufficient details were reported so that the results could be recalculated in terms of the modern constants listed in Chapter 8. Enthalpy of formation references preceded by a "B" indicate that the result listed is judged to be of reasonable accuracy but suffers from failure in reporting important details, insufficient characterization of the sample, or the fact that values for other compounds given in the same reference do not agree well with other investigations. When not enough details are reported and no comparisons are available, the reference is given a "C" rating, indicating that no judgment has been made. The "C" rating thus amounts to a *caveat emptor*. An "E" preceding the reference indicates that the listed enthalpy of formation was estimated. Many of the "E" values are of good accuracy, for example, those given by Prosen and Rossini (1203). In general, however, no judgment is implied for estimated values, and the concerned reader should consult the original reference.

A reference giving an entropy value derived from heat capacity measurements is preceded by "M", while references listing entropy values calculated from spectroscopic parameters are preceded by "S". A reference giving estimates is preceded by an "E". For the present purpose estimated entropy values include those calculated from spectroscopic data, which are in turn largely estimated.

In a very few instances enthalpy and entropy values were derived from the temperature dependence of measured equilibrium constants. Such data are denoted by the letter "F" preceding the references.

Compounds for which ideal gas tables are derived in Chapters 8 through 13 are referenced by the gas table number, which is preceded by a "T". Table numbers 10–48 are found in Chapter 8, 101–476 in Chapter 9, 501–569 in Chapter 10, 601–633 in Chapter 11, 701–786 in Chapter 12, and 800–918 in Chapter 13. The reader is referred to the discussion accompanying the ideal gas tables for data sources. Occasionally selection of condensed state values involves a compromise, and a table number is given as reference for discussion and/or for data sources.

If several references all described work on a specific compound, several courses were followed, depending on the situation. If one of the references involved work of obviously superior accuracy, it was selected. If the references were judged about equal in quality, the most recent reference was cited. In nearly all cases the cited publication lists sources of earlier efforts. Finally, if two or more references were judged about equal in quality and were published simultaneously or nearly so, values from all references were cited as separate entries.

Selected data for a particular compound were punched on cards and fed into a computer. Entropy and enthalpy data for the elements taken from Chapter 8 were stored in the computer. Appropriate programming then yielded the printed output reproduced here, which includes the Gibbs energy of formation when the data given are sufficient for its calculation. The limitations of the computer printer are evident, but are offset by the higher speed, accuracy, and economy of computer operation. Special symbols used to describe compounds are listed here.

ANG	angular structure
C	Celsius degrees
CALC	calculated
CIS	*cis* structural isomer
D	dextrorotatory optical isomer
ENDO	*endo* structural isomer
EXO	*exo* structural isomer
(G)	gas state
(HB)	high boiling isomer
(LB)	low boiling isomer
L	levorotatory optical isomer
(L)	liquid state
LIN	linear structure
M	*meta* structural isomer
M.P.	melting point
O	*ortho* structural isomer
P	*para* structural isomer
(S)	solid crystalline state
SYM	symmetrical structure
TRANS	*trans* structural isomer
(U)	unspecified structure

Selected Values for Inorganic Compounds from Chapter 8

Formula	Name	State	Mol Wt	$\Delta H f^\circ_{298}$ (kcal/mole)	Source	S°_{298} (cal/mole °K)	Source	$\Delta G f^\circ_{298}$ (kcal/mole)	Log $K p$
BRH	HYDROGEN BROMIDE	(G)	80.924	-8.66	10) TC	47.44	10) TC	-12.73	9.329
CLFO3	PERCHLORYL FLUORIDE	(G)	102.457	-6.49	24) TC	66.65	24) TC	-10.72	-7.858
CLH	HYDROGEN CHLORIDE	(G)	36.465	-22.06	25) TC	44.64	25) TC	-22.77	16.692
CLNO2	NITROSYL CHLORIDE	(G)	65.465	12.57	26) TC	62.53	26) TC	16.00	-11.727
CLNO2	NITRYL CHLORIDE	(G)	81.465	-3.12	27) TC	65.01	27) TC	6.87	-5.039
CL2OS	THIONYL CHLORIDE	(G)	118.980	-50.60	28) TC	73.23	28) TC	-46.97	34.425
CL2O2S	SULFURYL CHLORIDE	(G)	134.980	-85.40	29) TC	74.37	29) TC	-74.80	54.827
CL2S2	SULFUR MONOCHLORIDE	(G)	135.046	-4.66	30) TC	76.35	30) TC	-6.99	5.120
FH	HYDROGEN FLUORIDE	(G)	20.008	-64.80	31) TC	41.51	31) TC	-65.30	47.865
FNO	NITROSYL FLUORIDE	(G)	49.008	-15.70	32) TC	59.27	32) TC	-12.02	8.811
FNO2	NITRYL FLUORIDE	(G)	65.008	-19.00	33) TC	62.24	33) TC	-8.90	6.525
F4S	SULFUR TETRAFLUORIDE	(G)	108.066	-174.10	34) TC	69.58	34) TC	-163.68	119.976
HI	HYDROGEN IODIDE	(G)	127.918	6.30	35) TC	49.35	35) TC	0.38	-0.276
HNO3	HYDROGEN NITRATE	(G)	63.016	-32.02	36) TC	63.68	36) TC	-17.62	12.912
H2O	WATER	(G)	18.016	-57.80	37) TC	45.11	37) TC	-54.64	40.049
H2O2	HYDROGEN PEROXIDE	(G)	34.016	-32.53	38) TC	55.66	38) TC	-25.21	18.478
H2O4S	SULFURIC ACID (ANHYDROUS)	(L)	98.082	-194.45	39) TC	37.49	39) TC	-164.83	120.815
H2S	HYDROGEN SULFIDE	(G)	34.082	-4.82	40) TC	49.18	40) TC	-7.90	5.792
H3N	AMMONIA	(G)	17.032	-10.92	41) TC	46.03	41) TC	-3.86	2.831
H4N2	HYDRAZINE	(G)	32.048	22.75	42) TC	57.41	42) TC	37.89	-27.773
NO	NITRIC OXIDE	(G)	30.008	21.60	43) TC	50.35	43) TC	20.72	-15.184
NO2	NITROGEN DIOXIDE	(G)	46.008	8.09	44) TC	57.35	44) TC	12.42	-9.107
N2O	NITROUS OXIDE	(G)	44.016	19.49	45) TC	52.56	45) TC	24.77	-18.156
O2S	SULFUR DIOXIDE	(G)	64.066	-70.95	46) TC	59.30	46) TC	-71.74	52.588
O3	OZONE	(G)	48.000	34.00	47) TC	57.05	47) TC	38.91	-28.517
O3S	SULFUR TRIOXIDE	(G)	80.066	-94.47	48) TC	61.19	48) TC	-88.52	64.886

Selected Values for Organic Compounds

Formula	Name	State	Mol Wt	$\Delta H f^\circ_{298}$ (kcal/mole)	Source	S°_{298} (cal/mole °K)	Source	$\Delta G f^\circ_{298}$ (kcal/mole)	Log $K p$
CBRCLF2	BROMOCHLORODIFLUOROMETHANE	(G)	165.383	-112.7	E(938)	76.14	SC(465)	-107.18	78.563
CBRCLO	CARBONYL BROMIDE CHLORIDE	(G)	143.383			72.04	S(1094)		
CBRCL2F	BROMODICHLOROFLUOROMETHANE	(G)	181.840	-64.4	E(938)	78.87	SC(484)	-58.98	43.228
CBRCL3	BROMOTRICHLOROMETHANE	(G)	198.297	-8.9	B(104)	79.55	SC(912)	-2.96	2.167
CBRF3	BROMOTRIFLUOROMETHANE	(G)	148.926	-155.1	A(883)	71.19	SC(975)	-148.83	109.089
CBRN	CYANOGEN BROMIDE	(S)	105.934	32.5	B(883)				
CBRN	CYANOGEN BROMIDE	(G)	105.934	43.35	T(11)	59.07	T(11)	38.39	-28.140
CBR2CLF	DIBROMOCHLOROFLUOROMETHANE	(G)	226.299	-55.4	E(938)	81.89	SC(484)	-53.40	39.138
CBR2CL2	DIBROMODICHLOROMETHANE	(G)	242.756	-7.0	E(938)	83.23	SC(312)	-4.67	3.425
CBR2F2	DIBROMODIFLUOROMETHANE	(G)	209.842	-102.7	E(938)	77.66	SC(317)	-100.16	73.413
CBR2O	CARBONYL BROMIDE	(L)	187.842	-30.0	C(1249)				
CBR2O	CARBONYL BROMIDE	(G)	187.842	-23.0	C(1249)	73.82	S(1094)	-26.45	19.388
CBR3CL	TRIBROMOCHLOROMETHANE	(G)	287.215	3.0	E(938)	85.36	SC(465)	2.17	-1.592
CBR3F	TRIBROMOFLUOROMETHANE	(G)	270.758	-45.4	E(938)	82.65	SC(465)	-46.14	33.821
CBR4	CARBON TETRABROMIDE	(G)	331.674	12.0	C(1249)	85.53	SC(465)	8.60	-6.304
CCLFO	CARBONYL CHLORIDE FLUORIDE	(G)	82.467			66.18	SC(888)		
CCLF3	CHLOROTRIFLUOROMETHANE	(G)	104.467	-166.0	B(833)	68.28	SC(8)	-156.34	114.595
CCLN	CYANOGEN CHLORIDE	(G)	61.475	31.60	T(12)	56.28	T(12)	29.99	-21.984
CCLN3O6	TRINITROCHLOROMETHANE	(L)	185.491	-5.57	A(1647)				
CCL2F2	DICHLORODIFLUOROMETHANE	(G)	120.924	-115.0	B(833)	71.91	SC(8)	-105.70	77.477
CCL2O	PHOSGENE	(G)	98.924	-52.80	T(13)	67.82	T(13)	-49.42	36.225
CCL2S	THIOPHOSGENE	(G)	114.990			70.19	SC(1487)		
CCL3F	TRICHLOROFLUOROMETHANE	(G)	137.381	-68.0	B(833)	74.13	SC(8)	-58.64	42.983
CCL4	CARBON TETRACHLORIDE	(L)	153.838	-31.75	T(725)	51.67	MC(598)	-14.97	10.975
CCL4	CARBON TETRACHLORIDE	(G)	153.838	-24.00	T(725)	74.12	T(725)	-13.92	10.201
CCL4S	TRICHLOROMETHANESULFENYL CHLORIDE	(G)	185.904			88.43	A(412)		
CFN	CYANOGEN FLUORIDE	(L)	45.018			53.97	SC(890)		
CFN3O6	TRINITROFLUOROMETHANE	(L)	169.034	-53.4	A(1648)				
CF2O	CARBONYL FLUORIDE	(G)	66.010	-153.00	T(14)	61.84	T(14)	-149.28	109.421
CF3I	TRIFLUOROIODOMETHANE	(G)	195.920	-141.60	A(833)	73.50	SC(975)	-136.70	100.201
CF4	CARBON TETRAFLUORIDE	(G)	88.010	-223.00	T(704)	62.50	T(704)	-212.34	155.641
CF4O	TRIFLUOROMETHYL HYPOFLUORITE	(G)	104.010	-187.6	C(1435)				
CHBRCLF	BROMOCHLOROFLUOROMETHANE	(G)	147.391	-70.5	E(938)	72.88	SC(465)	-66.58	48.802
CHBRCL2	BROMODICHLOROMETHANE	(G)	163.848	-14.0	E(938)	75.56	S(1174)	-10.16	7.445
CHBRF2	BROMODIFLUOROMETHANE	(G)	130.934	-110.8	E(938)	70.51	S(1100)	-106.90	78.353

Formula	Name	State	Mol Wt	$\Delta H f^\circ_{298}$ (kcal/mole)	Source	S°_{298} (cal/mole °K)	Source	$\Delta G f^\circ_{298}$ (kcal/mole)	Log Kp
CHBR2CL	DIBROMOCHLOROMETHANE	(G)	208.307	−5.0	E(938)	78.31	S(1179)	−4.50	3.297
CHBR2F	DIBROMOFLUOROMETHANE	(G)	191.850	−53.4	E(938)	75.70	S(1037)	−52.84	38.732
CHBR2I	DIBROMOIODOMETHANE	(G)	299.760	20.0	E(938)				
CHBR3	TRIBROMOMETHANE	(L)	252.766	−4.8	C(1249)				
CHBR3	TRIBROMOMETHANE	(G)	252.766	6.0	C(1249)	79.03	S(465)	3.77	−2.762
CHCLF2	CHLORODIFLUOROMETHANE	(G)	86.475	−119.9	E(938)	67.13	M(1049)	−112.47	82.437
CHCL2F	DICHLOROFLUOROMETHANE	(G)	102.932	−71.4	E(938)	69.99	S(1585)	−64.10	46.983
CHCL3	CHLOROFORM	(L)	119.389	−31.6	T(724)	48.5	S(1249)	−17.17	12.585
CHCL3	CHLOROFORM	(G)	119.389	−24.2	T(724)	70.66	T(724)	−16.38	12.004
CHFI2	FLUORODIIODOMETHANE	(G)	285.838	−22.4	E(938)				
CHF2I	DIFLUOROIODOMETHANE	(G)	177.928	−94.8	E(938)				
CHF3	TRIFLUOROMETHANE	(G)	70.018	−166.71	T(703)	62.04	T(703)	−158.48	116.165
CHF3S	TRIFLUOROMETHANETHIOL	(G)	102.084			73.72	S(329)		
CHI3	TRIIODOMETHANE	(S)	393.748	33.60	C(1249)				
CHI3	TRIIODOMETHANE	(G)	393.748	50.40	T(777)	85.00	T(777)	42.53	−31.174
CHN	HYDROCYANIC ACID	(G)	27.026	31.20	T(15)	48.21	T(15)	28.71	−21.042
CHNO	ISOCYANIC ACID	(G)	43.026			56.92	S(892)		
CHNS	ISOTHIOCYANIC ACID	(G)	59.092	30.50	T(917)	59.28	T(917)	26.98	−19.777
CHN3O6	TRINITROMETHANE	(L)	151.042	−9.2	A(1193)				
CHN3O6	TRINITROMETHANE	(L)	151.042	−6.2	C(1640)				
CHN3O6	TRINITROMETHANE	(G)	151.042	18.6	C(613)				
CH2BRCL	BROMOCHLOROMETHANE	(G)	129.399	−12.0	E(938)	68.67	S(1582)	−9.39	6.886
CH2BRF	BROMOFLUOROMETHANE	(G)	112.942	−60.4	E(938)	65.97	S(465)	−57.71	42.302
CH2BRI	BROMOIODOMETHANE	(G)	220.852	12.0	E(938)	73.49	S(465)	9.36	−6.862
CH2BR2	DIBROMOMETHANE	(G)	173.858	−1.0	E(1249)	70.08	T(345)	−1.34	0.979
CH2CLF	CHLOROFLUOROMETHANE	(G)	68.483	−69.5	E(938)	63.16	S(465)	−63.45	46.510
CH2CLI	CHLOROIODOMETHANE	(G)	176.393	3.0	E(938)	70.78	S(465)	3.69	−2.705
CH2CL2	DICHLOROMETHANE	(L)	84.940	−29.7	T(723)	42.7	S(1249)	−16.83	12.337
CH2CL2	DICHLOROMETHANE	(G)	84.940	−22.80	T(723)	64.59	T(723)	−16.46	12.063
CH2F2	DIFLUOROMETHANE	(G)	52.026	−108.24	T(702)	58.94	T(702)	−101.66	74.513
CH2I2	DIIODOMETHANE	(L)	267.846	15.89	C(1249)				
CH2I2	DIIODOMETHANE	(G)	267.846	28.20	T(776)	73.88	T(776)	24.16	−17.709
CH2N2	CYANAMIDE	(S)	42.042	14.65	A(1275)				
CH2N2	CYANAMIDE	(G)	42.042			56.64	S(1041)		
CH2N2	DIAZOMETHANE	(G)	42.042	46.0	C(847)	58.03	S(1024)	52.06	−38.156
CH2N4	TETRAZOLE	(S)	70.058	56.7	B(974)				

Formula	Name	State	Mol Wt	$\Delta H f^\circ_{298}$ (kcal/mole)	Source	S°_{298} (cal/mole °K)	Source	$\Delta G f^\circ_{298}$ (kcal/mole)	Log Kp
CH2N4O	TETRAZOL-5-OL	(S)	86.058	1.5	B(1616)				
CH2N6O2	1-AZIDO-N NITROFORMAMIDINE	(S)	130.074	71.3	C(974)				
CH2N6O2	5-NITRAMINOTETRAZOLE	(S)	130.074	60.3	C(974)				
CH2O	FORMALDEHYDE	(G)	30.026	-27.70	T(544)	52.29	T(544)	-26.27	19.258
CH2O	PARAFORMALDEHYDE	(S)	30.026	-40.20	T(544)				
CH2O2	FORMIC ACID	(L)	46.026	-101.52	T(559)	30.82	T(559)	-86.39	63.320
CH2O2	FORMIC ACID	(G)	46.026	-90.49	T(559)	59.45	T(559)	-83.89	61.492
CH2S3	TRITHIOCARBONIC ACID	(L)	110.224	6.0	B(460)	53.3	M(460)	6.65	-4.871
CH3BR	BROMOMETHANE	(G)	94.950	-9.00	T(759)	58.75	T(759)	-6.73	4.932
CH3CL	CHLOROMETHANE	(G)	50.491	-20.63	T(722)	56.04	T(722)	-15.03	11.017
CH3F	FLUOROMETHANE	(G)	34.034	-55.90	T(701)	53.25	T(701)	-50.19	36.789
CH3I	IODOMETHANE	(L)	141.944	-3.29	T(775)	38.9	S(1249)	3.61	-2.649
CH3I	IODOMETHANE	(G)	141.944	3.34	T(775)	60.71	T(775)	3.74	-2.742
CH3NO	FORMAMIDE	(L)	45.042	-61.6	C(1249)				
CH3NO	FORMAMIDE	(G)	45.042	-44.5	C(82)	59.38	S(82)	-33.71	24.710
CH3NO2	METHYL NITRITE	(G)	61.042	-15.30	T(629)	67.95	T(629)	0.24	-0.174
CH3NO2	NITROMETHANE	(L)	61.042	-27.03	A(1193)	41.05	M(702)	-3.47	2.545
CH3NO2	NITROMETHANE	(G)	61.042	-17.86	T(623)	65.73	T(623)	-1.66	1.217
CH3NO3	METHYL NITRATE	(L)	77.042	-36.9	T(630)	51.86	M(528)	-9.26	6.787
CH3NO3	METHYL NITRATE	(G)	77.042	-28.80	T(630)	72.15	T(630)	-7.21	5.284
CH3N3O3	NITROUREA	(S)	105.058	-67.2	B(981)				
CH3N5	5-AMINOTETRAZOLE	(S)	85.074	49.7	B(981)				
CH4	METHANE	(G)	16.042	-17.89	T(101)	44.52	T(101)	-12.15	8.903
CH4N2O	UREA	(S)	60.058	-79.63	A(1265)	25.00	M(1265)	-47.12	34.535
CH4N2O	UREA	(G)	60.058	-58.7	B(1450)	59.6	M(1450)	-36.50	26.756
CH4N2S	THIOUREA	(G)	76.124	-21.5	C(91)	27.67	M(1596)	5.19	-3.802
CH4N2S	THIOUREA	(S)	76.124			72.44	S(909)		
CH4N4O2	NITROGUANIDINE	(S)	104.074	-21.5	B(981)				
CH4N4O2	NITROGUANIDINE	(S)	104.074	-22.2	B(974)				
CH4N4O4	N,N'-DINITROMETHANEDIAMINE	(S)	136.074	-13.8	B(1039)				
CH4O	METHANOL	(L)	32.042	-57.02	T(517)	30.3	M(732)	-39.73	29.123
CH4O	METHANOL	(G)	32.042	-48.08	T(517)	57.29	T(517)	-38.84	28.468
CH4S	METHANETHIOL	(L)	48.108	-11.18	A(511)				
CH4S	METHANETHIOL	(G)	48.108	-5.49	T(890)	60.96	T(890)	-2.37	1.739
CH5N	METHYLAMINE	(L)	31.058	-11.31	A(678)				
CH5N	METHYLAMINE	(G)	31.058	-5.5	T(601)	57.98	T(601)	7.71	-5.648

Formula	Name	State	Mol Wt	ΔHf°_{298} (kcal/mole)	Source	S°_{298} (cal/mole °K)	Source	ΔGf°_{298} (kcal/mole)	Log K_p
CH5N3	GUANIDINE	(S)	59.074	-17.0	C(1249)				
CH5N3O4	UREA NITRATE	(S)	123.074	-134.5	B(981)				
CH5N3S	HYDRAZINIUM THIOCYANATE	(S)	91.140	9.2	B(1446)				
CH5N3S	THIOSEMICARBAZIDE	(S)	91.140	5.7	B(1446)				
CH5N5O2	AMINONITROGUANIDINE	(S)	119.090	5.3	B(974)				
CH6ClN	METHYLAMINE HYDROCHLORIDE	(S)	67.523	-71.2	A(47)	33.13	M(47)	-37.99	27.845
CH6N2	METHYLHYDRAZINE	(L)	46.074	12.7	B(41)	39.66	M(34)	42.84	-31.403
CH6N2	METHYLHYDRAZINE	(G)	46.074	20.4	B(41)	66.61	M(34)	42.51	-31.158
CH6N2O2	AMMONIUM CARBAMATE	(S)	78.074	-154.21	C(1249)	39.70	F(1249)	-109.47	80.238
CH6N2O3	METHYLAMINE NITRATE	(S)	94.074	-84.3	C(983)				
CH6N2O3	METHYLAMINE NITRATE	(S)	94.074	-80.6	C(277)				
CH8N6O3	DIAMINOGUANIDINE NITRATE	(S)	152.122	-37.6	B(974)				
CIN	IODINE CYANIDE	(S)	152.928	38.30	A(883)	30.80	S(1249)	40.48	-29.674
CIN	IODINE CYANIDE	(G)	152.928	53.80	T(16)	61.33	TC 16)	46.88	-34.363
CI4	TETRAIODOMETHANE	(S)	519.650			93.61	S(1278)		
CI4	TETRAIODOMETHANE	(G)	519.650			93.60	SC 539)		
CN4O8	TETRANITROMETHANE	(L)	196.042	8.9	B(1252)				
CN4O8	TETRANITROMETHANE	(L)	196.042	8.8	BC 459)				
CO	CARBON MONOXIDE	(G)	28.010	-26.42	T(17)	47.30	T(17)	-32.81	24.050
COS	CARBONYL SULFIDE	(G)	60.076	-33.08	T(18)	55.33	TC 18)	-39.59	29.017
CO2	CARBON DIOXIDE	(G)	44.010	-94.05	T(19)	51.07	TC 19)	-94.26	69.091
CS	CARBON MONOSULFIDE	(G)	44.076	55.0	C(1435)	50.30	S(1435)	42.68	-31.287
CS2	CARBON DISULFIDE	(L)	76.142	21.37	A(511)	36.10	M(175)	15.56	-11.407
CS2	CARBON DISULFIDE	(G)	76.142	27.98	T(20)	56.83	T(20)	15.99	-11.722
C2BrClF2	1-BROMO-1-CHLORO-2,2-DIFLUOROETHYLENE	(G)	177.393			80.86	S(1483)		
C2BrClF2	BROMOTRIFLUOROETHYLENE	(G)	160.936			80.02	SC 918)		
C2BrF3	1,2-DIBROMO-1-CHLORO-1,2,2-TRIFLUOROETHANE	(G)	276.309	-153.2	C(823)				
C2Br2ClF3	1,1-DIBROMO-2,2-DIFLUOROETHYLENE	(G)	221.852			83.62	S(1076)		
C2Br2F2	1,1-DIBROMO-1,2,2,2-TETRAFLUOROETHANE	(G)	259.852	-188.5	B(823)				
C2Br2F4	TETRABROMOETHYLENE	(G)	343.684			92.49	S(923)		
C2Br4	CHLORODIFLUOROACETONITRILE	(G)	111.485			76.08	S(682)		

Formula	Name	State	Mol Wt	$\Delta H f^\circ_{298}$ (kcal/mole)	Source	S°_{298} (cal/mole °K)	Source	$\Delta G f^\circ_{298}$ (kcal/mole)	Log $K p$
C2CLF3	CHLOROTRIFLUOROETHYLENE · · · ·	(G)	116.477	-130.2	C(755)	77.12	S(917)	-122.77	89.989
C2CLF3	CHLOROTRIFLUOROETHYLENE · · · ·	(G)	116.477	-133.9	C(797)	77.12	S(917)	-126.47	92.701
C2CLF3	CHLOROTRIFLUOROETHYLENE · · · ·	(G)	116.477	-124.2	C(1574)	77.12	S(917)	-116.77	85.591
C2CLF5	CHLOROPENTAFLUOROETHANE · · · ·	(G)	154.477			85.47	M(45)		
C2CL2FN	DICHLOROFLUOROACETONITRILE · · ·	(G)	127.942			79.33	S(682)		
C2CL2F2	1,1-DICHLORO-2,2-								
C2CL2F2	DIFLUOROETHYLENE	(G)	132.934	-80.2	A(833)				
C2CL2F4	1,1,2,2-TETRAFLUOROETHANE	(G)	170.934	-218.6	C(755)				
C2CL2F4	1,2-DICHLORO-								
C2CL2F4	1,1,2,2-TETRAFLUOROETHANE	(G)	170.934	-214.6	B(831)				
C2CL2O2	OXALYL CHLORIDE · · · ·	(G)	126.934			79.27	S(1650)		
C2CL3F	TRICHLOROFLUOROETHYLENE · · · ·	(G)	149.391			81.86	S(924)		
C2CL3F	1,1,2-TRICHLORO-								
C2CL3F3	1,1,2-TRIFLUOROETHANE	(G)	187.391	-178.2	C(831)				
C2CL3F3	2,2,2-TRIFLUOROETHANE	(G)	187.391	-12.8	B(1384)	86.37	S(891)		
C2CL3F3	TRICHLOROACETONITRILE · · · ·	(G)	144.399	-3.40	T(751)	79.38	S(682)		
C2CL3N	TETRACHLOROETHYLENE · · · ·	(L)	165.848						
C2CL4	TETRACHLOROETHYLENE · · · ·	(G)	165.848			81.46	T(751)	4.90	-3.592
C2CL4	1,1,1,2-TETRACHLORO-								
C2CL4F2	2,2-DIFLUOROETHANE	(G)	203.848	-121.1	B(833)	90.85	S(1074)	-101.15	74.145
C2CL4F2	TRICHLOROACETYL CHLORIDE · · ·	(G)	181.848	-67.7	B(1187)				
C2CL4O	PENTACHLOROFLUOROETHANE · · · ·	(G)	220.305	-75.8	E(833)				
C2CL5F	HEXACHLOROETHANE · · · ·	(S)	236.762	-49.3	B(1384)	93.54	S(1074)	-55.93	40.999
C2CL6	HEXACHLOROETHANE · · · ·	(G)	236.762	-33.80	T(732)	94.77	T(732)	-13.58	9.954
C2CL6	TRIFLUOROACETONITRILE · · · ·	(G)	95.028			71.22	T(682)		
C2F3N	TRIFLUOROACETONITRILE · · · ·	(G)	95.028			71.27	M(1098)		
C2F3N	TETRAFLUOROETHYLENE · · · ·	(G)	100.020	-157.40	T(714)	71.69	T(714)	-149.07	109.269
C2F4	HEXAFLUOROETHANE · · · ·	(G)	138.020	-321.00	T(708)	79.37	T(708)	-300.52	220.275
C2F6	1-BROMO-2-CHLORO-								
C2HBRCLF3	1,1,2-TRIFLUOROETHANE	(G)	197.401	-159.9	C(830)				
C2HBRCLF3	2-BROMO-2-CHLORO-								
C2HBRCLF3	1,1,1-TRIFLUOROETHANE	(G)	197.401			86.77	S(1482)		
C2HBRCL2F2	1-BROMO-1,1DIFLUORO-								
C2HBRCL2F2	2,2-DICHLOROETHANE	(G)	213.858	-111.7	B(833)				

639

Formula	Name	State	Mol Wt	ΔHf°_{298} (kcal/mole)	Source	S°_{298} (cal/mole °K)	Source	ΔGf°_{298} (kcal/mole)	Log Kp
C2HBRF2	2-BROMO-1,1-DIFLUOROETHYLENE	(G)	142.944			75.24	S(1481)		
C2HBRF4	1-BROMO-								
C2HBRF4	1,1,2,2-TETRAFLUOROETHANE	(G)	180.944	-198.9	B(830)	85.56	S(393)		
C2HBR3	TRIBROMOETHANE	(G)	264.776	-24.9	C(1189)				
C2HBR3O	TRIBROMOACETALDEHYDE	(L)	280.776	-79.2	B(833)	72.28	S(1072)	-72.90	53.433
C2HCLF2	2-CHLORO-,1-DIFLUOROETHYLENE	(G)	98.485						
C2HCL2F	1,1-DICHLORO-2-								
C2HCL2F	FLUOROETHYLENE	(G)	114.942			74.95	S(1033)		
C2HCL2F3	1,2-DICHLORO-								
C2HCL2F3	1,1,2-TRIFLUOROETHANE	(G)	152.942	-167.3	B(832)				
C2HCL2F3	2,2-DICHLORO-								
C2HCL2F3	1,1,1-TRIFLUOROETHANE	(G)	152.942			84.10	S(891)		
C2HCL3	TRICHLOROETHYLENE	(G)	131.399	-9.6	B(1384)	77.63	T(750)	4.75	-3.482
C2HCL3	TRICHLOROETHYLENE	(L)	131.399	-1.40	T(750)				
C2HCL3O	DICHLOROACETYL CHLORIDE	(G)	147.399	-68.2	B(1187)				
C2HCL3O	TRICHLOROACETALDEHYDE	(L)	147.399	-56.4	C(1189)				
C2HCL3O2	TRICHLOROACETIC ACID	(S)	163.399	-121.7	B(1384)				
C2HCL5	PENTACHLOROETHANE	(L)	202.313	-45.3	B(1384)				
C2HCL5	PENTACHLOROETHANE	(G)	202.313	-34.00	T(731)	90.95	T(731)	-15.93	11.678
C2HF3	TRIFLUOROETHYLENE	(G)	82.028	-118.50	T(713)	69.94	T(713)	-112.22	82.257
C2HN5	TETRAZOLE-5-CARBONITRILE	(S)	95.068	96.1	B(1616)				
C2H2	ACETYLENE	(G)	26.036	54.19	T(314)	48.00	T(314)	50.00	-36.646
C2H2BRF3	2-BROMO-1,1,1-TRIFLUOROETHANE	(G)	162.952			80.22	S(1075)		
C2H2BR2	1,2-DIBROMOETHYLENE, CIS	(G)	185.868			74.38	S(346)		
C2H2BR2	1,2-DIBROMOETHYLENE, TRANS	(G)	185.868			74.90	S(346)		
C2H2CLF	1-CHLORO-1-FLUOROETHYLENE	(G)	80.493			67.71	S(921)		
C2H2CLF3	2-CHLORO-1,1,1-TRIFLUOROETHANE	(G)	118.493			78.05	S(1073)		
C2H2CL2	1,1-DICHLOROETHYLENE	(L)	96.950	-6.00	A(1359)	48.17	M(599)	5.64	-4.136
C2H2CL2	1,1-DICHLOROETHYLENE	(G)	96.950	0.30	T(747)	68.85	T(747)	5.78	-4.235
C2H2CL2	1,2-DICHLOROETHYLENE, CIS	(L)	96.950	-6.7	B(1384)				
C2H2CL2	1,2-DICHLOROETHYLENE, CIS	(G)	96.950	0.45	T(748)	69.20	T(748)	5.82	-4.268
C2H2CL2	1,2-DICHLOROETHYLENE, TRANS	(L)	96.950	-6.1	B(1384)				
C2H2CL2	1,2-DICHLOROETHYLENE, TRANS	(G)	96.950	1.00	T(749)	69.29	T(749)	6.35	-4.652
C2H2CL2O	CHLOROACETYL CHLORIDE	(L)	112.950	-68.5	B(1187)				
C2H2CL2O2	DICHLOROACETIC ACID	(L)	128.950	-119.7	B(1384)				
C2H2CL3NO	TRICHLOROACETAMIDE	(S)	162.415	-86.6	B(1384)				

Formula	Name	State	Mol Wt	$\Delta H^{\circ}_{f\,298}$ (kcal/mole)	Source	S°_{298} (cal/mole °K)	Source	$\Delta G^{\circ}_{f\,298}$ (kcal/mole)	Log K_p
C2H2CL4	1,1,2,2-TETRACHLOROETHANE	(G)	167.864	-36.50	T(730)	86.69	T(730)	-20.45	14.992
C2H2CL4	1,1,1,2-TETRACHLOROETHANE	(G)	167.864			85.07	S(1071)		
C2H2F2	1,1-DIFLUOROETHYLENE	(G)	64.036	-82.50	T(712)	63.38	T(712)	-76.84	56.319
C2H2F2O2	DIFLUOROACETIC ACID	(L)	96.036	-207.5	C(1249)				
C2H2I2	DIIODOETHYLENE, CIS	(G)	279.856			79.81	S(1205)		
C2H2I2	DIIODOETHYLENE, TRANS	(G)	279.856			80.11	S(1205)		
C2H2O	KETENE	(G)	42.036	-14.60	T(557)	57.79	T(557)	-14.41	10.561
C2H2O2	GLYOXAL	(L)	58.036	-83.7	C(1249)				
C2H2O4	OXALIC ACID	(S)	90.036	-198.36	A(1612)	28.70	M(1108)	-167.58	122.833
C2H3BR	BROMOETHYLENE	(G)	106.960	-18.73	T(772)	65.83	T(772)	19.30	-14.144
C2H3BRO	ACETYL BROMIDE	(L)	122.960	-53.2	A(198)				
C2H3BR3O2	TRIBROMOACETALDEHYDE MONOHYDRATE	(S)	298.792						
C2H3CL	CHLOROETHYLENE	(G)	62.501	8.40	T(746)	63.08	T(746)	12.31	-9.021
C2H3CLF2	1-CHLORO-1,1-DIFLUOROETHYLENE	(G)	100.501			73.61	S(979)		
C2H3CLO	ACETYL CHLORIDE	(L)	78.501	-65.5	A(1187)				
C2H3CLO	ACETYL CHLORIDE	(G)	78.501	-58.30	T(758)	70.47	T(758)	-49.29	36.130
C2H3CLO	CHLOROACETALDEHYDE	(L)	78.501	-61.3	C(1249)				
C2H3CLO2	CHLOROACETIC ACID	(S)	94.501	-122.6	B(1384)				
C2H3CL2F	1,1-DICHLORO-1-FLUOROETHANE	(S)	116.958			76.49	S(1376)		
C2H3CL3	1,1,1-TRICHLOROETHANE	(L)	133.415			54.37	M(1261)		
C2H3CL3	1,1,1-TRICHLOROETHANE	(L)	133.415			77.09	S(776)		
C2H3CL3	1,1,2-TRICHLOROETHANE	(L)	133.415	-42.7	B(1354)				
C2H3CL3	1,1,2-TRICHLOROETHANE	(G)	133.415	-33.10	T(729)	80.57	T(729)	-18.52	13.575
C2H3CL3O2	TRICHLOROACETALDEHYDE MONOHYDRATE	(S)	165.415	-132.2	C(1249)				
C2H3FO	ACETYL FLUORIDE	(L)	62.044	-111.4	A(1188)				
C2H3FO	ACETYL FLUORIDE	(G)	62.044	-105.2	A(1188)				
C2H3FO2	FLUOROACETIC ACID	(S)	78.044	-162.3	C(1249)				
C2H3F2NO	DIFLUOROACETAMIDE	(S)	95.052	-167.8	C(1249)				
C2H3F3	1,1,1-TRIFLUOROETHANE	(G)	84.044	-178.20	T(707)	68.66	T(707)	-162.23	118.915
C2H3F3	1,1,2-TRIFLUOROETHANE	(G)	84.044	-158.9	A(833)				
C2H3F3O	2,2,2-TRIFLUOROETHANOL	(L)	100.044	-207.6	A(798)				
C2H3I	IODOETHYLENE	(G)	153.954			68.04	S(1224)		
C2H3IO	ACETYL IODIDE	(L)	169.954	-39.0	A(198)				
C2H3N	ACETONITRILE	(L)	41.052	12.7	C(1249)	35.76	M(1209)	23.63	-17.321

641

Formula	Name	State	Mol Wt	$\Delta H f^\circ_{298}$ (kcal/mole)	Source	S°_{298} (cal/mole °K)	Source	$\Delta G f^\circ_{298}$ (kcal/mole)	Log Kp
C2H3N	ACETONITRILE	(G)	41.052	21.00	T(617)	58.19	T(617)	25.24	-18.503
C2H3N	METHYL ISOCYANIDE	(L)	41.052	28.1	C(1249)				
C2H3N	METHYL ISOCYANIDE	(G)	41.052	35.9	C(1249)	58.98	S(1148)	39.91	-29.252
C2H3NO	GLYCOLONITRILE	(L)	57.052	-34.1	C(1249)				
C2H3NO	METHYL ISOCYANATE	(L)	57.052	-21.5	C(1249)				
C2H3NO3	OXAMIC ACID	(S)	89.052	-161.4	C(1249)				
C2H3NS	METHYL ISOTHIOCYANATE	(S)	73.118	18.7	A(1443)	69.29	S(909)		
C2H3NS	METHYL ISOTHIOCYANATE	(G)	73.118						
C2H3NS	METHYL THIOCYANATE	(L)	73.118	25.1	C(1249)				
C2H3N3O6	1,1,1-TRINITROETHANE	(L)	165.068	-27.53	A(1193)				
C2H3N5O2	3-NITRAMINO-1,2,4-TRIAZOLE	(S)	129.084	26.9	B(1616)				
C2H4	ETHYLENE	(G)	28.052	12.50	T(261)	52.45	T(261)	16.28	-11.936
C2H4BR2	1,2-DIBROMOETHANE	(L)	187.884	-19.3	A(1249)	53.37	M(1151)	-4.94	3.622
C2H4BR2	1,2-DIBROMOETHANE	(G)	187.884	-9.30	T(761)	78.81	T(761)	-2.53	1.852
C2H4CLNO	CHLOROACETAMIDE	(S)	93.517	-81.1	B(1384)				
C2H4CL2	1,1-DICHLOROETHANE	(L)	98.966	-38.42	T(727)	50.61	M(862)	-18.20	13.339
C2H4CL2	1,1-DICHLOROETHANE	(G)	98.966	-31.05	T(727)	72.89	T(727)	-17.47	12.806
C2H4CL2	1,2-DICHLOROETHANE	(L)	98.966	-39.44	T(728)	49.84	M(1151)	-18.99	13.919
C2H4CL2	1,2-DICHLOROETHANE	(G)	98.966	-31.00	T(728)	73.66	T(728)	-17.65	12.938
C2H4FNO	2-FLUOROACETAMIDE	(S)	77.060	-117.7	C(1249)				
C2H4F2	1,1-DIFLUOROETHANE	(G)	66.052	-118.00	T(706)	67.52	T(706)	-104.26	76.424
C2H4F2N2O2	2,2-DIFLUORO-N-NITRO-ETHYLAMINE	(L)	126.068	-118.9	C(1249)				
C2H4F2O	2,2-DIFLUOROETHANOL	(L)	82.052	-164.2	C(1249)				
C2H4I2	1,2-DIIODOETHANE	(S)	281.872	0.2	B(1)				
C2H4I2	1,2-DIIODOETHANE	(G)	281.872	15.90	T(779)	83.30	T(779)	18.76	-13.753
C2H4N2O	2-URETIDINONE	(S)	72.068	-77.7	B(1472)				
C2H4N2O2	FORMYLUREA	(L)	88.068	-117.5	C(1249)				
C2H4N2O2	GLYOXIME	(S)	88.068	-21.1	B(1013)				
C2H4N2O2	OXAMIDE	(S)	88.068	-121.2	C(1249)	28.23	M(374)	-81.94	60.059
C2H4N2O2	OXAMIDE	(G)	88.068			72.56	M(374)		
C2H4N2O4	1,1-DINITROETHANE	(L)	120.068	-34.67	A(1193)				
C2H4N2O4	1,2-DINITROETHANE	(S)	120.068	-43.9	B(984)				
C2H4N2O6	ETHYLENE DINITRATE	(L)	152.068	-58.02	C(1230)				
C2H4N4	3-AMINO-1,2,4-TRIAZOLE	(S)	84.084	18.4	B(1616)				
C2H4N4	CYANOGUANIDINE	(S)	84.084	7.14	A(1275)	30.90	M(1411)	44.64	-32.721

642

Formula	Name	State	Mol Wt	$\Delta H_{f\,298}^{\circ}$ (kcal/mole)	Source	S_{298}° (cal/mole °K)	Source	$\Delta G_{f\,298}^{\circ}$ (kcal/mole)	Log K_p
C2H4N4O	5-METHOXYTETRAZOLE	(S)	100.084	16.5	B(1616)				
C2H4N4O2	1,1'-AZOBISFORMAMIDE	(S)	116.084	-69.9	B(1616)				
C2H4N4O2	URAZINE	(S)	116.084	-63.0	B(1640)				
C2H4N10	5,5'-HYDRAZODITETRAZOLE	(S)	168.132	135.0	C(974)				
C2H4O	ACETALDEHYDE	(L)	44.052	-45.9	A(245)				
C2H4O	ACETALDEHYDE	(G)	44.052	-39.76	T(545)	63.15	T(545)	-31.86	23.353
C2H4O	ETHYLENE OXIDE	(L)	44.052	-18.5	T(513)	36.76	M(997)	-2.73	2.003
C2H4O	ETHYLENE OXIDE	(G)	44.052	-12.58	T(513)	57.94	T(513)	-3.13	2.292
C2H4OS	THIOACETIC ACID	(L)	76.118	-52.39	A(1447)				
C2H4OS	THIOACETIC ACID	(G)	76.118	-43.49	T(918)	74.86	T(918)	-36.81	26.979
C2H4O2	ACETIC ACID	(L)	60.052	-115.7	A(389)	38.20	M(1108)	-93.06	68.209
C2H4O2	ACETIC ACID	(G)	60.052	-103.93	T(560)	67.52	T(560)	-90.03	65.989
C2H4O2	METHYL FORMATE	(L)	60.052	-90.4	C(1249)				
C2H4O2	METHYL FORMATE	(G)	60.052	-83.6	T(561)	72.0	T(561)	-71.03	52.067
C2H4O3	GLYCOLIC ACID	(S)	76.052	-158.1	C(1249)				
C2H4S	THIACYCLOPROPANE	(L)	60.118	12.41	A(1447)	38.84	E(552)	22.53	-16.512
C2H4S	THIACYCLOPROPANE	(G)	60.118	19.65	T(882)	61.01	T(882)	23.16	-16.974
C2H5BR	BROMOETHANE	(L)	108.976	-21.9	T(760)				
C2H5BR	BROMOETHANE	(G)	108.976	-15.30	T(760)	68.71	T(760)	-6.29	4.608
C2H5CL	CHLOROETHANE	(L)	64.517	-32.5	T(726)	45.60	M(518)	-14.08	10.318
C2H5CL	CHLOROETHANE	(G)	64.517	-26.70	T(726)	65.93	T(726)	-14.34	10.509
C2H5CLO	2-CHLOROETHANOL	(L)	80.517	-70.3	C(1249)				
C2H5F	FLUOROETHANE	(G)	48.060	-62.50	T(705)	63.32	T(705)	-50.08	36.709
C2H5FO	2-FLUOROETHANOL	(L)	64.060	-111.1	C(1249)				
C2H5F2N	2,2-DIFLUOROETHYLAMINE	(L)	81.068	-118.3	C(1249)				
C2H5I	IODOETHANE	(L)	155.970	-9.7	T(778)				
C2H5I	IODOETHANE	(G)	155.970	-2.00	T(778)	70.82	T(778)	5.10	-3.737
C2H5N	ETHYLENIMINE	(L)	43.068	21.96	A(1052)				
C2H5N	ETHYLENIMINE	(G)	43.068	29.50	T(611)	59.90	T(611)	42.54	-31.180
C2H5NO	ACETALDEHYDE OXIME	(S)	59.068	-18.2	C(1249)				
C2H5NO	ACETAMIDE	(L)	59.068	-75.9	B(188)				
C2H5NO2	ETHYL NITRITE	(G)	75.068	-24.8	C(1249)				
C2H5NO2	GLYCINE	(S)	75.068	-128.4	A(666)	24.74	M(665)	-90.27	66.165
C2H5NO2	GLYCINE	(S)	75.068	-126.33	A(653)	24.74	M(665)	-88.20	64.647
C2H5NO2	NITROETHANE	(L)	75.068	-34.33	A(1193)				
C2H5NO2	NITROETHANE	(G)	75.068	-24.20	T(624)	75.39	T(624)	-1.17	0.857

Formula	Name	State	Mol Wt	$\Delta H f^\circ_{298}$ (kcal/mole)	Source	S°_{298} (cal/mole·°K)	Source	$\Delta G f^\circ_{298}$ (kcal/mole)	Log K_p
C2H5NO3	ETHYL NITRATE	(L)	91.068	-45.5	A(405)	59.08	M(529)	-10.30	7.551
C2H5NO3	ETHYL NITRATE	(G)	91.068	-36.80	T(631)	83.25	T(631)	-8.81	6.456
C2H5NO3	2-NITROETHANOL	(L)	91.068	-83.6	B(983)				
C2H5NO3	2-AZIDOETHANOL	(L)	87.084	22.47	A(401)				
C2H5N3O6	1,1,1-TRINITROPROPANE	(L)	179.094	-28.06	A(1193)				
C2H5N5	5-AMINO-1H-METHYLTETRAZOLE	(S)	99.100	46.2	B(1616)				
C2H5N5	5-AMINO-2H-METHYLTETRAZOLE	(S)	99.100	50.4	B(1616)				
C2H5N5	AMINO-5-METHYLTETRAZOLE	(S)	99.100	48.4	B(1616)				
C2H5N5O3	3-AMINO-1,2,4-TRIAZOLE NITRATE	(S)	147.100	-40.9	B(1616)				
C2H5N5O3	1-(NITROAMIDINO)SEMICARBAZIDE	(S)	147.100	-35.1	B(1616)				
C2H5N7	(TETRAZOL-5-YL)GUANIDINE	(S)	127.116	40.6	B(974)				
C2H6	ETHANE	(G)	30.068	-20.24	T(102)	54.85	T(102)	-7.87	5.765
C2H6N2O	N-NITROSODIMETHYLAMINE	(L)	74.084	0.7	C(1249)				
C2H6N2O2	N-NITROETHYLAMINE	(L)	90.084	-20.6	B(1039)				
C2H6N2O2	N-NITRODIMETHYLAMINE	(L)	90.084	-19.9	B(1039)				
C2H6N2O2	N-NITRODIMETHYLAMINE	(S)	90.084	-17.0	B(1621)				
C2H6N2O5	GLYCINE NITRATE	(S)	138.084	-173.6	C(277)				
C2H6N4O2	BIUREA	(S)	118.100	-119.2	B(1616)				
C2H6N4O2	OXALYL DIHYDRAZIDE	(S)	118.100	-70.5	B(747)				
C2H6N4O4	N,N'-DINITROETHYLENEDIAMINE	(S)	150.100	-24.7	B(985)				
C2H6N6O3	1-(NITROAMIDINO)SEMICARBAZIDE	(S)	162.116	-80.1	B(204)				
C2H6O	ETHYL ALCOHOL	(L)	46.068	-66.20	A(215)	38.40	M(733)	-41.62	30.504
C2H6O	ETHYL ALCOHOL	(G)	46.068	-56.12	T(518)	67.54	T(518)	-40.22	29.483
C2H6O	METHYL ETHER	(G)	46.068	-43.99	T(501)	63.83	T(501)	-26.99	19.781
C2H6OS	METHYL SULFOXIDE	(L)	78.134	-47.2	B(908)				
C2H6OS	METHYL SULFOXIDE	(G)	78.134	-34.6	B(908)				
C2H6O2	ETHYLENE GLYCOL	(L)	62.068	-108.73	A(1121)	39.9	M(1108)	-77.29	56.651
C2H6O2	ETHYLENE GLYCOL	(G)	62.068	-93.05	T(542)	77.33	T(542)	-72.77	53.337
C2H6O2	ETHYL HYDROPEROXIDE	(L)	62.068	-58.0	C(1406)				
C2H6O2	METHYL PEROXIDE	(G)	62.068	-30.0	B(64)				
C2H6O2S	METHYL SULFONE	(L)	94.134	-107.5	B(908)	34.77	M(1595)	-72.25	52.960
C2H6O2S	METHYL SULFONE	(G)	94.134	-89.1	B(908)				
C2H6O4	BIS(HYDROXYMETHYL) PEROXIDE	(S)	94.068	-163.7	B(685)				
C2H6S	ETHANETHIOL	(L)	62.134	-17.60	A(958)	49.48	A(958)	-1.35	0.989
C2H6S	ETHANETHIOL	(G)	62.134	-11.02	T(891)	70.77	T(891)	-1.12	0.818
C2H6S	METHYL SULFIDE	(L)	62.134	-15.63	A(958)	46.94	M(1092)	1.38	-1.010

Formula	Name	State	Mol Wt	$\Delta H f^\circ_{298}$ (kcal/mole)	Source	S°_{298} (cal/mole °K)	Source	$\Delta G f^\circ_{298}$ (kcal/mole)	Log $K p$
C2H6S	METHYL SULFIDE	(G)	62.134	−8.97	T(800)	68.32	T(800)	1.66	−1.220
C2H6S2	1,2-ETHANEDITHIOL	(L)	94.200	−12.83	A(925)				
C2H6S2	1,2-ETHANEDITHIOL	(G)	94.200	−2.15	A(925)				
C2H6S2	METHYL DISULFIDE	(L)	94.200	−14.96	T(627)	56.26	M(1302)	1.54	−1.132
C2H6S2	METHYL DISULFIDE	(G)	94.200	−5.77	T(872)	80.46	T(872)	3.52	−2.580
C2H7CLO	METHYL ETHER HYDROCHLORIDE .	(G)	82.533	−71.1	C(1249)	89.3	C(1249)	−49.09	35.986
C2H7N	DIMETHYLAMINE	(L)	45.084	−10.50	A(678)				
C2H7N	DIMETHYLAMINE	(G)	45.084	−4.50	T(607)	65.24	T(607)	16.25	−11.913
C2H7N	ETHYLAMINE	(L)	45.084	−17.71	A(678)				
C2H7N	ETHYLAMINE	(G)	45.084	−11.00	T(602)	68.08	T(602)	8.91	−6.528
C2H7NO	2-AMINOETHANOL	(L)	61.084	−65.6	B(1354)				
C2H7NO3S	2-AMINOETHANESULFONIC ACID .	(S)	125.150	−187.8	C(1249)	36.8	M(652)	−134.38	98.497
C2H7N9O2	GUANIDINE SALT OF 5-NITRAMINOTETRAZOLE	(S)	189.148	26.6	B(974)				
C2H8N2	1,1-DIMETHYLHYDRAZINE	(L)	60.100	11.9	A(336)	47.86	MC	49.31	−36.143
C2H8N2	1,1-DIMETHYLHYDRAZINE	(G)	60.100	20.3	(CALC)	72.82	MC	50.27	−36.846
C2H8N2	1,2-DIMETHYLHYDRAZINE	(L)	60.100	12.1	A(336)	47.60	MC	49.59	−36.347
C2H8N2	1,2-DIMETHYLHYDRAZINE	(G)	60.100	21.5	A(41)	74.39	MC	51.00	−37.382
C2H8N2	ETHYLENEDIAMINE	(L)	60.100	−6.36	C(1249)				
C2H8N2O3	DIMETHYLAMINE NITRATE	(S)	108.100	−79.0	C(277)				
C2H8N2O3	ETHYLAMINE NITRATE	(S)	108.100	−86.9	C(277)				
C2H8N2O4	2-AMINOETHANOL NITRATE . . .	(S)	124.100	−137.6	C(277)				
C2H8N2O4	AMMONIUM OXALATE	(S)	124.100	−268.7	C(1249)				
C2H8N10O	1-AMINO-4-(5-TETRAZOLYL)-TETRAZENE HYDRATE .	(S)	188.164	45.2	B(1616)				
C2H8N10O	ETHYLENEDIAMINE MONOHYDRATE	(S)	78.116	−77.1	C(1249)				
C2H10N2O	ETHYLENEDIAMINE DINITRATE .	(S)	186.132	−155.8	B(981)				
C2H10N4O6	DIAMMONIUM SALT OF N,N-DINITROETHYLENEDIAMINE	(S)	184.164	−80.1	B(986)				
C2H12N6O4	DIIODOACETYLENE	(G)	277.840	73.0	C(1249)				
C2I2	TETRAIODOETHYLENE	(S)	531.660	73.84	T(21)	57.90	T(21)	71.03	−52.067
C2I4	CYANOGEN	(G)	52.036						
C2N2	SULFUR DICYANIDE	(G)	84.102			71.81	S(1144)		
C2N2S	HEXANITROETHANE	(S)	300.068	28.6	B(1079)				
C3CL2F6	1,2-DICHLORO-								

Formula	Name	State	Mol Wt	$\Delta H f°_{298}$ (kcal/mole)	Source	$S°_{298}$ (cal/mole °K)	Source	$\Delta G f°_{298}$ (kcal/mole)	Log Kp
C3CL2F6	1,1,2,3,3,3-HEXAFLUOROPROPANE	(G)	220.944	-313.3	C(832)				
C3CL3N3	CYANURIC CHLORIDE	(S)	184.425	21.9	A(663)				
C3F6	HEXAFLUOROPROPENE	(G)	150.030	-266.2	C(360)				
C3F8	OCTAFLUOROPROPANE	(G)	188.030	-417.9	E(833)				
C3H2N2	MALONONITRILE	(S)	66.062	44.6	C(744)				
C3H2N2	MALONONITRILE	(G)	66.062	59.7	(CALC)	69.05	S(567)	63.28	-46.384
C3H2N2O3	IMIDAZOLIDINETRIONE	(S)	114.062	-138.0	C(744)				
C3H2O	PROPYNAL	(G)	54.046			65.66	S(1211)		
C3H3BR	1-BROMOPROPYNE	(G)	118.970			70.61	S(1649)		
C3H3BR	3-BROMOPROPYNE	(G)	118.970			72.24	S(396)		
C3H3CL	1-CHLOROPROPYNE	(G)	74.511			67.89	S(1649)		
C3H3CL	3-CHLOROPROPYNE	(G)	74.511			69.46	S(396)		
C3H3CL3O2	METHYL TRICHLOROACETATE	(L)	177.425	-111.8	B(1384)				
C3H3I	1-IODOPROPYNE	(G)	165.964			72.42	S(1649)		
C3H3N	ACRYLONITRILE	(L)	53.062	35.9	A(314)				
C3H3N	ACRYLONITRILE	(G)	53.062	44.20	T(618)	65.47	T(618)	46.68	-34.215
C3H3NO2	CYANOACETIC ACID	(S)	85.062	-85.6	C(1549)				
C3H3N3	S-TRIAZINE	(S)	81.078	40.2	e(1549)				
C3H3N3O3	CYANURIC ACID	(S)	129.078	-165.1	e(1549)				
C3H3N5O4	4-METHYL-3,5-DINITROTRIAZOLE	(S)	173.094	29.35	A(1193)				
C3H4	ALLENE	(G)	40.062	45.92	T(304)	58.30	T(304)	48.37	-35.452
C3H4	CYCLOPROPENE	(G)	40.062	66.0	A(1608)	58.38	S(1607)	68.42	-50.153
C3H4	PROPYNE	(G)	40.062	44.32	T(315)	59.30	T(315)	46.47	-34.060
C3H4CL2	DICHLOROCYCLOPROPANE	(U,L)	110.976	-71.0	C(744)				
C3H4CL2O2	METHYL DICHLOROACETATE	(L)	142.976	-116.1	B(1384)				
C3H4N2	IMIDAZOLE	(S)	68.078	14.53	A(96)				
C3H4N2	IMIDAZOLE	(G)	68.078	30.5	B(96)				
C3H4N2	PYRAZOLE	(S)	68.078	28.32	A(96)				
C3H4N2	PYRAZOLE	(G)	68.078	43.3	B(96)				
C3H4N2O	CYANOACETAMIDE	(S)	84.078	-42.3	C(744)				
C3H4N2OS	THIOHYDANTOIN	(S)	116.144	-59.6	C(744)				
C3H4N2OS	2-THIOHYDANTOIN	(S)	116.144	-224.95	C(744)				
C3H4N2O2	HYDANTOIN	(S)	100.078	-106.9	C(744)				
C3H4N2O4	OXALURIC ACID	(S)	132.078	-211.2	C(744)				
C3H4N4O	N-ACETYLTETRAZOLE	(S)	112.094	-19.6	B(1557)				
C3H4O	ACROLEIN	(L)	56.062	-29.17	C(744)				

Formula	Name	State	Mol Wt	$\Delta H_{f\,298}^{\circ}$ (kcal/mole)	Source	S_{298}° (cal/mole °K)	Source	$\Delta G_{f\,298}^{\circ}$ (kcal/mole)	Log K_p
C3H4O	ACROLEIN	(L)	56.062	−25.09	CC 744				
C3H4O	2-PROPYN-1-OL	(L)	56.062	−5.2	CC 843				
C3H4O	2-PROPYN-1-OL	(G)	56.062	10.1	CC 744				
C3H4O2	ACRYLIC ACID	(L)	72.062	−91.8	CC 556				
C3H4O2	ACRYLIC ACID	(G)	72.062	−80.36	TC 564	75.29	TC 564	−68.37	50.113
C3H4O2	METHYLGLYOXAL	(L)	72.062	−73.9	B(1031)				
C3H4O2	BETA-PROPIOLACTONE	(L)	72.062	−78.85	AC 146				
C3H4O2	BETA-PROPIOLACTONE	(G)	72.062	−67.61	AC 146				
C3H4O3	ETHYLENE CARBONATE	(S)	88.062	−138.8	A(1349)				
C3H4O3	HYDROXYMALONALDEHYDE	(S)	88.062	−120.7	BC 385				
C3H4O3	PYRUVIC ACID	(L)	88.062	−139.7	A(1612)	42.9	F(1634)	−110.75	81.176
C3H4O4	MALONIC ACID	(S)	104.062	−212.97	A(1612)				
C3H4O4	TARTRONIC ACID	(S)	120.062	−253.4	CC 744				
C3H4O6	DIHYDROXYMALONIC ACID	(S)	136.062	−290.6	CC 744				
C3H5BR	3-BROMOPROPENE	(G)	120.986	11.80	TC 773	75.80	TC 773	19.11	−14.004
C3H5CL	3-CHLOROPROPENE	(G)	76.527	−0.15	TC 752	73.29	TC 752	10.42	−7.641
C3H5CLO	1-CHLORO-2,3-EPOXYPROPANE	(L)	92.527	−35.6	AC 141				
C3H5CLO	2-CHLORO-2-PROPEN-1-OL	(S)	92.527	−88.3	CC 843				
C3H5CLO2	2-CHLOROPROPIONIC ACID	(L)	108.527	−125.4	B(1384)				
C3H5CLO2	3-CHLOROPROPIONIC ACID	(L)	108.527	−131.5	B(1384)				
C3H5CLO2	METHYL CHLOROACETATE	(S)	108.527	−111.4	B(1384)				
C3H5CL3	1,2,3-TRICHLOROPROPANE	(L)	147.441	−55.65	AC 141				
C3H5CL3	1,2,3-TRICHLOROPROPANE	(G)	147.441	−44.40	TC 738	91.52	TC 738	−23.37	17.132
C3H5I	3-IODO-1-PROPENE	(L)	167.980	14.0	BC 466				
C3H5I	3-IODO-1-PROPENE	(G)	167.980	22.90	TC 785	76.46	TC 785	28.72	−21.053
C3H5IO2	2-IODOPROPIONIC ACID	(L)	199.980	−109.8	A(1251)				
C3H5N	ETHYL ISOCYANIDE	(L)	55.078	28.0	CC 744				
C3H5N	PROPIONITRILE	(L)	55.078	3.5	CC 744	45.25	M(1583)	21.31	−15.622
C3H5N	PROPIONITRILE	(G)	55.078	12.10	TC 619	68.50	TC 619	22.98	−16.844
C3H5NO	ETHYL ISOCYANATE	(L)	71.078	−28.2	CC 744				
C3H5NO	LACTONITRILE	(L)	71.078	−31.6	CC 744				
C3H5NO2	PYRUVALDEHYDE OXIME	(S)	87.078	−50.0	C(1056)				
C3H5NO3	METHYL OXAMATE	(S)	103.078	−147.4	CC 744				
C3H5NO4	METHYL NITROACETATE	(L)	119.078	−110.9	B(1640)				
C3H5NS	ETHYL ISOTHIOCYANATE	(L)	87.144	17.0	CC 744				
C3H5NS	ETHYL THIOCYANATE	(L)	87.144						

647

Formula	Name	State	Mol Wt	$\Delta Hf°_{298}$ (kcal/mole)	Source	$S°_{298}$ (cal/mole °K)	Source	$\Delta Gf°_{298}$ (kcal/mole)	Log Kp
C3H5N3O9	NITROGLYCERINE	(L)	227.094	-89.0	B(1474)				
C3H5N5O	5-ACETAMIDOTETRAZOLE	(S)	127.110	-1.8	BC 974)				
C3H5N5O2	3-NITRAMINO-5-METHYL- 1,2,4-TRIAZOLE								
C3H6	CYCLOPROPANE	(G)	42.078	12.74	TC 338)	56.75	TC 338)	24.95	-18.291
C3H6	PROPENE	(G)	42.078	4.88	TC 262)	63.80	TC 262)	14.99	-10.989
C3H6BR2	1,2-DIBROMOPROPANE	(L)	201.910	-17.40	TC 764)	89.90	TC 764)	-4.22	3.094
C3H6CL2	1,2-DICHLOROPROPANE	(L)	112.992	-48.3	TC 735)				
C3H6CL2	1,2-DICHLOROPROPANE	(G)	112.992	-39.60	TC 735)	84.80	TC 735)	-19.86	14.558
C3H6CL2	1,3-DICHLOROPROPANE	(L)	112.992	-48.3	BC(1384)				
C3H6CL2	1,3-DICHLOROPROPANE	(G)	112.992	-38.60	TC 736)	87.76	TC 736)	-19.74	14.472
C3H6CL2	2,2-DICHLOROPROPANE	(L)	112.992	-49.65	BC(1384)				
C3H6CL2	2,2-DICHLOROPROPANE	(G)	112.992	-42.00	TC 737)	77.92	TC 737)	-20.21	14.813
C3H6CL2O	1,3-DICHLORO-2-PROPANOL	(L)	128.992	-92.4	AC 141)				
C3H6CL2O	2,3-DICHLORO-1-PROPANOL	(L)	128.992	-91.4	AC 141)				
C3H6I2	1,2-DIIODOPROPANE	(G)	295.898	8.60	TC 782)	94.60	TC 782)	17.81	-13.051
C3H6N2O2	ACETYLUREA	(S)	102.094	-126.0	CC 744)				
C3H6N2O2	MALONAMIDE	(S)	102.094	-130.0	B(1471)				
C3H6N2O2	MALONIMIDE	(S)	102.094	-128.0	CC 744)				
C3H6N2O2	METHYLGLYOXIME	(S)	102.094	-29.6	B(1013)				
C3H6N2O2S	N-CARBAMOYLTHIOGLYCINE	(S)	134.160	-132.5	CC 744)				
C3H6N2O3	HYDANTOIC ACID	(S)	118.094	-178.0	CC 744)				
C3H6N2O4	1,1-DINITROPROPANE	(L)	134.094	-40.04	AC(1193)				
C3H6N2O4	1,1-DINITROPROPANE	(L)	134.094	-40.8	BC 613)				
C3H6N2O4	1,3-DINITROPROPANE	(L)	134.094	-53.5	BC 613)				
C3H6N2O4	2,2-DINITROPROPANE	(L)	134.094	-45.71	AC(1193)				
C3H6N2O4	2,2-DINITROPROPANE	(S)	134.094	-44.9	BC 613)				
C3H6N4	1,5-DIMETHYLTETRAZOLE	(S)	98.110	45.1	BC 974)				
C3H6N4O	1,4-DIMETHYL- 2-TETRAZOLIN-5-ONE	(S)	114.110	6.7	B(1616)				
C3H6N4O	HEXAHYDRO-1,3,5-TRINITROSO- S-TRIAZINE	(S)	126.126	-15.35	AC(1275)	35.63	M(1411)	44.10	-32.323
C3H6N6O3	HEXAHYDRO-1,3,5-TRINITROSO- S-TRIAZINE	(S)	174.126	69.1	BC 320)				
C3H6N6O3	HEXAHYDRO-1,3,5-TRINITROSO-	(S)	174.126	67.8	BC 981)				

Formula	Name	State	Mol Wt	$\Delta H f^\circ_{298}$ (kcal/mole)	Source	S°_{298} (cal/mole °K)	Source	$\Delta G f^\circ_{298}$ (kcal/mole)	Log Kp
C3H6N6O3	HEXAHYDRO-1,3,5-TRINITRO-S-TRIAZINE	(S)	174.126	66.0	B(1640)				
C3H6N6O6	HEXAHYDRO-1,3,5-TRINITRO-S-TRIAZINE (RDX)	(S)	222.126	13.6	B(1640)				
C3H6N6O6	HEXAHYDRO-1,3,5-TRINITRO-S-TRIAZINE (RDX)	(S)	222.126	16.8	B(184)				
C3H6N6O6	HEXAHYDRO-1,3,5-TRINITRO-S-TRIAZINE (RDX)	(S)	222.126	16.9	B(320)				
C3H6O	ACETONE	(S)	222.126	14.71	A(1193)				
C3H6O	ACETONE	(L)	58.078	-59.3	T(554)	47.9	M(734)	-37.14	27.225
C3H6O	ALLYL ALCOHOL	(G)	58.078	-52.00	T(554)	70.49	T(554)	-36.58	26.811
C3H6O	ALLYL ALCOHOL	(L)	58.078	-44.6	C(744)				
C3H6O	PROPIONALDEHYDE	(G)	58.078	-31.55	T(541)	73.51	T(541)	-17.03	12.481
C3H6O	PROPIONALDEHYDE	(L)	58.078	-52.95	A(1508)				
C3H6O	PROPYLENE OXIDE	(G)	58.078	-45.90	T(546)	72.83	T(546)	-31.18	22.851
C3H6O	PROPYLENE OXIDE	(L)	58.078	-28.84	A(1356)	46.91	M(1083)	-6.39	4.682
C3H6O	PROPYLENE OXIDE	(G)	58.078	-22.17	T(514)	68.53	T(514)	-6.16	4.517
C3H6O	TRIMETHYLENE OXIDE	(G)	58.078	-19.25	A(1133)	65.46	S(1652)	-2.33	1.706
C3H6O2	1,3-DIOXOLANE	(L)	74.078	-80.7	A(431)				
C3H6O2	ETHYL FORMATE	(L)	74.078	-95.44	C(744)				
C3H6O2	ETHYL FORMATE	(G)	74.078	-88.74	C(744)				
C3H6O2	GLYCIDOL	(L)	74.078	-71.3	A(141)				
C3H6O2	LACTALDEHYDE	(S)	74.078	-105.0	C(249)				
C3H6O2	METHYL ACETATE	(L)	74.078	-105.5	C(262)				
C3H6O2	METHYL ACETATE	(G)	74.078	-97.9	B(940)				
C3H6O2	PROPIONIC ACID	(L)	74.078	-122.1	A(853)				
C3H6O2S	2-MERCAPTOPROPIONIC ACID	(L)	106.144	-111.9	B(646)	54.70	M(646)	-82.19	60.243
C3H6O3	DIMETHYL CARBONATE	(L)	90.078	-146.6	C(744)				
C3H6O3	GLYCERALDEHYDE, D,L	(S)	90.078	-148.9	C(1057)				
C3H6O3	LACTIC ACID, D	(S)	90.078	-165.89	A(1279)	34.3	M(649)		
C3H6O3	LACTIC ACID, L	(S)	90.078	-161.2	C(744)	34.00	M(649)	-124.98	91.607
C3H6O3	LACTIC ACID, D,L	(L)	90.078	-124.4	B(320)	45.9	M(1117)	-123.84	90.769
C3H6O3	TRIOXYMETHYLENE, S	(S)	90.078		A(1447)				
C3H6S	1,2-EPITHIOPROPANE	(L)	74.144	2.77	A(1447)				
C3H6S	1,2-EPITHIOPROPANE	(G)	74.144	11.02	A(1447)				
C3H6S	THIACYCLOBUTANE	(L)	74.144	6.04	T(883)	44.72	M(1304)	24.12	-17.677

Formula	Name	State	Mol Wt	$\Delta H f^\circ_{298}$ (kcal/mole)	Source	S°_{298} (cal/mole °K)	Source	$\Delta G f^\circ_{298}$ (kcal/mole)	Log Kp
C3H6S	THIACYCLOBUTANE	(G)	74.144	14.61	T(883)	68.17	T(883)	25.69	-18.834
C3H7BR	1-BROMOPROPANE	(L)	123.002	-28.8	T(762)				
C3H7BR	1-BROMOPROPANE	(G)	123.002	-21.00	T(762)	79.08	T(762)	-5.37	3.934
C3H7BR	2-BROMOPROPANE	(L)	123.002	-30.7	T(763)				
C3H7BR	2-BROMOPROPANE	(G)	123.002	-23.20	T(763)	75.53	T(763)	-6.51	4.771
C3H7CL	1-CHLOROPROPANE	(L)	78.543	-38.0	T(733)				
C3H7CL	1-CHLOROPROPANE	(G)	78.543	-31.10	T(733)	76.27	T(733)	-12.11	8.876
C3H7CL	2-CHLOROPROPANE	(L)	78.543	-41.5	T(734)				
C3H7CL	2-CHLOROPROPANE	(G)	78.543	-35.00	T(734)	72.70	T(734)	-14.94	10.954
C3H7CLO2	2-CHLORO-1,3-PROPANEDIOL	(L)	110.543	-123.8	A(141)				
C3H7CLO2	3-CHLORO-1,2-PROPANEDIOL	(L)	110.543	-125.6	A(141)				
C3H7F	1-FLUOROPROPANE	(G)	62.086	-67.20	T(709)	72.71	T(709)	-47.87	35.087
C3H7F	2-FLUOROPROPANE	(G)	62.086	-69.00	T(710)	69.82	T(710)	-48.81	35.775
C3H7I	1-IODOPROPANE	(L)	169.996	-15.9	T(780)				
C3H7I	1-IODOPROPANE	(G)	169.996	-7.30	T(780)	80.32	T(780)	6.68	-4.894
C3H7I	2-IODOPROPANE	(L)	169.996	-18.7	T(781)				
C3H7I	2-IODOPROPANE	(G)	169.996	-10.00	T(781)	77.55	T(781)	4.80	-3.521
C3H7N	ALLYLAMINE	(L)	57.094	3.93	C(744)				
C3H7N	ALLYLAMINE	(G)	57.094	6.88	C(744)				
C3H7NO	ACETONE OXIME	(S)	73.094	-30.80	C(744)				
C3H7NO	PROPIONAMIDE	(S)	73.094	-81.1	C(744)				
C3H7NO2	ALANINE, D	(S)	89.094	-134.03	A(650)	31.6	M(644)	-88.23	64.672
C3H7NO2	ALANINE, D,L	(S)	89.094	-134.72	A(653)	31.6	M(648)	-88.92	65.178
C3H7NO2	ALANINE, L	(S)	89.094	-134.5	A(666)	30.88	M(665)	-88.49	64.860
C3H7NO2	ETHYL CARBAMATE	(S)	89.094	-123.8	C(744)				
C3H7NO2	1-NITROPROPANE	(L)	89.094	-40.15	T(625)				
C3H7NO2	1-NITROPROPANE	(G)	89.094	-29.80	T(625)	85.00	T(625)	0.08	-0.056
C3H7NO2	2-NITROPROPANE	(L)	89.094	-43.3	T(626)				
C3H7NO2	2-NITROPROPANE	(G)	89.094	-33.50	T(626)	83.10	T(626)	-3.06	2.240
C3H7NO2	SARCOSINE	(S)	89.094	-121.3	B(164)				
C3H7NO2S	CYSTINE, L	(S)	121.160	-127.6	B(646)				
C3H7NO3	ISOPROPYL NITRATE	(L)	105.094	-54.92	A(405)	40.6	M(647)	-82.21	60.258
C3H7NO3	ISOPROPYL NITRATE	(G)	105.094	-45.65	T(633)	89.20	T(633)	-9.72	7.125
C3H7NO3	ISOSERINE	(S)	105.094	-177.0	C(744)				
C3H7NO3	PROPYL NITRATE	(L)	105.094	-51.27	A(405)				
C3H7NO3	PROPYL NITRATE	(G)	105.094	-41.60	T(632)	92.10	T(632)	-6.53	4.790

Formula	Name	State	Mol Wt	$\Delta H_f^\circ{}_{298}$ (kcal/mole)	Source	S°_{298} (cal/mole °K)	Source	$\Delta G_f^\circ{}_{298}$ (kcal/mole)	Log K_p
C3H7NO3	SERINE, L	(S)	105.094	-173.6	C(668)	35.65	M(668)	-121.70	89.207
C3H7N3S	ACETALDEHYDE THIOSEMICARBAZONE	(S)	117.176	15.0	B(1446)				
C3H7N5	5-(DIMETHYLAMINO)TETRAZOLE	(S)	113.126	43.7	B(1616)				
C3H7N5	1-METHYL-5-(METHYL-AMINO)TETRAZOLE	(S)	113.126	47.8	B(1616)				
C3H7N5O3	1-ACETAMIDO-2-NITROGUANIDINE	(S)	161.126	-46.3	B(1616)				
C3H7N5O3	3-AMINO-5-METHYL-1,2,4-TRIAZOLE NITRATE	(S)	161.126	-54.6	B(1616)				
C3H7N5O5	1-(2-NITROXYETHYL)-3-NITROGUANIDINE	(S)	193.126	-48.9	B(1040)				
C3H7N5O5	2-AMINO-1-NITROIMIDAZOLINE NITRATE	(S)	193.126	-59.3	B(1040)				
C3H8	PROPANE	(G)	44.094	-24.82	T(103)	64.51	T(103)	-5.61	4.115
C3H8N2O	ETHYLUREA	(S)	88.110	-83.2	C(744)				
C3H8N2O2	PROPYL NITRAMINE	(L)	104.110	-25.8	B(1039)				
C3H8N2O3	1,3-BIS(HYDROXYMETHYL)UREA	(S)	120.110	-170.78	A(1275)				
C3H8N4O2	MALONYL DIHYDRAZIDE	(S)	132.126	-79.4	B(747)				
C3H8N6O3	1,3-DIMETHYL-5-IMINOTETRAZOLE NITRATE	(S)	176.142	-1.3	B(1616)				
C3H8O	ETHYL METHYL ETHER	(G)	60.094	-51.73	T(502)	74.24	T(502)	-28.12	20.611
C3H8O	ISOPROPYL ALCOHOL	(L)	60.094	-76.0	T(520)	43.16	M(20)	-43.12	31.609
C3H8O	ISOPROPYL ALCOHOL	(G)	60.094	-65.15	T(520)	74.07	T(520)	-41.49	30.411
C3H8O	PROPYL ALCOHOL	(L)	60.094	-72.79	A(215)	46.1	M(159)	-40.79	29.898
C3H8O	PROPYL ALCOHOL	(G)	60.094	-61.55	T(519)	77.63	T(519)	-38.95	28.550
C3H8O2	DIMETHOXYMETHANE	(L)	76.094	-92.3	C(744)	58.32	M(973)	-56.64	41.515
C3H8O2	DIMETHOXYMETHANE	(G)	76.094			80.24	M(973)		
C3H8O2	2-METHOXYETHANOL	(L)	76.094	-114.8	C(1351)				
C3H8O2	1,2-PROPANEDIOL	(L)	76.094	-119.3	B(1032)				
C3H8O2	PROPYL HYDROPEROXIDE	(L)	76.094	-76.1	C(1406)				
C3H8O2S	ETHYL METHYL SULFONE	(L)	108.160	-116.2	B(186)				
C3H8O2S	ETHYL METHYL SULFONE	(G)	108.160	-97.9	B(186)				
C3H8O3	GLYCEROL	(L)	92.094	-159.8	A(1121)	48.87	M(4)	-114.02	83.572
C3H8S	ETHYL METHYL SULFIDE	(L)	76.160	-21.89	A(638)	57.14	M(1308)	1.79	-1.311
C3H8S	ETHYL METHYL SULFIDE	(G)	76.160	-14.25	T(801)	79.62	T(801)	2.73	-1.998
C3H8S	1-PROPANETHIOL	(L)	76.160	-23.87	A(638)	57.96	M(1136)	-0.44	-0.320
C3H8S	1-PROPANETHIOL	(G)	76.160	-16.22	T(892)	80.40	T(892)	0.52	-0.384

651

Formula	Name	State	Mol Wt	$\Delta H_f^\circ{}_{298}$ (kcal/mole)	Source	S°_{298} (cal/mole °K)	Source	$\Delta G_f^\circ{}_{298}$ (kcal/mole)	Log K_p
C3H8S	2-PROPANETHIOL	(L)	76.160	-25.30	A(638)	55.82	M(955)	-1.23	0.900
C3H8S	2-PROPANETHIOL	(G)	76.160	-18.22	T(893)	77.51	T(893)	-0.61	0.451
C3H8S2	1,3-PROPANEDITHIOL	(L)	108.226	-18.84	A(925)				
C3H8S2	1,3-PROPANEDITHIOL	(G)	108.226	-6.97	A(925)				
C3H9N	PROPYLAMINE	(L)	59.110	-24.8	T(603)				
C3H9N	PROPYLAMINE	(G)	59.110	-17.30	T(603)	77.48	T(603)	9.51	-6.974
C3H9N	TRIMETHYLAMINE	(L)	59.110	-10.94	A(678)				
C3H9N	TRIMETHYLAMINE	(G)	59.110	-5.70	T(609)	69.02	T(609)	23.64	-17.325
C3H9NO4	ACETIC ACID 2-AMIDINOHYDRAZIDE NITRATE	(S)	179.142	-118.0	B(1616)				
C3H9N5O4	TRINITROSOTRIMETHYLENETRIAMINE	(S)	177.150	-68.0	B(320)				
C3H9N6O3	TRINITROTRIMETHYLENETRIAMINE	(S)	225.150	-15.7	B(320)				
C3H10N2	TRIMETHYLHYDRAZINE	(L)	74.126			56.34	M(48)		
C3H10N2	TRIMETHYLHYDRAZINE	(G)	74.126			80.00	M(48)		
C3H10N2O3	TRIMETHYLAMINE NITRATE	(S)	122.126	-73.1	C(277)				
C3H12N6O3	GUANIDINE CARBONATE	(S)	180.174	-232.13	A(641)	70.59	M(649)	-133.27	97.688
C3H12N6O3	GUANIDINE CARBONATE	(S)	180.174	-230.8	B(985)	70.59	M(649)	-131.94	96.713
C3N2O	CARBONYL CYANIDE	(L)	80.046	49.5	B(493)				
C3O2	CARBON SUBOXIDE	(L)	68.030	-28.03	A(819)				
C3O2	CARBON SUBOXIDE	(G)	68.030	-22.38	T(22)	66.05	T(22)	-26.25	19.237
C4CL2F6	1,2-DICHLORO-1,2,3,3,4,4-HEXAFLUOROCYCLOBUTANE	(G)	232.954	-289.4	C(50)				
C4CL2F6	HEXAFLUOROCYCLOBUTANE	(G)	232.954	-289.4	C(50)				
C4F6	HEXAFLUORO-1,3-BUTADIENE	(G)	162.040	-240.0	C(50)				
C4F6	OCTAFLUOROCYCLOBUTANE	(G)	162.040	-365.20	T(711)	95.69	T(711)	-334.33	245.057
C4F8	DECAFLUOROBUTANE	(G)	200.040	-516.0	E(833)				
C4F10	DECAFLUOROISOBUTANE	(G)	238.040	-524.1	E(833)				
C4F10	2,3,4,5-TETRAIODOPYRROLE	(S)	570.696	92.81	C(744)				
C4HI4N	1,3-BUTADIYNE	(G)	50.056	113.00	T(316)	59.76	T(316)	106.11	-77.778
C4H2	4-THIAZOLECARBONITRILE	(S)	110.138	52.64	A(927)				
C4H2N2S	4-THIAZOLECARBONITRILE	(G)	110.138	70.31	A(927)				
C4H2N2S	MALEIC ANHYDRIDE	(S)	98.056	-112.43	A(1612)				
C4H2O3	ACETYLENEDICARBOXYLIC ACID	(S)	114.056	-138.1	C(432)				
C4H2O4	1-BUTEN-3-YNE	(G)	52.072	72.80	T(317)	66.77	T(317)	73.13	-53.601
C4H4	N-BROMOSUCCINIMIDE	(S)	177.996	-80.35	A(618)				
C4H4BRNO2									

Formula	Name	State	Mol Wt	$\Delta H_f^\circ{}_{298}$ (kcal/mole)	Source	$S^\circ{}_{298}$ (cal/mole °K)	Source	$\Delta G_f^\circ{}_{298}$ (kcal/mole)	Log K_p
C4H4CLNO2	N-CHLOROSUCCINIMIDE	(S)	133.537	-85.58	A(618)				
C4H4N2	PYRAZINE	(S)	80.088	33.4	A(1507)				
C4H4N2	PYRAZINE	(G)	80.088	46.86	A(1507)				
C4H4N2	PYRIDAZINE	(L)	80.088	53.7	A(1507)				
C4H4N2	PYRIDAZINE	(G)	80.088	66.53	A(1507)				
C4H4N2	PYRIMIDINE	(L)	80.088	35.0	A(1507)				
C4H4N2	PYRIMIDINE	(G)	80.088	46.99	A(1507)				
C4H4N2	SUCCINONITRILE	(S)	80.088	33.24	C(744)	45.79	M(1633)	53.47	-39.191
C4H4N2	SUCCINONITRILE	(G)	80.088	20.1	C(1628)	79.04	M(1633)	30.41	-22.293
C4H4N2O2	5-METHYLHYDANTION	(S)	112.088	-55.6	B(1471)				
C4H4N2O3	BARBITURIC ACID	(S)	128.088	-153.9	C(744)				
C4H4N2O4	N-NITROSUCCINIMIDE	(S)	144.088	-77.1	B(1640)				
C4H4N2O5	ALLOXAN MONOHYDRATE	(S)	160.088	-239.19	A(1421)	44.6	M(1422)	-182.08	133.463
C4H4N8O14	BIS(TRINITROETHYL)NITRAZINE	(S)	388.136	-32.7	B(1640)				
C4H4O	FURAN	(L)	68.072	-14.90	A(551)	42.22	M(551)	0.05	-0.038
C4H4O	FURAN	(G)	68.072	-8.29	T(515)	63.86	T(515)	0.21	-0.154
C4H4O2	3-BUTYNOIC ACID	(L)	84.072	-55.2	B(432)				
C4H4O2	METHYL PROPIOLATE	(L)	84.072	-33.9	B(432)				
C4H4O2	TETROLIC ACID	(S)	84.072	-60.1	C(744)				
C4H4O3	SUCCINIC ANHYDRIDE	(S)	100.072	-142.9	B(1538)				
C4H4O4	FUMARIC ACID, TRANS	(S)	116.072	-193.85	A(1612)	39.7	M(1104)	-156.23	114.515
C4H4O4	MALEIC ACID, CIS	(S)	116.072	-188.96	A(1612)	38.1	M(1104)	-150.86	110.581
C4H4S	THIOPHENE	(L)	84.138	19.37	A(1444)	43.30	M(1548)	28.97	-21.234
C4H4S	THIOPHENE	(G)	84.138	27.66	T(887)	66.65	T(887)	30.30	-22.208
C4H5CL3O2	ETHYL TRICHLOROACETATE	(L)	191.451	-115.8	B(1384)				
C4H5N	ALLYL ISOCYANIDE	(L)	67.088	62.6	C(744)				
C4H5N	3-BUTENENITRILE	(L)	67.088	28.4	C(744)				
C4H5N	CROTONONITRILE	(L)	67.088	25.3	C(744)				
C4H5N	CROTONONITRILE, TRANS	(G)	67.088	35.0	C(744)	71.31	S(1635)		
C4H5N	CYCLOPROPANE CARBONITRILE	(L)	67.088						
C4H5N	CYCLOPROPANE CARBONITRILE	(G)	67.088			79.78	S(304)		
C4H5N	ISOCROTONONITRILE	(L)	67.088	27.3	C(744)				
C4H5N	PYRROLE	(L)	67.088	21.2	C(744)				
C4H5NO	2-METHYLISOXAZOLE	(L)	83.088	-5.6	C(1467)				
C4H5NO	4-METHYLISOXAZOLE	(L)	83.088	-4.1	C(1467)				
C4H5NO2	METHYL CYANOACETATE	(L)	99.088	-75.1	C(744)				

Formula	Name	State	Mol Wt	$\Delta H_{f\,298}^{\circ}$ (kcal/mole)	Source	S_{298}° (cal/mole °K)	Source	$\Delta G_{f\,298}^{\circ}$ (kcal/mole)	Log K_p
C4H5NO2	SUCCINIMIDE	(S)	99.088	−109.72	A(247)				
C4H5NS	ALLYL ISOTHIOCYANATE	(L)	99.154	42.0	C(744)				
C4H5NS	ALLYL ISOTHIOCYANATE	(G)	99.154	55.0	C(744)				
C4H5NS	4-METHYLTHIAZOLE	(L)	99.154	16.32	A(927)				
C4H5NS	4-METHYLTHIAZOLE	(G)	99.154	26.80	A(927)				
C4H5N3	TMINODIACETONITRILE	(G)	95.104	44.0	C(744)				
C4H5N3O3	AMINOBARBITURIC ACID	(U,S)	143.104	−167.7	C(744)				
C4H6	1,2-BUTADIENE	(L)	54.088	33.2	(CALC)	50.10	M(44)	47.80	−35.038
C4H6	1,2-BUTADIENE	(G)	54.088	38.77	T(305)	70.03	T(305)	47.43	−34.766
C4H6	1,3-BUTADIENE	(L)	54.088	20.4	(CALC)	47.58	M(1327)	35.75	−26.207
C4H6	1,3-BUTADIENE	(G)	54.088	26.33	T(306)	66.62	T(306)	36.01	−26.393
C4H6	1-BUTYNE	(L)	54.088	33.8	(CALC)	49.08	M(39)	48.71	−35.701
C4H6	1-BUTYNE	(G)	54.088	39.48	T(318)	69.51	T(318)	48.30	−35.400
C4H6	2-BUTYNE	(L)	54.088	28.65	(CALC)	46.63	M(1638)	44.29	−32.462
C4H6	2-BUTYNE	(G)	54.088	34.97	T(319)	67.71	T(319)	44.32	−32.487
C4H6	CYCLOBUTENE	(G)	54.088	31.00	T(344)	62.98	T(344)	41.76	−30.611
C4H6CL2O2	ETHYL DICHLOROACETATE	(L)	157.002	−119.6	B(1384)				
C4H6F2O2	DIFLUOROETHYL ACETATE	(U,L)	124.088	−211.5	C(744)				
C4H6F2O2	ETHYL DIFLUOROACETATE	(L)	124.088	−230.5	C(744)				
C4H6N2O2	5-METHYLHYDANTOIN	(S)	114.104	−105.0	B(1471)				
C4H6N2O2	5-METHYLHYDANTOIN	(S)	114.104	−105.3	B(1471)				
C4H6N2O2	2,5-PIPERAZINEDIONE	(S)	114.104	−105.8	C(744)				
C4H6N4O2	(ETHANEDIYLIDENE)DIUREA	(S)	142.120	−116.8	B(1472)				
C4H6N4O3	ALLANTOIN	(S)	158.120	−171.50	A(1421)	46.6	M(1421)	−106.65	78.171
C4H6N4O6	DIMETHYL DINITROXAMIDE	(S)	206.120	−73.1	B(985)				
C4H6N4O7	N-(1,1-DINITROETHYL)-1-NITROACETALDOXIME	(S)	222.120	−38.8	B(1640)				
C4H6N4O7	(2-HYDROXYETHYL)-N'-NITRO-OXAMIDE NITRATE	(S)	222.120	−132.7	B(984)				
C4H6N4O8	2,2,3-TETRANITROBUTANE	(S)	238.120	1.4	B(1001)				
C4H6N4O11	NITROISOBUTANETRIOLTRINITRATE	(S)	286.120	−54.6	B(981)				
C4H6N8	5,5'-ETHYLENEBISTETRAZOLE	(S)	166.152	106.2	B(1616)				
C4H6N10	5,5'-AZOBIS-(1-METHYLTETRAZOLE), CIS	(S)	194.168	188.6	B(1616)				
C4H6N10	5,5'-AZOBIS-(1-METHYLTETRAZOLE),TRANS	(S)	194.168	189.3	B(1616)				

Formula	Name	State	Mol Wt	$\Delta Hf°_{298}$ (kcal/mole)	Source	$S°_{298}$ (cal/mole °K)	Source	$\Delta Gf°_{298}$ (kcal/mole)	Log Kp
C4H6N10	5,5'-AZOBIS(2-METHYLTETRAZOLE)	(S)	194.168	180.40	B(1616)				
C4H6O	CROTONALDEHYDE	(L)	70.088	-34.45	A(1506)				
C4H6O	METHYL 2-PROPYNYL ETHER	(G)	70.088	19.7	C(744)				
C4H6O	VINYL ETHER	(G)	70.088	-3.03	A(1146)				
C4H6O2	ALLYL FORMATE	(G)	86.088	-55.4	C(744)				
C4H6O2	2,3-BUTANEDIONE	(L)	86.088	-87.60	A(1112)				
C4H6O2	2,3-BUTANEDIONE	(L)	86.088	-87.35	A(1067)				
C4H6O2	2,3-BUTANEDIONE	(G)	86.088	-86.36	A(1403)				
C4H6O2	2-BUTYNE-1,4-DIOL	(S)	86.088	-78.2	(CALC)				
C4H6O2	CROTONIC ACID	(S)	86.088	-54.3	C(843)				
C4H6O2	CYCLOPROPANECARBOXYLIC ACID	(S)	86.088	-103.1	C(744)				
C4H6O2	CYCLOPROPANECARBOXYLIC ACID	(S)	86.088	-101.1	C(744)				
C4H6O2	4-HYDROXYBUTYRIC ACID, GAMMA LACTONE	(L)	86.088	-97.5	C(744)				
C4H6O2	METHACRYLIC ACID	(L)	86.088	-99.20	C(843)				
C4H6O2	VINYL ACETATE	(L)	86.088	-103.9	C(1364)				
C4H6O2S	2,5-DIHYDROTHIOPHENE 1,1-DIOXIDE	(L)	118.154	-83.6	B(334)				
C4H6O2S	2,5-DIHYDROTHIOPHENE 1,1-DIOXIDE	(G)	118.154	-76.2	B(906)				
C4H6O3	ACETIC ANHYDRIDE	(L)	102.088	-149.14	A(1558)				
C4H6O3	ACETIC ANHYDRIDE	(G)	102.088	-137.60	TC 562)	64.2	MC 968)	-116.83	85.632
C4H6O4	ACETYL PEROXIDE	(L)	118.088	-127.9	B(679)	93.20	TC 562)	-113.93	83.511
C4H6O4	DIMETHYL OXALATE	(S)	118.088	-179.9	C(744)				
C4H6O4	METHYLMALONIC ACID	(S)	118.088	-218.8	C(744)				
C4H6O5	SUCCINIC ACID	(S)	118.088	-224.88	A(1612)	42.0	M(1104)	-178.64	130.942
C4H6O5	MALIC ACID, L	(S)	134.088	-264.27	A(1612)				
C4H6O6	MALIC ACID, D,L	(S)	134.088	-263.78	A(1612)				
C4H6O6	ACETYLENEDICARBOXYLIC ACID DIHYDRATE	(S)	150.088	-278.4	A(432)				
C4H6O6	TARTARIC ACID, D	(S)	150.088	-300.4	B(358)				
C4H6O6	TARTARIC ACID, D,L (RACEMIC)	(S)	150.088	-305.5	B(358)				
C4H6O6	TARTARIC ACID, MESO	(S)	150.088	-301.0	B(358)				
C4H6S	2,3-DIHYDROTHIOPHENE	(L)	86.154	12.82	A(309)				
C4H6S	2,3-DIHYDROTHIOPHENE	(G)	86.154	21.84	A(309)				

Formula	Name	State	Mol Wt	$\Delta H_{f\,298}^{\circ}$ (kcal/mole)	Source	S_{298}° (cal/mole °K)	Source	$\Delta G_{f\,298}^{\circ}$ (kcal/mole)	Log K_p
C4H6S	2,5-DIHYDROTHIOPHENE	(L)	86.154	11.37	AC 309				
C4H6S	2,5-DIHYDROTHIOPHENE	(G)	86.154	20.92	AC 309				
C4H7CLO2	2-CHLOROBUTYRIC ACID	(L)	122.553	−137.9	BC1384				
C4H7CLO2	3-CHLOROBUTYRIC ACID	(L)	122.553	−133.3	BC1384				
C4H7CLO2	4-CHLOROBUTYRIC ACID	(L)	122.553	−135.8	BC1384				
C4H7CLO2	ETHYL CHLOROACETATE	(L)	122.553	−120.8	BC1384				
C4H7CLO2	METHYL 2-CHLOROPROPIONATE	(L)	122.553	−117.9	BC1384				
C4H7CLO2	METHYL 3-CHLOROPROPIONATE	(L)	122.553	−118.9	BC1384				
C4H7FO2	ETHYL FLUOROACETATE	(L)	106.096	−155.2	CC 744				
C4H7FO2	FLUOROETHYL ACETATE	(U,L)	106.096	−158.3	CC 744				
C4H7F4N	TETRAFLUORODIETHYLAMINE	(U,L)	145.104	−220.6	CC 744				
C4H7N	BUTYRONITRILE	(L)	69.104	−1.39	AC 388				
C4H7N	BUTYRONITRILE	(G)	69.104	8.14	TC 620	77.78	TC 620	25.97	−19.032
C4H7N	ISOBUTYRONITRILE	(L)	69.104	−2.92	AC 388				
C4H7N	ISOBUTYRONITRILE	(G)	69.104	6.07	TC 621	74.88	TC 621	24.76	−18.149
C4H7NO	2-PYRROLIDINONE	(S)	85.104			32.7	MC 794		
C4H7NO3	ETHYL OXAMATE	(S)	117.104	−158.0	CC 744				
C4H7NO4	ASPARTIC ACID, L	(S)	133.104	−232.64	AC 664	40.66	MC 664	−174.53	127.925
C4H7NO4	DIGLYCOLIAMIDIC ACID	(S)	133.104	−219.0	CC 744				
C4H7N3O	CREATININE	(S)	113.120	−57.01	AC 650	40.0	MC 644	−6.97	5.109
C4H7N3O8	2-METHYL-2-NITRO-1,3-PROPANEDIOL DINITRATE	(S)	225.120	−93.1	BC 984				
C4H7N5	1-ALLYL-5-AMINOTETRAZOLE	(S)	125.136	63.4	BC1616				
C4H7N5	2-ALLYL-5-AMINOTETRAZOLE	(S)	125.136	67.6	BC1616				
C4H7N5O2	ETHYL 5-TETRAZOLECARBAMATE	(S)	157.136	−52.6	BC 974				
C4H8	1-BUTENE	(G)	56.104	−0.03	TC 263	73.04	TC 263	17.04	−12.489
C4H8	2-BUTENE, CIS	(G)	56.104	−1.67	TC 264	71.90	TC 264	15.74	−11.536
C4H8	2-BUTENE, TRANS	(G)	56.104	−2.67	TC 265	70.86	TC 265	15.05	−11.030
C4H8	CYCLOBUTANE	(L)	56.104	0.74	AC 267	43.44	MC1215	26.63	−19.522
C4H8	CYCLOBUTANE	(G)	56.104	6.37	TC 339	63.43	TC 339	26.30	−19.280
C4H8	2-METHYLPROPENE	(G)	56.104	−4.04	TC 266	70.17	TC 266	13.88	−10.177
C4H8BR2	1,2-DIBROMOBUTANE	(L)	215.936	−34.5	TC 768				
C4H8BR2	1,2-DIBROMOBUTANE	(G)	215.936	−23.70	TC 768	97.70	TC 768	−3.14	2.299
C4H8BR2	2,3-DIBROMOBUTANE	(G)	215.936	−24.40	TC 769	94.40	TC 769	−2.85	2.091
C4H8I2	1,2-DIIODOBUTANE	(G)	309.924	2.85	TC 784	101.80	TC 784	19.62	−14.381
C4H8N2O2	DIMETHYLGLYOXIME	(S)	116.120	−41.7	BC1013				

Formula	Name	State	Mol Wt	$\Delta H f^\circ_{298}$ (kcal/mole)	Source	S°_{298} (cal/mole °K)	Source	$\Delta G f^\circ_{298}$ (kcal/mole)	Log Kp
C4H8N2O2	SUCCINAMIDE	(S)	116.120	-140.0	C(744)	41.7	M(644)	-126.73	92.889
C4H8N2O3	ASPARAGINE, L	(S)	132.120	-188.70	A(650)	45.4	M(642)	-117.25	85.942
C4H8N2O3	GLYCYLGLYCINE	(S)	132.120	-178.12	A(643)				
C4H8N2O4	TARTARAMIDE, D	(L)	148.120	-222.3	C(744)				
C4H8N2O4	TARTARAMIDE, MESO	(L)	148.120	-222.7	C(744)				
C4H8N2S	1-ALLYL-2-THIOUREA	(S)	116.186	-1.5	C(744)				
C4H8N4O8	2,2'-(NITROIMINO)- DIETHANOL DINITRATE	(S)	240.136	-72.1	B(1471)				
C4H8N4O8	2,2'-(NITROIMINO) DIETHANOL DINITRATE	(S)	240.136	-78.1	B(984)				
C4H8N8O8	OCTAHYDRO-1,3,5,7-TETRANITRO-1,3,5,7-TETRAZOCINE (HMX)	(S)	296.168	17.93	A(1193)				
C4H8O	ALLYL METHYL ETHER	(G)	72.104	-25.5	C(744)	57.08	M(538)	-36.18	26.518
C4H8O	2-BUTANONE	(L)	72.104	-65.31	C(538)	80.81	T(555)	-34.91	25.591
C4H8O	2-BUTANONE	(G)	72.104	-56.97	T(555)	59.00	M(1109)	-28.50	20.890
C4H8O	BUTYRALDEHYDE	(L)	72.104	-57.06	T(547)	82.44	T(547)	-27.43	20.105
C4H8O	BUTYRALDEHYDE	(G)	72.104	-49.00	T(547)				
C4H8O	ETHYL VINYL ETHER	(G)	72.104	-33.50	A(1146)				
C4H8O	2-METHYLPROPIONALDEHYDE	(L)	72.104	-59.8	A(1508)				
C4H8O	TETRAHYDROFURAN	(L)	72.104	-51.68	A(1133)				
C4H8O	TETRAHYDROFURAN	(G)	72.104	-44.03	A(1133)				
C4H8OS	ETHYL THIOACETATE	(L)	104.170	-64.0	A(1554)				
C4H8O2	2-BUTENE-1,4-DIOL	(L)	88.104	-93.3	C(843)				
C4H8O2	BUTYRIC ACID	(L)	88.104	-127.6	A(853)	54.1	M(1105)	-90.27	66.170
C4H8O2	M-DIOXANE	(L)	88.104	-92.1	B(1391)				
C4H8O2	M-DIOXANE	(G)	88.104	-83.6	B(1391)				
C4H8O2	P-DIOXANE	(L)	88.104	-84.50	A(1391)	46.67	M(677)	-44.96	32.954
C4H8O2	P-DIOXANE	(G)	88.104	-75.30	T(516)	71.65	T(516)	-43.21	31.670
C4H8O2	ETHYL ACETATE	(L)	88.104	-114.49	A(1555)	62.0	M(1106)	-79.52	58.287
C4H8O2	ETHYL ACETATE	(G)	88.104	-105.86	T(563)	86.70	T(563)	-78.25	57.359
C4H8O2	3-HYDROXYBUTYRALDEHYDE	(L)	88.104	-102.9	C(744)				
C4H8O2	ISOBUTYRIC ACID	(L)	88.104	-132.1	C(744)				
C4H8O2	2-METHYL-1,3-DIOXOLANE	(L)	88.104	-90.3	C(843)				
C4H8O2	METHYL PROPIONATE	(L)	88.104	-113.04	C(186)				
C4H8O2S	ALLYL METHYL SULFONE	(L)	120.170	-91.5	B(186)				
C4H8O2S	ALLYL METHYL SULFONE	(G)	120.170	-72.6	B(186)				

Formula	Name	State	Mol Wt	$\Delta H^\circ_{f\,298}$ (kcal/mole)	Source	S°_{298} (cal/mole °K)	Source	$\Delta G^\circ_{f\,298}$ (kcal/mole)	Log K_p
C4H8O3	2-HYDROXYBUTYRIC ACID	(S)	104.104	-161.7	C C 744				
C4H8O3	HYDROXY-2-METHYLPROPIUNIC-								
C4H8O3	ACID	(U,S)	104.104	-177.7	C C 744				
C4H8S	METHYL LACTATE	(L)	104.104	-152.3	C C 744				
C4H8S	1,2-EPITHIO-2-METHYLPROPANE	(L)	88.170	-5.79	A C1447				
C4H8S	1,2-EPITHIO-2-METHYLPROPANE	(G)	88.170	2.76	A C1447				
C4H8S	2,3-EPITHIOBUTANE, CIS	(L)	88.170	-5.79	A C1447				
C4H8S	2,3-EPITHIOBUTANE, CIS	(G)	88.170	2.76	A C1447				
C4H8S	2,3-EPITHIOBUTANE, TRANS	(L)	88.170	-7.03	A C1447				
C4H8S	2,3-EPITHIOBUTANE, TRANS	(G)	88.170	0.92	A C1447				
C4H8S	THIACYCLOPENTANE	(L)	88.170	-17.34	A C 309	49.67	M C 628	8.97	-6.576
C4H8S	THIACYCLOPENTANE	(G)	88.170	-8.08	T C 884	73.94	T C 884	11.00	-8.059
C4H9BR	1-BROMOBUTANE	(L)	137.028	-34.47	A C 138				
C4H9BR	1-BROMOBUTANE	(G)	137.028	-25.65	T C 765	88.39	T C 765	-3.08	2.259
C4H9BR	2-BROMOBUTANE	(L)	137.028	-37.12	A C 138				
C4H9BR	2-BROMOBUTANE	(G)	137.028	-28.70	T C 766	88.50	T C 766	-6.16	4.518
C4H9BR	2-BROMO-2-METHYLPROPANE	(G)	137.028	-32.00	T C 767	79.34	T C 767	-6.73	4.935
C4H9CL	1-CHLOROBUTANE	(L)	92.569	-43.2	B C1384				
C4H9CL	1-CHLOROBUTANE	(G)	92.569	-35.20	T C 739	85.58	T C 739	-9.27	6.797
C4H9CL	2-CHLOROBUTANE	(L)	92.569	-46.2	B C1384				
C4H9CL	2-CHLOROBUTANE	(G)	92.569	-38.60	T C 740	85.94	T C 740	-12.78	9.368
C4H9CL	1-CHLORO-2-METHYLPROPANE	(L)	92.569	-45.8	B C1384				
C4H9CL	1-CHLORO-2-METHYLPROPANE	(G)	92.569	-38.10	T C 741	84.56	T C 741	-11.87	8.700
C4H9CL	2-CHLORO-2-METHYLPROPANE	(L)	92.569	-50.8	T C 742				
C4H9CL	2-CHLORO-2-METHYLPROPANE	(G)	92.569	-43.80	T C 742	77.00	T C 742	-15.32	11.226
C4H9F	2-FLUORO-1,1-DIMETHYL-	(G)	76.112	-82.0	C C 843				
C4H9FN2O2	N-FLUORO-N-NITROETHYLAMINE	(L)	136.128	-31.3	A C1646				
C4H9FN2O2	N-FLUORO-N-METHYL-								
C4H9FN2O2	N-NITROPROPYLAMINE	(L)	136.128	-38.7	A C1646				
C4H9FN2O2	N-FLUORO-N-NITROBUTYLAMINE	(L)	136.128	-39.9	A C1646				
C4H9I	2-IODO-2-METHYLPROPANE	(G)	184.022	-17.60	T C 783	81.79	T C 783	5.65	-4.142
C4H9N	PYRROLIDINE	(L)	71.120	-9.84	A C 948	48.76	M C 948	25.94	-19.016
C4H9N	PYRROLIDINE	(G)	71.120	-0.86	T C 612	73.97	T C 612	27.41	-20.089
C4H9NO	BUTYRAMIDE	(S)	87.120	-87.2	C C 744				
C4H9NO	2-BUTANONE OXIME	(L)	87.120	-37.14	C C 744				

658

Formula	Name	State	Mol Wt	$\Delta H f^\circ_{298}$ (kcal/mole)	Source	S°_{298} (cal/mole °K)	Source	$\Delta G f^\circ_{298}$ (kcal/mole)	Log K_p
C4H9NO	ISOBUTYRAMIDE	(S)	87.120	-87.3	C(744)				
C4H9NO2	ISOBUTYL NITRITE	(G)	103.120	-39.10	C(744)				
C4H9NO2	1-NITROBUTANE	(L)	103.120	-46.0	B(613)				
C4H9NO2	1-NITROBUTANE	(G)	103.120	-34.40	T(627)	94.28	T(627)	2.42	-1.775
C4H9NO2	2-NITROBUTANE	(L)	103.120	-49.6	B(613)				
C4H9NO2	2-NITROBUTANE	(G)	103.120	-39.10	T(628)	91.62	T(628)	-1.49	1.089
C4H9NO2	PROPYL CARBAMATE	(S)	103.120	-132.1	B(1640)				
C4H9NO3	2-METHYL-2-NITRO-1-PROPANOL	(S)	119.120	-98.1	B(300)				
C4H9NO3	3-NITRO-2-BUTANOL	(L)	119.120	-93.3	B(300)				
C4H9NO4	2-METHYL-2-NITRO-1,3-PROPANEDIOL	(S)	135.120	-138.6	B(300)				
C4H9NO5	2-(HYDROXYMETHYL)-2-NITRO-1,3-PROPANEDIOL	(S)	151.120	-176.5	B(983)				
C4H9NO6	AMMONIUM BITARTRATE, D	(S)	167.120	-341.9	C(744)				
C4H9NO6	AMMONIUM BIRACEMATE	(S)	167.120	-343.1	C(744)				
C4H9NO6	AMMONIUM BITARTRATE, MESO	(S)	167.120	-342.4	C(744)				
C4H9NS2	DIHYDRO-5-METHYL-4H-1,3,5-DITHIAZINE	(L)	135.252	-6.8	C(744)				
C4H9NS2	DIMETHYL METHYLDITHIO-IMIDOCARBONATE	(L)	135.252	-2.1	C(744)				
C4H9NS2	METHYL DIMETHYLDITHIOCARBAMATE	(S)	135.252	-17.4	C(744)				
C4H9N3O2	CREATINE	(S)	131.136	-128.39	A(650)	45.3	M(644)	-63.32	46.412
C4H9N3S	ACETONE THIOSEMICARBAZONE	(S)	131.202	4.6	B(1446)				
C4H9N5O5	2-AMINO-1-NITRO-1,3-DIAZA-CYCLOPENTENE METHYL NITRATE	(S)	207.152	-62.7	B(1040)				
C4H9N5O5	2-AMINO-1-NITRO-5-METHYL-1,3-DIAZACYCLOPENTENE NITRATE	(S)	207.152	-70.4	B(1040)				
C4H9N5O5	1-(2-NITROXYPROPYL)-3-NITROGUANIDINE	(S)	207.152	-57.7	B(1040)				
C4H9N5O5	1-(3-NITROXYPROPYL)-3-NITROGUANIDINE	(S)	207.152	-56.0	B(1040)				
C4H10	BUTANE	(L)	58.120	-35.29	A(1248)	55.2	M(40)	-3.60	2.637
C4H10	BUTANE	(G)	58.120	-30.15	T(104)	74.12	T(104)	-4.10	3.004
C4H10	2-METHYLPROPANE	(L)	58.120	-37.87	A(1248)	52.09	M(38)	-5.25	3.848
C4H10	2-METHYLPROPANE	(G)	58.120	-32.15	T(105)	70.42	T(105)	-4.99	3.661

Formula	Name	State	Mol Wt	$\Delta H f^\circ_{298}$ (kcal/mole)	Source	S°_{298} (cal/mole °K)	Source	$\Delta G f^\circ_{298}$ (kcal/mole)	Log $K p$
C4H10N2	PIPERAZINE	(S)	86.136	-10.93	A(93)				
C4H10N2O	TRIMETHYLUREA	(S)	102.136	-77.9	B(1472)				
C4H10N2O2	NITROBUTYLAMINE	(L)	118.136	-27.7	B(1039)				
C4H10N2O2	N-NITRODIETHYLAMINE	(L)	118.136	-25.6	B(1039)				
C4H10N2O2	N-NITRODIETHYLAMINE	(L)	118.136	-25.4	B(1621)				
C4H10N2O4	ASPARAGINE, L MONOHYDRATE	(S)	150.136	-259.7	A(664)	50.10	M(664)	-183.62	134.591
C4H10N2O6	ETHYL METHYLCARBAMATE NITRATE	(S)	182.136	-271.8	B(1471)				
C4H10N4O2	SUCCINYL DIHYDRAZIDE	(S)	146.152	-87.6	B(747)				
C4H10N4O9	BIS(2-HYDROXYETHYLNITRATE) AMMONIUM NITRATE	(S)	258.152	-139.4	B(984)				
C4H10O	BUTYL ALCOHOL	(L)	74.120	-77.87	T(521)	53.95	M(280)	-38.50	28.219
C4H10O	BUTYL ALCOHOL	(G)	74.120	-65.59	T(521)	86.80	T(521)	-36.01	26.397
C4H10O	SEC-BUTYL ALCOHOL	(L)	74.120	-81.9	T(522)				
C4H10O	SEC-BUTYL ALCOHOL	(G)	74.120	-69.86	T(522)	85.81	T(522)	-39.99	29.311
C4H10O	TERT-BUTYL ALCOHOL	(S)	74.120	-87.46	CALC	40.84	M(1082)	-44.18	32.383
C4H10O	TERT-BUTYL ALCOHOL	(L)	74.120	-85.87	A(1365)	46.16	M(1082)	-44.18	32.381
C4H10O	TERT-BUTYL ALCOHOL	(G)	74.120	-77.87	T(523)	77.98	T(523)	-45.66	33.471
C4H10O	ETHYL ETHER	(L)	74.120	-66.8	T(503)	60.5	M(1105)	-29.38	21.537
C4H10O	ETHYL ETHER	(G)	74.120	-60.28	T(503)	81.90	T(503)	-29.24	21.434
C4H10O	ISOBUTYL ALCOHOL	(L)	74.120	-79.73	A(215)				
C4H10O	ISOPROPYL METHYL ETHER	(L)	74.120	-60.24	T(505)	80.86	T(505)	-28.89	21.178
C4H10O	METHYL PROPYL ETHER	(G)	74.120	-56.82	T(504)	83.52	T(504)	-26.27	19.252
C4H10OS	ETHYL SULFOXIDE	(L)	106.186	-64.0	B(907)				
C4H10OS	ETHYL SULFOXIDE	(G)	106.186	-49.1	B(907)				
C4H10O2	ACETALDEHYDE DIMETHYL ACETAL	(L)	90.120	-97.4	C(744)				
C4H10O2	1,2-BUTANEDIOL	(L)	90.120	-124.9	C(1032)				
C4H10O2	1,3-BUTANEDIOL	(L)	90.120	-122.9	C(1032)				
C4H10O2	2,3-BUTANEDIOL	(L)	90.120	-129.2	C(1032)				
C4H10O2	ETHYL PEROXIDE	(L)	90.120	-53.4	B(64)				
C4H10O2	ETHYL PEROXIDE	(G)	90.120	-46.1	B(64)				
C4H10O2	2-METHYL-1,2-PROPANEDIOL	(L)	90.120	-128.8	C(1032)				
C4H10O2	TERT-BUTYL HYDROPEROXIDE	(L)	90.120	-70.2	B(807)				
C4H10O2S	ETHYLSULFONE	(L)	122.186	-123.1	B(904)				
C4H10O2S	ETHYLSULFONE	(G)	122.186	-102.9	B(904)				
C4H10O2S	ISOPROPYL METHYL SULFONE	(L)	122.186	-120.4	B(186)				
C4H10O2S	ISOPROPYL METHYL SULFONE	(G)	122.186	-103.3	B(186)				

Formula	Name	State	Mol Wt	ΔH_{298}° (kcal/mole)	Source	S_{298}° (cal/mole °K)	Source	ΔG_{298}° (kcal/mole)	Log K_p
C4H10O3	DIETHYLENE GLYCOL	(L)	106.120	-149.7	C(1032)				
C4H10O3	TRIMETHYL ORTHOFORMATE	(G)	106.120	-121.1	CC(744)				
C4H10O4	ERYTHRITOL	(S)	122.120	-217.6	A(1121)			-152.12	111.505
C4H10O6S3	TRIS(METHYLSULFONYL)METHANE	(S)	250.318						
C4H10S	1-BUTANETHIOL	(L)	90.186	-29.79	A(1309)	39.9	MC(1105)		
C4H10S	1-BUTANETHIOL	(G)	90.186	-21.05	A(894)	72.8	MC(311)	0.97	-0.711
C4H10S	2-BUTANETHIOL	(L)	90.186	-31.13	AC(956)	65.96	MC(1309)	2.64	-1.934
C4H10S	2-BUTANETHIOL	(G)	90.186	-23.00	TC(895)	89.68	MC(894)	-0.04	-0.033
C4H10S	ETHYL SULFIDE	(L)	90.186	-28.52	AC(631)	64.87	MC(956)	1.29	-0.948
C4H10S	ETHYL SULFIDE	(G)	90.186	-19.95	TC(802)	87.65	TC(895)	2.72	-1.992
C4H10S	ISOPROPYL METHYL SULFIDE	(L)	90.186	-29.79	AC(631)	64.36	MC(1305)	4.25	-3.116
C4H10S	ISOPROPYL METHYL SULFIDE	(G)	90.186	-21.61	TC(803)	87.96	TC(802)	1.89	-1.384
C4H10S	2-METHYL-1-PROPANETHIOL	(L)	90.186	-31.54	AC(614)	62.88	MC(953)	3.21	-2.356
C4H10S	2-METHYL-1-PROPANETHIOL	(G)	90.186	-23.24	TC(896)	85.87	TC(803)	-0.09	0.069
C4H10S	2-METHYL-2-PROPANETHIOL	(L)	90.186	-33.56	AC(963)	63.66	MC(614)	1.33	-0.973
C4H10S	2-METHYL-2-PROPANETHIOL	(G)	90.186	-26.17	TC(897)	86.73	TC(896)	-0.69	0.509
C4H10S	METHYL PROPYL SULFIDE	(L)	90.186	-28.21	A(1309)	58.90	MC(963)	-0.17	-0.124
C4H10S	METHYL PROPYL SULFIDE	(G)	90.186	-19.54	TC(804)	80.79	TC(897)	2.79	-2.048
C4H10S2	1,4-BUTANEDITHIOL	(L)	122.252	-25.12	AC(925)	65.14	A(1309)	4.40	-3.224
C4H10S2	1,4-BUTANEDITHIOL	(G)	122.252	-11.90	AC(925)	88.84	TC(804)		
C4H10S2	ETHYL DISULFIDE	(L)	122.252	-28.69	AC(1301)	72.90	MC(1307)	2.28	-1.668
C4H10S2	ETHYL DISULFIDE	(G)	122.252	-17.84	TC(873)	99.07	TC(873)	5.32	-3.902
C4H11N	BUTYLAMINE	(L)	73.136	-30.53	AC(387)				
C4H11N	BUTYLAMINE	(G)	73.136	-22.00	TC(604)	86.76	TC(604)	11.76	-8.619
C4H11N	DIETHYLAMINE	(L)	73.136	-24.79	AC(678)				
C4H11N	DIETHYLAMINE	(G)	73.136	-17.30	TC(608)	84.18	TC(608)	17.23	-12.628
C4H11N	ISOBUTYLAMINE	(L)	73.136	-37.8	CC(744)				
C4H11N	SEC-BUTYLAMINE	(L)	73.136	-32.88	AC(387)				
C4H11N	SEC-BUTYLAMINE	(G)	73.136	-24.90	TC(605)	83.90	TC(605)	9.71	-7.118
C4H11N	TERT-BUTYLAMINE	(L)	73.136	-35.95	AC(387)				
C4H11N	TERT-BUTYLAMINE	(G)	73.136	-28.65	TC(606)	80.76	TC(606)	6.90	-5.056
C4H11NO2	2,2'-IMINODIETHANOL	(L)	105.136	-112.6	B(1354)				
C4H11N3O3	CREATINE MONOHYDRATE	(S)	149.152	-198.6	CC(744)				
C4H12BRN	TETRAMETHYLAMMONIUM BROMIDE	(S)	154.060			56.0	MC(652)	-120.11	88.038
C4H12CLN	TETRAMETHYLAMMONIUM CHLORIDE	(S)	109.601			47.99	MC(211)		
C4H12IN	TETRAMETHYLAMMONIUM IODIDE	(S)	201.054			45.58	MC(211)		
						49.68	CC(843)		

Formula	Name	State	Mol Wt	ΔHf°_{298} (kcal/mole)	Source	S°_{298} (cal/mole °K)	Source	ΔGf°_{298} (kcal/mole)	Log Kp
C4H12N2O3	DIETHYLAMINE NITRATE	(S)	136.152	-98.7	C C(277)				
C4H12N2O3	TETRAMETHYLAMMONIUM NITRATE	(S)	136.152	-82.6	B C(984)				
C4H12N4	TETRAMETHYLTETRAZENE	(L)	116.168	55.34	B C(336)				
C4H13CL2N	TETRAMETHYLAMMONIUM HYDROGEN DICHLORIDE	(S)	146.066			60.63	M(212)		
C4N2	ACETYLENEDICARBONITRILE	(L)	76.056	120.6	A C(26)				
C4N2	ACETYLENEDICARBONITRILE	(G)	76.056	127.50	T C(23)	69.31	T(23)	122.10	-89.500
C5F11N	UNDECAFLUOROPIPERIDINE	(L)	283.058	-496.3	A C(515)	94.02	M C(515)	-436.04	319.607
C5F11N	UNDECAFLUOROPIPERIDINE	(G)	283.058	-489.1	A C(515)	116.10	M C(515)	-435.42	319.155
C5F12	DODECAFLUOROPENTANE	(G)	288.050	-613.6	E C(833)				
C5HN3	ETHENETRICARBONITRILE	(S)	103.082	105.0	A C(154)				
C5HN3	ETHENETRICARBONITRILE	(G)	103.082	123.9	A C(154)				
C5H4N4O	HYPOXANTHENE	(S)	136.114	-26.47	A C(1421)	34.8	M(1421)	18.39	-13.480
C5H4N4O	PURINE-8(7H)-ONE	(S)	136.114	-14.7	C C(744)				
C5H4N4O2	XANTHINE	(S)	152.114	-90.70	A C(1421)	38.5	M(1421)	-39.64	29.054
C5H4O2	2-FURALDEHYDE	(L)	96.082	-46.6	A C(1421)	52.10	M(1006)	-26.88	19.705
C5H4O2S	2-THIOPHENECARBOXYLIC ACID	(S)	128.148	-104.6	C C(744)				
C5H4O3	CITRACONIC ANHYDRIDE	(L)	112.082	-120.6	C C(258)				
C5H4O3	2-FUROIC ACID	(L)	112.082	-120.6	A C(1116)				
C5H4O3	ITACONIC ANHYDRIDE	(S)	112.082	-125.1	C C(744)				
C5H5CL3O2	ALLYL TRICHLOROACETATE	(L)	203.461	-95.8	B C(1384)				
C5H5N	PYRIDINE	(L)	79.098	23.9	A C(949)	42.52	M C(949)	43.34	-31.766
C5H5N	PYRIDINE	(G)	79.098	33.50	T C(613)	67.59	T(613)	45.46	-33.324
C5H5NO	2-PYRROLECARBOXALDEHYDE	(S)	95.098	-24.0	C C(1414)				
C5H5NO2S	4-METHYLTHIAZOLE	(L)	143.164	16.32	A C(927)				
C5H5NO2S	4-METHYLTHIAZOLE	(G)	143.164	26.80	A C(927)				
C5H5N4O3	URIC ACID	(S)	169.122	-147.91	A C(1421)	41.4	M(1421)	-85.75	62.857
C5H5N5	ADENINE	(S)	135.130	22.94	A C(1421)	36.1	M(1421)	71.58	-52.469
C5H5N5O	GUANINE	(S)	151.130	-43.96	A C(1421)	38.3	M(1421)	11.33	-8.306
C5H5N7	5-(GUANYLAMINO) TETRAZOLE	(S)	163.146	40.6	B C(974)				
C5H6	CYCLOPENTADIENE	(L)	66.098	25.1	B C(1580)				
C5H6	CYCLOPENTADIENE	(G)	66.098	32.00	A C(762)	64.00	(CALC)	42.86	-31.419
C5H6	3-PENTEN-1-YNE, CIS	(L)	66.098	54.2	A C(1364)				
C5H6	3-PENTEN-1-YNE, TRANS	(L)	66.098	54.6	A C(1364)				
C5H6CL2O2	ALLYL DICHLOROACETATE	(L)	169.012	-93.4	B C(1384)				

Formula	Name	State	Mol Wt	$\Delta H f^\circ_{298}$ (kcal/mole)	Source	S°_{298} (cal/mole °K)	Source	$\Delta G f^\circ_{298}$ (kcal/mole)	Log $K p$
C5H6N2	GLUTARONITRILE	(L)	94.114	24.9	C(744)	57.23	M(235)	51.43	-37.696
C5H6N2	GLUTARONITRILE	(G)	94.114	40.9	C(1628)	88.1	M(235)	58.22	-42.677
C5H6N2O	1-ACETYLIMIDAZOLE	(S)	110.114	-28.0	A(1557)				
C5H6N2O	2-PYRROLECARBOXALDEHYDE OXIME	(S)	110.114	4.5	C(1414)				
C5H6N2O2	5,5-DIMETHYLHYDANTOIN	(S)	126.114	-67.9	B(1471)				
C5H6N2O2	4-METHYLURACIL	(S)	126.114	-108.2	C(744)				
C5H6N2O2	THYMINE	(S)	126.114	-109.5	C(744)				
C5H6N2O3	DIMETHYLIMIDAZOLIDINETRIONE	(S)	142.114	-136.6	C(744)				
C5H6N4O4	PSEUDOURIC ACID	(U,S)	186.130	-220.8	C(744)				
C5H6N8O13	BIS(TRINITROETHYL)UREA	(S)	386.162	-76.9	C(1640)				
C5H6O	2-METHYLFURAN	(L)	82.098	-40.4	C(744)	51.12	M(190)		
C5H6O2	ETHYL PROPIOLATE	(L)	98.098	-66.05	A(1116)				
C5H6O2	FURFURYL ALCOHOL	(L)	98.098	-67.8	A(432)	51.50	M(1006)	-36.85	27.010
C5H6O2	3-PENTYNOIC ACID	(S)	98.098	-146.9	C(744)				
C5H6O3	GLUTARIC ANHYDRIDE	(S)	114.098	-147.6	B(1538)				
C5H6O3	METHYLSUCCINIC ANHYDRIDE	(S)	114.098	-145.4	B(1538)				
C5H6O4	1,2-CYCLOPROPANEDICARBOXYLIC ACID, CIS	(S)	130.098	-191.0	C(744)				
C5H6O4	1,1-CYCLOPROPANEDICARBOXYLIC ACID	(S)	130.098	-192.0	C(744)				
C5H6O4	CITRACONIC ACID, CIS	(S)	130.098	-197.06	A(1611)				
C5H6O4	ITACONIC ACID	(S)	130.098	-201.05	A(1611)				
C5H6O4	MESACONIC ACID, TRANS	(S)	130.098	-198.5	C(744)				
C5H6O5	2-OXOGLUTARIC ACID	(S)	146.098	-245.37	A(1611)				
C5H6S	2-METHYLTHIOPHENE	(L)	98.164	10.70	A(1134)	52.22	M(1134)	27.35	-20.048
C5H6S	2-METHYLTHIOPHENE	(G)	98.164	20.00	T(888)	76.62	T(888)	29.38	-21.532
C5H6S	3-METHYLTHIOPHENE	(L)	98.164	10.34	M(965)	52.19	M(965)	27.00	-19.791
C5H6S	3-METHYLTHIOPHENE	(G)	98.164	19.79	T(889)	76.79	T(889)	29.12	-21.341
C5H7CLO2	ALLYL CHLOROACETATE	(L)	134.563	-92.2	B(1384)				
C5H7CL3O2	ISOPROPYL TRICHLOROACETATE	(L)	205.477	-129.2	B(1384)				
C5H7CL3O2	PROPYL TRICHLOROACETATE	(L)	205.477	-123.9	B(1384)				
C5H7NO	3,5-DIMETHYLISOXAZOLE	(L)	97.114	-14.1	C(1467)				
C5H7NO2	ETHYL CYANOACETATE	(L)	113.114	-78.9	C(744)				
C5H7NO4	N-METHYLTARTRIMIDE, DL	(S)	145.114	-192.6	C(744)				
C5H7N3	2,6-DIAMINOPYRIDINE	(S)	109.130	-1.5	B(1640)				

Formula	Name	State	Mol Wt	$\Delta Hf°_{298}$ (kcal/mole)	Source	$S°_{298}$ (cal/mole °K)	Source	$\Delta Gf°_{298}$ (kcal/mole)	Log Kp
C5H8	CYCLOPENTENE	(L)	68.114	1.02	A(821)	48.10	M(645)	25.93	-19.006
C5H8	CYCLOPENTENE	(G)	68.114	7.87	T(345)	69.23	T(345)	26.48	-19.410
C5H8	2-METHYL-1,3-BUTADIENE	(L)	68.114	11.80	A(1248)	54.81	M(99)	34.71	-25.442
C5H8	2-METHYL-1,3-BUTADIENE	(G)	68.114	18.10	T(313)	75.44	T(313)	34.86	-25.551
C5H8	2-METHYL-1,2-BUTADIENE	(G)	68.114	31.00	T(312)	76.40	T(312)	47.47	-34.797
C5H8	3-METHYL-1-BUTYNE	(G)	68.114	32.60	T(322)	76.23	T(322)	49.12	-36.006
C5H8	1,2-PENTADIENE	(G)	68.114	34.80	T(307)	79.70	T(307)	50.29	-36.861
C5H8	1,3-PENTADIENE, CIS	(G)	68.114	18.70	T(308)	77.50	T(308)	34.84	-25.540
C5H8	1,3-PENTADIENE, TRANS	(G)	68.114	18.60	T(309)	76.40	T(309)	35.07	-25.708
C5H8	1,4-PENTADIENE	(L)	68.114	18.06	A(1119)	58.2	M(1119)	39.96	-29.289
C5H8	1,4-PENTADIENE	(G)	68.114	25.20	T(310)	79.70	T(310)	40.69	-29.824
C5H8	2,3-PENTADIENE	(G)	68.114	33.10	T(311)	77.60	T(311)	49.21	-36.074
C5H8	1-PENTYNE	(G)	68.114	34.50	T(320)	78.82	T(320)	50.25	-36.833
C5H8	2-PENTYNE	(G)	68.114	30.80	T(321)	79.30	T(321)	46.41	-34.016
C5H8	SPIROPENTANE	(L)	68.114	37.67	A(1303)	46.29	M(1303)	63.12	-46.266
C5H8	SPIROPENTANE	(G)	68.114	44.27	T(469)	67.45	T(469)	63.41	-46.479
C5H8	VINYLCYCLOPROPANE	(L)	68.114	18.5	C(1372)				
C5H8BR2	1,2-DIBROMOCYCLOPENTANE	(L)	227.946	-25.0	A(878)				
C5H8BR4	1,3-DIBROMO-								
C5H8BR4	2,2-BIS(BROMOMETHYL)PROPANE	(S)	387.778			69.58	M(1132)		
C5H8CL2O2	ISOPROPYL DICHLOROACETATE	(L)	171.028	-129.1	B(1384)				
C5H8CL2O2	PROPYL DICHLOROACETATE	(L)	171.028	-124.9	B(1384)				
C5H8CL4	1,3-DICHLORO-								
C5H8CL4	2,2-BIS(CHLOROMETHYL)PROPANE	(S)	209.942			61.54	M(1132)		
C5H8I4	1,3-DIIODO-								
C5H8I4	2,2-BIS(IODOMETHYL)PROPANE	(S)	575.754			75.70	M(1132)		
C5H8N2O2	5,5-DIMETHYLHYDANTOIN	(S)	128.130	-126.1	B(1471)				
C5H8N2O2	4-METHYLHYDROURACIL	(S)	128.130	-124.3	C(744)				
C5H8N2O5	GLYCYLGLYCINECARBOXYLIC ACID	(CU,S)	176.130	-271.5	C(744)				
C5H8N4O12	PENTAERYTHRITOL TETRANITRATE	(S)	316.146	-127.2	B(184)				
C5H8N6O	3-(5-TETRAZOLYLAZO)-2-BUTANONE	(S)	168.162	40.0	B(1616)				
C5H8O	ACETYLTRIMETHYLENE	(U,L)	84.114	-52.1	C(744)				
C5H8O	CYCLOPENTANONE	(L)	84.114	-56.62	A(1331)				
C5H8O	DIHYDROPYRAN	(L)	84.114	-37.6	B(202)				
C5H8O	3-PENTEN-2-ONE	(L)	84.114	-17.5	C(238)				
C5H8O	TIGLALDEHYDE	(L)	84.114	-54.5	C(238)				

Formula	Name	State	Mol Wt	$\Delta H^\circ_{f\,298}$ (kcal/mole)	Source	S°_{298} (cal/mole °K)	Source	$\Delta G^\circ_{f\,298}$ (kcal/mole)	Log K_p
C5H8O2	ALLYL ACETATE	(L)	100.114	-87.7	C(744)				
C5H8O2	ANGELIC ACID, CIS	(S)	100.114	-108.3	C(744)				
C5H8O2	CYCLOBUTANECARBOXYLIC ACID	(S)	100.114	-102.9	C(744)				
C5H8O2	ISOPROPENYL ACETATE	(L)	100.114	-85.0	A(1445)				
C5H8O2	METHYL CROTONATE	(L)	100.114	-91.6	C(1286)				
C5H8O2	METHYL METHACRYLATE	(L)	100.114	-88.7	C(1575)				
C5H8O2	2,4-PENTANEDIONE	(L)	100.114	-127.6	C(744)				
C5H8O2	2-PENTENOIC ACID	(S)	100.114	-106.9	B(1287)				
C5H8O2	3-PENTENOIC ACID	(L)	100.114	-104.0	B(1287)				
C5H8O2	4-PENTENOIC ACID	(L)	100.114	-103.1	B(1287)				
C5H8O2	TIGLIC ACID, TRANS	(S)	100.114	-116.7	C(744)				
C5H8O2S	TETRAHYDRO-2-THIOPHENE- CARBOXYLIC ACID	(S)	132.180	-132.6	C(744)				
C5H8O3	LEVULINIC ACID	(S)	116.114	-166.7	C(744)				
C5H8O3	METHYL ACETOACETATE	(L)	116.114	-149.6	C(744)				
C5H8O4	ETHYLMALONIC ACID	(S)	132.114	-224.9	B(1535)				
C5H8O4	DIMETHYLMALONIC ACID	(S)	132.114	-228.1	C(744)				
C5H8O4	DIMETHYL MALONATE	(L)	132.114	-188.9	C(744)				
C5H8O4	GLUTARIC ACID	(S)	132.114	-229.46	A(1612)				
C5H8O4	METHYLSUCCINIC ACID	(S)	132.114	-226.8	C(258)				
C5H8O7	2,3,4-TRIHYDROXYGLUTARIC ACID	(S)	180.114	-355.3	C(744)				
C5H9ClO2	ETHYL 2-CHLOROPROPIONATE	(L)	136.579	-128.1	B(1384)				
C5H9ClO2	ETHYL 3-CHLOROPROPIONATE	(L)	136.579	-125.2	B(1384)				
C5H9ClO2	ISOPROPYL CHLOROACETATE	(L)	136.579	-129.3	B(1384)				
C5H9ClO2	METHYL 2-CHLOROBUTYRATE	(L)	136.579	-136.1	B(1384)				
C5H9ClO2	METHYL 3-CHLOROBUTYRATE	(L)	136.579	-128.6	B(1384)				
C5H9ClO2	METHYL 4-CHLOROBUTYRATE	(L)	136.579	-129.0	B(1384)				
C5H9ClO2	PROPYL CHLOROACETATE	(L)	136.579	-123.3	A(141)				
C5H9FO	ETHYL FLUOROALLYL ETHER	(U,L)	104.122	-89.5	C(744)				
C5H9N	ISOBUTYL ISOCYANIDE	(L)	83.130	19.0	C(744)				
C5H9N	ISOVALERONITRILE	(L)	83.130	-5.0	C(744)				
C5H9N	1,2,5,6-TETRAHYDROPYRIDINE	(L)	83.130	7.96	A(93)				
C5H9NO2	PROLINE, L	(S)	115.130	-240.19	A(650)	39.21	M(241)		
C5H9NO4	GLUTAMIC ACID, D	(S)	147.130	-241.32	A(664)	45.7	M(644)	-173.87	127.443
C5H9NO4	GLUTAMIC ACID, L	(S)	147.130	-241.32	A(664)	44.98	M(664)	-174.78	128.113
C5H9N3	CYCLOPENTYL AZIDE	(S)	111.146	42.79	A(402)				

Formula	Name	State	Mol Wt	$\Delta Hf°_{298}$ (kcal/mole)	Source	$S°_{298}$ (cal/mole °K)	Source	$\Delta Gf°_{298}$ (kcal/mole)	Log K_p
C5H9N3O6	2-METHYL-2,3,3-TRINITROBUTANE	(S)	207.146	-79.6	B(1001)				
C5H9N3O8	2-ETHYL-2-NITRO-1,3- PROPANEDIOL DINITRATE	(S)	239.146	-86.2	B(983)				
C5H9N3O9	NITROPENTAGLYCERINE	(S)	255.146	-106.0	B(184)				
C5H9N3O10	TRINITRATE OF 3-(2-HYDROXY-ETHOXY)-1,2-PROPANEDIOL	(S)	271.146	-128.2	B(985)				
C5H10	CYCLOPENTANE	(L)	70.130	-25.31	A(1196)	48.82	M(341)	8.69	-6.370
C5H10	CYCLOPENTANE	(G)	70.130	-18.46	TC 340)	70.00	TC 340)	9.23	-6.763
C5H10	2-METHYL-1-BUTENE	(L)	70.130	-14.93	A(1248)	60.70	M(1511)	15.53	-11.382
C5H10	2-METHYL-1-BUTENE	(G)	70.130	-8.68	A(1248)	81.15	TC 270)	15.68	-11.495
C5H10	2-METHYL-2-BUTENE	(L)	70.130	-16.71	A(1248)	60.00	M(1511)	13.96	-10.231
C5H10	2-METHYL-2-BUTENE	(G)	70.130	-10.17	TC 272)	80.92	TC 272)	14.26	-10.453
C5H10	3-METHYL-1-BUTENE	(L)	70.130	-12.6	A(1248)	60.54	M(1511)	17.91	-13.125
C5H10	3-METHYL-1-BUTENE	(G)	70.130	-6.92	TC 271)	79.70	TC 271)	17.87	-13.101
C5H10	METHYLCYCLOBUTANE	(L)	70.130	-11.70	AC 662)				
C5H10	METHYLCYCLOBUTANE	(G)	70.130	-5.0	AC 662)				
C5H10	1-PENTENE	(L)	70.130	-11.16	A(1248)	62.75	M(1511)	18.69	-13.698
C5H10	1-PENTENE	(G)	70.130	-5.00	TC 267)	82.65	TC 267)	18.91	-13.864
C5H10	2-PENTENE, CIS	(L)	70.130	-13.18	A(1248)	61.81	M(1511)	16.95	-12.423
C5H10	2-PENTENE, CIS	(G)	70.130	-6.71	TC 268)	82.76	TC 268)	17.17	-12.587
C5H10	2-PENTENE, TRANS	(L)	70.130	-14.04	A(1248)	61.31	M(1511)	16.24	-11.902
C5H10	2-PENTENE, TRANS	(G)	70.130	-7.59	TC 269)	81.36	TC 269)	16.71	-12.248
C5H10BR2	2,3-DIBROMO-2-METHYLBUTANE	(G)	229.962	-33.20	TC 770)	98.60	TC 770)	-3.19	2.340
C5H10CLNO4	GLUTAMIC ACID HYDROCHLORIDE	(S)	183.595			59.3	M(649)		
C5H10N2O2	N,N'-DIMETHYLMALONAMIDE	(S)	130.146	-125.6	C(744)				
C5H10N2O3	ALANYLGLYCINE, D,L	(S)	146.146	-185.91	AC 643)	51.0	M(642)	-117.00	85.758
C5H10N2O3	ALANYLGLYCINE, L	(S)	146.146	-197.52	AC 664)	46.62	M(664)	-127.30	93.310
C5H10N2O4	2,2-DIMETHYL-1,3-DINITROPROPANE	(S)	162.146	-66.4	B(1001)				
C5H10N2S2	CARBOTHIALDINE	(U,L)	162.278	13.1	C(744)				
C5H10N2S2	DIHYDRO-5-(METHYLENEIMINO)-METHYL-4H-1,3,5-DITHIAZINE	(L)	162.278	-13.8	C(744)				
C5H10N2S2	DIMETHYL FORMOCARBOTHIALDINE	(U,L)	162.278	-1.2	C(744)				
C5H10N6O2	DINITROSOPENTA-METHYLENETETRAMINE	(S)	186.178	53.6	B(320)				

Formula	Name	State	Mol Wt	$\Delta H f^\circ_{298}$ (kcal/mole)	Source	S°_{298} (cal/mole °K)	Source	$\Delta G f^\circ_{298}$ (kcal/mole)	Log $K p$
C5H10N6O2	DINITROSOPENTA-								
C5H10N6O2	METHYLENETETRAMINE	(S)	186.178	54.8	B(986)				
C5H10O	CYCLOBUTANEMETHANOL	(L)	86.130	-64.0	C(744)				
C5H10O	CYCLOPENTANOL	(L)	86.130	-71.74	A(1109)	49.2	M(1109)	-30.55	22.391
C5H10O	2,3-EPOXY-2-METHYLBUTANE	(L)	86.130	-61.4	C(744)				
C5H10O	ISOPROPYL VINYL ETHER	(L)	86.130	-53.7	C(843)				
C5H10O	3-METHYL-2-BUTANONE	(L)	86.130	-77.9	C(744)				
C5H10O	2-PENTANONE	(L)	86.130	-71.7	(CALC)	65.11	M(1084)	-35.25	25.838
C5H10O	2-PENTANONE	(G)	86.130	-61.82	T(556)	89.91	T(556)	-32.76	24.016
C5H10O	3-PENTANONE	(L)	86.130	-71.7	(CALC)				
C5H10O	3-PENTANONE	(G)	86.130	-61.82	A(180)	88.44	M(180)	-32.33	23.695
C5H10O	1-PENTEN-3-OL	(L)	86.130	-58.9	C(744)				
C5H10O	TETRAHYDROPYRAN	(L)	86.130	-61.85	A(1133)				
C5H10O	TETRAHYDROPYRAN	(G)	86.130	-53.50	A(1133)				
C5H10O	VALERALDEHYDE	(L)	86.130	-63.5	(CALC)				
C5H10O	VALERALDEHYDE	(G)	86.130	-54.45	T(548)	91.53	T(548)	-25.88	18.968
C5H10OS	ALLYL ETHYL SULFOXIDE	(L)	118.196	-41.8	B(907)				
C5H10OS	ALLYL ETHYL SULFOXIDE	(G)	118.196	-24.7	B(907)				
C5H10OS	ISOPROPYL THIOACETATE	(L)	118.196	-71.3	A(1554)				
C5H10OS	PROPYL THIOACETATE	(L)	118.196	-70.3	A(1554)				
C5H10O2	1,1-CYCLOPROPANEDIMETHANOL	(L)	102.130	-104.2	C(744)				
C5H10O2	1,2-CYCLOPENTANEDIOL, CIS	(S)	102.130	-116.8	A(706)				
C5H10O2	1,2-CYCLOPENTANEDIOL, TRANS	(S)	102.130	-118.1	A(706)				
C5H10O2	ETHYL PROPIONATE	(L)	102.130	-114.12	B(1253)				
C5H10O2	ISOPROPYL ACETATE	(L)	102.130	-124.0	A(1555)				
C5H10O2	ISOVALERIC ACID	(L)	102.130	-134.0	B(569)				
C5H10O2	METHYL BUTYRATE	(L)	102.130	-119.0	C(744)				
C5H10O2	2-METHYLBUTYRIC ACID	(L)	102.130	-133.0	B(569)				
C5H10O2	METHYL ISOBUTYRATE	(L)	102.130	-117.3	C(744)				
C5H10O2	VALERIC ACID	(L)	102.130	-133.60	A(3)	62.10	M(970)	-88.95	65.197
C5H10O2	VALERIC ACID	(L)	102.130	-133.9	A(853)	62.10	M(970)	-89.25	65.417
C5H10O2	VALERIC ACID	(G)	102.130	-117.20	(CALC)	105.12	M(970)	-85.37	62.578
C5H10O2	PIVALIC ACID	(L)	102.130	-135.0	B(569)				
C5H10O2	TETRAHYDROFURFURYL ALCOHOL	(L)	102.130	-102.3	C(1006)				
C5H10O2S	ALLYL ETHYL SULFONE	(L)	134.196	-97.5	B(905)				
C5H10O2S	ALLYL ETHYL SULFONE	(G)	134.196	-77.6	H(905)				

Formula	Name	State	Mol Wt	$\Delta H f^\circ_{298}$ (kcal/mole)	Source	S°_{298} (cal/mole °K)	Source	$\Delta G f^\circ_{298}$ (kcal/mole)	Log Kp
C5H10O3	DIETHYL CARBONATE	(L)	118.130	-164.0	C (744)				
C5H10O3	ETHYL LACTATE	(L)	118.130	-158.6	C (744)				
C5H10O4	MONOACETIN	(L)	134.130	-216.3	B (1472)				
C5H10O5	ARABINOSE	(S)	150.130	-251.6	C (744)				
C5H10O5	RIBOSE, D	(S)	150.130	-253.9	B (1430)				
C5H10O5	XYLOSE, D	(S)	150.130	-250.8	C (1006)	34.30	M (1006)	-175.94	128.964
C5H10S	ALLYL ETHYL SULFIDE	(L)	102.196	-5.1	B (900)				
C5H10S	ALLYL ETHYL SULFIDE	(G)	102.196	4.3	B (900)				
C5H10S	CYCLOPENTANETHIOL	(L)	102.196	-21.35	A (111)	61.39	M (111)	11.18	-8.194
C5H10S	CYCLOPENTANETHIOL	(G)	102.196	-11.45	T (915)	86.38	T (915)	13.63	-9.989
C5H10S	2,3-EPITHIO-2-METHYLBUTANE	(L)	102.196	-14.46	A (1447)				
C5H10S	2,3-EPITHIO-2-METHYLBUTANE	(G)	102.196	-5.06	A (1447)				
C5H10S	THIACYCLOHEXANE	(L)	102.196	-25.32	A (951)	52.16	M (951)	9.96	-7.301
C5H10S	THIACYCLOHEXANE	(G)	102.196	-15.12	T (885)	77.26	T (885)	12.68	-9.292
C5H11BR	1-BROMOPENTANE	(L)	151.054	-40.72	A (138)				
C5H11BR	1-BROMOPENTANE	(G)	151.054	-30.87	T (771)	97.70	T (771)	-1.37	1.001
C5H11CL	1-CHLORO-3-METHYLBUTANE	(L)	106.595	-51.9	B (1384)				
C5H11CL	1-CHLORO-3-METHYLBUTANE	(G)	106.595	-43.10	T (744)	95.56	T (744)	-10.44	7.651
C5H11CL	2-CHLORO-2-METHYLBUTANE	(L)	106.595	-56.6	B (1384)				
C5H11CL	2-CHLORO-2-METHYLBUTANE	(G)	106.595	-48.40	T (745)	88.06	T (745)	-13.50	9.896
C5H11CL	1-CHLOROPENTANE	(L)	106.595	-50.9	T (743)				
C5H11CL	1-CHLOROPENTANE	(G)	106.595	-41.80	T (743)	94.89	T (743)	-8.94	6.551
C5H11N	1,1-AMINOCYCLOPROPYLETHANE	(U,L)	85.146	-22.6	C (744)				
C5H11N	PIPERIDINE	(L)	85.146	-21.10	A (93)				
C5H11N	PIPERIDINE	(G)	85.146	-11.71	A (93)				
C5H11NO	ISOVALERAMIDE	(S)	101.146	-94.0	C (744)				
C5H11NO	VALERAMIDE	(S)	101.146	-90.7	B (1640)				
C5H11NO2	VALINE, L	(S)	117.146	-147.7	A (666)	42.75	M (666)	-85.80	62.892
C5H11NO2S	METHIONINE, L	(S)	149.212	-181.3	B (667)	55.32	M (667)	-120.88	88.600
C5H11NO4	2-ETHYL-2-NITRO-1,3-PROPANEDIOL	(S)	149.146	-145.3	B (300)				
C5H11NO6	METHYLAMMONIUM BIRACEMATE	(S)	181.146	-340.0	C (744)				
C5H11NO6	METHYLAMMONIUM D-BITARTRATE	(S)	181.146	-338.0	C (744)				
C5H12	2,2-DIMETHYLPROPANE	(L)	72.146	-45.02	A (770)				
C5H12	2,2-DIMETHYLPROPANE	(G)	72.146	-39.67	T (108)	73.23	T (108)	-3.64	2.669
C5H12	2-METHYLBUTANE	(L)	72.146	-42.95	A (770)	62.24	M (550)	-3.64	2.671

Formula	Name	State	Mol Wt	$\Delta H f°_{298}$ (kcal/mole)	Source	$S°_{298}$ (cal/mole °K)	Source	$\Delta G f°_{298}$ (kcal/mole)	Log Kp
C5H12	2-METHYLBUTANE	(G)	72.146	-36.92	T(107)	82.12	T(107)	-3.54	2.596
C5H12	PENTANE	(L)	72.146	-41.40	A(1248)	62.78	M(992)	-2.26	1.653
C5H12	PENTANE	(G)	72.146	-35.00	T(106)	83.40	T(106)	-2.00	1.468
C5H12N2O2	ORNITHINE, D,L	(S)	132.162			46.2	M(652)		
C5H12O	TERT-BUTYL METHYL ALCOHOL	(G)	88.146	-70.00	T(506)	84.36	T(506)	-29.98	21.978
C5H12O	TERT-BUTYL METHYL ETHER	(L)	88.146	-77.2	A(1390)				
C5H12O	2-METHYL-1-BUTANOL	(L)	88.146	-85.24	A(215)				
C5H12O	2-METHYL-2-BUTANOL	(L)	88.146	-90.71	A(215)				
C5H12O	3-METHYL-1-BUTANOL	(L)	88.146	-85.18	A(215)				
C5H12O	3-METHYL-2-BUTANOL	(L)	88.146	-87.63	A(215)				
C5H12O	2-PENTANOL	(L)	88.146	-87.75	A(215)				
C5H12O	3-PENTANOL	(L)	88.146	-88.51	A(215)				
C5H12O	PENTYL ALCOHOL	(L)	88.146	-85.55	A(215)	60.9	M(1106)	-38.54	28.249
C5H12O	PENTYL ALCOHOL	(G)	88.146	-72.27	T(524)	96.21	T(524)	-35.79	26.232
C5H12O	TERT-PENTYL ALCOHOL	(L)	88.146	-90.71	A(215)	54.8	M(1106)	-41.88	30.698
C5H12O	TERT-PENTYL ALCOHOL	(G)	88.146	-78.65	T(525)	87.68	T(525)	-39.62	29.044
C5H12O2	ACETONE DIMETHYL ACETAL	(L)	104.146	-109.9	A(1416)				
C5H12O2	DIETHOXYMETHANE	(L)	104.146	-111.0	C(744)				
C5H12O2	1,5-PENTANEDIOL	(L)	104.146	-105.1	C(1006)	76.82	M(1006)	-55.53	40.704
C5H12O2S	BUTYL METHYL SULFONE	(L)	136.212	-128.0	B(904)				
C5H12O2S	BUTYL METHYL SULFONE	(L)	136.212	-109.2	B(904)				
C5H12O2S	TERT-BUTYL METHYL SULFONE	(L)	136.212	-132.8	B(186)				
C5H12O2S	TERT-BUTYL METHYL SULFONE	(G)	136.212	-111.5	B(186)				
C5H12O3	2-(2-METHOXYETHOXY)ETHANOL	(L)	120.146	-160.8	C(1351)				
C5H12O3	1,1,1-TRIS(HYDROXYMETHYL)-ETHANE	(S)	120.146	-177.6	B(984)				
C5H12O4	3-(2-HYDROXYETHOXY)-1,2-PROPANEDIOL	(L)	136.146	-206.7	B(985)				
C5H12O4	PENTAERYTHRITOL	(S)	136.146	-219.7	B(984)	47.34	M(1131)	-146.73	107.552
C5H12O5	ARABINITOL	(S)	152.146	-267.3	C(744)				
C5H12O5	XYLITOL	(S)	152.146	-267.3	B(1081)				
C5H12S	BUTYL METHYL SULFIDE	(L)	104.212	-34.15	A(952)	73.49	M(952)	4.08	-2.988
C5H12S	BUTYL METHYL SULFIDE	(G)	104.212	-24.42	T(805)	98.43	T(805)	6.37	-4.670
C5H12S	TERT-BUTYL METHYL SULFIDE	(L)	104.212	-37.49	A(1311)				
C5H12S	TERT-BUTYL METHYL SULFIDE	(G)	104.212	-28.93	A(1311)				
C5H12S	ETHYL PROPYL SULFIDE	(L)	104.212	-34.58	A(952)	73.98	M(952)	3.50	-2.566

Formula	Name	State	Mol Wt	$\Delta H f^\circ_{298}$ (kcal/mole)	Source	S°_{298} (cal/mole °K)	Source	$\Delta G f^\circ_{298}$ (kcal/mole)	Log $K p$
C5H12S	ETHYL PROPYL SULFIDE	(G)	104.212	-25.00	T(806)	98.97	T(806)	5.63	-4.126
C5H12S	ETHYL ISOPROPYL SULFIDE	(L)	104.212	-37.3	B(899)				
C5H12S	ETHYL ISOPROPYL SULFIDE	(G)	104.212	-28.0	B(899)				
C5H12S	2-METHYL-2-BUTANETHIOL	(L)	104.212	-38.90	A(1299)	69.34	M(1299)	0.56	-0.413
C5H12S	2-METHYL-2-BUTANETHIOL	(G)	104.212	-30.36	A(898)	92.48	T(898)	2.20	-1.616
C5H12S	3-METHYL-1-BUTANETHIOL	(L)	104.212	-27.44	A(957)				
C5H12S	1-PENTANETHIOL	(L)	104.212	-36.02	A(1443)	74.18	M(425)	2.00	-1.466
C5H12S	1-PENTANETHIOL	(L)	104.212	-35.74	A(957)	74.18	M(425)	2.28	-1.672
C5H12S	1-PENTANETHIOL	(G)	104.212	-25.91	T(899)	99.28	T(899)	4.63	-3.392
C5H12S2	1,5-PENTANEDITHIOL	(L)	136.278	-30.97	A(925)				
C5H12S2	1,5-PENTANEDITHIOL	(G)	136.278	-16.80	A(925)				
C5H13N	ISOPENTYLAMINE	(L)	87.162	-46.8	C(744)	63.74	M(1399)		
C5H14CLN	PENTYLAMINE HYDROCHLORIDE	(S)	123.627			70.25	M(649)		
C5H14CL2N2O2	ORNITHINE DIHYDROCHLORIDE	(S)	205.092						
C6CL4O2	TETRACHLOROBENZOQUINONE	(S)	245.888	-70.3	B(1384)	62.20	M(597)	0.25	-0.186
C6CL6	HEXACHLOROBENZENE	(S)	284.802	-31.30	A(1359)	105.45	T(757)	10.56	-7.740
C6CL6	HEXACHLOROBENZENE	(G)	284.802	-8.10	T(757)	66.90	M(279)	-211.43	154.973
C6F6	HEXAFLUOROBENZENE	(L)	186.060	-237.25	A(290)	91.59	T(719)	-210.18	154.058
C6F6	HEXAFLUOROBENZENE	(G)	186.060	-228.64	T(719)				
C6F10	DECAFLUOROCYCLOHEXENE	(L)	262.060	-470.6	A(290)				
C6F10	DECAFLUOROCYCLOHEXENE	(G)	262.060	-463.3	A(290)				
C6HCL3O2	TRICHLOROBENZOQUINONE	(S)	211.439	-65.5	B(1384)				
C6HCL5O	PENTACHLOROPHENOL	(S)	266.353	-70.6	A(1359)	60.21	M(597)	-34.44	25.243
C6H2CL2O2	2,3-DICHLOROBENZOQUINONE	(S)	176.990	-59.9	B(1384)				
C6H2CL2O2	2,5-DICHLOROBENZOQUINONE	(S)	176.990	-59.1	B(1384)				
C6H2CL2O2	2,6-DICHLOROBENZOQUINONE	(S)	176.990	-59.0	B(1384)				
C6H2CL2O4	2,5-DICHLORO-3,6-DIHYDROXY-P-BENZOQUINONE	(S)	208.990	-157.6	C(744)				
C6H2CL4O2	TETRACHLOROHYDROQUINONE	(S)	247.904	-109.6	B(1384)				
C6H3CLO2	CHLOROBENZOQUINONE	(S)	142.541	-52.7	B(1384)				
C6H3CL3O2	TRICHLOROHYDROQUINONE	(S)	213.455	-106.4	B(1384)				
C6H3FN2O4	1-FLUORO-2,4-DINITROBENZENE	(S)	186.100	-40.3	C(744)				
C6H3F2NO2	1,4-DIFLUORO-2-NITROBENZENE	(L)	159.092	-79.4	C(744)				
C6H3N3O6	1,2,4-TRINITROBENZENE	(S)	213.108	6.8	C(744)				
C6H3N3O6	1,3,5-TRINITROBENZENE	(S)	213.108	9.0	B(184)				

Formula	Name	State	Mol Wt	$\Delta H f^\circ_{298}$ (kcal/mole)	Source	S°_{298} (cal/mole °K)	Source	$\Delta G_f^\circ_{298}$ (kcal/mole)	Log Kp
C6H3N3O7	PICRIC ACID	(S)	229.108	-51.7	B(57)				
C6H3N3O8	2,4,6-TRINITRORESORCINOL	(S)	245.108	-103.9	B(981)				
C6H3N5O8	2,3,4,6-TETRANITROANILINE	(S)	273.124	-11.7	C(1231)				
C6H4BR2	P-DIBROMOBENZENE	(S)	235.924	13.0	C(1180)				
C6H4CL2	M-DICHLOROBENZENE	(L)	147.006	-5.12	A(635)				
C6H4CL2	M-DICHLOROBENZENE	(G)	147.006	6.32	T(754)	82.09	T(754)	18.78	-13.765
C6H4CL2	O-DICHLOROBENZENE	(L)	147.006	-4.40	T(755)				
C6H4CL2	O-DICHLOROBENZENE	(G)	147.006	7.16	T(755)	81.61	T(755)	19.76	-14.485
C6H4CL2	P-DICHLOROBENZENE	(L)	147.006	-10.0	T(756)				
C6H4CL2	P-DICHLOROBENZENE	(G)	147.006	5.50	T(756)	80.47	T(756)	18.44	-13.518
C6H4CL2O2	2,3-DICHLOROHYDROQUINONE	(S)	179.006	-100.0	B(1384)				
C6H4CL2O2	2,6-DICHLOROHYDROQUINONE	(S)	179.006	-101.7	B(1384)				
C6H4FNO2	1-FLUORO-2-NITROBENZENE	(L)	141.100	-19.4	C(744)				
C6H4FNO2	1-FLUORO-3-NITROBENZENE	(L)	141.100	-22.8	C(744)				
C6H4FNO2	1-FLUORO-4-NITROBENZENE	(L)	141.100	-73.0	C(744)				
C6H4FNO3	4-FLUORO-3-NITROPHENOL	(S)	157.100	-82.38	A(512)				
C6H4F2	M-DIFLUOROBENZENE	(G)	114.092	-74.09	T(716)	76.57	T(716)	-61.43	45.026
C6H4F2	M-DIFLUOROBENZENE	(L)	114.092	-79.04	A(512)	53.20	M(1322)	-59.41	43.547
C6H4F2	O-DIFLUOROBENZENE	(G)	114.092	-70.39	T(717)	76.94	T(717)	-57.84	42.395
C6H4F2	O-DIFLUOROBENZENE	(L)	114.092	-81.98	A(512)				
C6H4F2	P-DIFLUOROBENZENE	(G)	114.092	-73.43	T(718)	75.43	T(718)	-60.43	44.294
C6H4I2	M-DIIODOBENZENE	(S)	329.912	44.8	B(1382)				
C6H4I2	O-DIIODOBENZENE	(L)	329.912	41.3	B(1382)				
C6H4I2	O-DIIODOBENZENE	(S)	329.912	44.7	B(1382)				
C6H4I2	P-DIIODOBENZENE	(S)	329.912	38.5	B(1382)				
C6H4N2	NICOTINONITRILE	(S)	104.108	46.78	A(1275)				
C6H4N2	PICOLINONITRILE	(S)	104.108	62.0	E(680)	77.09	S(680)	73.71	-54.026
C6H4N2O4	M-DINITROBENZENE	(S)	168.108	-6.2	B(184)				
C6H4N2O4	O-DINITROBENZENE	(S)	168.108	2.7	C(744)				
C6H4N2O4	P-DINITROBENZENE	(S)	168.108	-9.05	B(57)				
C6H4N2O5	2,4-DINITROPHENOL	(S)	184.108	-55.5	B(57)				
C6H4N2O5	2,6-DINITROPHENOL	(S)	184.108	-49.4	B(57)				
C6H4N2O6	2,4-DINITRORESORCINOL	(S)	200.108	-104.9	B(984)				
C6H4N2O6	4,6-DINITRORESORCINOL	(S)	200.108	-110.7	B(984)				
C6H4N2S3	BIS(DIMETHYLTHIOCARBAMOYL)								

Formula	Name	State	Mol Wt	$\Delta H_{f\ 298}^{\circ}$ (kcal/mole)	Source	S_{298}° (cal/mole °K)	Source	$\Delta G_{f\ 298}^{\circ}$ (kcal/mole)	Log Kp
C6H4N2S3	BIS(DIMETHYLTHIOCARBAMOYL) SULFIDE	(S)	200.306	11.44	AC 510)				
C6H4N2S4	BIS(DIMETHYLTHIOCARBAMOYL) DISULFIDE	(S)	232.372	9.74	AC 510)				
C6H4N4O6	2,4,6-TRINITROANILINE	(S)	228.124	-17.8	BC 981)				
C6H4O2	P-BENZOQUINONE	(S)	108.092	-44.65	AC(1147)	38.5	(CALC)	-20.47	15.006
C6H4O2	P-BENZOQUINONE	(G)	108.092	-28.3	CC 913)	76.65	SC 89)	-15.50	11.359
C6H5BR	BROMOBENZENE	(L)	157.016	14.5	TC 774)	52.0	TC 774)	30.12	-22.077
C6H5BR	BROMOBENZENE	(G)	157.016	25.10	TC 774)	77.53	TC 774)	33.11	-24.267
C6H5CL	CHLOROBENZENE	(L)	112.557	2.58	TC 753)	50.0	TC 753)	21.32	-15.624
C6H5CL	CHLOROBENZENE	(G)	112.557	12.39	TC 753)	74.92	TC 753)	23.70	-17.368
C6H5CLO	M-CHLOROPHENOL	(S)	128.557	-49.7	B(1384)				
C6H5CLO	O-CHLOROPHENOL	(L)	128.557	-44.3	B(1384)				
C6H5CLO	P-CHLOROPHENOL	(S)	128.557	-47.6	B(1384)				
C6H5CLO2	CHLOROHYDROQUINONE	(S)	144.557	-91.5	B(1384)				
C6H5F	FLUOROBENZENE	(L)	96.100	-36.13	AC 514)	49.22	M(1318)	-17.88	13.109
C6H5F	FLUOROBENZENE	(G)	96.100	-27.86	TC 715)	72.33	TC 715)	-16.50	12.097
C6H5FN2O2	4-FLUORO-3-NITROANILINE	(S)	156.116	-41.0	CC 744)				
C6H5FO	M-FLUOROPHENOL	(L)	112.100	-81.3	CC 744)				
C6H5FO	O-FLUOROPHENOL	(L)	112.100	-72.1	CC 744)				
C6H5FO	P-FLUOROPHENOL (M.P. 48.2 C)	(S)	112.100	-79.9	CC 744)				
C6H5I	IODOBENZENE	(L)	204.010	27.0	TC 786)	53.3	TC 786)	40.95	-30.012
C6H5I	IODOBENZENE	(G)	204.010	38.85	TC 786)	79.84	TC 786)	44.88	-32.898
C6H5IO	M-IODOPHENOL	(S)	220.010	-22.5	B(1382)				
C6H5IO	O-IODOPHENOL	(L)	220.010	-22.8	B(1382)				
C6H5IO	P-IODOPHENOL	(S)	220.010	-22.7	B(1382)				
C6H5NO	NITROSOBENZENE	(S)	107.108	-7.0	CC 843)				
C6H5NO2	P-BENZOQUINONE OXIME	(C)	123.108	-21.1	CC 744)				
C6H5NO2	NITROBENZENE	(L)	123.108	3.8	C(1118)	53.6	M(1118)	34.95	-25.618
C6H5NO3	M-NITROPHENOL	(S)	139.108	-50.3	CC 744)				
C6H5NO3	O-NITROPHENOL	(S)	139.108	-46.4	CC 744)				
C6H5NO3	P-NITROPHENOL	(S)	139.108	-46.4	CC 744)				
C6H5N3	BENZOTRIAZOLE	(S)	119.124	59.66	AC 401)				
C6H5N3O3	BENZENEDIAZONIUM NITRATE	(S)	167.124	48.0	CC 744)				
C6H5N3O4	2,3-DINITROANILINE	(S)	183.124	-2.8	B 896)				
C6H5N3O4	2,4-DINITROANILINE	(S)	183.124	-22.4	BC 984)				
C6H5N3O4	2,5-DINITROANILINE	(S)	183.124	-10.6	CC 896)				

Formula	Name	State	Mol Wt	$\Delta H f°_{298}$ (kcal/mole)	Source	$S°_{298}$ (cal/mole °K)	Source	$\Delta G f°_{298}$ (kcal/mole)	Log $K p$
C6H5N3O4	2,6-DINITROANILINE	(S)	183.124	-12.1	C(896)				
C6H5N3O4	3,4-DINITROANILINE	(S)	183.124	-7.8	C(896)				
C6H5N3O4	3,5-DINITROANILINE	(S)	183.124	-9.3	C(896)				
C6H5N3O5	2-AMINO-4,6-DINITROPHENOL	(S)	199.124	-59.4	B(984)				
C6H5N5O6	2,4,6-TRINITROPHENYLHYDRAZINE	(L)	243.140	3.0	B(1248)				
C6H6	BENZENE	(L)	78.108	11.72	A(1248)	41.41	M(1085)	29.72	-21.788
C6H6	BENZENE	(G)	78.108	19.82	A(396)	64.34	T(396)	30.99	-22.714
C6H6	1,5-HEXADIYNE	(L)	78.108	91.9	A(1364)				
C6H6	2,4-HEXADIYNE	(G)	78.108	79.0	C(744)				
C6H6CL6	ALPHA-HEXACHLOROCYCLOHEXANE	(S)	290.850	-41.23	B(1294)				
C6H6CL6	BETA-HEXACHLOROCYCLOHEXANE	(S)	290.850	-42.99	B(1294)				
C6H6CL6	GAMMA-HEXACHLOROCYCLOHEXANE	(S)	290.850	-37.92	B(1294)				
C6H6CL6	DELTA-HEXACHLOROCYCLOHEXANE	(S)	290.850	-41.00	B(1294)				
C6H6FN	M-FLUOROANILINE	(L)	111.116	-35.0	C(744)				
C6H6FN	O-FLUOROANILINE	(L)	111.116	-33.9	C(744)				
C6H6FN	P-FLUOROANILINE	(L)	111.116	-29.9	C(744)				
C6H6IN	IODOANILINE	(U,L)	219.026	41.8	C(744)				
C6H6N2O2	M-NITROANILINE	(S)	138.124	-4.1	C(744)				
C6H6N2O2	O-NITROANILINE	(S)	138.124	-3.0	C(744)				
C6H6N2O2	P-NITROANILINE	(S)	138.124	-9.92	A(244)				
C6H6N2O3	N-(O-NITROPHENYL)HYDROXYL-AMINE	(L)	154.124	-3.2	C(744)				
C6H6N4	7-METHYLPURINE	(L)	134.140	52.3	C(744)				
C6H6N4	NITRILOTRIACETONITRILE	(L)	134.140	77.3	C(744)				
C6H6N4O	7-METHYLHYPOXANTHINE	(S)	150.140	-8.9	C(744)				
C6H6N4O4	2,4-DINITROPHENYLHYDRAZINE	(S)	198.140	6.4	B(984)				
C6H6N4O7	AMMONIUM PICRATE(RED)	(S)	246.140	-93.0	B(981)				
C6H6N4O7	AMMONIUM PICRATE(YELLOW)	(S)	246.140	-92.4	B(981)				
C6H6O	PHENOL	(S)	94.108	-39.44	A(285)	34.42	M(19)	-12.05	8.829
C6H6O	PHENOL	(L)	94.108	-23.03	T(565)	75.43	T(565)	-7.86	5.763
C6H6O	2-VINYLFURAN	(L)	94.108	1.52	C(844)				
C6H6O2	HYDROQUINONE	(S)	110.108	-87.51	A(1147)				
C6H6O2	HYDROQUINONE	(G)	110.108	-66.0	A(1147)				
C6H6O2	PYROCATECHOL	(S)	110.108	-83.9	C(744)				
C6H6O2	RESORCINOL	(S)	110.108	-85.7	C(744)				
C6H6O3	DIMETHYLMALEIC ANHYDRIDE	(S)	126.108	-136.2	C(1414)				

Formula	Name	State	Mol Wt	$\Delta H^\circ_{f\,298}$ (kcal/mole)	Source	S°_{298} (cal/mole °K)	Source	$\Delta G^\circ_{f\,298}$ (kcal/mole)	Log K_p
C6H6O3	PHLOROGLUCINOL	(S)	126.108	-151.5	C(744)				
C6H6O3	PYROGALLOL	(S)	126.108	-130.1	C(744)				
C6H6O3S	BENZENESULFONIC ACID	(G)	158.174			87.88	E(69)		
C6H6O6	ACONITIC ACID, CIS	(S)	174.108	-292.7	B(1611)				
C6H6O6	ACONITIC ACID, TRANS	(S)	174.108	-294.65	A(1611)				
C6H6S	BENZENETHIOL	(L)	110.174	15.27	A(1319)	53.25	M(1319)	32.02	-23.470
C6H6S	BENZENETHIOL	(G)	110.174	26.66	T(916)	80.51	T(916)	35.28	-25.861
C6H7N	ANILINE	(L)	93.124	20.76	T(616)	45.72	M(579)	35.63	-26.113
C6H7N	ANILINE	(G)	93.124	13.50	A(1315)	76.28	T(616)	39.84	-29.205
C6H7N	2-PICOLINE	(L)	93.124	23.65	T(1310)	52.07	M(1315)	39.80	-29.174
C6H7N	2-PICOLINE	(G)	93.124	14.75	A(1315)	77.68	T(614)	42.32	-31.017
C6H7N	3-PICOLINE	(L)	93.124	25.37	T(615)	51.70	M(1310)	41.16	-30.171
C6H7N	3-PICOLINE	(G)	93.124	13.57	A(288)	77.67	T(615)	44.04	-32.280
C6H7N	4-PICOLINE	(L)	93.124	24.43	A(288)				
C6H7N	4-PICOLINE	(G)	93.124	-42.8	C(744)				
C6H7NO	P-AMINOPHENOL	(S)	109.124	1.7	C(744)				
C6H7NO	N-PHENYLHYDROXYLAMINE	(L)	109.124	-118.0	C(744)				
C6H7NO3	METHYL 2-CYANOACETOACETATE	(S)	141.124	25.9	A(762)				
C6H8	1,3-CYCLOHEXADIENE	(G)	80.124	0.8	B(747)				
C6H8N2	P-PHENYLENEDIAMINE	(S)	108.140	33.99	A(244)				
C6H8N2	PHENYLHYDRAZINE	(L)	108.140	-42.5	C(277)				
C6H8N2O3	ANILINE NITRATE	(S)	156.140	25.2	B(490)				
C6H8N2S	3,3'-THIODIPROPIONITRILE	(G)	140.206						
C6H8N6O2	3-(5-TETRAZOLYLAZO)- 2,4-PENTANEDIONE	(S)	196.172	3.4	C(1616)				
C6H8N6O18	MANNITOL HEXANITRATE	(S)	452.172	-154.0	C(843)				
C6H8O2	2,4-PENTANEDIONE	(L)	112.124	-195.4	B(1064)				
C6H8O2	SORBIC ACID	(S)	112.124	-94.0	C(744)				
C6H8O3	2,2-DIMETHYLSUCCINIC ANHYDRIDE	(S)	128.124	-154.0	B(1538)				
C6H8O3	2,3-DIMETHYLSUCCINIC ANHYDRIDE, CIS	(S)	128.124	-155.6	B(1538)				
C6H8O3	2,3-DIMETHYLSUCCINIC ANHYDRIDE, TRANS	(S)	128.124	-157.4	B(1538)				
C6H8O3	2-ETHYLSUCCINIC ANHYDRIDE	(L)	128.124	-151.9	B(1538)				
C6H8O4	ALLYLMALONIC ACID	(S)	144.124	-199.4	C(744)				

Formula	Name	State	Mol Wt	$\Delta Hf°_{298}$ (kcal/mole)	Source	$S°_{298}$ (cal/mole °K)	Source	$\Delta Gf°_{298}$ (kcal/mole)	Log Kp
C6H8O4	1,1-CYCLOBUTANEDICARBOXYLIC ACID	(S)	144.124	-195.2	C(744)				
C6H8O4	1,2-CYCLOBUTANEDICARBOXYLIC ACID, CIS	(S)	144.124	-195.1	C(744)				
C6H8O4	1,3-CYCLOBUTANE-DICARROXYLIC ACID	(S)	144.124	-198.2	C(744)				
C6H8O4	DIMETHYL FUMARATE	(S)	144.124	-172.8	C(744)				
C6H8O4	DIMETHYL MALEATE	(S)	144.124	-168.4	C(744)				
C6H8O4	2-HEXENEDIOIC ACID	(S)	144.124	-208.4	C(744)				
C6H8O4	3-HEXENEDIOIC ACID	(S)	144.124	-208.1	C(744)				
C6H8O6	1,2,3-PROPANETRICARBOXYLIC ACID	(S)	176.124	-321.3	C(744)				
C6H8O7	CITRIC ACID	(S)	192.124	-369.0	A(1612)	58.49	M(191)		
C6H8S	2,5-DIMETHYLTHIOPHENE	(L)	112.190	-131.8	B(1384)				
C6H9CL3O2	BUTYL TRICHLOROACETATE	(L)	219.503	-133.5	B(1384)				
C6H9CL3O2	ISOBUTYL TRICHLOROACETATE	(L)	219.503	-28.7	B(432)				
C6H9NO	2-HEXYNAMIDE	(S)	111.140	-16.7	C(1467)				
C6H9NO	3,4,5-TRIMETHYLISOXAZOLE	(L)	111.140	-82.2	C(744)				
C6H9NO2	PROPYL CYANOACETATE	(L)	127.140	-200.2	C(744)				
C6H9NO4	N-ETHYLTARTRIMIDE, D,L	(S)	159.140	-200.1	C(744)				
C6H9NO4	N-ETHYLTARTRIMIDE, D,L	(S)	159.140	-199.4	C(744)				
C6H9NO4	N-ETHYLMESOTARTRIMIDE	(S)	159.140	-311.8	C(744)				
C6H9NO6	(2-(CARBAMOYLMETHOXY)ETHOXY)-ACETIC ACID	(S)	191.140	-200.7	B(1535)				
C6H9O4	PROPYLMALONIC ACID	(L)	145.132	-4.82	AC 821)				
C6H10	BICYCLO(3.1.0)HEXANE	(L)	82.140	-9.28	AC 821)				
C6H10	CYCLOHEXENE	(L)	82.140	-1.28	T(346)				
C6H10	CYCLOHEXENE	(G)	82.140	11.7	C(1374)				
C6H10	2-CYCLOPROPYLPROPENE	(L)	82.140	4.0	AC 298)				
C6H10	2,3-DIMETHYL-1,3-BUTADIENE	(L)	82.140	11.4	AC 335)				
C6H10	2,3-DIMETHYL-1,3-BUTADIENE	(G)	82.140	-7.7	C(744)				
C6H10	DIMETHYLMETHYLENECYCLOPROPANE	(U,L)	82.140	12.54	AC 268)				
C6H10	1,5-HEXADIENE	(L)	82.140	20.0	AC 762)	51.67	M(645)	24.28	-17.795
C6H10	1,5-HEXADIENE	(G)	82.140	29.55	AC 323)	74.27	T(346)	25.54	-18.720
C6H10	1-HEXYNE	(L)	82.140	-9.05	AC 821)				
C6H10	1-METHYLCYCLOPENTENE	(L)	82.140			88.13	T(323)	52.24	-38.288

Formula	Name	State	Mol Wt	$\Delta H_f^\circ_{298}$ (kcal/mole)	Source	S°_{298} (cal/mole °K)	Source	$\Delta G_f^\circ_{298}$ (kcal/mole)	Log K_P
C6H10	1-METHYLCYCLOPENTENE	(G)	82.140	-1.30	T(368)	78.00	T(368)	24.41	-17.890
C6H10	1-METHYLCYCLOPENTENE	(L)	82.140	-5.68	A(821)				
C6H10	3-METHYLCYCLOPENTENE	(G)	82.140	2.07	T(369)	79.00	T(369)	27.48	-20.141
C6H10	3-METHYLCYCLOPENTENE	(L)	82.140	-4.22	A(821)				
C6H10	4-METHYLCYCLOPENTENE	(G)	82.140	3.53	T(370)	78.60	T(370)	29.06	-21.299
C6H10BR2	1,2-DIBROMOCYCLOHEXANE	(L)	241.972	-38.37	A(137)				
C6H10BR2	1,2-DIBROMOCYCLOHEXANE	(G)	241.972	-26.27	A(137)				
C6H10CLN3	HISTIDINE HYDROCHLORIDE, L	(S)	191.621			65.99	M(242)		
C6H10CL2O2	BUTYL DICHLOROACETATE	(L)	185.054	-132.3	B(1384)				
C6H10CL2O2	ISOBUTYL DICHLOROACETATE	(L)	185.054	-133.1	B(1384)				
C6H10N2O2	ALANINE ANHYDRIDE	(S)	142.156	-118.2	C(744)				
C6H10N2O2	ALANINE ANHYDRIDE, D	(S)	142.156	-118.6	C(744)				
C6H10N4S2	5-ETHYL-5-METHYLHYDANTOIN	(S)	142.156	-133.5	B(1471)				
C6H10N4S2	TETRAHYDRO-3,7-DIMETHYL-S-TRIAZOLO[A]-S-TRIAZOLE-1,5-DITHIONE	(S)	202.304	32.7	B(1446)				
C6H10N6O6	HEXAMETHYLENETETRAMINE DINITRATE	(S)	262.188	92.8	B(320)				
C6H10N6O6	5,5'-AZOBIS(2-ETHYLTETRAZOLE)	(S)	222.220	156.6	B(1616)				
C6H10O	ALLYL ETHER	(G)	98.140	0.8	C(744)				
C6H10O	CYCLOBUTYL METHYL KETONE	(L)	98.140	-49.0	C(744)				
C6H10O	CYCLOHEXANONE	(L)	98.140	-65.1	T(558)	77.00	T(558)	-21.69	15.898
C6H10O	CYCLOHEXANONE	(G)	98.140	-55.00	T(558)				
C6H10O	1-HEXEN-4-ONE	(L)	98.140	-47.3	C(744)				
C6H10O	1-HEXEN-5-ONE	(L)	98.140	-47.7	C(744)				
C6H10O	3-METHYLCYCLOPENTANONE	(L)	98.140	-72.4	C(744)				
C6H10O	4-METHYL-3-PENTEN-2-ONE	(L)	98.140	-59.3	C(843)				
C6H10O	7-OXABICYCLO(2.2.1)HEPTANE	(L)	98.140	-53.58	A(94)				
C6H10O2	ACETYLACETONE-O-METHYL ETHER	(U,L)	114.140	-87.5	C(744)				
C6H10O2	ETHYL CROTONATE	(L)	114.140	-100.7	B(1286)				
C6H10O2	3-HEXENOIC ACID	(L)	114.140	-109.1	C(744)				
C6H10O2	METHYL ACETYLACETONE	(U,L)	114.140	-112.7	C(744)				
C6H10O2	METHYL CYCLOBUTANECARBOXYLATE	(L)	114.140	-96.8	C(744)				
C6H10O2	METHYL DIMETHYLACRYLATE	(L)	114.140	-100.1	C(744)				
C6H10O2	BIS(VINYLOXY)ETHANE	(L)	114.140	-66.2	C(843)				
C6H10O3	ETHYL ACETOACETATE	(L)	130.140	-152.3	C(744)				

Formula	Name	State	Mol Wt	$\Delta H f^\circ_{298}$ (kcal/mole)	Source	S°_{298} (cal/mole °K)	Source	$\Delta G f^\circ_{298}$ (kcal/mole)	Log Kp
C6H10O3	PROPIONIC ANHYDRIDE	(S)	130.140	-162.4	B(258)				
C6H10O4	ADIPIC ACID	(S)	146.140	-236.5	C(744)				
C6H10O4	DIETHYL OXALATE	(L)	146.140	-189.9	C(744)				
C6H10O4	DIMETHYL SUCCINATE	(L)	146.140	-202.2	C(744)				
C6H10O4	DIMETHYL SUCCINATE	(L)	146.140	-198.0	C(744)				
C6H10O4	2,2-DIMETHYLSUCCINIC ACID	(S)	146.140	-235.4	B(1538)				
C6H10O4	2,3-DIMETHYLSUCCINIC ACID, CIS	(S)	146.140	-232.9	B(1538)				
C6H10O4	2,3-DIMETHYLSUCCINIC ACID, TRANS	(S)	146.140	-234.4	B(1538)				
C6H10O4	2,3-DIMETHYLSUCCINIC ACID, L TRANS	(S)	146.140	-234.1	B(1538)				
C6H10O4	ERYTHRITEMETHYLAL	(U,S)	146.140	-160.4	C(744)				
C6H10O4	ETHYLMETHYLMALONIC ACID	(S)	146.140	-233.6	C(744)				
C6H10O4	ETHYLSUCCINIC ACID	(S)	146.140	-235.7	B(1538)				
C6H10O4	ISOPROPYLMALONIC ACID	(S)	146.140	-229.4	C(744)				
C6H10O4	2-METHYLGLUTARIC ACID	(S)	146.140	-234.9	C(744)				
C6H10O4	PROPIONYL PEROXIDE	(L)	146.140	-148.2	B(679)				
C6H10O4	PROPYLMALONIC ACID	(S)	146.140	-230.7	B(1535)				
C6H10O4S2	2,2'-DITHIODIPROPIONIC ACID	(S)	210.272	-231.7	C(646)	65.50 M(647)	-168.50	123.504
C6H10O5	LEVOGLUCOSAN	(S)	162.140	-227.8	C(744)				
C6H10O5	SACCHARINIC ACID LACTONE	(S)	162.140	-248.9	C(744)				
C6H10O6	DIMETHYL TARTRATE, D	(S)	178.140	-286.1	B(358)				
C6H10O6	DIMETHYL TARTRATE, RACEMIC	(S)	178.140	-287.5	B(358)				
C6H10O6	DIMETHYL TARTRATE, MESO	(S)	178.140	-286.3	B(358)				
C6H10O6	GLUCONOLACTONE, L	(S)	178.140	-291.3	C(744)				
C6H10O6	MANNOLACTONE, D	(L)	178.140	-287.3	C(744)				
C6H10O6	MANNOLACTONE, L	(S)	178.140	-289.7	C(744)				
C6H10O8	ALLOGALACTARIC ACID	(S)	210.140	-411.8	C(744)				
C6H10O8	CITRIC ACID MONOHYDRATE	(S)	210.140	-398.63	C(843)	67.74 M(386)	-311.42	228.268
C6H10O8	GALACTARIC ACID	(S)	210.140	-422.0	C(744)				
C6H11ClO2	BUTYL CHLOROACETATE	(L)	150.605	-128.7	A(141)				
C6H11ClO2	ETHYL 2-CHLOROBUTYRATE	(L)	150.605	-146.6	B(1384)				
C6H11ClO2	ETHYL 3-CHLOROBUTYRATE	(L)	150.605	-135.6	B(1384)				
C6H11ClO2	ETHYL 4-CHLOROBUTYRATE	(L)	150.605	-135.8	B(1384)				
C6H11ClO2	ISOBUTYL CHLOROACETATE	(L)	150.605	-132.1	B(1384)				
C6H11ClO2	ISOPROPYL 2-CHLOROPROPIONATE	(L)	150.605	-140.3	B(1384)				
C6H11ClO2	ISOPROPYL 3-CHLOROPROPIONATE	(L)	150.605	-132.9	B(1384)				

Formula	Name	State	Mol Wt	$\Delta H f^\circ_{298}$ (kcal/mole)	Source	S°_{298} (cal/mole °K)	Source	$\Delta G f^\circ_{298}$ (kcal/mole)	Log Kp
C6H11CL02	PROPYL 2-CHLOROPROPIONATE	(L)	150.605	-131.7	B(1384)				
C6H11CL02	PROPYL 3-CHLOROPROPIONATE	(L)	150.605	-113.5	B(1384)				
C6H11I	IODOCYCLOHEXANE	(L)	210.058	-23.0	B(1382)				
C6H11N	ISOPENTYL ISOCYANIDE	(L)	97.156	10.3	C(744)				
C6H11NO	HEXAHYDRO-2H-AZEPIN-2-ONE	(L)	113.156	-131.92	B(1269)				
C6H11N3	AZIDOCYCLOHEXANE	(L)	125.172	25.90	A(402)				
C6H11N3O6	DIGLYCYLGLYCINE	(S)	189.172	-229.2	C(744)				
C6H11N3O6	2-METHYL-2,3,3-TRINITROPENTANE	(S)	221.172	-70.0	B(1001)				
C6H11N3O9	2-(HYDROXYMETHYL)-2-METHYL-1,3-PROPANEDIOL TRINITRATE	(S)	269.172	-111.1	B(983)				
C6H12	CYCLOHEXANE	(L)	84.156	-37.34	A(1196)	48.84	M(1264)	6.37	-4.666
C6H12	CYCLOHEXANE	(G)	84.156	-29.43	T(341)	71.28	T(341)	7.59	-5.560
C6H12	2-CYCLOPROPYLPROPANE	(L)	84.156	-11.1	C(1375)				
C6H12	2,3-DIMETHYL-1-BUTENE	(L)	84.156	-22.85	A(80)				
C6H12	2,3-DIMETHYL-1-BUTENE	(G)	84.156	-13.32	T(287)	87.39	T(287)	18.89	-13.848
C6H12	3,3-DIMETHYL-1-BUTENE	(L)	84.156	-21.01	A(80)	61.3	M(741)	18.89	-13.847
C6H12	3,3-DIMETHYL-1-BUTENE	(G)	84.156	-10.31	T(288)	82.16	T(288)	23.46	-17.197
C6H12	2,3-DIMETHYL-2-BUTENE	(L)	84.156	-24.48	A(80)	64.58	M(1306)	14.53	-10.653
C6H12	2,3-DIMETHYL-2-BUTENE	(G)	84.156	-14.15	T(289)	87.15	T(289)	18.13	-13.292
C6H12	2-ETHYL-1-BUTENE	(L)	84.156	-20.82	A(80)				
C6H12	2-ETHYL-1-BUTENE	(G)	84.156	-12.32	T(286)	90.01	T(286)	19.11	-14.008
C6H12	1-HEXENE	(L)	84.156	-17.30	A(80)	70.55	M(950)	19.93	-14.611
C6H12	1-HEXENE	(G)	84.156	-9.96	T(273)	91.93	T(273)	20.90	-15.319
C6H12	2-HEXENE, CIS	(L)	84.156	-20.05	A(80)				
C6H12	2-HEXENE, CIS	(G)	84.156	-12.51	T(274)	92.37	T(274)	18.22	-13.353
C6H12	2-HEXENE, TRANS	(L)	84.156	-20.44	A(80)				
C6H12	2-HEXENE, TRANS	(G)	84.156	-12.88	T(275)	90.97	T(275)	18.27	-13.388
C6H12	3-HEXENE, CIS	(L)	84.156	-18.87	A(80)				
C6H12	3-HEXENE, CIS	(G)	84.156	-11.38	T(276)	90.73	T(276)	19.84	-14.540
C6H12	3-HEXENE, TRANS	(L)	84.156	-20.57	A(80)				
C6H12	3-HEXENE, TRANS	(G)	84.156	-13.01	T(277)	89.59	T(277)	18.55	-13.594
C6H12	METHYLCYCLOPENTANE	(L)	84.156	-33.08	A(1196)	59.26	M(341)	7.52	-5.512
C6H12	METHYLCYCLOPENTANE	(G)	84.156	-25.50	T(347)	81.24	T(347)	8.55	-6.264
C6H12	2-METHYL-1-PENTENE	(L)	84.156	-21.50	A(80)				
C6H12	2-METHYL-1-PENTENE	(G)	84.156	-12.49	T(278)	91.34	T(278)	18.55	-13.593

Formula	Name	State	Mol Wt	$\Delta H f^\circ_{298}$ (kcal/mole)	Source	S°_{298} (cal/mole °K)	Source	$\Delta G f^\circ_{298}$ (kcal/mole)	Log Kp
C6H12	3-METHYL-1-PENTENE	(L)	84.156	-18.68	A(80)				
C6H12	3-METHYL-1-PENTENE	(G)	84.156	-10.76	T(279)	90.06	T(279)	20.66	-15.141
C6H12	4-METHYL-1-PENTENE	(L)	84.156	-19.13	A(80)				
C6H12	4-METHYL-1-PENTENE	(G)	84.156	-10.54	T(280)	87.89	T(280)	21.52	-15.776
C6H12	2-METHYL-2-PENTENE	(L)	84.156	-23.55	A(80)				
C6H12	2-METHYL-2-PENTENE	(G)	84.156	-14.28	T(281)	90.45	T(281)	17.02	-12.476
C6H12	3-METHYL-2-PENTENE, CIS	(L)	84.156	-22.56	A(80)				
C6H12	3-METHYL-2-PENTENE, CIS	(G)	84.156	-13.80	T(282)	90.45	T(282)	17.50	-12.827
C6H12	3-METHYL-2-PENTENE, TRANS	(L)	84.156	-22.60	A(80)				
C6H12	3-METHYL-2-PENTENE, TRANS	(G)	84.156	-14.02	T(283)	91.26	T(283)	17.04	-12.489
C6H12	4-METHYL-2-PENTENE, CIS	(L)	84.156	-20.80	A(80)				
C6H12	4-METHYL-2-PENTENE, CIS	(G)	84.156	-12.03	T(284)	89.23	T(284)	19.63	-14.391
C6H12	4-METHYL-2-PENTENE, TRANS	(L)	84.156	-21.88	A(80)				
C6H12	4-METHYL-2-PENTENE, TRANS	(G)	84.156	-12.99	T(285)	88.02	T(285)	19.03	-13.952
C6H12F4N2	1,2-BIS(DIFLUOROAMINO)-4-METHYLPENTANE	(L)	188.172	-62.8	A(422)				
C6H12F4N2	1,2-BIS(DIFLUOROAMINO)-4-METHYLPENTANE	(G)	188.172	-52.3	A(422)				
C6H12F4N2	TRIETHYLENEDIAMINE	(S)	112.172	-3.4	B(1122)	37.67	M(1593)	57.28	-41.987
C6H12N2O3	ETHYL GLYCYLGLYCINATE	(S)	160.172	-169.8	C(744)				
C6H12N2O4	2,3-DIMETHYL-2,3-DINITROBUTANE	(S)	176.172	-83.6	B(1001)				
C6H12N2O4S2	CYSTINE, L	(S)	240.304	-251.4	C(646)	67.06	M(667)	-165.71	121.462
C6H12N2O4S2	CYSTINE, L	(S)	240.304	-249.6	C(667)	67.06	M(667)	-163.91	120.143
C6H12N2O5	SERINE ANHYDRIDE	(S)	192.172	-277.6	B(1177)				
C6H12N2O5	SERYLSERINE	(S)	192.172	-281.5	B(1176)				
C6H12N2S3	BIS(DIMETHYLTHIOCARBAMOYL) SULFIDE	(S)	208.370	11.35	A(510)				
C6H12N2S4	BIS(DIMETHYLTHIOCARBAMOYL) DISULFIDE	(S)	240.436	9.62	A(510)				
C6H12N4	HEXAMETHYLENETETRAMINE	(S)	140.188	30.0	B(320)	39.05	M(209)	103.92	-76.169
C6H12O	CYCLOHEXANOL	(L)	100.156	-83.22	A(1331)	47.7	M(735)	-31.87	23.359
C6H12O	CYCLOHEXANOL	(G)	100.156	-70.40	T(543)	78.32	T(543)	-28.18	20.654
C6H12O	3,3-DIMETHYL-2-BUTANONE	(L)	100.156	-82.3	C(744)				
C6H12O	1,2-EPOXY-2-ETHYLBUTANE	(L)	100.156	-61.7	C(843)				
C6H12O	3,4-EPOXYHEXANE	(L)	100.156	-59.6	C(744)				

Formula	Name	State	Mol Wt	$\Delta H^{\circ}_{f\,298}$ (kcal/mole)	Source	S°_{298} (cal/mole °K)	Source	$\Delta G^{\circ}_{f\,298}$ (kcal/mole)	Log K_p
C6H12O	HEXANAL	(G)	100.156	-59.37	T(549)	101.07	T(549)	-23.93	17.541
C6H12O	2-HEXANONE	(L)	100.156	-78.9	C(744)				
C6H12O	ISOBUTYL VINYL ETHER	(L)	100.156	-64.1	C(843)				
C6H12O	2-METHYLCYCLOPENTANOL	(L)	100.156	-86.5	C(744)				
C6H12O	4-METHYL-2-PENTANONE	(L)	100.156	-92.9	B(1472)				
C6H12O	2-METHYL-4-PENTEN-2-OL	(L)	100.156	-87.6	C(744)				
C6H12OS	BUTYL THIOACETATE	(L)	132.222	-76.1	A(1554)				
C6H12OS	TERT-BUTYL THIOACETATE	(L)	132.222	-78.2	A(1554)				
C6H12O2	BUTYL ACETATE	(L)	116.156	-126.1	A(1555)				
C6H12O2	1,2-CYCLOHEXANEDIOL, CIS	(S)	116.156	-132.7	B(1536)				
C6H12O2	1,2-CYCLOHEXANEDIOL, TRANS	(S)	116.156	-131.7	B(1536)				
C6H12O2	CYCLOHEXYL HYDROPEROXIDE	(L)	116.156	-65.3	B(1190)				
C6H12O2	ETHYL BUTYRATE	(L)	116.156	-123.0	C(744)				
C6H12O2	2-ETHYLBUTYRIC ACID	(L)	116.156	-143.4	C(744)				
C6H12O2	ETHYL ISOBUTYRATE	(L)	116.156	-128.5	C(744)				
C6H12O2	HEXANOIC ACID	(L)	116.156	-139.51	A(3)				
C6H12O2	HEXANOIC ACID	(L)	116.156	-140.0	A(853)				
C6H12O2	4-HYDROXY-4-METHYL-2-PENTANONE	(L)	116.156	-126.8	B(313)				
C6H12O2	1-METHYL-1,2CYCLOPENTANEDIOL- CIS	(S)	116.156	-126.6	B(1536)				
C6H12O2	1-METHYL-1,2CYCLOPENTANEDIOL- TRANS	(S)	116.156	-128.9	B(1536)				
C6H12O2	METHYL ISOVALERATE	(L)	116.156	-128.5	B(569)				
C6H12O2	METHYL 2-METHYLBUTYRATE	(L)	116.156	-127.4	B(569)				
C6H12O2	METHYL PIVALATE	(L)	116.156	-131.7	B(569)				
C6H12O2	PROPYL PROPIONATE	(L)	116.156	-122.95	B(1284)				
C6H12O2	METHYL VALERATE	(L)	116.156	-127.4	A(3)				
C6H12O2	4-METHYLVALERIC ACID	(L)	116.156	-136.2	C(744)				
C6H12O3	METALDEHYDE	(S)	132.156	-168.1	C(744)				
C6H12O3	PARALDEHYDE	(L)	132.156	-164.2	B(431)				
C6H12O5	1,2,3,4,5-CYCLOHEXANEPENTOL	(S)	164.156	-269.6	C(744)				
C6H12O5	FUCOSE	(S)	164.156	-261.9	C(744)				
C6H12O5	RHAMNOSE	(S)	164.156	-255.5	C(744)				
C6H12O6	ALPHA-GALACTOSE, D	(S)	180.156	-307.21	A(228)	49.1	M(676)	-219.75	161.073
C6H12O6	GLUCOSE, D (DEXTROSE)	(S)	180.156	-304.6	A(651)	50.7	M(651)	-217.62	159.510
C6H12O6	FRUCTOSE, L	(S)	180.156	-303.5	C(744)				

Formula	Name	State	Mol Wt	$\Delta Hf°_{298}$ (kcal/mole)	Source	$S°_{298}$ (cal/mole °K)	Source	$\Delta Gf°_{298}$ (kcal/mole)	Log Kp
C6H12O6	INOSITOL (INOSITE)	(S)	180.156	-311.7	C(744)				
C6H12O6	BETA-LEVULOSE, D	(S)	180.156	-303.5	A(228)				
C6H12O6	MANNOSE, D	(S)	180.156	-301.9	B(1430)				
C6H12O6	SORBOSE, D	(S)	180.156	-305.6	C(744)				
C6H12O6	SORBOSE, L	(S)	180.156	-303.68	A(228)	52.8	M(676)	-217.32	159.294
C6H12S	CYCLOHEXANETHIOL	(G)	116.222	-22.93	A(957)				
C6H12S	2,3-EPITHIO-2,3-DIMETHYLBUTANE	(L)	116.222	-19.84	A(1447)				
C6H12S	THIACYCLOHEPTANE	(L)	116.222	-26.96	A(1447)				
C6H12S	THIACYCLOHEPTANE	(G)	116.222	-14.66	T(886)	86.50	T(886)	20.09	-14.728
C6H13BR	1-BROMOHEXANE	(L)	165.080	-46.48	A(138)				
C6H13BR	1-BROMOHEXANE	(G)	165.080	-35.88	A(138)				
C6H13NO	N-BUTYLACETAMIDE	(L)	115.172	-91.1	A(1558)				
C6H13NO	HEXANAMIDE	(S)	115.172	-101.6	B(1640)				
C6H13NO2	ISOLEUCINE, L	(S)	131.172	-152.65	A(666)	49.71	M(666)	-82.97	60.814
C6H13NO2	LEUCINE, D	(S)	131.172	-152.36	A(653)	49.5	E(648)	-82.76	60.665
C6H13NO2	LEUCINE, L	(S)	131.172	-154.6	A(666)	50.62	M(666)	-85.34	62.552
C6H13NO2	LEUCINE, D,L	(S)	131.172	-153.14	A(653)	49.5	M(648)	-83.54	61.237
C6H13NS2	DIETHYL METHYLDITHIO- IMIDOCARBONATE	(L)	163.304	-6.7	C(744)				
C6H13NS2	DIHYDRO-2,4,6-TRIMETHYL- 4H-1,3,5-DITHIAZINE	(L)	163.304	-32.5	C(744)				
C6H13NS2	METHYL DIETHYLDITHIO- CARBAMATE	(L)	163.304	-24.4	C(744)				
C6H13NO4	2-NITRO-2-PROPYL- 1,3-PROPANEDIOL	(S)	163.172	-149.5	B(300)				
C6H13NO4	2-ISOPROPYL-2-NITRO- 1,3-PROPANEDIOL	(S)	163.172	-149.0	B(300)				
C6H13N3O3	N5-CARBAMOYLORNITHINE, D,L	(S)	175.188						
C6H14	2,2-DIMETHYLBUTANE	(L)	86.172	-51.00	A(1248)	60.8	M(652)	-2.81	2.059
C6H14	2,2-DIMETHYLBUTANE	(G)	86.172	-44.35	T(112)	65.01	M(342)	-2.30	1.689
C6H14	2,3-DIMETHYLBUTANE	(L)	86.172	-49.48	A(1248)	85.62	T(112)	-1.68	1.233
C6H14	2,3-DIMETHYLBUTANE	(G)	86.172	-42.49	T(113)	66.33	M(342)	-0.98	0.719
C6H14	HEXANE	(L)	86.172	-47.52	A(1248)	87.42	T(113)	-1.04	0.765
C6H14	HEXANE	(G)	86.172	-39.96	T(109)	70.76	M(342)	-0.06	0.047
C6H14	2-METHYLPENTANE	(L)	86.172	-48.82	A(1248)	92.83	T(109)	-1.95	1.431
C6H14	2-METHYLPENTANE	(G)	86.172	-41.66	T(110)	69.45	M(342)	-1.20	0.882
						90.95	T(110)		

Formula	Name	State	Mol Wt	$\Delta H f^\circ_{298}$ (kcal/mole)	Source	S°_{298} (cal/mole °K)	Source	$\Delta G f^\circ_{298}$ (kcal/mole)	Log Kp
C6H14	3-METHYLPENTANE	(L)	86.172	-48.28	A(1248)	69.22	E(1248)	-1.34	0.985
C6H14	3-METHYLPENTANE	(G)	86.172	-41.02	T(111)	90.77	T(111)	-0.51	0.373
C6H14N2	1,1'-DIMETHYLAZOETHANE	(S)	114.188	10.9	B(239)				
C6H14N2O2	N-NITRODIPROPYLAMINE	(L)	146.188	-41.4	B(1039)				
C6H14N4O2	ARGININE, D	(S)	174.204	-149.05	A(653)	59.9	M(648)	-57.43	42.098
C6H14N6O6	HEXAMETHYLENE TETRAMINEDINITRATE	(S)	266.220	-90.2	B(320)				
C6H14O	3,3-DIMETHYL-2-BUTANOL	(L)	102.172	-103.8	C(744)				
C6H14O	HEXYL ALCOHOL	(L)	102.172	-90.67	A(215)	68.6	M(733)	-36.24	26.567
C6H14O	HEXYL ALCOHOL	(G)	102.172	-76.39	T(526)	105.52	T(526)	-32.97	24.168
C6H14O	ISOPROPYL ETHER	(L)	102.172	-83.95	T(508)	70.4	M(1106)	-30.06	22.034
C6H14O	ISOPROPYL ETHER	(G)	102.172	-76.20	T(508)	93.27	T(508)	-29.13	21.352
C6H14O	3-METHYL-3-PENTANOL	(L)	102.172	-115.4	C(744)				
C6H14O	PROPYL ETHER	(L)	102.172	-78.59	A(252)				
C6H14O	PROPYL ETHER	(G)	102.172	-70.00	T(507)	100.98	T(507)	-25.23	18.492
C6H14OS	TERT-BUTYL ETHYL SULFOXIDE	(L)	134.238	-83.3	B(907)				
C6H14OS	TERT-BUTYL ETHYL SULFOXIDE	(G)	134.238	-65.5	B(907)				
C6H14OS	PROPYL SULFOXIDE	(L)	134.238	-78.7	B(907)				
C6H14OS	PROPYL SULFOXIDE	(G)	134.238	-60.9	B(907)				
C6H14O2	ACETALDEHYDE DIETHYL ACETAL	(L)	118.172	-117.3	C(744)				
C6H14O2	1,2-DIETHOXYETHANE	(L)	118.172	-117.2	C(744)				
C6H14O2	2,3-DIMETHYL-2,3-BUTANEDIOL	(S)	118.172	-144.9	C(744)				
C6H14O2S	TERT-BUTYL ETHYL SULFONE	(L)	150.238	-137.9	B(904)				
C6H14O2S	TERT-BUTYL ETHYL SULFONE	(G)	150.238	-116.8	B(904)				
C6H14O2S	PROPYL SULFONE	(L)	150.238	-131.0	B(904)				
C6H14O2S	PROPYL SULFONE	(G)	150.238	-113.3	B(904)				
C6H14O3	2-ETHYL-2-(HYDROXYMETHYL)-1,3-PROPANEDIOL	(S)	134.172	-177.8	B(983)				
C6H14O4	TRIETHYLENE GLYCOL	(L)	150.172	-191.4	C(1032)				
C6H14O6	GALACTITOL	(S)	182.172	-321.90	A(1121)	56.0	M(1105)	-227.19	166.528
C6H14O6	MANNITOL, D	(S)	182.172	-319.61	A(1121)	57.0	M(1105)	-225.20	165.068
C6H14O6	RHAMNOSE MONOHYDRATE	(S)	182.172	-330.6	C(744)				
C6H14O7	GLUCOSE HYDRATE	(S)	198.172	-375.80	A(228)				
C6H14S	BUTYL ETHYL SULFIDE	(L)	118.238	-41.1	T(807)				
C6H14S	BUTYL ETHYL SULFIDE	(L)	118.238	-40.7	T(807)				
C6H14S	BUTYL ETHYL SULFIDE	(G)	118.238	-29.92	T(807)	108.27	T(807)	7.65	-5.606

Formula	Name	State	Mol Wt	$\Delta H f^\circ_{298}$ (kcal/mole)	Source	S°_{298} (cal/mole °K)	Source	$\Delta G f^\circ_{298}$ (kcal/mole)	Log $K p$
C6H14S	BUTYL ETHYL SULFIDE	(G)	118.238	-30.3	B(899)				
C6H14S	TERT-BUTYL ETHYL SULFIDE	(L)	118.238	-44.7	B(899)				
C6H14S	TERT-BUTYL ETHYL SULFIDE	(G)	118.238	-35.3	B(899)				
C6H14S	1-HEXANETHIOL	(L)	118.238	-41.97	A(507)				
C6H14S	1-HEXANETHIOL	(G)	118.238	-30.83	T(900)	108.58	T(900)	6.65	-4.871
C6H14S	ISOPROPYL PENTYL SULFIDE	(G)	118.238	-33.76	T(808)	99.30	T(808)	6.48	-4.752
C6H14S	METHYL PENTYL SULFIDE	(L)	118.238	-39.9	B(899)				
C6H14S	METHYL PENTYL SULFIDE	(L)	118.238	-40.14	T(809)				
C6H14S	METHYL PENTYL SULFIDE	(G)	118.238	-29.34	T(809)	107.73	T(809)	8.39	-6.149
C6H14S	METHYL PENTYL SULFIDE	(G)	118.238	-29.1	B(899)				
C6H14S	PROPYL SULFIDE	(L)	118.238	-40.62	A(952)	80.85	M(952)	5.12	-3.755
C6H14S	PROPYL SULFIDE	(G)	118.238	-29.96	T(810)	107.16	T(810)	7.94	-5.819
C6H14S2	PROPYL DISULFIDE	(L)	150.304	-40.95	A(627)	89.28	T(627)	4.56	-3.339
C6H14S2	PROPYL DISULFIDE	(G)	150.304	-28.01	T(874)	118.30	T(874)	8.84	-6.482
C6H15CLN2	LYSINE, L HYDROCHLORIDE	(S)	182.653			63.21	M(242)		
C6H15CLN4	ARGININE, L HYDROCHLORIDE	(S)	210.669			68.43	M(242)		
C6H15N	HEXYLAMINE	(L)	101.188	-53.7	C(744)				
C6H15N	TRIETHYLAMINE	(L)	101.188	-32.07	A(678)				
C6H15N	TRIETHYLAMINE	(G)	101.188	-23.80	B(1354)	96.90	T(610)	26.36	-19.320
C6H15NO3	2,2',2''-NITRILOTRIETHANOL	(L)	149.188	-159.5	C(277)				
C6H16N2O3	TRIETHYLAMINE NITRATE	(S)	164.204	-97.3	B(983)				
C6H18N4O6	1,6-HEXANEDIAMINE DINITRATE	(S)	242.236	-181.4	A(154)				
C6N4	ETHENETETRACARBONITRILE	(S)	128.092	149.1	A(154)				
C6N4	ETHENETETRACARBONITRILE	(G)	128.092	168.5	A(154)				
C7F14	UNDECAFLUORO(TRIFLUOROMETHYL)-CYCLOHEXANE	(L)	350.070	-700.3	A(508)				
C7F14	UNDECAFLUORO(TRIFLUOROMETHYL)-CYCLOHEXANE	(G)	350.070	-692.2	A(508)	134.28	M(1086)	-739.24	541.850
C7F16	HEXADECAFLUOROHEPTANE	(L)	388.070	-817.6	A(508)				
C7F16	HEXADECAFLUOROHEPTANE	(G)	388.070	-808.9	A(508)	158.88	M(1086)	-737.87	540.849
C7HF5O2	PENTAFLUOROBENZOIC ACID	(S)	212.078	-289.66	A(290)				
C7H3N3O7	2,4,6-TRINITROBENZALDEHYDE	(S)	241.118	-31.5	C(1231)				
C7H3N3O8	TRINITROBENZUIC ACID	(S)	257.118	-96.3	B(981)				
C7H4CL2O	O-CHLOROBENZOYL CHLORIDE	(S)	175.016	-65.0	C(744)				
C7H4F3NO2	ALPHA,ALPHA,ALPHA-TRIFLUORO-								

Formula	Name	State	Mol Wt	$\Delta H f^\circ_{298}$ (kcal/mole)	Source	S°_{298} (cal/mole °K)	Source	$\Delta G f^\circ_{298}$ (kcal/mole)	Log $K p$
C7H4F3NO2	M-NITROTOLUENE	(L)	191.110	-152.3	C(744)				
C7H4F4	M,ALPHA,ALPHA,ALPHA-TETRAFLUOROTOLUENE	(L)	164.102	-198.7	A(508)				
C7H4F4	M,ALPHA,ALPHA,ALPHA-TETRAFLUOROTOLUENE	(G)	164.102	-189.7	A(508)				
C7H4I2O3	DIIODOSALICYLIC ACID	(U,S)	389.922	-94.9	C(744)				
C7H5BRO	BENZOYL BROMIDE	(L)	185.026	-25.5	A(197)				
C7H5CLO	BENZOYL CHLORIDE	(L)	140.567	-39.3	A(197)				
C7H5CLO	M-CHLOROBENZALDEHYDE	(L)	140.567	-30.5	B(1384)				
C7H5CLO	O-CHLOROBENZALDEHYDE	(L)	140.567	-28.7	B(1384)				
C7H5CLO	P-CHLOROBENZALDEHYDE	(S)	140.567	-35.4	B(1384)				
C7H5CLO2	CHLOROSALICYLALDEHYDE	(U,L)	156.567	-88.6	C(744)				
C7H5CLO2	M-CHLOROBENZOIC ACID	(S)	156.567	-101.5	B(1384)				
C7H5CLO2	O-CHLOROBENZOIC ACID	(S)	156.567	-95.6	B(1384)				
C7H5CLO2	P-CHLOROBENZOIC ACID	(S)	156.567	-102.1	A(141)				
C7H5FO2	M-FLUOROBENZOIC ACID	(S)	140.110	-139.21	A(514)				
C7H5FO2	O-FLUOROBENZOIC ACID	(S)	140.110	-135.75	A(514)				
C7H5FO2	P-FLUOROBENZOIC ACID	(S)	140.110	-140.32	A(514)				
C7H5FO2	P-FLUOROBENZOIC ACID	(S)	140.110	-138.47	A(290)				
C7H5F3	ALPHA,ALPHA,ALPHA-TRIFLUOROTOLUENE	(L)	146.110	-152.40	A(514)	64.89	M(1301)	-123.98	90.872
C7H5F3	ALPHA,ALPHA,ALPHA-TRIFLUOROTOLUENE	(G)	146.110	-143.42	T(720)	89.05	T(720)	-122.20	89.570
C7H5F3O	ALPHA,ALPHA,ALPHA-TRIFLUORO-O-CRESOL	(L)	162.110	-194.6	C(744)				
C7H5IO	BENZOYL IODIDE	(L)	232.020	-12.55	A(197)				
C7H5IO2	M-IODOBENZOIC ACID	(S)	248.020	-75.6	B(1382)				
C7H5IO2	O-IODOBENZOIC ACID	(S)	248.020	-72.2	B(1382)				
C7H5IO2	P-IODOBENZOIC ACID	(S)	248.020	-75.5	B(1382)				
C7H5IO3	IODOSALICYLIC ACID	(U,S)	264.020	-122.5	C(744)				
C7H5N	BENZONITRILE	(L)	103.118	39.0	A(388)				
C7H5N	BENZONITRILE	(G)	103.118	52.30	T(622)	76.73	T(622)	62.35	-45.702
C7H5NO3	M-NITROBENZALDEHYDE	(S)	151.118	-28.2	C(744)				
C7H5NO3	P-NITROBENZALDEHYDE	(S)	151.118	-36.1	C(1231)				
C7H5NO4	M-NITROBENZOIC ACID	(S)	167.118	-101.2	C(744)				
C7H5NO4	O-NITROBENZOIC ACID	(S)	167.118	-98.9	C(744)				

Formula	Name	State	Mol Wt	$\Delta H f^\circ_{298}$ (kcal/mole)	Source	S°_{298} (cal/mole °K)	Source	$\Delta G f^\circ_{298}$ (kcal/mole)	Log $K p$
C7H5NO4	P-NITROBENZOIC ACID	(S)	167.118	-100.6	C(744)				
C7H5NS	PHENYL ISOTHIOCYANATE	(L)	135.184	51.3	C(744)				
C7H5N3O6	2,3,4-TRINITROTOLUENE	(S)	227.134	3.6	C(744)				
C7H5N3O6	2,3,5-TRINITROTOLUENE	(S)	227.134	-5.5	C(744)				
C7H5N3O6	2,3,6-TRINITROTOLUENE	(S)	227.134	-4.0	C(744)				
C7H5N3O6	2,4,5-TRINITROTOLUENE	(S)	227.134	-3.7	C(744)				
C7H5N3O6	2,4,6-TRINITROTOLUENE	(S)	227.134	-15.0	C(55)				
C7H5N3O6	2,4,6-TRINITROTOLUENE	(S)	227.134	-14.2	B(1640)				
C7H5N3O6	3,4,5-TRINITROTOLUENE	(S)	227.134	-1.2	C(744)				
C7H5N3O7	2,4,6-TRINITRO-M-CRESOL	(S)	243.134	-60.3	B(57)				
C7H5N3O7	2,4,6-TRINITROANISOLE	(S)	243.134	-36.3	B(57)				
C7H5N5O8	N-METHYL-N,2,4,6-TETRANITRO-ANILINE	(S)	287.150	4.2	B(1640)				
C7H5N5O8	N-METHYL-N,2,4,6-TETRANITRO-ANILINE	(S)	287.150	7.6	B(184)				
C7H6F2	ALPHA,ALPHA-DIFLUOROTOLUENE	(L)	128.118	-82.3	C(744)				
C7H6N2O4	2,3-DINITROTOLUENE	(S)	182.134	-3.8	C(744)				
C7H6N2O4	2,4-DINITROTOLUENE	(S)	182.134	-16.3	C(744)				
C7H6N2O4	2,5-DINITROTOLUENE	(S)	182.134	-8.2	B(981)				
C7H6N2O4	2,6-DINITROTOLUENE	(S)	182.134	-10.5	C(744)				
C7H6N2O4	3,4-DINITROTOLUENE	(S)	182.134	-3.5	C(744)				
C7H6N2O4	3,5-DINITROTOLUENE	(S)	182.134	-10.4	B(57)				
C7H6N2O5	2,4-DINITROANISOLE	(S)	198.134	-44.4	B(981)				
C7H6N2O5	4,6-DINITRO-O-CRESOL	(S)	198.134	-67.2	B(981)				
C7H6N2O5	2,6-DINITRO-P-CRESOL	(S)	198.134	-57.6	B(974)				
C7H6N4	1-PHENYLTETRAZOLE	(S)	146.150	86.5	B(974)				
C7H6N4	5-PHENYLTETRAZOLE	(S)	146.150	69.9	B(974)				
C7H6N4O	1-PHENYL-1H-TETRAZOL-5-OL	(S)	162.150	26.7	B(974)				
C7H6N4O6	METHYL-2,4,N-TRINITRO-N-ANILINE	(S)	242.150	3.9	B(986)				
C7H6N4O6	METHYL-2,4,6-TRINITRO-N-ANILINE	(S)	242.150	-12.0	B(985)				
C7H6O	BENZALDEHYDE	(L)	106.118	-21.3	C(744)				
C7H6O	BENZALDEHYDE	(S)	106.118	-18.5	B(844)				
C7H6O2	BENZOIC ACID	(S)	122.118	-92.05	A(688)	40.05	M(454)	-58.62	42.970
C7H6O2	BENZOIC ACID	(G)	122.118	-69.36	T(569)	88.19	T(569)	-50.29	36.859

Formula	Name	State	Mol Wt	$\Delta H_{f\,298}^{\circ}$ (kcal/mole)	Source	S_{298}° (cal/mole °K)	Source	$\Delta G_{f\,298}^{\circ}$ (kcal/mole)	Log K_p
C7H6O2	1,3-BENZODIOXOLE	(L)	122.118	−44.0	A(201)				
C7H6O2	2-HYDROXY-2,4,6-CYCLOHEPTA-TRIEN-1-ONE	(S)	122.118	−57.23	A(632)				
C7H6O2	2-HYDROXY-2,4,6-CYCLOHEPTA-TRIEN-1-ONE	(G)	122.118	−37.23	A(632)				
C7H6O2	M-HYDROXYBENZALDEHYDE	(S)	122.118	−73.2	C(744)				
C7H6O2	P-HYDROXYBENZALDEHYDE	(S)	122.118	−70.6	C(744)				
C7H6O2	METHYL-X-BENZOQUINONE	(U,S)	122.118	−59.5	C(744)				
C7H6O2	SALICYLALDEHYDE	(S)	122.118	−67.2	C(744)				
C7H6O3	FURANACRYLIC ACID	(S)	138.118	−107.7	B(844)				
C7H6O3	M-HYDROXYBENZOIC ACID	(S)	138.118	−139.8	B(1110)	42.3	M(1110)	−99.74	73.107
C7H6O3	P-HYDROXYBENZOIC ACID	(S)	138.118	−139.7	B(1110)	42.0	M(1110)	−99.55	72.968
C7H6O3	SALICYLIC ACID	(S)	138.118	−139.8	B(1534)	42.6	M(1110)	−99.93	73.246
C7H6O4	2,4-DIHYDROXYBENZOIC ACID	(S)	154.118	−186.3	C(744)				
C7H6O5	GALLIC ACID	(S)	170.118	−229.2	C(744)				
C7H6O5	TRIHYDROXYBENZOIC ACID	(U,S)	170.118	−229.5	C(744)				
C7H6O8	1,1,2-CYCLOPROPANE-TETRACARBOXYLIC ACID	(S)	218.118	−380.4	C(744)				
C7H7BR	ALPHA-BROMOTOLUENE	(G)	171.042	20.0	F(107)	90.8	E(107)	33.76	−24.747
C7H7CL	ALPHA-CHLOROTOLUENE	(L)	126.583	−16.8	C(744)				
C7H7CL	M-CHLOROTOLUENE	(L)	126.583	−6.5	B(1384)				
C7H7CL	O-CHLOROTOLUENE	(L)	126.583	−4.9	B(1384)				
C7H7CL	P-CHLOROTOLUENE	(L)	126.583	−4.9	B(1384)				
C7H7F	O-FLUOROTOLUENE	(L)	110.126	−37.1	C(744)				
C7H7F	P-FLUOROTOLUENE	(L)	110.126	−44.80	A(512)	56.67	M(1321)	−19.06	13.974
C7H7F	P-FLUOROTOLUENE	(G)	110.126	−35.38	T(721)	81.15	T(721)	−16.94	12.419
C7H7I	ALPHA-IODOTOLUENE	(S)	218.036	30.4	A(1570)				
C7H7I	M-IODOTOLUENE	(L)	218.036	19.1	B(1382)				
C7H7I	O-IODOTOLUENE	(L)	218.036	18.9	B(1382)				
C7H7I	P-IODOTOLUENE	(L)	218.036	16.3	B(1382)				
C7H7N	2-VINYLPYRIDINE	(L)	105.134	75.1	A(67)				
C7H7NO	BENZALDEHYDE OXIME	(S)	121.134	9.8	C(744)				
C7H7NO	BENZAMIDE	(S)	121.134	−48.47	A(244)				
C7H7NO	FORMANILIDE	(S)	121.134	−35.8	C(744)				
C7H7NO2	P-AMINOBENZOIC ACID	(S)	137.134	−78.4	C(1207)				
C7H7NO2	M-NITROTOLUENE	(L)	137.134	−6.0	C(744)				

Formula	Name	State	Mol Wt	$\Delta H_f^{\circ}{}_{298}$ (kcal/mole)	Source	S°_{298} (cal/mole °K)	Source	$\Delta G_f^{\circ}{}_{298}$ (kcal/mole)	Log K_p
C7H7NO2	O-NITROTOLUENE	(L)	137.134	-0.4	C(744)				
C7H7NO2	P-NITROTOLUENE	(S)	137.134	-8.9	C(744)				
C7H7NO2	SALICYLALDEHYDE OXIME	(S)	137.134	-43.9	B(1640)				
C7H7N3O4	N-METHYL-2,4-DINITROANILINE	(S)	197.150	-22.6	B(984)				
C7H7N5	5-PHENYLAMINOTETRAZOLE	(S)	161.166	72.9	B(1616)				
C7H7N5	1-PHENYL-5-AMINOTETRAZOLE	(S)	161.166	74.3	B(1616)				
C7H8	1,3,5-CYCLOHEPTATRIENE	(L)	92.134	34.22	A(424)	51.30	M(424)	58.99	-43.237
C7H8	1,3,5-CYCLOHEPTATRIENE	(G)	92.134	43.47	T(470)	75.44	T(470)	61.04	-44.741
C7H8	2,5-NORBORNADIENE	(L)	92.134	-48.0	B(94)				
C7H8	TOLUENE	(L)	92.134	2.87	A(1248)	52.81	M(1314)	27.19	-19.928
C7H8	TOLUENE	(G)	92.134	11.95	T(397)	76.64	T(397)	29.16	-21.376
C7H8N2O	PHENYLUREA	(S)	136.150	-51.4	C(744)				
C7H8N2O2	N-METHYL-P-NITROANILINE	(S)	152.150	-7.4	C(744)				
C7H8N2O2	3-NITRO-P-TOLUIDINE	(S)	152.150	-17.2	B(1440)				
C7H8N2O2	5-NITRO-O-TOLUIDINE	(S)	152.150	-21.8	B(1440)				
C7H8N4O2	THEOBROMINE	(S)	180.166	-85.8	C(744)				
C7H8O	ANISOLE	(L)	108.134	-28.6	B(56)				
C7H8O	BENZYL ALCOHOL	(L)	108.134	-38.49	A(1112)	51.8	M(1118)	-6.57	4.813
C7H8O	M-CRESOL	(L)	108.134	-46.19	A(285)				
C7H8O	M-CRESOL	(G)	108.134	-31.63	T(566)	85.27	T(566)	-9.69	7.099
C7H8O	O-CRESOL	(L)	108.134	-48.84	A(285)				
C7H8O	O-CRESOL	(G)	108.134	-30.74	T(567)	85.47	T(567)	-8.86	6.491
C7H8O	P-CRESOL	(L)	108.134	-47.61	A(285)				
C7H8O	P-CRESOL	(G)	108.134	-29.97	T(568)	83.09	T(568)	-7.38	5.406
C7H8O2	O-HYDROXYBENZYL ALCOHOL	(S)	124.134	-86.2	C(744)				
C7H8O2	METHYLHYDROQUINONE	(S)	124.134	-95.9	C(744)				
C7H8O2	5-METHYLRESORCINOL	(S)	124.134	-106.7	C(744)				
C7H8O2S	METHYL PHENYL SULFONE	(L)	156.200	-82.5	B(905)				
C7H8O2S	METHYL PHENYL SULFONE	(G)	156.200	-62.6	B(905)				
C7H8O3S	P-TOLUENESULFONIC ACID	(S)	172.200	-298.4	C(1125)				
C7H8O3S	P-TOLUENESULFONIC ACID	(G)	172.200			94.87	E(69)		
C7H8S	METHYL PHENYL SULFIDE	(L)	124.200	11.5	B(898)				
C7H8S	METHYL PHENYL SULFIDE	(G)	124.200	23.6	B(898)				
C7H8S	ALPHA-TOLUENETHIOL	(L)	124.200	10.5	B(901)				
C7H8S	ALPHA-TOLUENETHIOL	(G)	124.200	22.8	B(901)				
C7H9N	BENZYLAMINE	(L)	121.158	2.0	C(744)				

Formula	Name	State	Mol Wt	$\Delta H_f^\circ{}_{298}$ (kcal/mole)	Source	S°_{298} (cal/mole °K)	Source	$\Delta G_f^\circ{}_{298}$ (kcal/mole)	Log K_p
C7H9N	2-ETHYLPYRIDINE	(L)	107.150	0.69	A(67)				
C7H9N	2,3-LUTIDINE	(L)	107.150	4.61	A(289)				
C7H9N	2,3-LUTIDINE	(G)	107.150	16.31	A(284)				
C7H9N	2,4-LUTIDINE	(L)	107.150	3.84	A(289)				
C7H9N	2,4-LUTIDINE	(G)	107.150	15.26	A(284)				
C7H9N	2,5-LUTIDINE	(L)	107.150	4.44	A(289)				
C7H9N	2,5-LUTIDINE	(G)	107.150	15.87	A(284)				
C7H9N	2,6-LUTIDINE	(L)	107.150	3.01	A(289)				
C7H9N	2,6-LUTIDINE	(G)	107.150	14.02	A(284)				
C7H9N	3,4-LUTIDINE	(L)	107.150	4.35	A(289)				
C7H9N	3,4-LUTIDINE	(G)	107.150	16.73	A(284)				
C7H9N	3,5-LUTIDINE	(L)	107.150	5.35	A(289)				
C7H9N	3,5-LUTIDINE	(G)	107.150	17.39	A(284)				
C7H9N	N-METHYLANILINE	(L)	107.150	7.7	C(1547)	53.6	E(1547)	43.26	-31.707
C7H9N	N-METHYLANILINE	(G)	107.150	20.4	C(1547)	81.6	E(1547)	47.61	-34.897
C7H9N	M-TOLUIDINE	(L)	107.150	-0.3	C(744)				
C7H9N	O-TOLUIDINE	(L)	107.150	-1.3	C(744)				
C7H9N	P-TOLUIDINE	(S)	107.150	-7.2	C(744)				
C7H9NO	P-ANISIDINE	(S)	123.150	-41.1	C(744)				
C7H9NO2	ETHYLMETHYLMALEIMIDE	(S)	139.150	-110.5	C(1414)				
C7H9NO3	ETHYL 2-CYANOACETOACETATE	(S)	155.150	-128.6	C(744)				
C7H10	1,3-CYCLOHEPTADIENE	(G)	94.150	22.3	A(261)				
C7H10N2	1-METHYL-1-PHENYLHYDRAZINE	(S)	122.166	-39.0	C(744)				
C7H10N2O3	BENZYLAMINE NITRATE	(S)	170.166	-57.2	C(277)				
C7H10O	3-METHYL-2-CYCLOHEXEN-1-ONE	(L)	110.150	-57.0	C(744)				
C7H10O	2-NORBORNANONE	(S)	110.150	-53.8	B(90)				
C7H10O2	1-CYCLOHEXENE-1-CARBOXYLIC ACID	(S)	126.150	-142.4	C(744)				
C7H10O2	2-CYCLOHEXENE-1-CARBOXYLIC ACID	(S)	126.150	-112.7	C(744)				
C7H10O2	ETHYL-2,4-PENTADIENOATE	(L)	126.150	-80.1	B(1288)				
C7H10O3	TRIMETHYLSUCCINIC ANHYDRIDE	(S)	142.150	-163.8	B(1538)				
C7H10O4	1,2-CYCLOPENTANEDICARBOXYLIC ACID, TRANS	(S)	158.150	-218.4	C(744)				
C7H10O4	DIMETHYLMETHYLENESUCCINIC ACID	(S)	158.150	-203.3	C(744)				

Formula	Name	State	Mol Wt	$\Delta Hf°_{298}$ (kcal/mole)	Source	$S°_{298}$ (cal/mole °K)	Source	$\Delta Gf°_{298}$ (kcal/mole)	Log Kp
C7H10O4	DIMETHYL TRIMETHYLENE-APLHA, ALPHA-DICARBOXYLATE	(U,L)	158.150	-171.7	C(744)				
C7H10O4	(1-HYDROXY-1-METHYLETHYL) SUCCINIC ACID GAMMA LACTONE	(L)	158.150	-221.7	C(744)				
C7H10O5	DIMETHYL ACETONYLMALONATE	(L)	174.150	-247.2	C(744)				
C7H10O10	3-HYDROXY-4-OXO-4H-PYRAN-DICARBOXYLIC ACID TRIHYDRATE	(S)	254.150	-304.3	C(744)				
C7H11CL3O2	ISOPENTYL TRICHLOROACETATE	(L)	233.529	-139.2	B(1384)				
C7H11N5	1-ALLYL-5-(ALLYLAMINO)TETRAZOLE	(S)	165.198	83.7	B(1616)				
C7H11N5	5-(DIALLYLAMINO)TETRAZOLE	(S)	165.198	83.9	B(1616)				
C7H12	CYCLOHEPTENE	(L)	96.166	-11.0	(CALC)				
C7H12	CYCLOHEPTENE	(G)	96.166	-2.23	A(261)				
C7H12	2-CYCLOPROPYL-1-BUTENE	(L)	96.166	19.9	C(1374)				
C7H12	2-CYCLOPROPYL-2-BUTENE	(L)	96.166	14.9	C(1374)				
C7H12	1,2-DIMETHYLCYCLOPENTENE	(G)	96.166			84.0	E(622)		
C7H12	1,3-DIMETHYLCYCLOPENTENE	(G)	96.166			86.4	E(622)		
C7H12	1,4-DIMETHYLCYCLOPENTENE	(G)	96.166			87.4	E(622)		
C7H12	1,5-DIMETHYLCYCLOPENTENE	(G)	96.166			86.4	E(622)		
C7H12	3,3-DIMETHYLCYCLOPENTENE	(G)	96.166			83.2	E(622)		
C7H12	3,4-DIMETHYLCYCLOPENTENE, CIS	(G)	96.166			86.8	E(622)		
C7H12	3,4-DIMETHYLCYCLOPENTENE, TRANS	(G)	96.166			86.6	E(622)		
C7H12	3,5-DIMETHYLCYLCOPENTENE, CIS	(G)	96.166			84.6	E(622)		
C7H12	3,5-DIMETHYLCYCLOPENTENE, TRANS	(G)	96.166			84.6	E(622)		
C7H12	1-ETHYLCYCLOPENTENE	(L)	96.166	-13.95	A(821)				
C7H12	ETHYLIDENECYCLOPENTANE	(L)	96.166	-13.58	A(821)				
C7H12	1-HEPTYNE	(L)	96.166	16.0	(CALC)				
C7H12	1-HEPTYNE	(G)	96.166	24.62	T(324)	97.44	T(324)	54.24	-39.759
C7H12	1-METHYLCYCLOHEXENE	(L)	96.166	-19.42	A(821)				
C7H12	4-METHYLCYCLOHEXENE	(L)	96.166	-24.4	C(744)				
C7H12	METHYLENECYCLOHEXANE	(L)	96.166	-17.63	A(1522)				
C7H12	NORBORNANE	(S)	96.166	-22.11	A(94)				
C7H12	VINYLCYCLOPENTANE	(L)	96.166	-8.00	A(821)				
C7H12CL2O2	ISOPENTYL DICHLOROACETATE	(L)	199.080	-138.1	B(1384)				
C7H12O	CYCLOHEPTANONE	(L)	112.166	-71.5	A(1371)				
C7H12O	2,5-DIMETHYLCYCLOPENTANONE	(L)	112.166	-78.3	C(744)				
C7H12O	ENDO-2-METHYL-	(L)							

Formula	Name	State	Mol Wt	$\Delta H f°_{298}$ (kcal/mole)	Source	$S°_{298}$ (cal/mole °K)	Source	$\Delta G f°_{298}$ (kcal/mole)	Log $K p$
C7H12O	7-OXABICYCLO(2.2.1)HEPTANE	(L)	112.166	-59.90	A(94)				
C7H12O	2-ETHYLCYCLOPENTANONE	(L)	112.166	-77.7	C(744)				
C7H12O	EXO-2-METHYL-								
C7H12O	7-OXABICYCLO(2.2.1)HEPTANE	(L)	112.166	-60.75	A(94)				
C7H12O	1,6-HEPTADIEN-4-OL	(L)	112.166	-39.9	C(744)				
C7H12O	3-METHYLCYCLOHEXANONE	(L)	112.166	-73.3	C(744)				
C7H12O2	ACETYLACETONE-O-ETHYL ETHER	(U,L)	128.166	-99.5	C(744)				
C7H12O2	CYCLOHEXANECARBOXYLIC ACID	(S)	128.166	-134.2	C(744)				
C7H12O2	ETHYL ANGELATE	(L)	128.166	-103.3	C(744)				
C7H12O2	ETHYL 2-PENTENOATE	(L)	128.166	-106.3	B(1287)				
C7H12O2	ETHYL 3-PENTENOATE	(L)	128.166	-104.1	B(1287)				
C7H12O2	ETHYL 4-PENTENOATE	(L)	128.166	-102.8	B(1287)				
C7H12O2	ETHYL TIGLATE	(S)	128.166	-113.1	C(744)				
C7H12O2	ETHYL TETRAMETHYLENE-								
C7H12O2	CARBOXYLATE	(U,L)	128.166	-103.2	C(744)				
C7H12O2	ISOPROPYL CROTONATE	(L)	128.166	-108.75	B(1286)				
C7H12O2	PROPYL CROTONATE	(L)	128.166	-105.55	B(1286)				
C7H12O3	ETHYL 2-ETHOXYACRYLATE	(S)	144.166	-141.5	C(744)				
C7H12O3	ETHYL 3-ETHOXYACRYLATE	(L)	144.166	-144.2	C(744)				
C7H12O4	BUTYLMALONIC ACID	(S)	160.166	-238.1	B(1535)				
C7H12O4	DIETHYL MALONATE	(L)	160.166	-207.9	C(744)				
C7H12O4	DIETHYLMALONIC ACID	(S)	160.166	-235.1	C(744)				
C7H12O4	DIMETHYL GLUTARATE	(L)	160.166	-204.5	C(744)				
C7H12O4	PIMELIC ACID	(S)	160.166	-242.8	A(1354)				
C7H12O4	2,4,8,10-TETRAOXASPIRO-								
C7H12O4	(5,5-UNDECANE)	(S)	160.166	-99.5	A(431)				
C7H12O4	TRIMETHYLSUCCINIC ACID	(S)	160.166	-238.4	B(1538)				
C7H12O5	DIACETIN	(S)	176.166	-266.5	B(1472)				
C7H12O7	GLUCOHEPTONIC ACID LACTONE	(S)	208.166	-342.1	C(744)				
C7H13ClO2	BUTYL 2-CHLOROPROPIONATE	(L)	164.631	-133.5	B(1384)				
C7H13ClO2	BUTYL 3-CHLOROPROPIONATE	(L)	164.631	-144.3	B(1384)				
C7H13ClO2	ISOBUTYL 2-CHLOROPROPIONATE	(L)	164.631	-137.1	B(1384)				
C7H13ClO2	ISOBUTYL 3-CHLOROPROPIONATE	(L)	164.631	-136.8	B(1384)				
C7H13ClO2	ISOPENTYL CHLOROACETATE	(L)	164.631	-154.8	B(1384)				
C7H13ClO2	ISOPROPYL 2-CHLOROBUTYRATE	(L)	164.631	-145.4	B(1384)				
C7H13ClO2	ISOPROPYL 3-CHLOROBUTYRATE	(L)	164.631		B(1384)				

Formula	Name	State	Mol Wt	$\Delta H f°_{298}$ (kcal/mole)	Source	$S°_{298}$ (cal/mole °K)	Source	$\Delta G f°_{298}$ (kcal/mole)	Log Kp
C7H13CLO2	ISOPROPYL 4-CHLOROBUTYRATE	(L)	164.631	-145.3	B(1384)				
C7H13CLO2	PROPYL 2-CHLOROBUTYRATE	(L)	164.631	-151.0	B(1384)				
C7H13CLO2	PROPYL 3-CHLOROBUTYRATE	(L)	164.631	-141.3	B(1384)				
C7H13CLO2	PROPYL 4-CHLOROBUTYRATE	(L)	164.631	-141.6	B(1384)				
C7H13N	1-AZABICYCLO(2,2,2)OCTANE	(S)	111.182						
C7H13NO	KSI-ENANTOLACTAM	(S)	127.182	-183.6	C(744)				
C7H13NO3	FORMYL LEUCINE, DL	(S)	159.182						
C7H14	CYCLOHEPTANE	(L)	98.182	-37.78	A(707)	49.47	M(1624)	12.92	-9.467
C7H14	CYCLOHEPTANE	(G)	98.182	-28.52	T(342)	45.4	M(794)	15.06	-11.042
C7H14	2-CYCLOPROPYLBUTANE	(L)	98.182	-17.9	C(1375)				
C7H14	1,1-DIMETHYLCYCLOPENTANE	(L)	98.182	-41.14	A(694)	57.97	M(424)	7.95	-5.831
C7H14	1,1-DIMETHYLCYCLOPENTANE	(G)	98.182	-33.05	T(349)	81.82	T(342)	9.33	-6.837
C7H14	1,1-DIMETHYL-2-ETHYLCYCLOPROPANE	(L)	98.182	-21.6	A(710)				
C7H14	1,2-DIMETHYLCYCLOPENTANE, CIS	(L)	98.182	-39.52	A(694)	63.34	M(544)	9.28	-6.802
C7H14	1,2-DIMETHYLCYCLOPENTANE, CIS	(G)	98.182	-30.96	T(350)	85.87	T(349)	10.93	-8.010
C7H14	1,2-DIMETHYLCYCLOPENTANE, TRANS	(L)	98.182	-40.94	A(694)	64.33	M(544)	7.70	-5.645
C7H14	1,2-DIMETHYLCYCLOPENTANE, TRANS	(G)	98.182	-32.67	T(351)	87.51	T(350)	9.17	-6.722
C7H14	1,3-DIMETHYLCYCLOPENTANE, CIS	(L)	98.182	-40.68	A(694)	64.86	M(655)	7.93	-5.812
C7H14	1,3-DIMETHYLCYCLOPENTANE, CIS	(G)	98.182	-32.47	T(352)	87.67	T(351)	9.37	-6.868
C7H14	1,3-DIMETHYLCYCLOPENTANE, TRANS	(L)	98.182	-40.19	A(694)	64.97	E(1248)	7.84	-5.749
C7H14	1,3-DIMETHYLCYCLOPENTANE, TRANS	(G)	98.182	-31.93	T(353)	66.90	M(544)	9.91	-7.264
C7H14	2,3-DIMETHYL-1-PENTENE	(L)	98.182	-19.5	E(1248)	87.67	T(353)		
C7H14	2,3-DIMETHYL-1-PENTENE	(G)	98.182	-21.1	E(1248)				
C7H14	2,4-DIMETHYL-1-PENTENE	(L)	98.182	-28.18	A(1236)				
C7H14	2,4-DIMETHYL-1-PENTENE	(G)	98.182	-20.27	A(1236)				
C7H14	2,4-DIMETHYL-2-PENTENE	(L)	98.182	-29.64	A(1236)				
C7H14	2,4-DIMETHYL-2-PENTENE	(G)	98.182	-21.44	A(1236)				
C7H14	3,3-DIMETHYL-1-PENTENE	(L)	98.182	-18.2	E(1248)				
C7H14	3,3-DIMETHYL-1-PENTENE	(G)	98.182	-17.6	E(1248)				
C7H14	3,4-DIMETHYL-2-PENTENE, CIS	(L)	98.182	-20.9	E(1248)				
C7H14	3,4-DIMETHYL-2-PENTENE, TRANS	(G)	98.182	-20.9	E(1248)				
C7H14	4,4-DIMETHYL-1-PENTENE	(L)	98.182	-26.65	A(1236)				
C7H14	4,4-DIMETHYL-1-PENTENE	(G)	98.182	-19.20	A(1236)				
C7H14	4,4-DIMETHYL-2-PENTENE, CIS	(L)	98.182	-25.39	A(1236)				
C7H14	4,4-DIMETHYL-2-PENTENE, CIS	(G)	98.182	-17.60	A(1236)				
C7H14	4,4-DIMETHYL-2-PENTENE, TRANS	(L)	98.182	-29.31	A(1236)				

Formula	Name	State	Mol Wt	$\Delta H_{f\,298}^{\circ}$ (kcal/mole)	Source	S_{298}° (cal/mole °K)	Source	$\Delta G_{f\,298}^{\circ}$ (kcal/mole)	Log K_p
C7H14	4,4-DIMETHYL-2-PENTENE, TRANS	(G)	98.182	-21.46	A(1236)				
C7H14	ETHYLCYCLOPENTANE	(L)	98.182	-39.08	A(1196)	67.00	M(544)	8.92	-6.541
C7H14	ETHYLCYCLOPENTANE	(G)	98.182	-30.37	T(348)	90.42	T(348)	10.65	-7.807
C7H14	2-ETHYL-1-PENTENE	(G)	98.182	-17.9	E(1248)				
C7H14	3-ETHYL-1-PENTENE	(G)	98.182	-15.3	E(1248)				
C7H14	3-ETHYL-2-PENTENE	(G)	98.182	-18.6	E(1248)				
C7H14	1-HEPTENE	(L)	98.182	-23.41	A(1236)	78.31	M(950)	21.22	-15.555
C7H14	1-HEPTENE	(G)	98.182	-14.89	T(290)	101.24	T(290)	22.90	-16.789
C7H14	2-HEPTENE, CIS	(G)	98.182	-16.5	E(1248)				
C7H14	2-HEPTENE, TRANS	(G)	98.182	-17.5	E(1248)				
C7H14	3-HEPTENE, CIS	(G)	98.182	-16.5	E(1248)				
C7H14	3-HEPTENE, TRANS	(G)	98.182	-17.5	E(1248)				
C7H14	METHYLCYCLOHEXANE	(L)	98.182	-45.45	A(1196)	59.26	M(341)	4.86	-3.563
C7H14	METHYLCYCLOHEXANE	(G)	98.182	-36.99	T(371)	82.06	T(371)	6.52	-4.781
C7H14	3-METHYL-2-ETHYL-1-BUTENE	(L)	98.182	-27.48	A(1236)				
C7H14	3-METHYL-2-ETHYL-1-BUTENE	(G)	98.182	-19.25	A(1236)				
C7H14	2-METHYL-1-HEXENE	(G)	98.182	-18.5	E(1248)				
C7H14	2-METHYL-2-HEXENE	(G)	98.182	-19.9	E(1248)				
C7H14	2-METHYL-3-HEXENE, CIS	(G)	98.182	-18.2	E(1248)				
C7H14	2-METHYL-3-HEXENE, TRANS	(G)	98.182	-19.2	E(1248)				
C7H14	3-METHYL-1-HEXENE	(G)	98.182	-16.0	E(1248)				
C7H14	3-METHYL-2-HEXENE, CIS	(G)	98.182	-19.3	E(1248)				
C7H14	3-METHYL-2-HEXENE, TRANS	(L)	98.182	-19.3	E(1248)				
C7H14	3-METHYL-3-HEXENE, CIS	(L)	98.182	-27.94	A(1236)				
C7H14	3-METHYL-3-HEXENE, CIS	(G)	98.182	-19.22	A(1236)				
C7H14	3-METHYL-3-HEXENE, TRANS	(L)	98.182	-27.16	A(1236)				
C7H14	3-METHYL-3-HEXENE, TRANS	(G)	98.182	-18.60	A(1236)				
C7H14	4-METHYL-1-HEXENE	(G)	98.182	-16.0	E(1248)				
C7H14	4-METHYL-2-HEXENE, CIS	(G)	98.182	-17.5	E(1248)				
C7H14	4-METHYL-2-HEXENE, TRANS	(G)	98.182	-18.5	E(1248)				
C7H14	5-METHYL-1-HEXENE	(L)	98.182	-24.36	B(268)				
C7H14	5-METHYL-1-HEXENE	(G)	98.182	-16.6	C(843)				
C7H14	5-METHYL-2-HEXENE, CIS	(G)	98.182	-18.2	E(1248)				
C7H14	5-METHYL-2-HEXENE, TRANS	(G)	98.182	-19.2	E(1248)				
C7H14	2,3,3-TRIMETHYL-1-BUTENE	(L)	98.182	-28.35	A(1236)				
C7H14	2,3,3-TRIMETHYL-1-BUTENE	(G)	98.182	-20.67	A(1236)				

Formula	Name	State	Mol Wt	$\Delta H f^\circ_{298}$ (kcal/mole)	Source	S°_{298} (cal/mole·°K)	Source	$\Delta G f^\circ_{298}$ (kcal/mole)	Log $K p$
C7H14BR2	1,2-DIBROMOHEPTANE	(L)	258.014	-37.6	B(878)				
C7H14BR2	1,2-DIBROMOHEPTANE	(G)	258.014	-25.2	B(878)				
C7H14N2	DIISOPROPYLCARBODIIMIDE	(S)	126.198	-4.37	A(1275)				
C7H14N2	DIISOPROPYLCYANAMIDE	(S)	126.198	-11.70	A(1275)				
C7H14N2O2	DIETHYLMALONAMIDE	(U,S)	158.198	-141.1	C(744)				
C7H14N2O3	GLYCYLVALINE	(S)	174.198	-199.6	B(1176)				
C7H14N2O3	VALINE GLYCINE ANHYDRIDE	(S)	174.198	-188.5	B(1178)				
C7H14O	CYCLOHEPTANOL	(L)	114.182	-86.3	B(844)				
C7H14O	CYCLOHEXANEMETHANOL	(L)	114.182	-88.2	B(844)				
C7H14O	1,3-DIMETHYLCYCLOPENTANOL	(L)	114.182	-102.5	C(744)				
C7H14O	2,5-DIMETHYLCYCLOPENTANOL	(L)	114.182	-105.9	C(744)				
C7H14O	2,4-DIMETHYL-3-PENTANONE	(L)	114.182	-90.9	C(744)				
C7H14O	1,3-EPOXY-3-ETHYLPENTANE	(L)	114.182	-76.6	C(843)				
C7H14O	2,3-EPOXY-3-ETHYLPENTANE	(L)	114.182	-71.1	C(843)				
C7H14O	2-ETHYLCYCLOPENTANOL	(L)	114.182	-97.5	C(744)				
C7H14O	HEPTANAL	(L)	114.182	-74.5	A(1065)	83.3	M(1109)	-24.05	17.629
C7H14O	HEPTANAL	(G)	114.182	-63.10	T(550)	110.34	T(550)	-20.71	15.183
C7H14O	4-HEPTANONE	(L)	114.182	-85.9	C(744)				
C7H14O	ISOPENTYL VINYL ETHER	(L)	114.182	-66.1	C(843)				
C7H14O	2-METHYLCYCLOHEXANOL, CIS	(L)	114.182	-93.3	C(1366)				
C7H14O	2-METHYLCYCLOHEXANOL, TRANS	(L)	114.182	-99.6	C(1366)				
C7H14O	3-METHYLCYCLOHEXANOL, CIS	(L)	114.182	-94.5	C(1366)				
C7H14O	3-METHYLCYCLOHEXANOL, TRANS	(L)	114.182	-99.6	C(1366)				
C7H14O	3-METHYL-5-HEXEN-3-OL	(L)	114.182	-86.4	C(744)				
C7H14O	4-METHYLCYCLOHEXANOL, CIS	(L)	114.182	-100.7	C(1366)				
C7H14O	4-METHYLCYCLOHEXANOL, TRANS	(L)	114.182	-103.9	C(1366)				
C7H14O2	ETHYL ISOVALERATE	(L)	130.182	-136.6	B(569)				
C7H14O2	ETHYL 2-METHYLBUTYRATE	(L)	130.182	-135.5	B(569)				
C7H14O2	ETHYL PIVALATE	(L)	130.182	-138.0	B(569)				
C7H14O2	ETHYL VALERATE	(L)	130.182	-132.0	B(569)				
C7H14O2	2-ETHYLVALERIC ACID	(L)	130.182	-141.3	C(744)				
C7H14O2	HEPTANOIC ACID	(L)	130.182	-145.5	A(3)				
C7H14O2	HEPTANOIC ACID	(L)	130.182	-146.2	A(853)				
C7H14O2	1-METHYL-1,2-CYCLOHEXANEDIOL, CIS	(S)	130.182	-144.2	B(1536)				
C7H14O2	1-METHYL-1,2-CYCLOHEXANEDIOL,								

Formula	Name	State	Mol Wt	$\Delta H^\circ_{f\,298}$ (kcal/mole)	Source	S°_{298} (cal/mole °K)	Source	$\Delta G^\circ_{f\,298}$ (kcal/mole)	Log K_p
C7H14O2	METHYL HEXANOATE TRANS	(S)	130.182	-141.8	B(1536)				
C7H14O2	METHYL HEXANOATE	(L)	130.182	-129.2	A(3)				
C7H14O6	ALPHA-METHYLGLUCOSIDE	(U,S)	194.182	-292.8	C(744)				
C7H14O7	GLUCOHEPTOSE	(S)	210.182	-353.2	C(744)				
C7H14S	ALLYL TERT-BUTYL SULFIDE	(L)	130.248	-21.7	B(900)				
C7H14S	ALLYL TERT-BUTYL SULFIDE	(G)	130.248	-11.1	B(900)				
C7H15BR	1-BROMOHEPTANE	(L)	179.106	-52.24	A(138)				
C7H15BR	1-BROMOHEPTANE	(G)	179.106	-40.69	A(138)				
C7H15F	1-FLUOROHEPTANE	(L)	118.190	-92.1	C(744)				
C7H15N	3-METHYLCYCLOHEXYLAMINE	(L)	113.198	-51.9	C(744)				
C7H16	2,2-DIMETHYLPENTANE	(L)	100.198	-57.05	A(1203)	71.77	MC(654)	-1.16	-0.853
C7H16	2,2-DIMETHYLPENTANE	(G)	100.198	-49.27	T(118)	93.90	TC(118)	0.02	-0.014
C7H16	2,3-DIMETHYLPENTANE	(L)	100.198	-55.81	A(1203)	76.27	E(1248)	-1.26	-0.927
C7H16	2,3-DIMETHYLPENTANE	(G)	100.198	-47.62	T(119)	98.96	TC(119)	0.16	-0.117
C7H16	2,4-DIMETHYLPENTANE	(L)	100.198	-56.17	A(1203)	72.46	MC(120)	-0.49	-0.358
C7H16	2,4-DIMETHYLPENTANE	(G)	100.198	-48.28	T(120)	94.80	TC(120)	0.74	-0.543
C7H16	3,3-DIMETHYLPENTANE	(L)	100.198	-56.07	A(1203)	73.44	E(1248)	-0.68	-0.499
C7H16	3,3-DIMETHYLPENTANE	(G)	100.198	-48.17	T(121)	95.53	TC(121)	0.63	-0.464
C7H16	3-ETHYLPENTANE	(L)	100.198	-53.77	A(1203)	75.18	MC(654)	-1.10	-0.806
C7H16	3-ETHYLPENTANE	(G)	100.198	-45.33	T(117)	98.35	TC(117)	2.63	-1.929
C7H16	HEPTANE	(L)	100.198	-53.63	A(1203)	78.53	MC(654)	0.24	-0.177
C7H16	HEPTANE	(G)	100.198	-44.88	T(114)	102.27	TC(114)	1.91	-1.402
C7H16	2-METHYLHEXANE	(L)	100.198	-54.93	A(1203)	77.28	MC(654)	-0.69	-0.503
C7H16	2-METHYLHEXANE	(G)	100.198	-46.59	T(115)	100.38	TC(115)	0.77	-0.562
C7H16	3-METHYLHEXANE	(L)	100.198	-54.35	A(1203)	78.23	E(1248)	-0.39	-0.285
C7H16	3-METHYLHEXANE	(G)	100.198	-45.96	T(116)	101.37	TC(116)	1.10	-0.807
C7H16	2,2,3-TRIMETHYLBUTANE	(L)	100.198	-56.63	A(1203)	69.85	MC(654)	-0.17	-0.125
C7H16	2,2,3-TRIMETHYLBUTANE	(G)	100.198	-48.95	T(122)	91.61	TC(122)	1.02	-0.749
C7H16O	TERT-BUTYL ISOPROPYL ETHER	(L)	116.198	-94.0	B(1390)				
C7H16O	TERT-BUTYL ISOPROPYL ETHER	(G)	116.198	-85.60	T(509)	99.89	T(509)	-30.79	22.570
C7H16O	3-ETHYL-3-PENTANOL	(L)	116.198	-124.8	E(744)				
C7H16O	HEPTYL ALCOHOL	(L)	116.198	-95.31	C(215)	77.9	M(1109)	-33.95	24.882
C7H16O	HEPTYL ALCOHOL	(G)	116.198	-80.03	T(527)	114.83	T(527)	-29.68	21.752
C7H16O2	ACETONE DIETHYL ACETAL	(L)	132.198	-128.2	A(1416)				
C7H16O2	ACETONE DIETHYL ACETAL	(G)	132.198	-120.6	A(1416)				
C7H16O2	DIPROPOXYMETHANE	(L)	132.198	-118.2	C(744)				

Formula	Name	State	Mol Wt	$\Delta H f°_{298}$ (kcal/mole)	Source	$S°_{298}$ (cal/mole °K)	Source	$\Delta G f°_{298}$ (kcal/mole)	Log $K p$
C7H16O4S2	2,2-BIS(ETHYLSULFONYL)PROPANE	(S)	228.330	-341.6	C(1441)				
C7H12O7	D-GLYCERO-D-GALACTOHEPTITOL	(S)	212.198	-368.6	C(744)				
C7H16S	BUTYL PROPYL SULFIDE	(G)	132.264	-34.88	TC(811)	117.90	TC(811)	9.53	-6.984
C7H16S	ETHYL PENTYL SULFIDE	(G)	132.264	-34.85	TC(812)	117.58	TC(812)	9.65	-7.076
C7H16S	1-HEPTANETHIOL	(L)	132.264	-47.85	AC(507)				
C7H16S	1-HEPTANETHIOL	(G)	132.264	-35.76	TC(901)	117.89	TC(901)	8.65	-6.341
C7H16S	HEXYL METHYL SULFIDE	(G)	132.264	-34.27	TC(813)	117.04	TC(813)	10.39	-7.619
C7H17N	HEPTYLAMINE	(L)	115.214	-59.2	CC(744)				
C8F16	UNDECAFLUORO(PENTAFLUORO-ETHYL)CYCLOHEXANE	(L)	400.080	-798.2	AC(508)				
C8F16	UNDECAFLUORO(PENTAFLUORO-ETHYL)CYCLOHEXANE	(G)	400.080	-789.0	AC(508)				
C8H4Cl2O2	PHTHALOYL CHLORIDE	(S)	203.026	-97.4	CC(744)				
C8H4N2	PHTHALONITRILE	(S)	128.128	66.35	AC(1275)				
C8H4N6	AMMONIUM SALT OF 1,1,2,3-PENTACYANOPROPENE	(S)	184.160	120.7	AC(154)				
C8H4O3	PHTHALIC ANHYDRIDE	(S)	148.112	-110.1	A(1116)	42.9	M(1118)	-79.12	57.992
C8H5F3O2	ALPHA,ALPHA,ALPHA-TRIFLUORO-M-TOLUIC ACID	(S)	190.120	-254.5	AC(512)				
C8H5NO	PHENYLGLYOXYLONITRILE	(S)	131.128	17.8	CC(744)				
C8H5NO2	INDOLE-2,3-DIONE	(S)	147.128	-54.7	CC(744)				
C8H5NO2	ISATIN	(S)	147.128	-61.6	CC(1414)				
C8H5NO2	PHTHALIMIDE	(S)	147.128	-73.0	CC(744)				
C8H6	ETHYNYLBENZENE	(L)	102.128	67.72	AC(432)				
C8H6	ETHYNYLBENZENE	(G)	102.128	78.22	TC(471)	76.88	TC(471)	86.46	-63.375
C8H6Br4	TETRABROMO-M-XYLENE	(S)	421.792	7.9	CC(1180)				
C8H6Br4	TETRABROMO-O-XYLENE	(S)	421.792	6.4	CC(1180)				
C8H6Br4	TETRABROMO-P-XYLENE	(S)	421.792	1.3	CC(1180)				
C8H6Cl2	2,5-DICHLOROSTYRENE	(L)	173.042	8.3	AC(1359)				
C8H6Cl4	2,3,5,6-TETRACHLORO-P-XYLENE	(S)	243.956	-42.05	AC(1387)				
C8H6N2O	PHENYLFURAZAN	(S)	146.144	58.5	CC(1012)				
C8H6N2O2	3-PHENYLOXADIAZOL-5-OL	(S)	162.144	-24.5	CC(1012)				
C8H6N2O2	5-PHENYLOXADIAZOL-3-OL	(S)	162.144	0.6	CC(1012)				
C8H6N2O4	M,BETA-DINITROSTYRENE	(S)	194.144	0.7	CC(744)				
C8H6N2O4	P,BETA-DINITROSTYRENE	(S)	194.144	5.9	CC(744)				

Formula	Name	State	Mol Wt	$\Delta H_{f\ 298}^{\circ}$ (kcal/mole)	Source	S_{298}° (cal/mole °K)	Source	$\Delta G_{f\ 298}^{\circ}$ (kcal/mole)	Log K_P
C8H6N4O2	2-PHENYL-5-TETRAZOLE-								
C8H6N4O2	CARBOXYLIC ACID	(S)	190.160	-9.9	B(974)				
C8H6N4O10	2,4,6-TRINITROPHENOXY-								
C8H6N4O10	ETHYL NITRATE	(S)	318.160	-66.3	B(983)				
C8H6O2	PHTHALIDE	(S)	134.128	-72.6	C(744)				
C8H6O3	METHYLENEPROTOCATECHUALDEHYDE	(S)	150.128	-85.9	C(744)				
C8H6O4	ISOPHTHALIC ACID	(S)	166.128	-191.96	A(1295)				
C8H6O4	PHTHALIC ACID	(S)	166.128	-186.96	A(1295)	49.7	M(1118)	-141.39	103.640
C8H6O4	PIPERONYLIC ACID	(S)	166.128	-153.3	C(744)				
C8H6O4	TEREPHTHALIC ACID	(S)	166.128	-195.10	A(1295)				
C8H6S	BENZOTHIOPHENE	(S)	134.194			42.33	M(419)		
C8H7ClO	TOLUOYL CHLORIDE	(S)	154.593	-52.4	C(744)				
C8H7FN2O3	4'-FLUORO-3'-NITROACETANILIDE	(L)	198.152	-98.1	C(744)				
C8H7F2NO	DIFLUOROACETANILIDE	(U,S)	171.144	-131.1	C(744)				
C8H7IO2	METHYL M-IODOBENZOATE	(S)	262.046	-66.3	B(1382)				
C8H7IO2	METHYL O-IODOBENZOATE	(L)	262.046	-58.1	B(1382)				
C8H7IO2	METHYL P-IODOBENZOATE	(S)	262.046	-68.4	B(1382)				
C8H7N	BENZYL ISOCYANIDE	(L)	117.144	55.9	C(744)				
C8H7N	INDOLE	(S)	117.144	30.3	C(1413)				
C8H7N	PHENYLACETONITRILE	(L)	117.144	32.9	C(744)				
C8H7N	O-TOLUNITRILE	(L)	117.144	39.7	C(744)				
C8H7NO2	INDOLEHYDROXY-2-INDOLINE	(S)	149.144	-75.1	C(744)				
C8H7NO2	OMEGA-NITROSTYRENE	(S)	149.144	6.1	B(1640)				
C8H7N3O	PHENYLAMINO FURAZAN	(S)	161.160	54.2	C(1012)				
C8H7N3O6	1-ETHYL-2,4,6-TRINITROBENZENE	(S)	241.160	-25.1	B(984)				
C8H7N3O6	2,4,6-TRINITRO-M-XYLENE	(S)	241.160	-24.0	B(55)				
C8H7N3O7	2,4,6-TRINITROPHENETOLE	(S)	257.160	-48.0	B(57)				
C8H7N3O8	2,4-DINITROPHENOXYETHYL								
C8H7N3O8	NITRATE	(S)	273.160	-66.4	B(983)				
C8H7N5O8	METHYLTETRYL	(U,S)	301.176	17.6	C(744)				
C8H7N5O8	AR,N-TETRANITROPHENETHYLAMINE	(S)	301.176	-4.8	B(981)				
C8H8	1,3,5,7-CYCLOOCTATETRAENE	(L)	104.144	60.93	A(1197)	52.65	M(1312)	85.70	-62.817
C8H8	1,3,5,7-CYCLOOCTATETRAENE	(G)	104.144	71.23	T(472)	78.10	T(472)	88.41	-64.805
C8H8	STYRENE	(L)	104.144	24.83	A(1248)	56.78	M(1160)	48.37	-35.454
C8H8	STYRENE	(G)	104.144	35.22	T(439)	82.48	T(439)	51.10	-37.453
C8H8FNO	3'-FLUOROACETANILIDE	(S)	153.152	-90.4	C(744)				

Formula	Name	State	Mol Wt	$\Delta H f^\circ_{298}$ (kcal/mole)	Source	S°_{298} (cal/mole °K)	Source	$\Delta G f^\circ_{298}$ (kcal/mole)	Log $K p$
C8H8FNO	4'-FLUOROACETANILIDE	(S)	153.152	-89.6	C(744)				
C8H8FNO3	4'-FLUORO-3-NITROPHENETOL	(S)	185.152	-87.2	C(744)				
C8H8N2O2	PHENYLGLYOXIME(ALPHA)	(S)	164.160	-3.2	B(1013)				
C8H8N2O2	PHENYLGLYOXIME(BETA)	(S)	164.160	11.8	B(1013)				
C8H8N2O3	2'-NITROACETANILIDE	(S)	180.160	-51.1	C(744)				
C8H8N2O3	3'-NITROACETANILIDE	(S)	180.160	-56.2	C(744)				
C8H8N2O3	4'-NITROACETANILIDE	(S)	180.160	-56.8	C(744)				
C8H8N2O4	BISUCCINIMIDE	(S)	196.160	-169.55	A(247)				
C8H8N2O4	2,4-DINITRO-M-XYLENE	(S)	196.160	-19.7	B(55)				
C8H8N2O4	4,6-DINITRO-M-XYLENE	(S)	196.160	-23.7	B(55)				
C8H8N2O4	1-ETHYL-2,4-DINITROBENZENE	(L)	196.160	-19.6	B(1471)				
C8H8N2O5	2,4-DINITROPHENETOLE	(S)	212.160	-53.4	B(57)				
C8H8N4	1-METHYL-5-PHENYLTETRAZOLE	(S)	160.176	69.8	B(1616)				
C8H8N4	1-PHENYL-5-METHYLTETRAZOLE	(S)	160.176	69.2	B(1616)				
C8H8N4	2-PHENYL-5-METHYLTETRAZOLE	(S)	160.176	65.6	B(974)				
C8H8N6	BENZAL-5-HYDRAZINOTETRAZOLE	(S)	188.192	105.4	B(1616)				
C8H8N6O6	AMMONIUM 5,5-NITRILO-DIBARBITURIC ACID	(S)	284.192	-289.4	C(744)				
C8H8O	ACETOPHENONE	(L)	120.144	-34.06	A(251)	59.62	M(1354)	-4.06	-2.977
C8H8O	ACETOPHENONE	(G)	120.144	-20.76	A(1434)	89.12	M(1434)	0.44	-0.324
C8H8OS	PHENYL THIOLACETATE	(L)	152.210	-32.4	A(1556)				
C8H8O2	M-ANISALDEHYDE	(L)	136.144	-64.9	C(928)				
C8H8O2	O-ANISALDEHYDE	(S)	136.144	-62.6	C(928)				
C8H8O2	P-ANISALDEHYDE	(L)	136.144	-63.8	C(928)				
C8H8O2	1,4-BENZODIOXAN-TETRAHYDRONAPHTHALENE	(L)	136.144	-61.0	A(201)				
C8H8O2	2-HYDROXYACETOPHENONE	(L)	136.144	-84.4	C(145)				
C8H8O2	3-HYDROXYACETOPHENONE	(S)	136.144	-87.1	C(145)				
C8H8O2	4-HYDROXYACETOPHENONE	(S)	136.144	-86.2	C(145)				
C8H8O2	METHYL BENZOATE	(L)	136.144	-81.4	C(744)				
C8H8O2	PHENYLACETIC ACID	(S)	136.144	-94.7	C(744)				
C8H8O2	M-TOLUIC ACID	(S)	136.144	-101.84	A(250)				
C8H8O2	O-TOLUIC ACID	(S)	136.144	-99.55	A(250)				
C8H8O2	P-TOLUIC ACID	(S)	136.144	-102.58	A(250)				
C8H8O3	P-ANISIC ACID	(S)	152.144	-130.6	C(744)				
C8H8O3	2,3-CRESOTIC ACID	(S)	152.144	-146.3	C(744)				

Formula	Name	State	Mol Wt	ΔH°_{298} (kcal/mole)	Source	S°_{298} (cal/mole °K)	Source	$\Delta G^\circ_{f\,298}$ (kcal/mole)	Log K_p
C8H8O3	2,4-CRESOTIC ACID	(S)	152.144	-147.2	C(744)				
C8H8O3	2,5-CRESOTIC ACID	(S)	152.144	-145.5	C(744)				
C8H8O3	2,6-CRESOTIC ACID	(S)	152.144	-142.2	C(744)				
C8H8O3	3-HYDROXY-P-ANISALDEHYDE	(S)	152.144	-107.3	C(928)				
C8H8O3	4-HYDROXY-M-ANISALDEHYDE	(S)	152.144	-110.9	C(744)				
C8H8O3	MANDELIC ACID, D,L	(S)	152.144	-138.5	B(1651)				
C8H8O3	MANDELIC ACID, L	(S)	152.144	-138.7	B(1651)				
C8H8O3	METHYL P-HYDROXYBENZOATE	(S)	152.144	-129.6	C(744)				
C8H8O3	METHYL SALICYLATE	(L)	152.144	-127.1	A(1354)				
C8H8O3	O-OXYMETHYLBENZOIC ACID	(L,S)	152.144	-137.7	C(744)				
C8H8O3	PHENOXYACETIC ACID	(S)	152.144	-122.2	C(744)				
C8H8O4	CYCLOHEXADIENE- 1,4-DICARBOXYLIC ACID	(S)	168.144	-182.5	C(744)				
C8H8O4	1,4-CYCLOHEXADIENE- 1,4-DICARBOXYLIC ACID	(S)	168.144	-189.5	C(744)				
C8H8O4	2,4-CYCLOHEXADIENE- 1,4-DICARBOXYLIC ACID	(S)	168.144	-183.0	C(744)				
C8H8O4	2,5-CYCLOHEXADIENE- 1,4-DICARBOXYLIC ACID	(S)	168.144	-180.2	C(744)				
C8H8O5	METHYL GALLATE	(S)	184.144	-223.6	C(744)				
C8H9CL	1-CHLORO-2-ETHYLBENZENE	(L)	140.609	-12.88	A(635)				
C8H9CL	1-CHLORO-4-ETHYLBENZENE	(L)	140.609	-12.33	A(635)				
C8H9FO	M-FLUOROPHENETOLE	(L)	140.152	-79.5	C(744)				
C8H9FO	P-FLUOROPHENETOLE	(L)	140.152	-82.7	C(744)				
C8H9NO	ACETANILIDE	(S)	135.160	-50.3	A(1558)				
C8H9NO	ACETOPHENONE OXIME	(S)	135.160	-5.3	C(744)				
C8H9NO2	2-OXYACETOPHENONE OXIME	(S)	151.160	-52.9	C(843)				
C8H9NO2	4-OXYACETOPHENONE OXIME	(S)	151.160	-50.9	C(843)				
C8H9NO2	PHENYLGLYCINE	(L,S)	151.160	-103.1	C(744)				
C8H9NO2	N-PHENYLGLYCINE	(L)	151.160	-93.0	C(744)				
C8H9NO3	M-NITROPHENETOLE	(L)	167.160	-49.9	C(744)				
C8H9NO3	O-NITROPHENETOLE	(L)	167.160	-38.6	C(744)				
C8H9NO3	P-NITROPHENETOLE	(L)	167.160	-53.1	C(744)				
C8H9N3O4	DIMETHYL-2,4-DINITRO- N,N-ANILINE	(S)	211.176	-6.9	B(204)				
C8H10	ETHYLBENZENE	(L)	106.160	-2.98	A(1248)	60.99	M(1325)	28.61	-20.971

Formula	Name	State	Mol Wt	$\Delta H f^\circ_{298}$ (kcal/mole)	Source	S°_{298} (cal/mole °K)	Source	$\Delta G f^\circ_{298}$ (kcal/mole)	Log $K p$
C8H10	ETHYLBENZENE	(G)	106.160	7.12	T(398)	86.15	T(398)	31.21	-22.875
C8H10	1,7-OCTADIYNE	(L)	106.160	80.0	A(433)				
C8H10	M-XYLENE	(L)	106.160	-6.07	A(1248)				
C8H10	M-XYLENE	(G)	106.160	4.12	T(399)	60.27	M(1248)	25.73	-18.863
C8H10	O-XYLENE	(L)	106.160	-5.84	A(1248)				
C8H10	O-XYLENE	(G)	106.160	4.54	T(400)	85.49	T(399)	28.41	-20.821
C8H10	P-XYLENE	(L)	106.160	-5.84	A(1248)	58.80	M(1166)	26.40	-19.353
C8H10	P-XYLENE	(G)	106.160	4.29	T(401)	84.31	T(400)	29.18	-21.386
C8H10N2O	N,N-DIMETHYL-P-NITROSOANILINE	(S)	150.176	31.6	C(744)	59.20	M(1166)	26.28	-19.265
C8H10N2O	N-ETHYL-N-NITROSOANILINE	(S)	150.176	25.3	C(744)	84.23	T(401)	28.95	-21.221
C8H10N4O2	CAFFEINE	(S)	194.192	-79.2	C(744)				
C8H10N4O10	ALLOXANTHINE DIHYDRATE	(U,S)	322.192	-475.9	C(744)				
C8H10O	M-ETHYLPHENOL	(L)	122.160	-51.29	A(128)				
C8H10O	M-ETHYLPHENOL	(G)	122.160	-35.01	A(128)				
C8H10O	O-ETHYLPHENOL	(L)	122.160	-49.99	A(128)				
C8H10O	O-ETHYLPHENOL	(G)	122.160	-34.82	A(128)				
C8H10O	P-ETHYLPHENOL	(S)	122.160	-53.71	A(128)				
C8H10O	P-ETHYLPHENOL	(G)	122.160	-34.55	A(128)				
C8H10O	M-METHYLANISULE	(L)	122.160	-37.6	B(56)				
C8H10O	PHENETOLE	(L)	122.160	-38.6	B(56)				
C8H10O	M-XYLENOL	(L)	122.160	-55.7	C(744)				
C8H10O	O-XYLENOL	(U,S)	122.160	-58.0	C(744)				
C8H10O	P-XYLENOL	(U,S)	122.160	-57.8	C(744)				
C8H10O2	O-DIMETHOXYBENZENE	(L)	138.160	-70.1	A(201)				
C8H10O2	P-DIMETHOXYBENZENE	(S)	138.160	-78.5	C(744)				
C8H10O2	P-DIMETHOXYBENZENE	(L)	138.160	-70.6	C(744)				
C8H10O2S	P-XYLENE-ALPHA,ALPHA'-DIOL	(S)	138.160	-94.23	A(1114)				
C8H10O2S	BENZYL METHYL SULFONE	(L)	170.226	-88.7	B(905)				
C8H10O2S	BENZYL METHYL SULFONE	(G)	170.226	-67.2	B(905)				
C8H10O2S	METHYL P-TOLYL SULFONE	(L)	170.226	-92.4	B(905)				
C8H10O2S	METHYL P-TOLYL SULFONE	(G)	170.226	-70.8	B(905)				
C8H10O3	1,2-CYCLOHEXANEDICARBOXYLIC ANHYDRIDE, CIS	(S)	154.160	-161.3	C(744)				
C8H10O3	1,2-CYCLOHEXANEDICARBOXYLIC ANHYDRIDE, TRANS	(S)	154.160	-162.0	C(744)				
C8H10O3S	P-ETHYLBENZENESULFONIC ACID	(G)	186.226			104.04	E(69)		

Formula	Name	State	Mol Wt	$\Delta H_f^\circ{}_{298}$ (kcal/mole)	Source	S°_{298} (cal/mole °K)	Source	$\Delta G_f^\circ{}_{298}$ (kcal/mole)	Log K_p
C8H10O4	1-CYCLOHEXENE-1,4-DICARBOXYLIC ACID	(S)	170.160	-211.1	C(744)				
C8H10O4	2-CYCLOHEXENE-1,4-DICARBOXYLIC ACID	(S)	170.160	-212.3	C(744)				
C8H10O4	DIETHYL ACETYLENEDICARBOXYLATE	(L)	170.160	-136.4	C(744)				
C8H10S	BENZYL METHYL SULFIDE	(L)	138.226	6.2	B(898)				
C8H10S	BENZYL METHYL SULFIDE	(G)	138.226	19.0	B(898)				
C8H10S	ETHYL PHENYL SULFIDE	(L)	138.226	5.2	B(898)				
C8H10S	ETHYL PHENYL SULFIDE	(G)	138.226	18.4	B(898)				
C8H11N	N,N-DIMETHYLANILINE	(L)	121.176	8.2	C(1547)	61.2	E(1547)	51.20	-37.531
C8H11N	N,N-DIMETHYLANILINE	(G)	121.176	20.10	C(1547)	87.5	E(1547)	55.26	-40.506
C8H11N	N-ETHYLANILINE	(L)	121.176	0.9	A(1547)	57.2	E(1547)	45.10	-33.054
C8H11N	N-ETHYLANILINE	(G)	121.176	13.40	B(1547)	84.1	E(1547)	49.58	-36.338
C8H11N	2-NORBORNANECARBONITRILE,ENDO	(L)	121.176			53.29	M(1334)		
C8H11N	2-NORBORNANECARBONITRILE, EXO	(L)	121.176			55.11	M(1334)		
C8H11N	2-OCTYNENITRILE	(L)	121.176	36.3	C(744)				
C8H11N	2,4-XYLIDINE	(L)	121.176	-19.2	C(744)				
C8H11NO2	ETHYL 2-METHYL-3-PYRROLECARBOXYLATE	(S)	153.176	-95.6	C(1414)				
C8H11NO2	METHYL 2,4-DIMETHYL-3-PYRROLECARBOXYLATE	(S)	153.176	-98.3	C(1413)				
C8H11NO2	METHYL 3,5-DIMETHYL-2-PYRROLECARBOXYLATE	(S)	153.176	-98.3	C(1413)				
C8H11NO3	ETHYL 4-HYDROXY-5-METHYL-3-PYRROLECARBOXYLATE	(S)	169.176	-143.5	C(1414)				
C8H12	BICYCLO(2,2,2)OCT-2-ENE	(S)	108.176	-13.8	G(744)	50.30	M(1624)		
C8H12	1,3-DIMETHYLDIHYDROBENZENE	(U,L)	108.176	-7.9	C(744)				
C8H12	1,4-DIMETHYL-1,3-CYCLOHEXADIENE	(L)	108.176						
C8H12	1-METHYL-3-METHYLENE-CYCLOHEXENE-1	(L)	108.176	-10.9	C(744)				
C8H12	VINYLCYCLOHEXENE	(G)	108.176	16.8	E(356)	96.4	E(356)	47.14	-34.551
C8H12N2O3	5,5-DIETHYLBARBITURIC ACID	(S)	184.192	-178.8	C(744)				
C8H12N2O6	1,2-BIS(3-CARBOXYPROPIONYL)-HYDRAZINE	(S)	232.192	-317.8	B(1640)				
C8H12N4	2,2'-AZOBIS-2-METHYL-PROPIONITRILE	(S)	164.208	54.6	B(974)				

700

Formula	Name	State	Mol Wt	$\Delta H^\circ_{f\,298}$ (kcal/mole)	Source	S°_{298} (cal/mole °K)	Source	$\Delta G^\circ_{f\,298}$ (kcal/mole)	Log Kp
C8H12O	BETA-DICYCLOOCTANONE, CIS	(L)	124.176	-67.8	A(74)				
C8H12O	BETA-DICYCLOOCTANONE, TRANS	(L)	124.176	-61.0	A(74)				
C8H12O	3,5-DIMETHYL-2-CYCLOHEXEN- 1-ONE	(L)	124.176	-59.9	C(744)				
C8H12O	ETHYLENECYCLOHEXANONE, ENDO	(L,S)	124.176	-65.5	A(90)	56.47	M(1624)		
C8H12O	3-OXABICYCLO(3,2,2)NONANE	(S)	124.176						
C8H12O2	1-CYCLOHEXENE-1-ACETIC ACID	(S)	140.176	-115.7	C(744)				
C8H12O2	DELTA-1-ALPHA-CYCLOHEXANE- ACETIC ACID	(S)	140.176	-118.5	C(744)				
C8H12O2	ETHYL SORBATE	(L)	140.176	-148.5	C(744)				
C8H12O2	2-OCTYNOIC ACID	(S)	140.176	-79.2	C(744)				
C8H12O3	2,2-DIETHYLSUCCINIC ANHYDRIDE	(L)	156.176	-164.0	B(1538)				
C8H12O3	DIETHYLSUCCINIC ANHYDRIDE,CIS	(L)	156.176	-165.1	B(1538)				
C8H12O3	DIETHYLSUCCINIC ANHYDRIDE,TRANS	(L)	156.176	-166.7	B(1538)				
C8H12O3	TETRAMETHYLSUCCINIC ANHYDRIDE	(S)	156.176	-169.8	B(1538)				
C8H12O4	CYCLOHEXANE-1,2- DICARBOXYLIC ACID, CIS	(S)	172.176	-229.8	B(1579)				
C8H12O4	CYCLOHEXANE-1,2- DICARBOXYLIC ACID, TRANS	(S)	172.176	-232.2	B(1579)				
C8H12O4	CYCLOHEXANE-1,4-DICARBOXYLIC ACID, CIS	(S)	172.176	-233.5	C(744)				
C8H12O4	CYCLOHEXANE-1,4-DICARBOXYLIC ACID, TRANS	(S)	172.176	-232.6	C(744)				
C8H12O4	DIMETHYL TETRAMETHYLENE- ALPHA,BETA-DICARBOXYLATE	(L)	172.176	-177.2	C(744)				
C8H12O4	ETHYL 2-ACETYLACETOACETATE	(L)	172.176	-190.2	C(744)				
C8H12S6	1,3,5,7-TETRAMETHYL-2,4,6,8,9, 10-HEXATHIAADAMANTANE	(S)	300.572						
C8H13NO	N-PENTYLPROPIOLAMIDE	(S)	139.192	-46.0	C(744)				
C8H14	BICYCLO(2,2,2)OCTANE	(S)	110.192	-16.7	B(878)	76.75	M(213)		
C8H14	CYCLOOCTENE	(L)	110.192	-6.76	A(261)	50.18	M(1624)		
C8H14	CYCLOOCTENE	(G)	110.192	-15.77	A(821)				
C8H14	3-CYCLOPENTYL-1-PROPENE	(L)	110.192	3.0	C(1373)				
C8H14	2-CYCLOPROPYL-3-METHYL-1-BUTENE	(L)	110.192	-7.0	C(1374)				
C8H14	2-CYCLOPROPYL-1-PENTENE	(L)	110.192	-2.0	C(1374)				
C8H14	2-CYCLOPROPYL-2-PENTENE	(L)	110.192						

Formula	Name	State	Mol Wt	$\Delta H f^\circ_{298}$ (kcal/mole)	Source	S°_{298} (cal/mole °K)	Source	$\Delta G f^\circ_{298}$ (kcal/mole)	Log $K p$
C8H14	2,4-DIMETHYLCYCLOHEXENE . . .	(L)	110.192	-35.8	C(744)				
C8H14	1-ETHYLCYCLOHEXENE	(L)	110.192	-25.53	A(821)				
C8H14	ETHYLIDENECYCLOHEXANE	(L)	110.192	-24.76	A(821)				
C8H14	2-METHYLNORBORNANE, ENDO . .	(L)	110.192			56.93	M(1333)		
C8H14	2-METHYLNORBORNANE, EXO . .	(L)	110.192			58.85	M(1333)		
C8H14	OCTAHYDROPENTALENE, CIS . . .	(L)	110.192	-30.9	B(75)				
C8H14	OCTAHYDROPENTALENE, TRANS . .	(L)	110.192	-24.9	B(75)				
C8H14	1-OCTYNE	(G)	110.192	19.70	T(325)	106.75	T(325)	56.26	-41.236
C8H14	1,2,3-TRIMETHYLCYCLOPENTENE	(L)	110.192	-37.9	C(744)				
C8H14	2,3,3-TRIMETHYLCYCLOPENTENE	(L)	110.192	-37.6	C(744)				
C8H14	VINYLCYCLOHEXANE	(L)	110.192	-21.22	A(821)				
C8H14BR2	1,2-DIBROMOCYCLOOCTANE . . .	(L)	270.024	-31.7	B(878)				
C8H14BR2	1,2-DIBROMOCYCLOOCTANE . . .	(G)	270.024	-18.7	B(878)				
C8H14N4	2,2'-HYDRAZOBIS- 2-METHYLPROPIONITRILE	(S)	166.224	28.5	B(974)				
C8H14N4O5	TRIGLYCYLGLYCINE	(U,S)	246.224	-282.6	C(744)				
C8H14O	CYCLOOCTANONE	(L)	126.192	-76.8	C(1272)				
C8H14O	2,2-DIMETHYLCYCLOHEXANONE . .	(L)	126.192	-76.4	C(744)				
C8H14O	2-ETHYL-2-HEXENAL	(L)	126.192	-62.46	A(1506)				
C8H14O	4-METHYL-1,6-HEPTADIEN-4-OL	(L)	126.192	-49.8	C(744)				
C8H14O	2-OCTYNE-1-OL	(L)	126.192	-38.6	C(744)				
C8H14O2	BUTYL CROTONATE	(L)	142.192	-112.9	B(1286)				
C8H14O2	SEC-BUTYL CROTONATE	(L)	142.192	-114.4	B(1286)				
C8H14O2	CYCLOHEPTANECARBOXYLIC ACID	(L)	142.192	-141.4	C(744)				
C8H14O2	CYCLOOCTANECARBOXYLIC ACID (ACTIVE)	(S)	142.192	-142.1	C(744)				
C8H14O2	2,5-DIMETHYL-3-HEXYNE-2,5-DIOL	(L)	142.192	-87.4	C(744)				
C8H14O2	ISOBUTYL CROTONATE	(L)	142.192	-115.6	B(1286)				
C8H14O2	ISOPROPYL 2-PENTENOATE . . .	(L)	142.192	-116.6	B(1287)				
C8H14O2	ISOPROPYL 3-PENTENOATE . . .	(L)	142.192	-114.2	B(1287)				
C8H14O2	ISOPROPYL 4-PENTENOATE . . .	(L)	142.192	-111.9	B(1287)				
C8H14O2	3-METHYLCYCLOHEXANECARBOXYLIC ACID	(S)	142.192	-144.4	C(744)				
C8H14O2	PROPYL 2-PENTENOATE	(L)	142.192	-114.2	B(1287)				
C8H14O2	PROPYL 3-PENTENOATE	(L)	142.192	-109.8	B(1287)				
C8H14O2	PROPYL 4-PENTENOATE	(L)	142.192	-108.3	B(1287)				

Formula	Name	State	Mol Wt	$\Delta H f^\circ_{298}$ (kcal/mole)	Source	S°_{298} (cal/mole °K)	Source	$\Delta G f^\circ_{298}$ (kcal/mole)	Log Kp
C8H14O3	ETHYL 2-ETHOXYCROTONATE	(L)	158.192	-154.7	A(996)				
C8H14O3	ETHYL 2-ETHYLACETOACETATE	(L)	158.192	-170.9	A(996)				
C8H14O3	ISOBUTYRIC ANHYDRIDE	(L)	158.192	-181.3	C(258)				
C8H14O4	BUTYRYL PEROXIDE	(L)	174.192	-161.0	A(679)				
C8H14O4	DIETHYL SUCCINATE	(L)	174.192	-223.4	C(744)				
C8H14O4	2,2-DIETHYLSUCCINIC ACID	(S)	174.192	-246.0	B(1538)				
C8H14O4	2,3-DIETHYLSUCCINIC ACID, CIS	(S)	174.192	-242.8	B(1538)				
C8H14O4	2,3-DIETHYLSUCCINIC ACID, TRANS	(S)	174.192	-244.5	B(1538)				
C8H14O4	DIMETHYL ADIPATE	(L)	174.192	-210.3	C(744)				
C8H14O4	DIMETHYLADIPIC ACID	(U,S)	174.192	-213.3	C(744)				
C8H14O4	ERYTHRITEDIACETAL	(U,S)	174.192	-180.4	C(744)				
C8H14O4	ETHYLPROPYLMALONIC ACID	(S)	174.192	-241.4	C(744)				
C8H14O4	PENTYLMALONIC ACID	(S)	174.192	-244.0	B(1535)				
C8H14O4	SUBERIC ACID	(S)	174.192	-246.6	C(744)				
C8H14O4	TETRAMETHYLSUCCINIC ACID	(S)	174.192	-241.2	B(1538)				
C8H14O6	DIETHYL TARTRATE, D	(L)	206.192	-299.5	C(744)				
C8H14O6	DIETHYL TARTRATE, MESO	(L)	206.192	-298.9	C(744)				
C8H14O6	DIMETHYLDIHYDROXYADIPIC ACID	(U,S)	206.192	-341.7	C(744)				
C8H14O6	DIMETHYL DIMETHOXY-SUCCINATE, DL	(S)	206.192	-265.3	B(358)				
C8H14O6	DIMETHYL MESO-DIMETHOXY-SUCCINATE	(S)	206.192	-275.7	B(358)				
C8H14O8	GLUCOOCTONOLACTONE	(S)	238.192	-394.1	C(744)				
C8H15CLO2	BUTYL 2-CHLOROBUTYRATE	(L)	178.657	-156.8	B(1384)				
C8H15CLO2	BUTYL 3-CHLOROBUTYRATE	(L)	178.657	-146.2	B(1384)				
C8H15CLO2	BUTYL 4-CHLOROBUTYRATE	(L)	178.657	-147.9	B(1384)				
C8H15CLO2	ISOBUTYL 2-CHLOROBUTYRATE	(L)	178.657	-158.3	B(1384)				
C8H15CLO2	ISOBUTYL 3-CHLOROBUTYRATE	(L)	178.657	-149.2	B(1384)				
C8H15CLO2	ISOBUTYL 4-CHLOROBUTYRATE	(L)	178.657	-151.0	B(1384)				
C8H15CLO2	ISOPENTYL 2-CHLOROPROPIONATE	(L)	178.657	-150.1	B(1384)				
C8H15CLO2	ISOPENTYL 3-CHLOROPROPIONATE	(L)	178.657	-142.0	B(1384)				
C8H15N	3-AZABICYCLO(3.2.2)NONANE	(S)	125.208						
C8H16	CYCLOOCTANE	(L)	112.208	-40.42	A(707)	56.14	M(73)	18.60	-13.634
C8H16	CYCLOOCTANE	(G)	112.208	-30.06	T(343)	62.62	M(424)	21.49	-15.755
C8H16	2-CYCLOPROPYL-3-METHYLBUTANE	(L)	112.208	-14.8	C(1373)	87.66	T(343)		
C8H16	2-CYCLOPROPYLPENTANE	(L)	112.208	-14.8	C(1375)				

703

Formula	Name	State	Mol Wt	$\Delta H_{f\,298}^{\circ}$ (kcal/mole)	Source	S_{298}° (cal/mole °K)	Source	$\Delta G_{f\,298}^{\circ}$ (kcal/mole)	Log K_p
C8H16	1,1-DIMETHYLCYCLOHEXANE	(L)	112.208	-52.31	A(1248)	63.87	MC 657)	6.34	-4.646
C8H16	1,1-DIMETHYLCYCLOHEXANE	(G)	112.208	-43.26	T(373)	87.24	T(373)	8.42	-6.172
C8H16	1,1-DIMETHYL-2-PROPYLCYCLOPROPANE	(L)	112.208	-27.8	AC 710)				
C8H16	1,2-DIMETHYLCYCLOHEXANE, CIS	(L)	112.208	-50.64	A(1248)	65.52	MC 657)	7.52	-5.509
C8H16	1,2-DIMETHYLCYCLOHEXANE, CIS	(G)	112.208	-41.15	T(374)	89.51	T(374)	9.85	-7.222
C8H16	1,2-DIMETHYLCYCLOHEXANE, TRANS	(L)	112.208	-52.19	A(1248)	65.30	MC 657)	6.03	-4.421
C8H16	1,2-DIMETHYLCYCLOHEXANE, TRANS	(G)	112.208	-43.02	T(375)	88.65	MC 375)	8.24	-6.040
C8H16	1,3-DIMETHYLCYCLOHEXANE, CIS	(L)	112.208	-53.30	A(1248)	65.16	MC 657)	4.96	-3.638
C8H16	1,3-DIMETHYLCYCLOHEXANE, CIS	(G)	112.208	-44.16	T(376)	88.54	T(376)	7.13	-5.228
C8H16	1,3-DIMETHYLCYCLOHEXANE, TRANS	(L)	112.208	-51.57	A(1248)	66.03	MC 657)	6.43	-4.716
C8H16	1,3-DIMETHYLCYCLOHEXANE, TRANS	(G)	112.208	-42.20	T(377)	89.92	T(377)	8.68	-6.363
C8H16	1,4-DIMETHYLCYCLOHEXANE, CIS	(L)	112.208	-51.55	A(1248)	64.80	MC 657)	6.82	-4.999
C8H16	1,4-DIMETHYLCYCLOHEXANE, CIS	(G)	112.208	-42.22	T(378)	88.54	T(378)	9.07	-6.650
C8H16	1,4-DIMETHYLCYCLOHEXANE, TRANS	(L)	112.208	-53.18	A(1248)	64.06	MC 657)	5.41	-3.966
C8H16	1,4-DIMETHYLCYCLOHEXANE, TRANS	(G)	112.208	-44.12	T(379)	87.19	T(379)	7.58	-5.552
C8H16	2,2-DIMETHYL-3-HEXENE, CIS	(L)	112.208	-30.63	A(1236)				
C8H16	2,2-DIMETHYL-3-HEXENE, CIS	(G)	112.208	-21.77	A(1236)				
C8H16	2,2-DIMETHYL-3-HEXENE, TRANS	(L)	112.208	-35.05	A(1236)				
C8H16	2,2-DIMETHYL-3-HEXENE, TRANS	(G)	112.208	-26.16	A(1196)				
C8H16	ETHYLCYCLOHEXANE	(L)	112.208	-50.72	A(1196)	67.14	MC 657)	6.95	-5.096
C8H16	ETHYLCYCLOHEXANE	(G)	112.208	-41.05	T(372)	91.44	T(372)	9.38	-6.874
C8H16	3-ETHYL-2-METHYL-1-PENTENE	(L)	112.208	-33.36	A(1236)				
C8H16	3-ETHYL-2-METHYL-1-PENTENE	(G)	112.208	-24.40	A(1236)				
C8H16	1-OCTENE	(L)	112.208	-29.52	A(1236)	86.15	MC 950)	22.49	-16.481
C8H16	1-OCTENE	(G)	112.208	-19.82	T(291)	110.55	T(291)	24.91	-18.259
C8H16	PROPYLCYCLOPENTANE	(L)	112.208	-45.22	A(1196)	74.29	MC 993)	10.33	-7.573
C8H16	PROPYLCYCLOPENTANE	(G)	112.208	-35.39	T(354)	99.73	T(354)	12.57	-9.211
C8H16	1,2,4-TRIMETHYLCYCLOPENTANE	(L)	112.208	-53.3	CC 744)				
C8H16	2,4,4-TRIMETHYL-1-PENTENE	(L)	112.208	-35.21	A(1236)	73.2	M(1119)	20.66	-15.141
C8H16	2,4,4-TRIMETHYL-1-PENTENE	(G)	112.208	-26.68	A(1236)				
C8H16	2,4,4-TRIMETHYL-2-PENTENE	(L)	112.208	-34.44	A(1236)	74.5	M(1119)	21.04	-15.421
C8H16	2,4,4-TRIMETHYL-2-PENTENE	(G)	112.208	-25.50	A(1236)				
C8H16N2O3	LEUCYLGLYCINE, D-L	(S)	188.224	-205.36	AC 643)	67.2	MC 642)	-112.14	82.200
C8H16N2O3	VALINE ALANINE ANHYDRIDE	(S)	188.224	-192.0	B(1177)				
C8H16N2O4	DIETHYLURETHAN OF								

Formula	Name	State	Mol Wt	$\Delta H_{f\,298}^{\circ}$ (kcal/mole)	Source	S_{298}° (cal/mole °K)	Source	$\Delta G_{f\,298}^{\circ}$ (kcal/mole)	Log Kp
C8H16N2O4	ETHYLENEDIAMINE	(CS)	204.224	-233.1	B(982)				
C8H16N2O4	TARTARICDIETHYLAMIDE, D	(U,L)	204.224	-234.2	C(744)				
C8H16N2O4	TARTARICDIETHYLAMIDE, D,L	(U,L)	204.224	-234.0	C(744)				
C8H16N2O4	TARTARICDIETHYLAMIDE, MESO	(U,L)	204.224	-233.2	C(744)				
C8H16O	1,2-DIMETHYLCYCLOHEXANOL	(L)	128.208	-100.3	C(744)				
C8H16O	1,3-DIMETHYLCYCLOHEXANOL	(L)	128.208	-106.3	C(744)				
C8H16O	2,6-DIMETHYLCYCLOHEXANOL	(L)	128.208	-102.8	C(744)				
C8H16O	3,5-DIMETHYLCYCLOHEXANOL								
C8H16O	CIS, CIS, CIS	(S)	128.208	-102.4	C(1367)				
C8H16O	3,5-DIMETHYLCYCLOHEXANOL								
C8H16O	CIS, CIS, TRANS	(L)	128.208	-116.9	C(1367)				
C8H16O	3,5-DIMETHYLCYCLOHEXANOL								
C8H16O	CIS, TRANS, CIS	(L)	128.208	-126.2	C(1367)				
C8H16O	3-ETHYL-5-HEXEN-3-OL	(L)	128.208	-91.7	C(744)				
C8H16O	4-METHYL-1-HEPTEN-4-OL	(L)	128.208	-96.9	C(744)				
C8H16O	OCTANAL	(L)	128.208	-80.0	B(463)				
C8H16O	OCTANAL	(G)	128.208	-69.23	T(551)	119.66	T(551)	-19.91	14.594
C8H16O	2-OCTANONE	(L)	128.208	-91.9	B(463)				
C8H16O	3-OCTANONE	(L)	128.208	-91.5	B(463)	89.35	M(1084)	-33.54	24.587
C8H16O	4-OCTANONE	(L)	128.208	-92.3	B(463)				
C8H16O2	SEC-BUTYL BUTYRATE	(L)	144.208	-141.5	B(1287)				
C8H16O2	METHYL HEPTANOATE	(L)	144.208	-135.62	A(3)				
C8H16O2	OCTANOIC ACID	(L)	144.208	-151.75	A(3)				
C8H16O2	OCTANOIC ACID	(L)	144.208	-152.2	A(853)				
C8H16O2	PROPYL VALERATE	(L)	144.208	-141.4	B(1287)				
C8H16O2	2-PROPYLVALERIC ACID	(L)	144.208	-146.5	C(744)				
C8H16O2	TETRAMETHYLBUTENEDIOL								
C8H16O2	FUMAROID	(U,S)	144.208	-122.3	C(744)				
C8H16O2	TETRAMETHYLBUTENEDIOL								
C8H16O2	MALEINOID	(U,S)	144.208	-125.7	C(744)				
C8H17BR	1-BROMOOCTANE	(L)	193.132	-58.61	A(138)				
C8H17BR	1-BROMOOCTANE	(G)	193.132	-46.26	A(138)				
C8H17F	1-FLUOROOCTANE	(L)	132.216	-105.0	C(744)				
C8H17N	CONIINE	(L)	127.224	-57.4	C(744)				
C8H17N	N-ISOBUTYLIDENEBUTYLAMINE	(L)	127.224	-31.76	A(96)				
C8H17N	N-ISOBUTYLIDENEBUTYLAMINE	(G)	127.224	-21.8	A(96)				

Formula	Name		State	Mol Wt	$\Delta Hf°_{298}$ (kcal/mole)	Source	$S°_{298}$ (cal/mole °K)	Source	$\Delta G°_{f\,298}$ (kcal/mole)	Log Kp
C8H17NO	OCTANAMIDE		(S)	143.224	-113.1	B(1640)			0.71	-0.523
C8H18	2,2-DIMETHYLHEXANE . .		(L)	114.224	-62.63	A(1248)	79.33	E(1248)	2.56	-1.876
C8H18	2,2-DIMETHYLHEXANE . .		(G)	114.224	-53.71	T(128)	103.06	T(128)	2.17	-1.594
C8H18	2,3-DIMETHYLHEXANE . .		(L)	114.224	-60.40	A(1248)	81.91	E(1248)	4.23	-3.100
C8H18	2,3-DIMETHYLHEXANE . .		(G)	114.224	-51.13	T(129)	106.11	T(129)	0.89	-0.655
C8H18	2,4-DIMETHYLHEXANE . .		(L)	114.224	-61.47	A(1248)	82.62	E(1248)	2.80	-2.053
C8H18	2,4-DIMETHYLHEXANE . .		(G)	114.224	-52.44	T(130)	106.51	T(130)	0.60	-0.438
C8H18	2,5-DIMETHYLHEXANE . .		(L)	114.224	-62.26	A(1248)	80.96	E(1248)	2.50	-1.833
C8H18	2,5-DIMETHYLHEXANE . .		(G)	114.224	-53.21	T(131)	104.93	T(131)	1.23	-0.902
C8H18	3,3-DIMETHYLHEXANE . .		(L)	114.224	-61.58	A(1248)	81.12	E(1248)	3.17	-2.324
C8H18	3,3-DIMETHYLHEXANE . .		(G)	114.224	-52.61	T(132)	104.70	T(132)	2.03	-1.487
C8H18	3,4-DIMETHYLHEXANE . .		(L)	114.224	-60.23	A(1248)	82.97	E(1248)	4.14	-3.034
C8H18	3,4-DIMETHYLHEXANE . .		(G)	114.224	-50.91	T(133)	107.15	T(133)	1.79	-1.311
C8H18	3-ETHYLHEXANE		(L)	114.224	-59.88	A(1248)	84.95	E(1248)	3.95	-2.892
C8H18	3-ETHYLHEXANE		(G)	114.224	-50.40	T(127)	109.51	T(127)	3.03	-2.224
C8H18	3-ETHYL-2-METHYLPENTANE		(L)	114.224	-59.69	A(1248)	81.41	E(1248)	5.08	-3.725
C8H18	3-ETHYL-2-METHYLPENTANE		(G)	114.224	-50.48	T(134)	105.43	T(134)	2.69	-1.974
C8H18	3-ETHYL-3-METHYLPENTANE		(L)	114.224	-60.46	A(1248)	79.97	E(1248)	4.76	-3.492
C8H18	3-ETHYL-3-METHYLPENTANE		(G)	114.224	-51.38	T(135)	103.48	T(135)	0.92	-0.677
C8H18	2-METHYLHEPTANE . . .		(L)	114.224	-60.98	A(1248)	84.16	E(1248)	3.05	-2.239
C8H18	2-METHYLHEPTANE . . .		(G)	114.224	-51.50	T(124)	108.81	T(124)	1.12	-0.819
C8H18	3-METHYLHEPTANE . . .		(L)	114.224	-60.34	A(1248)	85.66	E(1248)	3.28	-2.407
C8H18	3-METHYLHEPTANE . . .		(G)	114.224	-50.82	T(125)	110.32	T(125)	1.87	-1.367
C8H18	4-METHYLHEPTANE . . .		(L)	114.224	-60.17	A(1248)	83.72	E(1248)	4.00	-2.933
C8H18	4-METHYLHEPTANE . . .		(G)	114.224	-50.69	T(126)	108.35	T(126)	1.55	-1.134
C8H18	OCTANE		(L)	114.224	-59.74	A(1248)	86.23	M(420)	3.92	-2.872
C8H18	OCTANE		(G)	114.224	-49.82	T(123)	111.55	T(123)	2.21	-1.621
C8H18	2,2,3-TRIMETHYLPENTANE		(L)	114.224	-61.44	A(1248)	78.30	E(1248)	4.09	-2.997
C8H18	2,2,3-TRIMETHYLPENTANE		(G)	114.224	-52.61	T(136)	101.62	T(136)	1.65	-1.210
C8H18	2,2,4-TRIMETHYLPENTANE		(L)	114.224	-61.97	A(1248)	78.40	M(1152)	3.27	-2.396
C8H18	2,2,4-TRIMETHYLPENTANE		(G)	114.224	-53.57	T(137)	101.15	T(137)	2.54	-1.858
C8H18	2,3,3-TRIMETHYLPENTANE		(L)	114.224	-60.63	A(1248)	79.93	E(1248)	4.52	-3.309
C8H18	2,3,3-TRIMETHYLPENTANE		(G)	114.224	-51.73	T(138)	103.14	T(138)	2.55	-1.868
C8H18	2,3,4-TRIMETHYLPENTANE		(L)	114.224	-60.98	A(1248)	78.71	M(1165)	4.52	-3.315
C8H18	2,3,4-TRIMETHYLPENTANE		(G)	114.224	-51.97	T(139)	102.31	T(139)	3.26	-2.388
C8H18	2,2,3,3-TETRAMETHYLBUTANE		(S)	114.224	-64.23	A(1248)	65.43	M(1300)		

Formula	Name	State	Mol Wt	$\Delta H f^{\circ}_{298}$ (kcal/mole)	Source	S°_{298} (cal/mole °K)	Source	$\Delta G f^{\circ}_{298}$ (kcal/mole)	Log Kp
C8H18	2,2,3-TETRAMETHYLBUTANE	(G)	114.224	-53.99	TC 140	93.06	TC 140	5.26	-3.856
C8H18N2O2	N-NITRODIBUTYLAMINE	(L)	174.240	-50.2	BC1039				
C8H18N2O2	2-METHYL-1-NITROSOPROPANE DIMER, CIS	(L)	174.240	-46.2	AC 387				
C8H18N2O2	2-METHYL-1-NITROSOPROPANE DIMER, TRANS	(L)	174.240	-57.7	CC 387				
C8H18O	BUTYL ETHER	(L)	130.224	-90.3	TC 510				
C8H18O	BUTYL ETHER	(G)	130.224	-79.80	TC 510	119.60	TC 510	-21.16	15.508
C8H18O	SEC-BUTYL ETHER	(L)	130.224	-96.0	AC 252				
C8H18O	SEC-BUTYL ETHER	(G)	130.224	-86.20	TC 511	110.57	TC 511	-24.87	18.226
C8H18O	TERT-BUTYL ETHER	(L)	130.224	-96.1	BC1390				
C8H18O	TERT-BUTYL ETHER	(G)	130.224	-87.20	TC 512	102.12	TC 512	-23.35	17.112
C8H18O	2-ETHYL-1-HEXANOL	(L)	130.224	-103.46	AC1506				
C8H18O	4-METHYL-4-HEPTANOL	(L)	130.224	-134.4	CC 744				
C8H18O	OCTYL ALCOHOL	(L)	130.224	-101.62	AC 215				
C8H18O	OCTYL ALCOHOL	(G)	130.224	-85.34	TC 528	124.14	TC 528	-28.05	20.561
C8H18O2	TERT-BUTYL PEROXIDE	(L)	146.224	-91.1	BC 64				
C8H18O2	TERT-BUTYL PEROXIDE	(G)	146.224	-81.5	BC 64				
C8H18O2S	BUTYL SULFONE	(L)	178.290	-145.8	BC 904				
C8H18O2S	BUTYL SULFONE	(G)	178.290	-121.4	BC 904				
C8H18O2S	TERT-BUTYL SULFONE	(L)	178.290	-153.1	BC 904				
C8H18O2S	TERT-BUTYL SULFONE	(G)	178.290	-128.9	BC 904				
C8H18O2S	ISOBUTYL SULFONE	(L)	178.290	-149.4	BC 904				
C8H18O2S	ISOBUTYL SULFONE	(G)	178.290	-125.5	BC 904				
C8H18O5	TETRAETHYLENE GLYCOL	(L)	194.224	-233.6	BC1032				
C8H18O9	PARAFORMALDEHYDE MONOHYDRATE	(S)	258.224	-406.6	AC 320				
C8H18S	BUTYL SULFIDE	(L)	146.290	-52.74	AC 952	96.82	MC 952	7.66	-5.618
C8H18S	BUTYL SULFIDE	(G)	146.290	-39.99	TC 814	125.84	TC 814	11.76	-8.622
C8H18S	TERT-BUTYL SULFIDE	(L)	146.290	-60.3	BC 899				
C8H18S	TERT-BUTYL SULFIDE	(G)	146.290	-49.5	BC 899				
C8H18S	ETHYL HEXYL SULFIDE	(L)	146.290	-39.77	TC 815	126.89	TC 815	11.67	-8.553
C8H18S	HEPTYL METHYL SULFIDE	(G)	144.274	-39.19	TC 816	126.35	TC 816	3.10	-2.276
C8H18S	ISOBUTYL SULFIDE	(L)	146.290	-54.8	BC 899				
C8H18S	ISOBUTYL SULFIDE	(G)	146.290	-42.9	BC 899				
C8H18S	1-OCTANETHIOL	(L)	146.290	-40.68	TC 902	127.20	TC 902	10.67	-7.819
C8H18S	PENTYL PROPYL SULFIDE	(G)	146.290	-39.81	TC 817	127.21	TC 817	11.53	-8.454

Formula	Name	State	Mol Wt	$\Delta Hf°_{298}$ (kcal/mole)	Source	$S°_{298}$ (cal/mole °K)	Source	$\Delta Gf°_{298}$ (kcal/mole)	Log Kp
C8H18S2	BUTYL DISULFIDE	(L)	178.356	-53.1	B(902)				
C8H18S2	BUTYL DISULFIDE	(G)	178.356	-37.86	T(875)	136.91	T(875)	12.87	-9.431
C8H18S2	TERT-BUTYL DISULFIDE	(L)	178.356	-59.8	B(902)				
C8H18S2	TERT-BUTYL DISULFIDE	(G)	178.356	-47.1	B(902)				
C8H18S2	ISOBUTYL DISULFIDE	(L)	178.356	-55.5	B(902)				
C8H18S2	ISOBUTYL DISULFIDE	(G)	178.356	-40.7	B(902)				
C8H19N	N-ISOBUTYLBUTYLAMINE	(L)	129.240	-51.59	A(96)				
C8H19N	N-ISOBUTYLBUTYLAMINE	(G)	129.240	-41.8	A(96)				
C8H19N	DIISOBUTYLAMINE	(L)	129.240	-51.9	C(744)				
C8H20N2OS	N,N'-SULFINYL-BIS(DIETHYLAMINE)	(L)	192.322	-75.5	B(232)				
C8H20N2O2S	TETRAETHYLSULFAMIDE	(L)	208.322	-149.8	B(232)				
C8H20N2S2	N,N'-DITHIO-BIS(DIETHYLAMINE)	(L)	208.388	-29.4	B(232)				
C9H6N2O3	3-BENZOYLOXADIAZOL-5-OL	(S)	190.154	-62.8	C(1012)				
C9H6N2O3	5-BENZOYLOXADIAZOL-3-OL	(S)	190.154	-8.4	C(1012)				
C9H6O	PHENYLPROPIOL ALDEHYDE	(L)	130.138	29.8	C(744)				
C9H6O2	PHENYLPROPIOL ACID	(S)	146.138	-30.2	C(744)				
C9H6O6	TRIMESIC ACID	(S)	210.138	-283.9	C(744)				
C9H7FO2	ALPHA-FLUOROCINNAMIC ACID	(S)	166.146	-114.8	C(744)				
C9H7N	QUINOLINE	(L)	129.154	35.7	C(1118)	51.9	M(1118)	63.27	-46.376
C9H7NO	BENZOYLACETONITRILE	(S)	145.154	-21.0	C(744)				
C9H7NO	8-HYDROXYQUINOLINE	(S)	145.154	-21.0	B(1440)				
C9H7NO	ALPHA-PHENYLISOXAZOLE	(S)	145.154	21.1	C(1467)				
C9H7NO	GAMMA-PHENYLISOXAZOLE	(L)	145.154	19.9	G(1467)				
C9H7NO	3-PHENYLPROPIOLAMIDE	(S)	145.154	10.2	C(744)				
C9H7NO4	M-NITROCINNAMIC ACID	(S)	193.154	-89.3	C(744)				
C9H7NO4	O-NITROCINNAMIC ACID	(S)	193.154	-85.9	C(744)				
C9H7NO4	P-NITROCINNAMIC ACID	(S)	193.154	-88.4	C(744)				
C9H7N3O2	BENZOYLAMINOFURAZAN	(S)	189.170	20.8	C(1012)				
C9H8	INDENE	(L)	116.154	26.39	A(1438)	51.19	M(1438)	52.00	-38.117
C9H8F3NO	ALPHA,ALPHA,ALPHA-TRIFLUORO-M-ACETOTOLUIDIDE	(S)	203.162	-202.8	C(744)				
C9H8N2O	METHYLPHENYL-1,3,4-OXADIAZOLE	(S)	160.170	10.5	C(1012)				
C9H8N2O	3-METHYL-5-PHENYLOXADIAZOLE	(S)	160.170	26.4	C(1012)				

Formula	Name	State	Mol Wt	$\Delta Hf°_{298}$ (kcal/mole)	Source	$S°_{298}$ (cal/mole °K)	Source	$\Delta Gf°_{298}$ (kcal/mole)	Log Kp
C9H8N2O	5-METHYL-3-PHENYLOXADIAZOLE	(S)	160.170	24.9	C(10121)				
C9H8N2O	PYRRYL KETONE	(S)	160.170	-3.4	C(14131)				
C9H8N2O	P-TOLYLFURAZAN	(S)	160.170	40.6	C(10121)				
C9H8N2O2	METHYLPHENYLFURAZAN	(S)	176.170	31.8	C(10111)				
C9H8N2O2	5-METHYL-4-PHENYL-1,2,3,6-DIOXADIAZOLE	(S)	176.170	40.7	C(10111)				
C9H8O	CINNAMALDEHYDE	(L)	132.154	-6.4	C(744)				
C9H8O	6,6A-DIHYDRO-1A-OXIRENO-(A)INDENE	(L)	132.154	-1.9	C(744)				
C9H8O	1-INDANONE	(L)	132.154	-31.8	C(744)				
C9H8O	2-INDANONE	(L)	132.154	-30.5	C(744)				
C9H8O	3-PHENYL-2-PROPYN-1-OL	(L)	132.154	18.1	C(744)				
C9H8O2	ATROPIC ACID	(S)	148.154	-74.5	C(744)				
C9H8O2	CINNAMIC ACID, CIS(M.P.42 C)	(S)	148.154	-75.9	BC(376)				
C9H8O2	CINNAMIC ACID, CIS(M.P.58 C)	(S)	148.154	-74.5	BC(376)				
C9H8O2	CINNAMIC ACID, CIS(M.P.68 C)	(S)	148.154	-72.1	BC(376)				
C9H8O2	CINNAMIC ACID, TRANS	(S)	148.154	-80.60	A(1114)				
C9H8O3	P-HYDROXYCINNAMIC ACID, CIS (M.P. 126 TO 127 C)	(S)	164.154	-121.5	C(744)				
C9H8O3	P-HYDROXYCINNAMIC ACID, TRANS (M.P. 206 C)	(S)	164.154	-126.6	C(744)				
C9H8O4	UVITIC ACID	(U,S)	180.154	-190.8	C(744)				
C9H9ClO2	ETHYL O-CHLOROBENZOATE	(L)	184.619	-93.1	C(744)				
C9H9FO2	ETHYL P-FLUOROBENZOATE	(S)	168.162	-136.6	C(744)				
C9H9N	2-METHYLINDOLE	(S)	131.170	15.9	C(744)				
C9H9N	3-METHYLINDOLE	(L)	131.170	17.7	C(744)				
C9H9NO2	1-PHENYL-2-NITROPROPENE	(S)	163.170	-5.5	B(1640)				
C9H9NO3	HIPPURIC ACID	(S)	179.170	-145.54	AC(630)	57.2	M(642)	-88.33	64.744
C9H9N3O6	TRINITROMESITYLENE	(S)	255.186	-30.5	B(981)				
C9H9N9O12	N,N',N''-TRIMETHYL-N,N',N''-2,4,6-HEXANITRO-1,3,5-BENZENETRIAMINE	(S)	435.234	32.0	B(983)				
C9H10	INDANE	(L)	118.170	2.56	AC(1438)	56.01	M(1438)	36.04	-26.417
C9H10	ALPHA-METHYLSTYRENE	(G)	118.170	27.00	T(440)	91.7	T(440)	49.84	-36.531
C9H10	M-METHYLSTYRENE	(G)	118.170	27.60	T(443)	93.1	T(443)	50.02	-36.665
C9H10	O-METHYLSTYRENE	(G)	118.170	28.30	T(444)	91.7	T(444)	51.14	-37.484

Formula	Name	State	Mol Wt	$\Delta H f°_{298}$ (kcal/mole)	Source	$S°_{298}$ (cal/mole °K)	Source	$\Delta G f°_{298}$ (kcal/mole)	Log K_p
C9H10	P-METHYLSTYRENE	(G)	118.170	27.40	T(445)	91.7	T(445)	50.24	-36.825
C9H10	PHENYLCYCLOPROPANE	(L)	118.170	19.0	A(804)				
C9H10	PROPENYLBENZENE, CIS	(G)	118.170	29.00	T(441)	91.7	T(441)	51.84	-37.997
C9H10	PROPENYLBENZENE, TRANS	(G)	118.170	28.00	T(442)	90.9	T(442)	51.08	-37.439
C9H10N2	METHYLENEBISPYRROLE	(S)	146.186	32.9	C(1413)				
C9H10N2O2	P-TOLYLGLYOXIME(ALPHA)	(S)	178.186	-10.0	B(1013)				
C9H10N2O2	P-TOLYLGLYOXIME(BETA)	(S)	178.186	6.0	B(1013)				
C9H10N2O4	DINITROMESITYLENE	(U,S)	210.186	-12.4	C(744)				
C9H10N2O4	O-OMEGA-DINITROMESITYLENE	(U,S)	210.186	-33.0	C(744)				
C9H10O	2H-1,5-BENZODIOEPINE	(L)	134.170	-57.75	A(201)				
C9H10O	PHENYL-2-PROPANONE	(L)	134.170	-39.0	B(1112)				
C9H10O	PHENYL-2-PROPANONE	(L)	134.170	-36.5	B(1403)				
C9H10O	PROPIOPHENONE	(L)	134.170	-39.95	A(251)				
C9H10O2	2,3-DIMETHYLBENZOIC ACID	(S)	150.170	-107.65	A(253)				
C9H10O2	2,4-DIMETHYLBENZOIC ACID	(S)	150.170	-109.58	A(253)				
C9H10O2	2,5-DIMETHYLBENZOIC ACID	(S)	150.170	-109.02	A(253)				
C9H10O2	2,6-DIMETHYLBENZOIC ACID	(S)	150.170	-105.33	A(253)				
C9H10O2	3,4-DIMETHYLBENZOIC ACID	(S)	150.170	-112.04	A(253)				
C9H10O2	3,5-DIMETHYLBENZOIC ACID	(S)	150.170	-111.48	A(253)				
C9H10O2	ETHYL BENZOATE	(L)	150.170	-88.4	C(744)				
C9H10O2	HYDROCINNAMIC ACID	(S)	150.170	-102.2	C(744)				
C9H10O2	2-HYDROXY-3-METHYLACETOPHENONE	(L)	150.170	-95.3	C(843)				
C9H10O2	1,2-INDANDIOL, CIS	(S)	150.170	-89.2	B(1536)				
C9H10O2	1,2-INDANDIOL, TRANS	(S)	150.170	-91.0	B(1536)				
C9H10O2	M-TOLYL ACETATE	(L)	150.170	-89.2	C(744)				
C9H10O3	ETHYL P-HYDROXYBENZOATE	(S)	166.170	-144.5	A(1445)				
C9H10O3	ETHYL SALICYLATE	(L)	166.170	-136.0	C(744)				
C9H10O3	2-HYDROXY-4-METHOXYACETOPHENONE	(S)	166.170	-127.6	C(843)				
C9H10O3	METHYL ANISATE	(U,S)	166.170	-118.5	C(744)				
C9H11F	FLUOROPSEUDOCUMENE	(L)	138.178	-56.4	C(744)				
C9H11N	1,2,3,4-TETRAHYDROQUINOLINE	(L)	133.186	7.6	B(1640)				
C9H11NO	P-DIMETHYLAMINOBENZALDEHYDE	(S)	149.186	-32.8	C(843)				
C9H11NO	2'-METHYLACETOPHENONE OXIME	(S)	149.186	-57.3	C(843)				
C9H11NO	3'-METHYLACETOPHENONE OXIME	(S)	149.186	-60.7	C(843)				
C9H11NO	4'-METHYLACETOPHENONE OXIME	(S)	149.186	-59.2	C(843)				
C9H11NO	PROPIONANILIDE	(S)	149.186	-53.6	C(744)				

Formula	Name	State	Mol Wt	$\Delta Hf°_{298}$ (kcal/mole)	Source	$S°_{298}$ (cal/mole °K)	Source	$\Delta Gf°_{298}$ (kcal/mole)	Log Kp
C9H11NO2	ETHYL P-AMINOBENZOATE . . .	(S)	165.186	-100.1	B(1440)				
C9H11NO2	2-HYDROXY-3-METHYL-								
C9H11NO2	2-HYDROXY-2-METHYL- ACETOPHENONE OXIME	(S)	165.186	-115.4	C(843)				
C9H11NO2	ACETOPHENONE OXIME	(S)	165.186	-108.7	C(843)				
C9H11NO2	2-METHOXYACETOPHENONE OXIME	(S)	165.186	-87.2	C(843)				
C9H11NO2	NITROMESITYLENE	(U,S)	165.186	-17.4	G(744)				
C9H11NO2	ALPHA-NITROMESITYLENE	(S)	165.186	-27.8	C(744)				
C9H11NO2	PHENYLALANINE, L	(S)	165.186	-111.6	A(241)	51.06	M(241)	-50.56	37.058
C9H11NO2	PHENYLALANINE, D,L . . .	(S)	165.186	-109.9	E(164)				
C9H11NO3	2-HYDROXY-4-METHOXY- ACETOPHENONE OXIME	(S)	181.186	-141.9	C(843)				
C9H11NO3	2-HYDROXY-5-METHOXY- ACETOPHENONE OXIME	(S)	181.186	-139.5	C(843)				
C9H11NO3	TYROSINE, L	(S)	181.186	-160.5	B(241)	51.15	M(241)	-92.18	67.566
C9H11NO4	METHYL 3,5-DIMETHYL- 2,4-PYRROLEDICARBOXYLATE	(S)	197.186	-198.2	C(1413)				
C9H11NS2	DIMETHYL PHENYLDITHIO- IMIDOCARBONATE	(L)	197.318	34.5	C(744)				
C9H12	CUMENE	(L)	120.186	-9.85	A(692)	66.87	E(1248)	29.70	-21.768
C9H12	CUMENE	(G)	120.186	0.94	T(403)	92.87	T(403)	32.74	-23.995
C9H12	M-ETHYLTOLUENE . . .	(L)	120.186	-11.67	A(692)	69.90	E(1248)	26.97	-19.772
C9H12	M-ETHYLTOLUENE . . .	(G)	120.186	-0.46	T(404)	96.60	T(404)	30.22	-22.154
C9H12	O-ETHYLTOLUENE . . .	(L)	120.186	-11.11	A(692)	68.42	E(1248)	27.98	-20.506
C9H12	O-ETHYLTOLUENE . . .	(G)	120.186	0.29	T(405)	95.42	T(405)	31.33	-22.961
C9H12	P-ETHYLTOLUENE . . .	(L)	120.186	-11.92	A(692)	68.84	E(1248)	27.04	-19.820
C9H12	P-ETHYLTOLUENE . . .	(G)	120.186	-0.78	T(406)	95.34	T(406)	30.28	-22.195
C9H12	MESITYLENE . . .	(L)	120.186	-15.18	A(692)	65.38	M(1476)	24.81	-18.187
C9H12	MESITYLENE . . .	(G)	120.186	-3.84	T(409)	92.09	T(409)	28.19	-20.662
C9H12	PROPYLBENZENE . . .	(L)	120.186	-9.18	A(692)	68.78	M(994)	29.80	-21.842
C9H12	PROPYLBENZENE . . .	(G)	120.186	1.87	T(402)	95.76	T(402)	32.80	-24.045
C9H12	1,2,3-TRIMETHYLBENZENE . .	(L)	120.186	-14.01	A(692)	64.04	M(1475)	26.38	-19.337
C9H12	1,2,3-TRIMETHYLBENZENE . .	(G)	120.186	-2.29	T(407)	91.98	T(407)	29.77	-21.822
C9H12	1,2,4-TRIMETHYLBENZENE . .	(L)	120.186	-14.78	A(692)	67.73	M(1208)	24.51	-17.967
C9H12	1,2,4-TRIMETHYLBENZENE . .	(G)	120.186	-3.33	T(408)	94.59	T(408)	27.95	-20.489
C9H12N2O	1-METHYL-1-TOLYLUREA . .	(S)	164.202	-49.9	B(1472)				

Formula	Name	State	Mol Wt	$\Delta H f^\circ_{298}$ (kcal/mole)	Source	S°_{298} (cal/mole °K)	Source	$\Delta G f^\circ_{298}$ (kcal/mole)	Log Kp
C9H12O	P-METHYLPHENETOLE	(L)	136.186	-42.5	C(744)				
C9H12O	PHENYL PROPYL ETHER . . .	(L)	136.186	-42.2	C(744)				
C9H12O	3,5,5-TRIMETHYLHEXANOL . .	(L)	136.186	-109.2	B(1066)				
C9H12O	2,4,5-TRIMETHYLPHENOL . .	(S)	136.186	-64.1	C(744)				
C9H12O	M-XYLENOL METHYL ETHER . .	(U,L)	136.186	-41.9	C(744)				
C9H12O2	CUMENE HYDROPEROXIDE . . .	(L)	152.186	-35.5	C(807)				
C9H12O2	METACROLEIN	(U,S)	168.186	-86.7	C(744)				
C9H12O3S	P-PROPYLBENZENESULFONIC ACID	(G)	200.252			113.21	E(69)		
C9H12S	BENZYL ETHYL SULFIDE . . .	(L)	152.252	-1.2	B(898)				
C9H12S	BENZYL ETHYL SULFIDE . . .	(G)	152.252	12.4	B(898)				
C9H13N	N-ETHYLBENZYLAMINE	(L)	135.202	0.3	C(744)				
C9H13N	PSEUDOCUMIDINE	(U,S)	135.202	-23.5	C(744)				
C9H13NO	5-ALDEHYDO-2,4-DIMETHYL-3-ETHYLPYRROLE	(S)	151.202	-58.7	C(1255)				
C9H13NO2	ETHYL 2,3-DIMETHYLPYRROLE-4-CARBOXYLATE	(S)	167.202	-108.4	C(1255)				
C9H13NO2	ETHYL 2,3-DIMETHYLPYRROLE-5-CARBOXYLATE	(S)	167.202	-105.0	C(1255)				
C9H13NO2	ETHYL 2,4-DIMETHYLPYRROLE-3-CARBOXYLATE	(S)	167.202	-111.1	C(1255)				
C9H13NO2	ETHYL 2,4-DIMETHYLPYRROLE-5-CARBOXYLATE	(S)	167.202	-110.1	C(1255)				
C9H13NO2	ETHYL 2,5-DIMETHYLPYRROLE-3-CARBOXYLATE	(S)	167.202	-112.1	C(1255)				
C9H13NO3	ETHYL 5-HYDROXY-2,4-DIMETHYL-3-PYRROLECARBOXYLATE	(S)	183.202	-160.9	C(1413)				
C9H13N5O16	2,2,4,4-TETRAKIS(HYDROXY-METHYLNITRATE) PYRANOL NITRATE	(S)	447.234	-189.6	B(984)				
C9H14	1-ETHYL-4-METHYL-1,3-CYCLOHEXADIENE	(L)	122.202	-11.4	C(744)				
C9H14O	1,6-HEPTADIEN-4-ONE . . .	(L)	138.202	-41.6	C(744)				
C9H14O	HEXAHYDRO-2-INDANONE, CIS	(L)	138.202	-75.5	C(744)				
C9H14O	HEXAHYDRO-2-INDANONE, TRANS	(L)	138.202	-80.6	C(744)				
C9H14O	3,5,5-TRIMETHYL-3-CYCLOHEXEN-1-ONE	(L)	138.202	-76.1	C(744)				

Formula	Name	State	Mol Wt	$\Delta H f^\circ_{298}$ (kcal/mole)	Source	S°_{298} (cal/mole °K)	Source	$\Delta G f^\circ_{298}$ (kcal/mole)	Log Kp
C9H14O2	1-CYCLOHEXENE-1-PROPIONIC ACID	(S)	154.202	-123.0	C(744)				
C9H14O2	METHYL 2-CYCLOHEXENE-1-ACETATE	(L)	154.202	-112.4	C(744)				
C9H14O2	METHYL DELTA-1-ALPHA-CYCLOHEXANEACETATE	(L)	154.202	-106.0	C(744)				
C9H14O2	2-NONYNOIC ACID	(L)	154.202	-92.8	C(744)				
C9H14O4	1,2-CYCLOPENTYLENE DIACETATE, CIS	(L)	186.202	-210.2	B(1537)				
C9H14O4	1,2-CYCLOPENTYLENE DIACETATE, TRANS	(L)	186.202	-210.95	B(1537)				
C9H14O4	DIMETHYL PENTAMETHYLENE-ALPHA,BETA-DICARBOXYLATE	(L,L)	186.202	-206.2	C(744)				
C9H14O6	MANNITETRIMETHYLAL	(L,S)	218.202	-239.9	C(744)				
C9H14O6	TRIACETIN	(L)	218.202	-316.1	B(1472)				
C9H14O7	TRIMETHYL CITRATE	(S)	234.202	-341.0	C(1414)				
C9H14S	2-THIAADAMANTANE	(S)	154.268	34.33	A(835)				
C9H15N	3-ETHYL-2,4,5-TRIMETHYLPYRROLE	(S)	137.218	-18.5	C(1414)				
C9H15NO	N-HEXYLPROPIOLAMIDE	(S)	153.218	-50.7	C(744)				
C9H15NO	HEXYLMALONIC ACID	(S)	187.210	-216.7	B(1535)				
C9H16	2-CYCLOPROPYL-1-HEXENE	(L)	124.218	11.2	C(1374)				
C9H16	2-CYCLOPROPYL-2-HEXENE	(HB,L)	124.218	-3.8	C(1374)				
C9H16	HEXAHYDROINDAN, CIS	(L)	124.218	-41.41	A(176)				
C9H16	HEXAHYDROINDAN, TRANS	(L)	124.218	-42.15	A(176)				
C9H16	1-ISOPROPYLCYCLOHEXENE	(L)	124.218	-50.9	C(744)				
C9H16	1-NONYNE	(G)	124.218	14.77	T(326)	116.06	T(326)	58.26	-42.706
C9H16	SPIRO(4.4)NONANE	(L)	124.218	-34.5	A(805)				
C9H16N2O5	ETHYL-ALPHA-CARBETHOXY-GLYCYLGLYCINATE	(U,S)	232.234	-270.8	C(744)				
C9H16N2O5	ETHYL-BETA-CARBETHOXY-GLYCYLGLYCINATE	(U,S)	232.234	-299.4	C(744)				
C9H16O	CYCLOHEXYL ETHYL KETONE	(L)	140.218	-103.3	C(744)				
C9H16O	CYCLOHEPTYL METHYL KETONE	(L)	140.218	-114.5	C(744)				
C9H16O	METHYL 4-METHYLCYCLOHEXYL KETONE	(L)	140.218	-125.0	C(744)				
C9H16O	2-NONYN-1-OL	(L)	140.218	-52.4	C(744)				
C9H16O	2,4,6-TRIMETHYL-1-CYCLOHEXEN-1-OL	(L)	140.218	-98.1	C(744)				

713

Formula	Name	State	Mol Wt	$\Delta H f^\circ_{298}$ (kcal/mole)	Source	S°_{298} (cal/mole °K)	Source	$\Delta G f^\circ_{298}$ (kcal/mole)	Log Kp
C9H16O2	BUTYL 3-PENTENOATE	(L)	156.218	-160.0	B(1287)				
C9H16O2	BUTYL 4-PENTENOATE	(L)	156.218	-162.6	B(1287)				
C9H16O2	ISOBUTYL 3-PENTENOATE	(L)	156.218	-164.8	B(1287)				
C9H16O2	ISOBUTYL 4-PENTENOATE	(L)	156.218	-167.5	B(1287)				
C9H16O2	ISOPENTYL CROTONATE	(L)	156.218	-120.2	B(1286)				
C9H16O4	AZELAIC ACID	(S)	188.218	-250.5	B(744)				
C9H16O4	HEXYLMALONIC ACID	(S)	188.218	-210.2	B(1535)				
C9H16O4	DIMETHYL PIMELATE	(L)	188.218	-216.2	C(744)				
C9H16O4	DIPROPYLMALONIC ACID	(S)	188.218	-246.4	C(744)				
C9H17CLO2	ISOPENTYL 2-CHLOROBUTYRATE	(L)	192.683	-163.6	B(1384)				
C9H17CLO2	ISOPENTYL 3-CHLOROBUTYRATE	(L)	192.683	-154.1	B(1384)				
C9H17CLO2	ISOPENTYL 4-CHLOROBUTYRATE	(S)	192.683	-157.1	B(1384)				
C9H17NO	CYCLOHEPTYL METHYL KETONE OXIME	(S)	155.234	-63.5	C(744)				
C9H17NO	TETRAHYDRO-3,3,5,5-TETRAKIS-(HYDROXYMETHYL)-4-HYDROXYPYRAN	(S)	221.226	-274.37	B(986)				
C9H17O6	BUTYLCYCLOPENTANE	(L)	126.234	-51.22	A(1196)	82.18	M(993)	11.68	-8.561
C9H18	BUTYLCYCLOPENTANE	(G)	126.234	-40.22	T(355)	109.04	T(355)	14.67	-10.754
C9H18	CYCLONONANE	(L)	126.234	-43.7	A(707)				
C9H18	CYCLONONANE	(G)	126.234	-31.8	A(707)				
C9H18	CYCLOPROPYLHEXANE	(L)	126.234	-16.6	C(1375)				
C9H18	ETHYLCYCLOHEPTANE	(L)	126.234	-54.2	C(744)				
C9H18	METHYL-1-PROPYLCYCLOPENTANE	(U,L)	126.234	-59.6	C(744)				
C9H18	1-NONENE	(G)	126.234	-24.74	T(292)	119.86	T(292)	26.93	-19.736
C9H18	PROPYLCYCLOHEXANE	(L)	126.234	-56.98	A(1196)	74.54	M(423)	8.20	-6.009
C9H18	PROPYLCYCLOHEXANE	(G)	126.234	-46.20	T(380)	100.27	T(380)	11.31	-8.288
C9H18	1,1,3-TRIMETHYLCYCLOHEXANE	(L)	126.234	-66.3	C(744)				
C9H18	1,2,3-TRIMETHYLCYCLOHEXANE	(L)	126.234	-65.1	C(744)				
C9H18	1,3,5-TRIMETHYLCYCLOHEXANE, CIS	(G)	126.234	-51.48	T(381)	93.30	T(381)	8.10	-5.941
C9H18	1,3,5-TRIMETHYLCYCLOHEXANE, TRANS	(G)	126.234	-49.37	T(382)	95.60	T(382)	9.53	-6.985
C9H18O	CYCLOHEPTYLMETHYLCARBINOL	(U,L)	142.234	-118.9	C(744)				
C9H18O	1-ETHYL-3-METHYLCYCLOHEXANOL	(L)	142.234	-138.7	C(744)				
C9H18O	4-METHYL-1-OCTEN-4-OL	(L)	142.234	-96.2	C(744)				
C9H18O	NONANAL	(G)	142.234	-74.16	T(552)	128.97	T(552)	-17.91	13.124
C9H18O	2,2,3-TRIMETHYL-5-HEXEN-3-OL	(L)	142.234	-98.1	C(744)				
C9H18O2	BUTYL VALERATE	(L)	158.234	-147.9	B(1287)				

Formula	Name	State	Mol Wt	$\Delta H f°_{298}$ (kcal/mole)	Source	$S°_{298}$ (cal/mole °K)	Source	$\Delta G f°_{298}$ (kcal/mole)	Log Kp
C9H18O2	SEC-BUTYL VALERATE	(L)	158.234	-151.1	B(1287)				
C9H18O2	ISOBUTYL VALERATE	(L)	158.234	-150.0	B(1287)				
C9H18O2	METHYL OCTANOATE	(L)	158.234	-141.2	A(3)				
C9H18O2	NONANOIC ACID	(L)	158.234	-157.3	A(3)				
C9H18O2	NONANOIC ACID	(G)	158.234	-128.85	A(853)				
C9H20	3,3-DIETHYLPENTANE	(L)	128.250	-65.85	A(693)				
C9H20	3,3-DIETHYLPENTANE	(G)	128.250	-55.44	T(168)	110.31	T(168)	8.38	-6.141
C9H20	2,2-DIMETHYLHEPTANE	(L)	128.250	-69.11	E(820)				
C9H20	2,2-DIMETHYLHEPTANE	(G)	128.250	-59.00	T(147)	113.07	T(147)	4.00	-2.929
C9H20	2,3-DIMETHYLHEPTANE	(L)	128.250	-66.74	E(820)				
C9H20	2,3-DIMETHYLHEPTANE	(G)	128.250	-56.32	T(148)	116.79	T(148)	5.57	-4.080
C9H20	2,4-DIMETHYLHEPTANE	(L)	128.250	-67.74	E(820)				
C9H20	2,4-DIMETHYLHEPTANE	(G)	128.250	-57.48	T(149)	116.79	T(149)	4.41	-3.230
C9H20	2,5-DIMETHYLHEPTANE	(L)	128.250	-67.82	E(820)				
C9H20	2,5-DIMETHYLHEPTANE	(G)	128.250	-57.48	T(150)	116.79	T(150)	4.41	-3.230
C9H20	2,6-DIMETHYLHEPTANE	(L)	128.250	-68.52	E(820)				
C9H20	2,6-DIMETHYLHEPTANE	(G)	128.250	-58.17	T(151)	114.03	T(151)	4.54	-3.327
C9H20	3,3-DIMETHYLHEPTANE	(L)	128.250	-67.92	E(820)				
C9H20	3,3-DIMETHYLHEPTANE	(G)	128.250	-57.74	T(152)	115.25	T(152)	4.61	-3.376
C9H20	3,4-DIMETHYLHEPTANE	(L)	128.250	-66.05	E(820)				
C9H20	3,4-DIMETHYLHEPTANE	(G)	128.250	-55.63	T(153)	117.48	T(153)	6.05	-4.435
C9H20	3,5-DIMETHYLHEPTANE	(L)	128.250	-67.13	E(820)				
C9H20	3,5-DIMETHYLHEPTANE	(G)	128.250	-56.79	T(154)	116.10	T(154)	5.30	-3.887
C9H20	4,4-DIMETHYLHEPTANE	(L)	128.250	-67.83	E(820)				
C9H20	4,4-DIMETHYLHEPTANE	(G)	128.250	-57.74	T(155)	113.87	T(155)	5.02	-3.678
C9H20	3-ETHYLHEPTANE	(L)	128.250	-65.71	E(820)				
C9H20	3-ETHYLHEPTANE	(G)	128.250	-55.08	T(145)	118.52	T(145)	6.29	-4.611
C9H20	4-ETHYLHEPTANE	(L)	128.250	-65.62	E(820)				
C9H20	4-ETHYLHEPTANE	(G)	128.250	-55.08	T(146)	118.52	T(146)	6.29	-4.611
C9H20	3-ETHYL-2,2-DIMETHYLPENTANE	(L)	128.250	-66.92	E(820)				
C9H20	3-ETHYL-2,2-DIMETHYLPENTANE	(G)	128.250	-56.96	T(169)	109.96	T(169)	6.96	-5.104
C9H20	3-ETHYL-2,3-DIMETHYLPENTANE	(L)	128.250	-66.02	E(820)				
C9H20	3-ETHYL-2,3-DIMETHYLPENTANE	(G)	128.250	-55.82	T(170)	112.14	T(170)	7.45	-5.463
C9H20	3-ETHYL-2,4-DIMETHYLPENTANE	(L)	128.250	-66.30	E(820)				
C9H20	3-ETHYL-2,4-DIMETHYLPENTANE	(G)	128.250	-56.18	T(171)	112.30	T(171)	7.05	-5.164
C9H20	3-ETHYL-2-METHYLHEXANE	(L)	128.250	-65.96	E(820)				

Formula	Name	State	Mol Wt	$\Delta H f^{\circ}_{298}$ (kcal/mole)	Source	S°_{298} (cal/mole °K)	Source	$\Delta G f^{\circ}_{298}$ (kcal/mole)	Log Kp
C9H20	3-ETHYL-2-METHYLHEXANE	(G)	128.250	-55.63	T(156)	116.79	T(156)	6.26	-4.586
C9H20	3-ETHYL-3-METHYLHEXANE	(L)	128.250	-66.73	E(820)			5.87	-4.300
C9H20	3-ETHYL-3-METHYLHEXANE	(G)	128.250	-56.48	T(158)	115.25	T(158)		
C9H20	3-ETHYL-4-METHYLHEXANE	(L)	128.250	-65.35	E(820)			6.95	-5.092
C9H20	3-ETHYL-4-METHYLHEXANE	(G)	128.250	-54.94	T(159)	116.79	T(159)		
C9H20	4-ETHYL-2-METHYLHEXANE	(L)	128.250	-67.04	E(820)			5.51	-4.037
C9H20	4-ETHYL-2-METHYLHEXANE	(G)	128.250	-56.79	T(157)	115.41	T(157)		
C9H20	2-METHYLOCTANE	(L)	128.250	-67.18	E(820)			4.92	-3.607
C9H20	2-METHYLOCTANE	(G)	128.250	-56.45	T(142)	118.52	T(142)		
C9H20	3-METHYLOCTANE	(L)	128.250	-66.49	E(820)			5.19	-3.804
C9H20	3-METHYLOCTANE	(G)	128.250	-55.77	T(143)	119.90	T(143)		
C9H20	4-METHYLOCTANE	(L)	128.250	-66.40	E(820)			5.19	-3.804
C9H20	4-METHYLOCTANE	(G)	128.250	-55.77	T(144)	119.90	T(144)		
C9H20	NONANE		128.250	-65.84	A(693)	94.09	M(420)	2.81	-2.063
C9H20	NONANE		128.250	-54.74	T(141)	120.86	T(141)	5.93	-4.349
C9H20	2,2,3-TRIMETHYLHEXANE	(L)	128.250	-67.62	E(820)			5.86	-4.296
C9H20	2,2,3-TRIMETHYLHEXANE	(G)	128.250	-57.65	T(160)	111.34	T(160)		
C9H20	2,2,4-TRIMETHYLHEXANE	(L)	128.250	-67.86	E(820)			5.38	-3.945
C9H20	2,2,4-TRIMETHYLHEXANE	(G)	128.250	-58.03	T(161)	111.34	T(161)		
C9H20	2,2,5-TRIMETHYLHEXANE	(L)	128.250	-70.44	E(820)			3.21	-2.355
C9H20	2,2,5-TRIMETHYLHEXANE	(G)	128.250	-60.71	T(162)	109.96	T(162)		
C9H20	2,3,3-TRIMETHYLHEXANE	(L)	128.250	-67.13	E(820)			6.19	-4.539
C9H20	2,3,3-TRIMETHYLHEXANE	(G)	128.250	-57.08	T(163)	112.14	T(163)		
C9H20	2,3,4-TRIMETHYLHEXANE	(L)	128.250	-66.39	E(820)			6.43	-4.712
C9H20	2,3,4-TRIMETHYLHEXANE	(G)	128.250	-56.18	T(164)	114.37	T(164)		
C9H20	2,3,5-TRIMETHYLHEXANE	(L)	128.250	-68.08	E(820)			5.20	-3.808
C9H20	2,3,5-TRIMETHYLHEXANE	(G)	128.250	-57.37	T(165)	112.30	T(165)		
C9H20	2,4,4-TRIMETHYLHEXANE	(L)	128.250	-67.37	E(820)			5.71	-4.188
C9H20	2,4,4-TRIMETHYLHEXANE	(G)	128.250	-57.56	T(166)	112.14	T(166)		
C9H20	3,3,4-TRIMETHYLHEXANE	(L)	128.250	-66.52	E(820)			6.47	-4.744
C9H20	3,3,4-TRIMETHYLHEXANE	(G)	128.250	-56.39	T(167)	113.52	T(167)		
C9H20	2,2,3,3-TETRAMETHYLPENTANE	(L)	128.250	-66.54	A(693)			8.20	-6.009
C9H20	2,2,3,3-TETRAMETHYLPENTANE	(G)	128.250	-56.70	T(172)	106.69	T(172)		
C9H20	2,2,3,4-TETRAMETHYLPENTANE	(L)	128.250	-66.40	A(693)			7.80	-5.716
C9H20	2,2,3,4-TETRAMETHYLPENTANE	(G)	128.250	-56.64	T(173)	108.23	T(173)		
C9H20	2,2,4,4-TETRAMETHYLPENTANE	(L)	128.250	-66.95	A(693)				

Formula	Name	State	Mol Wt	$\Delta H_{f\,298}^{\circ}$ (kcal/mole)	Source	S_{298}° (cal/mole °K)	Source	$\Delta G_{f\,298}^{\circ}$ (kcal/mole)	Log K_p
C9H20	2,2,4-TETRAMETHYLPENTANE	(G)	128.250	-57.83	T(174)	103.13	T(174)	8.13	-5.959
C9H20	2,3,3,4-TETRAMETHYLPENTANE	(L)	128.250	-66.46	A(693)				
C9H20	2,3,3,4-TETRAMETHYLPENTANE	(G)	128.250	-56.46	T(175)	107.65	T(175)	8.15	-5.975
C9H20N2O	TETRAETHYLUREA	(L)	172.266	-130.1	B(1472)				
C9H20O	4-ETHYL-4-HEPTANOL	(L)	144.250	-109.75	A(215)				
C9H20O	NONYL ALCOHOL	(L)	144.250	-109.75	A(215)				
C9H20O	NONYL ALCOHOL	(G)	144.250	-92.47	T(529)	133.45	T(529)	-28.25	20.703
C9H20O	3,5,5-TRIMETHYL-1-HEXANOL	(L)	144.250	-109.2	B(1066)				
C9H20O2	DIISOBUTOXYMETHANE	(L)	160.250	-133.2	C(744)				
C9H20S	BUTYL PENTYL SULFIDE	(G)	160.316	-44.92	T(818)	136.52	T(818)	13.36	-9.792
C9H20S	ETHYL HEPTYL SULFIDE	(G)	160.316	-44.70	T(819)	136.20	T(819)	13.67	-10.023
C9H20S	HEXYL PROPYL SULFIDE	(G)	160.316	-44.73	T(820)	136.52	T(820)	13.55	-9.931
C9H20S	METHYL OCTYL SULFIDE	(G)	160.316	-44.12	T(821)	135.66	T(821)	14.42	-10.567
C9H20S	1-NONANETHIOL	(G)	160.316	-45.61	T(903)	136.51	T(903)	12.67	-9.289
C10H5N3O6	1,3,8-TRINITRONAPHTHALENE	(S)	263.164	6.9	B(54)				
C10H5N3O6	1,4,5-TRINITRONAPHTHALENE	(S)	263.164	9.9	B(54)				
C10H6N2O4	1,5-DINITRONAPHTHALENE	(S)	218.164	8.1	B(986)				
C10H6N2O4	1,8-DINITRONAPHTHALENE	(S)	218.164	11.1	B(986)				
C10H6N2O5	2,4-DINITRO-1-NAPHTHOL	(S)	234.164	-41.9	B(58)				
C10H6O	PHENYLCYCLOBUTADIENQUINONE	(S)	142.148	-20.12	A(809)				
C10H6O2	1,2-NAPHTHOQUINONE	(S)	158.148	-38.1	C(744)				
C10H6O2	1,4-NAPHTHOQUINONE	(S)	158.148	-43.9	C(913)				
C10H6O4	FURIL	(S)	190.148	-80.3	C(744)				
C10H6O8	PYROMELLITIC ACID	(S)	254.148	-368.2	C(744)				
C10H7Cl	1-CHLORONAPHTHALENE	(L)	162.613	12.7	B(1384)				
C10H7Cl	2-CHLORONAPHTHALENE	(S)	162.613	12.9	B(1384)				
C10H7I	1-IODONAPHTHALENE	(L)	254.066	38.6	B(1382)				
C10H7I	2-IODONAPHTHALENE	(S)	254.066	34.6	B(1382)				
C10H7NO2	1,2-NAPHTHOQUINONE 1-OXIME	(S)	173.164	-12.0	C(744)				
C10H7NO2	1,2-NAPHTHOQUINONE 2-OXIME	(S)	173.164	-11.3	C(744)				
C10H7NO2	1,4-NAPHTHOQUINONE OXIME	(S)	173.164	-15.4	C(744)				
C10H7NO2	1-NITRONAPHTHALENE	(S)	173.164	11.0	B(54)				
C10H7N3	NAPHTHOTRIAZULE, ANG	(S)	169.180	64.8	A(447)				
C10H7N3	NAPHTHOTRIAZULE, LIN	(S)	169.180	68.1	A(447)				
C10H8	AZULENE	(S)	128.164	50.7	A(803)				

Formula	Name	State	Mol Wt	$\Delta H^\circ_{f\,298}$ (kcal/mole)	Source	S°_{298} (cal/mole °K)	Source	$\Delta G^\circ_{f\,298}$ (kcal/mole)	Log $K p$
C10H8	AZULENE	(G)	128.164	66.90	T(473)	80.75	T(473)	84.10	-61.647
C10H8	NAPHTHALENE	(S)	128.164	18.66	T(446)	39.89	M(1398)	48.05	-35.218
C10H8	NAPHTHALENE	(G)	128.164	36.08	T(446)	80.22	T(446)	53.44	-39.172
C10H8N2O2	DIMETHYL-2,2'-BENZO-4,5,4',5'-BISOXAZOLE, ANG.	(S)	188.180	-48.1	C(446)				
C10H8N2O2	DIMETHYL-2,2'-BENZO-4,5,4',5'-BISOXAZOLE, LIN.	(S)	188.180	-49.4	C(446)				
C10H8N2O2	METHYLBENZOYLFURAZAN	(S)	188.180	-30.3	C(1012)				
C10H8N2O2	4-PHENYLURACIL	(S)	188.180	-80.8	C(744)				
C10H8N8	AMMONIUM SALT OF HEXACYANOISOBUTYLENE	(S)	240.228	121.4	A(154)				
C10H8O	1-NAPHTHOL	(S)	144.164	-26.8	B(855)				
C10H8O	2-NAPHTHOL	(S)	144.164	-29.8	B(855)				
C10H8O	3-PHENYLCYCLOBUTENONE	(S)	144.164	6.65	A(809)				
C10H8O	4-PHENYL-3-BUTYN-2-ONE	(L)	144.164	21.5	C(744)				
C10H8O2	METHYL PHENYLPROPIOLATE	(L)	160.164	-17.4	C(112)				
C10H8O2	1,4-NAPHTHALENEDIOL	(S)	160.164	-73.0	C(112)				
C10H8O3	2-PHENYLSUCCINIC ANHYDRIDE	(S)	176.164	-119.7	B(1538)				
C10H8O4	BENZYLIDENEMALONIC ACID	(S)	192.164	-157.0	C(744)				
C10H8O4	FUROIN	(S)	192.164	-98.9	C(744)				
C10H8O4	3,4-(METHYLENEDIOXY)CINNAMIC ACID, CIS (M.P. 99 TO 100 C)	(S)	192.164	-135.4	C(744)				
C10H8O4	3,4-(METHYLENEDIOXY)CINNAMIC ACID, TRANS (M.P. 238 C)	(S)	192.164	-144.5	C(744)				
C10H9FO2	METHYL ALPHA-FLUOROCINNAMATE	(S)	180.172	-103.2	C(744)				
C10H9N	1-NAPHTHYLAMINE	(S)	143.180	16.8	C(1618)				
C10H9N	2-NAPHTHYLAMINE	(S)	143.180	11.3	C(744)				
C10H9N	PHENYLPYRROLE	(S)	143.180	36.8	C(1413)				
C10H9N	ALPHA-PHENYLPYRROLE	(S)	143.180	35.9	C(1413)				
C10H9N	1-PHENYLPYRROLE	(L)	143.180	39.5	C(744)				
C10H9N	QUINALDINE	(L)	143.180	41.0	C(744)				
C10H9NO	5-METHYL-3-PHENYLISOXAZOLE	(S)	159.180	5.5	C(1467)				
C10H9NO	ALPHA-PHENYL-GAMMA-METHYLISOXAZOLE	(S)	159.180	6.6	C(1467)				
C10H9NO4	5,6-DIMETHOXYPHTHALDEHYDIC-ACID ANHYDRIDE, OXIME	(S)	207.180	-95.7	C(744)				

Formula	Name	State	Mol Wt	$\Delta H f^\circ_{298}$ (kcal/mole)	Source	S°_{298} (cal/mole °K)	Source	$\Delta G f^\circ_{298}$ (kcal/mole)	Log Kp
C10H9NO4	3,4-DIMETHOXYPHTHALIMIDE	(S)	207.180	-151.0	B(713)				
C10H9NO4	N-PHENYLTARTRIMIDE, D	(L)	207.180	-161.5	C(744)				
C10H9NO4	N-PHENYLTARTRIMIDE, D,L	(L)	207.180	-161.3	C(744)				
C10H9N3	NAPHTHO-(2',3',4,5)-1,2,3-TRIAZOLE	(S)	171.196	-0.8	A(447)				
C10H9N3	NAPHTHOTRIAZOLE, ANG	(S)	171.196	-4.2	A(447)				
C10H9N3O2	P-TOLUYLAMINOFURAZAN	(S)	203.196	12.1	C(1012)				
C10H10	1,2-DIHYDRONAPHTHALENE	(L)	130.180	16.5	C(744)				
C10H10	1,4-DIHYDRONAPHTHALENE	(S)	130.180	18.8	C(744)				
C10H10	4-PHENYL-1-BUTYNE	(L)	130.180	58.6	C(744)				
C10H10N2O2	METHYL-P-METHOXY-PHENYLFURAZANE	(S)	190.196	21.4	C(1011)				
C10H10N2O3	4-P-METHOXYPHENYL-5-METHYL-1,2,6-DIOXADIAZOLE	(S)	206.196	6.7	C(1011)				
C10H10N2O3	METHYL-P-METHOXY-PHENYLFUROXANE	(S)	206.196	-1.9	C(1011)				
C10H10N2O3	2-NAPHTHYLAMINE NITRATE	(S)	206.196	-42.3	C(1618)				
C10H10O	3,4-DIHYDRO-1(2H)-NAPHTHALENONE	(L)	146.180	-50.1	C(609)				
C10H10O	4-PHENYL-3-BUTEN-2-ONE	(S)	146.180	-23.6	C(744)				
C10H10O	3-PHENYLCYCLOBUTANONE	(L)	146.180	-13.52	A(809)				
C10H10O2	4-ALLYL-1,2-(METHYLENEDIOXY)-BENZENE	(L)	162.180	-37.0	C(744)				
C10H10O2	METHYL CINNAMATE	(S)	162.180	-69.5	C(744)				
C10H10O2	ALPHA-METHYLCINNAMIC ACID	(S)	162.180	-81.5	C(744)				
C10H10O2	BETA-METHYLCINNAMIC ACID	(S)	162.180	-83.0	C(744)				
C10H10O2	1,2-(METHYLENEDIOXY)-4-PROPENYLBENZENE	(L)	162.180	-47.2	C(744)				
C10H10O2	PHENYLISOCROTONIC ACID, CIS	(S)	162.180	-85.7	C(744)				
C10H10O3	HYDROXY-METHYLCINNAMIC ACID, CIS (M.P. 91 TO 92 C)	(U,S)	178.180	-112.9	C(744)				
C10H10O3	HYDROXY-METHYLCINNAMIC ACID, TRANS (M.P. 182 TO 183 C)	(U,S)	178.180	-119.1	C(744)				
C10H10O3	ALLO-P-METHOXYCINNAMIC ACID, CIS (M.P. 66 C)	(S)	178.180	-107.8	C(744)				
C10H10O3	P-METHOXYCINNAMIC ACID, TRANS	(S)	178.180	-117.1	C(744)				

Formula	Name	State	Mol Wt	$\Delta H f^{\circ}_{298}$ (kcal/mole)	Source	S°_{298} (cal/mole °K)	Source	$\Delta G f^{\circ}_{298}$ (kcal/mole)	Log $K p$
C10H1004	BENZYLMALONIC ACID	(S)	194.180	-198.8	C(744)				
C10H1004	2,4-DIACETYLRESORCINOL	(S)	194.180	-178.6	C(145)				
C10H1004	4,6-DIACETYLRESORCINOL	(S)	194.180	-184.2	C(145)				
C10H1004	5,6-DIMETHOXYPHTHALIDE	(S)	194.180	-145.4	C(744)				
C10H1004	DIMETHYL ISOPHTHALATE	(S)	194.180	-170.2	C(744)				
C10H1004	DIMETHYL PHTHALATE	(L)	194.180	-161.6	C(744)				
C10H1004	DIMETHYL TEREPHTHALATE	(S)	194.180	-169.7	C(744)				
C10H1004	PHENYLSUCCINIC ACID	(S)	194.180	-199.8	B(1538)				
C10H1005	5,6-DIMETHOXYPHTHALALDEHYDIC ACID	(S)	210.180	-192.4	C(744)				
C10H1006	4,5-DIMETHOXYPHTHALIC ACID	(S)	226.180	-257.1	C(744)				
C10H11N03	N-BENZOYLALANINE	(S)	193.196	-147.3	C(744)				
C10H11N03	N-BENZOYLSARCOSINE	(S)	193.196	-135.6	C(744)				
C10H11N03	N-(PHENYLACETYL)GLYCINE	(S)	193.196	-150.5	C(744)				
C10H11N03	N-(M-TOLUYL)GLYCINE	(S)	193.196	-148.4	C(744)				
C10H11N03	N-(O-TOLUYL)GLYCINE	(S)	193.196	-147.7	C(744)				
C10H11N03	N-(P-TOLUYL)GLYCINE	(S)	193.196	-147.9	C(744)				
C10H11N04	N-(P-ANISOYL)GLYCINE	(S)	209.196	-180.3	C(744)				
C10H11N5	BENZAL-3-HYDRAZINO-5-METHYL-1,2,4-TRIAZOLE	(S)	201.228	61.6	B(1616)				
C10H12	1-PHENYL-1-BUTENE	(L)	132.196	-1.7	C(744)				
C10H12	1-PHENYL-2-BUTENE	(L)	132.196	13.4	C(744)				
C10H12	2-PHENYL-2-BUTENE	(S)	132.196	9.4	C(744)				
C10H12	4-PHENYL-1-BUTENE	(L)	132.196	7.2	C(744)				
C10H12	1,2,3,4-TETRAHYDRONAPHTHALENE	(L)	132.196	-12.3	C(721)				
C10H12	1,2,3,4-TETRAHYDRONAPHTHALENE	(L)	132.196	-6.0	(CALC)	60.10	M(954)	35.97	-26.367
C10H12	1,2,3,4-TETRAHYDRONAPHTHALENE	(G)	132.196	6.6	B(1621)	89.2	F(1621)	39.90	-29.243
C10H12	3A,4,7,7A-TETRAHYDRO-4,7-METHANOINDENE	(S)	132.196	38.66	B(83)				
C10H120	P-ALLYLANISOLE	(L)	148.196	-14.6	B(744)				
C10H120	BUTYROPHENONE	(L)	148.196	-45.13	A(251)				
C10H120	ETHYL 1-PHENYLVINYL ETHER	(L)	148.196	-33.3	C(744)				
C10H120	P-PROPENYLANISOLE	(S)	148.196	-24.8	C(744)				
C10H120	1,2,3,4-TETRAHYDRO-1-NAPHTHOL	(S)	148.196	-68.2	B(163)				
C10H1202	4-ALLYL-2-METHOXYPHENOL	(L)	164.196	-62.7	C(744)				
C10H1202	4-ALLYL-2-METHOXYPHENOL	(L)	164.196	-62.9	C(744)				

Formula	Name	State	Mol Wt	$\Delta H f°_{298}$ (kcal/mole)	Source	$S°_{298}$ (cal/mole °K)	Source	$\Delta G f°_{298}$ (kcal/mole)	Log $K p$
C10H12O2	ETHYL PHENYLACETATE	(L)	164.196	-95.0	C(843)				
C10H12O2	P-ISOPROPYLBENZOIC ACID	(S)	164.196	-111.3	C(744)				
C10H12O2	P-MENTHA-3,6-DIENE-2,5-DIONE	(S)	164.196	-78.1	C(744)				
C10H12O2	2-METHOXY-4-PROPENYLPHENOL	(L)	164.196	-71.7	C(744)				
C10H12O2	PROPYL BENZOATE	(L)	164.196	-94.9	C(744)				
C10H12O2	1,2,3,4-TETRAHYDRO- NAPHTHALENE-1,2-DIOL, CIS	(S)	164.196	-99.28	B(1536)				
C10H12O2	1,2,3,4-TETRAHYDRO- NAPHTHALENE-1,2-DIOL, TRANS	(S)	164.196	-100.63	B(1536)				
C10H12O2	1,2,3,4-TETRAHYDRO- NAPHTHALENE-2,3-DIOL, CIS	(S)	164.196	-99.26	B(1536)				
C10H12O2	1,2,3,4-TETRAHYDRO- NAPHTHALENE-2,3-DIOL, TRANS	(S)	164.196	-100.58	B(1536)				
C10H12O2	1,2,3,4-TETRAHYDRONAPHTHALENE- HYDROPEROXIDE	(S)	164.196	-31.1	C(609)				
C10H12O2	2,3,4-TRIMETHYLBENZOIC ACID	(S)	164.196	-116.30	A(255)				
C10H12O2	2,3,5-TRIMETHYLBENZOIC ACID	(S)	164.196	-116.79	A(255)				
C10H12O2	2,3,6-TRIMETHYLBENZOIC ACID	(S)	164.196	-113.68	A(255)				
C10H12O2	2,4,5-TRIMETHYLBENZOIC ACID	(S)	164.196	-118.46	A(255)				
C10H12O2	2,4,6-TRIMETHYLBENZOIC ACID	(S)	164.196	-114.20	A(255)				
C10H12O2	3,4,5-TRIMETHYLBENZOIC ACID	(S)	164.196	-119.71	A(255)				
C10H12O2	2,4-XYLYLACETIC ACID	(S)	164.196	-118.2	C(571)				
C10H12O3	PROPYL P-HYDROXYBENZOATE	(S)	180.196	-148.9	C(744)				
C10H12O3	PROPYL SALICYLATE	(S)	180.196	-143.8	C(744)				
C10H12O4	DIMETHYL 1,4-CYCLOHEXADIENE- 1,4-DICARBOXYLATE	(S)	196.196	-169.0	C(744)				
C10H13N	1,2,3,4-TETRAHYDROQUINALDINE	(L)	147.212	-1.2	C(744)				
C10H13NO	ETHYL N-METHYLCARBANILATE	(L)	179.212	-101.2	B(984)				
C10H13NO2	P-MENTHA-3,6-DIENE-2,5-DIONE OXIME	(S)	179.212	-51.7	C(744)				
C10H13NO2	PHENACETIN	(S)	179.212	-98.4	C(744)				
C10H13NO3	ETHYL 2,4-DIMETHYL-5-FORMYL- 3-PYRROLECARBOXYLATE	(S)	195.212	-150.3	C(1413)				
C10H13NO3	ETHYL 4-FORMYL-3,5-DIMETHYL- 2-PYRROLECARBOXYLATE	(S)	195.212	-150.1	C(1413)				
C10H13NO3	4'-METHOXY-2'-METHYL-								

Formula	Name	State	Mol Wt	$\Delta H f^\circ_{298}$ (kcal/mole)	Source	S°_{298} (cal/mole °K)	Source	$\Delta G f^\circ_{298}$ (kcal/mole)	Log $K p$
C10H13NO3	ACETOPHENONE OXIME	(S)	195.212	-95.2	C(843)				
C10H13NO3	4'-METHOXY-3'-METHYL- ACETOPHENONE OXIME	(S)	195.212	-100.0	C(843)				
C10H13NO3	2-METHYL-2-NITRO- 1-PHENYL-1-PROPANOL	(S)	195.212	-75.9	B(300)				
C10H13NO3	2-METHYL-2-NITRO- 3-PHENYL-1-PROPANOL	(S)	195.212	-83.1	B(300)				
C10H13NO4	DIMETHYL 3,5-DIMETHYL- 2,4-PYRROLEDICARBOXYLATE	(S)	211.212	-199.1	C(1413)				
C10H13NO6	PHENYLAMMONIUM D-BITARTRATE	(S)	243.212	-305.3	E(744)				
C10H13NO6	PHENYLAMMONIUM BIRACEMATE	(S)	243.212	-307.3	C(744)				
C10H14	BUTYLBENZENE	(L)	134.212	-15.28	A(1195)	76.77	M(994)	31.03	-22.743
C10H14	BUTYLBENZENE	(G)	134.212	-3.30	A(1195)	105.04	T(410)	34.58	-25.346
C10H14	SEC-BUTYLBENZENE	(L)	134.212	-15.89	A(1195)				
C10H14	SEC-BUTYLBENZENE	(G)	134.212	-4.17	A(1195)				
C10H14	TERT-BUTYLBENZENE	(L)	134.212	-16.92	A(1195)	66.6	M(656)	32.42	-23.763
C10H14	TERT-BUTYLBENZENE	(G)	134.212	-5.42	A(1195)				
C10H14	M-CYMENE	(L)	134.212	-18.69	E(1195)				
C10H14	M-CYMENE	(G)	134.212	-6.79	E(1195)				
C10H14	O-CYMENE	(L)	134.212	-18.19	E(1195)				
C10H14	O-CYMENE	(G)	134.212	-6.10	E(1195)	73.28	M(655)	28.65	-20.999
C10H14	P-CYMENE	(L)	134.212	-18.7	E(1195)				
C10H14	P-CYMENE	(G)	134.212	-6.9	E(1195)				
C10H14	M-DIETHYLBENZENE	(L)	134.212	-17.44	E(1195)				
C10H14	M-DIETHYLBENZENE	(G)	134.212	-5.22	T(411)	104.99	T(411)	32.67	-23.949
C10H14	O-DIETHYLBENZENE	(L)	134.212	-16.94	E(1195)				
C10H14	O-DIETHYLBENZENE	(G)	134.212	-4.53	T(412)	103.81	T(412)	33.72	-24.713
C10H14	P-DIETHYLBENZENE	(L)	134.212	-17.47	E(1195)				
C10H14	P-DIETHYLBENZENE	(G)	134.212	-5.32	T(413)	103.73	T(413)	32.95	-24.152
C10H14	2-ETHYL-M-XYLENE	(L)	134.212	-19.84	E(1195)				
C10H14	2-ETHYL-M-XYLENE	(G)	134.212	-7.11	E(1195)				
C10H14	2-ETHYL-P-XYLENE	(L)	134.212	-20.38	E(1195)				
C10H14	2-ETHYL-P-XYLENE	(G)	134.212	-7.91	E(1195)				
C10H14	3-ETHYL-O-XYLENE	(L)	134.212	-19.84	E(1195)				
C10H14	3-ETHYL-O-XYLENE	(G)	134.212	-7.11	E(1195)				
C10H14	4-ETHYL-O-XYLENE	(L)	134.212	-20.38	E(1195)				

Formula	Name	State	Mol Wt	$\Delta H_{f\,298}^{\circ}$ (kcal/mole)	Source	S_{298}° (cal/mole °K)	Source	$\Delta G_{f\,298}^{\circ}$ (kcal/mole)	Log K_p
C10H14	4-ETHYL-O-XYLENE	(G)	134.212	-7.91	E(1195)				
C10H14	4-ETHYL-M-XYLENE	(L)	134.212	-20.38	E(1195)				
C10H14	4-ETHYL-M-XYLENE	(G)	134.212	-7.91	E(1195)				
C10H14	5-ETHYL-M-XYLENE	(L)	134.212	-20.86	E(1195)				
C10H14	5-ETHYL-M-XYLENE	(G)	134.212	-8.50	E(1195)				
C10H14	HEXAHYDRONAPHTHALENE	(L)	134.212	0.93	C(744)				
C10H14	ISOBUTYLBENZENE	(L)	134.212	-16.70	A(1195)				
C10H14	ISOBUTYLBENZENE	(L)	134.212	-5.15	A(1195)				
C10H14	O-PROPYLTOLUENE	(L)	134.212	-17.52	E(1195)				
C10H14	O-PROPYLTOLUENE	(G)	134.212	-5.17	E(1195)				
C10H14	M-PROPYLTOLUENE	(L)	134.212	-18.02	E(1195)				
C10H14	M-PROPYLTOLUENE	(G)	134.212	-5.86	E(1195)				
C10H14	P-PROPYLTOLUENE	(L)	134.212	-18.06	E(1195)				
C10H14	P-PROPYLTOLUENE	(G)	134.212	-5.97	E(1195)				
C10H14	3A,4,7,7A-TETRAHYDRO-4,7-METHANOINDAN	(S)	134.212	-5.6	B(90)				
C10H14	1,2,3,4-TETRAMETHYLBENZENE	(L)	134.212	-23.0	E(1195)				
C10H14	1,2,3,4-TETRAMETHYLBENZENE	(G)	134.212	-10.02	E(1195)	69.45	M(655)	25.49	-18.684
C10H14	1,2,3,5-TETRAMETHYLBENZENE	(L)	134.212	-23.54	E(1195)	99.55	T(414)	29.50	-21.620
C10H14	1,2,3,5-TETRAMETHYLBENZENE	(G)	134.212	-10.71	T(415)	74.06	M(655)	23.58	-17.281
C10H14	1,2,4,5-TETRAMETHYLBENZENE	(S)	134.212	-30.7	B(71)	100.99	T(415)	28.38	-20.800
C10H14	1,2,4,5-TETRAMETHYLBENZENE	(L)	134.212	-23.58	B(1195)	58.70	M(655)	21.00	-15.389
C10H14	1,2,4,5-TETRAMETHYLBENZENE	(G)	134.212	-10.82	T(416)	71.83	M(655)	24.20	-17.739
C10H14N2	NICOTINE	(L)	162.228	10.2	C(744)	100.03	T(416)	28.55	-20.929
C10H14N2O	N-TOLYL-N'-ETHYLUREA	(S)	178.228	-70.6	B(1472)				
C10H14O	CARVACROL	(L)	150.212	-63.0	C(744)				
C10H14O	P-MENTHA-6,8-DIEN-2-ONE	(L)	150.212	-43.1	C(744)				
C10H14O	THYMOL	(S)	150.212	-67.8	C(744)				
C10H14O	THYMOL	(L)	150.212	-64.1	C(744)				
C10H14O	2,6,6-TRIMETHYL-2,4-CYCLOHEPTADIEN-1-ONE	(L)	150.212	-43.1	C(744)				
C10H14O2	P-XYLENOLETHYL ETHER	(U,L)	150.212	-48.9	C(744)				
C10H14O2	P-CYMENE-2,5-DIOL	(S)	166.212	-110.6	C(744)				
C10H14O3	CAMPHORIC ANHYDRIDE	(S)	182.212	-166.0	C(744)				
C10H14O3S	P-BUTYLBENZENESULFONIC ACID	(G)	214.278			122.38	E(69)		
C10H14O4	DIMETHYL CYCLOHEXENE-								

Formula	Name	State	Mol Wt	$\Delta H_{f\,298}^{\circ}$ (kcal/mole)	Source	S_{298}° (cal/mole °K)	Source	$\Delta G_{f\,298}^{\circ}$ (kcal/mole)	Log K_p
C10H14O4	1,4-DICARBOXYLATE	(S)	198.212	-191.6	C(744)				
C10H14O8	SYMMETRICAL-TETRAMETHYL ETHANETETRACARBOXYLATE	(S)	262.212	-373.0	C(744)				
C10H15N	N,N-DIETHYLANILINE	(L)	149.228	-3.9	E(1547)	63.5	E(1547)	57.84	-42.396
C10H15N	N,N-DIETHYLANILINE	(G)	149.228	9.6	E(1547)	92.4	E(1547)	62.72	-45.975
C10H15NO2	ETHYL 2,4,5-TRIMETHYL-3-PYRROLECARBOXYLATE	(S)	181.228	-114.1	E(1413)				
C10H15NO2	ALPHA-NITROCAMPHOR	(U,S)	197.228	-80.8	C(744)				
C10H15NO3	ADAMANTANE	(S)	136.228	-31.6	C(209)	46.81	M(209)	32.95	-24.149
C10H16	ALLOCIMENE, CIS	(L)	136.228	-5.9	A(580)				
C10H16	BORNEOCAMPHENE	(U,L)	136.228	-15.4	C(744)				
C10H16	CAMPHENE (CRYST)	(S)	136.228	-16.6	C(744)				
C10H16	3-DECEN-1-YNE, CIS	(L)	136.228	23.67	A(1364)				
C10H16	3-DECEN-1-YNE, TRANS	(L)	136.228	23.97	A(1364)				
C10H16	1,5-DIMETHYL-3-VINYL-CYCLOHEXENE	(L)	136.228	-28.6	C(744)				
C10H16	HEXAHYDRO-4,7-METHANOINDAN	(L)	136.228	-32.9	B(90)				
C10H16	1-ISOPROPYL-4-METHYL-1,3-CYCLOHEXADIENE	(L)	136.228	-13.5	C(744)				
C10H16	P-MENTHA-1,3-DIENE	(L)	136.228	-13.8	C(744)				
C10H16	P-MENTHA-1,8-DIENE, D	(L)	136.228	-13.8	A(580)				
C10H16	P-MENTHA-1,8-DIENE, DL	(L)	136.228	-12.4	A(580)				
C10H16	7-METHYL-3-METHYLENE-1,6-OCTADIENE	(L)	136.228	3.3	A(580)				
C10H16	1-(2-METHYLPROPENYL)CYCLOHEXENE	(L)	136.228	-22.4	C(744)				
C10H16	OCTAHYDRONAPHTHALENE	(L)	136.228	-34.7	C(303)				
C10H16	2-PINENE	(L)	136.228	-4.1	A(580)				
C10H16	2(10)-PINENE	(L)	136.228	-2.0	A(580)				
C10H16	BETA-PINOLENE	(U,L)	136.228	-14.9	C(744)				
C10H16	SYLVESTRENE	(U,L)	136.228	-19.5	C(744)				
C10H16	TERECAMPHENE (INACTIVE)	(L)	136.228	-18.9	C(744)				
C10H16	1,7,7-TRIMETHYLTRICYCLO-(2.2.1.0-2,6)-HEPTANE	(S)	136.228	-16.9	C(744)				
C10H16N6O19	DIPENTAERYTHRITOL HEXANITRATE	(S)	524.276	-234.0	B(986)				
C10H16N6O19		(S)	152.228	-78.1	B(1461)				
C10H16O	CAMPHOR								

724

Formula	Name	State	Mol Wt	$\Delta H f°_{298}$ (kcal/mole)	Source	$S°_{298}$ (cal/mole °K)	Source	$\Delta G f°_{298}$ (kcal/mole)	Log Kp
C10H16O	CARONE	(U,L)	152.228	-89.7	C(744)				
C10H16O	CARVENONE	(U,L)	152.228	-76.6	C(744)				
C10H16O	DIHYDREUCARVONE	(U,L)	152.228	-58.2	C(744)				
C10H16O	3,7-DIMETHYL-2,6-OCTADIENAL	(L)	152.228	-47.4	C(744)				
C10H16O	P-MENTH-8-EN-2-ONE	(L)	152.228	-74.6	C(744)				
C10H16O	P-MENTH-4(8)-EN-3-ONE . . .	(L)	152.228	-72.3	C(744)				
C10H16O	P-MENTH-8-EN-3-ONE	(L)	152.228	-68.0	C(744)				
C10H16O	OCTAHYDRO-2(1H)- NAPHTHALENONE, CIS .	(L)	152.228	-81.4	C(744)				
C10H16O	OCTAHYDRO-2(1H)- NAPHTHALENONE, TRANS	(L)	152.228	-83.7	C(744)				
C10H16O	3-THUJANONE	(L)	152.228	-54.0	C(744)				
C10H16O2	CAMPHOLENIC ACID(M.P. 50 C)	(U,S)	168.228	-121.2	C(744)				
C10H16O2	2,7-DIMETHYL-2,6- OCTADIENOIC ACID	(L)	168.228	-105.6	C(744)				
C10H16O2	ETHYL CYCLOHEXYLIDENE- ACETATE	(L)	168.228	-117.6	C(744)				
C10H16O2	ETHYLCYCLOHEXENE-1 ACETATE	(L)	168.228	-124.2	G(744)				
C10H16O2	ETHYL 3,5-DIMETHYLSORBATE .	(L)	168.228	-133.3	C(744)				
C10H16O2	ETHYL 2-OCTYNOATE	(L)	168.228	-92.9	C(744)				
C10H16O2	ISOCAMPHOLYTIC ACID	(U,S)	168.228	-123.9	C(744)				
C10H16O2	METHYL 2-CYCLOHEX-1-ENE- PROPIONATE	(L)	168.228	-135.4	C(744)				
C10H16O2	METHYL 2-CYCLOHEXYLIDENE- PROPIONATE	(L)	168.228	-110.0	C(744)				
C10H16O2	METHYL 4-METHYLCYCLOHEX-1-ENE- ACETATE	(L)	168.228	-122.6	C(744)				
C10H16O2	METHYL 4-METHYLCYCLOHEXYLIDENE- ACETATE	(L)	168.228	-110.7	C(744)				
C10H16O3	3-(1-HYDROXY-1-METHYLETHYL)- 6-OXOHEPTANOIC ACID, GAMMA- LACTONE	(S)	184.228	-182.7	C(571)				
C10H16O3	OCTAHYDRONAPHTHALENE OZONIDE	(S)	184.228	-141.9	C(303)				
C10H16O3	ALPHA-TANACETONEKETOCARBOXYLIC ACID	(S)	184.228	-160.0	C(744)				
C10H16O3	TRIETHYLSUCCINIC ANHYDRIDE	(S)	184.228	-177.6	B(1538)				

Formula	Name	State	Mol Wt	$\Delta H f^\circ_{298}$ (kcal/mole)	Source	S°_{298} (cal/mole °K)	Source	$\Delta G f^\circ_{298}$ (kcal/mole)	Log Kp
C10H16O4	CAMPHORIC ACID, D	(S)	200.228	-242.3	C(744)				
C10H16O4	1,2-CYCLOHEXYLENE DIACETATE, CIS	(S)	200.228	-225.0	B(1537)				
C10H16O4	1,2-CYCLOHEXYLENE DIACETATE, TRANS	(S)	200.228	-223.4	B(1537)				
C10H16O4	DIMETHYL 1,3-CYCLOHEXANE-DICARBOXYLATE, CIS	(L)	200.228	-189.4	C(1368)				
C10H16O4	DIMETHYL 1,3-CYCLOHEXANE-DICARBOXYLATE, TRANS	(L)	200.228	-191.3	C(1368)				
C10H16O4	DIMETHYL 1,4-CYCLOHEXANE-DICARBOXYLATE	(S)	200.228	-212.6	C(744)				
C10H16O4	DIMETHYL 2,3-DIMETHYL 1,3-CYCLO-BUTANEDICARBOXYLATE, CIS	(L)	200.228	-197.4	C(744)				
C10H16S4	1,3,5,7-TETRAMETHYL-2,4,6,8-TETRATHIAADAMANTANE	(S)	264.492			71.90	M(213)		
C10H17CL	CAMPHENE HYDROCHLORIDE	(L)	172.693	-57.1	C(744)				
C10H17CL	2(10)-PINENE HYDROCHLORIDE	(L)	172.693	-57.7	C(744)				
C10H17NO	CAMPHOR OXIME	(S)	167.244	-39.2	C(744)				
C10H18	BICYCLOPENTANE	(L)	138.244	-41.9	A(806)				
C10H18	DECAHYDRONAPHTHALENE, CIS	(L)	138.244	-52.45	A(1401)	63.34	M(954)	16.47	-12.074
C10H18	DECAHYDRONAPHTHALENE, CIS	(G)	138.244	-40.38	T(474)	90.28	T(474)	20.51	-15.034
C10H18	DECAHYDRONAPHTHALENE, TRANS	(L)	138.244	-55.14	A(1401)	63.32	M(954)	13.79	-10.107
C10H18	DECAHYDRONAPHTHALENE, TRANS	(G)	138.244	-43.57	T(475)	89.52	T(475)	17.55	-12.862
C10H18	1-DECYNE	(G)	138.244	9.85	T(327)	125.36	T(327)	60.28	-44.186
C10H18	2-ETHYL-4,4-DIMETHYL-CYCLOHEXENE	(L)	138.244	-48.0	C(744)				
C10H18	P-MENTH-3-ENE	(L)	138.244	-30.7	C(744)				
C10H18	THUJANE	(L)	138.244	-46.1	C(744)				
C10H18	1,3,3-TRIMETHYLNORBORNANE	(L)	138.244	-52.2	C(744)				
C10H18	SPIRO(4.5)DECANE	(L)	138.244	-48.0	A(805)				
C10H18CL2	P-MENTHA-1,3-DIENE-DIHYDROCHLORIDE	(L)	209.158	-99.1	C(744)				
C10H18CL2	BORNEOL	(S)	154.244	-66.2	C(1636)				
C10H18O	BORNEOL, D	(S)	154.244	-88.9	C(744)				
C10H18O	BORNEOL, L	(S)	154.244	-88.3	C(744)				
C10H18O	BORNEOL (BORNEOL CAMPHOR)	(L)	154.244	-85.5	C(744)				

Formula	Name	State	Mol Wt	$\Delta H f^\circ_{298}$ (kcal/mole)	Source	S°_{298} (cal/mole °K)	Source	$\Delta G f^\circ_{298}$ (kcal/mole)	Log $K p$
C10H18	BORNEOL (SYNTHETIC). . . .	(L)	154.244	−87.9	C(744)				
C10H18	DECAHYDRO-1-NAPHTHOL, CIS .	(S)	154.244	−101.9	C(843)				
C10H18	DECAHYDRO-1-NAPHTHOL, TRANS	(S)	154.244	−105.9	C(843)				
C10H18	DECAHYDRO-2-NAPHTHOL, CIS .	(S)	154.244	−104.7	C(843)				
C10H18	DECAHYDRO-2-NAPHTHOL, TRANS	(S)	154.244	−106.2	C(843)				
C10H18	9-HYDROXYDECAHYDRO- NAPHTHALENE, TRANS . . .	(S)	154.244	−106.0	B(163)				
C10H18	ISOBORNEOL	(S)	154.244	−82.9	C(1636)				
C10H18	4-PROPYL-1,6-HEPTADIEN-4-OL	(L)	154.244	−83.0	C(744)				
C10H18	TERPINEOL	(S)	154.244	−85.6	C(744)				
C10H18	3-THUJANOL	(L)	154.244	−75.1	C(744)				
C10H18O2	4-ISOPROPYLCYCLOHEXANE- CARBOXYLIC ACID	(S)	170.244	−158.8	C(744)				
C10H18O2	9-PEROXYDECAHYDRO- NAPHTHALENE, TRANS	(S)	170.244	−83.2	B(163)				
C10H18O2	1,2,2,3-TETRAMETHYLCYCLOPENTANE- CARBOXYLIC ACID	(S)	170.244	−143.2	C(744)				
C10H18O4	DIETHYL DIMETHYLSUCCINATE,MESO	(S)	202.244	−230.3	C(744)				
C10H18O4	DIETHYL DIMETHYLSUCCINATE, RACEMIC	(L)	202.244	−231.5	C(744)				
C10H18O4	DIETHYL DIMETHYLSUCCINATE,SYM	(L)	202.244	−235.8	C(744)				
C10H18O4	DIMETHYL SUBERATE	(L)	202.244	−221.2	C(744)				
C10H18O4	HEPTYLMALONIC ACID	(S)	202.244	−257.7	B(1535)				
C10H18O4	SEBACIC ACID	(S)	202.244	−254.1	C(744)				
C10H18O4	TRIETHYLSUCCINIC ACID . .	(S)	202.244	−254.1	B(1538)				
C10H19N	CAMPHYLAMINE	(U+L)	153.260	−55.1	C(744)				
C10H19N3O4	LEUCYLGLYCYLGLYCINE . . .	(U+S)	245.276	−254.3	A(1196)	82.45	M(423)	9.62	−7.052
C10H20	BUTYLCYCLOHEXANE	(L)	140.260	−62.91	T(383)	109.58	T(383)	13.49	−9.890
C10H20	BUTYLCYCLOHEXANE	(G)	140.260	−50.95	T(383)				
C10H20	CYCLODECANE	(L)	140.260	−49.55	A(271)				
C10H20	1-CYCLOPENTYLPENTANE . . .	(G)	140.260	−45.15	T(356)	118.35	T(356)	16.68	−12.224
C10H20	1-DECENE	(L)	140.260	−41.73	A(1236)	101.58	M(950)	25.10	−18.396
C10H20	1-DECENE	(G)	140.260	−29.67	T(293)	129.17	T(293)	28.93	−21.206
C10H20	1,1-DIMETHYL-2-PENTYL- CYCLOPROPANE	(L)	140.260	−40.0	A(710)				
C10H20	1-ISOPROPYL-4-METHYLCYCLOHEXANE	(L)	140.260	−108.7	C(744)				

Formula	Name	State	Mol Wt	$\Delta H f^\circ_{298}$ (kcal/mole)	Source	S°_{298} (cal/mole °K)	Source	$\Delta G f^\circ_{298}$ (kcal/mole)	Log $K p$
C10H20	1-METHYL-3-PROPYLCYCLOHEXANE	(L)	140.260	-120.5	C(744)				
C10H20	2,2,5,5-TETRAMETHYL-								
C10H20	3-HEXENE, CIS	(L)	140.260	-44.18	A(1236)				
C10H20	2,2,5,5-TETRAMETHYL-								
C10H20	3-HEXENE, TRANS	(L)	140.260	-33.67	A(1236)				
C10H20O	DECANAL	(G)	156.260	-79.09	T(553)	138.28	T(553)	-15.90	11.654
C10H20O	MENTHOL	(S)	156.260	-114.7	C(744)				
C10H20O	4-PROPYL-1-HEPTEN-4-OL	(L)	156.260	-73.9	C(744)				
C10H20O	4-PROPYL-1-HEPTEN-4-OL	(L)	156.260	-105.4	C(744)				
C10H20O2	DECANOIC ACID	(S)	172.260	-170.70	A(3)				
C10H20O2	DECANOIC ACID	(L)	172.260	-163.68	A(3)				
C10H20O2	METHYL NONANOATE	(L)	172.260	-147.40	A(3)				
C10H22	DECANE	(L)	142.276	-71.95	A(1200)	101.79	M(420)	4.12	-3.020
C10H22	DECANE	(G)	142.276	-59.67	T(176)	130.17	T(176)	7.94	-5.819
C10H22	2,2-DIMETHYLOCTANE	(L)	142.276	-75.25	E(820)				
C10H22	2,2-DIMETHYLOCTANE	(G)	142.276	-63.93	T(183)	122.29	T(183)	6.03	-4.419
C10H22	2,3-DIMETHYLOCTANE	(L)	142.276	-72.87	E(820)				
C10H22	2,3-DIMETHYLOCTANE	(G)	142.276	-61.25	T(184)	126.01	T(184)	7.60	-5.570
C10H22	2,4-DIMETHYLOCTANE	(L)	142.276	-73.83	E(820)				
C10H22	2,4-DIMETHYLOCTANE	(G)	142.276	-62.41	T(185)	126.01	T(185)	6.44	-4.720
C10H22	2,5-DIMETHYLOCTANE	(L)	142.276	-73.87	E(820)				
C10H22	2,5-DIMETHYLOCTANE	(G)	142.276	-62.41	T(186)	126.01	T(186)	6.44	-4.720
C10H22	2,6-DIMETHYLOCTANE	(L)	142.276	-73.97	E(820)				
C10H22	2,6-DIMETHYLOCTANE	(G)	142.276	-62.41	T(187)	126.01	T(187)	6.44	-4.720
C10H22	2,7-DIMETHYLOCTANE	(L)	142.276	-74.67	E(820)				
C10H22	2,7-DIMETHYLOCTANE	(G)	142.276	-63.10	T(188)	123.25	T(188)	6.57	-4.817
C10H22	3,3-DIMETHYLOCTANE	(L)	142.276	-74.05	E(820)				
C10H22	3,3-DIMETHYLOCTANE	(G)	142.276	-62.67	T(189)	124.47	T(189)	6.64	-4.866
C10H22	3,4-DIMETHYLOCTANE	(L)	142.276	-72.14	E(820)				
C10H22	3,4-DIMETHYLOCTANE	(G)	142.276	-60.56	T(190)	126.70	T(190)	8.08	-5.925
C10H22	3,5-DIMETHYLOCTANE	(L)	142.276	-73.17	E(820)				
C10H22	3,5-DIMETHYLOCTANE	(G)	142.276	-61.72	T(191)	126.70	T(191)	6.92	-5.075
C10H22	3,6-DIMETHYLOCTANE	(L)	142.276	-73.27	E(820)				
C10H22	3,6-DIMETHYLOCTANE	(G)	142.276	-61.72	T(192)	125.32	T(192)	7.33	-5.376
C10H22	4,4-DIMETHYLOCTANE	(L)	142.276	-73.91	E(820)				
C10H22	4,4-DIMETHYLOCTANE	(G)	142.276	-62.67	T(193)	124.47	T(193)	6.64	-4.866

Formula	Name	State	Mol Wt	ΔH°_{298} (kcal/mole)	Source	S°_{298} (cal/mole °K)	Source	ΔG°_{298} (kcal/mole)	Log K_p
C10H22	4,5-DIMETHYLOCTANE · · · ·	(L)	142.276	−72.07	E(820)			8.49	−6.227
C10H22	4,5-DIMETHYLOCTANE · · · ·	(G)	142.276	−60.56	T(194)	125.32	T(194)	9.81	−7.189
C10H22	3,3-DIETHYLHEXANE · · · · ·	(L)	142.276	−71.60	E(820)				
C10H22	3,3-DIETHYLHEXANE · · · · ·	(G)	142.276	−60.15	T(222)	122.29	T(222)	10.50	−7.698
C10H22	3,4-DIETHYLHEXANE · · · · ·	(L)	142.276	−70.67	E(820)				
C10H22	3,4-DIETHYLHEXANE · · · · ·	(G)	142.276	−59.17	T(223)	123.25	T(223)	11.40	−8.353
C10H22	3,3-DIETHYL-2-METHYLPENTANE	(L)	142.276	−70.91	E(820)				
C10H22	3,3-DIETHYL-2-METHYLPENTANE	(G)	142.276	−59.49	T(245)	119.18	T(245)		
C10H22	3-ETHYL-2,2-DIMETHYLHEXANE	(L)	142.276	−72.93	E(820)				
C10H22	3-ETHYL-2,2-DIMETHYLHEXANE	(G)	142.276	−61.89	T(224)	120.56	T(224)	8.58	−6.292
C10H22	3-ETHYL-2,3-DIMETHYLHEXANE	(L)	142.276	−72.01	E(820)				
C10H22	3-ETHYL-2,3-DIMETHYLHEXANE	(G)	142.276	−60.75	T(226)	122.74	T(226)	9.07	−6.651
C10H22	3-ETHYL-2,4-DIMETHYLHEXANE	(L)	142.276	−71.71	E(820)				
C10H22	3-ETHYL-2,4-DIMETHYLHEXANE	(G)	142.276	−60.43	T(228)	123.59	T(228)	9.14	−6.700
C10H22	3-ETHYL-2,5-DIMETHYLHEXANE	(L)	142.276	−73.39	E(820)				
C10H22	3-ETHYL-2,5-DIMETHYLHEXANE	(G)	142.276	−62.27	T(230)	122.90	T(230)	7.51	−5.502
C10H22	3-ETHYL-3,4-DIMETHYLHEXANE	(L)	142.276	−71.26	E(820)				
C10H22	3-ETHYL-3,4-DIMETHYLHEXANE	(G)	142.276	−60.06	T(232)	122.74	T(232)	9.76	−7.157
C10H22	4-ETHYL-2,2-DIMETHYLHEXANE	(L)	142.276	−73.22	E(820)				
C10H22	4-ETHYL-2,2-DIMETHYLHEXANE	(G)	142.276	−62.37	T(225)	119.18	T(225)	8.52	−6.242
C10H22	4-ETHYL-2,3-DIMETHYLHEXANE	(L)	142.276	−71.74	E(820)				
C10H22	4-ETHYL-2,3-DIMETHYLHEXANE	(G)	142.276	−60.43	T(227)	122.90	T(227)	9.35	−6.851
C10H22	4-ETHYL-2,4-DIMETHYLHEXANE	(L)	142.276	−72.43	E(820)				
C10H22	4-ETHYL-2,4-DIMETHYLHEXANE	(G)	142.276	−61.23	T(229)	121.36	T(229)	9.01	−6.601
C10H22	4-ETHYL-3,3-DIMETHYLHEXANE	(L)	142.276	−71.86	E(820)				
C10H22	4-ETHYL-3,3-DIMETHYLHEXANE	(G)	142.276	−60.63	T(231)	121.36	T(231)	9.61	−7.041
C10H22	3-ETHYL-2-METHYLHEPTANE	(L)	142.276	−72.04	E(820)				
C10H22	3-ETHYL-2-METHYLHEPTANE	(G)	142.276	−60.56	T(197)	126.01	T(197)	8.29	−6.076
C10H22	3-ETHYL-3-METHYLHEPTANE	(L)	142.276	−72.81	E(820)				
C10H22	3-ETHYL-3-METHYLHEPTANE	(G)	142.276	−61.41	T(200)	124.47	T(200)	7.90	−5.789
C10H22	3-ETHYL-4-METHYLHEPTANE	(L)	142.276	−71.37	E(820)				
C10H22	3-ETHYL-4-METHYLHEPTANE	(G)	142.276	−59.87	T(203)	126.01	T(203)	8.98	−6.582
C10H22	3-ETHYL-5-METHYLHEPTANE	(L)	142.276	−72.48	E(820)				
C10H22	3-ETHYL-5-METHYLHEPTANE	(G)	142.276	−61.04	T(202)	126.01	T(202)	7.81	−5.724
C10H22	4-ETHYL-2-METHYLHEPTANE	(L)	142.276	−73.03	E(820)				
C10H22	4-ETHYL-2-METHYLHEPTANE	(G)	142.276	−61.72	T(198)	126.01	T(198)	7.13	−5.226

Formula	Name	State	Mol Wt	$\Delta H_{f\,298}^{\circ}$ (kcal/mole)	Source	S_{298}° (cal/mole·°K)	Source	$\Delta G_{f\,298}^{\circ}$ (kcal/mole)	Log K_p
C10H22	4-ETHYL-3-METHYLHEPTANE . .	(L)	142.276	-71.34	E(820)			8.77	-6.431
C10H22	4-ETHYL-3-METHYLHEPTANE . .	(G)	142.276	-59.87	T(201)	126.70	T(201)		-5.789
C10H22	4-ETHYL-4-METHYLHEPTANE . .	(L)	142.276	-72.70	E(820)			7.90	-5.527
C10H22	4-ETHYL-4-METHYLHEPTANE . .	(G)	142.276	-61.41	T(204)	124.47	T(204)		
C10H22	5-ETHYL-2-METHYLHEPTANE . .	(L)	142.276	-73.17	E(820)			7.54	-6.101
C10H22	5-ETHYL-2-METHYLHEPTANE . .	(G)	142.276	-61.72	T(199)	124.63	T(199)		
C10H22	3-ETHYLOCTANE	(L)	142.276	-71.82	E(820)			8.32	-5.799
C10H22	3-ETHYLOCTANE	(G)	142.276	-60.01	T(181)	127.74	T(181)		
C10H22	4-ETHYLOCTANE	(L)	142.276	-71.68	E(820)			7.91	-8.116
C10H22	4-ETHYLOCTANE	(G)	142.276	-60.01	E(182)	129.12	T(182)		
C10H22	2-ETHYL-2,3,4-TRIMETHYLPENTANE	(L)	142.276	-71.33	E(820)			11.07	-8.290
C10H22	2-ETHYL-2,3,4-TRIMETHYLPENTANE	(G)	142.276	-60.09	T(248)	118.25	T(248)		
C10H22	2-ETHYL-2,4,4-TRIMETHYLPENTANE	(L)	142.276	-71.40	E(820)			11.31	-8.121
C10H22	2-ETHYL-2,4,4-TRIMETHYLPENTANE	(G)	142.276	-60.55	T(247)	115.91	T(247)		
C10H22	3-ETHYL-2,2,3-TRIMETHYLPENTANE	(L)	142.276	-71.46	E(820)			11.08	-7.399
C10H22	3-ETHYL-2,2,3-TRIMETHYLPENTANE	(G)	142.276	-60.37	T(246)	117.29	T(246)		
C10H22	3-ETHYL-2,3,4-TRIMETHYLPENTANE	(L)	142.276	-72.66	E(820)			10.10	-6.773
C10H22	3-ETHYL-2,3,4-TRIMETHYLPENTANE	(G)	142.276	-61.67	T(244)	116.23	T(244)		
C10H22	4-ISOPROPYLHEPTANE	(L)	142.276	-71.39	E(820)			9.24	-6.654
C10H22	4-ISOPROPYLHEPTANE	(G)	142.276	-60.02	T(196)	124.63	T(196)		-1.337
C10H22	3-ISOPROPYL-2-METHYLHEXANE .	(L)	142.276	-72.29	E(820)			9.08	-5.097
C10H22	3-ISOPROPYL-2-METHYLHEXANE .	(G)	142.276	-61.11	T(221)	121.52	T(221)	1.82	-2.461
C10H22	4-ISOPROPYL-2-METHYLHEXANE .	(L)	142.276	-74.66	A(1025)	100.4	M(1120)	6.95	-5.293
C10H22	2-METHYLNONANE	(L)	142.276	-61.38	T(177)	127.74	T(177)	3.36	-2.629
C10H22	2-METHYLNONANE	(G)	142.276	-72.62	E(820)	102.1	M(1120)	7.22	-5.293
C10H22	3-METHYLNONANE	(L)	142.276	-60.70	T(178)	129.12	T(178)	3.59	-2.629
C10H22	3-METHYLNONANE	(G)	142.276	-72.51	E(820)	101.7	M(639)	7.22	-5.293
C10H22	4-METHYLNONANE	(L)	142.276	-60.70	T(179)	129.12	T(179)	3.59	-1.566
C10H22	4-METHYLNONANE	(G)	142.276	-74.08	A(1025)	101.3	M(1120)	2.14	-5.595
C10H22	5-METHYLNONANE	(L)	142.276	-60.70	T(180)	127.74	T(180)	7.63	
C10H22	5-METHYLNONANE	(G)	142.276	-69.99	E(820)			13.71	-10.047
C10H22	2,2,3,3,4-PENTAMETHYLPENTANE	(L)	142.276	-59.08	T(249)	112.80	T(249)		
C10H22	2,2,3,3,4-PENTAMETHYLPENTANE	(G)	142.276	-69.56	E(820)			14.40	-10.553
C10H22	2,2,3,4,4-PENTAMETHYLPENTANE	(L)	142.276	-59.04	T(250)	110.62	T(250)		
C10H22	2,2,3,4,4-PENTAMETHYLPENTANE	(G)	142.276	-71.58	E(820)			8.97	-6.577
C10H22	4-PROPYLHEPTANE	(L)	142.276	-60.01	T(195)	125.56	T(195)		
C10H22	4-PROPYLHEPTANE	(G)	142.276							

Formula	Name	State	Mol Wt	$\Delta H f^\circ_{298}$ (kcal/mole)	Source	S°_{298} (cal/mole°K)	Source	$\Delta G f^\circ_{298}$ (kcal/mole)	Log $K p$
C10H22	2,2,3,3-TETRAMETHYLHEXANE	(L)	142.276	-72.57	EC 820				
C10H22	2,2,3,3-TETRAMETHYLHEXANE	(G)	142.276	-61.63	TC 233	115.91	TC 233	10.23	-7.499
C10H22	2,2,3,4-TETRAMETHYLHEXANE	(L)	142.276	-71.50	EC 820				
C10H22	2,2,3,4-TETRAMETHYLHEXANE	(G)	142.276	-60.55	TC 234	118.14	TC 234	10.65	-7.803
C10H22	2,2,3,5-TETRAMETHYLHEXANE	(L)	142.276	-75.08	EC 820				
C10H22	2,2,3,5-TETRAMETHYLHEXANE	(G)	142.276	-64.29	TC 235	117.45	TC 235	7.11	-5.212
C10H22	2,2,4,4-TETRAMETHYLHEXANE	(L)	142.276	-71.92	EC 820				
C10H22	2,2,4,4-TETRAMETHYLHEXANE	(G)	142.276	-61.50	TC 236	115.91	TC 236	10.36	-7.594
C10H22	2,2,4,5-TETRAMETHYLHEXANE	(L)	142.276	-74.27	EC 820				
C10H22	2,2,4,5-TETRAMETHYLHEXANE	(G)	142.276	-63.61	TC 237	117.45	TC 237	7.79	-5.711
C10H22	2,2,5,5-TETRAMETHYLHEXANE	(L)	142.276	-78.54	EC 820				
C10H22	2,2,5,5-TETRAMETHYLHEXANE	(G)	142.276	-68.18	TC 238	112.35	TC 238	4.74	-3.476
C10H22	2,3,3,4-TETRAMETHYLHEXANE	(L)	142.276	-71.83	EC 820				
C10H22	2,3,3,4-TETRAMETHYLHEXANE	(G)	142.276	-60.66	TC 239	119.63	TC 239	10.09	-7.397
C10H22	2,3,3,5-TETRAMETHYLHEXANE	(L)	142.276	-72.68	EC 820				
C10H22	2,3,3,5-TETRAMETHYLHEXANE	(G)	142.276	-61.83	TC 240	118.25	TC 240	9.33	-6.841
C10H22	2,3,4,5-TETRAMETHYLHEXANE	(L)	142.276	-71.02	EC 820				
C10H22	2,3,4,5-TETRAMETHYLHEXANE	(G)	142.276	-59.98	TC 241	119.63	TC 241	10.77	-7.895
C10H22	3,3,4,4-TETRAMETHYLHEXANE	(L)	142.276	-72.79	EC 820				
C10H22	3,3,4,4-TETRAMETHYLHEXANE	(G)	142.276	-61.67	TC 242	119.10	TC 242	9.24	-6.772
C10H22	3,3,4,4-TETRAMETHYLHEXANE	(L)	142.276	-71.50	EC 820				
C10H22	3,3,4,4-TETRAMETHYLHEXANE	(G)	142.276	-60.37	TC 243	116.71	TC 243	11.25	-8.247
C10H22	2,2,3-TRIMETHYLHEPTANE	(L)	142.276	-73.73	EC 820				
C10H22	2,2,3-TRIMETHYLHEPTANE	(G)	142.276	-62.58	TC 205	120.56	TC 205	7.89	-5.786
C10H22	2,2,4-TRIMETHYLHEPTANE	(L)	142.276	-73.92	EC 820				
C10H22	2,2,4-TRIMETHYLHEPTANE	(G)	142.276	-63.06	TC 206	120.56	TC 206	7.41	-5.434
C10H22	2,2,5-TRIMETHYLHEPTANE	(L)	142.276	-75.91	EC 820				
C10H22	2,2,5-TRIMETHYLHEPTANE	(G)	142.276	-64.95	TC 207	120.56	TC 207	5.52	-4.049
C10H22	2,2,6-TRIMETHYLHEPTANE	(L)	142.276	-76.64	EC 820				
C10H22	2,2,6-TRIMETHYLHEPTANE	(G)	142.276	-65.64	TC 208	119.18	TC 208	5.25	-3.845
C10H22	2,3,3-TRIMETHYLHEPTANE	(L)	142.276	-73.22	EC 820				
C10H22	2,3,3-TRIMETHYLHEPTANE	(G)	142.276	-62.01	TC 209	121.36	TC 209	8.23	-6.029
C10H22	2,3,4-TRIMETHYLHEPTANE	(L)	142.276	-72.43	EC 820				
C10H22	2,3,4-TRIMETHYLHEPTANE	(G)	142.276	-61.11	TC 210	123.59	TC 210	8.46	-6.201
C10H22	2,3,5-TRIMETHYLHEPTANE	(L)	142.276	-73.53	EC 820				
C10H22	2,3,5-TRIMETHYLHEPTANE	(G)	142.276	-62.27	TC 211	123.59	TC 211	7.30	-5.351

Formula	Name	State	Mol Wt	$\Delta H f^\circ_{298}$ (kcal/mole)	Source	S°_{298} (cal/mole °K)	Source	$\Delta G f^\circ_{298}$ (kcal/mole)	Log $K p$
C10H22	2,3,6-TRIMETHYLHEPTANE	(L)	142.276	-74.22	E(820)				
C10H22	2,3,6-TRIMETHYLHEPTANE	(G)	142.276	-62.96	T(212)	122.90	T(212)	6.82	-4.996
C10H22	2,4,4-TRIMETHYLHEPTANE	(L)	142.276	-73.37	E(820)				
C10H22	2,4,4-TRIMETHYLHEPTANE	(G)	142.276	-62.49	T(213)	121.36	T(213)	7.75	-5.677
C10H22	2,4,5-TRIMETHYLHEPTANE	(L)	142.276	-73.49	E(820)				
C10H22	2,4,5-TRIMETHYLHEPTANE	(G)	142.276	-62.27	T(214)	123.59	T(214)	7.30	-5.351
C10H22	2,4,6-TRIMETHYLHEPTANE	(L)	142.276	-71.58	E(820)				
C10H22	2,4,6-TRIMETHYLHEPTANE	(G)	142.276	-60.52	T(215)	121.52	T(215)	9.67	-7.086
C10H22	2,5,5-TRIMETHYLHEPTANE	(L)	142.276	-75.40	E(820)				
C10H22	2,5,5-TRIMETHYLHEPTANE	(G)	142.276	-64.38	T(216)	121.36	T(216)	5.86	-4.292
C10H22	3,3,4-TRIMETHYLHEPTANE	(L)	142.276	-72.56	E(820)				
C10H22	3,3,4-TRIMETHYLHEPTANE	(G)	142.276	-61.32	T(217)	122.74	T(217)	8.50	-6.233
C10H22	3,3,5-TRIMETHYLHEPTANE	(L)	142.276	-72.81	E(820)				
C10H22	3,3,5-TRIMETHYLHEPTANE	(G)	142.276	-61.80	T(218)	122.74	T(218)	8.02	-5.881
C10H22	3,4,4-TRIMETHYLHEPTANE	(L)	142.276	-72.52	E(820)				
C10H22	3,4,4-TRIMETHYLHEPTANE	(G)	142.276	-61.32	T(219)	122.74	T(219)	8.50	-6.233
C10H22	3,4,5-TRIMETHYLHEPTANE	(L)	142.276	-71.81	E(820)				
C10H22	3,4,5-TRIMETHYLHEPTANE	(G)	142.276	-60.43	T(220)	123.59	T(220)	9.14	-6.700
C10H22O	DECYL ALCOHOL	(L)	158.276	-114.66	A(215)				
C10H22O	DECYL ALCOHOL	(G)	158.276	-96.38	T(530)	142.76	T(530)	-25.22	18.486
C10H22O	PENTYL ETHER	(L)	158.276	-82.5	C(744)				
C10H22O2	1,10-DECANEDIOL	(S)	174.276	-165.86	A(1114)				
C10H22O3	P-MENTHANE-1,8-DIOL	(S)	190.276	-241.7	C(744)				
C10H22O7	DIPENTERYTHRITOL	(S)	254.276	-376.2	B(986)				
C10H22S	BUTYL HEXYL SULFIDE	(L)	174.342	-49.84	T(822)				
C10H22S	1-DECANETHIOL	(L)	174.342	-66.10	A(507)				
C10H22S	1-DECANETHIOL	(G)	174.342	-50.54	T(904)	145.83	T(822)	15.37	-11.270
C10H22S	ETHYL OCTYL SULFIDE	(G)	174.342	-49.63	T(823)	145.82	T(904)	14.68	-10.759
C10H22S	HEPTYL PROPYL SULFIDE	(G)	174.342	-49.66	T(824)	145.51	T(823)	15.68	-11.493
C10H22S	ISOPENTYL SULFIDE	(L)	174.342	-67.4	B(899)		145.83	T(824)	15.55	-11.402
C10H22S	ISOPENTYL SULFIDE	(G)	174.342	-53.0	B(899)				
C10H22S	METHYL NONYL SULFIDE	(G)	174.342	-49.05	T(825)	144.97	T(825)	16.42	-12.037
C10H22S	PENTYL SULFIDE	(L)	174.342	-63.7	B(899)				
C10H22S	PENTYL SULFIDE	(G)	174.342	-49.84	T(826)	144.45	T(826)	15.79	-11.571
C10H22S	PENTYL SULFIDE	(G)	174.342	-49.0	B(899)				
C10H22S2	PENTYL DISULFIDE	(G)	206.408	-47.71	T(876)	155.53	T(876)	16.89	-12.379

Formula	Name	State	Mol Wt	$\Delta H_{f\ 298}^{\circ}$ (kcal/mole)	Source	S_{298}° (cal/mole °K)	Source	$\Delta G_{f\ 298}^{\circ}$ (kcal/mole)	Log Kp
C10H23N	DIISOPENTYLAMINE	(L)	157.292	-64.3	C(744)				
C11H7N	1-NAPHTHONITRILE	(S)	153.174	53.8	C(744)				
C11H7N	2-NAPHTHONITRILE	(S)	153.174	48.6	C(744)				
C11H8O2	2-METHYL-1,4-NAPHTHOQUINONE	(S)	172.174	-54.8	C(112)				
C11H8O2	1-NAPHTHOIC ACID	(S)	172.174	-75.0	C(744)				
C11H8O2	2-NAPHTHOIC ACID	(S)	172.174	-79.2	C(744)				
C11H8O3	3-HYDROXY-2-NAPHTHOIC ACID	(S)	188.174	-130.2	B(1640)				
C11H10	1-METHYLNAPHTHALENE	(L)	142.190	21.5	C(606)	60.90	MC 954)	54.33	-39.826
C11H10	1-METHYLNAPHTHALENE	(L)	142.190	13.43	A(1401)	60.90	MC 954)	46.26	-33.911
C11H10	1-METHYLNAPHTHALENE	(G)	142.190	27.93	TC 447)	90.21	TC 447)	52.03	-38.134
C11H10	2-METHYLNAPHTHALENE	(S)	142.190	10.72	A(1401)	52.58	MC 954)	46.03	-33.743
C11H10	2-METHYLNAPHTHALENE	(L)	142.190	13.55	A(1401)				
C11H10	2-METHYLNAPHTHALENE	(G)	142.190	27.75	TC 448)	90.83	TC 448)	51.66	-37.866
C11H10O	1-PHENYL-1-PENTYN-3-ONE . .	(L)	158.190	9.7	C(744)				
C11H10O	1,4,4A,8A-TETRAHYDRO-1,4- METHANONAPHTHALENE-5,8-DIONE	(S)	158.190	-39.2	C(843)				
C11H10O2	ETHYL PHENYLPROPIOLATE . .	(L)	174.190	-37.4	C(744)				
C11H10O2	2-METHYL-1,4-NAPHTHO- HYDROQUINONE	(S)	174.190	-99.0	C(112)				
C11H10O2	5-PHENYL-2,4-PENTADIENOIC ACID (M.P. 165 C)	(S)	174.190	-65.4	C(744)				
C11H10O2	5-PHENYL-2,4-PENTADIENOIC ACID (M.P. 138 C)	(S)	174.190	-56.6	C(744)				
C11H10O4	ACETYLCOUMARIC ACID, CIS (M.P. 85 C)	(U,S)	206.190	-162.4	C(744)				
C11H10O4	ACETYLCOUMARIC ACID, TRANS (M.P. 154 TO 155 C)	(U,S)	206.190	-166.5	C(744)				
C11H10O4	(HYDROXYMETHYL)PHENYLSUCCINIC ACID GAMMA-LACTONE	(S)	206.190	-179.7	C(744)				
C11H11NO4	N-BENZYLTARTRIMIDE, D . . .	(S)	221.206	-171.7	C(744)				
C11H11NO4	N-BENZYLTARTRIMIDE, D,L . .	(S)	221.206	-171.7	C(744)				
C11H11NO4	N-BENZYLTARTRAMIDE, MESO .	(S)	221.206	-168.7	C(744)				
C11H12N2O	TRYPTOPHAN, L	(S)	204.222	-99.2	A(241)	60.00	MC 241)	-28.54	20.917
C11H12N2O4	HIPPURYLGLYCINE	(S)	236.222	-199.09	A(643)	75.2	MC 642)	-118.35	86.747
C11H12O	3-METHYL-4-PHENYL-								

733

Formula	Name	State	Mol Wt	$\Delta H^\circ_{f\,298}$ (kcal/mole)	Source	S°_{298} (cal/mole °K)	Source	$\Delta G^\circ_{f\,298}$ (kcal/mole)	Log K_p
C11H12O	3-BUTEN-2-ONE 3-METHYL-4-PHENYL-	(S)	160.206	-29.7	C(744)				
C11H12O	3-BUTEN-2-ONE	(L)	160.206	-25.5	C(744)				
C11H12O3	ETHYL-O-HYDROXYCINNAMIC ACID, CIS, (M.P. 101 TO 102 C)	(S)	192.206	-119.2	C(744)				
C11H12O3	ETHYL-HYDROXYCINNAMIC ACID, TRANS (M.P. 133 TO 134 C)	(U,S)	192.206	-125.7	C(744)				
C11H12O5	METHYL 5,6-DIMETHOXYPHTHAL- ALDEHYDATE	(S)	224.206	-181.5	C(744)				
C11H13NO3	N-(O-TOLUOYL)ALANINE	(S)	207.222	-155.9	C(744)				
C11H13NO3	N-(P-TOLUOYL)ALANINE	(S)	207.222	-158.1	C(744)				
C11H13NO4	5-CARBOMETHOXY-2,4-DIMETHYL-3-PYRROLEACRYLIC ACID	(S)	223.222	-189.8	C(1413)				
C11H14	1-PHENYL-2-PENTENE	(L)	146.222	0.2	C(744)				
C11H14	1,2,3,4-TETRAHYDRO-1-METHYLNAPHTHALENE	(L)	146.222	-14.1	C(721)				
C11H14	1,2,3,4-TETRAHYDRO-5-METHYLNAPHTHALENE	(L)	146.222	-14.1	C(606)				
C11H14N2O2	ETHYL 4-(CYANOMETHYL)-3,5-DI-METHYL-2-PYRROLECARBOXYLATE	(S)	206.238	-89.1	C(1414)				
C11H14N2O3	GLYCINE PHENYLALANINE ANHYDRIDE	(S)	222.238	-150.9	B(1177)				
C11H14N2O3	GLYCINE TYROSINE ANHYDRIDE	(S)	222.238	-190.8	B(1177)				
C11H14N2O3	GLYCYLPHENYLALANINE	(S)	222.238	-163.6	B(1176)				
C11H14O	ISOBUTYLACETOPHENONE	(L)	162.222	-52.62	A(251)				
C11H14O	METHYL-1,2,3,4-TETRAHYDRO-NAPHTHOL	(L)	162.222	-58.4	C(609)				
C11H14O	TERT-BUTYLACETOPHENONE	(L)	162.222	-49.91	A(251)				
C11H14O	2,4,5-TRIMETHYLACETOPHENONE	(L)	162.222	-61.39	B(65)				
C11H14O	2,4,6-TRIMETHYLACETOPHENONE	(L)	162.222	-64.96	B(65)				
C11H14O2	(1,2,3,4-TETRAHYDRO-1-METHYL-NAPHTHYL) HYDROPEROXIDE	(L)	178.222	-37.6	C(609)				
C11H14O2	2,3,4,5-TETRAMETHYL BENZOIC ACID	(S)	178.222	-123.74	A(254)				
C11H14O2	ISOBUTYL BENZOATE	(L)	178.222	-100.2	C(744)				
C11H14O2	METHYLEUGENOL	(U,L)	178.222	-53.0	C(744)				

Formula	Name	State	Mol Wt	$\Delta H^\circ_{f\,298}$ (kcal/mole)	Source	S°_{298} (cal/mole °K)	Source	$\Delta G^\circ_{f\,298}$ (kcal/mole)	Log K_p
C11H14O2	METHYLISOEUGENOL	(L,L)	178.222	-64.2	C(744)				
C11H14O2	2,3,4,6-TETRAMETHYL- BENZOIC ACID	(S)	178.222	-121.33	A(254)				
C11H14O2	2,3,5,6-TETRAMETHYL- BENZOIC ACID	(S)	178.222	-120.95	A(254)				
C11H14O3	ISOBUTYL SALICYLATE	(L)	194.222	-146.2	C(744)				
C11H14O8	TETRAMETHYL ALPHA,ALPHA,BETA- BETA,TRIMETHYLENE- TETRACARBOXYLATE	(L,S)	274.222	-342.4	C(744)				
C11H15NO	CYANOCAMPHOR	(L,S)	177.238	-50.3	C(744)				
C11H15NO2	ETHYL 3,5-DIMETHYL-4-VINYL- 2-PYRROLECARBOXYLATE	(S)	193.238	-109.7	C(1413)				
C11H15NO2	ETHYL N-ETHYLCARBANILATE .	(S)	193.238	-110.8	B(984)				
C11H15NO3	ETHYL 3-ACETYL-2,4-DIMETHYL- PYRROLE-5-CARBOXYLATE	(S)	209.238	-153.7	C(1255)				
C11H15NO6	BENZYLAMMONIUM BITARTRATE, MESO	(S)	257.238	-317.0	C(744)				
C11H15NO6	BENZYLAMMONIUM BIRACEMATE . .	(S)	257.238	-315.4	C(744)				
C11H16	PENTAMETHYLBENZENE	(S)	148.238	-32.33	A(1121)	70.22	M(655)	25.64	-18.795
C11H16	PENTAMETHYLBENZENE	(G)	148.238	-17.80	T(418)	106.09	T(418)	29.48	-21.606
C11H16	PENTYLBENZENE	(G)	148.238	-8.23	T(417)	114.47	T(417)	36.55	-26.790
C11H16O	METHYL THYMYL ETHER . . .	(L,L)	164.238	-55.4	C(744)				
C11H16O2	ETHYL 1-METHYLCYCLOHEXENE- 1-METHENE-3-CARBOXYLATE	(L,L)	180.238	-101.4	C(744)				
C11H16O3S	P-PENTYLBENZENESULFONIC ACID	(G)	228.304			131.55	E(69)		
C11H16O4	DIMETHYL SPIROHEPTANE- DICARBOXYLATE . .	(L,L)	212.238	-171.0	C(744)				
C11H16O8	TETRAMETHYL 1,1,3,3- PROPANETETRACARBOXYLATE	(S)	276.238	-378.9	C(744)				
C11H17NO2	ETHYL 4-ETHYL-3,5-DIMETHYL- PYRROLE-2-CARBOXYLATE	(S)	195.254	-125.0	C(1255)				
C11H18	1,5-DIMETHYL-3-ISOPROPENE- CYCLOHEXENE-1	(L,L)	150.254	-31.3	C(744)				
C11H18O2	ETHYL 2-NONYNOATE	(L)	182.254	-98.4	C(744)				
C11H18O2	UNDECYNOIC ACID	(L,S)	182.254	-110.3	C(744)				
C11H18O3	ETHYL METHYL-1-ETHYL-4-								

735

Formula	Name	State	Mol Wt	$\Delta H°_{298}$ (kcal/mole)	Source	$S°_{298}$ (cal/mole °K)	Source	$\Delta G f°_{298}$ (kcal/mole)	Log Kp
C11H18O3	CYCLOPENTANONE CARBOXYLATE	(U,L)	198.254	-183.5	C(744)				
C11H18O3	METHYL-3-ACETYL-2,2-DIMETHYL-								
C11H18O3	CYCLOBUTANE ACETATE	(L)	198.254	-169.6	C(744)				
C11H18O4	DIMETHYL 3-CARBOXY-2,2-DIMETHYL-								
C11H18O4	CYCLOBUTANEACETATE	(L)	214.254	-206.8	C(744)				
C11H18O4	DIMETHYL ALPHA-TANACETONE-								
C11H18O4	DICARBOXYLATE	(U,L)	214.254	-195.6	C(744)				
C11H20	DICYCLOPENTYLMETHANE	(L)	152.270	-55.0	A(806)				
C11H20	DECAHYDRO-2-METHYL-								
C11H20	NAPHTHALENE, TRANS	(L)	152.270	-63.3	A(806)				
C11H20	DECAHYDRO-9-METHYL-								
C11H20	NAPHTHALENE, CIS	(L)	152.270	-58.31	A(307)				
C11H20	DECAHYDRO-9-METHYL-								
C11H20	NAPHTHALENE, TRANS	(L)	152.270	-59.70	A(307)				
C11H20	SPIRO(5.5)UNDECANE	(L)	152.270	-59.5	A(805)				
C11H20	1-UNDECYNE	(G)	152.270	4.92	T(328)	134.67	T(328)	62.29	-45.656
C11H20O2	UNDECENOIC ACID	(U,S)	184.270	-136.7	C(744)				
C11H20O4	DIMETHYL AZELATE	(L)	216.270	-228.3	C(744)				
C11H20O4	OCTYLMALONIC ACID	(S)	216.270	-264.9	B(1535)				
C11H20O4	UNDECANEDIOIC ACID	(S)	216.270	-261.1	C(744)				
C11H22	1-TERT-BUTYL-2-METHYL-								
C11H22	CYCLOHEXANE	(L)	154.286	-61.4	C(1338)				
C11H22	1-TERT-BUTYL-3-METHYL-								
C11H22	CYCLOHEXANE (HB,L)		154.286	-68.4	C(1338)				
C11H22	1-TERT-BUTYL-3-METHYL-								
C11H22	CYCLOHEXANE (LB,L)		154.286	-68.4	C(1338)				
C11H22	1-TERT-BUTYL-4-METHYL-								
C11H22	CYCLOHEXANE (HB,L)		154.286	-67.4	C(1338)				
C11H22	1-TERT-BUTYL-4-METHYL-								
C11H22	CYCLOHEXANE (LB,L)		154.286	-73.4	C(1338)				
C11H22	1-CYCLOPENTYLHEXANE	(G)	154.286	-50.07	T(357)	127.66	T(357)	18.69	-13.702
C11H22	CYCLOUNDECANE	(L)	154.286	-57.45	A(271)				
C11H22	1,1-DIMETHYL-								
C11H22	2-HEXYLCYCLOPROPANE	(L)	154.286	-46.2	A(710)				
C11H22	PENTYLCYCLOHEXANE	(G)	154.286	-55.88	T(384)	118.89	T(384)	15.50	-11.360
C11H22	1-UNDECENE	(L)	154.286			109.3	M(950)		

Formula	Name	State	Mol Wt	$\Delta H f^\circ_{298}$ (kcal/mole)	Source	S°_{298} (cal/mole °K)	Source	$\Delta G f^\circ_{298}$ (kcal/mole)	Log $K p$
C11H22	1-UNDECENE	(G)	154.286	-34.60	T(294)	138.48	T(294)	30.94	-22.676
C11H22N2O3	VALINE LEUCINE ANHYDRIDE	(S)	230.302	-218.0	B(1177)				
C11H22O	4-METHYL-1-DECEN-4-OL	(L)	170.286	-119.1	C(744)				
C11H22O2	METHYL DECANOATE	(L)	186.286	-153.20	A(3)				
C11H22O2	UNDECANOIC ACID	(L)	186.286	-176.00	A(3)				
C11H22O2	UNDECANOIC ACID	(L)	186.286	-169.83	A(686)				
C11H24	UNDECANE	(L)	156.302	-78.05	T(531)	109.49	M(420)	5.44	-3.984
C11H24	UNDECANE	(G)	156.302	-64.60	T(531)	139.48	T(251)	9.94	-7.289
C11H24O	UNDECYL ALCOHOL	(G)	172.302	-100.91	T(531)	152.07	T(531)	-22.81	16.722
C11H24O2	BIS(ISOPENTYLOXY)METHANE	(L)	188.302	-144.3	C(744)				
C11H24S	BUTYL HEPTYL SULFIDE	(G)	188.368	-54.77	T(827)	155.14	T(827)	17.38	-12.740
C11H24S	DECYL METHYL SULFIDE	(G)	188.368	-53.97	T(828)	154.28	T(828)	18.44	-13.514
C11H24S	ETHYL NONYL SULFIDE	(G)	188.368	-54.55	T(829)	154.82	T(829)	17.70	-12.971
C11H24S	OCTYL PROPYL SULFIDE	(G)	188.368	-54.56	T(830)	155.14	T(830)	17.59	-12.894
C11H24S	1-UNDECANETHIOL	(G)	188.368	-55.46	T(905)	155.13	T(905)	16.69	-12.236
C12H4N4	2,5-CYCLOHEXADIENE-DELTA-								
C12H4N4	1,ALPHA:4-ALPHA'- DIMALONONITRILE	(S)	204.184	158.9	A(154)				
C12H4N4	2,5-CYCLOHEXADIENE-DELTA-								
C12H4N4	1,ALPHA:4-ALPHA'- DIMALONONITRILE	(G)	204.184	184.0	A(154)	105.0	E(748)	142.15	-104.192
C12H5N7O12	HEXANITRODIPHENYLAMINE	(S)	439.216	9.9	B(981)				
C12H6O3	NAPHTHALIC ANHYDRIDE	(S)	198.168	-75.8	C(744)				
C12H6O12	MELLITIC ACID	(S)	342.168	-545.6	A(156)				
C12H8	ACENAPHTHYLENE	(S)	152.184	44.7	A(156)				
C12H8	ACENAPHTHYLENE	(G)	152.184	61.7	A(95)				
C12H8	BIPHENYLENE	(S)	152.184	84.4	A(95)				
C12H8	BIPHENYLENE	(G)	152.184	115.2	A(1386)				
C12H8CL2	2,2'-DICHLOROBIPHENYL	(S)	223.098	7.34	B(1386)				
C12H8CL2	2,2'-DICHLOROBIPHENYL	(G)	223.098	30.3	A(1386)				
C12H8CL2	4,4'-DICHLOROBIPHENYL	(S)	223.098	3.91	A(1386)				
C12H8CL2	4,4'-DICHLOROBIPHENYL	(G)	223.098	28.7	B(1386)				
C12H8F2	2,2'-DIFLUOROBIPHENYL	(S)	190.184	-70.8	A(1386)				
C12H8F2	2,2'-DIFLUOROBIPHENYL	(G)	190.184	-48.1	B(1386)				
C12H8F2	4,4'-DIFLUOROBIPHENYL	(S)	190.184	-71.0	A(1386)				

Formula	Name	State	Mol Wt	$\Delta H f^\circ_{298}$ (kcal/mole)	Source	S°_{298} (cal/mole °K)	Source	$\Delta G f^\circ_{298}$ (kcal/mole)	Log K_p
C12H8F2	4,4'-DIFLUOROBIPHENYL	(G)	190.184	-49.2	B(1386)				
C12H8N2	PHENAZINE	(S)	180.200	57.6	C(1618)				
C12H8N5	AMMONIUM 2,5-CYCLOHEXADIENE-								
C12H8N5	DELTA-1,ALPHA:4-ALPHA-								
C12H8N5	DIMALONONITRILE ION-RADICAL	(S)	222.224	103.2	A(154)				
C12H8O	DIBENZOFURAN	(S)	168.184	-1.4	A(202)				
C12H8O4	NAPHTHALIC ACID	(S)	216.184	-157.7	C(744)				
C12H8S2	THIANTHRENE	(S)	216.316	43.65	A(1447)				
C12H9N	CARBAZOLE	(S)	167.200	40.4	C(744)				
C12H9NO	2-METHYL-1',2'-NAPHTHO-OXAZOLE	(S)	183.200	-19.2	A(447)				
C12H9NO	2-METHYL-2',1'-NAPHTHO-OXAZOLE	(L)	183.200	-10.5	A(447)				
C12H9NO	2-METHYL-2',3'-NAPHTHO-OXAZOLE	(S)	183.200	-10.8	A(447)				
C12H9NO2	3-NITROBIPHENYL	(S)	199.200	17.7	C(177)				
C12H9NO2	4-NITROBIPHENYL	(S)	199.200	12.3	C(177)				
C12H10	ACENAPHTHENE	(S)	154.200	16.8	A(156)				
C12H10	ACENAPHTHENE	(G)	154.200	37.4	A(156)				
C12H10	BIPHENYL	(S)	154.200	24.02	A(246)	49.2	M(656)	60.75	-44.527
C12H10	BIPHENYL	(L)	154.200	28.5	(CALC)	59.8	M(656)	62.07	-45.495
C12H10	BIPHENYL	(G)	154.200	43.52	T(476)	93.85	T(476)	66.94	-49.063
C12H10N2	AZOBENZENE, CIS	(S)	182.216	86.59	A(244)				
C12H10N2	AZOBENZENE, TRANS	(S)	182.216	76.49	A(244)				
C12H10N2O	AZOXYBENZENE	(S)	198.216	63.4	C(744)				
C12H10N2O	N-NITROSODIPHENYLAMINE	(S)	198.216	43.4	B(984)				
C12H10N2O	P-NITROSODIPHENYLAMINE	(S)	198.216	50.9	B(986)				
C12H10N2O	P-(PHENYLAZO)PHENOL	(S)	198.216	33.1	C(744)				
C12H10N4O4	4,4'-DINITROHYDRAZOBENZENE	(S)	274.232	26.2	A(244)				
C12H10O	PHENYL ETHER	(S)	170.200	-7.57	A(453)	55.91	M(453)	34.46	-25.261
C12H10O	PHENYL ETHER	(L)	170.200	-3.48	A(453)	69.62	M(453)	34.47	-25.262
C12H10O	O-PHENYLPHENOL	(S)	170.200	12.3	C(177)				
C12H10OS	PHENYL SULFOXIDE	(L)	202.266	2.3	B(907)				
C12H10OS	PHENYL SULFOXIDE	(G)	202.266	25.5	B(907)				
C12H1002	METHYL 2-NAPHTHOATE	(S)	186.200	-67.5	C(744)				
C12H1002	1-NAPHTHYL ACETATE	(S)	186.200	-66.8	B(855)				
C12H1002	2-NAPHTHYL ACETATE	(S)	186.200	-70.3	B(855)				
C12H10O2S	PHENYL SULFONE	(L)	218.266	-53.3	B(905)				
C12H10O2S	PHENYL SULFONE	(G)	218.266	-27.9	B(905)				

Formula	Name	State	Mol Wt	$\Delta H f^\circ_{298}$ (kcal/mole)	Source	S°_{298} (cal/mole·°K)	Source	$\Delta G f^\circ_{298}$ (kcal/mole)	Log $K p$
C12H10O4	CINNAMYLIDENEMALONIC ACID	(S)	218.200	-150.7	C(744)				
C12H10O4	QUINHYDRONE	(S)	218.200	-134.7	C(744)				
C12H10O4S2	PHENYL DISULFONE	(L)	282.332	-153.62	A(910)				
C12H10O4S2	PHENYL DISULFONE	(G)	282.332	-114.9	B(910)				
C12H10S2	PHENYL DISULFIDE	(L)	218.332	35.6	B(900)				
C12H10S2	PHENYL DISULFIDE	(G)	218.332	58.3	B(900)				
C12H11N	2-BIPHENYLAMINE	(S)	169.216	29.1	C(177)				
C12H11N	4-BIPHENYLAMINE	(S)	169.216	20.8	C(177)				
C12H11N	DIPHENYLAMINE	(S)	169.216	27.93	A(244)				
C12H11N	DIPHENYLAMINE	(L)	169.216	31.5	C(1547)	67.5	E(1547)	74.25	-54.422
C12H11N	DIPHENYLAMINE	(G)	169.216	48.20	E(843)	97.5	E(843)	82.00	-60.107
C12H11N3	N-(PHENYLAZO)ANILINE	(S)	197.232	79.03	A(401)				
C12H11N3	P-(PHENYLAZO)ANILINE	(S)	197.232	70.96	C(744)				
C12H12	1,2-DIMETHYLNAPHTHALENE	(G)	156.216	19.97	T(451)	97.23	T(451)	51.68	-37.883
C12H12	1,3-DIMETHYLNAPHTHALENE	(G)	156.216	19.55	T(452)	97.86	T(452)	51.08	-37.438
C12H12	1,4-DIMETHYLNAPHTHALENE	(G)	156.216	19.72	T(453)	95.86	T(453)	51.84	-37.999
C12H12	1,5-DIMETHYLNAPHTHALENE	(G)	156.216	19.55	T(454)	95.86	T(454)	51.67	-37.875
C12H12	1,6-DIMETHYLNAPHTHALENE	(G)	156.216	19.72	T(455)	97.86	T(455)	51.25	-37.562
C12H12	1,7-DIMETHYLNAPHTHALENE	(G)	156.216	19.55	T(456)	97.86	T(456)	51.08	-37.438
C12H12	2,3-DIMETHYLNAPHTHALENE	(G)	156.216	19.97	T(457)	98.22	T(457)	51.39	-37.667
C12H12	2,6-DIMETHYLNAPHTHALENE	(G)	156.216	19.72	T(458)	97.68	T(458)	51.30	-37.602
C12H12	2,7-DIMETHYLNAPHTHALENE	(G)	156.216	19.72	T(459)	97.68	T(459)	51.30	-37.602
C12H12	1-ETHYLNAPHTHALENE	(L)	156.216	-0.4	C(606)				
C12H12	1-ETHYLNAPHTHALENE	(G)	156.216	23.10	T(449)	99.94	T(449)	54.01	-39.585
C12H12	2-ETHYLNAPHTHALENE	(G)	156.216	22.92	T(450)	100.56	T(450)	53.64	-39.318
C12H12N2	BENZIDINE	(S)	184.232	16.9	B(1175)				
C12H12N2	2,4-BIPHENYLDIAMINE	(L)	184.232	24.7	C(744)				
C12H12N2	HYDRAZOBENZENE	(S)	184.232	52.9	A(244)				
C12H12N4	1,3-DIPHENYLTETRAZENE	(S)	212.248	98.4	B(974)				
C12H12N4	4-(PHENYLAZO)- M-PHENYLENEDIAMINE	(S)	212.248	60.2	C(744)				
C12H12O	1-PHENYL-1-HEXYN-3-ONE	(L)	172.216	-0.3	C(744)				
C12H12O	HINOKITIOL	(S)	188.216	-81.0	C(1465)				
C12H12O3	BETA-BENZALLEVULINIC ACID	(U,S)	204.216	-124.0	C(744)				
C12H12O3	DELTA-BENZALLEVULINIC ACID	(U,S)	204.216	-121.3	C(744)				
C12H12O6	TRIMETHYL TRIMESATE	(S)	252.216	-245.9	C(744)				

Formula	Name	State	Mol Wt	$\Delta H^\circ_{f\,298}$ (kcal/mole)	Source	S°_{298} (cal/mole °K)	Source	$\Delta G^\circ_{f\,298}$ (kcal/mole)	Log K_p
C12H12O12	1,2,3,4,5,6-CYCLOHEXANEHEXA- CARBOXYLIC ACID	(S)	158.232	-614.8	C(744)				
C12H12O12	HEXACYCLO-(7.2.1.0(2.5).0-(3.10)-								
C12H14	.0(4.8).0(6.12))-DODECANE	(S)	158.232	12.06	A(629)				
C12H14	PHENYLCYCLOHEXENE	(L)	158.232	-2.8	B(177)				
C12H14I2N2	BENZIDINE DIHYDRIODIDE	(S)	440.068	-29.8	C(1175)				
C12H14I2N2	HYDRAZOBENZENE DIHYDRIODIDE	(S)	440.068	38.0	C(1175)				
C12H14N4O6	DESOXYAMALIC ACID	(U,S)	310.264	-283.9	C(744)				
C12H14N4O8	TETRAMETHYLALLOXANTHINE .	(U,S)	342.264	-366.8	C(744)				
C12H14O2	ETHYL ALPHA-METHYLCINNAMATE	(L)	190.232	-77.3	C(744)				
C12H14O2	ETHYL BETA-METHYLCINNAMATE	(L)	190.232	-76.6	C(744)				
C12H14O2	PROPYL CINNAMATE	(L)	190.232	-78.1	C(744)				
C12H14O3	4-ALLYL-2-METHOXYPHENYL ACETATE	(S)	206.232	-107.8	C(744)				
C12H14O3	ETHYL O-HYDROXYMETHYLCINNAMATE	(L)	206.232	-110.0	C(744)				
C12H14O3	O-HYDROXYPROPYLCINNAMIC ACID, CIS (M.P. 83 TO 84 C)(U,S)		206.232	-127.9	C(744)				
C12H14O3	HYDROXYPROPYLCINNAMIC ACID, TRANS (M.P. 105 TO 106 C)(U,S)		206.232	-133.8	C(744)				
C12H14O3	2-METHOXY-4-PROPENYLPHENYL ACETATE	(S)	206.232	-117.4	C(744)				
C12H14O3	METHYL ETHYL-O- HYDROXYCINNAMATE, CIS	(L)	206.232	-109.2	C(744)				
C12H14O3	METHYL ETHYL-O- HYDROXYCINNAMATE, TRANS	(L)	206.232	-113.9	C(744)				
C12H14O4	1-ALLYL-2,5-DIMETHOXY- 3,4-(METHYLENEDIOXY)BENZENE	(S)	222.232	-106.9	C(744)				
C12H14O4	DIETHYL PHTHALATE	(L)	222.232	-183.67	B(982)				
C12H14O4	1,4-DIMETHOXY-2,3-(METHYLENE- DIOXY)-5-PROPENYLBENZENE	(S)	222.232	-117.5	C(744)				
C12H16	DICYCLOHEXADIENE	(U,L)	160.248	6.3	B(90)				
C12H16	2-ETHYL-1-PHENYL-1-BUTENE	(L)	160.248	-7.0	C(744)				
C12H16	1-ETHYL- 1,2,3,4-TETRAHYDRONAPHTHALENE(L)		160.248	-21.0	C(721)				
C12H16	5-ETHYL- 1,2,3,4-TETRAHYDRONAPHTHALENE(L)		160.248	-21.0	C(606)				

740

Formula	Name	State	Mol Wt	$\Delta H f°_{298}$ (kcal/mole)	Source	$S°_{298}$ (cal/mole °K)	Source	$\Delta G f°_{298}$ (kcal/mole)	Log $K p$
C12H16N2	2,5,2',5'-TETRAMETHYL-								
C12H16N2	1,1'-DIPYRIDYL	(S)	188.264	31.60	A(247)				
C12H16N2O3	ALANINE PHENYLALANINE ANHYDRIDE	(S)	236.264	-157.3	B(1177)				
C12H16N2O3	ALANYLPHENYLALANINE	(S)	236.264	-169.8	B(1176)				
C12H16N2O3	ETHYL-4-PROPENYLGUAICOL . . .	(U,S)	192.248	-71.5	C(744)				
C12H16O2	PENTAMETHYLBENZOIC ACID	(S)	192.248	-128.12	A(254)				
C12H16O2	PENTYL BENZOATE	(L)	192.248	-104.3	C(744)				
C12H16O2	1-PHENYL-1,2- CYCLOHEXANEDIOL, CIS	(S)	192.248	-111.80	B(1536)				
C12H16O2	1-PHENYL-1,2- CYCLOHEXANEDIOL, TRANS	(S)	192.248	-110.94	B(1536)				
C12H16O3	1,2,4-TRIMETHOXY- 5-PROPENYLBENZENE	(L)	208.248	-97.7	C(744)				
C12H17NO3	ETHYL 2,4-DIMETHYL-5-PROPIONYL- PYRROLE-3-CARBOXYLATE	(S)	223.264	-159.2	C(1413)				
C12H17NO3	ETHYL 2,4-DIMETHYL-5-PROPIONYL- PYRROLE-3-CARBOXYLATE	(S)	223.264	-111.3	C(1255)				
C12H17NO3	ETHYL 3,5-DIMETHYL-4-PROPIONYL- 2-PYRROLECARBOXYLATE	(S)	223.264	-158.8	C(1413)				
C12H17NO4	DIMETHYL 5-CARBOXY-2,4- DIMETHYLPYRROLE-3-PROPIONATE	(S)	239.264	-199.2	C(1413)				
C12H17NO4	ETHYL 3-(2-CARBOXYETHYL)- 2,4-DIMETHYLPYRROLE- 5-CARBOXYLATE	(S)	239.264	-220.6	C(1255)				
C12H17NO4	ETHYL 2,4-DIMETHYLPYRROLE- 3,5-DICARBOXYLATE	(S)	239.264	-216.0	C(1255)				
C12H17NO5	DIETHYL(HYDROXYMETHYL)-3-METHYL- PYRROLE-2,4-DICARBOXYLATE	(S)	255.264	-248.8	C(1413)				
C12H18	3,9-DODECADIYNE	(L)	162.264	47.00	A(433)				
C12H18	5,7-DODECADIYNE	(L)	162.264	43.09	A(433)				
C12H18	2-ETHYL-1-PHENYLBUTANE	(L)	162.264	-28.8	C(839)				
C12H18	HEXAMETHYLBENZENE	(S)	162.264	-39.19	A(1121)	71.66	M(441)	28.06	-20.570
C12H18	HEXAMETHYLBENZENE	(G)	162.264	-25.26	T(423)	108.12	T(423)	31.12	-22.813
C12H18	HEXYLBENZENE	(G)	162.264	-13.15	T(419)	123.78	T(419)	38.56	-28.267
C12H18	3-PHENYLHEXANE	(L)	162.264	-33.8	C(839)				

741

Formula	Name	State	Mol Wt	$\Delta H f^\circ_{298}$ (kcal/mole)	Source	S°_{298} (cal/mole °K)	Source	$\Delta G f^\circ_{298}$ (kcal/mole)	Log $K p$
C12H18	1,2,3-TRIETHYLBENZENE	(G)	162.264	-16.25	T(420)	121.23	T(420)	36.22	-26.552
C12H18	1,2,4-TRIETHYLBENZENE	(G)	162.264	-16.99	T(421)	123.84	T(421)	34.71	-25.439
C12H18	1,3,5-TRIETHYLBENZENE	(G)	162.264	-17.86	T(422)	121.34	T(422)	34.58	-25.348
C12H18N2O4	DIETHYL 1-AMINO-2,5-DIMETHYL- PYRROLE-3,4-DICARBOXYLATE	(S)	254.280	-177.7	C(1414)				
C12H18O	ETHYL THYMYL ETHER	(L)	178.264	-62.0	C(744)				
C12H18O2	ETHYL 1,3-DIMETHYL-4-CYCLOHEXENE- 3-METHENE-5-CARBOXYLATE	(U,L)	194.264	-109.1	C(744)				
C12H18O3S	P-HEXYLBENZENESULFONIC ACID	(G)	242.330						
C12H18O8	RHAMNOSE TRIACETATE	(S)	290.264	-393.1	C(744)				
C12H19NO2	ETHYL 2,4-DIMETHYL-3-PROPYL- PYRROLE-5-CARBOXYLATE	(S)	209.280	-131.4	C(1413)				
C12H20	2,3-DIMETHYLBUTADIENE, DIMER	(L)	164.280	7.4	C(786)				
C12H20	TETRAHYDRODICYCLOHEXADIENE	(U,S)	164.280	-50.9	B(90)				
C12H20O2	PROPYL 2-NONYNOATE	(L)	196.280	-94.3	C(744)				
C12H20O3	TETRAETHYLSUCCINIC ANHYDRIDE	(S)	212.280	-190.4	B(1538)				
C12H20O4	DIETHYLHEXAHYDROISOPHTHALATE, CIS	(L)	228.280	-223.65	C(1368)				
C12H20O4	DIETHYLHEXAHYDROISOPHTHALATE, TRANS	(L)	228.280	-225.50	C(1368)				
C12H20O6	MANNITETRIACETAL	(U,S)	260.280	-271.7	C(744)				
C12H20O7	TRIETHYL CITRATE	(L)	276.280	-352.8	C(744)				
C12H22	CYCLOPENTYLCYCLOHEPTANE	(L)	166.296	-54.1	A(806)				
C12H22	DICYCLOHEXYL	(L)	166.296	-77.57	B(177)				
C12H22	1-DODECYNE	(G)	166.296	-0.01	T(329)	143.98	T(329)	64.29	-47.126
C12H22	SPIRO(5.6)DODECANE	(L)	166.296	-60.7	A(805)				
C12H22N2O2	3,6-DIISOBUTYL-2,5- PIPERAZINEDIONE	(S)	226.312	-153.4	C(744)				
C12H22N2O2	PENTYLPROPIOLIC ACETAL	(U,L)	198.296	-87.9	C(744)				
C12H22O3	DIETHYLACETIC ANHYDRIDE	(U,S)	214.296	-210.9	C(744)				
C12H22O4	ACETYLENEDICARBOXALDEHYDE- BIS(DIETHYL ACETAL)	(S)	230.296	-153.9	C(744)				
C12H22O4	DIMETHYL SEBACATE	(S)	230.296	-243.1	C(744)				
C12H22O4	DIMETHYL SEBACATE	(L)	230.296	-234.2	C(744)				
C12H22O4	DODECANEDIOIC ACID	(S)	230.296	-268.2	C(744)				
C12H22O4	NONYLMALONIC ACID	(S)	230.296	-272.0	B(1535)				

Formula	Name	State	Mol Wt	$\Delta H^{\circ}_{f\,298}$ (kcal/mole)	Source	S°_{298} (cal/mole °K)	Source	$\Delta G^{\circ}_{f\,298}$ (kcal/mole)	Log K_p
C12H22O4	TETRAETHYLSUCCINIC ACID	(S)	230.296	-261.3	B(1538)				
C12H22O11	CELLOBIOSE	(S)	342.296	-529.4	C(744)				
C12H22O11	BETA-LACTOSE	(S)	342.296	-534.59	AC(228)	92.3	MC 15)	-374.52	274.519
C12H22O11	MALTOSE	(S)	342.296	-528.0	C(744)				
C12H22O11	SUCROSE	(S)	342.296	-531.1	AC(228)	86.1	M(1106)	-369.18	270.605
C12H22O11	TREHALOSE	(S)	342.296	-529.9	C(744)				
C12H24	CYCLODODECANE	(S)	168.312	-73.48	AC(271)				
C12H24	1-CYCLOHEXYL-2-ETHYLBUTANE	(L)	168.312	-82.2	C(839)				
C12H24	3-CYCLOHEXYLHEXANE	(L)	168.312	-82.2	C(839)				
C12H24	1-CYCLOHEXYLHEXANE	(G)	168.312	-60.80	T(385)	128.20	T(385)	17.51	-12.837
C12H24	1-CYCLOPENTYLHEPTANE	(G)	168.312	-55.00	T(358)	136.96	T(358)	20.70	-15.174
C12H24	1-DODECENE	(L)	168.312			117.07	MC(950)		
C12H24	1-DODECENE	(G)	168.312	-39.52	T(295)	147.78	T(295)	32.96	-24.156
C12H24	2-METHYLPROPENE, TRIMER	(L)	168.312	-89.7	C(744)				
C12H24O2	T-BUTYL OCTANOATE	(L)	200.312	-167.1	A(1454)				
C12H24O2	LAURIC ACID	(S)	200.312	-185.28	AC 3)				
C12H24O2	LAURIC ACID	(L)	200.312	-176.51	AC 3)				
C12H24O2	METHYL UNDECANOATE	(L)	216.312	-159.11	AC 3)				
C12H24O3	PEROXYLAURIC ACID	(L)	216.312	-162.7	A(1454)				
C12H24O12	ALPHA-LACTOSE MONOHYDRATE	(S)	360.312	-593.6	AC(226)	99.1	MC 15)	-418.95	307.083
C12H24O12	BETA-LACTOSE MONOHYDRATE	(S)	360.312	-587.8	AC(228)	99.8	MC 15)	-413.36	302.984
C12H26	DODECANE	(L)	170.328	-84.02	A(1200)	117.26	MC(420)	6.86	-5.029
C12H26	DODECANE	(G)	170.328	-69.52	T(252)	148.78	T(252)	11.96	-8.769
C12H26O	DODECYL ALCOHOL	(L)	186.328	-105.84	T(532)	161.38	T(532)	-20.81	15.252
C12H26O3S	1-DODECANESULFONIC ACID	(G)	250.394	-360.1	C(1254)				
C12H26O4	SEC-BUTYLIDENE-BIS(TERT-BUTYLPEROXIDE)	(L)	234.328	-147.93	C(843)				
C12H26O13	TREHALOSE DIHYDRATE	(S)	378.328	-674.5	C(744)				
C12H26S	BUTYL OCTYL SULFIDE	(G)	202.394	-59.84	T(831)	164.45	T(831)	19.40	-14.217
C12H26S	DECYL ETHYL SULFIDE	(G)	202.394	-64.13	T(832)	164.13	T(832)	19.34	-14.177
C12H26S	1-DODECANETHIOL	(G)	202.394	-60.39	T(906)	164.44	T(906)	18.70	-13.706
C12H26S	HEXYL SULFIDE	(G)	202.394	-59.69	T(833)	163.07	T(833)	19.81	-14.519
C12H26S	METHYL UNDECYL SULFIDE	(G)	202.394	-58.90	T(834)	163.59	T(834)	20.44	-14.984
C12H26S	NONYL PROPYL SULFIDE	(G)	202.394	-59.51	T(835)	164.45	T(835)	19.58	-14.349
C12H26S2	HEXYL DISULFIDE	(G)	234.460	-57.56	C(877)	174.15	T(877)	20.91	-15.326
C12H27N	TRIISOBUTYLAMINE	(L)	185.344	-75.6	C(744)				

Formula	Name	State	Mol Wt	$\Delta H_{f\,298}^{\circ}$ (kcal/mole)	Source	S_{298}° (cal/mole °K)	Source	$\Delta G_{f\,298}^{\circ}$ (kcal/mole)	Log K_p
C12F22	DOCOSAFLUOROBICYCLOHEXYL	(S)	562.120	−1047.2	A(291)				
C13H9N	ACRIDINE	(S)	179.210	46.2	C(1618)				
C13H9N	2-BIPHENYLCARBONITRILE	(S)	179.210	57.4	C(177)				
C13H10	FLUORENE	(S)	166.210	21.1	C(744)				
C13H10N2	3-AMINOACRIDINE	(S)	194.226	37.9	C(1618)				
C13H10N2	5-AMINOACRIDINE	(S)	194.226	36.1	C(1618)				
C13H10N2O	ACRIDINE NITRATE	(S)	242.242	−12.2	C(1618)				
C13H10N4	1,5-DIPHENYLTETRAZOLE	(S)	222.242	99.3	B(974)				
C13H10N4	2,5-DIPHENYLTETRAZOLE	(S)	222.242	94.4	B(974)				
C13H10O	BENZOPHENONE	(S)	182.210	−8.3	A(256)				
C13H10O	BENZOPHENONE	(L)	182.210	−4.3	A(256)				
C13H10O	DIBENZOPYRAN	(S)	182.210	−15.26	A(202)				
C13H10O2	PHENYL BENZOATE	(S)	198.210	−52.4	C(744)				
C13H10O2	O-PHENYLBENZOIC ACID	(S)	198.210	−81.8	B(177)				
C13H10O3	DIPHENYL CARBONATE	(S)	214.210	−95.93	A(1357)	66.54	M(1357)	−42.05	30.822
C13H10O3	PHENYL SALICYLATE	(S)	214.210	−104.2	C(185)				
C13H11Cl	CHLORODIPHENYLMETHANE	(S)	202.675	14.1	C(744)				
C13H11N	N-BENZYLIDENEANILINE	(S)	181.226	41.0	C(239)				
C13H11NO	BENZANILIDE	(S)	197.226	−21.5	C(744)				
C13H11NO	BENZOPHENONE OXIME	(S)	197.226	28.7	C(744)				
C13H11N2O3	3-AMINOACRIDINE NITRATE	(S)	243.234	−17.6	C(1618)				
C13H11N2O3	5-AMINOACRIDINE NITRATE	(S)	243.234	−35.9	C(1618)				
C13H11N3	2,8-DIAMINOACRIDINE	(S)	209.242	28.4	C(1618)				
C13H12	DIPHENYLMETHANE	(S)	168.226	16.81	A(1115)	57.2	M(656)	60.86	−44.613
C13H12	DIPHENYLMETHANE	(L)	168.226	21.25	A(1115)	72.1	M(656)	60.86	−44.611
C13H12	M-METHYLBIPHENYL	(L)	168.226	21.81	B(177)				
C13H12	O-METHYLBIPHENYL	(L)	168.226	27.28	B(177)				
C13H12	P-METHYLBIPHENYL	(L)	168.226	14.59	B(177)				
C13H12N2O	CARBANILIDE	(S)	212.242	−19.1	C(744)				
C13H12N2O	1,1-DIPHENYLUREA	(S)	212.242	−26.7	B(982)				
C13H12N4	FORMAZANE	(S)	224.258	109.3	B(974)				
C13H12N4O3	2,8-DIAMINOACRIDINE NITRATE	(S)	272.258	−39.3	C(1618)				
C13H12O	BENZHYDROL	(S)	184.226	−25.16	A(1112)				
C13H12O	2,7-DIMETHYL-4,5-BENZTROPONE	(S)	184.226	−17.40	A(1290)				
C13H13N	4-METHYLDIPHENYLAMINE	(S)	183.242	14.2	B(1472)				

Formula	Name	State	Mol Wt	$\Delta H_{f\,298}^{\circ}$ (kcal/mole)	Source	S_{298}° (cal/mole °K)	Source	$\Delta G_{f\,298}^{\circ}$ (kcal/mole)	Log K_p
C13H13N	N-METHYLDIPHENYLAMINE	(S)	183.242	31.4	B(1472)				
C13H14	2-ETHYL-3-METHYLNAPHTHALENE	(G)	170.242	15.72	T(462)	109.33	T(462)	53.54	-39.242
C13H14	2-ETHYL-6-METHYLNAPHTHALENE	(G)	170.242	14.65	T(463)	108.79	T(463)	52.63	-38.576
C13H14	2-ETHYL-7-METHYLNAPHTHALENE	(G)	170.242	14.65	T(464)	108.79	T(464)	52.63	-38.576
C13H14	1-PROPYLNAPHTHALENE	(L)	170.242	2.8	C(606)				
C13H14	1-PROPYLNAPHTHALENE	(G)	170.242	17.85	T(460)	109.55	T(460)	55.60	-40.755
C13H14	2-PROPYLNAPHTHALENE	(G)	170.242	17.65	T(461)	110.18	T(461)	55.21	-40.471
C13H14N2O	TETRAHYDROCARBAZOLE CARBOXAMIDE	(S)	214.258	-256.0	B(1472)				
C13H14N2O	5-METHYL-1-PHENYL-1-HEXYN- 3-ONE	(L)	186.242	20.0	C(744)				
C13H14O	7-PHENYL-4-HEPTYN-3-ONE	(L)	186.242	-14.1	C(744)				
C13H14O4	1,2-INDANYLENE DIACETATE, CIS	(S)	234.242	-185.1	B(1537)				
C13H14O4	1,2-INDANYLENE DIACETATE, TRANS	(L)	234.242	-181.6	B(1537)				
C13H16O3	BUTYL O-HYDROXYCINNAMIC ACID, CIS (M.P. 53 TO 54 C)	(U,S)	220.258	-129.3	C(744)				
C13H16O3	BUTYL-HYDROXYCINNAMIC ACID, TRANS (M.P. 89 TO 90 C)	(U,S)	220.258	-135.7	C(744)				
C13H17NO5	ETHYL 5-CARBETHOXY-2,4- DIMETHYLPYRROLE-3-GLYOXYLATE	(S)	267.274	-238.6	C(1413)				
C13H19NO5	DIETHYL 5-(ALPHA-HYDROXY- ETHYL)-3-METHYL-2,4- PYRROLEDICARBOXYLATE	(S)	269.290	-266.8	C(1413)				
C13H20	2,4-DIMETHYL-1-PHENYLPENTANE	(L)	176.290	-45.7	C(839)				
C13H20	2,4-DIMETHYL-3-PHENYLPENTANE	(L)	176.290	-40.7	C(839)				
C13H20	1-PHENYLHEPTANE	(G)	176.290	-18.08	T(424)	133.09	T(424)	40.57	-29.737
C13H20O	6,10-DIMETHYL-3,5,9- UNDECATRIEN-2-ONE	(L)	192.290	-51.2	C(744)				
C13H20O	4-(2,6,6-TRIMETHYL-1- CYCLOHEXEN-1-YL)-3-BUTEN-2-ONE	(L)	192.290	-62.5	C(744)				
C13H20O	4-(2,6,6-TRIMETHYL-2- CYCLOHEXEN-1-YL)-3-BUTEN-2-ONE	(L)	192.290	-67.0	C(744)				
C13H20O3S	P-HEPTYLBENZENESULFONIC ACID	(G)	256.356	-71.3	C(1335)	149.89	E(69)		
C13H24	DICYCLOHEXYLMETHANE	(L)	180.322	-71.3	C(517)				
C13H24	2-METHYLBICYCLOHEXYL	(HB,L)	180.322	-46.3	C(517)				
C13H24	2-METHYLBICYCLOHEXYL	(LB,L)	180.322	-61.3	C(517)				

Formula	Name	State	Mol Wt	ΔHf°_{298} (kcal/mole)	Source	S°_{298} (cal/mole °K)	Source	ΔGf°_{298} (kcal/mole)	Log Kp
C13H24	1-TRIDECYNE	(G)	180.322	-4.93	T(330)	153.29	T(330)	66.31	-48.603
C13H2402	HEXYLPROPIOLIC ACETAL	(U,L)	212.322	-94.4	C(744)				
C13H2404	DECYLMALONIC ACID	(S)	244.322	-279.1	B(1535)				
C13H2404	TRIDECANEDIOIC ACID	(S)	244.322	-272.5	C(744)				
C13H26	3-CYCLOHEXYL-2,4-DIMETHYL-PENTANE	(L)	182.338	-84.1	C(839)				
C13H26	1-CYCLOHEXYLHEPTANE	(L)	182.338	-84.51	A(1113)	106.8	M(1113)	9.90	-7.253
C13H26	1-CYCLOHEXYLHEPTANE	(G)	182.338	-65.73	T(386)	137.51	T(386)	19.52	-14.307
C13H26	1-CYCLOPENTYLOCTANE	(G)	182.338	-59.92	T(359)	146.27	T(359)	22.72	-16.651
C13H26	CYCLOTRIDECANE	(L)	182.338	-74.21	A(271)				
C13H26	1-TRIDECENE	(G)	182.338	-44.45	T(296)	157.09	T(296)	34.96	-25.626
C13H2602	METHYL DODECANOATE	(L)	214.338	-165.82	A(3)				
C13H2602	TRIDECANOIC ACID	(S)	214.338	-192.97	A(3)				
C13H2602	TRIDECANOIC ACID	(L)	214.338	-182.67	A(3)				
C13H28	TRIDECANE	(L)	184.354	-90.27	E(1203)	124.97	M(420)	8.02	-5.881
C13H28	TRIDECANE	(G)	184.354	-74.45	T(253)	158.09	T(253)	13.97	-10.239
C13H280	1-TRIDECANOL	(G)	200.354	-110.77	T(533)	170.37	T(533)	-18.71	13.712
C13H28S	BUTYL NONYL SULFIDE	(G)	216.420	-64.62	T(836)	173.76	T(836)	21.40	-15.687
C13H28S	DECYL PROPYL SULFIDE	(G)	216.420	-64.44	T(837)	173.76	T(837)	21.58	-15.819
C13H28S	DODECYL METHYL SULFIDE	(G)	216.420	-63.82	T(838)	172.90	T(838)	22.46	-16.461
C13H28S	ETHYL UNDECYL SULFIDE	(G)	216.420	-64.40	T(839)	173.44	T(839)	21.72	-15.918
C13H28S	1-TRIDECANETHIOL	(G)	216.420	-65.31	T(907)	173.75	T(907)	20.71	-15.183
C14H8N10016	BIS(TRINITROPHENYL)-	(S)	572.284	32.7	B(981)				
C14H8N10016	N,N'-DINITROETHYLENEDIAMINE	(S)	208.204	-49.72	C(913)				
C14H802	ANTHRAQUINONE	(S)	208.204	-55.30	C(913)				
C14H802	PHENANTHRENEQUINONE	(S)	208.204	-107.0	C(744)				
C14H803	HYDROXYANTHRAQUINONE	(S)	224.204	-139.0	C(744)				
C14H804	1,2-DIHYDROXYANTHRAQUINONE	(S)	240.204	-186.8	C(744)				
C14H805	1,2,4-TRIHYDROXYANTHRAQUINONE	(L)	256.204						
C14H808	1,2,3,5,6,7-HEXAHYDROXYANTHRA-QUINONE	(S)	304.204	-340.0	C(744)				
C14H9NO4	4-NITROBENZIL	(S)	255.220	-22.0	C(744)				
C14H10	ANTHRACENE	(S)	178.220	30.87	A(246)	49.58	M(655)	68.30	-50.060
C14H10	ANTHRACENE	(G)	178.220	55.17	A(246)				
C14H10	DIPHENYLACETYLENE	(S)	178.220	76.4	A(432)	58.6	M(1388)	111.14	-81.462

Formula	Name	State	Mol Wt	$\Delta H_{f\,298}^{\circ}$ (kcal/mole)	Source	S_{298}° (cal/mole °K)	Source	$\Delta G_{f\,298}^{\circ}$ (kcal/mole)	Log K_p
C14H10	PHENANTHRENE	(S)	178.220	27.76	AC 246)	50.63	M(655)	64.87	-47.551
C14H10	PHENANTHRENE	(G)	178.220	49.46	AC 246)				
C14H10N2O	DIPHENYLFURAZAN	(S)	222.236	79.9	C(1012)				
C14H10N2O	DIPHENYL-1,2,4-OXADIAZOLE	(S)	222.236	51.9	C(1012)				
C14H10N2O	DIPHENYL-1,3,4-OXADIAZOLE	(S)	222.236	41.9	C(1012)				
C14H10N2O4	4,4'-DINITROSTILBENE, CIS	(S)	270.236	17.00	AC 16)				
C14H10N2O4	4,4'-DINITROSTILBENE, TRANS	(S)	270.236	12.40	AC 16)				
C14H10O2	ANTHRACENE-9,10-PEROXIDE	(S)	210.220	19.7	AC 102)				
C14H10O2	BENZIL	(S)	210.220	-36.91	A(1114)				
C14H10O2	9,10-PHENANTHRENE-2,5-DIOL	(S)	210.220	-56.8	CC 744)				
C14H10O3	BENZOIC ANHYDRIDE	(S)	226.220	-103.1	BC 163)				
C14H10O3	DISALICYLIC ALDEHYDE	(U,S)	226.220	-69.1	CC 744)				
C14H10O4	BENZOYL PEROXIDE	(S)	242.220	-93.8	BC 163)				
C14H10O5	SALICYLOSALICYLIC ACID	(S)	258.220	-249.7	C(1340)				
C14H11NO2	4-NITROSTILBENE, CIS	(S)	225.236	30.73	AC 16)				
C14H11NO2	4-NITROSTILBENE, TRANS	(S)	225.236	23.73	AC 16)				
C14H12	9,10-DIHYDROANTHRACENE	(S)	180.236	16.0	BC 914)				
C14H12	1,1-DIPHENYLETHENE	(L)	180.236	40.34	AC 263)				
C14H12	STILBENE, CIS	(L)	180.236	43.00	AC 263)				
C14H12	STILBENE, TRANS	(S)	180.236	32.27	AC 263)	60.0	M(1104)	75.90	-55.630
C14H12N2	BENZALDEHYDE AZINE	(S)	208.252	81.0	CC 239)				
C14H12N2O2	DIBENZOYLHYDRAZINE	(S)	240.252	-51.23	A(1114)				
C14H12N2O2	DIPHENYLGLYOXIME, ALPHA	(S)	240.252	11.7	BC1013)				
C14H12N2O2	DIPHENYLGLYOXIME, BETA	(S)	240.252	6.0	BC1013)				
C14H12N2O2	DIPHENYLGLYOXIME, GAMMA	(S)	240.252	14.4	BC1013)				
C14H12O	P-METHYLBENZOPHENONE	(L)	196.236	-118.6	AC 256)				
C14H12O	2-PHENYLACETOPHENONE	(S)	196.236	-17.11	A(1114)				
C14H12O2	BENZOIN	(S)	212.236	-59.35	A(1114)				
C14H12O2	DIPHENYLACETIC ACID	(S)	212.236	-73.7	CC 744)				
C14H12O2	P-TOLYL BENZOATE	(S)	212.236	-65.1	CC 744)				
C14H12O3	BENZILIC ACID	(S)	228.236	-107.1	CC 744)				
C14H12O4	O-METHOXYPHENYL SALICYLATE	(S)	244.236	-135.6	CC 744)				
C14H13NO	N-ACETYLDIPHENYLAMINE	(S)	211.252	-7.6	B(1472)				
C14H14	BIBENZYL	(L)	182.252	12.31	AC 246)	64.6	M(656)	63.87	-46.815
C14H14	1,1-DIPHENYLETHANE	(L)	182.252	11.7	AC 270)	80.28	M(1388)	58.58	-42.941
C14H14	BI-M-TOLYL	(L)	182.252	6.2	BC 177)				

Formula	Name	State	Mol Wt	$\Delta Hf°_{298}$ (kcal/mole)	Source	$S°_{298}$ (cal/mole °K)	Source	$\Delta Gf°_{298}$ (kcal/mole)	Log K_p
C14H14	BI-P-TOLYL	(S)	182.252	4.8	B(177)				
C14H14N2O	N-ACETYLHYDRAZOBENZENE	(S)	226.268	-2.0	C(1175)				
C14H14N2O	4,4'-(P-AMINOPHENYL)ACETANILIDE	(S)	226.268	-38.8	C(1175)				
C14H14N2O	4,4'-AZODIANISOLE	(S)	226.268	2.9	C(744)				
C14H14N2O	1-METHYL-1,3-DIPHENYLUREA	(S)	226.268	-22.7	B(982)				
C14H14O2	1,2-DIPHENYL-1,2-ETHANEDIOL	(S)	214.252	-70.4	C(744)				
C14H14O2	1,2-DIPHENYL-1,2-ETHANEDIOL	(S)	214.252	-65.6	C(744)				
C14H14O2S	BENZYL SULFONE	(L)	246.318	-63.3	B(905)				
C14H14O2S	BENZYL SULFONE	(G)	246.318	-36.0	B(905)				
C14H14O2S	P-TOLYL SULFONE	(L)	246.318	-74.4	B(905)				
C14H14O2S	P-TOLYL SULFONE	(G)	246.318	-48.2	B(905)				
C14H14O4	ALLYL PHTHALATE	(L)	246.252	-131.5	B(985)				
C14H14S	BENZYL SULFIDE	(L)	246.384	23.7	B(900)				
C14H14S	BENZYL SULFIDE	(G)	246.384	46.0	B(900)				
C14H15N	DIBENZYLAMINE	(S)	197.268	25.7	C(744)				
C14H15N	DI-P-TOLYLAMINE	(S)	197.268	-0.1	B(1472)				
C14H16	1-BUTYLNAPHTHALENE	(G)	184.268	12.68	T(465)	118.83	T(465)	57.38	-42.056
C14H16	2-BUTYLNAPHTHALENE	(G)	184.268	12.50	T(466)	119.46	T(466)	57.01	-41.786
C14H16N4O	M-AZOXYTOLUIDINE	(U,S)	256.300	41.6	C(744)				
C14H16O	1-PHENYL-1-OCTYN-3-ONE	(L)	200.268	-29.5	C(744)				
C14H16O4	1,2,3,4-TETRAHYDRO-1,2-NAPHTHYLENE DIACETATE, CIS	(S)	248.268	-194.2	B(1537)				
C14H16O4	1,2,3,4-TETRAHYDRO-1,2-NAPHTHYLENE DIACETATE, TRANS	(S)	248.268	-195.8	B(1537)				
C14H18O	1-PHENYL-2-OCTYN-1-OL	(L)	202.284	-39.7	C(744)				
C14H18O3	O-HYDROXY-ISOPENTYLCINNAMIC ACID, CIS(M.P. 80 TO 80.5 C)(U,S)	(S)	234.284	-137.1	C(744)				
C14H18O3	HYDROXY-ISOPENTYLCINNAMIC ACID, TRANS (M.P. 79 TO 79.5 C)(U,S)		234.284	-138.7	C(744)				
C14H19NO	N-ACETYLCYCLOHEXYLPHENYLAMINE	(S)	217.300	-69.5	B(1472)				
C14H19N3O4	GLYCYLALANYLPHENYLALANINE	(S)	293.316	-221.5	B(1176)				
C14H20	1,8-CYCLOTETRADECADIYNE	(S)	188.300	35.23	A(449)				
C14H20	1,8-CYCLOTETRADECADIYNE	(G)	188.300	74.91	B(449)				
C14H20N2O3	VALINE PHENYLALANINE ANHYDRIDE	(S)	264.316	-162.2	B(1177)				
C14H20N2O3	VALYLPHENYLALANINE	(S)	264.316	-183.1	B(1178)				

Formula	Name	State	Mol Wt	$\Delta H f^°_{298}$ (kcal/mole)	Source	$S^°_{298}$ (cal/mole °K)	Source	$\Delta G f^°_{298}$ (kcal/mole)	Log Kp
C14H20O8	TETRAETHYL ETHENETETRACARBOXYLATE	(S)	316.300	-372.2	C(744)				
C14H20O8	2,4-DIETHYL-3,5-DIPROPIONYL-								
C14H21NO2	PYRROLE	(S)	235.316	-110.7	C(1413)				
C14H21NO5	DIETHYL 5-(1-HYDROXY- PROPYL)-3-METHYL-								
C14H21NO5	2,4-PYRROLEDICARBOXYLATE	(S)	283.316	-254.7	C(1414)				
C14H22	1-PHENYLOCTANE	(G)	190.316	-23.00	T(425)	142.40	T(425)	42.59	-31.214
C14H22	1,2,3,4-TETRAETHYLBENZENE	(G)	190.316	-29.46	T(426)	138.55	T(426)	37.27	-27.321
C14H22	1,2,3,5-TETRAETHYLBENZENE	(G)	190.316	-29.36	T(427)	139.99	T(427)	36.94	-27.079
C14H22	1,2,4,5-TETRAETHYLBENZENE	(G)	190.316	-29.46	T(428)	139.03	T(428)	37.13	-27.216
C14H22O3S	P-OCTYLBENZENESULFONIC ACID	(G)	270.382			159.06	E(69)		
C14H22O8	TETRAETHYL ETHANETETRACARBOXYLATE, SYM	(S)	318.316	-405.9	C(744)				
C14H24	PERHYDROANTHRACENE TRANS-ANTI-TRANS	(S)	192.332	-70.31	A(931)				
C14H24	PERHYDROANTHRACENE TRANS-ANTI-TRANS	(G)	192.332	-52.93	B(931)				
C14H24	PERHYDROANTHRACENE TRANS-SYN-TRANS	(S)	192.332	-79.21	A(931)				
C14H24	PERHYDROANTHRACENE TRANS-SYN-TRANS	(G)	192.332	-58.32	B(931)				
C14H26	BICYCLOHEPTANE	(L)	194.348	-67.5	A(806)				
C14H26	1,2-DICYCLOHEXYLETHANE	(L)	194.348	-90.1	B(177)				
C14H26	3,3-DIMETHYLBICYCLOHEXYL	(L)	194.348	-98.9	C(744)				
C14H26	1-TETRADECYNE	(G)	194.348	-9.86	T(331)	162.60	T(331)	68.31	-50.073
C14H26O3	HEPTANOIC ANHYDRIDE	(S)	242.348	-219.2	C(744)				
C14H26O4	UNDECYLMALONIC ACID	(S)	258.348	-286.2	B(1535)				
C14H28	1-CYCLOHEXYLOCTANE	(G)	196.364	-70.65	T(387)	146.82	T(387)	21.53	-15.784
C14H28	1-CYCLOPENTYLNONANE	(G)	196.364	-64.85	T(360)	155.58	T(360)	24.72	-18.121
C14H28	CYCLOTETRADECANE	(S)	196.364	-89.15	A(449)				
C14H28	CYCLOTETRADECANE	(G)	196.364	-56.94	B(449)				
C14H28	1-TETRADECENE	(G)	196.364	-49.36	T(297)	166.40	T(297)	36.99	-27.111
C14H28O2	TERT-BUTYL DECANOATE	(L)	228.364	-180.1	A(1454)				
C14H28O2	METHYL TRIDECANOATE	(L)	228.364	-171.73	A(3)				
C14H28O2	MYRISTIC ACID	(S)	228.364	-199.7	A(1454)				

Formula	Name	State	Mol Wt	$\Delta H f^\circ_{298}$ (kcal/mole)	Source	S°_{298} (cal/mole °K)	Source	$\Delta G f^\circ_{298}$ (kcal/mole)	Log $K p$
C14H28O2	MYRISTIC ACID	(S)	228.364	-199.39	A(3)				
C14H28O2	MYRISTIC ACID	(L)	228.364	-188.70	A(3)				
C14H28O3	TERT-BUTYL PEROXYCAPRATE	(L)	244.364	-164.7	A(1454)				
C14H28O3	PEROXYMYRISTIC ACID	(S)	244.364	-179.4	A(1454)				
C14H30	TETRADECANE	(L)	198.380	-96.38	E(1203)	132.74	M(420)	9.31	-6.823
C14H30	TETRADECANE	(G)	198.380	-79.38	T(254)	167.40	T(254)	15.97	-11.709
C14H30O	1-TETRADECANOL	(G)	214.380	-115.70	T(534)	179.68	T(534)	-16.70	12.242
C14H30S	BUTYL DECYL SULFIDE	(G)	230.446	-69.55	T(840)	183.07	T(840)	23.41	-17.157
C14H30S	DODECYL ETHYL SULFIDE	(G)	230.446	-69.33	T(841)	182.75	T(841)	23.72	-17.388
C14H30S	HEPTYL SULFIDE	(G)	230.446	-69.54	T(842)	181.69	T(842)	23.83	-17.466
C14H30S	METHYL TRIDECYL SULFIDE	(G)	230.446	-68.75	T(843)	182.21	T(843)	24.46	-17.931
C14H30S	PROPYL UNDECYL SULFIDE	(G)	230.446	-69.36	T(844)	183.07	T(844)	23.60	-17.296
C14H30S	1-TETRADECANETHIOL	(G)	230.446	-70.24	T(908)	183.06	T(908)	22.72	-16.654
C14H30S2	HEPTYL DISULFIDE	(G)	262.512	-67.41	T(878)	192.77	T(878)	24.93	-18.274
C15H10O	3-PHENYLPROPIOLOPHENONE	(S)	206.230	35.3	C(744)				
C15H11NO	ALPHA,GAMMA-DIPHENYLISOXAZOLE	(S)	221.246	37.5	C(1467)				
C15H12O2	1,3-DIPHENYL-1,3-PROPANEDIONE	(S)	224.246	-53.26	A(1114)				
C15H14	1,1-DIPHENYLCYCLOPROPANE	(L)	194.262	44.3	A(804)				
C15H14	1,2-DIPHENYLCYCLOPROPANE, CIS	(L)	194.262	42.7	A(804)				
C15H14	1,2-DIPHENYLCYCLOPROPANE, TRANS	(L)	194.262	39.7	A(804)				
C15H14	ALPHA,BETA-METHYLPHENYLSTYRENE	(U,S)	194.262	40.2	C(744)				
C15H14O	1,3-DIPHENYL-2-PROPANONE	(S)	210.262	-20.3	B(1403)				
C15H14O	P-ETHYLBENZOPHENONE	(L)	210.262	-15.40	A(256)				
C15H14O2	O-XYLENYL BENZOATE	(U,S)	226.262	-73.1	C(744)				
C15H15NO2	ETHYL DIPHENYLCARBAMATE	(S)	241.278	-80.8	B(984)				
C15H16	1,1-DIPHENYLPROPANE	(L)	196.278	11.8	C(1335)				
C15H16	1,2-DIPHENYLPROPANE	(L)	196.278	6.8	C(1335)				
C15H16	1,3-DIPHENYLPROPANE	(L)	196.278	11.8	C(1335)				
C15H16	2,2-DIPHENYLPROPANE	(L)	196.278	21.8	C(1335)				
C15H16	2-ETHYLDIPHENYLMETHANE	(L)	196.278	6.8	C(841)				
C15H16	3-ETHYLDIPHENYLMETHANE	(L)	196.278	1.8	C(841)				
C15H16	4-ETHYLDIPHENYLMETHANE	(L)	196.278	1.8	C(841)				
C15H16	(M-ETHYLPHENYL)PHENYLMETHANE	(L)	196.278	1.8	C(841)				
C15H16	(O-ETHYLPHENYL)PHENYLMETHANE	(L)	196.278	6.8	C(841)				
C15H16	(P-ETHYLPHENYL)PHENYLMETHANE	(L)	196.278	1.8	C(841)				

Formula	Name	State	Mol Wt	$\Delta H f^\circ_{298}$ (kcal/mole)	Source	S°_{298} (cal/mole °K)	Source	$\Delta G f^\circ_{298}$ (kcal/mole)	Log Kp
C15H16	ALPHA-METHYLBIBENZYL	(L)	196.278	6.8	C(1335)				
C15H16	2-ISOPROPYLBIPHENYL	(L)	196.278	11.8	C(516)				
C15H16	2-PROPYLBIPHENYL	(L)	196.278	11.8	C(516)				
C15H16N2O	1,3-DIMETHYL-1,3-DIPHENYLUREA	(S)	240.294	-14.6	B(982)				
C15H16N2O	3-ETHYL-1,1-DIPHENYLUREA . .	(S)	240.294	-33.5	B(982)				
C15H16N2O2	3-(2-HYDROXYETHYL)-								
	1,1-DIPHENYLUREA . .	(S)	256.294	-72.9	B(1471)				
C15H16N2O2	4,4'-ISOPROPYLIDENEDIPHENOL .	(S)	228.278	-88.2	A(634)				
C15H18	CADALENE	(L)	198.294	-32.2	C(1139)				
C15H18	GUAIAZULENE	(S)	198.294	-14.4	A(802)				
C15H18	1-PENTYLNAPHTHALENE	(L)	198.294	-11.0	A(606)				
C15H18	1-PENTYLNAPHTHALENE	(G)	198.294	7.75	T(467)	128.26	T(467)	59.35	-43.500
C15H18	2-PENTYLNAPHTHALENE	(G)	198.294	7.57	T(468)	128.89	T(468)	58.98	-43.230
C15H18	ALPHA-TRICYCLOPENTADIENE . .	(S)	198.294	30.2	B(90)				
C15H20	DIHYDRO-ALPHA-								
	TRICYCLOPENTADIENE	(U,S)	200.310	-2.3	B(90)				
C15H20	DIHYDRO-BETA-								
	TRICYCLOPENTADIENE	(U,S)	200.310	-4.5	B(90)				
C15H20	1,2,3-PROPANETRIYL								
C15H20O6	TRICYCLOBUTANECARBOXYLATE .	(L)	296.310	-249.0	C(744)				
C15H20O6	1,2,3,4-TETRAHYDRO-								
	1-PENTYLNAPHTHALENE	(L)	202.326	-36.5	C(721)				
C15H22	TETRAHYDRO-ALPHA-								
	TRICYCLOPENTADIENE	(U,S)	202.326	-26.4	B(90)				
C15H22	TETRAHYDRO-BETA-								
	TRICYCLOPENTADIENE	(U,S)	202.326	-30.0	B(90)				
C15H24	1-PHENYLNONANE	(G)	204.342	-27.93	T(429)	151.71	T(429)	44.59	-32.685
C15H24O3S	P-NONYLBENZENESULFONIC ACID	(G)	284.408			168.23	E(69)		
C15H26O6	1,2,3-PROPANETRIYL								
	TRIBUTYRATE	(L)	302.358	-357.8	C(744)				
C15H28	CYCLOHEXYL(2-ETHYLCYCLOHEXYL)-								
	METHANE .	(HB,L)	208.374	-84.9	C(841)				
C15H28	CYCLOHEXYL(2-ETHYLCYCLOHEXYL)-								
	METHANE .	(LB,L)	208.374	-69.9	C(841)				
C15H28	CYCLOHEXYL(3-ETHYLCYCLOHEXYL)-								
	METHANE .	(L)	208.374	-69.9	C(841)				

751

Formula	Name	State	Mol Wt	$\Delta H_{f\,298}^{\circ}$ (kcal/mole)	Source	S_{298}° (cal/mole °K)	Source	$\Delta G_{f\,298}^{\circ}$ (kcal/mole)	Log K_p
C15H28	CYCLOHEXYL(4-ETHYLCYCLOHEXYL)-								
C15H28	METHANE (HB,L)		208.374	-84.9	C(841)				
C15H28	CYCLOHEXYL(4-ETHYLCYCLOHEXYL)-								
C15H28	METHANE (LB,L)		208.374	-69.9	C(841)				
C15H28	1,1-DICYCLOHEXYLPROPANE	(L)	208.374	-79.9	C(1335)				
C15H28	1,2-DICYCLOHEXYLPROPANE	(L)	208.374	-54.9	C(1335)				
C15H28	1,3-DICYCLOHEXYLPROPANE	(L)	208.374	-64.9	C(1335)				
C15H28	2,2-DICYCLOHEXYLPROPANE	(L)	208.374	-69.9	C(1335)				
C15H28	4-ETHYLDICYCLOHEXYLMETHANE(HB,L)		208.374	-85.0	C(841)				
C15H28	4-ETHYLDICYCLOHEXYLMETHANE(LB,L)		208.374	-70.0	C(841)				
C15H28	2-ISOPROPYLBICYCLOHEXYL (HB,L)		208.374	-54.9	C(517)				
C15H28	2-ISOPROPYLBICYCLOHEXYL (LB,L)		208.374	-79.9	C(517)				
C15H28	1-PENTADECYNE	(G)	208.374	-14.78	TC(332)	171.91	TC(332)	70.33	-51.550
C15H28	2-PROPYLBICYCLOHEXYL (HB,L)		208.374	-59.9	C(517)				
C15H28	2-PROPYLBICYCLOHEXYL (LB,L)		208.374	-74.9	C(517)				
C15H28O	CYCLOPENTADECANONE	(S)	224.374	-107.2	C(1272)				
C15H28O4	DODECYLMALONIC ACID	(S)	272.374	-293.8	B(1535)				
C15H30	1-CYCLOHEXYLNONANE	(G)	210.390	-75.58	TC(388)	156.12	TC(388)	23.54	-17.257
C15H30	CYCLOPENTADECANE	(S)	210.390	-90.10	AC(271)				
C15H30	1-CYCLOPENTYLDECANE	(L)	210.390	-87.52	AC(880)	128.71	MC(993)	19.78	-14.495
C15H30	1-CYCLOPENTYLDECANE	(G)	210.390	-69.78	TC(361)	164.89	TC(361)	26.73	-19.591
C15H30	1-PENTADECENE	(G)	210.390	-54.31	TC(298)	175.71	TC(298)	38.97	-28.566
C15H30O2	METHYL TETRADECANOATE	(L)	242.390	-177.98	AC(3)				
C15H30O2	PENTADECANOIC ACID	(S)	242.390	-206.16	AC(3)				
C15H30O2	PENTADECANOIC ACID	(L)	242.390	-194.21	AC(3)				
C15H32	PENTADECANE	(L)	212.406	-102.51	E(1248)	140.41	MC(420)	10.60	-7.771
C15H32	PENTADECANE	(G)	212.406	-84.31	TC(255)	176.71	TC(255)	17.98	-13.179
C15H32N2O	1,3-DIHEPTYL UREA	(S)	256.422	-150.60	B(218)				
C15H32O	1-PENTADECANOL	(G)	228.406	-120.62	TC(535)	188.99	TC(535)	-14.69	10.765
C15H32S	BUTYL UNDECYL SULFIDE	(G)	244.472	-74.47	TC(845)	192.38	TC(845)	25.42	-18.634
C15H32S	DODECYL PROPYL SULFIDE	(G)	244.472	-74.29	TC(846)	192.38	TC(846)	25.60	-18.766
C15H32S	ETHYL TRIDECYL SULFIDE	(G)	244.472	-74.26	TC(847)	192.06	TC(847)	25.73	-18.858
C15H32S	METHYL TETRADECYL SULFIDE	(G)	244.472	-73.68	TC(848)	191.52	TC(848)	26.47	-19.401
C15H32S	1-PENTADECANETHIOL	(G)	244.472	-75.17	TC(909)	192.37	TC(909)	24.73	-18.124
C15H33N	TRIISOPENTYLAMINE	(L)	227.422	-76.5	C(744)				
C15H34N2O2	HEPTYLAMMONIUM								

Formula	Name	State	Mol Wt	$\Delta H^{\circ}_{f\,298}$ (kcal/mole)	Source	S°_{298} (cal/mole °K)	Source	$\Delta G^{\circ}_{f\,298}$ (kcal/mole)	Log Kp
C15H34N2O2	HEPTYL CARBAMATE	(S)	274.438	−222.0	B(218)				
C16H10	DIPHENYLBUTADIYNE	(S)	202.240	123.7	A(266)				
C16H10	FLUORANTHENE	(S)	202.240	45.35	A(1600)	55.09	M(1600)	81.95	−60.064
C16H10	FLUORANTHENE	(S)	202.240	46.0	A(156)	55.09	M(1600)	82.60	−60.541
C16H10	FLUORANTHENE	(G)	202.240	70.4	A(156)				
C16H10	PYRENE (DELTA−2,2′,−BIINDOLINE)−	(S)	202.240	27.41	A(1600)	53.75	M(1600)	64.40	−47.208
C16H10N2O2	3,3′−DIONE	(S)	262.256	−29.9	C(744)				
C16H10N2O3	DIBENZOYLFURAZAN	(S)	278.256	−28.0	C(1012)				
C16H10O3	DIPHENYLMALEIC ANHYDRIDE	(S)	250.240	−76.0	C(744)				
C16H11NO2	BENZALHIPPURIC ACID LACTONE(U,L)		249.256	−25.1	C(744)				
C16H11NO5	P−NITROACETYLBENZOIN	(U,S)	297.256	−14.8	C(744)				
C16H12O2	1,4−DIPHENYL−2−BUTENE−1,4−DIONE	(S)	236.256	−27.55	A(1112)	76.3	M(1104)	26.64	−19.525
C16H12O3	2,3−DIPHENYLSUCCINIC ANHYDRIDE (RACEMIC)	(S)	252.256	−97.3	C(744)				
C16H12O3	2,3−DIPHENYLSUCCINIC ANHYDRIDE, TRANS	(S)	252.256	−98.8	B(1538)				
C16H13N	PHENYL−ALPHA−NAPHTHYLAMINE	(U,S)	219.272	56.1	C(744)				
C16H13N	PHENYL−BETA−NAPHTHYLAMINE	(U,S)	219.272	51.0	C(744)				
C16H13N	N−PHENYL−2−NAPHTHYLAMINE		219.272	38.2	A(1234)				
C16H13NO3	BENZALHIPPURIC ACID	(U,S)	267.272	−97.2	C(744)				
C16H14	2,7−DIMETHYLPHENANTHRENE	(S)	206.272	8.50	A(448)				
C16H14	4,5−DIMETHYLPHENANTHRENE	(S)	206.272	21.05	A(448)				
C16H14	1,4−DIPHENYL−1,3−BUTADIENE, CIS−CIS	(S)	206.272	47.30	A(265)				
C16H14	1,4−DIPHENYL−1,3−BUTADIENE, TRANS−TRANS	(S)	206.272	42.52	A(265)				
C16H14	DI−O−TOLYLACETYLENE	(S)	206.272	55.9	A(265)				
C16H14	DI−P−TOLYLACETYLENE	(S)	206.272	53.8	A(266)				
C16H14O2	1,4−DIPHENYL−1,4−BUTANEDIONE	(S)	238.272	−61.24	A(1112)	77.6	M(1104)	1.87	−1.367
C16H14O3	O−TOLUIC ACID ANHYDRIDE	(S)	254.272	−127.6	B(163)				
C16H14O3	P−TOLUIC ACID ANHYDRIDE	(S)	254.272	−124.6	B(163)				
C16H14O4	2,3−DIPHENYLSUCCINIC ACID, CIS	(S)	270.272	−174.0	B(1538)				

Formula	Name	State	Mol Wt	$\Delta H^°_{298}$ (kcal/mole)	Source	$S^°_{298}$ (cal/mole °K)	Source	$\Delta G f^°_{298}$ (kcal/mole)	Log Kp
C16H14O4	2,3-DIPHENYLSUCCINIC ACID, TRANS	(S)	270.272	-175.3	A(1538)				
C16H14O4	O-TOLYL PEROXIDE	(S)	270.272	-119.7	B(163)				
C16H14O4	P-TOLYL PEROXIDE	(S)	270.272	-108.0	B(163)				
C16H15NO2	N,N-DIPHENYLACETOACETAMIDE	(S)	253.288	-51.2	B(1472)				
C16H15NO3	BENZOYLPHENYLALANINE	(U,S)	269.288	-123.7	C(744)				
C16H16	1,1-DI-O-TOLYLETHYLENE	(S)	208.288	21.0	A(265)				
C16H16	1,1-DI-P-TOLYLETHYLENE	(S)	208.288	20.3	A(265)				
C16H16	1,2-DI-O-TOLYLETHYLENE, CIS	(S)	208.288	20.3	A(265)				
C16H16	1,2-DI-O-TOLYLETHYLENE, TRANS	(S)	208.288	17.7	A(265)				
C16H16	1,2-DI-P-TOLYLETHYLENE, CIS	(S)	208.288	23.2	A(265)				
C16H16	1,2-DI-P-TOLYLETHYLENE, TRANS	(S)	208.288	13.6	A(265)				
C16H16	2,2-PARACYCLOPHANE	(S)	208.288	36.9	A(155)				
C16H16	2,2-PARACYCLOPHANE	(G)	208.288	59.9	B(155)				
C16H16N2	DIBENZYLIDENEETHYLENEDIAMINE	(S)	236.304	65.0	C(239)				
C16H16N2	N-ALLYLCARBANILIDE	(S)	252.304	-6.1	B(1472)				
C16H16N2O2	N,N'-DIACETYLBENZIDINE	(S)	268.304	-113.0	C(1175)				
C16H16N2O2	N,N'-DIACETYLHYDRAZOBENZENE	(S)	268.304	-45.1	C(1175)				
C16H16N2O2	SUCCINANILIDE, SYM	(S)	268.304	-79.3	C(744)				
C16H16O	P-ISOPROPYLBENZOPHENONE	(L)	224.288	-28.4	A(256)				
C16H16O	8,9,10,11-TETRAHYDRO-6,12-METHANO-7H-BENZOCYCLO-UNDECEN-14-ONE	(S)	224.288	14.4	A(1290)				
C16H16O2	BETA-BENZYLHYDROCINNAMIC ACID	(S)	240.288	-95.3	C(744)				
C16H16O2	4,4'-DIMETHOXYSTILBENE	(S)	240.288	-34.4	C(744)				
C16H16O2	PSEUDOCUMENYL BENZOATE	(U,S)	240.288	-81.8	C(744)				
C16H18	2-BUTYLBIPHENYL	(L)	210.304	5.0	C(516)				
C16H18	(P-CUMENYL)PHENYLMETHANE	(L)	210.304	20.0	C(840)				
C16H18	ALPHA,ALPHA'-DIMETHYL-BIBENZYL, DL								
C16H18	ALPHA,ALPHA'-DIMETHYL-BIBENZYL, MESO	(S)	210.304	15.0	C(1337)				
C16H18	1,1-DIPHENYLBUTANE	(S)	210.304	10.0	C(1337)				
C16H18	1,2-DIPHENYLBUTANE	(L)	210.304	10.0	C(1337)				
C16H18	1,2-DIPHENYLBUTANE	(L)	210.304	5.0	C(1337)				
C16H18	1,3-DIPHENYLBUTANE	(L)	210.304	10.0	C(1337)				
C16H18	1,4-DIPHENYLBUTANE	(S)	210.304	-2.58	A(264)				

Formula	Name	State	Mol Wt	$\Delta H^{\circ}_{f\,298}$ (kcal/mole)	Source	S°_{298} (cal/mole °K)	Source	$\Delta G^{\circ}_{f\,298}$ (kcal/mole)	Log K_p
C16H18	1,4-DIPHENYLBUTANE	(L)	210.304	10.0	C(1337)				
C16H18	2,2-DIPHENYLBUTANE	(L)	210.304	10.0	C(1337)				
C16H18	2,3-DIPHENYLBUTANE, MESO	(L)	210.304	10.0	C(1337)				
C16H18	2,3-DIPHENYLBUTANE, DL	(L)	210.304	15.0	C(1337)				
C16H18	1,3-DIPHENYL-2-METHYLPROPANE	(L)	210.304	20.0	C(206)				
C16H18	1,2-DI-O-TOLYLETHANE	(L)	210.304	-9.27	A(264)				
C16H18	1,1-DI-O-TOLYLETHANE	(S)	210.304	-5.93	A(264)				
C16H18	1,2-DI-P-TOLYLETHANE	(S)	210.304	-9.74	A(264)				
C16H18	1,1-DI-P-TOLYLETHANE	(L)	210.304	-7.60	A(264)				
C16H18	ALPHA-ETHYLBIBENZYL	(L)	210.304	5.0	C(1337)				
C16H18	4-ISOPROPYLDIPHENYLMETHANE	(L)	210.304	20.0	C(840)				
C16H18	PHENYL(P-PROPYLPHENYL)METHANE	(L)	210.304	0.0	C(840)				
C16H18	4-PROPYLDIPHENYLMETHANE	(L)	210.304	0.0	C(840)				
C16H18N2O	1-ETHYL-3-METHYL-1,3-DIPHENYLUREA	(S)	254.320	-24.6	B(982)				
C16H18N2O	4,4'-AZODIPHENETOLE	(S)	270.320	-17.9	C(744)				
C16H18N2O2	2,2'-AZOXYDIPHENETOLE	(S)	286.320	-31.9	C(744)				
C16H18N2O3	4,4'-AZOXYDIPHENETOLE	(S)	286.320	-19.6	C(744)				
C16H20N2O5	DIETHYL 3-HYDROXY-5,5'-DIMETHYL(2,3'-BIPYRROLE)-4,4'-DICARBOXYLATE	(S)	320.336	-236.0	C(1414)				
C16H20O4	1-PHENYL-1,2-CYCLOHEXYLENE DIACETATE, CIS	(S)	276.320	-203.7	B(1537)				
C16H22O4	DIBUTYL PHTHALATE	(L)	278.336	-198.3	B(12)				
C16H22O4	ALPHA-GLUCOSE PENTAACETATE, D	(S)	278.336	-537.7	A(230)				
C16H22O4	BETA-GLUCOSE PENTAACETATE, D	(S)	278.336	-533.7	A(230)				
C16H22O11	GALACTOSE PENTAACETATE	(S)	390.336	-531.0	C(744)				
C16H26	PENTAETHYLBENZENE	(G)	218.368	-41.87	T(431)	154.84	T(431)	39.43	-28.901
C16H26	1-PHENYLDECANE	(L)	218.368	-51.58	A(880)				
C16H26	1-PHENYLDECANE	(G)	218.368	-32.86	T(430)	161.02	T(430)	46.60	-34.155
C16H26O3S	P-DECYLBENZENESULFONIC ACID	(G)	298.434			177.40	E(69)		
C16H30	2-BUTYLDICYCLOHEXYL	(LB,L)	222.400	-42.9	C(517)				
C16H30	2-BUTYLDICYCLOHEXYL	(HB,L)	222.400	-42.9	C(517)				
C16H30	1,1-DICYCLOHEXYLBUTANE	(L)	222.400	-57.9	C(1337)				
C16H30	1,2-DICYCLOHEXYLBUTANE	(L)	222.400	-47.9	C(1337)				
C16H30	1,3-DICYCLOHEXYLBUTANE	(L)	222.400	-57.9	C(1337)				

Formula	Name	State	Mol Wt	$\Delta H f^\circ_{298}$ (kcal/mole)	Source	S°_{298} (cal/mole °K)	Source	$\Delta G f^\circ_{298}$ (kcal/mole)	Log Kp
C16H30	1,4-DICYCLOHEXYLBUTANE	(L)	222.400	-57.9	C(1337)				
C16H30	2,2-DICYCLOHEXYLBUTANE	(L)	222.400	-52.9	C(1337)				
C16H30	2,3-DICYCLOHEXYLBUTANE, DL	(L)	222.400	-57.9	C(1337)				
C16H30	2,3-DICYCLOHEXYLBUTANE, MESO	(L)	222.400	-47.9	C(1337)				
C16H30	1,3-DICYCLOHEXYL-2-METHYLPROPANE								
C16H30		(L)	222.400	-42.9	C(206)				
C16H30	1-HEXADECYNE	(G)	222.400	-19.71	T(333)	181.22	T(333)	72.34	-53.020
C16H30	4-ISOPROPYL-								
C16H30	DICYCLOHEXYLMETHANE	(LB,L)	222.400	-62.9	C(840)				
C16H30	4-ISOPROPYL-								
C16H30	DICYCLOHEXYLMETHANE	(HB,L)	222.400	-67.9	C(840)				
C16H3OO4	TRIDECYLMALONIC ACID	(S)	286.400	-300.8	A(1535)				
C16H32	CYCLOHEXADECANE	(S)	224.416	-96.66	A(271)	129.10	M(423)	16.86	-12.358
C16H32	1-CYCLOHEXYLDECANE	(L)	224.416	-100.03	A(880)	165.43	T(389)	25.55	-18.727
C16H32	1-CYCLOHEXYLDECANE	(G)	224.416	-80.51	T(389)	174.20	T(362)	28.74	-21.069
C16H32	1-CYCLOPENTYLUNDECANE	(G)	224.416	-74.70	T(362)	147.9	M(950)	33.95	-24.881
C16H32	1-HEXADECENE	(L)	224.416	-77.34	T(880)	185.02	T(299)	40.99	-30.043
C16H32	1-HEXADECENE	(G)	224.416	-59.23	T(299)				
C16H3202	TERT-BUTYL DODECANOATE	(L)	256.416	-195.6	A(1454)				
C16H3202	METHYL PENTADECANOATE	(L)	256.416	-184.45	A(3)	108.12	M(1622)	-75.54	55.373
C16H3202	PALMITIC ACID	(S)	256.416	-213.30	A(3)	108.12	M(1622)	-73.54	53.907
C16H3202	PALMITIC ACID	(S)	256.416	-211.3	A(853)	108.12	M(1622)	-75.54	55.373
C16H3202	PALMITIC ACID	(S)	256.416	-213.3	A(1454)				
C16H3202	PALMITIC ACID	(G)	256.416	-201.0	A(853)				
C16H3203	TERT-BUTYL PEROXYLAURATE	(L)	256.416	-200.53	A(3)				
C16H3203	PEROXYPALMITIC ACID	(L)	272.416	-176.6	A(1454)				
C16H34	HEXADECANE	(S)	272.416	-191.8	A(1454)	148.09	M(420)	11.46	-8.403
C16H34	HEXADECANE	(L)	226.432	-109.07	A(443)	186.02	T(256)	20.00	-14.656
C16H340	1-HEXADECANOL	(G)	226.432	-89.23	T(256)	108.0	M(1109)	-23.76	17.414
C16H340	1-HEXADECANOL	(S)	242.432	-163.55	A(1116)	145.0	M(1109)	-23.10	16.932
C16H340	1-HEXADECANOL	(L)	242.432	-151.86	A(1116)	198.30	T(536)	-12.67	9.288
C16H340	1-HEXADECANOL	(G)	242.432	-125.54	T(536)	201.69	T(849)	27.43	-20.105
C16H34S	BUTYL DODECYL SULFIDE	(G)	258.498	-79.40	T(849)	201.36	T(850)	27.75	-20.338
C16H34S	ETHYL TETRADECYL SULFIDE	(G)	258.498	-79.18	T(850)	201.67	T(910)	26.74	-19.603
C16H34S	1-HEXADECANETHIOL	(G)	258.498	-80.09	T(910)	200.82	T(851)	28.49	-20.881
C16H34S	METHYL PENTADECYL SULFIDE	(G)	258.498	-78.60	T(851)				

Formula	Name	State	Mol Wt	$\Delta H_{f\,298}^{\circ}$ (kcal/mole)	Source	S_{298}° (cal/mole °K)	Source	$\Delta G_{f\,298}^{\circ}$ (kcal/mole)	Log K_p
C16H34S	OCTYL SULFIDE	(G)	258.498	-79.39	T(852)	200.31	T(852)	27.85	-20.413
C16H34S	PROPYL TRIDECYL SULFIDE	(G)	258.498	-79.22	T(853)	201.69	T(853)	27.61	-20.236
C16H34S2	OCTYL DISULFIDE	(G)	290.564	-77.27	T(879)	211.39	T(879)	28.94	-21.214
C17H12O2	2-NAPHTHYL BENZOATE	(S)	248.266	-44.0	B(855)				
C17H13NO2	3-ANILINO-2-NAPHTHOIC ACID	(S)	263.282	-62.9	B(1640)				
C17H14O	1,5-DIPHENYL-1,4-PENTADIENE-3-ONE	(S)	234.282	12.8	C(744)				
C17H16O3	ALLYL-3,4-GUAIACOL BENZOATE	(U,S)	268.298	-79.3	C(744)				
C17H16O3	4-ALLYL-2-METHOXYPHENYL BENZOATE	(S)	268.298	-79.5	C(744)				
C17H16O3	2-METHOXY-4-PROPENYLPHENYL BENZOATE	(S)	268.298	-88.7	C(744)				
C17H16O3	BETA-TOLYLMETHOXYCINNAMIC ACID (LABILE)	(U,S)	268.298	-102.5	C(744)				
C17H16O3	BETA-TOLYLMETHOXYCINNAMIC ACID (STABLE)	(U,S)	268.298	-106.5	C(744)				
C17H18O	P-TERT BUTYLBENZOPHENONE	(L)	238.314	-32.4	A(256)				
C17H18N2O2	N,N-DIPHENYL-4-MORPHOLINECARBOXAMIDE	(S)	282.330	-48.6	B(1471)				
C17H18O2	THYMYL BENZOATE	(S)	254.314	-84.2	C(744)				
C17H18O2	THYMYL BENZOATE	(L)	254.314	-79.5	C(744)				
C17H20	2-BENZYL-1-PHENYLBUTANE	(L)	224.330	2.6	C(206)				
C17H20	(O-BUTYLPHENYL)PHENYLMETHANE	(L)	224.330	-17.4	C(840)				
C17H20	(O-SEC-BUTYLPHENYL)PHENYLMETHANE	(L)	224.330	-17.4	C(840)				
C17H20	(P-BUTYLPHENYL)PHENYLMETHANE	(L)	224.330	-12.4	C(840)				
C17H20	(P-SEC-BUTYLPHENYL)PHENYLMETHANE	(L)	224.330	-22.4	C(840)				
C17H20	1,1-DIPHENYLPENTANE	(L)	224.330	2.6	C(1336)				
C17H20	1,5-DIPHENYLPENTANE	(L)	224.330	-7.4	C(1336)				
C17H21NO4	MORPHINE MONOHYDRATE	(S)	303.346	-169.7	C(744)				
C17H22N2O2	ETHYL 5-(ANILINOMETHYL)-3-ETHYL-4-METHYL-PYRROLE-2-CARBOXYLATE	(S)	286.362	-85.8	C(1414)				

Formula	Name	State	Mol Wt	ΔH°_{298} (kcal/mole)	Source	S°_{298} (cal/mole °K)	Source	ΔGf°_{298} (kcal/mole)	Log Kp
C17H24N2	4-ETHYL-2-(4-ETHYL-3,5-DIMETHYL-2-PYRRYLMETHYLENE)-3,5-DIMETHYL-ISOPYRROLE	(S)	256.378	-1.9	C(1414)				
C17H24N2O	BIS(4-ETHYL-3,5-DIMETHYL-2-PYRRYL)KETONE	(S)	272.378	-55.4	C(1413)				
C17H25NO6	TRIETHYL 4-(2,2-DICARBOXYETHYL)-3,5-DIMETHYL-PYRROLE-2-CARBOXYLATE	(S)	339.378	-330.1	C(1414)	170.32	T(432)	48.62	-35.634
C17H25NO6		(G)	232.394	-37.78	T(432)	186.57	E(69)		
C17H28O3S	1-PHENYLUNDECANE P-UNDECYLBENZENESULFONIC ACID	(G)	312.460	-106.6	C(1272)				
C17H30O	CYCLOHEPTADECENONE	(S)	250.410						
C17H32	1,3-DICYCLOHEXYL-2-ETHYLPROPANE	(L)	236.426	-89.2	C(206)				
C17H32	1,1-DICYCLOHEXYLPENTANE	(L)	236.426	-89.2	C(1336)				
C17H32	1,5-DICYCLOHEXYLPENTANE	(L)	236.426	-89.2	C(1336)				
C17H32	1-HEPTADECYNE	(G)	236.426	-24.64	T(334)	190.53	T(334)	74.34	-54.490
C17H32O	CYCLOHEPTADECANONE	(S)	252.426	-118.9	C(1272)				
C17H32O4	TETRADECYLMALONIC ACID	(S)	300.426	-307.9	B(1535)				
C17H34	CYCLOHEPTADECANE	(S)	238.442	-103.13	A(271)				
C17H34	1-CYCLOPENTYLDODECANE	(G)	238.442	-80.28	T(363)	183.51	T(363)	30.10	-22.062
C17H34	1-CYCLOHEXYLUNDECANE	(G)	238.442	-85.43	T(390)	174.74	T(390)	27.56	-20.204
C17H34	1-HEPTADECENE	(G)	238.442	-64.15	T(300)	194.33	T(300)	43.00	-31.521
C17H34O2	HEPTADECANOIC ACID	(L)	270.442	-221.17	A(3)				
C17H34O2	HEPTADECANOIC ACID	(L)	270.442	-207.12	A(3)				
C17H34O2	METHYL PALMITATE	(S)	270.442	-114.69	E(1203)	118.33	M(1622)	13.25	-9.710
C17H36	HEPTADECANE	(L)	240.458	-94.15	T(257)	155.83	E(1248)	22.01	-16.133
C17H36	HEPTADECANE	(G)	240.458	-170.0	B(218)	195.33	T(257)		
C17H36N2O	1,3-DIOCTYL UREA	(S)	284.474	-130.47	T(537)	207.61	T(537)	-10.67	7.818
C17H36O	1-HEPTADECANOL	(G)	256.458	-84.32	T(854)	210.99	T(854)	29.45	-21.584
C17H36S	BUTYL TRIDECYL SULFIDE	(G)	272.524	-84.11	T(855)	210.67	T(855)	29.75	-21.808
C17H36S	ETHYL PENTADECYL SULFIDE	(G)	272.524	-85.02	T(911)	210.98	T(911)	28.75	-21.073
C17H36S	1-HEPTADECANETHIOL	(G)	272.524	-83.53	T(856)	210.13	T(856)	30.49	-22.351
C17H36S	HEXADECYL METHYL SULFIDE	(G)	272.524	-84.14	T(857)	210.99	T(857)	29.63	-21.716
C17H36S	PROPYL TETRADECYL SULFIDE								
C17H38N2O2	OCTYLAMMONIUM OCTYL CARBAMATE	(S)	302.490	-243.0	B(218)				

Formula	Name	State	Mol Wt	$\Delta Hf°_{298}$ (kcal/mole)	Source	$S°_{298}$ (cal/mole °K)	Source	$\Delta Gf°_{298}$ (kcal/mole)	Log Kp
C18H10O2	BENZANTHRAQUINONE	(S)	258.260	-55.6	C(913)				
C18H10O2	TETRACENEQUINONE	(S)	258.260	-34.3	C(913)				
C18H12	1,2-BENZOANTHRACENE	(S)	228.276	40.9	B(914)				
C18H12	3,4-BENZOPHENANTHRACENE	(S)	228.276	44.3	B(914)				
C18H12	CHRYSENE	(S)	228.276	34.8	B(914)				
C18H12	NAPHTHACENE	(S)	228.276	38.0	B(914)	51.48	M(1624)	85.79	-62.882
C18H12	TRIPHENYLENE	(S)	228.276	33.69	A(1600)	60.87	M(1600)	78.68	-57.670
C18H14N2O2S	BENZONAPHTHOQUINONETHIAZINE	(S)	322.374	-200.7	C(744)				
C18H14N2O3	DI-P-TOLUYLFURAZAN	(S)	306.308	8.3	C(1012)				
C18H14O3	CINNAMIC ANHYDRIDE	(S)	278.292	-83.2	B(163)				
C18H15N	TRIPHENYLAMINE	(S)	245.308	58.7	C(1105)	73.0	C(1105)	120.85	-88.584
C18H16	DIPHENYLHEXATRIENE	(S)	232.308	50.5	C(744)				
C18H16O2	ISOPROPYL-X-METHYL-PHENANTHRENEQUINONE	(U,S)	264.308	-85.1	C(744)				
C18H16O4	DIMETHYL DIPHENYLMALEATE	(C,S)	296.308	-125.8	C(744)				
C18H16O4	ALPHA-TRUXILLIC ACID	(U,S)	296.308	-155.7	C(744)				
C18H14	5,12-DIHYDROTETRACENE	(S)	230.292	24.5	B(914)				
C18H14O3	CINNAMOYL PEROXIDE	(S)	298.292	-85.2	B(163)				
C18H18	CYCLOOCTADECANONAENE	(S)	234.324	39.0	A(98)				
C18H18	1,6-DIPHENYL-1,5-HEXADIENE	(S)	234.324	39.0	C(744)				
C18H18	1,6-DIPHENYL-2,4-HEXADIENE	(S)	234.324	35.6	C(744)				
C18H18	7-ISOPROPYL-1-METHYL-PHENANTHRENE	(S)	234.324	1.3	C(744)				
C18H18	2,4,5,7-TETRAMETHYL-PHENANTHRENE	(S)	234.324	2.98	A(720)				
C18H18	3,4,5,6-TETRAMETHYL-PHENANTHRENE	(S)	234.324	5.56	A(720)				
C18H18O2	CINNAMOIN	(L)	266.324	-30.9	C(1332)				
C18H18O12	HEXAMETHYL MELLITATE	(S)	426.324	-482.3	C(744)				
C18H20	3,6-DIPHENYL-3-HEXENE	(L)	236.340	1.3	C(744)				
C18H20N2O	N-PIPERIDYL-N,N-DIPHENYLFORMAMIDE	(S)	280.356	-22.5	B(1472)				
C18H20N2O3	PHENYLALANINE ANHYDRIDE	(S)	312.356	-137.1	B(1178)				
C18H20O2	HYDROCINNAMOIN	(L)	268.340	-44.7	C(1332)				
C18H22	2-BENZYL-1-PHENYLPENTANE	(L)	238.356	16.3	C(206)				

Formula	Name	State	Mol Wt	$\Delta H f^{\circ}_{298}$ (kcal/mole)	Source	S°_{298} (cal/mole °K)	Source	$\Delta G f^{\circ}_{298}$ (kcal/mole)	Log Kp
C18H22	1,1-DIPHENYLHEXANE	(L)	238.356	1.3	C(1336)				
C18H22	1,6-DIPHENYLHEXANE	(L)	238.356	1.3	C(1336)				
C18H23NO4	CODEINE MONOHYDRATE	(S)	317.372	-150.7	CC 744)				
C18H26O4	DIPENTYL PHTHALATE	(L)	306.388	-219.0	BC 12)				
C18H26O6	1,2,3-PROPANETRIYL TRICYCLOPENTANECARBOXYLATE	(L)	338.388	-272.3	CC 744)				
C18H30	HEXAETHYLBENZENE	(G)	246.420	-53.60	TC 434)	166.62	TC 434)	43.61	-31.965
C18H30	1-PHENYLDODECANE	(G)	246.420	-42.71	TC 433)	179.63	TC 433)	50.62	-37.104
C18H30O3S	P-DODECYLBENZENESULFONIC ACID	(G)	326.486			195.74	EC 69)		
C18H32O2	LINOLEIC ACID	(L)	280.436	-161.1	BC 728)				
C18H32O2	9-OCTADECYNOIC ACID	(S)	280.436	-155.1	CC 744)				
C18H32O16	RAFFINOSE (MELITOSE)	(S)	504.436	-759.3	CC 744)				
C18H34	1-CYCLOHEXYL-2-(CYCLOHEXYLMETHYL)PENTANE	(L)	250.452	-85.5	CC 206)				
C18H34	1,1-DICYCLOHEXYLHEXANE	(L)	250.452	-90.5	C(1336)				
C18H34	1,6-DICYCLOHEXYLHEXANE	(L)	250.452	-90.5	C(1336)				
C18H34	1-OCTADECYNE	(G)	250.452	-29.56	TC 335)	199.84	TC 335)	76.36	-55.968
C18H34O2	ELAIDIC ACID	(S)	282.452	-187.8	CC 744)				
C18H34O2	OLEIC ACID	(L)	282.452	-191.8	C(1439)				
C18H34O17	MELEZITOS MONOHYDRATE	(S)	522.452	-812.3	CC 744)				
C18H36	1-CYCLOHEXYLDODECANE	(L)	252.468	-113.0	A(1025)	147.1	M(1113)	17.95	-13.154
C18H36	1-CYCLOHEXYLDODECANE	(G)	252.468	-90.36	TC 391)	184.05	TC 391)	29.57	-21.674
C18H36	1-CYCLOPENTYLTRIDECANE	(G)	252.468	-84.55	TC 364)	192.89	TC 364)	32.74	-24.001
C18H36	1-OCTADECENE	(G)	252.468	-69.08	TC 301)	203.64	TC 301)	45.01	-32.991
C18H36O2	TERT-BUTYL TETRADECANOATE	(L)	284.468	-206.5	A(1454)				
C18H36O2	HEXADECYL ACETATE	(S)	284.468	-192.2	CC 744)				
C18H36O2	STEARIC ACID	(S)	284.468	-226.80	AC 853)				
C18H36O2	STEARIC ACID	(S)	284.468	-229.0	A(1454)				
C18H36O2	STEARIC ACID	(L)	284.468	-211.74	AC 3)				
C18H36O2	STEARIC ACID	(L)	284.468	-213.1	AC 853)				
C18H36O3	TERT-BUTYL PEROXYMYRISTATE	(L)	300.468	-190.4	A(1454)				
C18H36O3	PEROXYSTEARIC ACID	(S)	300.468	-205.1	A(1454)				
C18H38	OCTADECANE	(S)	254.484	-135.92	A(1113)	118.7	M(1113)	12.80	-9.382
C18H38	OCTADECANE	(L)	254.484	-121.54	A(1113)	166.5	M(1113)	12.93	-9.476
C18H38	OCTADECANE	(G)	254.484	-99.08	TC 258)	204.64	TC 258)	24.02	-17.604

Formula	Name	State	Mol Wt	$\Delta H_{f\,298}^{\circ}$ (kcal/mole)	Source	S_{298}° (cal/mole °K)	Source	$\Delta G_{f\,298}^{\circ}$ (kcal/mole)	Log Kp
C18H38O	1-OCTADECANOL	(G)	270.484	-135.39	T(538)	216.92	T(538)	-8.65	6.340
C18H38S	BUTYL TETRADECYL SULFIDE	(G)	286.550	-89.25	T(858)	220.30	T(858)	31.45	-23.054
C18H38S	ETHYL HEXADECYL SULFIDE	(G)	286.550	-89.03	T(859)	219.98	T(859)	31.77	-23.285
C18H38S	HEPTADECYL METHYL SULFIDE	(G)	286.550	-88.45	T(861)	219.44	T(860)	32.51	-23.828
C18H38S	NONYL SULFIDE	(G)	286.550	-89.25	T(861)	218.92	T(861)	31.86	-23.356
C18H38S	PENTADECYL PROPYL SULFIDE	(G)	286.550	-89.07	T(862)	220.30	T(862)	31.63	-23.186
C18H38S	1-OCTADECANETHIOL	(G)	286.550	-89.94	T(912)	220.29	T(912)	30.77	-22.551
C18H38S2	NONYL DISULFIDE	(G)	318.616	-87.12	T(880)	230.00	T(880)	32.97	-24.163
C18H42O21	RAFFINOSE PENTAHYDRATE	(S)	594.516	-1108.7	C(744)				
C19H13N3O6	TRINITROTRIPHENYLMETHANE	(U,S)	379.318	40.5	C(744)				
C19H13N3O7	TRINITROTRIPHENYLCARBINOL	(U,S)	395.318	-14.4	C(744)				
C19H15	TRITYL	(S)	243.310	79.9	C(744)				
C19H15CL	CHLOROTRIPHENYLMETHANE	(S)	278.767	43.4	C(744)				
C19H16	TRIPHENYLMETHANE	(S)	244.318	38.7	A(1121)	74.6	M(656)	98.61	-72.281
C19H16	TRIPHENYLMETHANE	(S)	244.318	39.7	A(269)	74.6	M(656)	99.61	-73.014
C19H16N4	ALPHA-(PHENYLAZO)-ALPHA-(PHENYLHYDRAZONO)TOLUENE	(S)	300.350	129.5	B(974)				
C19H16O	TRIPHENYLMETHANOL	(S)	260.318	-0.80	A(1112)	78.7	M(1388)	65.20	-47.787
C19H17N	N-BENZYLDIPHENYLAMINE	(S)	259.334	47.9	B(1472)				
C19H18O4	1,2-CYCLOPENTYLENE DIBENZOATE, CIS	(S)	310.334	-159.5	B(1537)				
C19H18O4	1,2-CYCLOPENTYLENE DIBENZOATE, TRANS	(S)	310.334	-161.3	B(1537)				
C19H19N3O	TRIAMINOTRIPHENYLCARBINOL	(U,S)	305.366	47.9	C(744)				
C19H21NO3	THEBIANE	(S)	311.366	-62.7	C(744)				
C19H24N2O4	DIETHYL 2-(4-CARBOXY-3,5-DIMETHYL-2-PYRRYLMETHYLENE)-3,5-DIMETHYLISOPYRROLE-4-CARBOXYLATE	(S)	344.398	-191.1	C(1413)				
C19H26N2O4	DIETHYL METHYLENE-BIS(2,4-DIMETHYLPYRROLE-3-CARBOXYLATE)	(S)	346.414	-201.8	C(1413)				
C19H26N2O4	DIETHYL METHYLENEBIS(2,5-DIMETHYLPYRROLE-4-CARBOXYLATE)	(S)	346.414	-203.2	C(1414)				
C19H26N2O4	DIETHYL METHYLENEBIS(2,4-DI								

Formula	Name	State	Mol Wt	$\Delta H^\circ_{f\,298}$ (kcal/mole)	Source	S°_{298} (cal/mole °K)	Source	$\Delta G^\circ_{f\,298}$ (kcal/mole)	Log K_p
C19H26N2O4	METHYLPYRROLE-5-CARBOXYLATE	(S)	346.414	-199.7	C(1414)				
C19H32O3S	P-TRIDECYL- BENZENESULFONIC ACID								
C19H32O3S	BENZENESULFONIC ACID					204.91	E(69)		
C19H32	1-PHENYLTRIDECANE	(G)	340.512	-47.63	TC(435)	188.94	TC(435)	52.64	-38.582
C19H36	1-NONADECYNE	(G)	260.446	-34.49	TC(336)	209.15	TC(336)	78.36	-57.438
C19H36O2	METHYL ELAIDATE	(L)	264.478	-174.6	BC(730)				
C19H36O2	METHYL OLEATE	(L)	296.478	-172.9	BC(730)				
C19H36O4	HEXADECYLMALONIC ACID	(S)	328.478	-307.2	CC(744)				
C19H38	1-CYCLOHEXYLTRIDECANE	(G)	266.494	-95.28	TC(392)	193.36	TC(392)	31.59	-23.151
C19H38	1-CYCLOPENTYLTETRADECANE	(G)	266.494	-89.48	TC(365)	202.13	TC(365)	34.77	-25.486
C19H38	1-NONADECENE	(G)	266.494	-74.00	TC(302)	212.95	TC(302)	47.02	-34.468
C19H38O2	METHYL STEARATE	(S)	298.494	-208.5	C(1439)				
C19H38O2	NONADECANOIC ACID	(S)	298.494	-235.42	AC(3)				
C19H38O2	NONADECANOIC ACID	(L)	298.494	-219.24	AC(3)				
C19H38O4	2,3-DIHYDROXYPROPYL PALMITATE	(S)	330.494	-306.2	AC(229)				
C19H38O4	2-HYDROXY-1-(HYDROXYMETHYL)- ETHYL PALMITATE								
C19H40	NONADECANE	(S)	330.494	-296.7	AC(229)	171.35	E(1248)	15.83	-11.605
C19H40	NONADECANE	(L)	268.510	-126.9	E(1248)	213.95	TC(259)	26.03	-19.081
C19H40O	1-NONADECANOL	(L)	268.510	-114.00	TC(259)	226.23	TC(539)	-6.64	4.870
C19H40S	BUTYL PENTADECYL SULFIDE	(G)	284.510	-140.32	TC(539)	229.61	TC(863)	33.46	-24.524
C19H40S	ETHYL HEPTADECYL SULFIDE	(G)	300.576	-94.18	TC(863)	229.29	TC(864)	33.77	-24.755
C19H40S	HEXADECYL PROPYL SULFIDE	(G)	300.576	-93.96	TC(864)	229.61	TC(865)	33.65	-24.663
C19H40S	METHYL OCTADECYL SULFIDE	(G)	300.576	-93.99	TC(865)	228.75	TC(866)	34.51	-25.298
C19H40S	1-NONADECANETHIOL	(G)	300.576	-94.87	TC(913)	229.60	TC(913)	32.77	-24.021
C20H10O2	1,12-PERYLENEDIONE	(S)	282.280	-10.44	CC(843)				
C20H12	PERYLENE	(S)	252.296	43.66	AC(1600)	63.23	M(1600)	88.76	-65.057
C20H14O4	M-PHENYLENE DIBENZOATE	(S)	318.312	-120.3	CC(744)				
C20H16	1',9-DIMETHYLBENZANTHRACENE	(S)	256.328	32.96	AC(448)				
C20H16	3',6-DIMETHYLBENZANTHRACENE	(S)	256.328	17.96	AC(448)				
C20H16	1,12-DIMETHYLBENZO(C)- PHENANTHRENE	(S)	256.328	36.67	AC(448)				
C20H16	5,8-DIMETHYLBENZO(C) PHENANTHRENE	(S)	256.328	25.62	AC(448)				
C20H16	TRIPHENYLETHYLENE	(S)	256.328	56.0	BC(263)	78.7	M(1388)	115.10	-84.363

Formula	Name	State	Mol Wt	$\Delta Hf°_{298}$ (kcal/mole)	Source	$S°_{298}$ (cal/mole °K)	Source	$\Delta Gf°_{298}$ (kcal/mole)	Log Kp
C20H18	1,1,1-TRIPHENYLETHANE	(S)	258.344	37.3	A(270)	78.7	M(1388)	105.70	-77.477
C20H18	1,1,2-TRIPHENYLETHANE	(S)	258.344	30.9	A(270)				
C20H20O4	1,2-CYCLOHEXYLENE DIBENZOATE, CIS	(S)	324.360	-171.7	B(1537)				
C20H20O4	1,2-CYCLOHEXYLENE DIBENZOATE, TRANS	(S)	324.360	-175.5	B(1537)				
C20H20O4	DIMETHYL BETA-TRUXILLATE	(U,S)	324.360	-140.5	C(744)				
C20H20O4	1-METHYL-1,2-CYCLOPENTYLENE DIBENZOATE, TRANS	(S)	324.360	-168.0	B(1537)				
C20H21NO4	PAPAVERINE	(S)	339.376	-120.0	C(744)				
C20H22O4	DIETHYL DIPHENYLSUCCINATE, MESO	(S)	326.376	-181.1	C(744)				
C20H22O4	DIETHYL DIPHENYLSUCCINATE RACEMIC	(S)	326.376	-179.6	C(744)				
C20H24	TETRACYCLOPENTADIENE	(S)	264.392	30.0	B(90)				
C20H26	DIHYDROTETRACYCLOPENTADIENE	(S)	266.408	-3.0	B(90)				
C20H27NO11	AMYGDALIN	(S)	457.424	-453.4	C(744)				
C20H28N2O4	DIETHYL 5,5'-ETHYLIDENE-BIS(2,4-DIMETHYL-PYRROLE-3-CARBOXYLATE)	(S)	360.440	-207.6	C(1413)				
C20H28N2O4	1-PHENYLTETRADECANE	(G)	274.472	-52.56	T(436)	198.25	T(436)	54.64	-40.052
C20H34O3S	P-TETRADECYL BENZENESULFONIC ACID	(G)	342.528			214.08	E(69)	80.38	-58.915
C20H34O3S		(G)				218.46	T(337)		
C20H38	1-EICOSYNE	(G)	278.504	-39.41	T(337)				
C20H38O2	ETHYL ELAIDATE	(L)	310.504	-183.3	B(730)				
C20H38O2	ETHYL OLEATE	(L)	310.504	-183.9	B(730)				
C20H40	1-CYCLOHEXYLTETRADECANE	(G)	280.520	-100.21	T(393)	202.67	T(393)	33.59	-24.621
C20H40	1-CYCLOPENTYLPENTADECANE	(G)	280.520	-94.41	T(366)	211.44	T(366)	36.78	-26.956
C20H40	1-EICOSENE	(G)	280.520	-78.93	T(303)	222.26	T(303)	49.03	-35.938
C20H40O2	EICOSANOIC ACID	(S)	312.520	-242.07	A(3)				
C20H40O2	EICOSANOIC ACID	(L)	312.520	-224.87	A(3)				
C20H42	EICOSANE	(S)	282.536	-149.7	E(1248)	133.5	M(1107)	14.03	-10.283
C20H42	EICOSANE	(L)	282.536	-133.0	E(1248)	179.1	E(1248)	17.13	-12.559
C20H42	EICOSANE	(G)	282.536	-108.93	T(260)	223.26	T(260)	28.04	-20.551
C20H42O	1-EICOSANOL	(L)	298.536	-145.25	T(540)	235.54	T(540)	-4.64	3.400
C20H42S	BUTYL HEXADECYL SULFIDE	(G)	314.602	-99.10	T(867)	238.92	T(867)	35.47	-26.002
C20H42S	DECYL SULFIDE	(G)	314.602	-99.10	T(868)	237.54	T(868)	35.89	-26.303

763

Formula	Name	State	Mol Wt	$\Delta H f°_{298}$ (kcal/mole)	Source	$S°_{298}$ (cal/mole °K)	Source	$\Delta G f°_{298}$ (kcal/mole)	Log $K p$
C20H42S	1-EICOSANETHIOL	(G)	314.602	-99.80	T(914)	238.91	T(914)	34.78	-25.491
C20H42S	ETHYL OCTADECYL SULFIDE	(G)	314.602	-98.89	T(869)	238.60	T(869)	35.78	-26.225
C20H42S	HEPTADECYL PROPYL SULFIDE	(G)	314.602	-98.92	T(870)	238.92	T(870)	35.65	-26.133
C20H42S	METHYL NONADECYL SULFIDE	(G)	314.602	-98.31	T(871)	238.06	T(871)	36.52	-26.769
C20H42S2	DECYL DISULFIDE	(G)	346.668	-96.97	T(881)	248.62	T(881)	36.99	-27.111
C21H13NO5	P-NITROBENZOYLBENZOIN	(U,S)	359.322	26.3	C(744)				
C21H16O	3,3-DIPHENYLACRYLOPHENONE	(L)	284.338	24.3	C(744)				
C21H16O	1,1,3-TRIPHENYL-2-PROPYN-1-OL	(L)	284.338	51.4	C(744)				
C21H16O2	BETA-DIOXYDINAPHTHYLMETHANE	(U,S)	300.338	-45.3	C(744)				
C21H18O2	BETA-NAPHTHOLFORMAL	(U,S)	300.338	-20.6	C(744)				
C21H18O2	3-HYDROXY-3,3-DIPHENYL-PROPIOPHENONE	(L)	302.354	-51.4	C(744)				
C21H21N	TRIBENZYLAMINE	(S)	287.386	38.8	B(1472)				
C21H22N2O2	STRYCHNINE	(S)	334.402	-38.5	C(744)				
C21H22O4	1-METHYL-1,2-CYCLOHEXYLENE DIBENZOATE, TRANS	(L)	338.386	-181.4	B(1537)				
C21H36	1-PHENYLPENTADECANE	(G)	288.498	-57.49	T(437)	207.56	T(437)	56.65	-41.522
C21H40O2	PROPYL ELAIDATE	(L)	324.530	-189.1	B(730)				
C21H40O2	PROPYL OLEATE	(L)	324.530	-187.6	B(730)				
C21H42	1-CYCLOHEXYLPENTADECANE	(G)	294.546	-105.14	T(394)	211.98	T(394)	35.60	-26.092
C21H42	1-CYCLOPENTYLHEXADECANE	(G)	294.546	-99.33	T(367)	220.75	T(367)	38.79	-28.434
C21H44N2O	1,3-DIDECYL UREA	(S)	340.578	-209.0	B(218)				
C21H46N2O2	DECYLAMMONIUM DECYL CARBAMATE	(S)	358.594	-282.0	B(218)				
C22H12O2	PENTACENEQUINONE	(S)	308.316	-17.7	C(913)				
C22H18O	METHYL 1,1,3-TRIPHENYL-2-PROPYNYL ETHER	(L)	298.364	56.1	C(744)				
C22H23NO7	NARCOTINE	(S)	413.412	-210.2	C(744)				
C22H38	1-PHENYLHEXADECANE	(G)	302.524	-62.41	T(438)	216.87	T(438)	58.66	-42.999
C22H40O2	DOCOSYNOIC ACID	(U,S)	336.540	-177.7	C(744)				
C22H42O2	BRASSIDIC ACID	(S)	338.556	-224.3	B(729)				
C22H42O2	BUTYL ELAIDATE	(L)	338.556	-195.0	B(730)				
C22H42O2	BUTYL OLEATE	(L)	338.556	-193.4	B(730)				
C22H42O2	13-DOCOSENOIC ACID, CIS	(S)	338.556	-204.2	C(744)				

Formula	Name	State	Mol Wt	$\Delta H f^\circ_{298}$ (kcal/mole)	Source	S°_{298} (cal/mole °K)	Source	$\Delta G f^\circ_{298}$ (kcal/mole)	Log Kp
C22H44	1-CYCLOHEXYLHEXADECANE	(G)	308.572	-110.06	T(395)	221.29	T(395)	37.61	-27.569
C22H44O2	DOCOSANOIC ACID	(S)	340.572	-254.1	C(843)				
C22H44O4	DIHYDROXYDOCOSANOIC ACID	(U,S)	372.572	-333.7	C(744)				
C23H18N2O	N-(1-NAPHTHYL)CARBANILIDE	(S)	338.390	22.9	B(1472)				
C23H18N2O	N-(2-NAPHTHYL)CARBANILIDE	(S)	338.390	19.4	B(1471)				
C23H18O4	1,2-INDANYLENE DIBENZOATE,CIS	(S)	358.374	-132.7	B(1537)				
C23H18O4	1,2-INDANYLENE DIBENZOATE, TRANS	(S)	358.374	-130.7	B(1537)				
C23H20O	ETHYL 1,1,3-TRIPHENYL-2-PROPYNYL ETHER	(L)	312.390	45.8	C(744)				
C23H22	TRI-O-TOLYLETHYLENE	(S)	298.406	29.6	A(265)				
C23H22	TRI-P-TOLYLETHYLENE	(S)	298.406	28.7	A(265)				
C23H24	1,1,2-TRI-O-TOLYLETHANE	(S)	300.422	5.5	A(264)				
C23H24	1,1,2-TRI-P-TOLYLETHANE	(S)	300.422	3.6	A(264)				
C23H26N2O4	BRUCINE	(S)	394.454	-115.9	C(744)				
C23H30O	2,7-DODECAMETHYLENE-4,5-BENZOTROPON	(U,S)	322.470	-60.4	A(1290)				
C23H31NO10	NARCEINE DIHYDRATE	(S)	481.486	-419.1	C(744)				
C23H44O2	PENTYL ELAIDATE	(L)	352.582	-201.3	B(730)				
C24H12O2	BENZO(RST)PENTAPHENE-5,8-DIONE	(S)	332.336	-60.3	C(913)				
C24H12O2	3,9-DIACETYLPERYLENE	(S)	336.368	-22.4	C(843)				
C24H16O2	1,3,5-TRIPHENYLBENZENE	(S)	306.384	52.3	A(1226)	87.8	M(1118)	119.61	-87.673
C24H18	TETRAPHENYLHYDRAZINE	(S)	336.416	109.2	A(244)				
C24H20N2	1,2,3,4-TETRAHYDRONAPHTHALENE-1,2-DIBENZOATE, CIS	(S)	372.400	-141.0	B(1537)				
C24H20O4	1,2,3,4-TETRAHYDRONAPHTHYLENEDIBENZOATE, CIS	(S)	372.400	-142.5	B(1537)				
C24H20O4	1,2,3,4-TETRAHYDRO-2,3-NAPHTHYLENEDIBENZOATE, CIS	(S)	372.400	-142.6	B(1537)				
C24H20O4	1,2,3,4-TETRAHYDRO-2,3-NAPHTHYLENEDIBENZOATE, TRANS	(S)	372.400	-140.3	B(1537)				
C24H20O6	1,2,3-PROPANETRIYL TRIBENZOATE	(S)	404.400	-219.3	C(744)				

Formula	Name	State	Mol Wt	$\Delta H^\circ_{f\,298}$ (kcal/mole)	Source	S°_{298} (cal/mole °K)	Source	$\Delta G^\circ_{f\,298}$ (kcal/mole)	Log $K p$
C24H20O8	CINNAMYLIDENEMALONIC ACID (EX-		436.400	-301.6	C(744)				
C24H20O8	POSED TO THE ACTION OF LIGHT)	(S)							
C24H22O	PROPYL 1,1,3-TRIPHENYL-2- PROPYNYL ETHER	(L)	326.416	39.6	C(744)				
C24H34	1,1-DIPHENYLDODECANE	(L)	322.512			163.70	M(716)		
C24H38O4	BIS(ETHYLHEXYL) PHTHALATE . .	(L)	390.544	-257.1	B(1471)				
C24H40	1-CYCLOHEXYL-1-PHENYLDODECANE	(L)	328.560			166.33	M(716)		
C24H42O21	STACHYOSE.	(S)	666.576	-981.6	C(744)				
C24H46	1,1-DICYCLOHEXYLDODECANE . .	(S)	334.608		(CALC)	130.42	M(716)	25.50	-18.689
C24H50	TETRACOSANE	(S)	338.640	-170.5		155.56	M(1113)		
C25H20	TETRAPHENYLMETHANE	(S)	320.410	57.8	A(268)	91.6	M(1388)	133.69	-97.992
C25H20N2	TRIPHENYL(PHENYLAZO)METHANE .	(S)	348.426	137.0	C(1610)				
C25H38N2	2,2-(DIISOBUTENYLMETHYLENE)-								
C25H38N2	BIS(4-ETHYL-3,5-DIMETHYL- PYRROLE)	(S)	366.570	-31.9	C(1413)				
C25H52	PENTACOSANE	(S)	352.666	-186.0	C(1105)	160.4	M(1107)	18.27	-13.389
C26H18	9,10-DIPHENYLANTHRACENE . . .	(S)	330.404	73.9	B(914)				
C26H20	DIANTHRACENE	(S)	332.420	255.4	C(744)				
C26H20	TETRAPHENYLETHYLENE	(S)	332.420	76.1	A(263)				
C26H20N2O	ALPHA-(TRIPHENYLMETHYLAZO)- BENZALDEHYDE	(S)	376.436	110.0	C(1610)				
C26H20O	BETA-BENZOPINACOLONE	(S)	348.420	52.0	C(1610)				
C26H20O2	3,9-DIPROPIONYLPERYLENE . . .	(S)	364.420	-47.3	C(843)				
C26H22	1,1,1,2-TETRAPHENYLETHANE . .	(S)	334.436	54.1	A(270)	96.9	M(1388)	138.12	-101.240
C26H22	1,1,2,2-TETRAPHENYLETHANE . .	(S)	334.436	52.4	A(270)	99.6	M(1388)	135.62	-99.404
C26H22N4O2	1,2-BIS(DIPHENYLCARBAMOYL)- HYDRAZINE	(S)	422.468	18.5	B(1472)				
C26H46	3-PHENYLEICOSANE	(S)	358.628	-131.6	A(768)				
C26H46	9-PHENYLEICOSANE	(L)	358.628	-114.9	A(768)				
C26H52	3-CYCLOHEXYLEICOSANE	(L)	364.676	-159.6	A(768)				
C26H52	9-CYCLOHEXYLEICOSANE	(L)	364.676	-161.5	A(768)				
C26H52	11-CYCLOPENTYLHENEICOSANE . .	(L)	364.676	-155.5	A(768)				
C26H54	5-BUTYLDOCOSANE	(L)	366.692	-171.0	A(768)				
C26H54	11-BUTYLDOCOSANE	(L)	366.692	-171.5	A(768)				

Formula	Name	State	Mol Wt	$\Delta H_f^{\circ}{}_{298}$ (kcal/mole)	Source	S°_{298} (cal/mole °K)	Source	$\Delta G_f^{\circ}{}_{298}$ (kcal/mole)	Log Kp
C27H48	11-PHENYLHENEICOSANE . . .	(L)	372.654	-120.8	A(1113)	207.4	M(1113)	51.65	-37.861
C27H54	11-CYCLOHEXYLHENEICOSANE . .	(L)	378.702	-165.1	A(1113)	211.3	M(1113)	34.11	-25.000
C28H18	9,9'-BIANTHRYL	(S)	354.424	78.1	B(914)				
C28H18	9,9'-BIPHENANTHRYL	(S)	354.424	51.0	B(914)				
C28H20	DELTA-9,9'-(10H,10'H)- BIANTHRACENE	(S)	356.440	73.2	A(102)				
C28H22	1,1,4,4-TETRAPHENYL- 1,3-BUTADIENE	(S)	358.456	78.5	A(265)				
C28H22	1,2,3,4-TETRAPHENYLBUTADIENE, TRANS,TRANS	(S)	358.456	85.4	A(265)				
C28H22O3	DIPHENYLACETIC ANHYDRIDE	(S)	406.456	-76.2	C(744)				
C28H24O2	3,9-DIBUTYRYLPERYLENE . . .	(S)	392.472	-55.9	C(843)				
C28H26	1,1,4,4-TETRAPHENYLBUTANE .	(S)	362.488	38.65	A(264)				
C28H26N4O2	N,N'-BISDIPHENYLCARBAMOYL)- ETHYLENEDIAMINE	(S)	450.520	-41.1	B(1472)				
C28H26N4O2	CELLOBIOSE OCTAACETATE . .	(S)	678.584	-896.9	C(744)				
C28H38O19	LACTOSE OCTAACETATE . . .	(S)	678.584	-900.2	C(744)				
C28H38O19	MALTOSE OCTAACETATE . . .	(S)	678.584	-898.9	C(744)				
C28H38O19	SUCROSE OCTAACETATE . . .	(S)	678.584	-896.2	C(744)				
C28H44O	ERGOSTEROL	(S)	396.632	-189.9	C(130)				
C30H28	TETRA-P-TOLYLETHYLENE . . .	(S)	388.524	40.6	A(264)				
C30H30	1,1,2,2-TETRA-P-TOLYLETHANE	(S)	390.540	17.21	A(264)				
C30H56O2	1,6-CYCLOTRIACONTANEDIONE .	(S)	448.748	-232.7	C(1272)				
C30H60	CYCLOTRIACONTANE	(S)	420.780	-212.2	C(1272)				
C31H56	13-PHENYLPENTACOSANE . . .	(S)	428.758	-164.7	A(1121)				
C31H62	13-CYCLOHEXYLPENTACOSANE .	(L)	434.806	-189.9	A(1121)				
C31H64	11-DECYLHENEICOSANE . . .	(L)	436.822	-201.7	A(1121)	259.6	M(429)	31.26	-22.911
C32H3O03	DIBENZYLACETIC ANHYDRIDE (GLASSY)	(U,S)	462.560	-102.2	C(744)				
C32H36N4O2	METHYLPYROPORPHYRINATE	(XV)	(S)	508.640	-83.4	C(1415)			
C32H38N4	ETIOPORPHYRIN	(S)	478.656	9.8	C(1415)				

767

Formula	Name	State	Mol Wt	$\Delta H f^\circ_{298}$ (kcal/mole)	Source	S°_{298} (cal/mole °K)	Source	$\Delta G f^\circ_{298}$ (kcal/mole)	Log Kp
C32H64O2	HEXADECYL PALMITATE	(S)	480.832	-318.6	C(744)				
C32H66	DOTRIACONTANE	(S)	450.848	-231.8	A(1121)	203.5	M(1113)	27.59	-20.227
C32H66	DOTRIACONTANE	(L)	450.848	-205.4	A(1113)	280.0	M(1113)	31.19	-22.859
C33H38N4O2	MONOMETHYL-GAMMA-PHYLLOPORPHYRINATE	(S)	522.666	-81.8	C(1415)				
C33H42N4O2	5-(4-ACETYL-3,5-DIMETHYL-2-PYRRYLMETHYL)-2-[5-(4-ACETYL-								
C33H42N4O2	3,5-DIMETHYL-2-PYRRYLMETHYL)-								
C33H42N4O2	4-ETHYL-3-METHYL-2-PYRRYL-								
C33H42N4O2	METHYLENE]-4-ETHYL-3-METHYLISOPYRROLE	(S)	526.698	-61.8	C(1414)	209.8	M(1107)		
C33H68	TRITRIACONTANE	(S)	464.874						
C34H34N4O4	PROTOPORPHYRIN	(S)	562.644	-115.3	G(1415)				
C34H36N4O3	MONOMETHYLPHYLLOERYTHRINATE	(S)	548.660	-77.9	C(1415)				
C34H36N4O3	MONOMETHYLPYROPHEOPHORBIDE	(S)	548.660	-79.5	G(1415)				
C34H36N4O4	DIMETHYLVERDOPORPHYRINATE , .	(S)	564.660	-138.5	C(1415)				
C34H36N4O5	MONOMETHYL-18-PHEOPURPURINATE	(S)	580.660	-225.0	C(1415)				
C34H38N4O2	MONOMETHYL-DESOXY-PHYLLOERYTHRINATE	(S)	534.676	-33.3	C(1415)				
C34H38N4O4	DIMETHYLRHODOPORPHYRINATE-XV	(S)	566.676	-116.7	C(1415)				
C34H38N4O4	DIMETHYLRHODOPORPHYRINATE (XXI)	(S)	566.676	-118.4	C(1415)				
C35H38N4O5	DIMETHYLCHLOROPORPHYRIN-E5-ATE SAMPLE (A)	(S)	594.686	-172.5	C(1415)				
C35H38N4O5	DIMETHYLCHLOROPORPHYRIN-E5-ATE SAMPLE (B)	(S)	594.686	-197.6	C(1415)				
C35H40N4O4	DIMETHYLCHLOROPORPHYRIN-E4-ATE	(S)	580.702	-148.5	C(1415)				
C35H40N4O4	DIMETHYLCHLORIN-E4-ATE . . .	(S)	580.702	-139.2	C(1415)				
C36H24O2	3,9-DI-2-TOLUOYLPERYLENE . .	(S)	488.552	9.1	C(843)				
C36H38N4O4	DIMETHYLPROTOPORPHYRINATE . .	(S)	590.696	-116.9	C(1415)				

Formula	Name	State	Mol Wt	$\Delta Hf°_{298}$ (kcal/mole)	Source	$S°_{298}$ (cal/mole °K)	Source	$\Delta Gf°_{298}$ (kcal/mole)	Log Kp
C36H38N4O5	DIMETHYLPHEOPORPHYRIN-A5-ATE	(S)	606.696	-158.7	C(1415)				
C36H38N4O5	METHYLPHEOPHORBID SAMPLE (A)	(S)	606.696	-153.8	C(1415)				
C36H38N4O5	METHYLPHEOPHORBID SAMPLE (B)	(S)	606.696	-147.2	C(1415)				
C36H40N4O6	TRIMETHYLCHLORIN-P6-ATE	(S)	624.712	-286.4	C(1415)				
C36H42N4O4	DIMETHYLMESOPORPHYRINATE(IX)	(S)	594.728	-190.3	C(1415)				
C36H46N4	OCTAETHYLPORPHYRIN	(S)	534.760	-33.5	C(1415)				
C37H40N4O7	DIMETHYLPHEOPURPURIN-7	(S)	652.722	-239.2	C(1415)				
C37H42N4O6	TRIMETHYLCHLORIN-E6-ATE	(S)	638.738	-215.4	C(1415)				
C37H42N4O6	TRIMETHYLCHLOROPORPHYRIN-E6-ATE	(S)	638.738	-224.3	C(1415)				
C39H74O6	1,2,3-PROPANETRIYL TRILAURATE	(S)	638.982	-484.6	C(744)				
C40H46N4O8	TETRAMETHYLKOPROPORPHYRINATE	(S)	710.800	-341.8	C(1415)				
C42H28	TETRAPHENYLNAPHTHACENE	(S)	532.644	134.4	C(378)				
C45H86O6	1,2,3-PROPANETRIYL TRIMYRISTATE	(L)	723.138	-514.0	C(744)	297.8	M(216)	-140.56	103.027
C47H88O5	1,2,3-PROPANETRIYL DIBRASSIDATE	(S)	733.174	-467.3	C(744)				
C47H88O5	1,2,3-PROPANETRIYL DIERUCATE	(S)	733.174	-441.3	C(744)				
C48H38O12	MANNITOL HEXABENZOATE	(S)	806.784	-449.6	C(744)				
C48H54N4O16	OCTAMETHYL-ISO-UROPORPHYRIN (II)	(S)	942.944	-612.7	C(1415)				
C51H98O6	1,2,3-PROPANETRYL TRIPALMITATE	(S)	807.294			331.6	M(216)		
C57H110O6	1,2,3-PROPANETRYL TRISTEARATE	(S)	891.450			366.8	M(216)		
C69H128O6	1,2,3-PROPANETRIYL TRIBRASSIDATE	(S)	1053.714	-624.2	C(744)				
C69H128O6	1,2,3-PROPANETRIYL TRIERUCATE	(S)	1053.714	-595.3	C(744)				

APPENDICES

APPENDIX A-1

Glossary of Symbols[a]

Primary Thermodynamic Quantities[b]

A	Helmholtz energy
Cp	molal heat capacity at constant pressure
Cv	molal heat capacity at constant volume
E	energy
ΔEc	energy increment for a combustion process
G	Gibbs energy
ΔGf	Gibbs energy of formation of a compound from the elements
$(G - H_{298})/T$	Gibbs energy function at temperature T based on 298.15°K
H	enthalpy
$H - H_{298}$	enthalpy increment between T°K and 298.15°K
ΔHc	enthalpy increment for a combustion process
ΔHf	enthalpy of formation of a compound from the elements
ΔHm	enthalpy increment of melting
ΔHr	enthalpy increment of reaction
ΔHs	enthalpy increment of sublimation
ΔHt	enthalpy increment of solid-state transition
ΔHv	enthalpy increment of vaporization
Kf	equilibrium constant of a reaction (expressed in fugacities)
Kn	equilibrium constant (expressed in number of moles)
Kp	equilibrium constant of a reaction (expressed in atmospheric pressure units)
P	pressure
Q	quantity of heat
S	entropy
V	volume

Other Quantities in Italic Letters

a	activity
a, b, c	constants in heat capacity equations
$B_P(C_P, D_P, \ldots)$ $B_V(C_V, D_V, \ldots)$	second and higher virial coefficients
c	velocity of light
d	interatomic distance in centimeters

Appendix A-1

dc	critical density
f	fugacity
g	gaseous state
g_i	degeneracy of ith level
h	Planck constant
I_A, I_B, I_C	principal moments of inertia of a molecule
k	Boltzmann constant
l	liquid state
M	molecular weight
m	mass of atoms (in grams)
N	Avogadro's constant; mole fraction
n	number of potential energy maxima per revolution of rotating group
Pc	critical pressure
Q_f	partition function for free rotation (parameter used in internal rotation functions)
R	gas constant
s	solid or crystalline state
T	temperature in degrees Kelvin (degrees Celsius + 273.15°)
t	temperature in degrees Celsius
Tb	boiling point at 1 atm pressure
Tc	critical temperature
Tm	melting temperature at 1 atm pressure
Tt	transition temperature
Ttp	triple-point temperature
u	$h\nu/kT = hc\omega/kT = 1.4386\omega/T$
V	potential energy barrier
x	$h\nu_{max}/kT$ (or for otherwise specified ν)
Z	compressibility factor

Greek Letters

γ	fugacity coefficient
Δ	indicates the increment in a given property for a process or reaction
ν_i	stoichiometric coefficient of ith component
σ	symmetry number
ω_i	wave number ("frequency"), per centimeter

Superscripts

°	indicates the standard state of unit fugacity or unit activity on any thermodynamic quantity
*	thermodynamic property excluding symmetry effects (e.g. $S^* = \Delta S^\circ + R \ln \sigma_t$); also used to designate quantities for a reference substance
′	property (e.g., f', fugacity, and Kf', equilibrium constant) in terms of pure component at total pressure of mixture

Subscripts

Numerical (e.g., $_{298}$)	denotes temperature in degrees Kelvin (note: subscript 298 is to be interpreted throughout as 298.15)
D	Debye, as in θ_D, Debye characteristic temperature, or f_D, Debye function
E	Einstein, as in θ_E
el	electronic
ir	internal rotational
or	over-all rotational
r	rotational
tot, t	total
tr	translational
vib	vibrational
w	whole

---

[a] In Chapter 14 limitations in the computer printer necessitated exclusive use of capital letters in characterizing the substances. Although most of these abbreviations will be readily recognized, definitions appear in the introduction to that chapter.

[b] The thermodynamic symbolism is further detailed and defined in Chapter 8.

Appendix Table A-2 Einstein Heat Capacity Function, $Cv°/R$, as a Function of $x = \theta_E/T$[a]

x	0.0	0.1	0.2	0.3	0.4	0.5	0.6	0.7	0.8	0.9
0.0	1.0000	0.9992	0.9967	0.9925	0.9868	0.9794	0.9705	0.9602	0.9483	0.9352
1.0	0.9207	0.9050	0.8882	0.8703	0.8515	0.8318	0.8114	0.7903	0.7687	0.7466
2.0	0.7241	0.7013	0.6783	0.6552	0.6320	0.6089	0.5859	0.5631	0.5405	0.5182
3.0	0.4963	0.4747	0.4536	0.4330	0.4129	0.3933	0.3743	0.3558	0.3380	0.3207
4.0	0.3041	0.2881	0.2726	0.2578	0.2436	0.2300	0.2170	0.2046	0.1928	0.1815
5.0	0.1707	0.1605	0.1508	0.1416	0.1329	0.1246	0.1168	0.1094	0.1025	0.09588
6.0	0.08968	0.08383	0.07833	0.07315	0.06828	0.06371	0.05942	0.05539	0.05162	0.04808
7.0	0.04476	0.04166	0.03876	0.03605	0.03351	0.03115	0.02894	0.02687	0.02495	0.02316
8.0	0.02148	0.01993	0.01848	0.01713	0.01587	0.01471	0.01362	0.01261	0.01168	0.01081
9.0	0.01000	0.00925	0.00855	0.00791	0.00731	0.00676	0.00624	0.00577	0.00533	0.00492
10.0	0.00454		0.00387		0.00329		0.00280		0.00238	
11.0	0.00202		0.00172		0.00145		0.00123		0.00104	
12.0	0.00088		0.00075		0.00063		0.00054		0.00045	
13.0	0.00038		0.00032		0.00027		0.00023		0.00019	
14.0	0.00016		0.00014		0.00012		0.00010		0.00008	
15.0	0.00007		0.00006		0.00005		0.00004		0.00003	

[a] Abstracted from the more detailed tables of Hilsenrath and Ziegler (605).

Appendix Table A-3 Einstein Energy Function, $(E° - E_0°)/RT$, in Terms of $x = \theta_E/T$ [a]

x	0.0	0.1	0.2	0.3	0.4	0.5	0.6	0.7	0.8	0.9
0.0	0.9995	0.9508	0.9033	0.8575	0.8133	0.7708	0.7298	0.6905	0.6528	0.6166
1.0	0.5820	0.5489	0.5172	0.4870	0.4582	0.4308	0.4048	0.3800	0.3565	0.3342
2.0	0.3130	0.2930	0.2741	0.2563	0.2394	0.2236	0.2086	0.1945	0.1813	0.1689
3.0	0.1572	0.1462	0.1360	0.1264	0.1174	0.1090	0.1011	0.09380	0.08695	0.08057
4.0	0.07463	0.06909	0.06394	0.05915	0.05469	0.05055	0.04671	0.04314	0.03983	0.03676
5.0	0.03392	0.03128	0.02885	0.02659	0.02450	0.02257	0.02078	0.01914	0.01761	0.01621
6.0	0.01491	0.01371	0.01261	0.01159	0.01065	0.00979	0.00899	0.00826	0.00758	0.00696
7.0	0.00639	0.00586	0.00538	0.00493	0.00453	0.00415	0.00381	0.00349	0.00320	0.00293
8.0	0.00268	0.00246	0.00225	0.00206	0.00189	0.00173	0.00158	0.00145	0.00133	0.00121
9.0	0.00111	0.00102	0.00093	0.00085	0.00078	0.00071	0.00065	0.00059	0.00054	0.00050
10.0	0.00045		0.00038		0.00032		0.00026		0.00022	
11.0	0.00018		0.00015		0.00013		0.00011		0.00009	
12.0	0.00007		0.00006		0.00005		0.00004		0.00004	
13.0	0.00003		0.00002		0.00002		0.00002		0.00001	
14.0	0.00001		0.00001		0.00001		0.00001		0.00001	
15.0	0.00001		0.00000		0.00000		0.00000		0.00000	

[a] Abstracted from the more detailed tables of Hilsenrath and Ziegler (605).

Appendix Table A-4 Einstein Entropy Function, $S°/R$, in Terms of $x = \theta_E/T$[a]

x	0.0	0.1	0.2	0.3	0.4	0.5	0.6	0.7	0.8	0.9
0.0	7.9078	3.3030	2.6111	2.2077	1.9229	1.7035	1.5257	1.3768	1.2494	1.1384
1.0	1.0406	0.9536	0.8756	0.8052	0.7414	0.6833	0.6303	0.5817	0.5371	0.4962
2.0	0.4584	0.4237	0.3916	0.3619	0.3346	0.3092	0.2858	0.2641	0.2440	0.2254
3.0	0.2083	0.1923	0.1776	0.1640	0.1513	0.1396	0.1288	0.1188	0.1096	0.1010
4.0	0.09311	0.08580	0.07905	0.07281	0.06705	0.06172	0.05681	0.05228	0.04809	0.04424
5.0	0.04068	0.03740	0.03438	0.03159	0.02903	0.02666	0.02449	0.02249	0.02065	0.01895
6.0	0.01739	0.01596	0.01464	0.01343	0.01231	0.01129	0.01035	0.00949	0.00870	0.00797
7.0	0.00730	0.00669	0.00613	0.00561	0.00514	0.00470	0.00431	0.00394	0.00361	0.00330
8.0	0.00302	0.00276	0.00253	0.00231	0.00211	0.00193	0.00177	0.00162	0.00148	0.00135
9.0	0.00123	0.00113	0.00103	0.00094	0.00086	0.00079	0.00072	0.00066	0.00060	0.00055
10.0	0.00050		0.00042		0.00035		0.00029		0.00024	
11.0	0.00020		0.00017		0.00014		0.00012		0.00010	
12.0	0.00008		0.00007		0.00006		0.00005		0.00004	
13.0	0.00003		0.00003		0.00002		0.00002		0.00001	
14.0	0.00001		0.00001		0.00001		0.00001		0.00001	
15.0	0.00001		0.00000		0.00000		0.00000		0.00000	

[a] Abstracted from the more detailed tables of Hilsenrath and Ziegler (605).

Appendix Table A-5 Debye Heat Capacity Function, $Cv°/3R$, in Terms of θ_D/T [a]

$\frac{\theta_D}{T}$	0.0	0.1	0.2	0.3	0.4	0.5	0.6	0.7	0.8	0.9
0.0	1.0000	0.9995	0.9980	0.9955	0.9920	0.9876	0.9822	0.9759	0.9687	0.9606
1.0	0.9517	0.9420	0.9315	0.9203	0.9085	0.8960	0.8828	0.8692	0.8550	0.8404
2.0	0.8254	0.8100	0.7943	0.7784	0.7622	0.7459	0.7294	0.7128	0.6961	0.6794
3.0	0.6628	0.6461	0.6296	0.6132	0.5968	0.5807	0.5647	0.5490	0.5334	0.5181
4.0	0.5031	0.4883	0.4738	0.4595	0.4456	0.4320	0.4187	0.4057	0.3930	0.3807
5.0	0.3686	0.3569	0.3455	0.3345	0.3237	0.3133	0.3031	0.2933	0.2838	0.2745
6.0	0.2656	0.2569	0.2486	0.2405	0.2326	0.2251	0.2177	0.2107	0.2038	0.1972
7.0	0.1909	0.1847	0.1788	0.1730	0.1675	0.1622	0.1570	0.1521	0.1473	0.1426
8.0	0.1382	0.1339	0.1297	0.1257	0.1219	0.1182	0.1146	0.1111	0.1078	0.1046
9.0	0.1015	0.09847	0.09558	0.09280	0.09011	0.08751	0.08500	0.08259	0.08025	0.07800
10.0	0.07582	0.07372	0.07169	0.06973	0.06783	0.06600	0.06424	0.06253	0.06087	0.05928
11.0	0.05773	0.05624	0.05479	0.05339	0.05204	0.05073	0.04946	0.04823	0.04705	0.04590
12.0	0.04478	0.04370	0.04265	0.04164	0.04066	0.03970	0.03878	0.03788	0.03701	0.03617
13.0	0.03535	0.03455	0.03378	0.03303	0.03230	0.03160	0.03091	0.03024	0.02959	0.02896
14.0	0.02835	0.02776	0.02718	0.02661	0.02607	0.02553	0.02501	0.02451	0.02402	0.02354
15.0	0.02307	0.02262	0.02218	0.02174	0.02132	0.02092	0.02052	0.02013	0.01975	0.01938

[a] If $\theta_D/T \geq 16$, $Cv/3R = 77.927(T/\theta_D)^3$; after Pitzer and Brewer (1158).

Appendix Table A-6 Debye Energy Function, $(E° - E_0°)/3RT$, in Terms of θ_D/T [a]

$\frac{\theta_D}{T}$	0.0	0.1	0.2	0.3	0.4	0.5	0.6	0.7	0.8	0.9
0.0	1.0000	0.9630	0.9270	0.8920	0.8580	0.8250	0.7929	0.7619	0.7318	0.7026
1.0	0.6744	0.6471	0.6208	0.5954	0.5708	0.5471	0.5243	0.5023	0.4811	0.4607
2.0	0.4411	0.4223	0.4042	0.3868	0.3701	0.3541	0.3388	0.3241	0.3100	0.2965
3.0	0.2836	0.2712	0.2594	0.2481	0.2373	0.2269	0.2170	0.2076	0.1986	0.1900
4.0	0.1817	0.1739	0.1664	0.1592	0.1524	0.1459	0.1397	0.1338	0.1281	0.1227
5.0	0.1176	0.1127	0.1080	0.1036	0.09930	0.09524	0.09137	0.08768	0.08415	0.08079
6.0	0.07758	0.07452	0.07160	0.06881	0.06615	0.06360	0.06118	0.05886	0.05664	0.05453
7.0	0.05251	0.05057	0.04873	0.04696	0.04527	0.04366	0.04211	0.04063	0.03921	0.03786
8.0	0.03656	0.03532	0.03413	0.03298	0.03189	0.03084	0.02983	0.02887	0.02794	0.02705
9.0	0.02620	0.02538	0.02459	0.02384	0.02311	0.02241	0.02174	0.02109	0.02047	0.01987
10.0	0.01930	0.01874	0.01821	0.01769	0.01720	0.01672	0.01626	0.01581	0.01538	0.01497
11.0	0.01457	0.01418	0.01381	0.01345	0.01311	0.01277	0.01245	0.01213	0.01183	0.01153
12.0	0.01125	0.01098	0.01071	0.01045	0.01020	0.00996	0.00973	0.00950	0.00928	0.00907
13.0	0.00886	0.00866	0.00846	0.00827	0.00809	0.00791	0.00774	0.00757	0.00741	0.00725
14.0	0.00710	0.00695	0.00680	0.00666	0.00652	0.00639	0.00626	0.00613	0.00601	0.00589
15.0	0.00577	0.00566	0.00555	0.00544	0.00533	0.00523	0.00513	0.00503	0.00494	0.00485

[a] If $\theta_D/T \geq 16$, $(E - E_0)/3RT = 19.482(T/\theta_D)^3$; after Pitzer and Brewer (1158).

Appendix Table A-7 Debye Entropy Function, $S°/3R$, in Terms of θ_D/T[a]

$\dfrac{\theta_D}{T}$	0.0	0.1	0.2	0.3	0.4	0.5	0.6	0.7	0.8	0.9
0.0	∞	3.6362	2.9438	2.5396	2.2536	2.0327	1.8531	1.7022	1.5723	1.4587
1.0	1.3579	1.2676	1.1861	1.1120	1.0442	0.9820	0.9246	0.8714	0.8222	0.7763
2.0	0.7336	0.6937	0.6564	0.6214	0.5886	0.5578	0.5289	0.5017	0.4761	0.4519
3.0	0.4292	0.4077	0.3875	0.3683	0.3503	0.3332	0.3171	0.3018	0.2874	0.2737
4.0	0.2608	0.2486	0.2370	0.2260	0.2156	0.2057	0.1964	0.1875	0.1791	0.1711
5.0	0.1636	0.1564	0.1496	0.1431	0.1369	0.1311	0.1255	0.1203	0.1152	0.1105
6.0	0.1059	0.1016	0.09750	0.09358	0.08986	0.08631	0.08293	0.07971	0.07664	0.07371
7.0	0.07092	0.06826	0.06572	0.06329	0.06097	0.05876	0.05665	0.05463	0.05270	0.05085
8.0	0.04908	0.04739	0.04578	0.04423	0.04274	0.04132	0.03996	0.03866	0.03741	0.03621
9.0	0.03506	0.03395	0.03289	0.03187	0.03090	0.02996	0.02905	0.02818	0.02735	0.02655
10.0	0.02577	0.02503	0.02431	0.02362	0.02296	0.02232	0.02170	0.02111	0.02053	0.01998
11.0	0.01944	0.01893	0.01843	0.01795	0.01749	0.01704	0.01660	0.01618	0.01578	0.01539
12.0	0.01501	0.01464	0.01428	0.01394	0.01361	0.01328	0.01297	0.01267	0.01237	0.01209
13.0	0.01181	0.01155	0.01129	0.01103	0.01079	0.01055	0.01032	0.01010	0.00988	0.00967
14.0	0.00946	0.00926	0.00907	0.00888	0.00870	0.00852	0.00834	0.00818	0.00801	0.00785
15.0	0.00770	0.00754	0.00740	0.00725	0.00711	0.00697	0.00684	0.00671	0.00659	0.00646

[a] If $\theta_D/T \geq 16$, $S/3R = 25.976(T/\theta_D)^3$; after Pitzer and Brewer (1158).

Appendix Table A-8 Heat Capacity, C, for Restricted Rotation[a]

$\dfrac{V}{RT}$	\multicolumn{20}{c}{$1/Q_f$}																		
	0.0	0.05	0.10	0.15	0.20	0.25	0.30	0.35	0.40	0.45	0.50	0.55	0.60	0.65	0.70	0.75	0.80	0.85	0.90	0.95
0.0	0.994	0.994	0.994	0.994	0.994	0.994	0.994	0.994	0.994	0.994	0.994	0.994	0.994	0.994	0.994	0.994	0.994	0.994	0.994	0.994
0.2	1.0035	1.003	1.003	1.002	1.001	1.000	0.999	0.998	0.998	0.998	1.000	1.000	1.000	1.000	1.000	1.000	1.000	0.999	0.999	0.999
0.4	1.0328	1.033	1.032	1.030	1.028	1.025	1.024	1.021	1.019	1.017	1.018	1.017	1.015	1.013	1.010	1.010	1.008	1.007	1.005	1.004
0.6	1.0801	1.080	1.079	1.076	1.073	1.068	1.065	1.060	1.056	1.051	1.049	1.046	1.041	1.036	1.031	1.026	1.021	1.017	1.014	1.011
0.8	1.1435	1.143	1.141	1.138	1.133	1.128	1.121	1.114	1.106	1.099	1.092	1.084	1.075	1.067	1.058	1.049	1.040	1.031	1.025	1.020
1.0	1.2203	1.219	1.217	1.212	1.206	1.199	1.190	1.180	1.169	1.157	1.144	1.131	1.118	1.105	1.091	1.078	1.065	1.052	1.040	1.031
1.5	1.4508	1.449	1.444	1.435	1.423	1.408	1.391	1.370	1.348	1.324	1.299	1.273	1.247	1.218	1.192	1.165	1.141	1.115	1.090	1.070
2.0	1.6778	1.695	1.687	1.673	1.655	1.632	1.606	1.574	1.541	1.505	1.465	1.424	1.382	1.341	1.300	1.258	1.218	1.180	1.146	1.113
2.5	1.9213	1.917	1.908	1.888	1.866	1.840	1.801	1.756	1.717	1.670	1.619	1.562	1.504	1.448	1.393	1.341	1.289	1.238	1.190	1.146
3.0	2.0989	2.095	2.082	2.062	2.033	1.996	1.952	1.900	1.846	1.794	1.732	1.663	1.597	1.532	1.466	1.401	1.337	1.276	1.217	1.164
3.5	2.2226	2.218	2.204	2.180	2.146	2.106	2.054	1.995	1.934	1.869	1.803	1.727	1.654	1.580	1.506	1.432	1.361	1.293	1.226	1.165
4.0	2.2989	2.294	2.276	2.249	2.213	2.168	2.110	2.048	1.980	1.907	1.834	1.754	1.674	1.593	1.513	1.435	1.359	1.286	1.215	1.148
4.5	2.3358	2.330	2.312	2.280	2.238	2.190	2.129	2.062	1.990	1.911	1.832	1.749	1.664	1.578	1.496	1.413	1.333	1.259	1.185	1.115
5.0	2.3447	2.338	2.318	2.285	2.241	2.186	2.120	2.056	1.972	1.890	1.808	1.718	1.631	1.543	1.457	1.373	1.292	1.214	1.140	1.068
6.0	2.3158	2.307	2.283	2.245	2.192	2.130	2.059	1.979	1.893	1.803	1.711	1.614	1.520	1.429	1.342	1.255	1.173	1.096	1.022	0.954
7.0	2.2650	2.256	2.228	2.185	2.126	2.055	1.973	1.883	1.787	1.688	1.588	1.487	1.390	1.296	1.207	1.120	1.040	0.962	0.890	0.826
8.0	2.2160	2.205	2.174	2.125	2.058	1.979	1.888	1.788	1.684	1.576	1.468	1.366	1.262	1.164	1.074	0.988	0.908	0.834	0.765	0.704
9.0	2.1762	2.164	2.130	2.074	1.999	1.909	1.808	1.699	1.587	1.474	1.362	1.250	1.144	1.048	0.956	0.869	0.789	0.717	0.652	0.593
10.0	2.1457	2.133	2.094	2.033	1.951	1.854	1.745	1.630	1.507	1.382	1.262	1.151	1.045	0.943	0.850	0.765	0.688	0.618	0.556	0.499
12.0	2.1053	2.089	2.043	1.972	1.877	1.763	1.636	1.502	1.365	1.233	1.107	0.989	0.877	0.774	0.682	0.600	0.528	0.463	0.407	0.358
14.0	2.0813	2.063	2.009	1.923	1.814	1.686	1.546	1.400	1.254	1.112	0.978	0.855	0.744	0.644	0.554	0.479	0.411	0.352	0.303	0.262
16.0	2.0657	2.044	1.983	1.887	1.764	1.622	1.468	1.311	1.156	1.009	0.873	0.749	0.639	0.542	0.457	0.387	0.324	0.272	0.229	0.194
18.0	2.0547	2.031	1.961	1.853	1.717	1.562	1.397	1.232	1.070	0.919	0.780	0.657	0.549	0.456	0.378	0.312	0.259	0.215	0.175	0.144
20.0	2.0465	2.020	1.944	1.827	1.678	1.510	1.333	1.158	0.991	0.837	0.701	0.580	0.477	0.389	0.316	0.256	0.208	0.168	0.135	0.109

[a] After Pitzer and Brewer (1158); values in cal/(mole °K).

Appendix Table A-9   Energy Function, $(E_T^\circ - E_0^\circ)/T$, for Restricted Rotation[a]

$\dfrac{V}{RT}$	0.0	0.05	0.10	0.15	0.20	0.25	0.30	0.35	0.40	0.45	$1/Q_f$ 0.50	0.55	0.60	0.65	0.70	0.75	0.80	0.85	0.90	0.95
0.0	0.994	0.994	0.994	0.994	0.994	0.994	0.994	0.994	0.994	0.994	0.994	0.994	0.994	0.994	0.994	0.994	0.994	0.994	0.994	0.994
0.2	1.1824	1.142	1.106	1.074	1.050	1.032	1.022	1.015	1.008	1.004	1.000	0.996	0.994	0.994	0.994	0.992	0.992	0.991	0.990	0.989
0.4	1.3515	1.300	1.249	1.200	1.151	1.106	1.073	1.051	1.036	1.025	1.015	1.006	0.999	0.994	0.992	0.990	0.988	0.988	0.986	0.985
0.6	1.5013	1.437	1.374	1.311	1.251	1.190	1.138	1.099	1.072	1.049	1.030	1.014	1.004	0.995	0.990	0.987	0.984	0.982	0.980	0.979
0.8	1.6326	1.556	1.482	1.411	1.340	1.272	1.211	1.157	1.114	1.077	1.048	1.026	1.009	0.996	0.984	0.980	0.976	0.974	0.972	0.971
1.0	1.7463	1.660	1.576	1.495	1.418	1.344	1.275	1.211	1.155	1.106	1.065	1.038	1.014	0.996	0.982	0.972	0.965	0.962	0.960	0.959
1.5	1.9610	1.856	1.753	1.654	1.561	1.472	1.385	1.306	1.230	1.164	1.103	1.059	1.019	0.987	0.962	0.945	0.932	0.922	0.916	0.915
2.0	2.0937	1.971	1.854	1.742	1.636	1.536	1.440	1.350	1.265	1.190	1.120	1.057	1.005	0.962	0.928	0.904	0.886	0.873	0.864	0.860
2.5	2.1660	2.031	1.900	1.779	1.662	1.550	1.448	1.351	1.260	1.179	1.104	1.032	0.972	0.922	0.882	0.850	0.827	0.811	0.801	0.796
3.0	2.1974	2.049	1.909	1.777	1.651	1.535	1.426	1.321	1.224	1.140	1.060	0.988	0.924	0.870	0.828	0.791	0.763	0.744	0.732	0.728
3.5	2.2033	2.043	1.893	1.753	1.621	1.497	1.382	1.275	1.176	1.088	1.006	0.933	0.868	0.811	0.765	0.727	0.697	0.676	0.663	0.659
4.0	2.1947	2.024	1.864	1.715	1.577	1.448	1.329	1.221	1.121	1.030	0.947	0.872	0.806	0.749	0.701	0.661	0.630	0.609	0.595	0.590
4.5	2.1791	1.998	1.829	1.673	1.529	1.394	1.273	1.162	1.061	0.968	0.884	0.810	0.744	0.687	0.638	0.599	0.567	0.545	0.531	0.526
5.0	2.1610	1.971	1.794	1.631	1.481	1.344	1.218	1.104	1.002	0.909	0.824	0.750	0.685	0.628	0.580	0.540	0.508	0.485	0.470	0.465
6.0	2.1264	1.918	1.727	1.552	1.392	1.247	1.115	0.999	0.893	0.799	0.714	0.644	0.580	0.523	0.476	0.437	0.406	0.383	0.368	0.361
7.0	2.0987	1.875	1.670	1.484	1.315	1.164	1.029	0.908	0.802	0.708	0.624	0.554	0.491	0.437	0.392	0.354	0.324	0.302	0.286	0.279
8.0	2.0784	1.840	1.623	1.427	1.251	1.095	0.955	0.833	0.725	0.631	0.549	0.480	0.420	0.368	0.326	0.290	0.261	0.239	0.223	0.215
9.0	2.0637	1.811	1.583	1.379	1.196	1.035	0.892	0.768	0.661	0.569	0.488	0.421	0.363	0.312	0.273	0.240	0.211	0.191	0.176	0.168
10.0	2.0529	1.787	1.548	1.335	1.147	0.982	0.838	0.715	0.608	0.515	0.437	0.370	0.314	0.269	0.231	0.200	0.174	0.154	0.140	0.132
12.0	2.0385	1.749	1.492	1.264	1.067	0.896	0.745	0.624	0.519	0.431	0.356	0.296	0.244	0.202	0.170	0.143	0.121	0.104	0.091	0.084
14.0	2.0295	1.717	1.441	1.202	0.997	0.823	0.672	0.551	0.450	0.365	0.297	0.240	0.195	0.158	0.127	0.103	0.084	0.072	0.062	0.056
16.0	2.0232	1.690	1.401	1.150	0.937	0.760	0.613	0.493	0.394	0.314	0.249	0.198	0.157	0.127	0.098	0.076	0.061	0.051	0.044	0.038
18.0	2.0185	1.666	1.363	1.102	0.886	0.707	0.561	0.443	0.347	0.271	0.211	0.164	0.128	0.099	0.077	0.060	0.047	0.036	0.029	0.026
20.0	2.0150	1.646	1.329	1.061	0.841	0.660	0.515	0.399	0.307	0.236	0.181	0.138	0.105	0.080	0.061	0.047	0.036	0.028	0.022	0.018

[a] After Pitzer and Brewer (1158); values in cal/(mole °K).

Appendix Table A-10 Entropy $S°$, for Restricted Rotation[a]

$\dfrac{V}{RT}$	\multicolumn{15}{c}{$1/Q_f$}
	0.25	0.30	0.35	0.40	0.45	0.50	0.55	0.60	0.65	0.70	0.75	0.80	0.85	0.90	0.95
0.0	3.748	3.386	3.079	2.814	2.580	2.371	2.182	2.009	1.850	1.703	1.567	1.438	1.316	1.203	1.097
0.2	3.743	3.382	3.076	2.811	2.578	2.369	2.180	2.003	1.848	1.701	1.563	1.433	1.312	1.196	1.091
0.4	3.730	3.370	3.065	2.801	2.568	2.359	2.170	1.996	1.837	1.691	1.555	1.428	1.307	1.193	1.085
0.6	3.709	3.347	3.043	2.780	2.547	2.340	2.151	1.980	1.823	1.677	1.541	1.415	1.295	1.184	1.076
0.8	3.679	3.318	3.013	2.750	2.519	2.315	2.125	1.957	1.800	1.654	1.523	1.399	1.284	1.171	1.068
1.0	3.638	3.279	2.974	2.714	2.485	2.279	2.094	1.928	1.774	1.629	1.499	1.377	1.262	1.153	1.052
1.5	3.512	3.156	2.854	2.600	2.376	2.173	1.997	1.833	1.685	1.552	1.428	1.310	1.201	1.094	1.000
2.0	3.355	3.004	2.709	2.458	2.241	2.048	1.874	1.718	1.578	1.450	1.332	1.224	1.122	1.024	0.936
2.5	3.180	2.836	2.548	2.303	2.091	1.907	1.739	1.589	1.456	1.335	1.224	1.126	1.031	0.942	0.860
3.0	3.008	2.667	2.380	2.138	1.933	1.756	1.576	1.456	1.330	1.217	1.114	1.021	0.936	0.855	0.779
3.5	2.838	2.500	2.218	1.978	1.782	1.610	1.458	1.323	1.206	1.100	1.004	0.919	0.841	0.769	0.703
4.0	2.678	2.343	2.069	1.834	1.643	1.475	1.328	1.199	1.087	0.988	0.901	0.821	0.748	0.683	0.623
4.5	2.528	2.199	1.926	1.698	1.511	1.348	1.209	1.086	0.978	0.884	0.804	0.730	0.662	0.607	0.551
5.0	2.396	2.068	1.798	1.579	1.392	1.233	1.097	0.982	0.881	0.794	0.716	0.648	0.588	0.535	0.486
6.0	2.166	1.844	1.585	1.370	1.192	1.040	0.915	0.808	0.715	0.637	0.568	0.509	0.457	0.412	0.372
7.0	1.983	1.665	1.411	1.204	1.033	0.891	0.774	0.672	0.588	0.516	0.453	0.401	0.357	0.319	0.285
8.0	1.830	1.519	1.272	1.071	0.906	0.770	0.660	0.566	0.486	0.422	0.366	0.320	0.281	0.248	0.220
9.0	1.703	1.397	1.156	0.962	0.804	0.674	0.570	0.483	0.407	0.350	0.300	0.258	0.223	0.195	0.171
10.0	1.593	1.295	1.060	0.872	0.719	0.596	0.496	0.414	0.348	0.293	0.248	0.211	0.180	0.154	0.134
12.0	1.417	1.125	0.904	0.728	0.588	0.476	0.388	0.315	0.255	0.213	0.176	0.146	0.122	0.101	0.084
14.0	1.275	0.994	0.783	0.620	0.492	0.388	0.309	0.247	0.196	0.157	0.126	0.100	0.084	0.069	0.056
16.0	1.157	0.890	0.688	0.533	0.414	0.322	0.251	0.196	0.155	0.119	0.092	0.075	0.059	0.048	0.038
18.0	1.058	0.801	0.609	0.464	0.353	0.270	0.205	0.158	0.121	0.093	0.072	0.056	0.042	0.034	0.026
20.0	0.975	0.727	0.542	0.405	0.303	0.228	0.170	0.129	0.097	0.073	0.056	0.042	0.032	0.024	0.018

[a] After Pitzer and Brewer (1158); values in cal/(mole °K).

Appendix Table A-11   Entropy Decrease from Free Rotation, $S_f - S$ [a]

$\dfrac{V}{RT}$	\multicolumn{12}{c}{$1/Q_f$}
	0.0	0.05	0.10	0.15	0.20	0.25	0.30	0.35	0.40	0.45	0.50	0.55
0.0	0.0000	0.000	0.000	0.000	0.000	0.000	0.000	0.000	0.000	0.000	0.000	0.000
0.2	0.0049	0.005	0.004	0.004	0.004	0.004	0.004	0.003	0.003	0.002	0.002	0.002
0.4	0.0198	0.020	0.018	0.018	0.018	0.018	0.016	0.014	0.013	0.012	0.012	0.010
0.6	0.0440	0.044	0.043	0.043	0.040	0.039	0.039	0.036	0.034	0.033	0.031	0.028
0.8	0.0771	0.077	0.077	0.075	0.072	0.069	0.068	0.066	0.064	0.061	0.056	0.053
1.0	0.1185	0.118	0.117	0.115	0.112	0.110	0.107	0.105	0.100	0.095	0.092	0.086
1.5	0.2527	0.252	0.250	0.248	0.242	0.236	0.230	0.225	0.214	0.204	0.198	0.189
2.0	0.4182	0.417	0.415	0.410	0.402	0.393	0.382	0.370	0.356	0.339	0.323	0.308
2.5	0.6001	0.599	0.594	0.585	0.577	0.568	0.550	0.531	0.511	0.489	0.464	0.440
3.0	0.7856	0.783	0.777	0.768	0.757	0.740	0.719	0.699	0.676	0.647	0.615	0.581
3.5	0.9660	0.964	0.957	0.944	0.929	0.910	0.886	0.861	0.836	0.798	0.761	0.722
4.0	1.1356	1.133	1.126	1.111	1.094	1.070	1.043	1.011	0.980	0.937	0.896	0.855
4.5	1.2918	1.289	1.280	1.265	1.244	1.220	1.187	1.153	1.116	1.069	1.023	0.977
5.0	1.4339	1.431	1.421	1.404	1.380	1.352	1.318	1.281	1.235	1.188	1.138	1.086
6.0	1.6781	1.674	1.662	1.643	1.616	1.582	1.542	1.494	1.444	1.388	1.331	1.268
7.0	1.8783	1.874	1.860	1.837	1.807	1.765	1.721	1.668	1.610	1.547	1.480	1.411
8.0	2.0447	2.040	2.024	1.998	1.962	1.918	1.867	1.807	1.743	1.674	1.601	1.525
9.0	2.1864	2.180	2.163	2.134	2.095	2.045	1.989	1.923	1.852	1.776	1.697	1.612
10.0	2.3095	2.303	2.284	2.252	2.208	2.155	2.091	2.019	1.942	1.861	1.775	1.686
12.0	2.5155	2.508	2.485	2.447	2.394	2.331	2.261	2.175	2.086	1.992	1.895	1.793
14.0	2.6847	2.676	2.650	2.607	2.547	2.473	2.392	2.296	2.194	2.088	1.983	1.872
16.0	2.8289	2.819	2.788	2.740	2.674	2.591	2.496	2.391	2.281	2.166	2.049	1.930
18.0	2.9545	2.943	2.910	2.855	2.781	2.690	2.585	2.470	2.350	2.227	2.101	1.976
20.0	3.0659	3.054	3.017	2.956	2.872	2.773	2.659	2.537	2.409	2.277	2.143	2.011

[a] $S_f = R(1/2 + \ln Q_f)$; after Pitzer and Brewer (1158); values in cal/(mole °K).

Appendix Table A-12 Comparative Thermodynamic Properties of the Restricted Rotator and the Harmonic Oscillator[a]

$\dfrac{V}{RT}$	$S - S_h$	$\dfrac{-(G-G_h)}{T}$	$\dfrac{(H-H_h)}{T}$	$C - C_h$
0.6	−0.45	0.03	−0.48	−0.90
1.0	−0.02	0.22	−0.24	−0.76
2.0	0.37	0.27	0.10	−0.29
3.0	0.41	0.20	0.21	0.10
4.0	0.35	0.14	0.21	0.30
6.0	0.21	0.08	0.13	0.32
9.0	0.11	0.03	0.08	0.19
12.0	0.06	0.02	0.04	0.12
16.0	0.04	0.01	0.03	0.07
20.0	0.02	0.0	0.02	0.05

[a] The subscript h refers to the harmonic oscillator. $\dfrac{h\nu}{kT} = (1.74/Q_f)\,(V/RT)^{1/2}$. After Halford (563).

Appendix Table A-13 Harmonic Oscillator Contributions to the Heat Capacity Tabulated for Vibrational Frequencies ω [a]

ω(cm^{-1})	298.15°K	300°K	400°K	500°K	600°K	700°K	800°K	900°K	1000°K
				$\dfrac{C°}{R}$					
100	0.9808	0.9811	0.9893	0.9931	0.9952	0.9965	0.9973	0.9979	0.9983
110	0.9769	0.9771	0.9871	0.9917	0.9942	0.9957	0.9967	0.9974	0.9979
120	0.9725	0.9729	0.9846	0.9901	0.9931	0.9949	0.9961	0.9969	0.9975
130	0.9678	0.9682	0.9820	0.9884	0.9919	0.9940	0.9955	0.9964	0.9971
140	0.9628	0.9633	0.9791	0.9866	0.9907	0.9931	0.9947	0.9958	0.9966
150	0.9575	0.9580	0.9761	0.9846	0.9893	0.9921	0.9940	0.9952	0.9961
160	0.9518	0.9523	0.9729	0.9825	0.9878	0.9910	0.9931	0.9946	0.9956
170	0.9458	0.9464	0.9694	0.9803	0.9863	0.9899	0.9922	0.9939	0.9950
180	0.9394	0.9401	0.9658	0.9780	0.9846	0.9887	0.9913	0.9931	0.9944
190	0.9328	0.9336	0.9620	0.9755	0.9829	0.9874	0.9903	0.9923	0.9938
200	0.9259	0.9267	0.9580	0.9729	0.9811	0.9860	0.9893	0.9915	0.9931
210	0.9187	0.9196	0.9538	0.9701	0.9791	0.9846	0.9882	0.9907	0.9924
220	0.9112	0.9120	0.9494	0.9673	0.9770	0.9831	0.9871	0.9898	0.9917
230	0.9034	0.9045	0.9449	0.9643	0.9750	0.9816	0.9859	0.9888	0.9909
240	0.8954	0.8966	0.9402	0.9612	0.9729	0.9800	0.9846	0.9878	0.9901
250	0.8871	0.8884	0.9353	0.9580	0.9706	0.9783	0.9833	0.9868	0.9893
260	0.8786	0.8800	0.9302	0.9546	0.9682	0.9765	0.9820	0.9857	0.9884
270	0.8698	0.8713	0.9250	0.9512	0.9658	0.9747	0.9806	0.9846	0.9875
280	0.8608	0.8624	0.9196	0.9476	0.9633	0.9729	0.9791	0.9835	0.9866
290	0.8517	0.8533	0.9141	0.9440	0.9607	0.9709	0.9776	0.9823	0.9856
300	0.8423	0.8440	0.9084	0.9402	0.9580	0.9686	0.9761	0.9811	0.9846
310	0.8327	0.8345	0.9025	0.9363	0.9553	0.9669	0.9745	0.9798	0.9836
320	0.8229	0.8249	0.8966	0.9323	0.9524	0.9647	0.9729	0.9785	0.9826
330	0.8130	0.8151	0.8904	0.9281	0.9494	0.9625	0.9712	0.9771	0.9814
340	0.8030	0.8051	0.8842	0.9239	0.9464	0.9603	0.9694	0.9758	0.9803
350	0.7928	0.7950	0.8778	0.9196	0.9433	0.9580	0.9676	0.9743	0.9791
360	0.7824	0.7847	0.8713	0.9152	0.9402	0.9556	0.9658	0.9729	0.9780
370	0.7719	0.7743	0.8646	0.9107	0.9369	0.9532	0.9639	0.9714	0.9767
380	0.7613	0.7638	0.8579	0.9061	0.9336	0.9507	0.9620	0.9698	0.9755
390	0.7510	0.7532	0.8510	0.9014	0.9302	0.9481	0.9600	0.9682	0.9742

Appendix Table A-13 *continued*

$$\frac{C°}{R}$$

ω(cm⁻¹)	298.15°K	300°K	400°K	500°K	600°K	700°K	800°K	900°K	1000°K
400	0.7398	0.7425	0.8440	0.8966	0.9268	0.9456	0.9580	0.9666	0.9729
410	0.7290	0.7317	0.8369	0.8917	0.9232	0.9429	0.9559	0.9650	0.9715
420	0.7180	0.7208	0.8297	0.8867	0.9196	0.9402	0.9538	0.9633	0.9701
430	0.7070	0.7100	0.8224	0.8816	0.9159	0.9374	0.9516	0.9616	0.9687
440	0.6960	0.6990	0.8151	0.8765	0.9122	0.9346	0.9494	0.9598	0.9673
450	0.6849	0.6880	0.8076	0.8713	0.9084	0.9317	0.9472	0.9580	0.9658
460	0.6738	0.6769	0.8000	0.8660	0.9043	0.9287	0.9449	0.9561	0.9643
470	0.6626	0.6658	0.7924	0.8606	0.9005	0.9258	0.9425	0.9543	0.9628
480	0.6514	0.6548	0.7847	0.8551	0.8967	0.9227	0.9402	0.9524	0.9612
490	0.6403	0.6437	0.7769	0.8495	0.8925	0.9196	0.9378	0.9504	0.9596
500	0.6291	0.6326	0.7691	0.8440	0.8884	0.9165	0.9353	0.9484	0.9580
510	0.6179	0.6215	0.7612	0.8384	0.8844	0.9133	0.9328	0.9464	0.9563
520	0.6068	0.6104	0.7532	0.8326	0.8799	0.9100	0.9302	0.9444	0.9546
530	0.5957	0.5993	0.7452	0.8268	0.8756	0.9067	0.9276	0.9423	0.9529
540	0.5846	0.5883	0.7371	0.8210	0.8713	0.9034	0.9250	0.9402	0.9512
550	0.5736	0.5773	0.7290	0.8151	0.8669	0.9000	0.9223	0.9380	0.9494
560	0.5626	0.5664	0.7209	0.8091	0.8624	0.8966	0.9196	0.9358	0.9476
570	0.5517	0.5555	0.7127	0.8031	0.8579	0.8931	0.9169	0.9336	0.9458
580	0.5408	0.5447	0.7045	0.7970	0.8533	0.8896	0.9141	0.9314	0.9439
590	0.5300	0.5340	0.6962	0.7909	0.8487	0.8860	0.9113	0.9291	0.9421
600	0.5193	0.5233	0.6879	0.7847	0.8440	0.8824	0.9084	0.9268	0.9402
610	0.5087	0.5127	0.6797	0.7785	0.8393	0.8787	0.9055	0.9244	0.9382
620	0.4981	0.5021	0.6716	0.7722	0.8345	0.8750	0.9025	0.9220	0.9363
630	0.4876	0.4917	0.6631	0.7659	0.8297	0.8713	0.8996	0.9196	0.9343
640	0.4773	0.4814	0.6547	0.7596	0.8249	0.8675	0.8966	0.9172	0.9323
650	0.4670	0.4712	0.6467	0.7532	0.8200	0.8637	0.8935	0.9147	0.9302
660	0.4569	0.4610	0.6381	0.7468	0.8151	0.8598	0.8905	0.9122	0.9282
670	0.4468	0.4510	0.6298	0.7404	0.8101	0.8559	0.8873	0.9097	0.9261
680	0.4369	0.4410	0.6215	0.7339	0.8051	0.8520	0.8842	0.9071	0.9239
690	0.4271	0.4312	0.6132	0.7274	0.8000	0.8480	0.8810	0.9045	0.9218

Appendix Table A-13 *continued*

$$\frac{C°}{R}$$

ω(cm⁻¹)	298.15°K	300°K	400°K	500°K	600°K	700°K	800°K	900°K	1000°K
700	0.4174	0.4215	0.6048	0.7209	0.7950	0.8440	0.8778	0.9019	0.9196
710	0.4079	0.4120	0.5966	0.7142	0.7898	0.8400	0.8746	0.8992	0.9174
720	0.3983	0.4025	0.5883	0.7078	0.7847	0.8359	0.8713	0.8966	0.9152
730	0.3890	0.3932	0.5801	0.7012	0.7795	0.8318	0.8680	0.8939	0.9130
740	0.3798	0.3840	0.5719	0.6946	0.7743	0.8277	0.8646	0.8911	0.9107
750	0.3708	0.3750	0.5637	0.6880	0.7691	0.8234	0.8613	0.8884	0.9084
760	0.3619	0.3660	0.5555	0.6813	0.7638	0.8193	0.8579	0.8856	0.9061
770	0.3531	0.3572	0.5474	0.6747	0.7585	0.8151	0.8545	0.8828	0.9037
780	0.3444	0.3486	0.5393	0.6681	0.7532	0.8108	0.8510	0.8800	0.9014
790	0.3359	0.3401	0.5313	0.6614	0.7479	0.8065	0.8475	0.8771	0.8990
800	0.3275	0.3317	0.5233	0.6548	0.7425	0.8022	0.8440	0.8742	0.8966
810	0.3193	0.3234	0.5153	0.6481	0.7371	0.7979	0.8405	0.8713	0.8941
820	0.3112	0.3153	0.5074	0.6414	0.7317	0.7935	0.8369	0.8684	0.8917
830	0.3033	0.3074	0.4995	0.6348	0.7263	0.7891	0.8334	0.8654	0.8892
840	0.2955	0.2995	0.4917	0.6284	0.7209	0.7847	0.8297	0.8624	0.8867
850	0.2879	0.2919	0.4839	0.6214	0.7154	0.7803	0.8261	0.8594	0.8842
860	0.2803	0.2843	0.4762	0.6148	0.7100	0.7758	0.8224	0.8564	0.8816
870	0.2730	0.2769	0.4686	0.6082	0.7045	0.7713	0.8188	0.8533	0.8791
880	0.2657	0.2697	0.4610	0.6015	0.6990	0.7668	0.8151	0.8502	0.8765
890	0.2587	0.2625	0.4535	0.5949	0.6935	0.7623	0.8114	0.8471	0.8739
900	0.2517	0.2556	0.4460	0.5883	0.6880	0.7578	0.8076	0.8440	0.8713
910	0.2449	0.2487	0.4386	0.5817	0.6824	0.7532	0.8038	0.8409	0.8686
920	0.2383	0.2420	0.4312	0.5751	0.6769	0.7487	0.8001	0.8377	0.8660
930	0.2317	0.2355	0.4240	0.5686	0.6714	0.7441	0.7964	0.8346	0.8633
940	0.2253	0.2290	0.4167	0.5620	0.6658	0.7395	0.7924	0.8314	0.8606
950	0.2191	0.2228	0.4096	0.5555	0.6603	0.7348	0.7886	0.8281	0.8579
960	0.2130	0.2166	0.4025	0.5490	0.6548	0.7302	0.7847	0.8249	0.8551
970	0.2070	0.2106	0.3955	0.5425	0.6492	0.7255	0.7808	0.8216	0.8524
980	0.2012	0.2047	0.3886	0.5361	0.6436	0.7209	0.7769	0.8184	0.8496
990	0.1955	0.1990	0.3817	0.5297	0.6381	0.7162	0.7730	0.8151	0.8468

Appendix Table A-13 *continued*

	$\dfrac{C°}{R}$								
ω(cm⁻¹)	298.15°K	300°K	400°K	500°K	600°K	700°K	800°K	900°K	1000°K
1000	0.1899	0.1933	0.3750	0.5233	0.6326	0.7115	0.7691	0.8118	0.8440
1010	0.1844	0.1878	0.3683	0.5169	0.6270	0.7068	0.7651	0.8084	0.8412
1020	0.1791	0.1825	0.3616	0.5105	0.6214	0.7021	0.7612	0.8051	0.8384
1030	0.1739	0.1772	0.3550	0.5042	0.6159	0.6974	0.7572	0.8017	0.8355
1040	0.1689	0.1721	0.3488	0.4980	0.6104	0.6927	0.7532	0.7984	0.8326
1050	0.1639	0.1671	0.3422	0.4917	0.6048	0.6880	0.7492	0.7950	0.8297
1060	0.1591	0.1622	0.3358	0.4855	0.5993	0.6832	0.7452	0.7916	0.8268
1070	0.1544	0.1575	0.3306	0.4793	0.5938	0.6785	0.7412	0.7881	0.8239
1080	0.1498	0.1528	0.3234	0.4732	0.5883	0.6738	0.7372	0.7847	0.8210
1090	0.1453	0.1483	0.3173	0.4671	0.5828	0.6690	0.7331	0.7813	0.8180
1100	0.1411	0.1439	0.3113	0.4610	0.5773	0.6643	0.7290	0.7778	0.8151
1110	0.1367	0.1396	0.3055	0.4550	0.5718	0.6595	0.7250	0.7743	0.8121
1120	0.1326	0.1354	0.2995	0.4490	0.5664	0.6548	0.7209	0.7708	0.8091
1130	0.1285	0.1313	0.2938	0.4430	0.5609	0.6500	0.7168	0.7673	0.8061
1140	0.1246	0.1274	0.2881	0.4371	0.5555	0.6452	0.7127	0.7638	0.8031
1150	0.1208	0.1235	0.2825	0.4312	0.5501	0.6405	0.7086	0.7603	0.8000
1160	0.1170	0.1197	0.2769	0.4254	0.5447	0.6357	0.7045	0.7568	0.7970
1170	0.1134	0.1160	0.2715	0.4196	0.5393	0.6310	0.7004	0.7532	0.7939
1180	0.1099	0.1125	0.2661	0.4139	0.5339	0.6262	0.6963	0.7497	0.7909
1190	0.1065	0.1090	0.2608	0.4082	0.5286	0.6215	0.6921	0.7461	0.7878
1200	0.1031	0.1056	0.2556	0.4025	0.5232	0.6167	0.6880	0.7425	0.7847
1210	0.0999	0.1023	0.2504	0.3969	0.5179	0.6120	0.6838	0.7389	0.7816
1220	0.0967	0.0991	0.2454	0.3914	0.5127	0.6072	0.6797	0.7354	0.7785
1230	0.0937	0.0960	0.2404	0.3859	0.5074	0.6025	0.6756	0.7318	0.7754
1240	0.0907	0.0930	0.2355	0.3804	0.5021	0.5978	0.6714	0.7281	0.7722
1250	0.0878	0.0900	0.2306	0.3750	0.4969	0.5930	0.6672	0.7245	0.7691
1260	0.0850	0.0872	0.2259	0.3696	0.4917	0.5883	0.6631	0.7209	0.7659
1270	0.0823	0.0844	0.2212	0.3643	0.4865	0.5836	0.6589	0.7172	0.7628
1280	0.0796	0.0817	0.2166	0.3590	0.4814	0.5789	0.6548	0.7136	0.7596
1290	0.0770	0.0791	0.2121	0.3537	0.4762	0.5742	0.6506	0.7100	0.7564

Appendix Table A-13 *continued*

$$\frac{C°}{R}$$

ω(cm⁻¹)	298.15°K	300°K	400°K	500°K	600°K	700°K	800°K	900°K	1000°K
1300	0.0745	0.0765	0.2076	0.3486	0.4711	0.5695	0.6464	0.7063	0.7532
1310	0.0721	0.0741	0.2032	0.3434	0.4661	0.5649	0.6423	0.7027	0.7500
1320	0.0697	0.0717	0.1989	0.3384	0.4610	0.5603	0.6381	0.6990	0.7468
1330	0.0675	0.0693	0.1947	0.3333	0.4560	0.5555	0.6340	0.6953	0.7436
1340	0.0653	0.0671	0.1906	0.3284	0.4510	0.5509	0.6298	0.6917	0.7404
1350	0.0631	0.0649	0.1865	0.3234	0.4460	0.5462	0.6256	0.6880	0.7371
1360	0.0610	0.0628	0.1825	0.3185	0.4410	0.5416	0.6215	0.6843	0.7339
1370	0.0590	0.0607	0.1785	0.3137	0.4361	0.5370	0.6173	0.6806	0.7307
1380	0.0570	0.0587	0.1746	0.3089	0.4312	0.5324	0.6132	0.6769	0.7274
1390	0.0551	0.0568	0.1708	0.3042	0.4264	0.5278	0.6090	0.6732	0.7241
1400	0.0533	0.0549	0.1671	0.2995	0.4215	0.5233	0.6049	0.6696	0.7209
1410	0.0512	0.0531	0.1634	0.2949	0.4167	0.5187	0.6007	0.6659	0.7176
1420	0.0498	0.0513	0.1598	0.2903	0.4120	0.5142	0.5966	0.6622	0.7143
1430	0.0481	0.0496	0.1563	0.2858	0.4072	0.5096	0.5924	0.6585	0.7111
1440	0.0465	0.0479	0.1528	0.2813	0.4025	0.5052	0.5883	0.6548	0.7078
1450	0.0449	0.0463	0.1494	0.2769	0.3979	0.5007	0.5842	0.6511	0.7045
1460	0.0434	0.0447	0.1461	0.2725	0.3932	0.4962	0.5801	0.6474	0.7012
1470	0.0419	0.0432	0.1428	0.2682	0.3886	0.4917	0.5760	0.6437	0.6979
1480	0.0405	0.0418	0.1396	0.2640	0.3840	0.4873	0.5719	0.6400	0.6946
1490	0.0391	0.0403	0.1364	0.2597	0.3794	0.4829	0.5678	0.6364	0.6913
1500	0.0377	0.0390	0.1334	0.2556	0.3750	0.4785	0.5637	0.6326	0.6880
1510	0.0364	0.0376	0.1303	0.2514	0.3705	0.4741	0.5596	0.6289	0.6847
1520	0.0352	0.0363	0.1273	0.2474	0.3660	0.4697	0.5555	0.6252	0.6814
1530	0.0340	0.0351	0.1244	0.2434	0.3616	0.4653	0.5515	0.6217	0.6780
1540	0.0328	0.0339	0.1216	0.2394	0.3572	0.4610	0.5474	0.6178	0.6747
1550	0.0316	0.0327	0.1188	0.2355	0.3529	0.4568	0.5434	0.6141	0.6714
1560	0.0306	0.0316	0.1160	0.2316	0.3486	0.4524	0.5393	0.6100	0.6681
1570	0.0295	0.0305	0.1133	0.2278	0.3443	0.4481	0.5353	0.6067	0.6647
1580	0.0285	0.0294	0.1107	0.2240	0.3400	0.4439	0.5313	0.6030	0.6614
1590	0.0275	0.0284	0.1081	0.2203	0.3358	0.4396	0.5273	0.5993	0.6581

Appendix A-13

Appendix Table A-13 *continued*

$$\frac{C°}{R}$$

ω(cm⁻¹)	298.15°K	300°K	400°K	500°K	600°K	700°K	800°K	900°K	1000°K
1600	0.0265	0.0274	0.1056	0.2166	0.3317	0.4354	0.5233	0.5957	0.6548
1610	0.0255	0.0265	0.1031	0.2130	0.3275	0.4313	0.5193	0.5920	0.6513
1620	0.0246	0.0255	0.1007	0.2094	0.3234	0.4271	0.5153	0.5883	0.6481
1630	0.0238	0.0246	0.0983	0.2058	0.3193	0.4229	0.5114	0.5847	0.6448
1640	0.0229	0.0238	0.0960	0.2024	0.3153	0.4188	0.5074	0.5810	0.6414
1650	0.0221	0.0229	0.0936	0.1989	0.3113	0.4147	0.5035	0.5773	0.6381
1660	0.0213	0.0221	0.0915	0.1956	0.3073	0.4106	0.4995	0.5737	0.6348
1670	0.0206	0.0214	0.0893	0.1922	0.3034	0.4066	0.4956	0.5700	0.6314
1680	0.0198	0.0206	0.0872	0.1889	0.2995	0.4026	0.4917	0.5664	0.6281
1690	0.0191	0.0199	0.0850	0.1857	0.2957	0.3985	0.4879	0.5628	0.6248
1700	0.0184	0.0192	0.0831	0.1825	0.2918	0.3945	0.4840	0.5592	0.6215
1710	0.0178	0.0185	0.0810	0.1793	0.2881	0.3916	0.4801	0.5555	0.6181
1720	0.0171	0.0178	0.0791	0.1762	0.2843	0.3866	0.4763	0.5519	0.6148
1730	0.0165	0.0172	0.0772	0.1732	0.2806	0.3827	0.4724	0.5483	0.6115
1740	0.0159	0.0166	0.0753	0.1701	0.2769	0.3788	0.4686	0.5447	0.6082
1750	0.0154	0.0160	0.0735	0.1671	0.2732	0.3750	0.4648	0.5411	0.6048
1760	0.0148	0.0154	0.0717	0.1642	0.2696	0.3711	0.4610	0.5375	0.6015
1770	0.0143	0.0149	0.0699	0.1613	0.2661	0.3673	0.4572	0.5340	0.5982
1780	0.0138	0.0143	0.0682	0.1584	0.2625	0.3635	0.4535	0.5304	0.5949
1790	0.0134	0.0138	0.0665	0.1556	0.2590	0.3598	0.4497	0.5268	0.5916
1800	0.0128	0.0133	0.0649	0.1528	0.2555	0.3560	0.4460	0.5233	0.5883
1810	0.0123	0.0128	0.0633	0.1501	0.2521	0.3523	0.4423	0.5197	0.5853
1820	0.0118	0.0124	0.0617	0.1474	0.2487	0.3486	0.4386	0.5162	0.5817
1830	0.0114	0.0119	0.0602	0.1448	0.2453	0.3449	0.4349	0.5127	0.5784
1840	0.0110	0.0115	0.0587	0.1422	0.2420	0.3413	0.4312	0.5092	0.5751
1850	0.0106	0.0111	0.0572	0.1396	0.2387	0.3377	0.4276	0.5057	0.5719
1860	0.0102	0.0107	0.0558	0.1371	0.2355	0.3341	0.4240	0.5022	0.5686
1870	0.0098	0.0103	0.0544	0.1346	0.2322	0.3305	0.4204	0.4987	0.5653
1880	0.0095	0.0099	0.0531	0.1321	0.2290	0.3270	0.4168	0.4952	0.5620
1890	0.0091	0.0095	0.0517	0.1297	0.2259	0.3234	0.4132	0.4917	0.5588

Appendix Table A-13 *continued*

$$\frac{C°}{R}$$

ω(cm⁻¹)	298.15°K	300°K	400°K	500°K	600°K	700°K	800°K	900°K	1000°K
1900	0.0088	0.0092	0.0504	0.1273	0.2227	0.3199	0.4096	0.4883	0.5555
1910	0.0084	0.0088	0.0492	0.1250	0.2196	0.3165	0.4061	0.4848	0.5523
1920	0.0081	0.0085	0.0479	0.1227	0.2166	0.3130	0.4026	0.4814	0.5490
1930	0.0078	0.0082	0.0467	0.1205	0.2136	0.3096	0.3990	0.4780	0.5458
1940	0.0075	0.0079	0.0455	0.1182	0.2106	0.3063	0.3955	0.4745	0.5425
1950	0.0073	0.0076	0.0443	0.1160	0.2076	0.3029	0.3921	0.4711	0.5393
1960	0.0070	0.0073	0.0432	0.1139	0.2047	0.2995	0.3886	0.4678	0.5361
1970	0.0067	0.0070	0.0421	0.1118	0.2018	0.2962	0.3852	0.4644	0.5329
1980	0.0065	0.0068	0.0410	0.1097	0.1989	0.2929	0.3819	0.4610	0.5297
1990	0.0062	0.0065	0.0400	0.1076	0.1961	0.2897	0.3784	0.4577	0.5264
2000	0.0060	0.0063	0.0390	0.1056	0.1933	0.2865	0.3750	0.4543	0.5233
2010	0.0058	0.0061	0.0380	0.1036	0.1905	0.2833	0.3716	0.4510	0.5201
2020	0.0056	0.0058	0.0370	0.1017	0.1878	0.2801	0.3683	0.4477	0.5169
2030	0.0054	0.0056	0.0360	0.0998	0.1851	0.2769	0.3649	0.4444	0.5137
2040	0.0052	0.0054	0.0351	0.0979	0.1825	0.2738	0.3616	0.4410	0.5105
2050	0.0050	0.0052	0.0342	0.0942	0.1798	0.2707	0.3583	0.4378	0.5074
2060	0.0048	0.0050	0.0333	0.0924	0.1772	0.2676	0.3551	0.4345	0.5042
2070	0.0046	0.0048	0.0324	0.0906	0.1746	0.2646	0.3518	0.4312	0.5011
2080	0.0044	0.0046	0.0316	0.0889	0.1721	0.2615	0.3486	0.4280	0.4980
2090	0.0043	0.0045	0.0308	0.0872	0.1696	0.2585	0.3454	0.4248	0.4948
2100	0.0041	0.0043	0.0300	0.0855	0.1671	0.2556	0.3422	0.4216	0.4917
2110	0.0039	0.0041	0.0292	0.0839	0.1647	0.2526	0.3390	0.4184	0.4886
2120	0.0038	0.0040	0.0284	0.0822	0.1622	0.2496	0.3359	0.4152	0.4855
2130	0.0036	0.0038	0.0277	0.0806	0.1598	0.2468	0.3327	0.4120	0.4824
2140	0.0035	0.0037	0.0269	0.0791	0.1575	0.2439	0.3296	0.4088	0.4793
2150	0.0033	0.0035	0.0262	0.0776	0.1551	0.2412	0.3265	0.4057	0.4762
2160	0.0032	0.0034	0.0255	0.0760	0.1528	0.2383	0.3234	0.4026	0.4732
2170	0.0031	0.0032	0.0249	0.0746	0.1506	0.2355	0.3204	0.3995	0.4701
2180	0.0030	0.0031	0.0242	0.0731	0.1483	0.2327	0.3174	0.3963	0.4671
2190	0.0029	0.0030	0.0236	0.0717	0.1461	0.2300	0.3143	0.3932	0.4640

Appendix Table A-13 *continued*

$$\frac{C°}{R}$$

ω(cm⁻¹)	298.15°K	300°K	400°K	500°K	600°K	700°K	800°K	900°K	1000°K
2200	0.0028	0.0029	0.0229	0.0703	0.1439	0.2272	0.3113	0.3901	0.4610
2210	0.0027	0.0028	0.0224	0.0689	0.1417	0.2246	0.3084	0.3871	0.4580
2220	0.0026	0.0027	0.0217	0.0675	0.1396	0.2219	0.3054	0.3840	0.4550
2230	0.0025	0.0026	0.0212	0.0662	0.1375	0.2192	0.3025	0.3810	0.4520
2240	0.0024	0.0025	0.0206	0.0649	0.1354	0.2166	0.2995	0.3780	0.4490
2250	0.0023	0.0024	0.0201	0.0636	0.1333	0.2140	0.2967	0.3750	0.4460
2260	0.0022	0.0023	0.0195	0.0624	0.1313	0.2114	0.2938	0.3720	0.4430
2270	0.0021	0.0022	0.0190	0.0611	0.1293	0.2089	0.2909	0.3690	0.4401
2280	0.0020	0.0021	0.0185	0.0599	0.1273	0.2064	0.2881	0.3660	0.4371
2290	0.0019	0.0021	0.0180	0.0587	0.1254	0.2039	0.2853	0.3631	0.4342
2300	0.0018	0.0020	0.0175	0.0575	0.1235	0.2014	0.2825	0.3602	0.4312
2310	0.0018	0.0019	0.0170	0.0564	0.1216	0.1990	0.2797	0.3572	0.4283
2320	0.0017	0.0018	0.0166	0.0552	0.1197	0.1965	0.2769	0.3543	0.4254
2330	0.0017	0.0018	0.0162	0.0541	0.1179	0.1941	0.2742	0.3515	0.4225
2340	0.0016	0.0017	0.0157	0.0531	0.1160	0.1918	0.2715	0.3486	0.4196
2350	0.0015	0.0016	0.0153	0.0520	0.1142	0.1894	0.2688	0.3457	0.4167
2360	0.0015	0.0016	0.0149	0.0509	0.1125	0.1871	0.2661	0.3429	0.4139
2370	0.0014	0.0015	0.0144	0.0499	0.1107	0.1848	0.2634	0.3401	0.4110
2380	0.0014	0.0014	0.0140	0.0489	0.1090	0.1825	0.2608	0.3373	0.4082
2390	0.0013	0.0014	0.0137	0.0479	0.1073	0.1802	0.2582	0.3344	0.4054
2400	0.0013	0.0013	0.0133	0.0469	0.1056	0.1780	0.2556	0.3317	0.4025
2410	0.0012	0.0013	0.0129	0.0460	0.1039	0.1758	0.2530	0.3289	0.3997
2420	0.0012	0.0012	0.0126	0.0450	0.1023	0.1736	0.2504	0.3262	0.3969
2430	0.0011	0.0012	0.0122	0.0441	0.1007	0.1714	0.2479	0.3234	0.3941
2440	0.0011	0.0011	0.0119	0.0431	0.0991	0.1692	0.2454	0.3207	0.3914
2450	0.0010	0.0011	0.0116	0.0423	0.0975	0.1671	0.2429	0.3180	0.3886
2460	0.0010	0.0011	0.0113	0.0415	0.0960	0.1650	0.2404	0.3153	0.3858
2470	0.0010	0.0010	0.0109	0.0406	0.0945	0.1629	0.2379	0.3127	0.3831
2480	0.0009	0.0010	0.0106	0.0398	0.0930	0.1609	0.2354	0.3100	0.3804
2490	0.0009	0.0009	0.0104	0.0390	0.0915	0.1588	0.2331	0.3074	0.3777

Appendix Table A-13 *continued*

	$\dfrac{C°}{R}$								
ω(cm^{-1})	298.15°K	300°K	400°K	500°K	600°K	700°K	800°K	900°K	1000°K
2500	0.0008	0.0009	0.0101	0.0382	0.0900	0.1568	0.2306	0.3047	0.3750
2510	0.0008	0.0009	0.0098	0.0374	0.0886	0.1548	0.2283	0.3021	0.3723
2520	0.0008	0.0008	0.0095	0.0366	0.0872	0.1529	0.2259	0.2995	0.3696
2530	0.0007	0.0008	0.0093	0.0358	0.0858	0.1509	0.2236	0.2970	0.3669
2540	0.0007	0.0008	0.0090	0.0351	0.0844	0.1490	0.2212	0.2944	0.3643
2550	0.0007	0.0007	0.0088	0.0344	0.0830	0.1471	0.2189	0.2919	0.3616
2560	0.0007	0.0007	0.0085	0.0337	0.0817	0.1452	0.2166	0.2893	0.3590
2570	0.0006	0.0007	0.0083	0.0330	0.0804	0.1433	0.2143	0.2868	0.3564
2580	0.0006	0.0007	0.0080	0.0323	0.0791	0.1414	0.2121	0.2843	0.3539
2590	0.0006	0.0006	0.0078	0.0316	0.0778	0.1396	0.2099	0.2819	0.3511
2600	0.0006	0.0006	0.0076	0.0309	0.0765	0.1378	0.2076	0.2794	0.3486
2610	0.0005	0.0006	0.0074	0.0303	0.0753	0.1360	0.2054	0.2769	0.3460
2620	0.0005	0.0006	0.0072	0.0297	0.0741	0.1342	0.2033	0.2745	0.3435
2630	0.0005	0.0005	0.0070	0.0290	0.0729	0.1325	0.2011	0.2721	0.3409
2640	0.0005	0.0005	0.0068	0.0284	0.0717	0.1308	0.1990	0.2697	0.3383
2650	0.0005	0.0005	0.0066	0.0278	0.0705	0.1291	0.1968	0.2673	0.3358
2660	0.0004	0.0005	0.0064	0.0272	0.0693	0.1274	0.1947	0.2649	0.3333
2670	0.0004	0.0005	0.0062	0.0267	0.0682	0.1257	0.1926	0.2626	0.3308
2680	0.0004	0.0004	0.0060	0.0261	0.0671	0.1240	0.1906	0.2602	0.3284
2690	0.0004	0.0004	0.0059	0.0255	0.0660	0.1224	0.1885	0.2579	0.3259
2700	0.0004	0.0004	0.0057	0.0250	0.0649	0.1208	0.1865	0.2556	0.3234
2710	0.0004	0.0004	0.0056	0.0245	0.0638	0.1192	0.1845	0.2533	0.3210
2720	0.0003	0.0004	0.0054	0.0240	0.0628	0.1176	0.1825	0.2510	0.3185
2730	0.0003	0.0004	0.0052	0.0235	0.0617	0.1160	0.1805	0.2487	0.3161
2740	0.0003	0.0003	0.0051	0.0229	0.0607	0.1145	0.1785	0.2465	0.3137
2750	0.0003	0.0003	0.0050	0.0224	0.0597	0.1130	0.1766	0.2443	0.3113
2760	0.0003	0.0003	0.0048	0.0220	0.0587	0.1115	0.1747	0.2420	0.3082
2770	0.0003	0.0003	0.0046	0.0215	0.0577	0.1100	0.1727	0.2398	0.3066
2780	0.0003	0.0003	0.0045	0.0210	0.0568	0.1085	0.1708	0.2377	0.3042
2790	0.0003	0.0003	0.0044	0.0206	0.0558	0.1070	0.1690	0.2355	0.3019

Appendix Table A-13 *continued*

	$\dfrac{C^\circ}{R}$							
$\omega(\text{cm}^{-1})$ 298.15°K	300°K	400°K	500°K	600°K	700°K	800°K	900°K	1000°K
2800 0.0003	0.0003	0.0043	0.0202	0.0549	0.1056	0.1671	0.2333	0.2995
2810 0.0002	0.0003	0.0042	0.0197	0.0540	0.1042	0.1653	0.2312	0.2972
2820 0.0002	0.0003	0.0041	0.0193	0.0531	0.1028	0.1635	0.2290	0.2949
2830 0.0002	0.0002	0.0039	0.0189	0.0522	0.1014	0.1616	0.2269	0.2926
2840 0.0002	0.0002	0.0038	0.0185	0.0513	0.1000	0.1598	0.2248	0.2903
2850 0.0002	0.0002	0.0037	0.0181	0.0504	0.0987	0.1581	0.2228	0.2881
2860 0.0002	0.0002	0.0036	0.0177	0.0496	0.0973	0.1563	0.2207	0.2858
2870 0.0002	0.0002	0.0035	0.0173	0.0487	0.0960	0.1546	0.2186	0.2836
2880 0.0002	0.0002	0.0034	0.0169	0.0479	0.0947	0.1529	0.2166	0.2813
2890 0.0002	0.0002	0.0033	0.0166	0.0471	0.0934	0.1511	0.2146	0.2791
2900 0.0002	0.0002	0.0032	0.0162	0.0462	0.0921	0.1494	0.2126	0.2769
2910 0.0002	0.0002	0.0031	0.0159	0.0455	0.0909	0.1478	0.2106	0.2747
2920 0.0002	0.0002	0.0030	0.0155	0.0447	0.0896	0.1461	0.2086	0.2725
2930 0.0002	0.0002	0.0029	0.0152	0.0440	0.0884	0.1445	0.2067	0.2704
2940 0.0001	0.0002	0.0029	0.0148	0.0432	0.0872	0.1428	0.2047	0.2682
2950 0.0001	0.0002	0.0028	0.0145	0.0425	0.0860	0.1412	0.2028	0.2661
2960 0.0001	0.0001	0.0027	0.0142	0.0417	0.0848	0.1396	0.2009	0.2639
2970 0.0001	0.0001	0.0026	0.0139	0.0410	0.0836	0.1380	0.1990	0.2618
2980 0.0001	0.0001	0.0025	0.0136	0.0403	0.0825	0.1365	0.1971	0.2597
2990 0.0001	0.0001	0.0025	0.0133	0.0396	0.0813	0.1349	0.1952	0.2576
3000 0.0001	0.0001	0.0024	0.0130	0.0390	0.0802	0.1334	0.1933	0.2556
3010 0.0001	0.0001	0.0023	0.0127	0.0383	0.0791	0.1318	0.1915	0.2535
3020 0.0001	0.0001	0.0023	0.0124	0.0376	0.0780	0.1303	0.1897	0.2514
3030 0.0001	0.0001	0.0022	0.0122	0.0370	0.0769	0.1288	0.1878	0.2494
3040 0.0001	0.0001	0.0021	0.0119	0.0363	0.0758	0.1274	0.1860	0.2474
3050 0.0001	0.0001	0.0021	0.0116	0.0357	0.0748	0.1259	0.1842	0.2454
3060 0.0001	0.0001	0.0020	0.0114	0.0351	0.0737	0.1244	0.1825	0.2434
3070 0.0001	0.0001	0.0020	0.0111	0.0345	0.0727	0.1230	0.1807	0.2414
3080 0.0001	0.0001	0.0019	0.0109	0.0339	0.0717	0.1216	0.1790	0.2394
3090 0.0001	0.0001	0.0019	0.0106	0.0333	0.0707	0.1202	0.1772	0.2374

Appendix Table A-13 *continued*

$$\frac{C°}{R}$$

ω(cm⁻¹)	298.15°K	300°K	400°K	500°K	600°K	700°K	800°K	900°K	1000°K
3100	0.0001	0.0001	0.0018	0.0104	0.0327	0.0697	0.1189	0.1755	0.2355
3110	0.0001	0.0001	0.0017	0.0102	0.0321	0.0687	0.1174	0.1738	0.2335
3120	0.0001	0.0001	0.0017	0.0100	0.0316	0.0677	0.1161	0.1721	0.2316
3130	0.0001	0.0001	0.0016	0.0097	0.0310	0.0668	0.1148	0.1704	0.2297
3140	0.0001	0.0001	0.0016	0.0095	0.0305	0.0658	0.1134	0.1688	0.2278
3150	0.0001	0.0001	0.0015	0.0093	0.0300	0.0649	0.1120	0.1671	0.2259
3160	0.0001	0.0001	0.0015	0.0091	0.0294	0.0640	0.1107	0.1655	0.2240
3170	0.0001	0.0001	0.0015	0.0089	0.0289	0.0631	0.1094	0.1639	0.2221
3180	0.0000	0.0001	0.0014	0.0087	0.0284	0.0622	0.1081	0.1622	0.2203
3190	0.0000	0.0001	0.0014	0.0085	0.0279	0.0613	0.1069	0.1606	0.2184
3200	0.0000	0.0000	0.0013	0.0083	0.0274	0.0604	0.1056	0.1590	0.2166
3210	0.0000	0.0000	0.0013	0.0081	0.0269	0.0596	0.1044	0.1575	0.2148
3220	0.0000	0.0000	0.0012	0.0080	0.0265	0.0587	0.1031	0.1559	0.2130
3230	0.0000	0.0000	0.0012	0.0078	0.0260	0.0579	0.1019	0.1544	0.2112
3240	0.0000	0.0000	0.0012	0.0076	0.0255	0.0570	0.1007	0.1528	0.2094
3250	0.0000	0.0000	0.0012	0.0074	0.0251	0.0562	0.0995	0.1513	0.2076
3260	0.0000	0.0000	0.0011	0.0073	0.0246	0.0554	0.0984	0.1498	0.2059
3270	0.0000	0.0000	0.0011	0.0071	0.0242	0.0546	0.0972	0.1483	0.2041
3280	0.0000	0.0000	0.0011	0.0069	0.0238	0.0538	0.0960	0.1468	0.2024
3290	0.0000	0.0000	0.0010	0.0068	0.0234	0.0531	0.0949	0.1454	0.2007
3300	0.0000	0.0000	0.0010	0.0066	0.0229	0.0523	0.0937	0.1439	0.1989
3310	0.0000	0.0000	0.0010	0.0065	0.0225	0.0515	0.0926	0.1425	0.1972
3320	0.0000	0.0000	0.0009	0.0063	0.0221	0.0508	0.0915	0.1410	0.1956
3330	0.0000	0.0000	0.0009	0.0062	0.0217	0.0501	0.0904	0.1396	0.1939
3340	0.0000	0.0000	0.0009	0.0061	0.0213	0.0493	0.0893	0.1383	0.1922
3350	0.0000	0.0000	0.0009	0.0059	0.0210	0.0486	0.0882	0.1368	0.1905
3360	0.0000	0.0000	0.0008	0.0058	0.0206	0.0479	0.0872	0.1354	0.1889
3370	0.0000	0.0000	0.0008	0.0056	0.0202	0.0472	0.0861	0.1340	0.1873
3380	0.0000	0.0000	0.0008	0.0055	0.0199	0.0465	0.0851	0.1327	0.1857
3390	0.0000	0.0000	0.0008	0.0054	0.0195	0.0458	0.0841	0.1313	0.1841

Appendix Table A-13 *continued*

$$\frac{C°}{R}$$

ω(cm⁻¹)	298.15°K	300°K	400°K	500°K	600°K	700°K	800°K	900°K	1000°K
3400	0.0000	0.0000	0.0007	0.0053	0.0192	0.0452	0.0831	0.1300	0.1825
3410	0.0000	0.0000	0.0007	0.0052	0.0188	0.0445	0.0821	0.1287	0.1809
3420	0.0000	0.0000	0.0007	0.0050	0.0185	0.0439	0.0811	0.1274	0.1793
3430	0.0000	0.0000	0.0007	0.0049	0.0181	0.0432	0.0801	0.1261	0.1777
3440	0.0000	0.0000	0.0007	0.0048	0.0178	0.0426	0.0791	0.1248	0.1762
3450	0.0000	0.0000	0.0006	0.0047	0.0175	0.0420	0.0781	0.1235	0.1746
3660	0.0000	0.0000	0.0006	0.0046	0.0172	0.0413	0.0772	0.1222	0.1731
3470	0.0000	0.0000	0.0006	0.0045	0.0169	0.0407	0.0762	0.1210	0.1716
3480	0.0000	0.0000	0.0006	0.0044	0.0166	0.0401	0.0753	0.1197	0.1701
3490	0.0000	0.0000	0.0006	0.0043	0.0163	0.0396	0.0744	0.1185	0.1686
3500	0.0000	0.0000	0.0005	0.0042	0.0160	0.0390	0.0735	0.1173	0.1671
3510	0.0000	0.0000	0.0005	0.0041	0.0157	0.0384	0.0726	0.1160	0.1656
3520	0.0000	0.0000	0.0005	0.0040	0.0154	0.0378	0.0717	0.1148	0.1642
3530	0.0000	0.0000	0.0005	0.0039	0.0151	0.0373	0.0708	0.1137	0.1627
3540	0.0000	0.0000	0.0005	0.0038	0.0148	0.0367	0.0699	0.1125	0.1613
3550	0.0000	0.0000	0.0005	0.0038	0.0146	0.0362	0.0691	0.1113	0.1599
3560	0.0000	0.0000	0.0005	0.0037	0.0143	0.0356	0.0682	0.1101	0.1584
3570	0.0000	0.0000	0.0004	0.0036	0.0141	0.0351	0.0674	0.1090	0.1570
3580	0.0000	0.0000	0.0004	0.0035	0.0138	0.0346	0.0665	0.1079	0.1556
3590	0.0000	0.0000	0.0004	0.0034	0.0135	0.0341	0.0657	0.1067	0.1542
3600	0.0000	0.0000	0.0004	0.0033	0.0133	0.0336	0.0649	0.1056	0.1528
3610	0.0000	0.0000	0.0004	0.0033	0.0131	0.0331	0.0641	0.1045	0.1515
3620	0.0000	0.0000	0.0004	0.0032	0.0128	0.0326	0.0633	0.1034	0.1501
3630	0.0000	0.0000	0.0004	0.0031	0.0126	0.0321	0.0625	0.1023	0.1488
3640	0.0000	0.0000	0.0004	0.0030	0.0124	0.0316	0.0617	0.1012	0.1474
3650	0.0000	0.0000	0.0004	0.0030	0.0121	0.0311	0.0610	0.1002	0.1461
3660	0.0000	0.0000	0.0003	0.0029	0.0119	0.0307	0.0602	0.0991	0.1448
3670	0.0000	0.0000	0.0003	0.0028	0.0117	0.0302	0.0594	0.0981	0.1435
3680	0.0000	0.0000	0.0003	0.0028	0.0115	0.0297	0.0587	0.0970	0.1422
3690	0.0000	0.0000	0.0003	0.0027	0.0113	0.0293	0.0580	0.0960	0.1409

[a]Calculated from the table of Johnston, Savedoff, and Belzer (695) $hc/k = 1.4386$.

REFERENCES

(1) A. Abrams and T. W. Davis, *J. Am. Chem. Soc.*, 76, 5993–5995 (1954).
(2) R. Adams, H. B. Bramlet, and F. H. Tendick, *J. Am. Chem. Soc.*, 42, 2369–2374 (1920).
(3) N. Adriaanse, H. Dekker, and J. Coops, *Rec. Trav. Chim.*, 84, 393–407 (1965).
(4) J. E. Ahlberg, E. R. Blanchard, and W. O. Lundberg, *J. Chem. Phys.*, 5, 539–551 (1937).
(5) B. V. Aibazov, S. M. Petrov, V. R. Khairullina, and V. G. Yaprintseva, in B. V. Aibazov (Ed.), *Physicochemical Constants of Sulfur Organic Compounds*, Publishing House "Khimiya," Moscow, 1964.
(6) P. B. Aitken, H. L. Boxall, and L. G. Cook, *Rev. Sci. Instr.*, 25, 967–970 (1954).
(7) P. A. Akishin, L. V. Vilkov, and Y. I. Vesnin, *Dokl. Akad. Nauk SSSR*, 126, 310–313 (1959).
(8) L. F. Albright, W. C. Galegar, and K. K. Innes, *J. Am. Chem. Soc.*, 76, 6017–6019 (1954).
(9) G. Allen and H. J. Bernstein, *Can. J. Chem.*, 32, 1044–1046 (1954).
(10) T. L. Allen, *J. Chem. Phys.*, 31, 1039–1049 (1959).
(11) A. Amador, J. R. Lacher, and J. D. Park, "Heats of Hydrogenation of Alkyl Chlorides," presented at the 20th Calorimetry Conference, Ames, Iowa, 1965.
(12) H. R. Ambler, *J. Soc. Chem. Ind. (London)*, 55, 291T–292T (1936).
(13) D. Ambrose, J. D. Cox, and R. Townsend, *Trans. Faraday Soc.*, 56, 1452–1459 (1960).
(14) D. Ambrose and R. Townsend, *J. Chem. Soc.*, 3614–3625 (1963).
(15) A. G. Anderson and G. Stegeman, *J. Am. Chem. Soc.*, 63, 2119–2121 (1941).
(16) C. M. Anderson, L. G. Cole, and E. C. Gilbert, *J. Am. Chem. Soc.*, 72, 1263–1264 (1950).
(17) C. M. Anderson and E. C. Gilbert, *J. Am. Chem. Soc.*, 64, 2369–2372 (1942).
(18) J. W. Anderson, G. H. Beyer, and K. M. Watson, *Natl. Petrol. News*, 36, R476–R484 (1944).
(19) R. J. L. Andon, J. F. Counsell, E. F. G. Herington, and J. F. Martin, *Trans. Faraday Soc.*, 59, 830–835 (1963).
(20) R. J. L. Andon, J. F. Counsell, and J. F. Martin, *Trans. Faraday Soc.*, 59, 1555–1558 (1963).
(21) R. J. L. Andon, J. D. Cox, E. F. G. Herington, and J. F. Martin, *Trans. Faraday Soc.*, 53, 1074–1082 (1957).
(22) D. N. Andreevskii and A. M. Rozhnov, *Neftekhimiya*, 2, 378–383 (1962).
(23) D. Andrychuk, *Can. J. Phys.*, 29, 151–58 (1951).
(24) C. Antoine, *Compt. Rend.*, 107, 681–684, 836–837, 1143–1145, (1888).
(25) W. F. Arendale and W. H. Fletcher, *J. Chem. Phys.*, 26, 793–797 (1957).
(26) G. T. Armstrong and S. Marantz, *J. Phys. Chem.*, 64, 1776–1778 (1960).
(27) S. J. Ashcroft, A. S. Carson, W. Carter, and P. G. Laye, *Trans. Faraday Soc.*, 61, 225–229 (1965).
(28) M. J. Astle, *Industrial Organic Nitrogen Compounds*, Reinhold, New York, 1961.
(29) J. G. Aston, *Ind. Eng. Chem.*, 34, 514–521 (1942).
(30) J. G. Aston, "The Third Law of Thermodynamics and Statistical Mechanics," in

H. S. Taylor and S. Glasstone (Eds.), *A Treatise on Physical Chemistry*, Vol. I, 3rd Ed., Van Nostrand, New York, 1942, Chap. IV.
(31) J. G. Aston, *Discussions Faraday Soc.*, *10*, 73–79 (1951).
(32) J. G. Aston, M. L. Eidinoff, and W. S. Forster, *J. Am. Chem. Soc.*, *61*, 1539–1543 (1939).
(33) J. G. Aston, H. L. Fink, A. B. Bestul, E. L. Pace, and G. J. Szasz, *J. Am. Chem. Soc.*, *68*, 52–57 (1946).
(34) J. G. Aston, H. L. Fink, G. J. Janz, and K. E. Russell, *J. Am. Chem. Soc.*, *73*, 1939–1945 (1951).
(35) J. G. Aston, H. L. Fink, and S. C. Schumann, *J. Am. Chem. Soc.*, *65*, 341–346 (1943).
(36) J. G. Aston, H. L. Fink, J. W. Tooke, and M. R. Cines, *Anal. Chem.*, *19*, 218–221 (1947).
(37) J. G. Aston, G. J. Janz, and K. E. Russell, *J. Am. Chem. Soc.*, *73*, 1943–1945 (1951).
(38) J. G. Aston, R. M. Kennedy, and S. C. Schumann, *J. Am. Chem. Soc.*, *62*, 2059–2063 (1940).
(39) J. G. Aston, S. V. R. Mastrangelo, and G. W. Moessen, *J. Am. Chem. Soc.*, *72*, 5287–5291 (1950).
(39a) J. G. Aston and G. H. Messerly, *J. Am. Chem. Soc.*, *58*, 2354–2361 (1936).
(40) J. G. Aston and G. H. Messerly, *J. Am. Chem. Soc.*, *62*, 1917–1923 (1940).
(41) J. G. Aston, E. J. Rock, and S. Isserow, *J. Am. Chem. Soc.*, *74*, 2484–2486 (1952).
(42) J. G. Aston, M. L. Sagenkahn, G. J. Szasz, G. W. Moessen, and H. F. Zuhr, *J. Am. Chem. Soc.*, *66*, 1171–1177 (1944).
(43) J. G. Aston, C. W. Siller, and G. H. Messerly, *J. Am. Chem. Soc.*, *59*, 1743–1751 (1937).
(44) J. G. Aston and G. J. Szasz, *J. Am. Chem. Soc.*, *69*, 3108–3114 (1947).
(45) J. G. Aston, P. E. Wills, and T. P. Zolki, *J. Am. Chem. Soc.*, *77*, 3939–3941 (1955).
(46) J. G. Aston, J. L. Wood, and T. P. Zolki, *J. Am. Chem. Soc.*, *75*, 6202–6204 (1953).
(47) J. G. Aston and C. W. Ziemer, *J. Am. Chem. Soc.*, *68*, 1405–1413 (1946).
(48) J. G. Aston, T. P. Zolki, and J. L. Wood, *J. Am. Chem. Soc.*, *77*, 281–284 (1955).
(49) P. G. Atherton, B. Appl. Sci. Thesis, University of Queensland, Brisbane, Australia, 1952.
(50) B. Atkinson and M. Stedman, *J. Chem. Soc.*, 512–519 (1962).
(51) B. Atkinson and A. B. Trenwith, *J. Chem. Phys.*, *20*, 754–755 (1952).
(52) W. Auer, in K. Schafer and E. Lax (Eds.), *Landolt-Börnstein: Zahlenwerte und Funktionen aus Physik, Chemie, Astronomie, Geophysik und Technik*, Band II, Teil 4, Springer-Verlag, Berlin, 1961, pp. 179–393.
(53) G. W. Ayers, Jr., and M. S. Agruss, *J. Am. Chem. Soc.*, *61*, 83–85 (1939).
(54) M. Badoche, *Bull. Soc. Chim. France*, *4*, 549–558 (1937).
(55) M. Badoche, *Bull. Soc. Chim. France*, *6*, 570–579 (1939).
(56) M. Badoche, *Bull. Soc. Chim. France*, *8*, 212–220 (1941).
(57) M. Badoche, *Bull. Soc. Chim. France*, *9*, 86–95 (1942).
(58) M. Badoche, *Bull. Soc. Chim. France*, 37–43 (1946).
(59) V. F. Baibuz, *Dokl. Akad. Nauk SSSR*, *140*, 1358–1360 (1961).
(60) B. Bak, *Kgl. Danske Videnskab. Selskab, Mat.-Fys. Medd.*, *24*, 3–14 (1948).
(61) B. Bak and F. A. Andersen, *J. Chem. Phys.*, *22*, 1050–1053 (1954).
(62) B. Bak, S. Brodersen, and L. Hansen, *Acta Chem. Scand.*, *9*, 749–762 (1955).
(63) A. W. Baker and R. C. Lord, *J. Chem. Phys.*, *23*, 1636–1643 (1955).
(64) G. Baker, J. H. Littlefair, R. Shaw, and J. C. J. Thynne, *J. Chem. Soc.*, 6970–6972 (1965).

(65) J. W. Baker and W. T. Tweed, *J. Chem. Soc.*, 796–802 (1941).
(66) A. A. Balandin, *Kataliz v Vysshei Shkole (Catalysis in College)*, Vol. 1, No. 5, Izd. MGU, Moscow State University, Moscow, 1962.
(67) A. A. Balandin, E. I. Klabunovskii, A. P. Oberemok-Yakubova, and I. I. Brusov, *Izv. Akad. Nauk SSSR, Otd. Khim. Nauk*, 5, 784–786 (1960); *Bull. Acad. Sci. USSR, Div. Chem. Sci.*, 5, 735–737 (1960).
(68) G. Bambach, *Kältetechnik*, 8, 334–339 (1956).
(69) S. C. Banerjee, D. C. Bigg, and L. K. Doraiswamy, *Brit. Chem. Eng.*, 9, 688–689 (1964).
(70) S. C. Banerjee and L. K. Doraiswamy, *Brit. Chem. Eng.*, 9, 311–313 (1964).
(71) H. Banse and G. S. Parks, *J. Am. Chem. Soc.*, 55, 3223–3226 (1933).
(72) T. M. Barakat, N. Legge, and A. D. E. Pullin, *Trans. Faraday Soc.*, 59, 1764–1772 (1963).
(73) C. M. Barber and E. F. Westrum, Jr., *J. Phys. Chem.*, 67, 2373–2376 (1963).
(74) J. W. Barrett and R. P. Linstead, *J. Chem. Soc.*, 436–442 (1935).
(75) J. W. Barrett and R. P. Linstead, *J. Chem. Soc.*, 611–616 (1936).
(76) G. M. Barrow, *J. Chem. Phys.*, 20, 1739–1744 (1952).
(77) G. M. Barrow and A. L. McClellan, *J. Am. Chem. Soc.*, 73, 573–575 (1951).
(78) G. M. Barrow and K. S. Pitzer, *Ind. Eng. Chem.*, 41, 2737–2740 (1949).
(79) L. S. Bartell and L. O. Brockway, *J. Chem. Phys.*, 23, 1860–1862 (1955).
(80) H. F. Bartolo and F. D. Rossini, *J. Phys. Chem.*, 64, 1685–1689 (1960).
(81) O. Bastiansen and L. Smedvik, *Acta Chem. Scand.*, 7, 652–656 (1953).
(82) A. Bauder and H. H. Günthard, *Helv. Chim. Acta*, 41, 670–673 (1958).
(83) E. Baur and S. Frater, *Helv. Chim. Acta*, 24, 768–782 (1941).
(84) G. P. Baxter, F. K. Bezzenberger, and C. H. Wilson, *J. Am. Chem. Soc.*, 42, 1386–1393 (1920).
(85) G. P. Baxter and M. R. Grose, *J. Am. Chem. Soc.*, 37, 1061–1072 (1915).
(86) G. P. Baxter, C. H. Hickey, and W. C. Holmes, *J. Am. Chem. Soc.*, 29, 127–136 (1907).
(87) G. V. Beard, *Univ. Microfilms Publ. No. 867*, 34 pp.; *Microfilm Abstr.*, 7 (2), 42–44 (1947).
(88) J. A. Beattie, G.-J. Su, and G. L. Simard, *J. Am. Chem. Soc.*, 61, 924–925 (1939).
(89) E. D. Becker, E. Charney, and T. Anno, *J. Chem. Phys.*, 42, 942–949 (1965).
(90) G. Becker and W. A. Roth, *Ber.*, 67B, 627–632 (1934).
(91) G. Becker and W. A. Roth, *Z. Phys. Chem. (Leipzig)*, A169, 287–296 (1934).
(92) C. W. Beckett, N. K. Freeman, and K. S. Pitzer, *J. Am. Chem. Soc.*, 70, 4227–4230 (1948).
(93) A. F. Bedford, A. E. Beezer, and C. T. Mortimer, *J. Chem. Soc.*, 2039–2043 (1963).
(94) A. F. Bedford, A. E. Beezer, C. T. Mortimer, and H. D. Springall, *J. Chem. Soc.*, 3823–3828 (1963).
(95) A. F. Bedford, J. G. Carey, I. T. Millar, C. T. Mortimer, and H. D. Springall, *J. Chem. Soc.*, 3895–3898 (1962).
(96) A. F. Bedford, P. B. Edmondson, and C. T. Mortimer, *J. Chem. Soc.*, 2927–2931 (1962).
(97) O. Beeck, *Discussions Faraday Soc.*, 8, 118–128 (1950).
(98) A. E. Beezer, C. T. Mortimer, H. D. Springall, F. Sondheimer, and R. Wolovsky, *J. Chem. Soc.*, 216–220 (1965).
(99) N. Bekkedahl and L. A. Wood, *J. Res. Natl. Bur. Std.*, 19, 551–558 (1937).
(100) N. Bekkedahl, L. A. Wood, and M. Wojciechowski, *J. Res. Natl. Bur. Std.*, 17, 883–894 (1936).

(101) H. E. Bellis and E. J. Slowinski, Jr., *Spectrochim. Acta, 15,* 1103–1117 (1959).
(102) P. Bender and J. Farber, *J. Am. Chem. Soc., 74,* 1450–1452 (1952).
(103) K. Bennewitz and W. Rossner, *Z. Phys. Chem. (Leipzig), B39,* 126–144 (1938).
(104) S. W. Benson, *J. Chem. Phys., 43,* 2044–2046 (1965).
(105) S. W. Benson and A. Amano, *J. Chem. Phys., 36,* 3464–3471 (1962).
(106) S. W. Benson and A. N. Bose, *J. Chem. Phys., 37,* 2935–2940 (1962).
(107) S. W. Benson and J. H. Buss, *J. Phys. Chem., 61,* 104–109 (1957).
(108) S. W. Benson and J. H. Buss, *J. Chem. Phys., 29,* 546–572 (1958).
(109) H. A. Bent, "Electron Correlation and Bond Properties in Some Selected Sulfur Compounds," in N. Kharasch and C. Y. Meyers (Eds.), *Chemistry of Organic Sulfur Compounds,* Vol. II, Pergamon Press, London, 1966, pp. 1–34.
(110) L. Berg, H. C. Carpenter, J. B. Daly, R. Dev, H. A. Herzel, P. R. Hippely, E. O. Kindschy, and D. C. Popovac, *Proc. Montana Acad. Sci., 7/8,* 85–125 (1947–1948).
(111) W. T. Berg, D. W. Scott, W. N. Hubbard, S. S. Todd, J. F. Messerly, I. A. Hossenlopp, A. Osborn, D. R. Douslin, and J. P. McCullough, *J. Phys. Chem., 65,* 1425–1430 (1961).
(112) E. Berliner, *J. Am. Chem. Soc., 68,* 49–51 (1946).
(113) N. S. Berman, C. W. Larkam, and J. J. McKetta, *J. Chem. Eng. Data, 9,* 218–219 (1964).
(114) N. S. Berman and J. J. McKetta, *J. Phys. Chem., 66,* 1444–1448 (1962).
(115) N. S. Berman and J. J. McKetta, private communication, June, 1962.
(116) H. J. Bernstein, *J. Chem. Phys., 24,* 911–912 (1956).
(117) H. J. Bernstein, *Trans. Faraday Soc., 58,* 2285–2306 (1962).
(118) H. J. Bernstein and G. Herzberg, *J. Chem. Phys., 16,* 30–39 (1948).
(119) H. J. Bernstein and D. A. Ramsay, *J. Chem. Phys., 17,* 556–565 (1949).
(120) D. Berthelot, *J. Phys., 8,* 263–271 (1899).
(121) M. Berthelot, *Thermochimie: Données et Lois Numériques,* Gauthier-Villars, Paris, 1897.
(122) M. Berthelot, *Ann. Chim. Phys., 22,* 322–326 (1901).
(123) M. P. E. Berthelot, G. Andre, and C. Matignon, *Compt. Rend., 111,* 6–9 (1890).
(124) E. T. Beynon, Jr., and J. J. McKetta, *J. Phys. Chem., 67,* 2761–2765 (1963).
(125) F. R. Bichowsky and F. D. Rossini, *The Thermochemistry of the Chemical Substances,* Reinhold, New York, 1936.
(126) D. P. Biddiscombe, R. R. Collerson, R. Handley, E. F. G. Herington, J. F. Martin, and C. H. S. Sprake, *J. Chem. Soc.,* 1954–1957 (1963).
(127) D. P. Biddiscombe, E. A. Coulson, R. Handley, and E. F. G. Herington, *J. Chem. Soc.,* 1957–1967 (1954).
(128) D. P. Biddiscombe, R. Handley, D. Harrop, A. J. Head, G. B. Lewis, J. F. Martin, and C. H. S. Sprake, *J. Chem. Soc.,* 5764–5768 (1963).
(129) D. P. Biddiscombe and J. F. Martin, *Trans. Faraday Soc., 54,* 1316–1322 (1958).
(130) C. E. Bills, F. G. McDonald, L. N. BeMiller, G. E. Steel, and M. Nussmeier, *J. Biol. Chem., 93,* 775–785 (1931).
(131) J. L. Binder, *J. Chem. Phys., 17,* 499–500 (1949).
(132) J. L. Binder, *J. Chem. Phys. 18,* 77–78 (1950).
(133) C. M. Birdsall, A. C. Jenkins, F. S. DiPaolo, J. A. Beattie, and C. M. Apt, *J. Chem. Phys., 23,* 441–452 (1955).
(134) L. Bjellerup, unpublished work, quoted by L. Smith, L. Bjellerup, S. Krook, and H. Westermark, *Acta Chem. Scand., 7,* 65–86 (1953).
(135) L. Bjellerup, *Acta Chem. Scand., 13,* 1511–1514 (1959).
(136) L. Bjellerup, *Acta Chem. Scand., 14,* 617–624 (1960).

(137) L. Bjellerup, *Acta Chem. Scand.*, *15*, 121–140 (1961).
(138) L. Bjellerup, *Acta Chem. Scand.*, *15*, 231–241 (1961).
(139) L. Bjellerup, *Svensk Kem. Tidskr.*, *73*, 144–152 (1961).
(140) L. Bjellerup, "Combustion in a Bomb of Organic Bromine Compounds," in H. A. Skinner, Ed., *Experimental Thermochemistry*, Vol. II, Interscience Publishers, New York, 1962, pp. 41–55, Chapter 3.
(141) L. Bjellerup and L. Smith, *Kgl. Fysiograf. Sallskap. Lund, Forh.*, *24*, 21–33(1954).
(142) L. Bjellerup, S. Sunner, and I. Wadsö, *Acta Chem. Scand.*, *11*, 1761–1765 (1957).
(143) H. H. Blau, Jr., and H. H. Nielsen, *J. Mol. Spectroscp.*, *1*, 124–132 (1957).
(144) M. Bodenstein, P. Günther, and F. Hoffmeister, *Z. Angew. Chem.*, *39*, 875–880 (1926).
(145) G. B. Bonino, R. Manzoni-Ansidei, and M. Rolla *Ric. Sci.*, *8*, 357–361 (1937).
(146) B. Borjesson, Y. Nakase, and S. Sunner, *Acta Chem. Scand.*, *20*, 803–810 (1966).
(147) M. Born and T. von Kármán, *Phys. Z.*, *13*, 297–309 (1912).
(148) H. Borsook and H. Huffman, "Some Thermodynamical Considerations of Amino Acids, Peptides, and Related Substances," in C. L. A. Schmidt (Ed.), *The Chemistry of the Amino Acids and Proteins*, 2nd ed., Charles C. Thomas, Springfield, Ill., 1944, pp. 822–870, Chapter XV.
(149) A. N. Bose and S. W. Benson, *J. Chem. Phys.*, *37*, 1081–1084 (1962).
(150) R. W. Bost and J. E. Everett, *J. Am. Chem. Soc.*, *62*, 1752–1754(1940).
(151) R. H. Boundy and R. F. Boyer (Eds.), *Styrene, Its Polymers, Copolymers and Derivatives (A.C.S. Monograph No. 115)*, Reinhold, New York, 1952.
(152) H. J. M. Bowen, A. Gilchrist, and L. E. Sutton, *Trans. Faraday Soc.*, *51*, 1341–1354 (1955).
(153) D. R. J. Boyd and H. W. Thompson, *Spectrochim. Acta*, *5*, 308–312 (1952).
(154) R. H. Boyd, *J. Chem. Phys.*, *38*, 2529–2535 (1963).
(155) R. H. Boyd, *Tetrahedron*, *22*, 119–122 (1966).
(156) R. H. Boyd, R. L. Christensen, and R. Pua, *J. Am. Chem. Soc.*, *87*, 3554–3559 (1965).
(157) J. C. Bradley, L. Haar, and A. S. Friedman, *J. Res. Natl. Bur. Std.*, *56*, 197–200 (1956).
(158) R. S. Bradley and T. G. Cleasby, *J. Chem. Soc.*, 1690–1692 (1963).
(159) J. C. D. Brand and T. M. Cawthon, *J. Am. Chem. Soc.*, *77*, 319–323 (1955).
(160) J. T. Brandenburg, *Univ. Microfilms L. C. Mic. 58–1553*, 55 pp.; *Dissertation Abstr.*, *18*, 1994–1995 (1958).
(161) J. L. Brandt, and R. L. Livingston, *J. Am. Chem. Soc.*, *78*, 3573–3576 (1956).
(162) J. von Braun and P. Engelbertz, *Ber.*, *56*, 1573–1577 (1923).
(163) J. W. Breitenbach and J. Derkosch, *Monatsh.*, *82*, 177–179 (1951).
(164) J. W. Breitenbach, J. Derkosch, and F. Wessely, *Nature*, *169*, 922 (1952).
(165) J. G. M. Bremner and G. D. Thomas, *Trans. Faraday Soc.*, *44*, 230–238 (1948).
(166) J. G. M. Bremner and G. D. Thomas, *Trans. Faraday Soc.*, *44*, 338–341 (1948).
(167) L. Brewer, L. A. Bromley, P. W. Gilles, and N. L. Lofgren, in L. L. Quill (Ed.), *The Chemistry and Metallurgy of Miscellaneous Materials–Thermodynamics*, McGraw-Hill, New York, 1950, Papers 6 and 8.
(168) F. G. Brickwedde, private communication, cited in R. H. Boundy and R. F. Boyer (Eds.), *Styrene, Its Polymers, Copolymers and Derivatives (A.C.S. Monograph No. 115)*, Reinhold, New York, 1952, p. 65.
(169) G. Briegleb, *Z. Elektrochem.*, *53*, 350–361 (1949).
(170) S. R. Brinkley, Jr., *J. Chem. Phys.*, *15*, 107–110 (1947).
(171) L. O. Brockway, J. Y. Beach, and L. Pauling, *J. Am. Chem. Soc.*, *57*, 2693–2704 (1935).

(172) G. H. J. Broers, J. A. A. Ketelaar, and P. F. Van Velden, *Rec. Trav. Chim.*, *69*, 1122–1126 (1950).
(173) L. G. Brouwer and J. P. Wibaut, *Rec. Trav. Chim.*, *53*, 1001–1010 (1934).
(174) J. C. Brown, *J. Chem. Soc.*, *83*, 987–994 (1903).
(175) O. L. I. Brown and G. G. Manov, *J. Am. Chem. Soc.*, *59*, 500–502 (1937).
(176) C. C. Browne and F. D. Rossini, *J. Phys. Chem.*, *64*, 927–931 (1960).
(177) L. Brüll, *Gazz. Chim. Ital.*, *65*, 19–28 (1935).
(178) W. M. D. Bryant, *J. Polymer Sci.*, *56*, 277–296 (1962).
(179) D. Bryce-Smith and K. E. Howlett, *J. Chem. Soc.*, 1141–1142 (1951).
(180) E. Buckley and E. F. G. Herington, *Trans. Faraday Soc.*, *61*, 1618–1625 (1965).
(181) P. R. Bunker and H. C. Longuet-Higgins, *Proc. Roy. Soc. (London), Ser. A*, *280*, 340–352 (1964).
(182) E. J. Burcik and D. M. Yost, *J. Chem. Phys.*, *7*, 1114–1115 (1939).
(183) G. K. Burgess, *J. Res. Natl. Bur. Std.*, *1*, 635–640 (1928).
(184) E. Burlot, M. Thomas, and M. Badoche, *Mem. Poudres*, *29*, 226–282 (1939).
(185) F. Burriel, *Anales Real Soc. Espan. Fis. Quim. (Madrid)*, *29*, 89–125 (1931).
(186) W. K. Busfield, K. J. Ivin, H. Mackle, and P. A. G. O'Hare, *Trans. Faraday Soc.*, *57*, 1064–1069 (1961).
(187) J. A. V. Butler, C. N. Ramchandani, and D. W. Thomson, *J. Chem. Soc.*, 280–285 (1935).
(188) E. Calvet, *J. Chim. Phys.*, *30*, 140–166 (1933).
(189) D. L. Camin and F. D. Rossini, *J. Phys. Chem.*, *59*, 1173–1179 (1955).
(190) H. G. Carlson and E. F. Westrum, Jr., *J. Chem. Eng. Data*, *10*, 134–135 (1965).
(191) H. G. Carlson and E. F. Westrum, Jr., *J. Phys. Chem.*, *69*, 1524–1530 (1965).
(192) R. A. Carney, E. A. Piotrowski, A. G. Meister, J. H. Braun, and F. F. Cleveland, *J. Mol. Spectrosc.*, *7*, 209–222 (1961).
(193) L. G. Carpenter and T. F. Harle, *Phil. Mag.*, *23*, 193–208 (1937).
(194) A. S. Carson, E. M. Carson, and B. Wilmshurst, *Nature*, *170*, 320–321 (1952).
(195) A. S. Carson, W. Carter, and J. B. Pedley, *Proc. Roy. Soc. (London), Ser. A*, *260*, 550–557 (1961).
(196) A. S. Carson, K. Hartley, and H. A. Skinner, *Trans. Faraday Soc.*, *45*, 1159–1167 (1949).
(197) A. S. Carson, H. O. Pritchard, and H. A. Skinner, *J. Chem. Soc.*, 656–659 (1950).
(198) A. S. Carson and H. A. Skinner, *J. Chem. Soc.*, 936–939 (1949).
(199) D. W. H. Casey and S. Fordham, *J. Chem. Soc.*, 2513–2516 (1951).
(200) R. C. Cass, S. E. Fletcher, C. T. Mortimer, P. G. Quincey, and H. D. Springall, *J. Chem. Soc.*, 958–963 (1958).
(201) R. C. Cass, S. E. Fletcher, C. T. Mortimer, P. G. Quincey, and H. D. Springall, *J. Chem. Soc.*, 2595–2597 (1958).
(202) R. C. Cass, S. E. Fletcher, C. T. Mortimer, H. D. Springall, and T. R. White, *J. Chem. Soc.*, 1406–1410 (1958).
(203) E. Catalano and K. S. Pitzer, *J. Am. Chem. Soc.*, *80*, 1054–1057 (1958).
(204) E. Catalano and K. S. Pitzer, *J. Phys. Chem.*, *62*, 838–840 (1958).
(205) E. Catalano and K. S. Pitzer, *J. Phys. Chem.*, *62*, 873–874 (1958).
(206) R. M. Caves, R. L. McLaughlin, and P. H. Wise, *J. Am. Chem. Soc.*, *76*, 522–524 (1954).
(207) C. Černý and E. Erdos, *Collect. Czech. Chem. Commun.*, *19*, 646–652 (1954).
(208) A. R. Challoner, H. A. Gundry, and A. R. Meetham, *Phil. Trans. Roy. Soc. London, Ser. A*, *247*, 553–580 (1955).
(209) S. S. Chang and E. F. Westrum, Jr., *J. Phys. Chem.*, *64*, 1547–1551 (1960).
(210) S. S. Chang and E. F. Westrum, Jr., *J. Phys. Chem.*, *64*, 1551–1552 (1960).

(211) S. S. Chang and E. F. Westrum, Jr., *J. Chem. Phys.*, *36*, 2420–2423 (1962).
(212) S. S. Chang and E. F. Westrum, Jr., *J. Chem. Phys.*, *36*, 2571–2577 (1962).
(213) S. S. Chang and E. F. Westrum, Jr., *J. Phys. Chem.*, *66*, 524–527 (1962).
(214) J. Chao, Dissertation, Carnegie Institute of Technology, Pittsburgh, 1961.
(215) J. Chao and F. D. Rossini, *J. Chem. Eng. Data*, *10*, 374–379 (1965).
(216) G. H. Charbonnet and W. S. Singleton, *J. Am. Oil Chemists' Soc.*, *24*, 140–142 (1947).
(217) N. C. S. Chari, *Univ. Microfilms L.C. Mic. 60–6850*, 188 pp. *Dissertation Abstr.*, *21*, 2220 (1961).
(218) T. V. Charlu and M. R. A. Rao, *Proc. Indian Acad. Sci.*, Sect. A, *60*, 31–35 (1964).
(219) D. C.-H. Cheng and J. C. McCoubrey, *J. Chem. Soc.*, 4993–4995 (1963).
(220) H. A. G. Chermin, *Petrol. Refiner*, *40* (2), 145–148 (1961).
(221) H. A. G. Chermin, *Petrol. Refiner*, *40* (3), 181–184 (1961).
(222) H. A. G. Chermin, *Hydrocarbon Process. Petrol. Refiner*, *40* (6), 179–182 (1961).
(223) H. A. G. Chermin, *Hydrocarbon Process. Petrol. Refiner*, *40* (9), 261–263 (1961).
(224) H. A. G. Chermin, *Hydrocarbon Process. Petrol. Refiner*, *40* (10), 145–147 (1961).
(225) C. L. Chernick, H. A. Skinner, and I. Wadsö, *Trans. Faraday Soc.*, *52*, 1088–1093 (1956).
(226) H. H. Claassen, *J. Chem. Phys.*, *18*, 543–551 (1950).
(227) H. H. Claassen, *J. Chem. Phys.*, *22*, 50–52 (1954).
(228) T. H. Clarke and G. Stegeman, *J. Am. Chem. Soc.*, *61*, 1726–1730 (1939).
(229) T. H. Clarke and G. Stegeman, *J. Am. Chem. Soc.*, *62*, 1815–1817 (1940).
(230) T. H. Clarke and G. Stegemen, *J. Am. Chem. Soc.*, *66*, 457–459 (1944).
(231) R. Clausius, *Ann. Phys. Chem.*, *125*, 353–400 (1865).
(232) A. P. Claydon and C. T. Mortimer, *J. Chem. Soc.*, 3212–3216 (1962).
(233) J. O. Clayton and W. F. Giauque, *J. Am. Chem. Soc.*, *54*, 2610–2626 (1932).
(234) H. E. Clements, K. V. Wise, and S. E. J. Johnsen, *J. Am. Chem. Soc.*, *75*, 1593–1595 (1953).
(235) H. L. Clever, C. A. Wulff, and E. F. Westrum, Jr., *J. Phys. Chem.*, *69*, 1983–1988 (1965).
(236) J. E. Cline and G. B. Kistiakowsky, *J. Chem. Phys.*, *5*, 990 (1937).
(237) J. K. Cline and D. H. Andrews, *J. Am. Chem. Soc.*, *53*, 3668–3673 (1931).
(238) J. A. Clopatt, *Soc. Sci. Fenn. Commentat. Phys.-Math.*, *6* (4) 1–15 (1932).
(239) G. E. Coates and L. E. Sutton, *J. Chem. Soc.*, 1187–1196 (1948).
(240) E. R. Cohen, K. M. Crowe, and J. W. M. Dumond, *The Fundamental Constants of Physics*, Interscience Publishers, New York, 1957.
(241) A. G. Cole, J. O. Hutchens, and J. W. Stout, *J. Phys. Chem.*, *67*, 1852–1855 (1963).
(242) A. G. Cole, J. O. Hutchens, and J. W. Stout, *J. Phys. Chem.*, *67*, 2245–2247 (1963).
(243) L. G. Cole, M. Farber, and G. W. Elverum, Jr., *J. Chem. Phys.*, *20*, 586–590 (1952).
(244) L. G. Cole and E. C. Gilbert, *J. Am. Chem. Soc.*, *73*, 5423–5427 (1951).
(245) C. F. Coleman and T. De Vries, *J. Am. Chem. Soc.*, *71*, 2839–2841 (1949).
(246) D. J. Coleman and G. Pilcher, *Trans. Faraday Soc.*, *62*, 821–827 (1966).
(247) D. J. Coleman and H. A. Skinner, *Trans. Faraday Soc.*, *62*, 2057–2062 (1966).
(248) D. K. Coles and R. H. Hughes, *Phys. Rev.*, *76*, 178 (1949).
(249) H. Collatz, *Cellulosechem.*, *17*, 128–129 (1936).
(250) M. Colomina, M. L. Boned Corral, and C. Turrion, *Anales Real Soc. Espan. Fis. Quim. (Madrid)*, Ser. B, *57*, 655–664 (1961).
(251) M. Colomina, C. Latorre, and R. Perez-Ossorio, *Pure Appl. Chem.*, *2*, 133–135 (1961).

(252) M. Colomina, A. S. Pell, H. A. Skinner, and D. J. Coleman, *Trans. Faraday Soc.*, *61*, 2641–2645 (1965).
(253) M. Colomina, R. Perez-Ossorio, M. L. Boned Corral, M. Panea, and C. Turrion, *Anales Real Soc. Espan. Fis. Quim. (Madrid), Ser. B*, *57*, 665–672 (1961).
(254) M. Colomina, R. Perez-Ossorio, C. Turrion, M. L. Boned Corral, and B. Pedraja, *Anales Real Soc. Espan. Fis. Quim. (Madrid), Ser. B*, *60*, 627–638 (1964).
(255) M. Colomina, C. Turrion, M. L. Boned Corral, and M. Panea, *Anales Real Soc. Espan. Fis. Quim. (Madrid), Ser. B*, *60*, 619–626 (1964).
(256) M. Colomina Barberá, M. Cambeiro, R. Perez-Ossorio, and C. Latorre, *Anales Real Soc. Espan. Fis. Quim. (Madrid), Ser. B*, *55*, 501–514 (1959).
(257) J. H. Colwell, E. K. Gill, and J. A. Morrison, *J. Chem. Phys.*, *39*, 635–653 (1963).
(258) J. B. Conn, G. B. Kistiakowsky, R. M. Roberts, and E. A. Smith, *J. Am. Chem. Soc.*, *64*, 1747–1752 (1942).
(259) J. B. Conn, G. B. Kistiakowsky, and E. A. Smith, *J. Am. Chem. Soc.*, *60*, 2764–2771 (1938).
(260) J. B. Conn, G. B. Kistiakowsky, and E. A. Smith, *J. Am. Chem. Soc.*, *61*, 216–217 (1939).
(261) J. B. Conn, G. B. Kistiakowsky, and E. A. Smith, *J. Am. Chem. Soc.*, *61*, 1868–1876 (1939).
(262) E. D. Coon and F. Daniels, *J. Phys. Chem.*, *37*, 1-12 (1933).
(263) J. Coops and G. J. Hoijtink, *Rec. Trav. Chim.*, *69*, 358–367 (1950).
(264) J. Coops, G. J. Hoijtink, and T. J. E. Kramer, *Rec. Trav. Chim.*, *72*, 793–797 (1953).
(265) J. Coops, G. J. Hoijtink, T. J. E. Kramer, and A. C. Faber, *Rec. Trav. Chim.*, *72*, 765–773 (1953).
(266) J. Coops, G. J. Hoijtink, T. J. E. Kramer, and A. C. Faber, *Rec. Trav. Chim.*, *72*, 781–784 (1953).
(267) J. Coops and S. Kaarsemaker, *Rec. Trav. Chim.*, *69*, 1364 (1950).
(268) J. Coops, D. Mulder, J. W. Dienske, and J. Smittenberg, *Rec. Trav. Chim.*, *65*, 128 (1946).
(269) J. Coops, D. Mulder, J. W. Dienske, and J. Smittenberg, *Rec. Trav. Chim.*, *66*, 153–160 (1947).
(270) J. Coops, D. Mulder, J. W. Dienske, and J. Smittenberg, *Rec. Trav. Chim.*, *72*, 785–792 (1953).
(271) J. Coops, H. van Kamp, W. A. Lambregts, B. J. Visser, and H. Dekker, *Rec. Trav. Chim.*, *79*, 1226–1234 (1960).
(272) J. Coops, K. Van Nes, A. Kentie, and J. W. Dienske, *Rec. Trav. Chim.*, *66*, 113–130 (1947).
(273) J. L. Copp and T. J. V. Findlay, *Trans. Faraday Soc.*, *56*, 13–22 (1960).
(274) R. J. Corruccini and D. C. Ginnings, *J. Am. Chem. Soc.*, *69*, 2291–2294 (1947).
(275) C. C. Costain, *J. Chem. Phys.*, *29*, 864–874 (1958).
(276) T. L. Cottrell, *The Strengths of Chemical Bonds*, 2nd Ed., Butterworths, London, 1958.
(277) T. L. Cottrell and J. E. Gill, *J. Chem. Soc.*, 1798–1800 (1951).
(278) J. P. Coughlin, *U.S. Bur. Mines Bull.*, *542*, 80 pp. (1954).
(279) J. F. Counsell, J. H. S. Green, J. L. Hales, and J. F. Martin, *Trans. Faraday Soc.*, *61*, 212–218 (1965).
(280) J. F. Counsell, J. L. Hales, and J. F. Martin, *Trans. Faraday Soc.*, *61*, 1869–1875 (1965).
(281) J. W. Coutts and R. L. Livingston, *J. Am. Chem. Soc.*, *75*, 1542–1547 (1953).
(282) M. Cowan and W. Gordy, *Phys. Rev.*, *104*, 551–552 (1956).

(283) A. P. Cox and A. S. Esbitt, *J. Chem. Phys.*, *38*, 1636–1643 (1963).
(284) J. D. Cox, *Trans. Faraday Soc.*, *56*, 959–964 (1960).
(285) J. D. Cox, *Pure Appl. Chem.*, *2*, 125–128 (1961).
(286) J. D. Cox, *Tetrahedron*, *18*, 1337–1350 (1962).
(287) J. D. Cox and R. J. L. Andon, *Trans. Faraday Soc.*, *54*, 1622–1629 (1958).
(288) J. D. Cox, A. R. Challoner, and A. R. Meetham, *J. Chem. Soc.*, 265–271 (1954).
(289) J. D. Cox and H. A. Gundry, *J. Chem. Soc.*, 1019–1022 (1958).
(290) J. D. Cox, H. A. Gundry, and A. J. Head, *Trans. Faraday Soc.*, *60*, 653–665 (1964).
(291) J. D. Cox, H. A. Gundry, and A. J. Head, *Trans. Faraday Soc.*, *61*, 1594–1600 (1965).
(292) J. D. Cox and A. J. Head, *Trans. Faraday Soc.*, *58*, 1839–1845 (1962).
(293) J. R. Cox, Jr., and J. D. Ray, *J. Chem. Phys.*, *34*, 1072–1073 (1961).
(294) A. E. Craver, U.S. Patent 1,851,754 (March 29, 1932); A. E. Craver and G. C. Bailey, U.S. Patent 1, 383, 059 (June 28, 1921).
(295) P. C. Cross, *J. Chem. Phys.*, *3*, 825–827 (1935).
(296) R. W. Crowe and C. P. Smyth, *J. Am. Chem. Soc.*, *72*, 4009–4015 (1950).
(297) A. H. Cubberley and M. B. Mueller, *J. Am. Chem. Soc.*, *68*, 1149–1151 (1946).
(298) G. A. M. Cummings and E. McLaughlin, *J. Chem. Soc.*, 1391–1392 (1955).
(299) R. F. Curl, Jr., *J. Chem. Phys.*, *30*, 1529–1536 (1959).
(300) R. M. Currie, C. O. Bennett, and D. E. Holcomb, *Ind. Eng. Chem.*, *44*, 329–331 (1952).
(301) G. R. Cuthbertson and G. B. Kistiakowsky, *J. Chem. Phys.*, *3*, 631–634 (1935).
(302) L. W. Daasch, C. Y. Liang, and J. R. Nielsen, *J. Chem. Phys.*, *22*, 1293–1303 (1954).
(303) E. Dallwick and E. Briner, *Helv. Chim. Acta*, *36*, 1166–1173 (1953).
(304) L. H. Daly and S. E. Wiberley, *J. Mol. Spectrosc.*, *2*, 177–186 (1958).
(305) M. D. Danford and R. L. Livingston, *J. Am. Chem. Soc.*, *81*, 4157–4159 (1959).
(306) A. Danti, *J. Chem. Phys.*, *27*, 1227 (1957).
(307) W. G. Dauben, O. Rohr, A. Labbauf, and F. D. Rossini, *J. Phys. Chem.*, *64*, 283–284 (1960).
(308) J. Davies, J. R. Lacher, and J. D. Park, "Heats of Hydrogenation," Presented at the 19th Calorimetry Conference, Washington, D. C., 1964.
(309) J. V. Davies and S. Sunner, *Acta Chem. Scand.*, *16*, 1870–1876 (1962).
(310) M. Davies and J. I. Jones, *Trans. Faraday Soc.*, *50*, 1042–1047 (1954).
(311) T. Davies and L. A. K. Staveley, *J. Chem. Soc.*, 2563 (1956).
(312) A. Davis, F. F. Cleveland, and A. G. Meister, *J. Chem. Phys.*, *20*, 454–459 (1952).
(313) G. L. Davis and G. H. Burrows, *J. Am. Chem. Soc.*, *58*, 311–312 (1936).
(314) H. S. Davis and O. F. Wiedeman, *Ind. Eng. Chem.*, *37*, 482–485 (1945).
(315) P. Debye, *Ann. Phys.* *39*, 789–839 (1912).
(316) C. E. Decker, A. G. Meister, and F. F. Cleveland, *J. Chem. Phys.*, *19*, 784–788 (1951).
(317) C. E. Decker, A. G. Meister, F. F. Cleveland, and R. B. Bernstein, *J. Chem. Phys.*, *21*, 1781–1783 (1953).
(318) R. F. Deese, Jr., *J. Am. Chem. Soc.*, *53*, 3673–3683 (1931).
(319) L. Deffet, *Répertoire des Composés Organiques Polymorphes*, Editions Desoer, Liege, 1942.
(320) M. Délépine and M. Badoche, *Compt. Rend.*, *214*, 777–781 (1942).
(321) K. G. Denbigh, *The Principles of Chemical Equilibrium*, Cambridge University Press, Cambridge, England, 1957.
(322) N. DeNevers and J. J. Martin, *Am. Inst. Chem. Eng. J.*, *6*, 43–49 (1960).

(323) T. De Vries and B. T. Collins, *J. Am. Chem. Soc.*, 63, 1343–1346 (1941).
(324) T. De Vries and B. T. Collins, *J. Am. Chem. Soc.*, 64, 1224–1225 (1942).
(325) M. J. S. Dewar and R. Pettit, *J. Chem. Soc.*, 1625–1634 (1954)
(326) M. J. S. Dewar and H. N. Schmeising, *Tetrahedron*, 5, 166–178 (1959); 11, 96–120 (1960).
(327) H. C. Dickinson, *Natl. Bur. Std. Bull.*, 11, 189–257 (1914).
(328) M. H. Dilke and D. D. Eley, *J. Chem. Soc.*, 2601–2612 (1949).
(329) R. E. Dininny and E. L. Pace, *J. Chem. Phys.*, 32, 805–809 (1960).
(330) W. B. Dixon and E. B. Wilson, Jr., *J. Chem. Phys.*, 35, 191–198 (1961).
(331) C. J. Dobratz, *Ind. Eng. Chem.*, 33, 759–762 (1941).
(332) R. E. Dodd, L. A. Woodward, and H. L. Roberts, *Trans. Faraday Soc.*, 52, 1052–1061 (1956).
(333) B. F. Dodge, *Brit. Chem. Eng.*, 7, 603–606, 683–688 (1962).
(334) M. A. Dolliver, T. L. Gresham, G. B. Kistiakowsky, E. A. Smith, and W. E. Vaughan, *J. Am. Chem. Soc.*, 60, 440–450 (1938).
(335) M. A. Dolliver, T. L. Gresham, G. B. Kistiakowsky, and W. E. Vaughan, *J. Am. Chem. Soc.*, 59, 831–841 (1937).
(336) T. M. Donovan, C. H. Shomate, and W. R. McBride, *J. Phys. Chem.*, 64, 281–282 (1960).
(337) D. E. Douglas and C. A. Winkler, *Can. J. Res.*, 25B, 381–386 (1947).
(338) T. B. Douglas and E. G. King, "High-Temperature Drop Calorimetry," in J. P. McCullough and D. W. Scott (Eds.), *Experimental Thermodynamics*, Butterworths, London, 1966, Chap. 7.
(339) D. R. Douslin, R. H. Harrison, R. T. Moore, and J. P. McCullough, *J. Chem. Phys.*, 35, 1357–1366 (1961).
(340) D. R. Douslin, R. H. Harrison, R. T. Moore, and J. P. McCullough, *J. Chem. Eng. Data*, 9, 358–363 (1964).
(341) D. R. Douslin and H. M. Huffman, *J. Am. Chem. Soc.*, 68, 173–176 (1946).
(342) D. R. Douslin and H. M. Huffman, *J. Am. Chem. Soc.*, 68, 1704–1708 (1946).
(343) D. R. Douslin, R. T. Moore, J. P. Dawson, and G. Waddington, *J. Am. Chem. Soc.*, 80, 2031–2038 (1958).
(344) D. R. Douslin, R. T. Moore, and G. Waddington, *J. Phys. Chem.*, 63, 1959–1966 (1959).
(345) J. M. Dowling and A. G. Meister, *J. Chem. Phys.*, 22, 1042–1044 (1954).
(346) J. M. Dowling, P. G. Puranik, A. G. Meister, and S. I. Miller, *J. Chem. Phys.*, 26, 233–240 (1957).
(347) A. A. Draeger, G. T. Gwin, C. J. G. Leesemann, and M. R. Morrow, *Petrol. Refiner*, 30, (8), 71–76 (1951).
(348) F. Drahowzal and D. Klamann, *Monatsh.*, 82, 594–599 (1951).
(349) L. G. Drayton and H. W. Thompson, *J. Chem. Soc.*, 1416–1419 (1948).
(350) R. R. Dreisbach, *Pressure-Volume-Temperature Relationship of Organic Compounds*, 3rd Ed., Handbook Publishers, Sandusky, Ohio, 1952.
(351) R. R. Dreisbach, *Physical Properties of Chemical Compounds*, Vol. I *(Advances in Chemistry Series, No. 15)*, American Chemical Society, Washington, D.C., 1955.
(352) R. R. Dreisbach, *Physical Properties of Chemical Compounds*, Vol. II *(Advances in Chemistry Series, No. 22)*, American Chemical Society, Washington, D.C., 1959.
(353) R. R. Dreisbach, *Physical Properties of Chemical Compounds*, Vol. III *(Advances in Chemistry Series, No. 29)*, American Chemical Society, Washington, D.C., 1961.
(354) H. Dreizler and H. D. Rudolph, *Z. Naturforsch.*, 17a, 712–732 (1962).
(355) G. M. S. Duff, *J. Appl. Chem. (London)*, 5, 642 (1955).
(356) N. E. Duncan and G. J. Janz, *J. Chem. Phys.*, 20, 1644–1645 (1952).

(357) N. E. Duncan and G. J. Janz, *J. Chem. Phys.*, 23, 434–440 (1955).
(358) H. Dunken and K. L. Wolf, *Z. Phys. Chem. (Leipzig)*, B38, 441–450 (1938).
(359) A. K. Dunlop, *J. Am. Chem. Soc.*, 77, 2016 (1955).
(360) H. C. Duus, *Ind. Eng. Chem.*, 47, 1445–1449 (1955).
(361) M. E. Dyatkina, *Zh. Fiz. Khim.*, 28, 377–388 (1954).
(362) E. D. Eastman and W. C. McGavock, *J. Am. Chem. Soc.*, 59, 145–151 (1937).
(363) W. F. Eberz and H. J. Lucas, *J. Am. Chem. Soc.*, 56, 1230–1234 (1934).
(364) J. R. Eckman and F. D. Rossini, *J. Res. Natl. Bur. Std.*, 3, 597–618 (1929).
(365) W. F. Edgell, *J. Am. Chem. Soc.*, 69, 660–661 (1947).
(366) W. F. Edgell, P. A. Kinsey, and J. W. Amy, *J. Am. Chem. Soc.*, 79, 2691–2693 (1957).
(367) W. F. Edgell, G. B. Miller, and J. W. Amy, *J. Am. Chem. Soc.*, 79, 2391–2393 (1957).
(368) W. F. Edgell and C. J. Ultee, *J. Chem. Phys.*, 22, 1983–1992 (1954)
(369) W. C. Edmister, *Petrol. Refiner*, 26 (7), 565–570; (8), 625–630; (9), 689–695; (10), 735–741; (11), 742–749; (12), 808–814 (1947); 27 (1), 6-14; (2), 74–81; (3), 113–119; (4), 213–221; (5), 228–235; (6), 314–321; (11), 609–615; (12), 656–663 (1948); 28 (1), 128–133; (2), 137–148; (3), 139–150; (4), 157–166; (5), 149–160; (6), 143–148; (8), 128–133; (9), 95–102; (10), 143–150; (11), 149–155; (12), 140–145 (1949).
(370) E. Eftring, Thesis, University of Lund, Lund, Sweden, 1938.
(371) C. J. Egan and W. C. Buss, *J. Phys. Chem.*, 63, 1887–1890 (1959).
(372) C. J. Egan and J. D. Kemp, *J. Am. Chem. Soc.*, 59, 1264–1268 (1937).
(373) C. J. Egan and J. D. Kemp, *J. Am. Chem. Soc.*, 60, 2097–2101 (1938).
(374) E. P. Egan, Jr., Z. T. Wakefield, and T. D. Farr, *J. Chem. Eng. Data*, 10, 138–140 (1965).
(375) A. Einstein, *Ann. Phys.* 22, 180–190 (1907).
(376) F. Eisenlohr and A. Metzner, *Z. Phys. Chem. (Leipzig)*, A178, 339–349 (1937).
(377) M. Z. El-Sabban, A. G. Meister, and F. F. Cleveland, *J. Chem. Phys.*, 19, 855–864 (1951).
(378) L. Enderlin, *Ann. Chim. (Paris)*, 10, 5–116 (1938).
(379) S. G. Entelis and N. M. Chirkov, *Dokl. Akad. Nauk SSSR*, 113, 1318–1320 (1957).
(380) M. B. Epstein, K. S. Pitzer, and F. D. Rossini, *J. Res. Natl. Bur. Std.*, 42, 379–382 (1949).
(381) G. Erlandsson, *Arkiv Fys.*, 8, 341–342 (1954).
(382) G. Erlandsson, *J. Chem. Phys.*, 25, 579–580 (1956).
(383) A. Eucken and E. U. Franck, *Z. Elektrochem.*, 52, 195–204 (1948).
(384) A. Eucken and E. Schröder, *Z. Phys. Chem. (Leipzig)*, B41, 307–319 (1938).
(385) H. von Euler and C. Martius, *Ann.*, 505, 73–87 (1933).
(386) D. M. Evans, F. E. Hoare, and T. P. Melia, *Trans. Faraday Soc.*, 58, 1511–1514 (1962).
(387) F. W. Evans, D. M. Fairbrother, and H. A. Skinner, *Trans. Faraday Soc.*, 55, 399–403 (1959).
(388) F. W. Evans and H. A. Skinner, *Trans. Faraday Soc.*, 55, 255–259 (1959).
(389) F. W. Evans, and H. A. Skinner, *Trans. Faraday Soc.*, 55, 260–261 (1959).
(390) J. C. Evans, *J. Chem. Phys.*, 30, 934–936 (1959).
(391) J. C. Evans, *Spectrochim. Acta*, 16, 428–442 (1960).
(392) J. C. Evans, *Spectrochim. Acta*, 16, 1382–1392 (1960).
(393) J. C. Evans and H. J. Bernstein, *Can. J. Chem.*, 33, 1171–1182 (1955).
(394) J. C. Evans and H. J. Bernstein, *Can. J. Chem.*, 34, 1083–1092 (1956).
(395) J. C. Evans and R. A. Nyquist, *Spectrochim. Acta*, 16, 918–928 (1960).

(396) J. C. Evans and R. A. Nyquist, *Spectrochim. Acta, 19,* 1153–1163 (1963).
(397) M. V. Evans and R. C. Lord, *J. Am. Chem. Soc., 82,* 1876–1882 (1960).
(398) W. H. Evans, T. R. Munson, and D. D. Wagman, *J. Res. Natl. Bur. Std., 55,* 147–164 (1955).
(399) W. H. Evans and D. D. Wagman, *J. Res. Natl. Bur. Std., 49,* 141–148 (1952).
(400) R. H. Ewell and J. F. Bourland, *J. Chem. Phys., 8,* 635–636 (1940).
(401) T. F. Fagley, E. Klein, and J. F. Albrecht, Jr., *J. Am. Chem. Soc., 75,* 3104–3106 (1953).
(402) T. F. Fagley and H. W. Myers, *J. Am. Chem. Soc., 76,* 6001–6003 (1954).
(403) J. Fahrenfort, L. L. van Reijen, and W. M. H. Sachtler, *Z. Elektrochem., 64,* 216–224 (1960).
(404) D. M. Fairbrother and H. A. Skinner, *Trans. Faraday Soc., 52,* 956–960 (1956).
(405) D. M. Fairbrother, H. A. Skinner, and F. W. Evans, *Trans. Faraday Soc., 53,* 779–783 (1957).
(406) K. Fajans, *Ber., 53B,* 643–665 (1920).
(407) W. G. Fateley and F. A. Miller, *Spectrochim. Acta, 17,* 857–868 (1961).
(408) W. G. Fateley and F. A. Miller, *Spectrochim. Acta, 18,* 977–993 (1962).
(409) W. G. Fateley and F. A. Miller, *Spectrochim. Acta, 18,* 1389 (1962).
(410) W. G. Fateley and F. A. Miller, *Spectrochim. Acta, 19,* 611–628 (1963).
(411) F. S. Fawcett, *J. Am. Chem. Soc., 68,* 1420–1422 (1946).
(412) F. Fehér and H. J. Berthold, *Z. Anorg. Allg. Chem., 284,* 60–68 (1956).
(413) G. Feick, *J. Am. Chem. Soc., 76,* 5858–5860 (1954).
(414) H. Feilchenfeld, *J. Phys. Chem., 61,* 1133–1135 (1957).
(415) W. A. Felsing and G. W. Drake, *J. Am. Chem. Soc., 58,* 1714–1717 (1936).
(416) W. A. Felsing and F. W. Jessen, *J. Am. Chem. Soc., 55,* 4418–4422 (1933).
(417) S. M. Ferigle and A. Weber, *J. Chem. Phys., 20,* 1657 (1952).
(418) J. D. Ferry and S. B. Thomas, *J. Phys. Chem., 37,* 253–255 (1933).
(419) H. L. Finke, M. E. Gross, J. F. Messerly, and G. Waddington, *J. Am. Chem. Soc., 76,* 854–857 (1954).
(420) H. L. Finke, M. E. Gross, G. Waddington, and H. M. Huffman, *J. Am. Chem. Soc., 76,* 333–341 (1954).
(421) H. L. Finke, G. B. Guthrie, S. S. Todd, J. F. Messerly, and J. P. McCullough, *Bull. Thermodyn. Thermochem., 10,* 28 (March 1967).
(422) H. L. Finke, I. A. Hossenlopp, and W. T. Berg, *J. Phys. Chem., 69,* 3030–3031 (1965).
(423) H. L. Finke, J. F. Messerly, and S. S. Todd, *J. Phys. Chem., 69,* 2094–2100 (1965).
(424) H. L. Finke, D. W. Scott, M. E. Gross, J. F. Messerly, and G. Waddington, *J. Am. Chem. Soc., 78,* 5469–5476 (1956).
(425) H. L. Finke, D. W. Scott, M. E. Gross, G. Waddington, and H. M. Huffman, *J. Am. Chem. Soc., 74,* 2804–2806 (1952).
(426) E. F. Fiock, D. C. Ginnings, and W. B. Holton, *J. Res. Natl. Bur. Std., 6,* 881–900 (1931).
(427) F. Fischer, *Ind. Eng. Chem., 17,* 547–576 (1925).
(428) J. Fischer and J. Bingle, *J. Am. Chem. Soc., 77,* 6511–6512 (1955).
(429) F. B. Fischl, B. F. Naylor, C. W. Ziemer, G. S. Parks, and J. G. Aston, *J. Am. Chem. Soc., 67,* 2075–2079 (1945).
(430) C. Flanagan and L. Pierce, *J. Chem. Phys., 38,* 2963–2969 (1963).
(431) S. E. Fletcher, C. T. Mortimer, and H. D. Springall, *J. Chem. Soc.,* 580–584, (1959).

(467) Y. I. Gerasimov, A. N. Krest
in Non-ferrous Metallurgy, V

(468) Y. I. Gerasimov, A. N. Krest
in Non-ferrous Metallurgy, V

(469) Y. I. Gerasimov, A. N. Krest
in Non-ferrous Metallurgy, V

(470) Y. I. Gerasimov, A. N. Krest
in Non-ferrous Metallurgy, V

(471) J. C. Ghosh and S. R. D. Guh

(472) J. C. Ghosh, S. R. D. Guha, a

(473) S. N. Ghosh, R. Trambarulo,

(474) W. F. Giauque and C. J. Egan,

(475) W. F. Giauque and J. Gordon

(476) W. F. Giauque and W. M. Jon

(477) W. F. Giauque and J. B. Ott,

(478) W. F. Giauque and R. A. Rue

(479) J. W. Gibbs, *Trans. Conn. Aca*
248 (1876); *4,* 343–524 (187

(480) P. A. Giguère, *Can. J. Chem.,*

(481) P. A. Giguère and I. D. Liu, *J.*

(482) P. A. Giguère, I. D. Liu, J. S.
128 (1954).

(483) P. A. Giguère, B. G. Morissett
804–820 (1955).

(484) R. L. Gilbert, E. A. Piotrowsk
31, 1633–1636 (1959).

(485) O. R. Gilliam, H. D. Edwards,

(486) D. C. Ginnings and R. J. Corru

(487) D. C. Ginnings and G. T. Furu

(488) A. Girelli and L. Burlamacchi,

(489) P. H. Given, *J. Chem. Soc.,* 58

(490) S. Gladstone and H.-Y. Chang,

(491) A. R. Glasgow, Jr., C. B. Willin
2297 (1949).

(492) S. Glasstone, *Thermodynamics*

(493) O. Glemser and V. Häusser, *Z.*

(494) G. Glockler, *J. Chem. Phys., 21*

(495) G. Glockler, *J. Phys. Chem., 61*

(496) G. Glockler, *J. Phys. Chem., 62*

(497) G. Glockler and W. F. Edgell, *J.*

(498) G. Glockler and W. F. Edgell, *In*

(499) V. P. Glushko, L. V. Gurvich, G
Medvedev, *Thermodynamic Pro*
lishing House of the Academy

(500) V. P. Glushko, L. V. Gurvich, G
Medvedev, *Thermodynamic Pro*
House of the Academy of Scien

(501) I. N. Godnev and A. S. Sverdlin,

(502) I. N. Godnev, A. S. Sverdlin, an
(1950).

(432) T. L. Flitcroft and H. A. Skinner, *Trans. Faraday Soc.*, *54*, 47–53 (1958).
(433) T. Flitcroft, H. A. Skinner, and M. C. Whiting, *Trans. Faraday Soc.*, *53*, 784–790 (1957).
(434) W. H. Flygare and J. A. Howe, *J. Chem. Phys.*, *36*, 440–443 (1962).
(435) P. Fowell, J. R. Lacher, and J. D. Park, *Trans. Faraday Soc.*, *61*, 1324–1327 (1965).
(436) R. H. Fowler and E. A. Guggenheim, *Statistical Thermodynamics*, Cambridge University Press, London, 1939.
(437) J. J. Fox and A. E. Martin, *J. Chem. Soc.*, 884–886 (1939).
(438) W. G. Frankenburg, "The Catalytic Synthesis of Ammonia from Nitrogen and Hydrogen," in P. H. Emmett (Ed.), *Catalysis*, Vol. III, Reinhold, New York, 1955, pp. 171–263.
(439) J. L. Franklin, *Ind. Eng. Chem.*, *41*, 1070–1076 (1949).
(440) J. L. Franklin and H. E. Lumpkin, *J. Am. Chem. Soc.*, *74*, 1023–1026 (1952).
(441) M. Frankosky and J. G. Aston, *J. Phys. Chem.*, *69*, 3126–3132 (1965).
(442) F. M. Fraser and E. J. Prosen, *J. Res. Natl. Bur. Std.*, *54*, 143–148 (1955).
(443) F. M. Fraser and E. J. Prosen, *J. Res. Natl. Bur. Std.*, *55*, 329–333 (1955).
(444) K. J. Frederick and J. H. Hildebrand, *J. Am. Chem. Soc.*, *60*, 1436–1439 (1938).
(445) A. S. Friedman and L. Haar, *J. Chem. Phys.*, *22*, 2051–2058 (1954).
(446) K. Fries and F. Beyerlein, *Ann.*, *527*, 71–83 (1936).
(447) K. Fries, R. Walter, and K. Schilling, *Ann.*, *516*, 248–285 (1935).
(448) M. A. Frisch, C. Barker, J. L. Margrave, and M. S. Newman, *J. Am. Chem. Soc.*, *85*, 2356–2357 (1963).
(449) M. A. Frisch, R. G. Bautista, J. L. Margrave, C. G. Parsons, and J. H. Wotiz, *J. Am. Chem. Soc.*, *86*, 335–336 (1964).
(450) C. G. Frye, *J. Chem. Eng. Data*, *7*, 592–595 (1962).
(451) W. Fuchs, K. H. Andres, J. Plenz, and O. Veiser, *Chem. Ingr. Tech.*, *29*, 768–771 (1957).
(452) G. T. Furukawa, T. B. Douglas, R. E. McCoskey, and D. C. Ginnings, *J. Res. Natl. Bur. Std.*, *57*, 67–82 (1956).
(453) G. T. Furukawa, D. C. Ginnings, R. E. McCoskey, and R. A. Nelson, *J. Res. Natl. Bur. Std.*, *46*, 195–206 (1951).
(454) G. T. Furukawa, R. E. McCoskey, and G. J. King, *J. Res. Natl. Bur. Std.*, *47*, 256–61 (1951).
(455) G. T. Furukawa, R. E. McCoskey, and G. J. King, *J. Res. Natl. Bur. Std.*, *49*, 273–78 (1952).
(456) G. T. Furukawa, R. E. McCoskey, and M. L. Reilly, *J. Res. Natl. Bur. Std.*, *51*, 69–72 (1953).
(457) G. T. Furukawa, R. E. McCoskey, and M. L. Reilly, *J. Res. Natl. Bur. Std.*, *52*, 11–16 (1954).
(458) B. W. Gamson and K. M. Watson, *Natl. Petrol. News*, *36*, R623–R635 (1944).
(459) D. M. Gardner and J. C. Grigger, *J. Chem. Eng. Data*, *8*, 73–74 (1963).
(460) G. Gattow and B. Krebs, *Z. Anorg. Allg. Chem.*, *322*, 113–128 (1963).
(461) A. G. Gaydon, *Dissociation Energies and Spectra of Diatomic Molecules*, 2nd Ed. (revised), Chapman and Hall, London, 1953.
(462) G. Geiseler and M. Rätzsch, *Z. Phys. Chem. (Frankfurt)*, *26*, 131–137 (1960).
(463) G. Geiseler and M. Rätzsch, *Ber. Bunsenges. Phys. Chem.*, *69*, 485–488 (1965).
(464) G. Geiseler and W. Thierfelder, *Z. Phys. Chem. (Frankfurt)*, *29*, 248–257 (1961).
(465) E. Gelles and K. S. Pitzer, *J. Am. Chem. Soc.*, *75*, 5259–5267 (1953).
(466) O. H. Gellner and H. A. Skinner, *J. Chem. Soc.*, 1145–1148 (1949).

(467) Y. I. Gerasimov, A. N. Krestovnikov, and A. S. Shakov, *Chemical Thermodynamics in Non-ferrous Metallurgy,* Vol. I, Moscow, 1960.
(468) Y. I. Gerasimov, A. N. Krestovnikov, and A. S. Shakov, *Chemical Thermodynamics in Non-ferrous Metallurgy,* Vol. II, Moscow, 1961.
(469) Y. I. Gerasimov, A. N. Krestovnikov, and A. S. Shakov, *Chemical Thermodynamics in Non-ferrous Metallurgy,* Vol. III, Moscow, 1962.
(470) Y. I. Gerasimov, A. N. Krestovnikov, and A. S. Shakov, *Chemical Thermodynamics in Non-ferrous Metallurgy,* Vol. IV, Moscow, 1966.
(471) J. C. Ghosh and S. R. D. Guha, *Petroleum (London), 14,* 261–264 (1951).
(472) J. C. Ghosh, S. R. D. Guha, and A. N. Roy, *Current Sci. (India), 14,* 269 (1945).
(473) S. N. Ghosh, R. Trambarulo, and W. Gordy, *J. Chem. Phys., 20,* 605–607 (1952).
(474) W. F. Giauque and C. J. Egan, *J. Chem. Phys., 5,* 45–54 (1937).
(475) W. F. Giauque and J. Gordon, *J. Am. Chem. Soc., 71,* 2176–2181 (1949).
(476) W. F. Giauque and W. M. Jones, *J. Am. Chem. Soc., 70,* 120–124 (1948).
(477) W. F. Giauque and J. B. Ott, *J. Am. Chem. Soc., 82,* 2689–2695 (1960).
(478) W. F. Giauque and R. A. Ruehrwein, *J. Am. Chem. Soc., 61,* 2626–2633 (1939).
(479) J. W. Gibbs, *Trans. Conn. Acad. Arts Sci., 2,* 309–342, 382–404 (1873); *3,* 108–248 (1876); *4,* 343–524 (1878).
(480) P. A. Giguère, *Can. J. Chem., 32,* 1161 (1954).
(481) P. A. Giguère and I. D. Liu, *J. Am. Chem. Soc., 77,* 6477–6479 (1955).
(482) P. A. Giguère, I. D. Liu, J. S. Dugdale, and J. A. Morrison, *Can J. Chem., 32,* 117–128 (1954).
(483) P. A. Giguère, B. G. Morissette, A. W. Olmos, and O. Knop, *Can. J. Chem., 33,* 804–820 (1955).
(484) R. L. Gilbert, E. A. Piotrowski, J. M. Dowling, and F. F. Cleveland, *J. Chem. Phys., 31,* 1633–1636 (1959).
(485) O. R. Gilliam, H. D. Edwards, and W. Gordy, *Phys. Rev., 75,* 1014–1016 (1949).
(486) D. C. Ginnings and R. J. Corruccini, *Ind. Eng. Chem., 40,* 1990–1991 (1948).
(487) D. C. Ginnings and G. T. Furukawa, *J. Am. Chem. Soc., 75,* 522–527 (1953).
(488) A. Girelli and L. Burlamacchi, *Riv. Combust., 15,* 121–130 (1961).
(489) P. H. Given, *J. Chem. Soc.,* 589 (1943).
(490) S. Gladstone and H.-Y. Chang, *J. Chem. Eng. Data, 11,* 238–239 (1966).
(491) A. R. Glasgow, Jr., C. B. Willingham, and F. D. Rossini, *Ind. Eng. Chem., 41,* 2292–2297 (1949).
(492) S. Glasstone, *Thermodynamics for Chemists,* Van Nostrand, New York, 1947.
(493) O. Glemser and V. Häusser, *Z. Naturforsch., 3b,* 159–163 (1948).
(494) G. Glockler, *J. Chem. Phys., 21,* 1242–1248, 1249–1254 (1953).
(495) G. Glockler, *J. Phys. Chem., 61,* 31–38 (1957).
(496) G. Glockler, *J. Phys. Chem., 62,* 1049–1054 (1958).
(497) G. Glockler and W. F. Edgell, *J. Chem. Phys., 9,* 527–529 (1941).
(498) G. Glockler and W. F. Edgell, *Ind. Eng. Chem., 34,* 532–534 (1942).
(499) V. P. Glushko, L. V. Gurvich, G. A. Khachkuruzov, I. V. Veits, and V. A. Medvedev, *Thermodynamic Properties of Individual Substances,* Vol. I, Publishing House of the Academy of Sciences of the U.S.S.R., Moscow, 1962.
(500) V. P. Glushko, L. V. Gurvich, G. A. Khachkuruzov, I. V. Veits, and V. A. Medvedev, *Thermodynamic Properties of Individual Substances,* Vol. II, Publishing House of the Academy of Sciences of the U.S.S.R., Moscow, 1962.
(501) I. N. Godnev and A. S. Sverdlin, *Zh. Fiz. Khim., 24,* 670–682 (1950).
(502) I. N. Godnev, A. S. Sverdlin, and M. S. Savogina, *Zh. Fiz. Khim., 24,* 807–812 (1950).

References

(503) D. M. Golden, K. W. Egger, and S. W. Benson, *J. Am. Chem. Soc., 86,* 5416–5420 (1964).
(504) D. M. Golden, H. E. O'Neal, and S. W. Benson, "Additivity Rules for the Estimation of Thermochemical Properties. I. Hydrocarbons," to be published.
(505) D. M. Golden, R. Walsh, and S. W. Benson, *J. Am. Chem. Soc., 87,* 4053–4057 (1965).
(506) G. I. Golodets and V. A. Roiter, *Intern. Chem. Eng., 4,* 632–642 (1964).
(507) W. D. Good and B. L. DePrater, *J. Phys. Chem., 70,* 3606–3609 (1966).
(508) W. D. Good, D. R. Douslin, D. W. Scott, A. George, J. L. Lacina, J. P. Dawson, and G. Waddington, *J. Phys. Chem., 63,* 1133–1138 (1959).
(509) W. D. Good, J. L. Lacina, and J. P. McCullough, *J. Am. Chem. Soc., 82,* 5589–5591 (1960).
(510) W. D. Good, J. L. Lacina, and J. P. McCullough, *J. Phys. Chem., 65,* 860–862 (1961).
(511) W. D. Good, J. L. Lacina, and J. P. McCullough, *J. Phys. Chem., 65,* 2229–2231 (1961).
(512) W. D. Good, J. L. Lacina, D. W. Scott, and J. P. McCullough, *J. Phys. Chem., 66,* 1529–1532 (1962).
(513) W. D. Good and D. W. Scott, "Combustion in a Bomb of Organic Fluorine Compounds," in H. A. Skinner (Ed.), *Experimental Thermochemistry,* Vol. II, Interscience Publishers, New York, 1962, Chap. 2.
(514) W. D. Good, D. W. Scott, and G. Waddington, *J. Phys. Chem., 60,* 1080–1089 (1956).
(515) W. D. Good, S. S. Todd, J. F. Messerly, J. L. Lacina, J. P. Dawson, D. W. Scott, and J. P. McCullough, *J. Phys. Chem., 67,* 1306–1311 (1963).
(516) I. A. Goodman and P. H. Wise, *J. Am. Chem. Soc., 72,* 3076–3078 (1950).
(517) I. A. Goodman and P. H. Wise, *J. Am. Chem. Soc., 73,* 850–851 (1951).
(518) J. G. Gordon and W. F. Giauque, *J. Am. Chem. Soc., 70,* 1506–1510 (1948).
(519) J. S. Gordon, *J. Chem. Eng. Data, 6,* 390–394 (1961).
(520) J. S. Gordon, *J. Chem. Eng. Data, 7,* 82, (1962).
(521) C. A. Goy and H. O. Pritchard, *J. Phys. Chem., 69,* 3040–3042 (1965).
(522) W. S. Graham, R. J. Nichol, and A. R. Ubbelohde, *J. Chem. Soc.,* 115–121 (1955).
(523) A. P. Gray and R. C. Lord, *J. Chem. Phys., 26,* 690–705 (1957).
(524) P. Gray and M. W. T. Pratt, *J. Chem. Soc.,* 2163–2168 (1957).
(525) P. Gray and M. W. T. Pratt, *J. Chem. Soc.,* 3403–3412 (1958).
(526) P. Gray, M. W. T. Pratt, and M. J. Larkin, *J. Chem. Soc.,* 210–212 (1956).
(527) P. Gray and L. W. Reeves, *J. Chem. Phys., 32,* 1878–1880 (1960).
(528) P. Gray and P. L. Smith, *J. Chem. Soc.,* 2380–2385 (1953).
(529) P. Gray and P. L. Smith, *J. Chem. Soc.,* 769–773 (1954).
(530) J. H. S. Green, *Chem. Ind. (London),* 1215–1216 (1960).
(531) J. H. S. Green, *Chem. Ind. (London),* 369 (1961).
(532) J. H. S. Green, *J. Appl. Chem. (London), 11,* 397–404 (1961).
(533) J. H. S. Green, *J. Chem. Soc.,* 2236–2241 (1961).
(534) J. H. S. Green, *J. Chem. Soc.,* 2241–2242 (1961).
(535) J. H. S. Green, *Spectrochim. Acta, 17,* 607–613 (1961).
(536) J. H. S. Green, *Trans. Faraday Soc., 57,* 2132–2137 (1961).
(537) J. H. S. Green, *Chem. Ind. (London),* 1575–1576 (1962).
(538) J. H. S. Green, *Trans. Faraday Soc., 59,* 1559–1563 (1963).
(539) J. H. S. Green and D. J. Holden, *J. Chem. Soc.,* 1513–1514 (1962).
(540) J. H. S. Green and D. J. Holden, *J. Chem. Soc.,* 1794–1801 (1962).

(541) J. H. S. Green, W. Kynaston, and H. M. Paisley, *J. Chem. Soc.*, 473–478 (1963).
(542) E. Greenberg and W. N. Hubbard, "The Enthalpy of Formation of CF_4 by Fluorination of Graphite; Use of a Two-Compartment Bomb and Quartz Thermometer," Presented at the 22nd Calorimetry Conference, Thousand Oaks, Calif., 1967.
(543) J. B. Greenshields and F. D. Rossini, *J. Phys. Chem.*, 62, 271–280 (1958).
(544) M. E. Gross, G. D. Oliver, and H. M. Huffman, *J. Am. Chem. Soc.*, 75, 2801–2804 (1953).
(545) F. T. Gucker, Jr., and K. H. Schminke, *J. Am. Chem. Soc.*, 54, 1358–1373 (1932).
(546) C. W. Gullikson and J. R. Nielsen, *J. Mol. Spectrosc.*, 1, 158–178 (1957).
(547) H. A. Gundry, A. J. Head, and G. B. Lewis, *Trans. Faraday Soc.*, 58, 1309–1312 (1962).
(548) H. Günthard and E. Heilbronner, *Helv. Chim. Acta*, 31, 2128–2132 (1948).
(549) H. H. Günthard and E. Kováts, *Helv. Chim. Acta*, 35, 1190–1191 (1952).
(550) G. B. Guthrie, Jr., and H. M. Huffman, *J. Am. Chem. Soc.*, 65, 1139–1143 (1943).
(551) G. B. Guthrie, Jr., D. W. Scott, W. N. Hubbard, C. Katz, J. P. McCullough, M. E. Gross, K. D. Williamson, and G. Waddington, *J. Am. Chem. Soc.*, 74, 4662–4669 (1952).
(552) G. B. Guthrie, Jr., D. W. Scott, and G. Waddington, *J. Am. Chem. Soc.*, 74, 2795–2800 (1952).
(553) L. Guttman and K. S. Pitzer, *J. Am. Chem. Soc.*, 67, 324–327 (1945).
(554) L. Guttman, E. F. Westrum, Jr., and K. S. Pitzer, *J. Am. Chem. Soc.*, 65, 1246–1247 (1943).
(555) A. Guyer, H. Schütze, and M. Weidenmann, *Helv. Chim. Acta*, 20, 936–949 (1937).
(556) W. D. Gwinn and K. S. Pitzer, *J. Chem. Phys.*, 16, 303–309 (1948).
(557) L. Haar, J. C. Bradley, and A. S. Friedman, *J. Res. Natl. Bur. Std.*, 55, 285–290 (1955).
(558) A. Hadni, *Compt. Rend.*, 239, 349–351 (1954).
(559) V. Haensel and C. V. Berger, *Petrol. Process.*, 6, 264–267 (1951).
(560) J. E. Haggenmacher, *J. Am. Chem. Soc.*, 68, 1633–1634 (1946).
(561) J. L. Hales, J. D. Cox, and E. B. Lees, *Trans. Faraday Soc.*, 59, 1544–1554 (1963).
(562) J. O. Halford, *J. Chem. Phys.*, 10, 582–584 (1942).
(563) J. O. Halford, *J. Chem. Phys.*, 15, 364–367 (1947).
(564) J. O. Halford and D. Brundage, *J. Am. Chem. Soc.*, 64, 36–40 (1942).
(565) J. O. Halford and G. A. Miller, *J. Phys. Chem.*, 61, 1536–1539 (1957).
(566) A. Halpern and A. C. Glasser, *J. Am. Pharm. Assoc.*, 38, 287–290 (1949).
(567) F. Halverson and R. J. Francel, *J. Chem. Phys.*, 17, 694–703 (1949).
(568) F. Halverson, R. F. Stamm, and J. J. Whalen, *J. Chem. Phys.*, 16, 808–816 (1948).
(569) C. K. Hancock, G. M. Watson, and R. F. Gilby, *J. Phys. Chem.*, 58, 127–129 (1954).
(570) D. V. N. Hardy, *J. Chem. Soc.*, 1335–1340 (1934).
(571) J. V. Harispe, *Ann. Chim. (Paris)*, 6, 249–347 (1936).
(572) D. Harrison and E. A. Moelwyn-Hughes, *Proc. Roy. Soc. (London), Ser. A*, 239, 230–246 (1957).
(573) R. H. Harrison and K. A. Kobe, *Chem. Eng. Progr.*, 49, 349–353 (1953).
(574) R. H. Harrison and K. A. Kobe, *J. Chem. Phys.*, 26, 1411–1415 (1957).
(575) K. Hartley, H. O. Pritchard, and H. A. Skinner, *Trans. Faraday Soc.*, 46, 1019–1025 (1950).

(576) K. Hartley, H. O. Pritchard, and H. A. Skinner, *Trans. Faraday Soc., 47,* 254–263 (1951).
(577) S. H. Hastings and D. E. Nicholson, *J. Phys. Chem., 61,* 730–735 (1957).
(578) S. H. Hastings and D. E. Nicholson, *J. Chem. Eng. Data, 6,* 1–4 (1961).
(579) W. E. Hatton, D. L. Hildenbrand, G. C. Sinke, and D. R. Stull, *J. Chem. Eng. Data, 7,* 229–231 (1962).
(580) J. E. Hawkins and W. T. Eriksen, *J. Am. Chem. Soc., 76,* 2669–2671 (1954).
(581) E. Heilbronner and K. Wieland, *Helv. Chim. Acta, 30,* 947–956 (1947).
(582) G. Heim, *Bull. Soc. Chim. Belges, 42,* 467–482 (1933).
(583) H. Heinemann, J. W. Schall, and D. H. Stevenson, *Petrol. Refiner, 30,* (11), 107–110 (1951).
(584) R. V. Helm, W. J. Lanum, G. L. Cook, and J. S. Ball, *J. Phys. Chem., 62,* 858–862 (1958).
(585) E. F. G. Herington and J. F. Martin, *Trans. Faraday Soc., 49,* 154–162 (1953).
(586) D. R. Herschbach, *J. Chem. Phys., 25,* 358–359 (1956).
(587) D. R. Herschbach, *J. Chem. Phys., 31,* 91–108 (1959).
(588) D. R. Herschbach and J. D. Swalen, *J. Chem. Phys., 29,* 761–776 (1958).
(589) G. Herzberg, *Molecular Spectra and Molecular Structure. II. Infrared and Raman Spectra of Polyatomic Molecules,* Van Nostrand, Princeton, N. J., 1945.
(590) G. Herzberg, *Molecular Spectra and Molecular Structure. I. Diatomic Molecules,* 2nd Ed., Van Nostrand, New York, 1950.
(591) G. Herzberg and L. Herzberg, *J. Chem. Phys., 18,* 1551–1561 (1950).
(592) G. H. Hess, *Ann. Phys. [2], 52,* 97–114 (1841).
(593) P. Hestermans and D. White, *J. Phys. Chem., 65,* 362–365 (1961).
(594) J. F. G. Hicks, J. G. Hooley, and C. C. Stephenson, *J. Am. Chem. Soc., 66,* 1064–1067 (1944).
(595) D. L. Hildenbrand, unpublished results, Dow Chemical Company, 1958.
(596) D. L. Hildenbrand, W. R. Kramer, R. A. McDonald, and D. R. Stull, *J. Am. Chem. Soc., 80,* 4129–4132 (1958).
(597) D. L. Hildenbrand, W. R. Kramer, and D. R. Stull, *J. Phys. Chem., 62,* 958–959 (1958).
(598) D. L. Hildenbrand and R. A. McDonald, *J. Phys. Chem., 63,* 1521–1522 (1959).
(599) D. L. Hildenbrand, R. A. McDonald, W. R. Kramer, and D. R. Stull, *J. Chem. Phys., 30,* 930–934 (1959).
(600) D. L. Hildenbrand, G. C. Sinke, R. A. McDonald, W. R. Kramer, and D. R. Stull, *J. Chem. Phys., 31,* 650–654 (1959).
(601) E. A. Hill, *J. Am. Chem. Soc., 22,* 478–494 (1900).
(602) T. L. Hill, *Statistical Mechanics, Principles and Selected Applications,* McGraw-Hill, New York, 1956.
(603) T. L. Hill, *An Introduction to Statistical Thermodynamics,* Addison-Wesley, Reading, Mass., 1960.
(604) J. Hilsenrath, C. W. Beckett, W. S. Benedict, L. Fano, H. J. Hoge, J. F. Masi, R. L. Nuttall, Y. S. Touloukian, and H. W. Woolley, *Tables of Thermal Properties of Gases (National Bureau of Standards Circular No. 564),* U.S. Government Printing Office, Washington, D.C., 1953.
(605) J. Hilsenrath and G. G. Ziegler, *Tables of Einstein Functions: Vibrational Contributions to the Thermodynamic Functions (National Bureau of Standards Monograph No. 49),* U.S. Government Printing Office, Washington, D.C., 1962.
(606) H. F. Hipsher and P. H. Wise, *J. Am. Chem. Soc., 76,* 1747–1748 (1954).
(607) E. Hirota, *J. Chem. Phys., 37,* 283–291 (1962).
(608) J. O. Hirschfelder, *J. Chem. Phys., 8,* 431 (1940).

(609) H. Hock and G. Knauel, *Ber.*, *84*, 1–4 (1951).
(610) C. W. W. Hoffman, *Univ. Microfilms Publ. No. 24161*, 89 pp.; *Dissertation Abstr.*, *18*, 420 (1958).
(611) C. W. W. Hoffman and R. L. Livingston, *J. Chem. Phys.*, *21*, 565 (1953).
(612) H. T. Hoffman, Jr., G. E. Evans, and G. Glockler, *J. Am. Chem. Soc.*, *73*, 3028–3030 (1951).
(613) D. E. Holcomb and C. L. Dorsey, Jr., *Ind. Eng. Chem.*, *41*, 2788–2792 (1949); *42*, 570 (1950).
(614) Y.-C. Hou, *Univ. Microfilms Publ. No. 12589*, 122 pp.; *Dissertation Abstr.*, *15*, 1579–1580 (1955).
(615) Y.-C. Hou and J. J. Martin, *Am. Inst. Chem. Eng. J.*, *5*, 125–129 (1959).
(616) E. W. Hough, D. M. Mason and B. H. Sage, *J. Am. Chem. Soc.*, *72*, 5775–5777 (1950).
(617) E. W. Hough, D. M. Mason, and B. H. Sage, *J. Am. Chem. Soc.*, *73*, 1363–1364 (1951).
(618) P. B. Howard and H. A. Skinner, *J. Chem. Soc. [A]*, 1536–1540 (1966).
(619) K. E. Howlett, *J. Chem. Soc.*, 1409–1412 (1951).
(620) K. E. Howlett, *J. Chem. Soc.*, 1784–1789 (1955).
(621) K. E. Howlett, *J. Chem. Soc.*, 2834–2836 (1957).
(622) J. J. Hrostowski and G. C. Pimentel, *J. Am. Chem. Soc.*, *75*, 539–542 (1953).
(623) K. Hrynakowski and A. Smoczkiewiczowa, *Roczniki Chem.*, *17*, 165–168 (1937).
(624) C. C. Hsu and J. J. McKetta, *Am. Inst. Chem. Eng. J.*, *9*, 794–796 (1963).
(625) J.-H. Hu, D. White, and H. L. Johnston, *J. Am. Chem. Soc.*, *75*, 1232–1236 (1953).
(626) J.-H. Hu, D. White, and H. L. Johnston, *J. Am. Chem. Soc.*, *75*, 5642–5645 (1953).
(627) W. N. Hubbard, D. R. Douslin, J. P. McCullough, D. W. Scott, S. S. Todd, J. F. Messerly, I. A. Hossenlopp, A. George, and G. Waddington, *J. Am. Chem. Soc.*, *80*, 3547–3554 (1958).
(628) W. N. Hubbard, H. L. Finke, D. W. Scott, J. P. McCullough, C. Katz, M. E. Gross, J. F. Messerly, R. E. Pennington, G. Waddington, *J. Am. Chem. Soc.*, *74*, 6025–6030 (1952).
(629) W. N. Hubbard, F. R. Frow, and G. Waddington, *J. Phys. Chem.*, *62*, 821–823 (1958).
(630) W. N. Hubbard, F. R. Frow, and G. Waddington, *J. Phys. Chem.*, *65*, 1326–1328 (1961).
(631) W. N. Hubbard, W. D. Good, and G. Waddington, *J. Phys. Chem.*, *62*, 614–617 (1958).
(632) W. N. Hubbard, C. Katz, G. B. Guthrie, Jr., and G. Waddington, *J. Am. Chem. Soc.*, *74*, 4456–4458 (1952).
(633) W. N. Hubbard, C. Katz, and G. Waddington, *J. Phys. Chem.*, *58*, 142–152 (1954).
(634) W. N. Hubbard, J. W. Knowlton, and H. M. Huffman, *J. Am. Chem. Soc.*, *70*, 3259–3261 (1948).
(635) W. N. Hubbard, J. W. Knowlton, and H. M. Huffman, *J. Phys. Chem.*, *58*, 396–402 (1954).
(636) W. N. Hubbard, D. W. Scott, F. R. Frow, and G. Waddington, *J. Am. Chem. Soc.*, *77*, 5855–5857 (1955).

(637) W. N. Hubbard, D. W. Scott, and G. Waddington, "Standard States and Corrections for Combustions in a Bomb at Constant Volume," in F. D. Rossini (Ed.), *Experimental Thermochemistry*, Vol. I, Interscience Publishers, New York, 1956, Chap. 5.
(638) W. N. Hubbard and G. Waddington, *Rec. Trav. Chem.*, *73*, 910–923 (1954).
(639) W. Hückel, A. Gercke, and A. Gross, *Ber.*, *66B*, 563–567 (1933).
(640) V. N. Huff, S. Gordon, and V. E. Morrell, *General Method and Thermodynamic Tables for Computations of Equilibrium Composition and Temperature of Chemical Reactions (National Advisory Committee for Aeronautics Report No. 1037)*, U.S. Government Printing Office, Washington, D.C., 1951.
(641) H. M. Huffman, *J. Am. Chem. Soc.*, *62*, 1009–1011 (1940).
(642) H. M. Huffman, *J. Am. Chem. Soc.*, *63*, 688–689 (1941).
(643) H. M. Huffman, *J. Phys. Chem.*, *46*, 885–891 (1942).
(644) H. M. Huffman and H. Borsook, *J. Am. Chem. Soc.*, *54*, 4297–4301 (1932).
(645) H. M. Huffman, M. Eaton, and G. D. Oliver, *J. Am. Chem. Soc.*, *70*, 2911–2914 (1948).
(646) H. M. Huffman and E. L. Ellis, *J. Am. Chem. Soc.*, *57*, 41–46 (1935).
(647) H. M. Huffman and E. L. Ellis, *J. Am. Chem. Soc.*, *57*, 46–48 (1935).
(648) H. M. Huffman and E. L. Ellis, *J. Am. Chem. Soc.*, *59*, 2150–2152 (1937).
(649) H. M. Huffman, E. L. Ellis, and H. Borsook, *J. Am. Chem. Soc.*, *62*, 297–299 (1940).
(650) H. M. Huffman, E. L. Ellis, and S. W. Fox, *J. Am. Chem. Soc.*, *58*, 1728–1733 (1936).
(651) H. M. Huffman and S. W. Fox, *J. Am. Chem. Soc.*, *60*, 1400–1403 (1938).
(652) H. M. Huffman and S. W. Fox, *J. Am. Chem. Soc.*, *62*, 3464–3465 (1940).
(653) H. M. Huffman, S. W. Fox, and E. L. Ellis, *J. Am. Chem. Soc.*, *59*, 2144–2150 (1937).
(654) H. M. Huffman, M. E. Gross, D. W. Scott, and J. P. McCullough, *J. Phys. Chem.*, *65*, 495–503 (1961).
(655) H. M. Huffman, G. S. Parks, and M. Barmore, *J. Am. Chem. Soc.*, *53*, 3876–3888 (1931).
(656) H. M. Huffman, G. S. Parks, and A. C. Daniels, *J. Am. Chem. Soc.*, *52*, 1547–1558 (1930).
(567) H. M. Huffman, S. S. Todd, and G. D. Oliver, *J. Am. Chem. Soc.*, *71*, 584–592, (1949).
(658) A. M. Hughes, R. J. Corruccini, and E. C. Gilbert, *J. Am. Chem. Soc.*, *61*, 2639–2642 (1939).
(569) J. A. C. Hugill, I. E. Coop, and L. E. Sutton, *Trans. Faraday Soc.*, *34*, 1518–1534 (1938).
(660) R. Hultgren, R. L. Orr, P. D. Anderson, and K. K. Kelley, *Selected Values of Thermodynamic Properties of Metals and Alloys*, Wiley, New York, 1963.
(661) R. Hultgren, R. L. Orr, and K. K. Kelley, *Selected Values of Thermodynamic Properties of Metals and Alloys*, University of California (Berkeley) Institute of Engineering Research, Minerals Research Laboratory, looseleaf since 1955.
(662) G. L. Humphrey and R. Spitzer, *J. Chem. Phys.* *18*, 902 (1950).
(663) A. R. Humphries and G. R. Nicholson, *J. Chem. Soc.*, 2429–2431 (1957).
(664) J. O. Hutchens, A. G. Cole, R. A. Robie, and J. W. Stout, *J. Biol. Chem.*, *238*, 2407–2412 (1963).
(665) J. O. Hutchens, A. G. Cole, and J. W. Stout, *J. Am. Chem. Soc.*, *82*, 4813–4815 (1960).

(666) J. O. Hutchens, A. G. Cole, and J. W. Stout, *J. Phys. Chem.*, *67*, 1128–1130 (1963).
(667) J. O. Hutchens, A. G. Cole, and J. W. Stout, *J. Biol. Chem.*, *239*, 591–595 (1964).
(668) J. O. Hutchens, A. G. Cole, and J. W. Stout, *J. Biol. Chem.*, *239*, 4194–4195 (1964).
(669) Y.-T. Hwang and J. J. Martin, *Am. Inst. Chem. Eng. J.*, *10*, 89–91 (1964).
(670) M. Igarashi, *Bull. Chem. Soc. Japan*, *34*, 369–373 (1961).
(671) R. G. Inskeep, J. M. Kelliher, P. E. McMahon, and B. G. Somers, *J. Chem. Phys.*, *28*, 1033–1036 (1958).
(672) T. Itoh, *J. Phys. Soc. Japan*, *11*, 264–271 (1956).
(673) E. V. Ivash and D. M. Dennison, *J. Chem. Phys.*, *21*, 1804–1816 (1953).
(674) E. V. Ivash, J. C. M. Li, and K. S. Pitzer, *J. Chem. Phys.*, *23*, 1814–1818 (1955).
(675) B. L. Iwanciow, Thesis, University of Wisconsin, Madison, 1950.
(676) G. W. Jack and G. Stegeman, *J. Am. Chem. Soc.*, *63*, 2121–2123 (1941).
(677) C. J. Jacobs and G. S. Parks, *J. Am. Chem. Soc.*, *56*, 1513–1517 (1934).
(678) I. Jaffe, Thesis, University of Maryland, College Park, 1958.
(679) L. Jaffe, E. J. Prosen, and M. Szwarc, *J. Chem. Phys.*, *27*, 416–420 (1957).
(680) G. J. Janz, *Can. J. Res.*, *25B*, 331–332 (1947).
(681) G. J. Janz, *Thermodynamic Properties of Organic Compounds: Estimation Methods, Principles and Practice*, revised ed. [Vol. 6 of E. Hutchinson and P. Van Rysselberghe (Eds.), *Physical Chemistry*], Academic Press, New York, 1967.
(682) G. J. Janz and S. C. Wait, Jr., *J. Chem. Phys.*, *26*, 1766–1768 (1957).
(683) S. K. K. Jatkar, *J. Indian Inst. Sci.*, *22A*, 39–58 (1939).
(684) M. Jen and D. R. Lide, Jr., *J. Chem. Phys.*, *36*, 2525–2526 (1962).
(685) A. D. Jenkins and D. W. G. Style, *J. Chem. Soc.*, 2337–2340 (1953).
(686) R. S. Jessup, *J. Res. Natl. Bur. Std.*, *18*, 115–128, (1937).
(687) R. S. Jessup, *J. Res. Natl. Bur. Std.*, *29*, 247–270 (1942).
(688) R. S. Jessup, *J. Res. Natl. Bur. Std.*, *36*, 421–423 (1946).
(689) R. S. Jessup, *(National Bureau of Standards Monograph No. 7)*, U.S. Government Printing Office, Washington, D.C., 1960.
(690) R. S. Jessup and C. B. Green, *J. Res. Natl. Bur. Std.*, *13*, 469–495 (1934).
(691) R. S. Jessup, R. E. McCoskey, and R. A. Nelson, *J. Am. Chem. Soc.*, *77*, 244–245 (1955).
(692) W. H. Johnson, E. J. Prosen, and F. D. Rossini, *J. Res. Natl. Bur. Std.*, *35*, 141–146 (1945).
(693) W. H. Johnson, E. J. Prosen, and F. D. Rossini, *J. Res. Natl. Bur. Std.*, *38*, 419–422 (1947).
(694) W. H. Johnson, E. J. Prosen, and F. D. Rossini, *J. Res. Natl. Bur. Std.*, *42*, 251–255 (1949).
(695) H. L. Johnston, L. Savedoff, and J. Belzer, "Contribution to Thermodynamic Functions by a Planck-Einstein Oscillator in One Degree of Freedom," Contract ONR, Ohio State University Research Foundation, ONR Department of the Navy, Washington, D.C., NAVEXOS P-646, July 1949.
(696) H. S. Johnston and H. J. Bertin, Jr., *J. Am. Chem. Soc.*, *81*, 6402–6404 (1959).
(697) A. H. Jones, *J. Chem. Eng. Data*, *5*, 196–200 (1960).
(698) A. V. Jones, *Proc. Roy. Soc. (London), Ser. A*, *211*, 285–295 (1952).
(699) E. Jones and G. G. Fowlie, *J. Appl. Chem. (London)*, *3*, 206–213 (1953).
(700) J. L. Jones and R. A. Ogg, Jr., *J. Am. Chem. Soc.*, *59*, 1943–1945 (1937).
(701) S. O. Jones and E. E. Reid, *J. Am. Chem. Soc.*, *60*, 2452–2455 (1938).
(702) W. M. Jones and W. F. Giauque, *J. Am. Chem. Soc.*, *69*, 983–987 (1947).

(703) T. E. Jordan, *Vapor Pressure of Organic Compounds*, Interscience Publishers, New York, 1954.
(704) R. M. Joshi, *Makromol. Chem.*, 55, 35–49 (1962).
(705) N. I. Joukovsky, *Bull. Soc. Chim. Belges*, 43, 397–445 (1934).
(706) G. Jung and J. Dahmlos, *Z. Phys. Chem. (Leipzig)*, A190, 230–240 (1942).
(707) S. Kaarsemaker and J. Coops, *Rec. Trav. Chim.*, 71, 261–276 (1952).
(708) R. L. Kabel and L. N. Johanson, *J. Chem. Eng. Data*, 6, 496–498 (1961).
(709) I. A. Kablukov and F. M. Perelman, *Compt. Rend. Acad. Sci. URSS [A]*, 519–522 (1930).
(710) O. N. Kachinskaya, S. K. Togoeva, A. P. Meshcheryakov, and S. M. Skuratov, *Dokl. Akad. Nauk SSSR*, 132, 119–122 (1960); *Proc. Acad. Sci. USSR, Chem. Sect.*, 132, 451–454 (1960).
(711) Y. Kanazawa and K. Nukada, *Bull. Chem. Soc. Japan*, 35, 612–618 (1962).
(712) H. J. Kandiner and S. R. Brinkley, Jr., *Ind. Eng. Chem.*, 42, 850–855 (1950).
(713) W. Kangro and R. Grau, *Z. Phys. Chem., Bodenstein-Festband*, 85–92 (1931).
(714) L. H. Kaplan, *Univ. Microfilms L. C. Mic 58–1788*, 133 pp.; *Dissertation Abstr.*, 19, 3130 (1959).
(715) A. F. Kapustinskii and R. T. Kan'kovskii, *Zh. Fiz. Khim.*, 32, 2810–2816 (1958).
(716) K. A. Karasharli and P. G. Strelkov, *Dokl. Akad. Nauk SSSR*, 131, 568–569 (1960); *Proc. Acad. Sci. USSR, Chem. Sect.*, 131, 267–268 (1960).
(717) I. L. Karle and J. Karle, *J. Chem. Phys.*, 18, 963–971 (1950).
(718) I. L. Karle and J. Karle, *J. Chem. Phys.*, 20, 63–65 (1952).
(719) K. J. Karlsson, Thesis, University of Lund, Lund, Sweden, 1941.
(720) H. A. Karnes, B. D. Kybett, M. H. Wilson, J. L. Margrave, and M. S. Newman, *J. Am. Chem. Soc.*, 87, 5554–5558 (1965).
(721) W. Karo, R. L. McLaughlin, and H. F. Hipsher, *J. Am. Chem. Soc.*, 75, 3233–3235 (1953).
(722) P. H. Kasai and R. J. Myers, *J. Chem. Phys.*, 30, 1096–1097 (1959).
(723) T. Kasuya, *J. Phys. Soc. Japan*, 15, 1273–1277 (1960).
(724) T. Kasuya and T. Oka, *J. Phys. Soc. Japan*, 15, 296–303 (1960).
(725) J. E. Katon and E. R. Lippincott, *Spectrochim. Acta*, 15, 627–650 (1959).
(726) T. J. Katz and J. L. Margrave, *J. Chem. Phys.*, 23, 983 (1955).
(727) W. B. Kay and W. E. Donham, *Chem. Eng. Sci.*, 4, 1–16 (1955).
(728) L. Keffler, *Bull. Soc. Chim. Belges*, 44, 425–434 (1935).
(729) L. J. P. Keffler, *J. Chem. Soc. Ind. (London)*, 55, 331–335T (1936).
(730) L. J. P. Keffler, *J. Phys. Chem.*, 41, 715–721 (1937).
(731) W. A. Keith and H. Mackle, *Trans. Faraday Soc.*, 54, 353–366 (1958).
(732) K. K. Kelley, *J. Am. Chem. Soc.*, 51, 180–187 (1929).
(733) K. K. Kelley, *J. Am. Chem. Soc.*, 51, 779–786 (1929).
(734) K. K. Kelley, *J. Am. Chem. Soc.*, 51, 1145–1151 (1929).
(735) K. K. Kelley, *J. Am. Chem. Soc.*, 51, 1400–1406 (1929).
(736) K. K. Kelley, *U.S. Bur. Mines Bull.*, 584 (1960).
(737) K. K. Kelley, *U.S. Bur. Mines Bull.*, 601, 154 pp. (1962).
(738) K. K. Kelley and E. G. King, *U.S. Bur. Mines Bull.*, 592 (1961).
(739) J. D. Kemp and W. F. Giauque, *J. Am. Chem. Soc.*, 59, 79–84 (1937).
(740) R. M. Kennedy, M. Sagenkahn, and J. G. Aston, *J. Am. Chem. Soc.*, 63, 2267–2272 (1941).
(741) W. D. Kennedy, C. H. Shomate, and G. S. Parks, *J. Am. Chem. Soc.*, 60, 1507–1509 (1938).
(742) J. A. A. Ketelaar and S. Kruyer, *Rec. Trav. Chim.*, 62, 550–552 (1943).

(743) J. A. A. Ketelaar, P. F. Van Velden, and P. Zalm, *Rec. Trav. Chim.*, 66, 721–732 (1947).
(744) M. S. Kharasch, *J. Res. Natl. Bur. Std.*, 2, 359–430 (1929).
(745) M. S. Kharasch and B. Sher, *J. Phys. Chem.*, 29, 625–658 (1925).
(746) M. S. Kharasch and A. F. Zavist, *J. Am. Chem. Soc.*, 73, 964–967 (1951).
(747) G. M. Kibler and H. Hunt, *J. Phys. Colloid Chem.*, 53, 955–956 (1949).
(748) J. Kielland, *Ber.*, 71B, 220–226 (1938).
(749) R. W. Kilb, *J. Chem. Phys.*, 23, 1736–1737 (1955).
(750) J. E. Kilpatrick and K. S. Pitzer, *J. Am. Chem. Soc.*, 68, 1066–1072 (1946).
(751) J. E. Kilpatrick and K. S. Pitzer, *J. Res. Natl. Bur. Std.*, 37, 163–171 (1946).
(752) J. E. Kilpatrick and K. S. Pitzer, *J. Chem. Phys.*, 17, 1064–1075 (1949).
(753) J. E. Kilpatrick, K. S. Pitzer, and R. Spitzer, *J. Am. Chem. Soc.*, 69, 2483–2488 (1947).
(754) F. W. Kirkbride, *J. Appl. Chem. (London)*, 6, 11–21 (1956).
(755) F. W. Kirkbride and F. G. Davidson, *Nature*, 174, 79–80 (1954).
(756) B. Kirtman, *J. Chem. Phys.*, 41, 775–788 (1964).
(757) G. B. Kistiakowsky and W. W. Rice, *J. Chem. Phys.*, 8, 610–618 (1940).
(758) G. B. Kistiakowsky and W. W. Rice, *J. Chem. Phys.*, 8, 618–622 (1940).
(759) G. B. Kistiakowsky, H. Romeyn, Jr., J. R. Ruhoff, H. A. Smith, and W. E. Vaughan, *J. Am. Chem. Soc.*, 57, 65–75 (1935).
(760) G. B. Kistiakowsky, J. R. Ruhoff, H. A. Smith, and W. E. Vaughan, *J. Am. Chem. Soc.*, 57, 876–882 (1935).
(761) G. B. Kistiakowsky, J. R. Ruhoff, H. A. Smith, and W. E. Vaughan, *J. Am. Chem. Soc.*, 58, 137–145 (1936).
(762) G. B. Kistiakowsky, J. R. Ruhoff, H. A. Smith, and W. E. Vaughan, *J. Am. Chem. Soc.*, 58, 146–153 (1936).
(763) G. B. Kistiakowsky and C. H. Stauffer, *J. Am. Chem. Soc.*, 59, 165–170 (1937).
(764) D. Kivelson, E. B. Wilson, Jr., and D. R. Lide, Jr., *J. Chem. Phys.*, 32, 205–209 (1960).
(765) F. Klages, *Ber.*, 82, 358–375 (1949).
(766) M. J. Klein, F. F. Cleveland, and A. G. Meister, *J. Chem. Phys.*, 19, 1068–1069 (1951).
(767) J. A. Knopp, W. S. Linnell, and W. C. Child, Jr., *J. Phys. Chem.*, 66, 1513–1516 (1962).
(768) J. W. Knowlton and H. M. Huffman, *J. Am. Chem. Soc.*, 66, 1492–1494 (1944).
(769) J. W. Knowlton and E. J. Prosen, *J. Res. Natl. Bur. Std.*, 46, 489–495 (1951).
(770) J. W. Knowlton and F. D. Rossini, *J. Res. Natl. Bur. Std.*, 22, 415–424 (1939).
(771) J. W. Knowlton and F. D. Rossini, *J. Res. Natl. Bur. Std.*, 43, 113–115 (1949).
(772) K. A. Kobe and H. R. Crawford, *Petrol. Refiner*, 37 (7), 125–130 (1958).
(773) K. A. Kobe, H. R. Crawford, and R. W. Stephenson, *Ind. Eng. Chem.*, 47, 1767–1772 (1955).
(774) K. A. Kobe and R. H. Harrison, *Petrol. Refiner*, 30, (11), 151–154 (1951).
(775) K. A. Kobe and R. H. Harrison, *Petrol. Refiner*, 33 (11), 161–164 (1954).
(776) K. A. Kobe and R. H. Harrison, *Petrol. Refiner*, 36 (10), 155–159 (1957).
(777) K. A. Kobe, R. H. Harrison, and R. E. Pennington, *Petrol. Refiner*, 30 (8), 119–122 (1951).
(778) K. A. Kobe and D. H. Kobe, *Petrol. Refiner*, 38 (12), 117–120 (1959).
(779) K. A. Kobe and E. G. Long, *Petrol. Refiner*, 29 (5), 89–92 (1950).
(780) K. A. Kobe and R. E. Lynn, Jr., *Chem. Rev.*, 52, 117–236 (1953).
(781) K. A. Kobe and R. E. Pennington, *Petrol. Refiner*, 29 (9), 135–138 (1950).
(782) K. A. Kobe and R. E. Pennington, *Petrol. Refiner*, 29 (12), 93–96 (1950).

(783) K. A. Kobe, A. E. Ravicz, and S. P. Vohra, *Chem. Eng. Data Ser.*, *1*, 50–56 (1956).
(784) K. A. Kobe and co-workers, *Petrol. Refiner*, *28* (1), 83–87; (2), 113–116; (3), 125–128; (5), 161–163; (7), 145–148; (10), 133–136; (11), 127–132 (1949); *29*(1), 126–130; (2), 124–128; (3), 157–160; (5), 89–92; (7), 129–133; (9), 135–138; (12), 93–96 (1950); *30* (4), 123–124; (6), 143–144; (8), 119–122; (11), 151–154; (12), 114–116 (1951); *33* (8), 109–110; (11), 161–164 (1954); *36* (10), 155–159; (12), 147–148 (1957); *37* (7), 125–130 (1958); *38* (12), 117–120 (1959).
(785) J. K. Koehler and W. F. Giauque, *J. Am. Chem. Soc.*, *80*, 2659–2662 (1958).
(786) P. N. Kogerman, *Sitzbar. Naturforsch.-Ges. Univ. Tartu*, *41* (3–4), 62 pp. (1934).
(787) K. W. F. Kohlrausch, *Monatsh. Chem.*, *68*, 349–358 (1936).
(788) T. Kojima, *J. Phys. Soc. Japan*, *15*, 284–287 (1960).
(789) T. Kojima, *J. Phys. Soc. Japan*, *15*, 1284–1291 (1960).
(790) T. Kojima and T. Nishikawa, *J. Phys. Soc. Japan*, *12*, 680–686 (1957).
(791) H. J. Kolb and R. L. Burwell, Jr., *J. Am. Chem. Soc.*, *67*, 1084–1088 (1945).
(792) V. P. Kolesov, A. M. Martynov, S. M. Shtekher, and S. M. Skuratov, *Zh. Fiz. Khim.*, *36*, 2078–2081 (1962).
(793) V. P. Kolesov, A. M. Martynov, and S. M. Skuratov, *Zh. Fiz. Khim.*, *39*, 435–437 (1965).
(794) V. P. Kolesov, I. E. Paukov, S. M. Skuratov, and E. A. Seregin, *Dokl. Akad. Nauk SSSR*, *128*, 130–132 (1959); *Proc. Acad. Sci. USSR*, *128*, 751–753 (1959).
(795) V. P. Kolesov, O. G. Talakin, and S. M. Skuratov, *Zh. Fiz. Khim.*, *38*, 1701–1703 (1964).
(796) V. P. Kolesov, I. D. Zenkov, and S. M. Skuratov, *Zh. Fiz. Khim.*, *36*, 89–92 (1962).
(797) V. P. Kolesov, I. D. Zenkov, and S. M. Skuratov, *Zh. Fiz. Khim.*, *37*, 224–225 (1963).
(798) V. P. Kolesov, I. D. Zenkov, and S. M. Skuratov, *Zh. Fiz. Khim.*, *39*, 2474–2476 (1965).
(799) N. De Kolossovsky and A. Alimov, *Bull. Soc. Chim. France [5]*, *1*, 877–880 (1934).
(800) R. Kopelman, *J. Chem. Phys.*, *41*, 1547–1553 (1964).
(801) V. N. Kostryukov, O. P. Samorukov, and P. G. Strelkov, *Zh. Fiz. Khim.*, *32*, 1354–1361 (1958).
(802) E. Kováts, H. H. Günthard, and P. A. Plattner, *Helv. Chim. Acta*, *38*, 1912–1919 (1955).
(803) E. Kováts, H. H. Günthard, and P. A. Plattner, *Helv. Chim. Acta*, *40*, 2008 (1957).
(804) M. P. Kozina, M. Y. Lukina, N. D. Zubareva, I. L. Safonova, S. M. Skuratov, and B. A. Kazanskii, *Dokl. Akad. Nauk SSSR*, *138*, 843–845 (1961); *Proc. Acad. Sci. USSR, Chem. Sect.*, *138*, 537–539 (1961).
(805) M. P. Kozina, A. K. Mirzaeva, I. E. Sosnina, N. V. Elagina, and S. M. Skuratov, *Dokl. Akad. Nauk SSSR*, *155*, 1123–1125 (1964).
(806) M. P. Kozina, S. M. Skuratov, S. M. Shtekher, I. E. Sosnina, and M. B. Turova-Polyak, *Zh. Fiz. Khim.*, *35*, 2316–2321 (1961).
(807) N. A. Kozlov and I. B. Rabinovich, *Tr. po Khim. i Khim. Tekhnol.*, 189–193 (1964).
(808) J. Kraitchman and B. P. Dailey, *J. Chem. Phys.*, *23*, 184–190 (1955).
(809) R. E. Krall and J. D. Roberts, *Am. Chem. Soc. Div. Petrol. Chem. Symp. 3 (4B)*, 63–68 (1958).

(810) R. Kraus, *Proc. 12th Intern. Congr. Acetylene, Oxyacetylene Welding and Allied Ind.*, London, *1936*, Vol. I, pp. 1422–1423, 1459–1460.
(811) C. B. Kretschmer and R. Wiebe, *J. Am. Chem. Soc., 76,* 2579–2583 (1954).
(812) O. Kubaschewski and E. L. Evans, *Metallurgical Thermochemistry*, 4th Ed., Pergamon Press, New York, 1967.
(813) N. V. Kul'kova and M. I. Temkin, *Zh. Fiz. Khim., 36,* 1731–1734 (1962).
(814) V. Y. Kurbatov, *J. Gen. Chem. USSR, 18,* 372–387 (1948).
(815) N. P. Kurin and M. S. Zakharov, *Izv. Vysshikh Ucheb. Zavedenii, Khim. i Khim. Tekhnol., 3,* 141–145 (1960).
(816) K. Kusano, *Nippon Kagaku Zasshi, 78,* 614–620 (1958).
(817) L. M. Kushner, R. W. Crowe, and C. P. Smyth, *J. Am. Chem. Soc., 72,* 1091–1098 (1950).
(818) N. von Kutepow, W. Himmele, and H. Hohenschutz, *Chem. Ingr. Tech., 37* (4), 383–388 (1965).
(819) B. D. Kybett, G. K. Johnson, C. K. Barker, and J. L. Margrave, *J. Phys. Chem., 69,* 3603–3606 (1965).
(820) A. Labbauf, J. B. Greenshields, and F. D. Rossini, *J. Chem. Eng. Data, 6,* 261–263 (1961).
(821) A. Labbauf and F. D. Rossini, *J. Phys. Chem., 65,* 476–480 (1961).
(822) J. R. Lacher, T. J. Billings, D. E. Campion, K. R. Lea, and J. D. Park, *J. Am. Chem. Soc., 74,* 5291–5292 (1952).
(823) J. R. Lacher, L. Casali, and J. D. Park, *J. Phys. Chem., 60,* 608–610 (1956).
(824) J. R. Lacher, E. Emery, E. Bohmfalk, and J. D. Park, *J. Phys. Chem., 60,* 492–495 (1956).
(825) J. R. Lacher, H. B. Gottlieb, and J. D. Park, *Trans. Faraday Soc., 58,* 2348–2351 (1962).
(826) J. R. Lacher, A. Kianpour, P. Montgomery, H. Knedler, and J. D. Park, *J. Phys. Chem., 61,* 1125–1126 (1957).
(827) J. R. Lacher, A. Kianpour, F. L. Oetting, and J. D. Park, *Trans. Faraday Soc., 52,* 1500–1508 (1956).
(828) J. R. Lacher, A. Kianpour, and J. D. Park, *J. Phys. Chem., 60,* 1454–1455 (1956).
(829) J. R. Lacher, A. Kianpour, and J. D. Park, *J. Phys. Chem., 61,* 1124–1125 (1957).
(830) J. R. Lacher, K. R. Lea, C. H. Walden, G. G. Olson, and J. D. Park, *J. Am. Chem. Soc., 72,* 3231–3234 (1950).
(831) J. R. Lacher, J. J. McKinley, C. M. Snow, L. Michel, G. Nelson, and J. D. Park, *J. Am. Chem. Soc., 71,* 1330–1334 (1949).
(832) J. R. Lacher, J. J. McKinley, C. Walden, K. Lea, and J. D. Park, *J. Am. Chem. Soc., 71,* 1334–1337 (1949).
(833) J. R. Lacher and H. A. Skinner, *J. Chem. Soc., (A), 1968,* 1034–1038 (1968).
(834) J. R. Lacher, C. H. Walden, K. R. Lea, and J. D. Park, *J. Am. Chem. Soc., 72,* 331–333 (1950).
(835) J. L. Lacina, W. D. Good, and J. P. McCullough, *J. Phys. Chem., 65,* 1026–1027 (1961).
(836) B. S. Lacy, R. G. Dunning, and H. H. Storch, *J. Am. Chem. Soc., 52,* 926–938 (1930).
(837) W. J. Lafferty, A. G. Maki, and E. K. Plyler, *J. Chem. Phys., 40,* 224–229 (1964).
(838) K. J. Laidler, *Can. J. Chem., 34,* 626–648 (1956).
(839) J. M. Lamberti and P. H. Wise, *J. Am. Chem. Soc., 75,* 4787–4789 (1953).
(840) J. H. Lamneck, Jr., and P. H. Wise, *J. Am. Chem. Soc., 76,* 3475–3476 (1954).

(841)　J. H. Lamneck, Jr., and P. H. Wise, *J. Am. Chem. Soc.*, 76, 5108–5110 (1954).
(842)　L. D. Landau and E. M. Lifshitz, *Statistical Physics*, Pergamon Press, London, 1958.
(843)　*Landolt-Börnstein Tabellen*, Sechste Auflage, Band II, Teil 4, "Kalorische Zustandsgrössen," Springer-Verlag, Berlin, 1961.
(844)　P. Landrieu, F. Baylocq, and J. R. Johnson, *Bull. Soc. Chim. France*, 45, 36–49 (1929).
(845)　M. R. Lane, J. W. Linnett, and H. G. Oswin, *Proc. Roy. Soc. (London), Ser. A*, 216, 361–374 (1953).
(846)　E. Lange, *Z. Phys. Chem. (Leipzig)*, 110, 343–362 (1924).
(847)　A. Langer, J. A. Hipple, and D. P. Stevenson, *J. Chem. Phys.*, 22, 1836–1844 (1954).
(848)　J. P. Larmann, D. E. Martire, and L. Z. Pollara, *J. Chem. Eng. Data*, 6, 330 (1961).
(849)　V. W. Laurie, *J. Chem. Phys.*, 31, 1500–1505 (1959).
(850)　V. W. Laurie, *J. Chem. Phys.*, 34, 1516–1519 (1961).
(851)　W. F. Lautsch, P. Erzberger, and A. Tröber, *Wiss. Z. Tech. Hochsch. Chem. Leuna-Merseburg*, 1, 31–33 (1958–1959).
(852)　R. B. Lawrance and M. W. P. Strandberg, *Phys. Rev.*, 83, 363–369 (1951).
(853)　N. D. Lebedeva, *Zh. Fiz. Khim.*, 38, 2648–2651 (1964).
(854)　J. A. Leermakers and H. C. Ramsperger, *J. Am. Chem. Soc.*, 54, 1837–1845 (1932).
(855)　A. Leman and G. Lepoutre, *Compt. Rend.*, 226, 1976–1978 (1948).
(856)　R. G. Lerner and B. P. Dailey, *J. Chem. Phys.*, 26, 678–680 (1957).
(857)　S. V. Levanova and D. N. Andreevskii, *Neftekhimiya*, 4, 477–480 (1964).
(858)　D. C. Lewis, M. A. Frisch, and J. L. Margrave, *Carbon*, 2, 431–432 (1965).
(859)　G. N. Lewis, *Proc. Am. Acad. (Daedalus)*, 37, 49–69 (1901); *Z. Phys. Chem. (Leipzig)*, 38, 205–226 (1901).
(860)　G. N. Lewis and M. Randall, *Thermodynamics and the Free Energy of Chemical Substances*, McGraw-Hill, New York, 1923.
(861)　G. N. Lewis and M. Randall, *Thermodynamics*, 2nd Ed., revised by K. S. Pitzer and L. Brewer, McGraw-Hill, New York, 1961.
(862)　J. C. M. Li and K. S. Pitzer, *J. Am. Chem. Soc.*, 78, 1077–1080 (1956).
(863)　J. C. M. Li and F. D. Rossini, *J. Chem. Eng. Data*, 6, 268–270 (1961).
(864)　K. Li, *J. Phys. Chem.*, 61, 782–785 (1957).
(865)　D. R. Lide, Jr., *J. Am. Chem. Soc.*, 74, 3548–3552 (1952).
(866)　D. R. Lide, Jr., *J. Chem. Phys.*, 29, 1426–1427 (1958).
(867)　D. R. Lide, Jr., *J. Chem. Phys.*, 30, 37–39 (1959).
(868)　D. R. Lide, Jr., and M. Jen, *J. Chem. Phys.*, 38, 1504–1557 (1963).
(869)　D. R. Lide, Jr., and D. E. Mann, *J. Chem. Phys.*, 25, 1128–1131 (1956).
(870)　D. R. Lide, Jr., and D. E. Mann, *J. Chem. Phys.*, 27, 868–873 (1957).
(871)　D. R. Lide, Jr., and D. E. Mann, *J. Chem. Phys.*, 27, 874–877 (1957).
(872)　D. R. Lide, Jr., and D. E. Mann, *J. Chem. Phys.*, 28, 572–576 (1958).
(873)　D. R. Lide, Jr., and D. E. Mann, *J. Chem. Phys.*, 29, 914–920 (1958).
(874)　C. C. Lin and R. W. Kilb, *J. Chem. Phys.*, 24, 631 (1956).
(875)　C. C. Lin and J. D. Swalen, *Rev. Mod. Phys.*, 31, 841–892 (1959).
(876)　J. W. Linnett, *J. Chem. Phys.*, 6, 692–702 (1938).
(877)　E. R. Lippincott and R. C. Lord, *J. Am. Chem. Soc.*, 73, 3889–3891 (1951).
(878)　M. W. Lister, *J. Am. Chem. Soc.*, 63, 143–149 (1941).
(879)　R. L. Livingston, C. N. R. Rao, L. H. Kaplan, and L. Rocks, *J. Am. Chem. Soc.*, 80, 5368–5371 (1958).
(880)　Sr. M. C. Loeffler and F. D. Rossini, *J. Phys. Chem.*, 64, 1530–1533 (1960).

(881) D. A. Long, F. S. Murfin, and R. L. Williams, *Proc. Roy. Soc. (London), Ser. A, 223*, 251–266 (1954).
(882) L. H. Long and R. G. W. Norrish, *Phil. Trans. Roy. Soc. London, Ser. A, 241*, 587–617 (1949).
(883) G. Lord and A. A. Woolf, *J. Chem. Soc.*, 2546–2551 (1954).
(884) R. C. Lord, Jr., and E. R. Blanchard, *J. Chem. Phys., 4*, 707–710 (1936).
(885) R. C. Lord and B. Nolin, *J. Chem. Phys., 24*, 656–658 (1956).
(886) R. C. Lord and D. G. Rea, *J. Am. Chem. Soc., 79*, 2401–2406 (1957).
(887) F. P. Lossing, K. U. Ingold, and I. H. S. Henderson, *J. Chem. Phys., 22*, 1489–1492 (1954).
(888) R. J. Lovell, C. F. Stephenson, and E. A. Jones, *J. Chem. Phys., 22*, 1953–1955 (1954).
(889) E. G. Lovering and K. J. Laidler, *Can. J. Chem., 38*, 2367–2372 (1960).
(890) N. W. Luft, *J. Chem. Phys., 21*, 1900–1901 (1953).
(891) N. W. Luft, *J. Chem. Phys., 22*, 155–156 (1954).
(892) N. W. Luft and O. P. Kharbanda, *J. Chem. Phys., 22*, 956–957 (1954).
(893) N. W. Luft and K. H. Todhunter, *J. Chem. Phys., 21*, 2225–2226 (1953).
(894) A. Lydersen, R. A. Greenkorn, and O. A. Hougen, *Univ. Wis. Coll. Eng. Expt. Sta. Rept., 4*, 99 pp. (1955).
(895) F. H. MacDougall, *J. Am. Chem. Soc., 58*, 2585–2591 (1936).
(896) K. Macharacek, A. I. Zakharov, and L. A. Aleshina, *Chem. Prumysl, 12*, 23–24 (1962).
(897) H. Mackle and R. G. Mayrick, "A Plastic Capsule Technique for the Combustion Calorimetry of Volatile or Chemically Reactive Compounds: The Heat of Combustion of Polythene," in *Proceedings of the Symposium on Thermodynamics, IUPAC Fritzens-Wattens, Austria, August 1959*, pp. 25–28.
(898) H. Mackle and R. G. Mayrick, *Trans. Faraday Soc., 58*, 33–39 (1962).
(899) H. Mackle and R. G. Mayrick, *Trans. Faraday Soc., 58*, 230–237 (1962).
(900) H. Mackle and R. G. Mayrick, *Trans. Faraday Soc., 58*, 238–243 (1962).
(901) H. Mackle and R. T. B. McClean, *Trans. Faraday Soc., 58*, 895–899 (1962).
(902) H. Mackle and R. T. B. McClean, *Trans. Faraday Soc., 60*, 669–672 (1964).
(903) H. Mackle and R. T. B. McClean, *Trans. Faraday Soc., 60*, 817–821 (1964).
(904) H. Mackle and P. A. G. O'Hare, *Trans. Faraday Soc., 57*, 1070–1074 (1961).
(905) H. Mackle and P. A. G. O'Hare, *Trans. Faraday Soc., 57*, 1521–1526 (1961).
(906) H. Mackle and P. A. G. O'Hare, *Trans. Faraday Soc., 57*, 1873–1876 (1961).
(907) H. Mackle and P. A. G. O'Hare, *Trans. Faraday Soc., 57*, 2119–2124 (1961).
(908) H. Mackle and P. A. G. O'Hare, *Trans. Faraday Soc., 58*, 1912–1915 (1962).
(909) H. Mackle and P. A. G. O'Hare, *Trans. Faraday Soc., 59*, 309–315 (1963).
(910) H. Mackle and P. A. G. O'Hare, *Trans. Faraday Soc., 60*, 506–509 (1964).
(911) H. Mackle and P. A. G. O'Hare, *Trans. Faraday Soc., 60*, 666–668 (1964).
(912) J. R. Madigan and F. F. Cleveland, *J. Chem. Phys., 19*, 119–123 (1951).
(913) A. Magnus, *Z. Phys. Chem. (Frankfurt), 9*, 141–161 (1956).
(914) A. Magnus, H. Hartmann, and F. Becker, *Z. Phys. Chem. (Leipzig), 197*, 75–91 (1951).
(915) S. Makishima, Y. Yoneda, and Y. Saito, *Actes 2nd Intern. Congr. Catalyse, Paris, 1960, 1*, 617–643 (1961).
(916) F. E. Malherbe and H. J. Bernstein, *J. Am. Chem. Soc., 74*, 4408–4410 (1952).
(917) D. E. Mann, N. Acquista, and E. K. Plyler, *J. Chem. Phys., 21*, 1949–1953' (1953).
(918) D. E. Mann, N. Acquista, and E. K. Plyler, *J. Chem. Phys., 22*, 1199–1202 (1954).

(919) D. E. Mann, N. Acquista, and E. K. Plyler, *J. Chem. Phys.*, 22, 1586–1592 (1954).
(920) D. E. Mann, N. Acquista, and E. K. Plyler, *J. Res. Natl. Bur. Std.*, 52, 67–72 (1954).
(921) D. E. Mann, N. Acquista, and E. K. Plyler, *J. Chem. Phys.*, 23, 2122–2126 (1955).
(922) D. E. Mann, L. Fano, J. H. Meal, and T. Shimanouchi, *J. Chem. Phys.*, 27, 51–59 (1957).
(923) D. E. Mann, J. H. Meal, and E. K. Plyler, *J. Chem. Phys.*, 24, 1018–1022 (1956).
(924) D. E. Mann and E. K. Plyler, *J. Chem. Phys.*, 23, 1989–1993 (1955).
(925) M. Månsson and S. Sunner, *Acta Chem. Scand.*, 16, 1863–1869 (1962).
(926) M. Månsson and S. Sunner, *Acta Chem. Scand.*, 17, 723–727 (1963).
(927) M. Månsson and S. Sunner, *Acta Chem. Scand.*, 20, 845–848 (1966).
(928) R. Manzoni-Ansidei and T. Storto, *Atti Accad. Ital., Rend. Classe Sci. Fis., Mat. Nat. [7]*, 1, 465–470 (1940).
(929) J. L. Margrave, *J. Chem. Phys.*, 24, 475–476 (1956).
(930) J. L. Margrave, in *Proceedings of the Symposium on High Temperature–A Tool for the Future, Berkeley, Calif., June, 1956*, Stanford Research Institute, Palo Alto, Calif.
(931) J. L. Margrave, M. A. Frisch, R. G. Bautista, R. L. Clarke, and W. S. Johnson, *J. Am. Chem. Soc.*, 85, 546–548 (1963).
(932) W. Maroney, *J. Am. Chem. Soc.*, 57, 2397–2398 (1935).
(933) C. H. Marshall, *Chem. Eng. Progr.*, 46, 313–318 (1950).
(934) J. J. Martin, *J. Chem. Eng. Data*, 7, 68–72 (1962).
(935) J. F. Masi, *J. Am. Chem. Soc.*, 74, 4738–4741 (1952); 75, 2276–2277 (1953).
(936) J. F. Masi, *J. Am. Chem. Soc.*, 75, 5082–5084 (1953).
(937) P. G. Maslov, *Khim. Sera-i Azotorgan. Soedin., Soderzhashch. v Neft. i Nefteprod., Akad. Nauk SSSR, Bashkirsk. Filial*, 3, 115–120 (1960).
(938) P. G. Maslov and Y. P. Maslov, *Khim. i Tekhnol. Topliv i Masel*, 3 (10), 50–55 (1958).
(939) J. F. Mathews and J. J. McKetta, *J. Phys. Chem.*, 65, 758–762 (1961).
(940) J. H. Mathews, *J. Am. Chem. Soc.*, 48, 562–576 (1926).
(941) J. H. Mathews and P. R. Fehlandt, *J. Am. Chem. Soc.*, 53, 3212–3217 (1931).
(942) J. E. Mayer and M. G. Mayer, *Statistical Mechanics*, Wiley, New York, 1940.
(943) J. Mazur, *Z. Phys.*, 113, 710–720 (1939).
(944) B. J. McBride and S. Gordon, *J. Chem. Phys.*, 35, 2198–2206 (1961).
(945) K. E. McCulloh and G. F. Pollnow, *J. Chem. Phys.*, 22, 1144 (1954).
(946) J. P. McCullough, *J. Chem. Phys.*, 29, 966–967 (1958).
(947) J. P. McCullough, *U.S. Bur. Mines Inform. Circ., 8034*, 49 pp. (1961).
(948) J. P. McCullough, D. R. Douslin, W. N. Hubbard, S. S. Todd, J. F. Messerly, I. A. Hossenlopp, F. R. Frow, J. P. Dawson, and G. Waddington, *J. Am. Chem. Soc.*, 81, 5884–5890 (1959).
(949) J. P. McCullough, D. R. Douslin, J. F. Messerly, I. A. Hossenlopp, T. C. Kincheloe, and G. Waddington, *J. Am. Chem. Soc.*, 79, 4289–4295 (1957).
(950) J. P. McCullough, H. L. Finke, M. E. Gross, J. F. Messerly, and G. Waddington, *J. Phys. Chem.*, 61, 289–301 (1957).
(951) J. P. McCullough, H. L. Finke, W. N. Hubbard, W. D. Good, R. E. Pennington, J. F. Messerly, and G. Waddington, *J. Am. Chem. Soc.*, 76, 2661–2669 (1954).
(952) J. P. McCullough, H. L. Finke, W. N. Hubbard, S. S. Todd, J. F. Messerly, D. R. Douslin, and G. Waddington, *J. Phys. Chem.*, 65, 784–791 (1961).

(953)   J. P. McCullough, H. L. Finke, J. F. Messerly, R. E. Pennington, I. A. Hossenlopp, and G. Waddington, *J. Am. Chem. Soc.*, 77, 6119–6125 (1955).
(954)   J. P. McCullough, H. L. Finke, J. F. Messerly, S. S. Todd, T. C. Kincheloe, and G. Waddington, *J. Phys. Chem.*, 61, 1105–1116 (1957).
(955)   J. P. McCullough, H. L. Finke, D. W. Scott, M. E. Gross, J. F. Messerly, R. E. Pennington, and G. Waddington, *J. Am. Chem. Soc.*, 76, 4796–4802 (1954).
(956)   J. P. McCullough, H. L. Finke, D. W. Scott, R. E. Pennington, M. E. Gross, J. F. Messerly, and G. Waddington, *J. Am. Chem. Soc.*, 80, 4786–4793 (1958).
(957)   J. P. McCullough and W. D. Good, *J. Phys. Chem.*, 65, 1430–1432 (1961).
(958)   J. P. McCullough, W. N. Hubbard, F. R. Frow, I. A. Hossenlopp, and G. Waddington, *J. Am. Chem. Soc.*, 79, 561–566 (1957).
(959)   J. P. McCullough, R. E. Pennington, J. C. Smith, I. A. Hossenlopp, and G. Waddington, *J. Am. Chem. Soc.*, 81, 5880–5883 (1959).
(960)   J. P. McCullough and D. W. Scott, *J. Am. Chem. Soc.*, 81, 1331–1334 (1959).
(961)   J. P. McCullough and D. W. Scott, private communication to D. R. Stull, Sept. 27, 1960.
(962)   J. P. McCullough, D. W. Scott, H. L. Finke, M. E. Gross, K. D. Williamson, R. E. Pennington, G. Waddington, and H. M. Huffman, *J. Am. Chem. Soc.*, 74, 2801–2804 (1952).
(963)   J. P. McCullough, D. W. Scott, H. L. Finke, W. N. Hubbard, M. E. Gross, C. Katz, R. E. Pennington, J. F. Messerly, and G. Waddington, *J. Am. Chem. Soc.*, 75, 1818–1824 (1953).
(964)   J. P. McCullough, D. W. Scott, R. E. Pennington, I. A. Hossenlopp, and G. Waddington, *J. Am. Chem. Soc.*, 76, 4791–4796 (1954).
(965)   J. P. McCullough, S. Sunner, H. L. Finke, W. N. Hubbard, M. E. Gross, R. E. Pennington, J. F. Messerly, W. D. Good, and G. Waddington, *J. Am. Chem. Soc.*, 75, 5075–5081 (1953).
(966)   K. G. McCurdy and K. J. Laidler, *Can. J. Chem.*, 41, 1867–1871 (1963).
(967)   H. J. McDonald, *J. Phys. Chem.*, 48, 47–50 (1944).
(968)   R. A. McDonald, Dow Thermal Research Laboratory, personal communication, Mar. 12, 1958.
(969)   R. A. McDonald, S. A. Shrader, and D. R. Stull, *J. Chem. Eng. Data*, 4, 311–313 (1959).
(970)   L. A. McDougall and J. E. Kilpatrick, *J. Chem. Phys.*, 42, 2307–2310 (1965).
(971)   L. A. McDougall and J. E. Kilpatrick, *J. Chem. Phys.*, 42, 2311–2321 (1965).
(972)   R. S. McDowell and F. H. Kruse, *J. Chem. Eng. Data*, 8, 547–548 (1963).
(973)   D. M. McEachern, Jr., and J. E. Kilpatrick, *J. Chem. Phys.*, 41, 3127–3131 (1964).
(974)   W. S. McEwan and M. W. Rigg, *J. Am. Chem. Soc.*, 73, 4725–4727 (1951).
(975)   P. R. McGee, F. F. Cleveland, A. G. Meister, C. E. Decker, and S. I. Miller, *J. Chem. Phys.*, 21, 242–246 (1953).
(976)   M. L. McGlashan, *J. Chem. Educ.*, 43, 226–232 (1966).
(977)   J. S. McKinley-McKee and E. A. Moelwyn-Hughes, *Trans. Faraday Soc.*, 48, 247–253 (1952).
(978)   J. McMorris and R. M. Badger, *J. Am. Chem. Soc.*, 55, 1952–1957 (1933).
(979)   W. H. Mears, R. F. Stahl, S. R. Orfeo, R. C. Shair, L. F. Kells, W. Thompson, and H. McCann, *Ind. Eng. Chem.*, 47, 1449–1454 (1955).
(980)   R. Mecke, *Z. Elektrochem.*, 36, 589–596 (1930); *Z. Phys. Chem. (Leipzig)*, B16, 409–420, 421–437; B17, 1–20 (1932); *Z. Phys.* 104, 291–302 (1936).
(981)   L. Médard and M. Thomas, *Mem. Poudres*, 31, 173–196 (1949).
(982)   L. Médard and M. Thomas, *Mem. Poudres*, 34, 421–442 (1952).

(983)   L. Médard and M. Thomas, *Mem. Poudres*, *35*, 155–173 (1953).
(984)   L. Médard and M. Thomas, *Mem. Poudres*, *36*, 97–127 (1954).
(985)   L. Médard and M. Thomas, *Mem. Poudres*, *37*, 129–138 (1955).
(986)   L. Médard and M. Thomas, *Mem. Poudres*, *38*, 45–63 (1956).
(987)   W. Mehl, *Beih. Z. Ges. Kälte-Ind.*, Ser. *1* (3), 5–35 (1933).
(988)   W. Mehl, *Z. Ges. Kälte-Ind.*, *41*, 152–153 (1934).
(989)   W. Mehl, *Z. Phys. Chem. (Leipzig)*, *A169*, 312–313 (1934).
(990)   A. G. Meister, J. M. Dowling, and A. J. Bielecki, *J. Chem. Phys.*, *25*, 941–942 (1956).
(991)   G. H. Messerly and J. G. Aston, *J. Am. Chem. Soc.*, *62*, 886–890 (1940).
(992)   G. H. Messerly and R. M. Kennedy, *J. Am. Chem. Soc.*, *62*, 2988–2991 (1940).
(993)   J. F. Messerly, S. S. Todd, and H. L. Finke, *J. Phys. Chem.*, *69*, 353–359 (1965).
(994)   J. F. Messerly, S. S. Todd, and H. L. Finke, *J. Phys. Chem.*, *69*, 4304–4311 (1965).
(995)   C. H. Meyers, *J. Res. Natl. Bur. Std.*, *9*, 807–813 (1932).
(996)   A. Michael and G. H. Carlson, *J. Am. Chem. Soc.*, *57*, 159–164 (1935).
(997)   A. Michels, S. R. de Groot, and M. Geldermans, *Appl. Sci. Res.*, Sect. A, *1*, 55–65 (1947).
(998)   A. Michels, A. Visser, R. J. Lunbeck, and G. J. Wolkers, *Physica*, *18*, 114–120 (1952).
(999)   G. Milazzo, *Gazz. Chim. Ital.*, *74*, 49–57 (1944).
(1000) C. B. Miles and H. Hunt, *J. Phys. Chem.*, *45*, 1346–1359 (1941).
(1001) A. J. Miller and H. Hunt, *J. Phys. Chem.*, *49*, 20–21 (1945).
(1002) F. A. Miller and R. B. Hannan, Jr., *J. Chem. Phys.*, *21*, 110–114 (1953).
(1003) F. A. Miller, R. B. Hannan, Jr., and L. R. Cousins, *J. Chem. Phys.*, *23*, 2127–2129 (1955).
(1004) F. A. Miller, D. H. Lemmon, and R. E. Witkowski, *Spectrochim. Acta*, *21*, 1709–1716 (1965).
(1005) G. A. Miller, *J. Chem. Eng. Data*, *8*, 69–72 (1963).
(1006) P. Miller, *Iowa State Coll. J. Sci.*, *10*, 91–93 (1936).
(1007) S. A. Miller, *Brit. Chem. Eng.*, *7*, 656–662 (1962).
(1008) S. L. Miller, L. C. Aamodt, G. Dousmanis, C. H. Townes, and J. Kraitchman, *J. Chem. Phys.*, *20*, 1112–1114 (1952).
(1009) D. E. Milligan, E. D. Becker, and K. S. Pitzer, *J. Am. Chem. Soc.*, *78*, 2707–2711 (1956).
(1010) R. C. Millikan and K. S. Pitzer, *J. Chem. Phys.*, *27*, 1305–1308 (1957).
(1011) M. Milone, *Gazz. Chim. Ital.*, *61*, 153–158 (1931).
(1012) M. Milone and S. Allavena, *Gazz. Chim. Ital.*, *61*, 75–90 (1931).
(1013) M. Milone and G. Venturello, *Gazz. Chim. Ital.*, *66*, 808–812 (1936).
(1014) J. E. Mitchell, Jr., *Trans. Am. Inst. Chem. Eng.*, *42*, 293–308 (1946).
(1015) S. S. Mitra and H. J. Bernstein, *Can. J. Chem.*, *37*, 553–562 (1959).
(1016) T. Miyazawa and K. S. Pitzer, *J. Am. Chem. Soc.*, *80*, 60–62 (1958).
(1017) E. A. Moelwyn-Hughes, *Physical Chemistry*, revised ed., Pergamon Press, London, 1961.
(1018) E. A. Moelwyn-Hughes, *States of Matter*, Oliver and Boyd, Edinburgh, 1961.
(1019) K. D. Möller, *Compt. Rend.*, *250*, 3977–3979 (1960).
(1020) M. Momotani, H. Suga, S. Seki, and I. Nitta, *Proc. Natl. Acad. Sci. India*, Sect. A, *25*, 74–82 (1956).
(1021) J. B. Montgomery and T. De Vries, *J. Am. Chem. Soc.*, *64*, 2375–2377 (1942).
(1022) R. B. Mooney and E. B. Ludlam, *Proc. Roy. Soc. Edinburgh*, *49*, 160–169 (1929).
(1023) C. B. Moore and G. C. Pimentel, *J. Chem. Phys.*, *38*, 2816–2829 (1963).

(1024) C. B. Moore and G. C. Pimentel, *J. Chem. Phys., 40,* 342–355 (1964).
(1025) G. E. Moore, M. L. Renquist, and G. S. Parks, *J. Am. Chem. Soc., 62,* 1505–1507 (1940).
(1026) Y. Morino and K. Kuchitsu, *J. Chem. Phys., 28,* 175–184 (1958).
(1027) Y. Morino, I. Miyagawa, and T. Haga, *J. Chem. Phys., 19,* 791–792 (1951).
(1028) J. C. Morris, W. J. Lanum, R. V. Helm, W. E. Haines, G. L. Cook, and J. S. Ball, *J. Chem. Eng. Data, 5,* 112–116 (1960).
(1029) C. T. Mortimer, *Reaction Heats and Bond Strengths,* Pergamon Press, New York, 1962.
(1030) C. T. Mortimer, H. O. Pritchard, and H. A. Skinner, *Trans. Faraday Soc., 48,* 220–228 (1952).
(1031) L. de V. Moulds and H. L. Riley, *J. Chem. Soc.,* 621–626 (1938).
(1032) H. Moureu and M. Dodé, *Bull. Soc. Chim. France, 4,* 637–647 (1937).
(1033) C. J. Muelleman, K. Ramaswamy, F. F. Cleveland, and S. Sundaram, *J. Mol. Spectrosc., 11,* 262–274 (1963).
(1034) E. F. Mueller, "Precision Resistance Thermometry," in *Temperature, Its Measurement and Control in Science and Industry,* Reinhold, New York, 1941, pp. 162–179, Chapter 2.
(1035) R. S. Mulliken, *J. Phys. Chem., 56,* 295–311 (1952).
(1036) A. Münster, *Statistische Thermodynamik,* Springer-Verlag, Berlin, 1956.
(1037) H. Murata, A. Palm, and A. G. Meister, *J. Chem. Phys., 23,* 702–703 (1955).
(1038) G. M. Murphy and J. E. Vance, *J. Chem. Phys., 18,* 1514–1515 (1950).
(1039) J. Murrin and S. Goldhagen, *Memorandum Report No. 88,* U.S. Naval Powder Factory, Research and Development Department, Indian Head, Md., October 19, 1954.
(1040) J. W. Murrin and S. Goldhagen, *Memorandum Report No. 121,* U.S. Naval Powder Factory, Research and Development Department, Indian Head, Md., May 18, 1956.
(1041) Y. I. Mushkin and A. I. Finkel'shtein, *Opt. i Spektrosk. 13,* 289–291 (1962).
(1042) R. J. Myers and W. D. Gwinn, *J. Chem. Phys., 20,* 1420–1427 (1952).
(1043) G. Nagarajan, *Bull. Soc. Chim. Belges, 71,* 65–72 (1962).
(1044) K. Naito, I. Nakagawa, K. Kuratani, I. Ichishima, and S. Mizushima, *J. Chem. Phys., 23,* 1907–1910 (1955).
(1045) T. Nakamura, *Shokubai (Sapporo), 7,* 17–20 (1951).
(1046) P. S. Nangia and S. W. Benson, *J. Am. Chem. Soc., 86,* 2770–2773 (1964).
(1047) N. A. Narasimham, J. R. Nielsen, and R. Theimer, *J. Chem. Phys., 27,* 740–745 (1957).
(1048) E. Neale and L. T. D. Williams, *J. Chem. Soc.,* 2156–2159 (1954).
(1049) E. F. Neilson and D. White, *J. Am. Chem. Soc., 79,* 5618–5621 (1957).
(1050) B. Nelander and S. Sunner, *J. Chem. Phys., 44,* 2476–2480 (1966).
(1051) E. W. Nelson and R. F. Newton, *J. Am. Chem. Soc., 63,* 2178–2182 (1941).
(1052) R. A. Nelson and R. S. Jessup, *J. Res. Natl. Bur. Std., 48,* 206–208 (1952).
(1053) W. Nernst, *Nachr. Ges. Wiss. Göttingen,* 1–40 (1906); see also *The New Heat Theorem,* Methuen, London, 1926.
(1054) W. Nernst, *Sitzber. Kgl. Preuss. Akad. Wiss., 12, 13,* 261–292 (1910).
(1055) W. Nernst and co-workers:
 (a) cf. (1054).
 (b) W. Nernst, *Ann. Phys. 36,* 395–439 (1911).
 (c) F. Pollitzer, *Z. Elektrochem., 17,* 5–14 (1911).
 (d) F. Pollitzer, *Z. Elektrochem., 19,* 513–518 (1913).
 (e) W. Nernst and F. Schwers, *Sitzber. Preuss. Akad. Wiss.,* 355–370 (1914).

(f) P. Gunter, *Ann. Phys. 51*, 828 (1916).
(g) F. Simon, *Ann. Phys., 68*, 241–280 (1922).
(h) F. Simon and F. Lange, *Z. Phys., 15*, 312–321 (1923).
(i) F. Lange, *Z. Phys. Chem. (Leipzig), 110*, 343–362 (1924).
(j) F. Simon and W. Zeidler, *Z. Phys. Chem. (Leipzig),* 123, 383–404 (1926).
(k) F. Simon and M. Ruhemann, *Z. Phys. Chem. (Leipzig), 129*, 321–328 (1927).

(1056) C. Neuberg and E. Hofmann, *Biochem. Z., 252*, 440–450 (1932).
(1057) C. Neuberg, E. Hofmann, and M. Kobel, *Biochem. Z., 234*, 341–344 (1931).
(1058) C. A. Neugebauer, *Univ. Microfilms Publ. No. 21224*, 131 pp.; *Dissertation Abstr., 17*, 1478–1479 (1957).
(1059) C. A. Neugebauer and J. L. Margrave, *J. Phys. Chem., 60*, 1318–1321 (1956).
(1060) C. A. Neugebauer and J. L. Margrave, *J. Am. Chem. Soc., 79*, 1338–1340 (1957).
(1061) C. A. Neugebauer and J. L. Margrave, *J. Phys. Chem., 62*, 1043–1048 (1958).
(1062) R. H. Newton and B. F. Dodge, *J. Am. Chem. Soc., 55*, 4747–4759 (1933).
(1063) R. H. Newton and B. F. Dodge, *Ind. Eng. Chem., 27*, 577–581 (1935).
(1064) G. R. Nicholson, *J. Chem. Soc.*, 2431–2432 (1957).
(1065) G. R. Nicholson, *J. Chem. Soc.*, 2377–2378 (1960).
(1066) G. R. Nicholson, *J. Chem. Soc.*, 2378–2379 (1960).
(1067) G. R. Nicholson, M. Szwarc, and J. W. Taylor, *J. Chem. Soc.*, 2767–2769 (1954).
(1068) J. K. Nickerson, K. A. Kobe, and J. J. McKetta, *J. Phys. Chem., 65*, 1037–1043 (1961).
(1069) E. Nicolini, *Ann. Chim. (Paris), 6*, 582–629 (1951).
(1070) J. R. Nielsen, H. H. Claassen, and D. C. Smith, *J. Chem. Phys., 18*, 1471–1476 (1950).
(1071) J. R. Nielsen, C. Y. Liang, and L. W. Daasch, *J. Opt. Soc. Am., 43*, 1071–1079 (1953).
(1072) J. R. Nielsen, C. Y. Liang, and D. C. Smith, *J. Chem. Phys., 20*, 1090–1094 (1952).
(1073) J. R. Nielsen, C. Y. Liang, and D. C. Smith, *J. Chem. Phys., 21*, 1060–1069 (1953).
(1074) J. R. Nielsen, C. Y. Liang, D. C. Smith, and M. Alpert, *J. Chem. Phys., 21*, 1070–1076 (1953).
(1075) J. R. Nielsen and R. Theimer, *J. Chem. Phys., 27*, 891–895 (1957).
(1076) J. R. Nielsen and R. Theimer, *J. Chem. Phys., 30*, 103–104 (1959).
(1077) T. Nishikawa, *J. Phys. Soc. Japan, 12*, 668–680 (1957).
(1078) T. Nishikawa, T. Itoh, and K. Shimoda, *J. Chem. Phys., 23*, 1735–1736 (1955).
(1079) P. Noble, Jr., W. L. Reed, C. J. Hoffman, J. A. Gallaghan, and F. G. Borgardt, *Am. Inst. Aeron. Astronaut. J., 1*, 395–397 (1963).
(1080) C. R. Noddings and G. M. Mullet, *Handbook of Compositions at Thermodynamic Equilibrium*, Interscience Publishers, New York, 1965.
(1081) A. P. Oberemok-Yakubova and A. A. Balandin, *Izv. Akad. Nauk SSSR, Ser. Khim.*, 2210–2211 (1963).
(1082) F. L. Oetting, *J. Phys. Chem., 67*, 2757–2761 (1963).
(1083) F. L. Oetting, *J. Chem. Phys., 41*, 149–153 (1964).
(1084) F. L. Oetting, *J. Chem. Eng. Data, 10*, 122–125 (1965).
(1085) G. D. Oliver, M. Eaton, and H. M. Huffman, *J. Am. Chem. Soc., 70*, 1502–1505 (1948).
(1086) G. D. Oliver and J. W. Grisard, *J. Am. Chem. Soc., 73*, 1688–1690 (1951).
(1087) H. Oosaka, *Bull. Chem. Soc. Japan, 15*, 31–36 (1940).

(1088) H. Oosaka, H. Sekine, and T. Saito, *Bull. Chem. Soc. Japan, 27,* 182–184 (1954).
(1089) E. I. Organick and W. R. Studhalter, *Chem. Eng. Progr., 44,* 847–854 (1948).
(1090) R. A. Oriani and C. P. Smyth, *J. Chem. Phys., 16,* 930 (1948).
(1091) A. G. Osborn and D. R. Douslin, *J. Chem. Eng. Data, 11,* 502–509 (1966).
(1092) D. W. Osborne, R. N. Doescher, and D. M. Yost, *J. Am. Chem. Soc., 64,* 169–172 (1942).
(1093) D. F. Othmer, *Ind. Eng. Chem., 34,* 1072–1078 (1942).
(1094) J. Overend and J. C. Evans, *Trans. Faraday Soc., 55,* 1817–1825 (1959).
(1095) J. Overend, R. A. Nyquist, J. C. Evans, and W. J. Potts, *Spectrochim. Acta, 17,* 1205–1218 (1961).
(1096) J. D. Overmars and S. M. Blinder, *J. Phys. Chem., 68,* 1801–1803 (1964).
(1097) E. L. Pace and J. G. Aston, *J. Am. Chem. Soc., 70,* 566–570 (1948).
(1098) E. L. Pace and R. J. Bobka, *J. Chem. Phys., 35,* 454–457 (1961).
(1099) A. Palm and M. Kilpatrick, *J. Chem. Phys., 23,* 1562–1563 (1955).
(1100) A. Palm, F. L. Voelz, and A. G. Meister, *J. Chem. Phys., 23,* 726–728 (1955).
(1101) W. A. Pardee and W. Weinrich, *Ind. Eng. Chem., 36,* 595–603 (1944).
(1102) G. S. Parks and J. A. Hatton, *J. Am. Chem. Soc., 71,* 2773–2775 (1949).
(1103) G. S. Parks and H. M. Huffman, *J. Am. Chem. Soc., 48,* 2788–2793 (1926).
(1104) G. S. Parks and H. M. Huffman, *J. Am. Chem. Soc., 52,* 4381–4391 (1930).
(1105) G. S. Parks and H. M. Huffman, *The Free Energies of Some Organic Compounds,* Chemical Catalog Co., New York, 1932.
(1106) G. S. Parks, H. M. Huffman, and M. Barmore, *J. Am. Chem. Soc., 55,* 2733–2740 (1933).
(1107) G. S. Parks, H. M. Huffman, and S. B. Thomas, *J. Am. Chem. Soc., 52,* 1032–1044 (1930).
(1108) G. S. Parks, K. K. Kelley, and H. M. Huffman, *J. Am. Chem. Soc., 51,* 1969–1973 (1929).
(1109) G. S. Parks, W. D. Kennedy, R. R. Gates, J. R. Mosley, G. E. Moore, and M. L. Renquist, *J. Am. Chem. Soc., 78,* 56–59 (1956).
(1110) G. S. Parks and D. W. Light, *J. Am. Chem. Soc., 56,* 1511–1513 (1934).
(1111) G. S. Parks and K. E. Manchester, *Thermochem. Bull., 2,* 8 (March 1956).
(1112) G. S. Parks, K. E. Manchester, and L. M. Vaughan, *J. Chem. Phys., 22,* 2089–2090 (1954).
(1113) G. S. Parks, G. E. Moore, M. L. Renquist, B. F. Naylor, L. A. McClaine, P. S. Fujii, and J. A. Hatton, *J. Am. Chem. Soc., 71,* 3386–3389 (1949).
(1114) G. S. Parks and H. P. Mosher, *J. Chem. Phys., 37,* 919–920 (1962).
(1115) G. S. Parks and J. R. Mosley, *J. Am. Chem. Soc., 72,* 1850 (1950).
(1116) G. S. Parks, J. R. Mosley, and P. V. Peterson, Jr., *J. Chem. Phys., 18,* 152–153 (1950).
(1117) G. S. Parks, S. B. Thomas, and D. W. Light, *J. Chem. Phys., 4,* 64–69 (1936).
(1118) G. S. Parks, S. S. Todd, and W. A. Moore, *J. Am. Chem. Soc., 58,* 398–401 (1936).
(1119) G. S. Parks, S. S. Todd, and C. H. Shomate, *J. Am. Chem. Soc., 58,* 2505–2508 (1936).
(1120) G. S. Parks, T. J. West, and G. E. Moore, *J. Am. Chem. Soc., 63,* 1133–1135 (1941).
(1121) G. S. Parks, T. J. West, B. F. Naylor, P. S. Fujii, and L. A. McClaine, *J. Am. Chem. Soc., 68,* 2524–2527 (1946).
(1122) M. Parris, P. S. Raybin, and L. C. Labowitz, *J. Chem. Eng. Data, 9,* 221–223 (1964).
(1123) E. J. Partington, J. S. Rowlinson, and J. F. Weston, *Trans. Faraday Soc., 56,* 479–485 (1960).

(1124) J. R. Partington, *An Advanced Treatise on Physical Chemistry,* Vol. II. Longmans Green, New York, 1951, pp. 226–274.
(1125) W. Pässler, Doctoral Dissertation, Dresden, 1930; cited by W. Auer (52), p. 655.
(1126) C. R. Patrick, *Tetrahedron, 4,* 26–35 (1958).
(1127) C. R. Patrick, "The Thermochemistry of Organic Fluorine Compounds," in M. Stacey, J. C. Tatlow, and A. G. Sharpe (Eds.), *Advances in Fluorine Chemistry,* Vol. 2, Butterworths, London, 1961, pp. 1–34.
(1128) C. R. Patrick and C. S. Prosser, *Trans. Faraday Soc., 60,* 700–704 (1964).
(1129) M. A. Paul, *Principles of Chemical Thermodynamics,* McGraw-Hill, New York, 1951.
(1130) L. Pauling, *The Nature of the Chemical Bond,* 3rd Ed., Cornell University Press, Ithaca, N. Y., 1960.
(1131) D. H. Payne, Dissertation, University of Michigan, 1954; *Univ. Microfilms Publ. No. 11336 (L-2105),* 131 pp.; *Dissertation Abstr., 15,* 723 (1955).
(1132) D. H. Payne and E. F. Westrum, Jr., *J. Phys. Chem., 66,* 748–751 (1962).
(1133) A. S. Pell and G. Pilcher, *Trans. Faraday Soc., 61,* 71–77 (1965).
(1134) R. E. Pennington, H. L. Finke, W. N. Hubbard, J. F. Messerly, F. R. Frow, I. A. Hossenlopp, and G. Waddington, *J. Am. Chem. Soc., 78,* 2055–2060 (1956).
(1135) R. E. Pennington and K. A. Kobe, *J. Am. Chem. Soc., 79,* 300–305 (1957).
(1136) R. E. Pennington, D. W. Scott, H. L. Finke, J. P. McCullough, J. F. Messerly, I. A. Hossenlopp, and G. Waddington, *J. Am. Chem. Soc., 78,* 3266–3272 (1956).
(1137) A. Perez Masiá and M. Diaz Peña, *Anales Real Soc. Espan. Fis. Quim. (Madrid), Ser. B, 54,* 661–668 (1958).
(1138) A. Perlick, *Bull. Intern. Inst. Refrig., 18* (4), A1-9 (1937).
(1139) E. Perrottet, W. Taub, and E. Briner, *Helv. Chim. Acta, 23,* 1260–1268 (1940).
(1140) W. B. Person and G. C. Pimentel, *J. Am. Chem. Soc., 75,* 532–538 (1953).
(1141) J. C. Philip and S. C. Waterton, *J. Chem. Soc.,* 2783–2784 (1930).
(1142) E. W. Phillips and R. R. Klimpel, *J. Chem. Eng. Data, 13,* 97–101 (1968).
(1143) L. Pierce and M. Hayashi, *J. Chem. Phys., 35,* 479–485 (1961).
(1144) L. Pierce, R. Nelson, and C. Thomas, *J. Chem. Phys., 43,* 3423–3431 (1965).
(1145) G. Pilcher, A. S. Pell, and D. J. Coleman, *Trans. Faraday Soc., 60,* 499–505 (1964).
(1146) G. Pilcher, H. A. Skinner, A. S. Pell, and A. E. Pope, *Trans. Faraday Soc., 59,* 316–330 (1963).
(1147) G. Pilcher and L. E. Sutton, *J. Chem. Soc.,* 2695–2700 (1956).
(1148) M. G. K. Pillai and F. F. Cleveland, *J. Mol. Spectrosc., 5,* 212–217 (1960).
(1149) M. G. K. Pillai and F. F. Cleveland, *J. Mol. Spectrosc., 6,* 465–471 (1961).
(1150) K. S. Pitzer, *Chem. Rev., 27,* 39–57 (1940).
(1151) K. S. Pitzer, *J. Am. Chem. Soc., 62,* 331–335 (1940).
(1152) K. S. Pitzer, *J. Am. Chem. Soc., 62,* 1224–1227 (1940).
(1153) K. S. Pitzer, *J. Chem. Phys., 8,* 711–720 (1940).
(1154) K. S. Pitzer, *J. Am. Chem. Soc., 63,* 2413–2418 (1941).
(1155) K. S. Pitzer, *Ind. Eng. Chem., 36,* 829–831 (1944).
(1156) K. S. Pitzer, *J. Chem. Phys., 14,* 239–243 (1946).
(1157) K. S. Pitzer, *Discussions Faraday Soc., 10,* 66–73 (1951).
(1158) K. S. Pitzer and L. Brewer, *Thermodynamics* (by G. N. Lewis and M. Randall), 2nd Ed., McGraw-Hill, New York, 1961.
(1159) K. S. Pitzer and E. Gelles, *J. Chem. Phys., 21,* 855–858 (1953).
(1160) K. S. Pitzer, L. Guttman, and E. F. Westrum, Jr., *J. Am. Chem. Soc., 68,* 2209–2212 (1946).
(1161) K. S. Pitzer and W. D. Gwinn, *J. Am. Chem. Soc., 63,* 3313–3316 (1941).

(1162) K. S. Pitzer, and W. D. Gwinn, *J. Chem. Phys.*, *10*, 428–440 (1942).
(1163) K. S. Pitzer and J. L. Hollenberg, *J. Am. Chem. Soc.*, *76*, 1493–1496 (1954).
(1164) K. S. Pitzer and J. E. Kilpatrick, *Chem. Rev.*, *39*, 435–447 (1946)
(1165) K. S. Pitzer and D. W. Scott, *J. Am. Chem. Soc.*, *63*, 2419–2422 (1941).
(1166) K. S. Pitzer and D. W. Scott, *J. Am. Chem. Soc.*, *65*, 803–829 (1943).
(1167) K. S. Pitzer and W. Weltner, Jr., *J. Am. Chem. Soc.*, *71*, 2842–2844 (1949).
(1168) M. Planck, *Ber.*, *45*, 5–23 (1912).
(1169) J. R. Platt, *J. Chem. Phys.*, *15*, 419–420 (1947).
(1170) J. R. Platt, *J. Phys. Chem.*, *56*, 328–336 (1952).
(1171) E. K. Plyler, *J. Res. Natl. Bur. Std.*, *64A*, 377–379 (1960).
(1172) E. K. Plyler and W. S. Benedict, *J. Res. Natl. Burl Std.*, *47*, 202–220 (1951).
(1173) E. Pohland and W. Mehl, *Z. Phys. Chem. (Leipzig)*, *A164*, 48–54 (1933).
(1174) S. R. Polo, A. Palm, F. L. Voelz, F. F. Cleveland, A. G. Meister, R. B. Bernstein, and R. H. Sherman, *J. Chem. Phys.*, *23*, 833–837 (1955).
(1175) A. Pongratz, S. Böhmert-Süss, and K. Scholtis, *Ber.*, *77B*, 651–661 (1944).
(1176) V. V. Ponomarev, T. A. Alekseeva, and L. N. Akimova, *Zh. Fiz. Khim.*, *36*, 872–873 (1962).
(1177) V. V. Ponomarev, T. A. Alekseeva, and L. N. Akimova, *Zh. Fiz. Khim.*, *36*, 1083–1084 (1962).
(1178) V. V. Ponomarev, T. A. Alekseeva, and L. N. Akimova, *Zh. Fiz. Khim.*, *37*, 227–228 (1963).
(1179) D. A. Pontarelli, A. G. Meister, F. F. Cleveland, F. L. Voelz, R. B. Bernstein, and R. H. Sherman, *J. Chem. Phys.*, *20*, 1949–1954 (1952).
(1180) M. M. Popov and P. K. Schirokich, *Z. Phys. Chem. (Leipzig)*, *A167*, 183–187 (1933).
(1181) S. P. S. Porto, *J. Mol. Spectrosc.*, *3*, 248–258 (1959).
(1182) R. L. Potter, *J. Chem. Phys.*, *26*, 394–397 (1957).
(1183) R. L. Potter, *J. Chem. Phys.*, *31*, 1100–1103 (1959).
(1184) R. L. Poynter, *J. Chem. Phys.*, *39*, 1962–1966 (1963).
(1185) J. M. Prausnitz and W. B. Carter, *Am. Inst. Chem. Eng. J.*, *6*, 611–614 (1960).
(1186) D. Price, *J. Chem. Phys.*, *9*, 807–815 (1941); *10*, 80 (1942).
(1187) H. O. Pritchard and H. A. Skinner, *J. Chem. Soc.*, 272–276 (1950).
(1188) H. O. Pritchard and H. A. Skinner, *J. Chem. Soc.*, 1099 (1950).
(1189) H. O. Pritchard and H. A. Skinner, *J. Chem. Soc.*, 1928–1931 (1950).
(1190) W. Pritzkow and K. A. Müller, *Ber.*, *89*, 2318–2321 (1956).
(1191) E. J. Prosen, "Combustion in a Bomb of Compounds Containing Carbon, Hydrogen, Oxygen, and Nitrogen," in F. D. Rossini (Ed.), *Experimental Thermochemistry*, Vol. I, Interscience Publishers, New York, 1956, Chap. 6.
(1192) E. J. Prosen, quoted by E. M. Otto, *J. Electrochem. Soc.*, *111*, 88–92 (1964).
(1193) E. J. Prosen, M. E. Hill, and I. Jaffe, unpublished work, National Bureau of Standards, Washington, D.C.
(1194) E. J. Prosen, R. Gilmont, and F. D. Rossini, *J. Res. Natl. Bur. Std.*, *34*, 65–71 (1945).
(1195) E. J. Prosen, W. H. Johnson, and F. D. Rossini, *J. Res. Natl. Bur. Std.*, *36*, 455–461 (1946).
(1196) E. J. Prosen, W. H. Johnson, and F. D. Rossini, *J. Res. Natl. Bur. Std.*, *37*, 51–56 (1946).
(1197) E. J. Prosen, W. H. Johnson, and F. D. Rossini, *J. Am. Chem. Soc.*, *72*, 626–627 (1950).
(1198) E. J. Prosen, F. W. Maron, and F. D. Rossini, *J. Res. Natl. Bur. Std.*, *42*, 269–277 (1949).

(1199) E. J. R. Prosen and F. D. Rossini, *J. Res. Natl. Bur. Std.*, *27*, 289–310 (1941).
(1200) E. J. Prosen and F. D. Rossini, *J. Res. Natl. Bur. Std.*, *33*, 255–272 (1944).
(1201) E. J. Prosen and F. D. Rossini, *J. Res. Natl. Bur. Std.*, *33*, 439–446 (1944).
(1202) E. J. Prosen and F. D. Rossini, *J. Res. Natl. Bur. Std.*, *34*, 163–174 (1945).
(1203) E. J. Prosen and F. D. Rossini, *J. Res. Natl. Bur. Std.*, *34*, 263–269 (1945).
(1204) E. J. Prosen and F. D. Rossini, *J. Res. Natl. Bur. Std.*, *36*, 269–275 (1946).
(1205) P. G. Puranik and L. Ramaswamy, *Proc. Indian Acad. Sci., Sect. A*, *52*, 135–142 (1960).
(1206) P. G. Puranik and E. V. Rao, *Indian J. Phys.*, *35*, 177–182 (1961).
(1207) Z. V. Pushkareva and Z. Y. Kokoshko, *J. Gen. Chem. USSR* (Eng. Trans.), *16*, 1269–1278 (1946).
(1208) W. E. Putnam and J. E. Kilpatrick, *J. Chem. Phys.*, *27*, 1075–1080 (1957).
(1209) W. E. Putnam, D. M. McEachern, Jr., and J. E. Kilpatrick, *J. Chem. Phys.*, *42*, 749–755 (1965).
(1210) W. E. Railing, *J. Am. Chem. Soc.*, *61*, 3349–3353 (1939).
(1211) K. Ramaswamy, K. Sathianandan, and F. F. Cleveland, *Spectrosc. Mol.*, *11*, 14–15 (1962).
(1212) W. Ramsay and S. Young, *J. Chem. Soc.*, *49*, 453–462 (1886).
(1213) D. H. Rank and W. M. Baldwin, *J. Chem. Phys.*, *19*, 1210–1211 (1951).
(1214) G. W. Rathjens, Jr., N. K. Freeman, W. D. Gwinn, and K. S. Pitzer, *J. Am. Chem. Soc.*, *75*, 5634–5642 (1953).
(1215) G. W. Rathjens, Jr., and W. D. Gwinn, *J. Am. Chem. Soc.*, *75*, 5629–5633 (1953).
(1216) J. D. Ray and A. A. Gershon, *J. Phys. Chem.*, *66*, 1750 (1962).
(1217) J. D. Ray and R. A. Ogg, Jr., *J. Chem. Phys.*, *31*, 168–171 (1959).
(1218) J. D. Ray and R. A. Ogg, Jr., *J. Phys. Chem.*, *63*, 1522–1523 (1959).
(1219) A. L. G. Rees, *J. Chem. Phys.*, *26*, 1567–1571 (1957).
(1220) R. C. Reid and T. K. Sherwood, *The Properties of Gases and Liquids, Their Estimation and Correlation*, McGraw-Hill, New York, 1958.
(1221) A. E. Reynolds and T. De Vries, *J. Am. Chem. Soc.*, *72*, 5443–5445 (1950).
(1222) G. Ribaud, *Publ. Sci. Tech. Min. Air (France)*, No. 266, 169 pp. (1952).
(1223) F. O. Rice and J. Greenberg, *J. Am. Chem. Soc.*, *56*, 2268–2270 (1934).
(1224) R. E. Richards, *J. Chem. Soc.*, 1931–1933 (1948).
(1225) W. T. Richards and J. H. Wallace, Jr., *J. Am. Chem. Soc.*, *54*, 2705–2713 (1932).
(1226) J. W. Richardson and G. S. Parks, *J. Am. Chem. Soc.*, *61*, 3543–3546 (1939).
(1227) E. K. Rideal, *Proc. Roy. Soc. (London), Ser. A*, *99*, 153–162 (1921).
(1228) L. Riedel, *Z. Ges. Kälte-Ind.*, *47*, 87 (1940).
(1229) L. Riedel, *Bull. Intern. Inst. Refrig.*, *22* (3), Annex No. C4, 1–3 (1941).
(1230) W. H. Rinkenbach, *Ind. Eng. Chem.*, *18*, 1195–1197 (1926).
(1231) W. H. Rinkenbach, *J. Am. Chem. Soc.*, *52*, 115–120 (1930).
(1232) O. Risgin and R. C. Taylor, *Spectrochim. Acta*, *15*, 1036–1050 (1959).
(1233) H. L. Ritter and J. H. Simons, *J. Am. Chem. Soc.*, *67*, 757–762 (1945).
(1234) D. E. Roberts and R. S. Jessup, *J. Res. Natl. Bur. Std.*, *40*, 281–283 (1948).
(1235) G. W. Robinson, *J. Chem. Phys.*, *21*, 1741–1745 (1953).
(1236) J. D. Rockenfeller and F. D. Rossini, *J. Phys. Chem.*, *65*, 267–272 (1961).
(1237) E. Rosenthal and B. P. Dailey, *J. Chem. Phys.*, *43*, 2093–2110 (1965).
(1238) F. D. Rossini, *J. Res. Natl. Bur. Std.*, *6*, 1–35 (1931).
(1239) F. D. Rossini, *J. Res. Natl. Bur. Std.*, *6*, 37–49 (1931).
(1240) F. D. Rossini, *J. Res. Natl. Bur. Std.*, *8*, 119–139 (1932).
(1241) F. D. Rossini, *J. Res. Natl. Bur. Std.*, *12*, 735–750 (1934).

(1242) F. D. Rossini, *J. Res. Natl. Bur. Std., 13,* 189–202 (1934).
(1243) F. D. Rossini, *Ind. Eng. Chem., 29,* 1424–1430 (1937).
(1244) F. D. Rossini, *J. Res. Natl. Bur. Std., 22,* 407–414 (1939).
(1245) F. D. Rossini, *Chemical Thermodynamics,* Wiley, New York, 1950.
(1246) F. D. Rossini, *Chemical Thermodynamics; Fractionating Processes; Hydrocarbons from Petroleum (Reilly Lectures,* Vol. III, 1949), University of Notre Dame, Notre Dame, Ind. 1950.
(1247) F. D. Rossini (Ed.), *Experimental Thermochemistry,* Vol. I. Interscience Publishers, New York, 1956.
(1248) F. D. Rossini, K. S. Pitzer, R. L. Arnett, R. M. Braun, and G. C. Pimentel, *Selected Values of Physical and Thermodynamic Properties of Hydrocarbons and Related Compounds,* Carnegie Press, Pittsburgh, 1953.
(1249) F. D. Rossini, D. D. Wagman, W. H. Evans, S. Levine, and I. Jaffe, *Selected Values of Chemical Thermodynamic Properties, (National Bureau of Standards Circular No. 500),* U.S. Government Printing Office, Washington, D.C., 1952.
(1250) F. D. Rossini and co-workers, *Selected Values of Properties of Chemical Compounds,* Manufacturing Chemists' Association Research Project, Semiannual, looseleaf.
(1251) W. A. Roth, *Ber., 77B,* 535–537 (1944).
(1252) W. A. Roth and K. Isecke, *Ber., 77B,* 537–539 (1944).
(1253) W. A. Roth and F. Müller, *Ber., 62B,* 1188–1194 (1929).
(1254) W. A. Roth and E. Rist-Schumacher, *Z. Elektrochem., 50,* 7–9 (1944).
(1255) P. Rothemund and H. Beyer, *Ann., 492,* 292–299 (1932).
(1256) J. S. Rowlinson, *Nature, 162,* 820–821 (1948).
(1257) J. S. Rowlinson, *Liquids and Liquid Mixtures,* Butterworths, London, 1959.
(1258) A. M. Rozhnov and D. N. Andreevskii, *Dokl. Akad. Nauk SSSR, 147,* 388–391 (1962).
(1259) A. M. Rozhnov and D. N. Andreevskii, *Neftekhimiya, 4,* 111–118 (1964).
(1260) T. R. Rubin and W. F. Giauque, *J. Am. Chem. Soc., 74,* 800–804 (1952).
(1261) T. R. Rubin, B. H. Levedahl, and D. M. Yost, *J. Am. Chem. Soc., 66,* 279–282 (1944).
(1262) D. M. Rudkovskii, A. G. Trifel, and A. V. Frost, *Ukr. Khem. Zh., 10,* 277–282 (1935).
(1263) R. A. Ruehrwein and W. F. Giauque, *J. Am. Chem. Soc., 61,* 2940–2944 (1939).
(1264) R. A. Ruehrwein and H. M. Huffman, *J. Am. Chem. Soc., 65,* 1620–1625 (1943).
(1265) R. A. Ruehrwein and H. M. Huffman, *J. Am. Chem. Soc., 68,* 1759–1761 (1946).
(1266) R. A. Ruehrwein and T. M. Powell, *J. Am. Chem. Soc., 68,* 1063–1066 (1946).
(1267) O. Ruff and S.-C. Li, *Z. Anorg. Allg. Chem., 242,* 272–276 (1939).
(1268) L. H. Ruiter, *Rec. Trav. Chim., 74,* 1467–1481 (1955).
(1269) F. Runge and P. Maass, *Chem. Tech. (Berlin), 5,* 421–424 (1953).
(1270) H. Russell, Jr., D. R. V. Golding, and D. M. Yost, *J. Am. Chem. Soc., 66,* 16–20 (1944).
(1271) H. Russell, Jr., D. W. Osborne, and D. M. Yost, *J. Am. Chem. Soc., 64,* 165–169 (1942).
(1272) L. Ruzicka and P. Schläpfer, *Helv. Chim. Acta, 16,* 162–168 (1933).
(1273) O. Sackur, *Ann. Phys., 36,* 958–980 (1911); *40,* 67–86 (1913).
(1274) G. Sage and W. Klemperer, *J. Chem. Phys., 39,* 371–376 (1963).
(1275) D. J. Salley and J. B. Gray, *J. Am. Chem. Soc., 70,* 2650–2653 (1948).
(1276) T. N. Sarachman, *J. Chem. Phys., 39,* 469–473 (1963).
(1277) K. V. L. N. Sastry and R. F. Curl, Jr., *J. Chem. Phys., 41,* 77–80 (1964).
(1278) K. Sathianandan, K. Ramaswamy, and F. F. Cleveland, *J. Mol. Spectrosc., 8,* 470–474 (1962).

References

(1279) G. Saville and H. A. Gundry, *Trans. Faraday Soc.*, 55, 2036–2038 (1959).
(1280) F. E. C. Scheffer and M. Voogd, *Rec. Trav. Chim.*, 45, 214–223 (1926).
(1281) J. R. Scherer and J. C. Evans, *Spectrochim. Acta*, 19, 1739–1775 (1963).
(1282) H. L. Schick, *Thermodynamics of Certain Refractory Compounds*, Vol. 1, Academic Press, New York, 1966.
(1283) H. L. Schick, *Thermodynamics of Certain Refractory Compounds*, Vol. 2, Academic Press, New York, 1966.
(1284) E. Schjånberg, Dissertation, University of Lund, Sweden, 1934.
(1285) E. Schjånberg, *Z. Phys. Chem. (Leipzig)*, A172, 197–233 (1935).
(1286) E. Schjånberg, *Z. Phys. Chem. (Leipzig)*, A175, 342–346 (1936).
(1287) E. Schjånberg, *Z. Phys. Chem. (Leipzig)*, A178, 274–281 (1937).
(1288) E. Schjånberg, *Z. Phys. Chem. (Leipzig)*, A181, 430–440 (1938).
(1289) P. von R. Schleyer, personal communication.
(1290) R. W. Schmid, E. Kloster-Jensen, E. Kovats, and E. Heilbronner, *Helv. Chim. Acta*, 39, 806–812 (1956).
(1291) W. Schottky and C. Wagner, cited by W. Schottky, H. Ulich, and C. Wagner, *Thermodynamik*, Springer-Verlag, Berlin, 1929, p. 377.
(1292) E. Schrödinger, *Statistical Thermodynamics*, 2nd Ed., Cambridge University Press, New York, 1952.
(1293) G.-M. Schwab, *Handbuch der Katalyse*, Vol. IV, Springer-Verlag, Vienna, 1943.
(1294) K. Schwabe, *Z. Elektrochem.*, 60, 151–157 (1956).
(1295) K. Schwabe and W. Wagner, *Z. Elektrochem.*, 65, 812–814 (1961).
(1296) R. H. Schwendeman and G. D. Jacobs, *J. Chem. Phys.*, 36, 1245–1250 (1962).
(1297) R. H. Schwendeman and F. L. Tobiason, *J. Chem. Phys.*, 43, 201–205 (1965).
(1298) D. W. Scott and G. A. Crowder, *J. Chem. Phys.*, 46, 1054–1062 (1967).
(1299) D. W. Scott, D. R. Douslin, H. L. Finke, W. N. Hubbard, J. F. Messerly, I. A. Hossenlopp, and J. P. McCullough, *J. Phys. Chem.*, 66, 1334–1341 (1962).
(1300) D. W. Scott, D. R. Douslin, M. E. Gross, G. D. Oliver, and H. M. Huffman, *J. Am. Chem. Soc.*, 74, 883–887 (1952).
(1301) D. W. Scott, D. R. Douslin, J. F. Messerly, S. S. Todd, I. A. Hossenlopp, T. C. Kincheloe, and J. P. McCullough, *J. Am. Chem. Soc.*, 81, 1015–1020 (1959).
(1302) D. W. Scott, H. L. Finke, M. E. Gross, G. B. Guthrie, and H. M. Huffman, *J. Am. Chem. Soc.*, 72, 2424–2430 (1950).
(1303) D. W. Scott, H. L. Finke, W. N. Hubbard, J. P. McCullough, M. E. Gross, K. D. Williamson, G. Waddington, and H. M. Huffman, *J. Am. Chem. Soc.*, 72, 4664–4668 (1950).
(1304) D. W. Scott, H. L. Finke, W. N. Hubbard, J. P. McCullough, C. Katz, M. E. Gross, J. F. Messerly, R. E. Pennington, and G. Waddington, *J. Am. Chem. Soc.*, 75, 2795–2800 (1953).
(1305) D. W. Scott, H. L. Finke, W. N. Hubbard, J. P. McCullough, G. D. Oliver, M. E. Gross, C. Katz, K. D. Williamson, G. Waddington, and H. M. Huffman, *J. Am. Chem. Soc.*, 74, 4656–4662 (1952).
(1306) D. W. Scott, H. L. Finke, J. P. McCullough, M. E. Gross, J. F. Messerly, R. E. Pennington, and G. Waddington, *J. Am. Chem. Soc.*, 77, 4993–4998 (1955).
(1307) D. W. Scott, H. L. Finke, J. P. McCullough, M. E. Gross, R. E. Pennington, and G. Waddington, *J. Am. Chem. Soc.*, 74, 2478–2483 (1952).
(1308) D. W. Scott, H. L. Finke, J. P. McCullough, M. E. Gross, K. D. Williamson, G. Waddington, and H. M. Huffman, *J. Am. Chem. Soc.*, 73, 261–265 (1951).
(1309) D. W. Scott, H. L. Finke, J. P. McCullough, J. F. Messerly, R. E. Pennington, I. A. Hossenlopp, and G. Waddington, *J. Am. Chem. Soc.*, 79, 1062–1068 (1957).
(1310) D. W. Scott, W. D. Good, G. B. Guthrie, S. S. Todd, I. A. Hossenlopp, A. G. Osborn, and J. P. McCullough, *J. Phys. Chem.*, 67, 685–689 (1963).

(1311) D. W. Scott, W. D. Good, S. S. Todd, J. F. Messerly, W. T. Berg, I. A. Hossenlopp, J. L. Lacina, A. Osborn, and J. P. McCullough, *J. Chem. Phys. 36*, 406–412 (1962).
(1312) D. W. Scott, M. E. Gross, G. D. Oliver, and H. M. Huffman, *J. Am. Chem. Soc., 71*, 1634–1636 (1949).
(1313) D. W. Scott, G. B. Guthrie, J. P. McCullough, and G. Waddington, *J. Chem. Eng. Data, 4*, 246–251 (1959).
(1314) D. W. Scott, G. B. Guthrie, J. F. Messerly, S. S. Todd, W. T. Berg, I. A. Hossenlopp, and J. P. McCullough, *J. Phys. Chem., 66*, 911–914 (1962).
(1315) D. W. Scott, W. N. Hubbard, J. F. Messerly, S. S. Todd, I. A. Hossenlopp, W. D. Good, D. R. Douslin, and J. P. McCullough, *J. Phys. Chem., 67*, 680–685 (1963).
(1316) D. W. Scott and J. P. McCullough, *U.S. Bur. Mines Bull., 595* (1961).
(1317) D. W. Scott and J. P. McCullough, U.S. Bur. Mines, Rept. Invest., *5930*, 27 pp. (1962).
(1318) D. W. Scott, J. P. McCullough, W. D. Good, J. F. Messerly, R. E. Pennington, T. C. Kincheloe, I. A. Hossenlopp, D. R. Douslin, and G. Waddington, *J. Am. Chem. Soc., 78*, 5457–5463 (1956).
(1319) D. W. Scott, J. P. McCullough, W. N. Hubbard, J. F. Messerly, I. A. Hossenlopp, F. R. Frow, and G. Waddington, *J. Am. Chem. Soc., 78*, 5463–5468 (1956).
(1320) D. W. Scott, J. P. McCullough, J. F. Messerly, R. E. Pennington, I. A. Hossenlopp, H. L. Finke, and G. Waddington, *J. Am. Chem. Soc., 80*, 55–59 (1958).
(1321) D. W. Scott, J. F. Messerly, S. S. Todd, I. A. Hossenlopp, D. R. Douslin, and J. P. McCullough, *J. Chem. Phys., 37*, 867–873 (1962).
(1322) D. W. Scott, J. F. Messerly, S. S. Todd, I. A. Hossenlopp, A. Osborn, and J. P. McCullough, *J. Chem. Phys., 38*, 532–539 (1963).
(1323) D. W. Scott, G. D. Oliver, M. E. Gross, W. N. Hubbard, and H. M. Huffman, *J. Am. Chem. Soc., 71*, 2293–2297 (1949).
(1324) D. W. Scott, G. Waddington, J. C. Smith, and H. M. Huffman, *J. Am. Chem. Soc., 71*, 2767–2773 (1949).
(1325) R. B. Scott and F. G. Brickwedde, *J. Res. Natl. Bur. Std., 35*, 501–512 (1945).
(1326) R. B. Scott, W. J. Ferguson, and F. G. Brickwedde, *J. Res. Natl. Bur. Std., 33*, 1–20 (1944).
(1327) R. B. Scott, C. H. Meyers, R. D. Rands, Jr., F. G. Brickwedde, and N. Bekkedahl, *J. Res. Natl. Bur. Std., 35*, 39–85 (1945).
(1328) G. W. Sears and E. R. Hopke, *J. Am. Chem. Soc., 71*, 1632–1634 (1949).
(1329) Z. Seha, *Chem. Listy, 49*, 1569–1570 (1955).
(1330) S. Sekino and T. Nishikawa, *J. Phys. Soc. Japan, 12*, 43–48 (1957).
(1331) P. Sellers and S. Sunner, *Acta Chem. Scand., 16*, 46–52 (1962).
(1332) G. Semerano and A. Chisini, *Gazz. Chim. Ital., 66*, 510–518 (1936).
(1333) E. A. Seregin, N. N. Goroshko, V. P. Kolesov, N. A. Belikova, S. M. Skuratov, and A. F. Plate, *Dokl. Akad. Nauk SSSR, 159*, 1381–1384 (1964).
(1334) E. A. Seregin, V. P. Kolesov, N. A. Belikova, S. M. Skuratov, and A. F. Plate, *Dokl. Akad. Nauk SSSR, 145*, 580–583 (1962).
(1335) K. T. Serijan and P. H. Wise, *J. Am. Chem. Soc., 73*, 4766–4769 (1951).
(1336) K. T. Serijan and P. H. Wise, *J. Am. Chem. Soc., 73*, 5191–5193 (1951).
(1337) K. T. Serijan and P. H. Wise, *J. Am. Chem. Soc., 74*, 365–368 (1952).
(1338) K. T. Serijan, P. H. Wise, and L. C. Gibbons, *J. Am. Chem. Soc., 71*, 2265–2266 (1949).
(1339) W. F. Seyer and C. W. Mann, *J. Am. Chem. Soc., 67*, 328–329 (1945).
(1340) A. N. Shchukarev and L. A. Shchukareva, *Russ. J. Phys. Chem., 3*, 169–174 (1932).

(1341) N. Sheppard, *J. Chem. Phys.*, *17*, 79–83 (1949).
(1342) N. Sheppard, *Trans. Faraday Soc.*, *45*, 693–697 (1949).
(1343) N. Sheppard, *Trans. Faraday Soc.*, *46*, 527–533 (1950).
(1344) N. Sheppard, *Trans. Faraday Soc.*, *46*, 533–539 (1950).
(1345) T. Shimanouchi, *J. Chem. Soc. Japan*, *62*, 1264–1269 (1941).
(1346) K. Shimoda, T. Nishikawa, and T. Itoh, *J. Phys. Soc. Japan*, *9*, 974–991 (1954).
(1347) D. A. Shirley and W. F. Giauque, *J. Am. Chem. Soc.*, *81*, 4778–4779 (1959).
(1348) F. K. Signaigo (to E. I. du Pont de Nemours & Co.), U.S. Patent 2,406,410 (August 27, 1946).
(1349) G. Silvestro and C. Lenchitz, *J. Phys. Chem.*, *65*, 694–695 (1961).
(1350) F. E. Simon, *Kältetechnik*, *9*, 58–61, 95–100 (1957).
(1351) M. Simonetta, *Chim. Ind. (Milan)*, *29*, 37–39 (1947).
(1352) A. D. Singh and N. W. Krase, *Ind. Eng. Chem.*, *27*, 909–14 (1935).
(1353) G. C. Sinke, *J. Phys. Chem.*, *63*, 2063 (1959).
(1354) G. C. Sinke, unpublished results, Dow Chemical Company.
(1355) G. C. Sinke and T. De Vries, *J. Am. Chem. Soc.*, *75*, 1815–1818 (1953).
(1356) G. C. Sinke and D. L. Hildenbrand, *J. Chem. Eng. Data*, *7*, 74 (1962).
(1357) G. C. Sinke, D. L. Hildenbrand, R. A. McDonald, W. R. Kramer, and D. R. Stull, *J. Phys. Chem.*, *62*, 1461–1462 (1958).
(1358) G. C. Sinke and F. L. Oetting, *J. Phys. Chem.*, *68*, 1354–1358 (1964).
(1359) G. C. Sinke and D. R. Stull, *J. Phys. Chem.*, *62*, 397–401 (1958).
(1360) K. M. Sinnott, *J. Chem. Phys.*, *34*, 851–861 (1961).
(1361) H. A. Skinner, *J. Chem. Soc.*, 4396–4408 (1962).
(1362) H. A. Skinner, Ed., *Experimental Thermochemistry*, Vol. II. Interscience Publishers, New York, 1962.
(1361) H. A. Skinner and G. Pilcher, *Quart. Rev. (London)*, *17*, 264–288 (1963).
(1364) H. A. Skinner and A. Snelson, *Trans. Faraday Soc.*, *55*, 404–407 (1959).
(1365) H. A. Skinner and A. Snelson, *Trans. Faraday Soc.*, *56*, 1776–1783 (1960).
(1366) A. Skita and W. Faust, *Ber.*, *64B*, 2878–2892 (1931).
(1367) A. Skita and W. Faust, *Ber.*, *72B*, 1127–1138 (1939).
(1368) A. Skita and R. Rössler, *Ber.*, *72B*, 265–272 (1939).
(1369) A. Skupinski, *Intern. Chem. Eng.*, *5*, 323–330 (1965).
(1370) S. M. Skuratov and M. P. Kozina, *Dokl. Akad. Nauk SSSR*, *122*, 109–110 (1958); *Proc. Acad. Sci. USSR*, *122*, 643–644 (1958).
(1371) S. M. Skuratov and S. M. Shtekher, *Khim. Nauka i Promysl*, *3*, 688 (1958).
(1372) V. A. Slabey, *J. Am. Chem. Soc.*, *74*, 4930–4932 (1952).
(1373) V. A. Slabey, *J. Am. Chem. Soc.*, *74*, 4963–4964 (1952).
(1374) V. A. Slabey and P. H. Wise, *J. Am. Chem. Soc.*, *74*, 1473–1476 (1952).
(1375) V. A. Slabey and P. H. Wise, *J. Am. Chem. Soc.*, *74*, 3887–3889 (1952).
(1376) D. C. Smith, G. M. Brown, J. R. Nielsen, R. M. Smith, and C. Y. Liang, *J. Chem. Phys.*, *20*, 473–486 (1952).
(1377) D. C. Smith, E. E. Ferguson, R. L. Hudson, and J. R. Nielsen, *J. Chem. Phys.*, *21*, 1475–1479 (1953).
(1378) D. C. Smith, J. R. Nielsen, and H. H. Claassen, *J. Chem. Phys.*, *18*, 326–331 (1950).
(1379) D. C. Smith, C. Y. Pan, and J. R. Nielsen, *J. Chem. Phys.*, *18*, 706–712 (1950).
(1380) D. C. Smith, R. A. Saunders, J. R. Nielsen, and E. E. Ferguson, *J. Chem. Phys.*, *20*, 847–859 (1952).
(1381) J. M. Smith, *Trans. Am. Inst. Chem. Eng.*, *42*, 983–988 (1946).
(1382) L. Smith, *Acta Chem. Scand.*, *10*, 884–886 (1956).
(1383) L. Smith and L. Bjellerup, *Acta Chem. Scand.*, *1*, 566–570 (1947).

(1384) L. Smith, L. Bjellerup, S. Krook, and H. Westermark, *Acta Chem. Scand.*, 7, 65–86 (1953).
(1385) L. Smith and S. Sunner, "Verbrennungen von Chlor-und Bromhaltigen Substanzen in Beweglicher Bombe," in A. Tiselius and K. O. Pedersen (Eds.), *The Svedberg*, Almquist and Wiksells, Uppsala, Sweden, pp. 352–369 (1944).
(1386) N. K. Smith, G. Gorin, W. D. Good, and J. P. McCullough, *J. Phys. Chem.*, 68, 940–946 (1964).
(1387) N. K. Smith, D. W. Scott, and J. P. McCullough, *J. Phys. Chem.*, 68, 934–939 (1964).
(1388) R. H. Smith and D. H. Andrews, *J. Am. Chem. Soc.*, 53, 3644–3660 (1931).
(1389) W. L. Smith and I. M. Mills, *J. Mol. Spectrosc.*, 11, 11–38 (1963).
(1390) E. J. Smutny and A. Bondi, *J. Phys. Chem.*, 65, 546–550 (1961).
(1391) A. Snelson and H. A. Skinner, *Trans. Faraday Soc.*, 57, 2125–2131 (1961).
(1392) N. Solimene and B. P. Dailey, *J. Chem. Phys.*, 22, 2042–2044 (1954).
(1393) N. Solimene and B. P. Dailey, *J. Chem. Phys.*, 23, 124–129 (1955).
(1394) G. R. Somayajulu, *J. Chem. Phys.*, 34, 1449–1450 (1961); *Univ. Calif. Rad. Lab. Rept. No. 9590* (1961).
(1395) G. R. Somayajulu, A. P. Kudchadker, and B. J. Zwolinski, *Ann. Rev. Phys. Chem.*, 16, 213–244 (1965).
(1396) M. Souders, Jr., C. S. Matthews, and C. O. Hurd, *Ind. Eng. Chem.*, 41, 1037–1048 (1949).
(1397) M. Souders, Jr., C. S. Matthews, and C. O. Hurd, *Ind. Eng. Chem.*, 41, 1048–1056 (1949).
(1398) J. C. Southard and F. G. Brickwedde, *J. Am. Chem. Soc.*, 55, 4378–4384 (1933).
(1399) J. C. Southard, R. T. Milner, and S. B. Hendricks, *J. Chem. Phys.*, 1, 95–102 (1933).
(1400) M. E. Spaght, S. B. Thomas, G. S. Parks, *J. Phys. Chem.*, 36, 882–888 (1932).
(1401) D. M. Speros and F. D. Rossini, *J. Phys. Chem.*, 64, 1723–1727 (1960).
(1402) H. D. Springall and T. R. White, *Research*, 2, 296 (1949).
(1403) H. D. Springall and T. R. White, *J. Chem. Soc.*, 2764–2766 (1954).
(1404) R. F. Stamm, F. Halverson, and J. J. Whalen, *J. Chem. Phys.*, 17, 104–105 (1949).
(1405) H. Stammreich and R. Forneris, *Spectrochim. Acta*, 8, 52–53 (1956).
(1406) E. C. Stathis and A. C. Egerton, *Trans. Faraday Soc.*, 36, 606 (1940).
(1407) L. A. K. Staveley, W. I. Tupman, and K. R. Hart, *Trans. Faraday Soc.*, 51, 323–343 (1955).
(1408) L. A. K. Staveley, J. B. Warren, H. P. Paget, and D. J. Dowrick, *J. Chem. Soc.*, 1992–2001 (1954).
(1409) D. Steele and D. H. Whiffen, *Trans. Faraday Soc.*, 55, 369–376 (1959).
(1410) L. E. Steiner, *Introduction to Chemical Thermodynamics*, 2nd Ed., McGraw-Hill, New York, 1948.
(1411) C. C. Stephenson and D. J. Berets, *J. Am. Chem. Soc.*, 74, 882–883 (1952).
(1412) C. V. Stephenson and E. A. Jones, *J. Chem. Phys.*, 20, 135–136 (1952).
(1413) A. Stern and G. Klebs, *Ann.*, 500, 91–108 (1932).
(1414) A. Stern and G. Klebs, *Ann.*, 504, 287–297 (1933).
(1415) A. Stern and G. Klebs, *Ann.*, 505, 295–306 (1933).
(1416) J. H. Stern and F. H. Dorer, *J. Phys. Chem.*, 66, 97–99 (1962).
(1417) R. K. Steunenberg and R. C. Vogel, *J. Am. Chem. Soc.*, 78, 901–902 (1956).
(1418) D. P. Stevenson, *J. Chem. Phys.*, 7, 171–174 (1939).
(1419) D. P. Stevenson and J. Y. Beach, *J. Chem. Phys.*, 6, 25–29, 108–110, 341 (1938).
(1420) J. E. Stewart, *J. Chem. Phys.*, 30, 1259–1265 (1959).

(1421) R. D. Stiehler and H. M. Huffman, *J. Am. Chem. Soc., 57,* 1734–1740 (1935).
(1422) R. D. Stiehler and H. M. Huffman, *J. Am. Chem. Soc., 57,* 1741–1743 (1935).
(1423) H. F. Stimson, *J. Res. Natl. Bur. Std., 42,* 209–217 (1949).
(1424) H. F. Stimson, *Am. J. Phys., 23,* 614–622 (1955).
(1425) F. Stitt, *J. Chem. Phys., 7,* 1115 (1939).
(1426) A. Stojiljkovic and D. H. Whiffen, *Spectrochim. Acta, 12,* 47–56 (1958).
(1427) J. W. Stout, "Isothermal Low-Temperature Calorimetry," in J. P. McCullough and D. W. Scott (Eds.), *Experimental Thermodynamics,* Vol. I, Butterworths, London, 1966, Chapter 5.
(1428) J. W. Stout and L. H. Fisher, *J. Chem. Phys., 9,* 163–168 (1941).
(1429) F. S. Stow, Jr., and J. H. Elliott, *Anal. Chem., 20,* 250–253 (1948).
(1430) H. H. Stroh and C. R. Fincke, *Z. Chem., 3,* 265–266 (1963).
(1431) D. R. Stull, *J. Am. Chem. Soc., 59,* 2726–2733 (1937).
(1432) D. R. Stull, *Ind. Eng. Chem., 39,* 517–550 (1947).
(1433) D. R. Stull, in R. H. Boundy and R. F. Boyer (Eds.), *Styrene: Its Polymers, Copolymers and Derivatives,* Reinhold, New York, 1952, Chapter 3.
(1434) D. R. Stull, "Thermodynamic Aspects," in G. A. Olah (Ed.), *Friedel-Crafts and Related Reactions,* Vol. I, Interscience Publishers, New York, 1963, pp. 937–997, Chapter 12.
(1435) D. R. Stull, I. Carr, J. Chao, T. E. Dergazarian, L. A. du Plessis, R. E. Jostad, S. Levine, F. L. Oetting, R. V. Petrella, H. Prophet, and G. C. Sinke, *JANAF Thermochemical Tables,* Clearinghouse for Federal Scientific and Technical Information, Springfield, Va., 1966.
(1436) D. R. Stull and F. D. Mayfield, *Ind. Eng. Chem., 35,* 639–645 (1943).
(1437) D. R. Stull and G. C. Sinke, *Thermodynamic Properties of the Elements (Advances in Chemistry Series No. 18),* American Chemical Society, Washington, D.C., 1956.
(1438) D. R. Stull, G. C. Sinke, R. A. McDonald, W. E. Hatton, and D. L. Hildenbrand, *Pure Appl. Chem., 2,* 315–322 (1961).
(1439) E. Suito and H. Aida, *J. Chem. Soc. Japan, Ind. Chem. Sect., 54,* 765–768 (1951).
(1440) M. V. Sullivan and H. Hunt, *J. Phys. Colloid Chem., 53,* 497–500 (1949).
(1441) S. Sunner, *Svensk Kem. Tidskr., 58,* 71–81 (1946).
(1442) S. Sunner, Thesis, University of Lund, Sweden, 1949.
(1443) S. Sunner, *Acta Chem. Scand., 9,* 837–846 (1955).
(1444) S. Sunner, *Acta Chem. Scand., 9,* 847–854 (1955).
(1445) S. Sunner, *Acta Chem. Scand., 11,* 1757–1760 (1957).
(1446) S. Sunner, *Acta Chem. Scand., 11,* 1766–1770 (1957).
(1447) S. Sunner, *Acta Chem. Scand., 17,* 728–730 (1963).
(1448) G. B. B. M. Sutherland, *Proc. Indian Acad. Sci., Sect. A, 8,* 341–344 (1938).
(1449) L. E. Sutton (Ed.), *Interatomic Distances (Special Publication No. 11),* Chemical Society, London, 1958.
(1450) K. Suzuki, S. Onishi, T. Koide, and S. Seki, *Bull. Chem. Soc. Japan, 29,* 127–131 (1956).
(1451) A. S. Sverdlin, *Zh. Fiz. Khim., 28,* 780–784 (1954).
(1452) A. S. Sverdlin, *Zh. Fiz. Khim., 32,* 659–665 (1958).
(1453) A. S. Sverdlin and I. N. Godnev, *Zh. Fiz. Khim., 27,* 1580–1585 (1953).
(1454) H. A. Swain, Jr., L. S. Silbert, and J. G. Miller, *J. Am. Chem. Soc., 86,* 2562–2566 (1964).
(1455) J. D. Swalen, *J. Chem. Phys., 23,* 1739–1740 (1955).
(1456) J. D. Swalen and C. C. Costain, *J. Chem. Phys., 31,* 1562–1574 (1959).

(1457) J. D. Swalen and D. R. Herschbach, *J. Chem. Phys.*, *27*, 100–108 (1957).
(1458) F. Swarts, *J. Chim. Phys.*, *17*, 3–70 (1919).
(1459) D. A. Swick and I. L. Karle, *J. Chem. Phys.*, *23*, 1499–1504 (1955).
(1460) D. A. Swick, I. L. Karle, and J. Karle, *J. Chem. Phys.*, *22*, 1242–1245 (1954).
(1461) W. Swietoslawski and J. Bobínska, *Bull. Intern. Acad. Polon. [A]*, 621–630 (1929).
(1462) M. Szwarc, *Discussions Faraday Soc.*, *10*, 336–338 (1951).
(1463) W. J. Tabor, *J. Chem. Phys.*, *27*, 974–975 (1957).
(1464) R. W. Taft, Jr., and P. Riesz, *J. Am. Chem. Soc.*, *77*, 902–904 (1955).
(1465) T. Tanaka and T. Watase, *Technol. Rept. Osaka Univ.*, *6*, 367–371 (1956).
(1466) E. Tannenbaum, R. D. Johnson, R. J. Myers, and W. D. Gwinn, *J. Chem. Phys.*, *22*, 949 (1954).
(1467) G. Tappi, *Gazz. Chim. Ital.*, *70*, 414–422 (1940).
(1468) V. M. Tatevskii, V. A. Benderskii, and S. S. Yarovoi, *Rules and Methods for Calculating the Physico-Chemical Properties of Paraffinic Hydrocarbons* (Translated by B. P. Mullins), Pergamon Press, Oxford, 1961.
(1469) V. M. Tatevskii and A. V. Frost, *Vestn. Moskov. Univ.*, *No. 3*, 65–83 (1947).
(1470) V. M. Tatevskii and Y. G. Papulov, *Russ. J. Phys. Chem.*, *34*, 115–123, 231–238, 335–338 (1960); *36*, 93–100 (1962).
(1471) P. Tavernier and M. Lamouroux, *Mem. Poudres*, *37*, 197–206 (1955).
(1472) P. Tavernier and M. Lamouroux, *Mem. Poudres*, *38*, 65–88 (1956).
(1473) A. H. Taylor, Jr., and R. H. Crist, *J. Am. Chem. Soc.*, *63*, 1377–1385 (1941).
(1474) J. Taylor and C. R. L. Hall, *J. Phys. Colloid Chem.*, *51*, 593–611 (1947).
(1475) R. D. Taylor, B. H. Johnson, and J. E. Kilpatrick, *J. Chem. Phys.*, *23*, 1225–1231 (1955).
(1476) R. D. Taylor and J. E. Kilpatrick, *J. Chem. Phys.*, *23*, 1232–1235 (1955).
(1477) W. J. Taylor, D. D. Wagman, M. G. Williams, K. S. Pitzer, and F. D. Rossini, *J. Res. Natl. Bur. Std.*, *37*, 95–122 (1946).
(1478) H. Teranishi and S. W. Benson, *J. Am. Chem. Soc.*, *85*, 2890–2892 (1963).
(1479) E. Terres and H. Wesemann, *Angew. Chem.*, *45*, 795–802 (1932).
(1480) H. Tetrode, *Ann. Phys.*, *38*, 434–442 (1912).
(1481) R. Theimer and J. R. Nielsen, *J. Chem. Phys.*, *27*, 264–268 (1957).
(1482) R. Theimer and J. R. Nielsen, *J. Chem. Phys.*, *27*, 887–890 (1957).
(1483) R. Theimer and J. R. Nielsen, *J. Chem. Phys.*, *30*, 98–102 (1959).
(1484) L. F. Thomas, J. S. Heeks, and J. Sheridan, *Z. Elektrochem.*, *61*, 935–937 (1957).
(1485) L. F. Thomas, E. I. Sherrard, and J. Sheridan, *Trans. Faraday Soc.*, *51*, 619–625 (1955).
(1486) W. J. Thomas and S. Portalski, *Ind. Eng. Chem.*, *50*, 967–970 (1958).
(1487) H. W. Thompson, *Trans. Faraday Soc.*, *37*, 251–260 (1941).
(1488) H. W. Thompson, *Trans. Faraday Soc.*, *37*, 344–352 (1941).
(1489) H. W. Thompson and W. T. Cave, *Trans. Faraday Soc.*, *47*, 951–957 (1951).
(1490) H. W. Thompson and F. S. Dainton, *Trans. Faraday Soc.*, *33*, 1546–1555 (1937).
(1491) H. W. Thompson and C. H. Miller, *Trans. Faraday Soc.*, *46*, 22–27 (1950).
(1492) H. W. Thompson and N. P. Skerrett, *Trans. Faraday Soc.*, *36*, 812–817 (1940).
(1493) H. W. Thompson and R. B. Temple, *J. Chem. Soc.*, 1428–1432 (1948).
(1494) H. W. Thompson and P. Torkington, *Proc. Roy. Soc. (London), Ser. A*, 21–41 (1945).
(1495) J. Thomsen, *Thermochemische Untersuchungen*, Barth, Leipzig, 1882–1886; *Thermochemistry* (Translated by K. A. Burke), Longmans, Green, London, 1908.
(1496) G. W. Thomson, *Chem. Rev.*, *38*, 1–39 (1946).
(1497) H. M. Thorne, W. I. R. Murphy, J. S. Ball, K. E. Stanfield, and J. W. Horne, *Ind. Eng. Chem.*, *43*, 20–27 (1951).

(1498) C. G. Thornton, *Univ. Microfilms Publ. No. 7746*, 78 pp.; *Dissertation Abstr., 14,* 604 (1954).
(1499) J. Timmermans, *Bull. Soc. Chim. Belges, 43,* 626–638 (1934).
(1500) J. Timmermans, *Bull. Soc. Chim. Belges, 44,* 17–40 (1935).
(1501) J. Timmermans, *Physico-Chemical Constants of Pure Organic Compounds,* Elsevier, New York, 1950.
(1502) J. Timmermans, *Bull. Soc. Chim. Belges, 61,* 393–402 (1952).
(1503) J. Timmermans, *Les Constantes Physiques des Composés Organiques Cristallisés,* Masson, Paris, 1953, p. 478.
(1504) J. Timmermans and Y. Delcourt, *J. Chim. Phys., 31,* 85–124 (1934).
(1505) J. Timmermans and H.-R. Roland, *J. Chim. Phys., 52,* 223–245 (1955).
(1506) J. Tjebbes, *Pure Appl. Chem., 2,* 129–132 (1961).
(1507) J. Tjebbes, *Acta Chem. Scand., 16,* 916–921 (1962).
(1508) J. Tjebbes, *Acta Chem. Scand., 16,* 953–957 (1962).
(1509) M. C. Tobin, *J. Am. Chem. Soc., 75,* 1788–1790 (1953).
(1510) M. C. Tobin, *Spectrochim. Acta, 16,* 1108–1110 (1960).
(1511) S. S. Todd, G. D. Oliver, and H. M. Huffman, *J. Am. Chem. Soc., 69,* 1519–1525 (1947).
(1512) S. S. Todd and G. S. Parks, *J. Am. Chem. Soc., 58,* 134–137 (1936).
(1513) R. C. Tolman, *The Principles of Statistical Thermodynamics,* Oxford University Press, New York, 1938.
(1514) L. K. J. Tong and W. O. Kenyon, *J. Am. Chem. Soc., 69,* 2245–2246 (1947).
(1515) E. E. Toops, *J. Phys. Chem., 60,* 304–306 (1956).
(1516) P. Torkington, *Trans. Faraday Soc., 46,* 894–900 (1950).
(1517) I. F. Trotter and H. W. Thompson, *J. Chem. Soc.,* 481–488 (1946).
(1518) F. Trouton, *Phil. Mag. [5], 18,* 54–57 (1884).
(1519) J. C. Trowbridge and E. F. Westrum, Jr., *J. Phys. Chem., 67,* 2381–2385 (1963).
(1520) E. Tschuikow-Roux, *J. Phys. Chem., 66,* 1636–1639 (1962).
(1521) E. Tschuikow-Roux, *J. Phys. Chem., 69,* 1075–1077 (1965).
(1522) R. B. Turner and R. H. Garner, *J. Am. Chem. Soc., 79,* 253 (1957).
(1523) T. E. Turner, V. C. Fiora, and W. M. Kendrick, *J. Chem. Phys., 23,* 1966 (1955).
(1524) K. Ueberreiter and H. J. Orthmann, *Z. Naturforsch., 5a,* 101–108 (1950).
(1525) T. Ukaji and R. A. Bonham, *J. Am. Chem. Soc., 84,* 3627–3630 (1962).
(1526) R. H. Valentine, G. E. Brodale, and W. F. Giauque, *J. Phys. Chem., 66,* 392–395 (1962).
(1527) W. N. Vanderkooi and T. De Vries, *J. Phys. Chem., 60,* 636–639 (1956).
(1528) D. W. Van Krevelen and H. A. G. Chermin, *Ingenieur (Holland), 62,* Ch65–74 (1950).
(1529) J. D. Vaughn and E. L. Muetterties, *J. Phys. Chem., 64,* 1787–1788 (1960).
(1530) V. I. Vedeneev, cited by N. N. Semenov, *Some Problems in Chemical Kinetics and Reactivity,* Vol. 1 (Translated by M. Boudart), Princeton University Press, Princeton, N. J., 1958, p. 55.
(1531) S. Venkateswaran, *Indian J. Phys., 5,* 219–236 (1930).
(1532) S. Venkateswaran, *Nature, 126,* 434 (1930).
(1533) P. Venkateswarlu, J. G. Baker, and W. Gordy, *J. Mol. Spectrosc., 6,* 215–228 (1961).
(1534) P. E. Verkade, *J. Chim. Phys., 29,* 297–301 (1932).
(1535) P. E. Verkade and J. Coops, Jr., *Rec. Trav. Chim., 52,* 747–767 (1933).
(1536) P. E. Verkade, J. Coops, Jr., C. J. Maan, and A. Verkade-Sandbergen, *Ann., 467,* 217–239 (1928).
(1537) P. E. Verkade, J. Coops, Jr., A. Verkade-Sandbergen, and C. J. Maan, *Ann., 477,* 279–297 (1930).

(1538) P. E. Verkade and H. Hartman, *Rec. Trav. Chim.*, *52*, 945–968 (1933).
(1539) R. D. Verma, *J. Chem. Phys.*, *32*, 738–749 (1960).
(1540) F. L. Voelz, F. F. Cleveland, A. G. Meister, and R. B. Bernstein, *J. Opt. Soc. Am.*, *43*, 1061–1064 (1953).
(1541) V. V. Voevodskii, *Dokl. Akad. Nauk SSSR*, *79*, 455–458 (1951).
(1542) A. I. Vogel and D. M. Cowan, *J. Chem. Soc.*, 16–24 (1943).
(1543) R. Vogel-Högler, *Acta Phys. Aust.*, *1*, 311–322 (1948).
(1544) S. P. Vohra and K. A. Kobe, *J. Chem. Eng. Data*, *4*, 329–330 (1959).
(1545) R. D. Vold, *J. Am. Chem. Soc.*, *57*, 1192–1195 (1935).
(1546) A. F. Vorob'ev and S. M. Skuratov, *Russ. J. Inorg. Chem.*, *5*, 679–681 (1960).
(1547) G. N. Vriens and A. G. Hill, *Ind. Eng. Chem.*, *44*, 2732–2735 (1952).
(1548) G. Waddington, J. W. Knowlton, D. W. Scott, G. D. Oliver, S. S. Todd, W. N. Hubbard, J. C. Smith, and H. M. Huffman, *J. Am. Chem. Soc.*, *71*, 797–808 (1949).
(1549) G. Waddington and H. A. Skinner, *Thermochem. Bull.* (March 1955).
(1550) G. Waddington, J. C. Smith, K. D. Williamson, and D. W. Scott, *J. Phys. Chem.*, *66*, 1074–1077 (1962).
(1551) G. Waddington, S. Sunner, and W. N. Hubbard, in F. D. Rossini (Ed.), *Experimental Thermochemistry*, Vol. I, Interscience Publishers, New York, 1956, Chapter 7.
(1552) G. Waddington, S. S. Todd, and H. M. Huffman, *J. Am. Chem. Soc.*, *69*, 22–30 (1947).
(1553) A. M. Wadsö and I. Wadsö, private communication, October 31, 1961.
(1554) I. Wadsö, *Acta Chem. Scand.*, *11*, 1745–1751 (1957).
(1555) I. Wadsö, *Acta Chem. Scand.*, *12*, 630–634 (1958).
(1556) I. Wadsö, *Acta Chem. Scand.*, *14*, 561–565 (1960).
(1557) I. Wadsö, *Acta Chem. Scand.*, *14*, 903–908 (1960).
(1558) I. Wadsö, *Acta Chem. Scand.*, *16*, 471–478 (1962).
(1559) D. D. Wagman and W. H. Evans, *Tables of Selected Values of Chemical Thermodynamic Properties*, Series III, U.S. National Bureau of Standards, Washington, D.C., 1947–1956, looseleaf.
(1560) D. D. Wagman, W. H. Evans, I. Halow, V. B. Parker, S. M. Bailey, and R. H. Schumm, "Selected Values of Chemical Thermodynamic Properties. Part 1. Tables for the First Twenty-Three Elements in the Standard Order of Arrangement," *Natl. Bur. Std. Tech. Note No. 270–1*, 1965.
(1561) D. D. Wagman, W. H. Evans, I. Halow, V. B. Parker, S. M. Bailey, and R. H. Schumm, "Selected Values of Chemical Thermodynamic Properties. Part 2. Tables for the Elements Twenty-Three Through Thirty-Two in the Standard Order of Arrangement," *Natl. Bur. Std. Tech. Note No. 270–2*, 1966.
(1562) D. D. Wagman, J. E. Kilpatrick, W. J. Taylor, K. S. Pitzer, and F. D. Rossini, *J. Res. Natl. Bur. Std.*, *34*, 143–161 (1945).
(1563) J. Wagner, *Z. Phys. Chem. (Leipzig)*, *B40*, 36–50 (1938).
(1564) J. Wagner, *Z. Phys. Chem. (Leipzig)*, *B40*, 439–449 (1938).
(1565) R. S. Wagner and B. P. Dailey, *J. Chem. Phys.*, *26*, 1588–1593 (1957).
(1566) R. S. Wagner, B. P. Dailey, and N. Solimene, *J. Chem. Phys.*, *26*, 1593–1596 (1957).
(1567) B. E. Walker, Jr., M. S. Brooks, C. T. Ewing, and R. R. Miller, *Chem. Eng. Data Ser.*, *3*, 280–282 (1958).
(1568) L. C. Walker, "NF_3 Calorimetry: The Enthalpy of Reaction of Cyanogen and Nitrogen Trifluoride," Presented at the 22nd Calorimetry Conference, Thousand Oaks, Calif., 1967.
(1569) P. N. Walsh and N. O. Smith, *J. Chem. Eng. Data*, *6*, 33–35 (1961).

(1570) R. Walsh, D. M. Golden, and S. W. Benson, *J. Am. Chem. Soc., 88,* 650–656 (1966).
(1571) W. Waring, *Chem. Rev., 51,* 171–183 (1952).
(1572) H. von Wartenberg, *Z. Anorg. Chem., 258,* 356–360 (1949).
(1573) H. von Wartenberg and B. Lerner-Steinberg, *Z. Angew. Chem., 38,* 591–592 (1925).
(1574) H. von Wartenberg and J. Schiefer, *Z. Anorg. Allg. Chem., 278,* 326–332 (1955).
(1575) H. von Wartenberg and R. Schütte, *Z. Anorg. Allg. Chem., 211,* 222–226 (1933).
(1576) H. von Wartenberg and H. Schütza, *Z. Phys. Chem. (Leipzig), A164,* 386–388 (1933).
(1577) E. W. Washburn (Ed.-in-Chief), *International Critical Tables of Numerical Data, Physics, Chemistry and Technology,* Vol. V, McGraw-Hill, New York, 1929.
(1578) E. W. Washburn, *J. Res. Natl. Bur. Std., 10,* 525–558 (1933).
(1579) A. Wasserman, *Z. Phys. Chem. (Leipzig), A151,* 113–128 (1930).
(1580) A. Wassermann, *J. Chem. Soc.,* 828–839 (1935).
(1581) G. A. Webb and B. B. Corson, *Ind. Eng. Chem., 39,* 1153–1156 (1947).
(1582) A. Weber, A. G. Meister, and F. F. Cleveland, *J. Chem. Phys., 21,* 930–933 (1953).
(1583) L. A. Weber and J. E. Kilpatrick, *J. Chem. Phys., 36,* 829–834 (1962).
(1584) H. B. Weissman, R. B. Bernstein, S. E. Rosser, A. G. Meister, and F. F. Cleveland, *J. Chem. Phys., 23,* 544–551 (1955).
(1585) H. B. Weissman, A. G. Meister, and F. F. Cleveland, *J. Chem. Phys., 29,* 72–77 (1958).
(1586) W. Weltner, Jr., *J. Am. Chem. Soc., 77,* 3941–3950 (1955).
(1587) W. Weltner, Jr., private communication, March 15, 1955.
(1588) W. Weltner, Jr., and K. S. Pitzer, *J. Am. Chem. Soc., 73,* 2606–2610 (1951).
(1589) H. Wenker, *J. Am. Chem. Soc., 57,* 2328 (1935).
(1590) E. D. West, *J. Am. Chem. Soc., 81,* 29–37 (1959).
(1591) E. D. West and D. C. Ginnings, *J. Res. Natl. Bur. Std., 60,* 309–316 (1958).
(1592) E. D. West and E. F. Westrum, Jr., "Adiabatic Calorimetry from 300 to 800°K," in J. P. McCullough and D. W. Scott (Eds.), *Experimental Thermodynamics,* Butterworths, London, 1968, Chap. 8.
(1593) E. F. Westrum, Jr., *J. Phys. Chem. Solids, 18,* 83–85 (1961).
(1594) E. F. Westrum, Jr., *J. Chem. Educ., 39,* 443–454 (1962).
(1595) E. F. Westrum, Jr., unpublished data.
(1596) E. F. Westrum, Jr., and E. Chang, *Colloq. Intern. Centre Natl. Rech. Sci. (Paris), 156,* 163–173 (1967).
(1597) E. F. Westrum, Jr., G. T. Furukawa, and J. P. McCullough, "Adiabatic Low-Temperature Calorimetry," in J. P. McCullough and D. W. Scott (Eds.), *Experimental Thermodynamics,* Butterworths, London, 1968, Chap. 4.
(1598) E. F. Westrum, Jr., and J. P. McCullough, "Thermodynamics of Crystals," in D. Fox, M. M. Labes, and A. Weissberger (Eds.), *Physics and Chemistry of the Organic Solid State,* Interscience Publishers, New York, 1963, pp. 1–178.
(1599) E. F. Westrum, Jr., and J. C. Trowbridge, to be published.
(1600) E. F. Westrum, Jr., W.-K. Wong, and S.-W. Wong, "Strain Energies and Thermal Properties Through Fusion for Polynuclear Aromatic Hydrocarbons," Presented at the American Chemical Society Meeting, Miami Beach, April, 1967.
(1601) W. H. Wheeler, H. Whittaker, and H. H. M. Pike, *J. Inst. Fuel, 20,* 137–156, 159 (1947).
(1602) G. W. Wheland, *Resonance in Organic Chemistry,* Wiley, New York, 1955.
(1603) D. H. Whiffen, *J. Chem. Soc.,* 1350–1356 (1956).

(1604) A. G. White, *J. Chem. Soc., 121,* 1244–1270 (1922).
(1605) W. B. White, S. M. Johnson, and G. B. Dantzig, *J. Chem. Phys., 28,* 751–755 (1958).
(1606) R. Whytlaw-Gray, C. G. Reeves, and G. A. Bottomley, *Nature, 181,* 1004 (1958).
(1607) K. B. Wiberg and W. J. Bartley, *J. Am. Chem. Soc., 82,* 6375–6380 (1960).
(1608) K. B. Wiberg, W. J. Bartley, and F. P. Lossing, *J. Am. Chem. Soc., 84,* 3980–3981 (1962).
(1609) E. Wichers, *J. Am. Chem. Soc., 80,* 4121–4124 (1958).
(1610) H. Wieland, K. Heymann, T. Tsatsas, D. Juchum, G. Varvoglis, G. Labriola, O. Dobbelatein, and H. S. Boyd-Barrett, *Ann., 514,* 145–181 (1934).
(1611) R. C. Wilhoit and I. Lei, *J. Chem. Eng. Data, 10,* 166–168 (1965).
(1612) R. C. Wilhoit and D. Shiao, *J. Chem. Eng. Data, 9,* 595–599 (1964).
(1613) J. Wilks, *The Third Law of Thermodynamics,* Oxford University Press, London, 1961.
(1614) G. C. Williams, *Ind. Eng. Chem., 40,* 340–341 (1948).
(1615) J. Q. Williams and W. Gordy, *J. Chem. Phys., 18,* 994–995 (1950).
(1616) M. M. Williams, W. S. McEwan, and R. A. Henry, *J. Phys. Chem., 61,* 261–267 (1957).
(1617) K. D. Williamson and R. H. Harrison, *J. Chem. Phys., 26,* 1409–1411 (1957).
(1618) J. B. Willis, *Trans. Faraday Soc., 43,* 97–102 (1947).
(1619) A. H. Wilson, *Thermodynamics and Statistical Mechanics,* Cambridge University Press, London, 1957.
(1620) E. B. Wilson, Jr., *Chem. Rev., 27,* 17–38 (1940).
(1621) T. P. Wilson, E. G. Caflisch, and G. F. Hurley, *J. Phys. Chem., 62,* 1059–1061 (1958).
(1622) H. E. Wirth, J. W. Droege, and J. H. Wood, *J. Phys. Chem., 60,* 917–919 (1956).
(1623) C. R. Witschonke, *Anal. Chem., 26,* 562–564 (1954).
(1624) W.-K. Wong, Doctoral Dissertation, University of Michigan, Ann Arbor, June 1966.
(1625) J. L. Wood, R. J. Lagow, and J. L. Margrave, *J. Chem. Eng. Data, 12,* 255–256 (1967).
(1626) R. E. Wood and D. P. Stevenson, *J. Am. Chem. Soc., 63,* 1650–1653 (1941).
(1627) W. W. Wood and V. Shomaker, *J. Chem. Phys., 20,* 555–560 (1952).
(1628) A. L. Woodman, W. J. Murbach, and M. H. Kaufman, *J. Phys. Chem., 64,* 658–660 (1960).
(1629) H. W. Woolley, *J. Res. Natl. Bur. Std., 40,* 163–168 (1948).
(1630) H. W. Woolley, R. B. Scott, and F. G. Brickwedde, *J. Res. Natl. Bur. Std., 41,* 379–475 (1948).
(1631) E. L. Wu, G. Zerbi, S. Califano, and B. Crawford, Jr., *J. Chem. Phys., 35,* 2060–2064 (1961).
(1632) C. A. Wulff, *J. Chem. Phys., 39,* 1227–1234 (1963).
(1633) C. A. Wulff and E. F. Westrum, Jr., *J. Phys. Chem., 67,* 2376–2381 (1963).
(1634) R. Wurmser and N. Mayer-Reich, *Compt. Rend., 196,* 612–614 (1933).
(1635) H.-R. Wyss and Hs. H. Günthard, *Helv. Chim. Acta, 44,* 625–631 (1961).
(1636) S. Yamada, *Bull. Chem. Soc. Japan, 16,* 187–196 (1941).
(1637) M. Yamaha, *Bull. Chem. Soc. Japan, 27,* 170–177 (1954).
(1638) D. M. Yost, D. W. Osborne, and C. S. Garner, *J. Am. Chem. Soc., 63,* 3492–3496 (1941).
(1639) D. M. Yost and W. E. Stone, *J. Am. Chem. Soc., 55,* 1889–1895 (1933).
(1640) J. A. Young, J. E. Keith, P. Stehle, W. C. Dzombak, and H. Hunt, *Ind. Eng. Chem., 48,* 1375–1378 (1956).
(1641) C. T. Zahn, *J. Chem. Phys., 2,* 671–680 (1934).
(1642) W. Zeil, M. Winnewisser, and K. Mueller, *Z. Nat*
(1643) H. Zeise, *Thermodynamik auf den Grundlagen statistik und Spektroskopie,* Band III/1, *Tabelle Darstellungen und Literatur,* Hirzel-Verlag, Leip
(1644) A. K. Zhdanov, *J. Gen. Chem. USSR, 15,* 895–
(1645) J. P. Zietlow, F. F. Cleveland, and A. G. Meister (1956).
(1646) M. F. Zimmer, E. E. Baroody, M. G. Graff, G. *Chem. Eng. Data, 11,* 579–581 (1966).
(1647) M. F. Zimmer, E. E. Baroody, M. Schwartz, and *Data, 9,* 527–529 (1964).
(1648) M. F. Zimmer, R. A. Robb, E. E. Baroody, and *Data, 11,* 577–579 (1966).
(1649) J. S. Ziomek and F. F. Cleveland, *J. Chem. Phy*
(1650) J. S. Ziomek, A. G. Meister, F. F. Cleveland, an *21,* 90–100 (1953).
(1651) N. D. Zubareva, A. P. Oberemok-Yakubova, Y. A. A. Balandin, *Izv. Akad. Nauk SSSR, Ser. Kh*
(1652) R. F. Zürcher and H. H. Günthard, *Helv. Chim.*
(1653) B. J. Zwolinski (Ed.), *Selected Values of Prope Compounds,* American Petroleum Institute Re Research Center Data Project, Texas A & M Un 1966.
(1654) B. J. Zwolinski et al., *Selected Values of Prope* Thermodynamics Research Center Data Projec Station, Tex., 1966.
(1655) Anonymous, *Chem. Eng. News, 43* (50), 33–3
(1656) IUPAC Recommendation on Symbols, *Compt. Chim. Pure Appl.,* Montréal, August 2–5, 196

COMPOUND INDEX
To Ideal Gas Tables

Note: For additional Compound Data at 298.15°K only, see Table beginning on page 634.

Acetaldehyde, 439
Acetic acid, 449
Acetic anhydride, 450
Acetone, 444
Acetonitrile, 473
Acetyl chloride, 536
Acetylene, 334
Acetylenedicarbonitrile, 222
Acrylic acid, 452
Acrylonitrile, 474
Allene, 330
Allyl alcohol, 436
Ammonia, 231
Aniline, 472
Azulene, 402

Benzene, 367
Benzenethiol, 625
Benzoic acid, 456
Benzonitrile, 476
Biacetylene, 335
Biphenyl, 403
Bromine, 210
Bromobenzene, 547
1-Bromobutane, 542
2-Bromobutane, 543
Bromoethane, 538
Bromoethylene, 546
Bromomethane, 537
2-Bromo-2-methylpropane, 543
1-Bromopentane, 546
1-Bromopropane, 540
2-Bromopropane, 541
3-Bromo-1-propene, 547
1,2-Butadiene, 330
1,3-Butadiene, 331
Butadiyne, 335
Butane, 245
1-Butanethiol, 614
2-Butanethiol, 614
2-Butanone, 445

1-Butene, 313
2-Butene, *cis,* 314
2-Butene, *trans,* 314
1-Buten-3-yne, 335
Butyl alcohol, 425
sec-Butyl alcohol, 426
tert-Butyl alcohol, 426
Butylamine, 464
sec-Butylamine, 464
tert-Butylamine, 465
Butylbenzene, 373
Butylcyclohexane, 362
Butylcyclopentane, 351
Butyl decyl sulfide, 585
Butyl disulfide, 602
Butyl dodecyl sulfide, 589
Butyl ether, 417
sec-Butyl ether, 418
tert-Butyl ether, 148
Butyl ethyl sulfide, 567
Butyl heptyl sulfide, 578
Butyl hexadecyl sulfide, 598
Butyl hexyl sulfide, 576
Butyl methyl sulfide, 566
1-Butylnaphthalene, 398
2-Butylnaphthalene, 398
Butyl nonyl sulfide, 583
Butyl octyl sulfide, 580
Butyl pentadecyl sulfide, 596
Butyl pentyl sulfide, 574
Butyl propyl sulfide, 570
Butyl sulfide, 572
Butyl tetradecyl sulfide, 594
Butyl tridecyl sulfide, 592
Butyl undecyl sulfide, 587
1-Butyne, 336
2-Butyne, 336
Butyraldehyde, 441
Butyronitrile, 475

Carbon (graphite), 209

Carbon dioxide, 220
Carbon disulfide, 220
Carbon monoxide, 219
Carbon suboxide, 222
Carbon tetrachloride, 511
Carbon tetrafluoride, 496
Carbonyl fluoride, 217
Carbonyl sulfide, 219
Chlorine, 210
Chlorobenzene, 532
1-Chlorobutane, 522
2-Chlorobutane, 523
Chloroethane, 512
Chloroethylene, 526
Chloroform, 510
Chloromethane, 508
1-Chloro-3-methylbutane, 525
2-Chloro-2-methylbutane, 526
1-Chloro-2-methylpropane, 523
2-Chloro-2-methylpropane, 524
1-Chloropentane, 524
1-Chloropropane, 518
2-Chloropropane, 519
3-Chloro-1-propene, 531
m-Cresol, 454
o-Cresol, 455
p-Cresol, 455
Cumene, 370
Cyanogen, 221
Cyanogen bromide, 215
Cyanogen chloride, 216
Cyanogen iodide, 218
Cyclobutane, 343
Cyclobutene, 346
Cycloheptane, 345
1,3,5-Cycloheptatriene, 400
Cyclohexane, 344
Cyclohexanol, 437
Cyclohexanone, 447
Cyclohexene, 347
1-Cyclohexyldecane, 364
1-Cyclohexyldodecane, 365
1-Cyclohexylheptane, 363
1-Cyclohexylhexadecane, 367
1-Cyclohexylhexane, 363
1-Cyclohexylnonane, 364
1-Cyclohexyloctane, 364
1-Cyclohexylpentadecane, 366
1-Cyclohexyltetradecane, 366
1-Cyclohexyltridecane, 366
1-Cyclohexylundecane, 365
Cyclooctane, 345

1,3,5,7-Cyclooctatetraene, 401
Cyclopentane, 344
Cyclopentanethiol, 625
Cyclopentene, 346
1-Cyclopentyldecane, 353
1-Cyclopentyldodecane, 354
1-Cyclopentylheptane, 352
1-Cyclopentylhexadecane, 355
1-Cyclopentylhexane, 352
1-Cyclopentylnonane, 353
1-Cyclopentyloctane, 352
1-Cyclopentylpentadecane, 355
1-Cyclopentylpentane, 351
1-Cyclopentyltetradecane, 355
1-Cyclopentyltridecane, 354
1-Cyclopentylundecane, 354
Cyclopropane, 343

Decahydronaphthalene, $trans$, 403
Decanal, 443
Decane, 276
1-Decanethiol, 620
1-Decene, 326
Decyl alcohol, 431
Decyl disulfide, 605
Decyl ethyl sulfide, 581
Decyl methyl sulfide, 579
Decyl propyl sulfide, 583
Decyl sulfide, 598
1-Decyne, 339
1,2-Dibromobutane, 544
2,3-Dibromobutane, 544
1,2-Dibromoethane, 539
2,3-Dibromo-2-methylbutane, 545
1,2-Dibromopropane, 541
m-Dichlorobenzene, 533
o-Dichlorobenzene, 534
p-Dichlorobenzene, 535
1,1-Dichloroethane, 513
1,2-Dichloroethane, 514
1,1-Dichloroethylene, 527
1,2-Dichloroethylene, cis, 528
1,2-Dichloroethylene, $trans$, 529
Dichloromethane, 509
1,2-Dichloropropane, 520
1,3-Dichloropropane, 520
2,2-Dichloropropane, 521
Diethylamine, 466
m-Diethylbenzene, 374
o-Diethylbenzene, 374
p-Diethylbenzene, 375
3,3-Diethylhexane, 296

3,4-Diethylhexane, 296
3,3-Diethyl-2-methylpentane, 305
3,3-Diethylpentane, 272
m-Difluorobenzene, 504
o-Difluorobenzene, 505
p-Difluorobenzene, 506
1,1-Difluoroethane, 497
1,1-Difluoroethylene, 502
Difluoromethane, 494
1,2-Diiodobutane, 555
1,2-Diiodoethane, 552
Diiodomethane, 550
1,2-Diiodopropane, 554
Dimethylacetylene, 336
Dimethylamine, 466
2,2-Dimethylbutane, 248
2,3-Dimethylbutane, 248
2,3-Dimethyl-1-butene, 323
2,3-Dimethyl-2-butene, 324
3,3-Dimethyl-1-butene, 324
1,1-Dimethylcyclohexane, 358
1,2-Dimethylcyclohexane, cis, 358
1,2-Dimethylcyclohexane, trans, 359
1,3-Dimethylcyclohexane, cis, 359
1,3-Dimethylcyclohexane, trans, 359
1,4-Dimethylcyclohexane, cis, 360
1,4-dimethylcyclohexane, trans, 360
1,1-Dimethylcyclopentane, 348
1,2-Dimethylcyclopentane, cis, 349
1,2-Dimethylcyclopentane, trans, 349
1,3-Dimethylcyclopentane, cis, 349
1,3-Dimethylcyclopentane, trans, 350
2,2-Dimethylhexane, 255
2,3-Dimethylhexane, 255
2,4-Dimethylhexane, 255
2,5-Dimethylhexane, 256
3,3-Dimethylhexane, 256
3,4-Dimethylhexane, 256
2,2-Dimethylheptane, 263
2,3-Dimethylheptane, 263
2,4-Dimethylheptane, 263
2,5-Dimethylheptane, 264
2,6-Dimethylheptane, 264
3,3-Dimethylheptane, 265
3,4-Dimethylheptane, 265
3,5-Dimethylheptane, 266
4,4-Dimethylheptane, 266
2,4-Dimethyl-3-isopropylpentane, 305
1,2-Dimethylnaphthalene, 392
1,3-Dimethylnaphthalene, 392
1,4-Dimethylnaphthalene, 392
1,5-Dimethylnaphthalene, 393

1,6-Dimethylnaphthalene, 393
1,7-Dimethylnaphthalene, 394
2,3-Dimethylnaphthalene, 394
2,6-Dimethylnaphthalene, 395
2,7-Dimethylnaphthalene, 395
2,2-Dimethyloctane, 279
2,3-Dimethyloctane, 279
2,4-Dimethyloctane, 280
2,5-Dimethyloctane, 280
2,6-Dimethyloctane, 281
2,7-Dimethyloctane, 281
3,3-Dimethyloctane, 282
3,4-Dimethyloctane, 282
3,5-Dimethyloctane, 282
3,6-Dimethyloctane, 283
4,4-Dimethyloctane, 283
4,5-Dimethyloctane, 284
2,2-Dimethylpentane, 251
2,3-Dimethylpentane, 251
2,4-Dimethylpentane, 251
3,3-Dimethylpentane, 252
2,2-Dimethylpropane, 246
p-Dioxane, 421
Dodecane, 308
1-Dodecanethiol, 621
1-Dodecene, 327
Dodecyl alcohol, 432
Dodecyl ethyl sulfide, 585
Dodecyl methyl sulfide, 584
Dodecyl propyl sulfide, 588
1-Dodecyne, 340

Eicosane, 312
1-Eicosanethiol, 624
1-Eicosanol, 435
1-Eicosene, 330
1-Eicosyne, 342
Ethane, 244
Ethanethiol, 612
Ethyl acetate, 451
Ethylacetylene, 336
Ethyl alcohol, 423
Ethylamine, 462
Ethylbenzene, 368
2-Ethyl-1-butene, 323
Ethylcyclohexane, 358
Ethylcyclopentane, 348
3-Ethyl-2,2-dimethylhexane, 297
3-Ethyl-2,3-dimethylhexane, 297
3-Ethyl-2,4-dimethylhexane, 298
3-Ethyl-2,5-dimethylhexane, 299

Compound Index

3-Ethyl-3,4-dimethylhexane, 300
4-Ethyl-2,2-dimethylhexane, 297
4-Ethyl-2,3-dimethylhexane, 298
4-Ethyl-2,4-dimethylhexane, 299
4-Ethyl-3,3-dimethylhexane, 300
3-Ethyl-2,2-dimethylpentane, 273
3-Ethyl-2,3-dimethylpentane, 273
3-Ethyl-2,4-dimethylpentane, 273
Ethyl disulfide, 601
Ethylene, 312
Ethylene glycol, 436
Ethylene oxide, 419
Ethylenimine, 468
Ethyl ether, 413
Ethyl heptadecyl sulfide, 596
3-Ethylheptane, 262
4-Ethylheptane, 262
Ethyl heptyl sulfide, 574
Ethyl hexadecyl sulfide, 594
3-Ethylhexane, 254
Ethyl hexyl sulfide, 572
Ethyl methyl ether, 413
3-Ethyl-2-methylheptane, 285
3-Ethyl-3-methylheptane, 286
3-Ethyl-4-methylheptane, 288
3-Ethyl-5-methylheptane, 287
4-Ethyl-2-methylheptane, 285
4-Ethyl-3-methylheptane, 287
4-Ethyl-4-methylheptane, 288
5-Ethyl-2-methylheptane, 286
3-Ethyl-2-methylhexane, 267
3-Ethyl-3-methylhexane, 267
3-Ethyl-4-methylhexane, 268
4-Ethyl-2-methylhexane, 267
2-Ethyl-3-methylnaphthalene, 397
2-Ethyl-6-methylnaphthalene, 397
2-Ethyl-7-methylnaphthalene, 397
3-Ethyl-2-methylpentane, 257
3-Ethyl-3-methylpentane, 257
Ethyl methyl sulfide, 564
1-Ethylnaphthalene, 391
2-Ethylnaphthalene, 391
Ethyl nitrate, 483
Ethyl nonyl sulfide, 579
Ethyl octadecyl sulfide, 599
3-Ethyloctane, 278
4-Ethyloctane, 278
Ethyl octyl sulfide, 576
Ethyl pentadecyl sulfide, 592
3-Ethylpentane, 250
Ethyl pentyl sulfide, 571

Ethyl propyl sulfide, 567
Ethyl sulfide, 564
Ethyl tetradecyl sulfide, 590
m-Ethyltoluene, 371
o-Ethyltoluene, 371
p-Ethyltoluene, 371
Ethyl tridecyl sulfide, 588
3-Ethyl-2,2,3-trimethylpentane, 306
3-Ethyl-2,2,4-trimethylpentane, 306
3-Ethyl-2,3,4-trimethylpentane, 307
Ethyl undecyl sulfide, 584
Ethyne, 334
Ethynylbenzene, 401

Fluorine, 211
Fluorobenzene, 504
Fluoroethane, 497
Fluoromethane, 494
1-Fluoropropane, 500
2-Fluoropropane, 500
p-Fluorotoluene, 507
Formaldehyde, 438
Formic acid, 448
Furan, 420

Heptadecane, 311
1-Heptadecanethiol, 623
1-Heptadecanol, 434
1-Heptadecene, 329
Heptadecyl methyl sulfide, 594
Heptadecyl propyl sulfide, 599
1-Heptadecyne, 341
Heptanal, 442
Heptane, 249
1-Heptanethiol, 618
1-Heptene, 324
Heptyl alcohol, 430
Heptyl disulfide, 604
Heptyl methyl sulfide, 573
Heptyl propyl sulfide, 577
Heptyl sulfide, 586
1-Heptyne, 338
Hexachlorobenzene, 536
Hexachloroethane, 517
Hexadecane, 310
1-Hexadecanethiol, 622
1-Hexadecanol, 434
1-Hexadecene, 328
Hexadecyl methyl sulfide, 593
Hexadecyl propyl sulfide, 597
1-Hexadecyne, 341

Hexaethylbenzene, 384
Hexafluorobenzene, 506
Hexafluoroethane, 499
Hexamethylbenzene, 379
Hexanal, 442
Hexane, 247
1-Hexanethiol, 618
1-Hexene, 318
2-Hexene, *cis,* 318
2-Hexene, *trans,* 319
3-Hexene, *cis,* 319
3-Hexene, *trans,* 319
Hexyl alcohol, 429
Hexylbenzene, 377
Hexyl disulfide, 603
Hexyl methyl sulfide, 571
Hexyl propyl sulfide, 575
Hexyl sulfide, 581
1-Hexyne, 338
Hydrazine, 231
Hydrogen, 209
Hydrogen bromide, 215
Hydrogen chloride, 223
Hydrogen cyanide, 217
Hydrogen fluoride, 226
Hydrogen iodide, 228
Hydrogen nitrate, 228
Hydrogen peroxide, 229
Hydrogen sulfide, 230

Iodine, 211
Iodobenzene, 556
Iodoethane, 551
Iodomethane, 549
2-Iodo-2-methylpropane, 554
1-Iodopropane, 552
2-Iodopropane, 553
3-Iodo-1-propene, 555
Isobutane, 245
Isobutyronitrile, 476
Isopentane, 246
Isoprene, 333
Isopropyl alcohol, 424
Isopropyl *tert*-butyl ether, 417
Isopropyl ether, 416
4-Isopropylheptane, 285
3-Isopropyl-2-methylhexane, 295
Isopropyl methyl sulfide, 565
Isopropyl nitrate, 485
Isopropyl sulfide, 568

Isothiocyanic acid, 626

Ketene, 446

Mesitylene, 373
Methane, 243
Methanethiol, 611
Methanol, 422
Methylacetylene, 334
Methylamine, 461
2-Methyl-1,3-butadiene, 333
3-Methyl-1,2-butadiene, 333
2-Methylbutane, 246
2-Methyl-2-butanethiol, 616
2-Methyl-1-butene, 316
2-Methyl-2-butene, 317
3-Methyl-1-butene, 317
Methyl *tert*-butyl ether, 415
3-Methyl-1-butyne, 337
Methylcyclohexane, 357
Methylcyclopentane, 347
1-Methylcyclopentene, 356
3-Methylcyclopentene, 356
4-Methylcyclopentene, 357
Methyl disulfide, 600
Methyl ether, 412
Methyl formate, 450
2-Methylheptane, 253
3-Methylheptane, 254
4-Methylheptane, 254
2-Methylhexane, 249
3-Methylhexane, 250
Methyl isopropyl ether, 415
1-Methylnaphthalene, 390
2-Methylnaphthalene, 390
Methyl nitrate, 482
Methyl nitrite, 481
Methyl nonadecyl sulfide, 600
2-Methylnonane, 276
3-Methylnonane, 277
4-Methylnonane, 277
5-Methylnonane, 278
Methyl nonyl sulfide, 577
Methyl octadecyl sulfide, 597
2-Methyloctane, 260
3-Methyloctane, 261
4-Methyloctane, 261
Methyl octyl sulfide, 575
Methyl pentadecyl sulfide, 590
2-Methylpentane, 247
3-Methylpentane, 248

2-Methyl-1-pentene, 320
2-Methyl-2-pentene, 321
3-Methyl-1-pentene, 320
3-Methyl-2-pentene, *cis,* 321
3-Methyl-2-pentene, *trans,* 322
4-Methyl-1-pentene, 320
4-Methyl-2-pentene, *cis,* 322
4-Methyl-2-pentene, *trans,* 322
Methyl pentyl sulfide, 569
2-Methylpropane, 245
2-Methyl-1-propanethiol, 615
2-Methyl-2-propanethiol, 615
2-Methylpropene, 315
Methyl propyl ether, 414
Methyl propyl sulfide, 565
m-Methylstyrene, 388
o-Methylstyrene, 388
p-Methylstyrene, 388
α-Methylstyrene, 387
β-Methylstyrene, *cis,* 387
β-Methylstyrene, *trans,* 387
Methyl sulfide, 563
Methyl tetradecyl sulfide, 589
2-Methylthiophene, 610
3-Methylthiophene, 610
Methyl tridecyl sulfide, 586
Methyl undecyl sulfide, 582

Naphthalene, 389
Neopentane, 246
Nitric oxide, 232
1-Nitrobutane, 480
2-Nitrobutane, 481
Nitroethane, 478
Nitrogen, 213
Nitrogen dioxide, 232
Nitromethane, 477
1-Nitropropane, 479
2-Nitropropane, 479
Nitrosyl chloride, 224
Nitrosyl fluoride, 227
Nitrous oxide, 232
Nitryl chloride, 224
Nitryl fluoride, 227
Nonadecane, 311
1-Nonadecanethiol, 624
1-Nonadecanol, 435
1-Nonadecene, 329
1-Nonadecyne, 342
Nonanal, 443
Nonane, 260

1-Nonanethiol, 619
1-Nonene, 326
Nonyl alcohol, 431
Nonyl disulfide, 604
Nonyl propyl sulfide, 582
Nonyl sulfide, 595
1-Nonyne, 339

Octadecane, 311
1-Octadecanethiol, 623
1-Octadecanol, 435
1-Octadecene, 329
1-Octadecyne, 342
Octafluorocyclobutane, 501
Octafluoropropane, 500
Octanal, 443
Octane, 253
1-Octanethiol, 619
1-Octene, 325
Octyl alcohol, 430
Octyl disulfide, 604
Octyl propyl sulfide, 580
Octyl sulfide, 591
1-Octyne, 338
Oxygen, 213
Ozone, 233

Pentachloroethane, 517
Pentadecane, 310
1-Pentadecanethiol, 622
1-Pentadecanol, 433
1-Pentadecene, 328
Pentadecyl propyl sulfide, 595
1-Pentadecyne, 341
1,2-Pentadiene, 331
1,3-Pentadiene, *cis,* 332
1,3-Pentadiene, *trans,* 332
1,4-Pentadiene, 332
2,3-Pentadiene, 332
Pentaethylbenzene, 383
Pentamethylbenzene, 377
2,2,3,3,4-Pentamethylpentane, 307
2,2,3,4,4-Pentamethylpentane, 308
Pentane, 246
1-Pentanethiol, 617
2-Pentanone, 446
1-Pentene, 315
2-Pentene, *cis,* 316
2-Pentene, *trans,* 316
Pentyl alcohol, 428
tert-Pentyl alcohol, 428

Compound Index

Pentylbenzene, 377
Pentylcyclohexane, 363
Pentyl disulfide, 603
1-Pentylnaphthalene, 399
2-Pentylnaphthalene, 399
Pentyl propyl sulfide, 573
Pentyl sulfide, 578
1-Pentyne, 336
2-Pentyne, 337
Perchloryl fluoride, 223
Phenol, 454
1-Phenyldecane, 383
1-Phenyldodecane, 384
1-Phenylheptane, 380
1-Phenylhexadecane, 386
1-Phenylnonane, 382
1-Phenylpentadecane, 385
1-Phenyloctane, 380
1-Phenyltetradecane, 385
1-Phenyltridecane, 385
1-Phenylundecane, 383
Phosgene, 216
2-Picoline, 471
3-Picoline, 471
Propadiene, 330
Propane, 244
1-Propanethiol, 612
2-Propanethiol, 613
Propene, 313
Propenylbenzene, *cis*, 387
Propenylbenzene, *trans*, 387
Propionaldehyde, 440
Propionitrile, 474
Propyl alcohol, 424
Propylamine, 463
Propylbenzene, 370
Propylcyclohexane, 361
Propylcyclopentane, 350
Propyl disulfide, 602
Propylene oxide, 420
Propyl ether, 416
4-Propylheptane, 284
1-Propylnaphthalene, 395
2-Propylnaphthalene, 396
Propyl nitrate, 484
Propyl sulfide, 570
Propyl tetradecyl sulfide, 593
Propyl tridecyl sulfide, 591
Propyl undecyl sulfide, 587
Propyne, 334
Pyridine, 470

Pyrrolidine, 469

Spiropentane, 400
Styrene, 386
Sulfur, 214
Sulfur dioxide, 233
Sulfuric acid anhydrous, 230
Sulfur monochloride, 225
Sulfur tetrafluoride, 227
Sulfur trioxide, 234
Sulfuryl chloride, 225

1,1,2,2-Tetrachloroethane, 516
Tetrachloroethylene, 530
Tetradecane, 309
1-Tetradecanethiol, 621
1-Tetradecanol, 433
1-Tetradecene, 328
1-Tetradecyne, 340
1,2,3,4-Tetraethylbenzene, 381
1,2,3,5-Tetraethylbenzene, 381
1,2,4,5-Tetraethylbenzene, 381
Tetrafluoroethylene, 503
1,2,3,4-Tetramethylbenzene, 375
1,2,3,5-Tetramethylbenzene, 376
1,2,4,5-Tetramethylbenzene, 376
2,2,3,3-Tetramethylbutane, 259
2,2,3,3-Tetramethylhexane, 300
2,2,3,4-Tetramethylhexane, 301
2,2,3,5-Tetramethylhexane, 301
2,2,4,4-Tetramethylhexane, 302
2,2,4,5-Tetramethylhexane, 302
2,2,5,5-Tetramethylhexane, 303
2,3,3,4-Tetramethylhexane, 303
2,3,3,5-Tetramethylhexane, 303
2,3,4,4-Tetramethylhexane, 304
2,3,4,5-Tetramethylhexane, 304
3,3,4,4-Tetramethylhexane, 305
2,2,3,3-Tetramethylpentane, 274
2,2,3,4-Tetramethylpentane, 274
2,2,4,4-Tetramethylpentane, 275
2,3,3,4-Tetramethylpentane, 275
Thiacyclobutane, 606
Thiacycloheptane, 608
Thiacyclohexane, 607
Thiacyclopentane, 607
Thiacyclopropane, 605
Thioacetic acid, 626
Thionyl chloride, 224
Thiophene, 609
Toluene, 368

1,1,2-Trichloroethane, 515
Trichloroethylene, 530
1,2,3-Trichloropropane, 521
Tridecane, 309
1-Tridecanethiol, 621
1-Tridecanol, 432
1-Tridecene, 327
1-Tridecyne, 340
Triethylamine, 468
1,2,3-Triethylbenzene, 378
1,2,4-Triethylbenzene, 378
1,3,5-Triethylbenzene, 379
1,1,1-Trifluoroethane, 498
Trifluoroethylene, 502
Trifluoromethane, 495
α,α,α-Trifluorotoluene, 507
Triiodomethane, 550
Trimethylamine, 467
1,2,3-Trimethylbenzene, 372
1,2,4-Trimethylbenzene, 372
1,3,5-Trimethylbenzene (mesitylene), 373
2,2,3-Trimethylbutane, 252
1,3,5-Trimethylcyclohexane, *cis, cis,* 361
1,3,5-Trimethylcyclohexane, *cis, trans,* 362
2,2,3-Trimethylheptane, 288
2,2,4-Trimethylheptane, 289
2,2,5-Trimethylheptane, 289
2,2,6-Trimethylheptane, 290
2,3,3-Trimethylheptane, 290
2,3,4-Trimethylheptane, 291
2,3,5-Trimethylheptane, 291
2,3,6-Trimethylheptane, 291
2,4,4-Trimethylheptane, 292
2,4,5-Trimethylheptane, 292

2,4,6-Trimethylheptane, 293
2,5,5-Trimethylheptane, 293
3,3,4-Trimethylheptane, 294
3,3,5-Trimethylheptane, 294
3,4,4-Trimethylheptane, 294
3,4,5-Trimethylheptane, 295
2,2,3-Trimethylhexane, 268
2,2,4-Trimethylhexane, 269
2,2,5-Trimethylhexane, 269
2,3,3-Trimethylhexane, 270
2,3,4-Trimethylhexane, 270
2,3,5-Trimethylhexane, 271
2,4,4-Trimethylhexane, 271
3,3,4-Trimethylhexane, 272
2,2,3-Trimethylpentane, 258
2,2,4-Trimethylpentane, 258
2,3,3-Trimethylpentane, 258
2,3,4-Trimethylpentane, 259

Undecane, 308
1-Undecanethiol, 620
1-Undecene, 326
Undecyl alcohol, 432
1-Undecyne, 339

Valeraldehyde, 441
Vinylacetylene, 335
Vinyl chloride, 526

Water, 229

m-Xylene, 369
o-Xylene, 369
p-Xylene, 369

SUBJECT INDEX

Aliphatic iodine compounds, tables, 549ff
Alkadienes, tables, 330ff
Alkanals, tables, 438ff
Alkanediol, table, 436, 437
Alkanes, tables, 243ff
Alkanethiols, cyclic, tables, 624ff
  tables, 611ff
Alkanoic acids, derivative ideal gas tables, 448ff
  tables, 448ff
Alkanols, enthalpy of formation, 407
  enthalpy of vaporization, 408
  tables, 422ff
Alkanones, tables, 444ff
Alkenes, tables, 312ff
Alkenoic acid, table, 452
Alkenol, table, 436
Alkenone, table, 446, 447
Alkylamines, primary, tables, 461ff
Alkylamines, secondary, tables, 466
Alkylamines, tertiary, tables, 467
Alkylbenzenes, tables, 367ff
Alkylcyclohexanes, tables, 357ff
Alkylcyclopentanes, tables, 347ff
Alkylcyclopentenes, tables, 356ff
Alkylnaphthalenes, tables, 389ff
Alkyl nitrates, tables, 481
Alkyl nitrites, tables, 481
Alkynes, tables, 334ff
Allen's method for estimating thermodynamic properties, 144, 241
American Petroleum Institute, 61, 62, 235
Amines, cyclic, tables, 468ff
Antimony point, 7, 8
Antoine equation, 14, 15, 58
Aromatic bromine compound, table, 547
Aromatic chlorine compounds, tables, 532ff
Aromatic fluorine compounds, tables, 504ff
Aromatic iodine compounds, table, 556
Aromatic thiol, table, 625
Arylamine, table, 472
Assumptions in thermodynamic calculations, 156
Atmosphere, standard, 203
Atomic weights, 202
Atomization, heat of, 242
Attainment of equilibrium, 160
Automatic shield control, 19ff
Avogadro's constant, 203

Barriers to internal rotation, *see* Rotation, internal
Benzaldehyde synthesis, 191
Benzene, enthalpy of vaporization of, 56
  thermodynamic functions, 206
Benzoic acid, calibration with, 67ff
  table, 456
Berthelot equation of state, 10, 90, 181
Biochemical processes, 458
Biological cycles, 405
Boltzmann constant, 203
Bomb calorimetry, 67ff
  calibration in, 67, 68
  of fluorine compounds, 70, 71
  of halogen compounds, 70
  of sulfur compounds, 70
  Washburn reduction in, 68, 69
Bond angles, table, 96, 97
Bond energy schemes, 144
  comparison between, 145
Bond lengths, table, 96, 97
Boyle's law, 6
Bromine compounds, enthalpy of formation, 491
  entropy, 491
  heat capacity, 491, 492
  methyl substitution constants, 491
  tables, 537ff
Bromomethane, calculation of heat capacity of gaseous, 35, 36
Butane, entropy, 111
  heat capacity, 49

Calculation, activity of acetyl chloride, 134

Subject Index

bomb combustion data, 71, 72
compositions by shortcut methods, 123ff
constants in Antoine equation, 14ff
enthalpy dependence on temperature, 65
enthalpy of melting of phenol, 57
entropy, change, liquid to vapor, 414, 417, 420, 422, 429, 437, 448, 453
  by Simpson's rule, 110ff
  by statistical methods, 93, 101ff
equation of state, 74ff
equilibrium, composition, 120ff, 136
  constants, 61, 120ff
  from Gibbs energy, 427, 438
  temperature dependence, 80ff
Gibbs energy of formation, 205
heat, of formation, 205
  of melting, 57
  of reaction, 63
  of sublimation, 57
  of transition, 50
  of vaporization, 57, 449
  of fluorobenzene, 59
heat capacity, 449
  bromomethane gas, 35, 36
  diatomic gas, 32
  dimethylacetylene gas, 37, 42, 43
  extrapolation to $0°K$, 27
  monatomic gas, 32
  polyatomic gas, 32
  1,1,1-trifluoroethane gas, 43, 44, 45
Hess' law, 77
Kandiner and Brinkley method, 162
moment of inertia, 97, 98, 102ff
Calculation by computer, enthalpy, 50, 51
  entropy, 110
  equilibria, 163
  tables, 204
Calculations, agreement of, 164, 206
  thermodynamic assumptions and, 156
Calibration, combustion bomb, 67
  flame calorimeter, 72
Calorie, thermochemical, 9, 12, 203
Calorimeter, adiabatic intermediate temperature, 22ff
  adiabatic low-temperature, 19ff
  enthalpy of vaporization, 28ff
  flame, 72ff
  flow, 28ff
  gaseou3 heat capacity, 28ff
  oxygen bomb, 67ff
Calorimetry Conference, 22

Carbon dioxide, enthalpy of formation, 203
Carbon disulfide molecule, 101ff
Carbonyl fluoride reaction, 80
Catalyst selection by thermodynamics, 192ff
Celsius temperature, 6
Charles' law, 6
Chemical Abstracts, 631, 632
  index system, 214
Chemical affinity, index to, 3
Chemical change, 5, 11
Chemical potential, *see* Gibbs energy
Chemical reactions, entropy of, 112
Chemicals from methane, 165
Chlorine compounds, enthalpy of formation, 490
  entropy, 490
  heat capacity, 490
  methyl substitution constants, 490
  tables, 508
Clapeyron equation, 55
Clausius-Clapeyron equation, 55
Coefficient, activity, *see* Activity and activity coefficient
Combustion, calculation of data, 71
  effect of 1% change, 157
  heat of, 64
  symbol, 200
Compilations of thermodynamic properties, elements, viii
  hydrocarbons, vii, viii, 235
  inorganic substances, viii
  metals, viii
  refractories, viii
  sulfur compounds, viii, 563
Complex equilibria, 135
Composition, determination by thermodynamic data, 171
  of petroleum aromatic fractions, 159
Compressibility factor, 10
Computer, calculation of entropy by, 110
  calculation of gaseous heat capacity by, 48
  calculation of tables by, 204ff
  solution of equilibrium compsition, 135, 163
Concentration of isomeric species, 161
Condensed phase equilibria, 133
Constant, universal gas, 7
Constants, physical, 203
Continuing use of Gibbs energies, 164
Correction for gas imperfection, 90
Cracking of ethylbenzene, 66
  enthalpy of, 173

Subject Index

Critical constants, 55, 58
Cryoscopic heat of melting, 54, 57
Cryostat, 19ff
Crystal structure, unstable, 53
Cycles, biological, 405
Cyclic amines, table, 468ff
Cyclic ethers, tables, 419ff
Cycloalkanes, tables, 343ff
Cycloalkanethiols, table, 625
Cycloalkanols, table, 437ff
Cycloalkanone, table, 447, 448
Cycloalkenes, tables, 546ff
Cyclohexane Houdriforming equilibria, 160

Data verification, 207
Datum level of enthalpy, 13
Decomposition equilibria of acetyl chloride, 134
Decomposition of carbonyl fluoride, 80, 86
Debye functions for solids, 25
  characteristic temperature, 25
  energy table, 780
  entropy table, 781
  extrapolation to $0°K$ by means of, 92
  heat capacity table, 779
Deformation frequencies, 142
Degeneracy, vibrational, 35
Degrees of freedom, 28ff
Dehydrogenation of 2-propanol, 78, 82
Depression of melting point, 54
Diluent effect of steam in formaldehyde process, 188
Dimethylacetylene, calculation of heat capacity, 37, 42, 43
  internal rotation in, 37, 42, 43
Disorder, effect on entropy, 91
Disproportionation equilibria of methyl benzenes, 165
Disproportionation of styrene, 174
Dithiaalkanes, ideal gas tables, 600ff
Driving force, measure of, 4
  of reaction, 115
Dynamic equilibrium, 5, 55
Dyne, 10

Effective catalysis, index to, 194
Effect of diluent in formaldehyde process, 188
Einstein function, energy table, 777
  entropy of vibration from, 101
  entropy table, 778

extrapolation to $0°K$ by means of, 92
heat capacity of solids from, 24, 26
heat capacity table, 776
Electrical calibration, combustion bomb, 67ff
Electromechanical work, 12
Electronic charge, 203
Elements, polyphase, 208
  reference state, 13, 112, 116, 205
  thermodynamic tables, 209ff
Energy, 10, 199
  of combustion, 71
  conservation of, 10
  Debye function table, 780
  Einstein function table, 777
  equipartition of, 32, 33
  Gibbs, 115, 205
  Helmholtz, 116
  table for restricted rotation, 783
Enthalpy, 12, 199, 200
  of atomization, 242
  of combustion, 64
    bomb calorimetric measurement, 67ff
    effect of 1% change, 157
    flame calorimetric measurement, 72
    of gas, 50
    of graphite, 203
    of halogen compounds, 70
    of sulfur compounds, 70, 559
  datum level of, 13
  dependence on temperature, 65
  of ethylbenzene cracking, 65, 173
  of formation, 61ff, 200
    of alcohols, 409
    estimation for sulfide, 152
    estimation for thiols, 152
    estimation of, 140ff
    from combustion data, 67, 72
    from equilibrium data, 78
    from heat of reaction data, 75
    regularities of, 238
    summary table, 631ff
  of hydrogenation, of acetone, 76
    of styrene, 65, 66
  increment, 13
  master table, 634
  of melting, 53, 54, 201
  of polymerization of styrene, 174
  of reaction, 62, 202
    from equilibrium data, 83ff, 137, 138
    temperature dependence of, 64ff
  reference temperature, 51

Subject Index

of sublimation, 55, 200, 202
of transition, 53, 200
of vaporization, 55, 200ff
  alcohols, 408
  of benzene, 56
  corresponding state method, 58
  Haggenmacher equation for, 59
  heptane, 59
  measurement of, 28ff
  Othmer method, 58
  Trouton's rule for, 57
Entropy, 88, 200
  calculation of by Simpson's rule, 110ff
  correction for gas imperfection, 90
  Debye function table, 781
  Einstein function table, 778
  estimation of, 140ff
  of furan, 94
  of gases by statistical mechanics, 93
  from heat capacity, 92ff
  increment for chemical reaction, 112ff
  internal energy contribution to, 95ff
  of internal rotation, 7, 100, 106ff
  master table, 631
  of mixing, 88
  pressure coefficient of, 89
  of reaction from equilibria, 113, 114
  rotational, 95
  spontaneous process and, 87
  standard reference state and, 112
  symbol for, 200
  symmetry and, 407ff
  tables for restricted rotation, 784, 785
  temperature coefficient of, 88, 108, 113
  third-law, 91ff, 108
  of transition, 89
  transitional, 94, 101, 102, 105
  vibrational, 101, 102, 104, 107
Equation, Antoine, 58, 144ff
  Berthelot, 10, 90, 101
  Clapeyron, 55
  Clausius-Clapeyron, 55, 56
  Debye, for heat capacity of solids, 25
  Einstein, for heat capacity of solids, 24
    for vibrational heat capacity of gases, 35
  Haggenmacher, 59
  for heat capacity of gases, 48
  Kirchoff, 65
  Sackur-Tetrode, 94, 95
  Van't Hoff, 79, 137, 181
  vapor pressure, 14

Equation of state, 8, 9, 74ff
  Berthelot, 10, 90, 181
  ideal or perfect gas, 9
  real or imperfect gas, 74, 76, 90
  virial, 9
Equilibrium, attainment of, 160
  dynamic, 5, 55
  and Gibbs energy, 118
  heterogeneous, 133
  homogeneous, 133
  metastable, 5, 118
  thermodynamic approach to, 167
  vapor pressure and, 133
Equilibrium composition, in acetic acid
    manufacture, 188, 189
  in acrylonitrile synthesis, 177
  calculation of, 123ff
    by computer, 135, 161ff
  in complex systems, 135, 161ff
  of cracked petroleum, 158
  in dehydrogenation, of cyclohexane, 160
    of ethylbenzene, 172
  effect of high pressure, 129, 157
  in formaldehyde manufacture, 181ff
  graphs for, 125ff
  in methanol synthesis, 180, 181
  in methylbenzenes, 161ff
  of petroleum aromatic fractions, 159
  relation to Gibbs energy, 122
  of shale oil naphtha, 159
  in systems with condensed phases, 133ff
Equilibrium constant, activity and, 120
  calculation of, 61
  for dehydrogenation of isopropanol, 78ff
  for disproportionation of CF_2O, 80ff
  enthalpy of reaction from, second-law
    method, 79, 137
  sigma plot method, 83
  third-law method, 83ff, 138
  of formation, master table, 634
  symbol for, 200
  temperature dependence of, 80, 113, 136
Equipartition of energy, 32
Erg, 10
Estimation of thermodynamic quantities,
    140ff
  group contribution methods, 146ff
  by method of increments, 241
  from parent hydrocarbons, 146, 147, 406ff,
    459, 460, 487ff, 561
  statistical approximation methods, 151

steric interaction energy and, 153
symmetry number in, 151
valence bond methods, 141ff
Ethers, aliphatic, tables, 412ff
  cyclic, tables, 418
  enthalpy of formation of, 406, 407
  thermodynamics of, 406, 412
Ethylbenzene, cracking of, 66
  enthalpy of dehydrogenation, 65, 66
Explosive reaction mixture in formaldehyde process, 184
Extensive properties, 4, 115
Extrapolation of heat capacity data to $0°K$, 26, 27, 92

Factors in catalysis, 196
Feasibility of reactions, 157ff, 171
First law of thermodynamics, 12
Flame calorimetry, 72ff
Flow method for gas heat capacity, 28
Fluorine compounds, bomb calorimetry of, 70, 71
  enthalpy of formation, 487
  entropy, 487
  methyl substitution constants, 487
  tables, 494ff
Fluorobenzene, enthalpy of vaporization, 59
Formaldehyde, thermodynamics of manufacture, 181ff
Formation, from elements, 200
  enthalpy of, 61ff, 200
Formula index system, 214
Franklin method of estimating thermodynamic properties, 153
Free energy, see Gibbs energy
Fugacity, 119, 129ff
  activity and, 119
  coefficient, 129, 131ff
  of gaseous mixtures, 131
  of gases, 129
Functions, Gibbs energy, 138
Fusion, symbol for, 200
Furan, entropy of, 93
  heat capacity of, 93

Gas, calorimeter, 28ff
  constants, 7, 9, 203
  degrees of freedom, 32
  energy of, 33, 101ff
  enthalpy of, 50

entropy of, 94
  statistical calculation, 93ff
equation of state, 9
equilibria at high pressure, 129ff
heat capacity, calculation of, 48, 49
  equations for, 48
  theoretical, 32ff
  imperfection correction, 90
Gatterman-Koch reaction, 190–192
Gibbs energy, 3, 115
  of acetic acid manufacture, 188, 189
  of acrylonitrile synthesis, 177
  bibliography of data, vii
  calculations, third law, 83, 138
  of chemicals from methane, 167–170
  continuing use of, 164
  definition, 115, 199, 200
  of dissociation of ethyl chloride, 116, 121, 131
  equations, additions, and integration, 136
  and equilibrium composition, 115ff
  of formaldehyde manufacture, 181ff
  of formation, 116, 205
  fugacity and, 120
  function, 83, 138, 200, 205
  as index to reaction feasibility, 116, 117, 118, 155ff
  internal rotation contribution to, 785
  as measure of driving force, 3, 4
  of methanol synthesis, 180ff
  of styrene manufacture, 171ff
  temperature coefficient of, 136ff
  of thiaalkanes, 561
  of vinyl chloride synthesis, 178ff
Glossary of terms, 773
Gold point, 7, 8
Graphite, enthalpy of combustion, 203
  reference state table, 209
Gravity, standard, 203
Group contributions in estimation, 145ff

Haggenmacher equation, 59
Halogen compounds, bomb calorimetry of, 70
  tables for, 486
Harmonic oscillator, compared to restricted rotator, 786
  contribution to heat capacity, 787ff
Heat, of combustion, formation, fusion, etc., see Enthalpy
  latent, see Enthalpy

see also Enthalpy; Energy
Heat capacity, 17, 18, 200
  calculation of, 451
  Debye function table, 779
  Debye theory of, 25
  of dimethylacetylene, 43
  Einstein function table, 776
  Einstein's equation for, 24
  empirical equation for, 48, 49
  estimates of, 140ff
  of gases, ideal or perfect, 32
    internal rotation contribution, 37, 48, 784
    measurement of, 28ff
    statistico-mechanical calculation of, 32ff
  of heptane vapor, 31
  of liquids, 28
  measurement, of condensed phases, 19
  of gases, 28ff
  of methyl bromide, 35, 36
  of solids, 19ff
    extrapolation to $0°K$, 26
    standards for, 22
  of 1,1,1-trifluoroethane, 45
Heat content, see Enthalpy
Heptane, enthalpy of vaporization, 59
  heat capacity of vapor, 31
Hess' law, 62, 77, 115
Heterogeneous equilibria, 133
Heterogeneous system, 4
High pressure equilibria, 129ff
Hill index system, 631
Hindered rotation, see Restricted rotation
Homogeneous equilibria, 133
Homogeneous system, 4
Houdriforming of cyclohexane, 160
Huffman-Ellis method, 559
Hydrocarbons, methylene increments in thermodynamic properties, 240ff
  regularities in properties, 238, 240ff
  as root compounds, 237
Hydrogen bromide, enthalpy of formation, 204
Hydrogen fluoride, enthalpy of formation, 204

Ice point, 8, 202, 203
Ideal solutions, 54, 135
Ideal gas, molar volume, 203
Ideal gas tables, aliphatic bromine compounds, 537ff
  aliphatic chlorine compounds, 508ff
  aliphatic ethers, 412ff
  aliphatic fluorine compounds, 494ff
  aliphatic iodine compounds, 549ff
  alkadienes, 330ff
  alkanals, 438ff
  alkanediol, 436, 437
  alkanes, 243ff
  alkanethiols, 611ff
  alkanoic acids, 448ff
  alkanols, 422ff
  alkanones, 444ff
  alkenes, 312ff
  alkenols, 436
  alkenone, 446, 447
  alkylbenzenes, 367ff
  alkylcyclohexanes, 357ff
  alkylcyclopentanes, 347ff
  alkylcyclopentenes, 356ff
  alkyl|naphthalenes, 389ff
  alkynes, 334ff
  aromatic bromine compound, 547
  aromatic chlorine compounds, 532ff
  aromatic fluorine compounds, 504ff
  aromatic iodine compound, 556
  aromatic thiols, 625
  arylamine, 472
  cyclic amines, 468ff
  cyclic ethers, 419ff
  cycloalkanes, 343ff
  cycloalkane thiols, 625
  cycloalkanols, 437ff
  cycloalkanone, 447, 448
  cycloalkenes, 346ff
  dithiaalkanes, 600ff
  monothiaalkanes, 563ff
  nitrates, 481ff
  nitriles, 473ff
  nitrites, 481ff
  nitroalkanes, 477ff
  phenols, 454ff
  primary alkylamines, 459ff
  secondary alkylamines, 466
  styrenes, 386ff
  tertiary alkylamines, 467, 468
  thiacycloalkanes, 605ff
  thiacycloalkenes, 609ff
Impurity determination by fractilnal melting, 55
Increment estimation method, 241
Index, to catalyst effectiveness, 194

to chemical reactivity, 3, 157, 158
Hill system, 631
system of tables, 214
to work of Petroleum Research Center, 236
Inertia, moments of, 95ff
Inorganic compounds, thermodynamic tables, 214ff
Intensive properties, 4
Internal consistency of table, 206
Internal rotation, see Rotation, internal
International temperature scale, 7, 8
Iodine compounds, enthalpy of formation, 493
entropy, 493
heat capacity, 493
methyl substitution constants, 493
Irreversible process, see Process; Reaction
Isobutane dehydrogenation, 166
Isomerism, rotational, 47
Intrinsic energy, 199

Joule, 10
Joule-Thomson effect, 30

Kandiner and Brinkley method, 162
Kelvin temperature scale, 6, 8
"Key" compound approach to thermodynamic properties, 236, 561
Kinetic energy, 10, 33
Kinetic factors in reaction, 156
Kirchoff equation, 50, 54, 64

Law, Boyle's, 6
Charles', 6
first, of thermodynamics, 12
Hess', 62, 77, 115
Raoult's, 135
second, of thermodynamics, 87, 137
third, of thermodynamics, 83, 90ff, 108ff, 138ff
Limitations of thermodynamics, 4, 155, 156
Liquids, degrees of freedom of, 28
entropy of vaporization, 414, 417, 420, 422, 429, 437, 438, 448, 453
heat capacity of, 28
Low temperature heat capacity, extrapolation to $0°K$, 26, 27, 92
measurement of, 19ff

Machine computation of thermodynamic tables, 204
Manufacture, of acetic acid, 188
of acrylonitrile, 175
of chemicals from methane, 165
of formaldehyde, 181
by Houdriforming, 160
of styrene, 171
of vinylchloride, 178
Measurement, of enthalpy of vaporization, 28ff
of heat capcity, of condensed phases, 19ff
of gases, 28ff
Melting, enthalpy of, 53, 201
entropy of, 89, 201
Melting point, 53
Melting point depression, enthalpy of melting from, 54
Metastable equilibrium, 5, 118
Methane, chemicals from, 165
table of reactions, 167ff
Methanol, pyrolysis, 183
thermodynamics of synthesis, 180, 181
Method, Allen, for estimating thermodynamic properties, 144, 241
flow, for gas heat capacity, 28
Franklin, for thermodynamic properties, 153
group contributions, in estimation, 145ff
Huffman-Ellis, for sulfur combustions, 559
increments, in estimation, 241, 561
Kandiner and Brinkley for equilibria, 162
limitations of thermodynamic, 155
methylene increments in estimation, 406, 411
Othmer, for enthalpy of vaporization, 58
parent hydrocarbon estimation from, 146, 147, 406, 407, 409, 410, 411, 459, 460, 461, 487ff
Pitzer's for hydrocarbon heat capacities, 151
second-law, for equilibria, 79, 87, 137
sigma plot, for equilibria, 83
Skinner, estimation of thermodynamic properties, 144
statistical approximation, 151
Souders, Matthew, and Hurd, estimation of thermodynamic properties, 153
third-law, for equilibria, 83ff, 138ff
of valence bonds, for thermodynamic properties, 141ff
Van Krevelin and Chermin for estimating

Subject Index

Method, (Cont'd), thermodynamic properties, 141
Methylbenzenes disproportionation, 161ff
Methyl bromide molecule, 35
Methylene increments in thermodynamic properties, 406, 411
Methyl substitution constants, bromine molecule, 35
Methyl substitution constants, bromine compounds, 491
  chlorine compounds, 489
  fluorine compounds, 488
  iodine compounds, 493
  nitrogen compounds, 459
Mixing, entropy of, 88
Molar volume of ideal gas, 203
Moments of inertia, 95ff
  calculation of, 97, 98, 101
  reduced, 46
Monothiaalkanes, tables, 563ff

Nernst heat theorem, 91
Nitrates, tables, 481ff
Nitriles, tables, 473ff
Nitrites, tables, 481ff
Nitroalkanes, ideal gas tables, 477ff
Nitrogen compounds, as explosives, 457
  methyl substitution constants, 460, 461
  resonance in, 458
  tables, 457ff
  thermodynamics of, 457ff
Normal boiling point, 55

Octane cracking equilibria, 158
Organic compound reference state, 13
Oxidation of methanol, 181
Oxygen bomb calorimetry, 67ff
Oxygen compound tables, 405ff
Oxygen point, 7, 8

Parent hydrocarbon estimation method, 146, 147, 406, 407, 409, 410, 411, 459, 460, 487ff, 561
Partition function, 94
  for internal rotation, 46
Petroleum industry problems, 158ff
Petroleum Thermodynamics Laboratory, 236
Phase discontinuities in reference states of elements, 208
Phase equilibria, condensed, 133

Phase transitions, entropy of, 89ff
Phenol, enthalpy of melting of, 57
  tables, 454ff
Physical change, 5, 11
Physical constants, 202, 203
Pitzer's method for heat capacity of hydrocarbons, 151
Platinum resistance thermometry, 21, 29, 30
Potential, chemical or thermodynamic, *see* Gibbs energy
Potential energy, 10
Potential energy barriers to internal rotation, numerical values of, 38ff
Practical entropy, 91, 200, 205
Pressure, 5
  critical, 58
  effect of, *see* specific property affected
  reduced, 131
  vapor, *see* Vapor pressure
Primary alkylamines, tables, 459ff
Process, 5
  irreversible or actual, 87; *see also* Reaction
  reversible, 87
1-Propanethiol, vapor pressure, 15
Propanol, dehydrogenation, 78, 82ff
Properties (state functions), extensive, 4
  intensive, 4
  prediction of, 237
  regularity of, 238
Pseudorotation, 458, 470
Pyrolysis of methanol, 181

Raoult's law, 135
Reaction, 4, 5, 200
  driving force, 115
  entropy of from equilibria, 113, 114
  explosive, 184
  Gatterman-Koch, 190
  Gibbs energy of, 115
  heat of, *see* Enthalpy of reaction
  isotherm, 120
  kinetic factors in, 156
  side, 135
  spontaneous, 4, 87
Reactivity, index to, 3
Real gas, *see* Gas
Reduced moment of inertia, 46
Reduced pressure, 58, 131
Reduced temperature, 58, 131
Reference state, 13, 75
  of elements, 13, 116, 205, 208ff

Subject Index

of organic compounds, 13
standard, 62, 200, 560
of sulfur, 560
Reference temperature for enthalpy, 51
Regularity of thermodynamic properties, 238
Resistance thermometer, 7
Resonance in nitrogen compounds, 458, 461
Restricted rotation, *see* Rotation
Reversible process, 87
Review, adiabatic cryogenic calorimetry, 22
Root compounds, hydrocarbons as, 237
Rotating combustion bomb, 70
Rotation, internal, 37ff
   barriers to, 38ff, 458, 487
   comparison with oscillation, 786
   in dimethylacetylene, 37, 42, 43
   energy function table, 783
   entropy function table, 784
   entropy of, 100, 105ff
   heat capacity of, 37
   heat capacity table, 782
   isomerism due to, 47
   partition function for, 46
   symmetry and, 47
   in 1,1,1-trifluoroethane, 43ff, 105ff
  molecular, 33
   entropy of, 95ff
Rotator, symmetry of, 45
Rule, Simpson's parabolic, 51ff
  Trouton's, 57

Sackur equation, 94
Sackur-Tetrode equation, 95
Schottky transition, 89
Secondary alkylamines, tables, 466
Second law of thermodynamics, 87
  enthalpy of reaction method, 137
Selection of catalyst, thermodynamics of, 192
Shale oil naphtha equilibria, 159
Shield control, automatic, 21, 22
Shortcut methods of calculating compositions, 123ff
Side reactions, effect of, 135
Sigma plot of equilibria, 83
Sign in energy changes, 11
Silver point, 7, 8
Simpson's parabolic rule, 51ff, 103, 110
Skinner's method for estimating thermodynamic properties, 144
Solid, degrees of freedom, 24ff
  solutions, effect on entropy, 91
  state, organic, viii
  state transition, 200
  theoretical heat capacity, 24ff
  transition, entropy of, 89ff
  transition temperature, 53
  vibration, 53
  vibrational frequencies, 24ff
Solution, ideal binary, 54
Souders, Matthew, Hurd correlation method of estimating thermodynamic properties, 153
Specific heat, 18
Spectroscopic data, ideal gas properties calculated from, 32, 93, 108
Speculation, intelligent, 155
Spontaneous process or reaction, 4, 87
Stable equilibrium, 118
Standard, atmosphere, 203
  benzoic acid for calibration, 68ff
  enthalpy, 200
  Gibbs energy of formation, 116
  gravity, 203
  reference state, 62, 75, 112, 116, 200, 560
  state, 13, 208
  state elements, 13
  thermocouple, 7
State function, *see* Properties
State of a system, 4, 8, 9
States, quantum, *see* Energy levels
  reference, 208
  standard, *see* Standard; Reference state
Stationary combustion bomb, 70
Statistical probability, distribution of states, *see* Partition function
Statistical treatment, of ideal gas, approximate estimation, 151
  entropy, 93ff
  heat capacity, 32ff
Steam point, 6, 8
Steric interaction energy, 153
Styrene, heat of polymerization, 174
  thermodynamics of manufacture, 171ff
Styrenes, tables, 386ff
Sublimation, enthalpy of, 55
  entropy of, 89
Successive use of Gibbs energy data, 164
Sulfur, reference state, 560

864 Subject Index

Sulfur compounds, 558
  in bomb calorimetry, 70
  compilations of, 563
  enthalpy of combustion, 559
  enthalpy of formation, 560
  Gibbs energy, 561
Sulfuric acid, heat of formation, 203
Sulfur point, 8
Summary table of thermodynamic properties at $298°K$, 634
Surroundings, definition of, 4
Symbol, glossary, 199ff, 773
Symmetric top, 45, 100
Symmetryless entropy, 407ff
Symmetry number, 46, 98, 99
  effect on estimation, 151
Synthesis, of acrylonitrile, 175
  of vinyl chloride, 178
System, definition, 4
  heterogeneous, 4
  homogeneous, 4
  macroscopic, 5
  microscopic, 5
Systematic program for thermodynamic investigations, 236

Table, bond angles, 96, 97
  bond lengths, 96, 97
  Debye function, energy, 780
  entropy, 781
  heat capacity, 779
  Einstein function, energy, 777
  entropy, 778
  heat capacity, 776
  elements, thermodynamic properties, 214
  entropy, compounds, 634
  Debye function, 781
  Einstein function, 778
  elements, 214ff
  harmonic oscillator, 786ff
  heat capacity, Debye function, 779
  Einstein function, 776
  harmonic oscillator, 786ff
  internal rotation contribution, 784ff
  internal consistency of, 206
  internal rotation contributions, 784ff
  internal rotation contribution, 784ff
  restricted rotation, 782
    comparative properties of restricted rotator and harmonic oscillator, 786
    decrease from free rotation, 785
  energy, 783
  entropy, 784
  heat capacity, 784
Temperature, 6, 8, 87, 202
  absolute, 6
  boiling, 55
  Celsius, 6
  Centigrade, 6
  conversion, 202
  critical, 55, 58
Debye characteristic, 25
  dependence, enthalpy of reaction, 64ff
    equilibrium constant, 136
  Einstein characteristic, 25
  enthalpy, reference, 51
  ice point, 202
  international scale, 7, 8
  Kelvin scale, 8
  melting, 53
  normal boiling, 55
  reduced, 58, 131
  scale, 7, 202
  solid transition, 53
  thermodynamic scale, 7
  triple point, 54
Tertiary alkylamines, tables, 467, 468
Tetrode equation, 95
Theoretical evaluation of heat capacity of solids, 24ff
Thermal properties, 3
  of inorganic compounds, 213ff
  in intermediate temperature range, measurement of, 19ff
Thermochemical calorie, 9, 203
Thermochemistry, 3, 61
  of halogen compounds, 486
Thermocouple, standard, 7
Thermodynamic, approach to reaction equilibria, 167
  assumptions in calculation, 156
  constants, 203
  functions, estimation of, 140ff
    regularities of, 238
  glossary of terms, 773
  ideal program, 236
  methods, limitations of, 155
  properties, estimation of, 140ff
  symbols, 199
  temperature scale, 7
  utility in modern research, 160
Thermodynamics, of acetic acid manufacture,

188, 189
of acrylonitrile synthesis, 175ff
application to problems, 155ff
of biochemical processes, 458
in catalyst selection, 192ff
of chemicals from methane, 165ff
of dehydrogenation, of cyclohexane, 160
of isobutane, 166
of disproportionation of methylbenzenes, 161
first law of, 12
of formaldehyde manufacture, 181ff
of hydrocarbons, 235ff
limitations of, 4, 155, 156
of methanol synthesisk 180ff
of petroleum cracking, 158
and the petroleum industry, 158ff
second law of, 87
of shale oil naphtha, 159
of styrene manufacture, 171ff
of sulfur compounds, 558ff
third law of, 90ff
of vinyl chloride synthesis, 178ff
Thermometer, platinum resistance, 7, 21, 29, 30
Thiaalkanes, Gibbs energies, 561
methylene increments, 561
substitution constants, 561
tables, 600
Thiacycloalkanes, tables, 605ff
Thiacycloalkenes, tables, 609ff
Thiol, aromatic, tables, 625ff
Third law of thermodynamics, 90
application of data, 83
entropy, 91ff, 108ff
equilibrium methods, 139
method for enthalpy of reaction, 138
Tops, symmetrical, 100ff
unsymmetrical, 47, 100ff
Transition, enthalpy of, 53ff
Schottky, 89
temperature of solid, 53
Translational, degrees of freedom, 33
energy of a gas, 32ff
entropy of a gas, 94
Triethylenediamine, heat capacity, 24
1,1,1-Trifluoroethane, molecule, 43ff
1,2,4-Trimethylbenzene, heat capacity, 26
Triple point, 201
temperature, 54
of water, 8
Trouton's rule, 57

Units, energy conversion factors, 9, 203
Unstable, crystal structure, 53
equilibrium, 118
Unsymmetric top, 47, 100
U.S. Bureau of Mines, Petroleum Research Center, 236
Utility of thermodynamics in modern research, 155
Utilization of methane, 167

Valence bond contributions, 141ff
Van Krevelin and Chermin method of estimating thermodynamic properties, 141
Vaporization, 200
enthalpy of, 55, 200
entropy of, 89
Vapor pressure, 14, 55ff
activity, 133ff, 181
Antoine equation, 14
calculation of, 15
Clapeyron equation, 55
equilibria, 133
fugacity, 133ff, 181
measurement, 28
1-propanethiol, 15
Variable, extensive and intensive, 4
Velocity of light, 203
Verification of data, 207
Vibration, gaseous entropy of, 101ff
in a solid, 53
Vibrational degeneracy, 35
Vibrational energy, 32ff
Vibrational entropy, 101
Vibrational frequencies of a solid, 24ff
Vibrational heat capacity, 24ff, 32
Vinyl chloride, thermodynamics of synthesis, 178ff
Virial coefficients, 9, 76, 77
Volume, 4, 5

Washburn reduction in bomb calori,etry, 68, 69
Water, heat of formation, 204
triple point, 8
Watt-second, 10
Wave numbers, 35
Weights, molecular, 634ff; *see also* individual compound or
Work, 11
electromechanical, 12
Work function, 116; *see also* Helmholtz energy